Handbook of Geochemistry

Vol. II/3

Executive Editor: K. H. Wedepohl

Editorial Board:
C. W. Correns · D. M. Shaw · K. K. Turekian
J. Zemann

Springer-Verlag Berlin · Heidelberg · New York 1972

ISBN-13: 978-3-642-65041-3   e-ISBN-13: 978-3-642-65039-0
DOI: 10.1007/978-3-642-65039-0

This work is subject to copyright. All rights are reserved, whether the whole or part of the material is concerned specifically those of translation, reprinting, re-use of illustrations, broadcasting, reproducing by photocopying machine or similar means, and storage in data banks.
Under § 54 of the German Copyright Law where copies are made for other than private use, a fee is payable to the publisher, the amount of the fee to be determined by agreement with the publisher.
© by Springer-Verlag Berlin · Heidelberg 1972. Library of Congress Catalog Card Number 78-85402.
Softcover reprint of the hardcover 1st edition 1972

The use of registered names, trademarks, etc. in this publication does not imply, even in the absence of a specific statement, that such names are exempt from the relevant protective laws and regulations and therefore free of general use.
Universitätsdruckerei H. Stürtz AG, Würzburg

LIST OF CHAPTERS
Chemical Elements Occuring in Nature

Atomic numbers (in heavy print) are identical with chapter numbers in Volume II of this handbook. Elements with very close relations, such as the noble gases, the lanthanides and the platinum elements, are combined and covered under one element of the respective group. Elements printed in italics will not be included in this book.

H	Hydrogen	1	Rb	Rubidium	37	Tb	Terbium	65
He	Helium	2	Sr	Strontium	38		(see 57)	
Li	Lithium	3	Y	Yttrium	39	Dy	Dysprosium	66
Be	Beryllium	4		(see 57)			(see 57)	
B	Boron	5	Zr	Zirconium	40	Ho	Holmium	67
C	Carbon	6	Nb	Niobium	41		(see 57)	
N	Nitrogen	7	Mo	Molybdenum	42	Er	Erbium	68
O	Oxygen	8	Ru	Ruthenium	44		(see 57)	
F	Fluorine	9		(see 78)		Tm	Thulium	69
Ne	Neon (see 2)	10	Rh	Rhodium	45		(see 57)	
Na	Sodium	11		(see 78)		Yb	Ytterbium	70
Mg	Magnesium	12	Pd	Palladium	46		(see 57)	
Al	Aluminium (Aluminum)	13		(see 78)		Lu	Lutetium (see 57)	71
Si	Silicon	14	Ag	Silver	47	Hf	Hafnium	72
P	Phosphorus	15	Cd	Cadmium	48	Ta	Tantalum	73
S	Sulfur	16	In	Indium	49	W	Tungsten (Wolfram)	74
Cl	Chlorine	17	Sn	Tin	50			
Ar	Argon (see 2)	18	Sb	Antimony	51	Re	Rhenium	75
K	Potassium	19	Te	Tellurium	52	Os	Osmium	76
Ca	Calcium	20	I	Iodine	53		(see 78)	
Sc	Scandium	21	Xe	Xenon (see 2)	54	Ir	Iridium (see 78)	77
Ti	Titanium	22						
V	Vanadium	23	Cs	Cesium	55	Pt	Platinum	78
Cr	Chromium	24	Ba	Barium	56	Au	Gold	79
Mn	Manganese	25	La	Lanthanum	57	Hg	Mercury	80
Fe	Iron	26	Ce	Cerium (see 57)	58	Tl	Thallium	81
Co	Cobalt	27	Pr	Praseodymium (see 57)	59	Pb	Lead	82
Ni	Nickel	28				Bi	Bismuth	83
Cu	Copper	29	Nd	Neodymium (see 57)	60	Po	*Polonium*	84
Zn	Zinc	30				Rn	*Radon*	86
Ga	Gallium	31	Sm	Samarium (see 57)	62	Ra	Radium	88
Ge	Germanium	32				Ac	*Actinium*	89
As	Arsenic	33	Eu	Europium (see 57)	63	Th	Thorium	90
Se	Selenium	34				Pa	*Protactinium*	91
Br	Bromine	35	Gd	Gadolinium (see 57)	64	U	Uranium	92
Kr	Krypton (see 2)	36						

STANDARD SECTIONS OF CHAPTERS
(Chapters have the atomic numbers of the elements)

A. Crystal chemistry
B. Isotopes in nature
C. Abundance in cosmos, meteorites, lunar materials and tektites
D. Abundance in rock-forming minerals (phase equilibria), minerals
E. Abundance in common igneous rock types
F. Behavior in magmatogenic processes (pegmatites, gas transport, ore deposition etc.)
G. Behavior during weathering and alteration of rocks
H. Solubilities of compounds which control concentrations of the element in natural waters; adsorption processes; valence states in natural environments
I. Abundance in natural waters and in the atmosphere
K. Abundance in common sediments and sedimentary rock types
L. Biogeochemistry
M. Abundance in common metamorphic rock types
N. Behavior in metamorphic processes
O. Relations to other elements, economic importance etc.

References (Common abbreviations: SB = Strukturbericht; SR = Structure Reports)

LIST OF SYMBOLS DESIGNATING ANALYTICAL METHODS

Optical emission spectrography or spectrometry: arc, spark etc.	S
Atomic absorption spectrometry	A
Optical emission spectrometry: flame photometry	F
X-ray spectrometry and X-ray fluorescence spectrometry	X
Chemical (wet conventional and rapid) methods etc.	W
Colorimetric, spectrophotometric methods	C
Polarographic methods	P
Neutron activation, radiometric measurements	N/R
Isotope dilution, mass spectrometric measurements	I
Microprobe analysis	M
Fluorimetric measurements	L
Spark source mass spectrometry etc.	Mass

Or combinations of these symbols

Lithium   3

A   G. Cocco, L. Fanfani and P. F. Zanazzi   (Istituto di Mineralogia dell'Università di Perugia, Perugia, Italy)

B—O   K. S. Heier   (Mineralogisk – Geologisk Museum, Oslo, Norway)

and

G. K. Billings   (Department of Geosciences, New Mexico Institute of Mining and Technology, Socorro, New Mexico, U.S.A.)

3-A. Crystal Chemistry

The most important common characteristic of the alkali elements is the single electron in the outermost energy level of their atoms. As shown by the spectroscopically determined ionization energies, these electrons can be easily detached. The cores remaining after the first ionization have the same electronic configuration as the immediately preceeding noble gases in the periodic table, and this fact explains the very high second ionization potentials. The aptitude of the alkali metals to be transformed into positively charged ions results in their extremely reactive chemical character. Lithium exhibits many similarities with the other alkali metals, although the small size of the Li atom and the Li^+ ion does lead to some remarkable differences; in many respects its behavior constitutes a transition between the group of the alkali metals and that of the alkaline earth metals, especially magnesium. Characteristic properties for lithium are: the low solubilities of various salts e.g. the fluoride,

Table 3-A-1. *Lithium minerals*

Compound		References
Cryolithionite	$Na_3Al_2[Li^{[4]}F_4]_3$	Menzer (SB 1930, 498)
Lithiophosphatite	$Li_3^{[4]}[PO_4]$	Zemann (SR 1960, 397)
Triphylite	$Li^{[6]}(Fe, Mn)[PO_4]$	Destenay (SR 1950, 319)
Amblygonite	$Li^{[5]}Al(F, OH)[PO_4]$	Baur (SR 1959, 429)
Eucryptite	$Li^{[6]}Al[SiO_4]$	Winkler (SR 1953, 562)
Spodumene	$Li^{[6]}Al[Si_2O_6]$	Clark et al. (1969)
Bikitaite	$Li[AlSi_2O_6] \cdot H_2O$	Hurlbut (SR 1957, 436)
Petalite	$Li^{[4]}[AlSi_4O_{10}]$	Liebau (SR 1961, 505)
Neptunite	$Li^{[6]}KNa_2Fe_2Ti_2O_2[Si_8O_{22}]$	Cannillo et al. (1966)
Holmquistite (Li-amphibole)	$Li_2^{[6]}Al_2(Mg, Fe)_3(OH)_2[Si_4O_{11}]_2$	Whittaker (1969)
Lepidolite (Li-mica)	$KLi_2Al(F, OH)_2[Si_4O_{10}]$	Radoslovich (1963)
Zinnwaldite (Li-mica)	$KLiFeAl(F, OH)_2[AlSi_3O_{10}]$	Radoslovich (1963)
Cookeite (Li-chlorite)	$LiAl_4(OH)_8[AlSi_3O_{10}]$	Brown and Bailey (1962)
Elbaite (Li-tourmaline)	$Na(Al, Li, Mg)_9^{[6]}(OH, F)_4[BO_3]_3[Si_6O_{18}]$	Ito and Sadanaga (SR 1951, 310)
Lithiophorite	$(Al, Li)^{[6]}MnO_2(OH)_2$	Wadsley (SR 1952, 266)

© Springer-Verlag Berlin · Heidelberg 1972

carbonate and phosphate, the tendency towards covalent bond formation, and solvation — all of which are due to the high polarizing power of Li$^+$, the highest of all the alkali ions.

Because of the small size of Li$^+$ with respect to that of the next alkaline ion Na$^+$, we can expect only rather limited substitution of lithium for sodium in crystals. A possible diadochy between Li$^+$ and Al^{3+}, Fe^{2+} and especially Mg^{2+}, owing to the radius similarity of these ions, may take place in the late phases of magmatic crystallization and affects the compositions of some minerals, e.g. clinopyroxenes and micas. There are, however, a few independent lithium minerals, which are listed in Tables 3-A-1 and 3-D-1.

With the exception of cryolithionite, $Na_3Al_2[LiF_4]_3$, and lithiophorite, $(Al, Li)MnO_2(OH)_2$, lithium occurs in nature mostly in silicates or phosphates. Therefore, the present crystallochemical review concerns only lithium-halogen and lithium-oxygen bonds.

I. Li Halides

From alkali halides a set of effective ionic radii for the alkali ions can be derived. The value for Li$^+$ is 0.78 Å (see this handbook, Vol. I, Table 12—8, p. 390).

In the Li-halides which show the NaCl structure type, the Li$^+$ ions is in 6-fold coordination with Li-halogen bond lengths as follows: Li—F, 2.014 Å; Li—Cl, 2.570 Å; Li—Br, 2.751 Å; Li—I, 3.000 Å (after SYSIÖ, 1969). Very similar bond lengths for Li octahedrally coordinated by fluorine are found in Li[SbF$_6$] (BURNS, 1962) namely Li—F, 2.032 Å. Distorted octahedral coordination occurs in Li$_6$BeZrF$_{12}$ (SEARS and BURNS, 1964) with average Li—F bond lengths of 2.08 Å and 2.04 Å respectively for the two crystallographically non-equivalent LiF$_6$ octahedra.

Other fluorides show Li in tetrahedral coordination. In Li$_2$[BeF$_4$] (BURNS and GORDON, 1966), whose structure is isotypic with phenakite, the average Li—F distances for the two independent LiF$_4$ tetrahedra are 1.875 and 1.861 Å; the mineral cryolithionite is isostructural with garnets, with LiF$_4$ replacing SiO$_4$ tetrahedra, (MENZER, SB 1930, 498). Li also has tetrahedral coordination in Li$_2$NiF$_4$, which belongs to the family of inverse spinels (RÜDORFF and KANDLER, SR 1957, 267).

Concerning distances between Li and the other halogens, JACOBI and BREHLER (1969) report two independent Li(H$_2$O)$_2$Cl$_4$ octahedra in Li$_2$ZnCl$_4 \cdot$ 2 H$_2$O with average Li—Cl bond lengths of 2.75 Å, significantly longer than those in LiCl.

II. Oxygen-containing Li Compounds

In structures where lithium is coordinated to oxygen, Li$^+$ usually exhibits the coordination numbers four or six, in the form of more or less distorted tetrahedra and octahedra.

The occurrence of such different coordination numbers arises from the fact that the crystal structure of most of the oxygenated compounds, containing no other large ion, can be explained by close packing of oxygen atoms. The lithium ions can be accomodated in the tetrahedral and octahedral sites of this packing.

Tables 3-A-2a and 3-A-2b list Li$^+$—O bond lengths for 4- and 6-coordination respectively.

The tetrahedral coordination occurs more frequently than the octahedral one. As with the fluorides the Li$^{[4]}$—O distances are shorter than the Li$^{[6]}$—O distances. The average value for Li$^{[4]}$—O is 1.97 Å, while the average Li$^{[6]}$—O bond length is 2.14 Å. These values can be compared with 1.82 Å reported by AKIŠIN and RAMBIDI (SR 1958, 282) for the Li—O bond length from electronic diffraction data on Li$_2$O vapour. Solid Li$_2$O has an "anti-fluorite" structure, with a Li$^{[4]}$—O bond length of 2.00 Å (ZINTL et al., SB 1934, 283).

An intermediate coordination number was found (BAUR, SR 1959, 429) in the mineral amblygonite, LiAl(OH, F)[PO$_4$], in which Li$^+$ is coordinated in the form of a distorted trigonal bipyramid by four oxygen atoms of PO$_4$ groups (at distances 2.12±3, 2.05±3, 2.11±3, and 2.14±3 Å) and one (OH, F) ion (at a distance of 1.95±3 Å). The average Li$^{[5]}$—O bond length is 2.07 Å, intermediate between the values found for tetrahedral and octahedral coordinations. In Li[BO$_2$] the Li$^+$ ion also binds four oxygen atoms at an average distance of 1.967 Å and one more oxygen atom at 2.473 Å (ZACHARIASEN, 1964). In Li[AsO$_3$] (HILMER and DORNBERGER-SCHIFF, SR 1957, 391) one lithium ion shows tetrahedral coordination, and a second one has four neighboring oxygen atoms at an average distance of 1.99 Å and two more oxygen atoms at 2.54 Å. In these last cases the lithium coordination number can be described as (4+1) and (4+2) respectively. The structure of Li[AsO$_3$] closely resembles the structure of diopside CaMg[Si$_2$O$_6$].

The diopside atomic arrangement is present also in LiNa[BeF$_3$]$_2$ (HAHN, SR 1953, 336) and in spodumene, α-LiAl[Si$_2$O$_6$] (WARREN and BISCOE, SB, 1931, 527).

Several structures containing LiO$_4$ tetrahedra can be related to SiO$_2$ structure types; in KLi[SO$_4$] (BRADLEY, SB 1925, 376), Li tetrahedra share the four vertices with different SO$_4$ tetrahedra to form a tridymite-like framework. In tetragonal γ-LiAlO$_2$ (MAREZIO, 1965a; BERTAUT et al., 1965), AlO$_4$-tetrahedra share corners to build a cristobalite-like framework. The Li$^+$ ions are accomodated in tetrahedral holes and each LiO$_4$ tetrahedron shares an edge with an AlO$_4$ tetrahedron.

Some Li-silicate structures can be considered to be derivatives of SiO$_2$ structures. In β-Li[AlSiO$_4$] (WINKLER, SR 1948, 474), the small Li$^+$ ions are accomodated in the empty channels of an atomic arrangement where Al, Si and oxygen atoms occupy the same sites as in the high-quartz. The hexagonal form of spodumene (γ-spodumene, Li[AlSi$_2$O$_6$]-III) (LI, 1968), also has the high-quartz structure with the Li$^+$ ions randomly situated in tetrahedral interstitial holes on equipoints of rank 3; additional tetrahedral sites are also possible for Li$^+$ in the structure.

The framework of β-spodumene, Li[AlSi$_2$O$_6$]-II (LI and PEACOR, 1968), is isotypic with keatite and the four lithium ions per unit cell are randomly distributed among four sets of paired eight-fold equipoints with $^1/_2$ occupancy in tetrahedral oxygen environment. Several additional distorted tetrahedral sites are available; furthermore, Li may occupy octahedral positions at higher temperature.

Tetrahedral lithium coordination occurs in some layer or chain silicates and in structurally related germanates; in these structures chains or layers form the structure and are linked together by Li ions in a three-dimensional framework. In Li[SiO$_3$] (SEEMANN, SR 1956, 404), Li tetrahedra share all vertices with other LiO$_4$ tetrahedra. In Li$_2$[Si$_2$O$_5$] (LIEBAU, SR 1961, 506) and Li$_2$[(Si$_{0.25}$Ge$_{075}$)$_2$O$_5$] (VÖLLEN-

KLE et al., 1968), LiO$_4$ tetrahedra are linked together in pairs by sharing edges; these doublets share vertices to form complex chains which run parallel to the c axes. In the first case these chains bind two adjacent Si sheets; in the second case they connect different chains built up by (Si, Ge)O$_4$ tetrahedra and running parallel to the Li chains. In the mineral petalite, Li[AlSi$_4$O$_{10}$] (LIEBAU, SR 1961; 506), Li ions share opposite edges with Al tetrahedra forming Li—Al—Li—Al chains running along the b axis and connecting two sheets of SiO$_4$ tetrahedra in a three-dimensional framework.

Several structures of artificial Li compounds consist of vertex-linked tetrahedra frameworks similar to that of phenakite; typical examples are Li$_2$[WO$_4$] (ZACHARIASEN and PLETTINGER, SR 1961, 391), LiAlGeO$_4$, LiGaGeO$_4$, LiZnVO$_4$ (BLASSE, 1963), eucryptite (α-Li[AlSiO$_4$]) (WINKLER, SR 1953, 562), and Li$_2$[BeF$_4$]. In LiGaO$_2$ (MAREZIO, 1965b), connections between LiO$_4$ and GaO$_4$ tetrahedra occur through vertices, each oxygen atom being shared by four tetrahedra, two Ga-centered and two Li-centered; LiGaO$_2$ has the arrangement of a wurtzite-like structure. In lithiophosphatite, low-temperature Li$_3$[PO$_4$] (ZEMANN, SR 1960, 397), LiO$_4$ tetrahedra share edges and vertices with each other and are linked by corners to PO$_4$ tetrahedra; the resulting structure can be described as a distorted hexagonal close-packing of oxygen atoms, where Li and P atoms occupy tetrahedral sites. Comparison of x-ray powder data and i.r. spectra of the low- and high-temperature forms of Li$_3$[PO$_4$] suggests that important structural rearrangements are not involved in the transition (TARTE, 1967).

In (NH$_4$)Li[SO$_4$] (DOLLASE, 1969) and in (N$_2$H$_5$)Li[SO$_4$] (BROWN, 1964), each LiO$_4$ tetrahedron shares the four corners with four different SO$_4$ tetrahedra and vice-versa to form a three-dimensional network with four- and six- and four- and eight-membered rings respectively. Rings of four and eight tetrahedra Li- and S-centered are present in the framework of Li$_2$[SO$_4$] · H$_2$O (LARSON, 1965) where each Li tetrahedron shares corners with different Li and S tetrahedra. In Li$_2$[C$_2$O$_4$] (BEAGLEY and SMALL, 1964), the vertices of the LiO$_4$ tetrahedra are connected by C$_2$O$_4^{2-}$ ions into sheets which form the crystal structure.

Coordination octahedra around Li$^+$ can occur either separately or linked together in various ways. Isolated octahedra, for example, occur in α-spodumene, where Li$^+$ is in sixfold coordination in contrast to the arrangement observed in β- and γ-spodumene. Recent x-ray studies (CLARK et al., 1969) have shown that Li-pyroxenes, due to the small size of the cation in position M2, do not maintain the symmetry of the space group C2/c (corresponding to M2 site occupied by Na or Ca), the true symmetry being C2. However, since deviation from C2/c occurs for a few weak reflections, refinement in this space group represents a reliable average structure. Coordination around the cation, which is irregularly eightfold when M2 is occupied by Na or Ca, becomes sixfold in Li-pyroxenes because the two most distant oxygen atoms are removed from coordination polyhedron. In the synthetic pyroxene, (Li$_{0.95}$Fe$^{2+}_{0.05}$)(Fe$^{3+}_{0.95}$Fe$^{2+}_{0.05}$)[Si$_2$O$_6$], Li is still in octahedral coordination but two Li-O distances are significantly longer (CLARK et al., 1969).

Isolated octahedra occur also in neptunite (CANNILLO et al., 1966). In LiMn[PO$_4$] (GELLER and DURAND, SR 1960, 399), the synthetic analogue of the manganous compound of the mineral triphyline, Li octahedra are linked by sharing edges to

Lithium

Table 3-A-2a. $Li^{[4]}$—O bond lengths

Compound	Distances (Å)				Mean	References
LiOH (by neutron diff.)	1.96 (4)				1.96	Dachs (SR 1959, 330)
LiOH · H$_2$O	1.96 (2)	1.98 (2)			1.97	Rabaud et al. (SR 1957, 240)
γ-LiAlO$_2$	2.06 ± 2 (2)	1.95 ± 2 (2)			2.00	Marezio (1965a)
LiGaO$_2$	2.00 ± 3	2.00 ± 2			1.99	Marezio (1965b)
	2.00 ± 2	1.95 ± 2				
Li$_2$[CO$_3$]	1.96	1.97			1.97	Zemann (SR 1957, 394)
	1.96	2.00				
Li[BO$_2$]	1.964 ± 8	1.952 ± 8			1.967	Zachariasen (1964)
	1.945 ± 8	2.007 ± 5				
Li$_2$[B$_4$O$_7$]	1.97 ± 2	2.02 ± 2			2.05	Krogh-Moe (1968)
	2.07 ± 2	2.14 ± 2				
Li$_2$[SO$_4$] · H$_2$O	1.917 ± 5	1.950 ± 8			1.959	Larson (1965)
	1.969 ± 6	2.001 ± 7				
	1.907 ± 6	1.942 ± 6			1.935	
	1.937 ± 9	1.953 ± 6				
LiNa[SO$_4$]	1.87 ± 4	1.88 ± 4			1.98	Morosin and Smith (1967)
	2.08 ± 4	2.09 ± 4				
Li(NH$_4$)[SO$_4$]	1.89 ± 10	1.89 ± 4			1.91	Dollase (1969)
	1.98 ± 3	1.90 ± 2				
Li(N$_2$H$_5$)[SO$_4$]	1.96 ± 4	1.95 ± 4			1.94	Brown (1964)
	1.97 ± 4	1.88 ± 4				
Li$_2$[WO$_4$]	1.87 ± 8	2.10 ± 8			1.89	Zachariasen and Plettinger (SR 1961, 391)
	1.80 ± 8	1.80 ± 8				
	1.88 ± 8	1.98 ± 8			1.97	
	2.00 ± 8	2.02 ± 8				
Li$_3$[PO$_4$]	2.00 ± 5	1.97 ± 5			1.96	Zemann (SR 1960, 397)
	1.90 ± 5	1.95 ± 5				
	1.98 ± 5 (2)	2.00 ± 5			1.99	
	1.98 ± 5					
LiK$_2$[P$_3$O$_9$] · H$_2$O	1.96 ± 3	1.94 ± 3			1.95	Eanes and Ondik (1962)
	1.93 ± 3	1.97 ± 3				
Li$_2$[Si$_2$O$_5$]	1.85	2.04			1.94	Liebau (SR 1961, 506)
	1.99	1.88				
β-Li[AlSiO$_4$]	2.01 (4)				2.01	Winkler (SR 1948, 474)
Li[AlSi$_4$O$_{10}$] petalite	1.92 (2)	1.99 (2)			1.95	Liebau (SR 1961, 506)
Li$_2$[(Si$_{0.25}$Ge$_{0.75}$)$_2$O$_5$]	2.01 ± 7	1.93 ± 7			1.94	Völlenkle et al. (1968)
	1.93 ± 7	1.90 ± 7				
Li$_4$[GeO$_4$]	1.94 ± 5 (2)	1.92 ± 5 (2)			1.93	Völlenkle and Wittmann (1969)
	2.02 ± 5 (2)	1.94 ± 5			2.03	
	2.14 ± 5					
Li$_2$[C$_2$O$_4$]	2.076 ± 4	2.033 ± 4			2.011	Beagley and Small (1964)
	1.999 ± 4	1.935 ± 4				

form linear strings, in a structure which is related to chrysoberyl. Similar connections occur in $Li_{1+x}V_3O_8$ (WADSLEY, SR 1957, 301); in this compound additional Li atoms are probably randomly distributed in tetrahedral sites. Connections through edges occur also in rhombohedral double oxides with the general formula $LiXO_2$ (where X is V, Ni, Co, Al, Cr, Rh, Nb) (WYCKOFF, 1964) with the structure of the $NaHF_2$ type, that is with each LiO_6 octahedron sharing six edges with neighboring octahedra to form planes having hexagonal packing. In $Li[IO_3]$ (ROSENZWEIG and MOROSIN, 1966) and $LiSbO_3$ (EDSTRAND and INGRI, SR 1954, 444), Li octahedra share opposite faces to form infinite chains. A similar arrangement is observed in $Li[ClO_4] \cdot 3\,H_2O$ (DATT et al., 1968) and $Li[MnO_4] \cdot 3\,H_2O$ (KETELAAR, SB 1935, 447); the most interesting feature of these structures is that Li is coordinated only by oxygen atoms of water molecules forming $[Li^{[6]}(H_2O)_3]$-chains linked to the framework by hydrogen bonds. Many synthetic Li compounds show the atomic arrangement of spinel (WYCKOFF, 1965). In all these structures Li^+ ions show octahedral coordination. $LiFe_5O_8$ (KATO, SR 1952, 325), $LiAl_5O_8$ and $LiGa_5O_8$ also have a spinel-like structure, but show an ordered low-temperature and a disordered high-temperature form. This feature occurs also in lithium ferrite $LiFeO_2$ (KATO, SR 1958, 325) which, after quenching from high temperature, has an NaCl structure in which Li^+ and Fe^{3+} ions are randomly distributed among the positions of the metallic atoms in NaCl. By annealing at 570° C lithium and iron become segregated into a fully ordered tetragonal structure.

The latter examples show that diadochy between Li^+ and other cations may occur only under special conditions. Replacement of octahedral Li^+ by Fe^{3+} is reported for a cubic Li tungstate with the crystallochemical formula $(Li_{0.92}Fe_{0.08})_{12}^{[6]}[(WO_4)_3W_4O_{16}]$ (BORISOV et al., 1969). From a crystal structure determination of the mineral lithiophorite, $(Al, Li)MnO_2(OH)_2$ (WADSLEY, SR 1952, 266), it appears that lithium ions randomly occupy one-third of the octahedral aluminium sites in a layer of the brucite type. In tourmalines, where Mg and Al in octahedral coordination are located at different crystallographic sites, a partial substitution of Mg ions by Li and Al ions is observed, one particular example being elbaite (ITO and SADANAGA, SR 1951, 310). In such Li-rich micas as lepidolite and zinnwaldite, Li in octahedral sites replaces Mg and Al (RADOSLOVICH, 1963). In petalite, spodumene and amblygonite, no random distribution of Al—Li has been observed. In holmquistite, $Li_2Al_2(Mg, Fe)_3(OH)_2[Si_4O_{11}]_2$ (WHITTAKER, 1969), Li^+ ions are confined to a single crystallographic position. In such a case, the preferential entry of Li into its particular site can be explained by electrostatic ordering effects. Holmquistite, isotypic with anthophyllite $Mg_7(OH)_2[Si_4O_{11}]_2$, is a typical example of the substitution of $(Li^+ + Al^{3+})$ for $2\,Mg^{2+}$ ions.

Triphylite, $Li(Mn, Fe)[PO_4]$ (DESTENAY, SR 1950, 319), exhibits a completely ordered arrangement of the cations. This mineral belongs to a series of natural compounds such as sicklerite and iron sicklerite, with the general formula $Li_{1-y}Mn_{1-x}^{2+}Fe_{x-y}^{2+}Fe_y^{3+}[PO_4]$, the other end of the series being represented by the mineral heterosite, $Fe^{3+}[PO_4]$. Close analogies in the unit cell parameters and x-ray powder data suggest an essentially similar structure. Probably the mechanism of the valence compensation during oxidation causes Li^+ ions to be removed from their octahedral sites, until in heterosite these sites are completely empty. This confirms the peculiar role of lithium as interstitial component in close-packed arrangements.

Lithium

Table 3-A-2b. $Li^{[6]}$—O bond lengths

Compound	Distances (Å)		Mean	References
$LiNiO_2$	2.04 (6)		2.04	Dyer et al. (SR 1954, 424)
$LiVO_2$	2.14 (6)		2.14	Rudörff and Becker (SR 1954, 424)
$LiSbO_3$	2.01 (2) 2.07 (2)	2.07 (2)	2.05	Edstrand and Ingri (SR 1954, 444)
$Li[IO_3]$	2.04 ± 6 (3)	2.22 ± 6 (3)	2.13	Rosenzweig and Morosin (1966)
$Li[ClO_4] \cdot 3 H_2O$ (neutron diffraction)	2.18 (3)	2.08 (3)	2.13	Datt et al. (1968)
$LiMn[PO_4]$	2.22 ± 1 (2) 2.16 ± 1 (2)	2.11 ± 1 (2)	2.17	Geller and Durand (SR 1960, 399)
$LiAl[Si_2O_6]$ α-spodumene	2.105 (2) 2.251 (2)	2.278 (2)	2.211	Clark et al. (1969)
$Li_{1+x}V_3O_8$	1.99 2.28 (2)	2.08 2.33 (2)	2.25	Wadsley (SR 1957, 301)
$LiKNa_2Fe_2^{2+}Ti_2O_2[Si_8O_{22}]$ neptunite	2.13 ± 2 (2) 2.04 ± 2 (2)	2.15 ± 2 (2)	2.11	Cannillo et al.

Revised manuscript received: June 1970

References: Section 3-A

BEAGLEY, B., SMALL, R. W. H.: The structure of lithium oxalate. Acta Cryst. 17, 783 (1964).
BERTAUT, E. F., DELAPALME, A., BASSI, G., DURIF-VARAMBON, A., JOUBERT, J. C.: Structure de γ-LiAlO$_2$. Bull. Soc. Franç. Minéral. Crist. 88, 103 (1965).
BLASSE, G.: The structures of ZnLiVO$_4$ and ZnLiNbO$_4$. J. Inorg. Nucl. Chem. 25, 136 (1963).
BORISOV, S. V., KLEVTSOVA, R. F., BELOV, N. V.: The structure of cubic lithium tungstate. Soviet Phys.-Cryst. 13, 852 (1969).
BROWN, B. E., BAILEY, S. W.: Chlorite polytypism: I-Regular and semirandom one-layer structures. Am. Mineralogist 47, 819 (1962).
BROWN, I. D.: The crystal structure of lithium hydrazinium sulfate. Acta Cryst. 17, 654 (1964).
BURNS, J. H.: The crystal structure of lithium fluoroantimonate (V). Acta Cryst. 15, 1098 (1962).
— GORDON, E. K.: Refinement of the crystal structure of Li$_2$BeF$_4$. Acta Cryst. 20, 135 (1966).
CANNILLO, E., MAZZI, F., ROSSI, G.: The crystal structure of neptunite. Acta Cryst. 21, 200 (1966).
CLARK, J. R., APPLEMAN, D. E., PAPIKE, J. J.: Crystal-chemical characterization of clinopyroxenes based on eight new structure refinements. Mineral. Soc. Am. Spec. Pap. 2, 31 (1969).
DATT, I. D., RANNEV, N. V., OZEROV, R. P.: Neutron-diffraction study of the structure of hydrated lithium salts. II. LiClO$_4 \cdot$ 3H$_2$O. Soviet Phys.-Cryst. 13, 192 (1968)
DOLLASE, W. A.: NH$_4$LiSO$_4$: a variant of the general tridymite structure. Acta Cryst. B 25, 2298 (1969).
EANES, E. D., ONDIK, H. M.: The structure of lithium dipotassium trimetaphosphate monohydrate. Acta Cryst. 15, 1280 (1962).
JACOBI, H., BREHLER, B.: Die Kristallstruktur des Li$_2$ZnCl$_4 \cdot$ 2H$_2$O. Z. Krist. 128, 390 (1969).
KROGH-MOE, J.: Refinement of the crystal structure of lithium diborate, Li$_2$O $\cdot$ 2B$_2$O$_3$. Acta Cryst. B 24, 179 (1968).
LARSON, A. C.: The crystal structure of Li$_2$SO$_4 \cdot$ H$_2$O. A three-dimensional refinement. Acta Cryst. 18, 717 (1965).
LI, C. T.: The crystal structure of LiAlSi$_2$O$_6$-III (high-quartz solid solution). Z. Krist. 127, 327 (1968).
— PEACOR, D. R.: The crystal structure of LiAlSi$_2$O$_6$-II ("β-spodumene"). Z. Krist. 126, 46 (1968).
MAREZIO, M.: The crystal structure and anomalous dispersion of γ-LiAlO$_2$. Acta Cryst. 19, 396 (1965a).
— The crystal structure of LiGaO$_2$. Acta Cryst. 18, 481 (1965b).
MOROSIN, B., SMITH, D. L.: The crystal structure of lithium sodium sulfate. Acta Cryst. 22, 906 (1967).
RADOSLOVICH, E. W.: The cell dimensions and symmetry of layer-lattice silicates. V. Composition limits. Am. Mineralogist 48, 348 (1963).
ROSENZWEIG, A., MOROSIN, B.: A reinvestigation of the crystal structure of LiIO$_3$. Acta Cryst. 20, 758 (1966).
SEARS, D. R., BURNS, J. H.: Crystal structure of Li$_6$BeF$_4$ZrF$_8$. J. Chem. Phys. 41, 3478 (1964).
SYSIÖ, P. A.: On the additivity of crystal radii in alkali halides. Acta Cryst. B 25, 2374 (1969).

© Springer-Verlag Berlin · Heidelberg 1972

Tarte, P.: Isomorphism and polymorphism of the compounds Li_3PO_4, Li_3AsO_4 and Li_3VO_4. J. Inorg. Nucl. Chem. **29**, 915 (1967).
Völlenkle, H., Wittmann, A.: Die Kristallstruktur von Li_4GeO_4. Z. Krist. **128**, 66 (1969).
— — Nowotny, H.: Die Kristallstruktur von $Li_2(Si_{0.25}Ge_{0.75})_2O_5$. Z. Krist. **126**, 37 (1968).
Whittaker, E. J. W.: The structure of the orthorhombic amphibole holmquistite. Acta Cryst. B **25**, 394 (1969).
Wyckoff, R. W. G.: Crystal Structures. (second edit.) vol. 2, p. 292. New York and London: Interscience Publisher 1964.
— Crystal Structures. (second edit.) vol. 3, p. 80. New York and London: Interscience Publisher 1965.
Zachariasen, W. H.: The crystal structure of lithium metaborate. Acta Cryst. **17**, 749 (1964).

Nitrogen

A   W. H. Baur   (Department of Geological Sciences, University of Illinois at Chicago Circle, Chicago, Illinois, U.S.A.)

B—M, O   F. Wlotzka   (Max-Planck-Institut für Chemie Mainz, Germany)

7-B. Isotopes in Nature

Two stable isotopes of nitrogen occur in nature: ^{14}N and ^{15}N. The existence of an isotope of mass 15 had been suspected by several authors in the beginning of our century, before it was finally discovered by NAUDÉ (1929) in the band spectra of NO. The literature on ^{15}N and on the determination of the $^{14}N/^{15}N$ abundance ratio from 1919 upto 1952 was compiled by CHAPMAN and BROIDA (1956). A review and a discussion of the geochemistry of the nitrogen isotopes has been given by RANKAMA in 1964.

I. Cosmos

From an analysis of the CN red band systems in the solar spectrum, LAMBERT (1966) derived a lower limit of about 30 for the $^{14}N/^{15}N$ abundance ratio in the sun. CAMERON (1968) adopted the value of 273 measured on terrestrial atmospheric N_2 as representative for the solar system. FOWLER (1954) calculated a value about ten times higher for stars in which the CNO bi-cycle is operative under equilibrium conditions. Deviations from this calculated value can occur, however, if equilibrium is not attained (CAUGHLAN, 1965; FOWLER, 1968).

A first analysis of nitrogen in a lunar rock from the Apollo 11 mission (breccia sample No. 10021-20) showed the $^{15}N/^{14}N$ ratio to be the same as in terrestrial air within the error limit of $\pm 3\%$ (HINTENBERGER et al., 1970).

Nitrogen isotope ratio measurements are usually reported in delta-values, which give the deviation of the $^{15}N/^{14}N$ ratio in the sample relative to standard air nitrogen in per mil, according to the equation:

$$\delta^{15}N = \frac{(^{15}N/^{14}N)_{sample} - (^{15}N/^{14}N)_{standard}}{(^{15}N/^{14}N)_{standard}} \times 1{,}000.$$

Only two meteorites have so far been analyzed (SCALAN, 1959) and the following $\delta^{15}N$ values were found:

	ppm N	$\delta^{15}N$
Gruver (Texas, U.S.A.) CH	8.4	+4.2
La Lande (N. Mex., U.S.A.) CL	5.6	+2.1

These isotope ratios are similar to those of nitrogen in terrestrial igneous rocks (see below). They are also close to those observed for the ammonium ion in rain (HOERING, 1957, see Section 7-I). Hence, as both meteorites are finds and contamination is possible, these values must be regarded with caution until more measurements become available.

II. Atmosphere

The $^{14}N/^{15}N$ ratio in atmospheric air was measured as 273 ± 1 (NIER, 1950) and 272 ± 0.3 (JUNK and SVEC, 1958). The latter value was recommended by NIER and has been adopted as the standard value for atmospheric air (FULLER, 1959).

The ratio was found to be the same for single measurements made at different locations of the United States over a period of six months (HOERING, 1956). A systematic variation with altitude above sea level was reported by McQUEEN (1950), but DOLE et al. (1954) found a constant ratio upto an altitude of 51.6 km.

III. Lithosphere

a) Minerals and Rocks

Recent measurements on nitrogen in minerals and rocks have been mady by HOERING (1955, 1956), MAYNE (1957), SCALAN (1959) and PILOT (1963) (Tables 7-B-1 and 7-B-2).

The data in these tables show the isotopic fractionation to be small, but that on the average there is an enrichment of ^{15}N relative to the atmosphere. Fig. 7-B-1 gives a histogram of the $\delta^{15}N$ values of SCALAN (1959). The isotope ratios measured by SCALAN in twelve minerals average $+1.9^0/_{00}$, while his values for 21 igneous rocks average $+4.7^0/_{00}$, and MAYNE's (1957) 11 igneous rocks average $+4.0^0/_{00}$. MAYNE's (1957) isotopic ratios are correlated with the nitrogen content and show that the less nitrogen a rock contains the more enriched it is in ^{15}N (Fig. 7-B-2). He concluded that this cannot be an artifact due to incomplete nitrogen extraction, because if that were the case one would expect the extracted nitrogen to be enriched

Table 7-B-1 and 7-B-2. *Nitrogen content and its isotopic constitution in minerals and rocks. The $\delta^{15}N$-value gives the difference between the $^{15}N/^{14}N$ ratios of the sample and the standard (atmospheric air) in per mil:*

$$\delta^{15}N = 1{,}000 \times \frac{(^{15}N/^{14}N)_{\text{sample}} - (^{15}N/^{14}N)_{\text{standard}}}{(^{15}N/^{14}N)_{\text{standard}}}$$

Table 7-B-1

Mineral	N, ppm	$\delta^{15}N$ [$^0/_{00}$]	Reference
Pitchblende, Great Bear Lake, Canada	25.0	−7.2	2b
Pitchblende, Great Bear Lake, Canada	33.1	+0.3	2b
Pitchblende, Great Bear Lake, Canada	—	−2.3	1
Pitchblende, Central City Colo., U.S.A.	—	0±3	4
Samarskite, Mitchell Co., N.C., U.S.A.	10.3	−4.3	2b
Monazite sand	20.0	+8.3	2b
Cordierite, Fort Victoria, Ndanga Dist., Africa	125.3	1.3	2b
Cordierite granulite, Geel-kip, Orange River Valley, Africa	720	+5.9	2b
Microcline, Keystone, S. Dak., U.S.A.	63.9	+9.3	2b
Orthoclase, Ft. Bayard, N. Mex., U.S.A.	38.6	−1.0	2a
	28.3	+1.5	2a
Biotite, Deep Creek, N.C., U.S.A.	50.0	−0.05	2b
	70	+3.5	2b
	73.3	+2.1	2b

Nitrogen

Table 7-B-2

Rock	N, ppm	$\delta^{15}N$ [‰]	Reference
Granite, Chelmsford, Mass., U.S.A.		− 0.2	1
Granite, Milford, Mass., U.S.A.		− 0.9	1
Granite, Zehren ü. Meißen, Germany	9	+ 8.0	3
		+ 8.1	3
Granite, Graniteville, Mo., U.S.A.	4.2	+11.0	2b
Granite, Westerly R. I., U.S.A.	1.4	± 0.0	2e
Rhyolite porphyry, Iron Mountain, Mo., U.S.A.	26.9	+ 4.5	2b
	26.3	+ 5.0	2b
Rhyolite, Lipari, Italy	61	−15.6	3
Obsidian, Monolake, Calif., U.S.A.	3.2	+30.9	3
Obsidian, Lake Co., Oreg., U.S.A.	6.9	+ 1.7	2b
	8.7	+ 1.7	2b
Andesite, Hoosac Mt., Eureka Co., Nev., U.S.A.	23	−11.8	3
Andesite, San Juan Co., Colo., U.S.A.	16.9	+ 3.2	2b
	17.6	+ 2.5	2b
Trachyte, Manna Hualalai, Hawaii, U.S.A.	20	− 1.2	3
Olivine gabbro, Cripple Creek, Colo., U.S.A.	3.0	+12.5	2e
	3.1	+ 4.9	2e
Hypersthene gabbro, Sudbury, Ont., Canada		+11.9	2e
Olivine basalt, Boulder Co., Colo., U.S.A.	22.1	+ 2.3	2a
Olivine basalt porphyry, Boulder Co., Colo., U.S.A.	31.3	+ 5.8	2b
	22.1	+ 2.3	2a
Amygdaloidal basalt, Keewenaw, Mich., U.S.A.	41.3	+ 5.7	2b
Basalt, Disko Island, Greenland	31.8	+ 7.6	2b
	19	− 0.2	3
Basalt, Holyoke, Mass., U.S.A.	20	+ 4.8	3
	22	+ 5.2	3
Olivine basalt, Mauna Loa, Hawaii, U.S.A.	3.1	+ 9.9	3
Dunite, Jackson Co., N. C., U.S.A.	2.2	+ 5.5	3
	22.5	− 1.9	2a
	11.9	+ 4.6	2a
Peridotite dunite, Jackson Co., N. C., U.S.A.	3.6	+ 8.2	2c
Pyroxenite, Webster, N. C., U.S.A.	8.4	+ 2.4	2b
	5.0	+ 2.4	2b
Shale, Wissenbacher Schiefer, Elbingerode, Germany		+ 3	4
Keratophyre, Elbingerode, Germany		+ 3	4

References: 1. HOERING (1955, 1956): estimated error ±0.5‰; 2. SCALAN (1959) for methods *a* through *e* see Table 7-D-3, only methods *a* and *b* occured without fractionation of the isotopes; 3. MAYNE (1957): error ±0.2‰; 4. PILOT (1963): error ±1.5‰.

in ^{14}N, if showing any fractionation at all. He explained the correlation by a loss of nitrogen from the crust of the earth to the atmosphere in such a way that the nitrogen remaining in the rocks is enriched in ^{15}N. Although the evidence drawn from the scattered points of Fig. 7-B-2 is more suggestive than conclusive, this suggestion is supported by SCALAN's (1959) high average of $\delta^{15}N +4.9‰$ for igneous rocks, showing a general enrichment of these rocks in ^{15}N relative to the atmosphere.

In Fig. 7-B-3, SCALAN's values of nitrogen content are plotted vs. their $\delta^{15}N$ value; data obtained on different samples of the same rock or mineral are connected

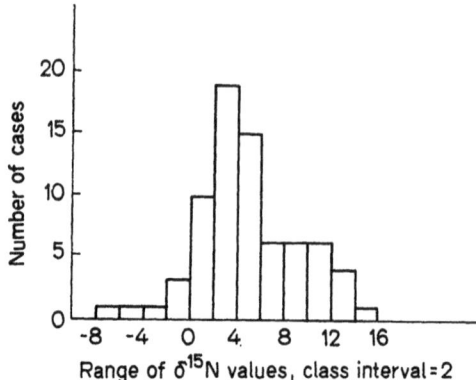

Fig. 7-B-1. Histogram of $\delta^{15}N$ values of minerals and rocks measured by Scalan (1959), relative to atmospheric air standard

by straight lines. Although these values, on the whole, do not follow the trend observed by Mayne (Fig. 7-B-2), several of the lines connecting identical rocks or minerals show this relation of high nitrogen content with a low $\delta^{15}N$ value. The opposite relation is displayed by only two samples.

Fractionation of nitrogen isotopes between igneous rocks and the atmosphere may result from isotopic exchange reactions between N_2, NH_3 and NH_4^+. The fractionation factors for the following isotope exchange equilibria at various temperatures have been calculated by Urey (1947) and since recalculated by Scalan (1959) using new data on the fundamental vibrational frequencies of the isotopic molecules:

$$^{15}NH_3 + ^{14}NH_4^+ = ^{14}NH_3 + ^{15}NH_4^+ \qquad (1)$$

$$^{14}NH_4^+ + ^{14}N^{15}N = ^{15}NH_4^+ + ^{14}N^{14}N \qquad (2)$$

$$^{15}NH_3 + ^{14}N^{14}N = ^{14}NH_3 + ^{14}N^{15}N \qquad (3)$$

Scalan's fractionation factors are tabulated in Table 7-B-3. Equilibrium constants greater than unity mean that the reaction products on the right side of Eq. (1) to (3) are preferred. Hence, for an enrichment of ^{14}N in the atmosphere relative to the crust of the earth, the reactions (1) and (2), involving the ammonium ion, offer a favourable possibility.

Table 7-B-3. *Equilibrium constants calculated for isotopic exchange equilibria (1), (2) and (3), page 7-B-4, from* Scalan *(1959)*

T, °K	Reaction (1)	Reaction (2)	Reaction (3)
273.1	1.034_1	1.018_9	1.015_9
298.1	1.031_1	1.016_7	1.014_3
303.1	1.029_2	1.015_9	1.013_1
400.0	1.021_6	1.011_7	1.009_9
600.0	1.010_4	1.005_6	1.004_8
800.0	1.007_7	1.004_9	1.002_8
1,200.0	1.003_5	1.003_2	1.000_8
1,600.0	1.001_4	1.001_4	1.000_6
2,000.0	1.001_5	1.001_2	1.000_4

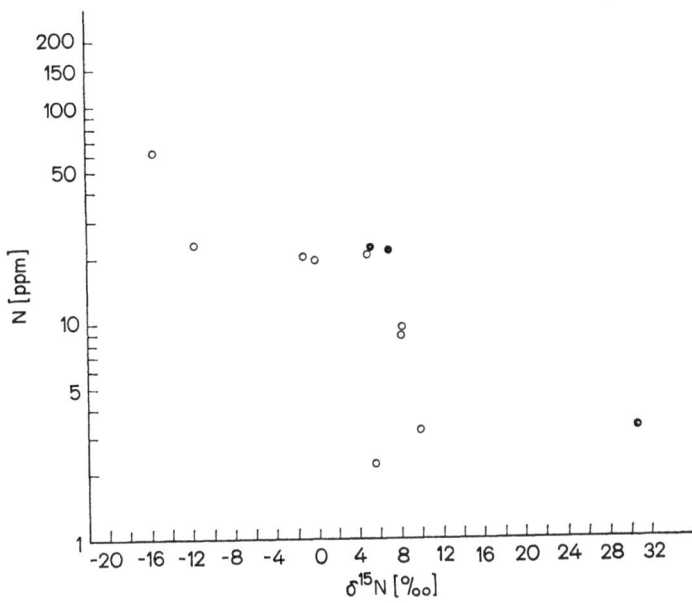

Fig. 7-B-2. Nitrogen content of rocks versus its isotopic constitution, as measured by MAYNE (1957), relative to atmospheric air standard

WHITE and YAGODA (1950) reported a high enrichment of ^{15}N in radioactive minerals. For pitchblende from Great Bear Lake, Canada, they found, e.g., $\delta^{15}N = +582$, and for samarskite from N. Carolina, U.S.A., $\delta^{15}N = +486$. These results were confirmed neither by SCALAN (1959), who found normal values of -7.2 and -4.3 respectively for these minerals from the same locations, nor by HOERING (1957) and PILOT (1963), see Table 7-B-1.

SCALAN (1959) measured the nitrogen content and its isotope ratio in 23 beryls and cordierites, which are known to contain high amounts of He and Ar (see Section 7-D). The amount of nitrogen found is higher than in other igneous minerals and rocks, whereas the isotopic composition shows on the average a similar enrichment of ^{15}N, $\delta^{15}N = +3.8$. No correlation between age of the minerals and their isotopic nitrogen ratio was found.

b) Volcanic Sublimates

Sal ammoniac from volcanoes shows a rather high enrichment of ^{15}N:

	$\delta^{15}N$, ‰	
Sal ammoniac, Paracutin, Mexico	+13.0	HOERING (1956)
Sal ammoniac, lava, Mt. Vesuvius, Italy, December 1886	+11.5	PARWEL et al. (1957)
Sal ammoniac, Mt. Etna, Sicily, eruption 1886	+11.0	PARWEL et al. (1957)

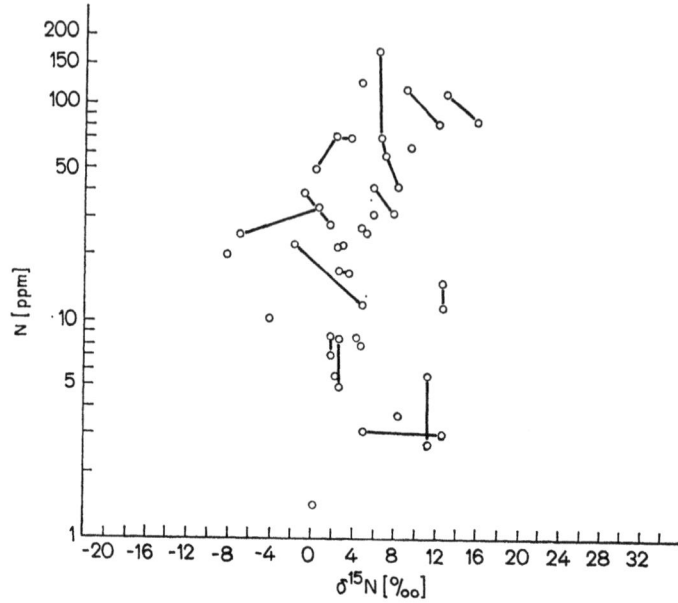

Fig. 7-B-3. Diagram of nitrogen content of minerals and rocks and its isotopic constitution as measured by SCALAN (1959), relative to atmospheric air standard. Values obtained on the same specimen are connected by straight lines

It is not possible to decide from these ratios whether the nitrogen has been derived from plants, soil or other biogenic sources, or was of truly magmatic origin. Both kinds of nitrogen may have high positive $\delta^{15}N$ values.

c) Nitrate Sediments

Nitrogen in caliche shows only small variations in the isotopic composition:

	$\delta^{15}N$, ‰	
Nitrate, Tarapaca, Chile	−2.6	HOERING (1956)
Caliche blanco, Tocopilla, Chile	−2.3	PARWEL et al. (1957)
Caliche colorado, Tocopilla, Chile	−1.2	PARWEL et al. (1957)

It is interesting to note that NO_3^- in rain shows a similar mean $\delta^{15}N$ value of ≈ -2 (HOERING, 1957). The measurements are only few in number, however, and the isotope ratio is not constant. The isotopic constitution of nitrogen contained in rain will be discussed below in Section 7-I.

d) Natural Gases, Oil and Coal

Table 7-B-4 gives the isotopic composition of nitrogen from natural gases, oil and coal. These ratios have been investigated in detail for clues as to the origin of these products.

Nitrogen

BOKHOVEN and THEEUWEN (1966) concluded from the enrichment of ^{15}N in natural gas and mine gas, relative to the associated coal, that the nitrogen could not have been derived from the coal, because a conversion of organic nitrogen to elemental nitrogen should discriminate in favour of the lighter isotope (Table 7-B-4).

Table 7-B-4. *Isotopic composition of nitrogen in natural gases, oil and coal (relative to atmospheric air standard)*

Sample	δ^{15}N [‰]		Reference
	Gas	Liquid	
Natural gas (82% methane, 15% nitrogen), Slochteren borehole, Netherlands	+18		6
Mine gas (98% methane, 1% nitrogen), State Mine Emma, South Limburg, Netherlands	+12		6
Coal, State Mine Maurits, South Limburg, Netherlands	+2		6
Natural gas and petroleum:			
Ella Well	− 8.1	—	1
Plaisted No. 1 Well, Marchand formation, Okla., U.S.A.	− 3.5	—	1
Fletcher No. 10 Well, same formation	− 7.6	+3.8	1
Steve No. 1 Well			
Upper Bradley formation, Okla., U.S.A.	− 8.2	+4.3	1
Lower Bradley formation, Okla., U.S.A.	+ 2.9	—	1
Bitt No. 1 Well, Hart formation, Okla., U.S.A.	−11.5	—	1
Washington Co., Ark., U.S.A.	− 5.9	—	1
Medrano Well No. 13, Medrano formation, Okla., U.S.A.	—	+3.5	1
Garrison No. 3 Well, Rowe formation, Okla., U.S.A.	—	+5.2	1
Shinn No. 5 Well, Bartlesville formation, Okla., U.S.A.	—	+6.3	1
Sylamore formation, Washington Co., Ark., U.S.A.	+ 2.7	—	7
Sells Sands formation, Franklin Co., Ark., U.S.A. range +1.3 to +14.4	—	av. +8.2	7
Peat, Eire	+1.9		1
Peat, Junius, N.Y., U.S.A.	−2.8		1
Lignite coal, Bowman, N.D., U.S.A.	−1.2		1
Bituminous coal, Pittsburgh, Pa., U.S.A.	−0.9		1
Cannel coal, Cannel City, Ky., U.S.A.	+1.6		1
Anthracite coal, Gunnison, Colo., U.S.A.	−1.2		1
Anthracite coal, Lehigh, Pa., U.S.A.	−1.4		1
Wealden coal, Northern Germany range +1.5 to +3.5 (8 samples)	av. +2.8		5
Carboniferous coal, Northwestern Germany range +0.5 to −2.5 (5 samples)	av. −1.0		5
Mine gas, borehole 167 Elbingerode, Germany (80% nitrogen, 20% methane)	−3.3		4
Mine gas, borehole 200b Elbingerode, Germany (65% nitrogen, 35% methane)	−2.2		4

References: 1. HOERING (1955, 1956); 4. PILOT (1963); 5. PARWEL, RYHAGE, and WICKMAN (1957), estimated error ±0.1% (these authors give delta-values for the ^{15}N^{14}N/^{14}N^{14}N ratio in the sample relative to the same ratio in standard air. This gives the same numerical values as the δ^{15}N usually reported, as long as the concentration of ^{15}N≪^{14}N); 6. BOKHOVEN and THEEUWEN (1966); 7. HOERING and MOORE (1958).

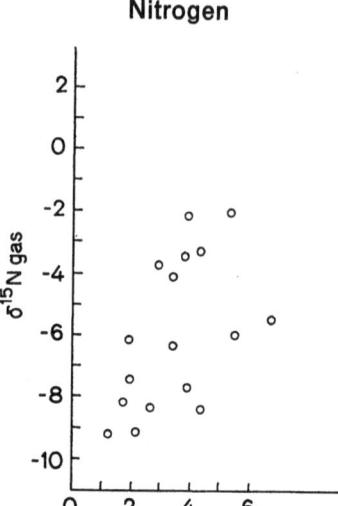

Fig. 7-B-4. Isotopic composition of nitrogen in crude oils and associated natural gases (relative to atmospheric air standard), as measured by HOERING and MOORE (1958)

HOERING and MOORE (1958) made a detailed study of the nitrogen isotopes in natural gases and the associated crude oils. Some of their values are given Table 7-B-4, the others are plotted in Fig. 7-B-4. It shows that the natural gases have negative $\delta^{15}N$ values, whereas the crude oil has positive values. There is also a tendency for low $\delta^{15}N$ values in the gas to be connected with low values in the crude oil. Furthermore the authors found a systematic trend in the $\delta^{15}N$ values across a gas field. They were able to show experimentally that an isotope fractionation can result from flow and migration of nitrogen through a porous structure like a sandstone.

These results were confirmed and expanded by measurements made by EICHMANN et al. (1971) in NW-Germany. They also found negative $\delta^{15}N$ values in oil gas and positive values in the crude oil. For dry natural gases they report a constant isotopic nitrogen composition for different horizons in a gas field, and a systematic trend in both the $\delta^{15}N$ values and the N_2-content from northwest ($\delta^{15}N + 45.7^0/_{00}$, N_2 89.5%) towards southeast ($\delta^{15}N - 8.0^0/_{00}$, N_2 5.0%). A systematic trend in the $\delta^{15}N$ values was also observed in the Staßfurt basin (Germany) by MÜLLER and WIENHOLZ (1968). HOERING and MOORE (1958) favour a genetic relationship between the crude oil and the nitrogen of the associated gases, while EICHMANN et al. (1970) conclude that such a relationship is improbable for the dry natural gases of NW-Germany. For a discussion of the origin of nitrogen in natural gases see Section 7-L.

IV. Biosphere

Relatively little is known about the fractionation of nitrogen isotopes by biological processes.

HOERING (1958) studied the fixation of atmospheric N_2 ($\delta^{15}N$ 0.0) by *Azobacter Vinelandii* and found an average $\delta^{15}N$ value of $-2.2^0/_{00}$ in a total amount of 14.3 mg nitrogen fixed during a period of 19 days. This shows a very small isotope effect with ^{14}N being fixed only 1.0022 ± 0.0010 times faster than ^{15}N.

Nitrogen

The denitrification of NO_3^- and NO_2^-, leading to N_2, was studied by WELLMAN et al. (1968). In experiments with *Pseudomonas stutzeri* they found a fractionation factor of 1.02, indicating that $^{14}NO_3^-$ and $^{14}NO_2^-$ are utilized 2% faster than the ^{15}N-compounds. The kinetic isotope effect resulting from a rupture of the N—O bond yields a theoretical fractionation factor still larger than the value measured (UREY, 1947).

Isotope ratio measurements of biological products are given in Table 7-B-5. The fractionations relative to atmospheric air are small. In plants they occur in both positive and negative directions, while in animal tissue the heavier isotope ^{15}N is favored.

Table 7-B-5. *Isotopic composition of nitrogen in organic material (relative to atmospheric air standard)*

Sample	$\delta^{15}N$ [⁰/₀₀]		Reference
Weeds, local	+4.3		1
Oats, unknown origin	+6.2		1
Seaweed, Tokyo Bay, Japan	+8.1		1
	April	August 1955	
Dandelion, local	−2.8	−0.4	1
White clover leaves, local	−6.5	−8.3	1
Red oak leaves, local	−0.9	−3.3	1
Cedar leaves, local	+1.3		1
American elm leaves, local	+1.9	−3.7	1
Chicken egg, local	+5.8		1
Clam flesh, Atlantic Ocean	+7.3		1
Lamb flesh, unknown origin	+5.0		1
Milk, local	+5.1		1
White rat, local, range: +0.3 to +12.1	av. +6.8		1
Fir, *Picea abies*, Ekolsund Sweden	−3.8		5
Birch, *Betula alba*, Ekolsund Sweden	−2.2		5
same, piece of bark	−5.0		5

References: 1. HOERING (1955, 1956), local = Arkansas, U.S.A.; 5. PARWEL, RYHAGE and WICKMAN (1957).

Revised manuscript received: July 1970

7-C. Abundance in Cosmos, Meteorites, Tektites and Lunar Materials

I. Cosmos

Nitrogen is one of the eight most abundant elements in the solar system, which have cosmic abundance units greater than 10^6 atoms per 10^6 Si atoms. It takes part in the CNO-cycle (VON WEIZSÄCKER, 1938) of nuclear energy production in stars.

The "cosmic" abundance of nitrogen is derived from measurements of the sun, because meteorites are strongly depleted in nitrogen as they are with respect to other volatiles. Table 7-C-1 lists a number of recent measurements of the sun's photosphere and corona. These values are normalized relative to the logarithm of the hydrogen abundance $= 12.00$. SUESS and UREY (1958) in tabulating cosmic abundances used the value of GOLDBERG, MÜLLER and ALLER (1960) from the sun, 6.48, in this case normalized to $\log (Si) = 6.00$, which is equivalent to $\log (H) = 10.50$. Si serves as a basis for comparison of meteoritic and solar abundance data. As Si is not accurately determined in the sun, CAMERON (1967) used the average of 10 non-volatile elements in meteorites relative to the sun to derive abundance figures for the volatile elements (Table 7-C-1).

Table 7-C-1 also gives the range of nitrogen abundances measured in "normal" stars taken from UNSÖLD (1967). These abundances are of the same order of magnitude as those of the sun. There are stars, however, which show a deficiency in nitrogen; M. and C. JASCHEK (1967) report a spectrum from two super-giants in which the N II-lines are extremely weak or absent. Anomalies in the abundance of nitrogen and carbon are also known from the Wolf-Rayet stars (ALLER, 1958).

Nitrogen has also been determined in other cosmic objects. ALLER and LILLER (1959) obtained an abundance value of 8.10 for the diffuse Orion Nebula. In solar cosmic radiation BISWAS and FICHTEL (1965) measured a C/N-ratio of 3.1 and a N/O-ratio of 0.19, which is not very different from the values of 2.8 and 0.17 respectively found in the solar photosphere. In comets the emission bands of CN, NH, NH_2 and NH_2^+ have been observed: these species are probably derived from the stable primary substances NH_4, C_2N_2 and N_2 (see WURM, 1959). In the ice grain theory of interstellar dust, it is assumed that oxygen, nitrogen and carbon combined with hydrogen as H_2O, NH_3 and CH_4, are the main constituents of these grains (GREENBERG, 1963). Recently, CHEUNG et al. (1968) were able to detect ammonia molecules in the interstellar medium in a region near the galactic center, using a radio telescope operating at a wavelength of 1.25 cm. They observed a line emission at a frequency corresponding to inversion transitions in the vibrational ground state of the NH_3 molecule. They estimate the density of NH_3 molecules to be 10^{-3} cm^{-3} for this interstellar gas cloud. With a hydrogen density of 10^3 cm^{-3} and a cosmic nitrogen/hydrogen ratio of 10^{-4}, it follows that about 1% of the total nitrogen here is in the form of NH_3.

Table 7-C-1. *Logarithm of nitrogen atomic abundance in the sun and stars, relative to log hydrogen abundance* $=12.00$, *except for the figures in brackets, which are normalized to log silicon* $=6.00$

Object	Log of abundance	Evidence	Reference
Sun photosphere	7.93 ± 0.10	Molecular band and neutral atomic spectra	Lambert (1968)
Sun photosphere	8.06	Neutral atomic spectra	Müller (1968)
Sun photosphere	7.88	Neutral atomic spectra	Goldberg, Müller and Aller (1960), corrected by Zwaan (1962)
Sun corona	7.85	Ultraviolet resonance lines	Pottasch (1967)
8 "normal" stars (main sequence stars, giants and supergiants)	8.01—9.40	Absorption spectrum	from Unsöld (1967)
Cosmic abundance	7.98 (6.48)	from solar abundance (normalized to log(Si) $=6.00$)	Suess and Urey (1958)
Solar system abundance	(6.38)	(normalized to log(Si) $=6.00$)	Cameron (1968)

II. Solar System

The sun has already been treated as a star in the preceding subsection, and its nitrogen content is listed in Table 7-C-1.

In the chemical evolution of the solar system as outlined by Urey (1952), nitrogen has played an important role. Thermodynamic calculations show that in the solar nebula the stable nitrogen molecule was ammonia, which was able to condense (as NH_4OH) in the region of Jupiter during the formation of the solar planetary system. However, some of it was also incorporated into the terrestrial protoplanets due to local variations in density and temperature. In the solar nebula, ammonia is closely related to water in behavior and in abundance, which is 160,000 for ammonia and 175,000 for water, relative to 10,000 atoms of Si (Urey, 1952). Thus, water and ammonia will be retained on Jupiter and the outer planets, whereas the earth and the other terrestrial planets will be heavily depleted. Condensing ammonia and water may also have served as a cementing agent during the initial aggregation of primordial dust grains, holding them together until a larger body was formed.

During a later heating state of the protoplanets, nitrogen may have been retained as an ammonium salt or, at higher temperatures and lower pressures, as a nitride. Ammonium silicates, particularly ammonium micas, may also have been formed, thus retaining nitrogen at first and later releasing it into the primitive atmosphere (Eugster and Munoz, 1966). This would again be a counterpart to the behavior of water being retained in the form of hydrated silicates.

Table 7-C-2. *Nitrogen compounds present in planetary atmospheres*

Planet	Nitrogen compound present	Amount	Evidence	Reference
Venus	N_2	2—5%, together with noble gases, rest CO_2	Measurements by Venera V and VI	Tass (1969)
	NH_4Cl?	in clouds	Model calculation	Lewis (1968)
Mars	N_2?	0—50%	Model calculations using Mariner occultation data	see Brandt and McElroy (1968)
	N_2	$\leq 5\%$	Ultraviolet spectrum measured by Mariner 6 and 7	Dalgarno and McElroy (1970)
Jupiter	NH_3	230 cm-atm	Calculated under the assumption of cosmic abundance of gases and chemical equilibrium	Greenspan and Owen (1967)
		12 m-atm	Estimate based on comparison of absorption spectrum with laboratory spectra	Owen (1969)
		0.2—5 m-atm		Cruikshank and Binder (1969)
		$\sim$0.05 cm-atm	ultraviolet albedo measurements	Greenspan and Owen (1967)
	HCN CH$_3$NH$_2$	<0.05 m-atm <0.02 m-atm	Absorption spectrum	Cruikshank and Binder (1969)
	NH_4SH NH_4-halides	in clouds	Model calculations	Lewis (1969b)
Saturn	NH_3	Mixing ratio 5×10^{-4} to 5×10^{-3}	Model calculations using radio emission spectra	Gulkis et al. (1969)

Today nitrogen will accumulate mainly in the atmospheres of the planets, due to its high volatility and the low chemical activity of the N_2 molecule. Table 7-C-2 lists the available data on the occurrence of nitrogen in the different planetary atmospheres.

For the planet Mercury a very tenuous atmosphere was deduced from light polarization measurements by Dollfus (1952). It may contain N_2 (Urey, 1959), although owing to the high exosphere temperature and the low escape velocity the main constituent should be radiogenic argon, see Field (1964) and Sagan and Morrison (1968).

In the past N_2 was thought to be the major constituent in the atmospheres of Mars and Venus, by analogy to Earth. For Mars the recent Mariner occultation experiments have led to the calculation of model atmospheres, which contain mainly CO_2 and zero to 50% of N_2 and/or Ar, the total pressure being about 10 mb. The occurrence of NO_2 in the atmosphere of Mars was deduced by KIESS et al. (1960) from absorption features in its spectrum. This result could not be confirmed by MARSHALL (1964), nor by OWEN (1966), who set an upper limit on the NO_2 abundance at 8 μ-atm.

For Venus the Venera V and VI measurements gave 93 to 97% of CO_2 and 2 to 5% of nitrogen together with noble gases (from TASS, 1969[1]; see also VINOGRADOV et al., 1968, for Venera IV results). This agrees well with model atmosphere calculations based on Mariner 5 data by McELROY (1969), who found an upper limit for the N_2 mixing ratio of 10%. If Venus still retains a primordial atmosphere, neon should be more abundant than nitrogen (SUESS, 1964; LEWIS, 1968). The clouds of Venus may contain NH_4Cl (LEWIS, 1968; HUNTEN and GOODY, 1969). For a detailed discussion of the Mars and Venus atmospheres see BRANDT and McELROY (1968), KLIORE et al. (1969) and McELROY (1969).

The giant outer planets Jupiter and Saturn contain large amounts of H_2 and He in their atmospheres. This supports the idea that they may have retained the elemental abundance of the primordial solar nebula. OWEN (1969) gives a H/N ratio of 1.6×10^4 in the Jovian atmosphere compared to 1.2×10^4 in the sun.

The two widely differing NH_3 abundance values in Table 7-C-2 reported by GREENSPAN and OWEN (1967) for the atmosphere of Jupiter can be reconciled, if the existence of clouds of condensed ammonia (WILDT, 1932) is taken into account. This means that the 230 cm-atm are present almost completely in condensed form, only 0.05 cm-atm being volatilized, which is consistent with the vapour pressure of ammonia under the P/T-conditions of Jupiter's atmosphere (GREENSPAN and OWEN, 1967). Detailed calculations by LEWIS (1969a) based on a solar-composition and adiabatic equilibrium model for the formation of clouds on Jupiter, show that they may consist mainly of aqueous ammonia solutions topped by a layer of solid NH_4SH and another layer of solid NH_3. Of the other nitrogen compounds, only those of the halides, NH_4Cl, NH_4Br and NH_4I, can be present in minor amounts under equilibrium conditions (LEWIS, 1969b).

III. Meteorites

a) Nitrogen Minerals

Two nitrogen-bearing minerals have been found in stony meteorites. The first, named osbornite, was discovered by MASKELYNE (1870) in the enstatite achondrite Bustee; BANNISTER (1941) later showed it to be titanium nitride, TiN. It has been observed also in the enstatite chondrites Hvittis and Atlanta by RAMDOHR (1969). The second one is the silicon oxynitride named sinoite, Si_2N_2O, found in several enstatite chondrites (ANDERSEN et al., 1964; KEIL and ANDERSEN, 1964; MASON, 1966). RYALL and MUAN (1969) established an approximate value of 10^{-15} for the oxygen to nitrogen pressure ratio required for the existence of sinoite under

---

[1] Added in proof: see also AVDUEVSKY et al. (1970).

equilibrium conditions at 1,400 to 1,500°C. Both minerals seem to be restricted to the most reduced types of stony meteorites, i.e., the enstatite achondrites and enstatite chondrites. They are not known to occur in terrestrial rocks.

b) Nitrogen Content

1. Carbonaceous Chondrites

Carbonaceous chondrites, which are noteworthy for their high content of water and other volatiles, also contain high amounts of nitrogen (WIIK, 1961; see Table 7-C-3). Some of this is present as organic compounds. MUELLER (1953) extracted 1.12% of organic material from Cold Bokkeveld (Cc2), and found this to contain 3.2% nitrogen, corresponding to 350 ppm N in the meteorite. Similar amounts were reported by BRIGGS (1962).

Recently, HAYATSU (1964) and HAYATSU et al. (1968) identified several organic nitrogen compounds in Orgueil (Cc_1). They found 20 ± 5 ppm adenine, 28 ± 7 ppm ammeline, 20 ± 5 ppm guanine, 270 ± 70 ppm guanylurea and 45 ± 10 ppm melamine. These compounds have no known biochemical significance; they form spontaneously, however, when CO, H_2 and NH_3 are allowed to react in the presence of iron meteorite powder (HAYATSU et al., 1968) supporting the suggestion of STUDIER et al. (1965) that the organic compounds in meteorites are not biogenic, but are formed in the solar nebula by chemical equilibrium processes. In Orgueil these organic compounds contain 205 ppm organic nitrogen, or only about one tenth of the total nitrogen present (Table 7-C-3). The bulk of the nitrogen may be present in other organic compounds, not yet found, as well as in ammonium salts or ammonium associated with the layer lattice silicates.

STUDIER et al. (1965) studied the gases evolved from samples of Orgueil, Cold Bokkeveld and Murray between 100°C and 140°C; of the nitrogen compounds, they found N_2 and NO but no free ammonia.

The presence of amino acids in the ppm range in carbonaceous and ordinary chondrites has been reported by several authors, e.g., KAPLAN et al. (1963), HAYATSU (1963) and VALLENTYNE (1965). The indigenous nature of these compounds has been questioned, however; see ANDERS et al. (1964). HAMILTON (1965) and ORÓ and SKEWES (1965) found human fingerprints to contain a similar distribution of amino acids as the chondrites. A review of this controversy and a comprehensive discussion of the organic compounds in meteorites is given by HAYES (1967); more recent measurements are given by SIMMONS et al. (1969).

2. Ordinary Chondrites

Nitrogen contents determined in ordinary chondrites are shown in Table 7-C-3, the range being similar to that for terrestrial igneous rocks. The (HL)-group chondrites which were measured have higher values, similar to the carbonaceous chondrites with which they were originally grouped as type III carbonaceous chondrites.

In Table 7-C-3 the nitrogen values determined by KÖNIG et al. (1961) and VINOGRADOV et al. (1963) refer to ammonium or nitride only, while the values of LIPMAN (1932) and BUDDHUE (1942) also include organic nitrogen. The agreement of both sets of values shows that most of the nitrogen in ordinary chondrites is in the ammonium (or nitride) form. KÖNIG et al. (1961) determined the nitrate nitrogen

Nitrogen

Table 7-C-3. *Nitrogen content of meteorites. All meteorites not marked as finds are observed falls. Nitrogen values of references (5) and (6) are ammonium and nitride nitrogen only, references (1), (3), and (4) give organic nitrogen also; references (2), (7), (8), and (9) are total nitrogen values including N_2*

Meteorite	N, ppm	Reference
Carbonaceous chondrites:		
Alais Cc1	2,900	1
Orgueil Cc1	2,400	1
Boriskino Cc2	700	1
Erakot Cc2	2,600	1
Al Rais Cc2	1,900	1
Renazzo Cc2	600	1
High iron — low metal (HL)-group chondrites:		
Grosnaja	172	10
Ornans	137	10
Vigarano	128	10
Karoonda	67, 60, 48	2
Mokoia	85, 125, 100	2
High iron (H)-group chondrites:		
Covert (find)	5	3
Morland (find)	16	3
Forest City	52	4
Forest City	75, 30, 46, 42	2
Gilgoin station (find)	48	4
Pultusk	34	4
Pantar	11	5
Zhovtnevyi Khutor	15	6
Gruver (find)	8.4	7
Allegan	25, 35	2
Richardton	48, 58	2
Saline (find)	50, 47, 53	2
Kesen	68, 40, 58	2
Beardsley	60, 69, 49	2
Low iron (L)-group chondrites:		
La Lande (find)	19	3
La Lande (find)	5.6	7
Ness County (find)	11	3
Holbrook	27	3
Holbrook	20	4
Long Island (find)	64	4
Tilden	36	4
L'Aigle	10	5
New Concord	10	5
New Concord	18, 24	2

References and methods: 1. Wiik (1962). Kjeldahl distillation. 2. Moore and Gibson (1969). Vacuum fusion of sample, gas chromatographic determination of nitrogen. 3. Buddhue (1942). Decomposition by caustic soda and gravimetric determination of ammonia. 4. Lipman (1932). Kjeldahl distillation. 5. König et al. (1961). HF-decomposition of sample, distillation, and colorimetric determination of ammonia. 6. Vinogradov et al. (1963). Phosphoric acid decomposition of sample, distillation, and colorimetric determination of ammonia. 7. Scalan (1959). Vacuum fusion of sample, volumetric determination of nitrogen as N_2. 8. Moore et al. (1969). Method as (2). 9. Moore et al. (1970a). Method as (2). 10. Gibson et al. (1971). Method as (2).

Table 7-C-3 (Continued)

Meteorite	N, ppm	Reference
Akaba	9	5
Alfianello	4	5
Mocs	3	5
Elenovka	27	6
Saratov	27	6
Kunashak	38	6
Bruderheim	43, 28, 29, 28	2
Leedey	32, 35	2
Marion (Iowa)	24, 34, 26	2
Farmington	26, 31, 27	2
Ergheo	73, 56, 65	2
Modoc	36, 52, 32	2
Low iron — low metal (LL)-group chondrites:		
Chainpur	57, 54, 55	2
Dhurmsala	42, 40, 42	2
Enstatite chondrites:		
Abee Ce1	295, 260, 264	2
Indarch Ce1	431, 430, 448, 413	2
Kota-Kota Ce1 (find)	293	8
St. Marks Ce1, 2	180	8
Saint Sauveur Ce1, 2	224	8
Atlanta Ce2 (find)	275, 224, 260, 228	2
Blithfield Ce2 (find)	397	8
Daniel's Kuil Ce2	174	8
Hvittis Ce2[a]	660, 650, 600	2
Jajh deh Kot Lalu Ce2[a]	668	8
Kairpur Ce2	60, 40, 59	2
Pillistfer Ce2[a]	783	8
Achondrites:		
Norton County Ae	53	6
Yurtuk Aor	32	6
Yurtuk Aor	61, 50	9
Haraiya Ap	36, 43	9
Pasamonte Ap	34, 45	9
Sioux County Ap	27, 33	9
Stannern Ap	19	6
Pallasites:		
Bragin (find)	16	6
Bragin, olivine fraction	22	6
Iron meteorites:		
Maldyak Om	107	6
Veliko-Nikolaevsky Priisk Og	59	6
Yardymly Og	68	6
Sikhote-Alin Ogg		
0—2 cm below surface, average of 3 dtm.	178	6
2—4 cm below surface	29	6
2—4 cm below surface	53	6
4—22 cm below surface	33	6
4—22 cm below surface	16	6
4—22 cm below surface	19	6

[a] Contains sinoite, Si_2N_2O.

content to be below 0.1 ppm, and LIPMAN (1932) and BUDDHUE (1942) also found no nitrate, nitride or cyanide. KÖNIG et al. (1961) found that only <1 ppm of the ammonium nitrogen could be extracted with water, in contrast to that of most terrestrial igneous rocks (see Section 7-E).

Another series of nitrogen determinations was made by MOORE and GIBSON (1968) by vacuum fusion and gas-chromatography; the results yielded total nitrogen values including elemental nitrogen. They are about a factor of two higher than the previous figures (Table 7-C-4), indicating that half of the nitrogen is probably present as N_2.

It is noteworthy that enstatite chondrites contain high amounts of nitrogen, similar to carbonaceous chondrites. Enstatite chondrites of type I are known to also have high contents of other volatile elements (ANDERS, 1964). The average nitrogen content in the type II enstatite chondrites measured is as high as in type I, however. The highest amounts are found in the meteorites containing sinoite, Si_2N_2O.

Table 7-C-4 gives the average nitrogen contents of the chondrites listed in Table 7-C-3. There exist no significant differences between finds and falls or between high iron- and low iron-group chondrites.

3. Achondrites

The achondrites measured by VINOGRADOV et al. (1963), and by MOORE et al. (1970), Table 7-C-3, have nitrogen contents similar to the chondrites.

4. Iron Meteorites

Table 7-C-3 gives the nitrogen contents of the iron meteorites measured by VINOGRADOV et al. (1963). Sikhote-Alin shows a very high content in a surface

Table 7-C-4. *Average nitrogen content of the stony meteorites listed in Table 7-C-3, and an additional achondrite average given by* MOORE (1969), *ref. 9, other references as in Table 7-C-3. Values by* MOORE *and* GIBSON (1969), *ref. 2, are averaged separately, because they include elemental* N_2

Meteorite class	Number analyzed	Nitrogen, ppm range	mean	Reference
Carbonaceous chondrites	6	600—2,900	1,850	1
Enstatite chondrites				
containing sinoite	3	600— 783	708	2, 8
without sinoite	6	40— 448	220	2, 8
(HL)-group chondrites	5	48— 172	114	2, 10
(H)-group chondrites	8	5— 52	24	3, 4, 5, 6, 7
	6	25— 75	50	2
(L)-group chondrites	15	3— 64	21	3, 4, 5, 6, 7
	6	24— 73	35	2
(LL)-group chondrites	2	40— 57	48	2
Achondrites	3	19— 53	35	6
	4	33— 61	41	9
Falls, (H)- and (L)-group	15	3— 52	22	3, 4, 5, 6, 7
Finds, (H)- and (L)-group	8	6— 64	23	3, 4, 5, 6, 7

Table 7-C-5. *Total nitrogen contents including N_2 in iron meteorites, from* GIBSON *and* MOORE (1971)

Chemical group	Structural class	Number analyzed	Nitrogen, ppm range	median
I	Off-Ogg	17	2—131	35
II a	H	11	4— 26	15
II b	Ogg-Oge	5	16— 27	18
II c	Opl	3	11— 16	14
II d	Of-Om	2	32— 44	38
III a	Om-Og	33	2— 80	25
III b	Of-Om	7	22— 70	46
IV a	Of	13	2— 34	4
IV b	D	4	2— 9	2
Median value		123		18

sample compared to samples more than 2 cm away from the surface. This suggests some external source of the nitrogen; the authors state, however, that the results need confirmation by further measurements.

GIBSON and MOORE's (1971) results (Table 7-C-5) are consistent with the measurements of VINOGRADOV et al. (1963). The iron meteorites display a large range of nitrogen contents, probably owing to the random occurrence of the recently discovered chromium nitride (SCOTT, 1970). There is no significant difference between dense irons, such as hexahedrites and ataxites, and irons having a Widmanstätten structure. A difference might be expected if the nitrogen had come from an external source, the Widmanstätten pattern offering a pathway for easier nitrogen diffusion and retention.

The solubility of nitrogen in a liquid Fe-Ni alloy with 10% Ni is 350 ppm N by weight, measured at 1,600°C and one atmosphere nitrogen pressure (BLOSSEY and PEHLKE, 1966), that is, about ten times higher than the average concentration found in iron meteorites.

IV. Tektites

MÜLLER and GENTNER (1968) determined the gases contained in bubbles of tektites and other natural glasses. They found nitrogen, oxygen and traces of carbon dioxide, the N_2/O_2 ratio being similar to that of atmospheric air. As other bubbles in the same tektites contained no gases (see also SUESS, 1951), inside diffusion of air during geologic time can be excluded. The gases found were therefore probably trapped during tektite formation. Their composition favours a terrestrial origin of tektites (MÜLLER and GENTNER, 1968). MÜLLER (1969) also determined the chemically bound nitrogen in a number of tektites and these results are shown in Table 7-C-6.

V. Lunar Materials

Nitrogen determinations on lunar material from the Apollo 11 landing site at Mare Tranquillitatis are listed in Table 7-C-7. The total nitrogen values of MOORE et al. (1970b) show no significant difference between fines, breccias or basaltic

Table 7-C-6. *Chemically bound nitrogen in tektites and other natural glasses, from* MÜLLER (1969)

	Number analyzed	Nitrogen, ppm range	mean
Indochinites	12	2—12	7
Australites	4	5—10	7
Philippinites	1		5
Ivory Coast tektites	3	6—14	9
Moldavites	2	3— 8	5
Bediasites	1		10
Bosumtwi crater glass	3	10—31	19
Ries Kessel glass	14	3—41	26

rocks. HINTENBERGER et al. (1970) also found about the same amount of N_2 in two breccias and a basaltic rock. There was a marked increase in ammonium and nitride nitrogen, however, in a sample of fines compared to a basaltic rock sample (see Table 7-C-7). A dissolution technique showed that this combined nitrogen is not water-soluble and seems to be contained in the surface layer of the mineral grains; this points to an origin by solar wind implantation (HINTENBERGER et al., 1970).

Table 7-C-7. *Nitrogen content of lunar material from the Apollo 11 landing site*

Sample No.	Rock type	Nitrogen compound	ppm	Reference
10086-A	fines	total nitrogen	102 ± 4	MOORE et al. (1970)
10086-B	fines	total nitrogen	153 ± 4	MOORE et al. (1970)
10002-54	C, breccia	total nitrogen	125 ± 4	MOORE et al. (1970)
10044-37	CB, coarse breccia	total nitrogen	98 ± 4	MOORE et al. (1970)
10049-23	A, basalt	total nitrogen	116 ± 4	MOORE et al. (1970)
10050-33	B, basalt	total nitrogen	30 ± 2	MOORE et al. (1970)
10021-20	C, breccia	N_2, heat extracted	130	HINTENBERGER et al. (1970)
10061-11	C, breccia	N_2, heat extracted	120	HINTENBERGER et al. (1970)
10057-40	A, basalt	ammonium and nitride	≤ 15	HINTENBERGER et al. (1970)
10084-18	fines	ammonium and nitride	80	HINTENBERGER et al. (1970)

Revised manuscript received: July 1970

7-D. Abundance in Rock-Forming Minerals

I. Introduction

The occurrence of nitrogen in the rocks and minerals of the earth's crust was already investigated early in the last century by several authors. One of the first phenomena studied was the sublimates of sal ammoniac found on fresh volcanic rocks. The origin of this ammonia has been widely discussed, with some authors ascribing it to nitrogen from burnt vegetation, e.g., BUNSEN (1853) and GOLDSCHMIDT (1954), whereas others held that the nitrogen was indigeneous to the magma, e.g., SARTORIUS VON WALTERSHAUSEN (1847) and VINOGRADOV et al. (1963). Today there can be little doubt that the magma itself contains nitrogen, as ammonia is present in the fumes of volcanic eruptions (CADLE and FRANK, 1968) and in magmatic minerals.

Analytical measurements of nitrogen in minerals and rocks were also made sometime ago by BRACANNOT (1838), ERDMANN (1896), GAUTIER (1901, 1906), and BRUN (1911). They found nitrogen in almost all rocks and minerals investigated in amounts varying between 10 and 200 ppm. Special reference should be made to the work of DELESSE (1860), who discovered the high nitrogen content of layered rocks (i.e. sediments) in contrast to massive ones (i.e. magmatic rocks). The first modern measurements of nitrogen in minerals and rocks were made by RAYLEIGH (1939). The results indicated values of 27 to 67 ppm total nitrogen in 13 igneous rocks (average 45 ppm) and showed that the nitrogen was mainly present in the rocks as ammonium or an ammonium-yielding compound. In more recent times four systematic studies have been published on which most of our knowledge of the geochemistry of nitrogen is based. These are the papers by SCALAN (1959), STEVENSON (1959, 1962), WLOTZKA (1961), and VINOGRADOV, FLORENSKII and VOLYNETS (1963).

II. Nitrogen Compounds

In the minerals of igneous rocks the main source of nitrogen seems to be present as ammonium. STEVENSON (1962) proposed to follow the convention of soil scientists and distinguished between "fixed" ammonium, which cannot be extracted by 1N KCl, and extractable ammonium. The latter may have existed as free ammonium salts or on broken edges of silicate minerals; it may also be a late contaminant introduced either by biological activity or in the laboratory. In a similar approach WLOTZKA (1961) determined the amount of ammonium extracted by shaking the powdered sample with 0.01N HCl. STEVENSON (1962) found the fixed nitrogen to be usually 80—90% of the total nitrogen of a given sample. WLOTZKA's determinations essentially confirmed these results, but furthermore indicated a difference between the average values for intrusive and volcanic rocks (see Section 7-E).

The radius of the NH_4^+ ion is 1.43 Å, similar to the radius of K^+, 1.33 Å (GOLDSCHMIDT, 1954). Hence, it may be assumed that ammonium can replace potassium in the lattice of silicate minerals. Experiments to check this possibility were performed by SCHNEIDER and CLARKE (1893), CLARKE and STEIGER (1902), and SCHACHTSCHABEL (1940). They observed a replacement of K^+ by NH_4^+ to a certain degree in feldspars treated with ammonium chloride gas or solution. The same replacement has been assumed by RANKAMA and SAHAMA (1950), SCALAN (1959), WLOTZKA (1961), and STEVENSON (1962) to occur in natural feldspar (see also MARSHALL, 1962). VOLYNETS and FRIDMAN (1965) determined nitrogen in a number of alkali rocks from the Kola peninsula and found a weak correlation, which however is statistically not significant, between nitrogen and potassium content in six rocks of the Lovozero Massif.

BARKER (1964) was able to synthesize an ammonium feldspar under hydrothermal conditions (550 to 650°C, 2,000 bars) out of alkali feldspars and aqueous NH_4Cl solutions. Here, ammonium substitutes for Na^+ and K^+ in the alkali feldspar structure, which is slightly expanded. At the same time, this feldspar takes up zeolitic water, so that the ammonium end member has the formula $NH_4AlSi_3O_8 \cdot 1/2H_2O$. This mineral has been found in nature by ERD et al. (1964) and was named buddingtonite. It occurs in pseudomorphs after plagioclase in andesitic rocks which have been hydrothermally altered by hot-spring waters high in ammonia.

Ammonium also enters the structure of layer lattice silicates. VEDDER (1964) and YAMAMOTO and NAKAHIRA (1966) have shown that NH_4^+ can replace K^+ in the interlayer of muscovites and sericites respectively; the ammonium ion replacement was established by the infrared absorption spectrum.

It is well known that ammonium takes part in the ion exchange processes occuring on clay minerals, see for example MORTLAND (1958). Ammonium penetrates into the interlayer space of montmorillonite and may be fixed there by chemisorption or physical adsorption. JAMES and HARWARD (1963) and RUSSELL (1965) have discussed and determined the different mechanisms for this process, which is important for the geochemical behavior of nitrogen in soils and sediments.

Another form for retaining nitrogen in minerals and rocks would be the nitride. Although the existence of nitrides in rocks has never been proven, GAUTIER (1901) and later geologists have assumed their occurrence. Most metal nitrides are not stable under magmatic conditions, nor under the high water and oxygen pressures of the earth's surface. Recently, it has been shown by MULFINGER and MEYER (1963) and KELEN and MULFINGER (1968) that glass melts take up appreciable amounts of nitrogen (upto 0.8%) when treated with atomic N or ammonia under reducing conditions. The same is known from silicate slags in the presence of reducing carbon and a N_2 gas phase (OELSEN et al., 1969). The nitrogen introduced in this way forms $-NH_2$ or $=NH$ groups which are chemically bound to the silicon atoms of the glass, replacing oxygen. In the case of low hydrogen or water vapour pressures, nitrogen may be bound in the form of nitride N_3^- ions to silicon. These results may be applied to the nitrogen content of natural minerals, see MEYER (1965).

WLOTZKA (1961) determined the amount of combined nitrogen in a number of co-existing minerals of different rocks, see Table 7-D-1. The results show that each mineral displays a wide range of nitrogen contents. In a given rock, however, quartz has usually the lowest, feldspar a medium, and biotite the highest content.

Table 7-D-1. *Ammonium and elemental nitrogen determined in co-existing minerals of rocks.*

NH$_3$-nitrogen from WLOTZKA (1961), method: see caption to Table 7-D-3. N$_2$-nitrogen from GOGUEL (1963), method: gas extraction by grinding in vacuum and volumetric determination of N$_2$

Rock		Nitrogen content in ppm					
		quartz	ortho-clase	plagio-clase	biotite		total rock
Granite, Wurmberg, Harz (Germany)	NH$_3$	24	25	48^a	112		35
Granite, Grimstad (Norway)	NH$_3$ N$_2$	10 1.0	12 1.5		24 1.9		12 1.4
Granite, Schluchsee (Germany)	NH$_3$ N$_2$	7 5.0	10 5.1	19 10	22 0.5		12 6.3
Granite, Weinsberg (Austria)	NH$_3$ N$_2$	2 1.0	40 2.5		34 1.0		30 1.9
Granite, Andlau I, Vosges (France)	NH$_3$ N$_2$	5 4.7	3 4.9	3 2.6	10 1.5		5 3.4
Granite, Andlau II, Vosges (France)	N$_2$	5.0	5.0	3.1	1.3		
Granite, Grafversfors-bolag (Sweden)	NH$_3$	0	16	25	30		15
Granite, Malsburg, Black Forest (Germany)	N$_2$	6.2	7.6	4.1	1.3		
Granodiorite, Aber Ildut (France)	NH$_3$	0	7	3	10		5
Granodiorite, Schönbach (Lusatia)	NH$_3$	24	19		53		27
Diorite, Neuntelstein (France)	NH$_3$	—	—	30	16	hornblende 12	17
Pegmatite, Malsburg (Germany)	NH$_3$ N$_2$	5 7.5	13 11.5	19 n.d.			
Pegmatite, Spessart (Germany)	NH$_3$	22	125	—	muscovite 44		—
Pegmatite, Schluchsee (Germany)	N$_2$	4.2	6.0	—	muscovite 1.4		—
Pegmatite, Landsverk (Norway)	N$_2$	1.7	0.6	1.5	amazonite 1.7		—
Pegmatite, unknown origin	NH$_3$	27	57	—	tourmaline 60		—
Basalt, Dransberg (Germany)	NH$_3$	—	—	20	pyroxene 11	olivine 0	13

^a Altered by sericitization.

Table 7-D-2. *Nitrogen content of rock-forming minerals*

Mineral	N, ppm	Reference
Quartz from pegmatite, Spittal, Austria	21	1
Adularia, Swiss Alps	17	1
Orthoclase, Tröstau, Fichtelgebirge, Germany	12	1
K-Feldspar, granite, Lengenfeld, Saxony	15	1
K-Feldspar, pegmatite, Caracol, Brazil	34	1
Sanidin, Serra d'Agua de Pan, the Azores	9	1
Sanidin, trachyte, Drachenfels, Germany	19	1
Labradorite, Auelandsfjord, Norway	12	1
Orthoclase, Ft. Bayard, N. Mex., U.S.A.	38.6	2a
Orthoclase, Ft. Bayard, N. Mex., U.S.A.	28.3	2a
Orthoclase, Ft. Bayard, N. Mex., U.S.A.	13.0	2a
Orthoclase, Ft. Bayard, N. Mex., U.S.A.	23.0	2d
Oligoclase, Mitchell Co., N.C., U.S.A.	15.0	2d
Microcline, Keystone, S. Dak., U.S.A.	63.9	2b
Orthoclase A., Illinois, U.S.A.	41	3
Orthoclase B, New Mexico, U.S.A.	20	3
Feldspar, Laven, Norway	19	1
Feldspar, granite from Wildenau, Saxony	2	1
Feldspar, Birkenfeld, Germany (40% feldspar, 40% quartz, 10% kaolinite)	20	5
Feldspar, Waidhaus, Germany	150	5
Feldspar, Amberg, Germany (74% feldspar, 12% quartz)	140	5
Plagioclase, from granodiorite, Opdal, Norway	23	1
Hornblende, black, granodiorite, Opdal, Norway	17	1
Hornblende, green, granodiorite, Opdal, Norway	42	1
Hornblende, Kragerö, Norway	15	1
Hornblende, Verona, Ontario, Canada	3.9	2b
Nepheline, Langesundfjord, Norway	18	1
Diallag, Crete	8	1
Bronzite, Kupferberg, Bavaria, Germany	20	1
Bronzite, Kraubath, Austria	9	1
Augite, Firmenich, Eifel, Germany	13	1
Spodumene, Keystone, S. Dak., U.S.A.	22.2	2b
Diopside, Rocca Nera, Piemont, Italy	17	1
Olivine, Hirzstein, Kassel, Germany	25	1
Epidot, Salzburg, Austria	3	1
Prehnite from spilite, Eckenberg, Germany	5	1
Calcite from spilite, Eckenberg, Germany	12	1
Ilmenite, unknown origin	13	1
Chromite, Ragusa, Italy	12	1
Biotite, Miask, U.S.S.R.	58	1
Biotite, Oslo, Norway	27	1
Biotite, Bodenmais, Bavaria, Germany	19	1
Biotite, Arendal, Norway	10	1

References: 1. WLOTZKA (1961); 2. SCALAN (1959); 3. STEVENSON (1962); 4. VINOGRADOV, FLORENSKII and VOLYNETS (1963); 5. MEYER (1965); 6. MÜLLER (1969). Method used in references 1, 2c, 2d, 3, 4a, 5, and 6: decomposition of sample by acids or NaOH-fusion, distillation of ammonia and determination by colorimetry or titration. This method gives ammonia and nitride nitrogen. Methods used in references 2a, 2b, 2e, 4b, and 7: vacuum fusion of sample, volumetric determination of nitrogen after conversion to N_2. This method gives total nitrogen including N_2.

Nitrogen

Table 7-D-2 (Continued)

Mineral	N, ppm	Reference
Biotite, Deep Creek, N.C., U.S.A.; average of 4 samples	61 (range 50—70 ppm)	2b
Biotite from granodiorite, Kaavi, Finland	94	2d
Biotite from granodiorite, Kalajoki, Finland	26.5	2d
Biotite, Canada	36	3
Biotite from granite, Ukraine, U.S.S.R., No. 1052/1	129	4
Biotite from pegmatite, Durulgui, Transbaikalia, U.S.S.R.	266	4
Muscovite, Bamle, Norway	25	1
Muscovite, Lenkersmühle, Oberpfalz, Germany	66	1
Muscovite, Spittal, Carinthia, Austria	105	1
Muscovite, Beryl Mt., N.H., U.S.A.	4	2b
Muscovite, India	107	3
Muscovite from pegmatite, Durulgui, Transbaikalia, U.S.S.R.	130	4
Lepidolite, Keystone, S. Dak., U.S.A. (mean of 9 dtm.)	133 (range 42 to 174 ppm)	2
Lepidolite, Gunnison, Colo., U.S.A.	8	2b
Lepidolite, South Rhodesia, Africa	13	2b
Fuchsite, Rutland, Vt., U.S.A.	80.3	2c
Fuchsite, Rutland, Vt., U.S.A.	22.1	2d
Fuchsite, Rutland, Vt., U.S.A.	4.8	2d
Pyrophyllite, Scania, Sweden	41	1
Chlorite, Chester Vt., U.S.A.	40	1
Talc, Zillertal, Austria	22	1
Topaz, Amelia Court House, Va., U.S.A.	17.3	2b
Tourmaline, San Diego Co., Calif., U.S.A.	11.9	2b
Tourmaline, unknown origin	13.4	2b
Pegmatite, Mitchell Co., N.C., U.S.A.	1.1	2a
Pegmatite, Mitchell Co., N.C., U.S.A.	6.1	2a
Rutile, pegmatite, Nelson Co., Va., U.S.A.	3.7	2a

Table 7-D-3. *Average nitrogen content of minerals listed in Tables 7-D-1 and 7-D-2*

Mineral	Number analyzed	N, ppm range	average
Pyroxene	6	9— 20	11
Quartz	12	0— 27	13
Hornblende	5	4— 42	18
Plagioclase	12	3— 48	22
Potassium feldspar	20	3—125	23
Biotite	19	10—266	55
Muscovite	7	25—130	68

Table 7-D-2 gives the nitrogen content of minerals from various sources. They display a range of nitrogen contents similar to that found above in co-existing minerals. The highest amounts are found in feldspars and micas; the same mineral species may have very low contents, however, indicating the importance of nitrogen availability during their formation.

The mean nitrogen contents of the rock-forming minerals listed in Tables 7-D-1 and 7-D-2, are given in Table 7-D-3. The spread of values is not large, but the general tendency, of quartz having low values and biotite high values, is still present.

III. Elemental Nitrogen

Gas or gas-liquid inclusions in minerals and rocks usually contain elemental nitrogen (CHAMBERLIN, 1908; RAYLEIGH, 1939; ELINSON, 1956; GOGUEL, 1963; CHAIGNEAU and MARINELLI, 1964; KRANZ, 1967). The amount of N_2 expressed as ppm by weight is of the same order of magnitude as the content of ammonium nitrogen. CHAIGNEAU (1964) found in the minerals of the magnetite ore body of Calamita, Italy, a nearly constant amount of 11 to 34 ppm N_2, the mean being 22 ppm. ELINSON (1956) reported large amounts (upto 1,200 ppm) of N_2 in quartz and other mineral samples, but GOGUEL (1963) has pointed out that the nitrogen may have been introduced from the grinding vessel used for grinding the minerals.

GOGUEL (1963) determined N_2 in a number of co-existing minerals out of the same rocks for which WLOTZKA (1961) determined NH_3. Thus, a direct comparison of these two forms of nitrogen is possible (Table 7-D-1). The results show that the N_2 content is usually lower than the NH_3 content. The distribution of N_2 between the rock-forming minerals is different from the distribution of ammonium nitrogen: N_2 is low in the first minerals to crystallize out of the magma, like biotite, and higher in the later ones, like quartz and orthoclase (GOGUEL, 1963). This tendency is also shown by the high content of N_2 in minerals of the Malsburg pegmatite as compared to their NH_3 content. DOLGOV and SHUGUROVA (1966) have found that in gas inclusions of late postmagmatic origin, N_2 is enriched relative to gases like H_2S and NH_3, in contrast to inclusions formed at earlier stages at higher temperatures.

IV. Nitrogen in Beryl and Cordierite

In connection with the work of DAMON and KULP (1958), who found large amounts of ^4He and ^{40}Ar in beryls and cordierites, SCALAN (1959) determined nitrogen in a series of these minerals. He found an average of 94 ppm N for 32 beryl samples (range 25 to 238 ppm) and 125 and 720 ppm N in a cordierite and a cordierite granulite respectively (see Table 7-B-1). These nitrogen contents are high compared to those for other minerals, but not as spectacular as the noble gas contents. An increase of nitrogen with the age of the beryls, as is the case with helium and argon, was not observed.

Revised manuscript received: July 1970

7-E. Abundance in Common Igneous Rock Types
I. Intrusive Rocks

The history of nitrogen determinations in rocks, especially the work of RAYLEIGH (1939), has been reviewed in Section D.

Table 7-E-1 shows the nitrogen content of a number of intrusive rocks which were determined by SCALAN (1959), WLOTZKA (1961), STEVENSON (1962), and VINOGRADOV, FLORENSKII and VOLYNETS (1963). It is apparent that nearly all intrusive rocks contain nitrogen, the amount varying from 2 to about 50 ppm by weight. The range of nitrogen contents and the mean values are given in Table 7-E-2. The mean values reported by various authors for granites do not vary much, except that SCALAN's figures are lower by a factor of 3 to 4; the same is true for his mean gabbro values. This may be due to the different analytical technique employed, i.e., vacuum fusion of the rock with or without flux and conversion of all nitrogen compounds to N_2, whereas the other authors used acid decomposition of the sample and determination of ammonia by chemical methods. SCALAN (1959) found the fusion method to have the advantage of avoiding fractionation of the nitrogen isotopes during extraction. VINOGRADOV et al. (1963), on the other hand, found that acid decomposition gave higher nitrogen yields.

The average of all 54 granites and granodiorites measured is 21 ppm and the average of 24 gabbros and diorites 11 ppm nitrogen. This lower value in the more basic rocks can be expected if ammonium is related to potassium or the alkalies in general, in the minerals of igneous rocks, as discussed in Section D. From the large spread of values, in the granites from 1.4 to 66 ppm and in the gabbros and diorites from 1.8 to 26 ppm, it is evident, however, that the more important factor is nitrogen availability during formation of the rock.

This fact is shown also by the exceptional high nitrogen amounts of 66 ppm in the lithionite-granite from Eibenstock, Saxony. This granite has been altered by a late autometamorphic process with participation of volatiles (TEUSCHER, 1936); it contains, for instance, 1.7% F (KORITNIG, 1951).

The average of all 103 intrusive rocks measured is 16 ppm of total combined nitrogen. The same average was found for 80 rocks in which fixed and total combined nitrogen were determined separately, with the average amount of fixed nitrogen in these rocks being 12 ppm or 75% of the total nitrogen.

Two series of granite samples from the same quarry were analysed by WLOTZKA (1961) to test the constancy of nitrogen content. In six samples of the Wurmberg granite the range was 9 to 44 ppm nitrogen (average 24 ppm) and in four samples of the Königskopf granite was 4 to 31 ppm (average 17 ppm). Fig. 7-E-1 shows these data plotted against the distance of the samples from the granite sediment contact. The nitrogen content increases towards the contact, probably owing to an enrichment of the volatiles in the outer parts of the magmatic body. The derivation

Nitrogen

Table 7-E-1. *Nitrogen content of intrusive rocks. Total combined N is ammonium and nitride nitrogen, except that figures from references 2a, 2b and 4b include also elemental nitrogen. Fixed N refers to the amount remaining after treatment of the sample with 0.01 N HCl (reference 1), or with 1 N KCl (reference 3), see subsection 7-D-II*

Rock	Total combined N ppm	Fixed N, ppm	Reference
Granite, G1 U.S. Geol. Survey standard	8	6	1
Granite, G1 U.S. Geol. Survey standard	2.4		6
Granite, Wurmberg, Harz, Germany	24	14	1
Granite, Königskopf, Harz, Germany	17	7	1
Granite, Schierke, Harz, Germany	8	8	1
Granite, Okertal, Harz, Germany	11	6	1
Granite, Hammerhus, Bornholm, Denmark	6	0	1
Granite, Eklara, Guajarat, India	22	20	1
Granite, Carlsbad, Czekoslovakia	24	17	1
Granite, Aaregranite, Gurtnellen, Switzerland	4	0	1
Granite, Mauthausen, Austria	27	23	1
Granite, Teufelsberg, Odenwald, Germany	16	8	1
Granite, Schluchsee, Germany	22	10	1
Granite, Grimstad, Norway	12	10	1
Granite, Grafversforsbolag, Sweden	13	10	1
Granite, porphyritic, Strehlen, Silesia	34	9	1
Granite, Graniteville, Mo., U.S.A.	2.7		2b
Granite, Graniteville, Mo., U.S.A.	5.6		2b
Granite, Westerly, R.I., U.S.A.	12.5		2b
Granite, Westerly, R.I., U.S.A.	1.4		2e
Granite, St. Cloud, Minn., U.S.A.	2.6		2a
Alkali granite, Quincy Mass., U.S.A.	8.8		2d
Granite, Desert Island, Maine, U.S.A.	34	19	3
Granite, Graniteville, Vt., U.S.A.	28	25	3
Granite, Vermilion Co., Ill., U.S.A.	25	21	3
Granite, composite of 13 German granites	28	18	1
Lithionite granite, Eibenstock, Germany	66	27	1
Muscovite biotite granite, Stone Mt., Ga., U.S.A.:			
unweathered	17	11	3
weathered	57	44	3
Granite, Rapakivi granite, Karelia U.S.S.R.	23		4a
Granite, Olga Bay, Sea of Japan	33		4a
Granite, Olga Bay, Sea of Japan	7		4b
Granite, Karelia, No. 787 U.S.S.R.	13		4a
Granite, Karelia, No. 787 U.S.S.R.	7		4b
Granite, Karelia, No. 1135 U.S.S.R.	9		4b
Granite, Karelia, No. 2197 U.S.S.R.	25		4a
Granite, Boguslavka, Ukraine, No. 754 U.S.S.R.	27		4a
Granite, Abkhazia, Caucasus, No. 678 U.S.S.R.	29		4a
Granite, Ukraine, No. 1052/1 U.S.S.R.	26		4a
Microcline granite, Chekan, S. Urals, depth 1,706 m U.S.S.R.	17		4b
Granitic gneiss, Karelia, No. 44/2 U.S.S.R.	22		4a
Granitic gneiss, Perm (Nuraminski granite) U.S.S.R.	51		4a
Two mica granite, Durulgui, Transbaikalia U.S.S.R.	13		4b

References: 1. Wlotzka (1961); 2. Scalan (1959); 3. Stevenson (1962); 4. Vinogradov, Florensky and Volynets (1963); 6. Müller (1969). For methods used see legend to Table 7-D-2.

Table 7-E-1 (Continued)

Rock	Total combined N ppm	Fixed N, ppm	Reference
Granodiorite, Andlau, Vosges, France	11	9	1
Granodiorite, Aber Ildut, Brest, France	6	4	1
Granodiorite, Schönbach, Lusatia	30	26	1
Granodiorite, Ulsberg, Norway	18	9	1
Granodiorite, Opdal, Norway	17	15	1
Quartz diorite, Adamello, Italy	25	15	1
Quartz diorite, Rossbach, Germany	31	22	1
Syenite, Zischewitz, Dresden, Germany	33	17	1
Nepheline syenite, composite of 22 rocks	9	5	1
Diorite, Leuthenberg, Bavaria, Germany	17	15	1
Diorite, Neuntelstein, Vosges, France	17	13	1
Gabbro, Harzburg, Germany	26	26	1
Gabbro, composite of 11 German gabbros	11	11	1
Gabbro, Vermilion Co., Ill., U.S.A.	21	17	3
Hypersthene gabbro, Sudbury, Ontario Canada	13.4		2e
Hypersthene gabbro, Sudbury, Ontario Canada	2.4		2e
Hypersthene gabbro, Essex Co., N.Y., U.S.A.	1.8		2e
Olivine gabbro, Cripple Creek, Colo., U.S.A.	5.3		2b
Olivine gabbro, Cripple Creek, Colo., U.S.A.	8.2		2c
Olivine gabbro, Cripple Creek, Colo., U.S.A.	3.0		2e
Olivine gabbro, Orybashevo, Urals, U.S.S.R. depth 2,030 m, No. 22/1	25		4b
Hornblende gabbro, Cheremshan, Urals, U.S.S.R., depth 1,818 m, No. 2—59a	4		4b

Table 7-E-2. *Mean nitrogen content of the intrusive rocks listed in Table 7-E-1*

Rock (number of samples analyzed), Reference	Total combined N, ppm		Fixed N, ppm	
	range	average	range	average
Granites (6), SCALAN (1959)	1.4—12.5	6		
Granites (7), STEVENSON (1962)	17 —57	30	11—44	23
Granites (8), VINOGRADOV et al. (1963)	7 —51	25		
Granites (28), WLOTZKA (1961)	4 —66	22	0—27	14
Granodiorites (5), WLOTZKA (1961)	6 —30	16	4—26	13
All granites and granodiorites (54)		21		
Gabbros (7), SCALAN (1959)	1.8—13.4	5		
Diorites and Gabbros (14), WLOTZKA (1961)	11 —26	13	11—26	12
All gabbros and diorites (24)		11		
All intrusive rocks (103)		16		
All intrusive rocks, where total and fixed N was determined (80)		16		12

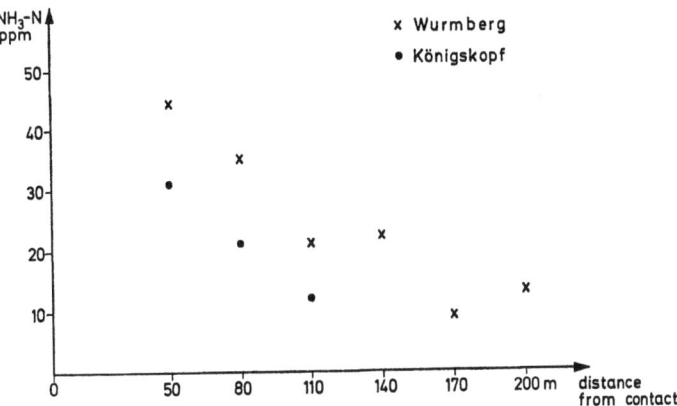

Fig. 7-E-1. Variation of nitrogen content of granite samples from Königskopf and Wurmberg, Harz, with distance from the contact between granite and sediments, from WLOTZKA (1961)

of the higher amounts of nitrogen out of the surrounding sediments seems less probable, because the nitrogen-enriched zone is still 50 to 80 m away from the contact. These rather high variations in nitrogen in the same magmatic body again show the dominant role of nitrogen availability for the nitrogen content and the minor role of the bulk rock chemistry.

II. Volcanic Rocks

The volcanic rocks listed in Table 7-E-3 show approximately the same spread in nitrogen content as the intrusive rocks. There are, however, in contrast to the intrusive rocks, quite a few which have high amounts (larger than 100 ppm) lying outside the general distribution (Fig. 7-E-2). Examples are the rhyolite from Eisenbach, Hungary, (119 ppm N), and the trachyte from Drachenfels, Germany, (102 ppm N). Both nitrogen determinations were checked with freshly ground rock samples and the same high amounts were found. An explanation may be incomplete degassing of the magma due to rapid cooling. Incomplete degassing was inferred by KORITNIG (1951) from the fluorine distribution in these two rocks; no F-bearing mineral was found for 80% of the fluorine, but numerous small inclusions in the matrix of the rocks were observed.

The andesite (porphyrite) of St. Andreasberg, Germany, also has a high nitrogen content, 170 ppm. This rock occurs concordantly in shales and has been hydrothermally altered (MATTIAT, 1958). The nitrogen may be derived from the sediments or from the late magmatic hydrothermal supply.

Alkali olivine basalts (22 ppm N) have a lower average nitrogen content than tholeiitic basalts (33 ppm), contrary to their higher content of potassium and alkalies. The statistics are too poor, however, to draw any conclusions from this negative correlation.

The average nitrogen content of all volcanic rocks measured (Table 7-E-4) is 37 ppm; i.e., it is more than twice as high as the 16 ppm N of intrusive rocks.

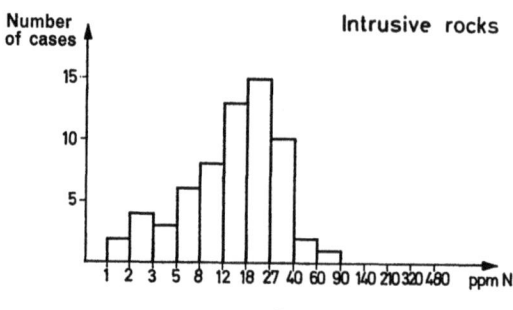

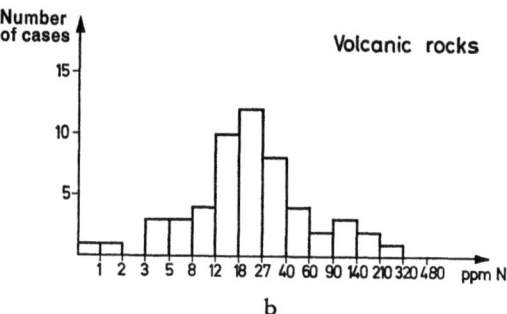

Fig. 7-E-2a and b. Histogram of nitrogen contents of intrusive (a) and volcanic (b) rocks, from Tables 7-E-1 and 7-E-3

Even if all rocks with nitrogen contents >100 ppm are disregarded for calculating the average, a value of 24 ppm N is obtained. A similar value of 22 ppm N is calculated as the median of all determinations, which may be chosen as more representative of volcanic rocks.

It is interesting to note that in volcanic rocks on the average only 51% of the average of 37 ppm total combined N are fixed, whereas in intrusive rocks fixed nitrogen was 75% of total combined nitrogen. This may be explained in the same way as the exceptional high nitrogen values in some volcanic rocks, that is by rapid cooling so that the volatile nitrogen compounds could not be fixed in the silicate mineral lattices. A drastic example of this effect is the andesite of Tokay, Hungary, where of the 144 ppm total combined N, none remains as fixed nitrogen.

III. Ultramafic Rocks

The nitrogen contents of ultramafic rocks are shown in Table 7-E-5. All values are low compared to other rock types, except for one peridotite containing 46 ppm N. The average value is 14 ppm total combined N, close to the 16 ppm value found for intrusive rocks.

IV. Elemental Nitrogen

The "total combined nitrogen" values of igneous rocks discussed in the preceding sections give the amount of ammonia and nitride nitrogen. They do not

Table 7-E-3. *Nitrogen content of volcanic rocks. For details see Table 7-E-1*

Rock	Total combined N, ppm	Fixed N, ppm	Reference
Rhyolite, Kemencepatak, Hungary	23	2	1
Rhyolite, Hrafntinnusker, Iceland	20	0	1
Rhyolite, Moskardshnjukal, Iceland	37	30	1
Rhyolite, Eisenbach, Hungary	119	43	1
Rhyolite, composite of 14 rhyolites from St. Pietro and St. Antioco, Italy	15	10	1
Rhyolite porphyry, Iron Mountain, Mo., U.S.A.	26.6		2b
Rhyolitic obsidian, Lake Co., Oreg., U.S.A.	7.8		2b
Rhyolitic obsidian, Rocce Rosse, Lipari, Italy	13	11	1
Rhyolitic obsidian, Fossa-Vulcano, Italy	12	7	1
Rhyolitic obsidian, composite of 8 rocks	5	5	1
Phonolite, composite of 10 German rocks	37	25	1
Phonolite, composite of 14 Bohemian rocks	43	21	1
Phonolite, Kempenich, Eifel, Germany	13	12	1
Phonolite, Schellkopf, Eifel, Germany	2	2	1
Phonolite, Hohentwiel, Singen, Germany	16	7	1
Trachyte, Hülsberg, Unterwesterwald, Germany	12	8	1
Trachyte, Drachenfels, Siebengebirge, Germany	102	66	1
Trachyte, Wölferlingen, Westerwald, Germany	8	5	1
Trachyte, Hahn, Westerwald, Germany	6	3	1
Trachyte, Monte Lentia, Italy	20	10	1
Andesite, Tokay, Hungary	144	0	1
Andesite, Sababurg, Reinhardswald, Germany	27	0	1
Andesite, composite of 14 rocks from Methana, Greece	26	7	1
Andesite, St. Andreasberg, Harz, Germany	170	141	1
Andesite, Easter Island	66		4a
Alkali olivine basalt, Dransberg, Germany	9	6	1
Alkali olivine basalt, Bramburg, Adelebsen, Germany	14	12	1
Alkali olivine basalt, Steinberg, Barlissen, Germany	12	9	1
Alkali olivine basalt, Steinberg, Meensen, Germany	14	7	1
Alkali olivine basalt, Güntersen, Germany, sample No. 4	68	34	1
Alkali olivine basalt, same, sample No. 5	34	21	1
Alkali olivine basalt, Gahrenberg, Reinhardswald, Germany	28	9	1
Alkali olivine basalt tuff, Ahrensberg, Habichtswald, Germany	25	11	1
Tholeiitic basalt, W1 U.S.Geol.Survey standard rock	14		6
	14	10	1
Tholeiitic basalt, Gufudalur, Iceland	19	10	1
Tholeiitic basalt, Brennisteinstindur, Iceland	35	20	1
Tholeiitic basalt, composite of 9, Deccan Peninsula, India	49	19	1
Tholeiitic basalt, Boulder Co., Colo., U.S.A.	22.1		2a
Tholeiitic basalt, Boulder Co., Colo., U.S.A.	31.3		2b
Tholeiitic basalt, Boulder Co., Colo., U.S.A.	18.0		2d

Table 7-E-3 (Continued)

Rock	Total combined N, ppm	Fixed N, ppm	Reference
Tholeiitic basalt, Tahiti	17		4a
Tholeiitic basalt, Karoo, Africa	40		4a
Tholeiitic basalt, Deccan Plateau, India	64		4a
Basalt, Mogda River, Siberia	26		4a
Basalt, Marquesas Island	42		4a
Basalt, Podkamennya, Tunguzka River, U.S.S.R.	105		4a
Basalt, Podkamennya, Tunguzka River, U.S.S.R.	24		4b
Basalt, Vermilion Co., Ill., U.S.A.	20	14	3
Amygdaloidal basalt, Keewenaw, Mich., U.S.A.	36.5		2b
Nepheline leucite tephrite, Niedermendig, Eifel, Germany	0	0	1
Nepheline leucite tephrite, Ettringer Feld, Eifel, Germany	5	5	1
Nepheline basanite, Druseltal, Kassel, Germany	36	29	1
Nepheline basanite, Mühlenberg, Eifel, Germany	4	1	1
Leucite basanite, Vulcanello, Italy	239	163	1
Picrite basalt, Eckenberg, Hessen, Germany	14	5	1

Table 7-E-4. *Mean nitrogen content of volcanic rocks, from Table 7-E-3. Figures in brackets are the averages for those samples for which total combined and fixed N were determined (see Table 7-E-1). No range of values is given for phonolithes, because two composite samples of 10 and 14 rocks, respectively, are included in the mean*

Rock	Number analyzed	Total combined N, ppm		Fixed N, ppm	
		range	mean	range	mean
Rhyolites and rhyolitic obsidians	10 (8)	5—119	28 (30)	0— 43	13.5
Phonolites	27		37		21
Trachytes	5	6—102	30	3— 66	18
Andesites	5 (4)	26—170	87 (92)	0—141	37
Alkali olivine basalts	7	9— 68	22	6— 34	14
Tholeiitic basalts	8 (4)	14— 64	33 (29)	10— 20	15
All basalts	20	9— 68	30		
Volcanic rocks, Scalan (1959)	6	8— 37	24		
Volcanic rocks, Vinogradov et al. (1963)	7	17—105	46		
Volcanic rocks, Wlotzka (1961)	40	0—239	37		
All volcanic rocks	53 (41)		37 (37)		19
Median of all volcanic rocks	53		22		

Table 7-E-5. *Nitrogen content of ultramafic rocks*

Rock	Total combined N, ppm	Reference
Pyroxenite, Webster, N.C., U.S.A.	6.7	2b
Dunite, Jackson Co., N.C., U.S.A.	7.6	2b
Dunite, Jackson Co., N.C., U.S.A. (average of 6 samples)	2.7 (range 1.9—3.6 ppm)	2c
Dunite, Jackson Co., N.C., U.S.A. (average of 2 samples)	3.0 (0.8 and 5.2 ppm)	2d
Dunite, Jackson Co., N.C., U.S.A.	17.2	2a
Dunite, Jackson Co., N.C., U.S.A.	11	3
Dunite, Jackson Co., N.C., U.S.A.	2.2	7
Dunite, Nan-Shan, China (average of 3 samples)	12 (range 11—14 ppm)	4a
Dunite, Monchegorsk Kola, U.S.S.R. (average of 2 samples)	10 (8 and 11 ppm)	4a
Dunite, Sobsko-Vaikarskii Massif, Urals, U.S.S.R. (average of 3 samples)	15 (range 7—22 ppm)	4a
Peridotite, Northern Urals, U.S.S.R.	46	4a
Meimechite, Gulinsk intrusive, Siberia, U.S.S.R.	12	4a
Average (10)	14	

References: 2. SCALAN (1959); 3. STEVENSON (1962); 4. VINOGRADOV et al. (1963); 7. MAYNE (1957). For methods used see legend to Table 7-D-2.

include elemental nitrogen, except for the values of SCALAN (1959) measured by method (a) or (b) (see legend to Table 7-D-2). Elemental nitrogen can be determined in two ways: either separately by vacuum grinding of the sample and determination of the liberated N_2, or together with the combined nitrogen by heat extraction.

The first method was employed by GOGUEL (1963) for a number of rock-forming minerals; these results were discussed in Section 7-D. They show that elemental nitrogen is usually lower than or about equal to combined nitrogen. In agreement with this BUACHIDZE and SIKHARULIDZE (1965), who used the same method, found amounts between 2.3 and 33 ppm N_2 (mean of 6 determinations: 15 ppm N_2) in deep seated basalts obtained from drillholes at depths of 200 to 1,300 m.

The heat extraction method employed by SHEPHERD (1938) gave very high amounts of total nitrogen. The range was 25 to 700 ppm, the average of 13 values being 220 ppm N. These high values are probably caused by air contamination (WAHLER, 1955). More recent heat extraction measurements, obtained by CHAIGNEAU and DEBRUNE (1961) as well as by CHAIGNEAU and BORDET (1962, 1963), agree better with the nitrogen values in the preceding section, see Table 7-E-6. Measurements by GIBSON and MOORE (1970), which were obtained by a vacuum fusion and gas chromatography technique and were carefully checked for reproducibility and accuracy, are included in Table 7-E-6. Here the U.S. Geological Survey standard rocks G-1 and W-1 show quite high amounts of total nitrogen (59 and 52 ppm N respectively) compared to the combined nitrogen reported in Tables 7-E-1 and 7-E-3: 8 and 2.4 ppm for G-1, and 14 and 14 ppm for W-1.

Table 7-E-6. *Total nitrogen content (including N_2) of rocks determined by heat extraction (at 1,000° C, CHAIGNEAU and co-workers) or vacuum fusion (at 2,400° C, GIBSON and MOORE)*

Rock (number of samples analyzed)	Nitrogen, ppm range	Nitrogen, ppm mean	Reference
Dacite, Montagne, Pelée, Martinique (4)	21—34	29	CHAIGNEAU and DEBRUNE (1961)
Obsidian, Colle-de-la-Motte, Esterél, France (2)	36—57	46	CHAIGNEAU and BORDET (1962)
Obsidian, Rocce Rosse, Lipari, Italy		16	CHAIGNEAU and BORDET (1962)
Pumice, Valley of Ten Thousand Smokes, Katmai, Alaska (4)	12—31	18	CHAIGNEAU and BORDET (1963)
Basalt, USGS rock standard W-1	51—53	52	GIBSON and MOORE (1970)
Basalt, USGS rock standard BCR-1	28—31	30	GIBSON and MOORE (1970)
Andesite, USGS rock standard AGV-1	40—49	44	GIBSON and MOORE (1970)
Granite, USGS rock standard G-1	57—61	59	GIBSON and MOORE (1970)
Granite, USGS rock standard G-2	50—65	56	GIBSON and MOORE (1970)
Granodiorite, USGS rock standard GSP-1	45—54	48	GIBSON and MOORE (1970)
Dunite, USGS rock standard DTS-1	24—31	27	GIBSON and MOORE (1970)
Peridotite, USGS rock standard PCC-1	37—49	43	GIBSON and MOORE (1970)

V. Nitrate Nitrogen

The occurrence of nitrate ions in igneous magmas is improbable because of the low oxidation potential of such magmas, which allows, for instance, the existence of ferrous iron. Accordingly, all attempts to detect nitrate nitrogen in igneous rocks have so far been negative (RAYLEIGH, 1939; WLOTZKA, 1961). WLOTZKA found no nitrate in the minerals and rocks analyzed for ammonium, the limit of detection lying at 5 ppm nitrate nitrogen. However, the situation is different in natural waters and sediments, see Sections 7-I and 7-K.

7-F. Behavior in Magmatogenic Processes

The physical solubility of N_2 in molten silicate glass was measured to be about 0.2 ppm by weight at 1 atm pressure and 1,400° C (MULFINGER, 1966); the amounts of N_2 found in inclusions of igneous rocks (GOGUEL, 1963; see Table 7-D-1) are of the same order of magnitude. The solubility of ammonia is much greater, as ammonia can be chemically bound in silicate melts under reducing conditions (upto 0.5% nitrogen; MULFINGER and MEYER, 1963). These results can be applied to natural igneous magmas; ammonia will behave in these situations as a volatile substance and concentrate in late magmatic minerals, as demonstrated by the relative high amounts found in zeolites (Table 7-F-1). Ammonium salts are also known from volcanic exhalation products [see e.g., BRUN (1911) and Table 7-F-1].

Table 7-F-1. *Nitrogen content of zeolites and of volcanic exhalation products, from* WLOTZKA (1961)

Mineral	N, ppm
Heulandite, Fassa-Tal, Tirol, Austria	370
Natrolite, Eikarholmen, Norway	22
Natrolite, Stempel, Marburg, Germany	27
Natrolite from phonolite, Hohentwiel, Germany	26
(in parent rock:	16)
Chabasite from granite, Königskopf, Harz, Germany	60
(in parent rock:	17)
Exhalation products, Vulcano, Italy	50
Sulfur, crater bottom, Vulcano, Italy	460

If ammonia escapes into the gas phase, it should dissociate almost completely into N_2 and H_2 under the conditions of high temperature and low nitrogen pressure (HABER, 1914) which are prevailing under magmatic conditions. Fig. 7-F-1 shows the NH_3 pressures in equilibrium with N_2 for magmatic vapors at different water and N_2 pressures at 600° C. Here the occurrence of NH_3 is mainly governed by the H_2 pressure from the $2H_2O = 2H_2 + O_2$ equilibrium, and only under very reducing conditions is some NH_3 formed. Nevertheless ammonia has been found in magmatic gases and volcanic particles (FOSHAG and HENDERSON, 1946; CADLE and FRANK, 1968). This is in accordance with experience from coal distillations, where the presence of water vapor prevents the decomposition of ammonia in the gas phase (KIRNER, 1945). Also it was found by VINOGRADOV et al. (1963) that N_2 and H_2 in contact with igneous rocks at temperatures between 350 and 400° C and at 1 atm pressure, can combine to give ammonia in amounts similar to those found in these rocks.

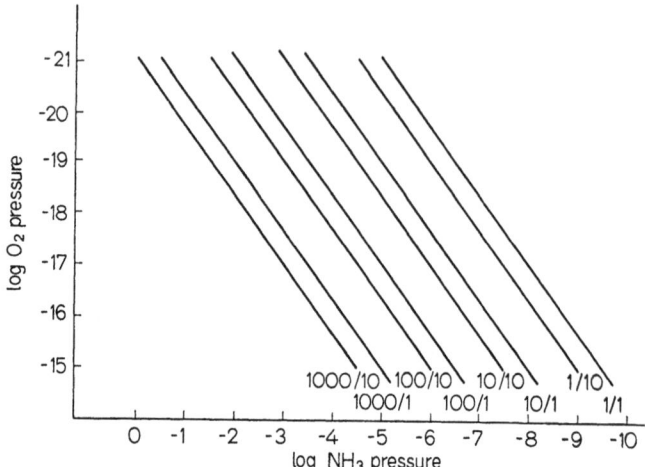

Fig. 7-F-1. NH_3 pressure in magmatic vapors at different O_2 pressures in equilibrium with H_2O and N_2 at 600°C. 1,000/10 etc. = $(H_2O)atm/(N_2)atm$. Calculated after a magmatic vapor model by KRAUSKOPF (1957). A material balance restriction of (NH_3) becomes effective only at $(H_2O)/(N_2)$ ratios of $< 1/100$

In volcanic gases nitrogen is a common component, ranging in amount from a few to more than 80 percent. Usually, however, all N_2, or the amount corresponding to the O_2 content found, is subtracted from the analysis as being due to air contamination. A review and discussion of the analytical data on volcanic emanations is given by WHITE and WARING (1963); for more recent data, see FINLAYSON et al. (1968). The amount of truly magmatic N_2 must be small as samples of magmatic gases from basaltic lavas at Surtsey, Iceland, had only 0.07 to 1.07 mole per cent N_2 (SIGVALDASON and ELISSON, 1968); these samples had been carefully selected and protected from air contamination.

A similar problem is encountered with thermal gases which usually contain large amounts of nitrogen. Groundwater equilibrated with air at 10°C has an Ar/N_2 ratio of 2.7/100 (HULSTON and MCCABE, 1962). If the ratio is smaller, an additional source of nitrogen, such as magmatic or organic nitrogen, may have been present. Another explanation for this lower ratio is the possibility of a later Rayleigh-type distillation (MAZOR and WASSERBURG, 1965; GUNTER and MUSGRAVE, 1966). The latter authors applied this model successfully to analyzed thermal gases from Yellowstone National Park and the Ar/N_2 ratios were interpreted to indicate a predominantly atmospheric origin of Ar and N_2.

7-G. Behavior during Weathering and Alteration of Rocks

Meteoric water and hydrothermal or pneumatolytic solutions can both carry high amounts of water-soluble nitrogen compounds. Thus, they may dissolve the nitrogen of rocks and carry it away, as well as add nitrogen derived either from biogenic sources or from the magmatic volatiles. The addition of nitrogen will occur whenever a fixation of nitrogen in layer lattice silicates or in feldspars is possible.

The higher amounts of nitrogen in the lithionite-granite and the sericitized feldspars of granites (Tables 7-E-1 and 7-D-2), described in Sections 7-E and 7-D, result from this nitrogen enrichment by late magmatic activity. The mechanism of

Table 7-G-1. *Nitrogen content of altered igneous rocks*

Rock	N, ppm	Reference
Granite, Brocken, Harz, Germany:		
fresh	19	1
highly weathered	48	1
Granite, Mt. Washington, New Hampshire, U.S.A.:		
highly weathered	78	3
Granite, Stone Mountain, Georgia, U.S.A.:		
fresh	17	3
weathered	57	3
Rhyolite, Moskardshnjukar, Iceland:		
sample M1, fresh	37	1
sample M6, altered by hydrothermal solutions	41	1
sample M7, altered by hydrothermal solutions	56	1
Rhyolite, Hrafntinnusker, Iceland:		
sample SL1, fresh	20	1
sample SL2, altered by hydrothermal solutions	47	1
Basalt, Bramburg, Germany:		
fresh	14	1
altered	13	1
altered	8	1
Basalt, Backenberg, Germany:		
fresh	26	1
altered	35	1
altered	21	1
Basalt, Steinberg, Germany		
fresh	5	1
slightly altered	6	1
altered	13	1
highly altered	2	1

References: 1. WLOTZKA (1961), 3. STEVENSON (1962). For analytical methods, see Table 7-D-3.

hydrothermal nitrogen fixation is demonstrated by the formation of the ammonium feldspar buddingtonite from plagioclase by the action of hot, ammonium-rich water (ERD et al., 1964).

Table 7-G-1 gives some nitrogen measurements of weathered and hydrothermally altered rocks. In the granites an increase in nitrogen due to weathering is apparent. The rhyolites from Iceland described by SIGVALDASON (1958, 1959) have been hydrothermally altered. They show a slight enrichment of nitrogen after alteration which is parallel to an increase in water (H_2O^-) content.

The basalts studied by BOLTER (1961) also show variations in nitrogen content with both increases and decreases occurring during alteration.

Revised manuscript received: July 1970

7-H. Solubility of Nitrogen in Water

Table 7-H-1 gives the solubility of molecular N_2 in sea water (DOUGLAS, 1965). Additional measurements are given by KÖNIG (1963). For the nitrogen solubility in distilled water compared to sea water, see BENSON and PARKER (1961). The solubility of oxygen in sea water (see Table 8-H-1) is about twice that of nitrogen, so that the ratio N_2/O_2 is around 2 compared to 4 in air — see, e.g., measurements by BENSON and PARKER (1961) or CRAIG et al. (1967).

Table 7-H-1. *Nitrogen solubility in sea water* (DOUGLAS, 1965). α *is given as volume of gas (STPD) absorbed by a unit volume of water when the pressure of the gas is 760 mm. Chlorinity is expressed as grams of chlorine per kilogram of sea water*

α-Nitrogen

Temp., °C	Chlorinity						
	15	16	17	18	19	20	21
0	0.01931	0.01904	0.01877	0.01850	0.01824	0.01797	0.01770
1	0.01886	0.01860	0.01835	0.01809	0.01783	0.01758	0.01732
2	0.01845	0.01820	0.01795	0.01770	0.01745	0.01720	0.01696
3	0.01804	0.01780	0.01756	0.01732	0.01708	0.01684	0.01660
4	0.01766	0.01742	0.01719	0.01696	0.01673	0.01650	0.01626
5	0.01727	0.01705	0.01683	0.01661	0.01639	0.01616	0.01594
6	0.01691	0.01669	0.01648	0.01627	0.01606	0.01584	0.01563
7	0.01656	0.01635	0.01615	0.01595	0.01574	0.01554	0.01533
8	0.01622	0.01602	0.01582	0.01562	0.01542	0.01522	0.01502
9	0.01587	0.01568	0.01549	0.01530	0.01511	0.01492	0.01473
10	0.01554	0.01536	0.01518	0.01500	0.01481	0.01463	0.01445
11	0.01523	0.01506	0.01488	0.01471	0.01453	0.01436	0.01418
12	0.01496	0.01478	0.01461	0.01444	0.01426	0.01409	0.01392
13	0.01469	0.01452	0.01435	0.01418	0.01401	0.01384	0.01367
14	0.01440	0.01424	0.01408	0.01392	0.01376	0.01360	0.01344
15	0.01419	0.01403	0.01386	0.01370	0.01354	0.01338	0.01321
16	0.01396	0.01380	0.01363	0.01346	0.01329	0.01312	0.01295
17	0.01373	0.01357	0.01341	0.01325	0.01309	0.01293	0.01277
18	0.01351	0.01335	0.01319	0.01303	0.01287	0.01271	0.01255
19	0.01329	0.01314	0.01298	0.01282	0.01266	0.01250	0.01235
20	0.01309	0.01293	0.01278	0.01263	0.01248	0.01232	0.01217
21	0.01288	0.01273	0.01259	0.01244	0.01299	0.01214	0.01199
22	0.01268	0.01254	0.01240	0.01226	0.01211	0.01197	0.01183
23	0.01249	0.01235	0.01222	0.01208	0.01194	0.01180	0.01166
24	0.01231	0.01218	0.01205	0.01192	0.01179	0.01165	0.01152
25	0.01214	0.01202	0.01189	0.01176	0.01164	0.01151	0.01138
26	0.01199	0.01187	0.01174	0.01162	0.01150	0.01137	0.01125
27	0.01185	0.01173	0.01160	0.01148	0.01136	0.01124	0.01112
28	0.01173	0.01160	0.01148	0.01136	0.01124	0.01112	0.01100
29	0.01159	0.01147	0.01135	0.01124	0.01112	0.01101	0.01089
30	0.01146	0.01134	0.01123	0.01111	0.01100	0.01088	0.01077

Revised manuscript received: July 1970

© Springer-Verlag Berlin · Heidelberg 1972

7-I. Abundance in Natural Waters and in the Atmosphere

I. Natural Waters

a) Molecular N_2

The bulk of the nitrogen in the ocean occurs in the form of N_2, the ratio molecular N to combined N being on the average about 25. In contrast to oxygen, dissolved molecular N_2 behaves in the ocean as a conservative element. A fixation of molecular nitrogen by blue-green algae takes place in the sea (DUGDALE et al., 1961), but the amount of nitrogen fixed seems to be negligible for the nitrogen balance or the distribution of dissolved nitrogen in the ocean. Measurements of KÖNIG et al. (1964) and LINNENBOM et al. (1965) showed the dissolved nitrogen to be in equilibrium with the atmosphere within the limits of error of the analytical technique employed.

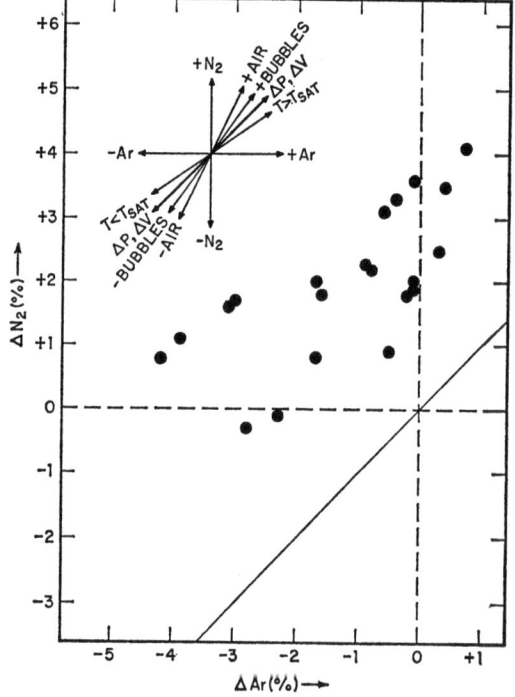

Fig. 7-I-1. Saturation anomalies ΔN_2 versus ΔAr in the Pacific Ocean. The vectors show approximate trajectories for addition of air to gas samples, interaction of waters with bubbles, surface pressure fluctuations (ΔP), volumetric errors in amount of water sampled or bulk loss of total gas (ΔV), and effects of heating or cooling of water masses after saturation at the surface. (From CRAIG et al., 1967)

More recently, measurements were made by Craig et al. (1967) in the equatorial and south Pacific Ocean. They found an average positive deviation for N_2 solubility and a negative one for Ar; both deviations are positively correlated (Fig. 7-I-1). This correlation can be explained by atmospheric pressure fluctuations, but the intercept of the correlation curve at positive values of ΔN_2 shows that N_2 is about 3.2% in excess of Ar, which suggests a production of N_2 in the ocean (Craig et al., 1967). A denitrification reaction in the sea is also postulated because of the nitrogen balance between atmosphere and ocean (see Section 7-O).

b) Inorganic Nitrogen Compounds

The inorganic nitrogen species ammonia, nitrite and nitrate play an important role in the biological cycle in natural waters. Their behavior has been widely studied: a comprehensive review for ocean waters is given, e.g., by Vaccaro (1965) and for inland waters by Feth (1966). These nitrogen compounds are non-conservative constituents and their amounts vary widely with locality, season and depth. In the ocean ammonium and nitrite ions are intermediate stages in the oxidative cycle of nitrogen and are found mainly in the photic surface layer of high biological activity. Here ammonium varies between 0 and 75 µg/l. For detailed studies of its distribution, see Beers and Kelly (1965). Nitrate is the end product of the nitrogen cycle. It is used by phytoplankton as the main nitrogen source so that its concentration can become very low in euphotic zones. Here a shortage of nitrate is the main growth-limiting factor for the phytoplankton. Fig. 7-I-2 shows the distribution of nitrate near the continental shelf in the Atlantic Ocean (Ketchum et al., 1958). In deeper layers of the ocean nitrate is the predominant nitrogen compound (except for N_2). Its measured

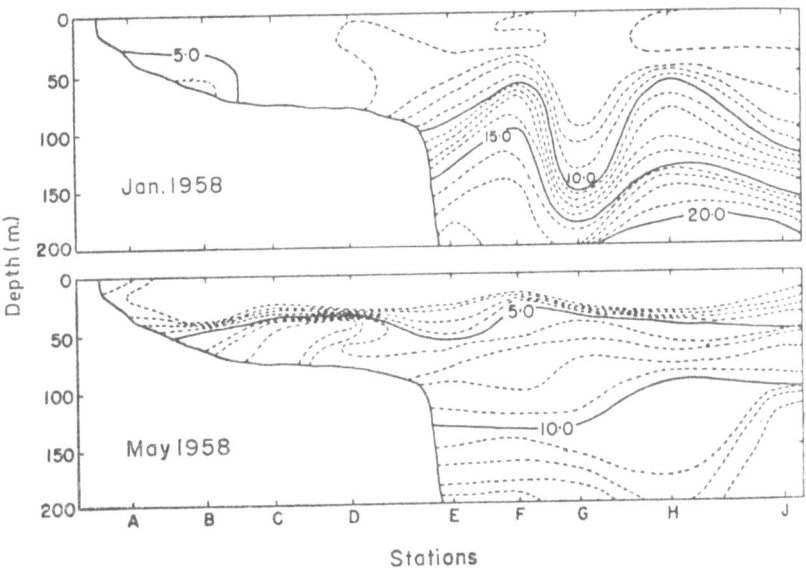

Fig. 7-I-2. Variations in nitrate nitrogen (µg-atoms N/l) at different times of the year across the Atlantic continental shelf south of Montauk, Long Island, U.S.A.
(From Ketchum et al., 1958)

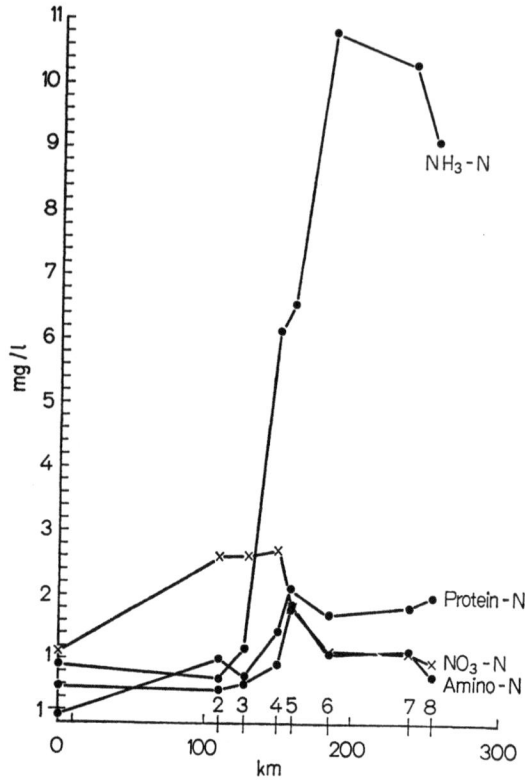

Fig. 7-I-3. Variation in nitrogen compound content along the Saale River, Germany. (After MEISSNER, 1965)

concentration was 2.5 mg/l in 2,000 to 6,000 m deep waters in the Pacific and 1.2 mg/l in the North Atlantic (CHOW and MANTYLA, 1965).

A similar nitrogen cycle is operative in rivers and lakes. Here artificial pollution and anoxic conditions favour the occurrence of ammonium nitrogen. As an example, Table 7-I-1 and Fig. 7-I-3 show the distribution of different forms of nitrogen in

Table 7-I-1. *Content of nitrogen compounds (mg N/l) in the Saale River, Germany, in 1962* (MEISSNER, 1965)

Station	NH_3—N	NO_2—N	NO_3—N	Protein-N	Amino-N	Total N
1 Eichicht	0.87	0.00	1.10	0.12	0.58	2.67
2 Naumburg	0.70	0.06	2.60	0.98	0.56	4.90
3 Weißenfels	1.18	0.09	2.60	0.74	0.61	5.22
4 Merseburg	6.10	0.14	2.70	1.23	0.93	7.10
5 Hohenweiden	6.55	0.14	1.80	2.10	1.83	12.60
6 Böllberg and Wettin	10.80	0.12	1.15	1.69	1.08	14.84
7 Bernburg	10.30	0.10	1.10	1.82	1.11	14.43
8 Nienburg and Groß-Rosenburg	9.10	0.08	0.95	1.96	0.74	12.56

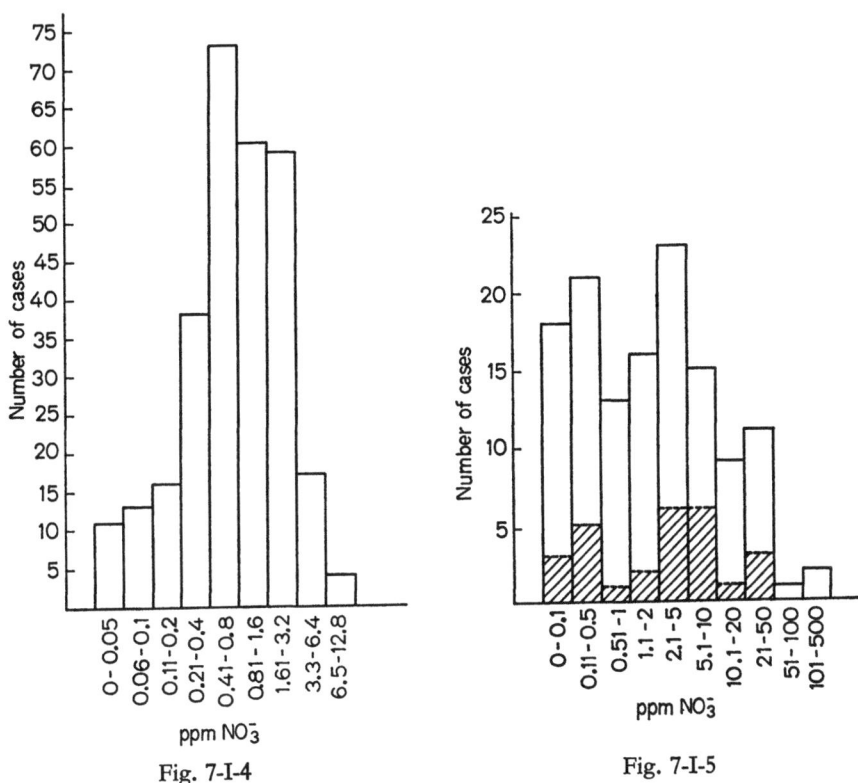

Fig. 7-I-4. Histogram of NO_3^- content (ppm NO_3^-) in rivers of the world. (After Livingstone, 1963)

Fig. 7-I-5. Histogram of NO_3^- contents (ppm NO_3^-) in subsurface waters. (After White et al., 1963). Shaded area: water from magmatic rocks; unshaded area: water from metamorphic and sedimentary rocks

the Saale River (Meissner, 1965). The large increase of ammonium ions between stations 3 and 5 is due to industrial sewage which leads to nearly anaerobic conditions connected with a decrease in nitrate nitrogen.

Fig. 7-I-4 gives a histogram of NO_3^--contents in rivers of the world as compiled by Livingstone (1963). He gives as average values 1 ppm NO_3^- for North America, 0.7 ppm for South America, 3.7 ppm for Europe, 0.7 ppm for Asia, 0.8 ppm for Africa, 0.005 ppm for Australia, and 1 ppm as the world average.

The nitrogen content of subsurface waters is shown in Fig. 7-I-5. No significant difference between ground waters from magmatic rocks and from sediments is found, which points to the secondary or biogenic origin of the nitrate nitrogen. Feth (1966) reported a tendency of groundwater from limestone areas to have a high nitrate content. This may be connected with the higher amounts of nitrate nitrogen occuring in limestones compared to other rocks (Wlotzka, 1961).

The nitrate content of ground waters is important for the estimation of biogenic pollution. To prevent infant methemoglobinemia, drinking water should not have more than 10 mg nitrate-N/l (United States Public Health Service, 1962). An investigation of water supplies in California (NAVONE et al., 1963) showed that 11% of the 781 sampling points studied exceeded this value. Similar results were obtained by ENGBERG (1967) in Nebraska, and by DARGENTA et al. (1967) in Roumania.

Ammonium is usually only a transitory constituent of subsurface waters. It is easily adsorbed on clay particles or oxidized to nitrate. Acid sulfate waters connected with volcanism may contain high amounts of ammonium — 1,400 ppm in "The Geysers", Sonoma Co., California — which is probably derived from magmatic gases (LIVINGSTONE, 1963). Oil field brines are usually high in ammonium: LIVINGSTONE (1963) gives an average of 40 ppm for Na—Cl-type brines, and 200 ppm for Na—Ca—Cl-types. These high values have been studied as a means for oil prospecting (YAKOVLEV, 1964; RAVIKOVICH, 1965; RAVIKOVICH and LUBYANSKAYA, 1969).

c) Organic Nitrogen Compounds

The organic nitrogen compounds in natural waters are mainly proteins and amino acids. The results of measurements of protein- and amino-N in the Saale River are shown in Table 7-I-1 (MEISSNER, 1965). Both compounds reach similar levels as nitrate-N. The amounts of dissolved and particulate organic nitrogen in the Indian Ocean (FRAGA, 1966) are given in Fig. 7-I-6; the maximum at a depth of 10 m corresponds to maximum photosynthetic activity. The C/N atomic ratios for particulate organic matter lie around 5 to 7, while the ratio for dissolved organic matter varies from 12 at the ocean surface to 8 at depths greater than 200 m. Additional data are found in the review by DUURSMA (1965) and the paper by GUSAROVA et al. (1966). For the organic nitrogen compounds in lakes, see HUTCHINSON (1957) and KOYAMA (1966).

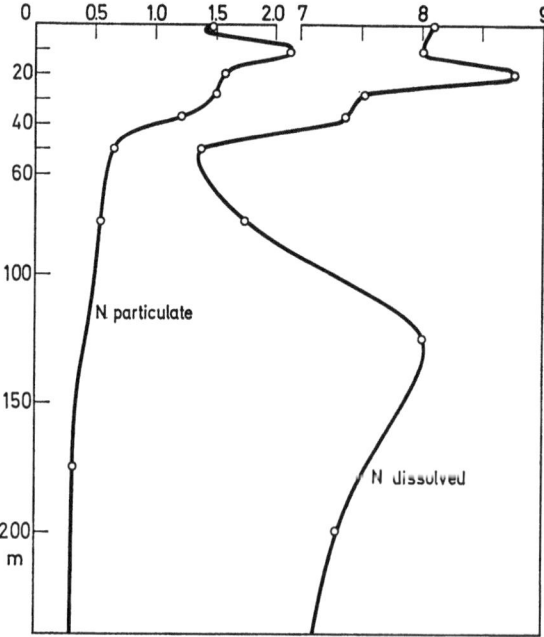

Fig. 7-I-6. Vertical distribution of particulate and dissolved organic nitrogen (μg-atoms N/l); average values from 40 stations in the western Indian Ocean. (From FRAGA, 1966)

II. Atmosphere and Precipitation

a) Molecular Nitrogen and Reaction Products

The terrestrial atmosphere contains 78.084 ± 0.004 per cent of N_2 by volume [GLUECKAUF (1951); U.S. Standard Atmosphere 1962] or 755 g nitrogen/cm² of earth surface. The atmospheric composition is constant up to a height of about 90 km. Number densities of N_2 between 120 and 200 km are given in Table 8-I-1 in the chapter on oxygen. Recent measurements of nitrogen and oxygen obtained with a rocket-borne mass spectrometer by ARNOLD et al. (1969) are shown in Fig. 7-I-7.

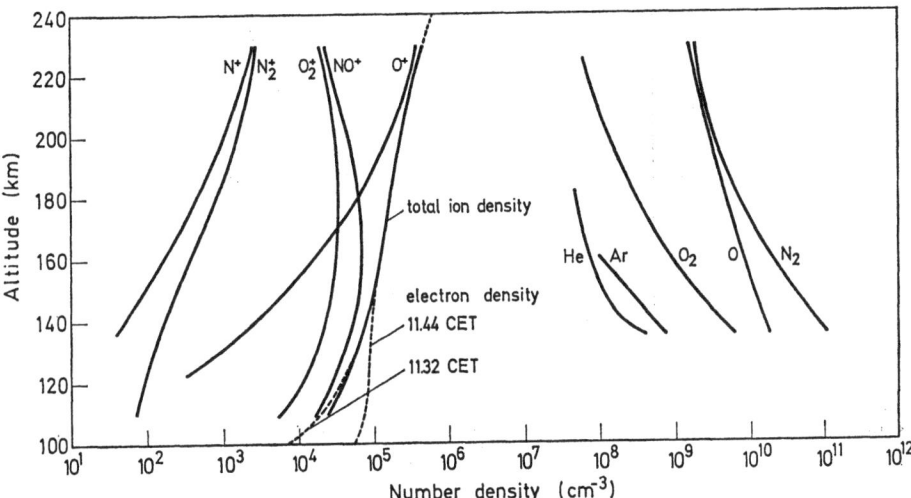

Fig. 7-I-7. Density versus altitude profiles of ionospheric and atmospheric constituents measured by a rocket-borne mass spectrometer. (From ARNOLD et al., 1969)

Photodissociation of N_2 to atomic nitrogen in the ionosphere occurs to a much lesser extent than in the case of oxygen. Rocket measurements of the N/N_2 ratio between 120 and 280 km were made by GHOSH et al. (1968) and are shown in Fig. 7-I-8.

A reaction of atomic nitrogen and oxygen in the ionosphere produces nitric oxide NO and the concentration of this species in the ionospheric D-region has been estimated by MITRA (1966) to be 4×10^5 atoms/cm³. For a review of the ionospheric composition and the reactions between nitrogen and oxygen, see DONAHUE (1968).

b) Nitrogen Oxides

In the troposphere both nitrous oxide, N_2O, and nitrogen dioxide, NO_2, have been found.

Nitrous oxide was first discovered by ADEL (1939) in the solar absorption spectrum. Measurements upto 1956, which gave a uniform mixing ratio of 0.5 ppm, were summarized by MILLER (1956). A detailed study by SCHÜTZ et al. (1970) showed the N_2O content at ground level to be 0.25—0.3 ppm and to decrease with height, see Fig. 7-I-10. They also found a concentration ten times this amount in

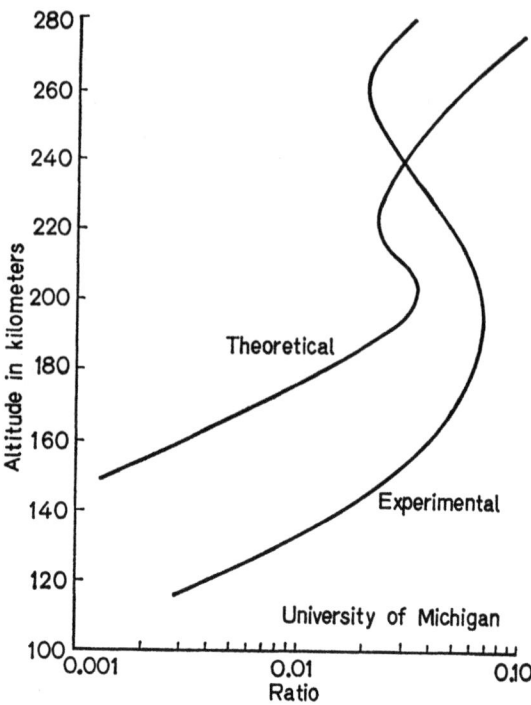

Fig. 7-I-8. Ratio of number densities N/N_2. Experimental profile from measurements by a mass spectrometer ejected from a rocket; theoretical profile calculated by GHOSH (1968). (From GHOSH et al., 1968)

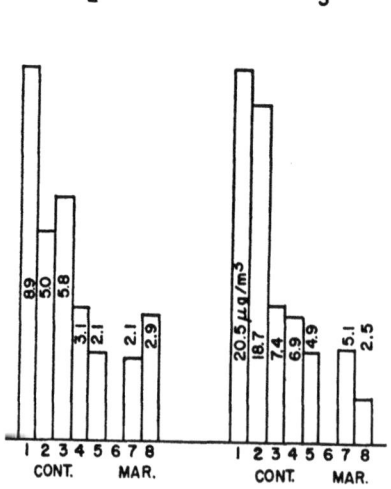

Fig. 7-I-9. Concentration of NO_2 and NH_3 gas in air. The concentrations are given in µg/m³ within the columns. The stations are arranged in the order polluted continental—unpolluted maritime areas. 1. Frankfurt/Main, Germany, winter; 2. same, summer; 3. Kleiner Feldberg, Taunus, Germany (Nov.—Jan.); 4. Zugspitze, Alps, Germany (August); 5. St. Moritz, Switzerland (August); 6. Round Hill, U.S.A.; 7. Florida, U.S.A.; 8. Hawaii. (From JUNGE, 1963)

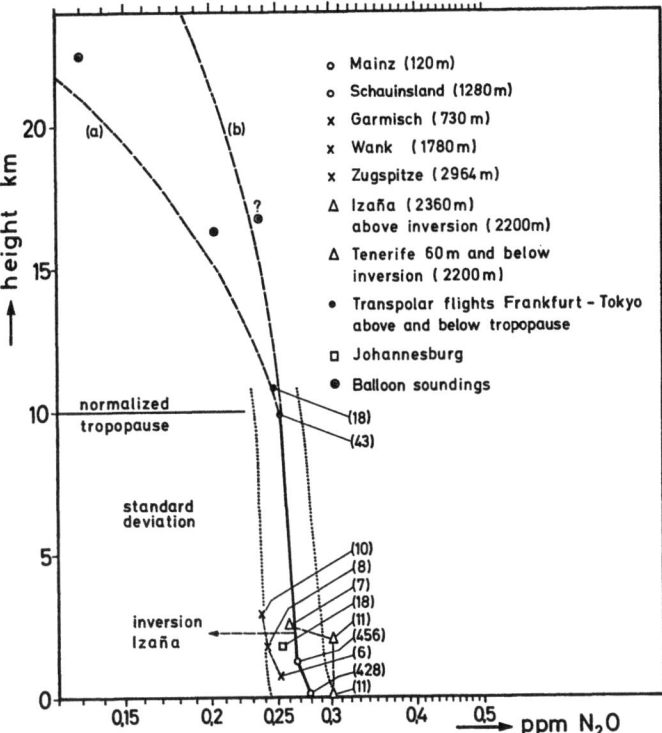

Fig. 7-I-10. Concentration of NO_2 (in ppm) in the atmosphere. Figures in brackets are number of measurements. The dotted lines give the standard deviation of the tropospheric series of measurements. The dashed lines a and b are calculated profiles of BATES and HAYS (1969) for stratospheric eddy diffusion coefficients of 10^3 and 10^4 cm²/sec, respectively. (From SCHÜTZ et al., 1970)

air inside a cultivated soil. This favours the view that atmospheric N_2O stems from the decomposition of nitrogen compounds by soil bacteria. Another possible source is the photochemical production of N_2O in the stratosphere (BATES and WITHERSPOON, 1952; HARTECK and DONDES, 1954). The vertical profile of the N_2O concentration (Fig. 7-I-10) shows, however, that most of the N_2O is released from the ground and destroyed in the stratosphere (SCHÜTZ et al., 1970).

Nitrogen dioxide in different localities was measured by JUNGE (1956), GEORGII (1960), and TANAEVSKY (1961) (see Fig. 7-I-9). A decrease in NO_2 from continental urban areas towards maritime localities is evident. This indicates the two main sources of NO_2: industrial pollution and decomposition of organic matter in soil (JUNGE, 1963). The contribution of molecular N_2 fixed by electric discharges, to the NO_2 content of the atmosphere and the NO_3^- content of rain has been discussed for a long time (LIEBIG, 1827). The general consensus now is that this contribution is small (HUTCHINSON, 1954; JUNGE, 1963), however, in certain regions it can be of importance. REITER (1966, 1968) has shown that a direct relationship exists between the NO_3^- ion concentration of precipitation in the Alps and the time-integral of the electric field strength. Industrial pollution can lead to high concen-

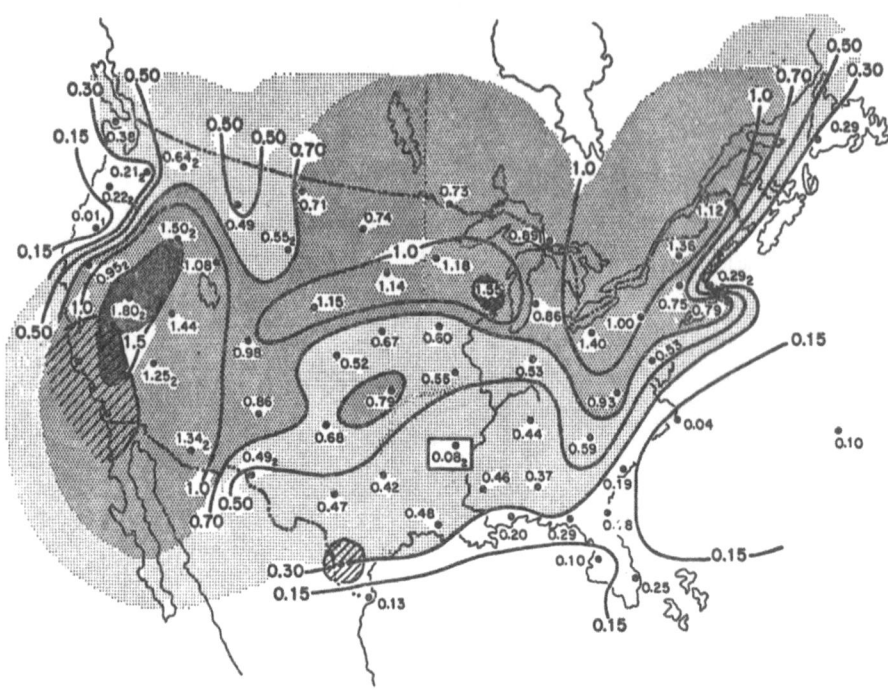

Fig. 7-I-11. Average NO_3^- concentration (mg/l) in rain over the U.S., July—Sept. 1955. (From JUNGE, 1958)

trations of NO_2 in the atmosphere. TEBBENS (1968) reports amounts between 0.25 and 2 ppm $NO+NO_2$ ($= 500$—$4{,}000$ $\mu g/m^3$) on smog days in the Los Angeles area.

The amounts of nitrogen compounds contained in rain were already being studied intensively during the last century when the main interest was the supply of nitrogen to the soil. The data have been reviewed and discussed by ERIKSSON (1952), HUTCHINSON (1954), and by JUNGE (1963).

The amount of nitrate found in rain varies widely with locality and season, the range of concentrations over land being 0.3 to 2.5 mg/l, whereas over sea or near the coast 0.15 to 0.5 mg/l is found. Usually an inverse relationship exists between the amount of precipitation and the concentration of nitrate, ammonium and other constituents (ÅNGSTRÖM and HÖGBERG, 1952). In unpolluted areas the seasonal maximum is found in spring or summer, which points to a source in the soil (JUNGE, 1963). The map of the average nitrate concentration in rain over the United States (JUNGE, 1958) shows the decrease towards the sea (Fig. 7-I-11), which seems to be at least a partial sink for NO_2. GAMBELL and FISHER (1964) found a strong positive correlation between SO_4^{2-} and NO_3^- in rain, indicating that both ions are aquired by similar means. Just as SO_4^{2-} is most likely derived from gaseous SO_2 (JUNGE and RYAN, 1958), NO_3^- stems from gaseous NO_2 and not from a particulate source. Table 7-I-2 gives some recent measurements of NO_3^- and NH_4^+ in rain.

Table 7-I-2. Recent measurements of nitrogen compounds in precipitation (mg N/l)

Locality	NO_3^-—N	NH_3—N, mg/l	Organic N	References
Dresden-Wahnsdorf, Germany:				MROSE (1966)
Summer	0.65	1.6	n.d.	
Winter	0.54	1.65	n.d.	
Oregon, U.S.A., forest opening 10 km from Pacific Ocean:				TARRANT et al. (1967)
June	0.10	<0.01	0.29	
August/Sept.	0.03	<0.01	0.00—0.03	
July	0.00	<0.01	0.22	
October-May	0.00	<0.01	0.00—0.09	
Yorkshire, England: Oct. 1961—Oct. 1962	0.33	0.72	n.d.	SYERS (1965)
Syktyvkar, U.S.S.R.	0.028—0.40 av. 0.14	0.024—3.68 av. 0.7	n.d.	CHEBYKINA (1964)
U.S.S.R.	n.d.	n.d.	0.004—0.02 amino-N 0.012—0.09 protein-N	SEMENOV et al. (1966)
U.S.S.R.:				SEMENOV et al. (1967)
rain	n.d.	n.d.	0.7 —5. av. 2.2	
rain			0.009—0.06 amino-N	
snow			0.7 —2.9, av. 1.8	
Menlo Park, Calif., U.S.A.:				WHITEHEAD and FETH (1964)
rain 1957—58	0.034	n.d.	n.d.	
rain 1958—59	0.036			
bulk precipitation	0.58			
South Pole, snow	0.005	n.d.	n.d.	WILSON and HOUSE (1965)

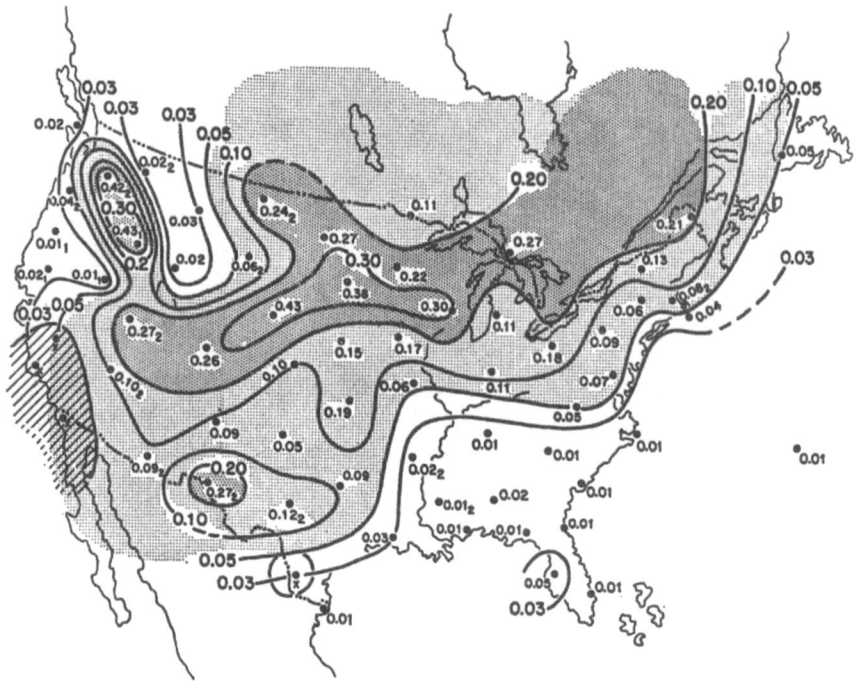

Fig. 7-I-12. Average NH_4^+ concentration (mg/l) in rain over the U.S., July—Sept. 1955. (From JUNGE, 1958)

Nitrite concentrations in rain are usually much lower than nitrate levels. ERIKSSON (1952) reports NO_2^-/NO_3^- ratios of 0.01 to 0.1, while REITER (1960) found a ratio of about 0.01 in rain over the Northern Alps.

c) Ammonia

The content of ammonia in air is shown in Fig. 7-I-9. The concentrations found in rain vary between 0.01 and 1 mg/l, most values lying around 0.1 to 0.2 mg/l (JUNGE, 1963). Fig. 7-I-12 gives the distribution of ammonium in precipitation over the United States (JUNGE, 1958). The concentrations are low over the sea and over the south-east land area which has laterite soils with low pH. This indicates the soil to be the most important source of ammonia in rain (JUNGE, 1963). This is further supported by the fact that the maximum ammonium concentrations are found in summer or fall. Data for Europe were collected by ERIKSSON (1952); the values are generally higher in Central Europe and show a maximum in winter, which was ascribed by ERIKSSON (1952) to the increased combustion of fuels. More recent measurements by LODGE and PATE (1966) on tropical air sampled on the Isthmus of Panama showed that aerosol particles on the coast contained high amounts of ammonium sulfate, whereas further inland free sulfuric acid predominates. The authors concluded that the Atlantic acts as a source for ammonia, which is oxidized to nitrogen oxide or dioxide over the land which thus becomes an ammonia sink.

Several authors, e.g., ÅNGSTRÖM and HÖGBERG (1952), VIRTANEN (1952) and SYERS (1966), have noted the almost constant ratio of 1:2 in which nitrate- and ammonia-N appear in precipitation in temperate regions; this would suggest a common, perhaps photochemical, origin of both compounds.

Ammonium is also a major constituent of aerosol particles in the troposphere and in the stratospheric aerosol layer above the tropopause (JUNGE, 1963; CADLE et al., 1968).

d) Organic Nitrogen

WILSON (1959) found 0.02 to 0.2 mg/l organic nitrogen in snow collected above the vegetation line in New Zealand, with most of the nitrogen occurring in this form in some samples. He considers this organic nitrogen to be derived from the sea. Similar amounts were found by TARRANT et al. (1967) in rain collected near the Pacific coast in Oregon (see Table 7-I-2); these samples were taken in a forest opening, however, and as on-shore storm activity was inversely related to the amount of organic nitrogen found, the authors think that the organic compounds stem from the forest itself. SEMENOV et al. (1966, 1967) found amino- and protein-nitrogen in rain and snow (see Table 7-I-2).

SIDLE (1967) found free amino acids in precipitation collected in Central London and its vicinity. Glycine with 7 µg/l was most abundant, followed by valine, alanine, glutamic acid, leucine/isoleucine, and aspartic acid. He advances the interesting theory of an abiotic origin in aerosol particles containing hydrochloric acid, under the influence of ultraviolet light.

The contribution of nitrogen compounds in rain to the nitrogen balance of soil, especially for cultivated land, is small (see Section 7-L).

Revised manuscript received: July 1970

7-K. Abundance in Common Sediments and Sedimentary Rock Types

I. Recent Sediments

Particulate organic matter is the main source of nitrogen in surface sediments. Thus nitrogen shows the same enrichment with decreasing grain size as organic carbon (CORRENS, 1939). EMERY (1960) gives 0.25% in the grain size fraction $>62\,\mu m$ and 0.48% in the fraction $<1\,\mu m$, as average nitrogen values for Recent basin sediments off Southern California.

Table 7-K-1 lists N contents of different types of Recent sediments. It shows that near-shore sediments contain more N than do deep sea sediments by upto a factor of ten. It is also worth noting that argillaceous and calcareous sediments contain about the same amount of nitrogen, in contrast to the amounts found in ancient sediments.

After deposition the organic matter undergoes a rapid decomposition, leading primarily to the production of ammonia. This results in a sharp decrease in N content within the first few inches of burial (REVELLE and SHEPARD, 1939; EMERY and RITTENBERG, 1952), while the interstitial water is highly enriched in ammonia compared to the overlying sea water (RITTENBERG et al., 1955; see Fig. 7-K-1).

The production of nitrate nitrogen occurs only at the surface of the sediment in aerobic environments. RITTENBERG et al. (1955) found nitrate-N only in the interstitial water of a sediment core with positive Eh-values (Catalina Basin, $Eh = +200$ to 300 mV). WLOTZKA (1961) reports 20 ppm nitrate-N for the upper 0.5 m only of a pelagic red clay from the Atlantic. Nitrate-N can be denitrified to N_2 and can escape into the overlying water, but only under anaerobic conditions.

Fig. 7-K-1 and Table 7-K-2 show that the bulk of the nitrogen in Recent sediments is present in organic form. Part of this is in the form of soluble amino acids and a large amount occurs as acid-insoluble compounds, i.e., asphaltics and kerogens (see FORSMAN and HUNT, 1958; MARLETT and ERDMAN, 1959; DEGENS et al., 1963; TILO, 1965).

II. Ancient Sediments

Fig. 7-K-2 gives the frequency distribution of the nitrogen contents of 10,423 North American ancient sediments as measured by TRASK and PATNODE (1942). Table 7-K-3 lists a few additional measurements in different rock types.

During diagenesis most of the organic nitrogen compounds in sediments are destroyed, mainly by deamination leading to the formation of ammonia (VALLENTYNE, 1957; ERDMAN, 1961). This may be lost or retained in the rock, the retention being favored by the presence of clay minerals capable of adsorbing ammonium. Hence, shales contain the highest amount of nitrogen (Table 7-K-3), and in grey-

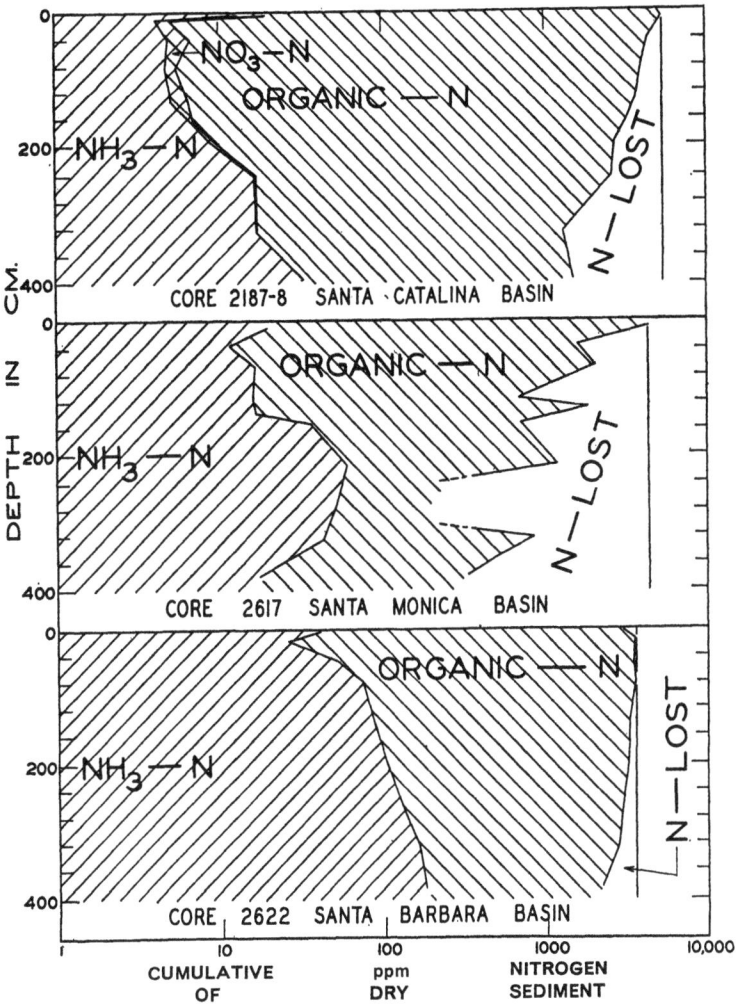

Fig. 7-K-1. Depth distribution of nitrogen compounds in interstitial water of sediments from three basins. Note logarithmic scale. (From RITTENBERG et al., 1955)

wacke, sandstone and limestone the N content varies in a manner proportional to the clay or mica content (Fig. 7-K-3).

In ancient sediments the main part of the nitrogen is contained in inorganic form, i.e., as ammonia. Exceptions are certain carbon-rich bituminous shales, such as the Kupferschiefer which is deposited under reducing conditions. WLOTZKA (1961) found in two Kupferschiefer samples 2,100 and 2,500 ppm total combined nitrogen, of which 85% and 76% respectively was organic nitrogen. Similarly, STEVENSON (1962) reports 1,700 and 4,000 ppm N in two carbon-rich shales, of which 86% was in organic form.

The amino acid content of ancient sediments reaches only a few ppm. MARLETT and ERDMAN (1959) found 0.7 to 2.1 ppm amino acid nitrogen in marine shales,

Table 7-K-1. Nitrogen content and C/N ratio of Recent sediments

Rock type and locality	Number analyzed	Nitrogen, %		C/N ratio		Reference
		range	mean	range	mean	
Basin sediments off Southern California,						
surface	13	0.18 —0.66	0.40	9.1—13.1	11.2	Emery and
60 cm below surface	13		0.19	10.6—14.4	12.3	Rittenberg (1952)
Near-shore sediments, Gulf of Maine and Puget Sound,						
surface samples	43	0.02 —0.58	0.19	5.1—25.5	9.9	Bader (1955)
core samples, surface	12	0.09 —0.58	0.33	5.1—11.4	8.6	
core samples, 8—250 cm below surface	22	0.08 —0.45	0.23	5.3—18.4	10.4	
Calcareous pelagic sediments	9	0.008—0.030	0.014	10—58	21	El Wakeel and Riley
Argillaceous pelagic sediments	12	0.006—0.028	0.016	5—54	15	(1961)
Siliceous pelagic sediments	6	0.002—0.059	0.016		26	
Atlantic pelagic clay	6	0.025—0.034	0.027			Wlotzka (1961)
Pelagic clay	3	0.03 —0.07	0.05	7 —14	10	Rittenberg et al. (1963)
Pelagic silt	1		<0.01		41	
Hemipelagic calcareous siliceous ooze	7	0.02 —0.07	0.04	16—140	61	
Sediments, Malacca Strait muddy sand	56	trace — 0.24		0.4—37	15.4	Keller and Richards (1967)
Lake sediments,						
Lake Kizaki-ko	16	0.26 —0.46	0.36	12.3—16.6	14.3	Koyama (1966)
Lake Nakatsuna-ko	4	0.49 —0.62	0.54	15.0—16.1	15.4	

Table 7-K-2. *Nitrogen content and nitrogen compounds of Experimental Mohole Sediments, from* TILO (1965). *For description of sediments see also* RITTENBERG *et al.* (1963)

Sample	Depth, m	Total N, ppm	Per cent of total N						C/N
			1	2	3	4	5	6	
EM 8-1	—	795	15.4	17.6	21.2	16.0	5.8	24.0	—
EM 8-9	83	637	31.0	12.4	2.4	18.0	2.7	33.5	140
EM 8-10	—	496	15.3	20.6	0	18.1	3.2	42.8	—
EM 8-11	101	456	20.4	16.5	4.0	27.6	3.1	28.4	35
EM 8-12	110	612	19.2	19.8	3.3	19.1	2.1	36.5	27
EM 8-13	—	645	10.9	21.5	0	15.7	1.2	50.7	—
EM 8-14	129	279	14.0	19.4	2.0	46.6	1.4	16.6	16
EM 8-15	138	627	34.2	9.9	2.3	15.3	1.1	37.2	95

1: Acid-insoluble N; 2: Fixed ammonium-N; 3: Ammonium-N; 4: Amino acid-N; 5: Aminosugar-N; 6: Unidentified N.

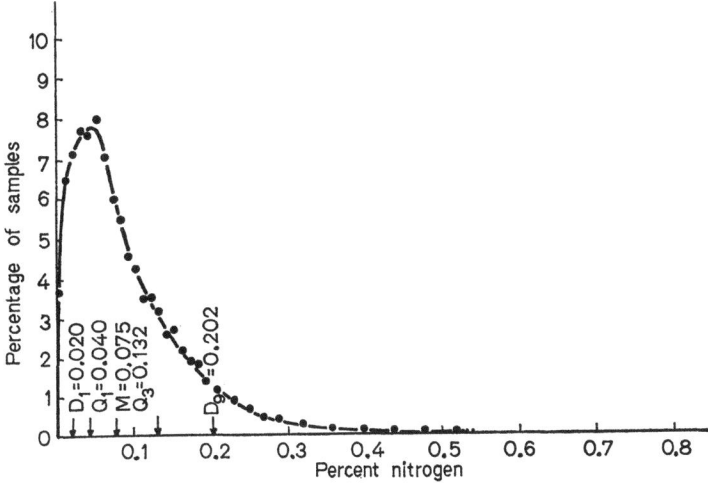

Fig. 7-K-2. Frequency distribution curve for nitrogen in 10,423 ancient North American sediments, from TRASK and PATNODE (1942). Points D, Q, and M give decil, quartil, and median values

accounting for only 0.2 to 0.4% of the total N; similar amounts were reported by DEGENS and BAJOR (1962). FORSMAN and HUNT (1958) found that about 20% of the nitrogen in shales occurs in the acid-insoluble kerogen, the rest being ammonium.

Nitrate nitrogen was found by WLOTZKA (1961) in only a few shales which occur in intermediate layers in salt deposits, in amounts between 5 and 15 ppm. In limestones, however, nitrate nitrogen occurs more frequently; of 37 rocks measured, 22 contained amounts between 5 and 20 ppm. Comparable amounts were reported by CHALK and KEENEY (1971). This may be due to a replacement of the CO_3^{2-} radical by NO_3^-, which has a similar size and configuration (WLOTZKA, 1961).

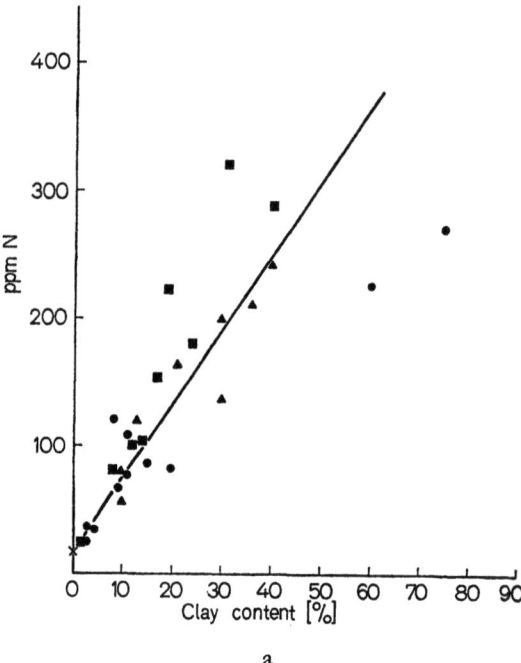

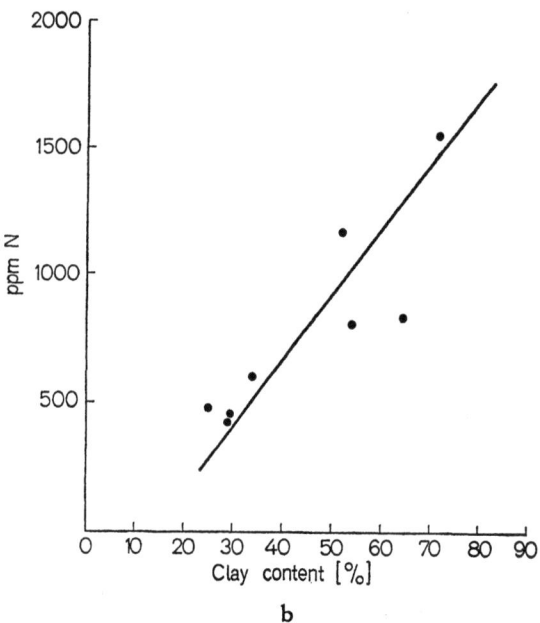

Fig. 7-K-3a and b. Nitrogen versus clay content in sediments. a Ancient sediments; ● limestones, ▲ greywackes, ■ sandstones (WLOTZKA, 1961); b Recent sediments (MARLETT and ERDMAN, 1959). The line drawn is the best fit straight line

Table 7-K-3. *Nitrogen content of ancient sediments*

Rock type	Number analyzed	Ammonia-N, ppm range	Ammonia-N, ppm mean	Reference
North American sediments	10,423	200—8,650	600	Trask and Patnode (1942)
Shales	13	640—4,800	1,870	Forsman and Hunt (1957)
Marine shales, Oligocene	6	190— 450	325	Marlett and Erdman (1959)
Shales	61	300—1,200	580	Wlotzka (1961)
Shales, Paleozoic	6	510—4,000	1,900	Stevenson (1962)
Argillaceous slate	9	900—3,000	1,500	Degens and Bajor (1962)
Greywackes	30	46— 240	180	Wlotzka (1961)
Sandstones	69	23— 300	120	Wlotzka (1961)
Limestones and dolomites	115	4— 200	73	Wlotzka (1961)
(nitrate-N):	(37)	(0— 20)	(5)	
Cherts	5	170— 350	210	Wlotzka (1961)

III. The C/N Ratio of Sediments

The C/N ratio of sediments depends on the nature of the organic material incorporated and on the process of decomposition after burial. Marine and lake planktons have C/N ratios of 5.7 (Sverdrup et al., 1942; Koyama, 1966), the mean value for plant material being 15.0 (Bowen, 1966). The main nitrogen-containing compounds, proteins and amino acids, are less stable than hydrocarbons under geologic conditions (Eglinton, 1969). Thus the relative and absolute amounts of amino acids in a Recent sediment decrease with depth (Rittenberg et al., 1963) and the C/N ratio increases with diagenesis. Trask and Patnode (1942) report an average C/N ratio of 10 for Recent sediments, and of 15 for 6,865 North American ancient sediments (10% of the values being below 8.5 and 10% larger than 25.2). Similar average values between 9 and 14 were found by Emery and Rittenberg (1952) and Emery (1960).

A study of the carbon to nitrogen relationships in distinct sedimentation areas shows a more complex pattern. Arrhenius (1952) found an increase of the C/N ratio with depth in calcareous sediment cores, but a decrease in North Pacific equatorial clays and siliceous sediments. Arrhenius (1950) and Bader (1955) noted a logarithmic linear relationship between carbon and nitrogen in subaquatic sediments of one area (Fig. 7-K-4). The slope of the C—N-correlation line is usually less than one, indicating an increase in the C/N ratio with decreasing organic content. Arrhenius (1950) explained this relationship by the dilution of the organics with mineral matter influencing the decomposition equilibrium. Bader (1955) found also several sedimentation areas showing slopes greater than one (Fig. 7-K-4). He thinks this is due to a high proportion of lignin in the organic matter, resulting in the formation of stable lignin-protein complexes.

Formerly it was common practice to convert the nitrogen content of a sediment into organic content by multiplying by a constant factor. Trask (1939) used a

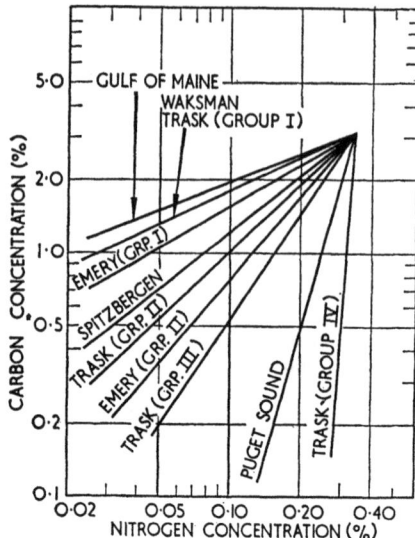

Fig. 7-K-4. Carbon-nitrogen relationship for sediments from various sources. Each line represents a core or group of cores. The point of origin represents the average surface carbon and nitrogen content of all core samples considered. (From BADER, 1955)

conversion factor of 18, based on an average content of 5.5% N in the organic matter of sediments, to obtain a "crude index" of the organic content. BADER (1954) has pointed out that large deviations occur from this mean value; he found nitrogen contents of between 2.0 and 19% in organic matter, leading to conversion factors between 5.2 and 50, with a mean at 20.5 and a coefficient of variation of 45%.

IV. Nitrate Deposits

Nitrate is found in normal desert soils to the extent of 0.1 to 0.2% nitrate nitrogen (MUELLER, 1968), and small nitrate deposits occur in caves and arid regions (MANSFIELD and BOARDMAN, 1932). CLARIDGE and CAMPBELL (1968) report an accumulation of nitrates and sulfates of Na, Ca and Mg, together with traces of iodate, in barren soils of Antarctica.

Large deposits of caliche occur in the North Chilean desert of Atacama. They contain 2 to 30% of NO_3^-, together with sulfates and chlorides of Na, K, Mg and Ca, and small amounts of perchlorates, iodates and borates. These deposits were of great economic value for the production of fertilizers, particularly in the second half of the last century. The production of niter was 2.5 million tons in 1925, but has declined since to 1.3 million tons in 1958 (GORHAM, 1949; HENGLEIN, 1958).

The common features of Antarctic soil and the North Chilean desert are the almost complete absence of biological activity and of leaching by natural waters, enabling nitrate salts to accumulate on a large scale. In the North Chilean desert a secondary concentration of nitrates and iodates, and a separation from sulfates and chlorates, can take place by capillary migration and evaporation (MUELLER, 1968). It does not seem necessary to invoke a special source of the nitrate nitrogen, as had been proposed in the older literature (STUTZER and WETZEL, 1932; GOLDSCHMIDT, 1954).

Revised manuscript received: July 1970

7-L. Biogeochemistry
I. The Biological Nitrogen Cycle

Nitrogen is one of the primary constituents of all living matter and is thus a strongly biophilic element. The average nitrogen content of plants is 3% and of animals 10% (dry weight), the C/N ratios being 15 and 4.5 respectively (BOWEN, 1966); this nitrogen is predominately contained in proteins. The cycle of nitrogen in the biosphere is one of the best studied biological cycles, mainly because of its importance in plant growth and soil fertility. The literature on nitrogen in soils is extensive and cannot be discussed here; the reader is referred to, e.g., the monograph on soil nitrogen edited by BARTHOLOMEW and CLARK (1965) or to SCHEFFER and SCHACHTSCHABEL (1966).

The N content of soils is highly variable. Desert soils contain about 0.1% N and highly organic soils upto 2% N, with most values in normal soils lying between 0.2 and 0.3% (STEVENSON, 1965). This nitrogen is derived from atmospheric N_2 by biological fixation. The most important factor for this process are bacteria living in symbiosis with legumes, but free living bacteria and blue-green algae are also able to fix nitrogen. The amount of nitrogen biologically fixed and added to the earth's surface has been estimated to be about 10^8 tons per year (STEWART, 1967), while the world production of fertilizers amounted to 3×10^7 tons N in 1969/70 (British Sulphur Corp., 1970). Leguminous plants may fix annually from a few to 200 kg N/ha, and nonsymbiotic bacteria 10 to 40 kg N/ha (ALLISON, 1965). The combined nitrogen contained in rain is highly variable, with a mean annual contribution to soil N being 5 kg N/ha, most of which is, however, cyclical (ERIKSSON, 1952; see also Section 7-I).

Fig. 7-L-1 gives sources and transformations of nitrogen in soils (ALLISON, 1965). The major part of nitrogen in surface soils is in organic form (BREMNER, 1965). In subsurface soils the amount of organic nitrogen decreases, and fixed ammonium contained in the lattice of clay minerals increases proportionately upto 60% of the total (STEVENSON and DHARIWAL, 1959).

The organic nitrogen compounds of animals and plants after their death are decomposed by bacteria, mainly to ammonia. This can again be used by plants as a nitrogen source, but most of it is oxidized to nitrite and finally to nitrate by soil bacteria (*Nitrosomonas* and *Nitrobacter*), which use this reaction as an energy source. Thus nitrate is the stable form of nitrogen in soils and is preferentially used as a nutrient by plants. Less common is the use of nitrate as a hydrogen acceptor by certain bacteria living under anaerobic conditions. Nitrate is then reduced to N_2, and to a lesser extent to N_2O. For details of the microbiological nitrogen cycle, see DELWICHE (1964).

Of geochemical importance are the processes which are able to remove nitrogen from the biological cycle. These are the fixation of ammonia in clay minerals, the

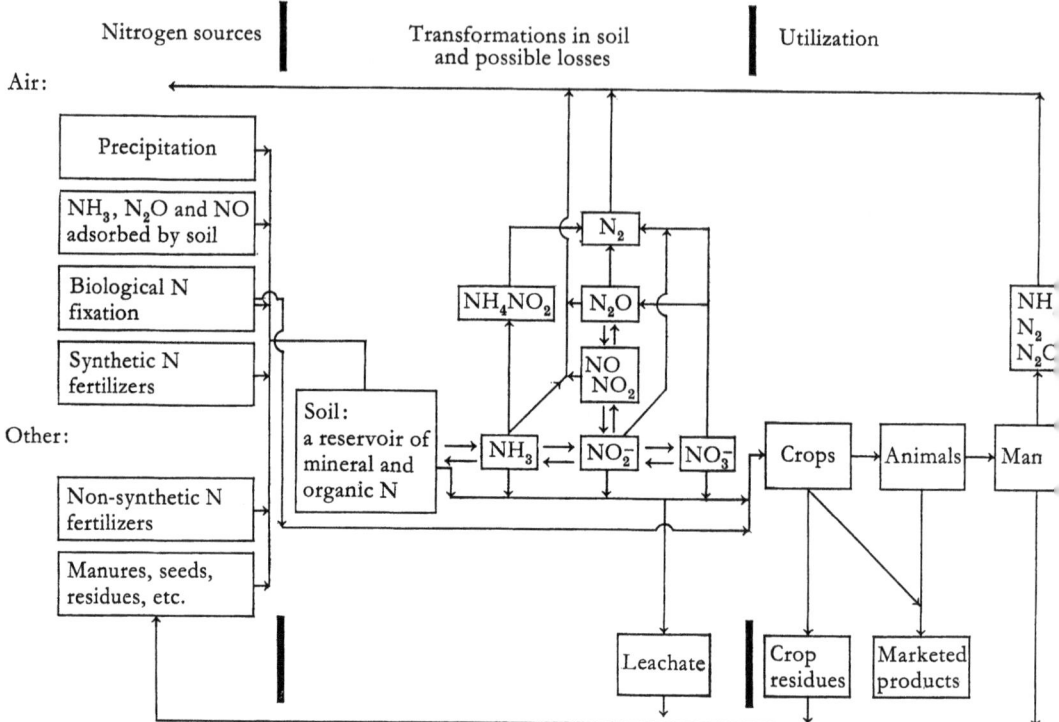

Fig. 7-L-1. Scheme of soil nitrogen sources, transformations and fate of the end products. (From ALLISON, 1965)

formation of stable and insoluble lignin or humic acid complexes (BREMNER, 1959; STEVENSON, 1960) and the leaching of nitrate by percolating waters. The consequences for the nitrogen content of sediments and for the geochemical nitrogen cycle are discussed in Sections 7-K and 7-O.

II. Coal, Petroleum and Natural Gases

Coal contains between 0.2 and 3% (dry weight) of nitrogen which is derived from animal and plant proteins (KIRNER, 1945; FRIEDEL and QUEISER, 1956). The transformation of these compounds and the fixation of nitrogen during coalification has been investigated by FLAIG (1968). During coal distillation this nitrogen is released mainly as ammonia and is used on a large scale for the production of fertilizers. The amount of nitrogen in the coal deposits of the world has been estimated to be 2×10^{11} tons (GORHAM, 1949).

Petroleum and crude oils contain 0.01 to 2% nitrogen (ERDMAN, 1965; SPEERS and WHITEHEAD, 1969). SNYDER (1969) determined the main types of nitrogen compounds in a Californian crude oil, finding that most of them were aromatics.

In natural gases associated with sediments nitrogen is the most frequent non-hydrocarbon component, the content lying usually between a few and 20% N_2 by volume, but sometimes going up to as much as 90% (ERDMAN, 1965; VOZNESENSKII and TERESHCHENKO, 1966). In certain gas fields a decrease of the N_2 content with depth has been found (MOORE, 1966; EICHMANN et al., 1970). A correlation between N_2 and He is found in certain occurrences (MOORE, 1966) but is lacking in others (EICHMANN et al., 1970). A constant Ar/N_2 ratio, which is less than that found in today's atmosphere is reported by SUGISAKI (1964) and EICHMANN et al. (1970).

The origin of this nitrogen is still open to speculation. For a review of the current theories see BEYER (1955) and TOMOR (1968). One possible source is the decomposition of organic matter (KREJCI-GRAF, 1930; BEYER, 1955); however, this is not sufficient to account for all the gases found. The other major source is atmospheric air, trapped in the sediment during its formation or introduced later dissolved in water. Apparently most of the nitrogen found in oil gases and dry natural gases has no genetic relationship to the hydrocarbons in these gases (EICHMANN et al., 1970). Nitrogen isotope measurements related to this problem are discussed in Subsection 7-B-IV.

7-M. Abundance in Common Metamorphic Rock Types

The main source of nitrogen in sediments is organic matter. During metamorphism this is decomposed and nitrogen is liberated, mainly as ammonia. This ammonia may be fixed in micas or other layer lattice silicates (Section 7-D), but at higher temperatures it will escape with other volatiles.

MILOVSKIY and VOLYNETS (1966, 1967) determined ammonia and organic carbon in a series of graphitic quartzites (75 to 90% quartz, 5 to 20% graphite) of known sedimentary origin. Their data are plotted in Fig. 7-M-1. Different samples

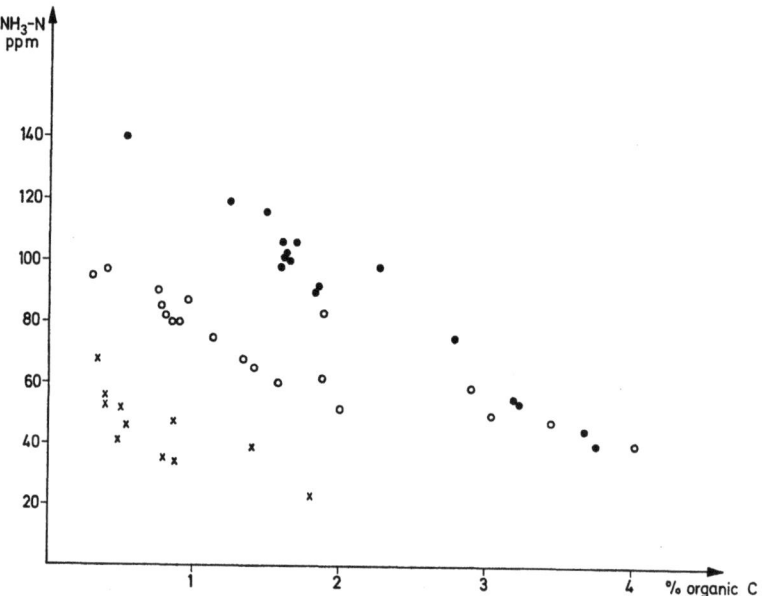

Fig. 7-M-1. Ammonium nitrogen content versus organic carbon content in three series of graphite quarzites, after MILOVSKIY and VOLYNETS (1967). × upper series; ○ middle series; ● lower series

of three formations were analyzed and within each formation an inverse relationship was found between organic carbon content and ammonium content. The organic nitrogen remained undetermined by the method employed. Comparison between formations, on the other hand, shows that the mean ammonium nitrogen content increases with increasing organic carbon content. This indicates that organic nitrogen is the predominant source of ammonia, which is liberated with increasing decomposition of the organic matter. This ammonia is fixed in the rock, while organic carbon is converted into graphite. Biotite seems to be most effective in

Table 7-M-1. *Ammonium nitrogen content in rocks of the Southern Mugodzhar Mountains, U.S.S.R., of progressive metamorphic facies, from* MILOVSKIY *and* VOLYNETS (1966)

Rock	Number analyzed	NH_3-nitrogen, ppm	
		range	average
Low rank metamorphic rocks, greenschist facies and lower	5	149—440	326
Schist formation, lower subfacies of amphibolite facies	5	73—163	114
Gneiss formation, highest subfacies of amphibolite facies	5	9— 58	36

fixing ammonia, as the above authors found the NH_3 content of three Volynian paragneisses could be related to their biotite content.

MILOVSKIY and VOLYNETS (1966) also studied a series of rocks showing different degrees of metamorphism (Table 7-M-1). The gneisses of the highest subfacies of the amphibolite facies had the lowest amount of ammonia N (mean 36 ppm), with schists of the intermediate series belonging to a lower subfacies of the amphibolite facies having higher contents (mean 114 ppm), and the upper series belonging to the greenschist facies showing the highest amounts (mean 326 ppm).

The same authors and also WLOTZKA (1961) studied several series of orthogneisses and paragneisses in order to find out whether a difference in nitrogen content exists between them. WLOTZKA (1961) found the following nitrogen amounts in a rock series from Lower Austria:

orthogneisses (4), range 9 to 38 ppm, average 18 ppm ammonium N
paragneisses (8), range 7 to 102 ppm, average 47 ppm ammonium N

The values of MILOVSKIY and VOLYNETS (1966) in rocks from Russia are quite similar:

orthogneisses (4), range 9 to 40 ppm, average 23 ppm ammonium N
paragneisses (15), range 20 to 106 ppm, average 48 ppm ammonium N

Table 7-M-2. *Nitrogen content of spilites, from* WLOTZKA (1961). *Total combined N is ammonium and nitride nitrogen in untreated sample. Fixed N is the same after shaking with 0.01 N HCl*

Rock	Total combined N ppm	Fixed N, ppm
Spilite, Wallenstein, Rheinisches Schiefergebirge, Germany	40	27
Spilite, Meschede, Rheinisches Schiefergebirge, Germany	36	26
Spilite, Remblinghausen, Rheinisches Schiefergebirge, Germany	47	29
Spilite, Heringhausen, Rheinisches Schiefergebirge, Germany	37	26
Spilite, Andreasberg, Harz, Germany	42	30
Spilite tuff, Langelsheim, Harz, Germany	13	8

This shows that the mean ammonium N content of paragneisses is about twice as high as that of orthogneisses, suggesting that the former derived their nitrogen from the parent sediments. An individual orthogneiss, however, cannot be distinguished from a paragneiss on its nitrogen content alone. Both ranges overlap and they are not significantly different from the range of nitrogen contents in igneous rocks.

The series of spilites analyzed by WLOTZKA (1961), (Table 7-M-2), has an average content of 38 ppm total N and 24 ppm fixed N, close to the average of normal unaltered volcanic rocks.

Revised manuscript received: July 1970

7-O. Nitrogen Cycle

The cycle of nitrogen has been investigated, mainly in its connection with the biosphere, by GILSON (1937), VIRTANEN (1952), HUTCHINSON (1954) and others. In Table 7-O-1 the amounts of nitrogen contained in the different units of the earth are listed. They are calculated from the results of the preceding sections. The ammonia N for igneous rocks of the earth's crust was obtained using a 1:1 ratio of granites (average 21 ppm N) and basalts (average 30 ppm N) as the most abundant rocks. The average molecular nitrogen content was taken to be 3 ppm N_2 from the 4 granite values of GOGUEL (1963) given in Table 7-D-1. For sediments, following WEDEPOHL (1969), a ratio of 77% shales (600 ppm N), 15% sandstones and greywackes (150 ppm N) and 8% limestones (70 ppm N) was used, resulting in an average of 490 ppm N. The mass of the earth's crust without sediments was taken as 2.3×10^{19} tons (POLDERVAART, 1955) and the mass of sediments as 1.5×10^{18} tons (WEDEPOHL, 1969). It should be noted that the result for ancient sediments (150 g N/cm²) is ten times higher than the value used by RANKAMA and SAHAMA (1950), which is the value most frequently quoted in the literature. They obtained

Table 7-O-1. *Nitrogen content of the earth*

	Nitrogen species	Nitrogen content			Reference
		in 10^{10} tons	g/cm²	%	
Atmosphere	N_2	386,000	755	73.1	GLÜCKAUF (1951)
	N_2O	0.2	4×10^{-4}		SCHÜTZ et al. (1970)
Igneous rocks in the crust	NH_3	58,000	110	10.6	This work
	N_2	7,000	14	1.4	This work
Ancient sediments	NH_3 and organic N	75,000	150	14.5	This work
Biosphere	Organic N	3	7×10^{-3}		BOWEN (1966)
Ocean	N_2	2,200	4.3	0.4	SVERDRUP et al. (1942)
	NO_3^-	57	0.11		SVERDRUP et al. (1942)
	Organic N	34	0.067		SVERDRUP et al. (1942)
Coal	NH_3	20	0.04		GORHAM (1949)
Nitrate deposits	NO_3^-	0.01	2×10^{-5}		HUTCHINSON (1954)
Total		528,300	1,033	100.0	
Earth's mantle		5,600,000			This work

this value from an indirect calculation using the carbon content of sediments and noted then that it appeared to be too low.

The 160 kg/cm² of igneous rocks weathered in the earth's history (CORRENS, 1948) gives only 4.0 g N as compared to the 83 g N trapped in the 170 kg/cm² of sediments formed. The difference has been withdrawn from the biosphere portion of the cycle and deposited as fossil nitrogen. The amount of 755 g N/cm² contained in the atmosphere today is much higher than the amount liberated during the weathering of rocks. Thus nitrogen belongs to the "excess volatiles" and has probably been introduced into the atmosphere by degassing of the earth's interior (RUBEY, 1955). The early atmosphere may have contained some of its nitrogen in the form of ammonia (UREY, 1952; HOLLAND, 1962), but no direct evidence for this is available from the nitrogen content of rocks. The presence of ammonia and other nitrogen compounds such as hydrogen cyanide may have played an important role in the formation of the first organic compounds and the origin of life (MILLER and UREY, 1959). BADA and MILLER (1968) have argued that the NH_4^+ concentration in the primitive ocean was regulated by its ion exchange equilibrium with clay, setting an upper limit of 0.01 M. They calculated the NH_3 pressure in equilibrium with this solution to be 7.3×10^{-6} atm at 25°C and 1 atm N_2.

The nitrogen content of the earth's mantle is not known. Here the average of 14 ppm N found for ultramafic igneous rocks (Table 7-E-5) was taken to calculate the nitrogen content of the mantle to be 11,000 g N/cm². This is about 15 times the amount contained in the atmosphere today. Evidence for the indigenous nature of nitrogen in deep-seated rocks is its occurrence in diamonds (KAISER and BOND, 1959).

For the cycle of nitrogen on earth, the exchange between atmosphere and biosphere is by far the most important process. The fixation of atmospheric N_2 has been estimated to be 8×10^8 tons (DELWICHE, 1964) and 10^8 tons (STUART, 1967) per year from the biomass production. A minimum value is obtained from the nitrogen discharge of rivers into the ocean. From LIVINGSTONE's (1963) summation of the annual discharge of the continents and their mean nitrogen content, an amount of 7.0×10^6 tons N/year results. The amount contained in the 10^{10} tons suspended material discharged is difficult to estimate. A value of 5×10^6 tons of N/year is obtained if an average nitrogen content of 500 ppm is assumed, as for ancient sediments. The fixation of 10^8 to 10^9 tons of N_2 per year will deplete the atmosphere in about 10^7 years. As SILLÉN (1966) has pointed out, under the present conditions of oxygen pressure, NO_3^- is the stable nitrogen compound so that all nitrogen converted into this form biologically, should end up in the ocean as NO_3^-. To maintain a steady state equilibrium a corresponding denitrification rate must be postulated. At present such reactions are known to occur only under anaerobic conditions on land (ALLISON, 1965) or in ocean water with oxygen concentrations below 10 ml/l (GOERING and DUGDALE, 1966). The missing link may be the production of N_2O in soil and in the ocean. The mixing ratio of 0.26 ppm N_2O together with a residence time of 10 to 70 years in the atmosphere (SCHÜTZ et al., 1970), yields a production rate of between 1.9×10^7 and 1.3×10^8 tons N per year. A production of N_2 in the ocean has also been inferred by CRAIG et al. (1967) from the nitrogen concentration in the Pacific, see Section 7-I.

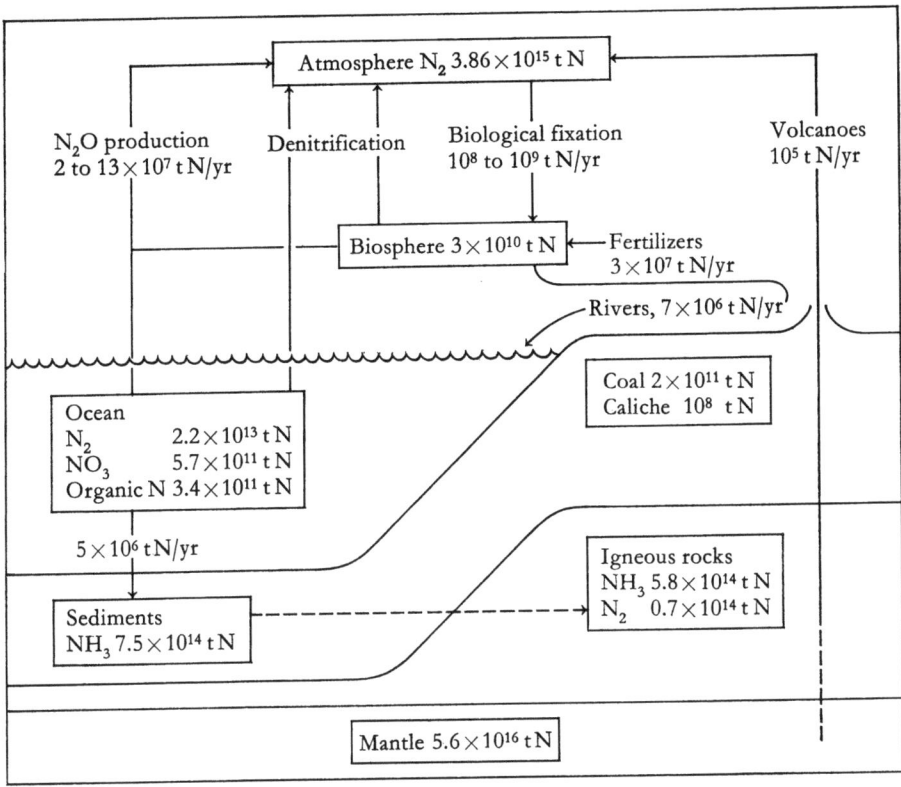

Fig. 7-O-1. Cycle of nitrogen. For references, see text and Table 7-O-1

The only other significant addition of N_2 to the atmosphere will be from volcanic emanations. WEDEPOHL (1969) estimated the amount of lava extruded each year to be 3×10^9 tons, expelling about 1% of water. If an average nitrogen content in these exhalations is taken as 0.1 to 1%, then a value of between 3×10^4 to 3×10^5 tons of N_2/year is obtained. An absolute upper limit for this value will be the amount of nitrogen in atmosphere, ocean and sediments (4.6×10^{15} tons) divided by the age of the earth, which gives 10^6 tons nitrogen per year.

Revised manuscript received: July 1970

References: Sections 7-B to 7-M, 7-O.

ADEL, A.: Note on the atmospheric oxides of nitrogen. Astrophys. J. **90**, 627 (1939).
ALLER, L. H.: The abundance of the elements in the sun and stars. In: Handbuch der Physik, Bd. 51, Astrophysik II. Berlin-Göttingen-Heidelberg: Springer 1958.
— LILLER, W.: Photoelectric spectrophotometry of gaseous nebulae. I. The Orion nebula. Astrophys. J. **130**, 45 (1959).
ALLISON, F. E.: Evaluation of incoming and outgoing processes that effect soil nitrogen. In: BARTHOLOMEW and CLARK (1965).
ANDERS, E.: Origin, age and composition of meteorites. Space Sci. Rev. **3**, 583 (1964).
— DUFRESNE, E. R., HAYATSU, R., CAVAILLÉ, A., DUFRESNE, A., FITCH, F. W.: Contaminated meteorite. Science **146**, 1157 (1964).
ANDERSEN, C. A., KEIL, K., MASON, B.: Silicon oxynitride: a meteoritic mineral. Science **146**, 256 (1964).
ÅNGSTRÖM, A., HÖGBERG, L.: On the content of nitrogen (NH_4—N and NO_3—N) in atmospheric precipitation. Tellus **4**, 31 (1952).
— — On the content of nitrogen in atmospheric precipitation in Sweden, II. Tellus **4**, 271 (1952).
ARNOLD, F., BERTHOLD, W., BETZ, B., LÄMMERZAHL, P., ZÄHRINGER, J.: Mass spectrometer measurements of positive ions and neutral gas between 100 and 233 km above Andöya. Norway. Space Res. **9**, 256 (1969).
ARRHENIUS, G.: Carbon and nitrogen in subaquatic sediments. Geochim. Cosmochim. Acta **1**, 15 (1950).
— Sediment cores from the East Pacific. Repts. Swedish Deep-Sea Expedition 1947—48, vol. V (1), 1—227 (1952).
AVDUEVSKY, V. S., MAROV, M. YA., ROZHDESTVENSKY, K. M.: A tentative model of the Venus atmosphere based on the measurements of Veneras 5 and 6. J. Atmospheric Sci. **27**, 561 (1970).
BADA, J. L., MILLER, S. L.: Ammonium ion concentration in the primitive ocean. Science **159**, 423 (1968).
BADER, R. G.: Use of factors for converting carbon or nitrogen to total sedimentary organics. Science **120**, 709 (1954).
— Carbon and nitrogen relations in surface and subsurface marine sediments. Geochim. Cosmochim. Acta **7**, 205 (1955).
BANNISTER, F. A.: Osbornite, meteoritic titanium nitride. Mineral. Mag. **26**, 36 (1941).
BARKER, D. S.: Ammonium in alkali feldspars. Am. Mineralogist **49**, 851 (1964).
BARTHOLOMEW, W. V., CLARK, F. E.: Soil Nitrogen. Madison, Wisconsin, USA; Amer. Soc. of Agronomy 1965.
BATES, D. R., HAYES, P. B.: Atmospheric nitrous oxide. Planet. Space Sci. **15**, 189 (1969).
— WITHERSPOON, A. E.: The photochemistry of some minor constituents of the earth's atmosphere. Monthly Notices Roy. Astron. Soc. **112**, 101 (1952).
BEERS, J. R., KELLY, A. C.: Short-term variation of ammonia in the Sargasso Sea off Bermuda. Deep-Sea Res. Oceanog. Abstr. **12**, 21 (1965).
BENSON, B. B., PARKER, P. D. M.: Relations among the solubilities of nitrogen, argon and oxygen in distilled water and sea water. J. Phys. Chem. **65**, 1489 (1961).
BEYER, A. K.: Der Ursprung des freien Stickstoffs in den Erdgasen. Wiss. Z. Univ. Greifswald **4**, 302 (1954/55).
BISWAS, S., FICHTEL, C. E.: Composition of solar cosmic rays. Space Sci. Rev. **4**, 709 (1965).
BLOSSEY, R. C., PEHLKE, R. D.: The solubility of nitrogen in liquid Fe-Ni-Co alloys. Trans. AIME **236**, 566 (1966).

© Springer-Verlag Berlin · Heidelberg 1972

BOKHOVEN, C., THEEUWEN, H. J.: Determination of the abundance of carbon and nitrogen isotopes in dutch coals and natural gases. Nature **211**, 927 (1966).
BOLTER, E.: Über Zersetzungsprodukte von Olivin-Feldspat-Basalten. Beitr. Mineral. Petrol. **8**, 111 (1961).
BOWEN, H. J. M.: Trace Elements in Biochemistry. London and New York: Academic Press 1966.
BRACANNOT, H.: Indice de débris organique dans les roches les plus annciennes du globe. Ann. Chim. Phys. **67**, 104 (1838).
BRANDT, J. C., MCELROY, M. B.: The Atmospheres of Venus and Mars. New York-London: Gordon and Breach, Publ. 1968.
BREMNER, J. M.: Organic nitrogen in soils. In: Soil Nitrogen (BARTHOLMEW and CLARK, eds.), p. 93. Madison, Wisconsin: Amer. Soc. of Agronomy 1965.
BRIGGS, M. H.: Organic extracts of some carbonaceous chondrites. Life Sci. **2**, 63 (1963).
British Sulphur Corporation: Statistical Supplement No. 2, The British Sulphur Corporation, Ltd. (London) (1970).
BRUN, A.: Recherches sur l'exhalaison volcanique. Genève: Libraire Kündig 1911.
BUACHIDZE, G. I., SIKHARULIDZE, G. G.: Occluded gases in diabases from the Barzhom area. Soobshch. Nauk. Gruz. SSR **39**, 349 (1965).
BUDDHUE, J. D.: Nitrogen and its compounds in meteorites. Popular Astronomy **50**, 561 (1942).
BUNSEN, M. R.: Recherches sur la formation des roches vulcanique en Islande. Ann. Chim. Phys. III, **38**, 289 (1853).
CADLE, R. D., FISCHER, W. H., FRANK, E. R., LODGE, J. P., JR.: Particles in the Antarctic atmosphere. J. Atmos. Sci. **25**, 100 (1968).
— FRANK, E. R.: Particles in the fume from the 1967 Kilauea eruption. J. Geophys. Res. **73**, 4780 (1968).
CAMERON, A. G. W.: A new table of abundances of the elements in the solar system. In: Origin and Distribution of the Elements (L. H. AHRENS, ed.). Oxford-London-New York: Pergamon Press 1968.
CAUGHLAN, G. R.: Approach to equilibrium in the CNO bicycle. Astrophys. J. **14**, 688 (1965).
CHAIGNEAU, M., BORDET, P.: Sur les teneurs en gaz occlus et eau des rétinites et obsidiennes. Compt. Rend. **255**, 3019 (1962).
— — Gaz occlus dans les verres de la vallée des Dix Mille Fumées (Katmai, Alaska). Compt. Rend. **256**, 3167 (1963).
— DEBRUNE, M.: Sur les gaz occlus dans diverses roches volcanique de la Montagne-Pelée (Martinique). Compt. Rend. **252**, 3842 (1961).
— MARINELLI, G.: Sur les gaz occlus dans les minéraux du gisement de magnétite de Calamita (Ile d'Elbe). Compt. Rend. **259**, 4299 (1964).
CHALK, P. M., KEENEY, D. R.: Nitrate and ammonium content of Wisconsin limestones. Nature **229**, 42 (1971).
CHAMBERLIN, R. T.: The gases in rocks. Carnegie Inst. Washington, Publ. 106 (1908).
CHAPMAN, M. W., BROIDA, H. P.: Bibliography on ^{15}N. Natl. Bur. Std. (U.S.) Circ. **575** (1956).
CHEBYKINA, N. V.: Content of bound nitrogen in atmospheric precipitation in the area of Syktyvkar City. Izv. Komi Filiala Vses. Geogr. Obshchestva **1964**, 83 (1964); cited after Chem. Abstr. 64:4812h.
CHEUNG, A. C., RANK, D. M., TOWNES, C. H., THORNTON, D. D., WELCH, W. J.: Detection of NH_3 molecules in the interstellar medium by their microwave emission. Physical Rev. Letters **21**, 1701 (1968).
CHOW, T. J., MANTYLA, A. W.: Inorganic nutrient anions in deep ocean waters. Nature **206**, 383 (1965).
CLARIDGE, G. G. C., CAMPBELL, I. B.: Origin of nitrate deposits. Nature **217**, 428 (1968).
CLARKE, F. W., STEIGER, G.: The action of ammonium chloride upon silicates. U.S. Geol. Surv. Bull. **207**, 57 (1902).

CORRENS, C. W.: Pelagic sediments of the north Atlantic Ocean. In: Recent Marine Sediments. Tulsa, Okla.: Assoc. Petrol. Geologists 1939.
— Die geochemische Bilanz. Naturwiss. **1**, 7 (1948).
CRAIG, H., WEISS, R. F., CLARKE, W. B.: Dissolved gases in the equatorial and south Pacific Ocean. J. Geophys. Res. **72**, 6165 (1967).
CRUIKSHANK, D. P., BINDER, A. B.: Minor constituents in the atmosphere of Jupiter. Astrophysics and Space Sci. **3**, 347 (1969).
DALGARNO, A., MCELROY, M. B.: Mars: Is nitrogen present? Science **170**, 167 (1970).
DAMON, P. E., KULP, J. L.: Excess helium and argon in beryl and other minerals. Am. Mineralogist **43**, 433 (1958).
DARGENTA, A., STANESCU, Z., ONAC, Z.: Nitrate content of ground water in the Dobrogea region. Igiena (Bukarest) **16**, 541 (1967); cit. after Chem. Abstr. **69**, 21787b.
DEGENS, E. T., BAJOR, M.: Die Verteilung der Aminosäuren in limnischen und marinen Schiefertonen des Ruhrkarbons. Fortschr. Geol. Rheinland-Westfalen **3** (2), 429 (1962).
— EMERY, K. O., REUTER, J. H.: Organic materials in Recent and ancient sediments, part III: Biochemical compounds in San Diego Trough, California. Neues Jahrb. Geol. Palaeontol. Monatsh. **1963**, 231 (1963).
DELESSE, A.: Recherches de l'azote et des matières organiques dans l'écorce terrestre. Ann. Min. Ser. 5, **18**, 151 (1860).
DELWICHE, C. C.: The cycling of carbon and nitrogen in the biosphere. In: Microbiology and Soil Fertility, Corvallis, Oregon: Oregon State Univ. Press 1964.
DOLE, M., LANE, G. A., RUDD, D. P., ZAUKELIS, D. A.: Isotopic composition of atmospheric oxygen and nitrogen. Geochim. Cosmochim. Acta **6**, 65 (1954).
DOLGOV, YU. A., SHUGUROVA, N. A.: O gazakh postmagmaticheskikh protsessov mineraloobrazovaniya. (The gases of postmagmatic minerogenetic processes.) Dokl. Akad. Nauk. SSSR **170**, 1422 (1966).
DOLLFUS, A.: Observation d'une atmosphère autour de la planète Mercure. Compt. Rend. **231**, 1430 (1950).
DONAHUE, T. M.: Ionospheric composition and reactions. Science **159**, 489 (1968).
DOUGLAS, E.: Solubilities of argon and nitrogen in sea water. J. Phys. Chem. **69**, 2608 (1965).
DUGDALE, R. C., MENZEL, D. W., RYTHER, J. H.: Nitrogen fixation in the Sargasso Sea. Deep-Sea Res. **7**, 297 (1961).
DUURSMA, E. K.: The dissolved organic constituents of sea water. In: Chemical Oceanography (J. P. RILEY and G. SKIRROW, eds.), p. 433. London and New York: Academic Press 1965.
EGLINTON, G.: Organic Geochemistry. The Organic Chemist's Approach. In: Organic Geochemistry—Methods and Results (G. EGLINTON and M. T. J. MURPHY, eds.) Berlin-Heidelberg-New York: Springer 1969.
EICHMANN, R., PLATE, A., BEHRENS, W., KROEPELIN, H.: Das Isotopenverhältnis des Stickstoffs in einigen Erdgasen, Erdölgasen und Erdölen Nordwestdeutschlands. Erdöl Kohle, **24**, 2 (1971).
ELINSON, M. M.: Gases which are occluded in rocks and minerals. Trudy Moskov. Geol. Razvedoch Inst. im. S Ordzhonikidze **29**, 195 (1956).
EL WAKEEL, S. K., RILEY, J. P.: Chemical and mineralogical studies of deep-sea sediments. Geochim. Cosmochim. Acta **25**, 110 (1961).
EMERY, K. O.: The sea off Southern California: A modern habitat of petroleum. New York: J. Wiley & Sons 1960.
— RITTENBERG, S. C.: Early diagenesis of California basin sediments in relation to origin of oil. Am. Ass. Petrol. Geol. Bull. **36**, 735 (1952).
ENGBERG, R. A.: The nitrate hazard in well water with special reference to Holt County Nebraska. Nebraska Water Surv. Paper No. 21, 1 (1967); cit. after Chem. Abstr. **68**, 33038a.
ERD, R. C., WHITE, D. C., FAHEY, J. J., LEE, D. E.: Buddingtonite, an ammonium feldspar with zeolitic water. Am. Mineralogist **49**, 831 (1964).
ERDMAN, J. G.: Petroleum- its origin in the earth. In: Fields in Subsurface Environments — a Symposium, Memoir 4, p. 20, Amer. Assoc. Petrol. Geol. (1965).

Nitrogen

ERDMAN, J. G.: Some chemical aspects of petroleum genesis as related to the problem of source bed recognition. Geochim. Cosmochim. Acta 22, 16 (1961).

ERDMANN, H.: Über das Vorkommen von Ammoniakstickstoff im Urgestein. Ber. Deut. Chem. Ges. 29, 1710 (1896).

ERIKSSON, E.: Composition of atmospheric precipitation, I. Nitrogen compounds. Tellus 4, 215 (1952).

EUGSTER, H. P., MUNOZ, J.: Ammonium micas: Possible sources of atmospheric ammonia and nitrogen. Science 151, 683 (1966).

FETH, J. H.: Nitrogen compounds in natural water — a review. Water Resources Research, Washington 2, 41 (1966).

FIELD, G. B.: The atmosphere of mercury. In: The Origin and Evolution of Atmospheres and Oceans (P. J. BRANCAZIO and A. G. W. CAMERON, eds.); New York-London: John Wiley & Sons 1964.

FINLAYSON, J. B., BARNES, I. L., NAUGHTON, J. J.: Developments in volcanic gas research in Hawai. In: The Crust and Upper Mantle of the Pacific Area (L. KNOPOFF et al., eds.). Washington, D. C : Amer. Geophys. Union 1968.

FLAIG, W.: Origin of nitrogen in coal. Chem. Geology 3, 485 (1968).

FORSMAN, J. P., HUNT, J. M.: Insoluble organic matter (kerogen) in sedimentary rocks. Geochim. Cosmochim. Acta 15, 170 (1958).

FOSHAG, W. F., HENDERSON, E. P.: Primary sublimates at Paricutin Volcano, Mexico. Am. Geophys. Union Trans. 27, 685 (1946).

FOWLER, W. A.: Experimental and theoretical results on nuclear reactions in stars. Mém. Soc. Roy. Sci. Liège 14, 88 (1964).

— The empirical foundations of nucleosynthesis. In: Origin and Distribution of the Elements (L. H. AHRENS, ed.). Oxford-London-New York: Pergamon Press 1968.

FRAGA, F.: Distribution of particulate and dissolved N in the western Indian Ocean. Deep-Sea Res. Oceanog. Abstr. 13, 413 (1966).

FRIEDEL, R. A., QUEISER, J. A.: Infrared analysis of bituminous coals and other carbonaceous materials. Anal. Chem. 28, 22 (1956).

FULLER, G. H.: Relative isotopic abundances. In: Nuclear Data Tables (K. WAY, ed.). Nat. Acad. Sci., Nat. Research Council, p. 66, Washington, D. C., 1959.

GAMBELL, A. W., FISHER, D. W.: Occurrence of sulfate and nitrate in rainfall. J. Geophys. Res. 69, 4203 (1964).

GAUTIER, A.: Sur l'existence d'azotures, argonures, arséniures et iodures dans les roches cristallins. Compt. Rend. 132, 932 (1901).

— La genèse des eaux thermales et ses rapports avec le vulcanisme. Ann. Mines, Ser. X, 9, 316 (1906).

GEORGII, H. W.: Untersuchungen über atmosphärische Spurenstoffe und ihre Bedeutung für die Chemie der Niederschläge. Geofis. Pura Appl. 47, 155 (1960).

GHOSH, S. N.: Distribution and lifetimes of N and NO between 100 and 280 km. J. Geophys. Res. 73, 309 (1968).

— HINTON, B. B., JONES, L. M., LEITE, R. J., MASON, C. J., SCHAEFER, E. J., WALTERS, M.: Atomic nitrogen in the upper atmosphere measured by mass spectrometers. J. Geophys. Res. 73, 4425 (1968).

GIBSON, E. K., MOORE, C. B.: The distribution of total nitrogen in iron meteorites. Geochim. Cosmochim. Acta 35, 877 (1971).

— — LEWIS, C. F.: Total nitrogen and carbon abundances in carbonaceous chondrites. Geochim. Cosmochim. Acta 35, 599 (1971).

GILSON, H. C.: The nitrogen cycle. The John Murray expedition 1933—34. Sci. Rep. 2, 21 (1937).

GLUECKAUF, E.: The composition of atmospheric air. In: Compendium of Meteorology. Boston, Mass.: Am. Meteorolog. Soc. 1951.

GEORING, J. J., DUGDALE, V. A.: Estimates of the rate of denitrification in a subarctic lake. Limnol. Oceanog. 11, 113 (1966).

GOGUEL, R.: Die chemische Zusammensetzung der in den Mineralen einiger Granite und ihrer Pegmatite eingeschlossenen Gase und Flüssigkeiten. Geochim. Cosmochim. Acta 27, 155 (1963).

GOLDBERG, L., MÜLLER, E. A., ALLER, L. H.: The abundance of the elements in the solar atmosphere. Astrophys. J., Suppl. Ser. 5, 1 (1960).

GOLDSCHMIDT, V. M.: Geochemistry. Oxford: Clarendon Press 1954.
GORHAM, H. R.: Nitrates and nitrogenous compounds. In: Industrial Minerals and Rocks (S. H. DOLBEAR and O. BOWLES, eds.), p. 643. New York: Am. Inst. Mining a. Metallurgical Engineers 1949.
GREENBERG, J. M.: Interstellar grains. Sci. Am. **217**, (4), 106 (1967).
GREENSPAN, J. A., OWEN, T.: Jupiter's atmosphere: its structure and composition. Science **156**, 1489 (1967).
GULKIS, S., MCDONOUGH, T. R., CRAFT, H.: The microwave spectrum of Saturn. Icarus **10**, 421 (1969).
GUNTER, B. D., MUSGRAVE, B. C.: Gas chromatographic measurements of hydrothermal emanations at Yellowstone National Park. Geochim. Cosmochim. Acta **30**, 1175 (1966).
GUSAROVA, A. N., KONNOV, V. A., SAPOZHNIKOV, V. V.: Significant regular characteristics of biogenic element distribution in the Pacific. Khim. Protsessy Moryakh Okeanakh, Akad. Nauk SSSR 119 (1966).
HAMILTON, P. B.: Amino acids on hands. Nature **205**, 284 (1965).
HARTECK, P., DONDES, S.: Origin of nitrous oxide in the atmosphere. Phys. Rev. **95**, 320 (1954).
HAYATSU, R.: Unpublished results (1953), reported in ANDERS *et al* (1964).
— Orgueil meteorite: organic nitrogen contents. Science **146**, 1291 (1964).
— STUDIER, M. H., ODA, A., FUSE, K., ANDERS, E.: Origin of organic matter in early solar system, II. Nitrogen compounds. Geochim. Cosmochim. Acta **32**, 175 (1968).
HAYES, J. M.: Organic constituents of meteorites — a review. Geochim. Cosmochim. Acta **31**, 1395 (1967).
HENGLEIN, F. A.: Chilesalpeter heute. Chemiker-Ztg. **82**, 287 (1958).
HINTENBERGER, H., WEBER, H. W., VOSHAGE, H., WÄNKE, H., BEGEMANN, F., VILCSEK, E., WLOTZKA, F.: Rare gases, hydrogen and nitrogen: concentrations and isotopic composition in lunar material. Science **167**, 543 (1970).
HOERING, T.: Variations of nitrogen-15 abundance in naturally occurring substances. Science **122**, 1233 (1955).
— Variations in the nitrogen isotope abundance. In: Nuclear Processes in Geologic Settings. Nat. Acad. Sci. — Nat. Res. Council Publ. **400**, 39 (1956).
— The isotopic composition of the ammonium and the nitrate ion in rain. Geochim. Cosmochim. Acta **12**, 97 (1957).
— Cosmological and geological implications of isotope ratio variations, Sect. V: Isotopic abundances of lighter elements. Nucl. Sci. Ser., Rep. 23, Nat. Acad. Sci. — Nat. Res. Council Publ. **572**, 161 (1958).
HOERING, T. C., MOORE, H. E.: The isotopic composition of the nitrogen in natural gases and associated crude oils. Geochim. Cosmochim. Acta **13**, 225 (1958).
HOLLAND, H. D.: Model for the evolution of the earth's atmosphere. In: Petrologic Studies, A volume to honour A. F. BUDDINGTON, p. 447. New York: Geol. Soc. Amer. 1962.
HULSTON, J. R., MCCABE, W. J.: Mass spectrometer measurements in the thermal areas of New Zealand. I. Carbon dioxide and residual gas analysis. Geochim. Cosmochim. Acta **26**, 383 (1962).
HUNTEN, D. M., GOODY, R. M.: Venus: the next phase of planetary exploration. Science **165**, 1317 (1969).
HUTCHINSON, G. E.: The biochemistry of the terrestrial atmosphere. In: The Earth as a Planet (G. P. KUIPER, ed.). Chicago: Chicago Univ. Press 1954.
— A Treatise on Limnology. I. Geography, Physics and Chemistry. New York: John Wiley & Sons 1957.
JAMES, D. W., HARWARD, M. E.: Mechanism of NH_3 adsorption by montmorillonite and kaolinite. In: Clays and Clay Minerals, vol. 11. Oxford-London-New York: Pergamon Press 1963.
JASCHEK, M., JASCHEK, C.: Nitrogen deficiency in two supergiants. Astrophys. J. **150**, 355 (1967).
JUNGE, C. E.: The distribution of ammonia and nitrate in rain water over the Unites States. Trans. Am. Geophys. Union **39**, 241 (1958).
— Air Chemistry and Radioactivity. New York-London: Academic Press 1963.

Junge, C. E.: Recent investigations in air chemistry. Tellus **8**, 127 (1956).
— Ryan, T. G.: Study of the SO_2 oxidation in solution and its role in atmospheric chemistry. Quart. J. Roy. Meteorol. Soc. **84**, 46 (1958).
Junk, G., Svec, H. S.: The absolute abundance of the nitrogen isotopes in the atmosphere and compressed gas from various sources. Geochim. Cosmochim. Acta **14**, 234 (1958).
Kaiser, W., Bond, W. L.: Nitrogen, a major impurity in common type I diamond. Phys. Rev. **115**, 857 (1959).
Kaplan, I. R., Degens, E. T., Reuter, J. H.: Organic compounds in stony meteorites. Geochim. Cosmochim. Acta **27**, 805 (1963).
Keil, K., Andersen, C. A.: Occurrences of sinoite, Si_2N_2O, in meteorites. Nature **207**, 745 (1965).
Kelen, T., Mulfinger, H. O.: Mechanismus der chemischen Auflösung von Stickstoff in Glasschmelzen. Glastech. Ber. **41**, 230 (1968).
Keller, G. H., Richards, A. F.: Sediments of the Malacca Strait, southeast Asia. J. Sediment. Petrol. **37**, 102 (1967).
Ketchum, B. H., Vaccaro, R. F., Corwin, N.: The annual cycle of phosphorus and nitrogen in New England coastal waters. J. Marine Res. **17**, 282 (1958).
Kiess, C. C., Karrer, S., Kiess, H. K.: A new interpretation of Martian phenomena. Publ. Astron. Soc. Pacific **72**, 256 (1960).
Kirner, W. R.: The occurrence of nitrogen in coal. In: Chemistry of Coal Utilization, vol. I (H. H. Lowry, ed.). New York: John Wiley 1945.
Kliore, A., Fjeldbo, G., Seidel, B. L., Rasool, S. I.: Mariners 6 and 7: Radio occultation measurements of the atmosphere of Mars. Science **166**, 1393 (1969).
König, H.: Über die Löslichkeit der Edelgase in Meerwasser. Z. Naturforsch. **18a**, 363 (1963).
— Keil, K., Hintenberger, H., Wlotzka, F., Begemann, F.: Untersuchungen an Steinmeteoriten mit extrem hohem Edelgasgehalt. I: Der Chondrit Pantar. Z. Naturforsch. **16a**, 1124 (1961).
— Wänke, H., Bien, G. S., Rakestraw, N. W., Suess, H. E.: Helium, neon and argon in the oceans. Deep-Sea Res. **11**, 243 (1964).
Koritnig, S.: Beitrag zur Geochemie des Fluor. Geochim. Cosmochim. Acta **1**, 89 (1951).
Koyama, T.: Ratios of organic carbon, nitrogen and hydrogen in Recent sediments. Advan. Chem. Ser. **55**, 43 (1966).
Kranz, R.: Stickstoffverbindungen in Silikatmineralien und deren Bedeutung für das technologische Verhalten der Feldspate. Ber. Deut. Keram. Ges. **44**, 430 (1967).
Krauskopf, K.: The heavy metal content of magmatic vapor at 600°C. Econ. Geol. **52**, 786 (1957).
Krejci-Graf, K.: Geochemie der Erdöllagerstätten. Abhandl. Prakt. Geol. Bergwirtschaftslehre, Bd. 20. Halle: W. Knapp 1930.
Lambert, D. L.: Abundance of C, N and O in the sun. Observatory **86**, 190 (1966).
— Abundance of the elements in the solar photosphere. I: C, N, O. Monthly Notices Roy. Astron. Soc. **138**, 143 (1968).
Lewis, J. S.: An estimate of the surface conditions of Venus. Icarus **8**, 434 (1968).
— The clouds of Jupiter and the NH_3—H_2O and NH_3—H_2S systems. Icarus **10**, 365 (1969a).
— Observability of the spectroscopically active compounds in the atmosphere of Jupiter. Icarus **10**, 393 (1969b).
Liebig, J. von: Ann. Chem. Phys. **35**, 329 (1827); cit. after Hutchinson (1954).
Linnenbom, V. J., Swinnerton, J. W., Cheek, C. H.: Evaluation of gas chromatography for the determination of dissolved gases in sea water. Oceanog. Sci. Eng., Trans. Joint Conf. Marine Tech. Soc.—Am. Soc. Oceanog., p. 1009 (1965).
Lipman, C. B.: Discovery of combined nitrogen in stony meteorites. Am. Museum Novitates **589**, 1 (1932).
Livingstone, D. A.: Chemical composition of rivers and lakes. U.S. Geol. Surv. Profess. Papers **440-G** (1963).
Lodge, J. P., Pate, J. B.: Atmospheric gases and particulates in Panama. Science **153**, 408 (1966).

MANSFIELD, G. R., BOARDMAN, L.: Nitrate deposits of the United States. U.S. Geol. Surv. Bull. **838**, 107 pp. (1932).
MARLETT, E. M., ERDMAN, J. G.: Carbon-nitrogen distribution and nitrogen-type relationships in Recent and ancient sediments. Paper presented before the Division of Petrol. Chem., Am. Chem. Soc., Boston Meeting, 1959.
MARSHALL, C. E.: Reactions of feldspars and micas with aqueous solutions. Econ. Geol. **57**, 1219 (1962).
MARSHALL, J. V.: Comm. Lunar and Planetary Lab. **2**, 167 (1964); cit. after OWEN (1966).
MASKELYNE, N. S.: On the mineral constituents of meteorites. Phil. Trans. Roy. Soc. London **160**, 189 (1870).
MASON, B.: The enstatite chondrites. Geochim. Cosmochim. Acta **30**, 23 (1966).
MATTIAT, B.: Beitrag zur Petrographie der Oberharzer Kulmgrauwacke. Beitr. Mineral. Petrog. **7**, 242 (1960).
MAYNE, K. I.: Natural variations in the nitrogen isotope abundance ratio in igneous rocks. Geochim. Cosmoschim. Acta **12**, 185 (1957).
MAZOR, E., WASSERBURG, G. J.: Helium, neon, argon, krypton and xenon in gas emanations from Yellowstone and Lassen Volcanic National Parks. Geochim. Cosmochim. Acta **29**, 443 (1965).
McELROY, M. B.: Structure of the Venus and Mars atmospheres. J. Geophys. Res. **74**, 29 (1969).
McQUEEN, J. H.: Isotopic separation due to settling in the atmosphere. Phys. Rev. **80**, 100 (1950).
MEISSNER, B.: Sauerstoff-, Stickstoff- und Kohlenstoffhaushalt der Saale zwischen Eichicht und Groß-Rosenburg. Fortschr. Wasserchemie Grenzgebiete **2**, 11 (1965).
MEYER, H.: Über den Einfluß von stickstoffhaltigen Verbindungen in Feldspaten auf die Blasenbildung. Ber. Deut. Keram. Ges. **42**, 248 (1965).
MILLER, L. E.: The chemistry and vertical distribution of the oxides of nitrogen in the atmosphere. Geophys. Res. Papers **39**, 1—135 (1956).
MILLER, S. L., UREY, H. C.: Organic compound synthesis on the primitive earth. Science **130**, 245 (1959).
MILOVSKIY, A. V., VOLYNETS, V. F.: Nitrogen in metamorphic rocks. Geochem. Internat. **3**, 752 (1966) [Translated from Geokhimiya **8**, 936 (1966)].
— — Voprosy̆ geokhimii azota (The problems of nitrogen geochemistry). Vestn. Mosk. Univ. Ser. IV, **22**, 125 (1967).
MITRA, A. P.: A ionospheric estimate of nitric oxide concentration in the D-region. J. Atmospheric Terrest. Phys. **28**, 945 (1966).
MOORE, C. A.: Occurrence of nitrogen in natural gases of the Oklahoma Panhandle. Proc. Gas Conditioning Conf., Univ. Oklahoma, p. E-1 (1966).
MOORE, C. B., GIBSON, E. K.: Nitrogen abundances in chondritic meteorites. Science **163**, 174 (1969).
— — KEIL, K.: Nitrogen abundances in enstatite chondrites. Earth Planetary Sci. Letters **6**, 457 (1969).
— — LARIMER, J. W., LEWIS, C. F., NICHIPORUK, W.: Total carbon and nitrogen abundances in Apollo 11 lunar samples and selected achondrites and basalts. Geochim. Cosmochim. Acta, Suppl. 1, Vol. 2, 1375 (1970a).
— LEWIS, C. F., GIBSON, E. K., NICHIPORUK, W.: Total carbon and nitrogen abundances in lunar samples. Science **167**, 495 (1970b).
MORTLAND, M. M.: Reactions of ammonia in soils. Advan. Agron. **10**, 325 (1958).
MROSE, M.: Measurements of pH, and chemical analyses of rain-, snow-, and fog-water. Tellus **18**, 266 (1966).
MÜLLER, E. A.: The solar abundances. In: Origin and Distribution of the Elements (L. H. AHRENS, ed.). Oxford-London-New York: Pergamon Press 1968.
MÜLLER, E. P., WIENHOLZ, E. P.: Isotopengeochemie und Erdölgenese. Z. Angew. Geol. **14**, 176 (1968).
MUELLER, G.: The properties and theory of genesis of the carbonaceous complex within the Cold Bokkeveld meteorite. Geochim. Cosmochim. Acta **4**, 1 (1953).
— Genetic histories of nitrate deposits from Antarctica and Chile. Nature **219**, 1131 (1968).

MÜLLER, O.: Personal communication, 1969.
— GENTNER, W.: Gas content in bubbles of tektites and other natural glasses. Earth Planetary Sci. Letters **4**, 406 (1968).
MULFINGER, H. O.: Physical and chemical solubility of nitrogen in glass melts. J. Am. Ceram. Soc. **49**, 462 (1966).
— MEYER, H.: Über die physikalische und chemische Löslichkeit von Stickstoff in Glasschmelzen. Glastech. Ber. **36**, 481 (1963).
NAUDÉ, S. M.: An isotope of nitrogen, mass 15. Phys. Rev. **34**, 1498 (1929).
NAVONE, R., HARMON, J. A., VOYLES, C. F.: Nitrogen content of ground water in Southern California. J. Am. Water Works Assoc. **55**, 615 (1963).
NIER, A. O.: A redetermination of the relative abundances of the isotopes of carbon, nitrogen, oxygen, argon and potassium. Phys. Rev. **77**, 789 (1950).
OELSEN, W., KELLER, H., SAUER, K. H.: Der Stickstoff der Gasphasen bei den Reaktionen zwischen Eisenschmelzen und Silikatschlacken im Graphittiegel. Arch. Eisenhüttenw. **40**, 911 (1969).
ORÓ, J., SKEWES, H. B.: Free amino acids on human fingers, the question of contamination in microanalysis. Nature **207**, 1042 (1965).
OWEN, T.: The composition and surface pressure of the Martian atmosphere: results from the 1965 opposition. Astrophys. J. **146**, 257 (1966).
— The spectra of Jupiter and Saturn in the photographic infrared. Icarus **10**, 355 (1969).
PARWEL, A., RYHAGE, R., WICKMAN, F. E.: Natural variations in the relative abundances of the nitrogen isotopes. Geochim. Cosmochim. Acta **11**, 165 (1957).
PILOT, J.: Über die massenspektrometrische Isotopenanalyse an Stickstoff aus Erdgasen und Gesteinen. Kernenergie **6**, 714 (1963).
POLDERVAART, A.: Chemistry of the earth's crust. Geol. Soc. Am., Spec. Papers **62**, Crust of the Earth (1955).
POTTASCH, S. R.: On the abundances in the solar corona. In: Origin and Distribution of the Elements (L. H. AHRENS, ed.). Oxford-London-New York: Pergamon Press 1968.
RAMDOHR, P.: Personal communication, 1969.
RANKAMA, K.: Progress in Isotope Geology. New York-London: Interscience Publ. 1963.
— SAHAMA, T. G.: Geochemistry. Chicago: Chicago Univ. Press 1950.
RAVIKOVICH, KH. A.: Some oil gas hydrochemical indexes. Tr. Sredneaz. Fil. Vses. Nauchn.-Issled. Inst. Prirodn. Gazov **1965**, 155 (1965); cit. after Chem. Abstr. **65**, 18352 g.
— LUBYANSKAYA, M. G.: Significance of the ammonium/sulfate hydrochemical index in evaluating potential resources in individual horizons. Uzb. Geol. Zh. **13**, 65 (1969); cit. after Chem. Abstr. **72**, 57700u.
RAYLEIGH, LORD R. J.: Nitrogen, argon and neon in the earth's crust with applications to cosmology. Proc. Roy. Soc. (London), Ser. A **170**, 451 (1939).
REITER, R.: Relationship between atmospheric electric phenomena and simultaneous meteorological conditions. Final Report, vol. I, Contract No. AF 61 (052)-55 (1960).
— Further experimental evidence for the importance, with respect to thunderstorm electrification, of NO_3^- ions contained in precipitation. J. Atmospheric Terrest. Phys. **28**, 1065 (1966); part II: **30**, 345 (1968).
REVELLE, R., SHEPARD, F. P.: Sediments of the California coast. In: Recent Marine Sediments. Am. Assoc. Petrol. Geologists, Tulsa, Okl. 1939.
RITTENBERG, S. C., EMERY, K. O., ORR, W. L.: Regeneration of nutrients in sediments of marine basins. Deep-Sea Res. **8**, 23 (1955).
— — HÜLSEMANN, J., DEGENS, E. T., FAY, R. C., REUTER, J. H., GRADY, J. R., RICHARDSON, S. H., BRAY, E. E.: Biogeochemistry of sediments in experimental Mohole. J. Sediment. Petrol. **33**, 140 (1963).
RUBEY, W. W.: Development of the hydrosphere and atmosphere with special reference to probable composition of the early atmosphere. In: Crust of the Earth. Geol. Soc. Am., Spec. Papers **62**, 631 (1955).
RUSSELL, J. D.: Infrared study of the reaction of ammonia with montmorillonite and saponite. Trans. Faraday Soc. **61**, 2284 (1965).
RYALL, W. R., MUAN, A.: Silicon oxynitride stability. Science **165**, 1363 (1969).

SAGAN, C., MORRISON, D.: The planet Mercury. Sci. J. **4**, 72 (1968).
SARTORIUS VON WALTERSHAUSEN: Physisch-geographische Skizze von Island. Göttingen: Vandenhoeck und Ruprecht 1847.
SCALAN, R. S.: The isotopic composition, concentration and chemical state of nitrogen in igneous rocks. Ph. D. Thesis, Univ. Arkansas, USA, 1959. University Microfilms, Ann Arbor, Michigan, Microfilm 59—1379. Most of the measurements are also reported in HOERING (1958).
SCHACHTSCHABEL, P.: Untersuchungen über die Sorption der Tonmineralien und organischen Bodenkolloide. Kolloid-Beihefte **51**, 199 (1940).
SCHEFFER, F., SCHACHTSCHABEL, P.: Lehrbuch der Bodenkunde. Stuttgart: Ferdinand Enke 1966.
SCHNEIDER, E. A., CLARKE, F. W.: Notes on the action of ammonium chloride upon silicates. U.S. Geol. Surv. Bull. **113**, 34 (1893).
SCHÜTZ, K., JUNGE, C. E., BECK, R., ALBRECHT, B.: Studies of atmospheric N_2O. J. Geophys. Res., **75**, 2230 (1970).
SCOTT, E. R. D.: Chromium nitride — a new meteoritic mineral. Paper presented at symposium "The Chemistry and Mineralogy of Meteorites and Extraterrestrial Matter". Mineral. Soc., London, April 6—8, 1970.
SEMENOV, A. D., NEMTSEVA, L. I., KISHKINOVA, T. S., PASHANOVA, A. P.: Content of individual groups of organic substances in atmospheric precipitations. Ghidrkhim. Mater. **42**, 17 (1966).
— — — — Organic substances of atmospheric precipitation. Dokl. Akad. Nauk. SSSR **173**, 1185 (1967).
SHEPERD, E. S.: The gases in rocks and some related problems. Am. J. Sci., 5th Ser., **35 A**, 311 (1938).
SIDLE, A. B.: Amino acid content of atmospheric precipitation. Tellus **19**, 128 (1967).
SIGVALDASON, G. E.: Das Liparitvorkommen der Móskardsnjúkar auf Island. Beitr. Mineral. Petrog. **6**, 100 (1958).
— Mineralogische Untersuchungen über Gesteinszersetzungen durch postvulkanische Aktivität in Island. Beitr. Mineral. Petrog. **6**, 405 (1959).
— ELISSON, G.: Collection and analysis of volcanic gases at Surtsey, Iceland. Geochim. Cosmochim. Acta **32**, 797 (1968).
SILLÉN, L. G.: Regulation of O, N, and CO_2 in the atmosphere; thoughts of a laboratory chemist. Tellus **18**, 198 (1966).
SIMMONS, P. G., BAUMANN, A. J., BOLLIN, E. M., GELPI, E., ORÓ, J.: Unextractable organic fraction of the Pueblito de Allende meteorite: evidence for its indigenous nature. Proc. Nat. Acad. Sci. U.S. **64**, 1027 (1969).
SNYDER, R. L.: Nitrogen and oxygen compound types in petroleum. Anal. Chem. **41**, 314 1085 (1969).
SPEERS, G. C., WHITEHEAD, E. V.: Crude petroleum. In: Organic Geochemistry. Berlin-Heidelberg-New York: Springer 1969.
STEVENSON, F. J.: On the presence of fixed ammonium in rocks. Science **130**, 221 (1959).
— Some aspects of the distribution of biochemicals in geologic environments. Geochim. Cosmochim. Acta **19**, 261 (1960).
— Chemical state of nitrogen in rocks. Geochim. Cosmochim. Acta **26**, 797 (1962).
— Origin and distribution of nitrogen in soil. In: Soil Nitrogen (BARTHOLOMEW and CLARK). Madison, Wisconsin: Am. Soc. of Agronomy 1965.
— DHARIWAL, A. P. S.: Distribution of fixed ammonium in soil. Soil Sci. Soc. Am. Proc. **23**, 121 (1959).
STEWART, W. D. P.: Nitrogen fixing plants. Science **158**, 1426 (1967).
STUDIER, M. H., HAYATSU, R., ANDERS, E.: Organic compounds in carbonaceous chondrites. Science **149**, 1455 (1965).
STUTZER, O., WETZEL, W.: Phosphat-Nitrat. In: Die wichtigsten Lagerstätten der Nichterze (O. STUTZER). Berlin: Bornträger 1932.
SUESS, H. E.: Gas content and age of tektites. Geochim. Cosmochim. Acta **2**, 76 (1951).
— Remarks concerning the chemical composition of the atmosphere of Venus. Z. Naturforsch. **19a**, 84 (1964).

SUESS, H. E., UREY, H. C.: Abundances of the elements. Rev. Mod. Phys. **28**, 53 (1956).
SUGISAKI, R.: Genetic relation of various types of natural gas deposits in Japan. Bull. Am. Assoc. Petrol. Geologists **48**, 85 (1964).
SVERDRUP, H. U., JOHNSON, M. W., FLEMING, R. H.: The Oceans, their Physics, Chemistry and General Biology. New York: Prentice Hall 1942.
SYERS, J. K.: The relationship between the concentration of nitrate- and ammonia-nitrogen in precipitation. Tellus **18**, 146 (1966).
TANAEVSKY, O.: Résultats de mesures de peroxyde d'azote atmosphérique. Compt. Rend. **252**, 2909 (1961).
TARRANT, R. F., LU, K. C., CHEN, C. S., BOLLEN, W. B.: Nitrogen content of precipitation in a coastal Oregon forest opening. Tellus **20**, 554 (1968).
TASS: International Service in English, Moscow, June 3rd, 1969; cit. after Icarus **11**, 139 (1969).
TEBBENS, B. D.: Gaseous pollutants in the air. In: Air Pollution (A. C. STERN, ed.), vol. I, p. 23. New York-London: Academic Press 1968.
TEUSCHER, E. O.: Umwandlungserscheinungen an Gesteinen des Granitmassivs von Eibenstock-Neudeck. Mineral. Petrog. Mitt. **47**, 273 (1936).
TILO, S. N.: Nitrogen compounds in the experimental Mohole sediments. Thesis, Univ. Illinois Urbana, Ill., 1965. Univ. Microfilms, Ann Arbor, Mich., No. 65-11, 881.
TOMOR, J.: Ungarns Stickstoff und Kohlendioxyd führende Gase und ihre Entstehung. Z. Angew. Geol. **14**, 14 (1968).
TRASK, P. D.: Organic content of Recent marine sediments. In: Recent Marine Sediments (P. D. TRASK, ed.). Tulsa, Okl.: Am. Assoc. Petroleum Geologists 1939.
— PATNODE, H. W.: Source beds of petroleum. Tulsa, Okl.: Am. Assoc. Petroleum Geologists 1942.
UNSÖLD, A.: Der Neue Kosmos. Berlin-Heidelberg-New York: Springer 1967.
UREY, H. C.: The thermodynamic properties of isotopic substances. J. Chem. Soc. **1947**, 562 (1947).
— The Planets, their Origin and Development. New Haven, Conn.: Yale Univ. Press 1952.
— The atmospheres of the planets. In: Handbuch der Physik, vol. 52, p. 363. Berlin-Göttingen-Heidelberg: Springer 1959.
U.S. Public Health Service: Drinking Water Standards, Federal Register, March 6, 1962, p. 2152.
U.S. Standard Atmosphere, Washington, D. C.: U.S. Government Printing Office 1962.
VACCARO, R. F.: Inorganic nitrogen in sea water. In: Chemical Oceanography, vol. I, p. 365 London and New York: Academic Press 1965.
VALLENTYNE, J. R.: Thermal degradation of amino acids. Carnegie Inst. Wash. Yearbook **56**, 185 (1957).
— Two aspects of the geochemistry of amino acids. In: The Origin of Prebiological Systems and of their Molecular Matrices (S. W. FOX, ed.). London-New York: Academic Press 1965.
VEDDER, W.: Ammonium in muscovite. Geochim. Cosmochim. Acta **29**, 221 (1965).
VINOGRADOV, A. P., FLORENSKII, K. P., VOLYNETS, V. F.: Ammiak v meteoritakh i izverzhennỹkh gornỹkh porodakh. (Ammonia in meteorites and igneous rocks). Geokhimiya **10**, 875 (1963). Translated in Geochemistry **10**, 905 (1963).
— SURKOV, U. A., FLORENSKY, C. P.: The chemical composition of the Venus atmosphere based on data of the interplanetary station Venera IV. J. Atmospheric Sci. **25**, 535 (1968).
VIRTANEN, A. J.: Molecular nitrogen fixation and nitrogen cycle in nature. Tellus **4**, 304 (1952).
VOLYNETS, V. F., FRIDMAN, A. I.: Sbyazannỹĭ azot v porodakh Khibinskogo shchelochnogo massiva. (Bound nitrogen in samples of the Khibinsk alkali massif.) Izv. Vysshikh Uchebn. Zavedenii, Geol. i Razvedka **7**, 32 (1965).
VOZNESENSKII, A. A., TERESHCHENKO, V. G.: Prirodnye gazy SSSR i ikh kharakteristiki (Natural gases of the USSR and their characteristics). Izv. Vyssikh Uchebn. Zavedenii, Energ. **9**, 107 (1966).
WAHLER, W.: Über die in Kristallen und Flüssigkeiten eingeschlossenen Flüssigkeiten und Gase. Geochim. Cosmochim. Acta **9**, 105 (1955).

Nitrogen

WEDEPOHL, K. H.: Einige Überlegungen zur Geschichte des Meerwassers. Fortschr. Geol. Rheinland Westfalen **10**, 129 (1963).
— Composition and abundance of common sedimentary rocks. In: Handbook of Geochemistry (K. H. WEDEPOHL, ed.), vol. I. Berlin-Heidelberg-New York: Springer 1969.
WEIZSÄCKER, C. F. VON: Über Elementumwandlungen im Inneren der Sterne, II. Z. Physik **39**, 633 (1938).
WELLMAN, R. P., COOK, F. D., KROUSE, H. R.: Nitrogen-15: microbiological alteration of abundance. Science **161**, 269 (1968).
WHITE, D. E., HEM, J. D., WARING, G. A.: Chemical composition of subsurface waters. U.S. Geol. Surv. Profess. Papers **440**-F (1963).
— WARING, G. A.: Volcanic emanations. U.S. Geol. Surv. Profess. Papers **440**-K (1963).
WHITE, W. C., YAGODA, H.: Abundance of ^{15}N in the nitrogen occluded in radioactive minerals. Science **111**, 307 (1950).
WHITEHEAD, H. C., FETH, J. H.: Chemical character of rain, dry fallout, and bulk precipitation at Menlo Park, California, 1957—1959. J. Geophys. Res. **69**, 3319 (1964).
WIIK, H. B.: Unpublished results (1962), reported in B. MASON: The carbonaceous chondrites. Space Sci. Rev. **1**, 621 (1962).
WILDT, R.: Veröffentl. Sternwarte Göttingen **2** (22), 171 (1932); cit. after GREENSPAN and OWEN (1967).
WILSON, A. T.: Organic nitrogen in New Zealand snows. Nature **183**, 318 (1959).
— HOUSE, D. A.: Chemical composition of South Polar snow. J. Geophys. Res. **70**, 5515 (1965).
WLOTZKA, F.: Untersuchungen zur Geochemie des Stickstoffs. Geochim. Cosmochim. Acta **24**, 106 (1961).
WURM, K.: The physics of comets. In: The Moon, Meteorites and Comets (B. M. MIDDLEHURST and G. P. KUIPER, eds.). Chicago: Chicago Univ. Press 1963.
YAKOVLEV, YU. I.: New method for prospecting for oil and gas deposits. Vklad. Molodykh Spetsialistov v. Gaz Prom. Moscow, Sb. **1964**, 3 (1964); cit. after Chem. Abstr. **63**, 17714.
YAMAMOTO, T., NAKAHIRA, M.: Ammonium ions in sericites. Am. Mineralogist **51**, 1775 (1966).
ZWAAN, C.: The continuous absorption coefficient in the violet and ultraviolet region. Bull. Astron. Inst. Neth. **16**, 225 (1962).

Revised manuscript received: July 1970

Fluorine   9

A   R. Allmann   (Mineralogisches Institut der Universität Marburg a.d. Lahn, Germany)

B—O   S. Koritnig   (Mineralogisch-Petrologisches Institut der Universität Göttingen, Germany)

9-B. Isotopes in Nature

Natural fluorine (atomic number 9) consists of a single isotope of mass number 19, with the isotopic mass of 18.9984046 ($^{12}C = 12.00000$), (D'Ans-Lax 1967; Handbook of Chemistry and Physics 1967/68; Strominger, Hollander and Seaborg, 1958).

Revised manuscript received: December 1971

© Springer-Verlag Berlin · Heidelberg 1972

9-C. Abundance in Cosmos, Meteorites, Tektites and Lunar Samples

I. Cosmos

Aller (1961) has summarized the astronomical data. Fluorine has been detected in only one gaseous nebula, NGC 7027 and one star, γ-*Pegasi*. The fluorine abundances in this nebula and star are about 5 to 10 times larger than the meteoritic value. In the sun, the earth's crust, and chondritic meteorites, the fluorine abundance has about the same magnitude (see Table 9-C-1).

Table 9-C-1. *Extraterrestrial fluorine-silicon ratios*

System	Abundance, atoms F/10^6 atoms Si	Reference
1. Cosmos	1.60×10^3	Suess (1965)
2. Pre-planetary medium	9×10^3	Brown (1950)
3. Gaseous nebula	14.0×10^3	Aller (1961)
4. γ-*Pegasi*	21.0×10^3	Aller (1961)
5. Chondrites[b]		
Pigeonite (1)	0.646×10^3	Reed (1964)
Hypersthene (5)	0.950×10^3	Reed (1964)
Bronzite (3)	1.22×10^3	Reed (1964)
Enstatite (12)	1.5×10^3	Larimer and Anders (1967)
Carbonaceous (1)	2.34×10^3	Reed (1964)
(14)	2.5×10^3	Larimer and Anders (1967)
6. Earth's crust	3.5×10^3	Reed (1964)[a]

[a] Using the data of Ahrens and Taylor (1961), Koritnig (1951) and Vinogradov (1956).

[b] Number of samples in brackets.

In achondrites and chondrites, fluorine concentrations are relatively variable, ranging from about 4 to 300 ppm; the mean values are about 10 to 200 ppm F (see Tables 9-C-2 and 9-C-3). According to Greenland and Lovering (1965), the chondrites described as "finds" by Prior (1953) show a slightly higher mean and larger standard deviation in fluorine content than do the "falls" (see Table 9-C-2). This could perhaps indicate a terrestrial contamination of the finds. Inter-element correlations between fluorine and other elements in chondrites indicate no significant trend and especially no correlation with phosphorus. Reed (1964) has found an inverse correlation between CaO and fluorine and a reasonable correlation between fluorine and pyroxene content.

Fluorine

II. Meteorites

a) Stones

Table 9-C-2. *Selected fluorine analyses of meteorites*

Class	Name	ppm F	Reference
Achondrites			
Enstatite	Bishopville	6.8 (N/R)	REED and JOVANOVIC (1969)
	Cumberland Falls	7.7 (N/R)	REED and JOVANOVIC (1969)
	Norton County	11.0 (N/R)	REED and JOVANOVIC (1969)
		10.7 (N/R)	REED (1964)
Hypersthene	Tatahouine	12.2 (N/R)	REED and JOVANOVIC (1969)
	Johnstown	3.5 (N/R)	REED and JOVANOVIC (1969)
Howardite	Frankfurt	57.1 (N/R)	REED and JOVANOVIC (1969)
	Sioux County	64 ± 5 (N/R)	REED (1964)
Eucrite	Moore County	91.5 (N/R)	REED and JOVANOVIC (1969)
		60 ± 10 (N/R)	REED (1964)
	Shergotty	50.1 (N/R)	REED and JOVANOVIC (1969)
	Stannern	19.9 (N/R)	REED and JOVANOVIC (1969)
Chondrites			
Enstatite	Indarch	220 (S)	GREENLAND and LOVERING (1965)
		136 (N/R) mean	FISHER (1963)
	Hvittis	250 (S)	GREENLAND and LOVERING (1965)
	Khaipur	180 (S)	GREENLAND and LOVERING (1965)
	Abee	280 (S)	GREENLAND and LOVERING (1965)
	St. Marks	140 (S)	GREENLAND and LOVERING (1965)
Olivine-bronzite (falls)	Allegan	170 (S)	GREENLAND and LOVERING (1965)
		114 (N/R)	REED (1964)
	Forest Vale	81 (S)	GREENLAND and LOVERING (1965)
	Mt. Browne	110 (S)	GREENLAND and LOVERING (1965)
Olivine-bronzite (finds)	Kaldoonera Hill	210 (S)	GREENLAND and LOVERING (1965)
	Morven	120 (S)	GREENLAND and LOVERING (1965)
	Cook	99 (S)	GREENLAND and LOVERING (1965)
	Nardoo	93 (S)	GREENLAND and LOVERING (1965)
Olivine-hypersthene (falls)	Bjurbole	100 (S)	GREENLAND and LOVERING (1965)
	Khohar	68 (S)	GREENLAND and LOVERING (1965)
	Farmington	250 (S)	GREENLAND and LOVERING (1965)
	Holbrook	130 (S)	GREENLAND and LOVERING (1965)
		189 (N/R) mean	FISHER (1963)
	Homestead	76 (S)	GREENLAND and LOVERING (1965)
	Mocs	160 (S)	GREENLAND and LOVERING (1965)
		147 (N/R) mean	FISHER (1963)
		119 (N/R)	REED (1964)
Olivine-hypersthene (finds)	Adelie Land	40 (S)	GREENLAND and LOVERING (1965)
	Barratta	78 (S)	GREENLAND and LOVERING (1965)

Table 9-C-2 (continued)

Class	Name	ppm F	Reference
	Cabell	180 (S)	GREENLAND and LOVERING (1965)
	Coolamon	200 (S)	GREENLAND and LOVERING (1965)
	Hermitage Plains	230 (S)	GREENLAND and LOVERING (1965)
	Silverton	280 (S)	GREENLAND and LOVERING (1965)
	Tenham	58 (S)	GREENLAND and LOVERING (1965)
Olivine-pigeonite	Mokoia	170 (S)	GREENLAND and LOVERING (1965)
	Karoonda	190 (S)	GREENLAND and LOVERING (1965)
	Warrenton	160 (S)	GREENLAND and LOVERING (1965)
	Lancé	66 (N/R)	REED (1964)
Carbonaceous			
	Orgueil (water-free)	190 (S)	GREENLAND and LOVERING (1965)
		237 (S)	
		390 (N/R)	FISHER (1963)
		420 (N/R)	FISHER (1963)
		206 (N/R)	REED (1964)

At present, very little is known about the minerals in meteorites in which fluorine occurs. In contrast to rocks in the earth's crust, the apatite in the meteorites (chondrites) is nearly pure chlorapatite (VAN SCHMUS and RIBBE, 1969; 0.4 wt.-% F, 5.4 wt.-% Cl and 0.4 wt.-% OH (M); see also SHANNON and LARSON, 1925) and apatites of silicate inclusions in iron meteorites have upto 1.3 wt.-% F. Recently OLSEN (1967) described the first occurrence of higher fluorine concentrations

Table 9-C-3. *Average fluorine concentrations in stony meteorites*

	Mean ppm	Range ppm	Reference
Chondrites			
Enstatite Type I	195	140—220	GREENLAND and LOVERING (1965)
Enstatite Type II	198		REED (1964); GREENLAND (1963)
High-iron (H)-group	130.4	81—210	GREENLAND and LOVERING (1965)
High-iron (L)-group	160.8	40—300	GREENLAND and LOVERING (1965)
High-iron-low-metal (HL)-group	172.5	160—190	GREENLAND and LOVERING (1965)
Carbonaceous Type I (water-free)	190 (237)		GREENLAND and LOVERING (1965)
Achondrites			
Enstatite	9	6.8—11	REED and JOVANOVIC (1969)
Hypersthene	8	3.5—12	REED and JOVANOVIC (1969)
Howardites	61.5	57—64	REED and JOVANOVIC (1969), REED (1964)
Eucrites	52	20—92	REED and JOVANOVIC (1969)

No fluorine-data is available for stone-iron meteorites.

in an amphibole (richterite) in a meteorite (coarse octahedrite). This amphibole contains 2.3 wt.-% (M) fluorine, an amount which occupies half the hydroxyl positions. No other minerals from meteorites with determined fluorine contents are known.

Even if the correlation between pyroxenes and fluorine (REED, 1964) is real, the substitution of fluorine in the structure of pyroxenes (for O^{2-}) is still not certain. For rock forming pyroxenes (diopsides) a small F content (0.1 to 0.6 wt.-%) has been reported in a few samples (DEER, HOWIE and ZUSSMAN, 1967), but the structural position is not known. KRANZ (1969) has found fluorine combined as CH_3F and F in alkali feldspars in inclusions. Fluorine could be similarly incorporated as inclusions in pyroxenes.

FUCHS (1962) has shown that much of the phosphate in ordinary chondrites is present as whitlockite $Ca_3(PO_4)_2$. Whitlockites from pegmatites contain 0.06 to 0.07% fluorine, but VAN SCHMUS and RIBBE (1969) found no fluorine in meteoritic whitlockite.

Recently BUNCH and KEIL (1970) found that the fluorine content in whitlockites of silicate inclusions from 6 iron meteorites is 0.05 wt.-% F (M). DOLGOV, POGREBNYAK and SHUGUROVA (1969) sometimes found small amounts of F in gas inclusions of tektites.

b) Iron Meteorites

For iron meteorites, no fluorine data is available. The reference to 1 ppm F in siderites by YAVNEL (1964) is a misinterpretation of the cited reference (NODDACK and NODDACK, 1930). This figure is an estimation for a mixture of stones plus irons, plus troilite.

If we take the Si-content of iron meteorites (FISHER, 1969; WAI and WASSON, 1969) with maximum, 20 to 40 ppm Si, and the relation of F content in chondrites with about 1.5×10^3 F per 10^6 atoms Si, this would correspond to an amount of about 0.05 to 0.1 ppb in iron meteorites. From OLSEN (1967) we also know that 2.3 wt.-% (M) fluorine is contained in richterite (an amphibole) from the Wichita County coarse octahedrite.

BUNCH and KEIL (1970) found small amounts of fluorine in apatites (0.3 to 1.3 wt.-% F) (M) and whitlockites (0.05 wt.-% F) (M) in silicate inclusions of iron meteorites.

III. Tektites

No quantitative fluorine data is available, but according to DOLGOV, POGREBNYAK and SHUGOROVA (1969) small amounts of F are sometimes detectable in inclusions of indochinites, phillipinites and australites.

IV. Lunar Rocks and Fines

In Table 9-C-4, the data of fluorine in lunar rock and fine (soil) samples of Apollo 11 and Apollo 12 missions are listed. The concentration of fluorine in the lunar material, according to MASON and MELSON (1970), for Apollo 11 material is comparable to that in W-1 and stony meteorites. The range of values (30 to 340 ppm F) is considerable and agreement of the determinations between two methods (R, N) for the same

Table 9-C-4. *Fluorine content in lunar materials*

Material	ppm F	Method[a]	Reference
Apollo 11 samples:			
Medium grained rock	202	R	REED et al. (1970)
Coarse grained rock	50	N	MORRISON et al. (1970a)
Vesicular rock	251	R	REED et al. (1970)
Fine grained vesicular rock	85	N	MORRISON et al. (1970a)
	70	N	MORRISON et al. (1970a)
	100	N	MORRISON et al. (1970a)
Vuggy rock	271	R	REED et al. (1970)
Microbreccia	342	R	REED et al. (1970)
Breccia	30	N	MORRISON et al. (1970a)
	80	N	MORRISON et al. (1970a)
Fines	144	R	REED et al. (1970)
Fines	66	N	MORRISON et al. (1970a)
Range, Apollo 11 samples	30—340	—	MASON and MELSON (1970)
Mean	140		
Apollo 12 samples:	wt.-%		
Rock (12004,46)	0.00	C	MAXWELL and WIIK (1971)
Rock (12051,34)	0.00	C	MAXWELL and WIIK (1971)
Regolith (12033,51)	0.00	C	MAXWELL and WIIK (1971)

[a] R: with photon activation analysis.

material is not good. The fluorine content of Apollo 12 material (MAXWELL and WIIK, 1971) seems to be distinctly lower than that of Apollo 11 (REED et al., 1970; MORRISON et al., 1970 a, b; MAXWELL, PECK and WIIK, 1970).

The only F-bearing mineral identified in the lunar material is apatite. This apatite contains less fluorine than pure F-apatite. The phosphorus content of the lunar samples could account for upto about 150 ppm fluorine in the form of apatite (MASON and MELSON, 1970).

9-D. Abundance in Rock-forming Minerals; Fluorine Minerals

Of the rock-forming minerals, only topaz and fluorite have fluorine as an essential constituent in the formula. The other minerals in which fluorine is an essential component, like villiaumite or cryolite, are accessory minerals. In all remaining F-bearing minerals, fluorine is an isomorphous replacement in the OH-position. Here we can distinguish two groups of minerals. In one, the fluorine content is in the OH-position as a constituent of a mixed crystal, varying from the pure OH- to the pure F-end member (i.e. apatite), sometimes with a smaller or larger miscibility gap. In the other group, fluorine is contained only in relatively small amounts, camouflaged in the OH-, rarely in the O^{2-}-, position. This group is very important for the geochemistry of rocks, because a great part of major OH-bearing minerals, such as mica and amphiboles, camouflage fluorine. Between the two groups there is no sharp boundary.

Two factors especially control the fluorine content in such minerals. One is the amount of fluorine available in the rock-forming environment, the other the physical properties of the lattice position of the OH-ion (or O-ion). The ionic sizes of the F^- ion (1.33 Å), OH^- ion (1.4 Å) and O^{2-} ion (1.32 Å) are similar, so that substitution should be easily possible. But the OH^- ion and the F^- ion have other, very different properties (polarizability, electronegativity, etc.), so that the coordination and kind of ligands of the OH site generally control the degree of replacement.

For micas and chlorites, FOSTER (1964) has also shown that the O fugacity of the environment may be a controlling factor. Also, whereas F^- cannot escape as an ion from the lattice, the OH^- ion may escape as H_2O simply by combining with a H^+ ion from a second nearby OH^- ion (BLOSS et al., 1959). This may also be an explanation for why O^{2-} is less often replaced by F than is the OH ion. In her very detailed investigation, RIMSAITE (1967) also studied the occupancy of the (OH, F) group of 48 micas.

Besides these rock-forming minerals, there are a great number of other fluorine-bearing minerals which occur mostly as rare accessory minerals especially in pegmatitic-pneumatolytic rocks (about 60% of all fluorine minerals) and in their altered wall rocks. Relatively few minerals belong to the hydrothermal cycle or other occurrences (compare Table 9-D-6).

According to GILLBERG (1964), biotites coexisting with muscovite (in pegmatites) contain more fluorine than the latter. Coexisting amphibole and biotite have fluorine contents of about the same magnitude (see Table 9-D-3).

The diagram of Fig. 9-D-1 shows the distribution of fluorine in biotites of igneous rocks. Fig. 9-D-2, from the data of TABORSZKY (1962), shows that the fluorine content of apatites, in gabbroic through granitic rocks, is distinctly dependent on

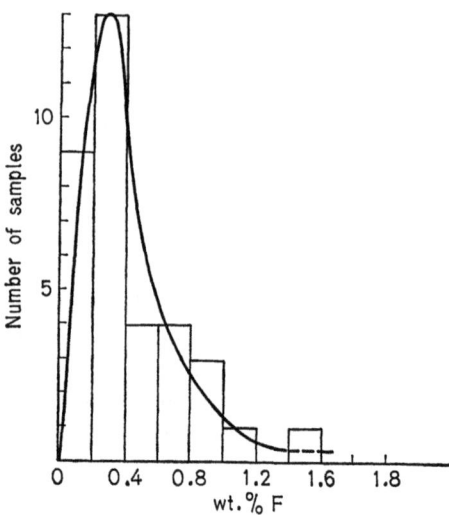

Fig. 9-D-1. Fluorine content in biotites (35 samples) of igneous rocks, 0.2% F intervals

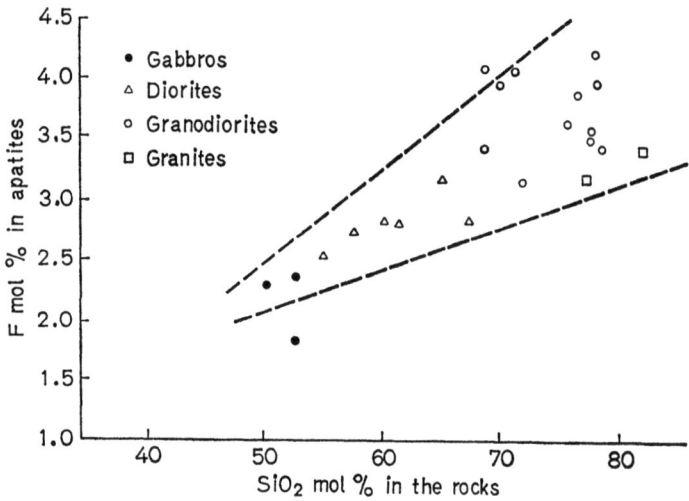

Fig. 9-D-2. Mol.-% fluorine in apatites of plutonic rocks with increasing SiO_2 content. (After TABORSZKY, 1962)

the SiO_2 content of the rocks. With increasing SiO_2, fluorine also increases. Fig. 9-D-3 shows the distribution of fluorine contents in apatites of plutonic rocks.

Free F and/or HF is reported from dark radioactive fluorites (antozonite) by HOFFMANN (1938). Recently KRANZ (1969) found sulforyl and thionyl fluorides as inclusions in such fluorites. In alkali feldspars from a pegmatite, he found CH_3F and HF (Method:I).

Fluorine

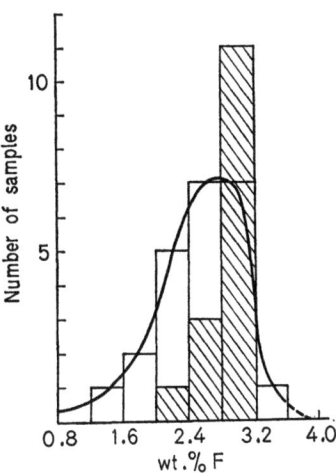

Fig. 9-D-3. Fluorine content in apatites from granitic to gabbroic rocks (23 samples), white areas (after TABORSZKY, 1962). Fluorine in apatites from granite pegmatites (15 samples), shaded (data of SHMAKIN and SHIRYAYEVA, 1968)

Table 9-D-1. *Fluorine content in igneous rock-forming minerals (Number of samples for average in brackets)*

	Range wt.-% F	Average wt.-% F	Method	Reference
Amphibole				
Alkali amphiboles	0.05—2.95	1.50 (20)	(W)	DEER et al. (1967)
Common hornblende	0.01—0.53	0.60 (11)	(W)	DEER et al. (1967)
— from lamprophyric tyke rocks	0.16—1.43		(C)	NĚMEC (1968)
Apatite				
— from eruptive rocks	1.35—2.56	1.99	(W)	KIND (1938)
— from gabbros	1.46—1.86		(W)	TABORSZKY (1962)
— from diorites	2.01—2.17		(W)	TABORSZKY (1962)
— from quartz diorites	2.23—2.50		(W)	TABORSZKY (1962)
— from granodiorites	2.49—3.31		(W)	TABORSZKY (1962)
— from pegmatites	2.37—3.36	2.97 (37)	(S, —)	SERAPHIM (1951); SHMAKIN and SHIRYAYEVA (1968)
Feldspar				
Plagioclase	0.069—0.089		(C)	KOKUBU (1956)
Microcline	0.03		(W)	DEER et al. (1967)
Mica				
Biotite				
— from gabbro	0.097		(S)	SERAPHIM (1951)
— from syenites	0.08—1.04	0.78 (13)	(W)	SOLOV'YEV et al. (1967)

Table 9-D-1 (continued)

	Range wt.-% F	Average wt.-% F	Method	Reference
— from syenites	0.52—1.16	0.88 (7)	(—)	Kostetskaya and Mordvinova (1968)
— from granites	0.08—0.415	0.24 (168)	(W)	Solov'yev et al. (1967)
— from granites	0.095—3.50	1.45 (6)	(S)	Seraphim (1951)
— from leucocratic granites	0.40—3.05	1.52 (4)	(—)	Kostetskaya and Mordvinova (1968)
— from pegmatites	0.22—3.05	0.27 (5)	(S, —)	Seraphim (1951); Kunitz (1924); Lokka (1943)
Lepidomelane	upto 4.36		(W)	Deer et al. (1967)
Lepidolite	0.62—9.19	5.41 (26)	(—, —)	Gmelin (1959); Heinrich et al. in Correns (1956) mean.
Muscovite				
— from granites	0.02—0.77	0.169 (6)	(S)	Seraphim (1951)
— from pegmatite	0.22—1.95	0.93 (8)	(—, C)	Kunitz (1924); Nemec (1969)
— from pegmatite	0.08—0.30	0.16 (11)	(—)	Glebov et al. (1968)
Phlogopite	0.05—6.74		(W, —)	Deer et al. (1967); Gmelin (1959)
Zinnwaldite	1.43—8.32		(W, —)	Deer et al. (1967); Gmelin (1959)
Olivine	0.045		(C)	Kokubu (1956)
Petalite	0.03		(W)	Deer et al. (1967)
Pyroxene	0.01—0.034		(S, C)	Seraphim (1951); Kokubu (1956)
Hypersthene	0.03		(?)	Lokka (1950) in Correns (1956)
Augite	0.01—0.1		(W)	Tröger (1935) in Correns (1956)
Aegirine	0.3		(?)	Vlasov et al. (1966)
Sphene (titanite)	0.03—0.76 (1.4)		(?, ?, S, C)	Sahama (1946); Jaffe (1947); Seraphim (1951); Kokubu (1956)
— from diorites	0.30—0.75		(?)	Kostetskaya and Mordvinova (1968)
— from subalkaline syenites	0.41—1.36		(?)	Kostetskaya and Mordvinova (1968)
— from leucocratic granites	0.36—0.82		(?)	Kostetskaya and Mordvinova (1968)
— from granitic rocks	0.28—1.36	0.67 (19)	(?)	Kostetskaya and Mordvinova (1968)
Topaz	13.01—20.43		(W, ?)	Deer et al. (1967); Gmelin (1959)
Tourmaline	0.07—1.27		(W, C)	Deer et al. (1967); Nĕmec (1969)

Table 9-D-2. *Fluorine content in selected igneous rock-forming minerals*

Mineral	Rock	Wt.-% F	Method	Reference
Amphibole				
Common hornblende	gabbro	0.23	(W)	Deer et al. (1967)
Common hornblende	hornblende gabbro	0.73	(W)	Deer et al. (1967)
Common hornblende	diorite pegmatites	0.53	(W)	Deer et al. (1967)
Hornblende	granite	1.50	(?)	Buddington (1953)
Basaltic hornblende	quartz latite	0.35	(W)	Deer et al. (1967)
Amphibole	malchite	0.16	(C)	Němec (1968)
Amphibole	vogesite	0.18	(C)	Němec (1968)
Amphibole	alkal. granite	1.43	(C)	Němec (1968)
Eckermannite	nepheline syenite	2.69	(W)	Deer et al. (1967)
Eckermannite	syenite	1.36	(W)	Solov'yev et al. (1967)
Arfvedsonite	nepheline syenite	2.95	(W)	Deer et al. (1967)
Ferrohastingite	hornblende granite	0.21	(W)	Deer et al. (1967)
Ferrihastingite	granite	0.96	(?)	Buddington (1953)
Ferrohastingite	rapakivi granite	1.06	(W)	Deer et al. (1967)
Holmquistite	pegmatite	0.24	(W)	Deer et al. (1967)
Hornblende	pumice	0.069	(C)	Kokubu (1956)
Kaersutite	camptonite	0.42	(W)	Deer et al. (1967)
Katophorite (taramite)	mariupolite	0.14	(W)	Deer et al. (1967)
Apatite				
Apatite	gabbro	1.83	(W)	Taborszky (1962)
Apatite	diorite	2.01	(W)	Taborszky (1962)
Apatite	quartz diorite	2.23	(W)	Taborszky (1962)
Apatite	quartz monzonite	2.68	(W)	Taborszky (1962)
Apatite	granodiorite	2.81	(W)	Taborszky (1962)
Apatite	granodiorite	3.31	(W)	Taborszky (1962)
Apatite	allivalite	0.089	(C)	Kokubu (1956)
Cookeite	Li-pegmatite	0.36	(?)	Sahama et al. (1968) in Cerny (1970)
Mica				
Biotite	gabbro diorite	0.097	(S)	Seraphim (1951)
Biotite	quartz diorite	0.26	(S)	Seraphim (1951)
Biotite	granite	0.415	(C)	Koritnig (1951)
Biotite	pegmatite	0.950	(S)	Seraphim (1951)
Biotite	pegmatite	1.40	(S)	Seraphim (1951)
Biotite	hornblende biotite granite	1.67	(C)	Němec (1968)
Biotite	tonalite	2.38	(?)	Nockolds and Mitchell (1948)
Biotite	pyroxene syenite	0.74	(W)	Solov'yev (1967)
Biotite	kersantite	0.42	(C)	Němec (1968)
Phlogopite	volcanic ejecta	2.57	(?)	Pieruccini (1950) in Correns (1956)
Phlogopite	kimberlite	0.27	(M)	Bouvier (1970)
Muscovite	granodiorite	0.16	(W)	Deer et al. (1967)
Muscovite	pegmatite	0.04	(W)	Jacob (1929)
Muscovite	pegmatite	0.09	(W)	Jacob (1929)
Muscovite	pegmatite	2.06	(W)	Deer et al. (1967)

Table 9-D-2 (continued)

Mineral	Rock	Wt.-% F	Method	Reference
Lepidolite	pegmatite	4.93	(W)	Deer et al. (1967)
Lepidolite	pegmatite	8.08	(W)	Deer et al. (1967)
Sphene (titanite)	gabbro	0.19	(?)	Kostetskaya and Mordvinova (1968)
Sphene	leuco granite	0.76	(?)	Kostetskaya and Mordvinova (1968)
Sphene	amphibole syenite	0.41	(?)	Kostetskaya and Mordvinova (1968)
Sphene	nepheline syenite	0.68	(?)	Kostetskaya and Mordvinova (1968)
Tourmaline				
Dravite	granite	0.07	(W)	Deer et al. (1967)
Schorl	pegmatite	0.18	(C)	Němec (1969)
Schorl	pegmatite	0.77	(C)	Němec (1969)

Table 9-D-3. *Fluorine content of coexisting biotites and hornblendes in plutonic rocks*

Rock	Biotite wt.-% F	Hornblende wt.-% F	Reference
Tonalite, Bonsall, Calif.	0.07	0.09	Larsen and Draisin (1950) in Correns (1956)
Tonalite, Green Val. Calif.	0.22	0.17	Larsen and Draisin (1950) in Correns (1956)
Tonalite, Lakeview, Calif.	1.37	0.18	Larsen and Draisin (1950) in Correns (1956)
Granodiorite, Bonsall, Calif.	0.58	0.23	Larsen and Draisin (1950) in Correns (1956)
Granite, Rubidoux, Calif.	0.21	1.20	Larsen and Draisin (1950) in Correns (1956)
Granitic rock, Calif.	0.19	0.08	Dodge and Ross (1971)
Granitic rock, Calif.	0.36	0.19	Dodge and Ross (1971)
Granitic rock, Calif.	0.45	0.22	Dodge and Ross (1971)
Granitic rock, Calif.	0.28	0.12	Dodge and Ross (1971)
Alkali syenite, Nordmarka, Norway	0.50	1.00	Butler (1954) in Correns (1956)

Table 9-D-4. *Fluorine content in metamorphic rock-forming minerals. (Number of samples in brackets)*

	Range, wt.-% F	Method	Reference
Amphibole			
Actinolite (2)	0.17—0.31	(W)	Deer et al. (1967)
Anthophyllite (4)	0.003—0.52	(W)	Deer et al. (1967)
Edenite (1)	0.20	(W)	Deer et al. (1967)
Ferrogedrite (1)	0.13	(W)	Deer et al. (1967)

Table 9-D-4 (continued)

	Range, wt.-% F	Method	Reference
Gedrite (1)	0.02	(W)	DEER et al. (1967)
Grunerite (2)	0.01—0.07	(W)	DEER et al. (1967)
Hornblende (1)	0.52	(W)	SOLOV'YEV (1967)
Pargasite (6)	0.06—1.90	(W)	DEER et al. (1967)
Richterite (8)	0.18—2.14	(W)	DEER et al. (1967)
Tremolite (1)	0.25	(W)	DEER et al. (1967)
Tschermakite (1)	0.22	(W)	DEER et al. (1967)
Amphibole (5)	0.36 aver.	(S)	SERAPHIM (1951)
Amphibole (14)	0.80 aver.	(?)	WARREN (1929)
Amphibole (17)	0.67 aver.	(?)	WITTELS (1951)
Amphibole (9)[a]	0.60—1.50 aver. 1.06	(C)	LEELANANDAM (1969)

Chlorite

Chlorite (1)	0.06	(?)	LOKKA (1943)
Chlorite (1)	0.13	(S)	SERAPHIM (1951)
Clinochlore (1)	0.02	(S)	SERAPHIM (1951)
Prochlorite (1)	0.02	(S)	SERAPHIM (1951)
Epidote (3)	0.02	(S)	SERAPHIM (1951)
Epidote (black)	0.17	(?)	LOKKA (1943)
Chondrodite	4.0—8.0	(?)	GMELIN (1959)
Clinohumite	1.70—5.0	(?)	GMELIN (1959)
Humite	3.08—5.43	(?)	GMELIN (1959)
Norbergite	12.78—13.70	(?)	SAHAMA (1946)
Lawsonite	0.02	(W)	DEER et al. (1967)

Mica

Biotite (1)	2.0	(S)	SERAPHIM (1951)
Biotite (18)	1.24 aver.	(S)	SERAPHIM (1951)
Biotite (4)	0.06—0.42	(W)	DEER et al. (1967)
Biotite[a] (12)	1.22—3.99, aver. 2.14	(C)	LEELANANDAM (1970)
Muscovite (1)	0.045	(W)	DEER et al. (1967)
Paragonite (2)	0.026, 0.08	(S, C)	SERAPHIM (1951); DEER et al. (1967)
Phlogopite	0.05—6.74	(?)	GMELIN (1959)
Phengite (1)	5.02	(?)	GMELIN (1959)
Margarite (1)	0.031	(S)	SERAPHIM (1951)

Pyroxene

Ferrosalite (1)	0.16	(W)	DEER et al. (1967)

Scapolite (5)	0.01—0.37	(W)	DEER et al. (1967)

Vesuvianite (4)	0.91—3.22	(W)	DEER et al. (1967)

[a] From charnockites.

CARMICHAEL (1970) investigated the fluorine contents of coexisting biotite, Ca-amphibole, epidote and sphene in a regionally metamorphosed calcareous rock from the Whetstone Lake area, Ontario; he determined (M): biotite 0.68, Ca-amphibole 0.30, epidote 0.00, and sphene 0.39 wt.-% F. EKSTRÖM (1972) reported a distribution coefficient of fluorine for coexisting Ca-amphibole and biotite of between 0.5 and 1.1. The coefficient is positively correlated with increasing metamorphic grade and titanium concentration in amphibole. NEMETS (1968) investigated the fluorine content in amphiboles from different mineral associations in skarns and found F concentrations from 0.13 to 1.53 wt.-% (C), mean 0.47 (24 samples).

Table 9-D-5. *Fluorine content in sedimentary rock-forming minerals*

	Range, wt.-% F	Method	Reference
Apatite			
Carbonate-ap.	0.2—5.6	(?, W)	DANA et al. (1958); DEER et al. (1967)
Francolite	2.8—5.60	(?)	GMELIN (1959)
Phosphorite (21)	3.8—4.2 aver. 3.25	(?)	GULBRANDSEN (1966)
Phosphorite recent	2.47—3.36	(?)	DIETZ, EMERY and SHEPARD (1942)
From sea floor (6)	aver. 3.08	(?)	DIETZ, EMERY and SHEPARD (1942)
Aragonite (7)	0.067—0.164	(?)	CARPENTER (1969)
Coral samples, Oolites	aver. 0.1074	(?)	CARPENTER (1969)
Muscovite (sandstone)	0.454	(C)	KORITNIG (1951)
Glauconite (continental samples)	0.0714—0.253 aver. 0.0156	(?)	CARPENTER (1969)
Kaolinite	0.026—0.15	(C, ?)	KORITNIG (1951); GRIGOREV (1939)
Halloysite (1)	0.018	(C)	KORITNIG (1951)
Vermiculite (1)	0.05	(W)	DEER et al. (1967)
Phillipsite (13) (Pacific samples)	0.020—0.100 aver. 0.0485	(?) (?)	CARPENTER (1969) CARPENTER (1969)
Opal (22) (all 3 oceans)	8—120 ppm aver. 50 ppm	(?) (?)	CARPENTER (1969) CARPENTER (1969)

Table 9-D-6. *Fluorine minerals. References:* Handbooks of mineralogy by DANA (1958); DOELTER (1912—1931); HINTZE (1904—1968); DEER et al. (1967); GMELIN (1926, 1959); and others

Mineral	Formula	F content wt.-%	
		Theoretical	Range observed
Halides			
Villiaumite	NaF	45.24	44.20—45.28
Carobbiite	KF	32.68	—
Sellaite	MgF_2	60.98	—
Fluorite	CaF_2	48.67	47.81—48.80
Yttrofluorite	$(Ca,Y)F_{2-2.33}$		41.64

Table 9-D-6 (continued)

Mineral	Formula	F content wt.-% Theoretical	F content wt.-% Range observed
Yttrocerite	$(Ca,Ce)F_{2-2.33}$		25.45
Gagarinite	$NaCaYF_6$	42.86	—
Fluocerite	$(Ce,La)F_3$	29.00	24.46—28.71
Fluellite	$AlF_3 \cdot H_2O$	55.89	56.25
Ferruccite	$Na[BF_4]$	69.22	69.22
Avogadrite	$K[BF_4]$	60.36	60.36
Prosopite	$Ca[Al(F,OH)_4]_2$ F:OH=1:1	31.92	29.95—35.01
Malladrite	$Na_2[SiF_6]$	60.62	60.62
Bararite	$(NH_4)_2[SiF_6]$	63.99	—
Hieratite	$K_2[SiF_6]$	51.76	51.90
Cryptohalite	$(NH_4)_2[SiF_6]$	64.00	63.29
Cryolithionite	$Na_3Li_3[AlF_6]_2$	61.33	60.79
Cryolite	$Na_3[AlF_6]$	54.29	53.48—54.37
Elpasolite	$K_2Na[AlF_6]$	47.08	46.98—47.90
Tikhonenkovite	$(Sr,Ca)[AlF_4OH \cdot H_2O]$	37.65	—
Jarlite	$NaSr_2[AlF_6 \cdot AlF_5H_2O]$	45.29	43.23—45.50
Gearksutite	$Ca[Al(F,OH)_5H_2O]$ F:OH=4:1	42.68	40.20—42.68
Creedite	$Ca_3[(Al(F,OH,H_2O)_6)_2/SO_4]$	25.85	30.30—30.68
Chukhrovite	$(Ca,Y,Ce)_{>3}[(AlF_6)_2/SO_4] \cdot 10H_2O$		27.88—29.55
Thomsenolite	$NaCa[AlF_6] \cdot H_2O$	51.34	50.08—50.65
Pachnolite	$NaCa[AlF_6] \cdot H_2O$	51.34	50.79—51.94
Chiolite	$Na_5[Al_3F_{14}]$	57.59	57.30—57.81
Weberite	$Na_2Mg[AlF_7]$	57.76	57.58
Ralstonite	$Al_2(F,OH)_6 \cdot H_2O$ (F:OH=2:1)	41.76	39.91, 57.68
Neighborite	$NaMgF_3$	54.65	—
(without name)	$KMgF_3$	47.33	—
(without name)	$KCoF_3$	36.76	—
Matlockite	$PbFCl$	7.21	7.11, 7.25
Zavaritskite	$BiOF$	7.79	—

Oxides

Mineral	Formula	F content wt.-% Theoretical	F content wt.-% Range observed
Romeite	$(Ca,NaH)Sb_2O_6(O,OH,F)$	max. 4.73	0—3.63
Pyrochlore-Microlite	$(Ca,Na)_2(Nb,Ta)_2O_6(O,OH,F)$		— ca. 5
Fersmite	$(Ca,Ce,Na)(Nb,Ti,Fe,Al)_2(O,OH,F)_6$		— 1.87

Carbonates

Mineral	Formula	F content wt.-% Theoretical	F content wt.-% Range observed
Stenonite	$Sr_2Al[F_5/CO_3]$	26.59	—
Bastnaesite	$Ce[F/CO_3]$	8.54	6.23—9.94
Parisite	$CaCe_2[F_2/(CO_3)_3]$	7.07	6.56, 6.82
Roentgenite	$Ca_2Ce_3[F_3/(CO_3)_5]$	6.65	—
Synchisite	$CaCe[F/(CO_3)_2]$	6.12	5.04, 5.82
Doverite	$CaY[F/(CO_3)_2]$	7.09	—
Huanghoite	$BaCe[F/(CO_3)_2]$	4.56	—
Cordylite	$Ba(Ce,La,Nd)_2[F_2/(CO_3)_3]$	5.99	4.87
Kettnerite	$CaBi[OF/CO_3]$	5.52	—
Schroeckingerite	$NaCa_3[UO_2/F/SO_4/(CO_3)_3] \cdot 10H_2O$	2.14	—

Table 9-D-6 (continued)

Mineral	Formula	F content wt.-%	
		Theoretical	Range observed
Borates			
Fluoborite	$Mg_3[(F,OH)_3/BO_3]$	30.19	9.30—20.94
Johachidolite	$\sim Ca_3Na_2Al_4H_4[(F,OH)/BO_3]_6$		12.21
Sulfates			
Sulfohalite	$Na_6[F/Cl/(SO_4)_2]$	4.94	4.71, 4.95
Galeite	$Na_3[(F,Cl)/SO_4]$ $F:Cl=4:1$	8.22	7.90
Schairerite	$Na_3[(F,Cl)/SO_4]$ $F:Cl=4:1$	8.22	—
Phosphates, Arsenates			
Griphite	$(Mn,Na,Ca)_3(Al,Mn)_2[PO_3(OH,F)]_3$		−3.03
Herderite	$CaBe[(F,OH)/PO_4]$	max. 11.65	traces −11.32
Amblygonite	$LiAl[(F,OH)/PO_4]$	max. 12.85	0.57—11.71
Natramblygonite	$(Na,Li)Al[(OH,F)/PO_4]$		0—5.6
Wagnerite	$Mg_2[F/PO_4]$	11.68	4.78—11.48
Zwieselite-Triplite	$(Fe^{\cdot\cdot},Mn)_2[F/PO_4]$—$(Mn,Fe^{\cdot\cdot})_2[F/PO_4]$ $Fe:Mn=1:1$	8.45	6.02—9.09
Isokite	$CaMg[F/PO_4]$	10.65	9.86
Tilasite	$CaMg[F/AsO_4]$	8.55	7.18—8.24
Durangite	$NaAl[F/AsO_4]$	9.14	7.67
Lacroixite	$Na_4Ca_2Al_3[(OH,F)_8/(PO_4)_3]$		6.53
Böggildite	$Na_2Sr_2Al_2[F_9/PO_4]$	31.60	31.70
Svabite	$Ca_5[F/(AsO_4)_3]$	2.99	1.41—2.80
Wilkeite	$Ca_5[(F,O)/(PO_4,SiO_4,SO_4)_3]$		0.9—2.1
Britholite	$(Na,Ce,Ca)_5[F/(SiO_4,PO_4)_3]$		1.33
Abukumalite	$(Y,Th,Ca)_5[(F,O)/(SiO_4,PO_4,AlO_4)_3]$		0.45
Morinite	$Ca_2NaAl_2[(F,OH)_5/(PO_4)_2]\cdot 2H_2O$	12.26	7.75—13.02
Richellite	$Ca_3Fe_{10}\cdots[(OH,F)_3/(PO_4)_2]_4\cdot nH_2O$		0.91—1.76
Silicates			
Topaz	$Al_2[F_2/SiO_4]$	20.7	13.01—20.43
Norbergite	$Mg_3[(OH,F)_2/SiO_4]$		12.78—13.70
Chondrodite	$Mg_5[(OH,F)_2/(SiO_4)_2]$	11.06	4.0—8.0
Humite	$Mg_7[(OH,F)_2/(SiO_4)_3]$		2.77—5.43
Clinohumite	$Mg_9[(OH,F)_2/(SiO_4)_4]$	6.08	0.06—5.0
Sonolithe	$Mn_9[(OH,F)_2/(SiO_4)_4]$		0.21
Bultfonteinite	$Ca_2[F/SiO_3OH]\cdot H_2O$	9.04	8.63—8.81
Fersmanite	$Na_4Ca_4Ti_4[(O,OH,F)_3/SiO_4]_3(?)$	—	3.09—3.61
Lessingite	$(Ca,Ce,La,Nd)_5[(O,OH,F)/(SiO_4)_3]$		0.54
Tritomite	$(Ce,La,Y,Th,Zr)_{3.5-3.7}$ $[(Si,B)_{2.8-3.0}(O,OH,F)_{13}]$	—	3.15—4.29
Spencite	$(Y,Ca,Ce,La,Th,Al)_{2.8-3.8}$ $[(Si,B)_{3.4-3.9}(O,OH,F)_{13}]$	—	0.44—1.02
Rowlandite	$\sim(Y,Fe,Ce)_3[(F,OH)/(SiO_4)_2]$	—	3.87
Melanocerite	$\sim Na_4Ca_{16}(Y,La)_3(Zr,Ce)_6$ $[F_{12}/(BO_3)_3/(SiO_4)_{12}]$	—	5.78
Caryocerite	Th-bearing melanocerite	—	5.63
Meliphanite	$(Ca,Na)_2(Be,Al)[Si_2O_6F]$	—	2.30—6.03

Fluorine

Table 9-D-6 (continued)

Mineral	Formula	F content wt.-%	
		Theoretical	Range observed
Leucophanite	$(Ca,NaH)_2Be[Si_2O_6(OH,F)](?)$	—	5.87—6.77
Cuspidine	$Ca_4[(F,OH)_2/Si_2O_7]$	10.37	9.05—10.05
Niobophyllite	$(K,Na)(Fe,Mn)_4(Nb,Ti)[(OH,F)/Si_2O_7]_2$	—	0.46
Woehlerite	$Ca_2NaZr[(F,OH,O)_2/Si_2O_7]$	4.78	2.80, 2.98
Lavenite	$(Na,Ca,Mn)_3Zr[(F,OH,O)_2/Si_2O_7]$	—	3.82—4.76
Niocalite	$Ca_3(Nb,Ca,Mg)[(O,F)_2/Si_2O_7]$	—	1.70
Guarinite	$Ca_2NaZr[(F,O)_2/Si_2O_7]$	4.78	1.28—5.83
Götzenite	$(Ca,Na)_3(Ti,Ce_{>1})[F_2/Si_2O_7]$	—	8.33
Rinkite	$(Na,Ca,Ce)_3(Ti,Ce)[(F,OH,O)_2/Si_2O_7]$	—	5.00—6.38
Mosandrite	$(Ca,Na,Y)_3(Ti,Zr,Ce)[(F,OH,O)_2/Si_2O_7]$	—	2.06
Rosenbuschite	$(Ca,Na)_6Zr(Ti,Mn,Nb)[(F,O)_2/Si_2O_7]_2$	—	5.83
Seidozerite	$Na_4MnTi(Zr_{1.5}Ti_{0.5})[O/(F,OH)/Si_2O_7]_2$	—	3.56
Lamprophyllite	$Na_3Sr_2Ti_3[(O,OH,F)_2/Si_2O_7]_2$	—	1.25—1.83
Innelite	$Ba_2(Na,K,Mn,Ti)_2Ti[(O,OH,F)_2/(S,Si)O_4/Si_2O_7]$	—	0.40
Zunyite	$Al_{12}[AlO_4/(OH,F)_{18}Cl/Si_5O_{16}]$	—	0.4—5.81
Canasite	$(Na,K)_5Ca_4[(OH,F)_3/Si_{10}O_{25}]$	—	2.17—2.21
Hyalotekite	$(Pb,Ca,Ba)_4B[Si_6O_{17}(F,OH)]$	—	0.99
Combeite	$Na_4Ca_3[Si_6O_{16}(OH,F)_2]$	—	1.87
Apophyllite	$KCa_4[F/(Si_4O_{10})_2]\cdot 8H_2O$	2.09	0.8—2.21
Phlogopite	$K Mg_3[(F,OH)_2/AlSi_3O_{10}]$	9.02	0.05—6.74
Lepidolite	$K(Li,Al)_3[(OH,F)_2(Al,Si)Si_3O_{10}]$	—	0.62—9.19
Zinnwaldite	$K(Li,Fe,Al)_3[(OH,F)_2(Al,Si)Si_3O_{10}]$	8.66	1.43—8.32
Taeniolite	$KLiMg_2[F_2/Si_4O_{10}]$	9.38	5.36—8.56
Zeophyllite	$Ca_4[F_2/(OH)_2/Si_3O_8]\cdot 2H_2O$	—	7.99—9.48
Hsianghualite	$\sim Ca_2[(Li,Be,Si)_6O_{12}]\cdot CaF_2$	7.32	
Leifite	$Na_2[(F,OH,H_2O)_{1-2}/(Al,Si)Si_5O_{12}]$	—	4.93, 4.98
Delhayelite	$\sim(Na,K)_{10}Ca_5[(Cl_2,F_2,SO_4)_3/O_4/Al_6Si_{32}O_{76}]\cdot 18H_2O$	—	0.33

9-E. Abundance in Common Igneous Rocks

If we compare the fluorine data for magmatic rocks (Tables 9-E-1 to 3) we observe a great variation even within rocks of a similar type. At first approximation we can recognize a positive correlation between the F- and SiO_2-content of rocks. But this trend is not always found. In general, ultramafic rocks (mean 100 ppm) and intermediate rocks (400 ppm) have a smaller F content than the rocks higher in SiO_2 (mean 800 ppm). If we neglect the pneumatolytic altered (greisenized) granitic rocks, alkalic rocks show the highest F content (mean 1,000 ppm). But alkalic rocks in general have a lower SiO_2 content.

The fluorine content in ultramafic rocks is the smallest of all magmatic rocks. The average concentration is about 20 ppm F, but in griquaites and kimberlites, two phlogopite-bearing ultramafic rocks, the fluorine content is about 10 times larger (average 252 ppm). Phlogopite is the principal carrier of fluorine in ultramafic rocks (RIMSAITE, 1970). Recently much higher fluorine contents (up to 660 ppm)

Table 9-E-1. *Average fluorine content of igneous rocks of various investigators*

Rock	Average ppm F	Number of samples	Method	Reference
A. Intrusive rocks				
Granites, leucocratic	800	(4)	(?)	KOSTETSKAYA et al. (1969)
Granites	1,900	(23)	(?)	JAHNS (1953)
Granites (Japan)	830	(5)	(C)	KOKUBU (1956)
Granites (Germany)	1,330	(13)	(C)	KORITNIG (1951)
Granites (Rapakivi)	2,950	(20)	(?)	SAHAMA (1945)
Granites, (New England, North America)	800	(40)	(S)	SERAPHIM (1951)
Granitic rocks (high Ca)	520			TUREKIAN and WEDEPOHL (1961)
Granitic rocks (low Ca)	850			TUREKIAN and WEDEPOHL (1961)
Granites and granodiorites	810	(93)		FLEISCHER and ROBINSON (1963)
Granites and granodiorites	810	(89)		FLEISCHER and ROBINSON (1963)
Granodiorites (Norway, Italy)	500	(2)	(C)	KORITNIG (1951)
Granodiorites	200		(W)	TRÖGER (1935)
Granodiorites and tonalites (Japan)	1,050	(19)	(C)	KOKUBU (1956)
Tonalites	670	(5)	(S)	SERAPHIM (1951)
Gabbros and basalts	420	(26)		FLEISCHER and ROBINSON (1963)

© Springer-Verlag Berlin · Heidelberg 1972.

Table 9-E-1 (continued)

Rock	Average ppm F	Number of samples	Method	Reference
Gabbros and basalts	440	(21)		Fleischer and Robinson (1963)
Gabbros (Japan)	480	(3)	(C)	Kokubu (1956)
Gabbros (Germany)	310	(11)	(C)	Koritnig (1951)
Gabbros	300		(W)	Tröger (1935)
Peridotites (inclusion in basalts)	21	(6)	(W/C)	Stueber et al. (1968)
Peridotites (intrusions)	19	(6)	(W/C)	Stueber et al. (1968)
Garnet peridotites and kimberlites	252	(6)	(W/C)	Stueber et al. (1968)
Basalts	450	(11)	(S)	Seraphim (1951)
Syenites	1,100	(7)	(?)	Saito (1950)
Syenites (U.S.A., Europe)	1,480	(16)	(S)	Seraphim (1951)
Syenites	1,200			Turekian and Wedepohl (1961)
Syenites	600		(W)	Tröger (1935)
Alkalic rocks	1,000	(65)		Fleischer and Robinson (1963)
Alkalic rocks	570	(6)		Fleischer and Robinson (1963)
Nepheline syenites (Germany)	1,190	(22)	(C)	Koritnig (1951)
Nepheline syenites (Korea)	500	(17)	(?)	Saito (1950)
Nepheline syenites	8,000		(W)	Tröger (1935)
Shonkinites	1,380	(6)	(W)	Shepherd (1940)

B. Extrusive rocks

Rhyolites, etc.	480	(78)		Fleischer and Robinson (1963)
Rhyolites, liparites	750	(8)	(?)	Comucci and Mazzi (1957)
Rhyolites	790	(65)		Fleischer and Robinson (1963)
Rhyolites, dacites	260	(6)	(C)	Kokubu (1956)
Rhyolites, liparites	280	(16)	(C)	Kokubu (1956)
Rhyolites, obsidians	700	(16)	(W)	Shepherd (1940)
Rhyolites	1,080	(41)	(W)	Shepherd (1940)
Rhyolites and rhyodacitic vitrophyres	820	(167)	(W/C)	Coats et al. (1963)
Andesites	210	(77)		Fleischer and Robinson (1963)
Andesites	470	(6)		Fleischer and Robinson (1963)
Andesites	260	(52)	(C)	Kokubu (1956)
Andesites	505	(2)	(C)	Koritnig (1951)
Basalts	360	(130)		Fleischer and Robinson (1963)
Basalts, Hawaii	340	(80)		Fleischer and Robinson (1963)
Basalts, Hawaii	320	(35)	(W)	Murata and Richter (1966)
Basalts	500	(58)		Fleischer and Robinson (1963)
Basalts (Iceland)	180	(5)	(W?)	Barth and Bruun (1945)
Basalts	480	(18)	(?)	Comucci and Mazzi (1957)
Basalts (Japan)	280	(21)	(C)	Kokubu (1956)
Basalts (Germany)	730	(3)	(C)	Koritnig (1951)
Basalts, Columbia River Plateau	540	(16)	(S)	Seraphim (1951)
Basaltic rocks	400			Turekian and Wedepohl (1961)

Fluorine

Table 9-E-2. *Abundance of fluorine in alkalic rocks*

Alkalic rocks	Location	ppm F	Method	Reference
Arfvedsonite granite	Oslo province, Norway	670	(?)	Barth and Bruun (1945)
Alkali granite	Münichreith, Austria	1,400	(W)	Tröger (1935)
Riebeckite granite	Northern Nigeria	2,700	(W/S)	Bowden (1966)
Riebeckite granite	Northern Nigeria	12,400	(W/S)	Bowden (1966)
Alkali granite	Pikes Peak, Colorado	3,600	(W)	Tröger (1935)
Alkalic hornblende biotite granite	Gniewoszów, C.S.S.R.	4,700	(C)	Němec (1968)
Nepheline syenite	Magnet Cove, Arkansas	200	(S)	Seraphim (1951)
Orthoclase nepheline syenite (lujavrite)	Lovozero massif, U.S.S.R.	220	(?)	Vlasov et al. (1966)
Analcime labradorite syenite	Big Bend Nat. Park, Texas	750	(S)	Seraphim (1951)
Larvikite (dark)	Tjølling, Oslo province	600	(?)	Barth and Bruun (1945)
Larvikite (light)	Tjølling, Oslo province	700	(?)	Barth and Bruun (1945)
Pulaskite (=alkali syenite)	Fourche Mt., Arkansas	1,950	(S)	Seraphim (1951)
Calcite alkali syenite	Laacher See district, Germany	2,500	(?)	Taylor et al. (1967)
Shonkinite	Shonkin Sag, Montana	1,410	(W)	Shepherd (1940)
Biotite augite foyaite (laerdalite)	Oslo, Norway	900	(?)	Barth (1947) in Seraphim (1951)
Essexite (platy)	Oslo province, Norway	500	(?)	Barth and Bruun (1945)
Essexite (light)	Oslo province, Norway	900	(?)	Barth and Bruun (1945)
Teschenite (analcime theralite)	Paskau, C.S.S.R.	1,000	(W)	Tröger (1935)
Nepheline teschenite	Bludowitz, C.S.S.R.	800	(W)	Tröger (1935)
Alkal. hornblende pyroxene minette	Stojków, C.S.S.R.	4,300	(C)	Němec (1968)
Nepheline basalt	Nagahama, Japan	640	(C)	Kokubu (1956)

Table 9-E-3. *Abundance of fluorine in standard igneous rocks*[a]. [*Sources in square brackets*]

Rock	Location	ppm F	Method	Reference
Granite GA	Andlau, Vosges [Nancy]	440	(W/C)	Huang and Johns (1967)
Granite G-1	Westerly, Rhode Island (U.S.A.) [U.S.G.S.]	705	(W/C)	Huang and Johns (1967)
Granite GR	Senones, Vosges [Nancy]	925	(W/C)	Huang and Johns (1967)

[a] See also data in Roubault et al. (1970).

Table 9-E-3 (continued)

Rock	Location	ppm F	Method	Reference
Granite G-2	Westerly, Rhode Island (U.S.A.) [U.S.G.S].	1,205	(W/C)	HUANG and JOHNS (1967)
Granite GH	Hoggar [Nancy]	4,550	(W/C)	HUANG and JOHNS (1967)
Granodiorite GSP-1	Silver Plume, Colorado [U.S.G.S.]	1,940	(W/C)	HUANG and JOHNS (1967)
Syenite Sy-1	Bancroft, Ontario [McGill, Canada]	1,770	(W/C)	HUANG and JOHNS (1967)
Quartz-diorite (tonalite) T-1	[Tanganyica Geol. Surv.]	390	(W/C)	HUANG and JOHNS (1967)
Andesite AGV-1	Guano Valley, Oregon [U.S.G.S.]	435	(W/C)	HUANG and JOHNS (1967)
Basalt BCR-1	Bridal Veil, Washington [U.S.G.S.]	500	(W)	FLANAGAN (1967)
Basalt ("diabase") W-1	Centerville, Virginia [U.S.G.S.]	208	(W/C)	HUANG and JOHNS (1967)
Peridotite PCC-1	Cozadero, Calif. [U.S.G.S.]	13	(W/C)	HUANG and JOHNS (1967)
Dunite DTS-1	Twin Sisters, Washington [U.S.G.S.]	12	(W/C)	HUANG and JOHNS (1967)

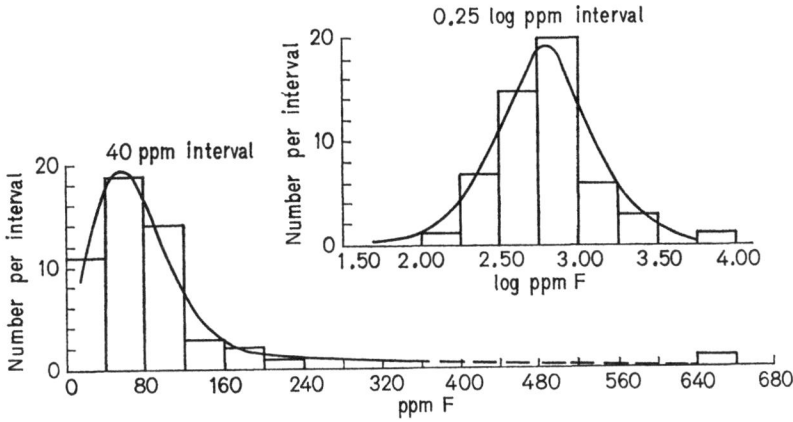

Fig. 9-E-1. Fluorine in Canadian and New England granite. (After AHRENS, 1954)

have been reported by STAVROV and UKHANOV (1971) in Russian kimberlites and eclogites.

According to CARSWELL and DAWSON (1970), South African kimberlites have 1 to 8 vol.-% phlogopite (mean 3.4, 10 samples). Using the F content of a Canadian phlogopite (0.27% F, BOUVIER, 1970) — no exact F data is available for South African phlogopites — we obtain about the same magnitude of F content which STUEBER et al. (1968) found for garnet peridotites.

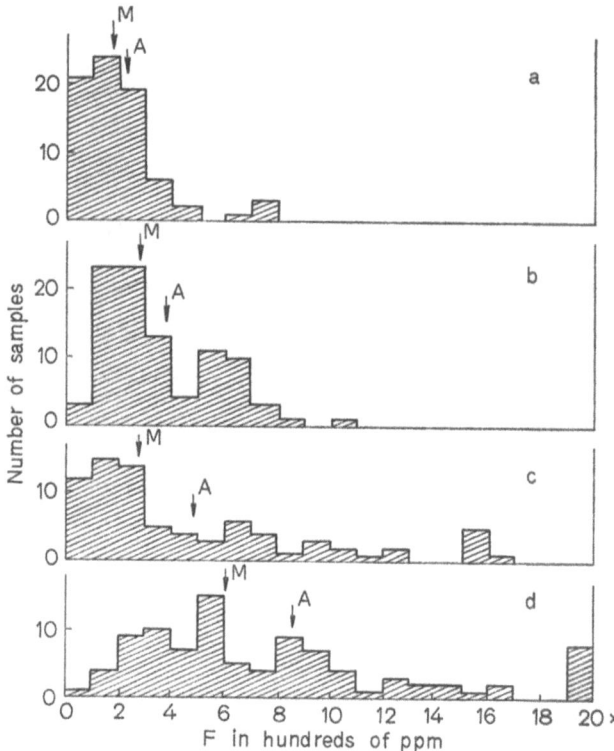

Fig. 9-E-2a—d. Distribution of fluorine in igneous rocks in ppm (after FLEISCHER and ROBINSON, 1963) (M=median, A=average). a) 77 andesites, median 175 ppm, average 210 ppm; b) 92 basalts, median 275 ppm, average 380 ppm; c) 99 obsidians, rhyolites etc., median 275 ppm, average 480 ppm; d) 44 granites and granodiorites, median 600 ppm, average 870 ppm

Small amounts of fluorine can also occur in accessory titanclinohumites (McGETCHIN et al., 1970).

If we compare effusive and plutonic rocks, it seems that plutonic rocks in general have higher fluorine contents than comparable effusive rocks. In Table 9-E-1 we find for instance the following means: gabbro/basalt, 420/210 ppm F; granite/rhyolite, 810/480 ppm F. But in detail we find many effusive rocks where the relation is quite the opposite.

Many factors overlap here, so that we cannot expect clear trends. These factors, which control magmatic rock-forming processes will be discussed in Section 9-F.

CLARKE (1924) thought that apatite is the major F-bearing component in all rocks, but from Table 9-E-4 we see that in most cases apatite contributes only a small amount (mostly 1 to 20%) of the total F content, whereas OH-bearing minerals often contribute up to 80%. In Table 9-E-4 some examples are given for the detailed distribution of fluorine in the different minerals in plutonic rocks.

Table 9-E-4. *Fluorine contribution of the different F-bearing minerals in igneous rocks*

Rock locality Reference	F-bearing mineral and wt.-% of this mineral in rock		F-content of F-bearing mineral, wt.-%	% F assigned to F-bearing minerals
1. Gabbro Harzburg; Germany KORITNIG (1951)	biotite apatite	8 0.13	0.46 2.5	97.5 2.5 100.0
2. Granodiorite (opdalite) Ulsberg; Norway KORITNIG (1951)	biotite apatite	12 0.6	0.44 2.2	80 20 100
3. Granite Brocken; Germany KORITNIG (1951)	biotite fluorite apatite	7.0 0.11 1.2	0.415 48.7 4.1	34 60 6 100
4. Syenite Dzhida; U.S.S.R. KOSTETSKAYA and MORDVINOVA (1968)	biotite amphibole sphene apatite	5.4 6.0 3.2 0.5	1.04 0.38 0.72 4.0	46 19 19 16 100
5. Monzonite ("syenite") Zitzschewig; Germany KORITNIG (1951)	amphibole apatite sphene	21 0.5 1.0	0.39 3.0 0.8	78 14 8 100
6. Quartz-diorite (tonalite) Adamello; Italy KORITNIG (1951)	amphibole biotite apatite	32 22 0.19	~0.05 ~0.055 3.0	47 35 18 100
7. Kimberlite Upper Canada Mine; Canada LEE and LAWRENCE (1968); BOUVIER (1970)	phlogopite apatite titanclino- humite	25 tr. tr.	0.27 ~3.0 ~0.06	~100 — — 100
8. Basalt Hoher Hagen; Germany KORITNIG (1951)	apatite	0.85	3.2	100 100
9. Rhyolite Eisenbach; Hungary KORITNIG (1951)	apatite biotite inclus. in matrix	0.08 2.0	4.0 0.5	4 13.3 82.7 100.0
10. Trachyte Drachenfels; Germany KORITNIG (1951)	apatite biotite remainder	0.33 0.5	3.2 0.4 ?	15 2 83 100

Carbonatites are somehow related to the alkalic rocks. PECORA (1956) has listed the ranges of fluorine contents in Alnö, Fen and East African carbonatites which are given in Table 9-E-5.

Table 9-E-5. *Fluorine content in carbonatites*

Province	Range wt.-%	Number of analyses	Reference
Alnö, Sweden	0.06—2.4	49[a]	PECORA (1956)
East Africa	0.02—0.3	21[b]	PECORA (1956), JEFFERY (1962)
Fen, Norway	0.08—0.56	10[c]	PECORA (1956)
Magnet Cove, U.S.A.	0.15	1	ERICKSON and BLADE (1963)
Oka, Canada	0.25	1	RIMSAITE (1969)

Analyses mentioned in: [a] ECKERMANN (1948); [b] KING (1949); SAGGERSON (1952); STRAUSS and TRUTER (1951); FOCKEMA (1953); FAWLEY and JAMES (1955); [c] BRÖGGER (1921), for reference see PECORA (1956).

In sövite from Laacher See district, TAYLOR et al. (1967) found 0.33 wt.-% and from Söve Fen area, TRÖGER (1935) found 0.28 wt.-% F.

The main F-bearing compounds of carbonatites besides F-apatite, biotite and phlogopite, are pyrochlore, chondrodite, rare-earth minerals, sellaite and fluorite.

9-F. Behavior in Magmatic Processes

I. Behavior during Crystallization and Differentiation of Magmas

Two kinds of magmas are especially important and of interest in rock-forming processes. These are the granitic and basaltic magmas. While the granitic magmas belong to the type of melts which are usually rich in volatiles (RUBEY, 1951, 1955), the basaltic melts are low in water (MOORE, 1970).

In general fluorine accumulates during magmatic crystallization and differentiation processes of the magma (TAUSON, 1967; RUBEY, 1951, 1955; GREENLAND and LOVERING, 1966).

One important factor for the concentration of fluorine in residual magmas during these processes is that the main rock-forming minerals are not able to accept appreciable amounts of F in their lattices. Only the accessory fluor-apatite has fluorine as a main constituent. With the start of crystallization of OH-bearing minerals in the late stages, fluorine is able to enter minerals in structural positions of OH⁻. However, the amount of this fluorine may be very limited, so that the residual magma is still further enriched in fluorine.

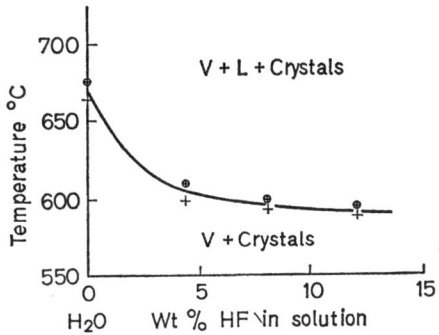

Fig. 9-F-1. (Perspective) TX projection for the "system" granite-H_2O-HF at 2,750 bars pressure, determined for a 1:1 weight ratio (approximately) of granite to total volatiles. The curve is a projection of the solidus, and it shows the effect of HF addition to H_2O, at constant pressure, on the temperature of the beginning of melting of granite, compared to the effect of H_2O alone. (After WYLLIE and TUTTLE, 1961)

Another important factor is that HF is more soluble than water in silicate melts, lowers the melting temperature more, and thus will be accumulated to a higher degree in the residual melts (WYLLIE and TUTTLE, 1961).

WYLLIE and TUTTLE (1961) in their experimental investigations in the system granite-H_2O-HF show that the addition of a small percentage of HF produces a marked depression of the melting temperature whereas further addition of HF produces a smaller effect (see Fig. 9-F-1).

In SiO_2-rich melts fluorine acts as a depolymerizer by forming Si-F bonds. In alkali-rich melts fluorine forms villiaumite. With increasing alkalinity the separation of fluorine into the gaseous phase decreases (DELITSYNA and MELENT'YEV, 1969; KOGARKO et al., 1968), thereby causing an enrichment of fluorine.

In magma chambers, water and HF migrate in the gradient of the external pressure and are concentrated in the upper parts of the magma chamber. Larger granitic bodies often show a more or less porphyritic and miarolitic development; and in this zone fluorite is a frequent mineral. This development leads to chamber and fissure pegmatites.

Similar pegmatitic and pneumatolytic developments are also found in alkalic rocks (e.g., nepheline syenites), but with villiaumite, NaF, as the fluorine mineral in the miarolitic cavities (e.g., Kola Peninsula, VLASOV et al., 1966).

TAUSON (1967) has shown that the distribution of fluorine during differentiation and crystallization of granitoids is determined by the composition, size, depth and tectonic setting of the intrusive. Fluorine ions in silicate melts have a high affinity for boron and beryllium (RINGWOOD, 1955; KOSTETSKAYA et al., 1969; KOSALS and MAZUROV, 1968), and some other elements, and tend to form complexes with them.

The average F contents in batholiths of different depths of intrusion, according to TAUSON (1967), are shown in Table 9-F-1.

Table 9-F-1. *Average fluorine content (wt.-%) in batholiths. (After TAUSON, 1967)*

Batholiths	Phase I	Phase II (principal)	Phase III	Aplite dikes
SUSAMYR (Centr.-Tien-Shan) (abyssal type)	0.070	0.100	0.125	0.025
DZHIDINSK (mesabyssal type)	0.082	0.098	0.079	0.015
KALBINSK (mesabyssal type)	0.070	0.150	0.066	0.018

In these rocks the main F-bearing minerals, besides apatite, are amphiboles and biotites.

GREENLAND and LOVERING (1966) have described the fractionation of fluorine during differentiation of a tholeiitic sill 529 m in thickness. Fluorine is reasonably constant in the lower-zone dolerites (240 to 260 ppm F) and increases steadily through the central-zone (300 to 480 ppm F) to a maximum in the granophyres (631 ppm F). The only F-bearing mineral that has been found in the sill is apatite, but this mineral accounts for not quite half of the fluorine content. The remainder is presumably concentrated in the mesostasis as a replacement for hydroxyl ions. In Hawaiian tholeiites MOORE (1970) found that the content of F increases directly with increasing H_2O content.

II. Pegmatites

In pegmatite veins, fluorine occurs mostly in apatite, topaz, Li-micas, tourmaline, phosphates, carbonates and fluorite. The majority of these fluorine minerals is formed during the pneumatolytic stage of the pegmatites. In general, fluorine is not especially enriched in vein pegmatites, and in aplitic veins it may be low. If fluorine is concentrated in the following hydrothermal stage and this fluid cannot escape, then high concentrations of fluorine minerals will be deposited in such pegmatites (e.g., the cryolite deposits at Ivigtut or Miask).

This process can, in some special cases, lead to the greisenization of the related granitic rocks by autometamorphism (compare Sections 9-M and N).

The distribution of fluorine around a chambered pegmatite (23 · 11 · 7 m) in the granitic rocks of Central Kazakhstan is described by BAZAROV et al. (1965). Fig. 9-F-2 is a summary of the results.

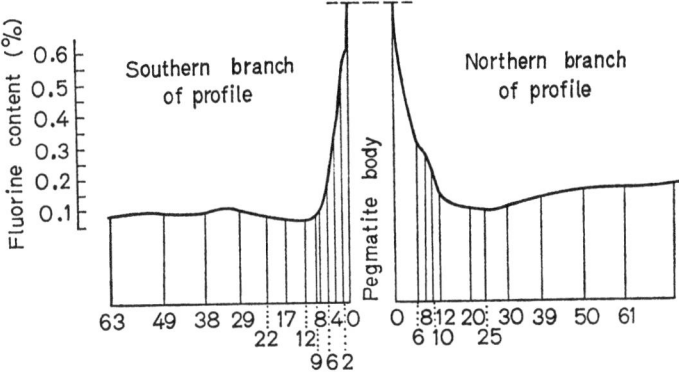

Fig. 9-F-2. Distribution of fluorine around a pegmatite body in Central Kazakhstan. (After BAZAROV et al., 1965)

III. Gas Transport

The degassing of a magma can occur in the pneumatolytic stage, which is limited generally to plutonic sources (with high temperature and pressure), or in the degassing of subvolcanic and volcanic magmas at low pressures and mostly lower temperatures.

Direct measurement of these gases at the earth's surface is possible only for volcanic emanations. However, such analyses of volcanic emanations, in general, can not give the true composition of the original gas since HF is a very active gas and can react with the country rock. LOVERING (1957) investigated altered rocks in the Valley of Ten Thousand Smokes and found that fluorine is highly concentrated in the altered rocks located closest to gas orifices. Similar enrichments are found in silicate rocks near fumaroles from other localities (YOSHIDA et al., 1969; HONDA, 1970; NABOKO, 1957). This may also be a reason for the fluorine fluctuations in the fumarolic gases since these gases can again leach fluorine from such altered rocks (HONDA, 1970).

The amounts of fluorine which are supplied by volcanoes can be very large. ZIES (1929) estimated the annual production of HF to be 200,000 tons from an area of 77 sq.km in the Valley of Ten Thousand Smokes. But the amount of fluorine, compared with the total gas emission of fumaroles, is relatively small. The principal component of these gases, H_2O, generally constitutes 99% of the total volume (rarely less than 80%). HF is generally only 1 to 2 vol.-% of the total gases. Both H_2O and HF are components of the "active gases" (WHITE and WARING, 1963). The total active gases, the HF content in % of the active gases, and the water gas content of volcanic gases are quoted in Table 9-F-2 (from the compilation of WHITE and WARING, 1963). See also TONANI (1957), SIGIURA et al. (1963) and MURATA et al. (1964). The energetics of HF in volcanic emanations is discussed by MUELLER (1970).

Table 9-F-2. *The HF content of volcanic gases in vol.-% according to the compilation of* WHITE *and* WARING *(1963)*

Location	Temperature °C	H₂O concentration	Active gases	HF as percent fraction of total active gases	Year of collection
Showa-shinzu, Usu Volcano (Japan) hypersthene dacite, extruded 1944–45.	700	99.39	0.592	3.54	1959
	464	99.10	0.859	0.88	1959
	203	~99.1	0.486	1.71	1959
Kliuchevskii Volcano, basalt Zavaritskii Crater (U.S.S.R.) erupted 1945 Apahonchich Crater, erupted 1946.	358	a	0.63	1	1949
	500	a	0.81	1	1949
	270	a	0.31	3	1949
	296	a	0.30	6	1949
Katmai Volcano (Alaska) rhyolite ash, erupted 1912.	400	99.97	0.03	11	1917
	400	99.96	0.04	17	1917
	350	99.87	0.13	23	1917
	250	99.90	0.10	32	1917
Mount Hekla (Iceland) basalt volcano, erupted 1845 and 1947.	620	—	7.0	trace	1951
Vesuvius (Italy), fumarole in crater, leucite tephritic volcano major eruptions 1906, 1944.	220—286	87.00	12.98	0.08	1917±

a Not determined.

Table 9-F-3. *HF content of fumarole condensates from volcanos, expressed in milligrams per liter water (From* WHITE *and* WARING, *1963)*

Location	T °C	HF mg per liter water	Year of collection
Sheveluch (Kamchatka, U.S.S.R.)	280	21.0	1953
	210	34.0	1953
	295	84.0	1953
	180	14.0	1953
Santa Maria (Guatemala), andesite	~300	20.4	1932
Hekla (Iceland), basalt, Shoulder Crater	520	13	1952
White Island (New Zealand) Big Donald fumarole, Schubert's Erl King Seven Dwarfs B/20	305	19	1939
	~340	290	1939
	~175	38	1939
Showa-shinzan, andesite, Hokkaido (Japan)	—	195 ppm	1958
Kilauea Iki, tholeiitic basalt (Hawaii)	—	20 ppm	

Table 9-F-4. *F content of sublimates of volcanic fumaroles and steams*

Location	T °C	F wt.-%	Year of collection	Reference
Paricutin (Mexico)	294	0.57	1943 (?)	WHITE and WARING (1963)
	120	0.53	1943 (?)	
Valley of 10,000 Smokes (Alaska)	350±	14.9	1919	ZIES (1929)
Phlegraean Fields, Naples (Italy)	—	0.195	1968	TADDEUCCI (1968)

Table 9-F-5. *Fluorine-gas included in rocks in volume percent (from the compilation of WHITE and WARING, 1963)*

Location	Volume of gas per gram rock (cc/g)	H_2O vol.-%	F_2 vol.-%	F_2-gas, recalculation to 100% gas, excluding H_2O
Obsidians				
Big Glass Mountain, Calif.	1.19	88.386	3.163	27.27
Coso Mountain, Calif.	1.45	89.246	7.795	54.54
Cerro Noagua (pitchstone)	214.8	98.551	—	78.51
Andesites, dacites				
Mount Pele, lava of 1912	4.8	82.525	3.317	—
Mount Pele, pumice glass (1903)	5.9	70.812	4.289	—
Lassen Peak (U.S.A.), bomb 1915	9.5	93.658	1.524	—
Andesites	—	97.20	0	—
Basalts				
Kilauea, molten lava 1916	6.9	80.834	3.609	18.83
Kilauea, pahoehoe 1923	5.0	74.953	14.115	56.37
Mauna Loa, pumice 1926	6.2	73.199	0.000	0.00
New Market basalt, Maryland (U.S.A.)	30.0	90.059	0.275	—
Granites				
Stone Mountain, Georgia (U.S.A.)	34.2	90.512	1.970	20.84
North Jay, Vermont (U.S.A.)	29.4	79.599	2.563	12.56
Granites and gneisses	—	95.54	0	—

IV. Hydrothermal Fluids

All fluorine which cannot be incorporated in crystalline phases during crystallization and differentiation of magmas will be accumulated in hydrothermal solutions. These fluids may form hydrothermal fluorite deposits and veins (e. g., BAZAROV et al., 1965). In some cases such fluids can react with the previously crystallized minerals to form greisen (SCHNEIDERHÖHN, 1961).

Fluorine transport in these aqueous solutions is controlled mainly by the solubility of CaF_2 (ELLIS and MAHON, 1967). Hydrothermal fluids low in fluorine can leach fluorine out of the country rocks which may be of magmatic, metamorphic or sedimentary origin (ELLIS and MAHON, 1967; WALKER and BUCHANAN, 1969). Experimental investigations have shown that the fluorine added by leaching is controlled by mineral solubility or ion exchange equilibria (ELLIS and MAHON, 1967).

The most abundant fluorine mineral in the hydrothermal cycle is fluorite, which often forms fluorite veins and is in many instances associated with sulfide ores. Recently sellaite, MgF_2, was also found in a low temperature fluorite vein along with barite and traces of sulfides (BORDET et al., 1964).

BROWNE and ELLIS (1970) have investigated the hot waters in a hydrothermal area of New Zealand. KRAYNOV et al. (1969) have studied water in deeper zones of the Lovozero Massif. In drillholes down to depths of 2134 m, BROWNE and ELLIS (1970) found 4 to 8 mg F/kg water. KRAYNOV et al. (1969) described highly alkaline (pH 12) water in deep mines in the alkalic Lovozero Massif which is strongly enriched in fluorine (up to 10 to 15 g F/liter). Fluorine-rich water with highest concentrations is found in rocks especially high (3%) in villiaumite.

WALKER and BUCHANAN (1969) have experimentally investigated the production of hydrothermal fluids from sedimentary rocks for processes in ore formation. Heating of sediments at temperatures of 300 to 400°C already releases a highly reactive mixture of gases which possibly contains F^+, HF^+ and SOF^+ with the mass numbers 19, 20 and 67 (I).

9-G. Behavior during Weathering and Alteration of Rocks

The behavior of fluorine during weathering in a humid climate was investigated by KORITNIG (1951) in a 5 m profile of weathered granite. The essentially down-circulating solutions in the marshy area are, in general, acidic. From Table 9-G-1 it can be seen that the fluorine was leached out very soon after the onset of weathering. Only in the uppermost parts is a small increase in fluorine to be observed. The most stable F-bearing component in this profile is apatite. The fluorine from micas has been leached, and the fluorite has been dissolved.

Table 9-G-1. *Fluorine content in a weathering profile of granite*

Sample, locality	Total F in sample in ppm	F not occurring in apatite as % of total F
Fresh, hard granite from Schierke, Harz (Germany)	860	95
from 5 m depth below surface; rock brittle and altered	40	1
from 2 m depth below surface; rock crumbled	30	1
from 1 m depth below surface; rock fine crumbled	20	0
from 30 cm depth below surface; earthy forest soil	50	3

Similar observations have been made by ROBINSON and EDGINGTON (1946) and MICHAEL and BLUME (1952) who found numerous soils (Table 9-G-2) in which fluorine decreased in the direction from the rock to soil surface. But in the case of these studies, micas are the major phases containing fluorine. The influence of weathering on the occupancy of the (OH,F)-structural position in micas has been investigated by RIMSAITE (1967, 1970b) in detail. If we compare the F content of clastic sediments (see Section 9-K), as well as from the weathering profile (where the fluorine is nearly entirely leached), we see that the mica proportion does not alone control the F content in clastic sediments.

This is also apparent from an investigation by KORITNIG (1951) of the fluorine content of a limestone and its micaceous natural residue. The limestone (Wellenkalk) has an average of 220 ppm F (C), whereas the micaceous residue has 800 ppm F.

HÜBNER (1969a, b) has shown, in his investigations of the F^-/OH^- exchange-adsorption of clay minerals, that the concentrations and pH of the circulating solutions have a very great influence on the ability to be leached and on the adsorption of fluorine. However, other factors can also control the fluorine content of the clay minerals in sediments.

Table 9-G-2. *Fluorine content of soils. (After* FLEISCHER *and* ROBINSON, *1963)*

Origin	F content, ppm		Number of samples	Reference
	Range	Average		
U.S.A.	10—1,500	250	(16)	STEINKOENIG (1919)
	10—7,070	290	(137)	ROBINSON and EDGINGTON (1946)
	130—5,600	980	(339)	MACINTIRE et al. (1958)
	104— 284	190	(9)	MACINTIRE et al. (1958)
Sweden, sandy	43— 198	90	(11)	NOMMIK (1953)
Sweden, clay-like	248— 657	450	(20)	NOMMIK (1953)
Germany	80—1,100	530	(19)	MICHAEL and BLUME (1952)
	120—1,100	390	(24)	JAHN-DEESBACH (1958)
U.S.S.R.	30— 320	200	(46)	VINOGRADOV (1957)
Japan	260— 520	370	(6)	KOKUBU (1956)
New Zealand	68— 540	200	(23)	GEMMELL (1946)

Among the clay minerals, the di- and tri-octahedral illites are especially important. An investigation of the F^-/OH^- exchange in these minerals was made by HÜBNER (1969b). He found that the exchange-adsorption reactions that occur during weathering are overlapped by processes of chemical decomposition of the layer silicates. In acidic environments, chemical decomposition and exchange-adsorption act in opposite directions. The adsorption of F prevails in the acidic range, whereas in the alkaline environment it is the F desorption.

Interpretation of fluctuations of the fluorine content during natural weathering processes is very difficult. Companion solvent ions, pH values, complex formation and special lattice properties of the clay minerals (e.g., see RIMSAITE, 1970b) are all important factors during weathering.

Since F-apatite is the least soluble of the apatites (see Section 9-H), fossil bones with their OH-apatite (KLEMENT, 1935) will be changed slowly into F-apatite in F-bearing ground water. This fact has even been used in an attempt to determine the age of fossil bones. However, the fluorine-hydroxyl exchange was found to depend on the chemical composition of the circulating solutions and cannot be easily correlated with the age.

NOBLE et al. (1967) found that during the natural hydration and devitrification of glassy rocks and glasses, about half of the F originally present was lost. In some cases significant amounts of fluorine may be removed from or added to the hydrated glass through contact with ground water.

Revised manuscript received: December 1971

9-H. Solubilities of Compounds which Control the Concentration of Fluorine in Natural Waters; Adsorption Processes

The solubility of $Ca_5F(PO_4)_3$ (fluorapatite) and CaF_2 (fluorite) are especially important for reactions in natural processes.

VALYASHKO et al. (1968) have calculated the thermodynamic stability of fluorapatite, chlorapatite and hydroxylapatite in aqueous solutions at temperatures from 25 to 350°C. STRÜBEL (1965) experimentally determined the solubility of fluorite in pure water from 20 to 600°C at pressures up to 3,020 bars and its phase relations in the ternary system CaF_2-NaCl-H_2O from 20°C to the boiling point. Both investigations are important for reactions in the sedimentary, magmatic and hydrothermal cycles.

According to the calculations of VALYASHKO et al. (1968), the solubility product of fluorapatite at room temperature is lower than the solubility product of the hydroxyl- and chlorapatites (see Table 9-H-1), but the absolute values of the calculated data are not in good agreement with the experimental data (Table 9-H-4). The calculated stability fields of F- and OH- apatites are shown in Fig. 9-H-1.

For an analysis of the conditions of hydrothermal formation of fluorite and of the formation of fluorite in the sedimentary cycle, the investigations of STRÜBEL (1965) are of interest. Tables 9-H-2 and 9-H-3 show his data for the solubility of CaF_2

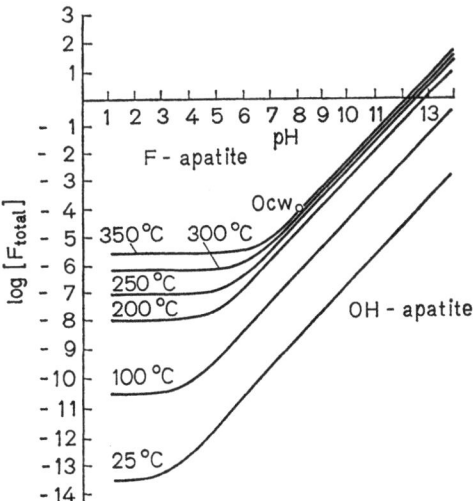

Fig. 9-H-1. Calculated stability fields of F- and OH-apatites (after VALYASHKO et al., 1968) (*Ocw*: ocean water)

© Springer-Verlag Berlin · Heidelberg 1972

in pure water and in water containing varying amounts of NaCl, respectively. These data are plotted in Fig. 9-H-2. Ocean water corresponds to a solution of about 0.6 n NaCl.

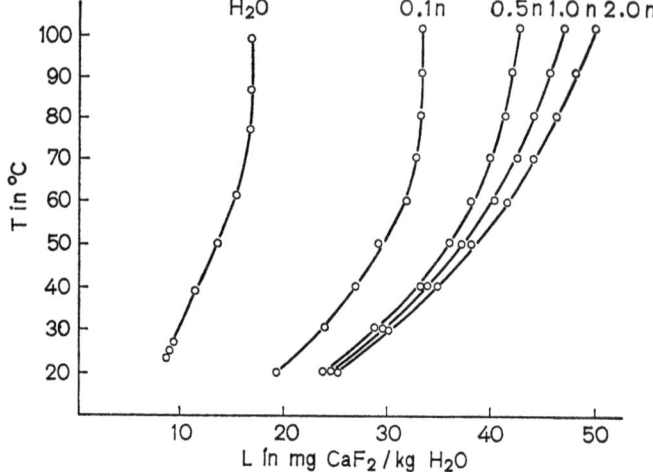

Fig. 9-H-2. Solubility of fluorite in the system CaF_2—NaCl—H_2O in solutions of constant NaCl concentrations. (After STRÜBEL, 1965)

Table 9-H-1. *Exponents of solubility products of hydroxyl-, fluor- and chlor-apatite (calculated).* (After VALYASHKO et al., 1968)

T °C	Exponents of solubility product		
	OH-apatite	F-apatite	Cl-apatite
25	−115.52	−120.87	−106.16
50	−118.04	−122.97	−108.93
100	−125.51	−129.68	−116.80

Table 9-H-2. *Solubility (L) of fluorite from 20° C to 350° C* (After STRÜBEL, 1965)

Temperature (°C)	Pressure (Bar)	L (mg CaF_2/kg H_2O)
23.0		8.76 ± 0.14
25.0		9.13 ± 0.16
26.5		9.24 ± 0.09
39.0		11.43 ± 0.10
50.0		13.70 ± 0.10
61.0		15.43 ± 0.16
76.5		16.70 ± 0.12
86.5		16.82 ± 0.11
98.5		16.78 ± 0.06
150.0	4.8	15.20
200.0	15.6	13.20
250.0	39.8	11.30
255.0	43.3	13.00 ± 9.00
300.0	85.9	7.85 ± 0.20
310.0	98.7	14.50 ± 4.50
350.0	165.4	6.50

Table 9-H-3. Solubility (L) of CaF_2 in the ternary system CaF_2-$NaCl$-H_2O. (After Strübel, 1965)

Temperature (°C)	Density (g·cm⁻³)	L (mg CaF_2/kg H_2O)			
		0.1 n NaCl	0.5 n NaCl	1.0 n NaCl	2.0 n NaCl
20.0	0.9982	19.35 ± 0.08	23.88 ± 0.11	24.58 ± 0.10	25.19 ± 0.11
30.0	0.9956	23.96 ± 0.06	28.84 ± 0.05	29.64 ± 0.06	29.88 ± 0.11
40.0	0.9922	27.13 ± 0.14	33.35 ± 0.08	33.81 ± 0.08	34.67 ± 0.32
50.0	0.9881	29.51 ± 0.11	36.08 ± 0.16	37.17 ± 0.00	38.66 ± 0.08
60.0	0.9832	31.78 ± 0.23	38.03 ± 0.08	40.22 ± 0.08	41.54 ± 0.15
70.0	0.9778	32.64 ± 0.15	40.15 ± 0.15	42.40 ± 0.24	44.20 ± 0.00
80.0	0.9718	33.27 ± 0.15	41.31 ± 0.08	44.04 ± 0.16	46.31 ± 0.23
90.0	0.9654	33.34 ± 0.08	41.86 ± 0.16	45.29 ± 0.32	48.03 ± 0.24
		33.58 ± 0.15	42.48 ± 0.15	46.31 ± 0.71	49.93 ± 0.11
Ts[a]		(100.5° C)	(99.5° C)	(100.1° C)	(101.5° C)

[a] Ts: Boiling temperature of the mixed phase.

Table 9-H-4. Solubility products (k_{sp}) of some F-bearing compounds

	k_{sp}		Reference
OH-apatite	2.6 × 10⁻⁴⁵	at 18° C, pH 7.0	1
OH-apatite	2.3 × 10⁻⁴¹	at 40° C, pH 7.4	1
Fluorite	3.4 × 10⁻¹¹	at 18° C	2
Sellaite	6.4 × 10⁻⁹	at 27° C	2

References: 1. Hayek, Mullner and Koller (1951) quoted in Linke and Seidell (1958). 2. D'Ans-Lax (1967).

Table 9-H-5. Solubility in mg per liter H_2O of some F-bearing compounds

Compound	Solubility	Temperature	Reference
Apatite	13 mg/l		1
Fluorite	8.67 ± 0.14 mg/kg	23° C	2
Sellaite	130 mg/l	25° C	3
Villiaumite	4,054 mg/l	25° C	3

References: 1. Hintze (1904—1968), vol. I, 4.1, p. 508. 2. Strübel (1965); for additional data see Tables 9-H-2 and 9-H-3. 3. Carter (1928).

Revised manuscript received: December 1971

9-I. Abundance in Natural Waters and in the Atmosphere

I. Natural Waters

a) Ground Water, Hot Spring Water

During weathering and circulation of water in rocks and soils, fluorine can be leached out and dissolved in ground water. The fluoride content of ground water varies greatly depending on the type of rocks from which they originate (BOND, 1945; WHITE et al., 1959; KÖPF et al., 1968; KRAINOV and PETROVA, 1962; VOROSHILOV, 1966). Table 9-I-1 summarizes data of BOND (1945) for South Africa and WHITE et al. (personal communication) for U.S.A. (as cited in FLEISCHER and ROBINSON, 1963). The latter have also summarized the maximum fluoride content of ground water of U.S.A. in a map, showing that there are enormous areas in which some of the water contains more than 1.5 ppm F; this is usually given as the concentration above which fluorine may be harmful. MÖSE and EXNER (1952) stated that in 110 samples of Styrian drinking waters more than 75 per cent had a fluorine content below 0.1 ppm (mg/l) and only one sample reached 0.8 ppm F.

Hot spring water (30°—95° C) from the Saga Pref., Japan, has concentrations from 0.15 to 8.0 ppm (mg/l) fluorine (MATUURA and KOKUBU, 1955). The fluorine content increases with the temperature of the springs, but the ratio of F/Cl in mol is approximately constant. The spring water is here diluted with cold ground water. SUGAWARA (1954, in CORRENS, 1956) gives values from 32.1 to 55.4 ppm fluorine for the Osama Hot Springs, Japan. For thermal and mineral springs of Japan, SUGAWARA (1967) found an average of 1.9 mg/l fluorine. Additional data and references for thermal and mineral springs are in GMELIN (1959).

b) Rivers, Lakes

The fluoride content of surface water also varies greatly depending on the fluoride content of ground water feeding a given stream (FLEISCHER and ROBINSON, 1963) and on the amounts of precipitation and run-off. The fluoride content is generally higher during dry periods. Most U.S. rivers contain less than 0.5 ppm fluoride, but even sizable rivers in areas such as western Texas may have fluoride contents upto 6.4 ppm F (e. g., Double Mountain Fork of the Brazos River and the Canadian River

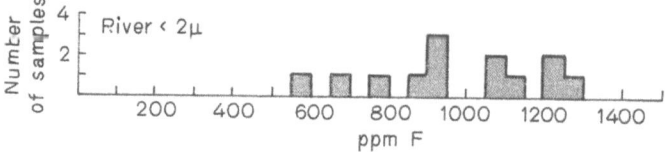

Fig. 9-I-1. Frequency distribution of fluorine in the $<2\mu$ size fraction of the suspended load of rivers (After CARPENTER, 1969)

Table 9-I-1. *Fluorine content of water from areas characterized by the predominance of one rock type, according to a list in* FLEISCHER *and* ROBINSON *(1963)*

Predominant rock type	Data of BOND (1945)			Data of WHITE, HEM, and WARING (personal communication)		
	F content ppm		Number of samples	F content ppm		Number of samples
	range	average		range	average	
Granite	0—9.0	1.4	78	0—3.4	0.9	14
Alkalic rocks	0.7—35.1	8.7	7	—	—	—
Basalt	0—0.5	0.1	44	0—0.4	0.2	11
Andesite	—	—	—	0—0.1	0.1	4
Sandstones and quartzite	0—2.7	0.1	69	0—1.9	0.4	16
Shale and clay	0—1.8	0.2	80	0—2.8	0.6	16
Limestone	—	—	—	0—0.9	0.3	14
Dolomite	0—0.3	0.02	21	0—1.7	0.5	5

near Amarillo; FLEISCHER and ROBINSON, 1963). LEVINSON et al. (1969) found for the Mackenzie River at Norma Wells, N.W.T., Canada, 0.09 to 0.20 ppm (average 0.14 ppm F). HEIDE and KAEDING (1954) investigated the water of the Saale river (Thuringia) by W/C method and found 0.15 to 0.35 ppm fluorine (also upto 0.90 ppm F, it is contaminated with sewages of potash industry). FLEISCHER and ROBINSON (1963) give an average of 0.2 ppm fluorine for stream water and SUGAWARA (1967), for 43 representation rivers in Japan, an average of 0.15 mg/l. SCHROLL (1959, 1965) gives the halogen content of rain, river, ground and lake water and the balance for a lake area (Neusiedlersee/Austria), which has only a very small flow off. The tributary river for the lake has 0.1 ppm F and the lake water 0.34 ppm F (enrichment by evaporation 3.4 ×). For additional data and references on subsurface water see WHITE et al. (1963), and GMELIN (1959). According to an investigation of KÖPF et al. (1968) on 30 different watersheds (27 wells, 3 tile drainage systems) in the State of Baden-Württemberg, Germany, water from limefree areas show lower F-contents (average 0.080 ppm F) than those in which limestone prevails (average 0.128 ppm F). The fluoride concentration in drainage water from a grey brown podzolic soil has a pronounced reciprocal relation with the fluorine content of the water and the run-off. For additional data on fluorine contents in rivers and lakes see GMELIN (1959).

With increasing use of fertilizers today, the fluorine content of surface water can also increase. Approximately 20 to 400 g F/hectare (2.47 English acres) are annually leached from soils, about the same amount which is added to the soil from the atmosphere, but fertilizing adds about 5 to 30 kg fluoride/hectare annually. This fluoride accumulates in the soils (KÖPF et al., 1968).

c) Seawater

According to BEWERS (1971) the mean concentration of fluoride in ocean water is 1.35 to 0.03 mg/l. In 152 samples from nine stations of the North Atlantic he could not find an increase in fluoride concentration with depth, as reported for some locations in the North Atlantic by RILEY (1965). Compare also KESTER (1971). For geologic history of seawater see RUBEY (1951) in BRANCAZIO and CAMERON (1963).

d) Rainwater

The main part of fluorine in rain water may originate from the sea, where tiny droplets of foam are caught up by the wind (CORRENS, 1956) and may be carried far from the ocean to continental areas. For the fluorine content of Japanese precipitation SUGAWARA (1967) gives an average of 0.089 mg/l. Amounts from 0.00 to 0.004 mg/l fluorine are given for rain water in U.S., far away from industry and dense population (BREDEMANN, McINTIRE, and others, for references see GMELIN, 1959). In the vicinity of cities and industrial areas, valves upto more than 1 mg/l (average 0.29 mg/l) fluorine can be found (GMELIN, 1959).

II. Atmosphere

The order of magnitude of the normal fluorine content in the air is $<0.01\ \mu g/m^3$ according to BOWEN (1966), about $0.3\text{—}0.4\ \mu g/m^3$ according to OELSCHLÄGER (1965) and in industrial areas up to $36\ \mu g/m^3$ (range $5\text{—}111\ \mu g/m^3$ in 26 samples) (OELSCHLÄGER et al., 1968). RADCZEWSKI (1968) investigated the dust (aerosol droplets) in the air over the North Sea, Germany, and identified K_2SiF_6 (hieratite) in the air in east wind and Na_2SiF_6 (malladrite) in west wind.

Fluorine plays an important role among toxic air pollutants, because even small amounts can be harmful for plant life and animals (see OELSCHLÄGER and RHEINWALD, 1968, and Section 9-L).

The main emitting sources of fluorine are plants producing hydrofluoric acid, aluminum, superphosphate and enamel, and also brickworks, metallurgical works and industrial plants consuming high amounts of low quality coal (OELSCHLÄGER, 1965). In exhaust gases from an aluminum-smelting furnace, RADCZEWSKI (1968) found $CaF_5 \cdot 5Al_2O_3$ in the dust.

SUGAWARA (1967) estimated the balance of fluorine of Japanese Islands. He found, for the total supply from the sky in the form of precipitation and dry fallout, $1.65 \cdot 10^{14}$ mg/year. Of this amount, $0.84 \cdot 10^{14}$ mg/year are discharged to the sea, leaving $0.79 \cdot 10^{14}$ mg/year on land.

Revised manuscript received: December 1971

9-K. Abundance in Common Sediments and Sedimentary Rock Types

Fluorine is the most abundant halogen in sedimentary rocks. Because of its similar ionic radius it is especially liable to become camouflaged in OH-bearing minerals.

The quantities of fluorine in sedimentary rocks and the mineral species in which fluorine occurs are listed in Tables 9-K-1 to 9-K-4.

Table 9-K-1. *Fluorine content in samples of different sedimentary rocks. (According to* KORITNIG, *1951, except data on two phosphatic sandstones reported by* PETTIJOHN, *1963)*

Rock	ppm F	Method
I. Clastic sediments		
a) Psammites		
Middle Bunter sandstone, Reinhausen near Göttingen (Germany)	80	(C)
Glauconitic sandstone (drill core 800 m), Ehra (Germany)	120	(C)
Calcareous sand, Meteor station 219, 220 m depth (Atlantic)	190	(C)
Greywackes, Culm, Lauterberg/Harz, (Germany)	80	(C)
Conglomerate, Zechstein (sandy facies), Albungen (Germany)	180	(C)
Loess, Hedemünden near Göttingen (Germany)	360	(C)
Phosphate sandstone, (Cenomanian), Kursk (U.S.S.R.)[a]	1.87%	(?)
Phosphatic sandstone, St. Pôt (France)[a]	2.86%	(?)
b) Pelites		
Shale, Culm, Waldenburg, Silesia (Poland)	500	(C)
Shale, Devonian, Rammelsberg, Harz (Germany)	960	(C)
Shale, (Dachschiefer) Raumland, Rothaargebirge (Germany)	310	(C)
Marl, (Kupferschiefer) (highly bituminous), Albungen (Germany)	650	(C)
Shale, Röt (gypsum bearing) near Göttingen (Germany)	50	(C)
Shale, Middle Bunter sandstone, Fredelsloh (Germany)	1,180	(C)
Shale, in marl, Upper Muschelkalk, Diemarden (Germany)	1,780	(C)
Shale, red Keuper, Friedland near Göttingen (Germany)	1,050	(C)
Shale, green Keuper, Friedland near Göttingen (Germany)	690	(C)
Glauconitic clay, Middle Oligocene, Uelsen, Lower Saxony (Germany)	550	(C)
Shale, blue (Estonia)	1,080	(C)
Clay, Sarmat, Zistersdorf (Austria)	410	(C)
Shale, Lias, Dobbertin, Mecklenburg (Germany)	760	(C)
Clay, (Middle Oligocene), Mallis, Mecklenburg (Germany)	990	(C)
Clay, Pleistocene, Papendorf near Rostock, Mecklenburg (Germany)	630	(C)
Clay, interglacial (vivianite bearing), Northeim (Germany)	300	(C)
Shale, gray (Salzton), Volkenroda, Thuringia (Germany)	2,750	(C)
Shale (Hauptlöser), Old. Potash bed, Volkenroda (Germany)	1,370	(C)
Marl (Bunte Zechsteinletten), Büdingen, Wetterau (Germany)	1,440	(C)
Shale, Posidonian, Reutlingen (Germany)	210	(C)
Pelagic clay, Meteor station 283, 5,729 m depth (Atlantic)	670	(C)
Pelagic, blue mud, Meteor station 229, 4,668 m depth (Atlantic)	730	(C)

Table 9-K-1 (continued)

Rock	ppm F	Method
II. Predominantly biogenic and chemical sediments		
a) Carbonates		
Limestone, Upper Muschelkalk, Diemarden (Germany)	510	(C)
Limestone, Upper Muschelkalk, Reckershausen (Germany)	20	(C)
Oolitic bed, Lower Muschelkalk (aver.), Auschnippetal (Germany)	190	(C)
Limestone, Lower Muschelkalk, Dransfeld (Germany)	220	(C)
Oolitic limestone, Malm, Freden, Lower Saxony (Germany)	60	(C)
Limestone, Cenomanian, Kaierde, Hils, Lower Saxony (Germany)	20	(C)
Chalk, Rügen (Germany)	280	(C)
Travertine, Recent, Lenglern near Göttingen (Germany)	100	(C)
Globigerina ooze, Meteor station 240, 2,271 m depth (Atlantic)	550	(C)
Dolomite, Zechstein, Scharzfeld, Harz (Germany)	110	(C)
Dolomite, Zechstein reef (fluorite bearing, average) Römerstein, Harz (Germany)	250	(C)
b) Siliceous rocks		
Chert, Culm, gray-black, Wallau (Germany)	500	(C)
Chert, Culm, yellowish, Wallau (Germany)	210	(C)
Chert, Gelstertal (Germany)	830	(C)
Diatomite, Beuren, Vogelsberg (Germany)	<5	(C)
Flint, Upper Chalk, Rügen (Germany)	40	(C)
Geyserite, Yellowstone Park (U.S.A.)	430	(C)
c) Rocks from potash salt deposits		
Anhydrite (drill core), Stassfurt (Germany)	800	(C)
Anhydrite (Middle Zechstein), Reyershausen near Göttingen (Germany)	810	(C)
Anhydrite, Upper Zechstein, Reyershausen (Germany)	130	(C)
Anhydrite (residue from rock salt), Reyershausen (Germany)	890	(C)
Gypsum, Niedersachswerfen (Germany)	870	(C)
Gypsum breccia, Hundelshausen (Germany)	130	(C)
Rock salt, Stassfurt (Germany)	2	(C)
Rock salt including residue, Reyershausen (Germany)	570	(C)
Rock salt excluding residue, Reyershausen (Germany)	7	(C)
Rock salt, Middle Zechstein, Reyershausen (Germany)	5	(C)
Rock salt, Upper Zechstein, Reyershausen (Germany)	6	(C)
Polyhalite, Stassfurt (Germany)	20	(C)
Potash bed, Reyershausen (Germany)	300	(C)

a PETTIJOHN (1963).

Table 9-K-2. *Average fluorine content in sedimentary rocks*

Rock	Average F, ppm	Number of samples	Method	Reference
I. Predominantly clastic sediments				
a) Psammites				
Sandstones and greywackes	180	(49)		FLEISCHER and ROBINSON (1963)
Sandstones	290	(24)	(C)	KOKUBU (1956)

Table 9-K-2 (continued)

Rock	Average F, ppm	Number of samples	Method	Reference
Sandstones Carboniferous, Germany (composite sample)	450	(11)	(C)	KORITNIG (1951)
Sandstones Triassic, Germany (composite sample)	320	(23)	(C)	KORITNIG (1951)
Sandstones Cretaceous, Germany (composite sample)	280	(11)	(C)	KORITNIG (1951)
Sandstones	270			TUREKIAN and WEDEPOHL (1961)
Sandstones	280	(11)	(W)	MICHAEL and BLUME (1952)
Greywackes, Germany (composite sample)	40	(17)	(C)	KORITNIG (1951)
Greywackes	330	(3)	(S)	SERAPHIM (1951)
Quartzites, Germany (composite sample)	200	(10)	(C)	KORITNIG (1951)
Volcanic ash	1,000	(5)	(C)	KOKUBU (1956)
Volcanic ash	720	(263)	(?)	POWERS[a], POWERS (1961)

b) Pelites

Rock	Average F, ppm	Number of samples	Method	Reference
Shales	800	(79)		FLEISCHER and ROBINSON (1963)
Shales	510	(7)	(?)	BARTH and BRUUN (1945)
Shales	290	(24)	(C)	KOKUBU (1956)
Shales	740			TUREKIAN and WEDEPOHL (1961)
Shales	750	(9)	(S)	SERAPHIM (1951)
Shales	680	(22)		TOURTELOT (1962)
Shales, Germany, Paleozoic (composite sample)	760	(36)	(C)	KORITNIG (1951)
Shales, Japan, Paleozoic (composite sample)	500	(14)	(C)	KORITNIG (1951)
Shales, Japan, Mesozoic (composite sample)	570	(10)	(C)	KORITNIG (1951)
Bentonites	5,950	(2)	(W)	ROBINSON and EDGINGTON (1946)

II. Predominantly biogenic and chemical sediments

a) Carbonates

Rock	Average F, ppm	Number of samples	Method	Reference
Limestones	220	(98)		FLEISCHER and ROBINSON (1963)
Limestones	227	(7)	(?)	DANILOVA (1949)
Limestones	340	(28)	(?)	JEFFRIES (1951)
Limestones	98	(30)	(C)	KOKUBU (1956)
Limestones, Germany, Devonian (composite sample)	210	(32)	(C)	KORITNIG (1951)
Limestones, Germany, Jurassic (composite sample)	550	(45)	(C)	KORITNIG (1951)

Table 9-K-2 (continued)

Rock	Average F, ppm	Number of samples	Method	Reference
Limestones, Germany, Cretaceous (composite sample)	270	(16)	(C)	Koritnig (1951)
Limestones	940	(7)	(W)	Michael and Blume (1952)
Limestones	90	(6)	(S)	Seraphim (1951)
Dolomites	260	(14)		Fleischer and Robinson (1963)
Dolomites	250	(10)	(?)	Danilova (1949)
Dolomites	390	(2)	(?)	Jeffries (1951)
Dolomites	180	(2)	(C)	Koritnig (1951)
III. Oceanic sediments				
Without specification	730	(79)		Fleischer and Robinson (1963)
Oceanic clays	700	(2)	(C)	Koritnig (1951)
Clays				
Pacific Ocean	600	(1)	(C)	Shishkina (1966)
Indian Ocean	580	(1)	(C)	Shishkina (1966)
Red clay, Pacific Ocean	540	(8)	(C)	Shishkina (1966)
Globigerina ooze	540	(17)	(S)	Seraphim (1951)
Carbonate sediments and calcareous clay:				
Pacific (6), Atlantic (1), Indian Ocean (2)	520	(9)	(C)	Shishkina (1966)
Black Sea	550	(35)	(C)	Shishkina (1966)
Baltic Sea	380	(1)	(C)	Shishkina (1966)
Cores from the Carnegie, Piggot and Thompson collection	430	(34)	(W)	Shepherd (1940)

[a] Personal communication to Fleischer and Robinson (1963).

Table 9-K-3. *Fluorine content (ranges and averages) of sedimentary rocks. (After Fleischer and Robinson, 1963)*

Rock	F ppm Range	F ppm Average	Number of samples
Limestones	upto 1,210	220	(98)
Dolomites	110—400	260	(14)
Carbonates	—	330[a]	
Sandstones and greywackes	10—1,100	200	(50)
Sandstones and greywackes	10— 880	180	(49)
Shales	10—7,600	940	(82)
Shales	10—7,600	800	(79)
Oceanic sediments (without specification)	100—1,600	730	(79)
Volcanic ashes and bentonites	100—2,900	750	(270)

[a] Average Turekian and Wedepohl (1961).

Table 9-K-4. *Fluorine distribution in the different F-bearing minerals in sedimentary rocks. (After* KORITNIG *1951, 1963)*

Rock Locality	F-bearing mineral and wt.-% of this mineral in rock		F content of F-bearing minerals wt.-%	Total F assigned to F-bearing minerals wt.-%
Mica-rich shale bed, Triassic Middle Bunter sandstone Fredelsloh, Lower Saxony (Germany)	mica apatite	55 0.2	0.20 3.5	94 6 — 100
Lias shale, Dobbertin, Mecklenburg (Germany)	illite apatite	52.6 0.44	0.11 3.5	79 21 — 100
Diluvial clay, Papendorf near Rostock (Germany)	mica montmorillo- nite apatite	36.9 7.0 0.23	0.14 0.03 3.5	84 3 13 — 100
Ceratites limestone, Upper Muschelkalk, Diemarden near Göttingen (Germany)	illite apatite fluorite	1.5 0.3 0.076	0.26 3.5 48.7	8.0 19.6 72.4 — 100.0
Chalk, Rügen (Germany)	apatite fluorite	0.21 0.043	3.5 48.7	25 75 — 100
Oolitic limestone, Malm, Freden, Lower Saxony (Germany)	apatite	0.18	3.5	100 — 100
Zechstein dolomite, Scharzfeld, Harz (Germany)	apatite fluorite	0.017 0.023	3.5 48.7	5 95 — 100
Chert (Kieselschiefer), Culm, Wallau near Marburg (Germany)	mica apatite	6 0.22	0.22 3.5	62 38 — 100
Anhydrite, Stassfurt (Germany)	apatite fluorite	0.000 0.164	3.5 48.7	0 100 — 100

I. Clastic Rocks

In clays approximately 80 to 90 per cent of fluorine is contained in muscovite, illite and related minerals of the mica group; the remainder is contained in montmorillonite, kaolinite and apatite.

The average F content of sedimentary micas and illites is about half as large as that of average igneous rocks. Montmorillonite and kaolinite contain only one tenth as much fluorine as sedimentary micas.

The factors controlling the marine geochemistry of fluorine and the behavior of fluorine in the present-day sedimentary cycle have been investigated by CARPENTER (1969). He found that calcium carbonate precipitation dominates the processes of removal of dissolved fluorine from sea water and that incorporation into calcium phosphates is apparently the next most important mechanism of removal. An enrichment of fluorine over chlorine by a factor of about 10^5 occurs during $CaCO_3$ precipitation. However, calcium carbonate and phosphate precipitation together can account for at most only 10 to 20% of the dissolved fluorine brought annually to the ocean by rivers. Very little of this dissolved fluorine is removed from solution by clays settling to the ocean floor (CARPENTER, 1969). SHISHKINA (1966) investigated the fluorine contents of oceanic sediments and their pore solutions with depth. F increases in concentration with depth, but the reverse is true in the Black Sea.

II. Carbonate Rocks and Evaporites

In carbonate rocks fluorine occurs in varying proportions of fluorite or apatite and clay minerals, depending on the prevalent chemical or biological mode of origin. In marine anhydrite fluorite represents nearly the total fluorine content. In rock salt deposits BRAITSCH (1960, 1962) also found, in addition to apatite, small amounts of the fluorine phosphates, isokite $CaMgFPO_4$ and wagnerite Mg_2FPO_4, possibly formed as secondary phases. HERRMANN and HOFFMANN (1961) analyzed fluorine in petroleum water and assumed that this element was introduced into the water during the formation of petroleum from primitive marine organisms.

The form in which fluorine is precipitated in carbonates has long been a matter of controversy. Many geological field observations indicated synsedimentary inorganic precipitation of CaF_2 in the carbonate rocks (SCHNEIDER, 1953, 1954; KRÜGER, 1962a, b, c). On the other hand, this was not in agreement with the known physicochemical data for the solubility of CaF_2. For instance, SILLEN (1959) assumed that there is a low probability of inorganic precipitation since the ionic activity product, for bulk sea water, $[Ca][F]^2$, is less than the K_{sp} of CaF_2. KAZAKOV and SOKOLOVA (1950), as reported by KÜHN (1968), studied the solubility relations of CaF_2 in distilled water and seawater and found that the solubility of CaF_2 increases to a certain degree with increasing concentrations of $MgSO_4$ or NaCl (see also STRÜBEL, 1965, Fig. 9-H-2). In evaluating these results, KÜHN (1968) states that as sea water is evaporated, these contrary influences become effective in a solubility minimum for fluorite of 4 mg F/l at a three- or four-fold concentration of seawater. This approximates the actual fluorine concentration or four-fold concentrated seawater. At concentrations slightly higher than this, calcium carbonate and calcium sulfate precipitate, thereby increasing the solubility of fluorite through removal of Ca. The overall result is a limited concentration range in which the precipitation of fluorite

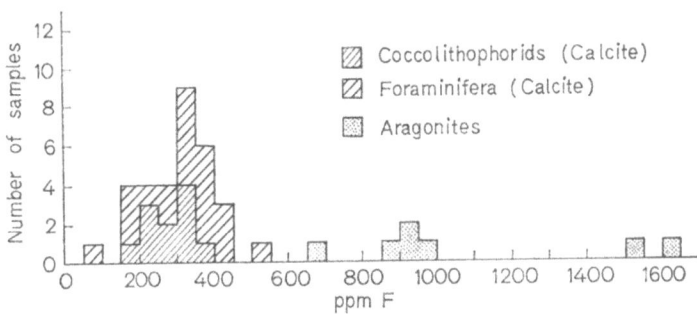

Fig. 9-K-1. Frequency distribution of the fluorine content in CaCO$_3$ samples. (After CARPENTER, 1969)

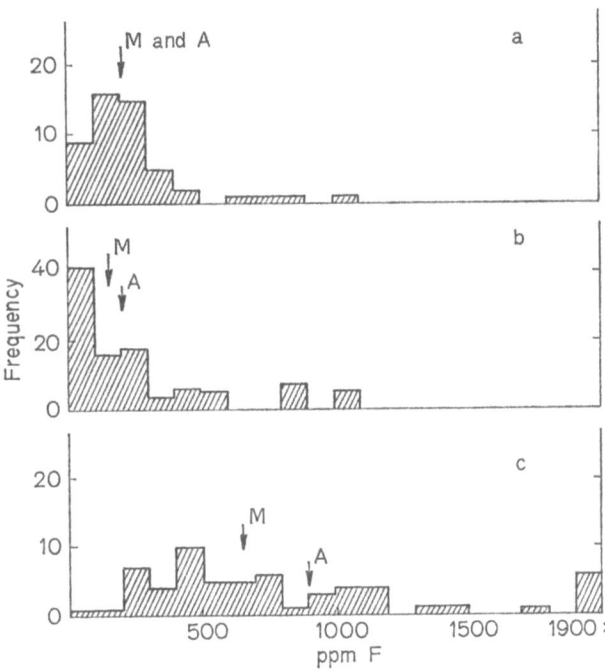

Fig. 9-K-2 a—c. Frequency distribution of fluorine in sedimentary rocks (after FLEISCHER and ROBINSON, 1963) (M median; A average). a 50 sandstones and greywackes; b 98 limestones; c 60 shales

in carbonate deposits is possible. However, if fresh river water periodically flows into the salt basin, the precipitation of fluorite will increase. Thus, the results of KAZAKOV and SOKOLOVA (1950) indicate that the discrepancy between the field observations and physico-chemical laws no longer exists and that the inorganic precipitation of CaF$_2$ in the environment of marine evaporites is possible.

The studies of WILBUR and WATABE (1963) indicate that coprecipitation of fluorite and CaCO$_3$ can also occur in open sea water within the cells of coccoliths,

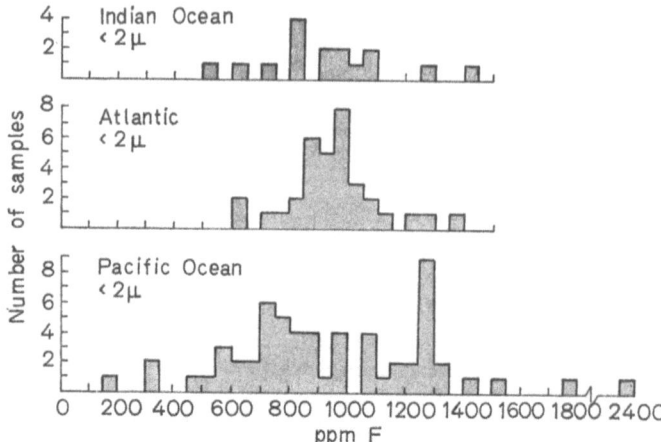

Fig. 9-K-3. Frequency distribution of fluorine of the $<2\,\mu$ size fraction of Recent oceanic sediments. (After CARPENTER, 1969)

where the activities of Ca^{2+} and F^- may be changed so that a precipitation of fluorite is possible. This could explain the observations of FÜCHTBAUER (1958) who found the abundance of CaF_2 in the Hauptdolomit to be correlated with the occurrence of calcareous algae (see also Fig. 9-K-1 from CARPENTER, 1969).

The average fluorine content of deposits of marine origin should be higher than that of sediments of the non-marine environments. In a study of carboniferous shales from the South Wales coalfield, BLOXAM and THOMAS (1969/70) have shown that the highest fluorine contents occur in the paleontologically defined offshore marine facies, and that the fluorine content generally decreases in the near-shore sediments. However, these results are influenced by other factors too. The fluorine content, therefore, is not a certain indication of marine or non-marine origin.

9-L. Biogeochemistry

A large amount of literature on fluorine in plants and animals has been published (reviews are given by BREDEMANN, 1956; UNDERWOOD, 1962; GMELIN, 1959; BOWEN, 1966; OELSCHLÄGER et al., 1967a, b, 1968).

The fluorine contents of plants — mostly cultivated plants — are in general low, except for tea (up to 400 ppm) and Dichapetalum cymosum (MARAIS, 1943). The fluorine content in the different parts of the plants is distinct. Roots and leaves generally have higher fluorine contents than fruits, seeds, stems, wood and barks (ROHOLM in HEFTER, 1938; UNDERWOOD, 1962).

Table 9-L-1 summarizes some data from plants and animal substances from areas with and without industrial emission of fluorine.

The magnitude of fluorine contents in leaves of different vegetables is about 3 to 20 ppm F in dry matter and in beef bones (cattle) 150 to 1,800 ppm F (ash). For additional data see MACINTIRE et al. (1942), OELSCHLÄGER et al., (1967a, b) and

Table 9-L-1. *Average fluorine content of plants and animal substances from areas with emission of fluorine (After OELSCHLÄGER et al., 1968), all determinations with W/C method*

From areas with fluorine emission				Average fluorine content from unpolluted areas, ppm F (dry weight)
Fluorine emitting source	Sample	Number of samples	Average ppm F (dry weight)	
Industrial area, Weisweiler/Germany	beet leaves	17	19	8.3
	grass	22	15	6.8
	red clover	6	20	6.7
City area, Mannheim/Germany	digitalis lanata	1	31	8.0
Brickyard, Baden/Germany	pine-needles 1 and 2 years old	4	53.9	3.3
Brickyard, Rheinland/Germany	pine-needles 1 year old	3	37.3	3.3
Brickyard, Rheinland/Germany	beet leaves	2	40	8.3
Glass factory, Baden/Germany	red clover	1	46	6.7
	pine-needles	3	85	3.3
Glass factory, Württemberg/Germany	hornbeam leaves	1	2,585	10.0
Enamel factory, Württemberg/Germany	red clover, unwashed	1	1,503	6.7
	red clover, washed	1	101	6.7

© Springer-Verlag Berlin · Heidelberg 1972

Table 9-L-1 (continued)

From areas with fluorine emission				Average fluorine content from unpolluted areas, ppm F (dry weight)
Fluorine emitting source	Sample	Number of samples	Average ppm F (dry weight)	
Enamel factory, Württemberg/Germany	linden leaves	7	317	14.2
	horse chestnut leaves	4	408	19.2
	maple leaves	3	211	7.0
	copper-beech (purpurea) leaves	2	316	10.6
	linden leaves	2	66	14.2
	horse chestnut leaves	1	120	19.2
	copper-beech (purpurea) leaves	1	40	10.6
	hornbeam leaves	1	58	10.0
Aluminum-smelting furnace, Baden/Germany	meadowgrass and hay	30	38.5	6.8
	pelvis bone (cattle)	70	2,730	150—1,800 (ash)
	bones (cattle)	124	5,724	150—1,800 (ash)
Aluminum-smelting furnace, France	vine leaves	2	1,733	11.9
	rib (cattle)	1	11,300	150—1,800 (ash)
	pelvis bone (cattle)	1	10,700	150—1,800 (ash)
Aluminum-smelting furnace, Westfalen/Germany	tail vertebra (cattle)	1	7,085	150—1,800 (ash)
	back vertebra (cattle)	1	7,848	150—1,800 (ash)
Hydrofluoric acid factory, Württemberg/Germany	hay	3	179	6.8
	clover	2	252	6.7
	grass	3	148	6.8
	summer barley	1	157	
	apple leaves	1	1,582	19.7
	pear leaves	2	1,189	19.8
	pear fruits	2	10.7	0.7
	strawberry leaves	1	232	2.3
	strawberry fruits	1	22	2.4
	red currant leaves	1	120	13.9
	red currant fruits	1	15	2.3
	sweet cherry leaves	1	934	4.9
	sweet cherry fruits	1	20	3.6
Hydrofluoric acid factory, Lower Saxony/Germany (time of sampling: 4 years)	meadowgrass	351	327	6.8
	hay	44	107	6.8
	straw	15	38	

OELSCHLÄGER and RHEINWALD (1968). In industrial areas with high emission of fluorine, the fluorine content in vegetables is 2 to 250 times and in beef bones 2 to 75 times larger (OELSCHLÄGER et al., 1968).

Land plants have 0.5 to 40 ppm F (BEESON, 1941; MITCHELL and EDMANN, 1945; BOWEN, 1966) and marine plants 4.5 ppm F (YOUNG and LANGILLE, 1958). Fossil plants (120 samples of coals from the British Commonwealth) contain 0 to

175 ppm fluorine (CROSSLEY, 1944, cited in CORRENS, 1956). The majority contains less than 80 ppm F. Apatite is the fluorine-bearing component.

The fluorine in bones has been converted to apatite. In extremely well preserved microsaur teeth (Hylerpeton dawsoni) from Carboniferous deposits at Joggins, Nova Scotia, F is found as francolite, a carbonate fluorapatite (STEVENSON and STEVENSON, 1966). BLOKH and KOCHENOV (1961) have proved that the fluorine content in fossil bones cannot be a measure for the geological age as proposed by MIDDLETON (1845, cited by BLOKH and KOCHENOV, 1961). Excess fluorine content over fluorapatite may be precipitated in the form of fluorite (BLOKH and KOCHENOV, 1961).

In Table 9-L-2 the fluorine contents of land, marine and fresh water animals are summarized according to KLEMENT (1935).

Table 9-L-2. *Average fluorine content in modern and fossil bones of animals. (Number of samples in brackets)*

Animals	Land	Marine	Fresh water	Method	Reference
Mammals	(14) 500 ppm	(6) 5,500 ppm		(W)	1
Birds	(9) 1,100 ppm	(14) 3,200 ppm		(W)	1
Fishes		(5) 4,300 ppm	(9) 300 ppm	(W)	1
Fossil fishes		(11) 22,100 ppm	(13) 19,900 ppm	(?)	2

References: 1. KLEMENT (1935). 2. BLOKH and KOCHENOV (1961).

Revised manuscript received: December 1971

9-M. Abundance in Common Metamorphic Rock Types

Data concerning the abundance of fluorine in metamorphic rocks are rare and do not represent all important metamorphic rock types. The majority of the data deals with fluorine metasomatism in and around granitic plutons and with skarn formation. In Table 9-M-1 the fluorine contents of some metamorphic rocks are listed. The frequency distribution for metamorphic rocks is given in Fig. 9-M-1 according to Kokubu (1956).

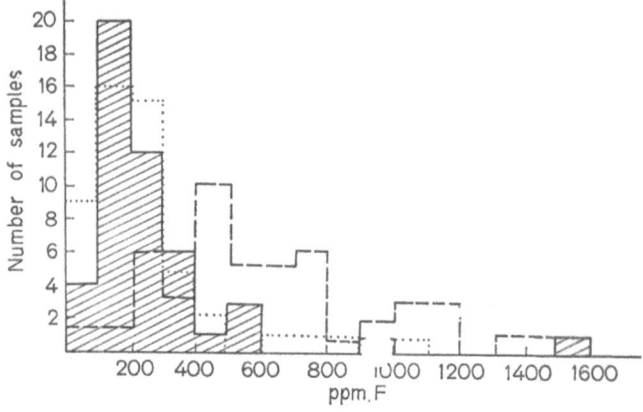

Fig. 9-M-1. Frequency distribution of fluorine in metamorphic rocks (shaded) according to Kokubu (1956), compared with the frequency distribution of fluorine in sandstones and greywackes (dotted line) and shales (dashed line) according to Fleischer and Robinson (1963), Fig. 9-K-2

Table 9-M-1. *Fluorine content in metamorphic rocks*

Rock, locality	ppm	Method	Reference
Regionally metamorphosed rocks			
Greenschist, Saganoseki-mati (Japan)	120	(C)	1
Greenschist, Mitutomo-mura (Japan)	160	(C)	1
Greenschist, Yosikawa-mura (Japan)	120	(C)	1
Greenschist, Sirataki bridge (Japan)	80	(C)	1
Greenschist, Kamai-tugawa (Japan)	160	(C)	1
Chlorite schist, Benverlodge (Canada)	250	(S)	2
Chlorite schist, Benverlodge (Canada)	450	(S)	2
Chlorite sericite schist, Benverlodge (Canada)	450	(S)	2
Chlorite sericite schist, Benverlodge (Canada)	450	(S)	2
Chloritized amphibolite, Benverlodge (Canada)	1,100	(S)	2
Chloritized amphibolite, Benverlodge (Canada)	1,400	(S)	2
Hornblende epidote greenschist, Tokuyama city (Japan)	200	(C)	1

Table 9-M-1 (continued)

Rock, locality	ppm	Method	Reference
Greenschist, Sugane-mura (Japan)	200	(C)	1
Greenschist, south of Nozi (Japan)	260	(C)	1
Greenschist, west of Sôten Mountain (Japan)	240	(C)	1
Hornblende schist, Sirataki mine (Japan)	220	(C)	1
Hornblende schist, Sirataki mine (Japan)	320	(C)	1
Hornblende schist, Sirataki mine (Japan)	60	(C)	1
Hornblende schist, Sirataki mine (Japan)	160	(C)	1
Amphibolite, Benverlodge (Canada)	800	(S)	2
Amphibolite, Benverlodge (Canada)	1,100	(S)	2
Banded amphibolite, Bessi mine (Japan)	140	(C)	1
Diopside amphibolite, Bessi mine (Japan)	140	(C)	1
Meta gabbro, Tadono-mura (Japan)	140	(C)	1
Meta gabbro, Tenzin-mura (Japan)	99	(C)	1
Meta basalt, Tukikawa (Japan)	160	(C)	1
Black schist, Notami Mountain (Japan)	360	(C)	1
Black schist, Nakanotani (Japan)	160	(C)	1
Black schist, Sukane-mura (Japan)	580	(C)	1
Black schist, Simizu-mura (Japan)	260	(C)	1
Biotite-bearing black schist, N. Tokuyamay city (Japan)	360	(C)	1
Black schist, Bessi mine (Japan)	300	(C)	1
Black schist, Sirataki mine (Japan)	300	(C)	1
Quartzite schist, Sagnoseki mati (Japan)	160	(C)	1
Quartzite schist, Hirose-mati (Japan)	240	(C)	1
Quartzite schist, Sirataki-guti (Japan)	100	(C)	1
Quartzite schist, Sirataki mine (Japan)	140	(C)	1
Quartzite schist, Sirataki mine (Japan)	200	(C)	1
Quartzite, Bessi mine (Japan)	200	(C)	1
Piedmontite quartz schist, Bessi mine (Japan)	340	(C)	1
Mica schist (biotite + muscovite) Riddarhyttan (Sweden)	1,600	(W)	6
Mica schist, Skyttgruvan mine (Sweden)	700	(W)	6

Contact metamorphic rocks
Hornfels, Oslo (Norway)	690	(?)	3
Hornfels, Yuu-mati (Japan)	260	(C)	1
Biotite hornfels, Sigino-mura (Japan)	500	(C)	1
Tremolite skarn, Persberg deposit (Sweden)	4,000		4
Actinolite pyroxene skarn, Persberg deposit (Sweden)	400		4
Talc skarn, Persberg deposit (Sweden)	4,700		4
Mica chlorite skarn, Persberg deposit (Sweden)	1,2100		4
Amphibole skarn, Yxiöfjeld (Sweden)	4,200		4
Garnet pyroxene skarn, Ubino (U.S.S.R.)	average 260	(C)	7
Idokrase skarn, Ubino (U.S.S.R.)	average 1,940	(C)	7
Garnet idokrase skarn, Talovo (U.S.S.R.)	average 430	(C)	7
Wollastonite skarn, Talovo (U.S.S.R.)	700	(C)	7

Metasomatic rocks
Greisenized granite, Eibenstock, Saxony (Germany)	1,7000	(C)	5
Greisenized granites, average of 24 German	2,0400	(C)	5

References: 1. KOKUBU (1956). 2. SERAPHIM (1951). 3. BARTH and BRUUN (1945). 4. LARSSON in CORRENS (1956). 5. KORITNIG (1951). 6. GEIJER (1960). 7. KOSALS (1968).

Two examples of the distribution of fluorine in greisenized granites are shown in Table 9-M-2. These are the only metamorphic rocks for which a detailed balance of fluorine distribution is given. In these extremely F-rich rocks the fluorine is contained in the minerals topaz, lithionite, zinnwaldite and fluorite.

Table 9-M-2. *Fluorine distribution in the different F-bearing minerals from greisenized granites. (After* KORITNIG, *1951)*

Rock Locality	F-bearing mineral and wt.-% of this mineral in rock		F content of F-bearing mineral wt.-%	F contributed to the rock wt.-%	Fraction of total F assigned to F-bearing minerals wt.-%
Greisenized granite (lithionite granite) Eibenstock, Saxony, Germany	topaz lithionite fluorite apatite	3.5 8 1.6 0.15	18 3.5 48.7 3.3	0.063 0.28 0.785 0.0055	37.0 16.5 46.2 0.3
				1.1335	100.0
Greisenized granite Zinnwald, C.S.S.R.	topaz zinnwaldite fluorite apatite	11.2 32.6 1.35 0.07	18 5.5 48.7 3.2	2.02 1.79 0.64 0.002	45 40 14.5 0.5
				4.452	100.0

In the common regional metamorphic rocks, as in sediments and magmatic rocks, fluorine occurs mainly in the OH-bearing minerals (especially micas, chlorites and amphiboles). All other minerals, including apatite, are generally unimportant sources of F in these rocks.

In skarns, where fluorine metasomatism is common (e.g., see GEIJER, 1960) we sometimes find especially fluorine-rich amphiboles (tremolite, actinolite and anthophyllite) in addition to the minerals of the humite group or vesuvianite.

SAHAMA (1948) and SERAPHIM (1951) investigated the variation of fluorine content as a function of distance from contacts with magmatic rocks.

Fig. 9-M-2 and Fig. 9-M-3 show the behavior of fluorine at the contacts of granites with siltstone and shale. Table 9-M-3 lists variations of F at the rapakivi granite contact with a mica gneiss at Ihovaara, Finland.

Table 9-M-3. *Fluorine content at the rapakivi granite contact of Ihovaara. (After* SAHAMA, *1947)*

	Rapakivi granite	Mica gneiss, fragment in rapakivi granite	Mica gneiss, immediate vicinity of the rapakivi contact	Mica gneiss, about 1 km E. of Ihovaara hill
ppm F	400	2,500	500	300

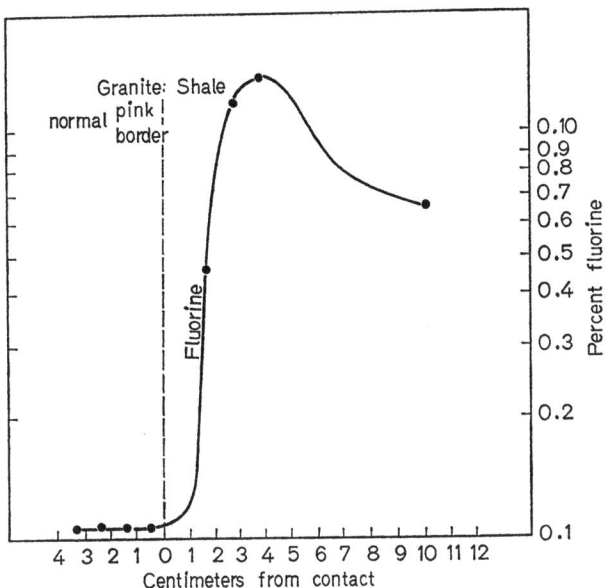

Fig. 9-M-2. Behavior of fluorine at a granite shale contact. (After Seraphim, 1951)

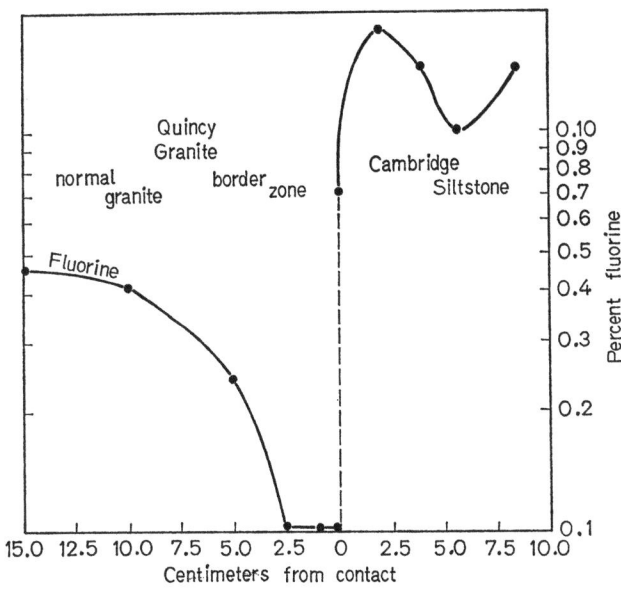

Fig. 9-M-3. Behavior of fluorine at the contact of a siltstone inclusion in granite. (Quincy Granite, Cambridge Siltstone, Mt. Ararat, Massach.) (After Seraphim, 1951)

Revised manuscript received: December 1971

9-N. Behavior in Metamorphic Reactions

Regional metamorphic equilibria are strongly affected by several "mobile" components of the intergranular phase. If an element is to be incorporated in a new phase growing during regional metamorphism, it must first be present in a mobile phase. In general this mobility is comparable to the transport in hydrothermal solutions or in the gas phase.

KORZHINSKY (1955) investigated the problem of geochemical mobility and proposed an average sequence of mobility in which fluorine is less mobile than H_2O, CO_2, S, SO_3, Cl, K_2O and Na_2O.

For regionally metamorphosed skarns of Western Moravia, NĚMETS (1968) found that fluorine now contained in the skarns was carried there together with water and other mobile components in the course of metamorphism. CARMICHAEL (1970) has treated the influence of fluorine on stability fields in metamorphic reactions. He found, for instance, that increasing activity of F^- extended the stability field of the assemblage, biotite-calcite-quartz, toward higher temperatures.

In pneumatolytic and hydrothermal processes fluorine can migrate into the country rocks from magmatic bodies and fluorite-bearing veins. For example, HÜBNER (1968) observed a maximum fluorine content 1 to 4 m away from the margins of hydrothermal fluorite veins in Saxony.

At the contacts of granite with siltstone and shale (see Fig. 9-M-2 and Fig. 9-M-3), the migration distances of fluorine are much shorter (SERAPHIM, 1951). At the contact between a pegmatite body and its enclosing granite, the fluorine maximum is right at the contact (see Fig. 9-F-2).

According to KOSALS (1968), during the formation of limestone skarns (from Gornyy Altay) fluorine does not form independent minerals. Instead, vesuvianite, the latest skarn mineral, is the chief host mineral for fluorine. In other instances of skarn formation, rare fluorite deposits are formed. The experimental results of AMES (1961) indicate that for the metasomatic replacement of limestone and dolostone with fluorite the altering solutions must be alkaline. A few ppm F in an alkaline solution, whose original Ca^{2+} and F^- concentrations never exceed the K_{sp} for CaF_2, can form a fluorite deposit by replacement because the anions of the replaced solid move only into the solution.

Fig. 9-M-1 shows the frequency distribution of fluorine in regional metamorphic and sedimentary rocks. If we compare the frequency distribution of metamorphic rocks with that of shales (dashed lines), it seems that metamorphic rocks have generally lost about half of their fluorine content during metamorphism. Comparing sandstones and greywackes (dotted lines), we find a good agreement. But the number of samples of Fig. 9-M-1 are too small to obtain a correct balance of the fluorine contents of sedimentary and regional metamorphic rocks.

© Springer-Verlag Berlin · Heidelberg 1972

Revised manuscript received: December 1971

9-O. Relations to Other Elements, Economic Importance etc.

I. Relation of Fluorine to Other Elements

During crystallization of acidic melts fluorine serves as a depolymerizer through the formation of Si-F bonds and acts as an effective flux in the liquid phase (WYLLIE and TUTTLE, 1961). Fig. 9-O-1, from KOKUBU (1956), seems to indicate that F is concentrated in Si-rich rocks. However, in detail many rocks are seen to deviate appreciably from this trend. SHEPHERD (1940) found no consistent relationships between F and other chemical components. KOKUBU (1956) mentions that we can not expect fluorine to display strong tendencies in solidified magmas or a close relationship between F and other elements, since an appreciable fraction of the fluorine in a magma escapes during crystallization.

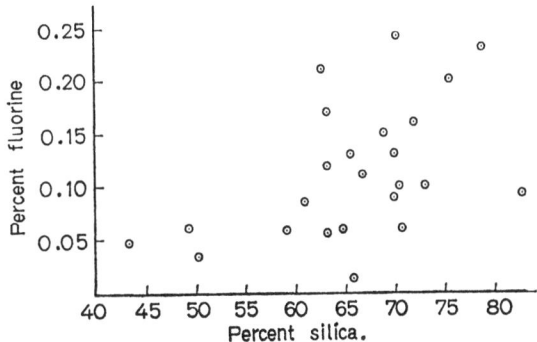

Fig. 9-O-1. Relation between fluorine and silica content of magmatic rocks. (After KOKUBU, 1956)

Recently, a correlation between fluorine and some other elements was found in some Russian granitic complexes. KOSTETSKAYA *et al.* (1969) found correlations between F and Be in the rocks and constituent minerals of the Dzhida granitic complex, especially within the leucocratic granites. In biotites from the complex, the correlation of fluorine with Be, Mn and Mg is particularly distinct. The strongest correlation observed was between Mn and F. TAUSON (1967) confirmed a close geochemical relationship between F and Li in the Kalbinsk, Dzhidinsk and Susamyr granitic series (see Fig. 9-O-2). In the rock-forming minerals which concentrate F and Li (e.g., biotites) the ratio F/Li remains constant.

Similar observations have been made by KOSALS and MAZUROV (1968) for fluorine with boron and rare alkalies in the Dzhida complex.

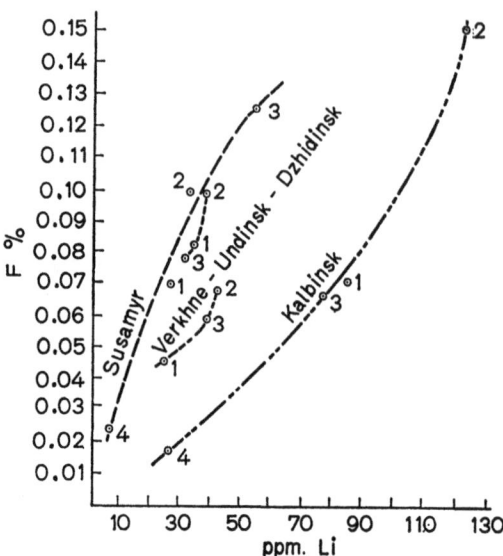

Fig. 9-O-2. Correlation between contents of lithium and fluorine in granite batholiths (after TAUSON, 1967). 1 first intrusive phase; 2 second (principal) intrusive phase; 3 third intrusive phase; 4 aplite dikes

In hydrothermal solutions fluorine forms stable complexes with some elements and thus can serve as a means of transport for less soluble elements. For instance, GLINKINA (1969) found experimentally that in the presence of fluorine, acidic hydrothermal solutions can transport tungsten and calcium in concentrations considerably higher than those found in the weakly alkaline, ore-forming solutions.

II. Economic Importance

For many years fluorite has been known and used as a flux in metallurgical processes. Depending upon its major application, fluorite concentrates are sold in three grades: metallurgical grade (85 to 95% CaF_2), ceramic grade (95 to 96% CaF_2) and acid grade (97 to 98% CaF_2) for the chemical industry.

Smelting works were formerly the largest users of fluorite, but since World War II their share of the total consumption, which is now three times that of 1937, has decreased. This is seen, for instance in the use of fluorite in U.S.A. (Table 9-O-1). The world production of fluorite is summarized in Table 9-O-2.

The main fluorite-producing countries of the world are plotted in Fig. 9-O-3 with the order of magnitude of production per year. For additional data about the development of the world production, and of world reserve see JACOBI (1971), US-Department of the Interior (1971), CHERMETTE (1970) and WELLS (1968).

Today fluorite is used only for special purposes (e.g., basic Siemens-Martin processes) in smelting works and foundaries. It is used in the ceramic industry and in the manufacturing of enamels and glass as a turbid substance ingredient and in cement industry. Small amounts (mostly synthetic fluorite) are used in the optical industry. The major proportion of fluorite is used today (about 50%) in the chemical industry for the preparation of numerous fluorine compounds (synthetic

Table 9-O-1. *Consumption of fluorite in U.S.A.* (After HILLER, 1962)

Steel industry (smelting)	270,000 t	40.7%
Aluminum production	110,000 t	16.7%
Chemical industry	210,000 t (U-prod.)	32.0%
	70,000 t (HF-a.o.)	10.6%
	Total 660,000 t	100.0%

Table 9-O-2. *World production of fluorite in short tons (1 sh.t. = 907.18 kg) according to U.S. Bureau of Mines (1967)*

Country	1962	1966[p, 2]
North America:		
Canada[e]	75,000	97,000
Mexico	553,642	799,602
United States (shipments)	206,026	253,068
South America:		
Argentina	13,799	10,472
Europe:		
France	154,06[3]	265,000[e]
Germany		
East[e]	80,000	90,000
West (marketable)	116,592	101,912
Italy	176,709	226,143
Spain (marketable)	165,356	264,004
Sweden (shales)	3,855	—
United Kingdom[4]	80,358[r]	203,927
Africa:		
Morocco	546	3,300[e]
Rhodesia, Southern	20	NA
South Africa, Republic of	111,683	90,266
Southwest Africa	240	—
Tunisia	—	6,600
Asia:		
China, mainland[e]	220,000	250,000
India	724	1,178
Japan	17,120	17,600[e]
Korea:		
North[e]	33,000	33,000
South	36,343	35,283
Mongolia	41,800	82,700[e]
Thailand	11,806	52,941
Turkey	640	1,659
U.S.S.R.[e, 5]	265,000	385,000
Australia	—	—
World total[e]	2,370,000	3,280,000

[e] Estimate. [p] Preliminary. [r] Revised. NA = Not available.
[2] Compiled mostly from data available July 1967.
[3] Marketable.
[4] Includes mineral obtained from some old mine dumps for which returns are available.
[5] U.S.S.R. in Europe included with U.S.S.R. in Asia, as the deposits are predominantly in Asiatic U.S.S.R.

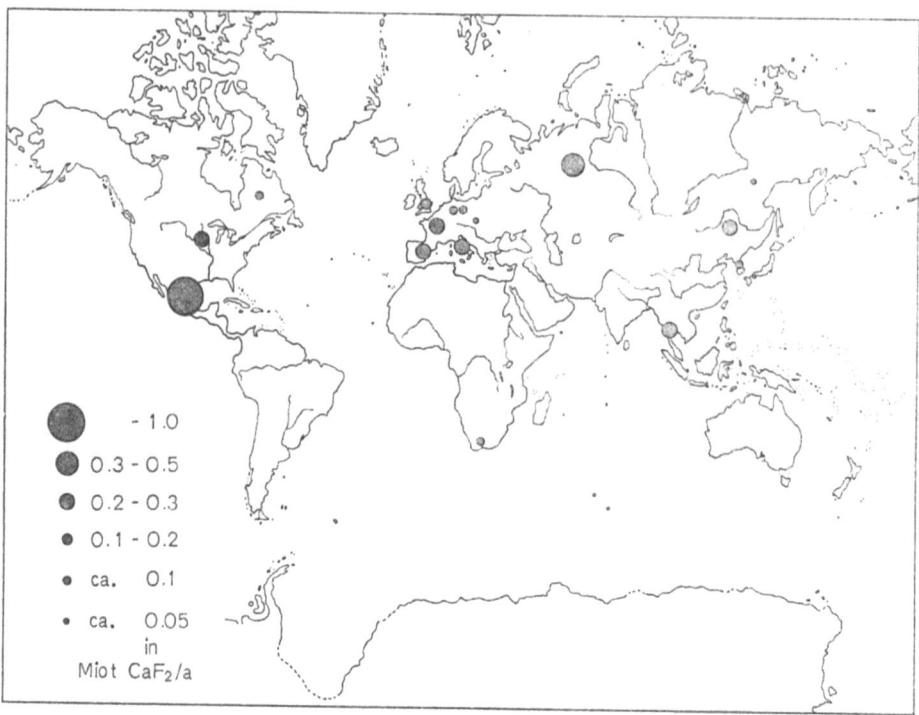

Fig. 9-O-3. The main fluorite producing countries of the world (after JACOBI, 1971)

cryolite, teflon etc.). Considerable advances in the industrial chemistry of fluorine have been made in the last few decades.

Fluorine is used in the purification and enrichment of uranium, for motor fuels with high octane ratings, and as an oxidizing agent in rocket engines.

The mineral cryolite, Na_3AlF_6, was formerly an important raw material for the production of aluminum. Today it is generally synthesized. It was only economically worth mining at Ivigtut, Greenland (10,000 to 30,000 ton/yr.).

Some organic fluorine compounds such as teflon have proved to be surprisingly stable and have become more and more important in the last 30 years. SCHERER (1970) has recently reviewed the technical uses of organic fluorine compounds.

Important applications of these compounds include refrigerants (Frigen®, Freon®, etc.), propellant gases in aerosol cans (world production in 1966: about 2.6 billion cans) and for such diverse uses as cosmetics, insecticides, disinfectants, and paints.

Organic fluorine compounds have also been used for fire-extinguishing substances, for the foaming of polyurethane and for its electrical and lubricating properties.

Revised manuscript received: December 1971

References: Sections 9-B to 9-O

AHRENS, L. H.: The lognormal distribution of elements (I—II). Geochim. Cosmochim. Acta: I: **5**, 49 (1954) and II: **6**, 121 (1954).
— TAYLOR, S. R.: Spectrochemical Analysis. New York: Addison-Wesley 1961.
ALLER, L. H.: The Abundance of the Elements. New York: Interscience 1961.
AMES, L. L., Jr.: The metasomatic replacement of limestone by alkaline, fluoride-bearing solutions. Econ. Geol. **56**, 730 (1961).
BARTH, T. F. W., BRUUN, B.: Fluorine in the Oslo petrographic province. Skrifter Norske Videnskaps-Akad. Oslo, I. Mat.-Naturvid. Kl. **8**, 5 (1945).
BAZAROV, L. SH., DOBRETSOVA, I. L., YUSUPOV, S. SH.: Pattern of distribution of fluorine around chambered pegmatite in granites. Dokl. Akad. Nauk. SSSR **157**, 140 (1965).
BEESON, K. C.: U.S. Department of Agriculture: Miscellaneous Publication 369 (1941).
BEWERS, J. M.: North Atlantic fluoride profiles. Deep-Sea Res. **18**, 237 (1971).
BLOKH, A. M., KOCHENOV, A. V.: The fluorine content in bone residue of fossil fishes. Dokl. Akad. Nauk. SSSR **135**, 1495 (1960). Transl. Geochemistry 1094 (1961).
BLOSS, F. D., SHEKARCHI, E., SHELL, H. R.: Hardness of synthetic and natural micas. Am. Mineralogist **44**, 33 (1959).
BLOXAM, T. W., THOMAS, R. L.: Fluorine in Carboniferous shales from the South Wales coalfield. Chem. Geol. **5**, 179 (1969/70).
BOND, G. W.: A geochemical survey of the underground water supplies of the Union of South Africa. Geol. Surv. South Africa Mem. **41**, 1 (1945).
BORDET, P., CHARRIER, J., GEFFROY, J.: La sellaite du gisment de fluorine de Font-Sante (Var). Bull. Soc. franç. **87**, 451 (1964).
BOUVIER, S. L.: Geol. Surv. Canada (person. commun. by RIMSAITE, J., 1970).
BOWDEN, P.: Lithium in younger granites of Northern Nigeria. Geochim. Cosmochim. Acta **30**, 555 (1966).
BOWEN, H. J. M.: Trace Elements in Biochemistry. London-New York: Academic Press 1966.
BRAITSCH, O.: Die Borate und Phosphate im Zechsteinsalz Südhannovers. Fortschr. Mineral. **38**, 190 (1960).
— Entstehung und Stoffbestand der Salzlagerstätten. Berlin-Göttingen-Heidelberg: Springer 1962.
BRANCAZIO, P. J., CAMERON, A. G. W. (eds.): The Origin and Evaluation of Atmospheres and Oceans. New York-London-Sydney: J. WILEY 1963.
BREDEMANN, C.: Biochemie und Physiologie des Fluors. 2. Aufl., Berlin: Akademie Verlag 1956.
BRÖGGER, W. C.: Die Eruptivgesteine des Kristianiagebietes, IV. Das Fengebiet in Telemark: Vidensk.-Skr. Kristiania, No. **9**, 408 (1921).
BROWN, H.: On the compositions and structures of the planets. Astrophys. J. **111**, 641 (1950).
BROWNE, P. R. L., ELLIS, A. J.: The Ohaki-Broadlands hydrothermal area, New Zealand: mineralogy and related geochemistry. Am. J. Sci. **269**, 97 (1970).
BUDDINGTON, A. F., LEONHARD, B. F.: Chemical petrology and mineralogy of hornblende in northwest Adirondack granitic rocks. Am. Mineralogist **38**, 891 (1953).
BUNCH, T. E., KEIL, K.: Mineralogy and petrology of silicate inclusions in iron meteorites. Contr. Mineral. and Petrol. **25**, 297 (1970).
CARMICHAEL, D. M.: Intersecting isograds in the Whetstone Lake Area, Ontario. J. Petrol. **11**, 147 (1970).

Fluorine

CARPENTER, R.: Factors controlling the marine geochemistry of fluorine. Geochim. Cosmochim. Acta **33**, 1153 (1969).

CARSWELL, D. A., DAWSON, J. B.: Garnet peridotite xenoliths in South African kimberlite pipes and their petrogenesis. Contr. Mineral. and Petrol. **25**, 163 (1970).

CARTER, R. H.: Solubilities of some inorganic fluorides in water at 25°C. Ind. Eng. Chem. **20**, 1195 (1928).

CERNY, P.: Compositional variations in cookeite. Canad. Mineralogist **10**, 636 (1970).

CHERMETTE, A.: Le spath-fluor Français en 1968. Mines et Metallurgie, March—Dec. (1969), Jan.—May (1970).

CLARKE, F. W.: The data of geochemistry. U.S. Geol. Surv. Bull. **770**, 1 (1924).

COATS, R. R., GOSS, W. D., RADER, L. F.: Distribution of fluorine in unaltered silicic volcanic rocks of the Western conterminous United States. Econ. Geol. **58**, 941 (1963).

COMUCCI, P., MAZZI, F.: Le vulcanite della Dancalia. Accad. Nazl. Lincei atti Mem. Classe Sci. Fis., Mat. e Nat., Ser. 8, 5: sect. 2a, **2**, 1 (1957).

CORRENS, C. W.: The geochemistry of the halogenes. In: Physics and Chemistry of the Earth, vol. 1, p. 181. London and New York: Pergamon Press 1956.

DANA, J., PALACHE, C., BERMAN, H., FRONDEL, C.: The System of Mineralogy, vol. 1—2. New York-London: J. Wiley & Sons, Inc., Chapman & Hall, Ltd. 1958.

DANILOVA, V. V.: On the content of fluorine in rocks. Akad. Nauk. SSSR, Biogeokhim. Lab., Trudy **9**, 129 (1949) [in Russian].

D'ANS-LAX: Taschenbuch für Chemiker und Physiker, 3. Aufl., vol. 1, p. 71. Berlin-Heidelberg-New York: Springer 1967.

DEER, W. A., HOWIE, R. A., ZUSSMAN, J.: Rock-forming Minerals. vol. I—V. London: Longmans 1967.

DELITSYNA, L. V., MELENT'YEV, B. N.: The coexistence of liquid phases at high temperatures: The system nepheline-villiaumite-lithium fluoride. Dokl. Acad. Sci. U.S.S.R. Earth scs. Sect. **189**, 215 (1969).

DIETZ, R. S., EMERY, K. O., SHEPARD, F. P.: Phosphorite deposits on the sea floor of southern California. Bull. Geol. Soc. Am. **53**, 815 (1942).

DODGE, F. C. W., ROSS, D. C.: Coexisting hornblendes and biotites from granitic rocks near the San Andreas fault, California. J. Geol. **79**, 158 (1971).

DOELTER, C.: Handbuch der Mineralchemie, Bd. 1—4. Dresden-Leipzig: Verlag Theodor Steinkopff 1912—1931.

DOLGOV, YU. A., POGREBNYAK, YU. F., SHUGUROVA, N. A.: Composition and pressure of gases in inclusions in tektites. Geochem. Internat. **6**, 525 (1969).

ECKERMANN, H. VON: The alkaline district of Alnö, Island: Sveriges Geol. Undersokn. **36**, 176 (1948).

EKSTRÖM, T. K.: The distribution of fluorine among some coexisting minerals. Contr. Mineral. and Petrol. **34**, 192 (1972).

ELLIS, A. J., MAHON, W. A. J.: Natural hydrothermal systems and experimental hot water/rock interaction (Part II) (Part I 1964). Geochim. Cosmochim. Acta **31**, 519 (1967).

ERICKSON, R. L., BLADE, L. V.: Geochemistry and petrology of the alkalic igneous complex at Magnet Cove, Arkansas. Geol. Surv. Profes. Papers **425** (1963).

FISHER, D. E.: The fluorine content of some chondritic meteorites. J. Geophys. Res. **68**, 6331 (1963).

— Silicon in iron meteorites and the earth's core. Nature **222**, 866 (1969).

FLANAGAN, F. J.: U.S. Geological Survey silicate rock standards. Geochim. Cosmochim. Acta **31**, 289 (1967).

FLEISCHER, M., ROBINSON, W. O.: Some problems of the geochemistry of fluorine in SHAW: D. M. (ed). Studies in Analytical Geochemistry. Toronto: University of Toronto Press 1963.

FOCKEMA, R. A. P.: Geology of the area around the confluence of the elands and crocodile rivers. Geol. Soc. South. Afr. Trans. **55**, 155 (1953).

FOSTER, M. P.: Water content of micas and chlorites. U.S. Geol. Surv. Profes. Papers **474-F**, 15 (1964).

FRIEDMAN, I., SMITH, R. L.: In MILLER, S. L., CRAIG, H., WASSERBURG, G. J.: Isotopic and Cosmic Chemistry. Amsterdam: North Holland Publ. Comp. 1964.

FUCHS, W.: Occurence of whitlockite in chondritic meteorites. Science **137**, 425 (1962).
FÜCHTBAUER, H.: Die petrographische Unterscheidung der Zechsteindolomite im Emsland durch ihren Säurerückstand. Erdoel Kohle **11**, 689 (1958).
GEIJER, P.: The distribution of halogens in skarn amphiboles in Central Sweden. Ark. Mineral. Geol. **2**, No. 32, 481 (1960).
GEMMELL, G. D.: Fluorine in New Zealand soils. New Zealand J. Sci. Technol. B **27**, 302 (1946).
GILLBERG, M.: Halogenes and hydroxyl contents of micas and amphiboles in Swedish granitic rocks. Geochim. Cosmochim. Acta **28**, 495 (1964).
GLEBOV, M. P., LEGEYDO, V. A., RYBAKOVA, M. M., SHIRYAYEVA, V. A.: Composition of muscovite from pegmatites of east Sayan formed under conditions of varying alkalinity. Geochem. Intern. **5**, 1016 (1968).
GLINKINA, M. I.: Behavior of tungsten and calcium in acid fluoride solutions at temperatures from 18 to 350°C. Geochem. Intern. **6**, 641 (1969).
GMELIN, L.: Handbuch der anorganischen Chemie. Fluor, Bd. 5. Leipzig-Berlin: Verlag Chemie, GmbH. 1926.
— Handbuch der anorganischen Chemie. Fluor Ergänzungsband, 8. Auflage. Weinheim: Verlag Chemie, GmbH 1959.
GRAF, D. L.: Geochemistry of carbonate sediments and sedimentary carbonate rocks: Illinois State Geol. Surv. Circ. 301, 297 (1960).
GREENLAND, L.: The enstatite chondrites (1963). In: MASON, B., Geochim. Cosmochim. Acta **30**, 23 (1966).
— LOVERING, J. F.: Minor and trace element abundance in chondritic meteorites. Geochim. Cosmochim. Acta **29**, 821 (1965).
— — Fractionation of fluorine, chlorine and other trace elements during differentiation of a tholeiitic magma. Geochim. Cosmochim. Acta **30**, 963 (1966).
GRIGOREV, D. P.: (1939) see GMELIN's Handbuch der anorganischen Chemie. Fluor Ergänzungsband, 8. Auflage. Weinheim: Verlag Chemie 1959.
GULBRANDSEN, R. A.: Chemical composition of phosphorites of the Phosphoria formation. Geochim. Cosmochim. Acta **30**, 769 (1966).
Handbook of Chemistry and Physics: Cleveland: Chemical Rubber 1967/68.
HEFTER, A.: Handbuch der experimentellen Pharmakologie, Ergänzungswerk Bd. 7. Berlin: Springer 1938.
HEIDE, F., KAEDING, J.: Der Halogengehalt des Saalewassers. Naturwissenschaften **41**, 256 (1954).
HERRMANN, A. G., HOFFMANN, R. O.: Zur Genese einiger Borate in den Salzablagerungen der Staßfurt-Serie des Südharzbezirkes einschließlich der Grube Königshall-Hindenburg. Neues Jahrb. Mineral. Monatsh. 52 (1961).
HILLER, J. E.: Die mineralischen Rohstoffe. Stuttgart: E. Schweizerbart'sche Verlagsbuchhandlung Nägele u. Obermiller 1962.
HINTZE, C.: Handbuch der Mineralogie, Bd. I, II und Ergänzungsbände, Berlin: De Gruyter 1904—1968.
HOFFMANN, J.: Über photolytisch neutralisierte Elementarstoffe der Fluorite. Chem. Erde **11**, 368 (1938).
HONDA, F.: A possible process for the fluctuation of halogene abundances in fumarolic gases. Geochem. J. **3**, 187 (1970) (Geochem. Soc. of Japan).
HUANG, W. H., JOHNS, W. D.: The chlorine and fluorine of geochemical standards. Geochim. Cosmochim. Acta **31**, 597 (1967).
HÜBNER, M.: Die Fluordispersion im Nebengestein hydrothermaler Flußspatgänge in Thüringen und im Harz. Z. Angew. Geol. **14**, 297 (1968).
— Untersuchung der F^-/OH^--Austauschadsorption an Mineralen der Illitgruppe und an Kaolinit. Freiberger Forschungsheft C 244 Mineralogie-Lagerstättenlehre (1969a).
— Geochemische Interpretation von Fluorid/Hydroxid-Austauschversuchen an Tonmineralen. Ber. Deutsch. Ges. geol. Wiss. B. Miner. Lagerstättenforsch. **14**, 5 (1969b).
JACOB, J.: Beiträge zur chemischen Konstitution der Glimmer. Z. Krist. **72**, 317 (1929).
JACOBI, K. H.: Nutzbare und potentielle Flußspatlagerstätten. Erzmetall **24**, H. 10, 486 (1971).

JAFFE, H. W.: Re-examination of sphene (titanite). Am. Mineralogist **32**, 637 (1947).
JAHN-DEESBACH, W.: Weitere Untersuchungen über Bindungszustand des Fluors in einigen landwirtschaftlich wichtigen Böden. Chem. Erde **19**, 308 (1958).
JAHNS, R. H.: The genesis of pegmatites. II. Quantitative analysis of lithium-bearing pegmatite, Mora County, New Mexico. Am. Mineralogist **38**, 1078 (1953).
JEFFERY, P. G.: The fluorine content of some standard samples. Geochim. Cosmochim. Acta **26**, 1355 (1962).
JEFFRIES, C. D.: Occurrence of fluorine in limestones and dolomites. Soil Sci. **71**, 287 (1951).
KAZAKOV, A. V., SOKOLOVA, E. J. (1950) reported by GRAF (1960).
KESTER, D. R.: Fluoride chlorinity ratio of sea water between the Grand Banks and the Mid-Atlantic Ridge. Deep-Sea Res. **18**, 1123 (1971).
KHITAROV, N. I. (ed.): Problems of Geochemistry. Jerusalem: Israel Program for Scientific Translations 1969.
KIND, A.: Der magmatische Apatit, seine chemische Zusammensetzung und seine physikalischen Eigenschaften. Chem. Erde **12**, 50 (1938).
KING, B. C.: The Napak area of Southern Karamoja, Uganda: Geol. Surv. Uganda Mem. **5**, 57 (1949).
KLEMENT, R.: Die Fluorgehalte der Knochen und Zähne. Ber. Deut. chem. Ges. Berlin **68** (2), 2012 (1935).
KÖPF, H., OELSCHLÄGER, W., BLEICH, K. E.: Fluorgehalte in boden- und gesteinsbürtigen Ursprungswässern. Z. Pflanzenernaehr. Dueng. Bodenk. **121**, 133 (1968).
KOGARKO, L. N., KRIGMAN, L. D., SHARUDILO, N. S.: Experimental investigations of the effect of alkalinity of silicate melts on the separation of fluorine into the gas phase. Geochem. Intern. **5**, 782 (1968).
KOKUBU, N.: Fluorine in rocks. Mem. Fac. Sci. Kyushu Univ., Ser. C **2**, No. 3, 95 (1956).
KORITNIG, S.: Ein Beitrag zur Geochemie des Fluor. Geochim. Cosmochim. Acta **1**, 89 (1951).
— Zur Geochemie des Fluors in den Sedimenten. Fortschr. Geol. Rheinld. Westf. **10**, 231 (1963).
KORZHINSKY, D. S.: Outline of Metasomatic Processes, 2nd ed. Moskwa: Akad. Nauk. U.S.S.R. 1955 [Russian].
KOSALS, YA. A.: Geochemistry of beryllium, boron, lithium and fluorine in the process of formation of limestone skarns (Gornyy Altay). Geochem. Intern. **5**, 131 (1968).
— MAZUROV, M. P.: Behavior of the rare alkalis, boron, fluorine and beryllium during the emplacement of the Bitu-Dzhida granitic batholith, Southwest Baykalia. Geochem. Intern. **5**, 1024 (1968).
KOSTETSKAYA, YE. V., MORDVINOVA, V. I.: Distribution of fluorine in the minerals of granitoids of the Dzhida complex, West Transbaykalia. Geochem. Intern. **5**, 519 (1968).
— PETROV, L. L., PETROVA, Z. I., MORDVINOVA, V. I.: Distribution of beryllium and fluorine and correlation between their contents in the minerals and rocks of the Dzhida paleozoic granitoid complex. Geochem. Intern. **6**, 72 (1969).
KRAINOV, S. R., PETROVA, N. G.: Trace elements in mineral waters of the Pamirs. Geokhimiya **1962**, 408.
KRANZ, R.: p. 525. In: EGLINTON, G., MURPHY, M. T. J. (eds.), Organic Geochemistry. Berlin-Heidelberg-New York: Springer 1969.
KRAYNOV, S. R., MER'KOV, A. N., PETROVA, N. G., BATURINSKAYA, I. V., ZHARIKOVA, V. M.: Highly alkaline (pH 12) fluosilicate waters in the deeper zones of the Lovozero massif. Geochem. Intern. **6**, 635 (1969).
KRUGER, P.: Über ein Vorkommen von syngenetisch-sedimentärem Fluorit im Plattendolomit des Geraer Beckens. Bergakademie **11**, 742 (1962a).
— Über ein Vorkommen von syngenetisch-sedimentärem Fluorit im Plattendolomit des Geraer Beckens. Bergakademie **11**, 1062 (1962b).
— Sedimentärer Fluorit—eine bis jetzt übersehene Gefahrenquelle für die Dolomitindustrie. Z. Angew. Geol. H. **8**, 400 (1962c).
KÜHN, R.: Geochemistry of German potash deposits. Geol. Soc. Am., Spec. Papers **88**, 427 (1968).

KUNITZ, W.: Die Beziehungen zwischen der chemischen Zusammensetzung und den physikalisch-optischen Eigenschaften innerhalb der Glimmergruppe. Neues Jahrb. Mineral. BB **50**, 365 (1924).

LARIMER, J. W., ANDERS, E.: Chemical fractionation in meteorites II. Abundance patterns and their interpretation. Geochim. Cosmochim. Acta **31**, 1239 (1967).

LEE, H. A., LAWRENCE, D. E.: A new occurrence of kimberlite in Gauthier Township, Ontario. Geol. Surv. Canada Paper **68—22**, 1 (1968).

LEELANANDAM, C.: Fluorine and chlorine in the charnockitic hornblendes from Kondapalli, India. Neues Jahrb. Mineral. Monatsh. 379 (1969).

— Chemical mineralogy of hornblendes and biotites from the charnockitic rocks of Kondapalli, India. J. Petrol. **11**, 475 (1970).

LEVINSON, A. A., (ed.): Proceedings of the Apollo 11 Lunar Science Conference. 2, Suppl. I, Geochim. Cosmochim. Acta (1970).

— HITCHON, B., REEDER, S. W.: Major element composition of the Mackenzie River at Norman Wells, N.W.T. Canada. Geochim. Cosmochim. Acta **33**, 133 (1969).

LINKE, W. F., SEIDELL, A.: Solubilities of Inorganic and Metal Organic Compounds. Fourth ed. vol. I. Toronto-New York-London: D. van Nostrand Company, Inc. 1958.

LOKKA, L.: (1943); see GMELIN, 1959.

LOVERING, T. S.: Halogen acid alteration of ash at fumarole No. 1. Valley of Ten Thousand Smokes, Alaska. Bull. Geol. Soc. Am. **68**, 1585 (1957).

MACINTIRE, W. H., HARDIN, L. J., BUEHLER, M. H.: Fluorine, Maury County, Tennessee. Univ. Tenn. Agric. Expt. Sta. Bull. **279**, 1 (1958).

— WINTERBERG, S. H., THOMPSON, J. G., HATCHER, B. W.: Fluorine content of plants. Ind. Eng. Chem. **34**, 1469 (1942).

MARAIS (1943): In BOWEN, H. J. M., p. 184 (1966).

MASON, B.: The enstatite chondrites. Geochim. Cosmochim. Acta **30**, 23 (1966).

— MELSON, W. G.: The Lunar Rocks. New York-London-Sydney-Toronto: Wiley Interscience 1970.

MATUURA, S., KOKUBU, N.: Fluorine content of the hot-spring water in Saga Prefecture. Chem. Fac. Sci. Kyusyu Univ. 1 (1955).

MAXWELL, J. A., PECK, L. C., WIIK, H. B.: Chemical composition of Apollo 11 lunar samples 10017, 10020, 10072 and 10084. p. 1369. In: LEVINSON, A. A. (ed.), Proceedings of the Apollo 11 Lunar Science Conference. Geochim. Cosmochim. Acta 2, Suppl. I (1970).

— WIIK, H. B.: Chemical composition of Apollo 12 lunar samples 12004, 12033, 12051, 12052 and 12065. Earth Planet. Sci. Letters **10**, 285 (1971).

MCGETCHIN, T. R., SILVER, L. T., CHODOS, A. A.: Titanclinohumite: A possible mineralogical site for water in the upper mantle. J. Geophys. Res. **75**, 255 (1970).

MICHAEL, G., BLUME, B.: Über den Fluorgehalt Thüringer Buntsandsteine und Muschelkalkböden. Chem. Erde **16**, 27 (1952).

MITCHELL, H. H., EDMANN, M.: Fluorine in the soil plants and animals. Soil Sci. **60**, 81 (1945).

MÖSE, J. R., EXNER, H.: (1952) see CORRENS (1956).

MOORE, J. G.: Water content of basalt erupted on the ocean floor. Contr. Mineral. and Petrol. **28**, 272 (1970).

— KOYANAGI, R. Y.: The October 1963 eruption of Kilauea Volcano Hawaii. U.S. Geol. Surv. Profess. Papers **614-C** (1969).

MORRISON, G. H., GERARD, J. T., KASHUBA, A. TH., GANGADHARAM, E. V., ROTHENBERG, A. M., POTTER, N. M., MILLER, G. B.: Elemental abundance of lunar soil and rocks. In: LEVINSON, A. A. (ed.), Proceedings of the Apollo 11 Lunar Science Conference 2, Suppl. I. Geochim. Cosmochim. Acta (1970a).

— — — — — — — Multielement analysis of lunar soil and rocks. Science **167**, 505 (1970b).

MUELLER, R. F.: Energetics of HCl and HF in volanic emanations. Geochim. Cosmochim Acta, **34**, 737 (1970).

MURATA, K. J., AULT, W. U., WHITE, D. E.: Halogen acids in fumarolic gases of Kilauea volcano. Bull. Volcanol. (Napoli) 17 (1964).

— RICHTER, D. H.: Chemistry of the lavas of the 1959—1960 eruption of Kilauea Volcano, Hawaii. U.S. Geol. Surv. Profes. Papers **537-A** (1966).

NABOKO, S. I.: A case of gaseous fluorine metasomatism of an active volcano. Geochemistry 5, 452 (1957).
NĚMEC, D.: Fluorine in lamprophyre and lamproid rocks. Geochim. Cosmochim. Acta 32, 523 (1968).
— Fluorine in tourmalines. Beitr. Mineral. Petrog. 20, 235 (1969).
NEMETS, D.: Fluorine in the regionally metamorphosed skarns of the Czech Massif (Czechoslovakia). Geochem. Intern. 5, 141 (1968).
NOBLE, D. C., SMITH, V., LEE, P.: Loss of halogens from crystallized and glassy silicic rocks. Geochim. Cosmochim. Acta 31, 215 (1967).
NOCKOLDS, S. R., MITCHELL, R. L.: The geochemistry of some Caledonian plutonic rocks. Trans. Roy. Soc. Edinburgh 61 (No. 20) (1948).
NODDACK, I., NODDACK, W.: Die Häufigkeit der chemischen Elemente. Naturwissenschaften 18, 757 (1930).
NOMMIK, H.: Fluorine in Swedish agricultural products, soil, and drinking water. Acta Polytech. 127, 1 (1953).
OELSCHLÄGER, W.: Die Verunreinigung der Atmosphäre durch Fluor. Z. Staub 12, 528 (1965).
— RHEINWALD, U.: Das Nahrungsfluor in toxikologischer und kariesprophylaktischer Hinsicht. Das öffentliche Gesundheitswesen 30, 11 (1968).
— WÖHLBIER, W., MENKE, K. H.: Über Fluor-Gehalte pflanzlicher, tierischer und anderer Stoffe aus Gebieten ohne und mit Fluoremission. I. Mitteilung (aus Gebieten ohne F-Emission). Landwirtsch. Forsch. 20, 199 (1967a).
— — — Über Fluorgehalte pflanzlicher, tierischer und anderer Stoffe aus Gebieten ohne und mit Fluoremission. II. Mitteilung (ohne F-Emission). Landwirtsch. Forsch. 20, 285 (1967b).
— — — Über Fluor-Gehalte pflanzlicher, tierischer und anderer Stoffe aus Gebieten ohne und mit Fluor-Emission. III. Mitteilung (aus Gebieten mit F-Emission). Landwirtsch. Forsch. 21, 82 (1968).
OLSEN, E.: Amphibole, first occurrence in a meteorite. Sciene 156, 61 (1967).
PALACHE, Ch., BERMAN, H., FRONDEL, Cl.: The System of Mineralogy, vol. II. New York-London: J. Wiley and Sons, Inc., Chapman and Hall. Ltd. 1957.
PECORA, W. T.: Carbonatites: A review. Geol. Soc. Am. Bull. 67, No. 11, 1537 (1956).
PETTIJOHN, F. J.: Chemical composition of sandstones-excluding carbonate and volcanic sands. Data of geochemistry, 6th ed., chapt. S. U.S. Geol. Surv. Profess. Papers 440-S, 1 (1963).
PIERRUCCINI, R.: La mica di un blocco pneumatollitico del Monte Somma ed i minerali che l'accompagnano. Atti Soc. Toscana Sci. Nat. Pisa, Proc. Verbali, Mem. Ser. A 57 (1950).
POWERS, H. A.: Chlorine and fluorine in silicic volcanic glass. U.S. Geol. Surv. Profess. Papers 424-B, 261 (1961).
PRIOR, G. T.: Catalogue of Meteorites. Brit. Mus. (Nat. Hist.) London: W. Clowes & Sons 1953.
RADCZEWSKI, O. E.: Feine Teilchen (Stäube) in Natur und Technik, ihre Bestimmung als Verunreinigungen und der Nachweis von Fluorverbindungen in der Luft. Ber. Deut. Keram. Ges. 45, 551 (1968).
RANKAMA, K., SAHAMA, Th. G.: Geochemistry. Chicago: University Press 1955.
REED, G. W. Jr.: Fluorine in stone meteorites. Geochim. Cosmochim. Acta 28, 1729 (1964).
— JOVANOVIC, S.: Some halogen measurements on achondrites. Earth and Planetary Sci. Letters 6, 316 (1969).
— — FUCHS, L. H.: Trace elements and accessory minerals in lunar samples. Science 167, 501 (1970).
RILEY, J. P.: The occurrence of anomalousy high fluoride concentrations in the North Atlantic. Deep-Sea Res. 12, 219 (1965).
RIMSAITE, J. H. Y.: Studies of rock forming micas. Geol. Surv. Canada Bull. 149, 1 (1967).
— Evolution of zoned micas and associated silicates in the Oka carbonatite. Contr. Mineral. and Petrol. 23, 340 (1969).
— Partly-fused phlogopite from eclogite nodules and its relation to diamond-bearing kimberlite and related rocks. Unpublished manuscript (1970a).

RIMSAITE, J. H. Y.: Structural formulae of oxidized and hydroxyl-deficient micas and decomposition of the hydroxyl-group. Beitr. Mineral. Petrog. **25**, 225 (1970b).

RINGWOOD, A. E.: The principles governing trace element behavior during magmatic crystallization. Part II: The role of complex formation. Geochim. Cosmochim. Acta **7**, 189 (1955).

ROBINSON, W. O., EDGINGTON, G.: Fluorine in soils. Soil Sci. **61**, 341 (1946).

ROUBAULT, M., DE LA ROCHE, H., GOVINDARAJU, K.: État actuel (1970) des études coopératives sur les standards géochimiques du centre de recherches pétrographiques et géochimiques. Sciences de la Terre **15**, 361 (1970).

RUBEY, W.: Geologic history of sea water. Bull. Geol. Soc. Am. **62**, 1111 (1951).

— Development of the hydrosphere and atmosphere, with special reference to probable composition of the early atmosphere. Geol. Soc. Am. Spec. Papers **62**, 631 (1955).

SAGGERSON, E. P.: Geology of the Kismu district. Kenya Geol. Surv. Rept. **21**, 86 (1952).

SAHAMA, Th. G.: Abundance relation of fluorite and sellaite in rocks. Ann. Acad. Sci. Fennicae, Ser. A, III **9**, (1945).

— On the chemistry of the mineral titanite. Compt. Rend. Soc. Géol. Finlande **19**, 88 (1946).

— Rapakivi amphibole from Uuksunjoki, Salmi area. Bull. Comm. Geol. Finlande No. 140 (1947) Helsinki.

— On the chemistry of the east Fennoscandian rapakivi granites. Bull. Comm. geol. Finlande No. 136 (1948).

SAITO, N.: Geochemistry of the Fukushinzan alkaline igneous complex. I. The distribution of chemical elements in alkaline rocks. Sci. Repts. Frac. Sci. Kyushu Univ. **1**, 121 (1950) [in Japanese, as quoted by KOKUBU, 1956].

SCHERER, O.: Technische organische Fluor-Chemie. Fortschr. Chem. Forsch. (Topics in Current Chemistry) **14**, 127 (1970)

SCHNEIDER, H. J.: Neue Ergebnisse zur Stoffkonzentration und Stoffwanderung in Blei-Zink-Lagerstätten der nördlichen Kalkalpen. Fortschr. Mineral. **32**, 26 (1953).

— Die sedimentäre Bildung von Flußspat im oberen Wettersteinkalk der nördlichen Kalkalpen. Abhandl. Bayer. Akad. Wiss., Math. Naturw. Kl., H. 66 (1954).

SCHNEIDERHÖHN, H.: Die Erzlagerstätten der Erde. Die Pegmatite, Bd. II. Stuttgart: Fischer 1961.

SCHROLL, E.: Zur Geochemie und Genese der Wässer des Neusiedlerseegebietes (Landschaft Neusiedler See) Wiss. Arb. aus dem Burgenland H. 23, 55 (1959).

— Zur Geochemie der Halogene in Wässern des Neusiedlerseegebietes und anderen mineralisierten Wässern des Burgenlandes. Wiss. Arb. a.d. Burgenland H. 30, 109 (1965).

SERAPHIM, R. H.: Some aspects of the geochemistry of fluorine. Thesis Mass. Inst. of Techn. (1951).

SHANNON, E. V., LARSON, E. S.: Merillite and chlorapatite from stony meteorites. Am. J. Sci. **209**, 250 (1925).

SHAW, D. M.: Studies in analytical geochemistry. Roy. Soc. Canada Spec. Publ. No. 6 Univ. of Toronto (1963).

SHEPHERD, E. S.: Note on the fluorine content of rocks and ocean bottom. Am. J. Sci. **238**, 117 (1940).

SHISHKINA, O. V.: Fluorine in oceanic sediments and their pore solutions. Geochem. Intern. **3**, 152 (1966).

SHMAKIN, B. M., SHIRYAYEVA, V. A.: Distribution of rare earth and some other elements in apatites of muscovite pegmatites, Eastern Siberia. Geochem. Intern. **5**, 796 (1968).

SILLEN, L. G.: The physical chemistry of sea water. In: Oceanography, (ed. M. SEARS). Am. Ass. Advanc. Sci. Publ. No. 67, 549 (1959).

SOLOV'YEV, A. T., CHUPROV, V. V., MOIZHES, I. B.: Geochemistry of fluorine in the alkalic rocks of Western Transbaikalia. Geochem. Intern. **4**, 278 (1967).

STAVROV, O. D., UKHANOV, A. V.: Alkali elements and fluorine in rocks and minerals of the mantle. Geochem. Intern. **8**, 168 (1971).

STEINKOENIG, L. A.: Relation of fluorine in soils, plants and animals. Ind. Eng. Chem. **11**, 463 (1919).

STEVENSON, J. S., STEVENSON, L. S.: Fluorine content of microsaur teeth from the Carboniferous rocks of Joggins, Nova Scotia. Science **154**, 1548 (1966).

STOIBER, R. E., ROSE, W. Jr.: The geochemistry of Central American volcanic gas condensates. Geol. Soc. Am. Bull. **81**, 2891 (1970).

STRAUSS, C. A., TRUTER, F. C.: The alkali complex at Spitzkop, Sekukuniland. Geol. Soc. South Africa Trans. **53**, 81 (1951a).
— — Post-Bushveld ultrabasic, alkali, and carbonatic eruptives at Magnet Heights, Sekukuniland, Eastern Transval. Geol. Soc. South Africa Trans. **53**, 169 (1951b).
STROMINGER, D., HOLLANDER, J. M., SEABORG, G. T.: Tables of isotopes. Rev. Mod. Phys. **30**, 585 (1958).
— — — Handbook of Chemistry and Physics, p. 380. Ohio (Cleveland): Chemical Rubber Co. (1967/68).
STRÜBEL, G.: Quantitative Untersuchungen über die hydrothermale Löslichkeit von Flußspat (CaF_2). Neues Jahrb. Mineral. Monatsh. 83 (1965).
STUEBER, A. M., HUANG, W. H., JOHNS, W. D.: Chlorine and fluorine in ultramafic rocks. Geochim. Cosmochim. Acta **32**, 353 (1968).
SUESS, H. E.: Abundance of the elements in the universe. In: LANDOLT-BOERNSTEIN, N. S., group VI: Astronomy and Astrophysics. I. Berlin-Heidelberg-New York: Springer 1965.
SUGAWARA, K.: Migration of elements through phases of the hydrosphere and atmosphere p. 501. In: VINOGRADOV, A. P., Chemistry of the Earth's Crust, vol. 2. Jerusalem: Israel Program for Scientific Translations 1967.
SUGIURA, T., MIZUTANI, Y., OANA, S.: Fluorine, bromine and iodine in volcanic gases. J. Earth Sci. Nagoya Univ. **11**, No. 2, (1963).
TABORSZKY, F. K.: Geochemie des Apatites in Tiefengesteinen am Beispiel des Odenwaldes. Beitr. Mineral. Petrog. **8**, 354 (1962).
TADDEUCCI, A.: Il boro ed il fluoro nei prodotti vulcanici dei Campi Flegrei. Periodico Min., Roma **37**, 845 (1968).
TAUSON, L. V.: Geochemical behavior of rare elements during crystallization and differentiation of granitic magmas. Geochem. Intern. **4**, No. 6, 1067 (1967).
TAYLOR, Jr., H. P., FRECHEN, J., DEGENS, E. T.: Oxygen and carbon isotope studies of carbonatites from the Laacher See District, West Germany and the Alnö District, Sweden. Geochim. Cosmochim. Acta **31**, 407 (1967).
TONANI, F.: Contributo alla conoscenza delle geochimica del fluoro. Il contenuto dell'isola di vulcano e del vesuvio. Atti Soc. Toscana Sci. Nat. Pisa Proc. Verbali Mem. Ser. A **64** (1957).
TOURTELOT, H. A.: Preliminary investigations of the geologic setting and chemical composition of the Pierre shale, Great Plains region. U.S. Geol. Surv. Profess. Papers **390**, 1 (1962).
TRÖGER, E.: Der Gehalt an seltenen Elementen bei Eruptivgesteinen. Chem. Erde **9**, 286 (1935).
TRÖGER, W. E.: Spezielle Petrographie der Eruptivgesteine. Ein Nomenklaturkompendium mit 1. Nachtrag. (1938) Berlin: Deutsch. Mineralog. Ges. 1935.
TUREKIAN, K. K., WEDEPOHL, K. H.: Distribution of elements in some major units of the earth's crust. Geol. Soc. Am. Bull. **72**, 175 (1961).
UNDERWOOD, E. J.: Trace Elements in Human and Animal Nutrition. New York-London: Academic Press 1962.
U.S. Bureau of Mines: Mineral Industry Survey, World Mineral Production in 1966. Government Printing Office, Wash. D. C. (1967).
U.S. Department of the Interior, Bureau of Mines: Commodity data summaries. Jan. (1971).
VALYASHKO, V. M., KOGARKO, L. N., KHODAKOVSKIY, I. L.: Stability of fluorapatite, chlorapatite and hydroxylapatite in aqueous solutions at different temperatures. Geochem. Intern. **5**, 21 (1968).
VAN SCHMUS, RIBBE, P. H.: Composition of phosphate minerals in ordinary chondrites. Geochim. Cosmochim. Acta **33**, 637 (1969).
VINOGRADOV, A. P. (ed.): Chemistry of the Earth's Crust, vol. 2. Jerusalem: Israel Program for Scientific Translations 1967.
VLASOV, K. A., KUZ'MENKO, M. Z., ES'KOVA, E. M.: The Lovozero Alkali Massif. Edinburgh and London: Oliver & Boyd 1966.
VOROSHILOV, Yu. I.: Geochemical behavior of fluorine in the ground waters of the Moscow region. Geochem. Intern. **3**, 261 (1966).

WAI, C. M., WASSON, J. T.: Silicon concentrations in the metal of iron meteorites. Geochim. Cosmochim. Acta **33**, 1465 (1969).
WALKER, A. L., BUCHANAN, A. S.: Geochemical processes in ore formation, pt. I. The production of hydrothermal fluids from sedimentary sequences. Econ. Geol. **64**, 919 (1969).
WARREN, B. K.: The crystal structure and chemical composition of the monoclinic amphiboles. Z. Krist. **72**, 493 (1929).
WELLS, J. R.: Fluorspar and Cryolite. In: U.S. Bureau of Mines, Minerals Yearbook, 1968.
WHITE, D. E., HEM, J. D., WARING, G. A.: Chemical composition of subsurface waters. In: Fleischer, M. (ed.): Data of Geochemistry, 6th ed., chapter F. U.S. Geol. Surv. Profess. Papers **440-F**, 1 (1963).
— WARING, G. A.: Volcanic emanations. In: FLEISCHER, M. (ed.): Data of Geochemistry, 6th ed. chapt. K., U.S. Geol. Surv. Profess. Papers **440-K**, 1 (1962).
WILBUR, K. M., WATABE, N.: Experimental studies on calcification in molluscs and the alga Coccolithus huxleyi. Ann. N.Y. Acad. Sci. **109**, 82 (1863).
WITTELS, M. C.: A thermo-chemical analysis of some amphiboles. Doctorate Thesis. Dept. of Geol. Mass. Inst. of Techn. 1951.
WYLLIE, P. J., TUTTLE, O. F.: Experimental investigation of silicate systems containing two volatile components; part II. Am. J. Sci. **259**, 128 (1961).
YAVNEL, A. A.: The abundance of elements in the metallic phase of iron meteorites and in chondrites. Geochem. Intern. **2**, 350 (1964).
YOSHIDA, M., OZAWA, T., OSSAKA, J.: A singular silica sublimate mineral found in Satuma-Iwo-Zima volcano. (1969) Cited by HONDA (1970).
YOUNG, E. G., LANGILLE, W. M.: Sea occurrence of inorganic elements in marine algae of the Atlantic Provinces of Canada. Can. J. Bot. **36**, 301 (1958).
ZIES, E. G.: The Valley of Ten Thousand Smokes. I. The acid gases contributed to the sea during volcanic activity. Nat. Geographic. Soc. Contributed Tech. Pap. I. Katmai Series No. 4, (1929).

Revised manuscript received: December 1971

Sodium

A	G. Cocco, L. Fanfani and P. F. Zanazzi	(Istituto di Mineralogia dell'Università di Perugia, Perugia, Italy)
B—G, **I—O**	K. S. Heier and	(Mineralogisk-Geologisk Museum, Oslo, Norway)
	G. K. Billings	(Department of Geosciences, New Mexico Institute of Mining and Technology, Socorro, New Mexico, U.S.A.)

11-A. Crystal Chemistry

The sodium atom in the ground state has the electron configuration $1s^2\,2s^2 2p^6 3s^1$, corresponding to a core with an electronic structure of the inert gas Ne and an additional single valence electron in the $3s$ orbital. This configuration explains its chemical reactivity and its occurrence only in the oxidation state $+1$ in ionic compounds. The usually accepted ionic radius of Na$^+$ is 0.98 Å (see Table 12-8 of vol. I of this handbook).

Because of the large differences in ionic radii, the mutual isomorphous replacement between sodium and the preceding and the following elements in the alkali series, Li and K, takes place only to a very limited extent. At high temperatures, the mutual solubility of Na—K compounds is much enhanced.

A close crystallochemical relationship exists between sodium and calcium; the mutual substitution is similar to that occurring between Li and Mg. As Na is more abundant than Li, this substitution affects a large number of natural compounds.

In nature, sodium occurs only in coordination with oxygen and halogen atoms; therefore, the present crystallochemical review is confined to these classes of compounds. Since the coordination around Na is essentially controlled by steric and electrostatic forces, Na$^+$ exhibits a variable stereochemistry in its compounds although the most frequent distribution of ligands around the cation is typically a more or less distorted octahedron. It is occasionally difficult to give the "coordination number" only on the basis of the bond lengths, as these sometimes exhibit a wide range; screen effects from nearer ligands and considerations of electrostatical equilibria should be taken into account.

In Table 11-A-1 there are listed several mineral structures from which reliable information concerning Na coordination can be obtained.

I. Halogen Compounds

Only flouride and chloride compounds of sodium have been found in nature, often in chemical combination with other cations. In most cases, the anionic group is represented by fluoride complexes such as [AlF$_6$]$^{3-}$, [BF$_4$]$^-$ and [SiF$_6$]$^{2-}$. The simple halogenides have the "NaCl" f.c.c. structure, with Na$^+$ in octahedral coordination; the sodium-halogen distances are: 2.317 Å in the fluoride; 2.820 Å in the chloride; 2.989 Å in the bromide and 3.237 Å in the iodide (SYSIÖ, 1969). Only a few reliable data are available for coordination and distances involving sodium in complex halogenides. Most of the crystal structures are inferred only by analogy with other substances. However, the most frequent coordination numbers seem to be six and eight. In cryolithionite, Na$_3$Al$_2$[LiF$_4$]$_3$, which is isostructural with garnet (MENZER, SB 1930, 498), Na$^+$ occurs in approximately eightfold coordination.

Sodium

Table 11-A-1

Mineral	Formula	Coordination	Range Å	Average Å	Reference
Acmite	NaFe[Si$_2$O$_6$]	Na[8]—O	2.398 —2.831	2.52	Clark et al. (1969)
Borax	Na$_2$[B[4]B$_2^{[3]}$O$_5$(OH)$_4$] · 8H$_2$O	Na[6]—O	2.39 —2.42	2.40	Morimoto (SR 1956, 376)
Buergerite (ferric tourmaline)	NaAl$_6$Fe$_3$O$_3$F[BO$_3$]$_3$[Si$_6$O$_{18}$]	Na[6]—O	2.40 —2.46	2.44	
Darapskite	Na$_3$[NO$_3$][SO$_4$] · H$_2$O	Na[9]—O	2.571—2.788	2.68	Barton (1969)
					Sabelli (1967)
Dawsonite	NaAl(OH)$_2$[CO$_3$]	Na[6]—O	2.338 —2.550	2.43	Frueh and Golightly (1967)
		Na[7]—O	2.300 —2.725	2.47	
Ferruccite	Na[BF$_4$]	Na[6]—F	2.39 —2.47	2.44	Brunton (1968)
Gaylussite	Na$_2$Ca[CO$_3$]$_2$ · 5H$_2$O	Na[6]—O	2.296 —2.609	2.47	Menchetti (1968)
Glauberite	CaNa$_2$[SO$_4$]$_2$	Na[6]—O	2.332 —2.612	2.43	Araki and Zoltai (1967)
Glaucophane	Na$_2$Mg$_3$Al$_2$(OH)$_2$[Si$_8$O$_{22}$]	Na[7]—O	2.335 —2.829	2.54	Papike and Clark (1968)
Halite	NaCl	Na[6]—Cl	2.337 —2.798	2.50	from Sysiö (1969)
Heidornite	$^2_\infty$Ca$_5$Na$_2$Cl[SO$_4$]$_2$[B$_5^{[4]}$B$_2^{[3]}$O$_8$(OH)$_2$]	Na[7]—1Cl	2.820	2.82	Burzlaff (1967)
		—6O	3.02		
Jadeite	NaAl[Si$_2$O$_6$]	Na[8]—O	2.41 —2.81	2.52	Prewitt and Burnham (1966)
Kernite	$^1_\infty$Na$_2$[B$_2^{[4]}$B$_2^{[3]}$O$_6$(OH)$_2$] · 3H$_2$O	Na[6]—O	2.357—2.741	2.47	Cialdi et al. (1967)
		Na[6]—O	2.275—2.488	2.41	
		Na[5]—O	2.288—2.431	2.39	
Kröhnkite	Na$_2$Cu[SO$_4$]$_2$ · 2H$_2$O	Na[7]—O	2.35 —2.59	2.49	Rao (SR 1961, 449)
Lecontite	Na(NH$_4$)[SO$_4$] · 2H$_2$O	Na[6]—O	2.34 —2.47	2.41	Corazza et al. (1967)
Leucophanite	CaNaBeF[Si$_2$O$_6$]	Na[6]—5O	2.35 —2.60	2.47	Cannillo et al. (1967)
		—1F	2.38		
Malladrite	Na$_2$[SiF$_6$]	Na[6]—F	2.30 —2.45	2.37	Zalkin et al. (1964)
		Na[6]—F	2.18 —2.31	2.27	
Mirabilite	Na$_2$[SO$_4$] · 10H$_2$O	Na[6]—O	2.34 —2.61	2.47	Cocco (1962)
		Na[6]—O	2.28 —2.60	2.41	
Nahcolite	NaH[CO$_3$]	Na[6]—O	2.385—2.464	2.44	Sharma (1965)
Narsarsukite	Na$_2$TiO[Si$_4$O$_{10}$]	Na[7]—O	2.39 —2.72	2.54	Peacor and Buerger (1962)
Natrochalcite	NaCu$_2$(OH)[SO$_4$]$_2$ · H$_2$O	Na[8]—O	2.57 —2.63	2.59	Rumanova and Volodina (SR 1958, 470)
Natrolite	Na$_2$[Al$_2$Si$_3$O$_{10}$] · 2H$_2$O	Na[6]—O	2.36 —2.62	2.44	Meier (SR 1960, 477)

Sodium

Mineral	Formula	Coordination	Distance	Distance 2	Reference
Natron	$Na_2[CO_3] \cdot 10 H_2O$	$Na_I^{[6]}$—O	2.31 —2.51	2.43	TAGA (1969)
		$Na_{II}^{[6]}$—O	2.39 —2.49	2.44	
Neptunite	$LiKNa_2Fe_2Ti_2O_2[Si_8O_{22}]$	$Na^{[6]}$—O	2.41 —2.73	2.53	CANNILLO et al. (1966)
Pectolite	$Ca_2NaH[Si_3O_9]$	$Na^{[6]}$—O	2.311—2.536	2.44	PREWITT (1967)
Pirssonite	$CaNa_2[CO_3]_2 \cdot 2H_2O$	$Na^{[6]}$—O	2.284—2.755	2.46	CORAZZA and SABELLI (1967)
Pollucite	$Cs_xNa_y[Al_{(x+y)}Si_{(48-x-y)}O_{96}] \cdot (16-x) H_2O$	$Na^{[6]}$—O	2.420—2.498	2.47	BEGER (1969)
Probertite	$NaCa[B^{[4]}_3B^{[3]}_2O_7(OH)_4] \cdot 3H_2O$	$Na^{[6]}$—O	2.34 —2.72	2.47	RUMANOVA et al. (1966)
Ramsayite	$Na_2Ti_2O_3[Si_2O_6]$	$Na^{[7]}$—O	2.26 —2.64	2.47	CHIN-KHAN et al. (1969)
Reedmergnerite	$Na[BSi_3O_8]$	$Na^{[5]}$—O	2.380—2.489	2.43	APPLEMAN and CLARK (1965)
Rinkite	$Na_2Ca(Ca, Ce)(Ti, Nb)(O, F)_2F_2[Si_2O_7]$	$Na_I^{[6]}$—2F	2.30 —2.40		SIMONOV and BELOV (1968)
		—2(O, F)	2.34 —2.60		
		—2O	2.30 —2.49		
		$Na_{II}^{[6]}$—2F	2.31 —2.32		
		—2(O, F)	2.43 —2.59		
		—2O	2.33 —2.35		
Searlesite	$NaB(OH)_2[Si_2O_5]$	$Na^{[6]}$—O	2.31 —2.74	2.52	KRAVCHENKO (1964)
Sodalite	$Na_8Cl_2[Si_6Al_6O_{24}]$	$Na^{[4]}$—3O	2.351	2.35	LONS and SCHULZ (1967)
		—1Cl	2.730		
Soda-alum	$NaAl[SO_4]_2 \cdot 12 H_2O$	$Na^{[6]}$—O	2.453	2.45	CROMER et al. (1967)
Soda-niter	$Na[NO_3]$	$Na^{[6]}$—O	2.42	2.42	SASS et al. (SR 1957, 567)
Tamarugite	$NaAl[SO_4]_2 \cdot 6 H_2O$	$Na^{[6]}$—O	2.26 —2.51	2.42	ROBINSON and FANG (1969)
Teepleite	$Na_2Cl[B(OH)_4]$	$Na^{[6]}$—4O	2.51	2.51	FORNASERI (SR 1949, 263)
		—2Cl	2.82	2.82	
Thomsenolite	$NaCa[AlF_6] \cdot H_2O$	$Na^{[8]}$—F	2.321—2.654	2.46	COCCO et al. (1967)
Tugtupite	$Na_8Al_2Be_2(Cl, S)_2[Si_8O_{24}]$	$Na^{[5]}$—4O	2.31 —2.57	2.41	after DANØ (1966)
		—1Cl	2.69		
Uklonskovite	$NaMg(OH)[SO_4] \cdot 2H_2O$	$Na^{[5]}$—O	2.29 —2.62	2.39	BORISOV et al. (1964)
Ureyite	$NaCr[Si_2O_6]$	$Na^{[8]}$—O	2.378—2.264	2.49	CLARK et al. (1969)
Vanthoffite	$Na_6Mg[SO_4]_4$	$Na_I^{[6]}$—O	2.34 —2.45	2.41	FISCHER (1964)
		$Na_{II}^{[6]}$—O	2.34 —2.52	2.41	
		$Na_{III}^{[7]}$—O	2.31 —2.75	2.53	
Villiaumite	NaF	$Na^{[6]}$—F	2.317	2.32	from SYSIÖ (1969)
Wardite	$NaAl_3(OH)_4[PO_4]_2 \cdot 2H_2O$	$Na^{[8]}$—O	2.38 —2.71	2.54	FANFANI et al. (1970)

The same coordination number is shown in neighborite, $NaMgF_3$ (CHAO et al., SR 1961, 304), which has a distorted perowskite-type structure, and in ferruccite, $Na[BF_4]$ (BRUNTON, 1968), a structure of the zircon-type with slight distortion to orthorhombic symmetry. The Na—F distances in this latter compound range between 2.296 and 2.609 Å with an average value of 2.47 Å. In weberite, $Na_2MgF[AlF_6]$ (BYSTRØM, SR 1949, 196) each of the two independent Na ions binds eight F ions with distances ranging from 2.24 to 2.70 Å. In thomsenolite (COCCO et al., 1967), sodium has coordination number of 8 in the form of a trigonal prism with two additional F ions above two prism faces; the average Na—F distance is 2.46 Å, with values in the range 2.321—2.654 Å. Sodium exhibits a six-fold coordination in rinneite, $K_3Na[FeCl_6]$ (BELLANCA, SR 1948, 415), in elpasolite, $K_2Na[AlF_6]$ (MENZER, SB 1932, 498) and in malladrite $Na_2[SiF_6]$ (ZALKIN et al., 1964); in this mineral the mean values are 2.37 and 2.27 Å respectively for Na—F distances in the two crystallographically independent Na-octahedra. NaF octahedra are present also in cryolite, $Na_3[AlF_6]$ (NÁRAY-SZABO and SASVARI, SB 1938, 120), where Na—F distances range from 2.21 to 2.68 Å. In teepleite $Na_2Cl[B(OH)_4]$ (FORNASERI, SR 1949, 263), the Na—Cl distances of the $Na(OH)_4Cl_2$ tetragonal bipyramids are 2.82 Å.

It is interesting to note that in the synthetic compound $Na[Sn_2^{[3+1]}F_5]$ (McDONALD et al., 1964), sodium has both a six- and an eight-fold coordination, according to an elongated octahedron and a distorted square antiprism; the average Na—F distances are 2.28 and 2.40 Å respectively.

II. Oxygen Compounds (Except Major Rock-forming Minerals)

In oxygen compounds, Na exhibits a larger variety of coordination numbers.

a) Sodium in *four-fold coordination* is unusual and it has been never found in minerals, with the exception of sodalite where Na^+ ions have three oxygen atoms and one Cl as nearest neighbors (LONS and SCHULZ, 1967). Na exhibits tetrahedral coordination in Na_2O, which has an "anti-fluorite" structure (ZINTL et al., SB 1934, 283), in β-$NaFeO_2$ (BERTAUT and BLUM, SR 1954, 422) and in related compounds such as $NaAlO_2$. In this latter compound the average $Na^{[4]}$—O distance is 2.25 Å. An approximately tetrahedral coordination occurs in $NaOH \cdot H_2O$ (WUNDERLICH, SR 1957, 240), with an average Na—O distance of 2.36 Å. In $Na_5[P_3O_{10}]$, phase I, (CORBRIDGE, SR 1960, 402), two Na^+ ions have a roughly octahedral coordination, while a third Na^+ ion has four short contacts in a rather "one-sided" arrangement; the mean Na—O value is 2.37 Å. In $Na_2Mn_2[Si_2O_7]$ (ASTAKHOVA and SIMONOV, 1969), $Na^{[4]}$—O distances range within the limits of 2.32—2.42 Å and 2.35—2.51 Å for the two different Na atoms with average values of 2.36 and 2.42 Å.

b) Some structures of minerals and inorganic synthetic compounds contain Na having a *five-fold coordination;* the resulting coordination polyhedron is a more or less distorted trigonal bipyramid or tetragonal pyramid. A typical example of this coordination has been found in the borate mineral kernite (CIALDI et al., 1967), where both independent sodium ions link five oxygen atoms with average distances of 2.39 and 2.41 Å. The Na(1) and Na(2) coordination polyhedra are connected at one corner and two Na(1) polyhedra share an edge which is running through an inversion center to form four-membered groups in the structure. In tugtupite $Na_8Al_2Be_2(Cl, S)_2[Si_8O_{24}]$, Na^+ binds four oxygen atoms and one chlorine atom

(DANØ, 1966); each chlorine is shared by four Na polyhedra. In uklonskovite NaMg(OH)[SO$_4$]·2H$_2$O, Na is coordinated to two oxygen atoms, two water molecules and one hydroxyl-group (BORISOV et al., 1964). In vanthoffite Na$_6$Mg[SO$_4$]$_4$ (FISCHER, 1964), only one of the three crystallographically different kinds of Na$^+$ ions is five-fold coordinated, with an average Na—O distance of 2.41 Å. In reedmergnerite Na[BSi$_3$O$_8$], the boron analogue of low-albite (APPLEMAN and CLARKE, 1965), Na is coordinated by five oxygen atoms within a range of 2.380—2.489 Å; two Na polyhedra are connected around an inversion center to form edge-joined doublets. In addition to five close oxygen atoms, there are two at longer distances (2.808 and 2.860 Å). Additional weak bonds are common when Na$^+$ exhibits a low coordination number. This is observed, for example, in probertite NaCa[B$_5$O$_7$(OH)$_4$]·3H$_2$O (RUMANOVA et al., 1966), in which in the Na polyhedron 5 Na—O distances lie very close together (2.34—2.49 Å) and an additional bond of 2.72 Å occurs between the cation and a water molecule. Reliable data on Na$^{[5]}$ are provided by some synthetic polyphosphates and silicates; in Na[PO$_3$] (McADAM et al., 1968), the average Na$^{[5]}$—O distances are 2.44 and 2.41 Å; in Na$_4$[P$_4$O$_{12}$]·4H$_2$O (ONDIK, 1964), 2.37 Å; in Na$_3$[P$_4$O$_{11}$(OH)] (JOST, 1968), 2.41 Å; in Na$_3$[P$_3$O$_9$] (ONDIK, 1965), 2.36 and 2.47 Å; in α-Na$_2$[Si$_2$O$_5$] (PANT and CRUICKSHANK, 1968), 2.40 Å; in β-Na$_2$[Si$_2$O$_5$] (PANT, 1968), 2.42 Å; in Na$_2$[SiO$_3$] (McDONALD and CRUICKSHANK, 1967), 2.38 Å; and in Na$_2$[SiO$_2$(OH)$_2$]·4H$_2$O (JOST and HILMER, 1966), 2.40 and 2.36 Å. The same coordination number for Na$^+$ occurs twice in Na$_3$Fe$_5$O$_9$ (ROMERS et al., 1967) with Na—O mean distances of 2.44 Å. The overall average Na$^{[5]}$—O bond length is 2.41 Å.

c) Examples of *six-fold coordination* around Na$^+$ are by far most numerous. Besides the minerals listed in Table 11-A-1, several synthetic inorganic crystals contain 6-coordinated sodium. The mean Na—O value resulting from 65 recent accurate determinations is 2.44 Å. The arrangement of oxygen atoms around Na is generally in the form of a distorted octahedron. In NaCo$_{2.31}$[MoO$_4$]$_3$ (IBERS and SMITH, 1969), however, the coordination polyhedron closely resembles a trigonal prism and the average Na$^{[6]}$—O distance is the shortest observed — 2.35 Å. Coordination octahedra around Na$^+$ can occur either isolated or linked together in various ways. Examples of isolated octahedra are reported for NaAl[SO$_4$]$_2$·12H$_2$O (CROMER et al., 1967) and for NaH$_3$[SeO$_3$]$_2$ (VIJAYAN, 1968). In the soda-alum, Na$^+$ binds oxygen atoms all belonging to water molecules; isolated Na(H$_2$O)$_6$ octahedra and SO$_4$ and Al(H$_2$O)$_6$ groups are linked by hydrogen bonds. In tamarugite NaAl[SO$_4$]$_2$·6H$_2$O (ROBINSON and FANG, 1969), two Na-octahedra share an edge around an inversion center. A similar arrangement is observed in the crystal structure of natron, Na$_2$[CO$_3$]·10H$_2$O (TAGA, 1969), which consists of pairs of sodium-water octahedra sharing an edge to form [Na$_2$(H$_2$O)$_{10}$]$^{2+}$ groups. The pairs of octahedra are linked by hydrogen bonds to form a three-dimensional framework in the crystal. In other structures the arrangement of coordination octahedra builds up infinite chains. In gaylussite Na$_2$Ca[CO$_3$]$_2$·5H$_2$O (MENCHETTI, 1968), Na-octahedra are connected alternately by sharing one corner or one edge to form zig-zag chains. Chains of octahedra sharing edges are present in borax Na$_2$[B$_4$O$_5$(OH)$_4$]·8H$_2$O (MORIMOTO, SR 1956, 376), in mirabilite Na$_2$[SO$_4$]·10H$_2$O (COCCO, 1962), and in julienite, Na$_2$Co[NCS]$_4$·8H$_2$O (PREISINGER, SR 1953, 462). In all these cases Na$^+$ is surrounded by six water molecules. Continuous chains, built by neighboring

Na-octahedra which share edges, are also observed in ulexite, $NaCa[B_5O_6(OH)_6] \cdot 5H_2O$. In this mineral Na is coordinated by four water molecules and two OH groups at an average distance of 2.42 Å (CLARK and APPLEMAN, 1964). In lecontite (CORAZZA et al., 1967), the Na ions share opposite faces of their coordination octahedra in order to form infinite straight chains; the Na—Na distance is very short (3.15 Å). In nahcolite (SHARMA, 1965), strings of two parallel chains are formed by edge-sharing. Examples of more complicated connections involving Na-octahedra are found in darapskite $Na_3[NO_3][SO_4] \cdot H_2O$ (SABELLI, 1967), and pirssonite $CaNa_2[CO_3]_2 \cdot 2H_2O$ (CORAZZA and SABELLI, 1967). In the former mineral, Na-octahedra share alternatively one face and one edge to form chains; the remaining Na^+ ions, in seven-fold coordination, are connected with Na-octahedra, NO_3 and SO_4 groups through edges and vertices to form thick layers linked by hydrogen bonds. In pirssonite, Na-octahedra are connected by edges to form two series of chains linked three-dimensionally at common corners.

d) The mean Na—O bond length at *coordination number 7* is 2.52 Å. This average value is the result of eight determinations; six are reported in Table 11-A-1. The other two concern the synthetic compounds $Na[BO_2]$ (MAREZIO et al., 1963) with $Na^{[7]}$—O mean value of 2.51 Å, and $Na_2[B_8O_{13}]$ (HYMAN et al., 1967), where one of the two independent Na^+ ions has seven oxygen neighbors with an average distance of 2.59 Å. Linkages between sodium-oxygen polyhedra occur through edges and corners to form either layers, as in kröhnkite $Na_2Cu[SO_4]_2 \cdot 2H_2O$ (RAO, SR 1961, 449), or three-dimensional frameworks, as in glauberite $CaNa_2[SO_4]_2$ (COCCO et al., 1965).

Higher coordination numbers for Na^+ are well known, but they do not occur frequently.

e) *Eight-coordinated* Na^+ is present in minerals and in other inorganic compounds. The coordination polyhedron is intermediate between a cube and a tetragonal antiprism. In the clinopyroxenes jadeite (PREWITT and BURNHAM, 1966), ureyite and acmite (CLARK et al., 1969), and in the clinoamphibole glaucophane (PAPIKE and CLARK, 1968), Na is coordinated to oxygen atoms by six short and two longer bonds. The average values are 2.38, 2.40, 2.41 and 2.40 Å for the shorter distances in the respective minerals and 2.74, 2.76, 2.83 and 2.80 Å for the longer distances. In natrochalcite $NaCu_2(OH)[SO_4]_2 \cdot H_2O$ (RUMANOVA and VOLODINA, SR 1958, 470) and in wardite $NaAl_3(OH)_4[PO_4]_2 \cdot 2H_2O$ (FANFANI et al., 1970), the mean $Na^{[8]}$—O values are 2.59 and 2.54 Å; in both these structures no connections occur between adjacent Na polyhedra. This coordination number is also observed in $NaFe[Ge_2O_6]$ (SOLOV'EVA and BAKAKIN, 1968), a structure isotypic with diopside; the average Na—O distance is 2.55 Å. Eight nearest neighboring oxygens at distances less than 2.74 Å (mean value 2.59 Å) surround one type of Na^+ in $NaNbO_3$ (SAKOWSKI-COWLEY et al., 1969) which has a distorted perowskite-type structure; the other Na^+ has 9 near neighbors at 2.80 Å or less, with a mean distance of 2.66 Å.

f) High Na^+ *coordination numbers* ($\geqslant 9$) in the $NaNbO_3$ structure can be explained by taking into account the fact that in the ideal (cubic) perowskite structure Na^+ would have 12-coordination. A crystallographic investigation of buergerite, a ferric tourmaline, (BARTON, 1969) has shown that nine-coordinated Na^+ is present in this mineral also. The average $Na^{[9]}$—O distance is 2.68 Å, in agreement with the preceding value.

High coordination numbers of 9 and above have been claimed for other minerals, such as astrophyllite (Woodrow, 1967), beryllonite (Golovastikov, SR 1961, 456), caracolite (Schneider, 1967), and swedenbergite (Aminoff, SB 1933, 381). In the case of swedenbergite, the coordination around Na is 12-fold with $Na^{[12]}$—O distances of 2.71 Å.

III. Major Rock-forming Minerals

In rock-forming minerals, such as feldspars, pyroxenes, amphiboles and zeolites, Na is often partially replaced by other ions, e.g., K^+ and Ca^{2+}; a typical example concerns the feldspars.

a) Feldspars

The structures of low-temperature and high-temperature albite, $Na[AlSi_3O_8]$, have been accurately refined by Ribbe et al. (1969). A common feature of albites is the marked "thermal" anisotropy observed for Na; the outstanding problem is whether the Na anisotropy represents true thermal vibration or a space average. The persistence of the anisotropy in low-albite even at $-180°C$ (Williams and Megaw, 1964) and the normal temperature factor found for Na in reedmergnerite, a boron-albite (Appleman and Clark, 1965), favors the hypothesis of a random spatial distribution of the cation in an irregular oxygen-surrounded cavity. Due to the strong "anisotropy" and its interpretation as probably being space-averaged, the coordination around Na can be described only approximately; assuming a single anisotropic-atom model Na has five nearest oxygen atoms, with distances ranging from 2.38 to 2.66 Å in low-albite, and from 2.34 to 2.71 Å in high-albite.

In sodium-potassium feldspars, the high-albite/high-sanidine series shows complete solid solution with a gradual symmetry change from triclinic to monoclinic. In low-temperature materials, segregation of the two phases occurs forming either a coarse or a fine perthitic structure. An explanation of these observations may be provided if one assumes that at high temperature Na and K are distributed in the holes of a continuous Si and Al tetrahedral framework, forming an average structure with a single diffraction pattern. With decrease of temperature solid solution becomes limited and the perthitic structure is the result of alternating Na-rich and K-rich regions throughout the crystal, caused by diffusion and segregation of the cations through the framework.

Since gradual changes in physical properties occur with change in composition, the plagioclase feldspars have long been considered to be a continuous series of solid solutions between the calcium- and sodium-feldspars. However, different members of the series show very complicated diffraction patterns referable to changes in structure. The complexity in this series is obvious in the difference in the crystal structures of the two end-members, anorthite and albite; the asymmetric unit of anorthite is four times that of albite (Megaw et al., 1962), though the atomic arrangement is roughly the same.

b) Pyroxenes

A series of solid solutions occurs between Na- and Ca-members in clinopyroxenes and in clinoamphiboles. An analysis of the coordination types of the metal atoms in clinopyroxenes containing Na and Ca indicates that these larger ions preferentially occupy the M2 positions in the structure, according to the crystallographic nomenclature proposed for clinopyroxenes by BURNHAM et al. (1967). Intermediate Ca—Na clinopyroxenes, named omphacites, are found to have the space group symmetries C2/c or P2, although this distinction cannot be made by examination of X-ray diffraction powder data. P2 omphacites are restricted to a narrow compositional range with $Na/(Na+Ca) \approx 0.5$ and $Al/(Al+Fe^{3+}) > 0.6$. Refinements of P2 omphacites (CLARK and PAPIKE, 1968; CLARK et al., 1969) show that the cation distribution is partially ordered, the large M2 polyhedra being alternatively occupied dominantly by either Na or Ca. These two cations exhibit similar M2—O distances.

c) Amphiboles

In clinoamphiboles Na and Ca occupy the M4 site, according to GHOSE (1965), a four-fold position of the space group. This site has a highly distorted octahedral coordination when it is occupied by Fe^{2+}, Mg^{2+} or Mn^{2+} in clinoamphiboles without alkali or alkaline earth cations; the coordination increases to eight if it is occupied by Ca^{2+} and Na^+. An additional site, named A, with a multiplicity of two occurs in the structure; in this site alkali ions (Na^+ or K^+) in excess of 4 (Ca+Na) ions per unit cell can be accomodated up to a maximum of two ions per cell (HERITSCH et al., SR 1960, 487; GHOSE, 1965). The ionic radii of the ions present in M4 sites seem to be closely related to the values of the β angle; the ionic radius of Na^+ (or Ca^{2+}) excludes the possibility of the existence of an orthorhombic amphibole with a composition rich in Na+Ca (WHITTAKER, SR 1960, 486). However, a partial occupancy of site A in orthoamphiboles by sodium has been observed in two gedrites (PAPIKE and ROSS, 1970). Sodium is coordinated to six oxygen atoms (average Na—O distance 2.55 Å) and shows little positional disorder, in contrast to sodium in the A site of clinoamphiboles.

d) Micas

Micas containing significant amounts of sodium are rare, the only examples being paragonite, a dioctahedral mica, and ephesite, a brittle mica. In the muscovite-paragonite system only a very limited solid solution occurs at normal temperatures (about 3% paragonite in muscovite and vice versa) (EUGSTER and YODER, 1955). Sodium analogues of the trioctahedral micas are not stable from a structural point of view (RADOSLOVICH, 1963).

e) Zeolites and Comparable Structures

In natural zeolites alkali and alkaline earth ions are located together with water molecules in holes of an open framework of tetrahedral groups of oxygen atoms around silicon and aluminum. In most cases the positions of the cations are only incompletely occupied. Replacement of cations in zeolites is rather complex because the structure of these minerals contains a large number of positions suitable for additional ions. In analcite (KNOWLES et al., 1965), 16 sodium ions in octahedral coordination (4 oxygen atoms +two water molecules at distances of about 2.5 Å) randomly occupy a 24-fold position. The ordering of the sodium atoms may be influenced by the aluminum atoms. Substitution of K and Ca for Na is observed in natural analcites.

The mineral pollucite has a structure similar to that of analcite. According to a recent determination (BEGER, 1969) Na ions randomly occupy a crystallographic position where they are octahedrally coordinated by four oxygen atoms and two water molecules (the mean value of the Na—O distance is 2.47 Å). A structure determination of the zeolite mineral faujasite, with a chemical composition $(Na_2Ca)_{0.075}[(Al_{0.3}Si_{0.7})O_2] \cdot 1.37 H_2O$ (BAUR, 1964), led to the conclusion that only 17 of the approximately 43 Na and Ca ions in the unit cell are located in a 32-fold position, the remaining cations being randomly distributed throughout the extremely open framework of this aluminosilicate. In natrolite (MEIER, SR 1960, 477), the Na content is only slightly less than the value corresponding to a chemical formula of $Na_2[Al_2Si_3O_{10}] \cdot 2H_2O$. Sodium, which completely occupies a 16-fold position, has four nearest oxygen atoms and two water molecules at an average distance of 2.44 Å.

Chemical analyses of natural samples of mordenite show a Na content close to 4 atoms per unit cell, additional cations being K and Ca. In an X-ray investigation performed on crystals treated with concentrated $NaNO_3$ solution, and having an approximate composition of $Na_8[Al_8Si_{40}O_{96}] \cdot 24 H_2O$, only four sodium ions occur on definite crystallographic positions. The remaining four sodium ions are distributed randomly (MEIER, SR 1961, 515).

Dachiardite has a structure related to that of mordenite (GOTTARDI and MEIER, 1963); sodium is partly replaced by calcium. Ferrierite (VAUGHAN, 1966) closely resembles mordenite and dachiardite; (Na, K) cations do not occupy definite crystallographic positions in this structure. In phillipsite (STEINFINK, 1962), Na and K atoms and water molecules could not be specifically located at the expected sites in the holes and it is assumed that they are statistically distributed. There exist additional Na-silicates with structures consisting of open three-dimensional networks, containing SO_4 groups or Cl or S atoms associated with the cations in the cavities. In sodalite, $Na_4Cl[Al_3Si_3O_{12}]$ (LÖNS and SCHULZ, 1967), Na^+ has an unusual four-fold coordination with one Cl^- at 2.730 Å and three oxygen atoms at 2.351 Å. Hauyne (SAALFELD, SR 1961, 537) and nosean structures (SCHULZ and SAALFELD, 1965) are frameworks similar to that of sodalite, containing the sodium ions and the sulfate groups in a disordered distribution. Cancrinite is another framework structure related to that of other sodalite-type minerals, with (Na, Ca) ions located in large channels; the CO_3 groups are disordered (JARCHOW, 1965).

f) Scapolites, Feldspathoids

In a scapolite of marialite composition with the chemical formula $(Na_{6.03}, Ca_{1.53}, K_{0.44})[Cl_{1.53}(CO_3)_{0.47}][Al_{1.22}Si_{2.78}O_8]_6$ (PAPIKE and ZOLTAI, 1965), the cation has one Cl at 3.018 Å and five oxygen atoms at distances ranging from 2.352 to 2.849 Å. Na—Ca substitution occurs in some recently determined structures. In meliphanite, closely related to melilite, with the chemical formula

$$Ca(Na_{0.63}Ca_{0.37})Be[(Si_{1.87}Al_{0.13})(O_{6.25}F_{0.75})]$$

(DAL NEGRO et al., 1967), (Na, Ca) cations have four oxygen and two fluorine atoms at an average distance of 2.44 Å and two oxygen atoms at distances of 2.72 and 2.82 Å. In delhayelite, $Ca(Na_3Ca)K_7Cl_2F_4[(Si_{14}Al_2)O_{38}]$ (CANNILLO et al., 1970),

(Na, Ca) is coordinated to two F atoms (2.27 Å) and to six oxygens with two of these latter ligands at a distance of 2.46 Å and four at a distance of 2.86 Å.

Another interesting Na mineral is the framework silicate nepheline, and the atomic arrangement is derived from the high-tridymite structure. Assuming an ideal cell content $K_2Na_6[Al_8Si_8O_{32}]$, Na occupies a six-fold position and has 7 nearest oxygen neighbors in fixed positions at 2.55—2.88 Å; one additional oxygen atom is statistically distributed on two of the three positions within the coordination range at distances of 2.51 and 2.83 Å (HAHN and BUERGER, SR 1955, 477). However, chemical analyses of several nephelines show always a cation deficiency and the presence of some substitutions; the actual formulae should be written as

$$K_xNa_yCa_z[Al_{x+y+2z}Si_{16-(x+y+2z)}O_{32}],$$

where $x+y+z<8$. In nature the most important substitution is that in which K replaces Na; according to DONNAY et al. (SR 1959, 488), the nepheline structure is retained up to an x-value of 4.73.

References: Section 11-A

APPLEMAN, D. E., CLARK, J. R.: Crystal structure of Reedmergnerite, a boron albite, and its relation to feldspar crystal chemistry. Am. Mineralogist 50, 1827 (1965).
ARAKI, T., ZOLTAI, T.: Refinement of the crystal structure of a glauberite. Am. Mineralogist 52, 1272 (1967).
ASTAKHOVA, L. P., SIMONOV, V. I.: Determination of the crystal structure of $Na_2Mn_2Si_2O_7$ by the superposition method. Soviet Phys.-Cryst. 14, 1 (1969).
BARTON, R., JR.: Refinement of the crystal structure of buergerite and the absolute orientation of tourmalines. Acta Cryst. B 25, 1524 (1969).
BAUR, W. H.: On the cation and water positions in faujasite. Am. Mineralogist 49, 697 (1964).
BEGER, R. M.: The crystal structure and chemical composition of pollucite. Z. Krist. 129, 280 (1969).
BORISOV, S. V., KLEVTSOVA, R. F., BELOV, N. V.: Crystal structure of uklonskovite $NaMg[SO_4](OH) \cdot 2H_2O$. Dokl. Akad. Nauk SSSR 158, 116 (1964).
BRUNTON, G.: Refinement of the structure of $NaBF_4$. Acta Cryst. B 24, 1703 (1968).
BURNHAM, C. W., CLARK, J. R., PAPIKE, J. J., PREWITT, C. T.: A proposed crystallographic nomenclature for clinopyroxene structures. Z. Krist. 125, 109 (1967).
BURZLAFF, H.: Die Struktur des Heidornit $Ca_5Na_2Cl(SO_4)_2B_5O_8(OH)_2$. Neues Jahrb. Mineral. Monatsh. 157 (1967).
CANNILLO, E., GIUSEPPETTI, G., TAZZOLI, V.: The crystal structure of leucophanite. Acta Cryst. 23, 255 (1967).
— MAZZI, F., ROSSI, G.: The crystal structure of neptunite. Acta Cryst. 21, 200 (1966).
— ROSSI, G., UNGARETTI, L.: The crystal structure of delhayelite. Rend. Soc. Ital. Miner. e Petrogr. 26, 63 (1970).
CHIN-KHAN, M. A., SIMONOV, V. I., BELOV, N. V.: The crystal structure of ramsayite $Na_2Ti_2Si_2O_9 \equiv Na_2Ti_2O_3[Si_2O_6]$. Dokl. Akad. Nauk SSSR 186, 820 (1969).
CIALDI, G., CORAZZA, E., SABELLI, C.: La struttura cristallina della kernite $Na_2B_4O_6(OH)_2 \cdot 3H_2O$. Rend. Accad. Naz. Lincei 42, 236 (1967).
CLARK, J. R., APPLEMAN, D. E.: Pentaborate polyanion in the crystal structure of ulexite, $NaCaB_5O_6(OH)_6 \cdot 5H_2O$. Science 145, 1295 (1964).
— — PAPIKE, J. J.: Crystal-chemical characterization of clinopyroxenes based on eight new structure refinements. Mineral. Soc. Amer. Spec. Pap. 2, 31 (1969).
— PAPIKE, J. J.: Crystal-chemical characterization of omphacites. Am. Mineralogist 53, 840 (1968).
COCCO, G.: Calcoli strutturistici con elaboratore elettronico. Appendice: Applicazione all'affinamento della struttura della mirabilite; Acc. Tosc. di Scienze e Lettere "La Colombaria" (1962).
— CASTIGLIONE, P. C., VAGLIASINDI, G.: The crystal structure of thomsenolite. Acta Cryst. 23, 162 (1967).
— CORAZZA, E., SABELLI, C.: The crystal structure of glauberite, $CaNa_2(SO_4)_2$. Z. Krist. 122, 175 (1965).
CORAZZA, E., SABELLI, C.: The crystal structure of pirssonite, $CaNa_2(CO_3)_2 \cdot 2H_2O$. Acta Cryst. 23, 763 (1967).
— — GIUSEPPETTI, G.: The crystal structure of lecontite, $NaNH_4SO_4 \cdot 2H_2O$. Acta Cryst. 22, 683 (1967).
CROMER, D. T., KAY, M. I., LARSON, A. C.: Refinement of the alum structures. II. X-ray and neutron diffraction of $NaAl(SO_4)_2 \cdot 12H_2O$, γ alum. Acta Cryst. 22, 182 (1967).
DAL NEGRO, A., ROSSI, G., UNGARETTI, L.: The crystal structure of meliphanite. Acta Cryst. 23, 260 (1967).

© Springer-Verlag Berlin · Heidelberg 1972

Sodium

DANØ, M.: The crystal structure of tugtupite — a new mineral, $Na_8Al_2Be_2Si_8O_{24}(Cl, S)_2$. Acta Cryst. 20, 812 (1966).

EUGSTER, H. P., YODER, H. S.: Carnegie Inst. Wash. Year Book No. 54, 123 (1955).

FANFANI, L., NUNZI, A., ZANAZZI, P. F.: The crystal structure of wardite. Mineral. Mag. 37, 598 (1970).

FISCHER, W.: Über die Struktur des Vanthoffits. Acta Cryst. 17, 1613 (1964).

FRUEH, A. J., JR., GOLIGHTLY, J. P.: The crystal structure of dawsonite $NaAl(CO_3)(OH)_2$. Can. Mineralogist 9, 51 (1967).

GHOSE, S.: A scheme of cation distribution in the amphiboles. Mineral. Mag. 35, 46 (1965).

GOTTARDI, G., MEIER, W. M.: The crystal structure of dachiardite. Z. Krist. 119, 53 (1963).

HYMAN, A., PERLOFF, A., MAUER, F., BLOCK, S.: The crystal structure of sodium tetraborate. Acta Cryst. 22, 815 (1967).

IBERS, A., SMITH, G. W.: Crystal structure of a sodium cobalt molybdate. Acta Cryst. 17, 190 (1964).

JARCHOW, O.: Atomanordnung und Strukturverfeinerung von Cancrinit. Z. Krist. 122, 407 (1965).

JOST, K. H.: Die Struktur des Polyphosphates $[Na_3H(PO_3)_4]_x$. Acta Cryst. B 24, 992 (1968).

— HILMER, W.: Die Struktur von $Na_2H_2SiO_4 \cdot 4H_2O$. Acta Cryst. 21, 583 (1966).

KNOWLES, C. R., RINALDI, F. F., SMITH, J. V.: Refinement of the crystal structure of analcime. Indian Mineralogist 6, 127 (1965).

KRAVCHENKO, V. B.: The crystal structure of searlesite, $NaBSi_2O_5(OH)_2$. Soviet Phys.- Cryst. 9, 143 (1964).

LÖNS, J., SCHULZ, H.: Strukturverfeinerung von Sodalit, $Na_8Si_6Al_6O_{24}Cl_2$. Acta Cryst. 23, 434 (1967).

MAREZIO, M., PLETTINGER, H. A., ZACHARIASEN, W. H.: The bond lengths in the sodium metaborate structure. Acta Cryst. 16, 594 (1963).

MCADAM, A., JOST, K. H., BEAGLEY, B.: Refinement of the structure of sodium Kurrol salt $(NaPO_3)_x$, type A. Acta Cryst. B 24, 1621 (1968).

MCDONALD, R. R., LARSON, A. C., CROMER, D. T.: The crystal structure of sodium pentafluorodistannate (II), $NaSn_2F_5$. Acta Cryst. 17, 1104 (1964).

MCDONALD, W. S., CRUICKSHANK, D. W. J.: A reinvestigation of the structure of sodium metasilicate, Na_2SiO_3. Acta Cryst. 22, 37 (1967).

MEGAW, H. D., KEMPSTER, C. J. E., RADOSLOVICH, E. W.: The structure of anorthite, $CaAl_2Si_2O_8$. II. Description and discussion. Acta Cryst. 15, 1017 (1962).

MENCHETTI, S.: The crystal structure of gaylussite. Rend. Accad. Naz. Lincei, Ser. VIII, 44, 183 (1968).

ONDIK, H. M.: The structure of the triclinic form of sodium tetrametaphosphate tetrahydrate. Acta Cryst. 17, 1139 (1964).

— The structure of anhydrous sodium trimetaphosphate, $Na_3P_3O_9$, and the monohydrate, $Na_3P_3O_9 \cdot H_2O$. Acta Cryst. 18, 226 (1965).

PANT, A. K.: A reconsideration of the crystal structure of β-$Na_2Si_2O_5$. Acta Cryst. B 24, 1077 (1968).

— CRUICKSHANK, D. W. J.: The crystal structure of α-$Na_2Si_2O_5$. Acta Cryst. B 24, 13 (1968).

PAPIKE, J. J., CLARK, J. R.: The crystal structure and cation distribution of glaucophane. Am. Mineralogist 53, 1156 (1968).

— ROSS, M.: Gedrites, crystal structures and intercrystalline cation distributions. Am. Mineralogist 55, 304 (1970).

— ZOLTAI, T.: The crystal structure of a marialite scapolite. Am. Mineralogist 50, 641 (1965).

PEACOR, D. R., BUERGER, M. J.: The determination of the structure of narsarsukite, $Na_2TiOSi_4O_{10}$. Am. Mineralogist 47, 539 (1962).

PREWITT, C. T.: Refinement of the structure of pectolite, $Ca_2NaHSi_3O_9$. Z. Krist. 125, 298 (1967).

— BURNHAM, C. W.: The crystal structure of jadeite $NaAlSi_2O_6$. Am. Mineralogist 51, 956 (1966).

RADOSLOVICH, E. W.: The cell dimensions and symmetry of layer-lattice silicates. V. Composition limits. Am. Mineralogist 48, 348 (1963).

RIBBE, P. H., MEGAW, H. D., TAYLOR, W. H.: The albite structures. Acta Cryst. B 25, 1503 (1969).
ROBINSON, P. D., FANG, J. H.: Crystal structure and mineral chemistry of double-salt hydrates: I. Direct determination of the crystal structure of tamarugite. Am. Mineralogist 54, 19 (1969).
ROMERS, C., ROOYMANS, C. J. M., DE GRAAF, R. A. G.: The preparation, crystal structure and magnetic properties of $Na_3Fe_5O_9$. Acta Cryst. 22, 766 (1967).
RUMANOVA, I. M., KURBANOV, KH. M., BELOV, N. V.: Determination of the crystal structure of probertite, $CaNa[B_5O_7(OH)_4] \cdot 3H_2O$. Soviet Phys.-Cryst. 10, 513 (1966).
SABELLI, C.: La struttura della darapskite. Rend. Accad. Naz. Lincei 42, 874 (1967).
SAKOWSKI-COWLEY, A. C., ŁUKASZEWICZ, K., MEGAW, H. D.: The structure of sodium niobate at room temperature and the problem of reliability in pseudosymmetric structures. Acta Cryst. B 25, 851 (1969).
SCHULZ, H., SAALFELD, H.: Zur Kristallstruktur des Noseans, $Na_8[SO_4(Si_6Al_6O_{24})]$. Tschermaks Mineral. Petrog. Mitt. 10, 225 (1965).
SHARMA, B. H.: Sodium bicarbonate and its hydrogen atom. Acta Cryst. 18, 818 (1965).
SIMONOV, V. I., BELOV, N. V.: Characteristics of the crystal structure of rinkite. Soviet Phys.-Cryst. 12, 740 (1968).
SOLOV'EVA, L. P., BAKAKIN, V. V.: X-ray study of the sodium-iron metagermanate $NaFeGe_2O_6$. Soviet Phys.-Cryst. 12, 517 (1968).
STEINFINK, H.: The crystal structure of the zeolite phillipsite. Acta Cryst. 15, 644 (1962).
SYSIÖ, P. A.: On the additivity of crystal radii in alkali halides. Acta Cryst. B 25, 2374 (1969).
TAGA, T.: Crystal structure of $Na_2CO_3 \cdot 10H_2O$. Acta Cryst. B 25, 2656 (1969).
VAUGHAN, P. A.: The crystal structure of the zeolite ferrierite. Acta Cryst. 21, 983 (1966).
VIJAYAN, M.: The crystal structure of sodium trihydrogen selenite. Acta Cryst. B 24, 1237 (1968).
WILLIAMS, P. P., MEGAW, H. D.: The crystal structure of low and high albites at $-180°C$. Acta Cryst. 17, 882 (1964).
WOODROW, P. J.: The crystal structure of astrophyllite. Acta Cryst. 22, 673 (1967).
ZALKIN, A., FORRESTER, J. D., TEMPLETON, D. H.: The crystal structure of sodium fluosilicate. Acta Cryst. 17, 1408 (1964).

Revised manuscript received: August 1970

Aluminum 13

A	P. B. Moore	(Department of Geophysical Sciences, The University of Chicago, Chicago, Ill., U.S.A.)
B—F	P. M. Bell	(Geophysical Laboratory Carnegie Institution, Washington, D. C., U.S.A.)
G—L	J.D. Hem	(U.S. Geological Survey, Menlo Park, Calif., U.S.A.)
M—O	P. M. Bell	(Geophysical Laboratory Carnegie Institution, Washington, D. C., U.S.A.)

13-A. Crystal Chemistry

I. Crystal Chemical Concepts and Properties of Al^{3+}

The Al^{3+} state ($1s^2 2s^2 2p^6$) is the only stable formal valence state in natural mineralogical systems. Only rarely and inconspicuously does it occur as a chalcogenide, replacing Mg^{2+} and Fe^{3+} in minor amounts in the hydroxide layer in valleriites $^2_\infty[(Mg^{[6]}_{0.68}Al^{[6]}_{0.32}(OH)_2] \cdot [CuFeS_2]$ (IIISHI et al., 1970). The highly soluble and unstable chlorides occur infrequently in fumarolic vents; the fluoroaluminates are built of $[AlF_6]$ octahedral linkages and comprise a substantial group of species, all of which occur in restricted environments, usually in association with pegmatite bodies. It is in the oxygen-bearing systems where Al^{3+} is particularly conspicuous as a cation.

Al^{3+} is known to occur in tetrahedral, distorted trigonal bipyramidal and octahedral oxygen coordination. The principal oxygen-bearing ligands (besides O^{2-}) are $(OH)^-$, $(H_2O)^0$ and O_T^{2-}, where O_T^{2-} is an oxide anion associated with a tetrahedral group such as $[SO_4]^{2-}$, $[PO_4]^{3-}$, $[SiO_4]^{4-}$, etc. The oxygen coordination about Al^{3+} is largely dictated by the geometrical relationship between the crystal radii, because Al^{3+} is predominantly an ionic cation with no strongly directed bonds. According to SMITH and BAILEY (1963), $Al^{[4]}$-O distances range, on the average, from 1.75 Å (for framework structures) to 1.80 Å (for insular tetrahedra). SHANNON and PREWITT (1969) have tabulated the Me—O distance dependence on oxygen coordination, which ranges from an $Al^{[4]}$—$O^{[2]}$ distance of 1.74 Å to an $Al^{[4]}$—$O^{[4]}$ distance of 1.77 Å in tetrahedral coordination, and an $Al^{[6]}$—$O^{[2]}$ distance of 1.88 Å to $Al^{[6]}$—$O^{[6]}$ distance of 1.93 Å in octahedral coordination. Unpublished results by the author reveal that the $[Al^{[6]}O_6^{2-}]$ average octahedral volumes in real crystals are strongly dependent on the number of edges (and faces) shared with other $[AlO_6]$ octahedra, and range from 8.9 Å^3 for five shared edges to approximately 10.0 Å^3 for no shared edges. The values for $[Al(H_2O)_6]$ and $[Al(OH)_6]$ octahedra with no shared edges tend to be lower than the anhydrous counterparts by 6 to 10%. Individual Al—O^{2-}, Al—$(OH)^-$ and Al—$(H_2O)^0$ distances, however, do not reveal averages indicative of systematic trends.

Al^{3+} exhibits possible solid solution with an extensive list of cations. Table 13-A-1 lists cations which, on the basis of their Me—Ø distances (where Me = cation and Ø = anion), should accommodate Al^{3+} in the crystal. Substitution of Al^{3+} for some other cation (Me) is written $Al^{3+} \rightarrow$ Me; substitution of the cation, Me, for Al^{3+} is written Me $\rightarrow Al^{3+}$; and both possibilities, $Al^{3+} \rightleftarrows$ Me. It is always difficult to certify the substitution of Al^{3+} in, for example, the tantalates on the basis of chemical analyses since coupled substitutions may occur, i.e., Fe^{3+} may also be present.

© Springer-Verlag Berlin · Heidelberg 1972

Table 13-A-1. $Al^{3+} \to Me$ and $Me \to Al^{3+}$ solid solutions (frequent examples are italicized)

Cation	Me—O^a distance (Å)	Examples	Comments
$Al^{3+[4]}$—O	1.75		
$Al^{3+[6]}$—O	1.92		
$As^{5+[4]}$—O	1.69	Hematolite?	$Al^{3+} \to As^{5+}$ limited, in arsenates
$B^{3+[4]}$—O	1.47	Kornerupine	$B^{3+} \to (Al^{3+}, Si^{4+})$ limited
$Be^{2+[4]}$—O	1.55	Bavenite	$Be^{2+} \rightleftarrows Al^{3+}$ limited
$Cr^{3+[6]}$—O	1.99	Spinels, garnets, etc.	$Cr^{3+} \rightleftarrows Al^{3+}$ extensive
$Fe^{3+[6]}$—O	2.03	Spinels, many silicates	$Fe^{3+} \rightleftarrows Al^{3+}$ extensive
$Fe^{3+[4]}$—O	1.88	Silicates	$Fe^{3+} \rightleftarrows Al^{3+}$, also involving Si^{4+}
$Ga^{3+[6]}$—O	2.00	Hydroxides	$Ga^{3+} \to Al^{3+}$, trace substituent
$Ge^{4+[6]}$—O	1.90	Silicates, etc.	$Ge^{4+} \to Al^{3+}$, trace substituent
$Li^{1+[6]}$—O	2.14	Lithiophorite, tourmaline	$Li^{1+} \to Al^{3+}$, few established examples
$Mg^{2+[6]}$—O	2.11	Sapphirine, staurolite, spinel, etc.	$Mg^{2+} \rightleftarrows Al^{3+}$, in high T,p assemblages
$Mn^{2+[6]}$—O	2.19	Mn silicates	$Al^{3+} \to Mn^{2+}$ limited
$Mn^{3+[6]}$—O	1.92	Henritermierite, diaspore, andalusite	$Mn^{3+} \rightleftarrows Al^{3+}$, limited by Jahn-Teller distortion of $Mn^{3+}\emptyset_6$?
$Mn^{4+[6]}$—O	1.93	Psilomelane, "wad"	$Al^{3+} \to Mn^{4+}$
$Mo^{6+[4]}$—O	1.78	Molybdates	$Al^{3+} \to Mo^{6+}$, very limited
$Nb^{5+[6]}$—O	2.02	Niobates, tantalates	$Al^{3+} \to Nb^{5+}$, limited
$Sc^{3+[6]}$—O	2.11	Al phosphates, silicates	$Sc^{3+} \to Al^{3+}$, see Frondel (1970)
$Si^{4+[4]}$—O	1.62	Zeolites, feldspars, sapphirine, etc.	$Si^{4+} \rightleftarrows Al^{3+}$ extensive, varying degrees of order
$Sn^{4+[6]}$—O	2.05	Complex oxides	$Al^{3+} \to Sn^{4+}$ very limited
$Ti^{4+[6]}$—O	1.99	Corundum, titanates, titanosilicates	$Al^{3+} \rightleftarrows Ti^{4+}$ limited
$V^{3+[6]}$—O	2.02	Roscoelite, hydroxides	$V^{3+} \rightleftarrows Al^{3+}$
$W^{6+[6]}$—O	1.94	Tungstates	$Al^{3+} \to W^{6+}$ very limited

^{a} Me—O averages from Shannon and Prewitt (1969).

II. Atomic Arrangements Involving Aluminum (3+)

It is illuminating to point out some of the outstanding properties of Al^{3+} coordinated by ligands in crystals, such as the halides, the oxyanions, the hydroxyl anion, and the water molecule. As seen from Table 13-A-1, Al^{3+} finds friendly consort with many other cations. One particular companion is ferric iron: the Al^{3+} compounds isostructural with Fe^{3+} equivalents are extensive among the sulfates, phosphates and silicates, including cases of isomorphism involving unusual polyhedral clusters. The picture changes when simple compositions are considered such as the aluminum oxides (Al_2O_3) and aluminum silicates (Al_2SiO_5); only infrequently do the ferric counterparts appear. Another remarkable feature of Al^{3+} is its appearance in extraordinarily complex polyhedral clusters, such as those found in sapphirine, $Mg_{3.5}Al_{9.00}Si_{1.5}O_{20}$ and kornerupine, $Mg_3(Al,Fe,Mg)_6[Si_2O_7][(Al, Si)_2SiO_{10}]O_3(O,OH)_2$. Yet another surprising feature is the frequency of (Al,Mg) solid solution among structures involving these cations in octahedral coordination. Along with the aforementioned sapphirine and kornerupine, we may include spinel, staurolite, hydrotalcite, possibly some of the amphiboles and sinhalite. The (Al,Si) *solid solutions* are well-established, particularly among the pyroxenes, amphiboles, sheet silicates, the feldspars and zeolites. Among the zeolites, a seeming contradiction exists. Although low temperature species, in the zeolites the Al^{3+} occurs in tetrahedral coordination, yet in oxide systems involving Al^{3+} a diminution in coordination generally occurs with increase in temperature on account of the necessity for increasing bond strengths to counteract the increase in oxygen thermal vibrations.

It is the plan of this section to tabulate and elaborate on those compounds whose crystal chemistry have been studied in detail. Predilections of the author's tastes color the tabulations. Although issue may be taken for including many close-packed Al-bearing silicates with the close-packed oxide structures, their interrelationships on topological grounds are in clearer light as silicate classification is not free from being arbitrary.

III. Fluoroaluminate Structures

All fluoroaluminate structures are erected from the $[AlF_6]$ octahedral module. Typical Al-F distances range from 1.80 to 1.86 Å, significantly shorter than Al-(OH)⁻ averages, implying possible degrees of order for mixed (OH,F) in crystals. Classification of their structures, on the basis of progressive condensations of the octahedra, was offered by PABST (1950) and Fig. 13-A-1 derives from his study. These compounds are listed in Table 13-A-2.

Insular octahedra occur in the cystal structures of cryolite, thomsenolite, and cryolithionite, the last possessing the garnet structure. Thallium fluoroaluminate,

Table 13-A-2 *Fluoroaluminate structures*

Species	Formula	Me-F average distance (Å)	Link type	Me:F ratio	Reference
Cryolite	$Na_3[AlF_6]$	1.81	Insular octahedron	1:6	Náray Szabó and Sasvári (SB 1938, 120)
Thomsenolite	$Na[Ca(H_2O)][AlF_6]$	1.801	Insular octahedron	1:6	Cocco et al. (1967)
Cryolithionite	$Na_3Li_3[AlF_6]_2$	1.807	Insular octahedron	1:6	Geller (1971)
Thallium fluoroaluminate	$\frac{1}{\infty}Tl_2[AlF_5]$	1.84	Corner-chains	1:5	Brosset (SB 1937, 104)
Chiolite	$\frac{2}{\infty}Na_5[Al_3F_{14}]$	1.85, 1.90	Open sheets	3:14	Brosset (SB 1938, 121)
Weberite	$\frac{3}{\infty}Na_2[MgAlF_7]$	1.83	Mg + Al-F open framework	2:7	Byström (SR 1949, 196)
Ralstonite	$\frac{3}{\infty}Na_x[Mg_xAl_{2-x}F_6]\cdot H_2O$	1.86	Corner-linked framework	1:3	Pabst (SB 1939, 127)
Aluminum fluoride	$\frac{3}{\infty}[AlF_3]$	1.80	Corner-linked framework	1:3	Ketelaar (SB 1933—35, 318)

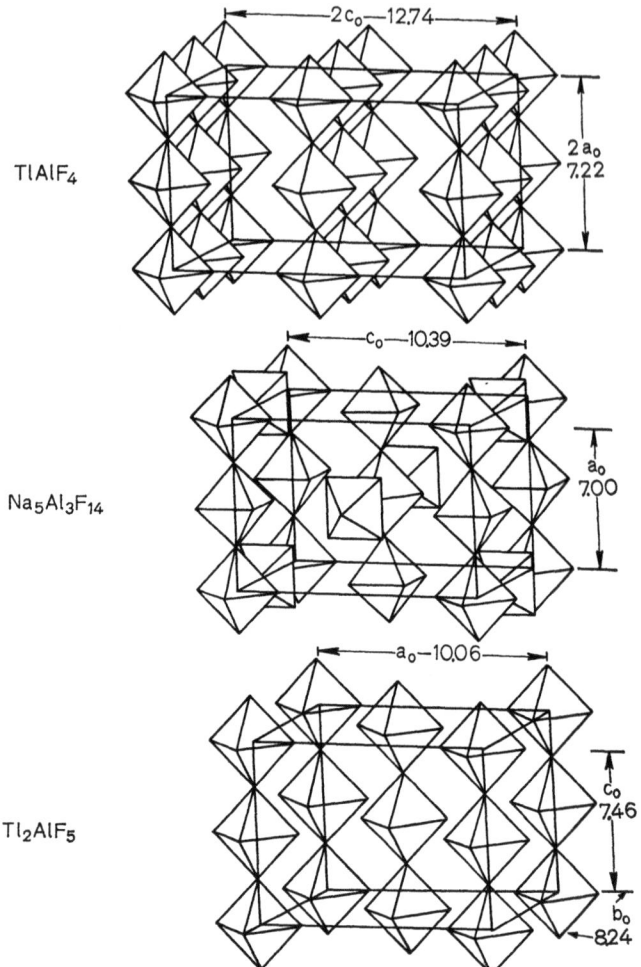

Fig. 13-A-1. Octahedral linkages in some fluoroaluminate structures. The diagrams are from PABST (1950)

$\frac{1}{\infty}Tl_2 [Al^{[6]}F_5]$, is based on corner-sharing octahedral chains (BROSSET, SB 1937, 104). Further condensation into open sheets leads to chiolite and the open framework of weberite, including the Mg^{2+} cation. Ralstonite and aluminum fluoride, $[AlF_3]$ link at all corners to form the infinite $\frac{3}{\infty} [AlF_3]$ framework.

IV. Octahedral Edge-Sharing Sheet Structures

It is appropriate to introduce the problem involving oxygen systems, and examples, with the arrangements derivative of planar $[AlØ_6]$ (where $Ø$ = octahedral vertex) octahedral linkages since many important structures and their derivatives belong here. Fig. 13-A-2 illustrates the most important octahedral sheets. The

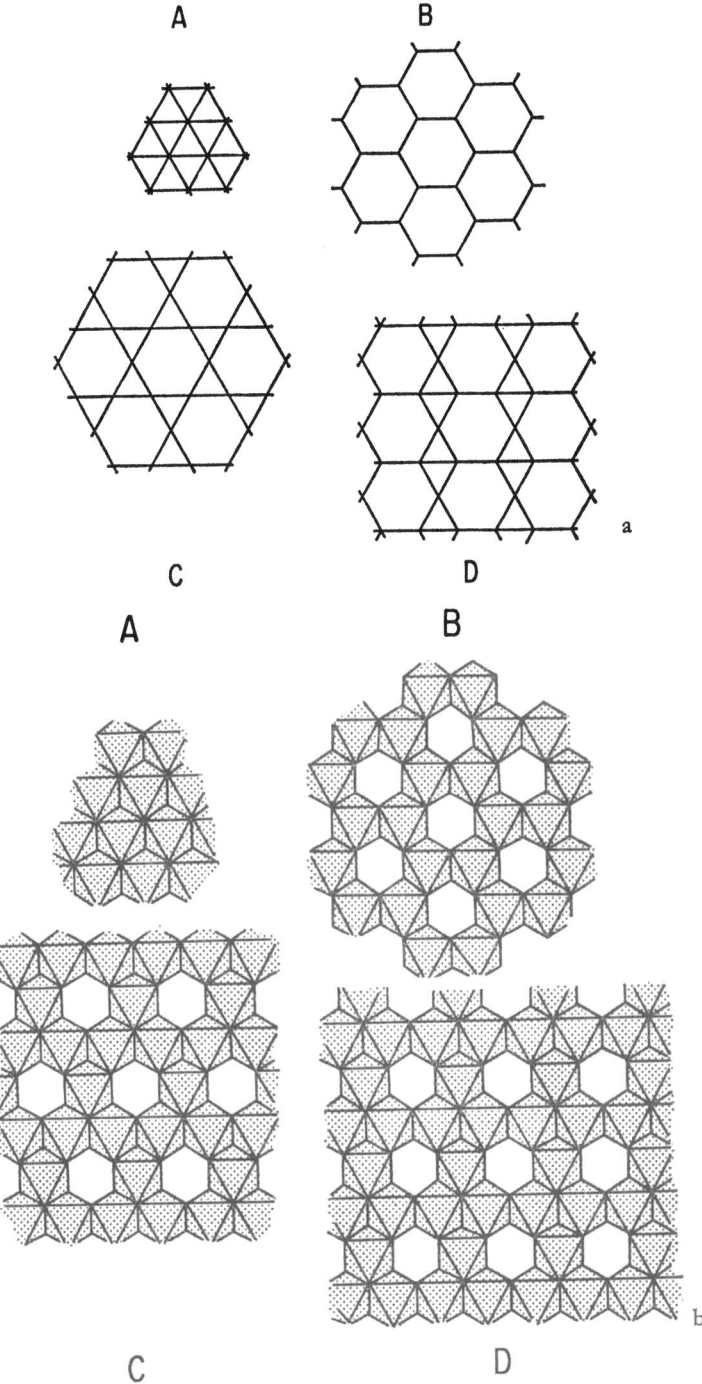

Fig. 13-A-2 a and b. Cation-cation linkages (a), and corresponding octahedral linkages (b) based on the planar sheets of triangular and hexagonal tessellations. A. $_\infty^2[M_3\varnothing_6]$ the triangular net. B. $_\infty^2[M_2\varnothing_6]$, the hexagonal net. C. $_\infty^2[M_3\varnothing_8]$, the mixed triangular-hexagonal net. D. The isomer of C.

Table 13-A-3. Octahedral layer structures

Species	Formula	Me—O distances	Me—O average	Reference
Bayerite	α-Al(OH)$_3$	Al-(OH)—	1.92	Rothbauer et al. (1967)
Gibbsite	γ-Al(OH)$_3$	Al$^{[6]}$(1)-(OH) 1.74, 1.90, 1.90, 1.92, 1.98, 2.06 Al$^{[6]}$(2)-(OH) 1.73, 1.73, 1.85, 1.91, 1.95, 1.98	1.86	Megaw (SB 1933—35, 322)
Nordstrandite	Al(OH)$_3$	Al(1)-, Al(2)-(OH) 1.80, 1.87, 1.90, 1.94, 1.97, 2.14	1.94 1.90	Saalfeld and Jarchow (1968)
Hydrotalcite	[Mg$_3$Al)$_8^{[6]}$(OH)$_{16}$]·CO$_3$·4H$_2$O	(Mg$_{0.67}$, Al$_{0.33}$)—(OH) 2.03 ($\times$6)	2.03	Allmann and Jepsen (1969)
Chloritoid	[(Fe^{2+},Mg)$_2$AlO$_2$(OH)$_4$]Si$_2$[Al$_3$O$_8$]			Harrison and Brindley (SR 1957, 463)

triangular net (Fig. 13-A-2a) is common among the sheet silicates, such as the trioctahedral micas, with average sheet stoichiometry $\overset{2}{\infty}$ [Me$_2^{2+}$Al^{3+}(OH)$_2$(O$_T$)$_4$].

The hexagonal net is the basis for the [Al$_2^{[6]}$(OH)$_4$O$_2$]-sheet in the dioctahedral micas and the $\overset{2}{\infty}$[Al(OH)$_3$] structures. The Kagomé net, with stoichiometry $\overset{2}{\infty}$[Al$_3$Ø$_8$] occurs parallel to {111} in the spinel structure. These sheets interpenetrate to form a three-dimensional edifice. In chloritoid, (Fe^{2+}, Mg)$_2$AlO$_2$(OH)$_4$Si$_2$ [Al$_3$O$_8$], the sheet occurs "isolated" and dictates, by simple electrostatic valence balances, how the insular [SiO$_4$] tetrahedra are disposed.

Table 13-A-3 lists the layer structures, most of which are hydroxides. Bayerite, gibbsite and nordstrandite are distinguished on the basis of the stackings of successive sheets of planar hexagonal nets. Lithiophorite consists of the triangular net and evidently involves the unusual Al^{3+} ⇌ Li^{1+} substitution. Hydrotalcite consists of a similar net, with random distribution of Mg^{2+} and Al^{3+} cations in the sheets.

V. Close-packed Oxygen Structures

Several important structure types involving complex [AlØ$_6$] octahedral and [AlØ$_4$] tetrahedral arrays are based on the dense packing of oxygen atoms. A list of important structures investigated in detail appears in Table 13-A-4. The corundum, α-Al$_2$O$_3$ structure, is based on the hexagonal close-packed (h.c.p.) stacking of sheets of the planar hexagonal net and is the only polymorph reported in nature; beause two-thirds of the octahedral voids are populated by Al^{3+}, [AlØ$_6$] octahedral face-sharing occurs. At least nine "polymorphs" of Al$_2$O$_3$ have been reported (see Table 13-A-5), but the synthesis of several depends on the nature of the starting materials, and their stability relationships are poorly

Table 13-A.4. *Close-packed oxygen structures*

Species	Formula	Me—O distances		Me—O average (Å)	Reference
Corundum	$\alpha\text{-}Al_2O_3$	$Al^{[6]}$—O	$1.86\,(\times 3),\ 1.97\,(\times 3)$	1.91	Newnham and DeHaan (1962)
Spinels	$Mg^{[4]}[Al_2^{[6]}O_4]$	$Al^{[6]}$—O	$1.929\,(\times 6)$	1.929	Fischer (1967)
Sapphirine	$(Al,Si)_3^{[4]}[(Al,Mg)_8^{[6]}O_{10}]$	$Al^{[6]}(1)$—O	$1.850,\ 1.915,\ 1.916,\ 1.920,\ 1.977,\ 1.977$	1.926	Moore (1969a)
		$Al^{[6]}(2)$—O	$1.842,\ 1.891,\ 1.918,\ 1.947,\ 1.975,\ 2.007$	1.930	
		$(Al_{0.5}Mg_{0.5})^{[6]}(3)$—O	$1.935,\ 1.964,\ 1.973,\ 1.986,\ 2.012,\ 2.057$	1.988	
		$Al^{[6]}(7)$—O	$1.863,\ 1.888,\ 1.889,\ 1.915,\ 1.956,\ 2.017$	1.921	
		$Al^{[6]}(8)$—O	$1.899,\ 1.913,\ 1.914,\ 1.924,\ 1.956,\ 1.973$	1.930	
		$Al^{[4]}(1)$—O	$1.738,\ 1.747,\ 1.790,\ 1.809$	1.771	
		$(Al_{0.25}Si_{0.75})^{[4]}(2)$—O	$1.617,\ 1.653,\ 1.665,\ 1.696$	1.658	
		$(Al_{0.5}Si_{0.5})^{[4]}(3)$—O	$1.687,\ 1.701,\ 1.706,\ 1.706$	1.700	
		$(Al_{0.75}Si_{0.25})^{[4]}(4)$—O	$1.663,\ 1.739,\ 1.752,\ 1.777$	1.733	
		$Al^{[4]}(5)$—O	$1.703,\ 1.759,\ 1.760,\ 1.797$	1.755	
		$(Al_{0.8}Si_{0.2})^{[4]}(6)$—O	$1.724,\ 1.736,\ 1.738,\ 1.744$	1.736	
Kornerupine	$(Si,Al)_5^{[4]}[(Al,Mg)_9^{[6]}(O,OH)_{22}]$	$Al^{[6]}(2)$—O	$1.80(\times 2),\ 1.89(\times 2),\ 2.11(\times 2)$	1.91	Moore and Bennett (1968)
		$(Al,Mg)^{[6]}(3)$—O	$1.84,\ 1.93,\ 1.93,\ 1.97,\ 2.09(\times 2)$	1.97	
		$Al^{[6]}(4)$—O	$1.80(\times 2),\ 1.95(\times 2),\ 1.98(\times 2)$	1.93	
		$(Al,Si)^{[4]}$—O	$1.61,\ 1.70,\ 1.71(\times 2)$	1.68	
Staurolite	$(Fe,Al)_2^{[4]}[(Si,Al)_4^{[4]}H_2[(Al,Mg)_9^{[6]}O_{24}]$	$Al^{[6]}(1a)$—O	$1.896(\times 2),\ 1.900(\times 2),\ 1.938(\times 2)$	1.91	Smith (1968)
		$Al^{[6]}(1b)$—O	$1.897(\times 2),\ 1.903(\times 2),\ 1.942(\times 2)$	1.91	
		$Al^{[6]}(2)$—O	$1.867,\ 1.872,\ 1.920,\ 1.921,\ 1.921,\ 1.930$	1.91	
		$(Al,Mg)^{[6]}(3a)$—O	$1.844(\times 2),\ 2.037(\times 4)$	1.97	
		$(Al,Mg)^{[6]}(3b)$—O	$1.855(\times 2),\ 2.060(\times 4)$	1.98	
Hibonite (synthetic)	$CaAl_2^{[4]}Al^{[5]}[Al_9^{[6]}O_{19}]$	$Al^{[6]}(1)$—O	$1.818(\times 2),\ 1.860,\ 1.970,\ 1.995(\times 2)$	1.91	Kato and Saalfeld (1968)
		$Al^{[6]}(3)$—O	$1.865(\times 3),\ 1.896(\times 3)$	1.88	
		$Al^{[6]}(4)$—O	$1.883(\times 6)$	1.88	
		$Al^{[5]}(5)$—O	$1.839(\times 3),\ 2.187(\times 2)$	disordered	
		$Al^{[4]}(2)$—O	$1.801(\times 3),\ 1.828(\times 1)$	1.81	

Aluminum

Table 13-A-4 (continued)

Species	Formula	Me—O distances	Me—O average (Å)	Reference
Kyanite	$Si^{[4]}[Al_2^{[6]}O_5]$	$Al^{[6]}(1)$—O 1.843, 1.853, 1.885, 1.892, 1.979, 1.992 $Al^{[6]}(2)$—O 1.894, 1.894, 1.915, 1.930, 1.936, 1.939 $Al^{[6]}(3)$—O 1.860, 1.891, 1.894, 1.920, 1.959, 1.971 $Al^{[6]}(4)$—O 1.814, 1.843, 1.883, 1.907, 1.939, 1.998	1.907 1.917 1.916 1.897	Burnham (1963a)
Chrysoberyl	$Be^{[4]}[Al_2^{[6]}O_4]$	$Al^{[6]}(1)$—O 1.861(×2), 1.892(×2), 1.917(×2) $Al^{[6]}(2)$—O 1.862, 1.895(×2), 1.940, 2.017(×2)	1.890 1.938	Farrell, et al. (1963)
Sinhalite	$B^{[4]}[Mg^{[6]}Al^{[6]}O_4]$	$Al^{[6]}$—O 1.850(×2), 1.880(×2), 1.967(×2)	1.899	Fang and Newnham (1965)
Diaspore	α-$Al^{[6]}O(OH)$	$Al^{[6]}$—O, (OH) 1.851(×2), 1.858, 1.975(×2), 1.980	1.915	Busing and Levy (SR 1958, 276)
Böhmite	γ-$Al^{[6]}O(OH)$	$Al^{[6]}$—O, (OH) 1.87(×2), 1.93(×2), 2.06(×2)	1.95	Reichertz and Yost (SR 1945—46, 99)
Topaz	$Si^{[4]}[Al_2^{[6]}F_2O_4]$	$Al^{[6]}$—{O 1.884, 1.896, 1.897, 1.901 F 1.790, 1.800}	1.895 1.795	Ribbe and Gibbs (1971)

Table 13-A-5. Outline of the alumina polymorphs

Polymorph		Reference
α (alpha)	α-Al(OH)$_3$, β-Al(OH)$_3$, > 1,000° C in air	Corundum (stable natural polymorph), $R\bar{3}c$, $a_h = 4.77$, $c_h = 13.04$ Å	Stumpf et al. (1950), Newnham and DeHaan (1962)
β (beta)	Corrosion zone in clinkers	$CaAl_{12}O_{19}$, isotypic to magneto-plumbite, $P6_3/m$ mc $a = 5.564$, $c = 21.892$ Å	Kato and Saalfeld (1968)
γ (gamma)	α-Al(OH)$_3$, in air or steam between 600 and 800° C.	Tetragonal?, similarities to η-Al$_2$O$_3$	Stumpf et al. (1950)
δ (delta)	α-Al(OH)$_3$, 1,000° C in steam	Tetragonal, $a = 7.94$, $c = 23.50$ Å	Stumpf et al. (1950), Lippens and DeBoer (1964)
η (eta)	β-Al(OH)$_3$, 750° C in air	Cubic, $a = 7.94$ Å, related to spinel	Stumpf et al. (1950)
θ (theta)	α-Al-(OH)$_3$, β-Al$^-$(OH)$_3$ at 1,150° C in air	Isotypic to β-Ga$_2$O$_3$. $A2/m$, $a = 5.72$, $b = 2.90$, $c = 11.8$ Å, $\beta = 104.5°$	Stumpf et al. (1950)
χ (chi)	Bayer process hydrate, 600° C in air; gibbsite at 470° to 800° C	Hexagonal, $a = 5.57$, $c = 8.64$ Å	Geller (SR 1960, 319) Stumpf et al. (1950), Brindley and Choe (SR 1961, 347)
ι (iota)	Rapid quenching of cryolite-alumina melt	Related to mullite	Foster (1959)
κ (kappa)	Gibbsite at 970—1,180° C (+ corundum)	Large cell. hexagonal, $a = 16.7$ Å	Brindley and Choe (SR 1961, 347)

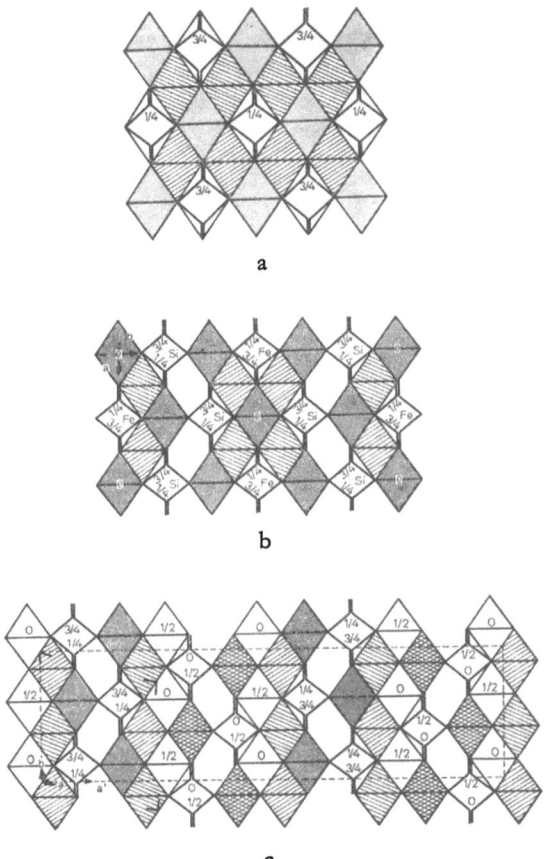

Fig. 13-A-3 a—c. Octahedral and tetrahedral linkages in spinel down [110] (a), kyanite (b) and staurolite (c). The Si tetrahedra are represented by spokes. Single octahedra at 1/4 are ruled NE-SW while those at 3/4 are NW-SE. Octahedral columns at 0 and 1/2 are stippled. The diagrams are from MOORE and SMITH (1970)

known. θ-Al$_2$O$_3$ is isomorphous to β-Ga$_2$O$_3$, therefore its formula should be written Al$^{[6]}$[Al$^{[4]}$O$_3$]. "β-Al$_2$O$_3$" has the crystal-chemical formula CaAl$_2^{[4]}$Al$^{[5]}$Al$_9^{[6]}$O$_{19}$ (hibonite).

The spinel structure type, Mg$^{[4]}$Al$_2^{[6]}$O$_4$, is among the most important in solid state science. As stated before, the structure is built of $_\infty^2$[Al$_8$O$_8$] Kagomé nets, four of which interpenetrate to form a three-dimensional structure. An alternative projection down [110], shown in Fig. 13-A-3, is the key to structural relationships with staurolite and kyanite (MOORE and SMITH, 1970). HAFNER (1962) has shown that only about 80% of the Al resides in the $16d$ site with $\bar{3}m$ point symmetry. On the basis of polyhedral volume calculations, the author suggests that limited (Al, Mg) solid solution occurs. The γ-Al$_2$O$_3$, better written $\square$ Al$_8$O$_{12}$, where $\square$ is a hole, is based on the spinel structure type.

Sapphirine is an extraordinarily complex structure and is based on cubic close packed (c.c.p.) oxygen atoms. It is built of [(Al, Si)$_6$O$_{18}$] pyroxene-like chains with

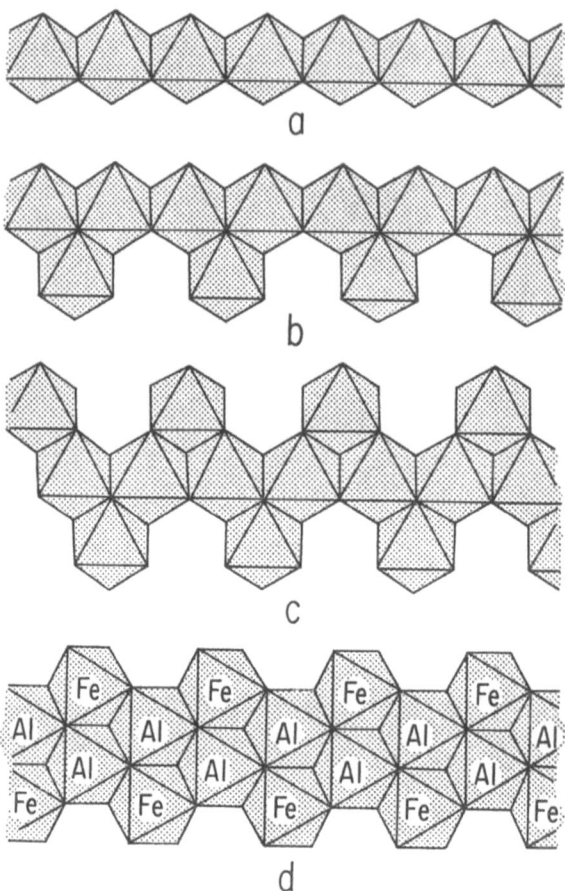

Fig. 13-A-4 a—g. Idealized octahedral chains and walls. a. The chain found in andalusite, sillimanite, etc. b. One of the chains found in clinozoisite. c. The serrated chain which is the basis for the olivine (chrysoberyl) structure. d. One of the chains found in ferrokarpholite. The Al distributions correspond to the M(1) positions in the clinopyroxene structures. e. The amphibole octahedral wall. f. The wall of Al-Ø octahedra found in kornerupine. g. The wall of Al-Ø octahedra found in sapphirine

"wings" and dense Al—O octahedral walls. These walls are featured in Fig. 13-A-4g. Table 13-A-4 shows that the interatomic Me—O averages for the eight non-equivalent octahedral and six non-equivalent tetrahedral sites reveal a range of (Al, Mg) solid solutions for the former and a range of (Al, Si) solid solutions for the latter. Similar solid solutions are revealed in kornerupine, a structure based on interrupted cubic close packed oxygen atoms. Kornerupine is constructed from $[T_3O_{10}]$ trimers and $[T_2O_7]$ dimers of tetrahedra, with T=(Si, Al), and (Al, Mg)-Ø octahedral walls (Fig. 13-A-4f). Staurolite reveals similar crystal-chemical phenomena, and Fig. 13-A-3 shows that it is based on bands of spinel-like octahedral occupancies. The staurolite data in Table 13-A-4 afford Me—O distances indicative of (Al, Mg) solution. Kyanite is a cubic close packed oxygen structure, whose relationship to spinel and staurolite can be seen in Fig. 13-A-3.

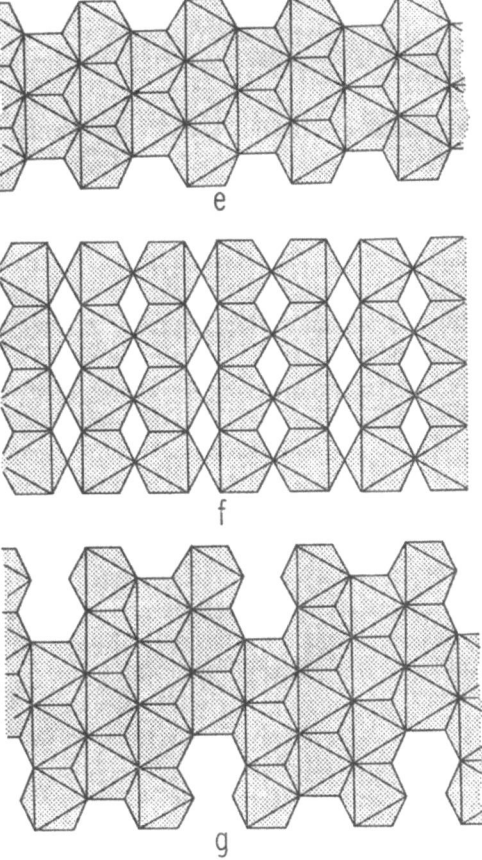

Fig. 13-A-4 e—g

Several structures occur which are based on hexagonal close packed oxygen arrays, including diaspore, α-Al$^{[6]}$O(OH), which is composed of bands of edge-sharing double-octahedra. Chrysoberyl possesses an octahedral, serrated chain of the olivine structure type, which appears in other Al^{3+}-bearing structures as well, including clinozoisite. This chain appears in Fig. 13-A-4c. The structurally related sinhalite may possess limited (Al, Mg) solid solution, according to the author's volume calculations.

Despite the fact that the atomic arrangement is not based on close-packed anions, the garnet, $X_3^{[8]}M_2^{[6]}[T^{[4]}O_4]_3$, structure conserves space efficiently as a consequence of extensive edge-sharing between the distorted cube (X) and the octahedron (M) and tetrahedron (T). The distorted cube shares two edges with the tetrahedra and six edges with the octahedra. A large number of compounds crystallize with the garnet structure. Refinement of pyrope by GIBBS and SMITH (1965) gave an Al$^{[6]}$-O distance of 1.886 Å and of grossularite by ABRAHAMS and GELLER (SR 1958, 500) yielded an Al$^{[6]}$-O distance of 1.945 Å, revealing large differences for Al—O distances as a result of compensatory effects of differing crystal radii between the two species.

Aluminum

Tetrahedrally coordinated $Al^{[4]}$ is relatively rare among natural garnets but many synthetic examples are known, such as the yttrium aluminates. The rare kimzeyite, $Ca_3^{[8]}Zr_2^{[6]}[AlO_4]_2[SiO_4]$, evidently possesses Al^{3+} in tetrahedral coordination.

VI. Anisodesmic Oxysalt $[AlØ_6]$ Octahedral Clusters

A remarkable list of compounds involves $[AlØ_6]$ octahedra which join to form finite dense clusters or infinite corner-sharing chains (Table 13-A-6). The role of H_2O molecules in low temperature systems is not simple; not only does it participate as a ligand, bonded directly to the Al^{3+} cation, but it may also be present solely as a hydrogen bonded system in the same crystal, or as weakly bonded "zeolitic" water.

$[Al(H_2O)_6]^{3+}$ insular octahedral groups occur in the crystal structures of tamarugite and Na-alum. The $Al-(H_2O)$ average distance (1.88 Å) is significantly shorter than typical $Al-O^{2-}$ distance averages (1.92 Å). In ettringite $Ca_6[Al(OH)_6]_2[SO_4]_3 \cdot 26\,H_2O$, the $[Al(OH)_6]$ insular octahedra bond to the weakly held Ca^{2+} cation. Setting O_P = oxygen associated with the $[PO_4]^{3-}$ ligand, the $[cis\text{-}Al(H_2O)_2(O_P)_4]$ insular octahedron occurs in clinovariscite and variscite (BORENSZTAJN, 1965). Throughout, "insularity" is defined as an octahedral cluster which does not bond to the same cation, but which may bond through ligands to other cations of significantly lower bond strength or ligand groups, such as $[PO_4]^{3-}$. Because of the condition of violent cation-cation repulsion effects across shared polyhedral elements other than corners, $[AlØ_6]$ octahedra are not observed to share edges with polyhedra of much greater bond strength such as $[PO_4]^{3-}$, $[SiO_4]^{4-}$, etc.

The presence of $(OH)^-$ groups and O_T anions allows for further condensation of the $[AlØ_6]$ octahedra. A large family of structures occurs based on the infinite corner-sharing $[AlØ_5]_\infty$ octahedral chains with the O_P and O_S (with the $[SO_4]^{2-}$ group) tetrahedra sharing corners like rungs in a ladder. Such an arrangement is shown in Fig. 13-A-5a. MOORE (1970a) showed that the varieties of structure can

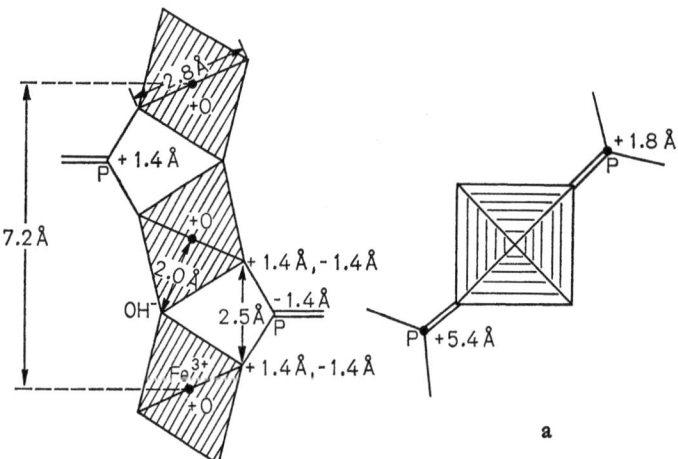

Fig. 13-A-5a—c. Some octahedral anisodesmic oxysalt clusters. a. The cluster of $[MeØ_5]_\infty$ corner-chains and associated tetrahedra in wavellite, eosphorite, etc. The diagram is from MOORE (1970a). b. The octahedral Al-Mg-Al face-sharing trimer found in lazulite. The diagram is from MOORE (1970b). c. The octahedral cluster found in zunyite. The diagram is from KAMB (1960)

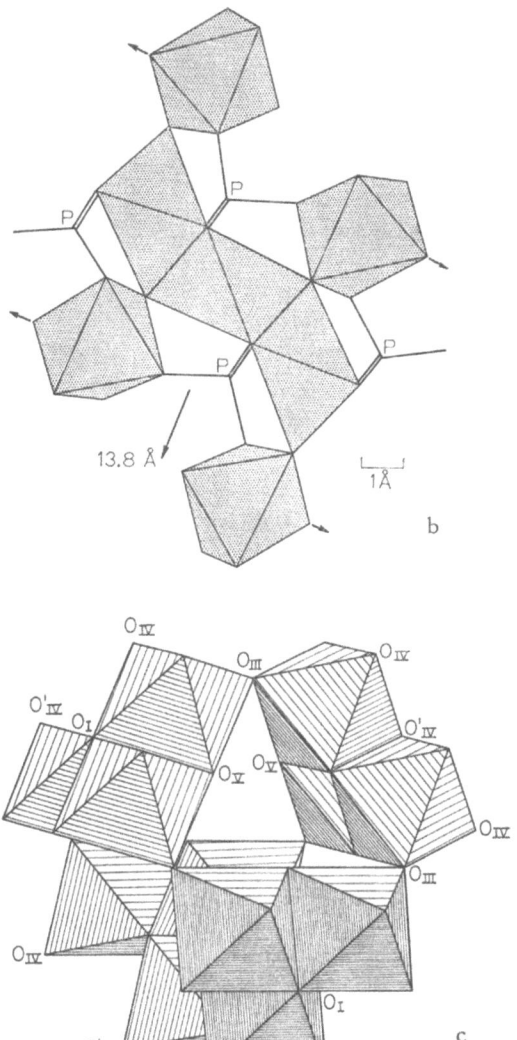

Fig. 13-A-5 b—c

be generated by ligand stereoisomerism, that is, by rearranging the (H$_2$O) and [TO$_4$] ligands around the octahedral backbone. Included here are wavellite, amblygonite, eosphorite, and fluellite.

The peculiar [Al$^{[6]}$Mg$^{[6]}$Al$^{[6]}$Ø$_{12}$] octahedral face-sharing trimer is the basis of the crystal structure of lazulite. This cluster (Fig. 13-A-5b) was investigated in detail by MOORE (1970b) who showed that it appears in several compounds involving Fe^{2+}, Fe^{3+}, Mg^{2+} and Al^{3+}. A highly condensed [Al$^{[6]}_{12}$O$_{16}$(OH)$_{30}$] cluster (Fig. 13-A-5c) occurs in the silicate mineral zunyite. It is placed with the anisodesmic oxysalts because of its similar crystal-chemical behavior. A curious finite cluster occurs in the augelite structure. It is a tetramer composed of an edge-sharing octahedral

Table 13-A-6. $[Al^{[6]}O_6]$-octahedral clusters: Anisodesmic oxysalts

Species	Formula	Me-O distances		Me-O average distance (Å)	Reference
Insular octahedra ($M:\varphi = 1:6$)					
Tamarugite	$NaAl(H_2O)_6[SO_4]_2$	$Al^{[6]}$—(H_2O)	1.845, 1.878, 1.880, 1.880, 1.897, 1.920	1.883	Robinson and Fang (1969)
Na-alum	$NaAl(H_2O)_6[SO_4]_2 \cdot 6H_2O$	$Al^{[6]}$—(H_2O)	$1.881(\times 6)$	1.881	Cromer et al. (1967)
Ettringite	$Ca_6[Al(OH)_6]_2[SO_4]_3 \cdot 26H_2O$	$Al^{[6]}(1)$—(OH) $Al^{[6]}(2)$—(OH)	$1.92(\times 3)$, $1.92(\times 3)$ $1.82(\times 3)$, $2.00(\times 3)$	1.92 1.91	Moore and Taylor (1970)
Clinovariscite	$Al(H_2O)_2[PO_4]$	$Al^{[6]}$—O, (OH)	1.80—1.94Å		Borensztajn (1965)
Infinite corner-chains ($M:\varphi = 1:5$)					
Wavellite	$Al_3^{[6]}(H_2O)_4(OH)_3[PO_4]_2 \cdot H_2O$	$Al^{[6]}(1)$—O, (OH), (H_2O) $Al^{[6]}(2)$—O, (OH), (H_2O)	1.777, 1.803, 1.834($\times 2$), 1.983($\times 2$) 1.879, 1.883, 1.895, 1.897, 1.927, 1.980	1.869 1.910	Araki and Zoltai (1968)
Eosphorite	$MnAl^{[6]}(OH)_2(H_2O)[PO_4]$	$Al^{[6]}$—O, (OH), (H_2O)	$1.87(\times 2)$, $1.90(\times 2)$, $1.97(\times 2)$	1.91	Hanson (1960)
Amblygonite	$LiAl^{[6]}(OH, F)[PO_4]$	$Al^{[6]}$—O, (OH)	$1.89(\times 2)$, $1.91(\times 2)$, $1.92(\times 2)$	1.91	Baur (SR 1959, 429)
Fluellite	$Al_2^{[6]}(H_2O, OH)_4F_2[PO_4] \cdot 4H_2O$	$Al^{[6]}$—O —F	$1.847(\times 2)$, $1.868(\times 2)$, $1.851(\times 2)$	1.857; 1.851	Guy and Jeffrey (1966)
Condensed face-sharing trimers ($M:\varphi = 3:10$)					
Lazulite	$MgAl_2^{[6]}(OH)_2[PO_4]_2$			1.95	Lindberg and Christ (SR 1959, 434)
Complex finite clusters:					
$Al_{12}^{[6]}O_{16}(OH)_{30} + Al^{[4]}O_4$					
Zunyite	$Al_{12}^{[6]}[Al^{[4]}O_4](OH)_{18}Cl[Si_5^{[4]}O_{16}]$	$Al^{[6]}$—O, (OH) $Al^{[4]}$—O	$1.78, 1.86(\times 2), 1.92(\times 2), 1.93$ $1.80(\times 4)$	1.88 1.80	Kamb (SR 1960, 474)

$Al_2^{[6]}Al_2^{[5]}(O, OH)_{14}$ tetramer			
Augelite	$Al^{[6]}Al^{[5]}(OH)_3[PO_4]$	$Al^{[6]}$—O, (OH) 1.826, ($\times 2$), 1.864($\times 2$), 1.983($\times 2$) $Al^{[5]}$—O, (OH) 1.750, 1.779($\times 2$), 1.798, 2.054	1.891 1.832	ARAKI et al. (1968)
$Al_4^{[6]}(O, OH)_{18}$ edge-sharing tetramer			
Axinite	$Ca_4Fe_2^{[6]}Al_4^{[6]}(OH)_2[B_2^{[4]}Si_8^{[4]}O_{30}]$			ITO et al. (1969)
Open sheets:			
$^2_\infty[Al(OH)_2O_2]$ fused corner-chains			

Wardite	$NaAl_3^{[6]}(H_2O)_2(OH)_4[PO_4]_2$	$Al^{[6]}(1)$—O, (OH), (H$_2$O)
		    1.84, 1.88, 1.88, 1.90, 1.90, 2.01
		$Al^{[6]}(2)$—O, (OH), (H$_2$O)
		    1.89($\times 2$), 1.93($\times 2$), 1.93($\times 2$)
			1.90
			
			1.92
				FANFANI et al. (1970)

$^2_\infty[AlO_4]$ fused corners			
Alunite	$KAl_3^{[6]}(OH)_6[SO_4]_2$	$Al^{[6]}$—O, (OH) 1.864($\times 4$), 1.963($\times 2$)	1.897	WANG et al. (1965)

dimer which further links by edge-sharing to two trigonal bipyramids, resulting in the cluster composition $[Al_2^{[6]}Al_2^{[5]}(OH)_6(O_P)_8]$ (ARAKI et al., 1968).

Further condensation of chains into corner-sharing sheets results in the structures of wardite and alunite. The sheets in these two structures have compositions $^2_\infty[Al^{[6]}(OH)_2(O_P)_2]$ and $^2_\infty[AlO_4]$ respectively, the latter representing a Kagomé net of corner-sharing octahedra.

VII. Silicate Structures

Complexities arise from the dual role of Al^{3+} in the rockforming silicate minerals: as an octahedrally coordinated cation and also as a tetrahedrally coordinated cation in important structures, such as the amphiboles, pyroxenes and phyllosilicates. Topological relationships among these structures, on the basis of assemblies of "I-beams", combined with permissible rotations of tetrahedral and octahedral components, has been discussed by THOMPSON (1970). Further difficulties are encountered in the nature of Al-Si ordering in specific structures; although the Smith-Bailey curves afford a reasonable estimate of the Al content in the individual tetrahedral site, there does not appear to be any consistency of Al-Si ordering even within one family of structures such as the zeolites. Consequently, the author has included $(Al_{0.x}Si_{1.0-0.x})$ distances in the tables, even though similar ground will be covered in the Si^{4+} chemistry (Section 14-A).

a) Octahedral Edge-sharing Chain Structures

Many structures are based on the octahedral edge-sharing $^1_\infty[AlO_4]$ chain as sketched in Fig. 13-A-4a. Included here are sillimanite, andalusite, and the struc-

Table 13-A-7. Octahedral edge-sharing chain structures

Species	Formula	Me-O distances		Me—O average distance (Å)	Reference
Sillimanite	$Al^{[6]}Al^{[4]}O[SiO_4]$	$Al^{[6]}$—O	$1.861(\times 2), 1.919(\times 2), 1.957(\times 2)$ Å	1.912	BURNHAM (1963b)
		$Al^{[4]}$—O	$1.721, 1.758, 1.800(\times 2)$	1.770	
Andalusite	$Al^{[6]}Al^{[5]}O[SiO_4]$	$Al^{[6]}$—O	$1.829(\times 2), 1.892(\times 2), 2.085(\times 2)$	1.935	BURNHAM and BUERGER (SR 1961, 502)
		$Al^{[5]}$—O	$1.816, 1.818(\times 2), 1.843, 1.886$	1.836	
Yoderite	$Mg_2Al_2^{[6]}Al_4^{[5]}(O,OH)_4[SiO_4]_4$	$(Al, Mg)^{[6]}$—O	$1.942, 1.949, 1.954, 1.955, 1.986, 2.011$	1.966	FLEET and MEGAW (1962)
		$(Al, Mg)^{[5]}$—O	$1.912, 1.928, 1.928, 1.935, 1.973$	1.935	
		$Al^{[5]}$—O	$1.824, 1.860, 1.862, 1.862, 1.944$	1.870	
Grandidierite	$(Mg, Fe)Al_2^{[6]}Al^{[5]}O_2[BO_3][SiO_4]$	$Al^{[6]}(1)$—O	$1.875(\times 2), 1.905(\times 2), 1.910(\times 2)$	1.897	STEPHENSON and MOORE (1968)
		$Al^{[6]}(2)$—O	$1.870(\times 2), 1.883(\times 2), 1.977(\times 2)$	1.910	
		$Al^{[5]}$—O	$1.800, 1.828, 1.828, 1.867(\times 2)$	1.838	
Epidote	$Ca_2Al_2^{[6]}(Fe^{3+}, Al)^{[6]}O(OH)[SiO_4][Si_2O_7]$	See diagram in reference			DOLLASE (1971)
Allanite	$Ca(R.E., Ca)(Al, Fe)Al(Fe, Al)O(OH)[SiO_4][Si_2O_7]^a$	See diagram in reference			DOLLASE (1971)
Hancockite	$Ca(Pb, Sr)(Al, Fe)Al(Fe, Al)O(OH)[SiO_4][Si_2O_7]$	See diagram in reference			DOLLASE (1971)
Zoisite	$Ca_2Al_3^{[6]}O(OH)[SiO_4][Si_2O_7]$	$Al^{[6]}(1,2)$—O	$1.839, 1.849, 1.849, 1.901, 1.927, 1.963$	1.888	DOLLASE (1968)
Clinozoisite	$Ca_2Al_3^{[6]}O(OH)[SiO_4][Si_2O_7]$	$Al^{[6]}(3)$—O	$1.784, 1.821, 1.964(\times 2), 2.133(\times 2)$	1.967	DOLLASE (1968)
		$Al^{[6]}(1)$—O	$1.850(\times 2), 1.930(\times 2), 1.937(\times 2)$	1.906	
		$Al^{[6]}(2)$—O	$1.852(\times 2), 1.859(\times 2), 1.923(\times 2)$	1.878	
Piemontite	$Ca(Ca, Sr)(Al, Mn^{3+})^{[6]}Al^{[6]}(Mn^{3+}, Fe^{3+}, Al)^{[6]}O(OH)[SiO_4][Si_2O_7]$	$Al^{[6]}(1)$—O	$1.872(\times 2), 1.941(\times 2), 1.985(\times 2)$	1.933	DOLLASE (1969)
		$Al^{[6]}(2)$—O	$1.857(\times 2), 1.881(\times 2), 1.934(\times 2)$	1.891	
Pumpellyite	$Ca_2Al_2^{[6]}(Al, Mg, Fe^{2+})^{[6]}(OH)_2(OH, O)[SiO_4][Si_2O_6(OH)]$	$Al^{[6]}$—O	$1.84, 1.87, 1.89, 1.93, 1.94, 2.04$	1.92	ALLMANN and DONNAY (1971)
Ardennite	$Mn_2^{2-}(Mn^{2+}, Ca)_2Al_4^{[6]}(Mg, Al, Fe^{3+})^{[6]}(OH)_6[(As, V)O_4][SiO_4]_2[Si_3O_{10}]$	$Al^{[6]}(1)$—O	$1.853(\times 2), 1.904(\times 2), 2.006(\times 2)$	1.921	DONNAY and ALLMANN (1968)
		$Al^{[6]}(2)$—O	$1.883(\times 2), 1.888(\times 2), 1.934(\times 2)$	1.902	

a: R. E.: Rare earth elements

turally related yoderite, and grandidierite. Al—O distances for these compounds are compiled in Table 13-A-7. DOLLASE (1968, 1969, 1971) has conducted extensive studies on the epidote group and related structures. A good example is the zoisite-clinozoisite dimorphous pair, in which three types of chains are represented (Fig. 13-A-4a, b, c). Elaborations on the $\frac{1}{\infty}[AlØ_4]$ chain appear in the complex structures of pumpellyite and ardennite.

b) Octahedral Wall Structures

The structures of the pyroxene and amphibole groups can be visualized in several ways, the most conventional of which is the articulation of the tetrahedral chains and ribbons. We shall find it more convenient to consider the octahedral walls, since here the role of Al^{3+} is most apparent. Ferrokarpholite, $FeAl_2^{[6]}(OH)_4[Si_2O_6]$ (MACGILLAVRY et al., SR 1956, 406) affords a good introduction, in several respects, to the pyroxene structures. It includes two types of octahedral linkages, one type with zig-zag, insular $\frac{1}{\infty}[AlØ_4]$, edge-sharing chains (the "M(1)" occupied pyroxene polyhedra) and another type, including additional edge-sharing Fe^{2+}—O octahedra (the "M(2)" occupied pyroxene polyhedra), forming a wall, shown in Fig. 13-A-4d. This wall is characteristic for the pyroxenes, although the additional polyhedra are severely distorted and approach 7- and 8-fold coordination in the clinopyroxenes. The silicate chains link above and below by corner-sharing with the central Al-Ø octahedra which, when seen end on, reveal a brickwork-like stacking. In the pyroxene structures, the M(1) polyhedron closely approximates an octahedron, and Al^{3+} preferentially substitutes at this site. The zig-zag edge-sharing chain occurs in the structures of spodumene, omphacite and jadeite.

In the amphiboles, $X^{[10]}Y_2^{[8]}M_5^{[6]}(OH)_2[T_8O_{22}]$, walls of edge-sharing octahedra, alternately two- and three-octahedra thick, appear as featured in Fig. 13-A-4e. The additional $Y^{[8]}Ø_8$ polyhedra ($Y = Na^+$, Ca^{2+}, etc.) approximate square antiprisms and share edges on either side of the wall, condensed in a manner akin to the M(2) polyhedra in the clinopyroxenes. The $\frac{1}{\infty}[T_8O_{22}]$ tetrahedral double band shares corners above and below, with the octahedra in the wall, the central apices of the latter occupied by $(OH)^-$ anions. The wall, its associated $Y^{[8]}$ and $X^{[10]}$ are rarely featured assembled together, but a good example appears in MOORE (1969b). Strong Al^{3+} site preference is apparent in the octahedral wall and is dictated by cation-cation repulsion effects and charge balance. In glaucophane, the Al^{3+} cations are ordered over the M(2) sites which are situated at the outer portion of the wall. In gedrites, Al^{3+} occurs in both octahedral and tetrahedral coordinations. The authors also present a detailed discussion (with diagrams) on amphibole polymorphism, utilizing the combinatorial concepts of THOMPSON (1970). Data on refined octahedral wall structures appear in Table 13-A-8.

c) Sheet Structures

The sheet structures include the micas, talc, clay minerals and chlorites, and the tetrahedral polymerization involves two-dimensional planar sheets of corner-sharing tetrahedra, forming an hexagonal outline. The free vertices of a tetrahedral sheet all point in the same direction. Except in kaolinites, octahedral sheets share corners with these tetrahedral apices and are sandwiched in between. In micas, individual

Table 13-A-8. Silicates: octahedral wall structures

Species	Formula	Me—O distances	Me—O average distance (Å)	Reference
Ferrokarpholite	$FeAl_2^{[6]}(OH)_4[Si_2O_6]$		1.92, 1.93	MacGillavry et al. (SR 1956, 406)
Spodumene	$LiAl^{[6]}[Si_2O_6]$	$Al^{[6]}$—O 1.818(×2), 1.943(×2), 1.997(×2)	1.919	Clark et al. (1969)
Jadeite	$NaAl^{[6]}[Si_2O_6]$	$Al^{[6]}$—O 1.856(×2), 1.933(×2), 1.996(×2)	1.928	Prewitt and Burnham (1966)
Omphacite	$(Na_{0.5}, Ca_{0.5})(Mg_{0.5}Al_{0.5})^{[6]}[Si_2O_6]$	$Al^{[6]}$—O 1.89(×2), 1.90(×2), 1.96(×2) (Calif.) $Al^{[6]}$—O 1.88(×2), 1.91(×2), 2.03(×2) (Venez.)	1.92 1.94	Clark et al. (1969)
Glaucophane	$Na_2Mg_3^{[6]}Al_2^{[6]}(OH)_2[Si_8O_{22}]$	$M(2) = (Al_{0.9}Fe_{0.1})^{[6]}$—O 1.849(×2), 1.943(×2), 2.038(×2)	1.943	Papike and Clark (1968)
Gedrite (sample 002)	ca. $Na_{0.5}(Fe, Mn)_{5.5}^{2+}(Al, Fe)_{1.5}^{3+}(OH)_2[Si_6Al_2O_{22}]$	$T(1A)^{[4]} = (Al_{0.27}Si_{0.73})$—O 1.641, 1.649, 1.653, 1.660	1.651	Papike and Ross (1970)
		$T(1B)^{[4]} = (Al_{0.44}Si_{0.56})$—O 1.666, 1.668, 1.677, 1.679	1.672	
		$T(2B)^{[4]} = (Al_{0.29}Si_{0.71})$—O 1.640, 1.660, 1.679, 1.683	1.666	
		$M(2)^{[6]} = (Al_{0.68}Mg_{0.23}Fe_{0.09})$—O 1.947(×2), 1.993(×2), 2.004(×2)	1.979	

sandwiches are held together electrostatically by large cations of low charge such as K^+, Na^+ and Ca^{2+}.

The Al^{3+} chemistry of the sheet silicates is further complicated by its appearance in both tetrahedral and octahedral coordination. Two kinds of octahedral layers are discerned: the "dioctahedral" sheets $^2_\infty[Me_2\emptyset_6]$, corresponding to the hexagonal net, and the "trioctahedral" sheets $^2_\infty[Me_3\emptyset_6]$, corresponding to the triangular net in Fig. 13-A-2a, b. Dioctahedral micas include margarite and muscovite, which possess $^2_\infty[Al_2^{[6]}O_4(OH)_2]$ sheets, with (Si, Al) solution in the tetrahedral positions. An example of a trioctahedral mica is xanthophyllite, with $^2_\infty[Mg_3AlO_4(OH)_2]$ sheets. $Mg \rightleftarrows Al$ and $Si \rightleftarrows Al$ solutions apparently occur in this structure. Mica polytypes are described by Ross et al. (1966). Experimental and theoretical studies on mica polymorphs are presented by SMITH and YODER (1956).

Chlorites are generated by the stacking of mica sandwiches and geometrically compatible brucite units (triangular nets). A detailed study on a prochlorite (STEINFINK, SR 1958a, 526), reveals complex partial ordering of Al^{3+} cations in both octahedral and tetrahedral layers. In the variety "kämmererite", $(Mg, Cr^{3+})_6(OH)_8$ $[Si_3AlO_{10}]$ (BROWN and BAILEY, 1963), $(Al_{0.4}Si_{0.6})$ and $(Al_{0.1}Si_{0.9})$ occupancies have been suggested. This paper presents an excellent diagram of tetrahedral Al-Si ordering relationships for the sheet structures. The variety "corundophilite" appears to possess a $(Si_{0.67}Al_{0.33})$ tetrahedral distribution (STEINFINK, SR 1958b, 525). Vermiculite is structurally related to the chlorites and appears to possess (Si, Al) disorder.

Detailed structural studies on the clay minerals are hindered by the rarity of suitable single crystals on account of their very soft nature resulting from the lack of strong bonds between the sandwiches. Dickite is composed of $^2_\infty[Al_2(OH)_4O_2]$ hexagonal sheets connected with tetrahedral sheets. BAILEY (1963) presents a general study on the polymorphism among the clay minerals.

Two noteworthy structures, distinct from the sheet silicates above, are included in this section. These are prehnite and the melilites, particularly gehlenite. In prehnite (PAPIKE and ZOLTAI, 1967) the T(2) tetrahedron is disordered, with average (Si, Al) = 1.67 Å. Varying degrees of (Si, Al) ordering apparently account for weak superstructure reflections which require cells of lower symmetry for some prehnites. The gehlenite structure is based on undulating sheets of tetrahedra, one of which has point symmetry $\bar{4}$ and shares all of its corners with other tetrahedra. LOUISNATHAN (1971) has shown that this tetrahedron is occupied by Al and the remaining tetrahedron has average population $(Al_{0.5}Si_{0.5})$.

d) Framework Silicates Containing Al^{3+}

The very important and frequently complicated framework structures are based on the polymerization of $[TO_4]$ tetrahedral units at each corner to form the infinite $^3_\infty[TO_2]$ framework. Such arrangements are difficult to classify, and detailed crystal structure refinements, often utilizing nuclear magnetic resonance techniques as well, are necessary to establish an approximate ordering scheme for the mixed cations over the tetrahedral positions. The introduction of Al^{3+} into tetrahedral positions results in an excess of negative charge which is balanced by the addition of cations incorporated into cavities in the framework, as in $Me^{n+}_{x/n}[Al^{3+}_x Si^{4+}_{1-x}O_2]$, where Me is

Table 13-A-9. Silicates: sheet structures

Species	Formula	Me—O distances	Me—O average distance (Å)	Reference
Xanthophyllite	$CaMg_2Al^{[6]}(OH)_2$ $[(Al_3Si)O_{10}]$	$(Mg, Al)^{[6]}$—O $(Al_{0.75}Si_{0.25})^{[4]}$—O 1.71, 1.71, 1.75, 1.77	1.98 1.74	Takéuchi and Sadanaga (SR 1959, 484)
Muscovite	$KAl_2^{[6]}(OH)_2[(Si_3Al)O_{10}]$	$Al^{[6]}$—O, OH 1.930, 1.932, 1.935, 1.939, 1.944, 2.048	1.955	Radoslovich (SR 1960, 483)
Prochlorite	$(Mg, Fe, Al)_6^{[6]}(OH)_8$ $[(Si, Al)_4O_{10}]$	$(Al_{0.75}Fe_{0.25})^{[6]}(4)$—O $2.04(\times2), 2.07(\times2), 2.17(\times2)$ $(Al_{0.75}Fe_{0.25})^{[6]}(5)$—O $2.04(\times2), 2.14(\times2), 2.17(\times2)$ $(Al_{0.7}Si_{0.3})^{[4]}(1)$—O 1.72, 1.73, 1.73($\times$2) $(Al_{0.2}Si_{0.8})^{[4]}(2)$—O 1.62, 1.63($\times$2), 1.68	2.09 2.12 1.73 1.64	Steinfink (SR 1958, 526)
"Kämmererite"	$(Mg, Cr^{2+})_6^{[6]}(OH)_8$ $[Si_3AlO_{10}]$	$(Al_{0.4}Si_{0.6})^{[4]}$—O 1.65, 1.66, 1.71, 1.72 $(Al_{0.1}Si_{0.9})^{[4]}$—O 1.61, 1.63, 1.64, 1.65	1.69 1.64	Brown and Bailey (1963)
"Corundophilite"	$(Mg, Fe, Al)_6^{[6]}(OH)_8$ $[(Si, Al)_4O_{10}]$	$(Si_{0.67}Al_{0.33})^{[4]}$—O	1.66	Steinfink (SR 1958, 525)
Vermiculite	$(Mg, Fe)_{3.3}^{[6]}(OH)_2(H_2O)_4$ $[(Si, Al)_4O_{10}]$	$(Si, Al)^{[4]}$—O	1.67	Mathieson (SR 1958, 529)
Dickite	$Al_2^{[6]}(OH)_4[Si_2O_5]$	$Al^{[6]}(1)$—O, OH 1.83, 1.86, 1.86, 1.93, 1.93, 2.01 $Al^{[6]}(2)$—O, OH 1.85, 1.86, 1.88, 1.90, 1.94, 1.98	1.90 1.90	Newnham (SR 1961, 504)
Prehnite	$Ca_2Al^{[6]}(OH)_2[Si_3AlO_{10}]$	$Al^{[6]}$—O $1.910(\times2), 1.930(\times4)$ $T(2) = (Si, Al)$—O $1.665(\times2), 1.683(\times2)$	1.923 1.674	Papike and Zoltai (1967)
Gehlenite	$Ca_2[Al^{[4]}(Al, Si)^{[4]}O_7]$	$Al^{[4]}$—O $1.785(\times4)$ $(Al_{0.5}Si_{0.5})^{[4]}$—O $1.681, 1.683(\times2), 1.718$	1.785 1.691	Louisnathan (1971)

typically Na^+, Li^+, K^+, Ca^{2+}, Ba^{2+} etc. Many such structures are stuffed derivatives of simpler arrangements.

A good example of the complexity of the problem includes nepheline, which is a stuffed derivative of the tridymite structure. FOREMAN and PEACOR (1970) refined three-dimensional single crystal data at 10°, 200°, 400°, 600° and 900°C, and concluded that no changes occur in the ordering scheme. They showed that $Si^{[4]}$—O average distances increase more readily than do $Al^{[4]}$—O. Two of the four independent tetrahedral sites are essentially occupied by Al^{3+}. Leucite, with a complicated structure involving 4-, 6-, 8-, and 12-membered rings, possesses $(Si_{0.67}Al_{0.33})$ solution at high temperature. The structurally related analcime often exhibits optical and X-ray deviation from the cubic $Ia3d$ space group, possibly a result of complex (Si, Al) ordering. KNOWLES et. al. (1965) report interatomic T—O averages suggestive of $(Si_{0.67}Al_{0.33})$ disorder of their isotropic crystal from the Cyclopean Islands. Pollucite possesses similar (Si, Al) disorder, and a complex coupling among large cation populations and Si:Al ratios.

The sodalite group of minerals is based on linked tetrahedra forming cages of packed Archimedean truncated octahedra. The large cavities accommodate anion complexes and Na^+ cations. The structure of cancrinite possesses 6-rings and 4-loops of tetrahedra in which the Al^{3+} cations are ordered.

Low cordierite is based on 6-membered rings and bridging tetrahedra. In the "ring", two tetrahedra are occupied essentially by Al^{3+}, and the ordered bridging tetrahedra are Al^{3+} and Si^{4+} in ratio 2:1 (GIBBS, 1966). In high cordierite (indialite), disorder of the (Si, Al) atoms in the ring and in the bridges leads to isotypy with beryl. Osumilite, often confused with cordierite, is based on hexagonal double rings. Between these double rings, which possess disordered (Si, Al) occupancies, are linking $[Al^{[4]}O_4]$-tetrahedra which form an aluminosilicate framework.

Petalite, in many respects, is transitional between the sheet and framework silicates since the structure consists of undulating $^2_\infty[Si_2O_5]$ sheets with bridging $[Al^{3+}O_4]$ tetrahedra resulting in the presence of 5- and 6-loops. The unusual bavenite has an incomplete framework of considerable complexity.

The feldspars and zeolites present an enormous problem. A review of zeolite structures, replete with structure diagrams, was presented by SMITH (1963). SMITH and RINALDI (1962) have shown that suitable projection of the feldspars (K, Na, Ca) $[(Si, Al)_4O_8]$, paracelsian, and harmotome structures reveals an underlying structural principle based on the linkage of parallel 4- and 8-membered rings. They retrieved seventeen simplest, topologically distinct arrangements based on this scheme. In general, the zeolites do not reveal any systematic trends in (Si, Al) ordering: some, such as gismondine possess essentially ordered tetrahedra, whereas many others do not. The most typical occurrence is partial (Si, Al) ordering over non-equivalent sites such as found epistibite, where SLAUGHTER and KANE (1969) proposed $(Al_{0.25}Si_{0.75})$, $(Al_{0.15}Si_{0.85})$ and $(Al_{0.40}Si_{0.60})$ over the nonequivalent tetrahedral positions. The degree of (Si, Al) ordering in zeolites may be related to the large cation occupancies in some way.

The feldspars possess problems in (Si, Al) order-disorder so enormous, that even a short review is quite meaningless. In the study of anorthite, $Ca[Al_2^{[4]}Si_2O_8]$, MEGAW et al. (1962) list 16 independent tetrahedral positions over the cell, half of which are Si-rich with the remaining half Al-rich, with small and large tetrahedra

Table 13-A-10. Framework silicates containing Al^{3+}

Species	Formula	Me—O distances		Me—O average distance (Å)	Reference
Bavenite	$Ca_4(OH)_2[BeAl_2^{[4]}Si_9O_{26}]$	$T(4)=Al^{[4]}$—O	1.725(×2), 1.732(×2)Å	1.728	CANNILLO et al. (1966)
Petalite	$Li[Al^{[4]}Si_4O_{10}]$		1.68(×2), 1.73(×2)	1.70	ZEMANN-HEDLIK and ZEMANN (1955)
Cordierite (low)	$Mg_2[Al_3^{[4]}(Al^{[4]}Si_5)O_{18}]·nH_2O$	$Al^{[4]}(1)$—O	1.749(×2), 1.756(×2)	1.753	GIBBS (1966)
		$Al^{[4]}(5)$—O	1.719, 1.720, 1.768(×2)	1.744	
Beryl	$Al_2^{[6]}[Be_3Si_6O_{18}]$	$Al^{[6]}$—O	1.903(×6)	1.903	GIBBS et al. (1968)
Osumilite	$K(Mg, Fe)[Al_3^{[4]}(Al_2Si_{10})^{[4]}O_{30}]·H_2O$	$Al^{[4]}$—O	1.762(×4)	1.762	BROWN and GIBBS (1969)
Nepheline	$KNa_3[Al^{[4]}SiO_4]_4$	$T(1)$—O	1.718(×3), 1.749(×1)	1.725	FOREMAN and PEACOR (1970)
		$T(4)$—O	1.709, 1.750, 1.755, 1.760	1.744	
Leucite (high)	$(K, Na)[Al^{[4]}Si_2O_6]$	$(Si_{0.67}Al_{0.33})^{[4]}$—O	1.611(×2), 1.627(×2)	1.619	PEACOR (1968)
Nosean	$Na_8[Al_6^{[4]}Si_6O_{24}]·(SO_4)$				SAALFELD (SR 1959, 491)
Cancrinite	$Na_6[Al_6^{[4]}Si_6O_{24}]·CaCO_3·2H_2O$	$Al(1)^{[4]}$—O	1.724, 1.761, 1.764, 1.764	1.753	JARCHOW (1965)
Analcite	$Na[Al^{[4]}Si_2O_6]·H_2O$	$(Si_{0.67}Al_{0.33})^{[4]}$—O	1.641(×2), 1.654(×2)	1.647	KNOWLES et al. (1965)
		$(Si_{0.67}Al_{0.33})^{[4]}$—O	1.644(×2), 1.648(×2)	1.646	
Pollucite	$Cs_{10-x}Na_x[Al_{16-x-y}^{[4]}Si_{32+x-y}O_{96}]·xH_2O$	$(Si_{0.63}Al_{0.37})^{[4]}$—O	1.630, 1.630, 1.655, 1.655	1.642	BEGER (1969)
Anorthite	$Ca[Al_2^{[4]}Si_2O_8]$	$(Si, Al)^{[4]}$—O	averages (lists 64 individual distances)	1.614	MEGAW et al. (1962)
		$(Al, Si)^{[4]}$—O		1.749	
Paracelsian	$Ba[Al_2^{[4]}Si_2O_8]$	$(Al_{0.5}Si_{0.5})^{[4]}(1)$—O	1.65, 1.69, 1.69, 1.70	1.68	SMITH (SR 1953, 556)
		$(Al_{0.5}Si_{0.5})^{[4]}(2)$—O	1.68, 1.70, 1.72, 1.76	1.71	
Mizzonite	$Ca_4[Al_6^{[4]}Si_6O_{24}]·CO_3$	$(Al_{0.3}Si_{0.7})^{[4]}$—O	1.623, 1.628, 1.670(×2)	1.648	PAPIKE and STEPHENSON (1966)
		$(Al_{0.5}Si_{0.5})^{[4]}$—O	1.677, 1.678, 1.679, 1.684	1.680	
Marialite	$Na_4[Al_3^{[4]}Si_9O_{24}]·Cl$	$(Si)^{[4]}$—O	1.584, 1.610(×2), 1.628	1.608	PAPIKE and ZOLTAI (1965)
		$(Al_{0.4}Si_{0.6})^{[4]}$—O	1.660, 1.660, 1.660, 1.682	1.665	
Zeolite structures					
Gismondine	$Ca[Al_2^{[4]}Si_2O_8]·4H_2O$	$Al^{[4]}(1)$—O	1.73, 1.73, 1.74, 1.75	1.74	FISCHER (1963)

Mordenite	$Na[Al^{[4]}Si_5O_{12}] \cdot 3H_2O$	$Al^{[4]}(2)$—O	1.74, 1.74, 1.76, 1.78	1.76	Meier (SR 1961, 515)
		T(1)—O		1.63	
		T(2)—O		1.58	
		T(3)—O		1.60	
		T(4)—O		1.59	
D'achiardite	$(K, Na)_5[Al_5^{[4]}Si_{19}O_{48}] \cdot 12H_2O$	T(1)—O		1.62	Gottardi and Meier (1963)
		T(2)—O		1.62	
		T(3)—O		1.63	
		T(4)—O		1.64	
Ferrierite	$NaMg_2[Al_5^{[4]}Si_{31}O_{72}] \cdot 18H_2O$	T(1)—O	1.589(×2), 1.629(×2)	1.609	Vaughan (1966)
		T(2)—O	1.618(×2), 1.619, 1.651	1.626	
		T(3)—O	1.567, 1.606(×2), 1.615	1.598	
		T(4)—O	1.591, 1.592, 1.599, 1.603	1.596	
Brewsterite	$(Sr, Ba)[Al_2^{[4]}Si_6O_{16}] \cdot 5H_2O$	$Si^{[4]}$—O	} averages	1.58	Perrotta and Smith (1964)
		$(Al_{0.33}Si_{0.67})^{[4]}$—O		1.66, 1.66, 1.67	
Yugawaralite	$Ca[Al_2^{[4]}Si_6O_{16}] \cdot 4H_2O$	$Si^{[4]}$—O	averages	1.60, 1.62	Leimer and Slaughter (1969)
		$(Si, Al)^{[4]}$—O		1.67, 1.68	
		$(Al, Si)^{[4]}$—O		1.70	
Epistilbite	$Ca_3[Al_6^{[4]}Si_{18}O_{48}] \cdot 15H_2O$	$(Si, Al)^{[4]}$—O	averages	1.61, 1.62, 1.64, 1.65	Slaughter and Kane (1969)
Chabazite	$Ca_2[Al_4^{[4]}Si_8O_{24}] \cdot 13H_2O$	$(Al_{0.33}Si_{0.67})^{[4]}$—O	1.64, 1.65, 1.66, 1.66	1.65	Smith et al. (1964)

alternating across each shared oxygen atom, obeying the aluminum avoidance rule of GOLDSMITH and LAVES (1955). Additional, more recent papers on feldspar refinements appear in the reference section (see WRIGHT, 1968 etc.).

The scapolite group of minerals resembles the feldspars in some respects. In projection down the c-axis, loops of 4- and 8-membered rings appear, which are reminiscent of the feldspars. These rings, however, are not parallel and do not belong to the structural scheme of SMITH and RINALDI (1962). Refinement of a mizzonite reveals $(Al_{0.3}Si_{0.7})$ and $(Al_{0.5}Si_{0.5})$ occupancies. For marialite $(Si_{1.0})$ and $(Al_{0.4}Si_{0.6})$ were found.

Crystal structure data for the framework structures appear in Table 13-A-10.

Revised manuscript received: December 1971

References: Section 13-A

ALLMANN, R., DONNAY, G.: Structural relations between pumpellyite and ardennite. Acta Cryst. B **27**, 1871 (1971).
— JEPSEN, H. P.: Die Struktur des Hydrotalkits. Neues Jahrb. Mineral., Monatsh. **1969**, 544 (1969).
ARAKI, T., FINNEY, J. J., ZOLTAI, T.: The crystal structure of augelite. Am. Mineralogist **53**, 1096 (1968).
— ZOLTAI, T.: The crystal structure of wavellite. Z. Krist. **127**, 21 (1968).
BAILEY, S. W.: Polymorphism of the kaolin minerals. Am. Mineralogist **48**, 1196 (1963).
BEGER, R. M.: The crystal structure and chemical composition of pollucite. Z. Krist. **129**, 280 (1969).
BORENSTAJN, J.: Structures cristallines de la métavariscite et de la métastrengite. Compt. Rend. (6) **261**, 376 (1965).
BROWN, B. E., BAILEY, S. W.: Chlorite polytypism. II. Crystal structure of a one-layer Cr-chlorite. Am. Mineralogist **48**, 42 (1963).
BROWN, G. E., GIBBS, G. V.: Refinement of the crystal structure of osumilite. Am. Mineralogist **54**, 101 (1969).
BURNHAM, C. W.: Refinement of the crystal structure of kyanite. Z. Krist. **118**, 337 (1963a).
— Refinement of the crystal structure of sillimanite. Z. Krist. **118**, 127 (1963b).
CANNILLO, E., CODA, A., FAGNANI, G.: The crystal structure of bavenite. Acta Cryst. **20**, 301 (1966).
CLARK, J. R., APPLEMAN, D. E., PAPIKE, J. J.: Crystal-chemical characterization of clinopyroxenes based on eight new structure refinements. Mineral. Soc. Am. Spec. Papers **2**, 31 (1969).
COCCO, G., CASTIGLIONE, P. C., VAGLIASINDI, G.: The crystal structure of thomsenolite. Acta Cryst. **23**, 162 (1967).
CROMER, D. T., KAY, M. I., LARSON, A. C.: Refinement of the alum structures. II. X-ray and neutron diffraction study of $NaAl(SO_4)_2 \cdot 12H_2O$, γ-alum. Acta Cryst. **22**, 182 (1967).
DOLLASE, W. A.: Refinement and comparison of the structures of zoisite and clinozoisite. Am. Mineralogist **53**, 1882 (1968).
— Crystal structure and cation ordering of piemontite. Am. Mineralogist **54**, 710 (1969).
— Refinement of the crystal structures of epidote, allanite and hancockite. Am. Mineralogist **56**, 447 (1971).
DONNAY, G., ALLMANN, R.: Si_3O_{10} groups in the crystal structure of ardennite. Acta Cryst. B **24**, 845 (1968).
FANFANI, L., NUNZI, A., ZANAZZI, P. F.: The crystal structure of wardite. Mineral. Mag. **37**, 598 (1970).
FANG, J. H., NEWNHAM, R. E.: The crystal structure of sinhalite. Mineral. Mag. **35**, 196 (1965).
FARRELL, E. F., FANG, J. H., NEWNHAM, R. E.: Refinement of the chrysoberyl structure. Am. Mineralogist **48**, 804 (1963).
FISCHER, K.: The crystal structure determination of the zeolite, gismondite, $CaAl_2Si_2O_8 \cdot 4H_2O$. Am. Mineralogist **48**, 664 (1963).
FISCHER, P.: Neutronbeugungsuntersuchung der Strukturen von $MgAl_2O_4$- und $ZnAl_2O_4$-Spinellen, in Abhängigkeit von der Vorgeschichte. Z. Krist. **124**, 275 (1967).
FLEET, S. G., MEGAW, H. D.: The crystal structure of yoderite. Acta Cryst. **15**, 721 (1962).
FOREMAN, N., PEACOR, D. R.: Refinement of the nepheline structure at several temperatures. Z. Krist. **132**, 45 (1970).

Aluminum

FOSTER, P. A.: The nature of alumina in quenched cryolite-alumina melts. J. Electrochem. Soc. 106, 971 (1959).

FRONDEL, C.: Scandium-rich minerals from rhyolite in the Thomas Range, Utah. Am. Mineralogist 55, 1058 (1970).

GELLER, S.: Refinement of the crystal structure of cryolithionite, $Na_3Al_2Li_3F_{12}$. Am. Mineralogist 56, 18 (1971).

GIBBS, G. V.: The polymorphism of cordierite. I. The crystal structure of low cordierite. Am. Mineralogist 51, 1068 (1966).

— BRECK, D. W., MEAGHER, E. P.: Structural refinement of hydrous and anhydrous synthetic beryl, $Al_2(Be_3Si_6)O_{18}$ and emerald, $Al_{1.9}Cr_{0.1}(Be_2Si_6)O_{18}$. Lithos 1, 275 (1968).

— SMITH, J. V.: Refinement of the crystal structure of synthetic pyrope. Am. Mineralogist 50, 2023 (1965).

GOLDSMITH, J. R., LAVES, F.: Cation order in anorthite ($CaAl_2Si_2O_8$) as revealed by gallium and germanium substitutions. Z. Krist. 106, 213 (1955).

GOTTARDI, G., MEIER, W. M.: The crystal structure of dachiardite. Z. Krist. 119, 54 (1963).

GUY, B. B., JEFFREY, G. A.: The crystal structure of fluellite, $Al_2PO_4F_2(OH) \cdot 7H_2O$. Am. Mineralogist 51, 1579 (1966).

HAFNER, S. S.: Die elektrische Quadrupolaufspaltung von Al^{27} in Spinell $MgAl_2O_4$ und Korund Al_2O_3. Z. Krist. 117, 38 (1962).

HANSON, A. W.: The crystal structure of eosphorite. Acta Cryst. 13, 384 (1960).

IIISHI, K., TOMISAKA, T., KATÔ, T., TAKENO, S.: Synthesis of valleriite. Am. Mineralogist 55, 2107 (1970).

ITO, T., TAKÉUCHI, Y., OZAWA, T., ARAKI, T., ZOLTAI, T., FINNEY, J. J.: The Crystal structure of axinite revised. Proc. Japan. Acad. 45, 490 (1969).

JARCHOW, O.: Atomanordnung und Strukturverfeinerung von Cancrinit. Z. Krist. 122, 407 (1965).

KAMB, W. B.: The crystal structure of zunyite. Acta Cryst. 13, 15 (1960).

KATO, K., SAALFELD, H.: Verfeinerung der Kristallstruktur von $CaO \cdot 6Al_2O_3$. Neues Jahrb. Mineral., Abhandl. 109, 192 (1968).

KNOWLES, C. R., RINALDI, F. F., SMITH, J. V.: Refinement of the crystal structure of analcime. Indian Mineralogist 6, 127 (1965).

LEIMER, H. W., SLAUGHTER, M.: The determination and refinement of the crystal structure of yugawaralite. Z. Krist. 130, 88 (1969).

LIPPENS, B. C., DEBOER, J. H.: Study of phase transformations during calcination of aluminum hydroxides by selected area electron diffraction. Acta Cryst. 17, 1312 (1964).

LOUISNATHAN, S. J.: Refinement of the crystal structure of a natural gehlenite, $Ca_2Al(Al, Si)_2O_7$. Can. Mineralogist 10, 822 (1971).

MEGAW, H. D., KEMPSTER, C. J. E., RADOSLOVICH, E. W.: The structure of anorthite, $CaAl_2Si_2O_8$. II. Description and discussion. Acta Cryst. 15, 1017 (1962).

MOORE, A. E., TAYLOR, H. F. W.: Crystal structure of ettringite. Acta Cryst. B 26, 386 (1970).

MOORE, P. B.: The crystal structure of sapphirine. Am. Mineralogist 54, 31 (1969a).

— Joesmithite. A novel amphibole crystal chemistry. Mineral. Soc. Am. Spec. Papers 2, 111 (1969b).

— Structural hierarchies among minerals containing octahedrally coordinating oxygen. I. Stereoisomerism among corner-sharing octahedral and tetrahedral chains. Neues Jahrb. Mineral., Monatsh. 163 (1970a).

— Crystal chemistry of the basic iron phosphates. Am. Mineralogist 55, 135 (1970b).

— BENNETT, J. M.: Kornerupine. Its crystal structure. Science 159, 524 (1968).

— SMITH, J. V.: Crystal structure of β-Mg_2SiO_4. Crystalchemical and geophysical implications. Phys. Earth Planet. Interiors 3, 166 (1970).

NEWNHAM, R. E., DEHAAN, Y. M.: Refinement of the α-Al_2O_3, Ti_2O_3, V_2O_3 and Cr_2O_3 structures. Z. Krist. 117, 235 (1962).

PABST, A.: A structural classification of fluoaluminates. Am. Mineralogist 35, 149 (1950).

PAPIKE, J. J., CLARK, J. R.: The crystal structure and cation distribution in glaucophane. Am. Mineralogist 53, 1156 (1968).

— ROSS, M.: Gedrites. Crystal structures and intracrystalline cation distributions. Am. Mineralogist 55, 1945 (1970).

PAPIKE, J. J., STEPHENSON, N. C.: The crystal structure of mizzonite, a calcium- and carbonate-rich scapolite. Am. Mineralogist 51, 1014 (1966).
— ZOLTAI, T.: The crystal structure of a marialite scapolite. Am. Mineralogist 50, 641 (1965).
— — Ordering of tetrahedral aluminum in prehnite, $Ca_2(Al, Fe^{+3})[Si_3AlO_{10}](OH)_2$. Am. Mineralogist 52, 974 (1967).
PEACOR, D. R.: A high temperature single crystal diffractometer study of leucite, (K, Na) $AlSi_2O_6$. Z. Krist. 127, 213 (1968).
PERROTTA, A. J., SMITH, J. V.: The crystal structure of brewsterite, $(Sr, Ba, Ca)(Al_2Si_6O_{16}) \cdot 5H_2O$. Acta Cryst. 17, 857 (1964).
PREWITT, C. T., BURNHAM, C. W.: The crystal structure of jadeite, $NaAlSi_2O_6$. Am. Mineralogist 51, 956 (1966).
RIBBE, P. H., GIBBS, G. V.: The crystal structure of topaz and its relation to physical properties. Am. Mineralogist 56, 24 (1971).
— — Statistical analysis and discussion of mean Al/Si-O bond distances and the aluminum content of tetrahedra in feldspars. Am. Mineralogist 54, 85 (1969).
ROBINSON, P. D., FANG, J. H.: Crystal structures and mineral chemistry of double salt hydrates. I. Direct determination of the crystal structure of tamarugite. Am. Mineralogist 54, 19 (1969).
ROSS, M., TAKEDA, H., WONES, D. R.: Mica polytypes. Systematic description and identification. Science 151, 191 (1966).
ROTHBAUER, R., ZIGAN, F., O'DANIEL, H.: Verfeinerung der Struktur des Bayerits, $Al(OH)_3$. Z. Krist. 125, 317 (1967).
SAALFELD, H., JARCHOW, O.: Die Kristallstruktur von Nordstrandit, $Al(OH)_3$. Neues Jahrb. Mineral., Abhandl. 109, 185 (1968).
SHANNON, R. D., PREWITT, C. T.: Effective ionic radii in oxides and fluorides. Acta Cryst. B 25, 925 (1969).
SLAUGHTER, M.: Crystal structure of stilbite. Am. Mineralogist 55, 387 (1970).
— KANE, W. T.: The crystal structure of a disordered epistilbite. Z. Krist. 130, 68 (1969).
SMITH, J. V.: Structural classification of zeolites. Mineral. Soc. Am. Spec. Papers 1, 281 (1963).
— The crystal structure of staurolite. Am. Mineralogist 53, 1139 (1968).
— BAILEY, S. W.: Second review of Al—O and Si—O tetrahedral distances. Acta Cryst. 16, 801 (1963).
— KNOWLES, C. R., RINALDI, F.: Crystal structures with a chabazite framework. III. Hydrated Ca-chabazite at +20 and −150°C. Acta Cryst. 17, 374 (1964).
— RINALDI, F.: Framework structures formed from parallel four- and eight-membered rings. Mineral. Mag. 33, 202 (1962).
— YODER, H.: Experimental and theoretical studies of mica polymorphs. Mineral. Mag. 31, 209 (1956).
STEPHENSON, D. A., MOORE, P. B.: The crystal structure of grandidierite, $(Mg, Fe)Al_3SiBO_9$. Acta Cryst B 24, 1518 (1968).
STUMPF, H. C., RUSSELL, A. S., NEWSOME, J. W., TUCKER, C. M.: Thermal transformations of aluminas and alumina hydrates. Ind. Eng. Chem. 42, 1398 (1950).
THOMPSON, J. B.: Geometrical possibilities for amphibole structures: model biopyriboles. Am. Mineralogist 55, 292 (1970).
VAUGHAN, P. A.: The crystal structure of the zeolite ferrierite. Acta Cryst. 21, 983 (1966).
WANG, R., BRADLEY, W. F., STEINFINK, H.: The crystal structure of alunite. Acta Cryst. 18, 249 (1965).
WRIGHT, T. L.: X-ray and optical study of alkali feldspar: II. An X-ray method for determining the composition and structural state from measurement of 20 values from three reflections. Am. Mineralogist 53, 88 (1968).
WRIGHT, T. L., STEWART, D. B.: X-ray and optical study of alkali feldspar: I. Determination of composition and structural state from refined unit cell parameters and 2V. Am. Mineralogist 53, 38 (1968).
ZEMANN-HEDLIK, A., ZEMANN, J.: Die Kristallstruktur von Petalit, $LiAlSi_4O_{10}$. Acta Cryst. 8, 781 (1955).

Revised manuscript received: December 1971

Silicon 14

A	F. Liebau	(Mineralogisch-Petrographisches Institut der Universität Kiel, Germany)
B	K. C. Condie	(Department of Geoscience, New Mexico Institute of Mining and Technology, Socorro, New Mexico, USA)
D I,	R. Siever	(Department of Geological Sciences, Harvard University, Cambridge, Mass., USA)
D II, E, F	K. C. Condie	
G—L	R. Siever	
M—O	K. C. Condie	

14-A. Crystal Chemistry[1]

I. Chemical Bond Character of Silicon

From the valency state of silicon, $1s^2 2s^2 2p^6 3s^1 3p_x^1 3p_y^1 3p_z^1 3d^0$, it follows that silicon atoms generally form bonds with four neighboring atoms, X, in a tetrahedral sp^3 hybrid. If the ligands, X, contain lone pair electrons, these may interact with the empty 3d orbitals of silicon resulting either in some π-bond character of the four Si—X bonds, in addition to the σ-bond character of the sp^3 hybrid, or in the formation of an octahedral sp^3d^2 hybrid with six σ-bonds. The degree of interaction between ligand electrons and the d orbitals of silicon depends on the electronegativity, χ_x, of the ligands (LIEBAU, 1971). The higher the χ_x the greater is the tendency of silicon to form $sp^3 d^2$ hybrids instead of the usual sp^3 hybrid. Thus with fluorine $(\chi=4.10)$[2] silicon usually forms $[SiF_6]^{2-}$ octahedra, while with oxygen $(\chi=3.50)$ $[SiO_4]^{4-}$ tetrahedra as well as $[SiO_6]^{8-}$ octahedra are known. Ligands with lower electronegativity, like $N(\chi=3.07)$, $Cl(\chi=2.83)$, $C(\chi=2.50)$, $S(\chi=2.44)$ and $Si(\chi=1.47)$, give only $[SiX_4]^{n-}$ tetrahedra, even if their radius is small.

The electronegativity, χ_A, of the next nearest neighbor, A, of silicon in phases with Si—X—A bond systems, causes a decrease in the tendency to form $[SiX_6]$ octahedra with decreasing χ_A.

This is clear from the fact that phases with $[SiO_6]$ octahedra are formed at normal pressure in chelates $(\chi_c=2.50)$ like $[(C_6H_4O_2)_3Si][C_5H_5NH]_2$ (FLYNN and BOER, 1969), in silicon diphosphate, $Si[P_2O_7]$ $(\chi_P=2.06)$ (LEVI and PEYRONEL, 1935; BISSERT and LIEBAU, 1970; LIEBAU and HESSE, 1971), and in thaumasite, $(X_H=2.1)$ $Ca_3[Si(OH)_6][CO_3][SO_4] \cdot 12H_2O$ (EDGE and TAYLOR, 1971). However, stishovite $SiO_2(\chi_{Si}=1.74)$ (STISHOV and BELOV, 1962), a garnet phase $Mg_3(AlSi_{0.5}Mg_{0.5})^{[6]}[Si^{[4]}O_4]_3$ (RINGWOOD, 1967) and polymorphs of the feldspars with hollandite structure $K[Si_3^{[6]}Al^{[6]}O_8]$ (RINGWOOD, REID and WADSLEY, 1967), $Sr_{1-2x}[(Si_{1-x}Al_x)_4^{[6]}O_8]$ and $Ba_{1-2x}[(Si_{1-x}Al_x)_4^{[6]}O_8]$ (REID and RINGWOOD, 1969) are synthesized only at very high pressures.

II. Silicon in Non-silicates

Elementary silicon is isostructural with diamond; carborundum, SiC, exists in the cubic sphalerite type, the hexagonal wurtzite type, and a large number of polytypes with higher periodicities. Cubic $Si^{[4]}C^{[4]}$, isostructural with sphalerite ZnS, and a hexagonal phase SiC—6H have been found in nature as β-moissanite and moissanite respectively. A survey of the numerous SiC polytypes has been made by WYCKOFF (1963). In the numerous silicides, either single silicon atoms are surrounded only by metal atoms or they form groups of two or four atoms,

---

[1] In this section reference is made to more recent publications rather than to earlier works so that references cited give no information about priority of a result.

[2] Electronegativity values according to ALLRED and ROCHOW (1958).

chains, layers or 3-dimensional frameworks with Si—Si distances similar to those in the element (ARONSSON, LUNDSTRÖM and RUNDQVIST, 1965; SCHÄFER, 1970). Since they are not found in nature, they will not be discussed in detail here. Of the two modifications of silicon nitride, $Si_3^{[4N]}N_4^{[3Si]}$, the β-phase is isostructural with phenacite, $Be_2[SiO_4]$ (BORGEN and SEIP, 1961). In both phases the silicon atoms are tetrahedrally coordinated by four nitrogen atoms which have planar threefold coordination (RUDDLESDEN and POPPER, 1958).

III. Silicon in Silicates

According to its place in the periodic table of the elements, silicon shows some amphoteric character; e.g., with very strong acids like $H_4P_2O_7$ or H_2SO_4, SiO_2 behaves as a basic oxide (LIEBAU, BISSERT and KÖPPEN, 1968; KRÜGER, BÜTTNER, OLIEW and THILO, 1968), with metal oxides it reacts as an acid anhydride. While monosilicic acid H_4SiO_4 exists only in diluted aqueous solutions, polysilicic acid $(H_2SiO_3)_n$ and phyllosilicic acid $(H_2Si_2O_5)_n$ have been synthesized in crystalline form (SCHWARZ and MENNER, 1924; LIEBAU, 1964; KALT and WEY, 1968). Acid salts of these and other hypothetical silicic acids like $Na_2[H_2SiO_4] \cdot nH_2O$ ($n = 4, 5, 8$) are well known (WILLIAMS and DENT GLASSER, 1971). The majority of silicates are neutral salts, but a number of basic salts like $Ca_3[SiO_4]O$, humite $Mg_7[SiO_4]_3(OH, F)_2$ (RIBBE and GIBBS, 1971) or $Ba_2[Si_2O_3(OH)_4](OH)_2$ (KRÜGER and WIEKER, 1965) exist as well. While the basic state of silicon is connected with the existence of $[SiO_6]$ octahedra in its compounds, phases with silicon as the acid component contain Si in tetrahedral coordination.

a) Silicon Oxygen Polyhedra, $[SiO_4]$ Tetrahedra[1]

In most cases silicon is in fourfold coordination with oxygen, giving slightly distorted $[SiO_4]$ tetrahedra. If there is no isomorphous replacement of silicon by other atoms, the mean distances within the $[SiO_4]$ tetrahedron are: Si—O = 1.62 Å, O—O = 2.64 Å, valence angle O—Si—O = 109.5°. There is a slight decrease in the average Si—O distance, with degree of condensation of the tetrahedra, from 1.63 Å for nesosilicates to 1.603 Å for tectosilicates (SMITH and BAILEY, 1963; JONES, 1968). The individual Si—O bond lengths vary from about 1.54 Å to about 1.70 Å, the lower values belonging to Si—O ($\rightarrow$ metal) bonds, the higher ones to Si—O ($\rightarrow$ Si) bridging bonds. Although this variation is usually explained by a shortening of the Si—O bond due to an increase in π-bond character (CRUICKSHANK, 1961; BROWN, GIBBS and RIBBE, 1969), this theory does not sufficiently consider the influence of *all* the coordinating cations of an oxygen atom. It should be supplemented by an extended electrostatic valence rule described by BAUR (1970, 1971). From this the individual Si—O bond length d(Si—O) can be computed under consideration of p_0 (the sum of the PAULING bond strengths), by the expression:

$$d(\text{Si—O}) = (1.44 + 0.09\, p_0) \text{ Å}.$$

---

[1] If not otherwise stated, $[SiO_4]$ will stand for $[SiO_4]$, $[AlO_4]$, $[FeO_4]$, $[TiO_4]$ etc. in cases of isomorphous replacement of Si by Al, Fe, Ti etc., that is when these tetrahedral ions are statistically distributed over the same sites of the structure, at least at high temperatures.

Isomorphous replacement of silicon by other tetrahedrally coordinated cations T causes a decrease (T = B^{3+}, Be^{2+}) or an increase (T = Al^{3+}, Ge^{4+}, Fe^{3+}, Ti^{4+}) of the T—O bond length, which is most accurately determined for [(Si, Al)O_4] tetrahedra in framework silicates. Here a linear increase has been found from 1.603 Å for pure [SiO_4], to 1.716 Å for pure [AlO_4] tetrahedra (JONES, 1968).

b) Silicon Oxygen Polyhedra, [SiO_6] Octahedra

In the few cases where [SiO_6] polyhedra have been found, they have been described as slightly distorted octahedra with the average dimensions Si—O 1.78 Å, O—O 2.54 Å, valence angle O—Si—O 90° and 180° respectively (LIEBAU, 1971). Here again, isomorphous replacement of silicon by aluminum causes an increase of the (Si, Al)—O bond (RINGWOOD, REID and WADSLEY, 1967).

IV. Classification of Phases with [SiO_n] Polyhedra

Since the days of MACHATSCHKI (1928), BRAGG (1930) and NÁRAY-SZABÓ (1930), the classification of silicates is usually based on the classification of silicate anions, i.e., on the silicon-oxygen polyhedra and their polymerisation. This classification is reasonable because it considers energy as well as genetic and crystal chemical relations among the different kinds of silicates (see Subsection 14-A-IV d).

In this section this scheme is also followed, beginning with a very rough division (Table 14-A-1) and proceeding to finer and finer subdivisions (Table 14-A-2). For each of the various silicate types one or two examples are mentioned.

The first, very rough division is into phases with [SiO_4] tetrahedra and those with [SiO_6] octahedra. Both these main groups are to be subdivided into phases with [SiO_n] polyhedra not connected to other [SiO_n] polyhedra (called isolated polyhedra), those with corner sharing, with edge sharing and with face sharing [SiO_n] polyhedra.

Table 14-A-1. *Broad division of silicon-oxygen complexes. The symbols used in the formulas are explained in the text*

	[SiO_4] tetrahedra	[SiO_6] octahedra
Isolated	olivine (Mg, Fe)$_2$ [SiO_4]	thaumasite Ca_3 [Si(OH)$_6$] [CO_3] [SO_4]·12 H_2O
Corner sharing	enstatite $Mg_2{}^1_\infty$ [Si_2O_6]	stishovite, ${}^3_\infty$ [SiO_2] $K{}^3_\infty$ [(Si$_{0.75}$Al$_{0.25}$)$_4$O$_8$] $Sr_{1-2x}{}^3_\infty$ [(Si$_{1-x}$Al$_x$)$_4$O$_8$] $\Big\} x=0.25$ $Ba_{1-2x}{}^3_\infty$ [(Si$_{1-x}$Al$_x$)$_4$O$_8$]
Edge sharing	fibrous${}^1_\infty$ [SiO_2]	
Face sharing	—	—

Due to increasing repulsion of the silicon atoms, the stability of anions with shared elements decreases rapidly from corner, via edge, to face sharing groups, and from octahedra to tetrahedra (PAULING's third rule). In agreement with this rule the majority of silicates contain isolated and corner shared [SiO_4] groups, while only one phase with edge shared [SiO$_4$] tetrahedra, fibrous SiO$_2$ (WEISS and WEISS, 1954) is known (Fig. 14-A-1). But of the few silicates with [SiO$_6$] octahedra, stishovite, SiO$_2$ with rutile type structure, and the high pressure phases of potassium, strontium and barium feldspars, with hollandite-type structure, show corner as well as edge sharing of these octahedra. Face sharing is unknown in both groups of silicates.

Fig. 14-A-1. Chain of edge shared tetrahedra: fibrous silica $^1_\infty$[SiO$_2$]

In the chemical formula, the silicon-oxygen complexes are written with square brackets, for example, [SiO$_4$] and [Si$_2$O$_7$]. Complexes with infinite extension have, in addition, an ∞ sign supplemented by the superscripts 1, 2 or 3 representing 1-dimension chains, 2-dimension layers or 3-dimension frameworks, for example $^1_\infty$[Si$_2$O$_6$], $^2_\infty$[Si$_4$O$_{10}$] or $^3_\infty$[Si$_4$O$_8$]. Silicate ring anions are marked with an index c (for cyclic) as in ^{c}[Si$_6$O$_{18}$], n-fold anions by an index n as $^{2\infty}$[Si$_4$O$_{11}$] for a double chain ($n=2$) or $^{3\infty}$[Si$_6$O$_{16}$] for a triple chain ($n=3$). Branched silicate anions are marked by an index b as in ^{b}[Si$_4$O$_{12}$], the branched chain of astrophyllite (see Subsection 14-A-IVa—6).

a) Silicates with [SiO$_4$] Tetrahedra

With few exceptions the silicate anions with isolated and corner shared [SiO$_4$] groups can be subdivided into a periodic table of anions (Table 14-A-2).

Table 14-A-2. *Broad classification of silicates with fourfold coordinated silicon atoms*

	Single	Double	Triple	Quadruple	Quintuple	Hexuple	...
Tetrahedra	+	+	+	−	−	−	
Chains	+	+	+	+	+	−	
Layers	+	+	−	−	−	−	
Frameworks	+						
Rings	+	+	−	−	−	−	

1. Nesosilicates and Sorosilicates

Single tetrahedra are very common among minerals like olivine, (Mg, Fe)$_2$[SiO$_4$] (BIRLE, GIBBS, MOORE and SMITH, 1968), and the garnets Me$_3^{2+}$Me$_2^{3+}$[SiO$_4$]$_3$ (NOVAK and GIBBS, 1971). *Double tetrahedra* [Si$_2$O$_7$] as found e.g. in barysilite, MnPb$_8$[Si$_2$O$_7$]$_3$ (LAJZEROWICZ, 1966; Fig. 14-A-2b), and hemimorphite, Zn$_4$[Si$_2$O$_7$](OH)$_2$ · H$_2$O

Fig. 14-A-2a—c. Low molecular silicate anions: a single tetrahedron; b double tetrahedron: barysilite MnPb$_8$[Si$_2$O$_7$]$_3$; c triple tetrahedron: aminoffite Ca$_3$(BeOH)$_2$[Si$_3$O$_{10}$]

(McDonald and Cruickshank, 1967b), are less frequent. *Triple tetrahedra* are rather rare and exist in aminoffite, $Ca_3(BeOH)_2[Si_3O_{10}]$ (Coda, Rossi and Ungaretti, 1967; Fig. 14-A-2c), and kinoite, $Ca_2Cu_2[Si_3O_{10}] \cdot 2H_2O$ (Laughon, 1971). Linear condensation of a small or medium number of $[SiO_4]$ tetrahedra larger than three has not yet been found in crystalline silicates.

2. Inosilicates

Silicates with Single Chains. The linear condensation product of a very large number of silicate tetrahedra is known as a single chain ("Einfachkette"). The various types of single chains differ in the number of tetrahedra in the identity period of the chain and are therefore called "Einer-Einfachkette", "Zweier-Einfachkette", "Dreier-Einfachkette" etc. Fig. 14-A-3 gives diagrams of the known types of silicate single chains.

"*Zweier-Einfachkette*": Chains of this type are most common as anions of the pyroxenes like enstatite, $Mg_2{}^1_\infty[Si_2O_6]$ (Morimoto and Koto, 1969), where they have an identity period of 5.2 to 5.3 Å. A shortened version of the "Zweier-Einfachkette" with a period of 4.5 to 4.9 Å has been found in some synthetic anhydrous silicates like $Na_4{}^1_\infty[Si_2O_6]$ (McDonald and Cruickshank, 1967a).

"*Dreier-Einfachkette*" exists in a number of calcium silicates of which wollastonite, $Ca_3{}^1_\infty[Si_3O_9]$ (Buerger and Prewitt, 1963; Trojer, 1968), is the most common. Other examples are foshagite, $Ca_4{}^1_\infty[Si_3O_9](OH)_2$ (Gard and Taylor, 1960), and pectolite, $Ca_2Na{}^1_\infty[Si_3O_8(OH)]$ (Buerger and Prewitt, 1963).

"*Vierer-Einfachkette*" has been found in haradaite, $Sr_2(VO)_2{}^1_\infty[Si_4O_{12}]$ (Takéuchi and Joswig, 1967), and in krauskopfite, $Ba_2{}^1_\infty[Si_4O_8(OH)_4] \cdot 4H_2O$ (Coda, Dal Negro and Rossi, 1967). In analogy to the two types of "Zweierkette" there is a bent "Viererkette" with a period of 7.06 Å in haradaite and a more stretched one with a 8.46 Å periodicity in krauskopfite.

"*Fünfer-Einfachkette*" has been described for rhodonite, $(Mn, Ca)_5{}^1_\infty[Si_5O_{15}]$ (Peacor and Niizeki, 1963), and has been proved to exist in babingtonite, $Ca_2Fe_2{}^1_\infty[Si_5O_{14}(OH)]$, and inesite, $Ca_2Mn_7{}^1_\infty[Si_5O_{14}(OH)]_2 \cdot 5H_2O$.

"*Sechser-Einfachkette*" of a helical kind is contained in the structure of stokesite $Ca_2Sn_2{}^1_\infty[Si_6O_{18}] \cdot 4H_2O$ (Vorma, 1963).

"*Siebener-Einfachkette*" is found in pyroxmangite, $(Mn, Fe, Ca)_7{}^1_\infty[Si_7O_{21}]$ (Liebau, 1959), and in pyroxferroite, $(Fe, Ca)_7{}^1_\infty[Si_7O_{21}]$, a mineral first discovered in lunar rocks (Burnham, 1971).

"*Neuner-Einfachkette*" occurs in the synthetic high-pressure polymorph ferrosilite III, $Fe_9{}^1_\infty[Si_9O_{27}]$ (Burnham, 1966).

"*Zwölfer-Einfachkette*" was found in alamosite, $Pb_{12}{}^1_\infty[Si_{12}O_{36}]$ (Boucher and Peacor, 1968). Although 12 tetrahedra is the highest periodicity known so far for silicates, the identity period in the chain direction of alamosite is only 19.63 Å due to its helical shape, which is less than the 22.61 Å for the "Neunerkette" in ferrosilite III.

"*Einer-Einfachkette*" and "*Achter-Einfachkette*" have not yet been discovered in silicates, but such chains have been verified in copper polygermanate, $Cu{}^1_\infty[GeO_3]$ (Völlenkle, Wittmann and Nowotny, 1967), and in copper dipotassium polyphosphate, $Cu_2K_4{}^1_\infty[P_8O_{24}]$ (Tordjman, Tran Qui and Laügt, 1970) respectively.

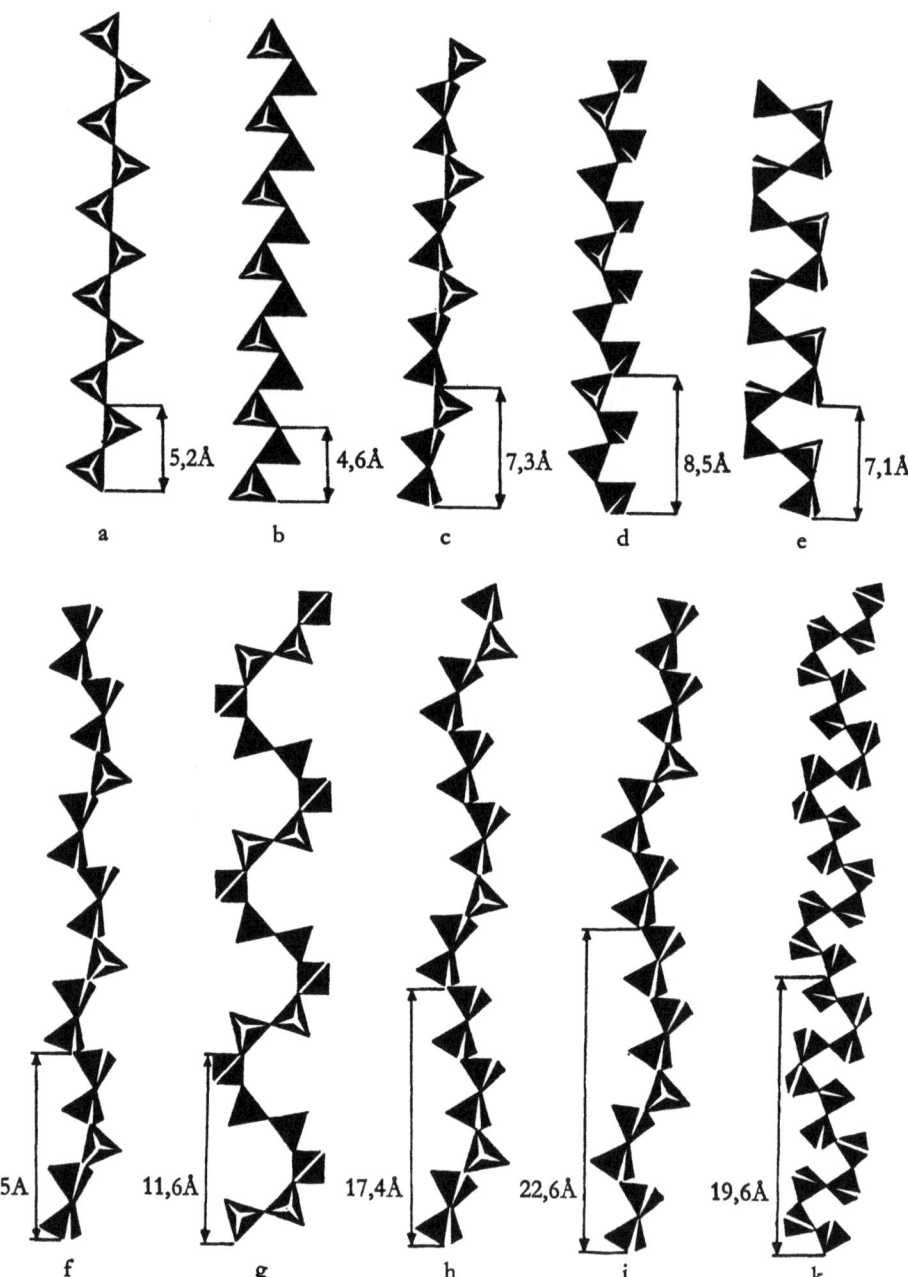

Fig. 14-A-3a—k. Single-chain types found in silicates: a extended "Zweier-Einfachkette": pyroxenes $A_2{}^1_\infty[Si_2O_6]$; b shortened "Zweier-Einfachkette": high-temperature $Ba_2{}^1_\infty[Si_2O_6]$; c "Dreier-Einfachkette": wollastonite $Ca_3{}^1_\infty[Si_3O_9]$; d extended "Vierer-Einfachkette": krauskopfite $Ba_2{}^1_\infty[Si_4O_8(OH)_4]\cdot 4H_2O$; e shortened "Vierer-Einfachkette": haradaite $Sr_2(VO)_2{}^1_\infty[Si_4O_{12}]$; f "Fünfer-Einfachkette": rhodonite $(Mn, Ca)_5{}^1_\infty[Si_5O_{15}]$; g "Sechser-Einfachkette": stokesite $Ca_2Sn_2{}^1_\infty[Si_6O_{18}]\cdot 4H_2O$; h "Siebener-Einfachkette": pyroxferroite $(Fe, Ca)_7{}^1_\infty[Si_7O_{21}]$; i "Neuner-Einfachkette": ferrosilite III high-pressure $Fe_9{}^1_\infty[Si_9O_{27}]$; k "Zwölfer-Einfachkette": alamosite $Pb_{12}{}^1_\infty[Si_{12}O_{36}]$

Silicates with Double Chains. Condensation of two single chains via corners of the tetrahedra results in the formation of double chains ("Doppelkette"). All tetrahedra of each chain can take part in the connection of the chains or only some of them. If p is the number of tetrahedra in the identity period of the single chains and q the number of bridges between the chains in one period, the possible double chains are described by the general formula

$$2^1_\infty [\mathrm{Si}_{2p} \mathrm{O}_{6p-q}]^{(4p-2q)-} \qquad 1 \leq q \leq p.$$

Of these numerous possible double chains only a few have been detected (Fig. 14-A-4).

Connection of two chains of "Zweier-Einfachkette" leads to "*Zweier-Doppelkette*" with every second tetrahedra shared, $2^1_\infty[\mathrm{Si}_4\mathrm{O}_{11}]^{6-}$ (Fig. 14-A-4a) or to those with all tetrahedra shared, $2^1_\infty[\mathrm{Si}_4\mathrm{O}_{10}]^{4-}$ (Fig. 14-A-4b). The former is quite frequent in the amphiboles like tremolite, $\mathrm{Ca_2Mg_5}\,2^1_\infty[\mathrm{Si}_4\mathrm{O}_{11}]_2(\mathrm{OH})_2$ (PAPIKE, ROSS and CLARK, 1969), while the latter has been confirmed as occurring only in synthetic $\mathrm{Li}_4\,2^1_\infty[\mathrm{SiGe}_3\mathrm{O}_{10}]$ (VÖLLENKLE, WITTMANN and NOWOTNY, 1968).

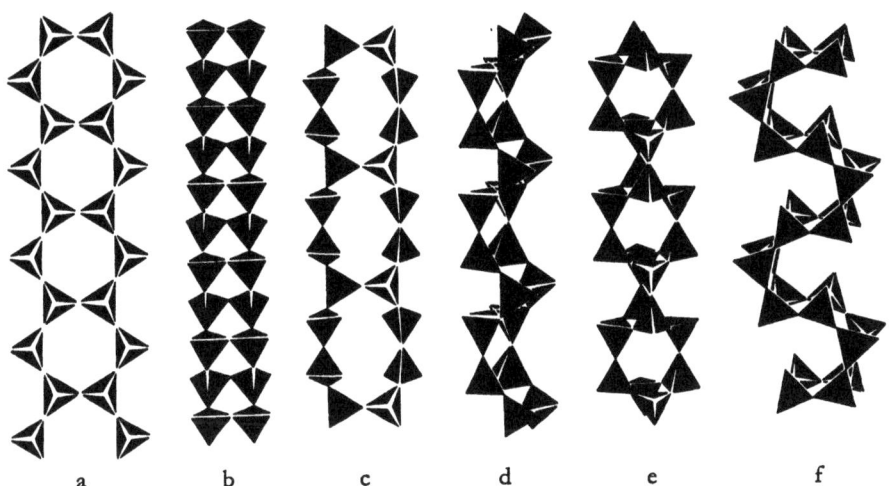

  a   b   c   d   e   f

Fig. 14-A-4 a—f. Double-chain types found in silicates: a "Zweier-Doppelkette" 2^1_∞ $[\mathrm{Si}_4\mathrm{O}_{11}]^{6-}$: amphiboles $\mathrm{A_2B_5}\,2^1_\infty[\mathrm{Si}_4\mathrm{O}_{11}]_2(\mathrm{OH})_2$; b "Zweier-Doppelkette" $2^1_\infty[\mathrm{Si}_4\mathrm{O}_{10}]^{4-}$: $\mathrm{Li}_4\,2^1_\infty[\mathrm{SiGe}_3\mathrm{O}_{10}]$; c "Dreier-Doppelkette" $2^1_\infty[\mathrm{Si}_6\mathrm{O}_{17}]^{10-}$: xonotlite $\mathrm{Ca}_6\,2^1_\infty[\mathrm{Si}_6\mathrm{O}_{17}](\mathrm{OH})_2$; d "Dreier-Doppelkette" $2^1_\infty[\mathrm{Si}_6\mathrm{O}_{15}]^{6-}$: epididymite $\mathrm{Na_2Be_2H}\,2^1_\infty[\mathrm{Si}_6\mathrm{O}_{15}]\mathrm{OH}$; e "Vierer-Doppelkette" $2^1_\infty[\mathrm{Si}_8\mathrm{O}_{20}]^{8-}$: narsarsukite $\mathrm{Na}_4(\mathrm{TiO})_2\,2^1_\infty[\mathrm{Si}_8\mathrm{O}_{20}]$; f "Sechser-Doppelkette" $2^1_\infty[\mathrm{Si}_{12}\mathrm{O}_{30}]^{12-}$: tuhualite $(\mathrm{Na},\mathrm{K})_2\mathrm{Fe}^{2+}_3\mathrm{Fe}^{3+}_3\,2^1_\infty[\mathrm{Si}_{12}\mathrm{O}_{30}]\cdot\mathrm{H_2O}$

"*Dreier-Doppelkette*", with every third tetrahedra shared between the single chains, has been found in xonotlite, $\mathrm{Ca}_6\,2^1_\infty[\mathrm{Si}_6\mathrm{O}_{17}](\mathrm{OH})_2$ (MAMEDOV and BELOV, 1956), those with two thirds of the tetrahedra shared have been suggested for devitrite, $\mathrm{Ca_3Na_2}\,2^1_\infty[\mathrm{Si}_6\mathrm{O}_{16}]$, and "Dreier-Doppelkette", with all tetrahedra shared, exists in epididymite, $\mathrm{Na_2Be_2H}\,2^1_\infty[\mathrm{Si}_6\mathrm{O}_{15}]\mathrm{OH}$ (ROBINSON and FANG, 1970).

"*Vierer-Doppelkette*" has only been verified, with all tetrahedra shared, in narsarsukite, $\mathrm{Na}_4(\mathrm{TiO})_2\,2^1_\infty[\mathrm{Si}_8\mathrm{O}_{20}]$ (PEACOR and BUERGER, 1962).

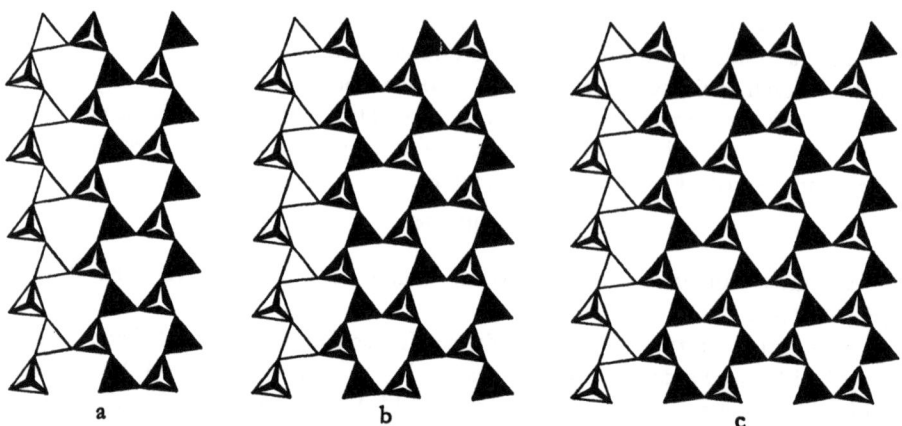

Fig. 14-A-5 a—c. n-fold chain types found in silicates: a "Zweier-Dreifachkette": $Ba_4 3^1_\infty[Si_6O_{16}]$; b "Zweier-Vierfachkette": $Ba_5 4^1_\infty[Si_8O_{21}]$; c "Zweier-Fünffachkette": $Ba_6 5^1_\infty[Si_{10}O_{26}]$

"*Sechser-Doppelkette*", with all tetrahedra shared between the two single chains, exists in tuhualite, $(Na, K)_2Fe^{2+}_2Fe^{3+}_2 2^1_\infty[Si_{12}O_{30}] \cdot H_2O$ (MERLINO, 1969).

Silicates with Triple and Higher-fold Chains. Condensation of three and more single chains to triple and higher-fold chains is very rare and has so far been found only for "Zweierkette" (Fig. 14-A-5). The "*Zweier-Dreifachkette*" (triple chain), in $Ba_4 3^1_\infty[Si_6O_{16}]$ (KATSCHER and LIEBAU, 1965), "*Zweier-Vierfachkette*" (quadruple chain), in $Ba_5 4^1_\infty[Si_8O_{21}]$, and "*Zweier-Fünffachkette*" (quintuple chain), in $Ba_6 5^1_\infty[Si_{10}O_{26}]$ (KATSCHER, 1969), are members of the homologous series of phases with

$$f^1_\infty[Si_{2f}O_{5f+1}]^{(2f+2)-},$$

where f is the number of single chains connected to form a band.

3. Phyllosilicates

Silicates with Single Layers. While condensation products of six and more single chains to bands have not yet been described, the condensation of a very large number of chains results in the formation of a single layer ("Einfachschicht"). The nomenclature of these layers is based on the chain type of lowest periodicity from which it is built by continuous connecting of one chain to the others. In the figures one of these chains is marked by open tetrahedra. So far, layers of the types "Zweier-Einfachschicht", "Dreier-Einfachschicht", "Vierer-Einfachschicht" and "Sechser-Einfachschicht" are known.

Another important difference between the various layers is the ratio between the numbers of unshared corners of the $[SiO_4]$ tetrahedra pointing to each side of the layer.

"*Zweier-Einfachschicht*" (Fig. 14-A-6). Anhydrous layer silicates like petalite, $LiAl^2_\infty[Si_4O_{10}]$ (LIEBAU, 1961), and sanbornite, $Ba_2^2_\infty[Si_4O_{10}]$ (DOUGLASS, 1958), have half the number of their unshared oxygen atoms of the $[SiO_4]$ tetrahedra pointing to one and half pointing to the other side of the layer alternately. The micas — e.g., muscovite, $KAl_2^2_\infty[Si_3AlO_{10}](OH, F)_2$ (BIRLE and TETTENHORST, 1968; SOBOLEVA and ZVYAGIN, 1969) — and most of the clay minerals — kaolinite,

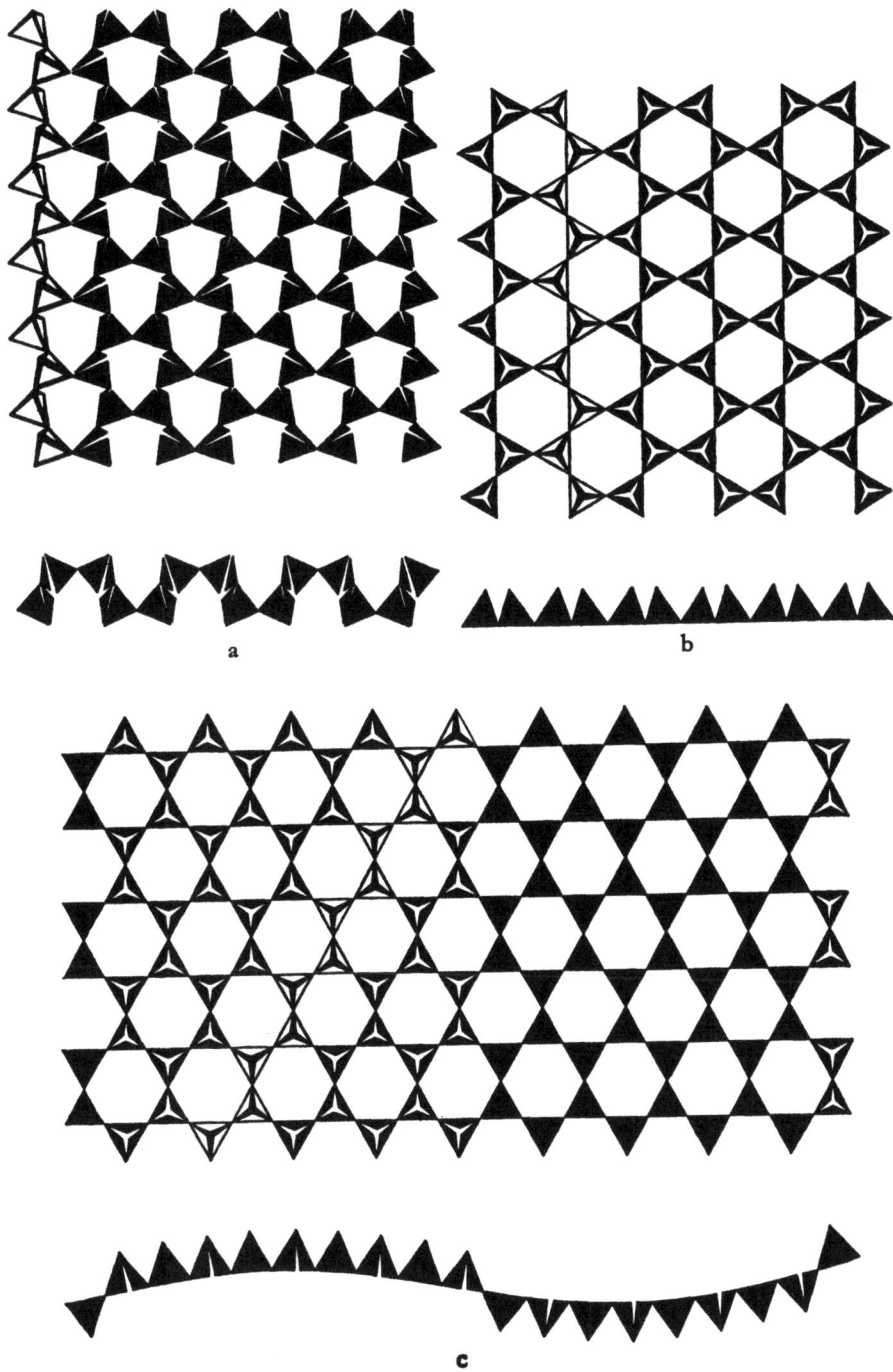

Fig. 14-A-6a—f. Types of layers of "Zweier-Einfachschicht" found in silicates: a sanbornite $Ba_2{}^2_\infty[Si_4O_{10}]$; b muscovite $KAl_2{}^2_\infty[Si_3AlO_{10}](OH, F)_2$; c antigorite $Mg_6{}^2_\infty[Si_4O_{10}](OH)_8$; d sepiolite $(Mg, Fe, Al)_8{}^2_\infty[Si_4O_{10}]_3(O, OH)_4 \cdot 8H_2O$; e palygorskite $Mg_5{}^2_\infty[Si_4O_{10}]_2(OH)_2 \cdot H_2O$; f $K_4{}^2_\infty[Si_8O_{18}]$

Fig. 14-A-6 d

$Al_4\overset{2}{\infty}[Si_4O_{10}](OH)_8$ (ZVYAGIN, 1960), montmorillonite like $(Al_{1.67}Mg_{0.33}Na_{0.33})\overset{2}{\infty}[Si_4O_{10}](OH)_2 \cdot 4H_2O$ (COWLEY and GOSWAMI, 1961), the chlorites like kammererite, $(Mg, Cr)_6\overset{2}{\infty}[Si_3AlO_{10}](OH)_8$ (BROWN and BAILEY, 1963), etc. — on the other hand, contain flat tetrahedral layers with all unshared oxygen atoms of the $[SiO_4]$ tetrahedra pointing to one side of the layer only (Fig. 14-A-6b).

In chrysotile, $Mg_6\overset{2}{\infty}[Si_4O_{10}](OH)_8$, the layers are rolled to form tubes (WHITTAKER, 1956; YADA, 1967). Antigorite, $Mg_6\overset{2}{\infty}[Si_4O_{10}](OH)_8$, has layers formed like corrugated iron where the unshared oxygen of one half-wave point to one side, those of the other half-wave point to the other side of the layer. Each half-wave is eight or nine tetrahedra wide (Fig. 14-A-6c) (KUNZE, 1959). In sepiolite, $(Mg, Fe, Al)_8\overset{2}{\infty}[Si_4O_{10}]_3(O, OH)_4 \cdot 8H_2O$, the unshared oxygen atoms of three adjacent "Zweierkette" chains point to one side of the "Zweierschicht", those of the next three chains to the other side, and so on (BRAUNER and PREISINGER, 1956). Palygorskite, $Mg_5\overset{2}{\infty}[Si_4O_{10}]_2(OH)_2 \cdot H_2O$, contains layers of "Zweier-Einfachschicht" in which the free corners of the tetrahedra of each of two "Zweierkette" chains point to one or the other side of the layer alternately (BRADLEY, 1940).

All these layers of "Zweierschicht" contain only rings of six $[SiO_4]$ tetrahedra, which result from the connection of the "Zweierkette" chains.

An unusual type of "Zweier-Einfachschicht" layers (Fig. 14-A-6f) exists in synthetic $K_4\overset{2}{\infty}[Si_8O_{18}]$ (SCHWEINSBERG and LIEBAU, 1971). It may be interpreted as a condensation product of "Zweier-Doppelkette" $2\overset{1}{\infty}[Si_4O_{10}]$ chains of the type found in $Li_4 2\overset{1}{\infty}[SiGe_3O_{10}]$ (VÖLLENKLE, WITTMANN and NOWOTNY, 1968).

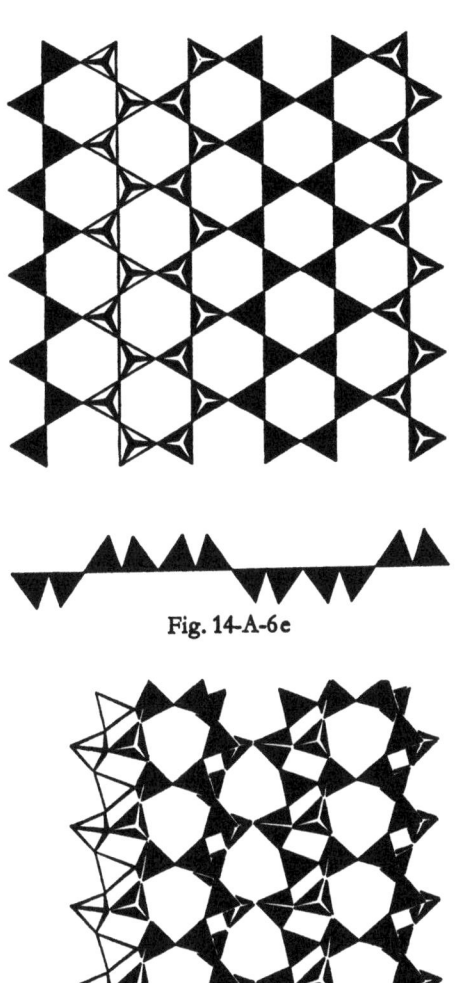

Fig. 14-A-6e

Fig. 14-A-6f

Layers of "*Dreier-Einfachschicht*" (Fig. 14-A-7a) formed by connection of "Dreierkette" chains have been described for dalyite, $K_2Zr_\infty^2[Si_6O_{15}]$ (FLEET, 1965).

Layers of the type "*Vierer-Einfachschicht*" have been found in apophyllite, $Ca_4K_\infty^2[Si_8O_{20}]F \cdot 8H_2O$ (Fig. 14-A-7b) (COLVILLE, ANDERSON and BLACK, 1971;

Fig. 14-A-7a—d. Types of single layers found in silicates: a "Dreier-Einfachschicht": dalyite $K_2Zr_\infty^2[Si_6O_{15}]$; b "Vierer-Einfachschicht": apophyllite $Ca_4K_\infty^2[Si_8O_{20}]F \cdot 8H_2O$; c "Vierer-Einfachschicht": prehnite $Ca_4(Al, Fe)_2{}_\infty^2[Si_6Al_2O_{20}](OH)_4$; d "Sechser-Einfachschicht": manganpyrosmalite $Mn_{16}{}_\infty^2[Si_{12}O_{30}](OH)_{18}Cl_2$

CHAO, 1971b; PRINCE, 1971), and in a number of anhydrous silicates isotypic with gillespite, $Ba_2Fe_2{}_\infty^2[Si_8O_{20}]$ (PABST, 1943). These tetragonal layers have rings of four and eight tetrahedra, respectively, with the free corners of half the four-membered rings pointing up, those of the other four-membered rings pointing down. The "Vierer-Einfachschicht" of prehnite, $Ca_4(Al, Fe)_2{}_\infty^2[Si_6Al_2O_{20}](OH)_4$ (Fig. 14-A-7c), is remarkable because it does not contain tetrahedra with three shared corners, but instead tetrahedra with four shared corners and two shared corners in the ratio 1:1 (PAPIKE and ZOLTAI, 1967).

Layers of the type *"Sechser-Einfachschicht"* (Fig. 14-A-7d) exist in manganpyrosmalite, $Mn_{16}{}^2_\infty[Si_{12}O_{30}](OH)_{18}Cl_2$ (TAKÉUCHI, KAWADA, IRIMAZIRI and SADANAGA, 1969). These layers contain rings of four, six and twelve tetrahedra. While the unshared oxygen atoms of one six-membered ring point to one side of the layer, those of the three adjacent six-membered rings point to the other side, and so forth.

Silicates with Double Layers. Condensation of two tetrahedral single layers results in the formation of a double layer (*"Doppelschicht"*). They are named after the component single layers, e.g., "Zweier-Doppelschicht", "Dreier-Doppelschicht" etc. Only a few examples of double layer silicates are known. In all of them the two single layers are related to each other by a mirror plane running through the oxygen atoms which join the two single layers to make a double layer.

Layers of the type *"Zweier-Doppelschicht"* built by connecting two single layers of the mica type (Fig. 14-A-6b), via all free corners (Fig. 14-A-8a), exist in the high-temperature phases of anorthite, $CaAl_2Si_2O_8$, and celsian, $BaAl_2Si_2O_8$, and should therefore be written as $Ca_2\,2^2_\infty[Al_4Si_4O_{16}]$ (TAKÉUCHI and DONNAY, 1959) and $Ba_2\,2^2_\infty[Al_4Si_4O_{16}]$ respectively (TAKÉUCHI, 1958).

Layers of the type *"Vierer-Doppelschicht"* have been found for delhayelite, $Ca_4(Na_3Ca)K_7\,2^2_\infty[Si_{14}Al_2O_{38}]Cl_2F_4$, (CANNILLO, ROSSI and UNGARETTI, 1969) and macdonaldite, $BaCa_4H_2\,2^2_\infty[Si_{16}O_{38}]\cdot 10.4H_2O$ (CANNILLO, ROSSI and UNGARETTI,

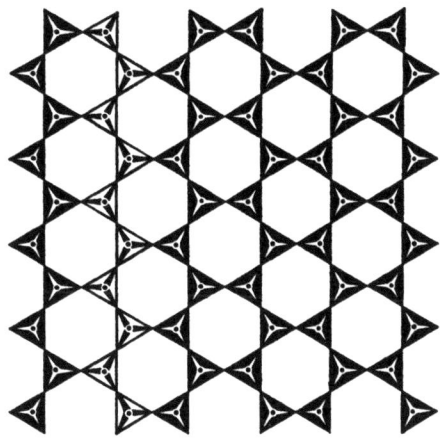

a

Fig. 14-A-8a—c. Types of double layers found in silicates. Two of the single layers shown with top view are joined via mirror planes through the tetrahedral corners marked with ●: a "Zweier-Doppelschicht": high-temperature $Ba_2\,2^2_\infty[Si_4Al_4O_{16}]$; b "Vierer-Doppelschicht": delhayelite $Ca_4(Na_3Ca)K_7\,2^2_\infty[Si_{14}Al_2O_{38}]Cl_2F_4$; c "Achter-Doppelschicht": carletonite $K_2Na_8Ca_8\,2^\infty_2[Si_{16}O_{36}][CO_3]_8(OH, F)_2\cdot 2H_2O$.
(For Fig. 14-A-8b and c see p. 14-A-14).

Silicon

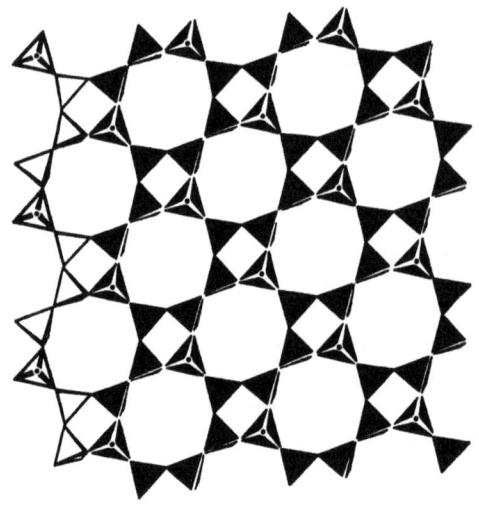

Fig. 14-A-8b

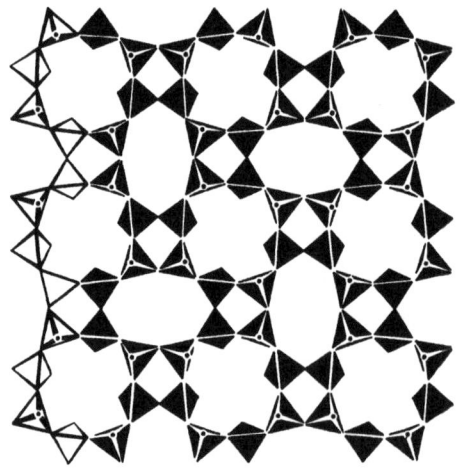

Fig. 14-A-8c

1968). In these double layers, two apophyllite-like single layers (Fig. 14-A-7b) are linked via the free corner of every fourth tetrahedron (Fig. 14-A-8b). Carletonite, $K_2Na_8Ca_8\, {}^2_\infty[Si_{16}O_{36}][CO_3]_8(OH, F)_2 \cdot 2H_2O$, contains "*Achter-Doppelschicht*" layers composed of two apophyllite-like single layers that are linked via the free corner of every second tetrahedron (Fig. 14-A-8c) (CHAO, 1971a).

4. Tectosilicates

In almost all tectosilicates, the tetrahedral framework (Gerüst) can be built by connecting very large numbers of single chains in all three dimensions. According to the chain types of lowest periodicity (number of tetrahedra in the chain period), from which the framework can completely be built by putting these chains together, the frameworks may be called "Zweiergerüst", "Dreiergerüst", etc. In general, the condensation takes place via all four corners of all tetrahedra, giving the formula ${}^3_\infty[Si_pO_{2p}]$, where p is the periodicity of the chain type from which the framework is built.

"*Zweiergerüst*" frameworks are found in cristobalite, ${}^3_\infty[Si_2O_4]$ (DOLLASE, 1965), and in tridymite, ${}^3_\infty[Si_2O_4]$ (Fig. 14-A-9a) (DOLLASE, 1967). In these silica phases pyroxene-type "*Zweier-Einfachkette*" chains (Fig. 14-A-3a) are linked to layers of

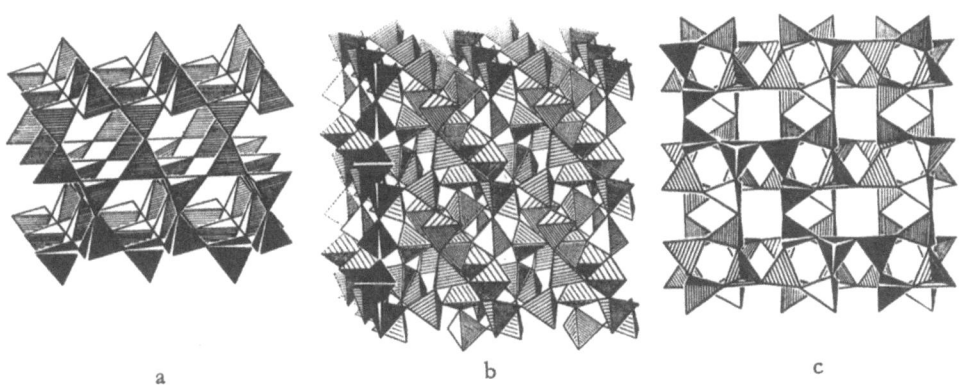

Fig. 14-A-9a—c. Framework types found in silicates: a "Zweiergerüst": tridymite ${}^3_\infty[Si_2O_4]$; b "Dreiergerüst": keatite ${}^3_\infty[Si_3O_6]$; c "Vierergerüst": orthoclase $K{}^3_\infty[(Si_{0.75}Al_{0.25})_4O_8]$

"Zweier-Einfachschicht" in such a way that the free corners of the tetrahedra of neighboring chains point to opposite sides of the layer. A very large number of those layers is linked via common corners to build the frameworks. Similarly a "*Dreiergerüst*", formed from wollastonite-like "Dreier-Einfachkette" chains, (Fig. 14-A-3c) is found in keatite, ${}^3_\infty[Si_3O_6]$ (Fig. 14-A-9b) (SHROPSHIRE, KEAT and VAUGHAN, 1959), while the "Dreiergerüst" of quartz, ${}^3_\infty[Si_3O_6]$, is built from spiral chains with three $[SiO_4]$ tetrahedra in the identity period of the chains (SMITH and ALEXANDER, 1963).

All the feldspars like orthoclase, $K\,{}^3_\infty[(Si_{0.75}Al_{0.25})_4O_8]$ (COLVILLE and RIBBE, 1968), or bytownite, $(Ca, Na)\,{}^3_\infty[(Si, Al)_4O_8]$ (FLEET, CHANDRASEKHAR and MEGAW, 1966), contain frameworks of "*Vierergerüst*", (Fig. 14-A-9c) as well as coesite, ${}^3_\infty[Si_4O_8]$, one of the high pressure phases of silica (ARAKI and ZOLTAI, 1969).

5. Cyclosilicates

Silicates with Single Rings. According to the number of [SiO$_4$] tetrahedra forming a single ring (Einfachring) these are called rings of type "Dreier-Einfachring", "Vierer-Einfachring", etc. All the silicate single rings known have the formula $^c[\text{Si}_p\text{O}_{3p}]^{2p-}$ and are shown in Fig. 14-A-10.

The type "*Dreier-Einfachring*", with the lowest possible number of tetrahedra, was described for high pressure wollastonite, Ca$_3$^{c}[Si$_3$O$_9$](p) (TROJER, 1969), as well as for benitoite, BaTi^c[Si$_3$O$_9$] (FISCHER, 1969).

"*Vierer-Einfachring*" is part of the structures of taramellite, Ba$_2$(Fe, Ti, Mg)$_2$ ^{c}[Si$_4$O$_{12}$](OH)$_2$ (MAZZI and ROSSI, 1965), and of a synthetic potassium silicate, K$_4$ ^{c}[Si$_4$O$_8$(OH)$_4$] (HILMER, 1964).

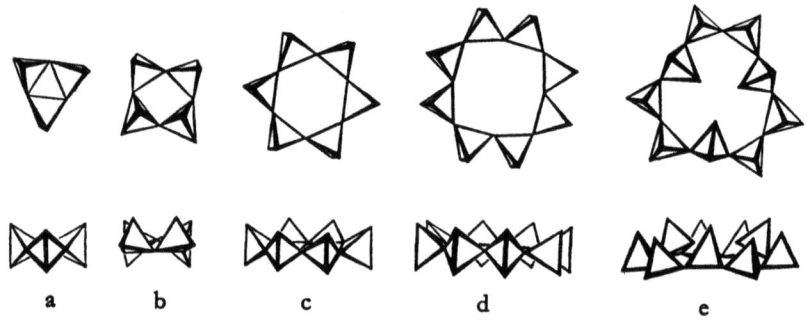

Fig. 14-A-10a—e. Types of single rings found in silicates two views each: a "Dreier-Einfachring": benitoite BaTi^c[Si$_3$O$_9$]; b "Vierer-Einfachring": taramellite Ba$_2$(Fe, Ti, Mg)$_2$ ^{c}[Si$_4$O$_{12}$](OH)$_2$; c "Sechser-Einfachring": beryl Al$_2$Be$_3$ ^{c}[Si$_6$O$_{18}$]; d "Achter-Einfachring": muirite, Ba$_{10}$(Ca, Mn, Ti)$_4$ ^{c}[Si$_8$O$_{24}$](Cl, O, OH)$_{12}$ · 4H$_2$O; e "Neuner-Einfachring": eudialyte (Fe, Mn, Mg)$_3$Zr$_3$(Zr, Nb)$_x$(Ca, R. E.)$_6$Na$_{12}$ ^{c}[Si$_9$O$_{27-y}$(OH)$_y$]$_2$ ^{c}[Si$_3$O$_9$]$_2$ Cl$_z$

While rings of "*Fünfer-Einfachring*" are still unknown, dioptase, Cu$_6$ ^{c}[Si$_6$O$_{18}$] · 6H$_2$O (HEIDE, BOLL-DORNBERGER, THILO and THILO, 1955), and beryl, Al$_2$Be$_3$ ^{c}[Si$_6$O$_{18}$] (GIBBS, BRECK and MEAGHER, 1968), have rings of "*Sechser-Einfachring*".

Two new types of silicate rings have been discovered recently: "*Achter-Einfachring*" in muirite, Ba$_{10}$(Ca, Mn Ti)$_4$ ^{c}[Si$_8$O$_{24}$](Cl, OH, O)$_{12}$ · 4H$_2$O (KHAN and BAUR, 1971), and "*Neuner-Einfachring*" in eudialyte, (Fe, Mn, Mg)$_3$Zr$_3$(Zr, Nb)$_x$(Ca, R.E.)$_6$-Na$_{12}$ ^{c}[Si$_9$O$_{27-y}$(OH)$_y$]$_2$ ^{c}[Si$_3$O$_9$]$_2$Cl$_z$, together with "Dreier-Einfachring" (GUISEPPETTI, MAZZI and TADINI, 1971) (see Subsection 14-A-IVc).

Silicates with Double Rings. A few silicates have been described in which two single rings are joined via all tetrahedra giving a double ring (Doppelring) of the formula $^{2c}[\text{Si}_{2p}\text{O}_{5p}]^{2p-}$.

"*Dreier-Doppelring*" exists in synthetic [Ni(H$_2$N · CH$_2$ · CH$_2$ · NH$_2$)$_3$]$_3$^{2c}[Si$_6$O$_{15}$] · 26H$_2$O (Fig. 14-A-11a) (SMOLIN, 1970), while ekanite K(Ca, Na)$_2$Th^{2c}[Si$_8$O$_{20}$] contains rings of "*Vierer-Doppelring*" (Fig. 14-A-11b) (MOKEEVA and GOLOVASTIKOW, 1966).

a   b   c

Fig. 14-A-11a—c. Types of double rings found in silicates: a "Dreier-Doppelring": [Ni(H$_2$N · CH$_2$ · CH$_2$ · NH$_2$)$_3$]$_3$^{2c}[Si$_6$O$_{15}$] · 26 H$_2$O; b "Vierer-Doppelring": ekanite K(Ca, Na)$_2$Th^{2c}[Si$_8$O$_{20}$]; c "Sechser-Doppelring": milarite K Ca$_2$Al Be$_2$ ^{2c}[Si$_{12}$O$_{30}$] · $^1/_2$H$_2$O

"*Sechser-Doppelring*" type of rings (Fig. 14-A-11c) was found for milarite, KCa$_2$AlBe$_2$^{2c}[Si$_{12}$O$_{30}$] · $^1/_2$H$_2$O (ITO, MORIMOTO and SADANAGA, 1952), and osumilite, (K, Na, Ca)(Fe, Mg)$_2$(Al, Fe)$_3$ ^{2c}[Si$_{10}$Al$_2$O$_{30}$] · H$_2$O (BROWN and GIBBS, 1969).

6. Other Silicates with [SiO$_4$] Tetrahedra

So far this classification is straightforward: [SiO$_4$] tetrahedra are linearly connected to oligo tetrahedra (sorosilicates) which may be condensed to form rings (cyclosilicates) or to form infinite chains (inosilicates with single chains). Putting together two or more chains results in the formation of double, triple, etc., chains (inosilicates with n-fold chains) and eventually 2-dimensional single layers (phyllosilicates). Connecting two or more tetrahedral layers leads to the formation of double layers and finally frameworks (tectosilicates).

Mainly in the last decade, a number of silicate structures has been found which at first sight does not seem to fit into this simple system of silicate anions. Some of them may be interpreted as links, intermediate between the different main types of silicates.

Silicates with Branched Anions. The ^{b}[Si$_5$O$_{16}$]$^{12-}$ groups (Fig. 14-A-12a) in zunyite, Al$_{12}$(AlO$_4$)(OH, F)$_{18}$ Cl^b[Si$_5$O$_{16}$] (KAMB, 1960), may be described as *branched triple tetrahedra*. In aenigmatite, Na$_2$Fe$_5$Ti b^1_∞[Si$_6$O$_{18}$]O$_2$ (CANNILLO, MAZZI, FANG, ROBINSON and OHYA, 1971), and in astrophyllite, K$_2$Na$_2$(Fe, Mn)$_5$Mg$_2$Ti$_2$ b^1_∞[Si$_4$O$_{12}$]$_2$ (O, OH, F) (WOODROW, 1967), *branched chains* (Fig. 14-A-11b, c) exist. Leucosphenite, Ba$_2$(Na, Ca)$_8$Ti$_4$B$_4$ b^2_∞[Si$_{20}$O$_{54}$]O$_6$ (SHUMYATSKAYA, VORONKOV and PYATENKO, 1968), is said to contain layers of "Sechser-Einfachschicht" with double tetrahedra attached to it, resulting in *branched layers* of composition b^2_∞[Si$_{20}$O$_{54}$]$^{28-}$. In these structures cations, showing no statistical replacement of silicon, complete the anion to form one of the anion types previously described. For example, in astrophyllite the branched "Zweierkette" b^1_∞[Si$_4$O$_{12}$] is supplemented by [TiO$_6$] octahedra to make a single layer, while the branched layer in leucosphenite is completed by [B$_2$O$_7$] double tetrahedra to form an ordinary "Sechser-Doppelschicht" 2^2_∞[Si$_{20}$B$_4$O$_{56}$]$^{20-}$.

In some cases the branching [SiO$_4$] groups have not only one corner but two corners shared with the larger silicate anion. Vlasovite, Na$_2$Zr b^1_∞[Si$_4$O$_{11}$] (FLEET and CANN, 1967), for example contains the chain type "Sechser-Einfachkette" with two additional [SiO$_4$] tetrahedra, each forming a bridge between two non-neighboring tetrahedra of the single chains (Fig. 14-A-12d).

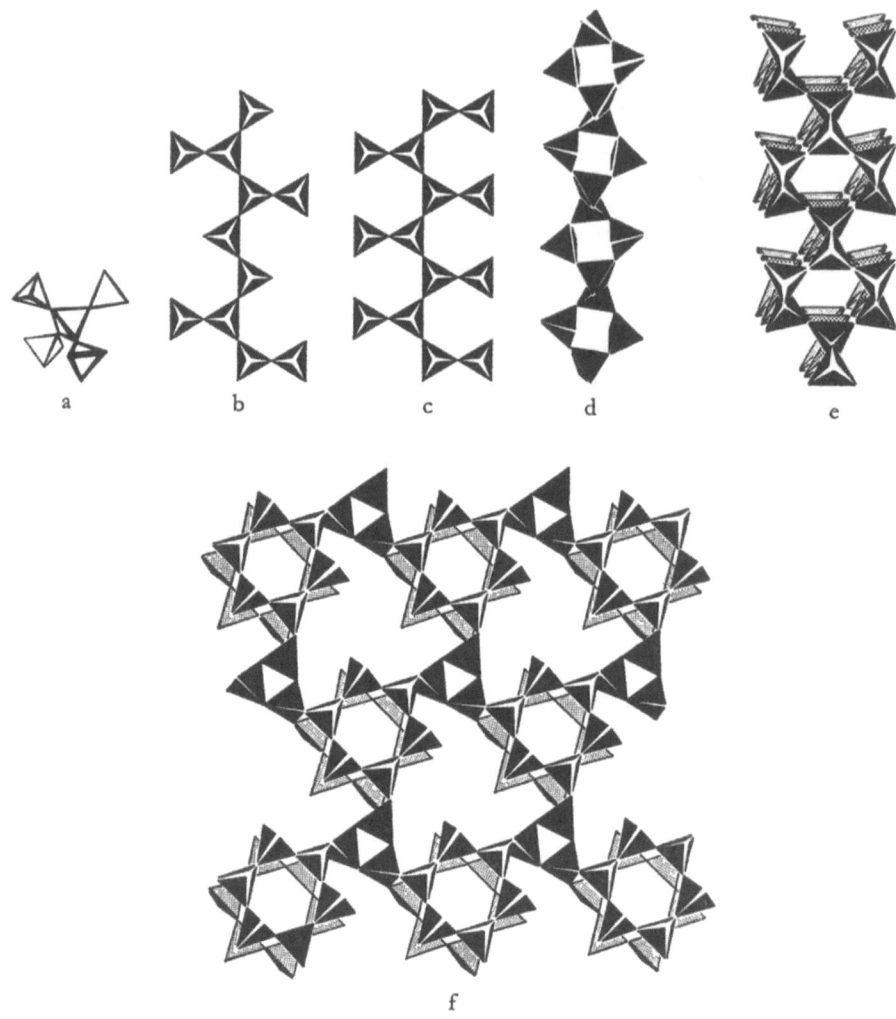

Fig. 14-A-12a—f. Unusual silicate anions: a branched triple tetrahedra: zunyite $Al_{12}(AlO_4)(OH, F)_{18}Clb[Si_5O_{16}]$; b branched chain: aenigmatite $Na_2Fe_5Ti b_\infty^1[Si_6O_{18}]O_2$; c branched chain: astrophyllite $K_2Na_2(Fe, Mn)_5Mg_2Ti_2 b_\infty^1[Si_4O_{12}]_2(O, OH, F)$; d branched chain: vlasovite $Na_2Zr\, b_\infty^1[Si_4O_{11}]$; e $1\tfrac{1}{2}$-fold layer: $Na_4\, b_\infty^2[Si_6O_{14}]$; f $1\tfrac{1}{2}$-fold layer: zussmanite $KFe_{11}(Mg, Mn)_2\, b_\infty^2[(Si, Al)_{18}O_{42}](OH)_{14}$

The silicate anion in synthetic $Na_4\, b_\infty^2[Si_6O_{14}]$ (JAMIESON, 1967) can be regarded as a "Zweier Einfachschicht" like that in $\alpha\text{-}Na_4\, \infty^2[Si_4O_{10}]$ (PANT and CRUICKSHANK, 1968) to which additional chains of "Zweier-Einfachkette" branches are linked. In this way, a silicate anion is formed which may just as well be interpreted as being a $1\tfrac{1}{2}$-fold layer (Fig. 14-A-12e).

Another silicate layer that lies between single and double layers is the anion of zussmanite, $KFe_{11}(Mg, Mn)_2\, b_\infty^2[(Si, Al)_{18}O_{42}](OH)_{14}$ (LOPES-VIEIRA and ZUSSMAN, 1969). It consists of a layer built of linked rings of "Sechserring" and "Dreierring"

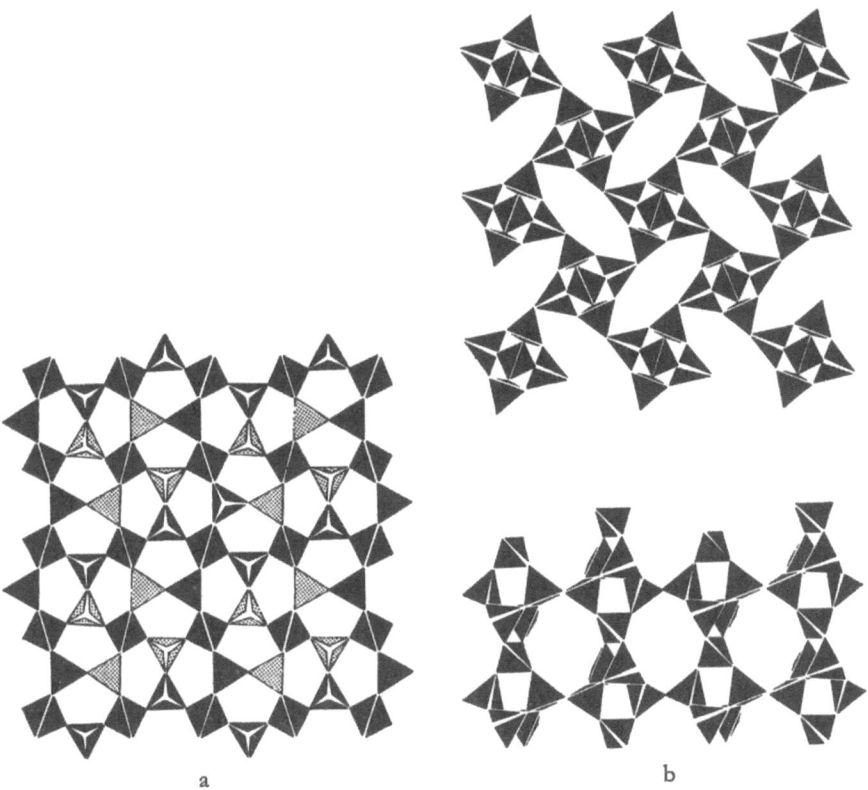

Fig. 14-A-13a and b. Infinite silicate anions which cannot be built completely by putting together single chains: a open single layer: meliphanite $Ca_4(Na, Ca)_4Be_4\overset{2}{\infty}[Si_7O_{20}][SiO_4]F_4$ (black tetrahedra $[SiO_4]$, dotted tetrahedra $[BeO_3F]$); b framework: natrolite $Na_4\overset{3}{\infty}[Si_6Al_4O_{20}] \cdot 4H_2O$ projected parallel and perpendicular to the strings of bridged "Dreierkette".

which can not be built completely by putting together single chains. To this single layer additional rings of the type "Sechserring" are linked giving again a $1^1/_2$-fold layer (Fig. 14-A-12f).

Silicates with infinite anion configurations not being formed completely by linking single chains. Meliphanite, $Ca_4(Na, Ca)_4Be_4\overset{2}{\infty}[Si_7O_{20}][SiO_4]F_4$ (DAL NEGRO, ROSSI and UNGARETTI, 1967), contains open layers consisting of chains of "Vierer-Einfachkette". Every fourth tetrahedron of a "Viererkette" is linked to another "Viererkette" by a triple-tetrahedron bridge (Fig. 14-A-13a). This structure is another border-line case showing that it is not always satisfactory to recognize only those tetrahedrally coordinated cations which are statistically distributed over the silicon positions as belonging to the silicate anion. In meliphanite this peculiar open silicate layer is supplemented by single $[BeO_3F]$ and $[SiO_4]$ tetrahedra, making an ordinary "Dreier-Einfachschicht". The open framework of neptunite, $LiNa_2K(Fe, Mg, Mn)_2(TiO)_2 \overset{3}{\infty}[Si_8O_{22}]$, is a similar case (CANNILLO, MAZZI and ROSSI, 1966).

As can be seen from Fig. 14-A-13b, in natrolite, $Na_4 \, {}^3_\infty[Si_6Al_4O_{20}] \cdot 4H_2O$, non-neighboring tetrahedra of a "Dreierkette" are bridged by pairs of tetrahedra, giving a string of composition ${}^1_\infty[Si_3Al_2O_{12}]$. Putting together those strings leads to the formation of the natrolite framework (MEIER, 1960).

b) Silicates with [SiO$_6$] Octahedra

The number of silicates with silicon in octahedral coordination is still rather small. A subdivision of these structures is therefore rather crude.

1. Phases with Single Octahedra

Thaumasite, $Ca_3[Si(OH)_6][SO_4][CO_3] \cdot 12H_2O$ (EDGE and TAYLOR, 1971), and at least three of the various phases of silicon diphosphate, $Si[P_2O_7]$ (LEVI and PEYRONEL, 1935; BISSERT and LIEBAU, 1970; LIEBAU and HESSE, 1971), contain [SiO$_6$] octahedra which have no corners in common with others. A number of chelates (see Subsection 14-A-I) may be mentioned here too.

2. Phases with Condensed Octahedra

So far, neither phases with a *limited* number of shared [SiO$_6$] octahedra nor those with chains or layers built from such octahedra have been described. Only 3-dimensional octahedral frameworks with edge- and corner-sharing of the octahedra are known. The hollandite-type high-pressure forms of the feldspars (RINGWOOD, REID and WADSLEY, 1967; REID and RINGWOOD, 1969) and stishovite SiO_2 (BAUR and KHAN, 1971) both have frameworks of corner- and edge-shared octahedra.

c) Silicates with More Than One Type of Silicate Anion

The classification of silicates as given in Subsections 14-A-IVa and 14-A-IVb is in fact a classification of the silicon-oxygen complexes. However, there are a few silicates which contain more than only one type of silicate anion. There are, for example, zoisite and clinozoisite, $Ca_2Al_3[SiO_4][Si_2O_7]O(OH)$, with single *and* double tetrahedra (DOLLASE, 1968), ardennite, $Mn_4(Mg, Al, Fe)_2Al_4[SiO_4]_2[Si_3O_{10}][(As, V)O_4](OH)_6$ (DONNAY and ALLMANN, 1968), and kilchoanite, $Ca_6[SiO_4][Si_3O_{10}]$ (TAYLOR, 1971), with both single and triple tetrahedra. Joesmithite, $(Pb, Ca, Ba)Ca_2Fe(Mg, Fe)_4[Si(O, OH)_4]_2 \, {}^1_\infty[Si_2O_6]_2(OH)_4$, is reported to contain single tetrahedra and chains of "Zweier-Einfachkette" (MOORE, 1968), while bavenite has triple tetrahedra in addition to $2{}^1_\infty[Si_6O_{16}]$ chains (CANNILLO, CODA and FAGNANI, 1966). Single chains as well as double chains have been found in vinogradovite, $Na_4(TiO)_4 \, {}^1_\infty[Si_2O_6]_2 \, 2{}^1_\infty[Si_4O_{10}] \cdot nH_2O$, by RASTSVETAEVA, SIMONOV and BELOV (1968), rings of "Dreier-Einfachring" and "Neuner-Einfachring" in eudialyte, $(Fe, Mn, Mg)_3Zr_3(Zr, Nb)_x(Ca, R.E.)_6Na_{12} \, {}^c[Si_9O_{27-y}(OH)_y]_2 \, {}^c[Si_3O_9]_2Cl_z$, by GUISEPPETTI, MAZZI and TADINI (1971). The structure of meliphanite with its open layers and isolated tetrahedra has been described in Subsection 14-A-IVa6, p.14-A-19. A garnet of composition $Mg_3^{[8]}(Al \, Si_{0.5}Mg_{0.5})^{[6]}[Si^{[4]}O_4]_3$, synthesized by RINGWOOD (1967) under very high pressure, even contains both single tetrahedra [SiO$_4$] and single octahedra [SiO$_6$].

d) Classification and Lattice Energy

Consciously or unconsciously, a classification based on the degree and mode of linking of $[SiO_n]$ polyhedra sometimes leads to an overestimation of the contribution of the bond energy of the silicate anion to the lattice energy of the crystal structure. On the other hand, a classification of the silicates based on the degree and mode of linking of the cation-oxygen polyhedra $[AO_n]$ would, apart from its greater inconvenience, lead to an unintended overemphasis of the bond energy of the cation polyhedra.

In fact, in tectosilicates the excess of $[SiO_4]$ tetrahedra over the $[AO_n]$ polyhedra results in a greater contribution of bond energy of the silicate framework to the lattice energy than of the cation polyhedra; in other words, the silicate anion of the tectosilicates is the rigid skeleton of the structure, and the cations more or less fill the voids between the oxygen atoms. However, in nesosilicates like forsterite, $Mg_2[SiO_4]$, the excess of $[AO_n]$ polyhedra over the $[SiO_4]$ tetrahedra gives the greater contribution to the lattice energy. The framework of cation-oxygen polyhedra is the prevailing skeleton of the structure. Therefore it must be recognized that the preference of the silicate anions as a basis for classification of silicates does not say anything about the significance of the bond energy of the silicate anion for the strength and stability of a silicate structure.

V. Factors Governing the Shape of the Silicate Anion

a) Electronegativity of Cations

As already shown in Subsection 14-A-I, octahedral coordination of silicon by oxygen in phases with Si—O—A bonds is favored by high electronegativity values of A, as well as by pressure, while the usual tetrahedral coordination is found in silicates with elements with medium or low electronegativites, χ_A.

b) Silicon to Oxygen Atomic Ratio

The ratio of SiO_2 to basic oxide in the material from which a silicate crystallizes is an important factor determining the degree of condensation of the $[SiO_4]$ tetrahedra. Each silicate anion has an atomic ratio, Si:O, determined by the mode of linking of its tetrahedra. Of the various possible condensation products for a given Si:O ratio (LAVES, 1932), those having the highest number of dimensions in space are the most stable (NOLL, 1962). For this reason, in the early days of silicate classification, it was often assumed that the type of silicate anion could be deduced simply from the atomic ratio Si:O of the particular silicate in the following way: the ratio 1:4 could only indicate a nesosilicate with $[SiO_4]^{4-}$ single tetrahedra, 1:3.5 only a sorosilicate with $[Si_2O_7]^{6-}$ double tetrahedra, 1:3 either a cyclosilicate with $^c[Si_pO_{3p}]^{2p-}$ rings or an inosilicate with $^1_\infty[SiO_3]_n^{2n-}$ chains, a ratio of 1:2.5 a phyllosilicate, and 1:2 a tectosilicate. From the examples mentioned in Subsection 14-A-IV of this review it can be seen that this is not true. For instance, despite its chemical composition, $Ca_3Si_2O_7$, kilchoanite does not contain $[Si_2O_7]^{6-}$ double tetrahedra but both single tetrahedra $[SiO_4]^{4-}$ and triple tetrahedra $[Si_3O_{10}]^{8-}$. A ratio Si:O = 1:2.5 has been found not only in phyllosilicates (e.g., sanbornite, $BaSi_2O_5$) but also in silicates with double chains, e.g., tuhualite $(Na, K)_2Fe_4(Si_2O_5)_6$.

H_2O, or with double rings, e.g., ekanite $K(Ca, Na)_2Th(Si_2O_5)_4$. On the other hand, silicate layers having Si:O values of 1:3 (leucosphenite), 1:2.5 (most of the phyllosilicates), 1:2.375 (madconaldite), 1:2.333 (zussmanite and synthetic $Na_4Si_6O_{14}$), 1:2.25 (synthetic $K_4Si_8O_{18}$) and 1:2 (hexagonal $BaSi_2Al_2O_8$) have so far been described. Therefore, although the Si:O ratio has an important influence on the kind of silicate anion, this is obviously not the only factor. The matter is even more complicated by the fact that many silicates contain oxygen atoms bonded not to silicon atoms, but to hydrogen or other atoms.

c) Size and Charge of Cations

The influence of cation size on the shape of silicate anions is most clearly demonstrated in case of the chain silicates (BELOV, 1960; LIEBAU, 1962). In order to form a cation-oxygen polyhedron in accordance with the radius ratio, $r_{cation}:r_{oxygen}$, the silicate chain has to adjust itself to the cations. Fig. 14-A-14 demonstrates the adjustment of the tetrahedral chains to columns of cation-oxygen octahedra in some

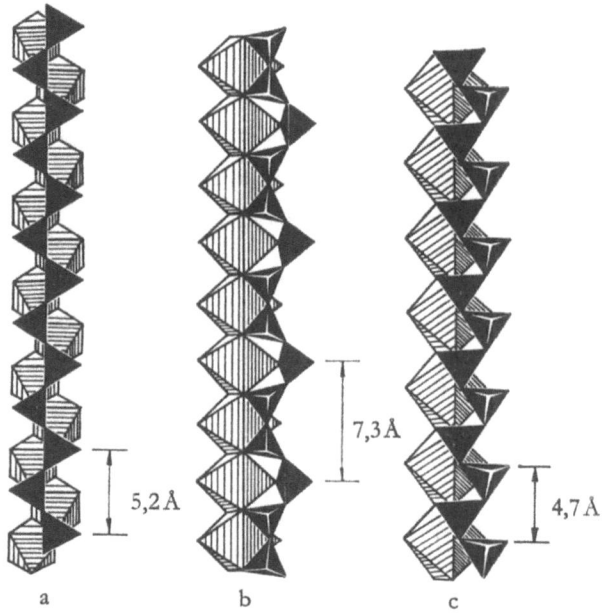

Fig. 14-A-14a—c. Adjustment of tetrahedral single chains to columns of $[AO_6]$ polyhedra in some silicates: a extended "Zweierkette": enstatite $Mg_{2\infty}^1[Si_2O_6]$; b "Dreierkette": wollastonite $Ca_{3\infty}^1[Si_3O_9]$; c shortened "Zweierkette": high-temperature $Ba_{2\infty}^1[Si_2O_6]$

silicates of general formula $A^{2+}SiO_3$: entstatite, $MgSiO_3$ ($r_{Mg^{2+}}=0.720$ Å)[1], wollastonite, $CaSiO_3$ ($r_{Ca^{2+}}=1.00$ Å), and the high-temperature phase $BaSiO_3$(h), ($r_{Ba^{2+}}=1.36$ Å). This figure as well as Table 14-A-3 clearly shows that the periodicity of the chain depends on the size of the cation A. In wollastonite, the length

---
[1] Ionic radii given are the effective ionic radii defined by SHANNON and PREWITT (1969).

Table 14-A-3. Relation between cation size and chain periodicity for silicates $A^{2+}SiO_3$ (hp: high-pressure phase; ht: high-temperature phase; lt: low-temperature phase)

	En-statite	Ferro-silite I	Ferro-silite III (hp)	Pyrox-ferro-ite	Rhodo-nite	$MnSiO_3$ (ht)	Wollas-tonite (lt)	Pseudo-wollas-tonite (ht)	$SrSiO_3$	$BaSiO_3$ (lt)	$BaSiO_3$ (ht)
Cation A^{2+}	Mg^{2+}		Fe^{2+}	$Fe^{2+}_{0.83}Ca^{2+}_{0.13}Mg^{2+}_{0.02}Mn^{2+}_{0.02}$		Mn^{2+}		Ca^{2+}	Sr^{2+}		Ba^{2+}
$r_{A^{2+}}$ [Å]	0.72		0.77	0.80		0.82		1.00	1.12		1.36
Periodicity of chain [tetrahedra]	2	2	9	7	5	3	3	rings			2
Chain period [Å]	5.2	5.3	22.7	17.4	12.5	7.2	7.3				4.7

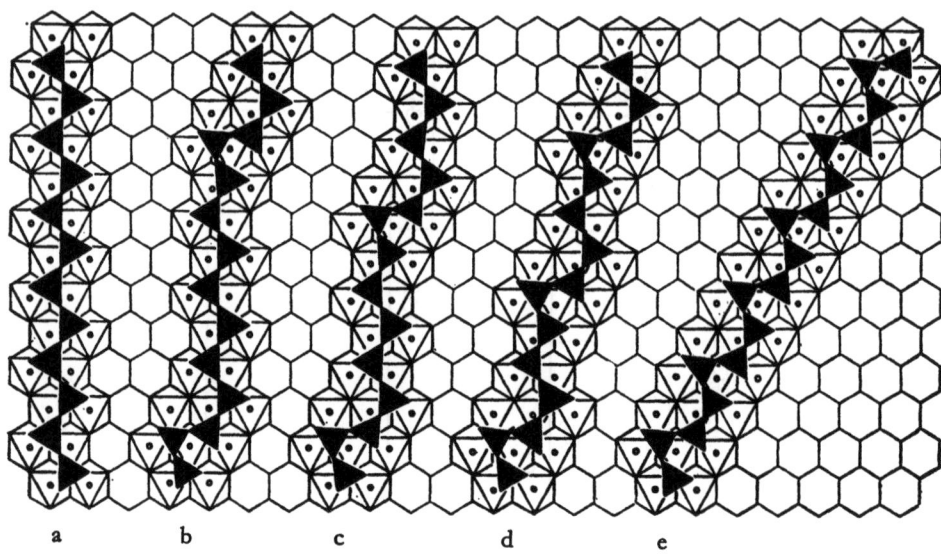

Fig. 14-A-15a—e. Adjustment of tetrahedral single chains to columns of [AO$_6$] octahedra in silicates with an odd number of [SiO$_4$] tetrahedra in the identity period of the chain. Schematic diagrams showing the orientation of the silicate chains relative to the columns of cation-oxygen octahedra: a "Zweierkette" in enstatite Mg$_2{}^1_\infty$[Si$_2$O$_6$]; b "Neunerkette" in ferrosilite III Fe$_9{}^1_\infty$[Si$_9$O$_{27}$](p); c "Siebenerkette" in pyroxferroite (Fe, Ca)$_7{}^1_\infty$[Si$_7$O$_{21}$]; d "Fünferkette" in rhodonite (Mn, Ca)$_5{}^1_\infty$[Si$_5$O$_{15}$]; e "Dreierkette" in wollastonite Ca$_3{}^1_\infty$[Si$_3$O$_9$]

of two edges of [CaO$_6$] octahedra is of about the same size as the period of three [SiO$_4$] tetrahedra forming a "Dreierkette" running parallel to the octahedral column. For the larger [BaO$_6$] polyhedra in BaSiO$_3$(h) the length of one octahedron is commensurable with the period 4.7 Å of the shortened version of the "Zweierkette".

In the pyroxenoids the adjustment of the tetrahedral chains to the columns of octahedra from Mn^{2+} to Mg^{2+}, with decreasing size, is mainly achieved by an increasing rotation of the chains in relation to the columns, resulting in chains of different periodicities (PREWITT and PEACOR, 1964; BURNHAM, 1966). In this regard, the "Zweierkette" of the pyroxenes is the end member of the homologous series of silicate chains with an odd number of tetrahedra in the period of the chain: "Dreierkette"-"Fünferkette"-"Siebenerkette"-"Neunerkette" — ··· — "Zweierkette" with a 5.3 Å periodicity (Fig. 14-A-15).

Table 14-A-4 contains the silicates for which chains with four, six or twelve, i.e., an even number of tetrahedra, have been found in the identity period.

Although only a rather limited number of such silicates is known, they seem to have in common that, besides cations with electronegativity values of 1.0 or less, they all also contain cations with appreciably higher electronegativities, giving rise to A—O bonds with considerably covalent character. In addition, most of these silicates have atomic ratios, $n_{\text{cations}} : n_{\text{silicon}} \neq 1$, i.e., the valence of the cations is an important factor for the determination of the chain type.

Table 14-A-4. *Silicates containing chains with four, six and twelve tetrahedra in the identity period of the chains*

Silicate	Formula	$n_{cat}:n_{Si}$	Chain period [Å]
a) "Viererkette"			
Krauskopfite	$Ba_2 \overset{1}{\infty}[Si_4O_8(OH)_4] \cdot 4\,H_2O$	1.5:1	8.46
Leukophane	$Ca_2Na_2Be_2 \overset{1}{\infty}[Si_4O_{12}]F_2$	1.5:1	7.42
Haradaite	$Sr_2(V(OH)_2)_2 \overset{1}{\infty}[Si_4O_{12}]$	1:1	7.06
Batisite	$BaNa_2(TiO)_2 \overset{1}{\infty}[Si_4O_{12}]$	1.25:1	8.08
Narsarsukite	$Na_4(Fe,Ti)_2(O,OH)_2 2\overset{1}{\infty}[Si_8O_{20}]$	0.75:1	8.0
b) "Sechserkette"			
Stokesite	$Ca_2Sn_2 \overset{1}{\infty}[Si_6O_{18}] \cdot 4\,H_2O$	0.67:1	11.63
Tuhualite	$(Na,K)_2Fe_2^{3+}Fe_2^{2+} 2\overset{1'}{\infty}[Si_{12}O_{30}] \cdot H_2O$	0.5:1	10.11
c) "Zwölferkette"			
Alamosite	$Pb_{12} \overset{1}{\infty}[Si_{12}O_{36}]$	1:1	19.63

These rules for the adjustment of 1-dimensionally infinite silicate chains to the system of cation-oxygen polyhedra also hold for tetrahedral layers. The phyllosilicates of Mg^{2+}, Fe^{2+}, Al^{3+}, e.g., contain layers of "Zweierschicht", while calcium phyllosilicates have chains of "Dreierkette" as chains of lowest periodicity, although it is not possible to build the layers by putting together these "Dreierkette" chains but "Viererkette" chains (Fig. 14-A-7b, 14-A-8b). For phyllosilicates this 1-dimension adjustment has to be completed by a fit between the tetrahedral and the polyhedral part of the structure in the second dimension. The smaller the cation of an *anhydrous* layer silicate, $A_2 \overset{2}{\infty}[Si_2O_5]$, the more the tetrahedral layer is folded (Fig. 14-A-16a—c). Of two cations of approximately equal size, that of higher valency gives rise to a stronger folding of the layer (Fig. 14-A-16b—d). Trivalent and small divalent cations would require so strong a folding of the layer, that no anhydrous phyllosilicates of these cations exist because the necessary structures are unstable.

In *hydrous* phyllosilicates the cations are not only coordinated by oxygen atoms of the tetrahedral layers, but also by additional OH groups or water. Therefore, the $[A(O,OH,H_2O)_n]$ polyhedra act as "effective cations" instead of the much smaller "naked" cations, A. In kaolinite the $[AlO_2(OH)_4]$ octahedra ($r_{Al^{3+}}=0.530$ Å) are the right size, so that a flat tetrahedral silicate layer fits nicely onto a layer of these octahedra (Fig. 14-A-17). For Mg^{2+} ($r_{Mg^{2+}}=0.720$ Å) the slightly larger size of $[Mg(O,OH)_6]$ results in a difference of mesh size of octahedral and tetrahedral layers. The tension between these two layers is reduced by rolling up the double layer to a tube (chrysotile, Fig. 14-A-18b), by corrugating the double layer (antigorite, Figs. 14-A-6c, 14-A-18c), or by breaking up the octahedral layer to laths of varying widths, compensated for by pointing the free corners of the tetrahedra of the $\overset{2}{\infty}[Si_4O_{10}]$ sheet to both sides of the sheet (sepiolite, Fig. 14-A-6d, 18d; palygorskite, Fig. 14-A-6e, 14-A-18e). With increasing size of the cation, the tension between tetrahedral and octahedral layers increases, giving rise to decreasing radii of the tubes from metahalloysite, $Al_4 \overset{2}{\infty}[Si_4O_{10}](OH)_8$, over chrysotile, $Mg_6 \overset{2}{\infty}[Si_4O_{10}](OH)_8$, and garnierite, $Ni_6 \overset{2}{\infty}[Si_4O_{10}](OH)_8$ ($r_{Ni^{2+}}=0.700$ Å), to synthetic cobalt

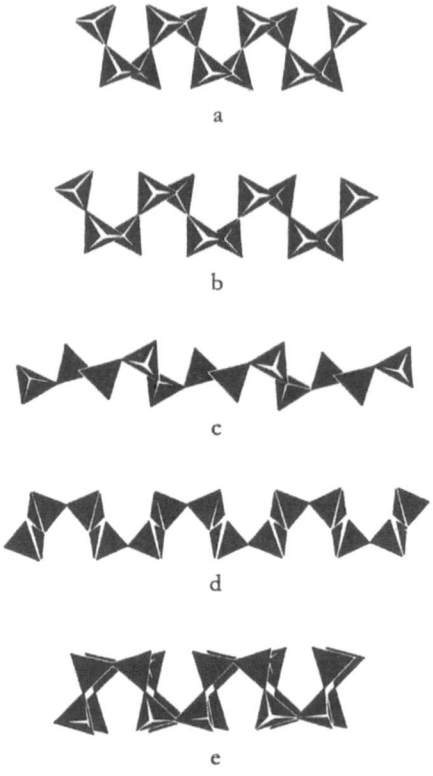

Fig. 14-A-16a—e. Folding of tetrahedral layers of some phyllosilicates. The projection is parallel to a direction within the layer: a $Li_4{}_\infty^2[Si_4O_{10}]$ ($r_{Li}^{[4]}=0.59$Å); b α-$Na_4{}_\infty^2[Si_4O_{10}]$ ($r_{Na}^{[6]}=1.00$ Å); c β-$Na_4{}_\infty^2[Si_4O_{10}]$ ($r_{Na}^{[6]}=1.02$ Å); d sanbornite $Ba_2{}_\infty^2[Si_4O_{10}]$ ($r_{Ba}^{[6]}=1.36$ Å); e gillespite $BaFe{}_\infty^2[Si_4O_{10}]$ ($r_{Ba}^{[6]}=1.36$ Å, $r_{Fe}^{[4]}=0.63$ Å)

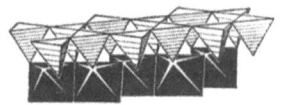

Fig. 14-A-17. Fit between tetrahedral and octahedral layers in kaolinite

chrysotile, $Co_6{}_\infty^2[Si_4O_{10}](OH)_8$ ($r_{Co^{2+}}=0.735$ Å), and iron chrysotile, $Fe_6{}_\infty^2[Si_4O_{10}](OH)_8$ ($r_{Fe^{2+}}=0.770$ Å). For even larger cations like Mn^{2+} ($r_{Mn^{2+}}=0.820$ Å), no smaller radius of curvature is possible; also, the breaking up of the octahedral layers into laths is no longer sufficient. Therefore these laths break up into small groups of cation-oxygen polyhedra which are bonded to tetrahedral layers in which neighboring rings (and not chains as in sepiolite and antigorite) point up and down alternately. In manganpyrosmalite, $Mn_{16}{}_\infty^2[Si_{12}O_{30}](OH)_{18}Cl_2$, this ring unit is a "Sechserring" $[Si_6O_{18}]$ (Fig. 14-A-7d), in apophyllite, $Ca_4K{}_\infty^2[Si_8O_{20}]F \cdot 8H_2O$ ($r_{Ca^{2+}}=1.07$ Å, seven coordinated; $r_{K^+}=1.51$ Å, eight coor-

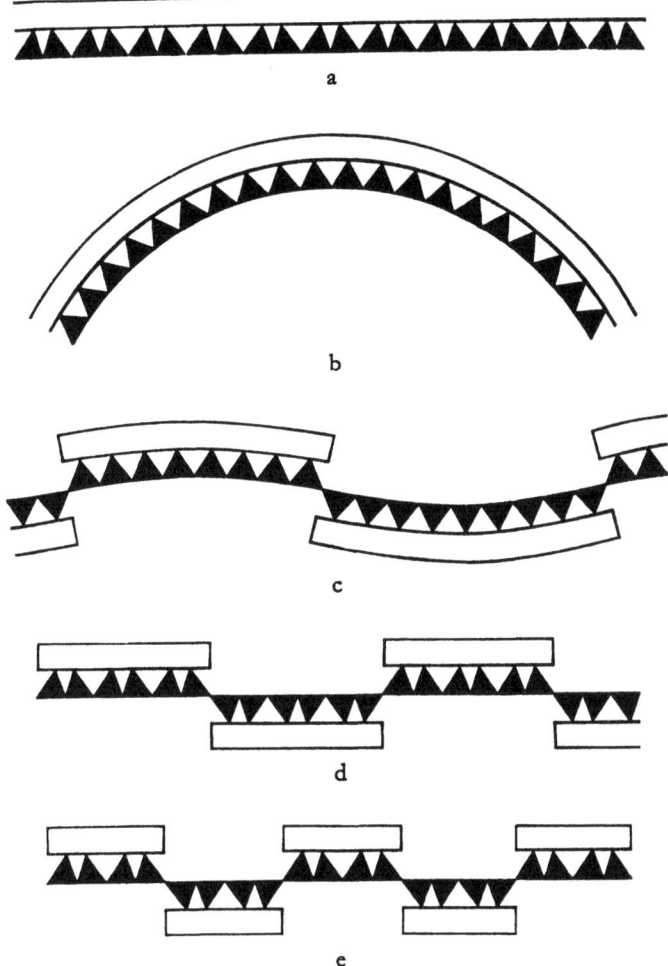

Fig. 14-A-18a—e. Adjustment between tetrahedral and octahedral layer in some hydrous layer silicates of aluminium and magnesium. (Black triangles: [SiO$_4$] tetrahedra; open bars: octahedral layer): a kaolinite Al$_4^{\frac{2}{\infty}}$[Si$_4$O$_{10}$](OH)$_8$; b chrysotile Mg$_6^{\frac{2}{\infty}}$[Si$_4$O$_{10}$](OH)$_8$; c antigorite Mg$_6^{\frac{2}{\infty}}$[Si$_4$O$_{10}$](OH)$_8$; d sepiolite (Mg, Fe, Al)$_8^{\frac{2}{\infty}}$[Si$_4$O$_{10}$]$_3$(O, OH)$_4 \cdot$ 8H$_2$O; e palygorskite Mg$_5^{\frac{2}{\infty}}$[Si$_4$O$_{10}$]$_2$(OH)$_2 \cdot$ H$_2$O

dinated), the curling unit is a "Viererring" [Si$_4$O$_{12}$] (Fig. 14-A-7b). For still larger cation-oxygen polyhedra, even smaller units would be necessary, but no examples of hydrous phyllosilicates with Sr^{2+} or Ba^{2+} are known.

A semiquantitative relation between size and charge of cations and the degree of folding or curling of the tetrahedral layers in hydrous and anhydrous phyllosilicates has been given by LIEBAU (1968).

In general, in tectosilicates the atomic ratio, cation:silicon, is low so that the influence of the cations on the tetrahedral framework is rather weak.

VI. Isotypism and Homeotypism of Silicates

Compounds which are isotypic with phosphates may be derived by
1. complete replacement of cations,
2. complete replacement of silicon atoms,
3. complete replacement of oxygen atoms, and
4. combinations of several of the above mentioned possibilities.

1. The adjustment of cation-oxygen polyhedra [MeO$_n$] to silicon-oxygen anions is most easily achieved with [SiO$_4$] single tetrahedra, but tensions between [MeO$_n$] polyhedra and silicate anions increase with increasing degree of polymerisation of the [SiO$_4$] tetrahedra. Therefore, isotypism by replacement of cations is most common for monosilicates, but rather rare for high-polymer silicates. For instance, the olivine structure has been found in A$_2^{2+}$[SiO$_4$] phases with A^{2+} = Ni^{2+}, Mg^{2+}, Co^{2+}, Fe^{2+}, Mn^{2+}, Cd^{2+}, Ca^{2+}, Sr^{2+} and Ba^{2+}, i.e., with cations of radii between 0.70 Å (Ni^{2+}) and 1.36 Å (Ba^{2+}). However, for inosilicates and phyllosilicates, the structure type varies strongly with the size of the cations (see 14-A-IV a 2, and 14-A-IV a 3).

2. Isotypism between silicates and germanates is quite common and extends from nesosilicates to tectosilicates. Detailed lists of isotypic silicates and germanates are given in recent reviews by STRUNZ (1961) and WITTMANN (1966). The higher tendency of germanium compared with silicon to accept octahedral coordination of oxygen atoms makes germanates especially valuable as model substances for high pressure studies of silicates.

Replacement of [SiO$_4$]$^{4-}$ by [BeO$_4$]$^{6-}$, [AlO$_4$]$^{5-}$, [GaO$_4$]$^{5-}$, [PO$_4$]$^{3-}$, [AsO$_4$]$^{3-}$, [VO$_4$]$^{3-}$ and sometimes [SO$_4$]$^{2-}$, [CrO$_4$]$^{2-}$, [WO$_4$]$^{2-}$ and [BO$_4$]$^{5-}$ leads to numerous phases which are isotypic with silicates, some of which are listed in Table 15-A-3.

Most remarkable is the substitution of Si by other atoms in the various phases of SiO$_2$. For tetrahedral [SiO$_4$] groups, the substitution of 2 Si^{4+} ↔ X^{3+} + X^{5+}, leads to ordered distribution of cations, i.e., to phases that are homeotypic instead of truly isotypic with the corresponding silica modifications. Analogues of low quartz are known of: BPO$_4$, AlPO$_4$, GaPO$_4$, FePO$_4$, MnPO$_4$, BAsO$_4$, AlAsO$_4$ and GaAsO$_4$, while high-quartz analogues have been described for AlPO$_4$, FePO$_4$ and MnPO$_4$. Low-tridymite structures have been found for AlPO$_4$ and FePO$_4$, high-tridymite structure only for AlPO$_4$. While low-cristobalite analogues have been verified for AlPO$_4$ and GaPO$_4$, high-cristobalite analogues are known for BPO$_4$, AlPO$_4$, GaPO$_4$, MnPO$_4$, BAsO$_4$ and AlAsO$_4$. The structure of BeSO$_4$ may be regarded as an ordered substitution 2 Si^{4+} ↔ Be^{2+} + S^{6+} in the high-cristobalite structure.

For stishovite, with [SiO$_6$] octahedra, the substitution 2 Si^{4+} ↔ X^{3+} + X^{5+} leads to random distribution of cations. A number of phases with X^{0+} = Al, Cr, Ga, Fe, Rh, Sc, In, and X^{5+} = V, Sb, Nb, Ta are known to be isotypic with stishovite. On the other hand, a homeotypic tri-rutile structure with ordered cation distribution has been found for the substitutions: 3 Si^{4+} ↔ X^{2+} + 2 X^{5+} and 3 Si^{4+} ↔ 2 X^{3+} + X^{6+} with X^{2+} = Ni, Mg, Co, Zn, Fe, X^{3+} = Cr, X^{5+} = Sb, Nb, Ta and X^{6+} = W, Te.

3. Some thiosilicates and selenosilicates like Ca$_2$SiS$_4$, Ca$_2$SiSe$_4$ (ROCKTÄSCHEL, RITTER and WEISS, 1964) and Li$_2$SiS$_3$ (WEISS and ROCKTÄSCHEL, 1960) have the

same structures as the corresponding silicates. SiS_2 and $SiSe_2$ are isotypic with fibrous SiO_2 (WEISS and WEISS, 1954).

4. Very often Si as well as the cations are replaced. This is the case for some of the phases given in Table 15-A-3. Most remarkable are the fluoroberyllates which were shown by GOLDSCHMIDT (1926) to be models of the silicates. The ratio $r_{Be^{2+}} : r_{F^-}$ is almost the same as $r_{Si^{4+}} : r_{O^{2-}}$. This causes isotypism of many silicates with their corresponding fluoroberyllates. In addition, the ionic charges of Be^{2+} and F^- are half those of Si^{4+} and O^{2-}, and therefore the weaker bonds between the atoms in the fluoroberyllates cause lower melting points and, in the case of polymorphous substances, even lower transformation temperatures compared with silicates.

The analogy is so striking, that in favorable cases, a binary phase diagram MeF_n—BeF_2 can be transformed into a silicate system MeO_n—SiO_2 by multiplying all temperatures measured in °K by a certain factor. For example, in the corresponding systems NaF—BeF_2 and CaO—SiO_2 this factor is 2.82 (THILO and SCHRÖDER, 1951). For $Ca_2[SiO_4]$, all four modifications have been found in the $Na_2[BeF_4]$ model.

The garnet structure $A_3^{[8]} B_2^{[6]} [X^{[4]} Z_4]_3$ is a structure type that is extremely variable in chemical composition. Not only the eight- and six-coordinated sites can be occupied by a whole series of cations, but the silicon atoms can also be replaced by Li^+, Zn^{2+}, Al^{3+}, Fe^{3+}, Ga^{3+}, Ge^{4+}, As^{5+}, V^{5+} (Table 14-A-5)[1]. The so-called hydrogarnet (ZÁBIŃSKI, 1966), $Ca_3Al_2[(OH)_4]_3$ (COHEN-ADDAD, DUCROS and BERTAUT, 1967), may formally be regarded as garnet with complete replacement of each Si^{4+} ion by four protons (see Subsection 14-A-VIIb). In $Na_3Al_2[LiF_4]_3$ even the oxygen atoms are replaced (GELLER, 1971). A review of recent work on garnets has been given by GELLER (1967).

VII. Isomorphism in Silicates

It was the extremely widespread isomorphous replacement of elements in silicates that prevented the deduction of their atomic structures by chemical methods until the development of X-ray methods for crystal structure analysis.

a) Partial Replacement of Cations

This is very common and follows the rules known from other compounds. It has been most extensively studied for the amphiboles (ERNST, 1968) and the olivine type phases. Although suitable ionic size is the main condition for a replacement, in the case of transition elements the stabilization energy due to crystal field effect also has an influence on the kind and degree of substitution (MATSUI and SYONO, 1968).

Regardless of the similarities in ionic radii of Mg^{2+} (0.720 Å), Zn^{2+} (0.745 Å) and Mn^{2+} (0.820 Å), only 16 mole% zinc ions are substituted for Mg^{2+} in olivine, $Mg_2[SiO_4]$, and 20 mole% Mg^{2+} ions in willemite, $Zn_2[SiO_4]$, at 850°C (SARVER and HUMMEL, 1962), and probably no more than 5 mole% ZnO for Mn^{2+} in tephroite, $Mn_2[SiO_4]$ (HURLBUT, 1961), under normal conditions. However, there is complete solid solution between $Mg_2[SiO_4]$ and $Mn_2[SiO_4]$.

---

[1] In contradiction to former statements, phosphate garnets $A_3B_2[PO_4]_3$ do not exist (SCHWARZ and SCHMIDT, 1971).

Table 14-A-5. $A_3^{[8]}B_2^{[6]}[X^{[4]}O_4]_3$ phases with garnet structure. Only a few of the numerous phases with mixed cations A, B and X respectively are included. Ionic radii in Å in parenthesis. The $B^{[6]}$ ions are tabulated in the sequence of increasing radii.

A[8] \ X[4]	Li+ (0.74)	Zn²⁺ (0.745)	Al³⁺ (0.530)	Ga³⁺ (0.620)	Fe³⁺ (0.645)	Si⁴⁺ (0.26)	Ge⁴⁺ (0.40)	As⁵⁺ (0.335)	V⁵⁺ (0.355)
Mg²⁺ (0.720)						Al³⁺ (0.530), Cr³⁺, Fe³⁺ (0.645)			
Co²⁺ (0.735)						Al³⁺ (0.530)			
Fe²⁺ (0.770)						Al³⁺ (0.530)			
Mn²⁺ (0.820)						Al³⁺ (0.530), Fe³⁺ (0.645)	Al³⁺ (0.530), Cr³⁺, Ga³⁺, V³⁺, Fe³⁺ (0.645)		(Li₀.₅⁺Mg₀.₅²⁺) M = Mg (0.720), Cu, Co, Zn (0.745)
R.E. ³⁺ (0.85—1.03)	Te⁶⁻		Al³⁺ (0.530) Fe³⁺ (0.645)	Ga³⁺ (0.620)	Fe³⁺ (0.645)				
Cd²⁺ (1.07)		Te⁶⁺							
Ca²⁺ (1.12)		Te⁶⁺		(Nb₀.₅⁵⁺M₀.₅⁴⁺), (Ta₀.₅⁵⁺M₀.₅⁴⁺) M = Ti (0.605), Sn, Hf, Zr (0.72)		Al³⁺ (0.530), Cr³⁺, Ga³⁺, V³⁺, Fe³⁺, Mn³⁺, Ti³⁺, Sc³⁺, In³⁺ (0.790)	Al³⁺ (0.530), Cr³⁺, Ga³⁺, V³⁺, Fe³⁺, Mn³⁺, Rh³⁺, (Zr₀.₅⁴⁺Ni₀.₅²⁺), (Sn₀.₅⁴⁺Co₀.₅²⁺), (Zr₀.₅⁴⁺Co₀.₅²⁺), Sc³⁺, In³⁺, R.E. (0.85—1.03)		(Li₀.₅⁺Co₀.₅²⁺) (0.74)
Na⁺₀.₃₃ Ca²⁺₀.₆₇				Sb⁵⁺ (0.61)				Ni²⁺ (0.700), Mg²⁺, Co²⁺, Mn²⁺ (0.820)	Cu²⁺ (0.73), Zn²⁺ (0.745)
Sr²⁺ (1.25)							Sc³⁺ (0.730), In³⁺, R.E.³⁺ (0.85—1.03)		

In addition to ionic size and crystal field effect, the degree of isomorphous replacement is governed by the availability of the various cations in the different geological environments in which silicates form. This may sometimes explain the observation of gaps in ionic replacement for natural minerals where complete miscibility has been found experimentally. The distribution of cations between a crystalline silicate and the phase from which it crystallizes depends strongly on temperature and also on pressure. If the p, T-diagram of a certain system for naturally occurring compounds is known, a silicate mineral can be used as geological thermometer and/or barometer. But there must be information concerning whether equilibrium was reached and whether the equilibrium state has been preserved.

The distribution of cations which can replace each other within a crystalline silicate is usually random at high temperatures. If the difference in size and/or crystal field stabilization energy of two cations is sufficient, they will adopt ordered distributions at lower temperatures. This is known from a number of recent, very accurate, structure determinations, studies of optical pleochroism, and Mössbauer studies, e.g., of orthopyroxenes (GHOSE, 1965; BURNS, 1966), amphiboles (PAPIKE, ROSS and CLARK, 1969; WHITFIELD and FREEMAN, 1967), and micas (TAKÉUCHI and SADANAGA, 1966).

b) Partial Replacement of Silicon Atoms

In many silicates silicon can be substituted, to some extent, by aluminum. For silicates found in nature the Al:Si ratio, in general, increases from nesosilicates to tectosilicates. This is at least partly due to an increase of aluminum content in the source from which the silicate crystallizes. According to LOEWENSTEIN's rule (1954) for silicates with Si/Al substitution in tetrahedral sites, the distribution is such that $Al^{[4]}$—O—$Al^{[4]}$ bonds are avoided, while $Si^{[4]}$—O—$Al^{[4]}$ and $Si^{[4]}$—O—$Si^{[4]}$ groups are stable. For silicates with tetrahedral chains, layers, or frameworks, this rule requires ratios $Al^{[4]}:Si^{[4]} \leq 1.0$, which is in good agreement with analytical data given by DEER, HOWIE and ZUSSMAN (1962, 1963). Only few exeptions are known, e.g., xanthophyllite (TAKÉUCHI and SADANAGA, 1966). The complete series of solid solution between the garnet structures of spessartine, $Mn_3Al_2^{[6]}[Si^{[4]}O_4]_3$, and of $Y_3Al_2^{[6]}[Al^{[4]}O_4]_3$ (YODER and KEITH, 1951) does not contradict this rule.

At high temperatures there is often a random distribution of silicon and aluminium over equivalent crystallographic sites in the structure. With decreasing temperature the tendency to adopt an ordered Si/Al distribution increases. These Si/Al order-disorder relations have been extensively studied within the feldspar group (TAYLOR, 1965; BAMBAUER, 1969). Similar relations seem to exist for nephelines, $(K, Na)_4 {}^3_\infty[Si_4Al_4O_{16}]$ (SAHAMA, 1962). In phyllosilicates partial Si/Al ordering has also been found (GÜVEN and BURNHAM, 1967), but is obviously less pronounced.

Replacement of silicon by other atoms is less pronounced but often of special geochemical interest. The substitution of Si^{4+} by P^{5+} has been described in Section 15-A, substitution by other ions such as Ti^{4+} (SAXENA, 1966; SCHRÖPFER, 1968; HARTMANN, 1969), Fe^{3+} (GELLER and MILLER, 1959; FAYE, 1969; HARTMANN, 1969), B^{3+} (CHRIST, 1965; KROGH-MOE, 1964), Be^{2+} (BEUS, 1956; ALEKSANDROVA and SIDORENKO, 1966) and As^{5+} (WUENSCH, 1960) have been proved by X-ray as well as by visible and IR spectroscopic methods, or have been suggested from crystal chemical consideration. Although several complete series of solid solutions between silicates and corresponding germanates have been synthesized (HAHN, 1970; GREBENSHCHIKOV, SHITOVA and TOROPOV, 1967), Si/Ge replacement in natural minerals is negligible due to the low abundance of germanium.

There has been much discussion about the substitution of $[SiO_4]$ tetrahedra by $[(OH)_4]$ groups in hydrogarnets, in minerals of the zircon groups, and a number of other silicates (e.g., McCONNEL, 1963). However, crystal structure determinations by combined X-ray diffraction, neutron diffraction, NMR and IR methods, have shown that at least the hydrogarnets, $Ca_3Al_2(OH)_{12}$ and $Ca_3Al_2[SiO_4]_{2.16}(OH)_{3.36}$, are more reasonably described as hydroxides with $[Al(OH)_6]$ octahedra than as silicates with $[(OH)_4]$ tetrahedra, because the protons substituting for silicon occupy places outside the "former" $[SiO_4]$ tetrahedra (COHEN-ADDAD, DUCROS and BERTAUT, 1967).

If silicon is substituted by non-tetravalent atoms, valency balance can be achieved in several ways:

(i) For higher charged tetrahedral ions like P^{5+} or As^{5+}, charge difference is compensated either by replacement of other cations of lower valency, as in the series lithiophilite $LiMn[PO_4]$ - tephroite $Mn_2[SiO_4]$ (BRADLEY, ENGEL and MUNRO, 1966), or by introduction of cation vacancies.

(ii) Charge deficiencies produced by replacement of Si^{4+} by Al^{3+}, Fe^{3+}, B^{3+} and Be^{2+} are counterbalanced either by replacement of higher valent cations, e.g. in plagioclase $(Na, Ca)_\infty^3[Si_{2+x}Al_{2-x}O_8]$ with substitution $Na^+ + Si^{4+} \leftrightarrow Ca^{2+} + Al^{3+}$, or by introduction of additional cations into voids in the structure. This latter case is verified in quartz where Al^{3+} impurities in tetrahedral sites are compensated by equivalent amounts of Na^+, Ca^{2+}, Cu^{2+}, Fe^{2+}, Mg^{2+}, etc., in hollow channels extending parallel to [0001] (KAMENTSEV, 1965).

As a rule in natural silicates, silicon atoms and cations are replaced by other atoms at the same time.

c) Partial Replacement of Oxygen

Partial replacement of oxygen bonded to silicon in crystalline silicates has not yet been definitely established, but in a number of minerals like topaz, $Al_2[SiO_4]$ (OH, F) (GEBERT and ZEMANN, 1965), amphiboles, or lithium micas, there is partial substitution of F^- for OH^- not bonded to silicon.

References: Section 14-A

ALEKSANDROVA, I. T., SIDORENKO, G. A.: Gadolinite Geol. Mestorozhd. Redkikh Elementov, Vses. Nauchn.-Issled. Inst. Mineral'n. Syrya No. 26, 66 (1966), cited after Chem. Abstr. **65**, 1968 (1966).

ALEONARD, S., LE FUR, Y.: Fluoberyllates de structure benitoite. Bull. Soc. Franç. Mineral. Crist. **89**, 425 (1966).

ALLRED, A. L., ROCHOW, E. G.: A scale of electronegativity based on electrostatic force. J. Inorg. Nucl. Chem. **5**, 264 (1958).

ARAKI, T., ZOLTAI, T.: Refinement of a coesite structure. Z. Krist. **129**, 381 (1969).

ARANSSON, B., LUNDSTRÖM, T., RUNDQVIST, S.: Borides, Silicides and Phosphides. London: Methuen 1965.

BAMBAUER, H. U.: Feldspat-Familie, p. 645. In: TRÖGER, W. E., Optische Bestimmung der gesteinsbildenden Minerale. Teil 2, Textband, 2. Aufl. Stuttgart: Schweizerbart 1969.

BAUR, W. H.: Bond length variation and distorted coordination polyhedra in inorganic crystals. Trans. Am. Cryst. Assoc. **6**, 129 (1970).

— The predictions of bond length variations in silicon-oxygen bonds. Am. Mineralogist **56**, 1573 (1971).

— KHAN, A. A.: Rutile-type compounds. IV, SiO_2, GeO_2 and a comparison with other rutile-type structures. Acta Cryst. B **27**, 2133 (1971).

BELOV, N. V.: Chapter B of the crystal chemistry of silicates. Fortschr. Mineral. **38**, 4 (1960).

BEUS, A. A.: Peculiarities of isomorph entry of beryllium into the crystalline structure of minerals. Geokhimiya 67 (1956).

BIRLE, J. D., GIBBS, G. V., MOORE, P. B., SMITH, J. V.: Crystal structures of natural olivines. Am. Mineralogist **53**, 807 (1968).

— TETTENHORST, R.: Refined muscovite structure. Mineral. Mag. **36**, 883 (1968).

BISSERT, G., LIEBAU, F.: Die Kristallstruktur von monoklinem Siliziumdiphosphat SiP_2O_7AIII: Eine Phase mit $[SiO_6]$-Oktaedern. Acta Cryst. B **26**, 233 (1970).

BORGEN, O., SEIP, H. M.: Crystal structure of β-Si_3N_4. Acta Chem. Scand. **15**, 1789 (1961).

BOUCHER, M. L., PEACOR, D. R.: The crystal structure of alamosite, $PbSiO_3$. Z. Krist. **126**, 98 (1968).

BRADLEY, R. S., ENGEL, P., MUNRO, D. C.: Subsolidus solubility between R_2SiO_4 and $LiR \cdot \cdot PO_4$: a hydrothermal investigation. Mineral. Mag. **35**, 742 (1966).

BRADLEY, W. F.: The structural schema of attapulgite. Am. Mineralogist **25**, 405 (1940).

BRAGG, W. L.: The structure of silicates. Z. Krist. **74**, 237 (1930).

BRAUNER, K., PREISINGER, A.: Struktur und Entstehung des Sepioliths. Tschermaks Mineral. Petrog. Mitt., III. F. **6**, 120 (1956).

BROWN, B. E., BAILEY, S. W.: Chlorite polytypism: II. Crystal structure of a one-layer Cr-chlorite. Am. Mineralogist **48**, 42 (1963).

BROWN, G. E., GIBBS, G. V.: Refinement of the crystal structure of osumilite. Am. Mineralogist **54**, 101 (1969).

— — RIBBE, P. H.: The nature and the variation in length of the Si—O and Al—O bonds in framework silicates. Am. Mineralogist **54**, 1044 (1969).

BUERGER, M. J., PREWITT, C. T.: Comparison of the crystal structures of wollastonite and pectolite. Miner. Soc. Am. Special Paper **1**, 293 (1963).

BURNHAM, C. W.: Ferrosilite III: A triclinic pyroxenoid-type polymorph of ferrous metasilicate. Science **154**, 513 (1966).

— The crystal structure of pyroxferroite from Mare Tranquillitatis. Proceedings of the Apollo 12 Lunar Science Conference. Cambridge: M.I.T. Press 1971.

Silicon

Burns, H., Gordon, E. K.: Refinement of the crystal structure of Li_2BeF_4. Acta Cryst. 20, 135 (1966).
Burns, R. G.: Origin of optical pleochroism in orthopyroxenes. Mineral. Mag. 35, 715 (1966).
Cannillo, E., Coda, A., Fagnani, G.: The crystal structure of bavenite. Acta Cryst. 20, 301 (1966).
— Mazzi, F., Fang, J. H., Robinson, P. D., Ohya, Y.: The crystal structure of aenigmatite. Am. Mineralogist 56, 427 (1971).
— — Rossi, G.: The crystal structure of neptunite. Acta Cryst. 21, 200 (1966).
Cannillo, E., Rossi, G., Ungaretti, L.: The crystal structure of macdonaldite. Accad. Naz. Lincei, Rend. Cl. Sci. fis., mat., natur., Ser. VIII 45, 399 (1968).
— — — The crystal structure of delhayelite. Rend. Soc. Ital. Mineral. Petrol. 26, 3 (1969).
Chao, G.: Refinement of the crystal structure of apophyllite. II. Determination of the hydrogen positions by X-ray diffraction. Am. Mineralogist 56, 1234 (1971b).
Chao, G. Y.: The crystal structure of carletonite, $KNa_4Ca_4Si_8O_{18}(CO_3)_4(OH, F) \cdot H_2O$. Am. Mineralogist 56, 1855 (1971a).
Christ, C. L.: Substitution of boron in silicate crystals. Norsk Geol. Tidsskr. 45, 423 (1965).
Coda, A., Dal Negro, A., Rossi, G.: The crystal structure of krauskopfite. Rend. Cl. Sc. fis. mat. nat. Acc. Naz. dei Lincei, Ser. VIII 42, 859 (1967).
— Rossi, G., Ungaretti, L.: The crystal structure of aminoffite. Rend. Cl. Sc. fis. mat. nat. Acc. Naz. dei Lincei, Ser. VIII 43, 225 (1967).
Cohen-Addad, C., Ducros, P., Bertaut, F.: Etude de la substitution du groupement SiO_4 par $(OH)_4$ dans les composes $Al_2Ca_3(OH)_{12}$ et $Al_2Ca_3(SiO_4)_{2.16}(OH)_{3.36}$ de type grenat. Acta Cryst. 22, 220 (1967).
Colville, A. A., Anderson, C. P., Black, P. M.: Refinement of the crystal structure of apophyllite. I. X-ray diffraction and physical properties. Am. Mineralogist 56, 1222 (1971).
— Ribbe, P. H.: The crystal structure of an adularia and a refinement of the structure of orthoclase. Am. Mineralogist 53, 25 (1968).
Cowley, J. M., Goswami, A.: Electron diffraction patterns from montmorillonite. Acta Cryst. 14, 1071 (1961).
Cruickshank, D. W. J.: The role of 3d-orbitals in π-bonds between (a) silicon, phosphorus, sulphur or chlorine and (b) oxygen or nitrogen. J. Chem. Soc. 5486 (1961).
Dachille, F., Roy, R.: High-pressure region of the silica isotypes. Z. Krist. 111, 451 (1959).
Dal Negro, A., Rossi, G., Ungaretti, L.: The crystal structure of meliphanite. Acta Cryst. 23, 260 (1967).
Deer, W. A., Howie, R. A., Zussman, J.: Rock-forming Minerals, vol. 2, 3 and 4. London: Longmans 1962, 1963.
Dollase, W. A.: Reinvestigation of the structure of low cristobalite. Z. Krist. 121, 369 (1965).
— The crystal structure at 220°C of orthorhombic tridymite from the Steinbach meteorite. Acta Cryst. 23, 617 (1967).
— Refinement and comparison of the structures of zoisite and clinozoisite. Am. Mineralogist 53, 1882 (1968).
Donnay, G., Allmann, R.: Si_3O_{10} groups in the crystal structure of ardennite. Acta Cryst. B 24, 845 (1968).
Douglass, R. M.: The crystal structure of sanbornite, $BaSi_2O_5$. Am. Mineralogist 43, 517 (1958).
Edge, R. A., Taylor, H. F. W.: Crystal structure of thaumasite, $[Ca_3Si(OH)_6 \cdot 12H_2O](SO_4)(CO_3)$. Acta Cryst. B 27, 594 (1971).
Ernst, W. G.: Amphiboles. Berlin-Heidelberg-New York: Springer 1968.
Faye, G. H.: Optical absorption spectrum of tetrahedrally bonded Fe^{3+} in orthoclase. Canad. Miner. 10, 112 (1969).
Fischer, K.: Verfeinerung der Kristallstruktur von Benitoit $BaTi[Si_3O_9]$. Z. Krist. 129, 222 (1969).
Fleet, S. G.: The crystal structure of dalyite. Z. Krist. 121, 349 (1965).

FLEET, S. G., CANN, J. R.: Vlasovite: a second occurrence and a triclinic to monoclinic inversion. Mineral. Mag. **36**, 233 (1967).
— CHANDRASEKHAR, S., MEGAW, H. D.: The structure of bytownite ('body centred anc orthite'). Acta Cryst. **21**, 782 (1966).
FLYNN, J. J., BOER, F. P.: Structural studies of hexacoordinate silicon. Tris(o-phenylenedioxy)siliconate. J. Am. Chem. Soc. **91**, 5756 (1969).
GARD, J. A., TAYLOR, H. F. W.: The crystal structure of foshagite. Acta Cryst. **13**, 785 (1960).
GEBERT, W., ZEMANN, J.: Der Pleochroismus der OH-Streckfrequenz in Topas. Neues Jahrb. Mineral., Monatsh. 380 (1965).
GELLER, S.: Crystal chemistry of the garnets. Z. Krist. **125**, 1 (1967).
— Refinement of the crystal structure of cryolithionite, $\{Na_3\}[Al_2](Li_3)F_{12}$. Am. Mineralogist **56**, 18 (1971).
— MILLER, C. E.: Silicate garnet — yttrium-iron garnet solid solutions. Am. Mineralogist **44**, 1115 (1959).
GHOSE, S.: Mg^{2+}—Fe^{2+} order in an orthopyroxene, $Mg_{0.93}Fe_{1.07}Si_2O_6$. Z. Krist. **122**, 81 (1965).
GIBBS, G. V., BRECK, D. W., MEAGHER, E. P.: Structural refinement of hydrous and anhydrous synthetic beryl, $Al_2(Be_3Si_6)O_{18}$ and emerald, $Al_{1.9}Cr_{0.1}(Be_3Si_6)O_{18}$. Lithos **1**, 275 (1968).
GIUSEPPETTI, G., MAZZI, F., TADINI, C.: The crystal structure of eudialyte. Tschermaks Mineral. Petrog. Mitt. **16**, 105 (1971).
GOLDSCHMIDT, V. M.: Geochemische Verteilungsgesetze der Elemente VII und VIII. Skr. Norske Vidensk.-Akad. Oslo I, Mat.-naturvidensk. Kl. (1926).
GREBENSHCHIKOV, R. G., SHITOVA, V. I., TOROPOV, N. A.: Phasendiagramm des Systems $BaSiO_3$—$BaGeO_3$. Dokl. Akad. Nauk SSSR **175**, 840 (1967).
GÜVEN, N., BURNHAM, C. W.: The crystal structure of 3T muscovite. Z. Krist. **125**, 163 (1967).
HAHN, T.: Modellbeziehungen zwischen Silikaten und Fluoberyllaten. Neues Jahrb. Mineral. Abhandl. **86**, 1 (1953).
— Solid solubility in the system Zn_2SiO_4—Zn_2GeO_4—Be_2SiO_4—Be_2GeO_4. Neues Jahrb. Mineral. Monatsh. 263 (1970).
HARTMAN, P.: Can Ti^{4+} replace Si^{4+} in silicates? Mineral. Mag. **37**, 366 (1969).
HEIDE, H. G., BOLL-DORNBERGER, K., THILO, E., THILO, E. M.: Die Struktur des Dioptas, $Cu_6(Si_6O_{18}) \cdot 6H_2O$. Acta Cryst. **8**, 425 (1955).
HILMER, W.: Die Kristallstruktur des sauren Kaliummetasilikates $K_4(HSiO_3)_4$. Acta Cryst. **17**, 1063 (1964).
HURLBUT, C. S.: Tephroite from Franklin, New Jersey. Am. Mineralogist **46**, 549 (1961).
ITO, T., MORIMOTO, N., SADANAGA, R.: The crystal structure of milarite. Acta Cryst. **5**, 209 (1952).
JAHN, W.: $NaLi[BeF_4]$, eine Modellsubstanz zu Monticellit $CaMg[SiO_4]$. Z. Anorg. Allgem. Chem. **276**, 113 (1954a).
— $Na_3Li[BeF_4]_2$, eine neue Verbindung im ternären System NaF—LiF—BeF_2 und ihre Beziehungen zum Merwinit $Ca_3Mg[SiO_4]_2$. Z. Anorg. Allgem. Chem. **277**, 274 (1954b).
— THILO, E.: Die Stellung der γ'-Modifikation im System der polymorphen Formen des Na_2BeF_4. Z. Anorg. Allgem. Chem. **274**, 72 (1953).
JAMIESON, P. B.: Crystal structure of $Na_2Si_3O_7$: a new type of silicate sheet. Nature **214**, 794 (1967).
JONES, J. B.: Al—O and Si—O tetrahedral distances in aluminosilicate framework structures. Acta Cryst. **B 24**, 355 (1968).
KALT, A., WEY, R.: Composes interfoliaires d'une silice hydratee cristallisee. Bull. Groupe Franc. Argiles **20**, 205 (1968).
KAMB, W. B.: The crystal structure of zunyite. Acta Cryst. **13**, 15 (1960).
KAMENTSEV, I. E.: Position of admixtures in the quartz structure. Zap. Vses. min. Obshch. **94**, 687 (1965), cited after Miner. Abstr. **17**, 657 (1966).
KATSCHER, H.: The structure of the quintuple chain in the barium silicate $Ba_6Si_{10}O_{26}$. Acta Cryst. **A 25**, S 107 (1969).
— LIEBAU, F.: Über die Kristallstruktur von $Ba_2Si_3O_8$. Ein Silikat mit Dreifachketten. Naturwissenschaften **52**, 512 (1965).

Silicon

KHAN, A. A., BAUR, W. H.: Eight-membered cyclosilicate rings in muirite. Science 173, 916 (1971).

KROGH-MOE, J.: Cross-linking of borate polymer chains by means of silicates. J. Am. Ceram. Soc. 47, 307 (1964).

KRÜGER, G., BÜTTNER, M., OLIEW, G., THILO, E.: Über Anhydride von Kieselsäure und Methyltrisilanol mit Schwefelsäure und deren Polymerisationsvermögen. Z. Anorg. Allgem. Chem. 360, 70 (1968).

— WIEKER, W.: Untersuchungen im System BaO—SiO_2—H_2O. Z. Anorg. Allgem. Chem. 340, 277 (1965).

KUNZE, G.: Fehlordnungen des Antigorits. Z. Krist. 111, 190 (1959).

LAJZÉROWICZ, J.: Étude par diffraction des rayons X et absorption infra-rouge de la barysilite, $MnPb_8 \cdot 3Si_2O_7$, et de composés isomorphes. Acta Cryst. 20, 357 (1966).

LAUGHON, R. B.: The crystal structure of kinoite. Am. Mineralogist 56, 193 (1971).

LAVES, F.: Zur Klassifikation der Silikate. Z. Krist. 82, 1 (1932).

LE FUR, Y., ALEONARD, S.: Orthofluoroberyllates $Me^I_2Me^{II}_2[BeF_4]_3$ with the langbeinite structure (Me^I=K, NH_4, Rb, Tl, Cs and Me^{II}=Mg, Ni, Co, Zn, Mn, Cd, Ca). Mater. Res. Bull. 4, 601 (1969).

LEVI, G. R., PEYRONEL, G.: Struttura cristallographica del gruppo isomorfo (Si^{4+}, Ti^{4+}, Zr^{4+}, Sn^{4+}, Hf^{4+})P_2O_7. Z. Krist. 92, 190 (1935).

LIEBAU, F.: Über die Kristallstruktur des Pyroxmangits (Mn, Fe, Ca, Mg)SiO_3. Acta Cryst. 12, 177 (1959).

— Zur Kristallchemie der Silicate, Germanate und Fluoberyllate des Formeltyps ABX_3. Neues Jahrb. Mineral. Abhandl. 94, 1209 (1960).

— Untersuchungen an Schichtsilikaten des Formeltyps $A_m(Si_2O_5)_n$. III. Zur Kristallstruktur von Petalit, $LiAlSi_4O_{10}$. Acta Cryst. 14, 399 (1961).

— Die Systematik der Silikate. Naturwissenschaften 49, 481 (1962).

— Über Kristallstrukturen zweier Phyllokieselsäuren, $H_2Si_2O_5$. Z. Krist. 120, 427 (1964).

— Ein Beitrag zur Kristallchemie der Schichtsilikate. Acta Cryst. B 24, 690 (1968).

— Zur Kristallchemie des sechsfach koordinierten Siliciums. Bull. Soc. Franç. Minéral. Crist. 94, 239 (1971).

— BISSERT, G., KÖPPEN, N.: Synthese und kristallographische Eigenschaften einiger Phasen im System SiO_2—P_2O_5. Z. Anorg. Allgem. Chem. 359, 113 (1968).

— HESSE, K. F.: Die Kristallstruktur einer zweiten monoklinen Siliziumdiphosphatphase, SiP_2O_7,AIV, mit oktaedrisch koordiniertem Silizium. Z. Krist. 133, 213 (1971).

LOEWENSTEIN, W.: The distribution of aluminium in the tetrahedra of silicates and aluminates. Am. Mineralogist 39, 92 (1954).

LOPES-VIEIRA, A., ZUSSMAN, J.: Further detail on the crystal structure of zussmanite. Mineral. Mag. 37, 49 (1969).

MACHATSCHKI, F.: Zur Frage der Struktur und Konstitution der Feldspate. Centralbl. Miner. Abt. A, 1928, 97 (1928).

MAMEDOV, K. S., BELOV, N. V.: Crystal structure of the minerals of the wollastonite group. I. Structure of xonotlite. Zap. Vses. Mineralog. Obshchestva 85, 13 (1956).

MATSUI, Y., SYONO, Y.: Unit cell dimensions of some synthetic olivine group solid solutions. Geochem. J. 2, 51 (1968).

MAZZI, F., ROSSI, G.: The crystal structure of taramellite. Z. Krist. 121, 243 (1965).

MCCONNEL, D.: Hydrogen ion incorporation in crystals. Science 141, 171 (1963).

MCDONALD, W. S., CRUICKSHANK, D. W. J.: A reinvestigation of the structure of sodium metasilicate, Na_2SiO_3. Acta Cryst. 22, 37 (1967a).

— — Refinement of the structure of hemimorphite. Z. Krist. 124, 180 (1967b).

MEIER, W. M.: The crystal structure of natrolite. Z. Krist. 113, 430 (1960).

MERLINO, S.: Tuhualite crystal structure. Science 166, 1399 (1969).

MOKEEVA, V. I., GOLOVASTIKOV, N. I.: The crystal structure of ekanite ThK(Ca, Na)$_2$ [Si_8O_{20}]. Dokl. Akad. Nauk SSSR 167, 1131 (1966).

MOORE, P. B.: The crystal structure of joesmithite: a preliminary note. Mineral. Mag. 36, 876 (1968).

Silicon

MORIMOTO, N., KOTO, K.: The crystal structure of orthoenstatite. Z. Krist. 129, 65 (1969).
NÁRAY-SZABÓ, ST.: Ein auf der Kristallstruktur basierendes Silicatsystem. Z. Physik. Chem., Abt. B 9, 356 (1930).
NOLL, W.: Kristallchemie der Silikate und Strukturchemie der Organopolysiloxane. Naturwissenschaften 49, 505 (1962).
NOVAK, G. A., GIBBS, G. V.: The crystal chemistry of the silicate garnets. Am. Mineralogist 56, 791 (1971).
O'DANIEL, H., TSCHEISCHWILI, L.: Zur Struktur von Gehlenit $Ca_2[AlSiAlO_7]$-Åkermanit $Ca_2[MgSi_2O_7]$ und ihrer Modellsubstanz $Na_2[LiBe_2F_7]$. Neues Jahrb. Mineral., Monatsh. A, 65 (1945a).
— — Zur Struktur von $NaBeF_3$ und β-$CaSiO_3$. Neues Jahrb. Mineral. Monatsh. A, 56 (1945b).
PABST, A.: Crystal structure of gillespite, $BaFeSi_4O_{10}$. Am. Mineralogist 28, 372 (1943).
PANT, A. K., CRUICKSHANK, D. W. J.: The crystal structure of α-$Na_2Si_2O_5$. Acta Cryst. B 24, 13 (1968).
PAPIKE, J. J., ROSS, M., CLARK, J. R.: Crystal chemical characterization of clinoamphiboles based on five new structure refinements. Pyroxenes and amphiboles: Crystal chemistry and phase petrology. Mineral. Soc. Am. Spec. Pap. 2, 117 (1969).
— ZOLTAI, T.: Ordering of tetrahedral aluminum in prehnite, $Ca_2(Al, Fe^{3+})[Si_3AlO_{10}]$ $(OH)_2$. Am. Mineralogist 52, 974 (1967).
PEACOR, D. R., BUERGER, M. J.: The determination and refinement of the structure of narsarsukite $Na_2TiOSi_4O_{10}$. Am. Mineralogist 47, 539 (1962).
— NIIZEKI, N.: The redetermination and refinement of the crystal structure of rhodonite, $(Mn, Ca)SiO_3$. Z. Krist. 119, 98 (1963).
PREWITT, C. T., PEACOR, D. R.: Crystal chemistry of the pyroxenes and pyroxenoids. Am. Mineralogist 49, 1527 (1964).
PRINCE, E.: Refinement of the crystal structure of apophyllite. III. Determination of the hydrogen positions by neutron diffraction. Am. Mineralogist 56, 1243 (1971).
RASTSVETAEVA, R. K., SIMONOV, V. I., BELOV, N. V.: Crystal structure of vinogradovite $Na_4Ti_4[Si_2O_6]_2[Si_4O_{10}]O_4 \cdot nH_2O$. Soviet. Phys. Doklady 12, 1090 (1968).
REID, A. F., RINGWOOD, A. E.: Six-coordinate silicon: High pressure strontium and barium aluminosilicates with the hollandite structure. J. Solid State Chem. 1, 6 (1969).
RIBBE, P. H., GIBBS, G. V.: Crystal structure of the humite minerals: III. Mg/Fe ordering in humite and its relation to other ferromagnesium silicates. Am. Mineralogist 56, 1155 (1971).
RINGWOOD, A. E.: The pyroxene-garnet transformation in the earth's mantle. Earth Planet. Sci. Letters 2, 255 (1967).
— REID, A. F., WADSLEY, A. D.: High pressure $KAlSi_3O_8$, an aluminosilicate with sixfold coordination. Acta Cryst. 23, 1093 (1967).
ROBINSON, P. D., and FANG, J. H.: The crystal structure of epididymite. Am. Mineralogist 55, 1541 (1970).
ROCKTÄSCHEL, G., RITTER, W., WEISS, A.: Ternäre Chalkogenide mit Elementen der 4. Hauptgruppe und Olivinstruktur. Z. Naturforsch. 19b, 958 (1964).
ROY, D. M., ROY, R., OSBORN, E. F.: The System NaF—BeF_2 and the polymorphism of Na_2BeF_4 and BeF_2. J. Am. Ceram. Soc. 36, 185 (1953).
RUDDLESDEN, S. N., POPPER, P.: On the crystal structures of the nitrides of silicon and germanium. Acta Cryst. 11, 465 (1958).
SAHAMA, T. G.: Order-disorder in natural nephelin solid solutions. J. Petrol. 3, 65 (1962).
SARVER, J. F., HUMMEL, F. A.: Solid solubility and eutectic temperature in the system Zn_2SiO_4—Mg_2SiO_4. J. Am. Ceram. Soc. 45, 304 (1962).
SAXENA, S. K.: Distribution of elements between coexisting muscovite and biotite and crystal chemical role of titanium in micas. Neues Jahrb. Mineral. Abhandl. 105, 1 (1966).
SCHÄFER, H.: Übergänge zwischen Metall- und Ionenbindung am Beispiel der Silicide der Alkali- und Erdalkalimetalle. Angew. Chem. 82, 959 (1970).
SCHRÖPFER, L.: Über den Einbau von Titan in Diopsid. Neues Jahrb. Mineral. Monatsh. 441 (1968).

SCHWARZ, H., SCHMIDT, L.: Arsenate des Typs $NaCa_2M_2^{II}(As_3)O_{12}$. Z. Anorg. Allgem. Chem. **382**, 257 (1971).

SCHWARZ, R., MENNER, E.: Zur Kenntnis der Kieselsäuren (I.). Ber. Deut. Chem. Ges. **57**, 1477 (1924).

SCHWEINSBERG, H., LIEBAU, F.: Ein neuer Schichttyp der Silikate. Naturwissenschaften **58**, 267 (1971).

SEIFERT, K. J., NOWOTNY, H., HAUSER, E.: Zur Struktur von Cristobalit GeO_2. Monatsh. Chem. **102**, 1006 (1971).

SHANNON, R. D., PREWITT, C. T.: Effective ionic radii in oxides and fluorides. Acta Cryst. **B 25**, 925 (1969).

SHROPSHIRE, J., KEAT, P. P., VAUGHAN, P. A.: The crystal structure of keatite, a new form of silica. Z. Krist. **112**, 409 (1959).

SHUMYATSKAYA, N. V., VORONKOV, A. A., PYATENKOV, YU. A.: The atomic structure of leucosphenite $Na_8Ba_2Ti_4O_4[B_4Si_{20}O_{56}]$. Soviet. Phys. Cryst. **13**, 130 (1968).

SMITH, G. S., ALEXANDER, L. E.: Refinement of the atomic parameters of α-quartz. Acta Cryst. **16**, 462 (1963).

SMITH, J. V., BAILEY, S. W.: Second review of Al—O and Si—O tetrahedral distances. Acta Cryst. **16**, 801 (1963).

SMOLIN, YU. I.: New silicon-oxygen radical. Ternary two-level Si_6O_{15} ring in the structure of $[Ni(en)_3]Si_2O_5 \cdot 8.7H_2O$. Soviet. Phys. Cryst. **15**, 23 (1970).

SOBOLEVA, S. V., ZVYAGIN, B. B.: Crystal structure of dioctahedral Al-mica 1 M. Soviet. Phys. Cryst. **13**, 516 (1969).

STISHOV, S. M., BELOV, N. V.: On the crystal structure of a new dense modification of silica SiO_2. Dokl. Akad. Nauk SSSR **143**, 951 (1962).

STRUNZ, H.: Kristallchemie der Germanate. Rend. Soc. Mineral. Ital. **17**, 537 (1961).

TAKÉUCHI, Y.: A detailed investigation of the structure of hexagonal $BaAl_2Si_2O_8$ with reference to its α—β inversion. Mineral. J. **5**, 311 (1958).

— DONNAY, G.: The crystal structure of hexagonal $CaAl_2Si_2O_8$. Acta Cryst. **12**, 465 (1959).

— JOSWIG, W.: The structure of haradaite and a note on the Si—O bond lengths in silicates. Mineral. J. **5**, 98 (1967).

— KAWADA, I., IRIMAZIRI, S., SADANAGA, R.: The crystal structure and polytypism of manganpyrosmalite. Mineral. J. **5**, 450 (1969).

— SADANAGA, R.: The structure of xanthophyllite refined. Mineral. J. **4**, 424 (1966).

TAYLOR, H. F. W.: The crystal structure of kilchoanite, $Ca_6[SiO_4][Si_3O_{10}]$, with some comments on related phases. Mineral. Mag. **38**, 26 (1971).

TAYLOR, W. H.: Framework silicates: The feldspars, Chapter 14. In: BRAGG, L., CLARINGBULL, G. F.: Crystal Structures of Minerals. London: Bell and Sons Ltd. 1965.

THILO, E., LIEBAU, F.: Die Modifikationen des Na_2BeF_4 und ihre Beziehungen zu denen des Ca_2SiO_4. Z. Physik. Chem. (Leipzig) **199**, 125 (1952).

— SCHRÖDER, H.: Über das System NaF—BeF_2 und seine Beziehungen zum System CaO—SiO_2. Z. Physik. Chem. (Leipzig) **197**, 39 (1951).

TORDJMAN, I., TRAN QUI, D., LAÜGT, M.: Structure cristalline du polyphosphate mixte de cuivre-potassium: $CuK_2(PO_3)_4$. Bull. Soc. Franç. Mineral. Crist. **93**, 160 (1970).

TROJER, F. J.: The crystal structure of parawollastonite. Z. Krist. **127**, 291 (1968).

— The crystal structure of a high-pressure polymorph of $CaSiO_3$. Z. Krist. **130**, 185 (1969).

VÖLLENKLE, H., WITTMANN, A., NOWOTNY, H.: Zur Kristallstruktur von $CuGeO_3$. Monatsh. Chem. **98**, 1352 (1967).

— — — Die Kristallstruktur von $Li_2(Si_{0.25}Ge_{0.75})_2O_5$. Z. Krist. **126**, 37 (1968).

VORMA, A.: Crystal structure of stokesite, $CaSnSi_3O_9 \cdot 2H_2O$. Mineral. Mag. **33**, 615 (1963)

WEISS, A., ROCKTÄSCHEL, G.: Zur Kenntnis von Thiosilikaten. Z. Anorg. Allgem. Chem. **307**, 1 (1960).

— WEISS, A.: Zur Kenntnis der faserigen Siliciumdioxyd-Modifikation. Z. Anorg. Allgem. Chem. **276**, 95 (1954).

WHITFIELD, H. J., FREEMAN, A. G.: Mössbauer study of amphiboles. J. Inorg. Nucl. Chem. **29**, 903 (1967).

WHITTAKER, E. J. W.: The structure of chrysotile. II. Clino-chrysotile. Acta Cryst. **9**, 855 (1956).

WILLIAMS, P. P., DENT GLASSER, L. S.: Sodium silicate hydrates. IV. Location of hydrogen atoms in $Na_2O \cdot SiO_2 \cdot 6H_2O$ by neutron diffraction. Acta Cryst. **B 27**, 2269 (1971).
WITTMANN, A.: Beiträge zur Strukturchemie der Germanate. Fortschr. Mineral. **43**, 230 (1966).
WOODROW, P. J.: The crystal structure of astrophyllite. Acta Cryst. **22**, 673 (1967).
WUENSCH, B. J.: The crystallography of mcgovernite, a complex arsenosilicate. Am. Mineralogist **45**, 937 (1960).
WYCKOFF, R. W. G.: Crystal Structures, 2nd edition, I, 113. New York and London: Interscience Publishers, 1963.
YADA, K.: Study of chrysotile asbestos by a high resolution electron microscope. Acta Cryst. **23**, 704 (1967).
YODER, H. S., KEITH, M. L.: Complete substitution of aluminum for silicon: The system $3MnO \cdot Al_2O_3 \cdot 3SiO_2 - 3Y_2O_3 \cdot 5Al_2O_3$. Am. Mineralogist **36**, 519 (1951).
ZÁBIŃSKI, W.: Hydrogarnets. Polska Akad. Nauk. Prace Mineral. **3**, 7 (1966).
ZVYAGIN, B. B.: Electron-diffraction determination of the structure of kaolinite. Soviet. Phys. Cryst. **5**, 32 (1960).

D. Abundance in Rock Forming Minerals
I. Silica and Silicates at Low Temperature
a) Detrital Silicates

The geochemistry of Si in the earth's crust is in reality the geochemistry of SiO_2 or of common silicates, for the oxidation potentials existing anywhere on (or presumably in) the earth forbid the persistence of metallic Si for any significant time period. One of the most abundant elements in the crust, Si is a major constituent of the most common rock-forming minerals in its combined form as the tetrahedral SiO_4-group. At low temperatures on earth, Si is found in metastably persisting detrital silicates, as dissolved SiO_2 in natural waters, as free silica in a great variety of sediments, and as silicates of low-temperature origin, primarily the clay minerals. The clay minerals are characteristic of soils, surface alteration products, and sediments of all types.

The variety of silicates found as detrital mineral grains in rocks of low temperature origin, sediments, is almost as large as the list of all silicate minerals known. All major rock-forming silicates of igneous and metamorphic rocks have been noted in sediments but not in the order of relative abundance in the original source rocks. The three silicates that persist as detritus in greatest abundance are quartz, feldspar and mica. All of the others make up only a few per cent of the total rock and many are found only rarely, usually as single grains. Undoubtedly this drastic change in relative abundance and wide differences in persistence as detrital minerals is related to the relative thermodynamic stabilities of the minerals in the low temperature, dilute aqueous environments of the earth's surface as well as to their original abundance.

The more common detrital silicates include, in addition to quartz, feldspar and mica: the clay minerals (smectites, kaolinite, chlorites, illites) and, chiefly as accessory heavy minerals, andalusite, augite, diopside, epidote, garnet, hornblende, hypersthene-enstatite, kyanite, sphene, staurolite, tourmaline and zoisite.

b) Free Silica

Free silica of low temperature origin, i.e., formed at temperatures less than 200° C, is common in sediments. It appears in the form of opal or other varieties of amorphous silica, as a variety of poorly crystallized α or low cristobalite, as chalcedony or microcrystalline quartz, and as macro-crystalline quartz crystals or overgrowths on detrital grains. It is now clear that truly amorphous silica in the form of gel, glass, siliceous sinter, etc., does not persist for geologically long periods of time even in low temperature environments. All of the forms of amorphous silica found in modern sediments recrystallize in nature, on a time scale of 10^7—10^8 years, to microcrystalline quartz and many will recrystallize in much shorter times. The abundant chert of ancient sediments, whose chalcedony is recrystallized to more or less coarsely microcrystalline quartz, ultimately originated as amorphous silica deposits, though is not always certain when or how the original material was precipitated, nor when it recrystallized.

Chert is found as a major component of thick, bedded, primary chemical precipitates characteristic of geosynclinal tracts. Some of the bedded cherts are pure SiO_2, others are mixed with varying proportions of carbonate or clay minerals. Chert is also found as nodules and concretions in limestones, where it can be shown to be of replacement (diagenetic) origin. Other kinds of occurrence of free silica are as cements binding together particulate rocks, as alteration products of volcanic glass, and as vug fillings and encrustations, the agates.

Macrocrystalline quartz of low temperature sedimentary origin is found only as a diagenetic product; there is only one report of crystalline quartz forming as a precipitate in modern sediments, that found in manganese nodules by HARDER and MENSCHEL (1967).

Quartz overgrowths and euhedral quartz crystals are common in sandstones, siltstones, shales, and all varieties of carbonate rock. Such quartz is extraordinarily pure; it contains few microscopically observable inclusions, other than carbonate when the host rock is a limestone, and has low concentrations of trace elements as determined by cathodo-luminescence and electron probe investigations in the author's laboratory.

c) Silicates

Silicates of low temperature origin are mainly the solid residues of the weathering of igneous and metamorphic silicates, with a much smaller proportion being directly precipitated from aqueous solution. The solid weathering residues are the clay minerals, hydrous aluminum silicates. The subject of clay mineralogy cannot be discussed here in detail, as it is a discipline in its own right.

It is important to note that the SiO_2 content of clays is highly variable and responsive to SiO_2 concentrations in the aqueous phases in which the clays are formed or altered. The stability of clays and other silicate minerals may be represented not only by standard phase diagrams but by plots of dissolved SiO_2 concentration (or activity) versus pH, activities of various alkali metals, or ratios of alkali metals to pH (GARRELS and CHRIST, 1965; HESS, 1966). The reactions of clays with aqueous solutions ranging from distilled water to artificial and natural sea waters have been investigated (MACKENZIE and GARRELS, 1965; MACKENZIE et al., 1967; SIEVER, 1968) and show that dissolved silica can either appear or increase as a result of clay dissolution or be depleted by clay mineral precipitation from a silica-rich solution. Dissolution or precipitation depends on the pH, type and concentration of cation, and silica concentration of the solution. Dissolution is favored by low pH, low silica and low cation concentration. The converse, a form of silica adsorption, favored by sea water or higher pH (8 or above), and higher concentrations of SiO_2 and cations, has been called "silicate reconstitution" (SIEVER, 1968) or "reverse weathering" (MACKENZIE and GARRELS, 1966). Kaolinite will dissolve in sea water at pH 8 at dissolved silica concentrations less than 12 ppm but will adsorb silica from more concentrated solutions. But kaolinite in distilled water at pH 5.5 will adsorb silica only from solutions more concentrated than 60 ppm SiO_2.

The zeolites have become increasingly recognized as minerals of low temperature origin that are quantitatively important as rock formers in some geological environments; this recognition followed the wide use of modern mineralogical identification methods. Zeolites have been found in abundance as weathering products of glassy material in many volcanic and mixed volcanic-sedimentary rocks. They are also abundant, particularly as phillipsite, in modern pelagic sediments (ARRHENIUS, 1963), and as primary precipitates in alkaline lakes and ancient lake deposits (HAY, 1966). The zeolites are included in thermodynamic stability diagrams that are plotted in terms of pH and activities of alkali metals and silica but thus far there have been few experimental studies at low temperatures in which chemical equilibrium could be regarded as established (APPS, 1968). But all of the zeolites, as would be deduced from their high silica content, are a response, in part, to high pH and high silica concentrations in the aqueous solutions in which they form. There is also little doubt that many of them are related to chemically reactive precursors, typically glasses.

Other silicates formed at low temperatures are known, the most abundant being zircon and tourmaline. Hitherto unreported as a sedimentary mineral, amphibole replacing carbonate shell material in a non-metamorphosed sediment has been noted recently (ENLOWS and OLES, 1966). Zircons and tourmalines occur in ancient rocks as clear, colorless overgrowths in optical continuity with the original detrital grain. It is obvious that they have been precipitated in void space from an aqueous solution but the chemical conditions necessary are not known.

Though there is some evidence of organic complexing of silica, and the biological evidence of silica secretion makes this obvious, there is no report of any organo-silicon compounds preserved in the rock column. Indeed there is no reason to believe that such compounds, either as solids or in solution are in existence long enough to be significant geologically, though such complexes may play an important role in segregating pure silica deposits.

Revised manuscript received: April 1970

G. Silica and Silicates in Weathering

Silicate minerals can be divided into three broad groups on the basis of their behavior during weathering: those that dissolve congruently, those that dissolve incongruently or alter, and those that are so slightly soluble that they are essentially resistant. The olivines, pyroxenes, and amphiboles dissolve congruently and are the first silicate minerals to disappear from a weathered soil profile. The feldspars and, to a lesser extent, the micas dissolve incongruently and alter to clay minerals by a process that is extremely complex in the details of its mechanism. The kinds of clay minerals formed are dependent on climate, topography, parent rock, and time. The K-feldspars are more resistant to weathering than the plagioclases and albitic feldspar is more resistant than anorthitic-rich varieties. The resistant group of silicates is typified by quartz, which remains in soil profiles after almost all else is altered or dissolved. This is so because of the extremely low solubility of quartz and its extremely slow kinetics of dissolution.

The behavior in weathering of the silicates corresponds well with their abundance as detritus. Those that dissolve quickly are rarely found as other than trace or accessory minerals. The feldspars and clay minerals are both abundant constituents, with the clays being far more abundant, an indication of the general efficiency of weathering. The main resistant silicate, quartz, is not only one of the most abundant minerals but is almost the sole constituent of the most mature detritus, the quartz-rich sandstones.

Solubility of silicates is discussed in subsection 14-H-V and the abundance of silica in soil waters in subsection 14-I-II.

H. Solubilities of Compounds which Control Concentrations in Natural Waters (I—V, VII). Adsorption Processes (VI)

I. Silica in Aqueous Solution

The equilibrium and colloid chemistry of silica has been explored extensively in laboratory investigations (ILER, 1955, gives a summary of the literature upto 1954). Much of this work has been directed to the chemistry of silica sols and gels in relation to commercial processes and is of little interest to geochemists. But another direction has been the measurement of solubilities of the various silica polymorphs and the elucidation of the nature of monomeric and polymeric species in solution. From this work it appears that definite solubility values can be given for the various forms of silica investigated and that those solubilities are more or less in harmony with thermodynamic properties arrived at by other means of experimentation.

As KRAUSKOPF (1959) has pointed out, much of the progress of the past decade has been achieved as a result of the demonstrated value of the silico-molybdate colorimetric method for the determination of dissolved silica (WEITZ, FRANCK and SCHUCHARD, 1950; ALEXANDER, HESTON and ILER, 1954). Using the yellow color developed by ammonium molybdate in strong acid solution, a rapidly developed color whose intensity is proportional solely to monomeric $Si(OH)_4$ in solution, as little as 5 mg/l of dissolved SiO_2 can be determined accurately and reproducibly. If a strong reducing agent such as p-amino-sulfonic acid is used, a strongly colored blue solution is developed after several hours, which extends the sensitivity of the determination by about one order of magnitude but includes the total of all species in solution, mono- and polymeric (MULLIN and RILEY, 1955). A comparison of the two types of analysis will reveal differences between the abundance of monomeric and polymeric species.

The dominant species in near neutral and acid solutions, as measured by lowering of freezing point determinations (WEITZ, FRANCK and SCHUCHARD, 1950) and diffusion rates (JANDER and JAHR, 1934), is monomeric silicic acid, H_4SiO_4 or $Si(OH)_4$. Hydration of the silicic acid was studied by ILER and DALTON (1956), who concluded that one layer of water molecules surrounds the $Si(OH)_4$ tetrahedron. In alkaline solutions, above pH 9, the ionization of H_4SiO_4 becomes appreciable and polymerization of charged and uncharged species begins. A thorough study of these species and an evaluation of the relevant equilibrium constants has been made by LAGERSTRÖM (1959), who studied the equilibria in $NaClO_4$ solutions. Though there is some discrepancy in the data there is little doubt that the first dissociation constant (for the formation of the $H_3SiO_4^-$ ion) is of the order of magnitude given by ROLLER and ERVIN (1940), $10^{-9.8}$; the second dissociation constant (for the formation of the $H_2SiO_4^{-2}$ ion) is about 10^{-12}. LAGERSTRÖM gives data in terms of the mono- and polynuclear species: $Si(OH)_4$, $SiO(OH)_3^-$, $SiO_2(OH)_2^{-2}$, $Si_4O_6(OH)_6^{-2}$, $Si_2O_3(OH)_4^{-2}$, and $Si_4O_8(OH)_4^{-4}$. Though most natural solutions are in the pH range 5—9, a range where such species complications need not be considered, geochemical work on more alkaline solutions such as those of some alkaline lakes or springs should take into account the much greater diversity of species in solution and their equilibrium with colloidal SiO_2 gels.

II. Solubility of Amorphous Silica

Silica gels will dissolve readily in aqueous solutions to reach an equilibrium value. The equilibrium solubility can be confirmed by precipitation of gel from supersaturated sols. Thus silica solubilities can be investigated by ordinary means and are similar to other compounds in every way. Equilibria are achieved rapidly, both between solid and solution and between various species in solution. It has been established by many investigators (ALEXANDER, HESTON and ILER, 1954; KRAUSKOPF, 1956; OKAMOTO, OKURA and GOTO, 1957) that the solubility of amorphous silica is independent of pH in the range 2—9 and the only significantly abundant species in solution is undissociated $Si(OH)_4$. At about pH 9 the solubility rises abruptly as a consequence of the formation of various silicate ions and poly-ions. The solubility in alkaline solutions is compatible with ROLLER and ERVIN's (1940) dissociation constants and also with those given by LAGERSTRÖM (1959) for $NaClO_4$ solutions. There seems to be little doubt, as KRAUSKOPF (1959) noted, that the data of CORRENS (1941) that showed a pronounced decrease in solubility at pH 3 must be anomalous.

In contrast to prevailing earlier ideas on the kinetics of the amorphous silica-water system and the presumed impossibility or difficulty of getting equilibrium results, the experimentation of the past decade has shown that reproducible equilibrium data can be obtained over a time scale of days. Some finely powdered silica gels will dissolve to saturation within hours if agitated strongly. Equilibrium can be approached from both under- and super-saturated concentrations of dissolved silica. The rate of dissolution is strongly affected by pH; at pH values below 5, the dissolution process is very slow and may take months, where at pH 8 the gel may dissolve to saturation overnight.

Similarly, the rate of gel formation is not uniform but is dependent on a number of factors. Gel rate is most rapid at intermediate to moderately low pH levels, between 4 and 7 (ILER, 1955), and very slow at high and low pH values. Gel rate is also strongly affected by degree of supersaturation, the greater the supersaturation, the faster the gel speed. A variety of impurities will also hasten gelling. It is important for geochemical processes to note that most natural waters, including both sea and many river and lake waters, are maintained at pH levels that promote rapid gelling; most natural waters also contain many of the impurities, both dissolved salts and particulate matter, that increase gelling speed. Thus it would seem that in spite of the success of laboratory chemists in making stable supersaturated silica sols, such as Ludox, such sols would be completely unexpected in nature.

Amorphous silica solubility increases regularly with increase in temperature (Fig. 14-H-1). At the same time, increase in temperature leads to much more rapid equilibration of the system. Though the solubility rises steadily and would be ex-

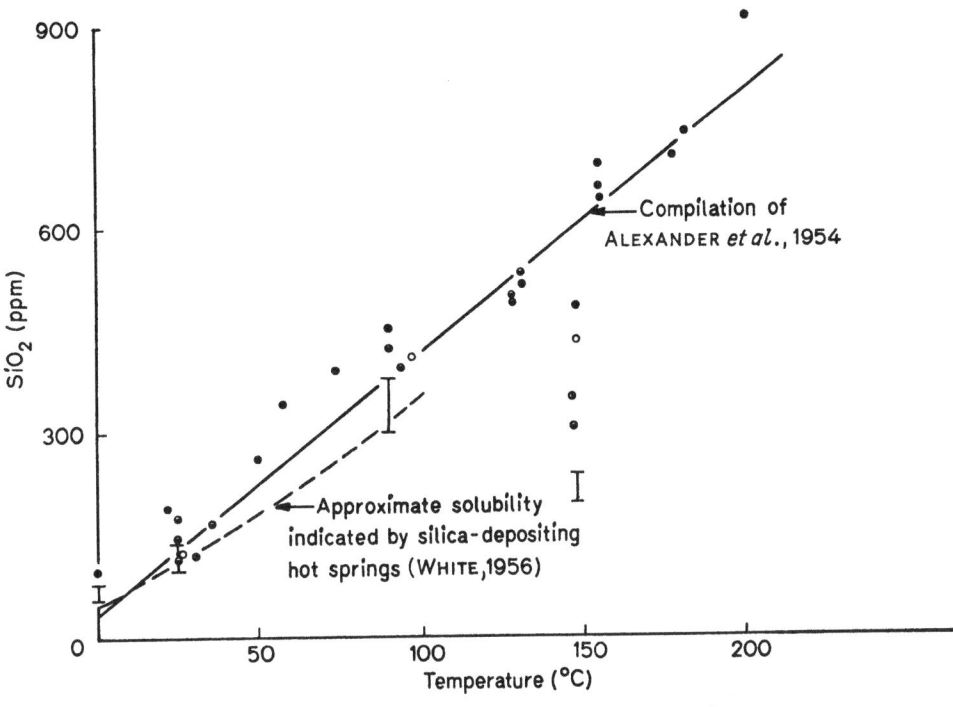

Fig. 14-H-1. Solubility of amorphous silica as a function of temperature (KRAUSKOPF, 1959, p. 5)

pected to continue doing so in subcritical water above 100° C, somewhere about 200° C the line is truncated as the amorphous silica starts to recrystallize to quartz and the solubility falls to that in equilibrium with the stable polymorph. Experiments at these temperatures show that the amorphous silica dissolves and then precipitates as quartz until all of the amorphous silica is converted. This behavior is utilized in the manufacture of artificial quartz crystals using high pressure vessels with a temperature gradient. At the hot end the amorphous silica dissolves and at the cool end quartz precipitates around a seed crystal. The temperature gradient is used to speed up and enhance the recrystallization that would take place isothermally in any case. It may be that this kind of recrystallization, with or without the temperature gradient, may be responsible for much of the secondary quartz that is found in sediments affected by diagenesis, the quartz having been "distilled" over from amorphous silica.

It is not clear whether the precise nature of the amorphous silica gel used has an effect on equilibrium values, though it certainly affects the kinetics. Most investigators agree that various kinds of gels and silica glasses have very nearly the same solubility but the precision of the solubility value is not good enough to distinguish the expected small differences between species that differ only in water content, total internal surface, bulk density, etc. It also appears that opals may have slightly slower kinetics than most gels and that its equilibrium value may be lower by some small amount but as yet the question is not settled. Diatom silica dissolves slowly but steadily, as reported by LEWIN (1961), and seems not to be significantly different in chemical properties from other varieties of amorphous silica. The solubility of various species of diatoms, radiolaria and sponge spicules may vary with the kinds and amounts of metal ions and organic materials bound to the silica secreted by the organisms. BERGER (1969) has shown this effect in the dissolution of radiolaria in sea water.

Dissolved impurities may affect the solubility of amorphous silica. It has been shown that addition of Al will reduce the solubility, presumably by the formation of some aluminosilicate (OKAMOTO, OKURA and GOTO, 1957). Recent work by MACKENZIE and GARRELS (1965), MACKENZIE et al. (1967), and SIEVER (1968) has demonstrated the lowering of silica concentrations in sea and other waters by reaction with clay minerals. It seems apparent that if any of the components of the very insoluble aluminosilicates are present, the long term equilibrium is concerned with those solids rather than amorphous silica. Presumably if enough Al and/or alkali metals or alkaline earths were provided, any solid amorphous silica would eventually be converted to aluminosilicate. Magnesium ions also affect silica in solution in the same way, forming a compound a magnesium silicate similar to sepiolite (WOLLAST, MACKENZIE and BRICKER, 1969). In sea water at its normal pH, about 8, silica solubility is reduced to about 25 ppm if the equilibrium is controlled by magnesium silicate alone.

III. Solubility of Quartz

Low-quartz is the stable modification of SiO_2 at low temperatures and pressures and would be expected to be the dominant precipitate from silica solutions if equilibrium thermodynamics were obeyed. But experiments with the simple silica-water system that results in precipitation of a silica phase at low temperatures, show that amorphous silica is produced rather than quartz. There is an inference of quartz precipitation from depletion of silica concentrations below the saturation level with respect to amorphous silica in several experiments reported by VAN LIER (1959) and MOREY, FOURNIER and ROWE (1962), but it was not possible to positively identify a quartz product. These laboratory experiments corroborate the experience of geologists who have long sought in vain for freshly precipitated quartz in modern sediments[1].

Because of the reluctance of quartz to precipitate it has proved impossible to establish an equilibrium solubility value at room temperature by precipitation from a supersaturated solution. Such solutions, undersaturated with respect to amorphous silica, will remain unchanged for years. On the other hand it is almost equally difficult to approach an equilibrium solubility from undersaturation. Even very finely powdered quartz, if it is washed free of any coating of amorphous silica, will not dissolve beyond a few parts per million over very long periods of time (SIEVER, 1962). Thus solubility values have been estimated by extrapolation from higher temperature

---

[1] Recently two papers have been published that show that quartz may be synthesized at low temperatures. HARDER and FLEHMIG (1970) have identified, by optical and x-ray diffraction methods, quartz and cristobalite coprecipitated with hydroxides of Al, Fe, and other metals after one month of reaction. MACKENZIE and GEES (1971) have published scanning electron photomicrographs of authigenic quartz crystals grown from seawater at 20° C. after three years of reaction time.

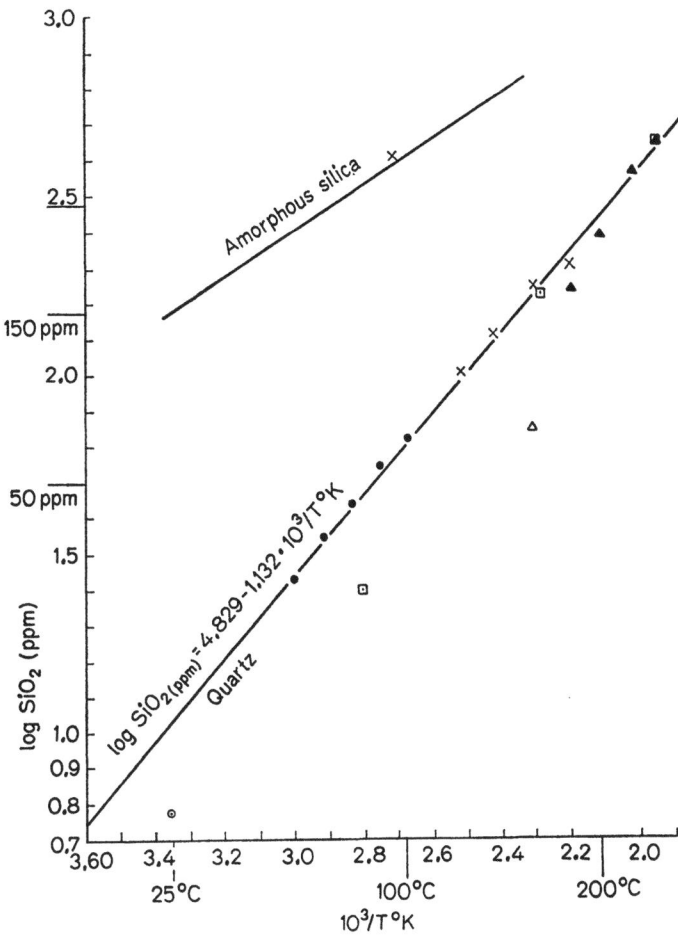

Fig. 14-H-2. Solubility of quartz as a function of temperature (SIEVER, 1962, p. 134)

data (VAN LIER, 1959; SIEVER, 1962) or by thermodynamic calculation (SIEVER, 1957). The extrapolations and calculations agree well on a value of 10—11 ppm at 25° C. A few direct determinations have been attempted but they may represent slightly undersaturated values; GARDNER (1938) reported 6 ppm and WHITE, in KRAUSKOPF (1959), reported 7 ppm or a little higher. The temperature dependence of the solubility of quartz is shown in Fig. 14-H-2.

Whatever the precise value of quartz solubility, the striking thing is that it is over an order of magnitude less than that of amorphous silica. This, coupled with the lack of quartz precipitation, makes possible a thorough study of the metastable equilibria of amorphous silica, and also the metastable persistence of amorphous silica and solutions supersaturated with respect to quartz in low temperature earth surface environments.

IV. Solubility of Other Forms of SiO$_2$

Chalcedony has been shown to be a form of microcrystalline quartz by FOLK and WEAVER (1952) and MIDGELY (1951). PELTO (1956) reports solubilities much greater than quartz but there is some question whether all amorphous silica from grinding and preparation procedures was removed. One may expect a very slight effect on solubility from extremely fine particle sizes in some cherts but the general level should be that of quartz.

Tridymite and cristobalite solubilities at low temperatures are not known. Because the structure of amorphous silica can be likened in some ways to a completely disordered cristobalite, one would expect crystalline cristobalite to have a solubility a little lower than that of amorphous silica but higher than that of quartz. Tridymite should have a value somewhere between quartz and cristobalite.

The equilibrium solubilities at low temperatures of the more recently discovered high pressure polymorphs, coesite and stishovite, are not accurately known. STÖBER (1967) has presented data on the kinetics of dissolution of these polymorphs as well as of quartz, tridymite, cristobalite and vitreous silica, but it is not considered that the final concentration values reached represented reversible saturation equilibria. The data suggest that stishovite may have a slightly higher solubility than amorphous silica and coesite a value between amorphous silica and quartz.

V. Solubility of Silicates

As silicates hydrolyze in water they release silica to the solution. The amount of silica in solution is, of course, a function of the composition and structure of the particular silicate. The feldspars have been most extensively investigated, starting with TAMM (1930), followed by CORRENS and VON ENGELHARDT (1938), ARMSTRONG (1940), GRUNER (1944), MOREY and CHEN (1955), NASH and MARSHALL (1956a, b) and more recently summarized by CORRENS (1961) and by GARRELS and CHRIST (1965). The latter show that the feldspars, micas, clay minerals and by implication all of the silicates have solubility diagrams that can be expressed in terms of the activities of dissolved SiO$_2$, pH, and the cations concerned. The solubility of amorphous silica is a barrier on one edge of the diagram, for any solutions with more dissolved silica than saturation with respect to amorphous silica will promptly precipitate a silica gel. On the other edge, the dissolution of quartz in solutions less concentrated than its solubility, implies a lower limit for most solutions that come to equilibrium with rocks in which quartz is an abundant constituent.

Thus far, information is available on alkali feldspars, some micas and clay minerals, sepiolite, analcime, phillipsite and a few others but most silicates have not been investigated. The difficulty of establishing equilibrium values is great, in most cases because the silicate being investigated cannot easily be precipitated at low temperatures, and in some because of the slowness of dissolution. For most of these minerals the dissolved silica values tend to be in the range from 10—30 ppm at 25° C. The extension of these data to the natural situation is beset with difficulties because the complete solubility diagram must be known together with the complete analysis of the pertinent ions in the natural waters under consideration. Even then one cannot be sure of equilibrium behavior in geologic situations.

VI. Adsorption of Silica

There are no thoroughly documented cases of dissolved silica being adsorbed on solid surfaces with characteristic adsorption isotherms. SIEVER (1968) has summarized the evidence for the formation of alkali aluminosilicates by reaction of clay minerals, dissolved silica, and alkali metal and alkaline earth ions. Recent experiments in the author's laboratories have shown the existence of a slight, very long term (periods of many months) adsorption of dissolved silica on kaolinite surfaces, and reversibility has been demonstrated. One would suspect that uncharged $Si(OH)_4$ tetrahedra, which would have a statistically excess negative charge distributed over the outside of the group, would be adsorbed only slightly by most negatively charged silicates at the pH of most natural environments.

VII. Organic-Silica Complexes

This is a virtually unknown area, though many have suspected the existence of such complexes. The fact that so many plants and some animals are able to transport dissolved silica effectively in body fluids and then precipitate amorphous silica through a membrane certainly suggests such complexes. The organic geochemistry of silica was reviewed in 1959 by SIEVER and SCOTT (1963). In that review it was pointed out that there were no naturally occurring organic silicon compounds in geologic deposits and that is still the case. JEFFREY and HOOD (1964) have reported the existence of a presumed silica-organic complex in sea water, but no other such studies have been reported. The possibility of such complexes remains attractive for the geochemist who could explain migration and segregation of free silica by such a mechanism.

Revised manuscript received: April 1970

I. Abundance in Natural Waters

The abundance of silica in natural waters is compatible with our knowledge of the chemistry of silica and silicates. Thus all waters except a very few are undersaturated with respect to amorphous silica and a great many are supersaturated with respect to quartz. Analyses of river and sea waters have shown that monomeric silicic acid is the species in solution (KRAUSKOPF, 1956), as would be expected from the pH of the waters. The concentration of dissolved silica varies widely in response to the amount released by surface weathering of silicate rocks, from hot springs and other groundwaters, the dilution by fresh rain water, and the removal of silica by chemical reactions or biological mechanisms.

I. Rivers

The most recent summary of dissolved silica in streams is that by DAVIS (1964) who statistically evaluated a great number of analyses and quotes medians, quartiles, and 5th and 95th percentile values by area (see Table 14-I-1). It is apparent that all streams are undersaturated with respect to amorphous silica and that the vast majority

Table 14-I-1. *Analyses of dissolved SiO_2 in some streams* (DAVIS, 1964). *Silica in stream water*

Location	Number of analyses	ppm silica at given percentiles					Source of data
		5	25	50	75	95	
United States	142	3	8	14	18	33	U.S. Geol. Survey Water-Supply Papers 1450, 1451, 1452 and 1453
Western United States	82	—	11	15	23	—	U.S. Geol. Survey Water-Supply Paper 1575
Alaska	63	—	4	7	10	—	U.S. Geol. Survey Water-Supply Paper 1500
California	545	8	12	17	23	36	California Dept. Water Resources Bull. 65
Sierra Nevada, California	21	—	—	16	—	—	California Dept. Water Resources Bull. 65
Puerto Rico	11	—	—	13	—	—	U.S. Geol. Survey Water-Supply Paper 1460-A
Gabon	17	—	—	7	—	—	Unpublished analyses by C. F. PARK, JR, 1958
Glacial "milk" various localities	17	—	—	1.5	—	—	see references (KELLER and REESMAN, 1963)
Greenland	8	—	—	2.8	—	—	see references (DAVIS, 1961)
Africa	53	—	16	24	35	—	U.S. Geol. Survey Prof. Paper 404 G
Venezuela	8	—	—	16	—	—	U.S. Geol. Survey Prof. Paper 404 G

are supersaturated with respect to quartz. The range in values is great, from as little as 2 or 3 ppm (parts per million, the common unit in which natural water concentrations are expressed) to over 80 ppm. The median value for world streams is approximately 13 ppm (LIVINGSTON, 1963). DAVIS' (1964) analyses of river water data indicates that silica concentrations are independent of the concentration of most dissolved ions and the total dissolved solids. DAVIS does suggest, however, that low dissolved silica concentrations correlate with high calcium concentrations. Dissolved silica concentrations appear to be independent of temperature or general climatic zones. DAVIS believes the time required for runoff to dissolve silica from its sources is very short, perhaps less than three days. The fact that silica concentrations do not change much with large increases in storm runoff that result in dilution of other dissolved solids, is possibly the result of fracturing and abrasion of quartz during the greatly increased turbulence of streams during storm runoff.

II. Soil Waters

Soil zones are the sites of active chemical weathering of silicate minerals as deduced from observations of etched and corroded minerals and progressive disappearance of silicate with leaching of the soil (MOHR and VAN BAREN, 1954). The pH of soil slurries and clarified waters extracted from soils show wide variations but typically are low, ranging from below 4 in some unusual circumstances to about 7.5 (GORHAM, 1960; EATON et al., 1968) and this acidity enhances the hydrolysis of silicates. Because chemical weathering of silicates depends in a complex way on climate, including temperature and rainfall, parent rock composition, and vegetation type, it is not surprising that dissolved silica concentrations of soil waters vary widely (MOHR and VAN BAREN, 1954; EATON et al., 1968).

Table 14-I-2 shows some of the analyses given by EATON et al. (1968). Concentrations are significantly higher than typical river water concentrations, as one would expect, but concentrations may be so greatly affected by dilution by fresh rain water

Table 14-I-2. *Analyses of dissolved SiO_2 in some soil waters* (EATON et al., 1968)

Location	SiO_2, ppm	
	range	mean
Middle Rio Grande Valley, New Mex.		
Surface	8—40	25
Subsoil	7—67	28
Elephant Butte Project, New Mex., Texas		
Subsoil	11—41	23

or evapo-transpiration that generalizations can be made only with caution. Some soils contain appreciable quantities of opal phytoliths (BAKER, 1959) and though no analyses of such soils are available it may be that they contain higher amounts of dissolved silica.

Direct observation of nonbiogenic opaline silica segregations in tropical soils is sufficient evidence to state that those soils must have periodic supersaturation levels with respect to amorphous silica, though no actual chemical analyses are known to the author. The nature of the sequential processes by which silica is leached from silicates and then concentrated to supersaturation so that amorphous silica can precipitate is not clear. Any equilibrium process between waters and hydrolyzing silicates will be poised at levels below amorphous silica saturation, so we must invoke some process by which soil waters of lower concentration are concentrated by evapo-transpiration during a dry period. The difficulty with this idea is that the cations leached from silicates might also be expected to concentrate and recombine with the silica to form clay minerals. Indeed this mode of formation may be characteristic for some montmorillonitic soils as well as for soils containing sepiolite and attapulgite. It is possible that plant roots act as temporary reservoirs for the silica and that the silica nodules and veinlets are simply redistributions of opal phytoliths and silica impregnated epidermal tissue. According to this view, it is not the plants that are responsible for increased weathering, as suggested by LOVERING (1959) but, as he also suggests, that they are accumulators.

III. Groundwaters

Analyses of dissolved silica concentrations in groundwaters, most recently reviewed by WHITE, HEM and WARING (1963) and DAVIS (1964), Table 14-I-3, show silica concentrations from just a few ppm in potable groundwaters to almost 100 ppm. The peculiar spring in California reported by FETH, ROGERS and

Table 14-I-3. *Analyses of dissolved SiO_2 in some subsurface waters* (DAVIS, 1964). Silica in saline water and brine

Location	Number of analyses	Ppm silica at given percentile					Source of data
		5	25	50	75	95	
Illinois	482	3	7	13	18	33	Illinois Geol. Survey, Petroleum, v. 66
North Dakota	504	6	13	21	28	36	U.S. Geol. Survey Water-Supply Paper 1428
By Formations:							
Hell Creek and Fort Union	226	6	10	15	22	35	U.S. Geol. Survey Water-Supply Paper 1428
Dakota	66	—	11	17	21	—	U.S. Geol. Survey Water-Supply Paper 1428
Pleistocene	212	13	21	26	31	38	U.S. Geol. Survey Water-Supply Paper 1428
Pennsylvania	37	—	7	9	15	—	Pennsylvania Geol. Survey Bull. M-47
Texas	61	—	11	16	23	—	U.S. Geol. Survey Water-Supply Paper 1365
New Mexico	130	8	14	20	26	55	U.S. Geol. Survey Water-Supply Paper 1601
By Eras:							
Paleozoic	44	—	15	19	23	—	U.S. Geol. Survey Water-Supply Paper 1601
Mesozoic	29	—	11	14	18	—	U.S. Geol. Survey Water-Supply Paper 1601
Cenozoic	57	—	18	23	37	—	U.S. Geol. Survey Water-Supply Paper 1601

ROBERSON (1961) has a level of almost 4,000 ppm. The upper zones of groundwater are typically dilute as a result of meteoric water infiltration and the deeper brines will typically contain much more dissolved SiO_2, though this is not always so. DAVIS (1964) points out that this is probably not a temperature effect. The silica concentrations in groundwaters are primarily a function of the petrographic nature of the containing aquifer; thus limestone aquifers, for example, are low in silica.

Thermal groundwaters issuing from hot springs have been studied by WHITE, BRANNOCK and MURATA (1956), Table 14-I-4; such waters tend to have high silica concentrations, upto almost 400 ppm. When these waters cool they become supersaturated and spontaneously precipitate siliceous and opaline sinters. The high silica content of such hot springs is probably chiefly attributable to the temperature effect on solubility but some may be due to the association with glassy volcanic rocks that are highly reactive and release much silica on hydrolysis.

Table 14-I-4. *Analyses of dissolved SiO_2 in hot springs* (WHITE, HEM and WARING, 1963)

Location	SiO_2, ppm
Upper Basin, Yellowstone Park, Wyo.	363
Norris Basin, Yellowstone Park, Wyo.	529
Steamboat Springs, Nevada	293
Geyser Bight, Umnak Is., Alaska	150
Hankadalur, Reykjavik, Iceland	359
Shumhaya, Kamchatka, U.S.S.R.	294
Wairakai, N. Island, New Zealand	386
Hot Creek, Mono County, Calif.	131
Jemez, Sandoval County, New Mex.	93
Agnano, Naples, Italy	152
Amedee Springs, Lassen County, Calif.	96
Boiling Springs, Valley County, Idaho	81
Hot Springs, Garland County, Arkansas	46
Plombiers, France	90

IV. Lakes

Average freshwater lakes in temperate to cold climates tend to have relatively small amounts of dissolved SiO_2, usually falling in the range of 2 to 10 ppm (LIVINGSTONE, 1963). Seasonal changes in the chemistry of lakes related to thermal stratification have an effect on the amount of dissolved SiO_2. During warm season algal blooms, the SiO_2 content of the surface waters may be greatly depleted. Release by bottom sediments of dissolved SiO_2 that significantly raises its concentration may occur under reducing conditions in the hypolimnion.

Lakes in more arid climates are generally higher in dissolved SiO_2, sometimes having as much as 70 ppm. Extreme aridity may lead to alkaline lakes in which dissolved silica may be at concentrations of several hundred ppm; these are lakes in which the lake water is essentially a sodium silicate solution with a pH of 10 or higher.

V. Oceans

The oceans are undersaturated with respect to amorphous silica at all depths, though the distribution in the water column and from place to place varies widely, primarily as a result of biological processes. Surface waters tend to be very depleted in silica (SVERDRUP, JOHNSON and FLEMING, 1942); in most places they contain one or less than one ppm. The concentration in deeper waters, as shown most recently by BERGER (1968), increases sharply at or below the oxygen minimum zone to about 6—9 ppm. BRUEVITCH (1953) reported concentrations up to 12 ppm in bottom waters of the Bering Sea. The mechanism responsible for this vertical distribution is the extraction of silica from sea water by diatoms, radiolaria, silico-flagellates, sponges, and other organisms, which then start to dissolve as they fall to the bottom after death. BERGER (1968) has observed the rates of dissolution of radiolaria in the Pacific and reports weight losses up to 15%. A certain amount of selectivity according to species was also revealed by BERGER's work; he concluded that deeper-living forms having more robust skeletons are enriched in bottom sediments as a result of this dissolution.

The distribution of dissolved silica in the different oceans seems to be a function of varying biological activity that is itself conditioned by the horizontal and vertical circulation and the petrologic character of terrigenous detritus and volcanic contributions. Thus the Pacific is slightly higher in dissolved silica concentrations in general than the Atlantic, and the Pacific is also covered by much more extensive areas of diatomaceous and radiolarian oozes. It appears that this is due to the much higher ratio of mafic volcanic materials to dominantly quartzose and felsic terrigenous detritus in the Pacific than in the Atlantic. The greater abundance of montomorillonitic and zeolitic clayey sediments in the Pacific than in the Atlantic is probably also attributable to this basic control.

Though the ultimate control on sea water composition, and therefore silica concentrations, is the steady state equilibrium (on a long term basis) between water and the minerals of the earth's crust and the atmosphere (SILLÉN, 1961; GARRELS, 1965), it appears that the short term control on silica is, at least at present, the biological activity of planktonic organisms that secrete silica (HARRISS, 1966; SCHINK, 1967; CALVERT, 1968). Though reactions of clay minerals with silica in sea water can deplete silica concentrations (MACKENZIE et al., 1967; SIEVER, 1968), it has not been established by unambiguous observations that this happens in the expected sites, estuaries; BIEN et al. (1958) indicated that for the Mississippi it does and STEFANSSON and RICHARDS (1963) that for the Columbia it does not. But it has been established that dissolved silica in surface waters will wax and wane in response to seasonal fluctuations in diatom populations (HARVEY, 1957).

VI. Interstitial Waters

Interstitial waters extracted from modern sediments have been analyzed for dissolved silica (EMERY and RITTENBERG, 1952; SIEVER, BECK and BERNER, 1965). The variation is great, ranging from 1 or 2 ppm to about 75 ppm (Table 14-I-5). The latter figure corresponds to approximate saturation with respect to amorphous silica at the low temperatures (about 2—5° C) characteristic of most bottom waters and sediments. The high silica concentrations as compared with the overlying sea

Table 14-I-5. *Analyses of dissolved SiO$_2$ in modern sediment interstitial waters*
(SIEVER *et al.*, 1965)

Location	Depth in core, cm	SiO$_2$, ppm
Cape Cod, Core 1	0—15	52
Cape Cod, Core 1	45—56	41
Cape Cod, Core 1	400—450	67
Cape Cod, Core 6	101—111	22
Cape Cod, Core 6	314—325	23
Equatorial Atlantic, Chain 17, Sta. 2	13—20	38
Equatorial Atlantic, Chain 17, Sta. 2	157—163	56
Equatorial Atlantic, Chain 17, Sta. 8	25—32	11
Equatorial Atlantic, Chain 17, Sta. 8	783—790	11
Gulf of California, Core L-154	18—25	54
Gulf of California, Core L-154	91—98	81
Southwest Atlantic, Core A5757	21—28	21
Southwest Atlantic, Core A5757	609—616	18
Northwest Atlantic, Core 1	0—40	4
Northwest Atlantic, Core 1	402—417	15

waters have been attributed by SIEVER, BECK and BERNER (1965) to the dissolution of diatom and other amorphous silica skeletal materials in the sediment after burial. The primary evidence for this is the low dissolved silica content of waters in sediments poor in siliceous skeletons and the high content of waters coming from diatomaceous sediments. The recent finding of lithified chert beds in a great variety of sediments from many parts of the deep ocean (in the first drilling reports of the JOIDES (1969) deep ocean drilling program of the United States National Science Foundation), is striking confirmation of the supersaturation levels that may be reached in interstitial waters.

Some dissolved silica in interstitial waters may come from hydrolysis of detrital feldspars in terrigenous sediments and from the hydrolysis and devitrification of volcanic glasses on the ocean floor. It is unlikely, however, that this is the major effect in most sediments of the oceans.

Revised manuscript received: April 1970

K. Abundance in Common Sediments and Sedimentary Rocks

The relative abundance of silica in modern sediments and ancient sedimentary rocks is highly variable, primarily because sediments represent such complex mixtures of widely assorted terrigenous detritus and chemical precipitates introduced at the time of sedimentation or later, during diagenesis. Thus the abundance of silica depends on the proportion of free silica as detrital quartz, amorphous silica in the form of siliceous skeletal materials added at the depositional site, and secondary quartz or amorphous silica added as cementing agents. Silica abundance also depends greatly on the kinds and relative abundance of the various silicates commonly found as terrigenous detritus and as diagenetic precipitates or alteration products. Finally, silica may be removed by dissolution of free silica or silicates during diagenesis or weathering. Though few sediments are exclusively detrital or wholly chemically precipitated, it is convenient to consider them in those major groups, recognizing that largely detrital rocks such as sandstones may contain appreciable quantities of secondary quartz, clay minerals, or feldspars, and that largely chemical rocks, such as limestones, may contain significant amounts of detrital silicate.

I. Detrital Sediments

Most sandstones and conglomerates show about the same range of silica proportions, from about 65 to 95 per cent (PETTIJOHN, 1963), see Table 8-4 in Volume I of this handbook. The most siliceous are the quartz sandstones and quartz pebble conglomerates, some of which are composed almost entirely of quartz and hence approach the 100 per cent silica composition. Naturally there are always some small amounts of clay and other minerals that reduce the amount of silica to less than 95 per cent. Most commonly, carbonate cement will reduce the silica in some sandstones to less than 50 per cent. Some phosphatic and glauconitic sandstones will likewise be very low in silica. But it is also true that some famous quartz sandstones used for glass making purposes, such as the Oriskany sandstone of the Eastern United States, are extraordinarily pure silica rocks.

Arkoses and graywackes, and their associated conglomerates, which contain intermediate amounts of feldspars, clay minerals, carbonates and oxides, will be much lower in silica, ranging from 40 to 85 per cent, depending on the amount of carbonate cement and other constituents. The graywackes, though largely composed of silicates, have the lowest silica amounts, most lying in the range 60—75 per cent, the remainder being made up largely of alumina, iron, alkali metals and alkaline earths but little carbonate. Many lithic sandstones and arkoses are cemented by carbonates or iron oxides, which reduce the silica content. The total amount of silica has little to do with the silica content of the clay mineral fraction, for it is the total clay content of the rock that is important. Thus it is common to find very aluminous clays, the kaolinites, in rocks that are very high in silica, the quartz sandstones.

Shales, too, very widely in silica content. The shales commonly include much silt size material and many rocks called shales are actually siltstones. Such siltstones will typically contain much quartz silt and so be higher in silica, but because practically no analytical data are accompanied by petrographic analysis that includes size distribution there is little firm information. The varieties of shale that contain unusually high amounts of chemically precipitated minerals, such as calcareous, phosphatic, ferruginous, or sulfidic shales, will tend to have silica contents reduced correspondingly. The siliceous shales show very high amounts of silica. Some of these are associated with bedded cherts and grade into them; many of these owe their high silica content to siliceous skeletal remains, the remnants of which can be seen in thin section. Others are less clearly of this origin and some may owe their high silica content to high contents of volcanic materials which have hydrolyzed and altered to a combination of silica and clays.

II. Chemical Sediments

The chemical sediments with the highest silica contents are the cherts, siliceous earths and sinters, and related rocks. The chemical composition of these rocks has been reviewed by CRESSMAN (1962), see Table 8-4 in Volume I of this handbook. Some are almost 100 per cent silica, others are as low as 64 per cent when admixed with other silicates and non-siliceous minerals. It has become increasingly clear that most cherts and siliceous shales derive their silica content from the remains of siliceous organisms; these even include the Paleozoic cherts that were hitherto not diagnosed as such but which are found to contain remains of radiolarians and sponge spicules (MCBRIDE and THOMSON, 1970). So far there is no evidence for such organic remains in Precambrian cherts, though presumed blue-green algae have been identified in a number of Precambrian cherts (BARGHOORN and TYLER, 1965). The Precambrian cherts may be the result of inorganic precipitation of amorphous silica in restricted environments. Recently some modern lake deposits of silica gels and sodium silicates have been described by EUGSTER and JONES (1968), who infer their origin from the interaction of hot alkaline springwaters with alkali volcanics and associated detritus.

Limestones normally contain some silica in the form of quartz silt or sand grains, secondary quartz crystals, or chert nodules. Some limestones are so chert-like that they may be very high in silica content and grade into what appear to be bedded calcareous cherts. There is no clear correlation with dolomite content or petrographic character of the limestone. Naturally, the argillaceous limestones contain moderately high amounts of silica.

The iron rich sediments normally contain silica in varying amounts, depending on the type of iron formation. The iron carbonate rocks contain the least, the iron silicate the most. The typical Precambrian bedded iron oxide formations are interbedded with chert and the formation as a whole may contain well over 50 per cent silica, though the individual laminae of iron oxide, whether hematite or limonite (almost always goethite), may be very pure and silica-free.

Other chemical sediments, such as the evaporites, may be relatively free of silica, the sole contribution coming from small amounts of quartz silt. On the other hand, phosphate rocks nearly always contain some silica, coming from admixed clays and other silicates.

In general, one can generalize from a petrographic analysis of a sediment what the silica content is likely to be. The higher the terrigenous detrital fraction or the greater the amount of chert, primary or secondary, the higher the silica content will be.

L. Biogeochemistry

Though there are many analyses of silica contents of organisms, usually performed by ashing the organic fraction, the biochemical mechanisms by which silica is secreted by organisms remains a matter of conjecture. Silica content in modern organisms was reviewed by Siever and Scott (1963), who pointed out that only a few groups of invertebrate aquatic animals secrete silica as skeletal material, the radiolarians, silicoflagellates, sponges and forminifera. Most animals seem to excrete silica rather than incorporate it to any significant extent in body tissue or hard parts. Plants, however, do typically contain silica, though silicon is not considered to be a nutritionally essential trace element except for the diatoms.

The diatoms are the quantitatively important organisms that use dissolved silica from waters to make their skeletons. The biochemistry of the process was considered by Rogall (1939) who theorized that the organic substrate membrane that secretes the silica has a configuration that acts as a template for the amorphous silica structures. Though it is possible that body fluids contain some silica in true solution as $Si(OH)_4$, and this would be the only species in solution at the slightly alkaline pH of typical fluids of this kind, there is no direct evidence that this is so. On the contrary, the fact that the diatoms are able to extract silica from very undersaturated waters is a strong argument for the existence of organic complexes by which the organism is able to sequester the silica, later to precipitate it through a membrane through which the complex is destabilized. This must be true not only for the diatoms but for many other species of silica-secreting plants as well.

Opaline silica is found in the tissues of many vascular plants. In many species they are localized as barbs of nettles or endocarps of palm fruits. Bamboo secretes an amorphous silica gel known as tabashir. Opal phytoliths, very tiny pure amorphous silica grains of a myriad of shapes and sizes, are formed by many plants. The well known scouring rush, *Equisetum*, contains unusual amounts of silica that account for its abrasive qualities. Lovering (1959) has pointed out the large amounts of silica accumulated by tropical plants.

Silicification of plant and animal tissues and hard parts is commonplace in sedimentary rocks. The silicified fossils range from impregnation of plant tissues that results in petrifactions such as petrified wood to structureless casts and replacements of delicate hard parts such as the spines of some brachiopods. Few soft bodied organisms, however, are preserved in this way. A detailed review of silicification has been given by Siever and Scott (1963). It is not certain what role, if any, the organic tissue of plants, or animals, plays in the silicification process. Silicification may be highly selective and dependent on the extent of degradation of organic tissue. Though the fine structure of the silica may reflect the nature of the replaced material, there is no firm evidence that the precise nature of the organic material has any relation to the amount or nature of the infiltration and precipitation of silica.

(The solubility of organic silica complexes is discussed under subsection 14-H-VII.)

References: Sections 14-DI, 14-G to 14-L

ALEXANDER, G. B., HESTON, W. M., ILER, R. K.: The solubility of amorphous silica in water. J. Phys. Chem. 58, 453 (1954).
ARMSTRONG, L. C.: Decomposition and alteration of feldspars and spodumene by water. Am. Mineralogist 25, 810—820 (1940).
ARRHENIUS, G.: Pelagic sediments. In: M. N. HILL, editor, The sea, vol. III, pp. 655—727. London: Interscience 1963.
BAKER, G.: Opal phytoliths in some Victorian soils and "Red Rain" residues. Australian J. Botany 7, 64 (1959).
BARGHOORN, E. S., TYLER, S. A.: Microorganisms from the gunflint chert. Science 147, 563 (1965).
BERGER, W. H.: Radiolarian skeletons: Solution at depths. Science 159, 1237 (1968).
BIEN, G. S., CONTOIS, D. E., THOMAS, W. H.: The removal of soluble silica from fresh water entering the sea. Geochim. Cosmochim. Acta 14, 35 (1958).
CALVERT, S. E.: Silica balance in the ocean and diagenesis. Nature 219, 919 (1968).
CORRENS, C. W.: Über die Löslichkeit von Kieselsäure in schwach sauren und alkalischen Lösungen. Chem. Erde 13, 92 (1941).
— The experimental chemical weathering of silicates. Clay Minerals Bull. 4, 249—265 (1961).
— VON ENGELHARDT, W.: Neue Untersuchungen über die Verwitterung des Kalifeldspates. Chem. Erde 12, 1 (1938).
CRESSMAN, E. R.: Nondetrital siliceous sediments. Geol. Surv. Profess. Paper 440-T (1962).
DAVIS, S.: Silica in streams and ground water. Am. J. Sci. 262, 870 (1964).
EATON, F. M., MCLEAN, G. W., BREDELL, G. S., DONER, H. E.: Significance of silica in the loss of magnesium from irrigation waters. Soil Sci. 105, 260 (1968).
EMERY, K. O., RITTENBERG, S. C.: Early diagenesis of California basin sediments in relation to origin of oil. Bull. Am. Assoc. Petrol. Geologists 36, 735 (1952).
ENLOWS, H. E., OLES, K. F.: Authigenic silicates in marine Spencer formation at Corvallis, Oregon. Bull. Am. Ass. Petrol. Geologists 50, 1918 (1966).
EUGSTER, H. P., JONES, B. F.: Gels composed of sodium-aluminum silicate, Lake Magadi, Kenya. Science 161, 160 (1968).
FETH, J. H., ROGERS, S. M., ROBERSON, C. E.: Aqua de Ney, California, a spring of unique chemical character. Geochim. Cosmochim. Acta 22, 75 (1961).
FOLK, R. L., WEAVER, C. E.: A study of the texture and composition of chert. Am. J. Sci. 250, 498 (1952).
GARDNER, L. U.: Reaction of the living body to different types of mineral dusts. Am. Inst. Mining Met. Eng., Tech. Pub. 929 (1938).
GARRELS, R. M.: Silica: Role in the buffering of natural waters. Science 148, 69 (1965).
— CHRIST, C. L.: Solutions, minerals, and equilibria. New York: Harper and Row 1965.
GORHAM, E.: pH of fresh soils. Ecology 41, 563 (1960).
GRUNER, J. W.: The hydrothermal alteration of feldspars in acid solutions between 300° and 400° C. Econ. Geol. 39, 578—589 (1944).
HARDER, H., FLEHMIG, W.: Quarzsynthese bei tiefen Temperaturen. Geochim. Cosmochim. Acta 34, 295 (1970).
— MENSCHEL, G.: Quarzbildung am Meeresboden. Naturwissensch. 54, 567 (1967).
HARRISS, R. C.: Biological buffering of oceanic silica. Nature 212, 275 (1966).
HARVEY, H. M.: The chemistry and fertility of sea waters. Cambridge: Cambridge University Press 1957.
HAY, R. L.: Zeolites and zeolitic reactions in sedimentary rocks. Geol. Soc. Am. Spec. Papers 85 (1966).

HESS, P. C.: Phase equilibria of some minerals in the $K_2O-Na_2O-Al_2O_3-SiO_2-H_2O$ system at 25° C and 1 atmosphere. Am. J. Sci. 289 (1966).

ILER, R. K.: The colloid chemistry of silica and silicates. Ithaca, N. Y.: Cornell University Press 1955.

— DALTON, R. L.: Degree of hydration of particles of colloidal silica in aqueous solution. J. Phys. Chem. 60, 955 (1956).

JANDER, G., JAHR, K. F.: Das System der Kieselsäuren. Kolloid-Beihefte 41, 48 (1934).

JEFFREY, L. M., HOOD, D. W.: Chemistry of organo-silicate complexes isolated from marine sediments. Geol. Soc. Am. Spec. Papers No. 82, 100 (1965).

JOIDES (Joint Oceanographic Institutions for Deep Earth Sampling): Initial reports of the deep sea drilling project, vol. I. Washington: U.S. Natl Sci. Foundation 1969.

KRAUSKOPF, K.: Dissolution and precipitation of silica at low temperatures. Geochim. Cosmochim. Acta 10, 1 (1956).

— The geochemistry of silica in sedimentary environments. In: Silica in sediments. Soc. Econ. Paleontologists Mineralogists, Spec. Publ. 7, 4—19 (1959).

LAGERSTRÖM, GÖSTA: Equilibrium studies of polyanions. III. Silicate ions in $NaClO_4$ medium. Acta Chem. Scand. 13, 722 (1959).

LEWIN, J. C.: The dissolution of silica from diatom walls. Geochim. Cosmochim. Acta 21, 182 (1961).

LIER, J. A. VAN: The solubility of quartz. Utrecht: Kemink en Zoon 1959.

LIVINGSTONE, D. A.: Chemical composition of rivers and lakes. Geol. Surv. Profess Paper 440-G (1963).

LOVERING, T. S.: Significance of accumulator plants in rock weathering. Geol. Soc. Am. Bull. 70, 781 (1959).

MACKENZIE, F. T., GARRELS, R. M.: Silicates: Reactivity with sea water. Science 150, 57 (1965).

— — Silica-bicarbonate balance in the ocean and early diagenesis. J. Sediment Petrol. 36, 1075 (1966).

— — BRICKER, O. P., BICKLEY, F.: Silica in sea water: Control by silica minerals. Science 155, 1404 (1967).

— GEES R.: Quartz-Synthesis at Earth-surface conditions. Science 173, 533 (1971).

MCBRIDE, E. F., THOMSON, A.: The Caballos Novaculite, Marathon Region, Texas. Geol. Soc. Am. Spec. Paper, 122 (1970).

MIDGELEY, H. G.: Chalcedony and flint. Geol. Mag. 88, 179 (1951).

MOHR, E. C. J., VAN BAREN, F. A.: Tropical soils. New York: Interscience Pub. 1954.

MOREY, G. W., CHEN, W. T.: The action of hot water on some feldspars. Am. Mineralogist 40, 996—1000 (1955).

— FOURNIER, R. O., ROWE, J. J.: The solubility of quartz in water in the temperature interval from 25—300° C. Geochim. Cosmochim. Acta 26, 1029 (1962).

MULLIN, J. B., RILEY, J. P.: The colorimetric determination of silicate with special reference to sea and natural waters. Anal. Chim. Acta 12, 162 (1955).

NASH, V. E., MARSHALL, C. E.: The surface reactions of silicate minerals. Part I. The reactions of feldspar surfaces with acidic solutions. Univ. Missouri Coll. Agr. Res. Bull. 613, 36 pp. (1956a).

— — The surface reactions of silicate minerals, Part II. Reactions of feldspar surfaces with salt solutions. Univ. Missouri Coll. Agr. Res. Bull. 614, 36 pp. (1956b).

OKAMOTO, G., OKURA, T., GOTO, K.: Properties of silica in water. Geochim. Cosmochim. Acta 12, 123 (1957).

PELTO, C. R.: A study of chalcedony. Am. J. Sci. 254, 32 (1956).

PETTIJOHN, F. J.: Chemical composition of sandstones — excluding carbonates and volcanic sands. Geol. Surv. Profess. Paper 440-S (1963).

ROGALL, E.: Über den Feinbau der Kieselmembrane der diatomeen. Planta 29, 27 (1939).

ROLLER, P. S., JR., ERVIN, GUY JR.: The system calcium oxide-silica-water at 30°. The Association of Silicate Ion in Dilute Alkaline Solution. J. Am. Chem. Soc. 62, 461 (1940).

SIEVER, R.: The silica budget in the sedimentary cycle. Am. Mineralogist 42, 821 (1957).

SIEVER, R.: Silica solubility, 0—200° C, and the diagenesis of siliceous sediments. J. Geol. 70, 127 (1962).
— Establishment of equilibrium between clays and sea water. Earth and Planet. Sci. Letters 5, 106 (1968).
— BECK, K. C., BERNER, R. A.: Composition of interstitial waters of modern sediments. J. Geol. 73, 39 (1965).
— SCOTT, R. A. Organic geochemistry of silica. In: Organic geochemistry. New York: Pergamon Press 1963.
SILLÉN, L. G.: The physical chemistry of sea water. In: Oceanography. Am. Assoc. Adv. Sci. Pub. 67, 549 (1961).
STEFANNSON, U., RICHARDS, F. A.: Processes contributing to the nutrient distributions off the Columbia river and the strait of Juan de Tica. Limnol. Oceanog. 8, 394 (1963).
STÖBER, W.: Formation of silicic acid in aqueous suspensions of different silica modifications. In: W. STUMM, editor, Equilibrium Concepts in Natural Water Systems. Am. Chem. Soc. Advances in Chemistry 67, Washington, D.C., p. 161—182 (1967).
SVERDRUP, H. U., JOHNSON, M. W., FLEMING, R. H.: The oceans. New York: Prentice-Hall 1942.
TAMM, OLOF: Experimentelle Studien über die Verwitterung und Tonbildung von Feldspäten. Chem. Erde 4, 420 (1930).
WEITZ, E., FRANCK, H., SCHUCHARD, M.: Silicic acid and silicates. Chemiker-Ztg. 74, 256 (1950).
WHITE, D. E., BRANNOCK, W. W., MURATA, K. J.: Silica in hot-spring waters. Geochim. Cosmochim. Acta 10, 27 (1956).
— HEM, J. D., WARING, G. A.: Chemical composition of subsurface waters. Geol. Surv. Profess. Paper 440-F (1963).
WOLLAST, R., MACKENZIE, F. T., BRICKER, O. P.: Experimental precipitation and genesis of sepiolite at earth-surface conditions (in press).

Revised manuscript received: April 1970

Sulfur   16

A	B. J. Wuensch	(Ceramics Division, Department of Metallurgy and Materials Science, Massachusetts Institute of Technology, Cambridge, Massachusetts, U.S.A.)
B	H. Nielsen	(Geochemisches Institut der Universität, Göttingen, Germany)
C—E	A. Schneider	(Geochemisches Institut der Universität, Göttingen, Germany)
F	H. E. Usdowski	(Sedimentpetrogr. Institut der Universität, Göttingen, Germany)
G	A. G. Herrmann	(Geochemisches Institut der Universität, Göttingen, Germany)
H	H. E. Usdowski	
I, K	A. G. Herrmann	
L	W. Orr	(Mobile Research and Development Corp. Dallas, Texas 75221, U.S.A.)
M, N	A. Schneider	
O	A. G. Herrmann	

16-A. Crystal Chemistry

I. Introduction

a) Bond Types and Coordination about Sulfur

The outer electrons of sulfur have the s^2p^4 configuration common to all Group VI elements. Elements of this series become progressively more metallic as their atomic number increases. Sulfur has an electronegativity of 2.5 which is moderate compared with the value of 3.5 assigned to oxygen. Further, in sulfur and subsequent elements, d orbitals are available for hybridization. The crystal chemistry of sulfur, and of the later Group VI elements, is therefore quite distinct from that of oxygen.

Several types of bonds may be formed to complete the s^2p^4 subshell of sulfur. Two electrons might be added to create S^{2-}; a single electron-pair bond might be formed and one electron added (SH^-, for example); two electron-pair bonds might be formed or, alternatively, a larger number of bonds involving one of a variety of possible hybrid orbitals. All of these modes of bond formation are known for sulfur. Its coordination number is therefore extremely variable, ranging from unity to unusually high values in disordered structures. Illustrations of this variability are presented in Table 16-A-1. The coordination polyhedra are often regular for the more common configurations (tetrahedral and octahedral coordination, for example); in complex sulfides, however, extremely distorted polyhedra are often found in which the interatomic distances are so irregular that a distinction between nearest- and second-nearest neighbors is somewhat arbitrary.

Table 16-A-1. *Coordination numbers of sulfur*

Coordination	Examples
1	S_2; thiocyanates (SCN)
2 po	SO_2; CS_2; S_6; S_8; orpiment, As_2S_3; stibnite Sb_2S_3
3	SO_3
3 po	molybdenite, MoS_2, and other sulfides with the CdI_2 structure type; stibnite, Sb_2S_3
3 po + S	covellite, CuS; pyrite and marcasite, FeS_2
4	wurtzite and sphalerite, ZnS, and many derivatives of these structures cooperite, PtS all sulfates
4 po	sulvanite, Cu_3VS_4
5	covellite, CuS; stromeyerite, CuAgS
6	galena, PbS; tetrahedrite, $Cu_{12}Sb_4S_{13}$; SF_6
6 pr	troilite, FeS, and other sulfides with NiAs structure type
7	bornite, Cu_5FeS_4 [a]
8	alkali metal sulfides with the antifluorite structure type
9	bornite, Cu_5FeS_4 [a]
17	high-chalcocite, β-Cu_2S [a]

[a] Statistical coordination in disordered structure.

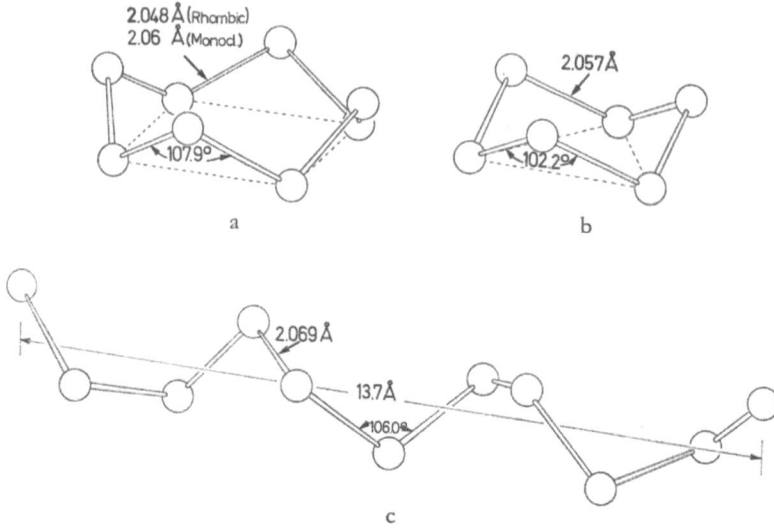

Fig. 16-A-1a—c. Structural groups in polymorphs of elemental sulfur. (Bond lengths and angles are averages for each molecule.) a S_8 ring in orthorhombic and monoclinic sulfur. b S_6 ring in rhombohedral sulfur. c Helical chain in fibrous forms of sulfur

b) Repeated S—S Bonds and the Structures of Elemental Sulfur

The tendency of sulfur to form covalent bonds is reflected by the ease with which repeated S—S bonds may form and be arranged to create chains or rings $S_n^{[2po]}$. Examples are provided by the polysulfide ions S_n^{2-} and the thionate ions $S_n O_6^{2-}$ ($n = 2$ to 6 for both). Further examples are provided by the many polymorphs of elemental sulfur.

At room temperature the stable form of sulfur is orthorhombic. This structure, Fig. 16-A-1a, contains molecules consisting of puckered, crown-like 8-membered rings (ABRAHAMS, SR 19, 312). The molecules are efficiently packed and intermolecular contacts range from 3.31 Å to 3.83 Å. The stable form of sulfur above 95.4° C is monoclinic and this structure (SANDS, 1965) contains similar 8-membered rings packed in a disordered fashion. A rhombohedral form of sulfur may be crystallized from solution; this phase contains 6-membered rings of symmetry $\bar{3}m$ (Fig. 16-A-1b) packed in a cubic close-packed sequence with the plane of the molecule normal to c (DONOHUE et al., SR 26, 288). Intermolecular S—S contacts range between 3.20 and 3.75 Å in this structure.

Three closely related phases may be synthesized under the influence of pressure (GELLER, 1966). These structures apparently consist of chains with three helical turns and ten atoms in a characteristic 13.7 Å period (Fig. 16-A-1c). These structures are related to the fibrous sulfur which may be prepared by stretching quenched amorphous sulfur (TUINSTRA, 1966; GELLER and LIND, 1969). Additional forms which have been reported include a purple paramagnetic form, condensed from the vapor, which is believed to consist of S_2 molecules (RICE and SPARROW, 1953), and a green form believed to contain S_8 chains (RICE and DITTER, SR 17, 314).

The S—S bond lengths in the ring molecules are all slightly less than the 2.08 Å separation assigned to a single bond by PAULING. This may reflect some double-bond character due to use of d orbitals. The length of S—S bonds, in general, show considerable ranges: from 1.887 Å in S_2 gas (double bond) to 2.39 Å in thionate ions. Typical bond angles about S are intermediate between the orthogonal bonds to be expected for p orbitals and the angle of 109.5° for tetrahedral coordination.

c) Occurrence of Sulfur in Minerals

Sulfur is a primary constituent in three major mineral groups: sulfides, sulfosalts, and sulfates. Each of these families is discussed below in a separate section. A few additional compounds occur in nature which are of interest because of their rarity as minerals. A thiocyanate, for example, is represented by the mineral julienite, $Na^{[6H_2O]}Co^{[4N]}[NCS]_4 \cdot 8H_2O$ (PREISINGER, SR 17, 462); S is bonded to C at one end of a linear NCS group. Oxysulfides other than sulfates are unusual in nature since SO_4^{2-} represents the final product of oxidation. An exception is kermesite, Sb_2S_2O, discussed below with Sb_2S_3. Voltzite, long described as Zn_5S_4O, has recently been shown to be a mixture of wurtzite and a zinc-containing organometallic compound (FRONDEL, 1967).

d) Derivative Structure and Superstructure in Sulfur Compounds

Certain structures bearing relationships to simpler atomic arrangements may be termed *derivative* structures (BUERGER, 1947). In such structures the symmetry of a simpler atomic array is degraded through mechanisms such as ordered substitution of one type of atom for another, ordered omission of atoms, addition of atoms to unoccupied sites ("stuffed" derivative structures), or distortion of the array. The symmetry of the derivative structure thus may be a subgroup of the simpler array. The resulting structure may also have a translational periodicity which is a multiple of that of the simpler structure. Such derivatives are termed *superstructures*. The latter relationship manifests itself in diffraction patterns as a subperiod of weak reflections added to a superperiod of intense reflections. Many complex sulfides and sulfosalts are derivatives of simpler structures and examples are provided in subsequent sections.

II. Sulfides

a) Ionic Structures

Sulfur forms purely ionic bonds only with ions of low charge of the most electropositive metals. These sulfides crystallize with typical ionic structures. The alkali metal sulfides assume the antifluorite structure, while the alkaline-earth sulfides have the sodium chloride structure.

b) Disulfides

Many dioxides form ionic structures. However, the moderate electronegativity of sulfur causes formation of S^{2-} to be energetically unfavorable in combination with highly charged M^{4+} ions. The disulfides therefore assume either the pyrite or marcasite structure, or one of several types of layer structures.

The structure of pyrite, $Fe^{[6]}[S_2]$ (BRAGG, SB 1, 215), is shown in Fig. 16-A-2a. The atomic arrangement is a derivative of the sodium chloride structure in which

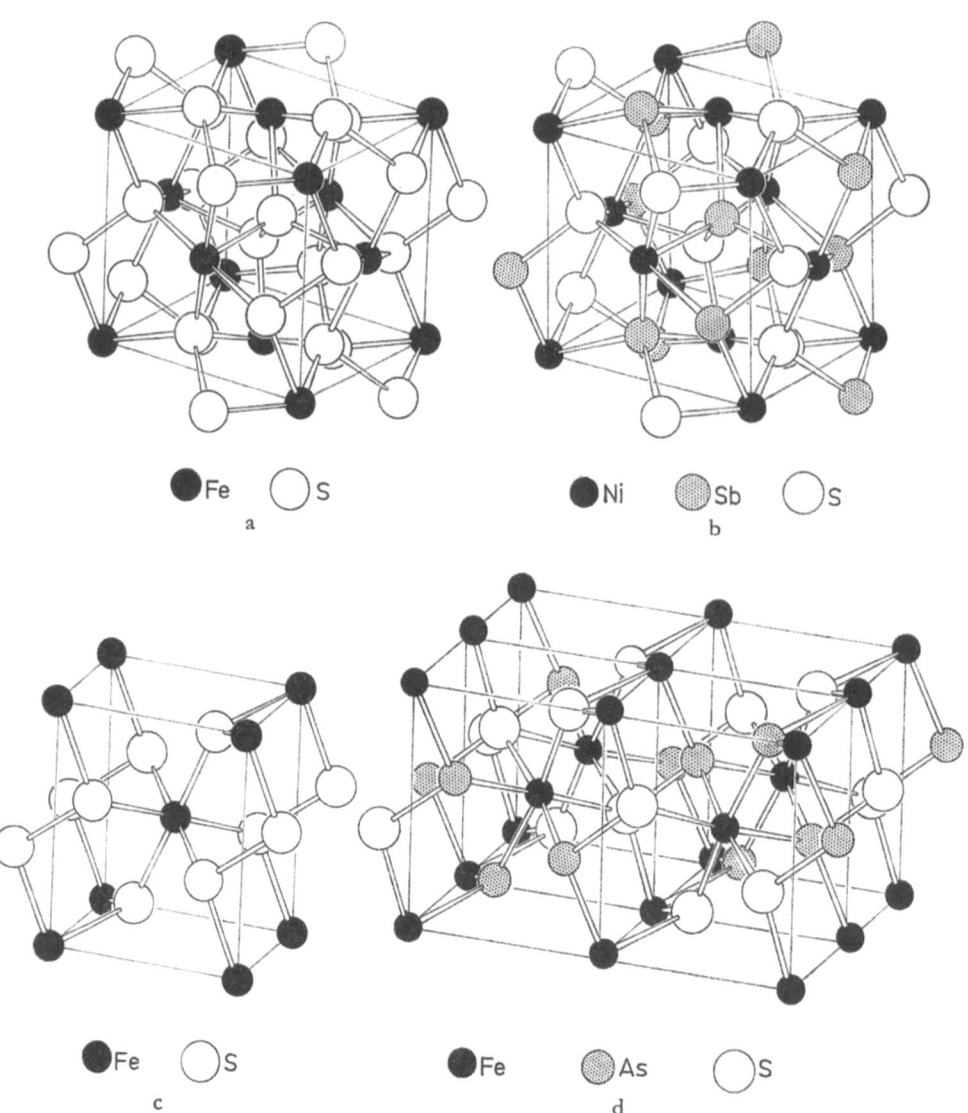

Fig. 16-A-2a—d. The pyrite and marcasite structures and derivatives. a Pyrite, FeS$_2$. b Ullmannite, NiSbS. c Marcasite, FeS$_2$. d Arsenopyrite, FeAsS

the anion has been replaced by a covalently-bonded pair of S separated by 2.14 Å. Each of the four S$_2$ pairs per cell is directed along a different body diagonal of the cell and each S is coordinated by 1S and 3Fe, the latter forming an octahedron about the S$_2$ pair. The orthorhombic marcasite modification of Fe$^{[6]}$[S$_2$], shown in Fig. 16-A-2c (BUERGER, SB 2, 272), has similar primary coordination; the separation of the S$_2$ pair is 2.21 Å.

Disulfides such as TiS$_2$, ZrS$_2$ and SnS$_2$ assume the CdI$_2$ structure type, Fig. 16-A-3a. Layers M$^{[6]}$S$_2$ are stacked in a simple-hexagonal sequence. The

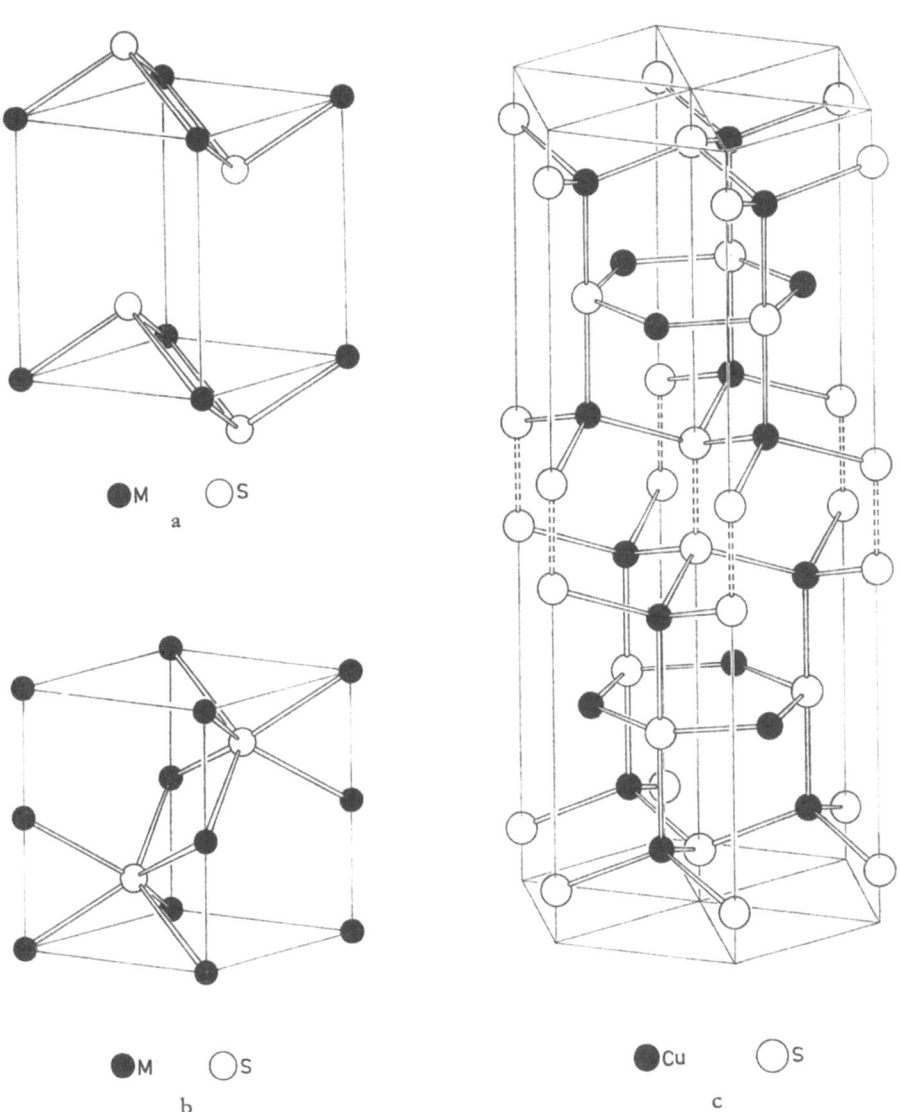

Fig. 16-A-3a—c. Hexagonal and layer structures assumed by sulfides. a CdI$_2$-type, MS$_2$. b NiAs-type, MS. c Covellite, CuS

structure may be viewed as a hexagonal close-packed array of S in which alternating layers of octahedral sites are unoccupied. Molybdenite, Mo$^{[6\ pr]}$S$_2$ (DICKINSON and PAULING, SB 1, 214), and tungstenite, W$^{[6\ pr]}$S$_2$, assume a different type of layer structure in which the coordination of the metal atom in the layer is prismatic. Stacking variations of these layers are known for molybdenite (TAKÉUCHI and NOWACKI, 1964).

c) Monosulfides

Monosulfides frequently assume one of several structure types characteristic of sulfides. Alternatively, a few unexpectedly complex structures occur which are assumed by only one particular mineral.

The NiAs structure type, $M^{[6]}S^{[6\,pr]}$, Fig. 16-A-3b, is assumed by most of the transition metal monosulfides. The prismatic coordination about S would not occur in a structure composed of ions. Furthermore, metal atoms in neighboring layers are close enough that bonds between them undoubtedly contribute to the stability of the structure. Non-stoichiometric compositions and derivative structures are common in sulfides of this general structure type. Comparison of Figs. 16-A-3a and 3b shows that the CdI_2 structure-type results when a layer of metal atoms is removed from the NiAs arrangement. Vacancies in this layer may result in complete solid solution between these two structure types, as in melonite, $NiTe$—$NiTe_2$. In other sulfur-rich structures the metal vacancies may be statistically distributed over all sites. Alternatively, metal atoms may be omitted in a regular, ordered fashion which gives rise to a superstructure. Examples are provided by pyrrhotite, Fe_7S_8 (BERTAUT, SR 17, 444), and a series of Cr sulfides whose compositions reflect different numbers and ordering schemes of vacancies: Cr_7S_8, Cr_5S_6, Cr_3S_4, and Cr_2S_3 (JELLINEK, SR 21, 95). Stoichiometric $Cr^{[4+2]}S$ is a distorted monoclinic derivative of the NiAs arrangement.

The Group IIB monosulfides crystallize with either the sphalerite-type structure of ZnS (BeS; CdS; metacinnabarite, HgS), Fig. 16-A-4a, or with the wurtzite form of ZnS (greenockite, CdS; MnS), Fig. 16-A-5a. The two arrangements differ in the manner in which the tetrahedra are stacked: a cubic close-packed sequence in the former and a hexagonal close-packed sequence in the latter. (Stacking polytypes are also known.) Many derivatives of these structure types occur. Chalcopyrite, $Cu^{[4]}Fe^{[4]}S_2$ (PAULING and BROCKWAY, SB 2, 346), is a tetragonal superstructure based upon a sphalerite-like arrangement with ordered substitution of Cu and Fe, Fig. 16-A-4b. Stannite, $Cu^{[4]}_2Fe^{[4]}Sn^{[4]}S_4$ (BROCKWAY, SB 3, 440), has a supercell of similar geometry in which layers of ordered Fe and Sn alternate with Cu. Examples of other derivatives are provided by sulfosalts described below. Structures in which all atoms have tetrahedral environment may form when the number of bonding electrons available in the structure is four times the number of atoms. The compositions for which this requirement is satisfied include many sulfides and have been explored in detail by PARTHÉ (1964).

Several additional monosulfides merit mention because of the unique nature of their structures. In cooperite, $Pt^{[4\,pl]}S$ (BANNISTER and HEY, SB 2, 234), S forms a square arrangement about Pt due to formation of dsp^2 bonds. The S—Pt—S angle is 97.5°, intermediate between the 90° angle required by $Pt^{[4\,pl]}$ and 109.5° for $S^{[4]}$. Cinnabar, $Hg^{[2+2+2]}S$ (AURIVILLUS, SR 13, 179), is a distorted hexagonal derivative of the NaCl structure type; two short Hg—S bonds reflect the tendency of Hg to form sp bonds. If nearest-neighbor bonds alone are considered, the structure consists of helical $Hg^{[2\,po]}S$ chains. The Hg—S—Hg bond angle is 105.2°, a value similar to that observed in S_n chains. Covellite, CuS (OFTEDAL, SB 2, 230; BERRY, SR 18, 380), provides another example of the unique and complex structures often assumed by sulfides of simple chemical composition. The structure contains two

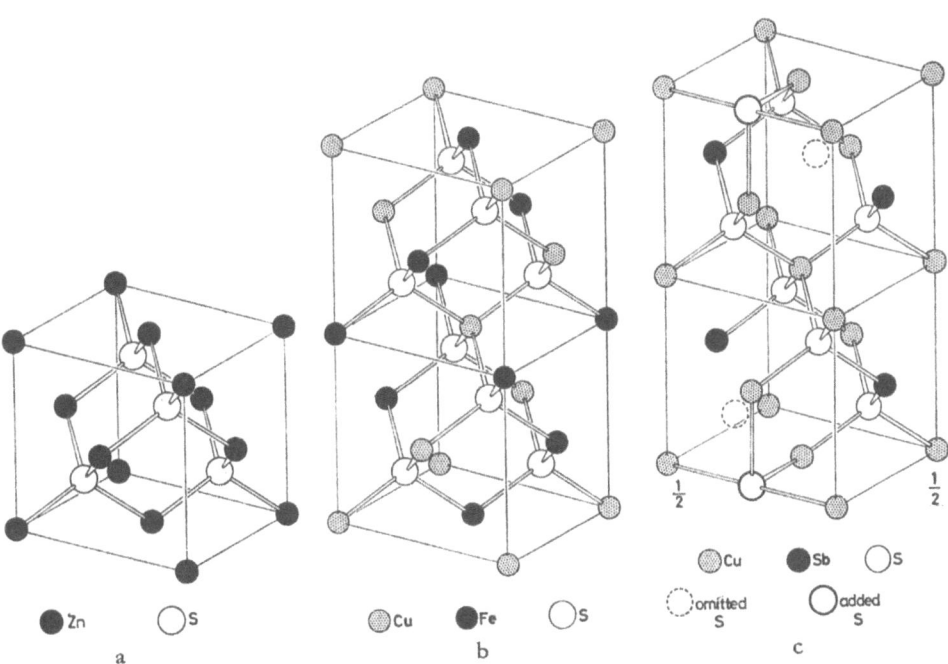

Fig. 16-A-4a—c. The sphalerite structure and derivatives. a Sphalerite, ZnS. b Chalcopyrite, $CuFeS_2$. c Tetrahedrite, $Cu_{12}Sb_4S_{13}$

types of Cu, one of which forms trigonal rings with S, while a second occupies a tetrahedral site, Fig. 16-A-3c. S—S bonds exist between neighboring layers. The structural formula must therefore be expressed as $Cu^{[3]}Cu_2^{[4]}[S|(S_2)]$. The single bonded S_2 pair also occurs in livingstonite, $HgSb_4S_8$ (NIIZEKI and BUERGER, SR 21, 347), and patronite, $V(S_2)_2$ (ALLMANN et al., 1964).

d) Other Chain and Layer Sulfides

A few sulfides form layer or chain structures as a result of the limited number of bonding electrons available from the metal. Group V metals, for example, often form three approximately orthogonal bonds with p electrons. Orpiment, $As_2^{[3\,p\text{o}]}S_3$ (MORIMOTO, SR 19, 405), thus forms a layer structure in which six AsS_3 pyramids share corners to form rings, Fig. 16-A-6a. The structures of stibnite, Sb_2S_3 (HOFMANN, SR 3, 349; ŠĆAVNIČAR, SR 24, 51), and the isostructural phase bismuthinite, Bi_2S_3 (HOFMANN, SR 3, 349), are more complex and the structures contain two types of Group V metals. In stibnite, Sb(1) forms a trigonal pyramid with S (Sb—S = 2.57 and 2.58 Å). Sb(2), however, forms one short (2.49 Å) and two longer bonds (2.68 Å) which share corners with the pyramids in the first chain, Fig. 16-A-6b. Two such double chains are joined such that each Sb(2) acquires two additional neighbors at longer distances (2.82 Å). The resulting structure consists of $Sb_2^{[3\,p\text{o}]}Sb_2^{[1+2+2]}S_2^{[2]}S_4^{[3\,p\text{o}]}$ quadruple chains, with Sb(2) in the base of a square pyramid, probably employing sp^3d^2 hybridization. The chains are joined by weaker Sb—S

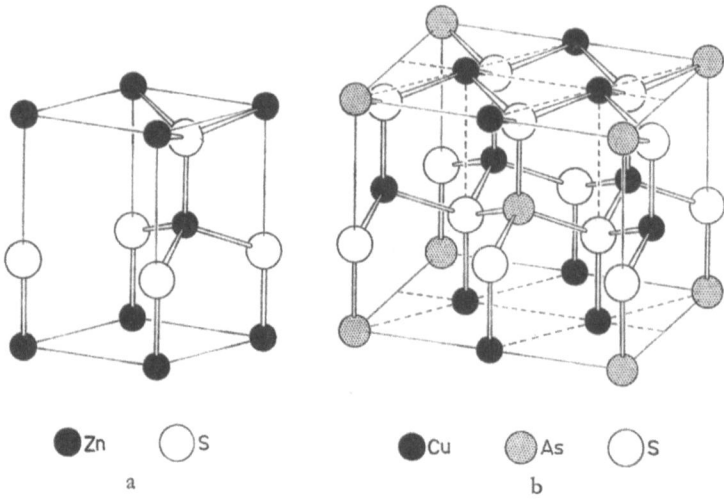

Fig. 16-A-5a and b. The wurtzite structure and derivatives. a Wurtzite, ZnS. b Enargite, Cu_3AsS_4

bonds (3.20 Å) to form sheets parallel to (010) which, in turn, are linked by still weaker bonds (3.33 Å) along b. The oxysulfide kermesite, $Sb^{[1+2+2S]}Sb^{[1 S+2O+O]}S_2O$ (KUPČÍK, 1967), has a structure closely related to that of stibnite. The 5-coordinated Sb form a double chain similar to the central portion of the stibnite quadruple-chain. The remaining Sb (corresponding to the outer $Sb^{[3Po]}$ in stibnite) again share an S with $Sb^{[5]}$, but have 3 oxygens as the additional neighbors. The bond to the additional oxygen neighbor links the quadruple chains into layers. Getchellite, $As^{[3Po]}Sb^{[3Po]}S_3^{[2Po]}$, a layer structure, bears no relation to orpiment (GUILLERMO and WUENSCH, 1969). Eight-membered rings are arranged normal to the plane of thick sheets.

The synthetic compound $Si^{[4]}S_2$ contains chains of tetrahedra sharing edges, in marked contrast to the invariable linkage of tetrahedral vertices in the more ionic SiO_2 minerals.

e) Phase Transitions and Disordered Sulfides

Order-disorder phase transformations which occur at low (ca. 100—300°C) temperatures are common in many sulfides important in ore mineralogy. Stromeyerite, $Cu^{[3]}Ag^{[2]}S$ (FRUEH, SR 19, 412), and chalcocite, $Cu_2^{[3]}S$ (EVANS, 1968), both have structures based upon hexagonal close-packed arrays of S. Half of the available triangular sites in every S layer are occupied by Cu in stromeyerite. These trigonal $Cu^{[3]}S$ sheets, similar to those present in covellite, are joined by zig-zag $Ag^{[2]}S$ chains. In chalcocite such sheets are formed with only alternate layers of S. The sheets are linked by Cu in triangular sites between layers. Both structures transform to similar disordered hexagonal forms in the neighborhood of 100°C. High chalcocite (BUERGER and WUENSCH, 1963) contains S in hexagonal close-packing; the Cu moves between three types of sites with 2-, 3-, and 4-fold coordination.

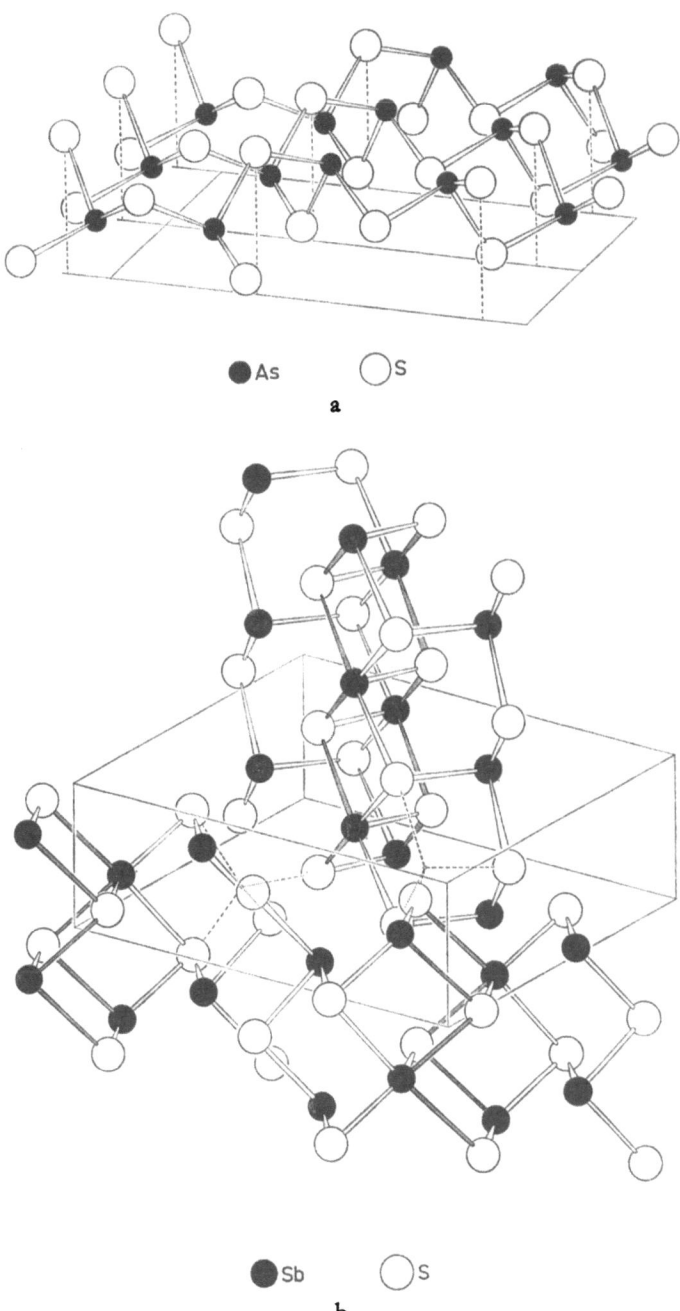

Fig. 16-A-6a and b. Chain and layer structures of Group V metal sulfides. a One sheet in the layer structure of orpiment, As_2S_3. b Quadruple-chain structure of stibnite, Sb_2S_3. One of the two chains per cell has been extended above the unit cell, the other below. Weak bonds joining two double chains $Sb_2^{[3\,po]}S_3^{[2\,po]}$ into a quadruple chain $Sb_2^{[3\,po]}Sb_2^{[3\,po+2]}$ $S_3^{[2\,po]}S_3^{[3\,po]}$ are shaded. Locations of Cu in the stuffed derivative structure of aikinite, $PbCuBiS_3$, are indicated by small circles

Similar statistical distributions of metals among sites in cubic closepacked arrays of S are known in the high-temperature polymorphs of digenite, Cu_9S_5 (MORIMOTO and KULLERUD, 1963), and bornite, Cu_5FeS_4 (MORIMOTO, 1964). Order-disorder transformations occur in many sulfides which are derivative structures, *e.g.*, chalcopyrite, pyrrhotite, and members of the cobaltite group.

Chalcogenides of Group V and IV metals readily form glasses which are of interest because of their semiconducting and infrared transmitting properties (PEARSON, 1964). Amorphous lead-arsenic sulfides are found in nature (AMSTUTZ *et al.*, 1957; MILTON and INGRAM, 1950; BAUMANN and AMSTUTZ, 1965; BURKART-BAUMANN *et al.*, 1966).

III. Sulfosalts

a) Introduction

Sulfosalts are a diverse group of minerals in which one or more metal atoms are combined with S and a Group V element such as As, Sb, or Bi. Ternary compounds are the most common and account for roughly 75% of the known sulfosalt minerals. Relatively few metals are present in these compounds: usually Pb, Cu, or Ag, or less frequently Zn or Hg. Separation of these minerals from the sulfides is based on early chemical concepts and is somewhat artificial. Many sulfosalts bear close structural relations to simpler sulfides.

b) Marcasite and Pyrite Derivatives

Arsenopyrite, $Fe^{[3As+3S]}As^{[3Fe+S]}S^{[3Fe+As]}$ (BUERGER, SB 4, 143), is a triclinic marcasite derivative in which As replaces one-half of the sulfur in the S_2 pairs, Fig. 16-A-2d. The As:S ratio may vary from 1.22 to 0.82; As-rich phases approach monoclinic symmetry and a cell which is dimensionally orthorhombic (MORIMOTO and CLARK, SR 26, 50). Gudmundite, FeSbS (BUERGER, SB 7, 122), has the same structure.

Ullmannite, NiSbS (TAKÉUCHI, SR 21, 34), Fig. 16-A-2b, and gersdorffite, NiAsS (BAYLISS and STEPHENSON, 1967), are derived from the pyrite structure in an analogous fashion. Cobaltite, CoAsS, has essentially the same structure (GIESE and KERR, 1965) but atomic displacements cause the dimensionally isometric structure to have orthorhombic symmetry. All three structures occur in disordered forms statistically isostructural with pyrite (PEACOCK and HENRY, SR 11, 357; GIESE and KERR, 1965; BAYLISS, 1968).

c) Sphalerite and Wurtzite Derivatives

A large group of sulfosalts are derivatives of the sphalerite structure. Lautite, $Cu^{[3S+1As]}As^{[1Cu+2As+1S]}S^{[3Cu+1As]}$, is an orthorhombic derivative with $[a, b, c] = [\frac{3}{2}\frac{3}{2}0/-\frac{1}{2}\frac{1}{2}0/0\,0\,1]\,[a_1, a_2, a_3]$ of sphalerite (KULPE, 1961; MARUMO and NOWACKI, 1964). The As—As separation is the same as that found in metallic As. Luzonite, $Cu_3^{[4]}As^{[4]}S_4$ (MARUMO and NOWACKI, 1967a), is a tetragonal derivative with $[a_1, a_2, c] = [100/010/002]\,[a_1, a_2, a_3]$ of sphalerite, in which layers of tetrahedrally coordinated Cu alternate with layers of Cu and As. Enargite, Fig. 16-A-5b, an orthorhombic polymorph of luzonite, provides an example of a derivative of the wurtzite arrangement (PAULING and WEINBAUM, SB 3, 438).

The above-mentioned derivatives, along with members of the cobaltite and arsenopyrite groups are atypical of the sulfosalts in that the Group V metal has tetrahedral coordination. (Only in the luzonite and enargite structures, however, is the metal bonded to 4 S.) In most sulfosalt structures the Group V metal forms either three nearly orthogonal single bonds with p electrons, or displays five-fold coordination with sp^3d^2 hybridization. The coordination group formed with S is usually a characteristic trigonal pyramid. The structures of some sulfosalts are derived from the sphalerite arrangement by omission of the fourth S which would have been bonded to the Group V metal. An example is provided by the rhombohedral mineral nowackiite, $Cu_6^{[4]}Zn_3^{[4]}As_4^{[3\,po]}]S_{12}$ (MARUMO, 1967).

Tennantite, $Cu_{12}As_4S_{13}$ (PAULING and NEUMAN, SB 3, 413; WUENSCH *et al.*, 1966), and tetrahedrite, $Cu_{12}Sb_4S_{13}$ (WUENSCH, 1964), are more complex derivatives with $a = 2a$ sphalerite, Fig. 16-A-4c. One-fourth of the metal atom sites are occupied by a Group V metal which forms but three bonds to S. One-fourth of the S atoms in the cell are therefore omitted. This, however, leaves half of the Cu with only 2 S neighbors, and an additional S is added at an interstitial site in the sphalerite array to complete a triangular coordination for these atoms. The structural formula for tetrahedrite must therefore be written as $Cu_6^{[4]}Cu_6^{[3]}Sb_4^{[3\,po]}[S_{12}^{[3\,Cu+Sb]}S^{[6\,Cu]}]$.

d) Rock-Salt Derivatives

As noted by HELLNER (1958), many sulfosalts are derivatives of the rock-salt structure or, alternatively, have structures composed of blocks of rocksalt-like structures. Such structures, however, invariably have highly irregular coordination polyhedra resulting in the S atoms being appreciably displaced from their ideal positions. An atom in a nominally octahedral interstice may have, in fact, only 3 or 4 nearest neighbors.

Miargyrite, $Ag^{[3+1+2]}Ag^{[3+3]}Sb_2^{[3\,po+3]}S_4$ (KNOWLES, 1964), is monoclinic with $[a, b, c] = [2\,1 - 1/0\,\tfrac{1}{2}\tfrac{1}{2}/1 - \tfrac{3}{2}\tfrac{3}{2}][a_1, a_2, a_3]$ NaCl. Two As analogues of myargyrite are also derivatives of the rock-salt structure: smithite, $Ag^{[3+1+2]}Ag^{[2+2+2]}As_2^{[3\,po+3]}S_4$ (HELLNER and BURZLAFF, 1964), is monoclinic with

$$[a, b, c] = [3 - \tfrac{1}{2}\tfrac{1}{2}/0\,1\,1/0\,2\,\bar{2}][a_1, a_2, a_3]\ \text{NaCl},$$

and trenchmannite, $Ag^{[3+1+2]}As^{[3\,po+2]}S_2$ (MATSUMOTO and NOWACKI, 1969), is rhombohedral with $a = \tfrac{1}{2}[\bar{1}4\bar{3}]$ NaCl and $c = [111]$ NaCl, but also has some unoccupied metal and sulfur sites. Both smithite and trenchmannite contain a 3-membered As_3S_6 ring. Freieslebenite, $PbAgSbS_3$ (HELLNER, SR 21, 32), and marrite, $Pb^{[6]}Ag^{[3+1+2]}As^{[3\,po+3]}S_3$ (WUENSCH and NOWACKI, 1967), are monoclinic with $[a, b, c] = [110/-\tfrac{3}{2}\tfrac{3}{2}0/001][a_1, a_2, a_3]$ NaCl. Diaphorite, $Pb_2Ag_3Sb_3S_8$ (HELLNER, SR 22, 376), is a similar monoclinic superstructure with

$$[a, b, c] = [2\,2\,0/-4\,4\,0/0\,0\,1][a_1, a_2, a_3]\ \text{NaCl}.$$

The lead sulfosalts, Table 16-A-2, form an extremely complex group of minerals which show interesting variations in structure as the Pb content of the minerals decreases. Gratonite, the As sulfosalt of highest Pb content, is a rhombohedral superstructure with $c = [111]$PbS (RÖSCH, 1963; RIBAR and NOWACKI, 1969). Jordanite is monoclinic, pseudohexagonal with $a \approx c \approx [110]$PbS, $b = 10/3\,[111]$PbS. The structure (WUENSCH and NOWACKI, 1966) is based upon 2 PbS-like slabs,

Table 16-A-2. *Lead-arsenic and lead-antimony sulfosalts* Pb_xM_yS

x	y	Mineral	Composition	Mineral	Composition
0.600	0.267	gratonite	$Pb_9As_4S_{15}$		
0.565	0.304	jordanite	$Pb_{13}As_7S_{23}(?)$	geocronite	$Pb_{13}(Sb, As)_8S_{23}(?)$
0.500	0.333			falkmanite	$Pb_3Sb_2S_6$
0.454	0.363			boulangerite	$Pb_5Sb_4S_{11}(?)$
0.444	0.370			sterryite	$Pb_{12}(Sb, As)_{10}S_{27}(?)$
0.429	0.381			semseyite[b]	$Pb_9Sb_8S_{21}$
0.425	0.550			sorbyite	$Pb_{17}(Sb, As)_{22}S_{40}(?)$
0.415	0.390			madocite	$Pb_{17}Sb_{16}S_{41}(?)$
0.400	0.400	dufrenoysite[a]	$Pb_2As_2S_5$	veenite	$Pb_2(Sb, As)_2S_5$
0.379	0.414			dadsonite	$Pb_{11}Sb_{12}S_{29}$
0.372	0.419			playfairite	$Pb_{16}Sb_{18}S_{43}(?)$
0.368	0.421			heteromorphite[b]	$Pb_7Sb_8S_{19}$
0.361	0.426			launayite	$Pb_{22}Sb_{26}S_{61}(?)$
0.350	0.450	rathite Ia[a]	$Pb_7As_9S_{20}$		
0.322	0.456	rathite II	$Pb_9As_{13}S_{28}$		
0.306	0.473	acentric baumhauerite	$Pb_{11}As_{17}S_{36}$	antimonian baumhauerite	$Pb_{11}(Sb, As)_{17}S_{36}$
0.300	0.500	rathite I[a]	$(Pb, Tl)_3As_5S_{10}$		
		rathite III[a]	$Pb_3As_5S_{10}$		
0.294	0.491	liveingite (?)	$Pb_5As_8S_{17}(?)$	plagionite[b]	$Pb_5Sb_8S_{17}$
0.280	0.480			robinsonite	$Pb_7Sb_{12}S_{25}(?)$
0.278	0.500	centric baumhauerite	$Pb_5As_9S_{18}$		
0.273	0.485			guettardite	$Pb_9(Sb, As)_{16}S_{33}(?)$
0.250	0.500	scleroclase	$PbAs_2S_4$	twinnite	$Pb(Sb, As)_2S_4$
0.222	0.519			zinkenite	$Pb_6Sb_{14}S_{27}(?)$
0.222	0.556	hutchinsonite	$(Pb, Tl)_2As_5S_9$		
0.200	0.553			fülöppite[b]	$Pb_3Sb_8S_{15}$

A composition followed by a question mark is the most satisfactory of several possible representations of an empirical analysis, and has not yet been confirmed by a structure determination. Solid solution is indicated in the composition when minerals have not yet been found which are reasonably close to the end-member of the solid solution series.

[a] Rathite group; 32 metal and 40 S atoms contained in virtually identical cells.

[b] Plagionite group; an apparent homologous series, $Pb_{3+2n}Sb_8S_{15+2n}$, in which two lattice dimensions remain invariant, while the third increases with n.

5 octahedra in thickness, rotated 180° about [111]PbS relative to one another. Minerals of higher As content break into regions of distorted PbS-like structure joined by layers with 9-coordinated Pb. The structure of dufrenoysite,

$$Pb^{[9]}Pb^{[6]}As_2^{[3\,po]}S_5$$

(MARUMO and NOWACKI, 1967b; RIBAR et al., 1969), is shown in Fig. 16-A-7. Other members of the rathite group have closely related structures. Their difference lies in the metal content of the PbS-like layer: Pb_4As_8 in dufrenoysite; $(Pb, Tl)_2AgAs_9$ in rathite I (MARUMO and NOWACKI, 1965); Pb_2As_{10} in rathite III (LEBIHAN, 1962); Pb_3As_9 in rathite Ia (LEBIHAN, 1962). Disordered crystals are known (MARUMO and NOWACKI, 1967b) in which the layering may be mixed. In minerals of still higher

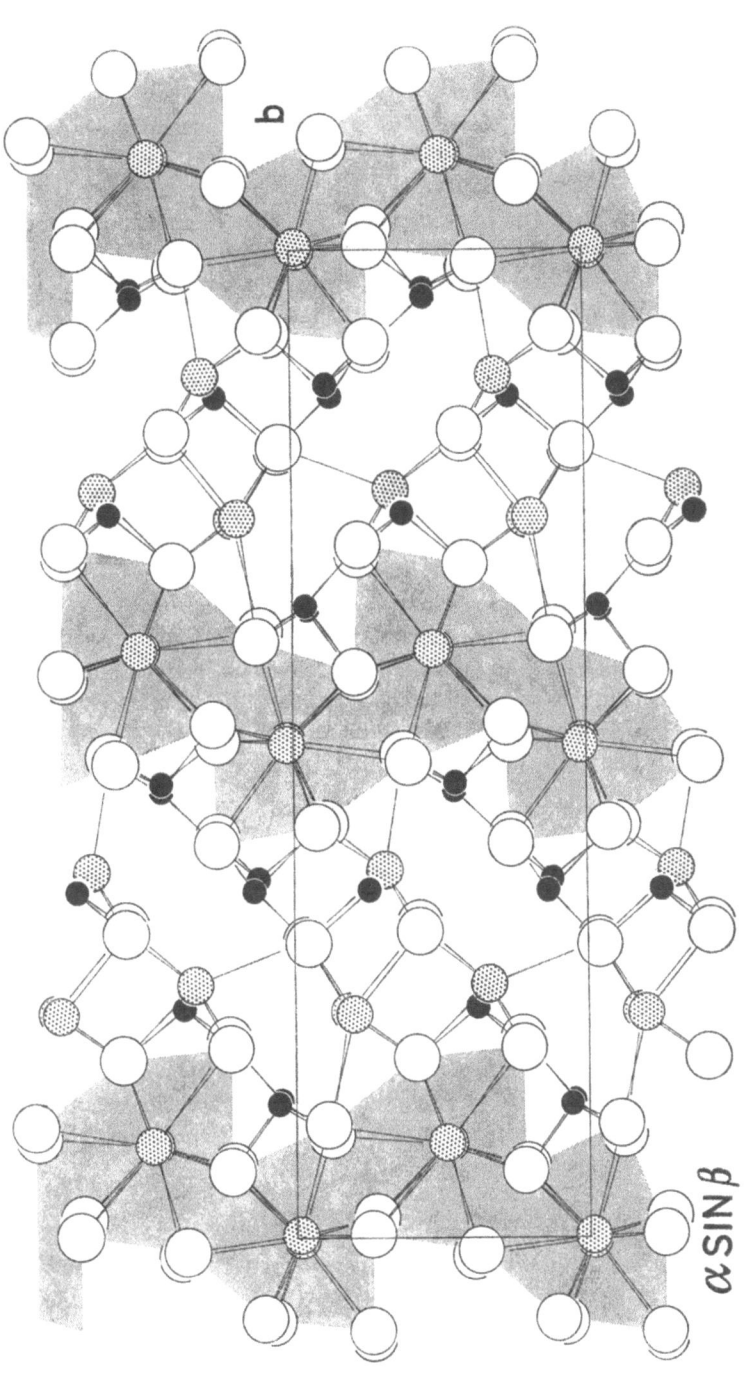

Fig. 16-A-7. Projection of the structure of dufrenoysite, $Pb_2As_2S_5$, an example of a sulfosalt of the rathite group. Layers of 9-coordinated Pb (shaded) alternate with layers of distorted rock salt-like structure. The composition of the latter layer differs in other sulfosalts of this family

As content, the thickness as well as the composition of the PbS-like slab varies. Minerals of this group include rathite II and baumhauerite (LeBihan, 1962; Engel and Nowacki, 1969), and scleroclase (Iitaka and Nowacki, SR 26, 426). Hutchinsonite, the mineral of highest As content, again has PbS-like layers (Takéuchi et al., 1965). The $Pb^{[9]}S_3$ layer, however, is replaced by a layer containing laterally-joined spiral chains of $As^{[3po]}S_3$ pyramids. Similar chains are found in lorandite, $TlAsS_2$ (Zemann and Zemann, SR 23, 42).

The Pb—Sb sulfosalts, Table 16-A-2, have equally complex crystal chemistry (see, e.g., Jambor, 1967). Interestingly, these minerals bear little resemblance to the As sulfosalts, but few structures are known. Plagionite, $Pb_5Sb_8S_{17}$ (Cho and Wuensch 1970), contains twisted rocksalt-like ribbons. Other members of the plagionite group probably differ in the number of Pb atoms in the rock-salt sequence and hence in the "pitch" of the ribbon. Jamesonite, $Fe^{[2+2+2]}Pb_2^{[6+1]}Pb_2^{[6+2]}Sb_6^{[3po]}S_{14}$ (Niizeki and Buerger, SR 21, 346), also contains PbS-like ribbons. Cosalite, $Pb_2Bi_2S_5$ (Weitz and Hellner, SR 24, 63), has large blocks of rocksalt-like structure.

e) Stibnite Derivatives and Related Structures

A number of sulfosalts are related to the stibnite structure or, alternatively, contain similar structural groups. Aikinite, $CuPbBiS_3$ (Wickman, SR 17, 451; Kohatsu and Wuensch, 1971), is a stuffed derivative in which Cu is added to a tetrahedral interstice in the stibnite array, Fig. 16-A-6b, and Pb substitutes for the 5-coordinated metal in the interior of the stibnite chain. The Cu tetrahedra share corners and form a chain parallel to the stibnite quadruple-chain. Several complex Cu—Pb—Bi sulfosalts of higher Bi content are superstructures based upon aikinite (Welin, 1966; Moore, 1967). Bournonite, $CuPbSbS_3$, and seligmannite, $CuPbAsS_3$ (Leineweber, SR 20, 30; Takéuchi and Haga, 1969), have

$$[a, b, c] = [\tfrac{1}{2} - \tfrac{1}{2} 0 / \tfrac{1}{2} \tfrac{1}{2} 0 / 0\, 0\, 2]\, [a, b, c]$$

of aikinite. Projections of these structures bear a resemblance to aikinite.

Group V metal chains are found in a number of additional sulfosalts. Berthierite, $Fe^{[6]}Sb^{[3po+2]}Sb^{[3po]}S_2$ (Buerger and Hahn, SR 19, 418), contains Sb with the two distinct coordinations of stibnite. The two different types of polyhedra, however, form separate chains which are linked by a distorted Fe octahedron. Galenobismutite, $PbBi_2S_4$ (Wickman, SR 15, 239; Iitaka and Nowacki, 1962), has a somewhat similar structure. In chalcostibite, $Cu^{[4]}Sb^{[1+2+2]}S_2$, and emplectite, $CuBiS_2$ (Hofmann, SB 3, 393), chains of Cu tetrahedra link double Group V metal chains which are similar to the central portion of the stibnite chain. All of the above chain structures are orthorhombic and are characterized by one short (ca. 4 Å) translation corresponding to the period of the chains.

f) Additional Sulfosalt Structures

The following are a few examples of unusual structure types which do not fit into the above categories. Pyrargyrite, $Ag_3^{[2po]}Sb^{[3po]}S_3$, and proustite, Ag_3AsS_3 (Harker, 1936; Engel and Nowacki, 1966), contain rows of Group V metal pyramids stacked along their axis of symmetry and linked by a trigonal spiral, $Ag^{[2po]}S$. Proustite has a monoclinic polymorph, xanthoconite, $Ag_2^{[3]}Ag^{[3po]}As^{[3po]}S_3$ (Engel and Nowacki, 1968). Sulvanite, $Cu_3^{[4]}V^{[4]}S_4$ (Pauling and Hultgren,

SB 3, 437; Trojer, 1966), is not a tetrahedral structure; the coordination about sulfur ($S^{[3\,Cu+Vpo]}$) is such that the S—V bond is in the inverse direction of that for tetrahedral coordination. An interpretation of this curious bonding has been given by Pauling (1965).

IV. Sulfates

a) Introduction

Sulfates are the final oxidation product of oxysulfide compounds and occur in nature as a large group of minerals. The bonding within the tetrahedral SO_4^{2-} group is covalent, with an S—O separation appreciably shorter than the sum of the single-bond radii. The multiple-bond character has been interpreted as $d\pi - p\pi$ bonding (Cruickshank, 1961). The tetrahedra are quite regular in hydrated sulfates. In well-refined structures the average S—O distance is 1.473 Å (Baur, 1964b) and the average bond angle is the same as that for a regular tetrahedron. The slight departures from regularity (usually 1.45—1.49 Å for the bond length and 108—112° for the bond angle) depend on the environment of the apical oxygen. A hydrogen atom attached to one of the oxygen, for example, results in a significantly longer bond length, and contraction of the other bonds; e.g., H_2SO_4: S—O = 1.426, S—O(H) = 1.535 (Pascard-Billy, 1965); $H_2SO_4 \cdot H_2O$: S—O = 1.448, S—O(H) = 1.560 (Taesler and Olovsson, 1968). The tetrahedron may be somewhat less regular in the network structures assumed by the anhydrous sulfates. S—O bonds ranging from 1.42 to 1.60 Å, and bond angles from 103° to 116° have been reported.

While the bonding within the sulfate group is covalent, the bond formed between this group and metal atoms is predominantly ionic. The structures assumed by the anhydrous sulfates are consequently determined by the relative sizes of the cations and the SO_4 tetrahedron. The resulting structures are often isostructural with a chlorate, chromate, selenate, vanadate or manganate. In hydrous sulfates the environment of the metal atoms may be either oxygen, water, or a combination of both; the coordination number of the cation is again determined by its size and the linkage of the polyhedra is in part dependent on the relative number of oxygen and H_2O neighbors. For convenience, the hydrates are subdivided below into "partially-hydrated", "fully-hydrated" or "over-hydrated" salts depending on whether the number of H_2O is less than, equal to, or in excess of the coordination number of the cation.

While the structures assumed by sulfates are often unexpectedly complex, their crystal chemistry is primarily a matter of packing considerations or hydrogen bonding, rather than any unique crystal-chemical feature of sulfur. The examples in subsequent sections indicate merely some of the great variety of structures which are formed, and the discussion is mainly confined to structures containing a single cation. A few more complex minerals with known structures are given in Table 16-A-3.

b) Anhydrous Sulfates

The decrease in coordination number with decreasing cation size is illustrated by the sulfates of the divalent metals, Table 16-A-4. (The tetrahedral coordination for Hg and Cd, however, is rather unexpected.)

Table 16-A-3 *Additional known structures of sulfate minerals containing more than one species of cation*

Mineral	Formula	Reference
I. Anhydrous		
apthitalite	$NaKSO_4$	Bellanca, SR 9, 232
palmierite	$K_2Pb(SO_4)_2$	Bellanca, SR 11, 390; SR 13, 326; Bachmann and Kleber, SR 17, 499
lanarkite	$Pb_2O(SO_4)$	Binnie, SR 15, 255
langbeinite	$K_2Mg_2(SO_4)_3$	Zemann and Zemann, SR 21, 362
linarite	$CuPb(OH)_2SO_4$	Araki, SR 22, 471; Bachmann and Zemann, SR 24, 384; SR 26, 451
brochantite	$Cu_4(OH)_6SO_4$	Cocco and Mazzi, SR 23, 451
antlerite	$Cu_3(OH)_4SO_4$	Araki, SR 26, 439
dolerophanite	$Cu_2O(SO_4)$	Flügel-Kahler (1963)
glauberite	$CaNa_2(SO_4)_2$	Cocco *et al.* (1965); Araki and Zoltai (1967)
II. Hydrous		
kröhnkite	$CuNa(SO_4)_2 \cdot 2 H_2O$	Dahlman, SR 16, 289; Rao, SR 26, 449
blödite	$Na_2Mg(SO_4)_2 \cdot 4 H_2O$	Rumanova, SR 22, 461; Giglio, SR 22, 464; Rumanova and Malickaja, SR 23, 449
natrochalcite	$Cu_2Na(SO_4)_2(OH) \cdot H_2O$	Rumanova and Volodina, SR 22, 470
manganleonite	$K_2Mn(SO_4)_2 \cdot 4 H_2O$	Schneider, SR 26, 447; Srikanta *et al.* (1968)
krausite	$FeK(SO_4)_2 \cdot H_2O$	Graeber *et al.* (1965)
schönite	$MgK_2(SO_4)_2 \cdot 6 H_2O$	Kannan and Viswamitra (1965)
syngenite	$CaK_2(SO_4)_2 \cdot H_2O$	Corazza and Sabelli (1967)
serpierite	$Ca(Cu, Zn)_4(SO_4)_2(OH)_6 \cdot 3 H_2O$	Sabelli and Zanazzi (1968)
jouravskite	$Ca_3Mn(SO_4)(CO_3)(OH)_6 \cdot 12 H_2O$	Granger and Protas (1969)
tamarugite	$AlNa(SO_4)_2 \cdot 6 H_2O$	Robinson and Fang (1969)
ettringite	$Ca_6[Al(OH)_6]_2(SO_4)_3 \cdot 26 H_2O$	Moore and Taylor (1970)
roemerite	$Fe^{2+}Fe_2^{3+}(SO_4)_4 \cdot 14 H_2O$	Fanfani *et al.* (1970)
loeweite	$Na_{12}Mg_7(SO_4)_2 \cdot 15 H_2O$	Fang and Robinson (1970)

In the acid sulfate mercallite, $K^{[9]}HSO_4$ (Loopstra and MacGillavry, SR 22, 458; Cruickshank, 1964), one type of SO_4 tetrahedra is linked to a second by two H bonds to form a dimer; a second type is linked by single H bonds to form an infinite chain. The potassium polyhedra are linked by edge- and face-sharing, as well as by SO_4 tetrahedra, to form a network.

c) Fully-hydrated Sulfates

In salts in which the number of available H_2O equals the coordination number of the cation, the metal is usually coordinated only by H_2O, and this polyhedron

Table 16-A-4. *Cation coordination in anhydrous sulfates of the divalent metals*

Coordination number of cation	Ionic radius of cation (Å)	Examples (reference)
12	1.35	barite, $BaSO_4$ (JAMES and WOOD, SB 1, 383; COLVILLE and STAUDHAMMER, 1967)
	1.21	anglesite, $PbSO_4$ (JAMES and WOOD, SB 1, 383; SAHL, 1963)
	1.17	$EuSO_4$ (MAYER et al., 1964)
	1.13	celestite, $SrSO_4$ (JAMES and WOOD, SB 1, 383; GARSKE and PEACOR, 1965)
	0.93	$SnSO_4$ (RENTZEPERIS, 1962)
8	0.99	anhydrite, $CaSO_4$ (WASASTJERNA, SB 1, 381; HOHNE, 1962; CHENG and ZUSSMAN, 1963)
6 ($CrVO_4$-type, SR 9, 181)	0.97	α-$CdSO_4$ (KOKKOROS and RENTZEPERIS, 1963)
	0.80	$MnSO_4$ ⎫
	0.75	$FeSO_4$ ⎬ (COING-BOYAT, SR 23, 444)
	0.72	β-$CoSO_4$ ⎭
	0.69	$NiSO_4$ (DIMARAS, SR 21, 365)
	0.65	$MgSO_4$ (RENTZEPERIS and SOLDATOS, SR 22, 448)
6	0.74	$Zn^{[2+2+2]}SO_4$ ⎫ (KOKKOROS and RENTZEPERIS, SR 22, 454)
	0.72	α-$CoSO_4$ ⎭
	0.72	chalcocyanite, $Cu^{[4+2]}SO_4$ (KOKKOROS and RENTZEPERIS, SR 22, 454; RAO, SR 26, 438)
4	1.10	$Hg^{[2+1+1]}SO_4$ ⎫ (KOKKOROS and RENTZEPERIS, 1963)
	0.97	β-$Cd^{[1+2+1]}SO_4$ ⎭
4 (BPO_4-type, SB 3, 92)	0.31	$BeSO_4$ (GRUND, SR 19, 445; KOKKOROS, SR 20, 347)

is linked to SO_4 tetrahedra by hydrogen bonds. $Be^{[4H_2O]}SO_4 \cdot 4H_2O$ (SCHONEFELD, and BEEVERS and LIPSON, SB 2, 427; DANCE and FREEMAN, 1969) contains a Be tetrahedron and a SO_4 group arranged in a CsCl-like array. Cations of medium size may form one of two types of hexahydrates with different schemes of hydrogen bonding: hexahydrite, $Mg^{[6H_2O]}SO_4 \cdot 6H_2O$ (ZALKIN et al., 1964), and $CoSO_4 \cdot 6H_2O$ (ZALKIN et al., 1962) are monoclinic, while $Ni^{[6H_2O]}SO_4 \cdot 6H_2O$ (BEEVERS and LIPSON, SB 2, 433) is tetragonal.

The alums are a large family of fully hydrated double salts with general composition $A^{+[6H_2O]}B^{3+[6H_2O]}(RO_4)_2 \cdot 12H_2O$. A^+ may be NH_4 or an alkali ion, B^{3+} may be Al, Ga, Cr, Fe, or V, and R is S, Se, or Te. All of these salts are cubic, but three distinct structure types are known. They are designated α, β, and γ in order of their discovery (LIPSON, SB 3, 455). For Al^{3+} alums the modifications are formed with intermediate, large, and small A^+ ions, respectively; other alums, such as Cr^{3+} compounds, do not always follow this scheme (LEDSHAM and STEEPLE, 1968; 1969). The coordination of B^{3+} is always a nearly-regular octahedron whose orientation relative to the cell edges differs in the three types. In γ alum, for which $NaAl(SO_4)_2 \cdot 12H_2O$ is the only known example, the B^+ octahedron is rotated about 40° about

[111] and the A^+ octahedron is regular (CROMER et al., 1967). In β alums the B^{3+} octahedron is parallel to the cell edges. The A^+ octahedron is severely compressed along [111] and acquires 6 additional oxygen neighbors from SO_4 groups (CROMER et al., 1966). In α alum, the most common form, the B^{3+} octahedron is slightly rotated, the A^+ octahedron is slightly compressed, and the orientation of the SO_4 groups is disordered (LARSON and CROMER, 1967; CROMER and KAY, 1967). The degree of disorder increases as the size of A^+ decreases.

d) Over-hydrated Sulfates

Cations are coordinated only by H_2O when the number of available H_2O exceeds their coordination number. Hydrogen bonds hold the additional water between SO_4 groups and the H_2O neighbors of the cation.

The heptahydrates of the divalent metals provide examples of such salts. Melanterite, $Fe^{[6H_2O]}SO_4 \cdot 7H_2O$ (BAUR, 1964a), is monoclinic; hydrogen bonds hold the one additional H_2O between the SO_4 group and one of the $6H_2O$ surrounding Fe. A similar isolated H_2O is present in the orthorhombic structures assumed by morenosite, $NiSO_4 \cdot 7H_2O$ (BEEVERS and SCHWARTZ, SB 3, 453), and epsomite, $MgSO_4 \cdot 7H_2O$ (BAUR, 1964b).

e) Partially Hydrated Sulfates

In many sulfates the number of available H_2O does not permit full coordination of the cations by H_2O. One might expect the cation to acquire all available H_2O as neighbors and then complete its coordination with oxygen. These structures would consist of polyhedra sharing certain oxygen atoms and linked in part by hydrogen bonds. A structure of this sort is assumed by gypsum, $Ca^{[6O,2H_2O]}SO_4 \cdot 2H_2O$ (WOOSTER, SB 4, 179; ATOJI and RUNDLE, SR 22, 449). The calcium ion has 6 O belonging to SO_4 as neighbors and together these polyhedra are linked to form a double layer. The H_2O coordinated to Ca are located at the surface of the sheet and the layers are linked by hydrogen bonds between these H_2O and the SO_4 group of a neighboring layer. In $Hg^{[1O+1H_2O+4O]}SO_4 \cdot H_2O$ (BONEFAČIĆ, SR 26, 443; TEMPLETON et al., 1964), octahedra share edges to form a chain; a SO_4 group links the chains by both sharing an oxygen and a hydrogen bond to H_2O.

Leonhardite, $Mg^{[2O,4H_2O]}SO_4 \cdot 4H_2O$, and rozenite, $FeSO_4 \cdot 4H_2O$ (BAUR, 1962), provide examples of ring structures. A pair of oxygens on each of two separate SO_4 groups are shared with two metal octahedra to form discrete four-membered rings of composition $M_2(SO_4)_2 \cdot 8H_2O$. The rings are linked only by hydrogen bonds. In kieserite, $Mg^{[4O+2H_2O]}SO_4 \cdot H_2O$ (LEONHARDT and WEISS, SR 21, 361), the cation coordinates oxygens from four different SO_4 groups as neighbors.

Tutton's salts constitute an extensive series of isomorphous monoclinic sulfates. Their general composition is $A_2^+B^{2+[6H_2O]}(SO_4)_2 \cdot 6H_2O$, where A^+ is K, NH_4, Rb, Cs or Tl, and B^{2+} may be any of a large collection of ions usually having a radius of the order of 0.7 Å. The structures of the ammonium salts have been extensively studied: detailed structures are known for $B^{2+}=$ Mg, V, Mn, Fe, Co, Ni, Cu, Zn and Cd (HOFMANN, SB 2, 436; MARGULIS and TEMPLETON, 1962; GRIMES et al., 1963; MONTGOMERY and LINGAFELTER, 1964—1967). Their structures consist of a packing of fully hydrated B^{2+} ions, SO_4 tetrahedra, and ammonium ions, all linked by hydrogen bonds. Each H_2O forms two hydrogen bonds to SO_4 oxygen. These oxygen form an icosohedron of second-nearest neighbors about B^{2+}. Three H on the ammonium ion form bonds to a SO_4 oxygen; the fourth forms a bifurcated bond. In schönite, a potassium-magnesium Tutton's salt, one H_2O and five SO_4 oxygen coordinate K (KANNAN and VISWAMITRA, 1965).

In a few sulfates extra H_2O are held between structural groups by hydrogen bonds as in the higher hydrates even though the cations in the structure have less than a full complement of H_2O neighbors. A fifth H_2O is held between an H_2O coordinated to Cu and the oxygen of a SO_4 group in chalcanthite, $Cu^{[4H_2O+2O]}SO_4 \cdot 5H_2O$ (BEEVERS and LIPSON, SB 3, 449; BACON and CURRY, 1962). Amarantite, $Fe_2(SO_4)_2O \cdot 7H_2O$ (SÜSSE, 1968a), and botryogen, $Fe^{3+}(Zn, Mn, Mg)(SO_4)_2OH \cdot 7H_2O$ (SÜSSE, 1968b), are examples of two more complex sulfates in which this situation occurs.

References: Section 16-A

ALLMANN, R., BAUMANN, I., KUTOGLU, A., RÖSCH, H., HELLNER, E.: Die Kristallstruktur des Patronits $V(S_2)_2$. Naturwissenschaften 11, 263 (1964).
AMSTUTZ, G. C., RAMDOHR, P., DE LAS CASAS, F.: A new low temperature mineral of hydrothermal origin from Cerro de Pasco. Soc. Geol. Peru 32, 25 (1957).
ARAKI, T., ZOLTAI, T.: Refinement of the crystal structure of a glauberite. Am. Mineralogist 52, 1272 (1967).
BACON, G. E., CURRY, N. A.: The water molecules in $CuSO_4 \cdot 5H_2O$. Proc. Roy. Soc. (London), Ser. A 266, 95 (1962).
BAUMANN, I. H., AMSTUTZ, G. C.: Natural X-ray amorphous lead-arsenic sulfides from Cerro de Pasco Mine, Peru. Naturwissenschaften 52, 585 (1965).
BAUR, W. H.: Zur Kristallchemie der Salzhydrate. Die Kristallstrukturen von $MgSO_4 \cdot 4H_2O$ (Leonhardit) und $FeSO_4 \cdot 4H_2O$ (Rozenit). Acta Cryst. 15, 815 (1962).
— On the crystal chemistry of salt hydrates. III. The determination of the crystal structure of $FeSO_4 \cdot 7H_2O$ (melanterite). Acta Cryst. 17, 1167 (1964a).
— On the crystal chemistry of salt hydrates. IV. The refinement of the crystal structure of $MgSO_4 \cdot 7H_2O$ (epsomite). Acta Cryst. 17, 1361 (1964b).
BAYLISS, P.: The crystal structure of disordered gersdorffite. Am. Mineralogist 53, 290 (1968).
— STEPHENSON, W. C.: The crystal structure of gersdorffite. Mineral. Mag. 36, 38 (1967).
BUERGER, M. J.: Derivative crystal structures. J. Chem. Phys. 15, 1 (1947).
— WUENSCH, B. J.: Distribution of atoms in high-chalcocite, Cu_2S. Science 141, 276 (1963).
BURKART-BAUMANN, I., OTTEMANN, J., AMSTUTZ, G. C.: Neue Beobachtungen an den röntgen-amorphen Sulfiden von Cerro de Pasco, Peru. Neues Jahrb. Mineral. Monatsh. 12, 353 (1966).
CHENG, G. C. H., ZUSSMAN, J.: The crystal structure of anhydrite ($CaSO_4$). Acta Cryst. 16, 767 (1963).
CHO, S.-A., WUENSCH, B. J.: Crystal chemistry of the plagionite group. Nature 225, 444 (1970).
COCCO, G., CORAZZA, E., SABELLI, C.: The crystal structure of glauberite, $CaNa_2(SO_4)_2$. Z. Krist. 122, 175 (1965).
COLVILLE, A. A., STAUDHAMMER, K.: A refinement of the structure of barite. Am. Mineralogist 52, 1877 (1967).
CORAZZA, E., SABELLI, C.: The crystal structure of syngenite, $K_2Ca(SO_4)_2 \cdot H_2O$. Z. Krist. 124, 398 (1967).
CROMER, D. T., KAY, M. I.: Refinement of the alum structures. IV. Neutron diffraction study of deuterated ammonium alum, $ND_4Al(SO_4)_2 \cdot 12H_2O$, an α-alum. Acta Cryst. 22, 800 (1967).
— — LARSON, A. C.: Refinement of the alum structures. I. X-ray and neutron diffraction study of $CsAl(SO_4)_2 \cdot 12H_2O$, a β-alum. Acta Cryst. 21, 383 (1966).
— — — Refinement of the alum structures. II. X-ray and neutron diffraction study of $NaAl(SO_4)_2 \cdot 12H_2O$, γ-alum. Acta Cryst. 22, 182 (1967).
CRUICKSHANK, D. W. J.: The role of 3d-orbitals in π-bonds between (a) silicon, phosphorus, sulfur or chlorine and (b) oxygen or nitrogen. J. Chem. Soc. 5486 (1961).
— Refinements of structures containing bonds between Si, P, S or Cl and O or N. VIII. $KHSO_4$ (mercallite). Acta Cryst. 17, 682 (1964).
DANCE, I. G., FREEMAN, H. C.: Refinement of the crystal structure of beryllium sulphate tetrahydrate. Acta Cryst. B 25, 304 (1969).

Sulfur

ENGEL, P., NOWACKI, W.: Die Verfeinerung der Kristallstruktur von Proustit, Ag_3AsS_3, und Pyrargyrite, Ag_3SbS_3. Neues Jahrb. Mineral. 6, 181 (1966).
— — Die Kristallstruktur von Xanthokon, Ag_3AsS_3. Acta Cryst. B 24, 77 (1968).
— — Die Kristallstruktur von Baumhauerit. Z. Krist. 129, 178 (1969).
EVANS, H. T.: A crystal structure study of low chalcocite. Program and Abstracts, Annual Meeting Geolog. Soc. Am. (1968).
FANFANI, L., NUNZI, A., ZANAZZI, F. F.: The crystal structure of roemerite. Am. Mineralogist 55, 78 (1970).
FANG, J. H., ROBINSON, P. D.: Crystal structures and mineral chemistry of double-salt hydrates: II. The crystal structure of loeweite. Am. Mineralogist 55, 378 (1970).
FLÜGEL-KAHLER, E.: Die Kristallstruktur von Dolerophanit, $Cu_2O(SO_4)$. Acta Cryst. 16, 1009 (1963).
FRONDEL, C.: Voltzite. Am. Mineralogist 52, 617 (1967).
GARSKE, D., PEACOR, D. R.: Refinement of the structure of celestite, $SrSO_4$. Z. Krist. 121, 204 (1965).
GELLER, S.: Pressure-induced phases of sulfur. Science 152, 644 (1966).
— LIND, M. D.: Indexing of the ψ-sulfur fiber pattern. Acta Cryst. B 25, 2166 (1969).
GIESE, R. F., KERR, P. F.: The crystal structures of ordered and disordered cobaltite. Am. Mineralogist 50, 1002 (1965).
GRAEBER, E. J., MOROSIN, B., ROSENZWEIG, A.: The crystal structure of krausite, $KFe(SO_4)_2 \cdot H_2O$. Am. Mineralogist 50, 1929 (1965).
GRANGER, M. M., PROTAS, J.: Détermination et étude de la structure cristalline de la jouravskite $Ca_3Mn^{IV}(SO_4)(CO_3)(OH)_6 \cdot 12H_2O$. Acta Cryst. B 25, 1943 (1969).
GRIMES, N. W., KAY, H. F., WEBB, M. W.: The crystal structure of ammonium nickel sulfate hexahydrate $(NH_4)_2Ni(SO_4)_2 \cdot 6H_2O$. Acta Cryst. 16, 823 (1963).
GUILLERMO, T. R., WUENSCH, B. J.; The crystal structure of getchellite, $AsSbS_3$. Program and Abstracts, Annual Meeting Geolog. Soc. Am. (1969).
HARKER, D.: The application of the three-dimensional Patterson method and the crystal structures of proustite, Ag_3AsS_3, and pyrargyrite, Ag_3SbS_3. J. Chem. Phys. 4, 381 (1936)
HELLNER, E : A structural scheme for sulfide minerals. J. Geol. 66, 503 (1958).
— BURZLAFF, H.: Die Struktur des Smithit $AgAsS_2$. Naturwissenschaften 51, 35 (1964).
HÖHNE, E.: Die Kristallstruktur des Anhydrit-$CaSO_4$. Monatsber. Deut. Akad. Wiss. Berlin 4, 72 (1962).
IITAKA, Y., NOWACKI, W.: A redetermination of the crystal structure of galenobismutite, $PbBi_2S_4$. Acta Cryst. 15, 691 (1962).
JAMBOR, J. L.: New lead sulfantimonides from Madoc, Ontario-Part I. Can. Mineralogist 9, 7 (1967)
— Part II. —Mineral Descriptions. Can. Mineralogist 9, 191 (1967).
KANNAN, K. K., VISWAMITRA, M. A.: Crystal structure of magnesium potassium sulfate hexahydrate $MgK_2(SO_4)_2 \cdot 6H_2O$. Z. Krist. 122, 161 (1965).
KNOWLES, C. R.: A redetermination of the crystal structure of miargyrite, $AgSbS_2$. Acta Cryst. 17, 847 (1964).
KOHATSU, I., WUENSCH, B. J.: The crystal structure of aikinite, $PbCuBiS_3$. Acta Cryst. B 27, 1245 (1971).
KOKKOROS, P. A., RENTZEPERIS, P. J.: The crystal structure of the anhydrous cadmium and mercuric sulfates. Z. Krist. 119, 234 (1963).
KULPE, S.: Die Kristallstruktur des Lautit. Fortschr. Mineral. 39, 332 (1961).
KUPČIK, V.: Die Kristallstruktur des Kermesits, Sb_2S_2O. Naturwissenschaften 54, 114 (1967).
LARSON, A. C., CROMER, D. T.: Refinement of the alum structures. III. X-ray study of the α-alums, K, Rb, and $NH_4Al(SO_4)_2 \cdot 12H_2O$. Acta Cryst. 22, 793 (1967).
LEBIHAN, M.-TH.: Étude structurale de quelques sulfures de plomb et d'arsenic naturels du gisement de Binn. Bull. Soc. Franc. Mineral. Crist. 85, 15 (1962).
LEDSHAM, A. H. C., STEEPLE, H.: The crystal structures of sodium chromium alum and caesium chromium alum. Acta Cryst. B 24, 1287 (1968).
— — The classification of the chromium alums. Acta Cryst. B 25, 398 (1969).

Sulfur

MARGULIS, T. N., TEMPLETON, D.: Crystal structure and hydrogen bonding of magnesium ammonium sulfate hexahydrate. Z. Krist. 117, 344 (1962).

MARUMO, F.: The crystal structure of nowackiite, $Cu_6Zn_3As_4S_{12}$. Z. Krist. 124, 352 (1967).

— NOWACKI, W.: The crystal structure of lautite and of sinnerite, a new mineral from the Lengenbach Quarry. Schweiz. Mineral. Petrog. Mitt. 44, 439 (1964).

— — The crystal structure of rathite I. Z. Krist. 122, 433 (1965).

— — A refinement of the crystal structure of luzonite, Cu_3AsS_4. Z. Krist. 124, 1 (1967a).

— — The crystal structure of dufrenoysite, $Pb_{16}As_{16}S_{40}$. Z. Krist. 124, 409 (1967b).

MATSUMOTO, T., NOWACKI, W.: The crystal structure of trenchmannite, $AgAsS_2$. Z. Krist. 129, 163 (1969).

MAYER, I., LEVY, E., GLASNER, A.: The crystal structure of $EuSO_4$ and $EuCO_3$. Acta Cryst. 17, 1071 (1964).

MONTGOMERY, H., LINGAFELTER, E. C.: The crystal structure of Tutton's salts. I. Zinc ammonium sulfate hexahydrate. Acta Cryst. 17, 1295 (1964).

— — II. Magnesium ammonium sulfate hexahydrate and nickel ammonium sulfate hexahydrate. Acta Cryst. 17, 1478 (1964).

— — III. Copper ammonium sulfate hexahydrate. Acta Cryst. 20, 659 (1966).

— — IV. Cadmium ammonium sulfate hexahydrate. Acta Cryst. 20, 728 (1966).

— — (With CHASTAIN, R. V.): V. Manganese ammonium sulfate hexahydrate. Acta Cryst. 20, 731 (1966).

— — (with CHASTAIN, R. V., NATT, J. J., WITKOWSKA, A. M.): VI. Vanadium (II), iron (II) and cobalt (II) ammonium sulfate hexahydrates. Acta Cryst. 22, 775 (1967).

MOORE, A. E., TAYLOR, H. F. W.: The crystal structure of ettringite. Acta Cryst. B 26, 386 (1970).

MOORE, P. B.: A classification of sulfosalt structures derived from the structure of aikinite. Am. Mineralogist 52, 1874 (1967).

MORIMOTO, N.: Structures of two polymorphic forms of Cu_5FeS_4. Acta Cryst. 17, 351 (1964).

— KULLERUD, G.: Polymorphism in digenite. Am. Mineralogist 48, 110 (1963).

PARTHÉ, E.: Crystal Chemistry of Tetrahedral Structures. New York: Gordon and Breach 1964.

PASCARD-BILLY, C.: Structure précise de l'acide sulfurique. Acta Cryst. 18, 827 (1965).

PAULING, L.: The nature of the chemical bonds in sulvanite, Cu_3VS_4. Tschermaks Mineral. Petrogr. Mitt. 10, 379 (1965).

PEARSON, A. D.: Sulphide, selenide and telluride glasses. Modern Aspects of the Vitreous State, vol. 3, chapt. 2, 29. London: Butterworths 1964.

RENTZEPERIS, P.: The crystal structure of the anhydrous stannous sulphate. Z. Krist. 117, 431 (1962).

RIRAR, B., NICCA, CH., NOWACKI, W.: Dreidimensionale Verfeinerung der Kristallstruktur von Dufrenoysit, $Pb_8As_8S_{20}$. Z. Krist. 130, 15 (1969).

— NOWACKI, W.: Neubestimmung der Kristallstruktur von Gratonit, $Pb_9As_4S_{15}$. Z. Krist. 128, 321 (1969).

RICE, F. O., SPARROW, C.: Purple sulfur, a new allotropic form. J. Am. Chem. Soc. 75, 848 (1953).

ROBINSON, P. D., FANG, J. H.: Crystal structures and mineral chemistry of double-salt hydrates: I. Direct determination of the crystal structure of tamarugite. Am. Mineralogist 54, 19 (1969).

RÖSCH, H.: Zur Kristallstruktur des Gratonits—$9PbS \cdot 2As_2S_3$. Neues Jahrb. Mineral. 99, 307 (1963).

SABELLI, C., ZANAZZI, P. F.: The crystal structure of serpierite. Acta Cryst. B 24, 1214 (1968).

SAHL, K.: Die Verfeinerung der Kristallstrukturen von $PbCl_2$ (Cotunnit), $BaCl_2$, $PbSO_4$ (Anglesit) and $BaSO_4$ (Baryt). Beitr. Mineral. Petrog. 9, 111 (1963).

SANDS, D. E.: The crystal structure of monoclinic (β) sulfur. J. Am. Chem. Soc., 87, 1395 (1965).

SRIKANTA, S., SEQUEIRA, A., CHIDAMBARAM, R.: Neutron diffraction study of the space group and structure of manganese-leonite, $K_2Mn(SO_4)_2 \cdot 4H_2O$. Acta Cryst. B 24, 1176 (1968).

Süsse, P.: The crystal structure of amarantite, $Fe_2(SO_4)_2O \cdot 7H_2O$. Z. Krist. **127**, 261 (1968a).
— Die Kristallstruktur des Botryogens. Acta Cryst. B **24**, 760 (1968b).
Taesler, I., Olovsson, I.: Hydrogen bond studies. XXI. The crystal structure of sulphuric acid monohydrate. Acta Cryst. B **24**, 299 (1968).
Takéuchi, Y., Ghose, S., Nowacki, W.: The crystal structure of hutchinsonite, $(Tl, Pb)_2As_5S_9$. Z. Krist. **121**, 321 (1965).
— Haga, N.: On the crystal structures of seligmannite, $PbCuAsS_3$, and related minerals. Z. Krist. **130**, 254 (1969).
— Nowacki, W.: Detailed crystal structure of rhombohedral MoS_2 and systematic deduction of possible polytypes of molybdenite. Schweiz. Mineral. Petrog. Mitt. **44**, 105 (1964).
Templeton, L. K., Templeton, D. H., Zalkin, A.: Refinement of the crystal structure of mercuric sulfate monohydrate. Acta Cryst. **17**, 933 (1964).
Trojer, F. J.: Refinement of the structure of sulvanite. Am. Mineralogist **51**, 890 (1966).
Tuinstra, F.: The structure of fibrous sulfur. Acta Cryst. **20**, 341 (1966).
Welin, E.: Notes on the mineralogy of Sweden 5. Bismuth-bearing sulphosalts from Gladhammar, a revision. Ark. Mineral. Geol. **4**, 377 (1966).
Wuensch, B. J.: The crystal structure of tetrahedrite, $Cu_{12}Sb_4S_{13}$. Z. Krist. **119**, 437 (1964).
— Nowacki, W.: The substructure of the sulfosalt jordanite. Schweiz. Mineral. Petrog. Mitt. **46**, 89 (1966).
— — The crystal structure of marrite, $PbAgAsS_3$. Z. Krist. **125**, 459 (1967).
— Takéuchi, Y., Nowacki, W.: Refinement of the crystal structure of binnite, $Cu_{12}As_4S_{13}$. Z. Krist. **123**, 1 (1966).
Zalkin, A., Ruben, H., Templeton, D. H.: The crystal structure of cobalt sulfate hexahydrate. Acta Cryst. **15**, 1219 (1962).
— — — The crystal structure and hydrogen bonding of magnesium sulfate hexahydrate. Acta Cryst. **17**, 235 (1964).

Potassium 19

A	G. Cocco, L. Fanfani and P. F. Zanazzi	(Istituto di Mineralogia dell'Università di Perugia, Perugia, Italy)
B—E, G, I—N	K. S. Heier	(Mineralogisk-Geologisk Museum, Oslo, Norway)
	and	
	G. K. Billings	(Department of Geosciences, New Mexico Institute of Mining and Technology, Socorro, New Mexico, U.S.A.)

19-A. Crystal Chemistry

Potassium, like the other alkali metals, has an inner electronic configuration corresponding to that of the preceding noble gas in the Periodic Table, and a single valence electron in an s orbital. The electronic configuration of potassium explains the existence of only one oxidation state ($+1$) and the strong affinity for oxygen and the halogens.

The ionic radius of potassium is 1.33 Å, as reported in Table 12-8 of Vol. I of this handbook. Despite their chemical similarity, substitution of sodium for potassium is not common at room temperature because their ionic radii are too dissimilar. However the ability to undergo isomorphous replacement increases at higher temperatures. K-feldspar and albite form a complete solid solution at 660°C but on cooling "exolution textures" (perthite and antiperthite intergrowths) are produced. The similarity between the ionic radii of K^+ and Rb^+ allows mutual substitution of the two elements; the observed range of this replacement in nature, however, is limited because of the scarcity of rubidium. The crystallochemical relationships between K and Ba are very similar to those between Na and Ca, but as Ba is much less abundant in nature, K-minerals usually contain Ba as a minor constituent only; a typical example of a major substitution product is the hyalophane series. Pb^{2+} also has an ionic radius close to that of K^+, hence traces of this element can be found in K-minerals. The size of K^+ is similar to that of H_3O^+ and NH_4^+; isomorphous replacement of these ions for K^+ is observed in many natural substances such as micas, zeolites, clay minerals, as well as in jarosite and torbernite minerals.

In connection with the geochemistry of sediment formation an essential feature of the behavior of K^+ ions is the ease of adsorption of this ion on colloids etc. (as aluminum and ferric hydroxides) and their preferential fixation in cation exchange reactions with clay minerals. This property seems to be related to the lower affinity for hydration of K^+ as compared with Na^+ and Ca^{2+} in competition with the surface charge of the mineral.

Potassium occurs in nature only in coordination with oxygen or the halogens. Therefore the present crystallochemical review concerns these substances only, but also includes some synthetic compounds as well as minerals.

I. Halogen Compounds

Simple K-halides have the NaCl type structure, $K^{[6]}X^{[6]}$. In nature, only the fluoride, carobbite, and the chloride, sylvite, are known. In the simple halogenides the following K-halogen bond distances are observed: K—F, 2.673 Å; K—Cl, 3.147 Å; K—Br, 3.298 Å; K—I, 3.533 Å (SYRSIÖ, 1969). A few K-compounds occur in nature as complex fluorides or chlorides, such as avogadrite $K[BF_4]$, hieratite

$K_2[SiF_6]$, elpasolite $K_2Na[AlF_6]$, chlorocalcite $KCaCl_3$, rinneite $K_3Na[FeCl_6]$ and chloromanganocalite $K_4[MnCl_6]$. Some other complex K-chlorides are hydrated, such as the important mineral carnallite, $KMgCl_3 \cdot 6H_2O$, and the fumarolic products mitscherlichite, $K_2CuCl_4 \cdot 2H_2O$, douglasite, $K_2FeCl_4 \cdot 2H_2O$, and erythrosiderite, $K_2FeCl_5 \cdot H_2O$. Reliable structural data on these minerals are lacking, but information on K-halogen bond distances can be obtained from some synthetic compounds. In KH_2F_3 (FORRESTER et al., 1963) two different kinds of K+ ions are coordinated by eight F atoms in the range 2.69—3.05 Å and 2.65—2.99 Å respectively; average values are 2.83 and 2.82 Å. Eight-fold coordinated K ions occur in $K_2[BeF_4]$ (MUSTAFAEV et al., 1966); K—F mean values are 2.91 and 2.76 Å. In K_2CuF_4 (KNOX, SR 1959, 290) nine F atoms surround potassium with distances from 2.59 to 2.94 Å (mean value 2.83 Å); this structure is typical of a larger number of compounds which include K_2MgF_4, K_2CoF_4 and K_2NiF_4. The same coordination number occurs in K_2NbF_7 (BROWN and WALKER, 1966), where K—F bond lengths are in the range 2.645—2.896 Å for one K+ and 2.641—2.935 Å for the other one (a common mean value is 2.78 Å). $KNbF_6$ and $KTaF_6$ are isostructural; in the first compound K+ is surrounded by 8F at 2.50 Å and by 4F at 2.90 Å (BODE and DÖHREN, SR 1958, 249). $KCuF_3$ has a tetragonal distorted perowskite structure; each K+ is surrounded by eight F atoms at 2.86 Å and by four F atoms at 2.93 Å (OKAZAKI and SUEMENE, SR 1961, 305). Probably $KCrF_4$ also has this structure-type. Many other complex K-fluorides with the chemical formula $KMeF_3$, where Me is Mn^{2+}, Fe^{2+}, Co, Ni, Zn, show cubic symmetry and perowskite-like arrangements (KNOX, SR 1961, 305); at low temperatures their symmetry changes to tetragonal. $K_2[TiCl_6]$ is isotypical with $K_2[PtCl_6]$ and other double chlorides; K+ ions are equally distant from 12 chlorine atoms at 3.46 Å (BLAND and FLENGAS, SR 1961, 325). In $K_3W_2Cl_9$ (WATSON and WASER, SR 1958, 253), two different K+ are present. K(1) has 3 Cl at 3.25 Å, 6 Cl at 3.37 Å and 3 Cl at 3.92 Å as nearest neighbors; K(2) is surrounded by 3 Cl at 3.31 Å, 3 Cl at 3.33 Å, 3 Cl at 3.54 Å and 3 Cl at 3.89 Å.

II. Oxygen-Compounds

In Table 19-A-1 some natural or synthetic compounds are listed from which reliable information concerning K—O bonds lengths can be obtained. Cryptomelane, $K_{2-x}Mn_8O_{16}$, has a tetragonal and a monoclinic modification. Its structure is a stuffed derivative of the α-MnO_2 arrangement; the introduction of large ions probably stabilizes the framework of the α-form of MnO_2 which does not occur in nature. K ions are located in the channels of the three-dimensional network of Mn-octahedra and are surrounded by 8 oxygen atoms forming a slightly distorted cube and by oxygen atoms at greater distances. The occupancy is about 0.5 so that the chemical formula can be written as KMn_8O_{16}. Replacement of K by Ba is observed, the barium-rich member being hollandite (BYSTRÖM and BYSTRÖM, SR 1950, 187).

Potassium bicarbonate occurs in nature as the mineral kalicine. K+ is coordinated by eight oxygen atoms in the range 2.69—3.13 Å (NITTA et al., SR 1952, 327).

The ionic size of the cation controls the structure in alkali nitrates. KNO_3, nitre, has a structure like aragonite, while $NaNO_3$ and $LiNO_3$ have structures like calcite.

Potassium

Table 19-A-1. K—O bond lengths

		Range, Å	Mean, Å	Reference
$KNbO_3$	$K^{[12]}$—O	2.79 —2.88	2.84	Katz and Megaw (1967)
KCr_3O_8	$K^{[10]}$—O	2.80 —3.21	2.99	Wilhelmi (SR 1958, 330)
K_3CrO_8	$K^{[8]}$—O	2.74	2.74	Stomberg and Brosset (SR 1960, 448)
	$K_I^{[8]}$—O		2.72	
	$K_{II}^{[8]}$—O		2.80	
$K_2Ti_6O_{13}$	$K^{[8]}$—O	2.66 —2.80	2.80	Cid-Dresdner and Buerger (1962)
$KTiNbO_5$	$K^{[8]}$—O	2.36 —3.14	2.91	Wadsley (1964)
$K_2[Al_2O(OH)_6]$	$K^{[6]}$—O	2.75 —3.13	2.85	Johansson (1966)
$K[ClO_3]$	$K^{[9]}$—O	2.76 —2.95	2.91	Aravindakshan (SR 1958, 484)
$K_4[I_2O_8(OH)_2]$	$K^{[8]}$—O	2.79 —3.08	2.88	Ferrari et al. (1965)
	$K_I^{[8]}$—O	2.653 —3.065	2.88	
	$K_{II}^{[8]}$—O	2.749 —3.101		
$^2_\infty K[B_5^{[3]}B^{[4]}O_8]$	$K^{[9]}$—O	2.90 —3.20	3.01	Krogh-Moe (1965)
$K[B_4^{[3]}B^{[4]}O_6(OH)_4] \cdot 2H_2O$	$K^{[8]}$—O	2.919 —3.034	2.90	Zachariasen and Plettinger (1963)
$^1_\infty K[B_2^{[3]}B_2^{[4]}O_5(OH)_4] \cdot 2H_2O$	$K^{[7]}$—O	2.772 —3.073	2.86	Marezio et al. (1963)
	$K_I^{[7]}$—O	2.771 —3.011	2.86	
$^2_\infty K_2[B_3^{[3]}B^{[4]}O_8(OH)] \cdot 2H_2O$	$K_I^{[8]}$—O	2.693 —2.890	2.82	Marezio (1969)
	$K_{II}^{[8]}$—O	2.654 —2.893	2.79	
$^1_\infty KHMg_2[B_6^{[3]}B_6^{[4]}O_{16}(OH)_{10}] \cdot 4H_2O$ kaliborite	$K^{[8]}$—O	2.909 —2.980	2.94	Corazza and Sabelli (1966)
$K[CO_2(OH)]$ kalicine	$K^{[8]}$—O	2.69 —3.13	2.84	Nitta et al. (SR 1952, 327)
$K[C_2O_3(OH)][C_2O_2(OH)_2]$	$K^{[9]}$—O	2.870 —3.250	2.95	Haas (1964)
$K[SO_3(OH)]$ mercallite	$K^{[9]}$—O	2.68 —3.06	2.89	Loopstra and McGillavry (SR 1958, 458)
	$K_I^{[9]}$—O	2.58 —3.02	2.82	
	$K_{II}^{[8]}$—O	2.983	2.98	
$KAl[SO_4]_2 \cdot 12H_2O$ α-alum				
$KCr[SO_4]_2 \cdot 12H_2O$ (by neutron diffraction)	$K^{[6]}$—O	2.97	2.97	Bacon and Gardiner (SR 1958, 467)
$KAl_3[SO_4]_2(OH)_6$ alunite	$K^{[12]}$—O	2.821 —2.871	2.85	Wang et al. (1965)
$KFe[SO_4]_2 \cdot H_2O$ krausite	$K^{[10]}$—O	2.755 —3.056	2.92	Graeber et al. (1965)

Table 19-A-1 (Continued)

		Range, Å	Mean, Å	Reference
$K_2Mg[SO_4]_2 \cdot 6H_2O$ picromerite	$K^{[7]}$—O	2.715—3.186	2.94	Kannan and Viswamitra (1965)
$K_2Mg_2[SO_4]_3$ langbeinite	$K_I^{[12]}$—O $K_{II}^{[9]}$—O $K^{[9]}$—O	2.88 —3.25 2.78 —3.11 2.63 —3.26	3.06 2.96 2.96	Zemann and Zemann (SR 1957, 362) Srikanta et al. (1968)
$K_2Mn[SO_4]_2 \cdot 4H_2O$ Mn-leonite				
$K_2Ca[SO_4]_2 \cdot H_2O$ syngenite	$K^{[8]}$—O	2.694—2.981	2.88	Corazza and Sabelli (1967)
$K_2[S_2O_7]$	$K^{[9]}$—O $K_I^{[8]}$—O $K_{II}^{[8]}$—O	2.71 —3.22 2.71 —3.20 2.70 —3.20	2.89 2.91 2.92	Lynton and Truter (SR 1960, 378) Koster et al. (1969)
$K_2[WO_4]$				
$KY[WO_4]_2$	$K^{[12]}$—O $K^{[7]}$—O $K^{[8]}$—O	2.80 —3.31 2.68 —3.08 2.64 —2.96	3.00 2.84 2.82	Borisov and Klevtsova (1968) Eanes and Ondik (1962)
$K_2Li[P_3O_9] \cdot H_2O$				
$K_2Zn_2[V_{10}^{[6]}O_{28}] \cdot 16H_2O$ Zn-hummerite	$K^{[10]}$—O	2.781—3.251	2.97	Evans (1966)
$K_4H_4[Si_6O_{12}]$	$K_I^{[6]}$—O $K_{II}^{[8]}$—O $K^{[9]}$—O	2.81 —3.06 2.64 —3.10 2.711—3.129	2.93 2.89 2.97	Hilmer (1964) Colville and Ribbe (1968)
$K[AlSi_3O_8]$ orthoclase				
$K[AlSi_3O_8]$ adularia	$K^{[9]}$—O	2.717—3.117	2.96	Colville and Ribbe (1968)
$K[AlSi_3O_8]$ microcline	$K^{[8]}$—O	2.750—3.136	2.92	Brown and Bailey (1964)
$K[AlSi_3O_8]$ microcline	$K^{[8]}$—O	2.743—3.139	2.93	Finney and Bailey (1964)
$K[AlSi_3O_8]$ sanidine	$K^{[9]}$—O	2.707—3.140	2.97	Ribbe (1963)
$(K_{0.85}Na_{0.15})[AlSi_2O_6]$ leucite	$K^{[12]}$—O	3.35 —3.54	3.45	Peacor (1968)
$KAl_2(OH)_2[AlSi_3O_{10}]$ muscovite (monoclinic)	$K^{[6+6]}$—O	2.82 —2.88 3.30 —3.51	2.85 3.38	Birle and Tettenhorst (1968)

Potassium

Kaliborite is a potassium borate mineral. Its chemical formula is

$$HKMg_2[B_{12}O_{16}(OH)_{10}] \cdot 4 H_2O$$

(CORAZZA and SABELLI, 1966). The potassium ions bind 8 oxygens each in the form of a distorted cube (average distance: 2.94 Å); four oxygen atoms belong to water molecules, two to hydroxyl groups and two to the complex borate polyanion.

The β-form of $K_2[SO_4]$ occurs in nature as the mineral arcanite and is isotypic with tarapacaite, $K_2[CrO_4]$; the two different K^+ ions in the structure bind 10 and 9 oxygens atoms respectively with mean K—O values of 2.98 and 2.86 Å (recalculated from ROBINSON, SR 1958, 447). The occurrence in nature of mixed crystals $(K, NH_4)_2[SO_4]$, named taylorite, is reported (after STRUNZ, 1966). In mercallite, $K[SO_3(OH)]$, both K^+ ions are surrounded in about the same way by 9 oxygen atoms; they form a strongly deformed tetragonal antiprism with one square face centered. Average K—O distances in the two polyhedra are 2.89 and 2.82 Å (LOOPSTRA and MACGILLAVRY, SR 1958, 458). Potassium is a constituent of a large number of double sulphates. In alunite $KAl_3(OH)_6[SO_4]_2$, K^+ is present in 12-coordination between 6 OH and 6 oxygen atoms with a mean distance of 2.85 Å (WANG et al., 1965). Mixed crystals within the whole range of Al^{3+}—Fe^{3+} substitution have been synthetized, although intermediate natural members are rare; the Fe-endmember is represented by the mineral jarosite (BROPHY and SHERIDAN, 1965). The following solid-solution series appears to exist in the jarosite-alunite mineral group: $A^+R_3^{3+}(OH)_6[SO_4]_2$ where A^+ is H_3O^+, NH_4^+, K^+ and Na^+, and R^{3+} is Al^{3+} and Fe^{3+} (ROSS and EVANS, 1965).

$KAl[SO_4]_2 \cdot 12 H_2O$ represents the prototype of the α-alums which include also the Rb- and NH_4-members. The univalent cation is surrounded by six water molecules at a distance of 2.983 Å forming an octahedron (LARSON and CROMER, 1967). This value is in agreement with that of 2.97 Å reported from a neutron diffraction analysis of $KCr[SO_4]_2 \cdot 12 H_2O$ (BACON and GARDINER, SR 1958, 467).

$KAl_2(OH)_2[AlSi_3O_{10}]$ muscovite (trigonal)	$K^{[6+6]}$—O	2.839—2.900 3.283—3.502	2.87 3.39
$(K_{0.9}Mn_{0.1})Mg_3(OH)_2[FeSi_3O_{10}]$ phlogopite	$K^{[6+6]}$—O	2.95 3.45	2.95 3.45
$KFe_3(OH)_2[FeSi_3O_{10}]$ synthetic mica	$K^{[6+6]}$—O	3.015—3.074 3.332—3.378	3.05 3.35
$K(Mg, Li)_3F_2[AlSi_3O_{10}]$ synthetic mica	$K^{[6+6]}$—O	2.990—2.997 3.276—3.279	2.99 3.28
$KFe_{13}(OH)_{14}[(Si, Al)_{18}O_{42}]$ zussmanite	$K^{[12]}$—O	3.00 —3.01	3.00
$KLiNa_2(Fe, Mg, Mn)_2(TiO)_2[Si_8O_{22}]$ neptunite	$K^{[10]}$—O	2.81 —3.21	2.96

GÜVEN and BURNHAM (1967)
STEINFINK (1962)
DONNAY et al. (1964)
TAKEDA and DONNAY (1966)
LOPES-VIEIRA and ZUSSMAN (1969)
CANNILLO et al. (1966)

In langbeinite, $K_2Mg_2[SO_4]_3$, the two kinds of K⁺ ions have different coordination numbers: one binds 12 oxygen atoms (average value: 3.06 Å), the other one is coordinated by 9 oxygen atoms (average K—O distance: 2.96 Å) (ZEMANN and ZEMANN, SR 1957, 362).

The structure of the synthetic Mn analogue of the mineral leonite, $K_2Mn[SO_4]_2 \cdot 4H_2O$, has been recently refined by neutron diffraction data (SRIKANTA et al., 1968); the mean value of the nine K—O bond lengths is 2.96 Å. In krausite, $KFe[SO_4]_2 \cdot H_2O$, each K⁺ is coordinated by ten nearest oxygen neighbors at an average distance of 2.92 Å (GRAEBER et al., 1965). Eightfold coordination of oxygen atoms around K⁺ occurs in syngenite, $K_2Ca[SO_4]_2 \cdot H_2O$ with an average K—O distance of 2.88 Å (CORAZZA and SABELLI, 1967). Cyanochroite, $K_2Cu[SO_4]_2 \cdot 6H_2O$, and picromerite, $K_2Mg[SO_4]_2 \cdot 6H_2O$, belong to the group of Tutton's salts. Coordination of K⁺ in picromerite is sevenfold: five oxygen atoms and two water molecules are linked to K ions at distances in the range 2.72—3.19 Å (KANNAN and VISWAMITRA, 1965). A refinement of synthetic $K_2Cu[SO_4]_2 \cdot 6H_2O$ shows that the K⁺ ions are surrounded by eight oxygen atoms forming an irregular polyhedron; the K—O distances range from 2.77 to 3.13 Å (CARAPEZZA and RIVA DI SANSEVERINO, 1968). Palmierite, $K_2Pb[SO_4]_2$, has an atomic arrangement similar to that of $Ba_3[PO_4]$. The analogous salts of NH_4^+, Rb⁺ and Tl⁺ are isotypic (MÖLLER, SR 1954, 475). In polyhalite, $K_2Ca_2Mg[SO_4]_4 \cdot H_2O$, K⁺ is surrounded by 11 oxygen atoms, ten belonging to SO_4 groups and one to a water molecule (SCHLATTI et al., 1970). In carnotite, $K_2(UO_2)_2[V_2O_8] \cdot nH_2O$, the K⁺ ions fit between the $^2_\infty[(UO_2)_2V_2O_8]$ sheets as in the case of the micas. The double salt $K_2Mg_2[V_{10}O_{28}] \cdot 16H_2O$ occurs in nature as the mineral hummerite; a structural determination of the analogous compound $K_2Zn_2[V_{10}O_{28}] \cdot 16H_2O$ (EVANS, 1966) shows that the K⁺ ion is irregularly coordinated to five oxygen atoms of three neighboring $[V_{10}O_{28}]^{6-}$ groups and to five water molecules. The average distance is 2.96 Å.

Abernathyte, $K(UO_2)[AsO_4] \cdot 3H_2O$, is a torbernite-like mineral characterized by infinite sheets of the type $^2_\infty[UO_2AsO_4]$ of the same kind as in autunite. One potassium ion and three water molecules are randomly distributed in a symmetrical square arrangement about the fourfold axes between the uranyl ions (Ross and EVANS, 1964). A solid solution series probably exists in the torbernite mineral group between abernathyte, troegerite $H_3O(UO_2)[AsO_4] \cdot 3H_2O$, and the compound $NH_4(UO_2)[AsO_4] \cdot 3H_2O$. A similar isomorphous series occurs among the analogous phosphates and includes the minerals meta-ankoleite $K(UO_2)[PO_4] \cdot 3H_2O$, uramphite $NH_4(UO_2)[PO_4] \cdot 3H_2O$, and the synthetic compound $H_3O(UO_2)[PO_4] \cdot 3H_2O$ (Ross and EVANS, 1965).

III. Silicates

K, together with Na and Ca, is an important cation in several silicate groups, like feldspars, micas and zeolites. The K-feldspars have monoclinic or very nearly monoclinic symmetry. Sanidine represents the high-temperature phase; orthoclase and microcline are lower temperature modifications. Sanidine is a monoclinic modification characterized by a random distribution of Al and Si over tetrahedral sites. There are 9 oxygen atoms around each K⁺ at distances between 2.707 and 3.140 Å (RIBBE, 1963). Microcline is triclinic; variations in interaxial angles range from values which are hardly distinguishable from rectangular ones to values $\alpha = 90°41'$ (combined with a $\gamma = 87°30'$) ("maximum microcline") (McKenzie, SR 1954, 550). Members of the microcline series, having interaxial angles roughly halfway between monoclinic symmetry and the maximum triclinic values are known as "intermediate microclines". Structural investigations were reported for an intermediate microcline (Bailey and Taylor, SR 1955, 474), for an igneous maximum microcline (Brown and Bailey, 1964) and for a sedimentary maximum microcline (Finney and Bailey, 1964). No significant structural differences occur in microclines; in all the structures the ordering of 4-coordinated Si, Al atoms is nearly complete. In intermediate microcline K^+ is surrounded by 8 oxygen atoms at distances in the range 2.758—3.105 Å with two more neighbors at 3.209 and 3.405 Å; in maximum microcline, K^+ has 8 oxygen neighbors. Distances range from 2.750 to 3.136 Å in igneous material and from 2.743 to 3.139 Å in sedimentary material. In both maximum-microcline structures the oxygen atoms of the Al-rich tetrahedra are closer to the potassium, on the average, than are the other oxygen atoms, whose negative charges are already satisfied by their bonds to silicon.

Orthoclase and adularia are structurally intermediate between sanidine and microcline. Adularia is characterized by a distinctive crystal habit but its validity as a true polymorphic form is still in question. Orthoclase is assumed as monoclinic but no final decision has been reached on its true symmetry — whether monoclinic or an intimate intergrowth of triclinic units — (Jones and Taylor, SR 1961, 524). A recent refinement of the adularia and orthoclase structures has shown that they are built of triclinic domains but are monoclinic on the average (Colville and Ribbe, 1968). The "average" structures show a partial ordering of (Si, Al) atoms and a similar coordination around K^+ ions; in orthoclase the average value of K—O distances for the nine nearest oxygen atoms is 2.97 Å, in adularia 2.96 Å. Orthoclase is transformed on heating into sanidine.

At high temperature complete solid solution exists between the monoclinic potassium-feldspars (sanidine and orthoclase) and the triclinic sodium-feldspar (albite). The intermediate feldspars are called anorthoclase. At lower temperatures anorthoclase exsolves into subparallel lamellae, alternately sodic and potassic in composition. Such an intergrowth is called perthite or antiperthite.

Because of the similarity in the size of potassium and barium ions, isomorphism exists between orthoclase and celsian, the Ba-feldspar, the members of the series being termed hyalophanes. According to Vermaas (SR 1953, 558) a transition point appears at a composition in the range $cn_{0.37}$ and $cn_{0.50}$ and indicates the existence of two isomorphous series cn_0—$cn_{0.40}$ and $cn_{0.45}$—$cn_{1.00}$. The structural complexity of intermediate terms can be related to the completely different ordering scheme (Al, Si) in orthoclase and celsian.

Synthetic ammonium (Barker, 1964) and rubidium (Barrer and McCallum, SR 1953, 563) feldspars have been prepared and appear to be closely related structurally to sanidine. The NH_4-feldspar is metastable but is stabilized by zeolitic water. The hydrated product is analogous to the mineral buddingtonite.

Synthetic $K[AlSi_3O_8]$ (sanidine) undergoes a polymorphic transition at 120 kilobars pressure and 900° C. The high-density modification (3.84 g · cm^{-3}) is tetragonal and has an hollandite-type structure with Al and Si randomly distributed in octahedral positions (Ringwood et al., 1967). Sanidine is transformed into leucite at pressures below 20 Kb and temperatures above 1150°C (LINDSLEY, 1966). Leucite, K[AlSi$_2$O$_6$], is cubic above 605°C and tetragonal at room temperature. The high-temperature leucite has a framework of completely disordered (Si, Al) tetrahedra, similar to analcite. Sixteen K$^+$ occupy the 16-fold positions which are occupied by water molecules in the analcite structure (WYART, SR 1940, 220). The refinement of the structure on a sample with the cation composition (Na$_{0.15}$K$_{0.85}$) shows the coordination of K$^+$ is twelve, with an average K—O distance of 3.45 Å (PEACOR, 1968). The room temperature form has an atomic arrangement related to pollucite (NARAY-SZABO, SR 1942, 256).

The structure of nepheline is a derivative of β-tridymite, the replacement of half of the silicon by aluminum being balanced by additional alkali cations in the voids of the framework. Chemical analyses of several nephelines show that the general formula of this mineral is K$_x$Na$_y$Ca$_z$[Al$_{x+y+2z}$Si$_{16-x-y-2z}$O$_{32}$] where $x+y+z<8$. The structure of nepheline is essentially retained up to a value of $x=4.73$ but the change of cell-parameters with the K content suggest the existence of three sub-phases: subpotassic ($0<x<0.25$), mediopotassic ($0.25<x<2.00$) and perpotassic ($2.00<x<4.73$) (DONNAY et al., SR 1959, 488). From a structural determination on material with ideal composition KNa$_3$[Al$_4$Si$_4$O$_{16}$] (HAHN and BUERGER, SR 1955, 477), coordination around K ions is ninefold with distances from 2.91 to 3.06 Å (average value 2.96 Å). The kalsilite structure consists also of a framework of high-temperature tridymite type. However the ordered pattern of Al and Si atoms reduces the symmetry making the c axis polar. Coordination around K$^+$ is uncertain because of a partial disorder affecting oxygen atoms (PERROTTA and SMITH, 1966).

In micas potassium ions connect negatively charged layers built of (Al, Si) tetrahedra and (Mg, Al) octahedra. The K$^+$ ions are in 12-fold coordination with six shorter and six longer K—O bonds. In a trigonal muscovite with a chemical composition close to the ideal formula KAl$_2$(OH)$_2$[Si$_3$AlO$_{10}$], the average values for the six shorter and longer distances are 2.87 and 3.39 Å respectively (GÜVEN and BURNHAM, 1967). The corresponding values for a monoclinic muscovite are 2.85 and 3.38 Å (BIRLE and TETTENHORST, 1968). A refinement of an iron-rich phlogopite shows the potassium ion has six near neighbors at a mean distance of 2.95 Å; six additional oxygen atoms are at a mean distance of 3.45 Å (STEINFINK, 1962). The crystal structure of a synthetic lithium fluor-mica, in composition between K(Mg$_2$Li)F$_2$[Si$_4$O$_{10}$] and KMg$_3$F$_2$[Si$_3$AlO$_{10}$] but potassium deficient, shows the vacant potassium sites are distributed randomly. Inner and outer K—O distances average 2.99 and 3.28 Å respectively (TAKEDA and DONNAY, 1966). For a synthetic iron mica, K—O average distances of 3.05 and 3.35 Å are observed (DONNAY et al., 1964a). A geometrical approach to relate K—O bond lengths in micas with the values of the b parameter and of the distances cation-oxygen in the tetrahedral and octahedral coordination is reported (DONNAY et al., 1964b). Concerning the possibility of replacement of K$^+$ in micas, EUGSTER and YODER (1955) revealed a very limited solid solution of the paragonite molecule (Na-mica) in muscovite (about 3% at room temperature). Ca is often present in chemical analyses of K-micas (DEER et al., 1966). Isomorphous replacement of H$_3$O$^+$ for K$^+$ is suggested by equilibrium studies in water (GARRELS and HOWARD, 1959).

In apophyllite, KCa$_4$F[Si$_8$O$_{20}$]·8H$_2$O, the sheets of linked tetrahedral groups are in rings of four and eight. The potassium ion lies in a large hole surrounded by eight water oxygen atoms situated at the apices of a square prism (IVLEV et al., 1969); K—O distances are 2.83 Å (TAYLOR and NARAY-SZABO, SB 1931, 545). K-zeolites are less common than Ca and Na zeolites. Two types of isomorphous replacement seem to effect the K content in natural zeolites. The first is a replacement similar to that in feldspars: K—Si$\rightleftharpoons$Ba—Al; the second type, Ba$\rightleftharpoons$K$_2$, changes the number of cations; this is possible because the framework contains a sufficient number of holes to include the additional ions. In phillipsite, $(K_xNa_{1-x})_5[Si_{11}Al_5O_{32}] \cdot 10H_2O$, K$^+$ is probably statistically distributed on sites with 12-fold coordination; K—O distances are in the range 2.92—3.57 Å (STEINFINK, 1962). Twelve-fold coordination is exhibited by K$^+$ ions in another framework silicate: osumilite, $(K, Na, Ca)(Fe, Mg, Mn)_2[(Si, Al, Fe)_{15}O_{30}] \cdot H_2O$. A recent X-ray study performed on crystals with a large amount of Na replacing K has shown that the alkali ions are at a distance of 3.10 Å from the twelve equivalent oxygen atoms (BROWN and GIBBS, 1969). A partial replacement of Na for K is observed also in the sheet silicate dalyite. $(K, Na)_2Zr[Si_6O_{15}]$; K/Na ions are in 8-fold coordination, the bond lengths being in a large range (2.83—3.24 Å) with an average (K, Na)—O distance of 3.02 Å. The very high anisotropy of these ions suggests the possibility of a space-averaged location (FLEET, 1965). In wadeite, $K_4Zr[Si_6O_{18}]$, where the site is nearly completely occupied by K$^+$, the mean K$^{[9]}$—O distance is 3.11 Å (HENSHAW, SR 1955, 463).

Revised manuscript received: August 1970

References: Section 19-A

BARKER, D. S.: Ammonium in alkali feldspars. Am. Mineralogist **49**, 851 (1964).
BIRLE, J. D., TETTENHORST, R.: Refined muscovite structure. Mineral. Mag. **36**, 883 (1968).
BORISOV, S. V., KLEVTSOVA, R. F.: Crystal structure of $KY(WO_4)_2$. Soviet Phys.-Cryst. **13**, 420 (1968).
BROPHY, G. P., SHERIDAN, M. F.: Sulfate studies. IV. The jarosite-natrojarosite-hydronium jarosite solid solution series. Am. Mineralogist **50**, 1595 (1965).
BROWN, B. E., BAILEY, S. W.: The structure of maximum microcline. Acta Cryst. **17**, 1391 (1964).
BROWN, G. E., GIBBS, G. V.: Refinement of the crystal structure of osumilite. Am. Mineralogist **54**, 101 (1969).
BROWN, G. N., WALKER, L. A.: Refinement of the structure of potassium heptafluoroniobate, K_2NbF_7, from neutron diffraction data. Acta Cryst. **20**, 220 (1966).
CANNILLO, E., MAZZI, F., ROSSI, G.: The crystal structure of neptunite. Acta Cryst. **21**, 200 (1966).
CARAPEZZA, M., RIVA DI SANSEVERINO, L.: Crystallography and genesis of double sulfates and their hydrates. II. Structure powder pattern and thermoanalysis of cyanochroite, $K_2Cu(SO_4)_2 \cdot 6H_2O$. Mineral. Petrog. Acta **14**, 23 (1968).
CID-DRESDNER, H., BUERGER, M. J.: The crystal structure of potassium hexatitanate $K_2Ti_6O_{13}$. Z. Krist. **117**, 411 (1962).
COLVILLE, A. A., RIBBE, P. H.: The crystal structure of an adularia and a refinement of the structure of orthoclase. Am. Mineralogist **53**, 25 (1968).
CORAZZA, E., SABELLI, C.: The crystal structure of kaliborite. Rend. Acc. Naz. Lincei, Ser. VIII **41**, 527 (1966).
— — The crystal structure of syngenite, $K_2Ca(SO_4)_2 \cdot H_2O$. Z. Krist. **124**, 398 (1967).
DEER, W. A., HOWIE, R. A., ZUSSMAN, J.: An Introduction to the Rockforming Minerals, p. 198. London: Longmans, Green & Co. 1966.
DONNAY, G., MORIMOTO, N., TAKEDA, H., DONNAY, J. O. H.: Trioctahedral one-layer micas. I. Crystal structure of a synthetic iron mica. Acta Cryst. **17**, 1369 (1964).
EANES, E. D., ONDIK, H. M.: The structure of lithium dipotassium trimetaphosphate monohydrate. Acta Cryst. **15**, 1280 (1962).
EVANS, JR., H. T.: The molecular structure of the isopoly complex ions decavanadate $(V_{10}O_{28})^{-6}$. Inorg. Chem. **5**, 967 (1966).
FERRARI, A., BRAIBANTI, A., TIRIPICCHIO, A.: The crystal structure of tetrapotassium dihydrogendecaoxodiiodate (VII) octahydrate. Acta Cryst. **19**, 629 (1965).
FINNEY, J. J., BAILEY, S. W.: Crystal structure of an authigenic maximum microcline. Z. Krist. **119**, 413 (1964).
FLEET, S. G.: The crystal structure of dalyite. Z. Krist. **121**, 349 (1965).
FORRESTER, J. D., SENKO, M. E., ZALKIN, A., TEMPLETON, D. H.: Crystal structure of KH_2F_3 and geometry of the H_2F_3 ion. Acta Cryst. **16**, 58 (1963).
GARRELS, R. M., HOWARD, P.: Reactions of feldspar and mica with water at low temperature and pressure. Nat. Conf. Clays Clay Minerals, 6th Proc., 68 (1959).
GRAEBER, E. J., MOROSIN, B., ROSENZWEIG, A.: The crystal structure of krausite, $KFe(SO_4)_2 \cdot H_2O$. Am. Mineralogist **50**, 1929 (1965).
GÜVEN, N., BURNHAM, C. W.: The crystal structure of 3T muscovite. Z. Krist. **125**, 163 (1967).
HAAS, D. J.: The crystal structure of potassium tetraoxalate, $K_2(HC_2O_4)(H_2C_2O_4) \cdot 2H_2O$. Acta Cryst. **17**, 1511 (1964).
HILMER, W.: Die Kristallstruktur des sauren Kaliummetasilikates, $K_4(HSiO_3)_4$. Acta Cryst. **17**, 1063 (1964).

© Springer-Verlag Berlin · Heidelberg 1972

IVLEV, V. F., ZAVERZINA, N. I., GABUDA, S. P.: Position of protons and diffusion of water molecules in apophyllite, $KFCa_4Si_8O_{20} \cdot 8H_2O$. Soviet Phys.-Cryst. 13, 705 (1969).
JOHANSSON, G.: The crystal structure of the potassium aluminate $K_2[Al_2O(OH)_6]$. Acta Chem. Scand. 20, 505 (1966).
KANNAN, K. K., VISWAMITRA, M. A.: Crystal structure of magnesium potassium sulfate hexahydrate $MgK_2(SO_4)_2 \cdot 6H_2O$. Z. Krist. 122, 161 (1965).
KATZ, L., MEGAW, H. D.: The structure of potassium niobate at room temperature: the solution of a pseudosymmetric structure by Fourier methods. Acta Cryst. 22, 639 (1967).
KOSTER, A. S., KOOLS, F. X. N. M., RIECK, G. D.: The crystal structure of potassium tungstate, K_2WO_4. Acta Cryst. B 25, 1704 (1969).
KROGH-MOE, J.: Least-squares refinement of the crystal structure of potassium pentaborate. Acta Cryst. 18, 1088 (1965).
LARSON, A. C., CROMER, D. T.: Refinement of the alum structure. III. X-ray study of the α-alums, K, Rb and $NH_4Al(SO_4)_2 \cdot 12H_2O$. Acta Cryst. 22, 793 (1967).
LINDSLEY, D. H.: Melting relations of $KAlSi_3O_8$: effect of pressures up to 40 kilobars. Am. Mineralogist 51, 1793 (1966).
LOPES-VIEIRA, A., ZUSSMAN, J.: Further details on the crystal structure of zussmanite. Mineral. Mag. 37, 49 (1969).
MAREZIO, M.: The crystal structure of $K_2[B_5O_8(OH)] \cdot 2H_2O$. Acta Cryst. B 25, 1787 (1969).
— PLETTINGER, H. A., ZACHARIASEN, W. H.: The crystal structure of potassium tetraborate tetrahydrate. Acta Cryst. 16, 975 (1963).
MUSTAFAEV, N. M., ILYUKHIN, V. V., BELOV, N. V.: The crystal structures of K and Rb orthofluoroberyllates and their relationship to compound of general formula M_2BX_4. Soviet Phys.-Cryst. 10, 676 (1966).
PEACOR, D. R.: A high temperature single crystal diffractometry study of leucite (K, Na)$AlSi_2O_6$. Z. Krist. 127, 213 (1968).
PERROTTA, A. J., SMITH, J. V.: The crystal structure of kalsilite, $KAlSiO_4$. Mineral. Mag. 35, 588 (1965).
RIBBE, P. H.: A refinement of the crystal structure of sanidinized orthoclase. Acta Cryst. 16, 426 (1963).
RINGWOOD, A. E., REID, A. F., WADSLEY, A. D.: High-pressure $KAlSi_3O_8$, an aluminosilicate with sixfold coordination. Acta Cryst. 23, 1093 (1967).
ROSS, M., EVANS, H. T., JR.: Studies of the torbernite minerals (I): the crystal structure of abernathyte and the structurally related compounds, $NH_4(UO_2AsO_4) \cdot 3H_2O$ and $K(H_3O)(UO_2AsO_4)_2 \cdot 6H_2O$. Am. Mineralogist 49, 1578 (1964).
— — Studies of the torbernite minerals (III): role of the interlayer oxonium, potassium, and ammonium ions, and water molecules. Am. Mineralogist 50, 1 (1965).
SCHLATTI, M., SAHL, K., ZEMANN, A., ZEMANN, J.: Kristallstruktur von Polyhalit, $K_2Ca_2Mg[SO_4]_4 \cdot 2H_2O$. Tschermaks Mineral. Petrog. Mitt. 14, 75 (1970).
SRIKANTA, S., SEQUEIRA, A., CHIDAMBARAM, R.: Neutron diffraction study of the space group and structure of manganese-leonite $K_2Mn(SO_4)_2 \cdot 4H_2O$. Acta Cryst. B 24, 1176 (1968).
STEINFINK, H.: Crystal structure of a trioctahedral mica: phlogopite. Am. Mineralogist 47, 886 (1962).
— The crystal structure of the zeolite, phillipsite. Acta Cryst. 15, 644 (1962).
STRUNZ, H.: Mineralogische Tabellen, p. 242. Leipzig: Akademische Verlagsgesellschaft 1966.
SYSIÖ, P. A.: On the additivity of crystal radii in alkali halides. Acta Cryst. B 25, 2374 (1969).
TAKEDA, H., DONNAY, J. D. H.: Trioctahedral one-layer micas. III. Crystal structure of a synthetic lithium fluormica. Acta Cryst. 20, 638 (1966).
WADSLEY, A. D.: Alkali titanoniobates: The crystal structure of $KTiNbO_5$ and KTi_3NbO_9. Acta Cryst. 17, 623 (1964).
WANG, R., BRADLEY, W. F., STEINFINK, H.: The crystal structure of alunite. Acta Cryst. 18, 249 (1965).
ZACHARIASEN, W. H., PLETTINGER, H. A.: Refinement of the structure of potassium pentaborate tetrahydrate. Acta Cryst. 16, 376 (1963).

Revised manuscript received: August 1970

Titanium   **22**

A   E. TILLMANNS   (Institut für Mineralogie, Universität Bochum, Germany)

B—O   C. W. CORRENS   (Geochemisches Institut der Universität, Göttingen, Germany)

22-A. Crystal Chemistry

Titanium occurs in minerals predominantly in the tetravalent oxidation state. Most minerals which contain titanium as a major constituent are oxides, titanates or silicates.

Warwickite, $(Mg, Ti, Fe)_2^{[6]}O[B^{[3]}O_3]$ (TAKÉUCHI et al., SR 1950, 350) seems to be the only natural titanium borate known. Osbornite, $Ti^{[6]}N^{[6]}$, with NaCl structure (BANNISTER, SR 1941, 103) has been found in meteorites (Ca poor achondrites).

The ionic radius for Ti^{4+} is 0.64 Å according to GOLDSCHMIDT (1926) and 0.68 Å according to PAULING (1960) and AHRENS (1952). For Ti^{3+}, values of 0.69 Å (GOLDSCHMIDT) and 0.76 Å (AHRENS) have been derived. Values for O^{2-} range between 1.32 and 1.40 Å. With a radius ratio of 0.52 (PAULING, 1960) for Ti^{4+} and O^{2-}, titanium is expected to be octahedrally coordinated by oxygen. Only a few exceptions to this rule are known: in fresnoite, $Ba_2^{[10]}Ti^{[5py]}O[Si_2O_7]$, (see no. 9 in Table 22-A-1); lamprophyllite, $(Ba, Sr, K)Na^{[6]}(Fe, Ti)^{[6]}Ti^{[5py]}(O, OH, F)_2[Si_2O_7]$, (WOODROW, 1964); innelite, $Na_2^{[6]}Ba_2^{[11]}Ba^{[10]}(Ba, K, Mn)^{[10]}(Ca, Na)^{[6]}Ti^{[6]}Ti_2^{[5py]}O_4(SO_4)_2[Si_2O_7]_2$, (see no. 13) as well as in the synthetic compounds, $K_2^{[8]}Ti_2^{[5py]}O_5$ (ANDERSSON and WADSLEY, SR 1961, 373) and $Y_2^{[7]}Ti^{[5py]}O_5$ (MUMME and WADSLEY, 1968), Ti^{4+} is coordinated by five oxygen ions at the corners of a square pyramid. Bariumorthotitanate, $Ba^{[10]}Ba^{[8]}Ti^{[4]}O_4$ (BLAND, SR 1961, 371), has been found to crystallize in the β-K_2SO_4 structure, the Ti^{4+} coordination polyhedron being a distorted tetrahedron with distances Ti—O = 1.63 to 1.82 Å and angles ranging from 94 to 129°.

Table 22-A-1 lists bond lengths and angles for a number of compounds with well determined crystal structures. Minerals in which Ti partly replaces other cations of comparable size have not been included in the table.

Table 22-A-1. *Structural data of titanium minerals*

No.	Mineral structural formula	Ti—O distances (Å)	O—Ti—O angles (degrees)	Reference
1	Rutile $Ti^{[6]}O_2$	4 × 1.944 2 × 1.988	80.8—99.2 180	BAUR (SR 1956, 263)
2	Anatase $Ti^{[6]}O_2$	4 × 1.937 2 × 1.964	77.6—102.3 155.3—180.0	CROMER and HERRINGTON (SR 1955, 361)
3	Brookite $Ti^{[6]}O_2$	1.87; 1.92 1.94; 1.99 2.00; 2.04	77—105 157—171	BAUR (SR 1961, 351)
4	Perovskite $Ca^{[12]}Ti^{[6]}O_3$	2 × 1.92 2 × 1.92 2 × 1.93	86.6—93.4 180	KAY and BAILEY (SR 1957, 317)

Titanium

Table 22-A-1 (Continued)

No.	Mineral structural formula	Ti—O distances (Å)	O—Ti—O angles (degrees)	Reference
5	Ilmenite $Fe^{[6]}Ti^{[6]}O_3$	3×1.93 3×2.15	75.4—102.9 159	Shirane et al. (SR 1959, 374)
6	Pseudobrookite $Fe_2^{[6]}Ti^{[6]}O_5$	2×1.96 2×1.98 2×2.08	76.6—120.7 140—163	Hamelin (SR 1957, 279)
7	Brannerite (syn.) $Th^{[6]}Ti_2^{[6]}O_5$	1.83; 1.84 2×1.94 2.06; 2.19	76.2—102.3 161.6—171.3	Ruh and Wadsley (1966)
8	Titanite $Ca^{[7]}Ti^{[6]}O[SiO_4]$	2×1.87 2×1.98 2×2.02	90.2—91.7 180	Mongiorgi and di Sanseverino (1968)
9	Fresnoite $Ba_2^{[10]}Ti^{[5py]}O[Si_2O_7]$	4×2.00 1×1.63	84.7—107.7 144.5	Moore and Louisnathan (1969)
10	Bafertisite $Ba^{[13]}Ba^{[12]}Fe_4^{[6]}$ $Ti_2^{[6]}O_2(OH)_4[Si_2O_7]_2$	1.82; 2.06 2×1.94 2×2.15 1.62; 1.80 1.91; 2.02 2.17; 2.43	79.6—103.6 160.8—180 68—114 143—169	Ya-Hsien et al. (1963)
11	Perrierite $Ce_4^{[10]}Fe^{2+[6]}(Fe^{3+},Ti)_2^{[6]}$ $Ti_2^{[6]}O_8[Si_2O_7]_2$	1.78; 2.33 2×1.89 2×2.03	78.4—99.8 160—177	Galli (1966)
12	Seidozerite $Na_4^{[6]}Mn^{[6]}(Zr,Ti)_2^{[6]}$ $Ti^{[6]}O_2(F,OH)[Si_2O_7]_2$	2×1.93 2×1.98 2×1.98	86.3—93.9 176—179	Simonov and Belov (SR 1959, 472)
13	Innelite $Na_2^{[6]}Ba_2^{[11]}Ba^{[10]}(Ca,Na)^{[6]}$ $(Ba,K,Mn)^{[10]}Ti^{[6]}Ti_2^{[5py]}O_4$ $(SO_4)_2[Si_2O_7]_2$	1.99; 2.00 2.05; 2.06 1.62 1.93; 1.96 1.98; 2.01 1.71 1.90; 1.92 2.00; 2.04 2.09; 2.16	82—110 148—149 85—103 154—156 78—100 163—176	Chernov et al. (1971)
14	Benitoite $Ba^{[6]}Ti^{[6]}[Si_3O_9]$	6×1.942	86.8—91.1 177	Fischer (1969)
15	Astrophyllite $Na^{[10]}K_2^{[12]}Fe_7^{[6]}Ti_2^{[6]}$ $O_7[Si_8O_{24}]$	1.81; 1.89 1.90; 1.94 2.04; 2.11	79.8—102.2 163—178	Woodrow (1967)
16	Aenigmatite $Na_2^{[8]}Fe_5^{[6]}Ti^{[6]}O_2$ $[Si_6O_{18}]$	1.84; 1.86 1.98; 2.00 2.08; 2.09	75—108 160—180	Cannillo et al. (1971)
17	Narsarsukite $Na_2^{[7]}Ti^{[6]}O[Si_4O_{10}]$	1.90; 2.07 4×1.97	83.3—96.7 166.6—180	Peacor and Buerger (1962)

The most important titanium minerals are: ilmenite, $Fe^{[6]}Ti^{[6]}O_3$, (see no. 5); titanomagnetite, a solid solution of magnetite, $(Fe^{3+}Fe^{2+})^{[6]}Fe^{3+[4]}O_4$, and ulvöspinel, $(Fe^{2+}Ti^{4+})^{[6]}Fe^{2+[4]}O_4$; rutile, $Ti^{[6]}O_2$, (see no. 1); and titanite, $Ca^{[7]}Ti^{[6]}O[SiO_4]$ (see no. 8).

I. Crystal Structures of Titanium Minerals

a) Oxides and Titanates

Titanium dioxide occurs in nature in three different modifications: rutile (see no. 1, Table 22-A-1), brookite (see no. 3) and anatase (see no. 2). In rutile, TiO_6 octahedra share two edges each to form chains along [001]. In brookite, zigzag chains of TiO_6 octahedra, parallel to [001], are linked to neighboring chains by common edges to form double nets parallel to (100). Each octahedron shares three edges with other octahedra. Both in rutile and brookite the infinite units formed by edge-sharing TiO_6 octahedra are connected by common corners to form a 3-dimensional network. The crystal structure of anatase corresponds to a 3-dimensional framework of TiO_6 octahedra with each octahedron sharing four edges with other octahedra. A fourth modification of TiO_2 has been synthezised by SIMONS and DACHILLE (1967) at 40 kb and 450°C. It is isostructural with α-PbO_2 having four Ti—O distances of 1.91 Å and two distances of 2.05 Å.

In ilmenite, $Fe^{[6]}Ti^{[6]}O_3$, oxygen ions form a hexagonal close-packed array with Ti^{4+} and Fe^{2+} in octahedral voids. TiO_6 octahedra share three edges with other TiO_6 octahedra and a face with a neighboring FeO_6 octahedron. Geikielite, $Mg^{[6]}Ti^{[6]}O_3$, and pyrophanite, $Mn^{[6]}Ti^{[6]}O_3$ (BARTH and POSNJAK, SB 1934, 69), are isostructural with ilmenite. Ulvöspinel, $(FeTi)^{[6]}Fe^{[4]}O_4$, has the inverse spinel structure. According to FORSTER and HALL (1965), some Ti^{4+} ions occupy tetrahedral sites.

The crystal structure of perovskite, $Ca^{[12]}Ti^{[6]}O_3$ (see no. 4, Table 22-A-1) has been determined by KAY and BAILEY (SR 1957, 317) for synthetic material. The idealized perovskite structure is isometric with the large cations and anions forming cubic close-packed layers. TiO_6 octahedra share all six corners with other octahedra. In perovskite this structure is distorted to orthorhombic symmetry. In the ferroelectric tetragonal modifications of $BaTiO_3$ (EVANS, SR 1961, 370) and $PbTiO_3$ (SHIRANE et al., SR 1956, 288), which also have distorted perovskite structures, Ti^{4+} is displaced from the center of the coordination octahedron by 0.155 Å in $BaTiO_3$, and by 0.30 Å in $PbTiO_3$. There are four symmetrically equivalent Ti—O distances of 2.00 and 1.98 Å respectively, and also two opposite bonds of different lengths: 1.86 and 2.17 Å in $BaTiO_3$ and 1.78 and 2.38 Å in $PbTiO_3$.

In pseudobrookite, $Fe_2^{[6]}Ti^{[6]}O_5$ (see no. 6), TiO_6 octahedra form chains via shared corners. Each TiO_6 octahedron shares six edges with FeO_6 octahedra. Isostructural with pseudobrookite are the synthetic compounds "tialite", Al_2TiO_5 (HAMELIN, SR 1957, 279), "anosovite", $Ti_2^{3+}TiO_5$, and "karrooite", $MgTiTiO_5$, (RUSAKOV and ŽDANOV, SR 1952, 233), and the mineral armalcolite, $(Fe,Mg)Ti_2O_5$, which occurs in lunar crystalline rocks (ANDERSON et al., 1970). The crystal structure of the low temperature modification of $Ti_2^{3+}TiO_5$ (ÅSBRINK and MAGNÉLI, SR 1959, 335) is related to that of pseudobrookite.

RUH and WADSLEY (1966) used synthetic $Th^{[6]}Ti^{[6]}_2O_6$ (see no. 7, Table 22-A-1) to determine the crystal structure of brannerite, $(U,Ca,Th,Y)(Ti,Fe)_2O_6$. TiO_6

octahedra share edges to form infinite zigzag sheets centered about the (001) plane. The sheets are bound together by thorium in octahedral interlayer positions.

While TiO_6 coordination octahedra are usually connected by common corners and edges, several synthetic compounds are known in which octahedra have a common face. In the hexagonal modification of barium titanate, $Ba^{[12]}Ti^{[6]}O_3$, one third of the octahedra share a face to form a group of composition $[Ti_2O_9]^{10-}$ (BURBANK and EVANS, SR 1948, 448). This group is linked to three other octahedra by common corners. The Ti—Ti distance across the shared face is 2.67 Å. In $Ba^{[12]}Ti_6^{[6]}O_{13}$ (TILLMANNS, 1972), the Ti—Ti distance across the shared face is 2.824 Å, while in $Ba_4^{[11]}Ba_2^{[12]}Ti_{17}^{[6]}O_{40}$ it is 2.857 Å (TILLMANNS and BAUR, 1970). The shortest Ti—Ti distance across a shared edge in $Ba_6Ti_{17}O_{40}$ is 2.838 Å. In $Ti_2^{3+[6]}O_3$, which has corundum structure, the Ti^{3+}—Ti^{3+} distance across the shared face is 2.59 Å (NEWNHAM and DE HAAN, 1962). ÅSBRINK and MAGNÉLI (SR 1959, 335) give formal valencies of 3, $3\,1/3$ and $3\,2/3$ for the three crystallographically different Ti ions in low Ti_2TiO_5. They found Ti—Ti distances across shared edges of 2.61, 2.76, 2.82 and 3.07 to 3.17 Å.

b) Silicates

The most important titanium silicate is titanite, $Ca^{[7]}Ti^{[6]}O[SiO_4]$ (see no. 8, Table 22-A-1). TiO_6 octahedra share four edges with CaO_7 polyhedra and corners with two TiO_6 octahedra and four SiO_4 tetrahedra. In fresnoite, $Ba_2^{[10]}Ti^{[5py]}O[Si_2O_7]$ (see no. 9), $[Si_2O_7]^{6-}$ groups are linked by common corners to TiO_5 square pyramids to give flat sheets parallel with (001). The same units, although arranged differently, form sheets in lamprophyllite, $(Ba, Sr, K)Na^{[6]}(Ti, Fe)^{[6]}Ti^{[5py]}_2(O, OH, F)_2[Si_2O_7]$ (WOODROW, 1964), and innelite, $Na_2^{[6]}Ba_2^{[11]}Ba^{[10]}(Ba, K, Mn)^{[10]}(Ca, Na)^{[6]}Ti^{[6]}Ti_2^{[5py]}O_4(SO_4)_2[Si_2O_7]_2$ (see no. 13).

The crystal structures of bafertisite, $Ba^{[13]}Ba^{[12]}Fe_4^{[6]}Ti^{[6]}O_2(OH)_4[Si_2O_7]_2$ (see no. 10), seidozerite, $Na_4^{[6]}Mn^{[6]}(Zr, Ti)_2^{[6]}Ti^{[6]}O_2(F, OH)_2[Si_2O_7]_2$ (see no. 12), murmanite, $Na_2^{[6]}Mn^{[6]}Ti_3^{[6]}(OH)_4[Si_2O_7]_2 \cdot 4H_2O$ (KHALILOV et al., 1965), and rosenbuschite, $(Ca, Na)_6^{[6]}Zr^{[6]}(Ti, Mn, Nb)^{[6]}F_2O(O, F)[Si_2O_7]_2$ (SHIBAEVA et al., 1963), are related to that of lamprophyllite. In these minerals $[Si_2O_7]^{6-}$ groups link sheets and ribbons formed via common edges and corners by the coordination polyhedra around the other cations.

In perrierite, $Ce_4^{[10]}Fe^{2+[6]}(Fe^{3+}, Ti)_2^{[6]}Ti_2^{[6]}O_8[Si_2O_7]_2$ (see no. 11, Table 22-A-1), TiO_6 and $(Fe, Ti)O_6$ octahedra share edges to form sheets parallel to (001), with $[Si_2O_7]^{6-}$ groups and CeO_{10} polyhedra arranged between two layers of octahedra. Similar sheets formed by $(Ti, Nb)O_6$ and $(Na, Ca)O_6$ octahedra and $(Na, Ca)O_8$ square antiprisms are the main feature of the crystal structure of rinkite, $(Na, Ca)_2^{[6]}(Na, Ca)^{[8]}(Ca, Ce)_4^{[7]}(Ti, Nb)^{[6]}(O, F)_4[Si_2O_7]_2$ (GALLI and ALBERTI, 1971). In leucosphenite, $Na_8^{[7]}Ba_2^{[10]}Ti_4^{[6]}O_4[B_4^{[4]}Si_{20}O_{56}]$ (SHUMYATSKAYA et al., 1968), groups of the composition $[B_4Si_{20}O_{56}]^{20-}$ form infinite double layers parallel to the (001) plane. Coordination polyhedra around Na and Ti connect two of the double layers. The crystal structure of astrophyllite, $Na^{[10]}K_2^{[13]}Fe_7^{[6]}Ti_2^{[6]}O_7[Si_8O_{24}]$ (see no. 15), is related to that of biotite with a layer of TiO_6 octahedra and SiO_4 tetrahedra in the ratio 1:4 replacing the biotite layer of SiO_4 tetrahedra. In benitoite, $Ba^{[6]}Ti^{[6]}[Si_3O_9]$ (see no. 14), $[Si_3O_9]^{6-}$ rings are linked to TiO_6 and BaO_6 octahedra via common corners. The TiO_6 octahedron shares three edges with BaO_6 octahedra. The arrangement of TiO_6

octahedra in ramsayite, $Na_2^{[7]}Ti_2^{[6]}O_3[Si_2O_6]$ (BELOV and BELYAEV, SR 1950, 371), is similar to that in brookite: the zigzag chains of edge-sharing TiO_6 octahedra share corners with the $[SiO_3]^{2-}$ chains. The crystal structure of narsarsukite, $Na_2^{[7]}Ti^{[6]}O[Si_4O_{10}]$ (see no. 17), contains chains of corner-sharing TiO_6 octahedra. In aenigmatite, $Na_2^{[8]}Fe_5^{[6]}Ti^{[6]}O_2[Si_6O_{18}]$ (see no. 16), continuous layers formed by TiO_6 and FeO_6 octahedra, and by NaO_8 polyhedra, alternate with layers formed by $[Si_6O_{18}]^{12-}$ chains and FeO_6 octahedra.

II. Replacement of Cations by Titanium

In addition to compounds with titanium as a major constituent, there is a large number of minerals in which Ti partly replaces other ions such as Al^{3+}, Fe^{3+}, Nb^{5+}, Ta^{5+}, Mn^{3+} etc. Titanium is incorporated, for example, in minerals of the pyrochlore group, in priderite, freudenbergite, crichtonite, davidite and senaite (cf. STRUNZ, 1970). Appreciable amounts of titanium can be present in biotites, $K(Mg,Fe,Mn)_3(OH,F)_2[Si_3AlO_{10}]$ (SAXENA, 1966; DAHL, 1970), in andradite garnets, $Ca_3Fe_2[Si_3O_{12}]$ (ISAACS, 1968; HOWIE and WOOLLEY, 1968), in amphiboles such as kaersutite, $Ca_2(Na,K)(Mg,Fe^{2+},Fe^{3+})_4Ti(O,OH,F)_2[Si_6Al_2O_{22}]$ (BORLEY and FROST, 1963), and in augites, $(Ca,Mg,Fe^{2+},Fe^{3+},Ti,Al)_2[(Si,Al)_2O_6]$ (LEWIS, 1967).

Usually titanium substitutes for Fe^{3+} or Al in octahedral positions, as reported by GHOSE (1965) for kaersutite, and by PEACOR (1967) for Ti-augite. In a review of earlier literature, HARTMANN (1969) discussed the question of whether Ti^{4+} can also replace Si^{4+} in tetrahedral positions. He assumes that Al, Fe^{3+}, and Ti^{4+} substitute for Si^{4+} in decreasing order of preference, i.e., Ti^{4+} enters tetrahedral positions only if there is not enough Al and Fe^{3+} present to compensate the tetrahedral deficiency. From lattice parameter changes and from the results of a least-squares refinement of site-occupancy factors for a Ti-containing synthetic diopside, SCHRÖPFER (1968) concluded that Ti^{4+} can partly replace Si^{4+} in diopside. In a study on the substitution of $Mg^{2+}Si_2^{4+}$ by $Ti^{4+}Al_2^{3+}$ in synthetic diopside, he found evidence for the substitution of Si by Ti even when there is sufficient Al present to occupy all vacant tetrahedral positions (SCHRÖPFER, 1971). For andradite garnet, MOORE and WHITE (1971) concluded that most of the titanium replaces Fe^{3+} in the octahedral sites, with some unknown proportion as Ti^{3+}. The silica deficiency of high-titanium garnets seems to be compensated by the substitution of Al and Fe^{3+} into the tetrahedral sites.

References: Section 22-A

AHRENS, L. H.: The use of ionization potentials. Part 1. Ionic radii of the elements. Geochim. Cosmochim. Acta **2**, 155 (1952).

ANDERSON, A. T., BUNCH, T. E., CAMERON, E. N., HAGGERTY, S. E., BOYD, F. R., FINGER, L. W., JAMES, O. B., KEIL, K., PRINZ, M., RAMDOHR, P., EL GORESY, A.: Armalcolite: A new mineral from the Apollo 11 samples. Proceedings of the Apollo 11 Lunar Science Conference **1**, 55 (1970).

BORLEY, G., FROST, M. T.: Some observations on igneous ferrohastingsites. Mineral. Mag. **33**, 646 (1963).

CANNILLO, E., MAZZI, F., FANG, J. H., ROBINSON, P. D., OHYA, Y.: The crystal structure of aenigmatite. Am. Mineralogist **56**, 427 (1971).

CHERNOV, A. N., ILYUKHIN, V. V., MAKSIMOV, B. A., BELOV, N. V.: Crystal structure of innelite. Kristallografiya **16**, 87 (1971).

DAHL, O.: Octahedral titanium and aluminium in biotite. Lithos **3**, 161 (1970).

FISCHER, K.: Verfeinerung der Kristallstruktur von Benitoid $BaTi[Si_3O_9]$. Z. Krist. **129**, 222 (1969).

FORSTER, R. H., HALL, E. O.: A neutron diffraction study of ulvöspinel, Fe_2TiO_4. Acta Cryst. **18**, 859 (1965).

GALLI, E.: Affinamento della struttura della perrierite. Mineral. Petr. Acta (Bologna) **11**, 39 (1966).

— ALBERTI, A.: The crystal structure of rinkite. Acta Cryst. B **27**, 1277 (1971).

GHOSE, S.: A scheme of cation distribution in the amphiboles. Mineral. Mag. **35**, 46 (1965).

GOLDSCHMIDT, V. M.: Geochemische Verteilungsgesetze der Elemente. Skrifter Norske Videnskaps-Akad. Oslo, 1: Mat.-Naturv. Kl. (1926).

HARTMANN, P.: Can Ti^{4+} replace Si^{4+} in silicates? Mineral. Mag. **37**, 366 (1969).

HOWIE, R. A., WOOLLEY, A. R.: The role of titanium and the effect of TiO_2 on the cell-size, refractive index, and specific gravity in the andradite-melanite-schorlomite series. Mineral. Mag. **36**, 775 (1968).

ISAACS, T.: Titanium substitution in andradites. Chem. Geol. **3**, 219 (1968).

KHALILOV, A. D., MAMEDOV, K. S., MAKAROV, E. S., PYANZINA, L. Y.: Crystal structure of murmanite. Dokl. Akad. Nauk SSSR **161**, 1409 (1965).

LEWIS, J. F.: Unit-cell dimensions of some aluminous natural clinopyroxenes. Am. Mineralogist **52**, 42 (1967).

MONGIORGI, R., DI SANSEVERINO, L. R.: A reconsideration of the structure of titanite, $CaTiOSiO_4$. Mineral. Petr. Acta (Bologna) **14**, 123 (1968).

MOORE, P. B., LOUISNATHAN, J.: The crystal structure of fresnoite, $Ba_2(TiO)[Si_2O_7]$. Z. Krist. **130**, 438 (1969).

MOORE, R. K., WHITE, W. B.: Intervalence electron transfer effects in the spectra of the melanite garnets. Am. Mineralogist **56**, 826 (1971).

MUMME, W. G., WADSLEY, A. D.: The structure of orthorhombic Y_2TiO_5, an example of mixed seven- and five-fold coordination. Acta Cryst. B **24**, 1327 (1968).

NEWNHAM, R. E., DE HAAN, Y. M.: Refinement of the α-Al_2O_3, Ti_2O_3, V_2O_3 and Cr_2O_3 structures. Z. Krist. **117**, 235 (1962).

PAULING, L.: The Nature of the Chemical Bond. (2nd edition.) Ithaca-New York: Cornell University Press 1960.

PEACOR, D. R.: Refinement of the crystal structure of a pyroxene of formula $M_I M_{II}$ $(Si_{1.5}Al_{0.5})O_6$. Am. Mineralogist **52**, 31 (1967).

— BUERGER, M. J.: The determination and refinement of the structure of narsarsukite, $Na_2TiO[Si_4O_{10}]$. Am. Mineralogist **47**, 539 (1962).

RUH, R., WADSLEY, A. D.: The crystal structure of $ThTi_2O_6$ (brannerite). Acta Cryst. 21, 974 (1966).
SAXENA, S. K.: Distribution of elements between coexisting muscovite and biotite and crystal chemical role of titanium in the micas. Neues Jahrb. Mineral., Abhandl. 105, 1 (1966).
SCHRÖPFER, L.: Über den Einbau von Titan in Diopsid. Neues Jahrb. Mineral., Monatsh. 11, 441 (1968).
— Über den gekoppelten Ersatz von $Mg^{2+}Si_2^{4+}$ durch $Ti^{4+}Al_2^{3+}$ im Diopsid. Neues Jahrb. Mineral., Abhandl. 116, 20 (1971).
SHIBAEVA, R. P., SIMONOV, V. I., BELOV, N. V.: Crystal structure of the Ca, Na, Zr, Ti silicate rosenbuschite $Ca_{3.5} Na_{2.5} Zr$ (Ti, Mn, Nb) $[Si_2O_7]_2$ F_2O (O, F). Kristallografiya 8, 506 (1963).
SHUMYATSKAYA, N. G., VORONKOV, A. A., PYATENKO, Y. A.: The atomic structure of leucosphenite $Na_6Ba_2Ti_4O_4$ $[B_4Si_{20}O_{56}]$. Kristallografiya 13, 165 (1968).
SIMONS, P. Y., DACHILLE, F.: The structure of TiO_2-II, a high pressure phase of TiO_2. Acta Cryst. 23, 334 (1967).
STRUNZ, H.: Mineralogische Tabellen. (5. Aufl.) Leipzig: Akademische Verlagsgesellschaft 1970.
TILLMANNS, E.: Bariumhexatitanate, $BaTi_6O_{13}$. Crystal Structure Communications 1, 1 (1972).
— BAUR, W. H.: The crystal structure of hexabarium 17-titanate. Acta Cryst. B 26, 1645 (1970).
WOODROW, P. J.: Crystal structure of lamprophyllite. Nature 204, 375 (1964).
— The crystal structure of astrophyllite. Acta Cryst. 22, 673 (1967).
YA-HSIEN, K., SIMONOV, V. I., BELOV, N. V.: Crystal structure of bafertisite. Dokl. Akad. Nauk SSSR 149, 1416 (1963).

Revised manuscript received: February 1972

Chromium 24

A   E. Matzat   (Mineralog.-Kristallogr. Institut der Universität, Göttingen, Germany)

B—O   K. Shiraki   (Department of Earth Sciences, Nagoya University, Nagoya, Japan)

24-A. Crystal Chemistry

Chromium is known to occur in nature in the hexavalent and trivalent oxidation states. Besides minerals with chromium as a major constituent, there is a large number of minerals where trivalent chromium replaces other elements, preferably aluminium, upto approximately one weight % Cr_2O_3.

I. Hexavalent Chromium

Hexavalent chromium is found in nature to occur in tetrahedral oxygen coordination as the isolated $[CrO_4]^{2-}$ ion or with corner-sharing as the $[Cr_2O_7]^{2-}$ ion. The tetrahedra are distorted with angles ranging from 104° to 114°, and the mean Cr—O distance ranges from 1.60 Å to 1.67 Å. Table 24-A-1 lists the bond lengths and angles in these minerals.

In crocoite, $Pb^{[10]}[CrO_4]$, each CrO_4-tetrahedron shares 3 edges with 3 polyhedra and the 4 corners with 4 additional polyhedra of 10-fold coordinated Pb. Chromatite, $Ca^{[8]}[CrO_4]$, is isostructural with zircon where tetrahedra share 2 edges with 2 polyhedra and the 4 corners with 4 additional polyhedra of 8-fold coordinated Ca. In tarapacaite, $K^{[9]}K^{[10]}[CrO_4]$, the tetrahedron shares faces, edges and corners with the polyhedra of 10-fold coordinated K, but only edges and corners with those of 9-fold coordinated K. In phoenicochroite, $Pb_2^{[9]}O[CrO_4]$, the tetrahedron shares edges and corners with the polyhedra of two crystallographically different Pb atoms in 9-fold coordination. In vauquelinite, $Cu^{[4+2]}Pb_2^{[9]}(OH)[CrO_4][PO_4]$, and fornacite, $(Cu,Fe)^{[4+2]}Pb^{[9]}Pb^{[10]}(OH)[CrO_4][(P, As)O_4]$, sheets of connected oxygen-polyhedra around Pb are linked by zig-zag chains built-up of polyhedra around (4+2)-fold coordinated Cu. Additional linking by CrO_4- and PO_4-tetrahedra, sharing edges and corners, results in a compact 3-dimensional array. A structural relationship also exists with hemihedrite, $Zn^{[6]}Pb_8^{[9]}Pb_2^{[11]}Pb_2^{[8]}F_2[CrO_4]_6[SiO_4]_2$. Bellite, $(Pb, Ag)_5Cl[(Cr, As, Si)O_4]_3$ (STRUNZ, 1958), is a mimetite-type mineral with isomorphous replacement of AsO_4-tetrahedra by CrO_4. It is not known which of the known modifications of synthetic $K_2[Cr_2O_7]$ is the mineral lopezite. The structural investigation of the triclinic modification (BRANDON and BROWN, 1968) shows that two nearly tetrahedral CrO_4 groups join to form Cr_2O_7 groups by a shared oxygen. Two different ions are almost in the eclipsed configuration (deviations ~5° and ~10°) and one of the tetrahedral angles deviates significantly from the ideal value. Cr—O—Cr bridging angles are 124° and 127.6°; bridging Cr—O distances measure 1.79 Å and the terminal ones 1.63 Å. The investigation of a monoclinic modification (ZHUKOVA and PINSKER, 1964) gives approximately 1.71 Å as a mean Cr—O distance. The shape and dimensions of the Cr_2O_7 group are similar in several additional synthetic dichromates.

Table 24-A-1. *Minerals with hexavalent chromium*

Mineral	Cr—O distance [Å] (σ)	Mean distance [Å]	O—Cr—O angles [°] (σ)	M[a]	R[b]	Reference
Pb[10][CrO₄] Crocoite	1.61 (2) 1.66 (2) 1.67 (2) 1.67 (2)	1.65	105.7— 113.1 (1.3)	N	8.5	1
Ca[8][CrO₄] Chromatite	1.64 (4×)	1.64	—	S	—	2
K[9]K[10][CrO₄] Tarapacaite	1.61 1.59 1.60 1.60	1.60	—	S	—	3
Pb₂[9]O[CrO₄] Phoenicochroite	1.70 1.71 1.64 (2×)	1.67	108.4— 110.3	N	6.3	4
Cu[4+2]Pb₂[9](OH)[CrO₄][PO₄] Vauquelinite	1.63 (7) 1.68 (7) 1.60 (4) 1.59 (4)	1.62	105 (4)— 114 (3)	N	8.9	5
(Cu, Fe)[4+2]Pb[9]Pb[10] (OH)[CrO₄][(Pb, As)O₄] Fornacite	1.64 (3) 1.64 (3) 1.57 (3) 1.68 (3)	1.63	104.6 (1.4)— 113.4 (1.6)	N	9.8	6
Zn[6]Pb₆[9]Pb₂[11]Pb₂[8] F₂[CrO₄]₆[SiO₄]₂ Hemihedrite	1.65 (3) 1.68 (4) 1.67 (3) 1.64 (3) 1.69 (3) 1.60 (3) 1.68 (2) 1.68 (2) 1.60 (2) 1.64 (2) 1.70 (3) 1.70 (3)	1.66 1.66 1.66	106.1 (1.3)— 111.8 (1.4) 105.3 (1.2)— 112.2 (1.3) 107.5 (1.4)— 111.3 (1.1)	N	4.1	7

References: 1. QUARENI and DE PIERI (1965). 2. CLOUSE (1932). 3. ZACHARIASEN and ZIEGLER (1931). 4. WILLIAMS, McLEAN and ANTHONY (1970). 5. FANFANI and ZANAZZI (1968). 6. COCCO, FANFANI and ZANAZZI (1967). 7. McLEAN and ANTHONY (1970).

[a] M: Material used for structure determination: N = natural, S = synthetic.
[b] R: Reliability index.

II. Trivalent Chromium

Trivalent chromium occurs in nature in slightly distorted octahedral coordination in simple or complex oxides and sulfides. The mean Cr—O distance ranges from 1.97 Å to 2.00 Å, and the Cr—S distance from 2.39 Å to 2.43 Å. Table 24-A-2 gives bond lengths and references.

Chromium

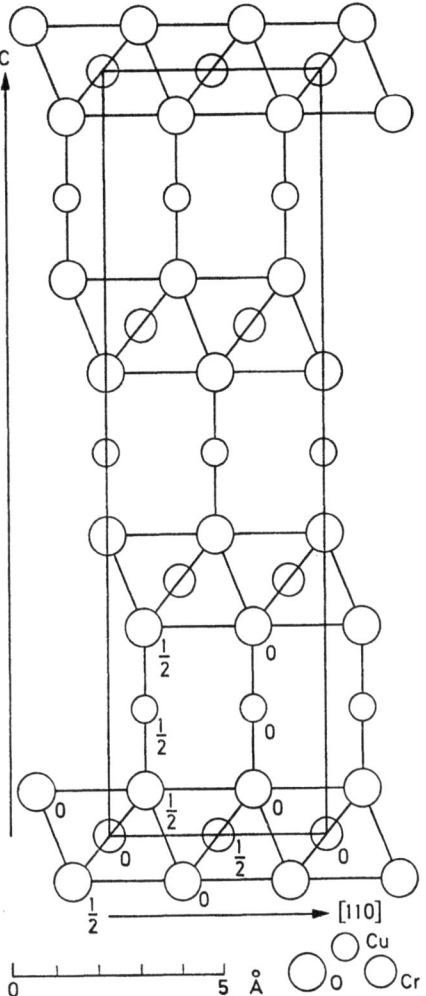

Fig. 24-A-1. Structure of mcconnellite projected on (1 $\bar{2}$·0). In grimaldiite, Cu is replaced by H, and S by O accompanied by relatively shorter O—H—O distances

a) Chromium Minerals

Eskolaite, $Cr_2^{[6]}O_3$, crystallizes isostructurally with α-Al_2O_3. The oxygen octahedra share edges and faces. Particularly notable is the short Cr—Cr distance of 2.65 Å. O—Cr—O octahedral angles range from 81.4° to 99.0°. Three modifications of $Cr^{[6]}OOH$ exist in nature (MILTON, APPLEMAN, CHAO et al., 1967). Guayanite is isostructural with InOOH (NØRLUND CHRISTENSEN, GRØNBÆK and RASMUSSEN, 1964), where octahedra share edges and corners, as in a deformed rutile-type structure, and are linked by hydrogen bonds. Bracewellite is isostructural with α-FeOOH being built up by corner-linked and hydrogen-bonded double bands of edge-sharing octahedra. In grimaldiite, as well as in mcconnellite, $Cu^{[2]}Cr^{[6]}O_2$, CrO_6-octahedra form sheets by sharing 6 of their edges with coplanar neighbors. Oxygen atoms of different

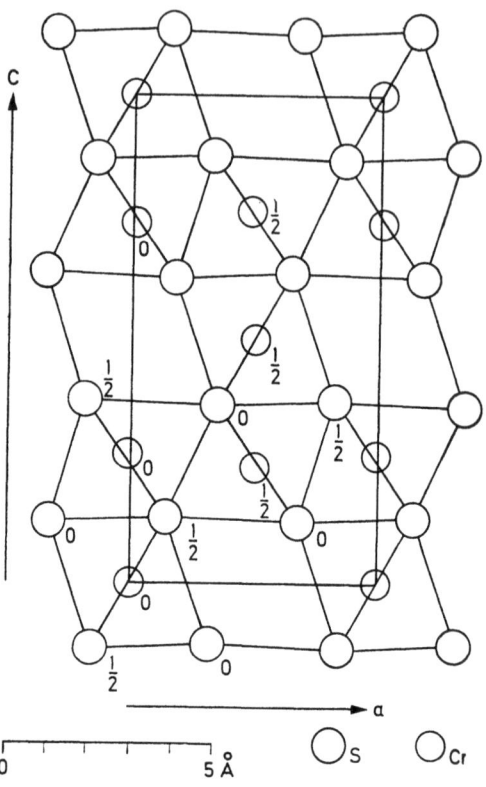

Fig. 24-A-2. Structure of brezinaite projected on (0 1 0) [related to NiAs structure projected on (1 2̄ 0)]

sheets are exactly superposed and are linked by hydrogen bonds in grimaldiite and by Cu in two-fold linear coordination in mcconnellite (Fig. 24-A-1). Chromite, $(Fe, Mg)^{[4]}Cr_2^{[6]}O_4$, characterized by a considerable replacement of Cr by Al, has the normal spinel structure, a framework of octahedra sharing edges with each other and corners with tetrahedra. The composition of meteoritic chromite is close to $FeCr_2O_4$ (cf. Table 24-A-2); in $MgCr_2O_4$ the Cr—O distance is 2.00 Å (VERWEY and HEILMANN, 1947). Ureyite, $Na^{[8]}Cr^{[6]}[Si_2O_6]$, is a clinopyroxene; Cr occupies the octahedral M_1 site which links 4 SiO_3 chains by corners. The octahedra share edges with 2 polyhedra of the 8-fold coordinated M_2 site occupied by Na. Stichtite and barbertonite, $Mg_6Cr_2(OH)_{16}CO_3 \cdot 4H_2O$, are isostructural with the stacking modifications pyroaurite (ALLMANN, 1968; INGRAM and TAYLOR, 1967) and sjögrenite, $Mg_6Fe_2(OH)_{16}CO_3 \cdot 4H_2O$ (idem; ALLMANN and LOHSE, 1966). The authors agree on the position of the cations as being situated on an octahedral site of a brucite-type layer, but disagree on the distribution of interlayer carbonate groups and water.

Natural chromium sulfides have been described only from meteoritic materials. Daubréelite, $Fe^{[4]}Cr_2^{[6]}S_4$, crystallizes in the normal spinel structure with 6-fold coordinated Cr. Brezinaite, $Cr_3^{[6]}S_4$ has a structure related to that of NiAs with ordered vacancies of metal positions; sheets of edge-sharing octahedra have faces

Table 24-A-2. *Minerals with trivalent chromium*

Mineral	Cr—O distance [Å] (σ)	Mean distance [Å]	Shortest distance Cr—Cr [Å]	R[a]	Reference
$Cr_2^{[6]}O_3$ Eskolaite	2.02 (2) (3 ×) 1.97 (1) (3 ×)	2.00	2.65	9.0	Newnham and De Haan (1962)
$Cr^{[6]}OOH$ Grimaldiite	1.97 (4) (6 ×)	1.97	2.984	6.0	Douglass (1957)
$Cu^{[2]}Cr^{[6]}O_2$ Mcconnellite	1.99 (2) (6 ×)	1.99	2.975	—	Dannhauser and Vaughan (1955)
$Fe^{[4]}Cr_2^{[6]}O_4$ Chromite	1.99 (6 ×)	1.99	2.96	—	Bacchella and Pinot (1964)
$Na^{[8]}Cr^{[6]}[Si_2O_6]$ Ureyite	2.038 (5) (2 ×) 2.011 (8) (2 ×) 1.946 (5) (2 ×)	2.00	—	5.0	Clark, Appleman and Papike (1968)
$Fe^{[4]}Cr_2^{[6]}S_4$ Daubréelite	2.39 (6 ×)	2.39	3.524	3.6	Broquetas Colominas, Ballestracci and Roult (1964)
$Cr_3^{[6]}S_4$ Brezinaite	1) 2.45 (4 ×) 2.39 (2 ×) 2) 2.44 2.43 (2 ×) 2.38 2.37 (2 ×)	2.43 2.40	2.97	—	Jellinek (1957)

[a] R: Reliability index; Synthetic material used for all structural determinations.

in common with additional octahedra as shown in Fig. 24-A-2. Compared with daubréelite, the short Cr—Cr distance of 2.97 Å indicates intermetallic bonding in this metalloid compound. Comparable distances and metal site vacancies have been observed in synthetic chromium sulfides (Jellinek, 1957).

b) Replacement of Cations by Trivalent Chromium

Spectroscopic absorption measurements are used to determine the position of chromium in compounds where it can replace cations of comparable size. Dunitz and Orgel (1957) calculated the octahedral site preference energy for Cr^{3+} to be 37.7 kcal. mole^{-1}. Comparable ionic radii are: Cr^{3+}, 0.69; Al^{3+}, 0.50; Fe^{3+}, 0.64; Fe^{2+}, 0.76; Mg^{2+}, 0.65; and Ti^{4+}, 0.68 Å (Pauling, 1960). Cr^{3+} with electronic d^3 configuration is in a ^4F ground state, which splits in an octahedral field into $^4A_{2g}$, $^4T_{2g}$ and $^4T_{1g}$ degenerate states. The two spin-allowed transitions $^4A_{2g} \rightarrow {^4T_{2g}}$ and $^4A_{2g} \rightarrow {^4T_{1g}}$ are always observed according to semi-quantitative calculations (Tanabe and Sugano, 1954). Further splitting of terms by lowering the crystal field symmetry is observed to be minimal within the broad absorption bands. Depending on the strength of the crystal field, the visible color of the chromium-containing compound

Table 24-A-3. *Minerals: isomorphous replacement of cations by trivalent chromium*

Investigated mineral (chromium variety name)	Mean cation-O distance [Å]	Reference (structure)	Reference (replacement)
$Mg_3Al_2[SiO_4]_3$ pyrope	Al: 1.886	GIBBS and SMITH (1965)	NEUHAUS (1960)
$Be_3Al_2[Si_6O_{18}]$ beryl (emerald)	Al: 1.903	GIBBS, BRECK and MEAGHER (1968)	WOOD (1965)
Al_2O_3 corundum (ruby)	Al: 1.915	NEWNHAM and DE HAAN (1962)	NEUHAUS (1960)
$K_2Al_2(OH)_2[AlSi_3O_{10}]$ muskovite (fuchsite)	Al: 1.925	BIRLE and TETTENHORST (1968)	NEUHAUS (1960)
$Ca_3Al_2[SiO_4]_3$ grossular (uvarovite)	Al: 1.927	PRANDL (1966)	MANNING (1969a)
$MgAl_2O_4$ spinel	Al: 1.93	STOLL, FISCHER, HÄLG and MAIER (1964)	NEUHAUS (1960)
$(Ca, Na)Mg_3Al_6(OH)_4[BO_3]_3[Si_6O_{18}]$ Tourmaline	Al: 1.93	BUERGER, BURNHAM and PEACOR (1962)	MANNING (1969b)
$Al_2[BeO_4]$ chrysoberyl (alexandrite)	Al_1: 1.890 Al_2: 1.938	FARREL, FANG and NEWNHAM (1963)	FARRELL and NEWNHAM (1965)
Fe_3O_4 magnetite	Fe: 2.05	O'REILLY (1968)	YEARIAN, KORTRIGHT and LANGENHEIM (1954)
TiO_2 rutil	Ti: 1.96	BAUR (1955)	LOHR and LIPSCOMP (1963)

varies continuously between green and red. ν_1 varies between $<15{,}400$ and $17{,}200$ cm^{-1}, ν_2 between $<21{,}800$ and $23{,}800$ cm^{-1} in a weak crystal field resulting in green visible colour, ν_1 between $17{,}500$ and $18{,}500$ cm^{-1} and ν_2 between $24{,}200$ and $25{,}600$ cm^{-1} in a strong crystal field resulting in red visible colour. The variation depends only to a small extent on concentration and crystal field symmetry; the strength of the crystal field is influenced mainly by distance and strength of ligands and contrapolarization by neighboring cations (NEUHAUS, 1960).

In complex oxides Cr has been proved to preferentially replace Al in 6-fold coordination. In some of these oxides this site is also partially occupied by Mg, Fe and Ti. Table 24-A-3 gives cation-oxygen distances of the octahedral site for several minerals where replacement takes place, and also references on structural investigations and replacement. The replacement of cations in the octahedral site has been proved also to occur in chlorite (kämmererite), $(Mg, Al)_{2-3}(OH)_6(Mg, Al)_{2-3}(OH)_2[(Si, Al)_4O_{10}]$, (where a position in the brucite-type layer is occupied), augite, $Ca(Mg, Fe^{2+}, Al)[(Si, Al)_2O_6]$, amphibole, $(Na, Ca)_2(Mg, Fe^{2+}, Al)_5(OH)_2[(Si, Al)_8O_{22}]$ (NEUHAUS, 1960), and in epidote (tawmawite) $Ca_2(Fe, Al)Al_2O(OH)[SiO_4][Si_2O_7]$ (BURNS and STRENS, 1967). According to investigations on synthetic spinels (VERWEY and HEILMANN, 1947; VERWEY, HAAYMAN and ROMEIJN, 1947), replacement of octahedral sites is to be expected in all spinel-type compounds.

Revised manuscript received: September 1970

References: Section 24-A

ALLMANN, R.: The crystal structure of pyroaurite. Acta Cryst. B **24**, 972 (1968).
— LOHSE, H. H.: Die Kristallstruktur des Sjögrenits und eines Umwandlungsproduktes des Koenenits (= Chlor-Manasseits). Neues Jahrb. Mineral., Monatsh. 1966, 161 (1966).
BACCHELLA, G. L., PINOT, M.: Étude sur la structure magnétique de $FeCr_2O_4$. J. Physique **25**, 537 (1964).
BAUR, W.: Atomabstände und Bindungswinkel im Rutil. Naturwissenschaften **42**, 295 (1955).
BIRLE, J. D., TETTENHORST, R.: Refined muskovite structure. Mineral. Mag. **36**, 883 (1968).
BRANDON, J. K., BROWN, I. D.: An accurate determination of the crystal structure of triclinic potassium dichromate $K_2Cr_2O_7$. Can. J. Chem. **46**, 933 (1968).
BROQUETAS COLOMINAS, C., BALLESTRACCI, R., ROULT, G.: Étude par diffraction neutronique du spinelle $FeCr_2S_4$. J. Physique **25**, 526 (1964).
BUERGER, M. J., BURNHAM, C. W., PEACOR, D. R.: Assessment of the several structures proposed for tourmaline. Acta Cryst. **15**, 583 (1962).
BURNS, R. G., STRENS, R. G. J.: Structural interpretation of polarized absorption spectra of the Al-Fe-Mn-Cr epidotes. Mineral. Mag. **36**, 204 (1967).
CLARK, J. R., APPLEMAN, D. E., PAPIKE, J. J.: Bonding in eight ordered clinopyroxenes isostructural with diopside. Contr. Mineral. Petrol. **20**, 81 (1968).
CLOUSE, J. H.: Investigations on the X-ray crystal structures of $CaCrO_4$, $CaCrO_4 \cdot H_2O$ and $CaCrO_4 \cdot 2H_2O$. Z. Krist. **83**, 161 (1932).
COCCO, G., FANFANI, L., ZANAZZI, P. F.: The crystal structure of fornacite. Z. Krist. **124**, 385 (1967).
DANNHAUSER, W., VAUGHAN, P. A.: The crystal structure of cuprous chromite. J. Am. Chem. Soc. **77**, 896 (1955).
DOUGLASS, R. M.: The crystal structure of $HCrO_2$. Acta Cryst. **10**, 423 (1957).
DUNITZ, J. D., ORGEL, L. E.: Electronic properties of transition element oxides. II. Cation distribution amongst octahedral and tetrahedral sites. J. Phys. Chem. Solids **3**, 318 (1957).
FANFANI, L., ZANAZZI, P. F.: The crystal structure of vauquelinite and the relationships to fornacite. Z. Krist. **126**, 433 (1968).
FARRELL, E. F., FANG, J. H., NEWNHAM, R. E.: Refinement of the chrysoberyl structure. Am. Mineralogist **48**, 804 (1963).
— NEWNHAM, R. E.: Crystal-field spectra of chrysoberyl, alexandrite, peridot and sinhalite. Am. Mineralogist **50**, 1972 (1965).
GIBBS, G. V., BRECK, D. W., MEAGHER, E. P.: Structural refinement of hydrous and anhydrous synthetic beryl, $Al_2(Be_3Si_6)O_{18}$ and emerald $Al_{1.9}Cr_{0.1}(Be_3Si_6)O_{18}$. Lithos **1**, 275 (1968).
— SMITH, J. V.: Refinement of the crystal structure of synthetic pyrope. Am. Mineralogist **50**, 2023 (1965).
INGRAM, L., TAYLOR, H. F. W.: The crystal structures of sjögrenite and pyroaurite. Mineral. Mag. **36**, 465 (1967).
JELLINEK, F.: The structures of the chromium sulfides. Acta Cryst. **10**, 620 (1957).
LOHR, L. L., LIPSCOMB, W. N.: Molecular orbital theory of spectra of Cr^{3+} ions in crystals. J. Chem. Phys. **38**, 1607 (1963).
MANNING, P. G.: Optical absorption studies of grossular, andradite (var. colophonite) and uvarovite. Can. Mineralogist **9**, 723 (1969a).
— Optical absorption spectra of chromium-bearing tourmaline, black tourmaline and buergerite. Can. Mineralogist **10**, 57 (1969b).

McLean, W. J., Anthony, J. W.: The crystal sturcture of hemihedrite. Am. Mineralogist 55, 1103 (1970).

Milton, C., Appleman, D., Chao, E. C. T., Cuttita, F., Dinnin, J. L., Dwornik, E. J., Hall, M., Ingram, B. L., Rose, H. J.: Mineralogy of merumite, a unique assembledge of Cr-minerals from Guayana. Geol. Soc. Am. Progr. 1967 Ann. Meet. 151 (1967).

Neuhaus, A.: Über die Ionenfarben der Kristalle und Minerale am Beispiel der Chromfärbungen. Z. Krist. 113, 195 (1960).

Newnham, R. E., DeHaan, Y. M.: Refinement of the α-Al_2O_3, Ti_2O_3, V_2O_3 and Cr_2O_3 structures. Z. Krist. 117, 235 (1962).

Nørlund Christensen, A., Grønbæk, R., Rasmussen, S. E.: The crystal structure of InOOH. Acta Chem. Scand. 18, 1261 (1964).

O'Reilly, W.: Weak reflexions in the X-ray diffraction pattern of magnetite, Fe_3O_4. Acta Cryst. B 24, 422 (1968).

Pauling, L.: The Nature of Chemical Bond. Ithaca, Cornell University Press 1960.

Prandl, W.: Verfeinerung der Kristallstruktur des Grossulars mit Neutronen- und Röntgenstrahlbeugung. Z. Krist. 123, 81 (1966).

Quareni, S., DePieri, R.: A three-dimensional refinement of the structure of crocoite, $PbCrO_4$. Acta Cryst. 19, 287 (1965).

Stoll, E., Fischer, P., Hälg, W., Maier, G.: Redetermination of the cation distribution of spinel ($MgAl_2O_4$) by means of neutron diffraction. J. Physique 25, 447 (1964).

Strunz, H.: Bellit, ein Chromapatit. Naturwissenschaften 45, 127 (1958).

Tanabe, Y., Sugano, S.: On the absorption spectra of complex ions. J. Phys. Soc. Japan 9, 753 (1954).

Verwey, E. J., Haayman, P. W., Romeijn, F. C.: Physical properties and cation arrangement of oxides with spinel structures II. J. Chem. Phys. 15, 181 (1947).

Verwey, E. J. W., Heilmann, E. L.: Physical properties and cation arrangement of oxides with spinel structures I. J. Chem. Phys. 15, 174 (1947).

Williams, S. A., McLean, W. J., Anthony, J. W.: A study of phoenicochroite—its structure and properties. Am. Mineralogist 55, 784 (1970).

Wood, D. L.: Absorption, fluorescence, and Zeeman effect in emerald. J. Chem. Phys. 42, 3404 (1965).

Yearian, H. J., Kortright, J. M., Langenheim, R. H.: Lattice parameters of the $FeFe_{2-x}Cr_xO_4$ system. J. Chem. Phys. 22, 1196 (1954).

Zachariasen, W. H., Ziegler, G. E.: The crystal structure of potassium chromate. Z. Krist. 80, 164 (1931).

Zhukova, L. A., Pinsker, Z. G.: An electron-diffraction study of the structure of potassium dichromate. Soviet Phys.-Cryst. 9, 31 (1964).

Revised manuscript received: September 1970

Manganese

A   D. R. Peacor   (Department of Geology and Mineralogy,
University of Michigan,
Ann Arbor, Mich., U.S.A.)

B—O   K. H. Wedepohl   (Geochemisches Institut der Universität,
Göttingen, Germany)

Manganese

25-A. Crystal Chemistry

I. Valence States

Manganese is found in naturally occurring minerals as the ions Mn^{2+}, Mn^{3+} or Mn^{4+}, more than one valence state occasionally occurring in a single phase. Some general factors affecting the crystal chemical characteristics of each of these ions in turn are briefly discussed in the following paragraphs (see COTTON and WILKINSON, 1967, as general reference).

Mn^{2+} has a $3d^5$ electron configuration and thus may theoretically occur in either a high-spin, $t_{2g}^3 e_g^2$, or low-spin, t_{2g}^5, state. For those coordinating ligands normally found in naturally occurring minerals, such as O^{2-}, OH^-, Cl^-, F^- or H_2O, the crystal field is relatively weak and only the high-spin state occurs. In octahedral coordination, the low-spin state is possible for some other ligands, but it should never be observed for tetrahedral coordination, since the tetrahedral crystal field splitting parameter, Δ_t, is only 4/9ths of that of the octahedral splitting parameter, Δ_0. Since all five d-orbitals are occupied in the high-spin state there is no crystal-field stabilization energy (CFSE), and this factor therefore does not influence the nature of the site coordination. The coordination polyhedron geometry is thus determined principally by relative ionic radius and local charge conditions. On this basis the coordination is predicted to be octahedral with ligands such as O^{2-}, OH^-, and H_2O. Since under most geological conditions Mn^{2+} is the stable ionic species, the most common state for manganese in minerals is therefore as high-spin, octahedrally coordinated Mn^{2+}. Table 25-A-1 is a compilation of bond lengths for Mn^{2+} compounds for which atom coordinates have been determined. This represents only a portion of the known representatives, including those phases which are isotypic with other compounds having well determined crystal structures. Although the coordination is octahedral or somewhat distorted octahedral for most of these compounds, tetrahedral, five-fold, eight-fold and other six-fold coordinations are not uncommon where factors other than relative radius are significant.

Mn^{3+} has a $3d^4$ electron configuration and therefore may also occur in a high-spin, $t_{2g}^3 e_g^1$, or low-spin, t_{2g}^4, state. As with Mn^{2+} only the high-spin state is normally found however. Mn^{3+} is predicted to have octahedral coordination on the basis of relative ionic radius. However, there is an additional factor favoring this coordination in that there is an octahedral site preference energy of $2/5\, \Delta_0$. Thus Mn^{3+} is normally found to be in octahedral coordination and although it has a smaller radius than Mn^{2+}, tetrahedral coordination is more common for the latter ion, as shown in part by the coordination data of Table 25-A-2. Tetrahedral coordination is found in some synthetic compounds such as $MnPO_4$ which has the quartz and cristobalite structures (SHAFER et al., SR 1956, 302). As for Mn^{2+}, only the high-spin state should occur for Mn^{3+} in tetrahedral coordination.

© Springer-Verlag Berlin · Heidelberg 1972

Table 25-A-1. Bond lengths for Mn^{2+}

Compound	Anion	Distance (Å)	Polyhedron	Reference
$Mn_3Fe_2[GeO_4]_3$ garnet	Mn—O	2.303 (4), 2.421 (4)	distorted cube	LIND and GELLER (1969)
Mn_2GeO_4	Mn—O	2.14, 2.14 (2), 2.20 (2), 2.29	trig. prism	WADSLEY and REID (1968)
$\tfrac{1}{2}Mn_4S[Be_3Si_3O_{12}]$ helvite	Mn(1)—O Mn(1)—S	2.074 (3) 2.432	tet.	HOLLOWAY (1969)
$MnZn_2(OH)_2[SiO_4]$ hodgkinsonite	Mn(1)—O Mn(1)—OH	2.250 2.179, 2.188, 2.190, 2.262, 2.265	oct.	RENTZEPIRIS (1963)
$MnCa[Si_2O_6]$ johannsenite	Mn—O	2.155 (2), 2.231 (2), 2.134 (2)	oct.	FREED and PEACOR (1967)
$MnCa[Si_2O_6]$ bustamite	Mn(1)—O Mn(2)—O	2.041, 2.144, 2.163, 2.286, 2.335, 2.499 2.154 (2), 2.215 (2), 2.241 (2)	oct. oct	PEACOR and BUERGER (1962)
$Mn_4(Ca, Mn)[Si_5O_{15}]$ rhodonite	Mn(1)—O Mn(2)—O Mn(3)—O Mn(4)—O	2.11, 2.15, 2.18, 2.25, 2.28, 2.35 2.07, 2.15, 2.22, 2.25, 2.27, 2.33 2.11, 2.14, 2.20, 2.28, 2.33, 2.41 1.98, 2.04, 2.12, 2.23, 2.39, 2.88	oct. oct. oct. distorted oct.	PEACOR and NIIZEKI (1963)
$MnPb_8[Si_2O_7]_3$ barysilite	Mn—O	2.26 (6)	oct.	LAJZÉROWICZ (1965)
$MnLi[PO_4]$ lithiophilite	Mn—O	2.130 (2), 2.139, 2.240, 2.283 (2)	oct.	GELLER and DURAND (SR 1960, 399)
$Mn_2[P_2O_7]$	Mn—O	2.08 (2), 2.08 (2), 2.21 (2)	oct.	LUKASZEWICZ (SR 1961, 466)
$(Mn_{0.85}Fe_{0.15})Be(OH)[PO_4]$ väyrynenite	(Mn, Fe)—OH (Mn, Fe)—O	2.37 2.08, 2.14, 2.18, 2.23, 2.24	oct.	MROSE and APPLEMAN (1962)

Manganese

Formula / Name	Bond	Distances	Coord.	Reference
$MnFe_2(OH)_2[PO_4]_2(H_2O)_6 \cdot 2H_2O$ laueite	Mn—(O, OH)	2.068 (2), 2.083 (2), 2.187 (2)	oct.	Moore (1965)
$Mn_7(OH)_8[AsO_4]_2$ allactite	Mn(1)—OH	2.300 (2), 2.314 (2)	oct.	Moore (1968)
	Mn(1)—O	2.185 (2)		
	Mn(2)—OH	2.085, 2.145, 2.173, 2.591	oct.	
	Mn(2)—O	2.141, 2.230		
	Mn(3)—OH	2.086, 2.149, 2.171, 2.181	oct.	
	Mn(3)—O	2.229, 2.288		
	Mn(4)—OH	2.065, 2.155, 2.211, 2.299	oct.	
	Mn(4)—O	2.115, 2.302		
$MnCa_2[AsO_4]_2 \cdot 2H_2O$ brandtite	Mn—O	2.11 (2), 2.19 (2), 2.27 (2)	oct.	Dahlman (SR 1952, 289)
$(Mn, Mg)_2Y(OH)_4[AsO_4]$ retzian	(Mn, Mg)—O	2.12 (2), 2.33 (2)	oct.	Moore (1967)
	(Mn, Mg)—OH	2.10 (2)		
$MnPb(OH)[VO_4]$ pyrobelonite	Mn—O	2.04 (2), 2.18 (2), 2.22 (2)	oct.	Donaldson and Barnes (SR 1955, 451)
$(Ca_{0.6}Mn_{0.4})(Mn_{0.7}Zn_{0.24}Mg_{0.04})Te_4O_{10}$ denningite	(Ca, Mn)—O	2.44 (4), 2.46 (4)	sq. antiprism	Walitzi (1965)
	(Mn, Zn)—O	2.02 (4), 2.39 (2)	distorted oct.	
$Mn[SO_4]$	Mn—O	2.11 (2), 2.25 (4)	oct.	Will (1965)
$MnK_2[SO_4]_2 \cdot 4H_2O$ manganese-leonite	Mn(1)—O	2.109 (2)	oct.	Sriktana et al. (1967)
	Mn(1)—O$_w$	2.188 (2), 2.227 (2)		
	Mn(2)—O	2.195 (2)		
	Mn(2)—O$_w$	2.193 (4)		
$Mn(NH_4)_2[SO_4]_2 \cdot 6H_2O$	Mn—O$_w$	2.150 (2), 2.193 (2), 2.200 (2)	oct.	Montgomeroy et al. (1966)
MnF_2	Mn—F	2.10 (2), 2.13 (4)	oct.	Baur (1958)
$BaMnF_4$	Mn—F	2.037—2.151	oct.	Keve et al. (1969)
$CsMnF_3$	Mn(1)—F	2.12 (6)	oct.	Zatkin et al. (1962)
	Mn(2)—F	2.12 (3), 2.16 (3)		
$K_2MnCl_4 \cdot 2H_2O$	Mn—Cl	2.531 (4)	oct.	Jensen (1968)
	Mn—O$_w$	2.175 (2)		

Manganese

Table 25-A-1 (continued)

Compound	Anion	Distance (Å)	Polyhedron	Reference
$MnCl_2 \cdot 4H_2O$	Mn—Cl Mn—O_w	2.475, 2.500 2.185, 2.206, 2.209, 2.224	oct.	Zalkin et al. (1964)
$MnCl_2 \cdot 2H_2O$	Mn—Cl Mn—O_w	2.515 (2), 2.592 (2) 2.150 (2)	oct.	Morosin and Graeber (1965)
$KMnCl_3 \cdot 2H_2O$	Mn—Cl Mn—O_w	2.482, 2.490, 2.570, 2.594 2.182, 2.187	oct.	Jensen (1968)
$Cs_2MnCl_4 \cdot 2H_2O$	Mn—Cl Mn—O_w	2.54 (2), 2.54 (2) 2.13 (2)	oct.	Jensen (1964)
MnO manganosite	Mn—O	2.222 (6)	oct.	U.S. Nat'l Bur. Stand. (1955)
$Mn[MoO_4]$	Mn(1)—O Mn(2)—O	2.098 (2), 2.148 (2), 2.252 (2) 2.091, 2.122 (2), 2.204 (2), 2.227	oct. oct.	Abrahams and Reddy (1965)
$MnTiO_3$ pyrophanite	Mn—O	2.10 (3), 2.26 (3)	oct.	Shirane et al. (SR 1959, 374)
MnS (α) alabandite	Mn—S	2.612 (6)	oct.	U.S. Nat'l Bur. of Stand. (1955)
MnS (β)	Mn—S	2.427 (4)	tet.	Corliss et al. (SR 1956, 150)
MnSe (α)	Mn—Se	2.718 (6)	oct.	U.S. Nat'l Bur. of Stand. (1960)
MnSe (β)	Mn—Se	2.52 (4)	tet.	Baroni (1938)
$Mn_2[SiS_4]$	Mn(1)—S Mn(2)—S	2.47 (2), 2.56 (2), 2.78 (2) 2.38 (2), 2.68, 2.82, 2.83 (2)	oct. oct.	Hardy et al. (1965)
$Mn_2[GeS_4]$	Mn(1)—S Mn(2)—S	2.47 (2), 2.49 (2), 2.89 (2) 2.68, 2.71 (2), 2.76 (2), 2.95	oct. oct.	Hardy et al. (1965)

Table 25-A-2. Bond lengths for Mn^{3+}

Compound	Anion	Distances (Å)	Polyhedron	Reference
MnO(OH) manganite	Mn—O Mn—OH	1.868, 1.878, 2.199 1.965, 1.981, 2.333	oct.	Dachs (1963)
MnO(OH) groutite	Mn—O Mn—OH	1.896 (2), 2.178 1.968 (2), 2.340	oct.	Dent, Glasser and Ingram (1968)
Mn_2O_3 (ortho.) bixbyite, synthetic	Mn(1)—O Mn(2)—O Mn(3)—O Mn(4)—O Mn(5)—O	1.955 (2), 1.989 (2), 2.058 (2) 1.964 (2), 1.991 (2), 2.067 (2) 1.893, 1.912, 1.957, 2.010, 2.192, 2.306 1.878, 1.918, 1.982, 1.984, 2.229, 2.274 1.875, 1.912, 1.962, 2.011, 2.215, 2.262	oct. oct. oct. oct. oct.	Norrestam (1967)
$(Mn, Fe)_2O_3$ (cubic) bixbyite	(Mn, Fe)(1)—O (Mn, Fe)(2)—O	2.01 (6) 1.90 (2), 1.92 (2), 2.24 (2)	oct.	Dachs (SR 1956, 273)
$CdMn_2O_4$	Mn—O	1.88 (4), 2.24 (2)	oct.	Sinha et al. (SR 1957, 329)
$PbMnO_2(OH)$ quenselite	Mn(1)—O Mn(2)—O	1.93 (2), 1.98 (2), 2.25 (2) 1.90 (2), 1.95 (2), 2.37 (2)	oct. oct.	Rouse (1969)
$CuMnO_2$ crednerite	Mn—O	1.92 (2), 2.28 (4)	oct.	Kondrasev (SR 1958, 334)
$LuMnO_3$	Mn—O	1.84, 1.93, 1.98, 2.06 (2)	trig. dipyr.	Yakel and Koehler (1963)
$Na_4Mn_4Ti_5O_{18}$	Mn(1)—O Mn(2)—O	1.948 (2), 1.985 (2), 2.171 1.940 (2), 1.970 (2), 2.032, 2.273	rect. pyr. oct.	Mumme (1968)
$Bi_2Mn_4O_{10}$	Mn(1)—O Mn(2)—O	1.83 (2), 1.90 (2), 1.98 (2) 1.91 (2), 2.04, 2.10 (2)	oct. sq. pyr.	Niizeki and Wachi (1968)
MnF_3	Mn—F	1.79 (2), 1.93 (2), 2.09 (2)	oct.	Hepworth and Jack (SR 1957, 207)

Table 25-A-2 (continued)

Compound	Anion	Distances (Å)	Polyhedron	Reference
MnAs	Mn—As	2.52, 2.56, 2.56, 2.58, 2.58, 2.61	oct.	WILSON and KASPER (1964)
$Ca_2Al(Al_{0.8}Mn^{3+}_{0.1}Fe^{3+}_{0.1}$$Mn^{3+}_{0.6}Fe_{0.2}Al_{0.2})(OH)_2[SiO_4][Si_2O_7]$ piemontite	Mn^{3+}—O	1.861, 1.900, 2.031 (2), 2.274 (2)	oct.	DOLLASE (1969)

Table 25-A-3. Bond lengths for Mn^{4+}, octahedral coordination

Compound	Anion	Distances (Å)	Reference
$BaMnO_3$	Mn—O	2.02 (6)	HARDY (1962)
Mg_6MnO_8	Mn—O	1.928 (6)	KASPER and PRENER (SR 1954, 524)
$ZnMn_3O_7 \cdot 3H_2O$ chalcophanite	Mn—O	1.94, 1.96	WADSLEY (SR 1955, 454)
γ-MnO_2 ramsdellite	Mn—O	1.85, 1.87, 1.93, 1.94	KONDRASEV and ZASLAVSKIJ (SR 1951, 182)
$(Ba, Pb, K, Na)(Mn, Fe, Al, Si)_{7.86}(O, OH)_{16}$ hollandite	Mn—(O, OH)	1.98	BYSTRÖM and BYSTRÖM (SR 1950, 186)
$(K_{0.54}H_2O_{1.46})_2(Mn^{4+}_{7.75}Mn^{2+}_{0.24})_8O_{16}$ cryptomelane	Mn—O	1.88, 1.89, 1.90, 1.90	KONDRASEV and ZASLAVSKIJ (SR 1951, 182)
$(Ba_{0.64}H_2O_{1.36})_2(Mn^{4+}_{4.25}Mn^{2+}_{0.56}R_{0.18})_5O_{10}$ psilomelane	Mn(1)—O Mn(2)—O Mn(3)—O	1.85, 1.95 1.92, 1.98, 2.01, 2.02 1.85, 1.86, 1.92, 1.94	WADSLEY (SR 1953, 429)
$(Al_{0.68}Li_{0.32})(Mn^{4+}_{0.83}Mn^{3+}_{0.17})O_3 \cdot 1H_2O$ lithiophorite	Mn—O	1.93 (4), 1.97 (2)	WADSLEY (SR 1952, 266)
$Ca_3Mn(OH)_6[SO_4]_2 \cdot 3H_2O$ despujolsite	Mn—OH	1.92 (6)	GAUDEFROY et al. (1968)

The Mn³⁺ coordination octahedron is subject to Jahn-Teller distortions. This may take the form of a contraction of two opposing bond lengths relative to the remaining four coplanar bonds such that the electron in the e_g state occupies the $d_{x^2-y^2}$ orbital; alternatively two opposing bonds are lengthened along a four-fold axis, so that the polyhedron becomes an elongated tetragonal dipyramid with one electron in the d_{z^2} orbital. The latter case is normally observed, as can be seen in the bond length values of Table 25-A-2. STRENS (1966) has shown that for isolated octahedra, or for compounds where the bonding is ionic, this polyhedron geometry is the most stable. Mn—O bond lengths are about 1.95 Å for the four short bonds, while the remaining two may be greater than 2.3 Å. The occurrence of such octahedron distortion, together with values of interatomic distances, may normally be used as confirmation of the occurrence of manganese in a 3+ oxidation state. As can be seen in the tabulated values of interatomic distances, such octahedron distortion does not occur for Mn²⁺ and Mn⁴⁺ despite the presence of other forms of octahedron irregularity. On the other hand, where structures containing Mn³⁺ have been refined using modern techniques, the Jahn-Teller distortion is usually observed. Where it is not, some doubt may exist as to the accuracy of structure determination or refinement.

The Jahn-Teller distortion is significant in relation to the substitution of Mn³⁺ for other ions which do not exhibit this effect. For example, STRENS (1965) has noted that the site symmetry ($\bar{3}$) of the octahedrally coordinated site of garnet is inconsistent with that of an octahedron distorted due to the Jahn-Teller effect. Therefore Mn³⁺ substitution for Al, for example, should be restricted. In structures where the octahedron distortion is inconsistent with site symmetry, or significant enough to cause

Table 25-A-4. Mn—O bond lengths for compounds with Mn in two principal valence states

Compound	Anion	Distances (Å)	Polyhedron	Reference
$Mn_2^{2+}Mn_3^{4+}O_8$	$Mn^{4+}(1)$—O	1.853 (4), 1.879 (2)	oct.	OSWALD and WAMPETICH (1967)
	$Mn^{4+}(2)$—O	1.849 (2), 1.893 (2), 1.919 (2)	oct.	
	$Mn^{2+}(3)$—O	2.047, 2.167 (2), 2.282, 2.314	trig. prism	
$DyMn^{3+}Mn^{4+}O_5$	Mn^{4+}—O	1.873 (2), 1.940 (2), 1.940 (2)	oct.	ABRAHAMS and BERNSTEIN (1967)
	Mn^{3+}—O	1.911 (2), 1.926 (2), 2.020	sq. pyr.	
$Mn_2^{2+}Mn^{3+}(OH)_4[AsO_4]$ flinkite	Mn^{3+}—O	1.91 (2), 1.80 (2), 2.29 (2)	oct.	MOORE (1967)
	Mn^{2+}—O	2.13, 2.14, 2.18, 2.21, 2.24, 2.26	oct.	
$Mn^{2+}Mn_2^{3+}Mg_3O_4[BO_3]_2$ pinakiolite	Mn^{2+}—O	2.00 (2), 2.06, 2.16 (2), 2.22	oct.	TAKÉUCHI et al. (SR 1950, 355)
	$Mn^{3+}(1)$—O	2.00 (2), 2.00 (2), 2.00 (2)	oct.	
	$Mn^{3+}(2)$—O	2.09 (2), 2.13 (2), 2.13 (2)	oct.	

distortion in other structural units, the overall structure symmetry may be lowered. Thus hausmannite, Mn_3O_4, has the normal spinel structure, but is a tetragonal, distorted derivative of cubic spinel.

Mn^{4+} has a $3d^3$ electron configuration and therefore has only one spin state if octahedrally coordinated. As with Mn^{2+} and Mn^{3+}, octahedral coordination by anions found in mineral structures is expected on the basis of relative ionic radii. Mn^{4+} has a large octahedral site preference energy ($4/5\ \Delta_0$). The coordination is therefore ubiquitously octahedral, although its effective ionic radius is significantly less than those of Mn^{2+} or Mn^{4+}.

II. Bond Lengths

Bond lengths for Mn^{2+}, Mn^{3+} and Mn^{4+} as coordinated by a variety of ligands are listed in Tables 25-A-1 to 25-A-4. Only where these ions are octahedrally coordinated by oxygen are there enough high-accuracy structure determinations to calculate reasonably accurate average values. These are 2.21, 2.01, and 1.90 Å for these ions, respectively. The average value for Mn^{3+}—O was determined by including bonds elongated due to the Jahn-Teller effect. The concept of ionic radius clearly has a restricted meaning for this ion. SHANNON and PREWITT (1969) derived average bond lengths of 2.22, 2.05, and 1.94 Å for these three ions, assuming values of "effective" radii of 1.40, 0.82, 0.65 and 0.54 Å for O^{2-}, Mn^{2+}, Mn^{3+} and Mn^{4+} respectively.

III. Sulfides, Tellurides, Selenides

MnS has been described as having the NaCl (alabandite), ZnS and ZnO structure types (NBS, 1955; CORLISS et al., SR 1956, 150) with Mn having tetrahedral coordination in the latter two structures. Similarly, MnSe occurs in all three structure types (NBS, 1960; BARONI, SB 1938, 58) while MnTe has the NaCl structure (JOHNSON and SESTRICH, SR 1961, 151). Hauerite, MnS_2, has the pyrite structure (GORDON, SR 1951, 237) as does $MnSe_2$ and $MnTe_2$ (HASTINGS et al., SR 1959, 180). A number of manganese compounds have the NiAs structure, including MnBi (HOCART and GUILLAUD, SB 1939, 210), MnSb, MnTe, MnS and MnAs (OFTEDAL, SB 1927, 765). MnAs and MnP have an orthorhombic structure and are distortionally derivative to NiAs (FYLKING, SB 1934, 263; WILSON and KASPER, 1964). Mn_2GeS_4 and Mn_2SiS_4 have been described as having the olivine structure, with Mn in octahedral coordination (HARDY et al., 1965).

IV. Oxides and Hydroxides

Mn^{2+} has no crystal field stabilization energy and thus should form normal spinels with Mn^{2+} on the tetrahedral sites, based on charge considerations. Examples are $MnTi_2O_4$ (LECARF and HARDY, SR 1961, 408), $MnFe_2O_4$, jacobsite (NOZIK and JAMZIN, SR 1961, 411), galaxite, $MnAl_2O_4$, $MnCr_2O_4$, and MnV_2O_4 (DUNN, MCCLURE and PEARSON, 1965). Mn^{3+} also should form normal spinels with Mn^{3+} on octahedral sites, both from ionic charge considerations and because there is an octahedral site preference energy. Thus Fe_3O_4 is an inverse spinel since there is a stabilization energy for Fe^{2+} favoring octahedral coordination, but none for

Fe^{3+}, while Mn_3O_4, hausmannite, has the normal spinel structure. The Mn^{3+} octahedron is subject to the Jahn-Teller distortion, however, and therefore hausmannite (SATOMI, SR 1961, 467), haeterolite, $ZnMn_2O_4$ (O'KEEFFE, SR 1961, 409), and $CdMn_2O_4$ (SINHA et al., SR 1957, 329) have structures which are tetragonal, distortional derivatives of the spinel structure. The compound γ-Mn_2O_3 has a cation-defect hausmannite structure (SINHA and SINHA, SR 1957, 330).

Bixbyite, $(Mn,Fe)_2O_3$, is cubic and has a defect structure derivative to that of fluorite, with Mn^{3+} in octahedral coordination. One Mn,Fe atom occupies a high symmetry site with six equivalent bonds, while the second site is subject to Jahn-Teller distortion (DACHS, SR 1956, 273; FERT, 1962; HASE, 1967). HASE and BRUCHNER (1969) note that when Mn:Fe is greater than about 3:2 the space group is different, although cubic, and that for compositions with less than 1% Fe_2O_3 there is a third form. GELLER et al. (1967) likewise noted that pure Mn_2O_3 is orthorhombic and that only 1 to 2% Fe_2O_3 results in a cubic structure, while NORRESTAM (1967) has refined the structure of orthorhombic Mn_2O_3. The two sites of the cubic bixbyite structure are replaced by five octahedrally coordinated sites, three of which are subject to major Jahn-Teller distortions.

A number of manganese oxides, hydroxides, chlorides, etc. have layer structures based on anion closest-packing with manganese in octahedral sites, and with octahedra sharing edges to form brucite or gibbsite-like sheets. Thus pyrochroite, $Mn(OH)_2$, is isotypic with brucite (AMINOFF, SB 1919, 195) as are $MnBr_2$ and MnI_2 (FERRARI and GIORGIO, SB 1929, 246—247). $MnCl_2$ has the $CdCl_2$ layer structure (FERRARI et al., SB 1929, 245). Several manganese compounds have the ilmenite structure with alternating gibbsite-like layers and with face-sharing of the octahedra between layers. These include $MnTiO_3$, pyrophanite (SHIRANE et al., SR 1959, 374), $NiMnO_3$ and $CoMnO_3$ (CLOUD, SR 1958, 332). Quenselite, PbMnOOH, crednerite, $CuMnO_2$, chalcophanite, $ZnMn_3O_7 \cdot 3H_2O$, and lithiophorite, $(Al,Li)MnO_2(OH)_2$ have layer structures with brucite-like manganese octahedral layers alternating with some other layered unit. In quenselite the layer sequence is Mn—O—Pb,OH—Pb,OH—O—Mn (ROUSE, unpublished), in crednerite it is Mn—O—Cu—O—Mn (KONDRASHEV, SR 1958, 334), in lithiophorite Mn—O—OH—Al,Li—OH—O—Mn (WADSLEY, SR 1952, 1966), and in chalcophanite Mn—O—Zn—H_2O—Zn—O—Mn (WADSLEY, SR 1955, 454). In the latter case, one of every seven octahedral sites of the brucite-like sheet is vacant.

Three well-defined structures have been described for MnO_2 or compounds substitutionally derivative to it, and these have related structures. Mn^{4+} octahedra share two opposite edges to form infinite chains, so that these compounds are characterized by a common translation or pseudotranslation of about 2.9 Å, corresponding to an octahedron edge parallel to the chain. In pyrolusite, β-MnO_2, which is isotypic with rutile, vertices are shared between chains to produce a tetragonal unit cell with $a=b\sim 4.4$ Å (FARRARI, SB 1926, 213). In the ramsdellite structure, γ-MnO_2 (isotypic with diaspore, α-AlOOH), two chains share edges to form bands of octahedra which in turn share vertices to form an orthorhombic structure with $a\sim 4.5$ and $c\sim 9.3$ Å (KONDRASEV and ZASLAVSKIJ, SR 1951, 182). Similar doubled chains crosslink through vertex-sharing in α-MnO_2 to yield a tetragonal or pseudotetragonal cell with $a\sim b\sim 9.9$ Å. This linkage provides 4-sided channels which may be occupied by H_2O or some relatively large cation such as Pb, Ba or K, provided that some appropriate substitution takes place on the octahedral sites to balance charge, such as Mn^{2+} for Mn^{4+}. Thus there is a series of solid solutions from α-MnO_2 to cryptomelane, $K_2Mn_8O_{16}$, hollandite, $Ba_2Mn_8O_{16}$, and coronadite, $Pb_2Mn_8O_{16}$ (BYSTRÖM and BYSTRÖM, SR 1950, 186). For example, KONDRASEV and ZALAVSKIJ (SR 1951, 182) determined the structure of a synthetic cryptomelane for which they reported the composition $(K_{0.54}H_2O_{1.48})$ $(Mn^{4+}_{7.75} Mn^{2+}_{0.24})O_{16}$. WADSLEY (SR 1953, 429) determined the structure of a fourth variant of this structure scheme having the composition $(Ba_{0.64}H_2O_{1.36})$ $(Mn^{4+}_{4.28}Mn^{2+}_{0.56}R_{0.18})O_{10}$ (approximately $A_2Mn_5O_{10}$). This structure has both double and triple edge-sharing chains of octahedra, in turn joined through vertex-sharing, to yield a monoclinic unit cell with $a{\sim}9.6$ and $c{\sim}13.9$ Å. WADSLEY referred to this material as psilomelane, but MUKHERJEE (1965) postulates that psilomelane has the composition $A_3X_6Mn_8O_{16}$, where A=Ba, Mn, etc. and X=(O, OH). However, he notes that such material has approximately the same lattice parameters as WADSLEY's phase, except that b is doubled. It should therefore have the same general type of octahedron linkage and formula. FAULRING (1965) concludes that the naturally occurring mineral nsutite, $Mn_{1-x}Mn_xO_{2-2x}(OH)_{2x}$, commonly called γ-MnO_2, is hexagonal with $a=9.65$ and $c=4.43$ Å, as contrasted with previous results which showed that it was orthorhombic with a unit cell dimensionally equivalent to that of ramsdellite (BYSTRÖM, SR 1949, 152).

Groutite, α-MnOOH, also has the γ-MnO_2 structure and is therefore isotypic with goethite and diaspore (DENT, GLASSER and INGRAM, 1968). Manganite, γ-MnOOH, has a structure which is derivative to that of marcasite (DACHS, 1963).

V. Complex Oxides

Manganese forms a large number of hydrated and anhydrous sulphates, carbonates, phosphates, arsenates, etc., in which Mn^{2+} is octahedrally coordinated by O, OH and/or H_2O such that there is extensive solid solution with cations such as Fe^{2+}, Mg, Co, and Ni. MOORE (1965) has proposed a classification of hydrated phosphates and arsenates containing isolated tetrahedra (all naturally occurring species) based on the nature of the vertex, edge and face linkages of Fe,Mn octahedra, and considering the varying roles of H_2O and OH in such structures. The structures of the manganese phosphates laueite, stewartite, strunzite, and eosphorite are considered relative to this general framework, as well as those of iron phosphates with no Mn.

The primary pegmatite minerals triphylite-lithiophilite, Li(Fe,Mn)PO_4, have the olivine structure (GELLER and DURAND, SR 1960, 399) and it is through the hydrothermal alteration and oxidation of these phases that a variety of Fe and Mn hydrated secondary phosphates are formed. Oxidation of iron and manganese with simultaneous leaching of Li produces the minerals sicklerite, $Li_{<1}(Mn^{2+}, Fe^{3+})PO_4$ to purpurite (Mn^{3+}, Fe^{3+})PO_4, which have the olivine structure defect in Li. MROSE and APPLEMAN (1961) have shown that the primary pegmatite phosphate sarcopside, (Fe,Mn,Ca)$_3$(PO$_4$)$_2$ also has a defect olivine structure. Natrophilite, NaMnPO$_4$, was shown to be isotypic with olivine, but with Na and Mn disordered over the octahedral M(1) and M(2) sites (BYSTRÖM, SR 1943, 236). In these and other phosphate minerals, Mn^{2+} has the usual octahedral coordination, but in graftonite,

(Fe, Mn, Ca, Mg)$_3$ (PO$_4$)$_2$, which is isostructural with Mn$_3$(PO$_4$)$_2$, Ca has pentagonal bipyramidal coordination, with Mn, Fe occupying sites with square pyramidal and trigonal bipyramidal coordination (CALVO, 1968).

VI. Silicates

Manganese occurs primarily as Mn^{2+} in the rock-forming silicates and, as noted above, since this ion has no CFSE its structural role is controlled principally by local electrostatic charge and ionic radius. Relative to oxygen, the radius ratio falls almost midway in the octahedral range. The structures of the common rock-forming olivines, pyroxenes, amphiboles and layer silicates are based on oxygen closest-packing with Si, Al in tetrahedral sites and Mg, Fe^{2+}, Fe^{3+}, Al, Ca, Mn^{2+}, Co, etc. on octahedral sites, and thus manganese occurs both as a major or minor substitutional constituent of these phases. The octahedral sites are somewhat distorted in these structures, producing polyhedra, such as M(2) of pyroxene, which may have eight-fold coordination. Relative to Mg and Fe^{2+}, on the basis of relative ionic radius, Mn^{2+} should occupy that site in these structures consistent with the largest average bond length. In olivine this is the M(2) site, in amphibole the M(4) site and in pyroxene the M(2) site. These are the sites also preferred by the large Ca ion relative to Mg and Fe^{2+}. BANCROFT et al. (1967) have shown that Mn^{2+} is in fact concentrated on the larger M(2) site of pyroxene. On the other hand, if only Ca and Mn^{2+} proxy for the various sites, Mn^{2+} should preferentially order on those sites with a shorter average bond length, such as M(1) of olivine and pyroxene. Thus glaucochroite, CaMnSiO$_4$, should have Mn on the M(1) site, while for johannsenite, CaMnSi$_2$O$_6$, FREED and PEACOR (1967) have verified the ordering of Mn on M(1). The situation is more complex, of course, if Mn is proxying with some cation which has a stabilization energy relative to one of the distorted sites, or where trivalent cations affect local charge balance.

The role of Mn^{2+} in silicates is exemplified by those pyroxenes and pyroxenoids with essential Mn. All of these phases have pyroxene-like structures; planes of octahedrally coordinated 2+ cations alternate with planes of tetrahedrally coordinated Si between planes of oxygen atoms arranged in distorted closest-packing. The octahedra share edges to form infinite bands parallel to SiO$_3$ chains, and these units have the tetrahedral apical oxygen atoms in common such that apical oxygen atoms from adjoining tetrahedra define an octahedron edge. The geometry of the chain of tetrahedra is thus controlled by the magnitudes of such octahedron edges and therefore by the relative radii and distributions of cations on the octahedral sites. In the pyroxene johannsenite, CaMnSi$_2$O$_6$, the chain translation-repeat unit is two tetrahedra; in bustamite, CaMnSi$_2$O$_6$, it is three tetrahedra (PEACOR and BUERGER, 1962); in rhodonite, Mn$_4$(Mn, Ca)Si$_5$O$_{15}$, it is five tetrahedra (PEACOR and NIIZEKI, 1963); in pyroxmangite, (Fe, Mn, Mg)SiO$_3$, it is seven tetrahedra (LIEBAU, SR 1959, 476). In all cases Mn^{2+} is ordered relative to Ca, and occupies central sites within the octahedron band with minimum octahedron distortion. Sites at the band edges are subject to distortion, however, and are preferred by the larger Ca atom. In rhodonite, M(5) is therefore principally Ca, but phases may be synthesized with the composition Mn$_5$Si$_5$O$_{15}$ (LIEBAU et al., SR 1958, 505). The limited substitution between Ca and Mn in these structures is shown by the fact that in bustamite, Ca and Mn are each ordered over two octahedral sites, and that wollastonite, $CaSiO_3$, which is structurally closely related, exhibits almost no solid solution with Mn in natural samples. On the other hand, serandite $(Mn\ Ca)_2NaH(SiO_3)_2$, is isotypic with pectolite, $Ca_2NaH(SiO_3)_2$, which also has a structure similar to that of bustamite.

Mn^{2+} may also have 4, 5, and 8 fold coordination in silicates. LIND and GELLER (1969) have verified that Mn occupies the dodecahedral, 8-fold site in synthetic $Mn_3Fe_2Ge_3O_{12}$, garnet, with an average Mn—O bond length of 2.362 Å, while MANNING (1967) has studied the optical absorption spectrum for Mn^{2+} in this site in the garnet spessartite, $Mn_3Al_2Si_3O_{12}$. GELLER (1967) noted in a review of garnet crystal chemistry, that depending on the nature of the other cations proxying for sites, Mn^{2+} may occupy either the dodecahedral or octahedral sites, while Mn^{3+} occupies the octahedral site. STRENS (1965) pointed out that the latter substitution is unlikely in view of the nature of the Jahn-Teller distortion for this ion. A high coordination number for Mn^{2+} is also required in the synthetic feldspar $MnAl_2Si_2O_8$ where Mn substitutes for Ca and Na (EBERHARD, 1962). Mn^{2+} is tetrahedrally coordinated in helvite, $Mn_4(BeSiO_4)_6S$, which is isotypic with sodalite, $Na_4(AlSiO_4)_6Cl$. Mn is on a triad axis with 3 Mn—O bonds of 2.074 Å and 1 Mn—S bond of 2.432 Å. The Mn site has approximately 20% Fe^{2+} and Zn, however, for the refined structure (HOLLOWAY, unpublished). The M(4) site of rhodonite approximates to five-fold coordination, rather than six, since the average of five bond lengths is 2.15 Å while a sixth is 2.88 Å. Some Mg and Fe are assigned to this site, however.

Mn^{3+} may substitute for other cations on octahedral sites in silicates, but the nature of the Jahn-Teller distortion is a determining factor. This ion will favor distorted sites due to a crystal field stabilization energy relative to regular octahedra, where such sites are defined by other cations. Thus epidote may have appreciable Mn^{3+} and, as such, is the mineral piemontite. Three of the five cation sites in epidote are Al octahedral sites. BURNS and STRENS (1967) showed that the piemontite absorption spectrum was consistent with most of the Mn^{3+} being on the M(3) site, which is the one with maximum distortion. DOLLASE (1969) verified this conclusion with a refinement of the piemontite structure, finding Mn,Al—O bond lengths of 1.861 to 2.274 Å. Contrary to the conclusion of BURNS and STRENS, he showed that the additional Mn^{3+} occupies M(1) rather than M(2) sites. STRENS (1966) has noted that the M(3) site distortion is an example of Jahn-Teller distortion with octahedron compression.

References: Section 25-A

ABRAHAMS, S. C., REDDY, J. M.: Crystal structure of the transition-metal molybdates. I. Paramagnetic alpha-$MnMoO_4$. J. Chem. Phys. 43, 2533 (1965).
— BERNSTEIN, J. L.: Crystal structure of paramagnetic $DyMn_2O_5$ at 298° K. J. Chem. Phys. 46, 3776 (1967).
BANCROFT, G. M., BURNS, R. G., HOWIE, R. A.: Determination of the cation distribution in the orthopyroxene series by Mossbauer effect. Nature 213, 1221 (1967).
BAUR, W. H.: Über die Verfeinerung der Kristallstrukturbestimmung einiger Vertreter des Rutiltyps. II. Die Difluoride von Mn, Fe, Co, Ni und Sn. Acta Cryst. 11, 488 (1958).
BURNS, R. G., STRENS, R. G. J.: Structural interpretation of polarized absorption spectra of the Al—Fe—Mn—Cr epidotes. Mineral. Mag. 36, 204 (1967).
CALVO, C.: The crystal structure of graftonite. Am. Mineralogist 53, 742 (1968).
COTTON, F. A., WILKINSON, G.: Advanced Inorganic Chemistry. New York: Interscience Publishers 1967.
DACHS, H.: Neutronen- und Röntgenuntersuchungen am Manganite, MnOOH. Z. Krist. 118, 303 (1963).
DOLLASE, W. A.: Crystal structure and cation ordering of piemontite. Am. Mineralogist 54, 711 (1969).
DORM, E., MARINDER, B.: Thortveitite-type structure of $Mn_2V_2O_7$. Acta Chem. Scand. 21, 590 (1967).
DUNN, T. M., McCLURE, D. S., PEARSON, R. G.: Some Aspects of Crystal Field Theory. New York: Harper and Row 1965.
EBERHARD, E.: The structure of $MnAl_2Si_2O_8$ (manganese feldspar). Fortschr. Mineral. 40, 52 (1962).
FAULRING, G. M.: Unit cell determination and thermal transformations of nsutite. Am. Mineralogist 50, 170 (1965).
FERT, A.: Structure de quelques oxydes de terres rares. Bull. Soc. Franc. Mineral. Crist. 85, 267 (1962).
FREED, R. L., PEACOR, D. R.: Refinement of the crystal structure of johannsenite. Am. Mineralogist 52, 709 (1967).
GAUDEFROY, C., GRANGER, M., PERMINGEAT, F. PROTAS, J.: La despujolsite, une nouvelle espèce minérale. Bull. Soc. Franc. Mineral. Crist. 91, 43 (1968).
GELLER, S.: Crystal chemistry of the garnets. Z. Krist. 125, 1 (1967).
—CAPE, J. A., GRANT, R. W., ESPINOSA, G. P.: Distortion in the crystal structure of α-Mn_2O_3. Phys. Letters 24A, 369 (1967).
GLASSER, L., DENT, S., INGRAM, L.: Refinement of the crystal structure of groutite, α-MnOOH. Acta Cryst. 24B, 1233 (1968).
HARDY, A.: Structures cristallines de deux variétés allotropiques de manganite de baryum. Nouvelle structure ABO_3. Acta Cryst. 15, 179 (1962).
— PEREZ, G., SERMENT, J.: Crystal structure of manganese orthothiosilicate and orthothiogermanate. Bull. Soc. Chim. France 1965, 2638 (1965).
HASE, W.: Kristallstrukturuntersuchungen an Sesquioxiden mit α-Mn_2O_3-Struktur. Z. Krist. 124, 428 (1967).
— BRÜCKNER, W.: Untersuchungen zur Kristallstruktur von $(Mn_{1-x}Fe_x)_2O_3$. Z. Krist. 129, 360 (1969).
HOLLOWAY, W.: Refinement of the crystal structure of helvite (Unpublished) (1969).
JENSEN, S. J.: The crystal structures of $Cs_2MnCl_4 \cdot 2H_2O$ and $Rb_2MnCl_4 \cdot 2H_2O$. Acta Chem. Scand. 18, 2085 (1964).

© Springer-Verlag Berlin · Heidelberg 1972

JENSEN, S. J.: The crystal structure of $KMnCl_3 \cdot 2H_2O$. Acta Chem. Scand. 22, 641 (1968).
— The crystal structure of $K_2MnCl_4 \cdot 2H_2O$. Acta Chem. Scand. 22, 647 (1968).
— ANDERSON, P., RASMUSSEN, S. E.: The crystal structure of $CsMnCl_3 \cdot 2H_2O$. Acta Chem. Scand. 16, 1890 (1962).
KEVE, E. T., ABRAHAMS, S. C., BERNSTEIN, J. L.: Crystal structure of piezoelectric paramagnetic barium manganese fluoride. (abs.) Prog. Ann. Meet. Am. Cryst. Assoc., 61 (1969).
LAJZÉROWICZ, J.: Étude par diffraction des rayons X et absorption infra-rouge de la barysilite, $MnPb_8 \cdot 3Si_2O_7$, et composes isomorphes. Acta Cryst. 20, 357 (1965).
LIND, M. D., GELLER, S.: Crystal structure of the garnet $Mn_3Fe_2Ge_3O_{12}$. Z. Krist. 129, 427 (1969).
MANNING, P. G.: The optical absorption spectra of the garnets almandine-pyrope, pyrope and spessartite and some structural interpretations of mineralogical significance. Can. Mineralogist 9, 237 (1967).
MONTGOMERY, H., CHASTAIN, R. V., LINGAFELTER, E. C.: The crystal structure of Tutton's Salts V. Manganese ammonium sulfate hexahydrate. Acta Cryst. 20, 731 (1966).
MOORE, P. B.: The crystal structure of laueite, $Mn^{2+}Fe_2^{3+}(OH)_2(PO_4)_2(H_2O)_6 \cdot 2H_2O$. Am. Mineralogist 50, 1884 (1965).
— A structural classification of Fe—Mn orthophosphate hydrates. Am. Mineralogist 50, 2052 (1965).
— Crystal chemistry of the basic manganese arsenate minerals. I. The crystal structures of flinkite, $Mn_2^{2+}Mn^{3+}(OH)_4(AsO_4)$, and retzian, $Mn_2^{2+}Y^{3+}(OH)_4(AsO_4)$. Am. Mineralogist 52, 1603 (1967).
— Crystal chemistry of the basic manganese arsenate minerals. II. The crystal structure of allactite. Am. Mineralogist 53, 733 (1968).
— The crystal structure of chlorophoenicite. Am. Mineralogist 53, 1110 (1968).
MOROSIN, M., GRAEBER, E. J.: Crystal structures of manganese (II) and iron (II) chlorine dihydrate. J. Chem. Phys. 42, 898 (1965).
MROSE, M. E., APPLEMAN, D. E.: Crystal structure and crystal chemistry of sarcopside $(Fe,Mn,Ca)_3(PO_4)_2$. Prog. Ann. Meet. Am. Cryst. Assoc., 18 (1961).
— — The crystal structures and crystal chemistry of väyrynenite, $(Mn,Fe)Be(PO_4)(OH)$, and euclase, $AlBe(SiO_4)(OH)$. Z. Krist. 117, 16 (1962).
MUKHERJEE, B.: Crystallography of psilomelane, $A_3X_6Mn_8O_{16}$. Mineral. Mag. 35, 643 (1965).
MUMME, W. G.: The structure of $Na_4Mn_4Ti_5O_{18}$. Acta Cryst. 24B, 1114 (1968).
NIIZEKI, N., WACHI, M.: The crystal structures of $Bi_2Mn_4O_{10}$, $Bi_2Al_4O_9$ and $Bi_2Fe_4O_9$. Z. Krist. 127, 173 (1968).
NORRESTAM, R.: α-Manganese (III) oxide — a C-type sesquioxide of orthorhombic symmetry. Acta Chem. Scand. 21, 2871 (1967).
OSWALD, H. R., WAMPETICH, M. J.: Die Kristallstrukturen von Mn_5O_8 und $Cd_2Mn_3O_8$. Helv. Chim. Acta 50, 2023 (1967).
PEACOR, D. R., BUERGER, M. J.: Determination and refinement of the crystal structure of bustamite, $CaMnSi_2O_6$. Z. Krist. 117, 331 (1962).
— NIIZEKI, N.: The redetermination and refinement of the crystal structure of rhodonite $(Mn,Ca)SiO_3$. Z. Krist. 119, 98 (1963).
RENTZEPERIS, P. J.: The crystal structure of hodgkinsonite $Zn_2Mn[(OH)_2/SiO_4]$. Z. Krist. 119, 117 (1963).
ROUSE, R.: Refinement of the crystal structure of quenselite (Unpublished) (1969).
SHANNON, R. D., PREWITT, C. T.: Effective ionic radii in oxides and fluorides. Acta Cryst. B25, 925 (1969).
SRIKTANA, S., SEQUEIRA, A., CHIDAMBARAM, R.: Neutron diffraction study of the space group and structure of manganese-leonite, $K_2Mn(SO_4)_2 \cdot 4H_2O$. Acta Cryst. B24, 1176 (1968).

STRENS, R. G. J.: Instability of the garnet $Ca_3Mn_2'''Si_3O_{12}$ and the substitution $Mn''' \rightleftharpoons Al$. Mineral. Mag. 35, 547 (1965).
— The axial-ratio-inversion effect in Jahn-Teller distorted ML_6 octahedra in the epidote and perovskite structures. Mineral. Mag. 35, 777 (1966).
WADSLEY, A. D., REID, A. F.: The high pressure form of Mn_2GeO_4, a member of the olivine group. Acta Cryst. B24, 740 (1968).
WALITZI, E. M.: Die Kristallstruktur von Denningit, $(Mn,Ca,Zn)Te_2O_5$. Ein Beispiel für die Koordination um vierwertiges Tellur. Mineral. u. Petrog. Mitt. 10, 241 (1965).
WILL, G.: The crystal structure of $MnSO_4$. Acta Cryst. 19, 854 (1965).
WILSON, R. H., KASPER, J. S.: The crystal structure of MnAs above 40 °C. Acta Cryst. 17, 95 (1964).
YAKEL, H. L., KOEHLER, W. C.: On the crystal structure of the manganese (III) trioxides of the heavy lanthanides and yttrium. Acta Cryst. 16, 957 (1963).
ZALKIN, A., FORRESTER, J. D., TEMPLETON, D.: The crystal structure of manganese dichloride tetrahydrate. Inorg. Chem. 3, 529 (1964).
— LEE, K., TEMPLETON, D.: Crystal structure of $CsMnF_3$. J. Chem. Phys. 37, 697 (1962).

Revised manuscript received: January 1970

Copper 29

A   J. Zemann   (Institut für Mineralogie und Kristallographie der Universität Wien, Austria)

B—O   K. H. Wedepohl   (Geochemisches Institut der Universität, Göttingen, Germany)

29-A. Crystal Chemistry

Due to its electron configuration the stereochemistry of copper shows many unusual features. As a consequence, Cu-compounds often crystallize in structure types which do not occur — or only rarely — with other elements. The more general aspects of the crystal chemistry of this element are found in the textbooks on crystal chemistry, inorganic chemistry, and chemical bonds. For Cu^{2+} see especially WELLS (1949), ORGEL and DUNITZ (1957), and PROUT (1969). For an earlier review article stressing the mineralogical aspects of the stereochemistry of this element the reader is referred to ZEMANN (1961).

I. Metallic Copper and Alloy-Like Phases

Native copper crystallizes in the cubic close-packed structure (BRAGG, SB 1913-28, 35) with a Cu—Cu distance of 2.56 Å (12 ×). Although the Cu-atoms in this structure can easily be replaced by Ag and Au, generally natural native copper is essentially free of these elements.

Of the alloy-like phases, those in the system Cu—As are of special interest here. In native copper, several percent of the atoms can be replaced by As (whitneyite). Cu_3As exists in two modifications of which one, α-domeykite, has not yet been synthesized. Considering Cu—As distances up to 2.50 Å and Cu—Cu distances up to 2.80 Å, the structural formulae read (according to the structures given by STEENBERG, SB 1938, 65; the limits of bond lengths used here are somewhat arbitrary; modern confirmations and refinements of the two structures are needed; SKINNER and LUCE, 1971, disagree with the space group given by STEENBERG for β-domeykite and for this phase give the composition $Cu_{2.7}As$):

α-domeykite: $Cu_3^{[1\,As,\,6\,Cu]}As^{[3\,Cu]}$ c;
β-domeykite: $Cu^{[6\,Cu]}Cu_2^{[3\,As,\,9\,Cu]}Cu_3^{[2\,As,\,12\,Cu]}As_2^{[12\,Cu]}$ h.

There exist some additional copper arsenide minerals generally with unknown structures. Very well known, however, is the atomic arrangement in lautite, $Cu^{[3\,S,\,1\,As]}As^{[Cu,\,2\,As,\,1\,S]}S^{[3\,Cu,\,1\,As]}$, with Cu—As = 2.42 Å (MARUMO and NOWACKI, 1964; see also HÖHNE and KULPE, 1959; CRAIG and STEPHENSON, 1965).

II. Sulfides and Sulfosalts

In sulfides and sulfosalts, Cu is usually coordinated by four S-atoms in the form of a tetrahedron (Cu—S $\sim$ 2.35 Å; cf. NOWACKI, 1969), or by three S-atoms to form the center of an equilateral triangle (with Cu—S $\sim$ 2.25 Å; cf. NOWACKI, 1969). Rather strongly distorted 4-coordinations with possible transitions to lower coordination numbers, have been found in anilite and hodrushite (cf. bond lengths around Cu(1) in both minerals, Table 29-A-2), and in wallisite, $PbTlCu[As_2S_5]$ (Cu—S = 2.22, 2.35, 2.41, 2.80 Å; TAKÉUCHI et al., 1968, 2-dim. structure determination). A 2-coordination possibly occurs in high-chalcocite, Cu_2S (SADANAGA et al., 1965, especially

p. 288) and in low-chalcocite (see below). Octahedrally coordinated copper has been reported for an artificially formed, high-pressure phase of CuS_2 with a pyrite-type structure (MUNSON, 1966), for fukuchilite, $(Cu_3, Fe)S_8$, also of pyrite-type (KAJIWARA, 1969), and for a part of the Cu-atoms in the very approximately known structure of digenite (DONNAY et al., SR 1958, 111). BELOV and POBEDIMSKAYA (1969) have postulated from theoretical reasoning that Cu in sulfides and sulfosalts is restricted to occur in tetrahedral 4-coordination and 2-coordination only; they have, however, not proved that all structures in which Cu has been reported to have another coordination (especially 3-coordination) are wrong.

There exist many Cu-sulfides (Table 29-A-1). Despite many efforts, only parts of their crystal structures are accurately known. At higher temperatures of more than about 100° C, Cu-rich phases have a strong tendency for structures with Cu occupying positions of statistically high multiplicity, and high mobility of Cu (cf. e.g. RAHLFS, SB 1936, 109). At room temperatures the structures are ordered.

The complicated structure of monoclinic low-chalcocite, Cu_2S, has been described in a first approximation as a hexagonal close packing of S-atoms, with all Cu-atoms in 3-coordinated interstices (EVANS, 1968). The detailed list of interatomic distances at an intermediate state of refinement (EVANS, private communication 1969) shows, however, that 80% of the Cu—S distances measure 2.31 ± 0.06 Å, while some are much larger. To obtain 3-coordination for all 24 species of crystallographically independent Cu-atoms it would be necessary to allow for two of them each one S-neighbor at a distance of 2.85 ± 0.04 Å. It seems therefore justified at this stage to attribute to these two Cu-atoms only two S-neighbours[1].

In the well known structure of anilite, Cu_7S_4, the arrangement of the S-atoms is approximately cubic close-packing. Neglecting Cu—S distances ≥ 2.36 Å (cf. Table 29-A-2), the chemical formula reads:

$$Cu^{[2]}Cu_5^{[3]}Cu^{[4]}S_3^{[5]}S^{[6]};$$

paying attention, however, to the Cu—S distances of 2.52 Å (then only Cu-distances follow with Cu—S ≥ 2.94 Å, while the smallest Cu—Cu distances measure 2.64 Å) it reads:

$$Cu^{[2+2]}Cu_5^{[3]}Cu^{[4]}S^{[5]}S_2^{[5+1]}S^{[6]}.$$

The coordination numbers given above consider only neighbors of the different species of atoms; it is to be stressed, however, that part of the Cu—Cu distances in anilite are only a few percent larger than in metallic copper.

Covellite, CuS, is very remarkable because, despite its simple bulk composition, Cu is partly in 3- and partly in 4-coordination; a high proportion of the S-atoms forms S_2-pairs (Fig. 29-A-1). The crystal chemical formula reads $Cu_2^{[4]}Cu^{[3]}[S_2]S$ h.

Of similar complexity are many of the ternary phases in the system Cu—Fe—S which has been carefully investigated up to 700°C (ROSEBOOM, 1966; YUND and KULLERUD, 1966). By far the most common copper sulfide in nature is chalcopyrite with the simple formula $Cu^{[4\,S]}Fe^{[4\,S]}S_2^{[2\,Cu,\,2\,Fe]}$ t, a structure which can be derived from that of sphalerite (see Table 29-A-2, for the magnetic structure see DONNAY et al., 1958). Chalcopyrite is an n-type semiconductor at room temperature (FRUEH, 1959). At temperatures $\geq$ ca. 550°C disorder occurs between Cu and Fe (HILLER and PROBSTHAIN, 1957). Phases with a composition close to chalcopyrite, but somewhat

---

[1] A refinement by EVANS [Nature. Phys. Sci. 232, 69 (1971)] has essentially confirmed the results given here.

Table 29-A-1. *Data on copper-sulfur phases*

Composition	Name	Approx. arrangement of S-atoms	Cu-coordination	Reference
Cu_2S	high-chalcocite ($>103°$ C)	hexagonal dense	4, 3, 2(?), partly occupied positions	BUERGER and WUENSCH (1963), WUENSCH and BUERGER (1963), SADANAGA et al. (1965)
	low-chalcocite ($<103°$ C)	hexagonal dense	essentially 3 (see text)	EVANS (1968)
$Cu_{1.97}S$	djurleite	hexagonal dense	unknown	TAKEDA et al. (1967)
$Cu_{1.96}S$	artificial	cubic dense	3	JÁNOSI (1964)
$Cu_{1.80}S$ ($=Cu_9S_5$)	high-digenite[a] ($>73°$ C)	cubic dense	4, 3(?), 2(?); disordered	MORIMOTO and KULLERUD (1963)
	metastable form[b]	cubic dense	more order than in high-digenite	MORIMOTO and KULLERUD (1963) (for a proposal for atomic arrangement based on a twinned rhombohedral cell see DONNAY et al., SR 1958, 111)
	low-digenite[c]	cubic dense	unknown	MORIMOTO and KULLERUD (1963)
$Cu_{1.75}S$ ($=Cu_7S_4$)	anilite	cubic dense	4, 3, (2+2)	KOTO and MORIMOTO (1970)
$Cu_{1+x}S$ ($x \leq 0.4$)	blaubleibender covellite	close to covellite	4, 3	FRENZEL (1960, 1961), SILLITOE and CLARK (1969), OTTEMANN and FRENZEL (1971)
CuS	covellite	(cf. Fig. 29-A-1)	4, 3	BERRY (SR 1954, 380)
	artificial, unstable modification (?)	cubic dense	4 (sphalerite-type)	KAZINETS (1970)
CuS_2	artificial, high pressure	as in pyrite	6	MUNSON (1966)

[a] According to RAHLFS (SB 1936, 109) and KAZINETS (1970), the high temperature phase of this composition forms a sphalerite-type CuS-framework with a little less than 1 Cu near the centers of the remaining tetrahedral holes.

[b] This is the phase into which high-digenite changes when cooled below 60—65° C (DONNAY et al., SR 1958, 111).

[c] According to MORIMOTO (1970), natural digenite contains ~1% Fe, and Fe-free digenite seems not to be stable at room temperature.

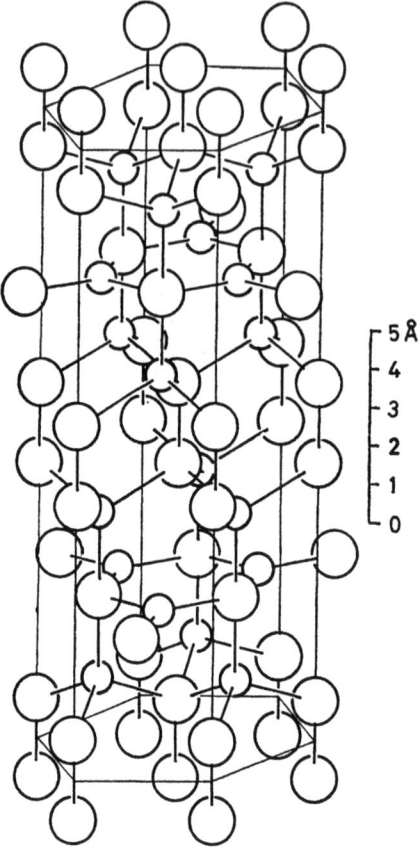

Fig. 29-A-1. Crystal structure of covellite, CuS. $a_0 = 3.80$ Å, $c_0 = 16.36$ Å; space group: P 6_3/m mc. Large circles: S, small circles: Cu

poorer in S have also been found [e.g. talnakhite, $Cu_9(Fe, Ni)_8S_{16}$, CABRI and HARRIS, 1971].

The atomic arrangement of cubanite, $Cu^{[4\,S]}Fe_2^{[4\,S]}S^{[2\,Cu,\,2\,Fe]}S_2^{[Cu,\,3\,Fe]}$, (Table 29-A-2) is partially similar to that in wurtzite, but two FeS_4-tetrahedra always share an edge (Fig. 29-A-2).

Highly remarkable, from the point of view of crystal chemistry, is valleriite, $\frac{2}{\infty}[Mg_{2/3}Al_{1/3}(OH)_2] \cdot \frac{2}{\infty}[(Cu, Fe)_2S_2]$, which — besides brucite-like layers — contains $\frac{2}{\infty}[Me_2^{[4]}S_2]$-layers with Cu:Fe ∼ 1:1 (Fig. 29-A-3) in which each $(Cu, Fe)S_4$-tetrahedron shares three edges with neighboring tetrahedra (EVANS and ALLMANN, 1968).

In the vast majority of natural pyrites Cu does not substitute Fe appreciably as can be explained by the JAHN-TELLER theorem (RADCLIFFE and McWEEN, 1969, 1970). However, the pyrite-like mineral Cu_3FeS_8, fukuchilite, has been found in nature (KAJIWARA, 1969) and has been synthesized at 225°C (SHIMIZAKI, 1969). Recently, SHIMIZAKI and CLARK (1971) have reported that the substitution of Fe by

Copper

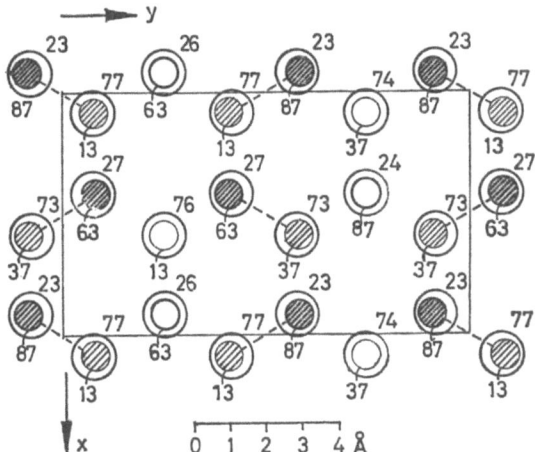

Fig. 29-A-2. Crystal structure of cubanite, $CuFe_2S_3$. $a_0=6.47$ Å, $b_0=11.12$ Å, $c_0=6.23$ Å; space group: P cmn. Projection parallel [001]. Large circles: S, small blank circles: Cu, small hatched circles: Fe. The height of the atoms is given in $c_0/100$. Broken lines give iron "pairs" (Fe—Fe = 2.82 Å) over common S—S edges

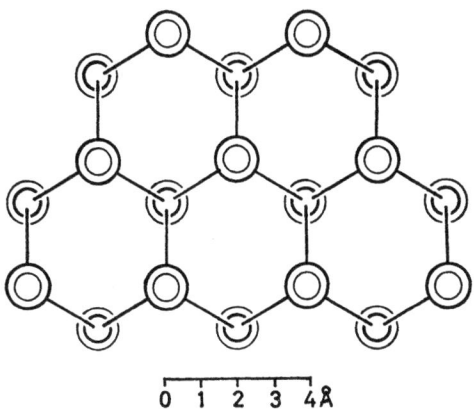

Fig. 29-A-3. $^2_\infty(Cu, Fe)^{[4]}_2 S^{[3]}_2$-layer of valleriite. Projection parallel [00.1]. Large circles: S, small circles: (Cu, Fe). The boldly and weakly drawn S-atoms differ in height by 3.14 Å; the (Cu, Fe)-atoms occupy the centers of the tetrahedra

Cu in pyrite is extensive at low temperatures (82% at 100°C) but that it decreases with rising temperatures to about 10% at 275°C.

Selected Cu—S bond lengths in sulfides and sulfosalts of well refined structures are given in Table 29-A-2.

III. Organic Complexes

In S-containing organic complexes, where the valence of copper is much better known than in the inorganic compounds previously discussed, Cu(I) shows a tetrahedral S-coordination [cf. tris (thiourea)copper(I)chloride, Table 29-A-3]. In

Table 29-A-2. *Selected Cu—S bond lengths (Å) in mineral structures*

Name, Formula	Cu—S bond lengths	Reference
Aikinite $CuPbBiS_3$	Cu—S = 2.31 = 2.35 (2 ×) = 2.40	OHMASA and NOWACKI (1970), see also KOHATSU and WUENSCH (1971) with very similar results
Anilite Cu_7S_4	Cu(1)—S = 2.29 = 2.32 = 2.52 (2 ×) Cu(2)—S = 2.30 = 2.31 (2 ×) Cu(3)—S = 2.30 = 2.33 (2 ×) = 2.35 Cu(4)—S = 2.28 = 2.33 (2 ×) Cu(5)—S = 2.24 (2 ×) = 2.35	KOTO and MORIMOTO (1970)
Binnite $Cu_{12}As_4S_{13}$	Cu(1)—S = 2.34 (4 ×) Cu(2)—S = 2.20 = 2.26 (2 ×)	WUENSCH et al. (1966)
Bournonite $CuPbSbS_3$	Cu—S = 2.29 = 2.34 (2 ×) = 2.41	EDENHARTER et al. (1970)
Chalcopyrite $CuFeS_2$	Cu—S = 2.32 (4 ×)	PAULING and BROCKWAY (SB 1928—32, 346), confirmation by BOON (SR 1945—46, 122) and DONNAY et al. (1958)
Cubanite $CuFe_2S_3$	Cu—S = 2.25 = 2.33 (2 ×) = 2.43	FLEET (1970)
Covellite CuS	Cu(1)—S = 2.19 (3 ×) Cu(2)—S = 2.30 (3 ×) = 2.34	BERRY (SR 1954, 380)
Enargite Cu_3AsS_4	Cu(1)—S = 2.30 = 2.35 = 2.38 (2 ×) Cu(2)—S = 2.31 = 2.32 (3 ×)	ADIWIDJAJA and LÖHN (1970)
Hodrushite $Cu_8Bi_{10}Bi_{2-x}Fe_xS_{22}$ (x~0.3—0.4)	Cu(1)—S = 2.27 = 2.33 (2 ×) = 2.58 Cu(2)—S = 2.28 = 2.34 (2 ×) = 2.46	KUPČÍK and MACKOVICKÝ (1968) for name and composition see KODĚRA et al. (1970)

Table 29-A-2 (Continued)

Name, formula	Cu—S bond lengths	Reference
Hodrushite (Continued)	Cu(3)—S = 2.30 = 2.35 = 2.36 (2 ×) Cu(4)—S = 2.26 = 2.34 (2 ×) = (2.94)	
Luzonite Cu$_3$($\underline{As}$, Sb)S$_4$	Cu(1)—S = 2.30 (4 ×) Cu(2)—S = 2.34 (4 ×)	Marumo and Nowacki (1967)
Seligmannite CuPbAsS$_3$	Cu—S = 2.26 = 2.33 = 2.34 = 2.38	Edenharter et al. (1970)
Sulvanite Cu$_3$VS$_4$	Cu—S = 2.30 (4 ×)	Trojer (1966)
Tetrahedrite Cu$_{12}$Sb$_4$S$_{13}$	Cu(1)—S = 2.34 (4 ×) Cu(2)—S = 2.23 = 2.27 (3 ×)	Wuensch (1964)

the same class of compounds, Cu(II) has been reported to be coordinated by four S in the form of a square (Cu—S $\sim$ 2.30 Å), apparently, as a rule, with a fifth S-neighbor at $\sim$2.8 Å, perpendicular to this plane [cf. copper(II)diethyldithiocarbamate, Bonamico et al., 1965]. This type of S-coordination has not yet been found in inorganic Cu-compounds, but it is very common for oxygen coordination of Cu^{2+}.

IV. Selenides and Tellurides

Copper-selenium minerals are sometimes similar to the corresponding sulfides; e.g. berzelianite, Cu$_2$Se, seems to have a structure similar to high-digenite. However, important differences also exist. For example no known copper sulfide corresponds to umangite, Cu$^{[4\,Se,\,4\,Cu]}$Cu$_2^{[4\,Se,\,3\,Cu]}$Se$_2^{[6\,Cu]}$ t (Cu-Se = 2.37—2.49 Å; Cu—Cu = 2.63—2.66 Å; Morimoto and Koto, 1966). The crystal structure of klockmannite, CuSe, is closely related to but not identical with that of covellite, CuS (Taylor and Underwood, SR 1960, 138; Lippmann, 1962; Elliott et al., 1969). Artificially formed CuSe$_2$ crystallizes with a marcasite-type structure (Gattow, 1965).

The stereochemistry of copper tellurides is, at least for the low temperature phases, radically different from that of the copper sulfides. For example, vulcanite, CuTe, has an orthorhombic, pseudo-tetragonal sheet structure $^2_\infty$Cu$^{[4\,Te,\,4\,Cu]}$ Te$^{[4\,Cu,\,2\,Te]}$ with Cu—Te = 2.63 Å (2 ×), 2.72 Å (2 ×), Cu—Cu = 2.64 Å (4 ×), Te—Te = 3.16 Å (2 ×) (Schubert et al., SR 1953, 158; cf. also SR 1954, 152). Rickardite, Cu$_{2-x}$Te (x $\sim$ 0.6), has an alloylike structure of the Cu$_2$Sb-type (Forman and Peacock, SR 1949, 158).

V. Oxides, Hydroxides, Oxygen-containing Salts etc.

a) Monovalent Copper

In halogenides, oxides, normal and basic salts of oxy-acids etc., the oxidation state of copper is much better known than in the chalcogenides. Contrary to artificial compounds, where 3-valent Cu is also known (e.g. in $^{1}_{\infty}$K[Cu$^{[4]}$O$_2$] with Cu^{3+} in planar 4-coordination; HESTERMANN and HOPPE, 1969), only mono- and divalent copper have been reported in minerals.

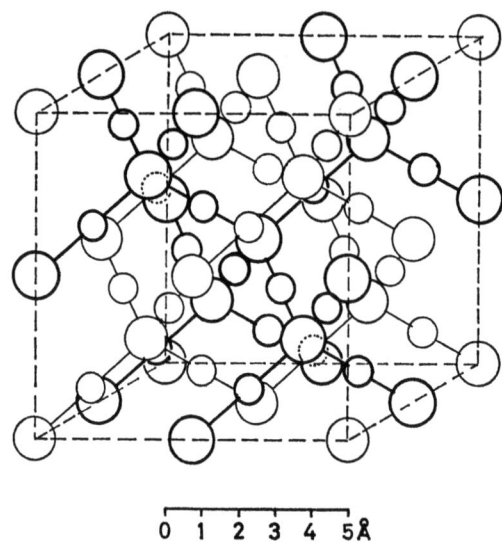

Fig. 29-A-4. Crystal structure of cuprite, Cu$_2$O. $a_0 = 4.27$ Å; space group: P n3m. Large circles: O, small circles: Cu. The bold and the weak parts of the structure have no close Cu—O bonds in common. The broken lines give a cube with edge length 2 a_0.

In cuprite, Cu$_2^{[2\,0]}$O$^{[4\,\text{Cu}]}$ (Fig. A-29-4), and some additional compounds, monovalent copper shows a linear 2-coordination; but a tetrahedral 4-coordination is also common (toward Cl etc.). In organic complexes 4-coordination and 3-coordination are known. Selected bond lengths for monovalent Cu are given in Table 29-A-3.

b) Divalent Copper

2-valent copper, which has nine 3d-electrons in the outer shell, shows in the vast majority of cases a square planar coordination (often slightly distorted), with Cu—O, Cu—OH, Cu—OH$_2$ distances of about 2.0 Å, whereas the Cu—Cl distance is about 2.3 Å etc.

Among minerals, e.g. tenorite, Cu$^{[4p]}$O$^{[4]}$ m (Table 29-A-4), and cuprorivaite, $^{2}_{\infty}$Ca$^{[8]}$Cu$^{[4p]}$[Si$_4$O$_{10}$] t, are examples of Cu in this coordination only. The same surrounding occurs for one third of the Cu-atoms in azurite, Cu$^{[4p2]}$Cu$_2^{[4pl+1]}$(OH)$_2$[CO$_3$]$_2$ m, with Cu(1)—O = 1.88 Å, Cu—OH = 1.98 Å (each 2×, error ~0.05 Å); additional neighbors are 2C at ca. 2.7 Å and 2O at ca. 2.9 Å (GATTOW and ZEMANN, SR 1958, 387; cf. Fig. 29-A-5). This kind of coordination can be explained by dsp^2 covalent bonds.

Table 29-A-3. *Coordination number and bond lengths around monovalent Cu (in Å) in minerals and selected artificial compounds (sulfides, selenides and tellurides excluded)*

Name, formula	Bond lengths	Reference
Cuprite $Cu_2^{[2]}O^{[4]}$	Cu—O = 1.85 (2 ×)	BRAGG and BRAGG (SB 1913—28, 222)
Crednerite $Cu^{[2]}Mn^{[4+2]}O_2$	Cu—O = 1.80 (2 ×)	KONDRAŠEV (SR 1958, 334)
Delafossite $Cu^{[2]}Fe^{[6]}O_2$	Cu—O = 1.90 (2 ×)	SOLLER and THOMPSON (SB 1933—35, 392), PABST (SR 1945—1946, 110)
Mcconnellite $Cu^{[2]}Cr^{[6]}O_2$	Cu—O = 1.85 (2 ×)	DANNHAUSER and VAUGHAN (SR 1955, 372)
$Cu^{[2]}La^{[6]}O_2$	Cu—O = 1.85 (2 ×)	HAAS and KORDES (1969)
Nantokite $Cu^{[4]}Cl^{[4]}$	Cu—Cl = 2.34 (4 ×)	WYCKOFF and POSNJAK (SB 1913—28, 110)
$Cu^{[4]}Br^{[4]}$	Cu—Br = 2.46 (4 ×)	WYCKOFF and POSNJAK (SR 1913—28, 110)
Marshite $Cu^{[4]}I^{[4]}$	Cu—I = 2.62 (4 ×)	WYCKOFF and POSNJAK (SB 1913—28, 110)
$^1_\infty K_2[Cu^{[4]}Cl_3]$	Cu—Cl = 2.31 = 2.32 = 2.43 (2 ×)	BRINK and MACGILLAVRY (SR 1949, 188)
$^1_\infty Cs[Cu_2^{[4]}Cl_3]$ [a]	Cu—Cl = 2.16 (2 ×) = 2.53 (2 ×)	BRINK et al. (SR 1954, 413)
$^1_\infty K[Cu^{[3]}(CN)_2]$	Cu—C = 1.92 (2 ×) —N = 2.05	CROMER (SR 1957, 260)
$^2_\infty K[Cu_2^{[3]}(CN)_3]$	Cu(1)—C = 1.89 (2 ×) —N = 1.98 Cu(2)—C = 1.87 —N = 1.96 = 2.02	CROMER and LARSON (1962)
$^2_\infty [Cu^{[4]}[NCCH_2CH_2CN)_2]\cdot ClO_4$	Cu—N = 1.99 (4 ×)	BLOUNT et al. (1969)
$Cu_2^{1+[18,30]}Cu^{2+[4+20]}[SO_3]_2\cdot 2H_2O$	Cu^{1+}—S = 2.14 —O = 2.11 = 2.13 = 2.14	KIERKEGAARD and NYBERG (1965)
$^1_\infty [Cu^{[4S]}(SCN_2H_4)_3]Cl$	Cu—S = 2.28 = 2.35 = 2.38 = 2.43	OKAYA and KNOBLER (1964)

[a] $^1_\infty[Cu^{[4]}Cl_3]^{1-}$-strings built from $[Cu^{[4]}Cl_4]^{3-}$-tetrahedra by edge sharing are linked by more shared edges to $^2_\infty[Cu_2^{[4]}Cl_3]^{1-}$-double strings. The Cu—Cl distances given are possibly not very accurate.

In a number of cases there occurs *one* additional neighbor at a larger distance approximately perpendicular to this plane. Azurite can again serve as an example. Two thirds of the Cu-atoms are here coordinated by 2O and 2OH at a distance of 1.97 ± 0.10 Å in the form of a square; one additional O-neighbor lies approximately perpendicular to this plane at 2.38 ± 0.05 Å, the next larger Cu-O distance measures 2.83 ± 0.05 Å (GATTOW and ZEMANN, loc. cit.; cf. Fig. 29-A-5). For further well established examples see Table 29-A-4. In many structures with this kind of Cu-coordination it has been found that Cu lies somewhat out of the "square" towards the fifth neighbor. A clear exception to this rule is $Cu_2^{[4pl+1]}[NO_3]_4 \cdot 5H_2O$ m, where Cu lies exactly in the plane of its four nearest neighbors (MOROSIN, 1970).

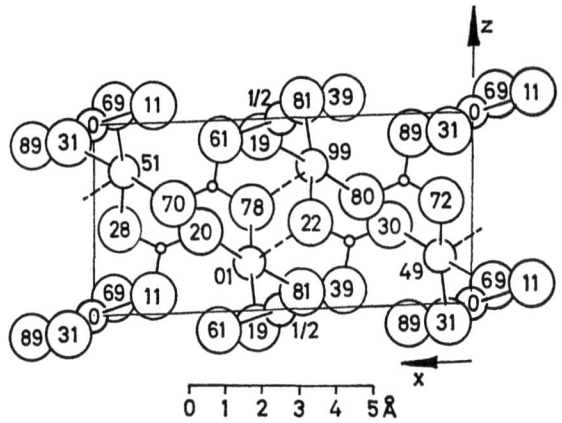

Fig. 29-A-5. Crystal structure of azurite, $Cu_3(OH)_2[CO_3]_2$. $a_0 = 10.35$ Å, $b_0 = 5.85$ Å, $c_0 = 5.00$ Å, $\beta = 92°20'$; space group: $P\,2_1/a$. Projection parallel [010]. Large circles: O (incl. OH), medium circles: Cu, small circles: C. The height of Cu and O is given in $b_0/100$. The Cu(2)—O distances of 2.38 Å are drawn as broken lines

Much more common, however, is a coordination in which *two* additional neighbors complete the square planar coordination to a coordination polyhedron in the form of an elongated tetragonal dipyramid, which is often slightly distorted. Examples are malachite, $Cu_2^{[4pl+2]}(OH)_2[CO_3]$ m, and chalcanthite, $Cu^{[4pl+2]}(H_2O)_4[SO_4] \cdot H_2O$ trcl; for bond lengths in these and other examples see Table 29-A-4. This coordination can be explained as a Jahn-Teller-distortion of an octahedron due to the electron configuration of Cu^{2+}. A deviation from the coordination "octahedron" around Cu^{2+} toward a somewhat flattened tetragonal dipyramid (also possible according to the Jahn-Teller theory) has been found in $K_2^{[9F]}Cu^{[2+4]}F_4$ t with Cu—F distances of 1.95 Å (2 ×) and 2.08 Å (4 ×) (KNOX, SR 1959, 290).

It seems worth mentioning that the hydroxide group has a strong tendency to enter the Cu-square coordination. No example of an hydroxysalt of Cu^{2+} is known where the oxygen of any OH-group does not belong to a square coordination around copper.

Examples with Cu^{2+} occurring in a trigonal dipyramidal 5-coordination are rare. A mineralogical example is dolerophanite, $Cu^{[4pl+2]}Cu^{[5]}O[SO_4]$ m, with Cu—O distances ranging between 1.87 and 2.14 Å for the 5-coordinated copper (FLÜGEL-KAHLER, 1963).

Table 29-A-4. *Selected Cu-O bond lengths (Å) in structures of minerals with divalent Cu*

Name, formula	Cu(II)-ligand distances	Reference
Antlerite[a] $Cu_3(OH)_4SO_4$	Cu(1)—OH = 1.98 (2 ×) = 1.99 (2 ×) Cu(2)—OH = 1.91 (2 ×) = 2.03 —O = 2.04 Cu(1)—OH = 2.32 —O = 2.57 Cu(2)—O = 2.36 = 2.41	Finney and Araki (1963)
Bonattite $CuSO_4 \cdot 3H_2O$	Cu—OH$_2$ = 1.96 = 1.97 = 1.98 —O = 1.94 Cu—O = 2.40 = 2.45	Zahrobsky and Baur (1968)
Chalcanthite $CuSO_4 \cdot 5H_2O$	Cu—OH$_2$ = 1.98 (2 ×) = 1.99 (2 ×) Cu—O = 2.38 (2 ×)	Bacon and Curry (1962)
Clinoclase $Cu_3(OH)_3AsO_4$	Cu(1)—OH = 1.91 (2 ×) —O = 1.97 = 2.07 Cu(2)—OH = 1.89 = 1.93 = 1.95 —O = 1.99 Cu(3)—OH = 1.90 —O = 1.98 = 1.99 = 2.02 Cu(1)—O = 2.31 Cu(2)—O = 2.51 Cu(3)—OH = 2.30	Ghose, Fehlmann and Sundaralingam (1965)
Cornetite $Cu_3(OH)_3PO_4$	Cu(1)—OH = 1.93 = 1.96 = 2.01 —O = 1.94 Cu(2)—OH = 1.93 = 1.96 = 2.04 —O = 1.96 Cu(1)—O = 2.49 Cu(2)—O = 2.66 = 2.74	Fehlmann, Ghose and Finney (1964)

[a] Distances recalculated from original parameters by the author.

Table 29-A-4 (Continued)

Name, formula	Cu(II)-ligand distances	Reference
Cornetite (Continued)	Cu(3)—OH = 1.95 = 1.97 —O = 1.93 = 2.06 Cu(3)—O = 2.28	
Eriochalcite $CuCl_2 \cdot 2H_2O$	Cu—OH_2 = 1.93 (2 ×) —Cl = 2.28 (2 ×) Cu—Cl = 2.91 (2 ×)	Peterson and Levy (SR 1957, 397)
Euchroite $Cu_2(OH)AsO_4 \cdot 3H_2O$	Cu(1)—OH = 1.94 = 2.04 —O = 1.96 = 1.98 Cu(2)—OH = 2.01 —O = 1.96 = 1.99 —OH_2 = 1.95 Cu(1)—OH_2 = 2.37 = 2.48 Cu(2)—O = 2.40 —OH_2 = 2.80	Finney (1966)
Kinoite $Cu_2Ca_2[Si_3O_{10}] \cdot 2H_2O$	Cu(1)—O = 1.94 (2 ×) —OH_2 = 1.94 (2 ×) Cu(2)—O = 1.97 (2 ×) —OH_2 = 1.96 (2 ×) Cu(1)—O = 2.45 (2 ×) Cu(2)—O = 2.24 (2 ×)	Laughon (1971)
Malachite $Cu_2(OH)_2CO_3$	Cu(1)—OH = 1.90 = 1.92 —O = 1.99 = 2.07 Cu(2)—OH = 1.92 (2 ×) —O = 2.04 = 2.12 Cu(1)—O = 2.52 = 2.63 Cu(2)—OH = 2.35 = 2.36	Süsse (1967)
Mitscherlichite $K_2CuCl_4 \cdot 2H_2O$	Cu—OH_2 = 1.97 (2 ×) —Cl = 2.29 (2 ×) Cu—Cl = 2.90 (2 ×)	Chidambaram et al. (1970)
Ransomite $CuFe_2[SO_4]_4 \cdot 6H_2O$	Cu—OH_2 = 1.96 (2 ×) = 2.01 (2 ×) Cu—O = 2.44 (2 ×)	Wood (1970)

Table 29-A-4 (Continued)

Name, formula	Cu(II)-ligand distances	Reference
Tenorite CuO	Cu—O = 1.95 (2 ×) —O = 1.96 (2 ×) Cu—O = 2.78 (2 ×)	Åsbrink and Norrby (1970)
Turquoise $CuAl_6(OH)_8[PO_4]_4 \cdot 4H_2O$	Cu—OH = 1.92 (2 ×) = 2.11 (2 ×) Cu—OH$_2$ = 2.42 (2 ×)	Cid-Dresdner (1965)

The coordination "octahedron" around Cu^{2+} in $(NH_4)_2Cu(H_2O)_6[SO_4]_2$ is relatively weakly distorted with Cu—(OH)$_2$ distances of 1.96 Å (2 ×), 2.08 Å (2 ×) and 2.21 Å (2 ×) (Montgomery and Lingafelter, 1966; Brown and Chidambaram, 1969). Undistorted octahedral coordination around Cu^{2+} has been reported for $PbCu_6O_8$ (murdochite, Christ and Clark, SR 1955, 398) and $K_2Pb[Cu(NO_2)_6]$ (Isaacs et al., 1967). Unfortunately, the R-values given for both structure determinations are not very low by present standards so that a careful confirmation of these interesting results seems to be desired.

Manuscript received: August 1971

References: Section 29-A

ADIWIDJAJA, G., LÖHN, J.: Strukturverfeinerung von Enargit, Cu_3AsS_4. Acta Cryst. B **26**, 1878 (1970).

ÅSBRINK, S., NORRBY, L.-J.: A refinement of the crystal structure of copper(II)oxide with a discussion of some exceptional E.s.d.'s. Acta Cryst. B **26**, 8 (1970).

BACON, G. E., CURRY, N. A.: The water molecules in $CuSO_4 \cdot 5H_2O$. Proc. Roy. Soc. (London), Ser. A **266**, 95 (1962).

BELOV, N. V., POBEDIMSKAYA, E. A.: Covellite (klockmannite) — chalcocite, (acanthite, stromeyerite, bornite) — fahlerz. Soviet Phys.-Cryst. **13**, 843 (1969).

BLOUNT, J. F., FREEMAN, H. C., HEMMERICH, P., SIGWART, C.: The crystal structure of bis (succinodinitrile)copper(I)perchlorate. Acta Cryst. B **25**, 1518 (1969).

BONAMICO, M., DESSY, G., MUGNOLI, A., VACIAGO, A., ZAMBONELLI, L.: Structural studies of metal dithiobarbamates. II. The crystal and molecular structure of copper diethyldithiocarbamate. Acta Cryst. **19**, 886 (1965).

BROWN, G. M., CHIDAMBARAM, R.: The structure of copper ammonium sulfate hexahydrate from neutron-diffraction data. Acta Cryst. B **25**, 676 (1969).

BUERGER, M. J., WUENSCH, B. J.: Distribution of atoms in high chalcocite, Cu_2S. Science **141**, 276 (1963).

CABRI, L. J., HARRIS, D. C.: New compositional data for talnakhite, $Cu_{18}(Fe, Ni)_{16}S_{32}$. Econ. Geol. **66**, 673 (1971).

CHIDAMBARAM, R., NAVARRO, Q. O., GRACIA, A., LINGGOATMODJO, L., SHI-CHIEN, L., IL-HWAN SUH, SEQUEIRA, A., SRIKANTA, S.: Neutron diffraction refinement of the crystal structure of potassium copper chloride dihydrate, $K_2CuCl_4 \cdot 2H_2O$. Acta Cryst. B **26**, 827 (1970).

CID-DRESDNER, H.: Determination and refinement of the crystal structure of turquois, $CuAl_6(PO_4)_4(OH)_8 \cdot 4H_2O$. Z. Krist. **121**, 87 (1965).

CRAIG, D. C., STEPHENSON, N. C.: The crystal structure of lautite, CuAsS. Acta Cryst. **19**, 543 (1965).

CROMER, DON T., LARSON, A. C.: The crystal structure of $KCu_2(CN)_3 \cdot H_2O$. Acta Cryst. **15**, 397 (1962).

DONNAY, G., CORLISS, L. M., DONNAY, J. D. H., ELLIOTT, N., HASTINGS, J. M.: Symmetry of magnetic structures: Magnetic structure of chalcopyrite. Phys. Rev. **112**, 1917 (1958).

EDENHARTER, A., NOWACKI, W., TAKÉUCHI, Y.: Verfeinerung von Bournonit [$(SbS_3)_2$|$Cu_2^{IV}Pb^{VII}Pb^{VIII}$] und von Seligmannit [$(AsS_3)_2$|$Cu_2^{IV}Pb^{VII}Pb^{VIII}$]. Z. Krist. **131**, 397 (1970).

ELLIOT, J. A., BICKNELL, J. A., COLLINGE, R. G.: Twinning in the superlattice of CuSe, synthetic klockmannite. Acta Cryst. B **25**, 2420 (1969).

EVANS, H. T., Jr.: A crystal structure study of low chalcocite. Abstr. 1968 Meetings Geol. Soc. Am. 92 (1968).

— ALLMANN, R.: The crystal structure and crystal chemistry of valleriite. Z. Krist. **127**, 73 (1968).

FEHLMANN, M., GHOSE, S., FINNEY, J. J.: Direct determination of the crystal structure of cornetite, $Cu_3PO_4(OH)_3$, by the Monte Carlo method. J. Chem. Phys. **41**, 1910 (1964).

FINNEY, J. J.: Refinement of the crystal structure of euchroite, $Cu_2(AsO_4)(OH) \cdot 3H_2O$. Acta Cryst. **21**, 437 (1966).

— ARAKI, T.: Refinement of the crystal structure of antlerite. Nature **197**, 70 (1963).

FLEET, M. E.: Refinement of the crystal structure of cubanite and polymorphism of $CuFe_2S_3$. Z. Krist. **132**, 276 (1970).

Copper

FLÜGEL-KAHLER, E.: Die Kristallstruktur von Dolerophanit, $Cu_2O(SO_4)$. Acta Cryst. 16, 1009 (1963).
FRENZEL, G.: Idait und "blaubleibender Covellin". Neues Jahrb. Mineral. Abhandl. 93, 87 (1960).
— Der Cu-Überschuß des blaubleibenden Covellins. Neues Jahrb. Mineral. Monatsh. 1961, 199 (1961).
FRUEH, A.: The use of zone theory in problems of sulfide mineralogy. Part II. The resistivity of chalcopyrite. Am. Mineralogist 44, 1010 (1959).
GATTOW, G.: Die Kristallstruktur von $CuSe_2$. Z. Anorg. Allgem. Chem. 340, 312 (1965).
GHOSE, S., FEHLMANN, M., SUNDARALINGAM, M.: The crystal structure of clinoclase, $Cu_3AsO_4(OH)_3$. Acta Cryst. 18, 777 (1965).
HAAS, H., KORDES, E.: Cu^{1+}-haltige Doppeloxide mit seltenen Erdmetallen. Z. Krist. 129, 259 (1969).
HESTERMANN, K., HOPPE, R.: Die Kristallstruktur von $KCuO_2$, $RbCuO_2$ und $CsCuO_2$. Z. Anorg. Allgem. Chem. 367, 249 (1969).
HILLER, J.-E., PROBSTHAIN, K.: Thermische und röntgenographische Untersuchungen am Kupferkies. Z. Krist. 108, 108 (1957).
HÖHNE, E., KULPE, S.: Zur Struktur des Lautit. Monatsber. Deut. Akad. Wiss. Berlin 1, 283 (1959).
ISAACS, N. W., KENNARD, C. H. L.: A neutron-diffraction study of $K_2Pb[Cu(NO_2)_6]$, an example of a regular octahedral copper(II)complex. Chem. Commun. 587 (1967).
JÁNOSI, A.: La structure du sulfure cuivreux quadratique. Acta Cryst. 17, 311 (1964).
KAJIWARA, Y.: Fukuchilite, Cu_3FeS_8, a new mineral from the Hanawa mine, Akita prefecture, Japan. Mineral. J. (Japan) 5, 399 (1969).
KAZINETS, M. M.: On the structure of certain phases in the copper-sulfur system. Soviet Phys.-Cryst. 14, 599 (1970).
KIERKEGAARD, P., NYBERG, B.: The crystal structure of $Cu_2SO_3 \cdot CuSO_3 \cdot 2H_2O$. Acta. Chem. Scand. 19, 2189 (1965).
KODĚRA, M., KUPČÍK, V., MACKOVICKÝ, E.: Hodrushite — a new mineral. Mineral. Mag. 37, 641 (1970).
KOHATSU, I., WUENSCH, B. J.: The crystal structure of aikinite, $PbCuBiS_3$. Acta Cryst. B 27, 1245 (1971).
KOTO, K., MORIMOTO, N.: The crystal structure of anilite. Acta Cryst. B 26, 915 (1970).
KUPČÍK, V., MACKOVICKÝ, E.: Die Kristallstruktur des Minerals $(Pb, Ag, Bi)Cu_4Bi_5S_{11}$. Neues Jahrb. Mineral. Monatsh. 1968, 236 (1968).
LAUGHON, R. B.: The crystal structure of kinoite. Am. Mineralogist 56, 193 (1971).
LIPPMANN, F.: Zur Deutung der Überstruktur von Klockmannit, CuSe. Neues Jahrb. Mineral. Monatsh. 1962, 99 (1962).
MARUMO, F., NOWACKI, W.: The crystal structure of lautite and of sinnerite, a new mineral from the Lengenbach quarry. Schweiz. Mineral. Petrog. Mitt. 44, 439 (1964).
— — A refinement of the structure of luzonite, Cu_3AsS_4. Z. Krist. 124, 1 (1967).
MONTGOMERY, H., LINGAFELTER, E. C.: The crystal structure of Tutton's salts. III. Copper ammonium sulfate hexahydrate. Acta Cryst. 20, 659 (1966).
MORIMOTO, N.: Crystal-chemical studies of the Cu—Fe—S system, p. 323. In: T. TATSUMI (ed.), Volcanism and Ore Genesis. Tokyo: Univ. of Tokyo Press 1970.
— KOTO, K.: Crystal structure of umangite, Cu_3Se_2. Science 152, 345 (1966).
— KULLERUD, G.: Polymorphism in digenite. Am. Mineralogist 48, 110 (1963).
MOROSIN, B.: The crystal structure of $Cu(NO_3)_2 \cdot 2 \cdot 5 H_2O$. Acta Cryst. B 26, 1203 (1970).
MUNSON, R. A.: The synthesis of copper disulfide Inorg. Chem. 5, 1296 (1966).
NOWACKI, W.: Zur Klassifikation und Kristallchemie der Sulfosalze. Schweiz. Mineral. Petrog. Mitt. 49, 109 (1969).
OHMASA, M., NOWACKI, W.: A redetermination of the crystal structure of aikinite $[BiS_2|S|-Cu^{IV}Pb^{VII}]$. Z. Krist. 132, 71 (1970).
OKAYA, Y., KNOBLER, C. B.: Refinement of the crystal structure of tris(thiourea)copper(I)-chloride. Acta Cryst. 17, 928 (1964).
ORGEL, L. E., DUNITZ, J. D.: Stereochemistry of cupric compounds. Nature 179, 462 (1957).

OTTEMANN, J., FRENZEL, G.: Neue Mikrosonden-Untersuchungen an Idait, Covellin und blaubleibendem Covellin. Neues Jahrb. Mineral. Monatsh. **1971**, 80 (1971).

PROUT, C. W.: Recent studies in copper(II) co-ordination chemistry. Izv. Jugoslav. Centra Krist. **4**, 23 (1969).

RADCLIFFE, D., McWEEN, H. Y.: Copper zoning in pyrite from Cerro de Pasco, Peru: A discussion. Am. Mineralogist **54**, 1216 (1969).

— — Copper zoning in pyrite from Cerro de Pasco Perú: Reply. Am. Mineralogist **55**, 527 (1970).

ROSEBOOM, E. H., Jr.: An investigation of the system Cu-S and some natural copper sulfides between 25° and 700°C. Econ. Geol. **61**, 641 (1966).

SADANAGA, R., OHMASA, M., MORIMOTO, N.: On the statistical distribution of copper ions in the structure of β-chalcocite. Mineral. J. (Japan) **4**, 275 (1965).

SHIMIZAKI, H.: Synthesis of a copper-iron disulfide phase. Can. Mineralogist **10**, 146 (1969).

— CLARK, L. A.: Synthetic FeS_2—CuS_2 solid solution and fukuchilite-like minerals. Can. Mineralogist **10**, 648 (1970).

SILLITOE, R. H., CLARK, A. H.: Copper and copper-iron sulfides as the initial products of supergene oxidation, Copiapó mining district, Northern Chile. Am. Mineralogist **54**, 1684 (1969).

SKINNER, B. J., LUCE, F. D.: Stabilities and composition of β-domeykite and algodonite. Econ. Geol. **66**, 133 (1971).

SÜSSE, P.: Verfeinerung der Kristallstruktur des Malachits, $Cu_2(OH)_2CO_3$. Acta Cryst. **22**, 146 (1967).

TAKEDA, H., DONNAY, J. D. H., ROSEBOOM, E. H., APPLEMAN, D. A.: The crystallography of djurleite, $Cu_{1.97}S$. Z. Krist. **125**, 404 (1967).

TAKÉUCHI, Y., OHMASA, M., NOWACKI, W.: The crystal structure of wallisite, $PbTlCuAs_2S_5$, the Cu analogue of hatchite, $PbTlAgAs_2S_5$. Z. Krist. **127**, 349 (1968).

TAYLOR, C. A., UNDERWOOD, F. A.: A twinning interpretation of "superlattice" reflexions in X-ray photographs of synthetic klockmannite CuSe. Acta Cryst. **13**, 361 (1960).

TROJER, F. J.: Refinement of the structure of sulvanite. Am. Mineralogist **51**, 890 (1966).

WELLS, A. F.: The crystal structure of atacamite and the crystal chemistry of cupric compounds. Acta Cryst. **2**, 175 (1949).

WOOD, M. M.: The crystal structure of ransomite. Am. Mineralogist **55**, 729 (1970).

WUENSCH, B. J.: The crystal structure of tetrahedrite, $Cu_{12}Sn_4S_{13}$. Z. Krist. **119**, 437 (1964).

— BUERGER, M. J.: The crystal structure of chalcocite, Cu_2S. Mineral. Soc. Am., Spec. Paper 1, 164 (1963).

— TAKÉUCHI, Y., NOWACKI, W.: Refinement of the crystal structure of binnite, $Cu_{12}As_4S_{13}$. Z. Krist. **123**, 1 (1966).

YUND, R. A., KULLERUD, G.: Thermal stability of assemblages in the Cu—Fe—S system. J. Petrol. **7**, 454 (1966).

ZAHROBSKY, R. F., BAUR, W. H.: On the crystal chemistry of salt hydrates. V. The determination of the crystal structure of $CuSO_4 \cdot 3H_2O$ (bonattite). Acta Cryst. B **24**, 508 (1968).

ZEMANN, J.: Die Kristallchemie des Kupfers. Fortschr. Mineral. **39**, 59 (1961).

30-B. Isotopes in Nature

HOLLANDER, PERLMAN and SEABORG (1953) list in their table of isotopes the following percent abundances: ^{64}Zn, 48.89; ^{66}Zn, 27.81; ^{67}Zn, 4.11; ^{68}Zn, 18.56; ^{70}Zn, 0.62.[1] In 18 samples of 7 different zinc minerals — mainly sphalerite from different environments — BLIX et al. (1957) could not detect any variation of the ratio ^{64}Zn/^{68}Zn outside the limits of analytical error. If fractionation occurs in nature, the variation in the mentioned ratio must be smaller than 0.1%. FILBY (1964) reports on the constancy of the ^{64}Zn/^{68}Zn ratio of one sphalerite, the standard granite G-1, the standard basalt W-1 and several artificial zinc compounds from neutron activation analyses. In this case the relative standard deviation of the method was approximately 0.3 percent.

---

[1] Note added in proof: K. J. R. ROSMAN (Geochim. Cosmochim. Acta **36**, 801, 1972) reports: ^{64}Zn, 48.63 ± 0.13; ^{66}Zn, 27.90 ± 0.08; ^{67}Zn, 4.10 ± 0.03; ^{68}Zn, 18.75 ± 0.16; ^{70}Zn, 0.62 ± 0.01.

30-C. Abundance in Cosmos, Meteorites, Tektites and Lunar Samples

I. Cosmic Abundance

The cosmic abundance of zinc relative to 10^6 atoms Si has been estimated to be 486 atoms by SUESS and UREY (1956). This is a value lower than the average for carbonaceous chondrites of type II (Table 30-C-5, converted into atoms per 10^6 atoms Si). Later, CAMERON (1968) published a higher value of 1500 atoms zinc per 10^6 atoms Si, based on data of type I carbonaceous chondrites from LARIMER and ANDERS (1967), a value which is comparable with the average of our Table 30-C-5: 1570 atoms per 10^6 atoms Si.

A recent value for the abundance in the solar photosphere is listed by MÜLLER (1968) as 113 (or alternatively 250) atoms Zn per 10^6 atoms Si. According to the same author silicon is as yet not an adequate reference element because of its questionable abundance determination in the solar photosphere.

II. Abundance in Meteorites

Due to the availability of samples for direct high precision analyses, meteoritic materials have been used to get information on cosmic abundances of the elements. The major problem connected with these samples is the fractionation of an element in cosmos. Numerous authors tend to take the carbonaceous chondrites of type I for reasonable samples of least fractionated material of the solar system, accumulated at low temperatures, with the exception of a few light gases (see GOLES: Chapter 5 in Volume I of this handbook). A high degree of fractionation among the different classes of meteorites is clearly observable from data listed in Tables 30-C-1, -2, -3 and -5. The analytical data for single meteorites are compiled in Table 30-C-5 as averages for classes. In addition, the ranges of zinc concentrations for each class are listed to give an indication of the heterogeneous or homogeneous distribution of the element within the class. The ordinary chondrites are rather homogeneous groups with respect to zinc. The irons have larger ranges of values than the stones, but zinc concentrations are lower in irons than in chondrites. The three different types of carbonaceous chondrites range high in zinc as in other readily volatile elements; averages decrease in order from type I to III: 399, 179, 116 ppm Zn. Only the highly reduced enstatite chondrites of type I are as high in Zn (and a few other elements) as the carbonaceous chondrites of type I (parts of Abee are even higher). Zinc is fractionated by a factor of about 2 between each of the successive chondrite groups following the order: Cc1, Cc2, Cc3, ordinary, Ce2.

Systematic differences, even between the most abundant classes of ordinary chondrites, are real. No systematic difference among data from different analytical

Table 30-C-1. *Zinc concentrations in chondrites*

Name	Class	ppm Zn	Analytical method	Reference
Abee	Ce1	730	C	1
Abee	Ce1	112	C	2, 3
Abee	Ce1	720	N/R	5
Abee (2 samples)	Ce1	210, 348	N/R	4
Indarch	Ce1	315	C	2, 3
Indarch (3 samples)	Ce1	316—540	N/R	4
St. Marks	Ce	112	C	3
Daniels Kuil	Ce2	8	N/R	4
Hvittis	Ce2	<5 (5)	C	2, 3
Hvittis	Ce2	16	C	1
Hvittis	Ce2	20	N/R	5
Hvittis	Ce2	29	N/R	4
Khairpur	Ce2	36.1	C	2, 3
Khairpur (2 samples)	Ce2	22, 30	N/R	4
Khairpur	Ce2	22	N/R	5
Pillistfer	Ce2	24.4	C	2, 3
Pillistfer	Ce2	10	N/R	4
Pillistfer	Ce2	7.8	N/R	5
Allegan	CH	42.9	C	3
Allegan	CH	48	N/R	4
Allegan	CH	43	N/R	5
Cook	CH	20.6	C	3
Ehole	CH	63	N/R	5
Ehole	CH	62	N/R	4
Estacado	CH	46	C	1
Forest Vale	CH	38.6	C	3
Gladstone	CH	62	C	1
Kaldoonera Hill	CH	24.8	C	3
Kanzaki	CH	61	C	1
Morven	CH	23.7	C	3
Mt. Brown	CH	28.4	C	3
Nardoo	CH	23.8	C	3
Ness County	CH	52	C	1
Ochansk	CH	36	C	1
Pantar	CH	52	N/R	5
Pipe Creek	CH	48.6	N/R	9
Plainview	CH	54	C	1
Plainview	CH	35	N/R	10
Pultusk	CH	69.0	N/R	9
Richardton	CH	65	C	1
Zhovtneryi	CH	88.5	C	3
Accalana	CL	18.7	C	3
Adelie Land	CL	80.9	C	3
Alfianello	CL	33	N/R	6
Alfianello	CL	58	N/R	9
Albareto	CL	290	N/R	6
Barratta	CL	76.3	C	3
Barratta	CL	61	N/R	8
Beenham	CL	59	C	1
Bjurböle	CL	67.6	C	3
Bjurböle	CL	51	N/R	9
Bluff	CL	48.6	N/R	9

Table 30-C-1 (Continued)

Name	Class	ppm Zn	Analytical method	Reference
Bruderheim	CL	51	C	1
Bruderheim	CL	50	N/R	4
Bruderheim	CL	55	N/R	8
Bruderheim	CL	56	N/R	5
Cabell	CL	29.2	C	3
Chantonnay	CL	53	N/R	9
Cocunda	CL	67	C	3
Cook	CL	39.7	C	3
Coolamon	CL	45.1	C	3
Dhurmsala	CL	55	N/R	9
Dimbola	CL	45.7	C	3
Elenovka	CL	81.3	C	3
Farmington	CL	102	C	3
Farmington	CL	59	N/R	8
Fukutomi	CL	73	N/R	8
Goodland	CL	66	N/R	8
Hermitage Plains	CL	51.5	C	3
Holbrook	CL	28.9	C	3
Holbrook	CL	39	C	1
Holbrook	CL	56	N/R	7
Holbrook	CL	41	N/R	10
Homestead	CL	63.8	C	3
Homestead	CL	57	N/R	8
Homestead	CL	54.3	N/R	9
Jelica	CL	51.2	N/R	9
Khohar	CL	157	C	3
Khohar	CL	57	N/R	8
Kingfisher	CL	70	C	1
Knyahinya	CL	35.2	N/R	7
Krymka	CL	65	N/R	8
Kulnine	CL	54.5	C	3
Kyushu	CL	56	C	1
Ladder Creek	CL	53.5	C	1
Lake Labyrinth	CL	76.6	C	3
Lake Brown	CL	45.5	C	3
L'Aigle	CL	47	N/R	9
Long Island	CL	48	N/R	7
Loongana	CL	89.6	C	3
Mangwendi	CL	85.4	C	3
Mezö Madaras	CL	66.5	N/R	8
Mezö Madaras	CL	54.2	N/R	9
Mocs	CL	58.4	C	3
Mocs	CL	58	N/R	4, 5
Mocs	CL	57.1	N/R	7
Modoc	CL	62	N/R	8
Narellan	CL	34.3	C	3
Nardo	CL	33.9	C	3
Ness County	CL	51.0	N/R	9
Olivenza	CL	51.6	C	3
Peace River	CL	56	N/R	4
Peace River	CL	52	N/R	5
Perpeti	CL	107	C	3

Table 30-C-1 (Continued)

Name	Class	ppm Zn	Analytical method	Reference
Rawlinna	CL	54.9	C	3
Silverton	CL	41	C	3
St. Michel	CL	8.7	C	3
Tenham	CL	18.7	C	3
Tennasilm	CL	135	N/R	8
Vincent	CL	27.9	C	3
Waconda	CL	44	N/R	7
Walters	CL	63	N/R	5
Yalgoo	CL	57	C	3
Ordinary chondrites (composite of 7)	CH, CL (4)	51	C	1
Ordinary chondrites (composite of 7)	CH, CL (3)	47	C	1
Ordinary chondrites (composite of 5)	CH, CL (3)	50	C	1
Hamlet	CLL	51	N/R	8
Soko Banja	CLL	33.6	N/R	7
Chainpur	CHL (Cc3)	186	N/R	4
Chainpur	CHL (Cc3)	58	N/R	5
Felix	CHL (Cc3)	110	N/R	5
Felix (2 samples)	CHL (Cc3)	110, 134	N/R	4
Karoonda	CHL (Cc3)	49.9	C	2, 3
Lancé	CHL (Cc3)	57.4	C	2, 3
Lancé	CHL (Cc3)	162	N/R	4
Lancé	CHL (Cc3)	110	N/R	5
Mokoia	CHL (Cc3)	80	N/R	5
Mokoia (2 samples)	CHL (Cc3)	128, 210	N/R	4
Mokoia	CHL (Cc3)	64.2	C	2, 3
Vigarano	CHL (Cc3)	174	N/R	4
Vigarano	CHL (Cc3)	85	N/R	5
Warrenton	CHL (Cc3)	98	N/R	5
Warrenton	CHL (Cc3)	158	N/R	4
Ivuna	Cc1	320	N/R	5
Ivuna	Cc1	566	N/R	4
Orgueil	Cc1	228	C	2, 3
Orgueil	Cc1	340	N/R	5
Orgueil (2 samples)	Cc1	390, 549	N/R	4
Mighei	Cc2	150	N/R	5
Mighei (3 samples)	Cc2	199—230 (210)	N/R	4
Murray	Cc2	200	N/R	5
Murray (3 samples)	Cc2	64—213 (156)	N/R	4

References: 1. NISHIMURA and SANDELL (1964). 2. GREENLAND (1963). 3. GREENLAND and LOVERING (1965). 4. GREENLAND and GOLES (1965). 5. GREENLAND (1967). 6. BRECCIA et al. (1967). 7. KIESL et al. (1967). 8. KEAYS et al. (1971). 9. SEITNER et al. (1971). 10. LAUL et al. (1970).

procedures could be observed. A gross average of zinc in ordinary chondrites is expected to be close to 55 ppm Zn, a value almost comparable with the mean of terrestrial peridotites. Achondrites are much lower in zinc than chondrites and especially lower than terrestrial basalts. The octahedrites exhibit a general trend of

Table 30-C-2. *Zinc concentrations in achondrites*

Name	Class	ppm Zn	Analytical method	Reference
Johnstown	Ab	3	C	1
Pasamonte	Aor[a]	2	C	1
Sioux County	Aor[a]	2	C	1
Angra dos Reis	Aa	10	Mass spec	2
Juvinas	Ap[a]	2.5	N/R	3

References: 1. NISHIMURA and SANDELL (1964). 2. SMALES et al. (1970). 3. MORRISON et al. (1970). Note added in proof: LAUL et al. (Geochim. Cosmochim. Acta 36, 329, 1972) report: 8 Ap: 9.2 ppm; 6 Aor: 5.8 ppm; 2 Ado 57 ppm Zn.

decreasing zinc concentration with decreasing coarseness of their structure or increasing Ni content. Hexahedrites are more related to fine octahedrites in zinc (in spite of the low Ni concentrations of the former).

GREENLAND (1967) has demonstrated that with the exception of Ce2 chondrites, the Se:Zn ratio is almost constant for the different chondrites. This ratio of elements with quite different behavior supports a suggestion that the fractionation of the elements occurred prior to the aggregation of primitive materials into chondrite parent bodies.

SMALES et al. (1967) failed to observe a simple correlation between zinc and sulfur in iron meteorites but proved the existence of correlations between zinc and germanium, zinc and gallium and zinc and silver.

Table 30-C-4 lists abundances of zinc in meteorite minerals, exhibiting different behavior of zinc in different environments of meteorite formation. The highly reduced meteorites, such as irons and enstatite chondrites, often have high concentrations of zinc in their troilite. In the case of Odessa (Og), a heterogeneous distribution is demonstrated even in troilite. The non-uniform distribution of Zn in troilite might be partly due to the occurrence of this element as sphalerite in intergrowth with the former. RAMDOHR (1963) has described sphalerite (high in iron) from highly reduced meteorites as the enstatite chondrites Pillistfer and Hvittis. EL GORESY (1965) reports about the occurrence of sphalerite in such iron meteorites as Canon Diablo, Toluca, Cape York and Odessa.

Daubreelite and niningerite are additional sulfides which can be high in zinc in highly reduced meteorites. Metallic iron is usually lower in zinc than silicates and sulfides. Chromite of ordinary chondrites can be a major zinc carrier, whereas the bulk zinc of those stones is about equally distributed in troilite and silicates. These observations do not support an easy classification of zinc as a chalcophilic or lithophilic element.

III. Tektites

The few available data on zinc in tektites are listed in Table 30-C-6. The regional averages range between 7 and 15 ppm Zn, almost as low as achondrites (the only class of stone meteorites very low in zinc). There are no abundant terrestrial crustal rocks as low in zinc either. If tektites are melting products of sediments produced during impact of cosmic matter, the primordial zinc content of the sediments must

Table 30-C-3. *Zn concentrations in stony iron meteorites, iron meteorites, and chondritic iron*

Name	Class	ppm Zn	Analytical method	Reference
Stony irons:				
Admire (metal phase)	P	<1	C	1
Admire (metal phase)	P	2.9	N/R	4
Hexahedrites:				
Braunau	H	0.5	N/R	5
Coahuila	H	<1	N/R	2
Coahuila I, II	H	1.0	N/R	5
La Primitiva	H	<1	N/R	2
Mount Joy	H	0.3	N/R	5
Nedagolla	H	<1	N/R	2
Otumpa	H	28	N/R	2
San Martin	H	<1	N/R	2
Tocopilla	H	<1	N/R	2
Tombigbee River	H	<1	N/R	2
Uwet	H	<1	N/R	2
Walker County	H	<1	N/R	2
Zacatecas	H	0.4	N/R	5
Octahedrites:				
Arispe	Ogg	7.3	N/R	2
Gladstone	Ogg	37	N/R	2
Nelson County	Ogg	0.1	N/R	5
Soa Juliao de Moreira	Ogg	<1	N/R	2
Seeläsgen	Ogg	27	N/R	2
Arva Magura	Og	17.3	N/R	3
Bendego	Og	19	N/R	2
Bennet County	Og	0.48	N/R	2
Billings	Og	2.0	N/R	2
Bischtübe	Og	42	N/R	2
Bohumilitz	Og	15.1	N/R	2
Canyon Diablo	Og	29	N/R	2
Canyon Diablo	Og	37 (32—41.4)	C	1
Canyon Diablo	Og	34 (33—35)	N/R	4
Cranbourne	Og	12	N/R	2
Odessa	Og	40.8	N/R	6
Pan de Azucar	Og	15.1	N/R	2
Wichita County	Og	25	N/R	2
Youndegin	Og	36	N/R	2
Bohumilitz	Om	1.8	N/R	5
Breece	Om	1.7	N/R	2
Campbellsville	Om	1.8	N/R	2
Canton	Om	2.8	N/R	2
Cape York	Om	4.2	N/R	2
Caperr	Om	<1	N/R	2
Chinautla	Om	<1	N/R	2
Clark County	Om	6.5	N/R	2
Colfax	Om	32	N/R	2
Cumpas	Om	<1	N/R	2
Descubridora	Om	2.8	N/R	2
Glorieta Mountain	Om	<1	N/R	2
Guw Creek	Om	1.9	N/R	2

Zinc

Table 30-C-3 (Continued)

Name	Class	ppm Zn	Analytical method	Reference
Henbury	Om	32 (27—38)	C	1
Henbury	Om	<1	N/R	2
Henbury	Om	19.4 (5.8—70)	N/R	4
Kenton County	Om	2.1	N/R	3
Kingston	Om	1.9	N/R	2
La Caille	Om	<1	N/R	2
Mount Edith	Om	1.2	N/R	2
Puquios	Om	3.3	N/R	2
Rhine Villa	Om	<1	N/R	2
Sams Valley	Om	<1	N/R	2
San Angelo	Om	3	C	1
Spearman	Om	8	C	1
Staunton	Om	3.1	N/R	5
Toluca	Om	<1	N/R	2
Toluca	Om	0.6	N/R	5
Trenton	Om	3	C	1
Xiquipilco	Om	33	N/R	2
Altonah	O_f	24	C	1
Boogaldi	O_f	<1	N/R	2
Carlton	O_f	21	N/R	2
Carlton	O_f	38.0	N/R	5
Duchesne	O_f	33	C	1
Gibeon	O_f	<1	N/R	2
Kamkas (Gibeon)	O_f	<1	N/R	2
Grand Rapids	O_f	3	N/R	2
Huizopa	O_f	<1	N/R	2
Lockport (Campria)	O_f	<1	N/R	2
Moonbi	O_f	<1	N/R	2
Obernkirchen	O_f	<1	N/R	2
Otchinjau	O_f	<1	N/R	2
Bacubirito	O_{ff}	<1	N/R	2
Ballinoo	O_{ff}	<1	N/R	2
Butler	O_{ff}	2.2	N/R	5
Mount Magnet	O_{ff}	<1	N/R	2
Salt River	O_{ff}	0.9	N/R	2
Barranca Blanca	O (Brecc.)	<1	N/R	2
Kopjes Vlei	(Granul. metabol.)	<1	N/R	2
Murnpeowie	(Granul. metabol.)	<1	N/R	2
Santa Rosa	O (Brecc.)	12.8	N/R	2
Weekero Station	O (Brecc.)	2.3	N/R	2
Babb's Mill	D	<1	N/R	2
Hoba	D	<1	N/R	2
Illinois Gulch	D	<1	N/R	2
San Cristobal	D	22	N/R	2
Santa Catharina	D	3	N/R	2
Smithland	D	0.28	N/R	2
Metals of chondrites:				
Holbrook	CL	4	C	1
Kyushu	CL	~1	C	1

References: 1. NISHIMURA and SANDELL (1964). 2. SMALES, MAPPER and FOUCHÉ (1967). 3. KIESL *et al.* (1967). 4. HAMAGUCHI *et al.* (1964). 5. SEITNER *et al.* (1971). 6. KIESL and WEINKE (1970).

Table 30-C-4. *Zinc concentrations in meteorite minerals*

Mineral (name of meteorite)	Class	ppm Zn	Analytical method	Reference
Troilite (5 samples)	irons	1,530	X	1
Troilite (3 samples)	irons	2,000	S	2
Troilite (3 samples Odessa)	Og	5, 19.5, 3,300	X	3, 9
Troilite (19 samples)	irons	50—521	X	3
Troilite (Trenton)	Om	800	C	4
Troilite (Holbrook)	CL	12	C	4
Troilite (Kyushu)	CL	10	C	4
Troilite (Brenham)	P	63	X	3
Troilite (Cullison)	CH	61	X	3
Troilite (Hvittis)	Ce2	40	C	4
Troilite (Kota-Kota; Abee)	Ce1	1,200, 1,700	M/X	5
Troilite (St. Mark's)	Ce	1,100	M/X	5
Troilite (12 samples)	Ce2, Ce1	900	M/X	5
Daubreelite (Kota-Kota)	Ce1	52,000	M/X	5
Daubreelite (St. Mark's)	Ce i	42,600	M/X	5
Daubreelite (8 samples)	Ce2	1,200	M/X	5
Daubreelite (Odessa)	Og	6,000—11,000	M/X	7
Chromite	P, M, Ap, Ao, CH	<200	M/X	10
Chromite (Pultusk)	CH	8,800	M/X	8
Chromite (Modoc, St. Severin)	CL, CLL	86, 1,800	Mass spec.	6
Graphite (Odessa)	Og	19.8	M/X	9
Schreibersite (Odessa)	Og	4.2—9.6	M/X	9
Niningerite (Abee)	Ce1	3,100	M/X	5
Olivine, pyroxenes (13 samples)	ordinary chondrites	3—48 (average: 15)	Mass spec	6
Plagioclase	ordinary chondrites	3—5	Mass spec	6

References: 1. I. and W. NODDACK (1930). 2. GOLDSCHMIDT (1937). 3. NICHIPORUK and CHODOS (1959). 4. NISHIMURA and SANDELL (1964). 5. KEIL (1968). 6. MASON and GRAHAM (1970). 7. EL GORESY (1967). 8. BUNCH et al. (1967). 9. KIESL and WEINKE (1970). 10. BUNCH and KEIL (1971).

have been partially lost due to its volatility. The abundance of the major elements in average tektites is almost comparable with a 75 percent shale, 25 percent quartz sandstone mixture. Such a mixture is expected to contain about 80 to 90 ppm Zn in contrast to less than 15 ppm Zn in tektites.

IV. Lunar Materials

Zinc data of Apollo 11, 12 and 14 samples from Mare Tranquillitatis, Oceanus Procellarum and Fra Mauro area are listed in Table 30-C-7. Because of discrepancies in data by different methods and investigators on the same basalt samples (for instance type A basalt sample 1072: 1.81, 7.0, 24, 34 ppm Zn by N/R, N/R, X, and A methods respectively), only the order of magnitude of 10 ppm Zn can be used. This is about ten times lower than zinc in terrestrial basalts (see Section

Table 30-C-5. *Ranges and averages of zinc concentrations in the different classes of meteorites (number of samples used for average in brackets)*

Class	Range in ppm Zn	Average in ppm Zn
Ce1 (10)	112—730	383
Ce2 (12)	5—36.1	19
CH (23)	20.6—88.5	47.3
CL (72)	8.7—290	60.4
CLL (2)	33.6—51	42.3
CHL (Cc3) (17)	49.9—210	116
Cc2 (4)	64—230	179
Cc1 (5)	228—566	399
Achondrites (5)	2—10	4
P (2)	<1—2.9	<2
H (13)	<1—28	<3 (0.5 ?)
Ogg (5)	<1—37	14.5
Og (14)	0.48—42	23.3
Om (30)	<1—70	$\leqslant 5.8$
O_f (13)	<1—38	<9.8
O_{ff} (5)	<1—2.2	<1.2
D (6)	<1—22	$\leqslant 4.7$

Table 30-C-6. *Zinc concentration in tektites (number of samples in brackets). (Spectrophotometric analyses by* GREENLAND *and* LOVERING, *1963)*

Type locality	Range in ppm Zn	Average in ppm Zn
Texas (2)	3.4, 16.0	14.7
Czechoslovakia (5)	3.2—10.0	7.6
Indochina (2)	3.2, 12.4	7.8
Philippines (3)	6.0—13.4	9.3
Java (9)	1.3—19.7	11.7
Australia (21)	1.7—19	9.6

30-E) but only slightly higher than zinc in achondrites (see Tables 30-C-2, 30-C-5). There is less scattering of zinc data on Apollo 12 than on Apollo 11 rocks.

VINOGRADOV (1971) reports concentrations of 10 to 33 ppm Zn (average 24 ppm Zn) in 4 samples of lunar regolith and 26 ppm Zn in lunar basalt obtained from the Russian automatic station "Luna 16" from Mare Crisium. Even considering the analytical discrepancies mentioned for zinc in Apollo 11 samples, there can be no doubt that this element is concentrated in lunar breccias and fines relative to the basaltic flows by at least a factor of 3.

Zn, Cd and non-radiogenic Pb are correlated in the lunar basalts. GANAPATHY et al. (1970) have demonstrated that about 2 percent of carbonaceous chondrite-like material can account for that portion of the elements, Zn, Cd, Ag, Bi, Tl, Br, ^{204}Pb etc., in which lunar soil and breccia exceed the basaltic rocks from Tranquillity Base.

Table 30-C-7. *Zinc concentration in lunar materials from Mare Tranquillitatis (Apollo 11), Oceanus Procellarum (Apollo 12), Fra Mauro area (Apollo 14) and Mare Crisium (Luna 16). (Number of samples in brackets.)*

Most references for Apollo 11 investigations are from: [P] Proceedings of the Apollo 11 Lunar Science Conference Vol. 2 — Geochim. Cosmochim. Acta 34, Supplement 1, 1970, and [S] Science 3918 Vol. 167, 1970; references for Apollo 12 are from: [P*] Proceedings of the Second Lunar Science Conference Vol. 2 — Geochim. Cosmochim. Acta Supplement 2, 1971.

Rock type	Range ppm Zn	Average ppm Zn	Analytical method	Reference
Apollo 11				
A, basalts, fine grained (4)	<4	<4	S	ANNELL, HELZ [P] [S]
A, basalts, fine grained (1)	—	34	X	COMPSTON et al. [P]
A, basalts, fine grained (3)	1.29—1.81	1.6	N/R	GANAPATHY et al. [P]
A, basalts, fine grained (1)	—	2.9	N/R	HASKIN et al. [P]
A, basalts, fine grained (2)	24—48	36	A	MAXWELL et al. [P]
A, basalts, fine grained (3)	2.1—7.0	4	N/R	MORRISON et al. [P] [S]
A, basalts, fine grained (3)	9.3—30	16.8	A	GAST, HUBBARD [S]
AB, A and B, basalts (7)	1.71—18.0	4.7	N/R	ANDERS et al. [P*]
B, basalts, medium grained (2)	<4	<4	S	ANNELL, HELZ [P] [S]
B, basalts, medium grained (3)	13—14	14	X	COMPSTON et al. [P]
B, basalts, medium grained (2)	1.75—5.76	3.75	N/R	GANAPATHY et al. [P]
B, basalts, medium grained (1)	—	2.9	N/R	HASKIN et al. [P]
B, basalts, medium grained (1)	—	26	A	MAXWELL et al. [P]
B, basalts, medium grained (2)	9.3—30	19.7	N/R	MORRISON et al. [P] [S]
B, basalts, medium grained (3)	7.4—10	9.1	A	GAST, HUBBARD [S]
C, breccias (7)	22—29	24.4	S	ANNELL, HELZ [P] [S]
C, breccias (2)	37—54	45.5	X	COMPSTON et al. [P]
C, breccias (2)	28.6—29.2	28.9	N/R	GANAPATHY et al. [P]
C, breccias (1)	—	30.2	N/R	HASKIN et al. [P]
C, breccias (2)	2.7—25	13.9	N/R	MORRISON et al. [P] [S]
C, breccias (1)	—	40	Mass spec	SMALES et al. [P] [S]
C, breccias (1)	—	24	A	GAST, HUBBARD [S]
D, fines (1)	—	19	S	ANNELL, HELZ [P] [S]
D, fines (6)	10.6—34.2	21.3	N/R	GANAPATHY et al. [P]
D, fines (1)	—	37	X	COMPSTON et al. [P]
D, fines (3)	18—21	20	N/R	LAUL et al. [P*]
D, fines (1)	—	22.8	N/R	HASKIN et al. [P]
D, fines (1)	—	41.5	A	MAXWELL et al. [P]
D, fines (1)	—	22	N/R	MORRISON et al. [P] [S]

Zinc

Table 30-C-7 (Continued)

Rock type	Range ppm Zn	Average ppm Zn	Analytical method	Reference
D, fines (1)	—	50	Mass spec	SMALES et al. [P] [S]
D, fines (2)	<5—14	<9.5	Mass spec	ORÓ et al. [P]
D, fines (1)	—	24	A	GAST, HUBBARD [S]
Anorthosite fragments (2)	1.4—2.0	1.7	N/R	LAUL et al. [P*]
Apollo 12				
AB, basalts (2)	0.52—0.54	0.53	N/R	GANAPATHY et al. (1970)
AB, B, basalts (11)	0.52—3.30	0.80	N/R	ANDERS et al. [P*]
A, B, basalts (6)	1.0—2.3	1.6	N/R	BAEDECKER et al. [P*]
AD, basalt and fines (1)	—	1.47	N/R	GANAPATHY et al. (1970)
AB, basalts (3)	1.2—4.5	2.4	X	WILLIS et al. [P*]
A, AB, B, basalts (4)	1.6—4.2	3.0	N/R	MORRISON et al. [P*]
AB, basalts (4)	6—16	9.5	Mass spec	BOUCHET et al. [P*]
AB, basalts (2)	—	2	N/R	SMALES et al. [P*]
AB, basalts (6)	1.2—7.2	3.6	N/R	BRUNFELT et al. [P*]
C, breccias (7)	1.94—4.10	2.77	N/R	ANDERS et al. [P*]
C, breccias (2)	0.82—2.0	1.4	N/R	LAUL et al. [P*]
C, breccias (1)	—	6.8	N/R	MORRISON et al. [P*]
D, fines (4)	5.1—6.9	5.4	N/R	GANAPATHY et al. (1970)
D, fines (1)	—	7.4	N/R	BRUNFELT et al. [P*]
D, fines (1)	—	11	N/R	SMALES et al. [P*]
D, fines (1)	—	14	Mass spec	BOUCHET et al. [P*]
D, fines (6)	<4—8.2	6.7	S	CUTTITA et al. [P*]
D, fines (3)	8.4—8.7	8.5	N/R	MORRISON et al. [P*]
D, fines (2)	7.1—8.9	8.0	N/R	BAEDECKER et al. [P*]
D, fines (2)	9.7—22	16.0	X	WILLIS et al. [P*]
D, fines (15)	1.5—6.9	5.3	N/R	LAUL et al. [P*]
Apollo 14				
Fines (1)		33	N/R	BRUNFELT et al. (1971)
Luna 16				
AB, basalt (1)	—	26	?	VINOGRADOV (1971)
Regolith (4)	10—33	24	?	—

Manuscript received: March 1972

30-D. Abundance in Rock-forming Minerals, Phase Equilibria, Zinc Minerals

I. Rock-forming Minerals

Analyses of rock-forming minerals for zinc show its tendency to be predominantly incorporated in certain structural positions of silicates and oxides. In these positions, zinc replaces ferrous iron and magnesium, favoring positions with coordination numbers 4 and 6; the latter only if certain ligands (OH⁻ etc.) are available (WEDEPOHL, 1953).

NEUMANN (1949) has already mentioned this tendency for tetracoordinated positions (for details see Section 30-D). He assumed that zinc occurs mainly as sphalerite in rocks. This could be disproved for numerous rocks by a combined spectrochemical and thermochemical method (WEDEPOHL, 1953), by ore microscopy (NEWHOUSE, 1936; RAMDOHR, 1940), by balance computations between zinc in total rock and its mineral constituents, and by microprobe analyses of basaltic magnetites. In contrast to these observations, DESBOROUGH (1963) reports the occurrence of sphalerite from 10 intrusive bodies and flows of gabbro and basalt. This is about half the number of specimens he has examined. Due to their intergrowth he assumes that most of these sphalerite grains are of primary origin.

In willemite, Zn_2SiO_4, zinc occurs in a tetracoordinated position contrary to the olivine structure. Table 30-A-13 lists several zinc silicates in which zinc has 4 oxygen ligands. The low pressure modification of ZnO has a wurtzite-type, the high pressure modification a sodium chloride-type (BATES, WHITE and ROY, 1962) lattice. The ε-$Zn(OH)_2$ polymorph is tetracoordinated, the α-$Zn(OH)_2$ hexacoordinated. Except in magnetite the major rock-forming minerals do not offer tetracoordinated ferrous iron or magnesium positions. The existence of a zincian biotite (hendricksite: FRONDEL, ITO, 1966a), zincian montmorillonite (sauconite: ROSS, 1946), zincian serpentine (BADALOW, 1958), zincian amphiboles (KLEIN, ITO, 1968), and zincian staurolite (JUURINEN, 1956) indicate the stability of structures having zinc in sixfold coordinated positions, if several ligands are hydroxide instead of oxygen. Hydroxide ligands will have a lower polarizability than oxygen ions. This might be the reason

Table 30-D-1. *Crystal chemical properties of zinc, ferrous iron and magnesium.* (Reference: HEYDEMANN: Chapter 12 of Vol. I of this handbook)

	Zn^{2+}	Fe^{2+}	Mg^{2+}
Effective ionic radii (GOLDSCHMIDT, 1926/34)	0.83 Å	0.83 Å	0.78 Å
Effective ionic radii (AHRENS, 1952)	0.74 Å	0.74 Å	0.67 Å
Ionization potentials (atomic state to X^{2+})	17.89 eV	16.16 eV	14.96 eV
Electronegativities (PAULING, 1960)	1.6	1.8	1.2

© Springer-Verlag Berlin · Heidelberg 1972

for a relatively higher proportion of ionic bonds between zinc and hydroxide ligands. In the case of several hydroxide ligands, better similarity between zinc bonds and ferrous iron or magnesium bonds in sixfold coordinated positions is to be expected than between zinc and oxygen ligands only.

For comparison of crystal chemical behavior, several properties of Zn^{2+}, Fe^{2+} and Mg^{2+} ions, which show closer relations between Zn^{2+} and Fe^{2+} than between Zn^{2+} and Mg^{2+}, are listed in Table 30-D-1.

a) Oxide Minerals

Data for this subsection are compiled in Table 30-D-2. The existence of the zinc spinels, franklinite $(Zn, Mn)Fe_2O_4$, donathite $(Fe^{2+}, Mg, Zn)(Cr, Fe^{3+})_2O_4$, gahnite $ZnAl_2O_4$ and hetaerolite $ZnMn_2O_4$, clearly demonstrates the good fit of Zn^{2+} into spinel structures. HAFNER (1960) mentions the importance of proportions of covalent bonding by sp^3 hybridization in these structures. In a regional survey considering the different types of basaltic magma, GRAMSE (1966) found a range of 1,000 to 4,000 ppm Zn in titanomagnetites from Tertiary basalts in northwest Germany. Zn concentrations in these magnetites increase in the sequence olivine nephelinite, olivine alkali basalt, tholeiitic basalt, with the decrease of the magnetite content of the rocks. In magnetites from granitic rocks, zinc concentrations are usually lower by more than a factor of ten (THEOBALD, THOMPSON, 1962). These authors assume that magnetites from granite intrusions, which are connected with sphalerite deposits, are higher in zinc than others. Besides magnetites from magmatic intrusions, they report alluvial magnetite deposits from five places in the U.S.A., ranging from <25 to >1,000 ppm Zn, with numerous samples in the 25 to 150 ppm Zn range. In the skarn deposits of Kondoma (western Siberia), zinc in magnetite ranges from 70 to 490 ppm (VAKHRUSHEV, 1959; 51 analyses; method: P).

Chromite can be much higher than several hundred ppm in zinc. THAYER et al. (1964) report that chromite from the Outokumpu ore deposit in Finland contains upto 5.8% Zn. SEELIGER and MÜCKE (1969) reexamined chromite from Hestmandö in North Norway (originally described by Donath) as having 2.3 and 2.7% ZnO. They named this tetragonal variety donathite.

Olivine from peridotites and basalts often contains tiny crystals of a chromium spinel. It has to be assumed that at least some zinc content of the olivine is connected with the intergrown spinel.

b) Silicate Minerals

The existence of some zincian silicates, containing hydroxide, mainly occurring in zinc rich skarn deposits, such as Franklin and Sterling Hill, New Jersey (U.S.A.), has already been mentioned in the introduction to Section 30-D. Besides these, zincian and manganoan aegirine-augite has been described from the same zinc and manganese skarn at Franklin. FRONDEL and ITO (1966b), analyzing this mineral with 6.9 to 8.8% ZnO, emphasize that the former name jeffersonite should be changed because of lack of varietal significance. Hendricksite (zincian biotite) from Franklin contains 0.3 to 23.8% ZnO (FRONDEL and ITO, 1966a) in addition to high manganese concentrations. The maximum zinc concentration of 3 zincian cummingtonites is 10.8% ZnO, of 9 zincian tremolites-actinolites is 9.5% Zn, and the content

Table 30-D-2. *Zinc concentrations in rock-forming oxides. (Number of samples in brackets)*

Mineral	Origin	Range ppm Zn	Average ppm Zn	Analytical method	Reference
Magnetite (14)	quartz monzonite, Front Range Colorado (U.S.A.)	150—2,500	1,500	W/C	1
Magnetite (10)	granodiorite, Cabarrus Co., N. Carolina (U.S.A.)	<25—50	<25	W/C	1
Magnetite (15)	granite, Cabarrus Co., N. Carolina (U.S.A.)	<25—25	25	W/C	1
Magnetite (10)	syenite, Cabarrus Co., N. Carolina (U.S.A.)	25—50	40	W/C	1
Magnetite (1)	granite, Eastern Ontario (Canada)	—	137	W/C	2
Magnetite (3)	granite, Susamir, Tian-Shan (U.S.S.R.)	100—230	173	W/C	3
Magnetite (30)	granodiorite, Butte Mont. (U.S.A.)	25—200	50	W/C	1
Magnetite (4)	mafic dike, Cabarrus Co., N. Carolina (U.S.A.)	50—4,000	1,500	W/C	1
Magnetite (1)	gabbro, Skaergaard (Greenland)	300—3,000	—	S	4
Magnetite (2)	tholeiitic basalt, (northwest Germany)	274—282	2,780	X	5
Magnetite (10)	olivine alkali basalt (northwest Germany)	1,080—1,490	1,240	X	5
Magnetite (4)	olivine nephelinite (northwest Germany)	990—1,270	1,110	X	5
Magnetite ilmenite concentrates	mafic rocks, Minnesota (U.S.A.)	300—400	—	W/C	6
Magnetite ilmenite concentrates (6)	pitchstone glasses, (Scotland, Iceland)	900—2,200	1,450	P	7
Chromite (4)	chromite deposits in serpentinite, Maryland, Pennsylvania (U.S.A.)	<200—5,000	1,550	S	8

References: 1. THEOBALD, THOMPSON (1962). 2. AZZARIA (1963). 3. TAUSON, KRAVCHENKO (1956). 4. WAGER, MITCHELL (1951). 5. GRAMSE (1966). 6. SANDELL, GOLDICH (1943). 7. CARMICHAEL, MCDONALD (1961). 8. PEARRE, HEYL (1960).

of one zincian magnesioriebeckite is 7.8% ZnO (KLEIN and ITO, 1968). The samples are again from Franklin from which two additional zinc silicates should be mentioned: hardystonite (zincian melilite) and willemite (Zn_2SiO_4), the latter as a separate structure type. High zinc concentrations in ore-forming media, such as those which have formed zincian silicates at Franklin and Sterling Hill, are apparently rare in nature. The occurrence of zinc in certain rock-forming ferrous iron and magnesium silicates, and oxides, is by far more important for the crustal abundance of this element than zinc in ore deposits.

In Table 30-D-3 zinc analyses of common rock-forming silicates are compiled.

Table 30-D-3. *Zinc concentrations in rock-forming silicates. (Number of samples in brackets)*

Mineral	Origin	Range ppm Zn	Average ppm Zn	Analytical method	Reference
Olivine (3)	peridotite inclusions in Tertiary basalt (northwest Germany)	50—65	56	S	1
Olivine (3)	peridotite inclusions, basaltic ejecta, Dreiser Weiher (Germany)	50—82	74	X	2
Olivine (1)	olivine alkali-basalt, Etna (Italy)	—	51	S	1
Garnet (almandite) (4)	amphibolites, Bamble district, Krageö (Norway)	1,480—5,275	3,316	A	3
Garnet (almandite) (2)	mica schists, Bamble district, Krageö (Norway)	230—560	400	A	3
Garnet (almandite) (25)	paragneisses, Emeryville Colton Region, New York (U.S.A.)	75—280	185	S(?)	4
Garnet (almandite) (21)	schists, mostly from Wyoming and western U.S.A.	<30—1,000	169	S	5
Staurolite (2)	schist, north of Idaho Batholith	2,000—6,000	4,000	S	6
Staurolite (1)	mica schist, (Finland)	—	7.44%	(?)	7
Pyroxene (aegirine) (3)	alkali granites (northern Nigeria)	850—2,250	1,470	W/P	8
Pyroxene (augite) (7)	trachytic ejecta, Eifel (Germany)	140—450	225	X	9
Pyroxene (diopside) (2)	peridotite inclusion, basalt, Dreiser Weiher (Germany)	26—130	78	X	2

Table 30-D-3. (Continued)

Mineral	Origin	Range ppm Zn	Average ppm Zn	Analytical method	Reference
Pyroxene (diopside) (13)	schists, mainly Wyoming (U.S.A.)	<30—200	53	S	10
Pyroxene (diopside) (3)	basalt and kimberlite	16—79	38	S	1
Amphibole (21)	amphibolite facies schists, mainly Wyoming (U.S.A.)	50—480	92	S	11
Amphibole (10)	granulites, Wyoming (U.S.A.)	120—400	220	S	11
Amphibole (8)	schists (mainly western U.S.A.)	34—680	200	S	10
Amphibole (12)	trachytic ejecta, Eifel (Germany)	257—690	355	X	9
Amphibole (riebeckite arfvedsonite) (7)	alkali granites, northern Nigeria	2,250—8,900	5,300	W/P	8
Amphibole (1)	granite, New England (U.S.A.)	—	1,600	W/C	12
Amphibole (1)	granite, Susamyr, Tien-Shan (U.S.S.R.)	—	710	W/C	13
Biotite (7)	granites, Susamyr, Tien-Shan (U.S.S.R.)	450—870	630	W/C	13
Biotite (19)	schists (mainly from Wyoming, U.S.A.)	50—700	270	S	11
Biotite (16)	schists (mainly western U.S.A.)	34—720	180	S	10
Biotite (151) (concentrates)	granites, California (U.S.A.)	—	440	S	14
Biotite (226) (concentrates)	granites, granodiorites, quartz monzonites, Arizona (U.S.A.)	900—4,000 (abundant values: 500—800)	720	S	15
Biotite (35)	gneisses (mainly Central Europe)	40—1,220	445	X	16
Biotite (29)	granites without muscovite (mainly Central Europe)	290—920	548	X	16
Biotite (16)	granites with equal amounts of muscovite and biotite (mainly Central Europe)	440—1,520	803	X	16
Biotite (8)	granites with more muscovite than biotite (mainly Central Europe)	520—1,350	1,040	X	16

Zinc

Table 30-D-3. (Continued)

Mineral	Origin	Range ppm Zn	Average ppm Zn	Analytical method	Reference
Biotite (95)	quartz monzonites, Basin Range (U.S.A.)	80—1,290	290	S	17
Biotite (10)	quartz monzonites, granodiorites, Central Sierra Nevada Batholith (U.S.A.)	420—1,000	584	A	18
Biotite (44)	granites (one mica- and two mica- granites, mineralized and unmineralized areas) (U.S.A., Great Britain)	37—919	350	X	19
Biotite (6)	alkali granite (northern Nigeria)	965—5,100	2,540	W/P	8
Biotite (10)	nepheline syenites, schists, etc. (Canada)	250—1,700	772	X	20
Phlogopite (6)	kimberlites, eclogites, carbonatites etc. (South Africa)	160—1,200	440	X	21
Phengites	gneisses, mica schists Western Alps (Switzerland)	29—211	93	X	2
Muscovite (9)	nepheline syenites, schists etc. (Canada)	40—200	85	X	20
Muscovite (8)	schists, pegmatites (mainly from western U.S.A.)	24—51	31	S	10
Muscovite (2)	pegmatites, (Finland, Brazil)	2—9	5	S	1
Illite (1)	Morris, Illinois (U.S.A.)	—	120	X	2
Chlorite (14)	schists (mainly from Wyoming, U.S.A.)	33—1,600	374	S	6
Chlorite (1)	glaucophane schist, California (U.S.A.)	—	700	S	22
Montmorillonite (2)	Georgia, Missouri (U.S.A.)	73—156	115	X	2
Kaolinite (50)	13 major kaolinite deposits (Europe, U.S.A.)	14—264	59	W/C	23
Plagioclase (3)	pitchstone glasses (Scotland, Iceland)	7.5—26	11	W/C	24
Plagioclase (1)	diorite porphyrite (Roumania)	—	16	W/C	25

Table 30-D-3. (Continued)

Mineral	Origin	Range ppm Zn	Average ppm Zn	Analytical method	Reference
Plagioclase concentrates (14)	anorthosites, ligth types (eastern Canada)	7—50	21	W/C	26
Plagioclase (5)	granite, gabbro etc. (Norway, Germany)	0.7—26	10	S	1
Feldspar (3)	granites, Susamyr, Tien-Shan (U.S.S.R.)	10—24	15	W/C	13
Feldspar (1)	granite, eastern Ontario (Canada)	—	15	W/C	27
Potassium feldspar (3)	granite, trachyte (Germany)	2—8	5	S	1
Quartz (3)	granite, Susamyr, Tien-Shan (U.S.S.R.)	7—11	8	W/C	13
Quartz	quartz sand, Dörentrup (Germany)	—	<5	S	1
Quartz (1)	granite, eastern Ontario (Canada)	—	4	W/C	27

References: 1. WEDEPOHL (1953). 2. WEDEPOHL (unpublished). 3. BURELL (1966). 4. ENGEL, ENGEL (1960). 5. DE VORE (1955a, b). 6. HIETANEN (1969). 7. JUURINEN (1956). 8. BUTLER, THOMPSON (1967). 9. JASMUND, SECK (1964). 10. DE VORE (1955b). 11. DE VORE (1955a). 12. SANDELL, GOLDICH (1943). 13. TAUSON, KRAVCHENKO (1956). 14. PUTMAN, ALFORS (1967). 15. PUTMAN, BURNHAM (1963). 16. HAACK (1969). 17. PARRY, NACKOWSKI (1963). 18. DODGE et al. (1969). 19. BLAXLAND (1970). 20. RIMSAITE (1968). 21. RIMSAITE (1971). 22. LEE et al. (1966). 23. KÖSTER (1969). 24. CARMICHAEL, MCDONALD (1961). 25. SAVUL et al. (1956a). 26. PAPEZIK (1965). 27. AZZARIA (1963).

Zinc abundances in the different minerals are a function of at least two parameters: the zinc concentration of the magma, premetamorphic rock etc., and the ability of the crystal structure to incorporate this element (at a given temperature and pressure). Because different rock types were used, Table 30-D-3 can only roughly inform about the tendency of the different mineral structures for zinc incorporation. Partition of zinc concentrations on pairs of coexisting minerals gives substantially more information on the crystal chemical behavior of this element. However, there are only a few data. JASMUND and SECK (1964) report that hornblende from trachytic ejecta has incorporated 1.75 times the zinc concentration of coexisting augite. RIMSAITE (1968) demonstrates that biotite, coexisting with muscovite in schists, gneisses and nepheline syenites, contains 26 to 5 times more zinc than muscovite (higher figure: syenites, quartz monzonite; lower figure: schists and gneisses).

AZZARIA (1963) and TAUSON and KRAVCHENKO (1956) investigated the distribution of zinc in the coexisting minerals of one granite from Canada and three granites from Tien-Shan (U.S.S.R.). They obtained the following ratios of zinc concentrations:

Canada:

$\dfrac{\text{biotite}}{\text{magnetite}}$: 2.15; $\dfrac{\text{biotite}}{\text{feldspar}}$: 19.5; $\dfrac{\text{feldspar}}{\text{quartz}}$: 3.8[1].

Tien-Shan:

$\dfrac{\text{biotite}}{\text{magnetite}}$: 2.7 to 8.7; $\dfrac{\text{biotite}}{\text{amphibole}}$: 1.05; $\dfrac{\text{biotite}}{\text{feldspar}}$: 26 to 87; $\dfrac{\text{feldspar}}{\text{quartz}}$: 1.1 to 3.4[1].

According to data published by SANDELL and GOLDICH (1943) for two granites from the U.S.A. the ratio of zinc concentrations is the following:

$\dfrac{\text{biotite}}{\text{feldspar}+\text{quartz}}$: ~100; $\dfrac{\text{amphibole}}{\text{feldspar}+\text{quartz}}$: ~300.

The information from subsections 30-D-Ia, b can be condensed in the statement that magnetite is quantitatively the most important zinc carrier in basaltic-gabbroic rocks, and biotite in granitic rocks. In coexisting biotite and magnetite, the former usually contains more zinc.

II. Phase Equilibria

a) Oxides

Complete mutual solubility has been established in $(Zn_2TiO_4)_x \cdot (ZnFe_2O_4)_{1-x}$ and $(Zn_2TiO_4)_x \cdot (Fe_3O_4)_{1-x}$ by SHCHEPETKIN et al. (1970). Relationships between composition and lattice parameters are listed by these authors.

Phase relations between the low and high pressure, modifications of ZnO (slightly distorted wurtzite type lattice at low pressure, NaCl-type lattice at high pressure) have been published by BATES et al. (1962). Their value for dp/dT is 42.5 atm/°C (two points of the equilibrium curve: 94 kilobars and 100°C; 105 kilobars and 350°C).

b) Silicates

The stable compound at normal pressure and temperature in the $ZnO-SiO_2$ system is Zn_2SiO_4, willemite (and not Zn-olivine). Increased pressure lowers the melting point of willemite from a one atm value of 1,510°C to below 1,425° at 30 kbar (TAYLOR et al., 1971). This is an unusual feature for a silicate mineral. Another group of investigators (HAYASHI et al., 1965) reports that even at 3.5 kbar and 750°C willemite decomposes into a monoclinic zinc pyroxene and zinc oxide. According to the latter authors, monoclinic zinc pyroxene and ZnO form orthorhombic zinc pyroxene and willemite at 37 kbar and 1,250°C. Subsequent transformation to a zinc olivine takes place above 42 kbar and 1,500°C. TAYLOR et al. (1971) mention the existence of a Zn_2SiO_4 spinel phase at high pressures.

Solid solubilities in the binary $Mg_2SiO_4-Zn_2SiO_4$ and $MgSiO_3-ZnSiO_3$ have been investigated by SEGNIT and HOLLAND (1965), the former system exclusively by SARVER and HUMMEL (1962).

The stability fields of sauconite, hemimorphite and willemite have been published by ROY and MUMPTON (1956) from an experimental study of the system $ZnO-SiO_2-H_2O$. From this investigation, it is to be expected that below 3 kbar,

---

[1] Quartz might have been contaminated by intergrown minerals higher in zinc.

sauconite is stable only upto 225°C and hemimorphite between 200 and 260°C. For the formation of willemite in the presence of water, a temperature of 250°C may serve as a minimum.

c) Carbonates

Decomposition temperatures and pressures of smithsonite have been investigated by HARKER and HUTTA (1956). At 310°C the equilibrium decomposition pressure is almost 800 atm and at 440°C about 2,800 atm. According to JAMBOR (1964) the dissociation temperatures of hydrozincite, $Zn_5(CO_3)_2(OH)_6$, and other basic carbonates are higher.

d) Sulfides

At normal pressures, transition from pure sphalerite to wurtzite occurs at $1,020 \pm 5°C$ (HANSEN, ANDERKO, 1958). The transition temperature is lowered by solid solution of O, Cd, Mn and Fe. The transition region can include stable polytypes. 56 mole percent FeS depresses the inversion temperature to approximately 875°C.

Most important in the use of natural impurities of sphalerite to determine the environment of formation, is a knowledge of the ternary system Fe—Zn—S.

The sulfur-bearing part of the Fe—Zn—S system is simple and contains only sphalerite, (wurtzite), pyrrhotite, pyrite and sulfur as phases in the range 200 to 700°C (BARTON et al., 1963; BARTON and SKINNER, 1967). At 750°C and low pressure, pyrite disappears as a stable compound of this system. The range of substitution of Fe for Zn is wide on the nearly binary join, FeS—ZnS. Sphalerite departs from a

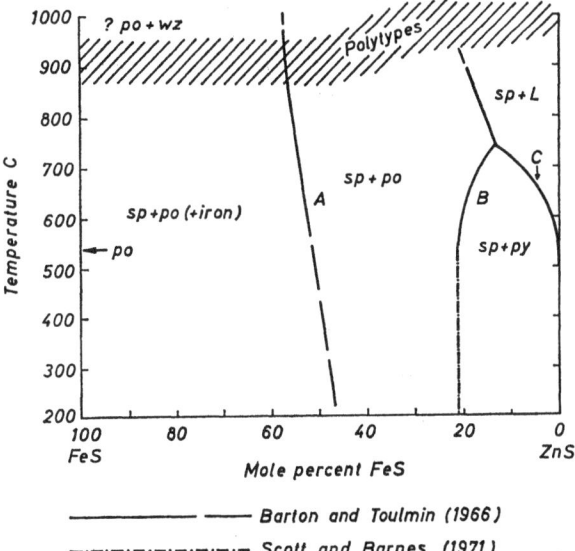

Fig. 30-D-1. Compositions of sphalerites in the system Fe-Zn-S in the temperature range 200 to 1,000° C after BARTON and SKINNER (1967) (with additions after SCOTT and BARNES, 1971). Vapor is present in all assemblages. Symbols: po = pyrrhotite, py = pyrite, sp = sphalerite, wz = wurtzite, L = liquid sulfur. Curves A, B and C give the compositions of sphalerites in equilibrium with pyrrhotite of FeS composition, with pyrite+pyrrhotite and with pyrite +liquid sulfur respectively

1:1 metal to sulfur ratio by less than 0.1 percent. Pyrrhotite and pyrite dissolve less than 0.1 percent ZnS. The FeS—ZnS solvus as shown in Fig. 30-D-1, represented by curve A, is unfortunately very steep. A change in temperature of 270°C (from 580 to 850°C) is related to only a slight change of equilibrium composition of sphalerite (in the FeS—ZnS system in which the activity of S_2 is buffered or controlled) from 52 to 56 percent FeS. Therefore, the mentioned solvus is hardly of any use as a geothermometer. The univariant composition curve for sphalerites equilibrated with coexistent pyrite plus pyrrhotite has a better slope between 550° and 700° for use as a geothermometer (curve B in Fig. 30-D-1 according to BARTON and SKINNER, 1967). For improved values on the phase boundary of ZnS+FeS and ZnS+FeS_2 below 550°C see SCOTT and BARNES (1971). This paper includes a discussion and references of data on a potential change of curve B published between 1967 and 1971 (BOORMAN, EINAUDI, CHERNYSHEV and ANFILOGOV).

The substitution of Fe for Zn in sphalerite decreases remarkably with increasing pressure (31.1 mole percent Fe at 5 kbar, 900°C; 2.5 mole percent at 60 kbar, 900°C; NEUHAUS and CEMIČ, 1970).

Information on the substitution of Zn in sphalerites by elements other than iron is compiled by FLEISCHER (1955). According to his report, sphalerites often contain Cd and Mn in the 1,000 to 5,000 ppm range. Ga, Ge and In concentrations range, in abundant cases, upto 100 ppm. Sb, As, Se, Ti and Hg are usually lower than 30 ppm.

III. Zinc Minerals

Zinc minerals are listed in Table 30-D-4. It is obvious that a high proportion of all zinc minerals (more than a third) contain hydroxide in structural positions. This observation is in line with the already mentioned importance of hydroxide groups as ligands of compounds in which zinc substitutes for divalent metals of about 0.8 Å ionic radius in sixfold coordinated structural positions.

Table 30-D-4. *Zinc minerals*

Metal:	
Zinc	Zn
Sulfides, selenides, sulfosalts:	
Sphalerite (blende)	ZnS (cubic)
Wurtzite	ZnS (hexag.)
Matraite	ZnS (rhombohedral)
Erythrozincite	(Zn, Mn) S
Stilleite	ZnSe
Guadalcazarite	(Hg, Zn)(S, Se)
Voltzite	Zn(S, As)
Koesterite	Cu_2ZnSnS_4
Renierite	$Cu_3(Fe, Ge, Zn)(S, As)_4$
Tetrahedrite-tennantite series	$(Cu, Ag)^+_{20}(Fe^{2+}, Zn, Hg^{2+}, Ge, Sn)_4 (As, Sb, Bi)^{3+}_8 S_{26}$
Oxides, hydroxides, etc.:	
Zincite	ZnO
Franklinite	$(Zn, Mn)Fe_2O_4$
Donathite	$(Fe^{2+}, Zn, Mg) (Cr, Fe^{3+})_2O_4$
Gahnite	$ZnAl_2O_4$

Table 30-D-4. (Continued)

Hetaerolite	$ZnMn_2O_4$
Nigerite	$(Al, Fe)_{12}(Sn, Zn, Mg, Fe)_3 \cdot O_{22}(OH)_2$
Woodruffite	$(Zn, Mn)_2Mn_5O_{12} \cdot 4H_2O$
Zincian dibraunite	$ZnMn_2O_3 \cdot 2H_2O$
Chalkophanite	$ZnMn_3O_7 \cdot 3H_2O$
Ordonezite	$ZnSb_2O_6$

Tellurites etc.:

Denningite	$(Mn, Ca, Zn) Te_2O_5$
Spiroffite	$(Mn, Zn)_2Te_3O_8$
Zemannite	$(Zn, Fe)_2(TeO_3)_3NaH_{2-x} \cdot H_2O$

Carbonates:

Smithsonite	$Zn[CO_3]$
(Ferrous-, manganous)	$(Zn, Fe^{2+}, Mn)[CO_3]$
Zincian dolomite	$Ca(Mg, Zn)[CO_3]$
Hydrozincite	$Zn_5(OH)_6[CO_3]_2$
Rosasite	$(Zn, Cu)_2(OH)_2[CO_3]$
Aurichalcite	$(Zn, Cu)_5(OH)_6[CO_3]_2$
Loseyite	$(Mn, Zn)_7(OH)_{10}[CO_3]_2$

Sulfates:

Zincosite	$Zn[SO_4]$
Poitevinite	$(Cu, Fe, Zn)[SO_4] \cdot H_2O$
Gunningite	$(Zn, Mn)[SO_4] \cdot H_2O$
Goslarite	$Zn[SO_4] \cdot 7H_2O$
Bianchite	$(Zn, Fe)[SO_4] \cdot 6H_2O$
Sommairite (zincian melanterite)	$(Fe, Cu, Zn)[SO_4] \cdot 7H_2O$
Zincian boothite	$(Cu, Zn)[SO_4] \cdot 7H_2O$
Zincian fauserite	$(Mn, Zn)[SO_4] \cdot 7H_2O$ (or $5H_2O$?)
Zincian roemerite	$(Fe^{2+}, Zn) Fe_2^{3+}[SO_4]_4 \cdot 14H_2O$
Dietrichite	$ZnAl_2[SO_4]_4 \cdot 22H_2O$
Ktenasite	$(Cu, Zn)_3(OH)_4[SO_4] \cdot 2H_2O$
Mooreite	$(Mg, Zn, Mn)_8(OH)_{14}[SO_4] \cdot 4H_2O$
Torreyite	$(Mg, Zn, Mn)_7(OH)_{12}[SO_4] \cdot 4H_2O$
Zincian aluminite	$Zn_3Al_3(OH)_{13}[SO_4] \cdot 2H_2O$
Glaukokerinite	$(Zn, Cu)_{10}Al_4(OH)_{30}[SO_4] \cdot 2H_2O$ (?)
Zincian botryogen	$(Zn, Mg, Mn, Fe^{2+}) Fe^{3+}(OH)[SO_4]_2 \cdot 7H_2O$
Serpierite	$Ca(Cu, Zn)_4(OH)_6[SO_4]_2 \cdot 3H_2O$
Zincian copiapite	$(Zn, Fe^{2+}, Mn)Fe_4^{3+}(OH)_2[SO_4]_6 \cdot 18H_2O$

Phosphates:

Hopeite	$Zn_3[PO_4]_2 \cdot 4H_2O$ (monoclin.)
Parahopeite	$Zn_3[PO_4]_2$ (tricl.)
Tarbuttite	$Zn_2(OH)[PO_4]$
Zincian rockbridgeite	$ZnFe_3^{4+}(OH)_5[PO_4]_3$
Scholzite	$CaZn_2[PO_4]_2 \cdot 2H_2O$
Phosphophyllite	$(Zn, Fe)_3[PO_4]_2 \cdot 4H_2O$
Spencerite	$Zn_4(OH)_2[PO_4]_2 \cdot 3H_2O$
Faustite	$(Zn, Cu)Al_6(OH)_8[PO_4]_4 \cdot 5H_2O$
Veszelyite	$(Cu, Zn)_3(OH)_3[PO_4] \cdot 2H_2O$
Kehoeite	$Zn_{5.5}Ca_{2.5}[AlP(H_3)]_{16}O_{96} \cdot 16H_2O$ (analcite type structure)

Table 30-D-4. (Continued)

Arsenites, arsenates, vanadates, tungstates etc.:

Reinerite	$Zn_3[AsO_3]_2$
Adamite	$Zn_2(OH)[AsO_4]$ (orthorhomb.)
Paradamite	$Zn_2(OH)[AsO_4]$ (triclin.)
Stranskiite	$CuZn_2[AsO_4]_2$
Chlorophoenicite	$(Zn, Mn)_5(OH)_7[AsO_4]$
Holdenite	$(Mn, Ca)_4(Zn, Mg, Fe^{2+})_2(OH)_5O_2[AsO_4]$
Austinite	$Ca\,Zn(OH)\,[AsO_4]$
Koettigite	$Zn_3[AsO_4]_2 \cdot 8H_2O$
Legrandite	$Zn_2(OH)[AsO_4] \cdot H_2O$
Zincian lavendulan	$(Ca, Na)_2(Zn, Cu)_5[AsO_4]_4Cl_4 \cdot 4{-}5H_2O$
Chudobaite	$(Na, K, Ca)(Mg, Zn, Mn)_2H[AsO_4]_2 \cdot 4H_2O$
Mixite	$(Cu, Zn, Fe, Ca)_{10}Bi(OH)_8[AsO_4]_5 \cdot 7H_2O$
Descloizite	$Pb(Zn, Cu)(OH)[VO_4]$
Mottramite	$Pb(Cu, Zn)\,(OH)[VO_4]$
Sanmartinite	$(Zn, Fe)[WO_4]$

Silicates:

Willemite	$Zn_2[SiO_4]$
Troostite	$(Zn, Mn)_2[SiO_4]$
Roepperite	$(Fe^{2+}, Mn, Zn, Mg)_2[SiO_4]$
Larsenite	$PbZn[SiO_4]$
Esperite	$(Ca, Pb)Zn[SiO_4]$
Hodgkinsonite	$MnZn_2(OH)_2[SiO_4]$
Zincian staurolite	$(Fe^{2+}, Mg, Zn)_2(Al, Fe^{3+})_9O_6(O,OH)_2[SiO_4]_4$
Fraipontite	$Zn_8Al_4(OH)_8[SiO_4]_5 \cdot 7H_2O$
Hemihedrite	$ZnPb_{10}[CrO_4]_6F_2[SiO_4]_2$
Macgovernite	$(Mn, Mg, Zn)_{15}(OH)_{14}O[AsO_4]_2[SiO_4]_2$
Yeatmanite	$(Mn, Zn)_{16}Sb_2O_{13}[SiO_4]_4$
Hardystonite	$Ca_2Zn[Si_2O_7]$
Klinohedrite	$Ca_2Zn_2(OH)_2[Si_2O_7] \cdot H_2O$
Hemimorphite	$Zn_4(OH)_2[Si_2O_7] \cdot H_2O$
Fowlerite	$Ca(Mn\,Zn)_4[Si_5O_{15}]$
Zincian chkalovite	$Na_2Zn[Si_2O_6]$
Zincian schefferite	$Ca(Mg, Zn, Mn)[Si_2O_6]$
Zincian aegirine augite ("jeffersonite")	$(Ca, Na)(Fe^{2+}, Zn, Mn, Mg, Fe^{3+})[(Si, Al)_2O_6]$
Zincian cummingtonite	$(Mg, Fe^{2+}, Zn, Mn)_7(OH)_2[Si_8O_{22}]$
Zincian tremolite (actinolite)	$Ca(Mg, Fe^{2+}, Zn, Mn)_5(OH)_2[Si_8O_{22}]$
Zincian magnesio-riebeckite	$Na_2(Mg, Zn, Mn)_3Fe_2^{3+}(OH)_2[Si_8O_{22}]$
Zinalsite	$(Zn, Al, Mg, Ca, Cu, Fe^{3+})_6(OH)_8[(Si,Al)_4O_{10}]$
Zincsilite	$Zn_3[Si_4O_{10}](OH)_2 \cdot 11H_2O$
(end member of the series montmorillonite-sauconite-zincsilite)	
Sauconite	$(Zn, Mg, Al, Ca, Na)_3(OH)_2[(Si,Al)_4O_{10}] \cdot 4H_2O$
Hendricksite	$K(Zn, Mn, Mg, Fe^{2+})_3(OH)_2[Si_3AlO_{10}]$
Zincian serpentine	$(Mg, Fe^{2+}, Zn)_3(OH)_4[Si_2O_5]$
Karpinskyite	$Na_2(Be, Zn, Mg)[Al_2Si_6O_{16}](OH)_2$
Helvite	$(Mn, Zn, Fe^{2+})_8S_2[BeSiO_4]_6$
Genthelvite	$(Zn, Fe, Mn)_8S_2[BeSiO_4]_6$

Manuscript received: March 1972

30-E. Abundance in Common Igneous Rock Types; Crustal Abundance

Information on the abundance of magmatic rock types in the upper continental earth's crust have been reported by the present author in Chapter 7 of Volume I of this handbook. According to these data, granites, granodiorites, quartz monzonites and quartz diorites are by far the most abundant magmatic rock types (about 85 percent) in the major part of the crust. Large proportions of metamorphic rocks are chemical equivalents of the so-called granitic rocks, with a tendency to granodiorites and quartz diorites. The average zinc concentration in granitic rocks will be generally representative for the zinc abundance in the continental crust. Mafic rocks contribute only 10 to 15 percent to the (upper) continental crust. They are the major constituents of the oceanic crust, however, but the oceanic crust represents only 22 percent by volume and mass of the total crust.

I. Standard Reference Rocks
(Analytical Precision and Accuracy of Zinc Determinations)

In recent years it became common practice to use several sets of standard reference rocks to check the accuracy of an analytical method. Series of new data on major and minor elements in rocks are often accompanied by data on the common reference standards for interlaboratory comparison and for the mentioned accuracy check. Because of the importance of the standard reference materials, data from literature are listed in Table 30-E-1, accompanied by the reference of the authors. Only those rocks were selected for which larger sets of data are available. The list is restricted to the more common analytical methods.

The smallest scattering of values around an average can be observed in the group of neutron activation data. The largest sets of neutron activation data are those on W-1 and BCR-1. Eight groups of data from different laboratories on each set scatter only from 76 to 92 ppm Zn (average: 83.5 ppm Zn) and 100 to 133 ppm Zn (average: 119.4 ppm Zn) respectively. Within the groups of X-ray data on the same rocks is a slightly larger scattering, but the averages (77.5 ppm Zn and 124 ppm Zn) are close to those from neutron activation. Because of the close similarity of the averages from such completely different methods as neutron activation and X-ray fluorescence, the accuracy of these methods, as applied, is high. The small scattering indicates high precision as well. There is usually appreciably more scattering within the groups of optical spectrographic and spectrometric methods (including atomic absorption). But with a larger set of twice as much data they give an average for W-1 which is again close to that from neutron activation (81.8 ppm Zn). The spectrographic average for BCR-1 is slightly higher (139.3 ppm Zn) than the neutron activation set because of one single high value. This makes a

Zinc

Table 30-E-1. *Zinc abundances in standard reference materials.* (*The abbreviations used for sample types are those introduced by the producer and reported in literature, see* FLANAGAN, GWYN, 1967) (*Reference numbers in brackets, recommended averages in boxes*)

Sample	Method			
	S/F/A	X	W/C	N/R and I
Dunite DTS-1, Twin Sisters, Washington (U.S.A.)	41 A (32), 140 A (18), 56 S (35), 53 S (36), 55 A (38), 38 S (21), 33.5 A (23), 66 S (44), 67 A (46), 39 A (47)	100 (33), 46 (34), 45 (25), 40 (37), 33 (39), 42 (42)		61 (29), 61 (45), 37.1 I (47), 53 (54)
	$\boxed{58.9}$			N/R $\boxed{58.3}$
Peridotite PCC-1 Cozadero, Calif. (U.S.A.)	52 A (18), 39 S (35), 50 S (36), 25 A (38), 31 S (21), 29 A (23), 57 S (44), 70 A (46), 37 A (47)	100 (33), 43 (34), 40 (25), 45 (37), 38 (39), 41 (42), 44 (40)		42 (29), 44 (45), 38.6 I (47), 45 (54)
	$\boxed{43.3}$			N/R $\boxed{43.7}$
Basalt ("diabase"), W-1 Centerville, Virginia (U.S.A.)	60 S (4), 68 S (5), 78 F (8), 91 A (9), 110 S (10), 100 S (11), 85 S (16), 45 S (17), 91 S (18), 95 (S 19), 86 S (20), 83 S (21), 69 A (22), 80 A (23), 82 S (44), 85 A (27)	78 (2), 69 (12), 80 (13), 84 (24), 76.5 (41), 78 (40), 69 (42), 85 (43)	78 (1), 85 (3), 66 (7), 78 (8), 64 (30)	85 (14), 82.8 (15), 82.8 (24), 86 (27), 76 (28), 83 (29), 80 (45), 89.3 I (47), 72 I (49), 90 I (50), 92 (54)
	$\boxed{81.8}$	$\boxed{77.5}$	$\boxed{74}$	N/R $\boxed{83.5}$
Basalt BCR-1, Bridal Veil, Oregon (U.S.A.)	106 A (32), 278 A (18), 125 S (35), 109 S (36), 120 A (38), 110 S (21), 123.5 A (23), 140 S (44), 153 A (46), 128 A (47)	180 (33), 113 (34), 125 (25), 100 (37), 120 (39), 102 (42), 130 (43), 121 (56), 125 (140)		115 (29), 117 (45), 129.4 I (47), 126.5 (48), 100 (51), 120 (52), 116 (53), 133 (54), 127.4 (55)
	$\boxed{124}$ (excluding: 278 A)	$\boxed{124}$		N/R $\boxed{119.4}$
Basalt BR, Essey-la-Côte Lorraine (France)	140 S (44), 192 A (46), 185 S (57), 175 A (58), 173 S (60),	120 (59)	80 (65), 160 (65)	167 (54)

Zinc

Tabel 30-E-1. (Continued)

Sample	Method			
	S/F/A	X	W/C	N/R and I
	154 S (61), 240 A (64), 155 S (65), 170 A (65), 155 S (66)			
	⎯169⎯			
Andesite AGV-1, Guano Valley, Oregon (U.S.A.)	75 S (31), 92 A (32), 304 A (18), 78 S (35), 85 S (36), 130 A (38), 64 S (21), 85.5 A (23), 90 S (44), 120 A (46), 86 A (47)	200 (33), 81 (34), 85 (25), 80 (37), 82 (34), 80 (39), 75 (42), 96 (43),		87 (29), 88 (45), 86.7 I (47), 96 (54)
	90.5 (excluding: 304 A)	82.7 (excluding: 200)		N/R ⎯90.3⎯
Granodiorite GSP-1, Silver Plume, Colorado (U.S.A.)	80 S (31), 107 A (32), 143 A (18), 110 S (35), 90 S (36) 110 A (38), 54 S (21), 100.5 A (23), 55 S (44), 128 A (46), 107 A (47)	340 (33), 102 (34), 100 (25), 100 (37), 90 (39), 95 (42), 110 (43), 106 (40)		108 (29), 116 (45), 105.1 I (47), 107 (45)
	⎯98.4⎯	99.5 (excluding: 340)		N/R ⎯110.3⎯
Granite G-1, Westerly, Rhode Island (U.S.A.)	20 S (4), 95 S (5), 42 F (8), 49 A (9), 55 S (10), 70 S (11), 55 S (19), 54 S (20), 43.5 A (23), 46 S (44)	26 (2), 29 (12), 40 (13), 45 (24), 40 (25), 44 (26), 33.5 (41), 38 (40), 51 (43)	41 (1), 42 (3), 52.5 (6), 31.5 (7), 41 (8), 48 (30)	44 (14), 45.7 (15), 45.7 (24), 44 (45), 17 I (49)
	⎯53⎯	⎯38.5⎯	⎯43⎯	N/R ⎯44.9⎯
Granite G-2, Westerly, Rhode Island (U.S.A.)	90 S (31), 42 A (32), 138 A (18), 86 S (35), 77 S (36), 105 A (38), 44 S (21), 82.5 (23), 76 S (44), 113 A (46), 81 A (47)	86 (34), 80 (25), 90 (37), 87 (34), 78 (39), 85 (42), 90 (43)		77 (29), 79 (45), 83.9 I (47), 93 (54)
	⎯85⎯	⎯85.2⎯		⎯83⎯

Zinc

Table 30-E-1. (Continued)

Sample	Method			
	S/F/A	X	W/C	N/R and I
Granite GA, Andlau, Vosges (France)	56 S (44), 121 A (46), 18 S (57), 94 A (58), 70 S (60), 88 S (61), 56 A (64), 33 S (65), 100 A (65) $\boxed{71.9}$	83 (59)		33 (65), 90 (65)
Granite GH, Hoggar (Algérie)	103 S (44), 30 S (57), 97 A (58), 90 S (60), 95 S (61), 96 A (64), 70 A (65) $\boxed{83}$	78 (59)	60 (65), 70 (65)	
Granite GR, Senones, Vosges (France)	52 S (44), 12 S (57), 56 A (58), 60 S (60), 70 S (61), 100 S (63), 112 A (64) $\boxed{67}$	58 (59), 83 (62)		

References for Table 30-E-1

1. Huffmann, C., Lipp, H. H., 1958 [in: U.S. Geol. Surv. Bull. **1113**, 103 (1960)].
2. Hower, J., Fancher, T. W., 1957 [in: U.S. Geol. Surv. Bull. **1113**, 103 (1960)].
3. Carmichael, I., McDonald, A., 1961 [in Fleischer, M., Stevens, R. E.: Geochim. Cosmochim. Acta **26**, 525 (1962)].
4. O'Neil, R. L., Suhr, N. H., 1960 [in Fleischer, M., Stevens, R. E.: Geochim. Cosmochim. Acta **26**, 525 (1962)].
5. Brooks, R. R., Ahrens, L. H., Taylor, S. R., 1960 [in Fleischer, M., Stevens, R. E.: Geochim. Cosmochim. Acta **26**, 525 (1962)].
6. Azzaria, L. M., 1963 [in Fleischer, M.: Geochim. Cosmochim. Acta **29**, 1263 (1965)].
7. Greenland, L., 1963 [in Fleischer, M.: Geochim. Cosmochim. Acta **29**, 1263 (1965)].
8. Huffmann, C., 1963 [in Fleischer, M.: Geochim. Cosmochim. Acta **29**, 1263 (1965)].
9. Belt, C. B. Jr., 1964 [in Fleischer, M.: Geochim. Cosmochim. Acta **29**, 1263 (1965)].
10. Ingamells, C. O., Suhr, N. H., 1963 [in Fleischer, M.: Geochim. Cosmochim. Acta **29**, 1263 (1965)].
11. Schroll, E., Weninger, M., 1965 [in Fleischer, M.: Geochim. Cosmochim. Acta **29**, 1263 (1965)].
12. Kalman, Z. H., Heller, L., 1962 [in Fleischer, M.: Geochim. Cosmochim. Acta **29**, 1263 (1965)].
13. Heller, L., 1953 [in Fleischer, M.: Geochim. Cosmochim. Acta **29**, 1263 (1965)].
14. Pierce, T. B., Peck, P. F., 1962 [in Fleischer, M.: Geochim. Cosmochim. Acta **29**, 1263 (1965)].
15. Filby, R. H., 1964 [in Fleischer, M.: Geochim. Cosmochim. Acta **29**, 1263 (1965)].
16. Ataman, G., Besnus, Y., 1965 [in Fleischer, M.: Geochim. Cosmochim. Acta **33**, 65 (1969)].
17. Moxham, R. L., 1965 [in Fleischer, M.: Geochim. Cosmochim. Acta **33**, 65 (1969)].
18. Kulbicki, G., 1966 [in Fleischer, M.: Geochim. Cosmochim. Acta **33**, 65 (1969)].
19. Giles, D. L., 1967 [in Fleischer, M.: Geochim. Cosmochim. Acta **33**, 65 (1969)].

20. TENNANT, W. C., 1967 [in FLEISCHER, M.: Geochim. Cosmochim. Acta **33**, 65 (1969)].
21. CHAMP, W. H., 1968 [in FLEISCHER, M.: Geochim. Cosmochim. Acta **33**, 65 (1969)].
22. BURRELL, D. C., 1965 [in FLEISCHER, M.: Geochim. Cosmochim. Acta **33**, 65 (1969)].
23. HUFFMANN, C., 1968 [in FLEISCHER, M.: Geochim. Cosmochim. Acta **33**, 65 (1969)].
24. BALL, T. K., FILBY, R. H., 1965 [in FLEISCHER, M.: Geochim. Cosmochim. Acta **33**, 65 (1969)].
25. CARMICHAEL, I. S. E., 1967 [in FLEISCHER, M.: Geochim. Cosmochim. Acta **33**, 65 (1969)].
26. GUM, B. M., 1967 [in FLEISCHER, M.: Geochim. Cosmochim. Acta **33**, 65 (1969)].
27. MATSUI, Y., BANNO, S., 1966 [in FLEISCHER, M.: Geochim. Cosmochim. Acta **33**, 65 (1969)].
28. ALIAN, A., SHABANA, R., 1967 [in FLEISCHER, M.: Geochim. Cosmochim. Acta **33**, 65 (1969)].
29. BRUNFELT, A. O., JOHANSEN, O., STEINNES, E., 1967 [in FLEISCHER, M.: Geochim. Cosmochim. Acta **33**, 65 (1969)].
30. MAH, D. C., MAH, S., TUPPER, W. M., 1965 [in FLEISCHER, M.: Geochim. Cosmochim. Acta **33**, 65 (1969)].
31. GOVINDARAJU, K., 1965 [in FLANAGAN, F. J.: Geochim. Cosmochim. Acta **33**, 81 (1969)].
32. LEWIS, C. L., 1966 [in FLANAGAN, F. J.: Geochim. Cosmochim. Acta **33**, 81 (1969)].
33. SINE, N. M., 1967 [in FLANAGAN, F. J.: Geochim. Cosmochim. Acta **33**, 81 (1969)].
34. GUNN, B. M., 1967 and 1967a [in FLANAGAN, F. J.: Geochim. Cosmochim. Acta **33**, 81 (1969)].
35. LE RICHE, H. H., 1967 [in FLANAGAN, F. J.: Geochim. Cosmochim. Acta **33**, 81 (1969)].
36. KULBICKI, G., SOURISSE, C., BARADAT, J., 1967 [in FLANAGAN, F. J.: Geochim. Cosmochim. Acta **33**, 81 (1969)].
37. PRICE, N. B., 1967 [in FLANAGAN, F. J.: Geochim. Cosmochim. Acta **33**, 81 (1969)].
38. QUESNEL, D. F., 1968 [in FLANAGAN, F. J.: Geochim. Cosmochim. Acta **33**, 81 (1969)].
39. WEDEPOHL, K. H., 1968 [in FLANAGAN, F. J.: Geochim. Cosmochim. Acta **33**, 81 (1969)].
40. WEDEPOHL, K. H. (unpublished).
41. KAYE, M. J.: [Geochim. Cosmochim. Acta **29**, 139 (1965)].
42. MURAD, E.: (Regensburg, Germany) (personal communication, 1968).
43. SCHNEIDER, G. [Neues Jahrb. Mineral. Monatsh. **11**, 504 (1969)].
44. HUBER-SCHAUSBERGER, I., JANDA, I., DOLEZEL, P., SCHROLL, E. [Tschermaks Mineral. Petrog. Mitt. **14**, 195 (1970)].
45. LAUL, J. C., CASE, D. R., WECHTER, M., SCHMIDT-BLEEK, F., LIPSCHUTZ, M. E. [J. Radioanal. Chem. **4**, 241 (1970)].
46. CIONI, R., INNOCENTI, F., MAZZUOLI, R. [Chem. Geology **7**, 19 (1971)].
47. ROSMAN, K. J. R., JEFFERY, P. M. [Chem. Geology **8**, 25 (1971)].
48. GANAPATHY, R., KEAYS, R. R., LAUL, J. C., ANDERS, E. [Geochim. Cosmochim. Acta, Suppl. 1, **34**, 1117 (1970)].
49. BROWN, R., WOLSTENHOLME, W. A., 1964 [in FLEISCHER, M.: Geochim. Cosmochim. Acta **33**, 65 (1969)].
50. NICHOLLS, G. D., GRAHAM, A. L., WILLIAMS, E., WOOD, M., 1967 [in FLEISCHER, M.: Geochim. Cosmochim. Acta **33**, 65 (1969)].
51. HASKIN, L. A., ALLEN, R. O., HELMKE, P. A., PASTER, T. P., ANDERSON, M. R., KOROTEV, R. L., ZWEIFEL, K. A. [Geochim. Cosmochim. Acta, Suppl. 1, **34**, 1213 (1970)].
52. MAXWELL, J. A., PECK, L. C., WIIK, H. B. [Geochim. Cosmochim. Acta, Suppl. 1, **34**, 1369 (1970)].
53. ANDERS, E., GANAPATHY, R., KEAYS, R. R., LAUL, J. C., MORGAN, J. W. [Proc. Sec. Lunar Sci. Conf. **2**, 1021 (1971)].
54. BAEDECKER, P. A., SCHANDY, R., ELZIE, J. L., KIMBERLIN, J., WASSON, J. T. [Proc. Sec. Lunar Conf. **2**, 1037 (1971)].
55. BRUNFELT, A. O., HEIER, K. S. [Proc. Sec. Lunar Conf. **2**, 1281 (1971)].
56. HOFMEYR, P. K. (PhD-Thesis Capetown, 1971).

57. Moxham, R. L., 1966—1968 [in Roubault, M. et al.: Sci. Terre 13, 379 (1968)].
58. Govindaraju, K., 1966—1968 [in Roubault, M. et al.: Sci. Terre 13, 379 (1968)].
59. Kaye, M., 1966—1968 [in Roubault, M. et al.: Sci. Terre 13, 379 (1968)].
60. Moal, J. Y., Béguinot, J., Rueil, G., Vannier, M., 1966—1968 [in Roubault, M. et al.: Sci. Terre 13, 379 (1968)].
61. Kulbicki, G. 1966—1968 [in Roubault, M. et al.: Sci. Terre 13, 379 (1968)].
62. Quintin, M., Martin, A., 1966—1968 [in Roubault, M. et al.: Sci. Terre 13, 379 (1968)].
63. Ingamells, C. O., Suhr, N. H., 1966—1968 [in Roubault, M. et al.: Sci. Terre 13, 379 (1968)].
64. Muller, V., 1966—1968 [in Roubault, M. et al.: Sci. Terre 13, 379 (1968)].
65. Branche, G., Agrinier, H., 1966—1968 [in Roubault, M. et al.: Sci. Terre 13, 379 (1968)].
66. Govindaraju, K. [in Roubault, M. et al.: Sci. Terre 11, 105 (1966)].

generally high accuracy (with a few exceptions) for spectrographic methods probable; but precision is expected to be appreciably less than that with neutron activation and X-ray methods. The sensitivity of the methods for zinc decreases in the sequence: N/R, A, W/C, X, S, having the optimum slightly lower than 1 ppm.

II. Peridotitic Rocks

Two large classes of peridotitic rocks can be distinguished depending on the pressure regime of their origin: garnet peridotites from more than 20 kbar pressure, and spinel peridotites formed at pressures lower than 20 kbar. The few available zinc analyses of peridotites are restricted to the latter type. They are listed in Table 30-E-2. As evidenced by the data of Table 30-D-3, the major minerals of peridotite as olivine, diopside and orthopyroxene usually contain zinc in the 40 to 80 ppm range. Chromium spinels of abundant peridotites are reported as having 800 ppm Zn (Carter, 1970). Garnets are usually higher in zinc than olivine and pyroxenes.

It cannot be concluded from our present information whether the major portion of zinc occurs in the olivine (and pyroxene) crystal structure or in inclusions of spinels, which are abundant in olivine.

Average zinc in spinel peridotites is expected to be 56 ppm Zn. This figure is close to the average zinc in ordinary chondrites (57.6 ppm Zn, of CH, CL classes: Table 30-C-5). There is no large difference expected in zinc between the upper mantle of the earth and ordinary chondrites.

Serpentinites contain about the same average zinc concentration as peridotites indicating no general loss or gain during serpentinization.

III. Gabbroic and Basaltic Rocks

The zinc content of basaltic and gabbroic rocks is mainly contained in their magnetite fraction. Desborough (1963) reports the occurrence of sphalerite in some mafic rocks from Missouri. According to Newhouse (1936), Ramdohr (1940) and Wedepohl (1953), sphalerite is a rare mineral in magmatic rocks.

Data on zinc abundances in mafic rocks are listed in Table 30-E-3.

Table 30-E-2. *Zinc abundances in periotitic rocks (Number of samples in brackets)*

Rock type	Origin	Range ppm Zn	Average ppm Zn	Analytical method	Reference
Lherozolite nodules from alkali olivine basalt (8)	Tertiary, northwest Germany	55—91	71	X	1
Lherzolite nodules from nephelinite ejecta (2)	Dreiser Weiher, Eifel (Germany)	60—85	68	X	1
Lherzolite nodules from alkali basalt[a]	Kilbourne Hole, New Mexico (U.S.A.)	—	67	A	2
Garnet lherzolite nodules from kimberlite	South Africa	25—69	42	A	6
Peridotite (PCC-1)	Cozadero, Calif. (U.S.A.)	—	43.5[b]	N/R, S	3
Dunite (DTS-1)	Twin Sisters, Washington (U.S.A.)	—	58.6[b]	N/R,(S)	3
Peridotite	Salt Lake, Honolulu (Hawaii)	—	52	N/R	4
Peridotites (2)	Finero, Ivrea area (Italy)	53—63	58	X	1
Peridotite	Sunmoere, Norway	—	41	X	1
Peridotites (2)	Iwate, Nagano Prefectures (Japan)	40—80	60	W/C	5

[a] Numerous nodules of the most abundant type having nine percent fayalite in olivine.
[b] Recommended after evaluation of Table 30-E-1.

References: 1. WEDEPOHL (unpublished). 2. CARTER (1970). 3. Table 30-E-1. 4. BAEDECKER, SCHAUDY et al. (1971). 5. MORITA (1955). 6. CHEN (1971).

Table 30-E-3. *Zinc concentrations in gabbroic and basaltic rocks (including spilites). (Number of samples in brackets)*

Rock type	Origin	Range ppm Zn	Average ppm Zn	Analytical method	Reference
Gabbros (4)	Honshu (Japan)	80—135	100	W/C	1
Gabbro	Precambrian, Duluth, Minnesota (U.S.A.)	—	100	W/C	2
Gabbros, norites (6)	Variscan, N. Tien-Shan (U.S.S.R.)	73—121	100	W/C	3
Gabbros (composite of 11)	Germany	—	102	X	4
Basalts, tholeiitic (19)	Keweenawan, Michigan (U.S.A.)	75—175	102	W/C	2

Table 30-E-3. (Continued)

Rock type	Origin	Range ppm Zn	Average ppm Zn	Analytical method	Reference
Basalts, tholeiitic (6)	Keweenawan etc., Minnesota (U.S.A.)	80—230	137	W/C	2
Basalts, tholeiitic etc. (163)	Precambrian, Coppermine River (Canada)	30—420	154	S	5
Basalts (mainly tholeiitic, olivine bearing) (16)	Japan, Korea, Manchuria	70—150	102	W/C	1
Basalts, tholeiitic ("diabases" etc.) (3)	Jurassic, Mount Wellington, Tasmania	75—171	109	P	6
Basalts, tholeiitic ("dolerites") (20)	Jurassic, Mount Wellington etc., Tasmania	42—89	67	X	7
Basalts, tholeiitic quartz normative ("dolerite" dikes) (18)	Mesozoic, Connecticut (U.S.A.)	—	88.2	A	8
Basalts, tholeiitic quartz normative ("dolerite" dikes) (9)	Mesozoic, Virginia (U.S.A.)	—	94	A	8
Basalt, W-1, tholeiitic, quartz normative ("dolerite" dike) (1)	Mesozoic, Centerville, Virginia (U.S.A.)	—	83.5	N/R	9
Basalts, quartz normative ("dolerite" dikes) (4)	Mesozoic, North Carolina	—	101	A	8
Basalts, tholeiitic quartz normative ("dolerite" dikes) (15)	Mesozoic, Nova Scotia (Canada), Connecticut, Pennsylvania (U.S.A.)	—	84	A	8
Basalts, tholeiitic quartz normative ("dolerite" dikes) (15)	Mesozoic, Georgia, Alabama (U.S.A.)	—	85	A	8
Basalts, tholeiitic olivine normative ("dolerite" dikes) (40)	Mesozoic, North and South Carolina, Virginia, Pennsylvania (U.S.A.)	—	84	A	8
Basalts, tholeiitic olivine normative ("dolerite" dikes) (7)	Mesozoic, Georgia, Alabama (U.S.A.)	—	85	A	8
Basalts, tholeiitic (31)	Mesozoic, Connecticut (U.S.A.)	64—120	90	W/C	10
Basalts, tholeiitic etc. (32)	Snake River area, Idaho (U.S.A.)	93—180	140	W/C	10

Table 30-E-3. (Continued)

Rock type	Origin	Range ppm Zn	Average ppm Zn	Analytical method	Reference
Basalts, tholeiitic etc. (28)	Mt. Hood area, Oregon (U.S.A.)	48—120	80	W/C	10
Basalts, tholeiitic etc. (20)	Mt. Lassen area, California (U.S.A.)	57—89	72	W/C	10
Basalt, tholeiitic (1)	Clear Lake area, California (U.S.A.)	—	75	W/C	2
Basalts, tholeiitic etc. (14)	Jemez Mountains, New Mexico (U.S.A.)	68—100	84	W/C	10
Basalts, tholeiitic etc. (34)	Hawaii (U.S.A.)	89—120	100	W/C	10
Basalts, tholeiitic (8)	Hawaii, Midatlantic Ridge (2)	82—123	107	N/R	11
Basalts, mainly tholeiitic (three composites with a total of 241)	America, Africa, India, Oceanic Islands (YCR-1, CB-I; YCR-2, CB-II; YCR-3, CB-III of Yale)	—	119	X	4
Basalts, tholeiitic (composite of 9)	Deccan Peninsula	—	108	X	4
Basalts, olivine tholeiitic (15)	Erta'Ale (Ethiopia)	42—130	89	P	12
Basalts, picritic (3)	Erta'Ale (Ethiopia)	48—67	60	P	12
Basalts, oceanic tholeiitic (4)	Midatlantic Ridge (22° N)	85—125	100		13
Basalts, olivine tholeiitic (25)	Tertiary, Baffin Island (West Greenland)	—	57	X	14
Basalts, olivine tholeiitic (17)	Tertiary, Svartenhuk (West Greenland)	—	70	X	14
Basalts, tholeiitic (14)	Tertiary, Svartenhuk (West Greenland)	—	81	X	14
Basalts, picritic (21)	Tertiary, Svartenhuk, Baffin Island (West Greenland)	—	64	X	14
Basalts (2)	Calemanian Mountains (Roumania)	—	80	W/C	15
Basalts, quartz tholeiitic (8)	Tertiary, Northern Hessia (Germany)	110—140	122	X	4
Basalts, alkali olivine (22)	Tertiary, Lower Saxony, Hessia (Germany)	81—142	108	X	4

Table 30-E-3. (Continued)

Rock type	Origin	Range ppm Zn	Average ppm Zn	Analytical method	Reference
Basalts (pillow lavas) (174)	Troodos (Cyprus)	mainly: 40—80	62 (90)[a]	A	16
Spilites (tholeiitic) (21)	Variskian geosyncline (northwest Germany)	62—220	104	X	17
Metabasalts (577)	East Amisk area, Saskatchewan (Canada)	mainly[b]: 65—125	94	X	18

[a] Centres of pillows (averages of four classes).
[b] For frequency distribution see Plate 2 in SMITH (1964).

References: 1. MORITA (1955). 2. SANDELL, GOLDICH (1943). 3. ZLOBIN et al. (1965). 4. WEDEPOHL (unpublished). 5. BARAGAR (1969). 6. SMYTHE, GATEHOUSE (1955). 7. TILLER (1959). 8. WEIGAND, RAGLAND (1970). 9. Recommended from Table 30-E-1. 10. RADER et al. (1960, 1963). 11. BAEDECKER, SCHAUDY et al. (1971). 12. TREUIL et al. (1971). 13. MELSON et al. (1968). 14. CLARKE (1970). 15. SAVUL et al. (1956a). 16. GOVETT, PANTAZIS (1971). 17. HERRMANN, WEDEPOHL (1970). 18. SMITH (1964).

Tholeiitic basalt is by far the most abundant basalt type. From the data listed in Table 30-E-3, it cannot be expected that there exists a systematic difference in zinc between gabbros and basalts, between quartz tholeiitic and olivine tholeiitic basalts, between continental and oceanic tholeiitic basalts, between tholeiitic and

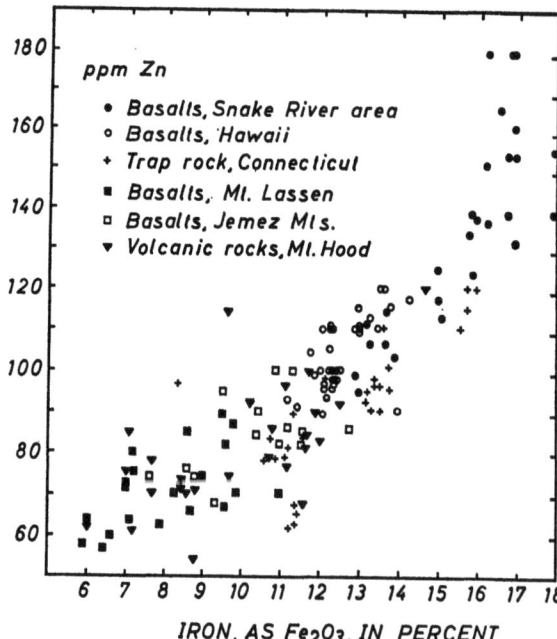

Fig. 30-E-1. Plot of zinc to total iron, calculated as Fe_2O_3 in basalts from six different areas in the United States. (After RADER et al., 1960)

alkali olivine basalts, between basalts and metabasalts (spilites etc.). Picritic basalts are slightly lower in zinc than the average because of the higher abundance of olivine. RADER *et al.* (1960, 1963) as well as WEIGAND and RAGLAND (1970) have clearly demonstrated a correlation between zinc and iron in basalts (see Fig. 30-E-1, after RADER *et al.* 1960, 1963).

The range of scattering of the averages from different authors and areas is not large, mainly between 80 and 120 ppm Zn. A gross average of the sample sets of Table 30-E-3 is close to 100 ppm Zn. The occurrence of regional differences can be concluded from reports such as that of RADER *et al.* (1960, 1963), considering different volcanic areas. A high precision and accuracy of the methods applied by the latter authors can be derived from the information they give.

IV. Alkalic Rocks
(Including Monzonites, Syenites and Trachytes)

We can only present zinc data on a few typical rocks but not on the whole variety of alkalic rocks. They are listed in Table 30-E-4. It is to be expected and has been proved in a few samples (GERASIMOVSKII, NESMEYANOVA, 1960; ZLOBIN, GORSHKOVA, 1961) that again the iron magnesium minerals (amphibole, biotite, aegirine etc.) are the major zinc carriers. Sphalerite occurs locally (Lovozero massiv). The most abundant alkalic rocks (including syenites, trachytes, monzonites) generally have less than half the total iron content of basaltic and gabbroic rocks. But the average zinc concentration in the former rocks is higher than 50 ppm Zn (half the average zinc of basalts). This indicates processes of melting which accumulate zinc relative to iron.

Assimilation of limestones and dolomites, which can locally form alkalic rocks, will dilute the zinc concentrations of magmatic melts because of low zinc concentrations in carbonate rocks.

Table 30-E-4. *Zinc abundances in alkalic rocks. (Number of samples in brackets)*

Rock type	Origin	Range ppm Zn	Average ppm Zn	Analytical method	Reference
Monzonites (2)	Iwate, Gifu (Japan)	40—100	70	W/C	1
Monzonites (6)	Sandyk Mountains, Northern Kirgiziya (U.S.S.R.)	—	65	W/C	2
Syenites (4)	Korea, Japan	35—280	171	W/C	1
Syenites (34)	Sandyk Mountains, Northern Kirgiziya (U.S.S.R.)	31—75	59	W/C	2
Syenites (27)	Kzyl Ompul Massif, Northern Tien-Shan (U.S.S.R.)	14—120	68	W/C	3

Table 30-E-4. (Continued)

Rock type	Origin	Range ppm Zn	Average ppm Zn	Analytical method	Reference
Syenites (8)	Matcha Massif, Southern Tien-Shan (U.S.S.R.)	35—178	81	W/C	4
Syenites (4)	Northern Tien-Shan (U.S.S.R.)	18—71	48	W/C	5
Trachytes (4)	Westerwald (Germany)	55—135	93	X	6
Trachytes (5)	Phlegrean Fields etc. Naples area (Italy)	62—165	91	X	6
Trachyte (1)	Ishikawa (Japan)	—	125	W/C	1
Trachytes (11)	Erta'Ale (Ethiopia)	48—155	122	P	7
Nepheline syenites (2)	Korea	50—160	105	W/C	1
Nepheline syenites (composite of 22)	Brazil, U.S.A., Canada, Portugal, Roumania, Norway	—	110	X	6
Nepheline syenites (15)	Sandyk Mountains, Northern Kirgiziya (U.S.S.R.)	36—84	53	W/C	2
Nepheline syenites (foyaites, khibinites, ijolites) (4)	Khibina Massif, Kola Peninsula (U.S.S.R.)	42—104	69	W/C	2
Nepheline syenites (5)	First intrusive stage, Lovozero Massif, Kola Peninsula (U.S.S.R.)	104—200	145	W/C	8
Nepheline syenites (foyaites, urtites etc.) (15)	Second (main) stage, Lovozero Massif (U.S.S.R.)	102—326	189	W/C	8
Nepheline syenites (lujavrites etc.) (10)	Third (major) stage, Lovozero Massif (U.S.S.R.)	115—1,070	260	W/C	8
Nepheline syenites (5)	Third substage, Kzyl Ompul Massif, Northern Tien-Shan (U.S.S.R.)	35—87	62	W/C	3
Nepheline syenites (8)	Matcha Massif, Southern Tien-Shan (U.S.S.R.)	42—231	105	W/C	4
Phonolites (2 composites of 24)	Czechoslovakia, Germany	—	98	X	6
Phonolites (3)	Eifel, Westerwald (Germany)	90—119	108	X	6

Table 30-E-4. (Continued)

Rock type	Origin	Range ppm Zn	Average ppm Zn	Analytical method	Reference
Nepheline tephrite (1)	Eifel (Germany)	—	81	X	6
Leucite tephrites (4)	Laach area, Eifel (Germany)	50—94	70	X	6
Leucite tephrites (17)	Vesuvius area (Italy)	60—90	80	X	9
Leucite tephrite (1)	Mount Somma (Italy)	—	75	W/C	1
Nephelinites (11)	Northern Hessia, Suebian Mountains (Germany)	54—134	90	X	6
Nephelinite (1)	Nagahama (Japan)	—	160	W/C	1
Leucitites (vesuvites) (21)	Vesuvius area (Italy)	70—105	80	X	9

References: 1. MORITA (1955). 2. ZLOBIN, GORSHKOVA (1961). 3. GAVRILIN, PEVTSOVA (1963). 4. GAVRILIN et al. (1965). 5. ZLOBIN et al. (1965). 6. WEDEPOHL (unpublished). 7. TREUIL et al. (1971). 8. GERASIMOVSKII, NESMEYANOVA (1960). 9. SAVELLI (1967).

The scattering of averages of zinc concentrations among sets of syenites (with an average zinc content of 70 ppm) and trachytes is comparable to that among basalts but less than that among sets of nepheline syenites. There seems to occur a regional difference in zinc among the so-called alkalic rock provinces as Khibina-, Lovozero-, etc. massifs. Within one massif an increase of zinc in the sequence of stages could be observed (GERASIMOVSKII, NESMEYANOVA, 1960).

V. Diorites, Andesites

The intermediate mineral composition of magmatic rocks between granite and gabbro is expected to result in intermediate zinc averages. Zinc data, as listed in Table 30-E-5, show this intermediate range of zinc concentrations.

For the 132 andesites and diorites listed in Table 30-E-5, an average of 69.9 ppm Zn can be computed. The scattering of the averages for the different sets around 70 ppm Zn is not large. No systematic difference in zinc between andesites and diorites can be observed.

VI. Granitic Rocks and Related Effusives

The zinc abundance in granitic rocks (see Table 30-E-6) is mainly controlled by their biotite (and/or amphibole) content as has been reported by WEDEPOHL (1953), TAUSON and KRAVCHENKO (1956), BELT (1960), PUTMAN and BURNHAM (1963), HAACK (1969) etc. Light granitic rocks often have lower zinc concentrations than dark varieties. The scattering of averages for sets of data from different localities and authors is mainly in the range from 30 to 70 ppm. Within the group of granites

Table 30-E-5. *Zinc abundances in intermediate rocks. (Number of samples in brackets)*

Rock type	Origin	Range ppm Zn	Average ppm Zn	Analytical method	Reference
Diorite (1)	Minnesota (U.S.A.)	—	100	W/C	SANDELL, GOLDICH (1943)
Diorites (11)	Payson, Arizona (U.S.A.)	30—100	65	X	PUTMAN, BURNHAM (1963)
Diorites (25)	East Amisk, Saskatchewan (Canada)	5—75	42	X	SMITH (1964)
Diorites (3)	Japan	65—95	77	W/C	MORITA (1955)
Diorites (27)	Bavarian Mountains, Spessart, Odenwald (Germany)	41—127	93	X	OKRUSCH, RICHTER (1969)
Diorites (3)	Variscan, Tien-Shan (U.S.S.R.)	65—126	96	W/C	ZLOBIN et al. (1965)
Andesites (26)	Japan	60—115	87	W/C	MORITA (1955)
Andesites (13)	Calimanian Mountains (Roumania)	—	63	W/C	SAVUL et al. (1956a)
Andesites (2 composites of 20)	Methana Island, Aegina Island (Greece)	—	60	X	WEDEPOHL (unpublished)
Andesites (2)	California (U.S.A.)	55—70	63	W/C	SANDELL, GOLDICH (1943)
Andesite (AGV-1)	Guano Valley, Oregon (U.S.A.)	—	90.4 (recommended)	N/R, S	Table 30-E-1

is a slightly larger scattering of averages than in the group of granodiorites (and quartz diorites). There exists a great number of data for these rock classes to give reliable arithmetic means.

In these averages there is no large systematic difference observable between the majority of granites and granodiorites: 482 granites have an average of 48.0 ppm Zn and 624 granodiorites an average of 52.4 ppm Zn. Because of this small difference the wrong classification of a few samples cannot influence the result. A frequency distribution of 883 granitic rocks of Table 30-E-6 (including all groups: granites, quartz monzonites granodiorites, and quartz diorites) has been plotted in Fig. 30-E-2. This is only slightly different from the curve representing 261 granites and quartz monzonites exclusively. The latter group has more of a tendency to lower zinc values. Higher values of 50 to 90 ppm Zn are not so rare in both groups of granitic rocks. There is no ready explanation for this observation.

Zinc

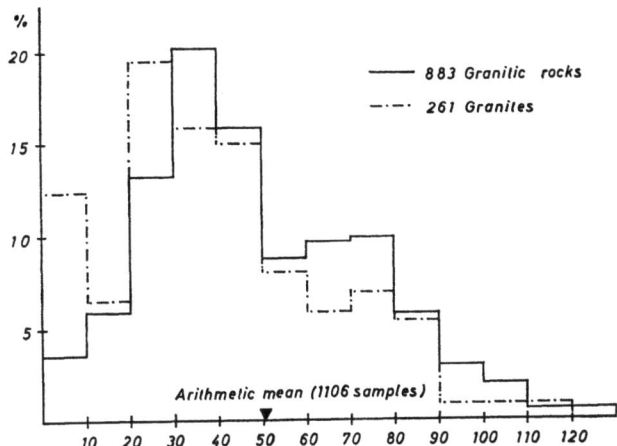

Fig. 30-E-2. Frequency distribution of 883 granitic rocks from all over the world, listed in Table 30-E-6 (heavy line) and of 261 granites and quartz monzonites (broken line)

The abnormally high zinc concentrations of granitic rocks from the alkaline province of Northern Nigeria have not been considered in the averages. Rhyolites and obsidians from this area are again higher than such rocks from other parts of the world. This might be due to a local anomaly. On the other side, it must be mentioned that the average of 234 *rhyolites, liparites, pitchstones and obsidians* (see Table 30-E-7), excluding rocks from northern Nigeria, is twice as high (97.8 ppm Zn) as that of granitic rocks. In their relatively high zinc concentration rhyolites and related glasses are closer to the basaltic rocks than to granitic types.

Besides this observation, the special accumulation of zinc in alkaline rhyolites as comendites and pantellerites must be mentioned; this is even higher than that of a group of nepheline syenites.

Table 30-E-6. *Abundance of zinc in granitic intrusive rocks. (Number of samples in brackets)*

Rock type	Origin	Range ppm Zn	Average ppm Zn	Analytical method	Reference
Granite (G-1)	Westerly, Rhode Island	—	44 (recommended)	N/R, W/C	1
Granite (G-2)	Westerly, Rhode Island (U.S.A.)	—	84 (recommended)	N/R, X, S	1
Granites (3)	Minnesota (U.S.A.)	50—90	70	W/C	2
Granites (4)	Missouri (U.S.A.)	15—85	35	W/C	2
Granites (6)	Texas (U.S.A.)	40—120	80	W/C	2
Granite (1)	New Hampshire (U.S.A.)	—	80	W/C	2
Granites (and granite gneisses) (composite of 50)	Precambrian, Western U.S.A. (Yale: YCR-5-CGII)	—	75	X	3

Table 30-E-6. (Continued)

Rock type	Origin	Range ppm Zn	Average ppm Zn	Analytical method	Reference
Granites, quartz monzonites (4)	Front Range, Colorado (U.S.A.)	48—94	70	W/C	4
Granites (35)	East Amisk, Saskatchewan (Canada)	25—75[a]	48	X	5
Granite (1)	Grenville, Ontario (Canada)	—	29	W/C	6
Granites, quartz monzonites (91)	northwest and central Arizona (U.S.A.)	6—210	35	X	7
Granites (5) adamellite (1)	Central Honshu, Shikoku, Kyushu (Japan)	20—70	46	W/C	8
Granite (1)	Eklara (India)	—	48	X	3
Granites, (alkaline, often riebeckite bearing) (23)	northern Nigeria	75—1,150	319[b]	P	9
Granites (composite of 14)	Germany	—	66	X	3
Granites (12)	Germany	21—103	68	X	3
Granites (17)	eastern and southern Alps	10—40	24	W/C	21
Granites (2)	Aare, Bergell (Switzerland)	20—41	30	X	3
Granites (8)	Spessart (Germany)	26—89	50	X	19
Granites (3) (GR, GA, GH)	Vosges, Hoggar (France)	67—83	74	S	1
Granites (19)	Bavarian Mountains, Black Forest (Germany), Upper Austria	9—108	49	X	10
Granites (4)	Caledonian, Susamyr, Tien-Shan (U.S.S.R.)	40—72	52	W/C	11
Granites (light) (5)	Susamyr, Tien-Shan (U.S.S.R.)	24—40	28	W/C	12
Granites (6)	Kzyl-Ompul, Tien-Shan (U.S.S.R.)	12—30	23	W/C	13
Granites (6)	Alay Range, Tien-Shan (U.S.S.R.)	45—179	89	W/C	14
Granites, granodiorites (36)	Paleozoic, East Transbaikalia (U.S.S.R.)	—	50	W/C	15

Table 30-E-6. (Continued)

Rock type	Origin	Range ppm Zn	Average ppm Zn	Analytical method	Reference
Adamellites, quartz diorites, syenites (31)	Variscan, Sonkul, Tien-Shan (U.S.S.R.)	16—69	41	W/C	16
Adamellites, granites (17)	Variscan, north Tien-Shan (U.S.S.R.)	10—65	39	W/C	17
Granites (22)	Paleozoic, east Transbaikalia (U.S.S.R.)	—	46	W/C	15
Granites, granodiorites (76)	Mesozoic, Transbaikalia (U.S.S.R.)	—	44	W/C	15
Granodiorite (GSP-1)	Silver Plume, Colorado (U.S.A.)	—	105 (recommended)	N/R, X	1
Granodiorites (2)	Minnesota (U.S.A.)	—	60	W/C	2
Tonalites (37)	Hanover, New Mexico (U.S.A.)	11—141	45	W/C	23
Granodiorites, quartz diorites (6)	northwest and central Arizona (U.S.A.)	17—70	56	X	7
Granodiorites (23)	Santa Rosa, Nevada (U.S.A.)	37—72	50	S	18
Granodiorites (hornblende bearing) (197)	East Amisk, Saskatchewan (Canada)	15—235[a]	73	X	5
Granodiorites (biotite bearing) (189)	East Amisk, Saskatchewan (Canada)	5—65[a]	36	X	5
Granodiorites (3)	Central Honshu (Japan)	65—75	70	W/C	8
Granodiorites (6)	Spessart (Germany)	57—75	68	X	19
Granodiorites (4)	Bavarian Mountains	62—83	74	X	20
Granodiorites, tonalites (86)	Eastern and southern Alps (Austria, Italy)	15—55	32	W/C	21
Quartz diorite (1)	Calimanian Mountains (Roumania)	—	41	W/C	22
Granodiorites, granites, granite gneisses (composite of 27)	Finland (Yale: YCR-4-CGI)	—	65	X	3
Granodiorites (20)	Susamyr, Tien-Shan (U.S.S.R.)	32—81	59	W/C	12

Table 30-E-6. (Continued)

Rock type	Origin	Range ppm Zn	Average ppm Zn	Analytical method	Reference
Granodiorite (1)	Kzyl-Ompul, Tien-Shan (U.S.S.R.)	—	60	W/C	13
Granodiorites, quartz diorites (9)	Variscan, north Tien-Shan (U.S.S.R.)	38—81	60	W/C	17

[a] For frequency distribution see Plate 2 in SMITH (1964).
[b] Excluded from computation of average.

References: 1. Table 30-E-1. 2. SANDELL, GOLDICH (1943). 3. WEDEPOHL (unpublished). 4. HUFF (1952). 5. SMITH (1964). 6. AZZARIA (1963). 7. PUTMAN, BURNHAM (1963). 8. MORITA (1955). 9. BUTLER, THOMPSON (1967). 10. HAACK (1969). 11. TAUSON, KRAVCHENKO (1956). 12. TAUSON, PEVTSOVA (1955). 13. GAVRILIN, PEVTSOVA (1963). 14. GAVRILIN, PEVTSOVA (1965). 15. TAUSON (1964). 16. ZLOBIN, PEVTSOVA (1964). 17. ZLOBIN et al. (1965). 18. WODZICKI (1971). 19. OKRUSCH, RICHTER (1967). 20. OKRUSCH, RICHTER (1969). 21. GUNDLACH et al. (1967). 22. SAVUL et al. (1956a). 23. BELT (1960).

Table 30-E-7. *Abundance of zinc in rhyolites, rhyodacites, obsidians, dacites etc. (Number of samples in brackets)*

Rock type	Origin	Range ppm Zn	Average ppm Zn	Analytical method	Reference
Rhyolites (2)	Minnesota (U.S.A.)	—	60	W/C	SANDELL, GOLDICH (1943)
Rhyolites, rhyodacites (6)	Missouri, California (U.S.A.)	30—210	75	W/C	SANDELL, GOLDICH (1943)
Rhyolites (6)	California, Oregon etc. (U.S.A.)	20—40	33	X	JACK, CARMICHAEL (1969)
Rhyolites (6)	Central Honshu (Japan) (5), Scotland	15—95	38	W/C	MORITA (1955)
Rhyolites (5)	Erta Ale (Ethiopia)	65—135	106	P	TREULL et al. (1971)
Rhyolites (5)	Northern Nigeria	215—400	297	P	BUTLER, THOMPSON (1967)
Volcanic rocks ("acidic") (11)	East Amisk, Saskatchewan (Canada)	5—115	45	X	SMITH (1964)
Pitchstones (20)	Iceland	80—200	123	W/C	CARMICHAEL, MCDONALD (1961)
Pitchstones (14)	Arran, Eigg (Scotland)	34—210	68	W/C	CARMICHAEL, MCDONALD (1961)
Obsidians (118)	Newberry Caldera, Oregon (U.S.A.)	—	120	A	LAIDLEY, MCKAY (1971)

Table 30-E-7 (Continued)

Rock type	Origin	Range ppm Zn	Average ppm Zn	Analytical method	References
Obsidians (40)	California, Oregon etc. (U.S.A.)	20—80	40	X	Jack, Carmichael (1969)
Obsidians (2)	Ascension Is., Caucasus (U.S.S.R.)	225—265	245	P	Butler, Thompson (1967)
Obsidians (composite of 8)	Iceland (4), Italy (3), U.S.A.	—	100	X	Wedepohl (unpublished)
Liparites (composite of 14)	Lipari (Italy)	—	84	X	Wedepohl (unpublished)
Comendites (5)	Nigeria, Sardinia, Greenland, New Zealand, Australia	200—570	345	P	Butler, Thompson (1967)
Comendites	New Zealand	—	290	X (?)	Nicholls, Carmichael (1969)
Pantellerites (6)	Pantelleria	350—520	440	X (?)	Nicholls, Carmichael (1969)
Pantellerite (1)	Pantelleria	—	500	P	Butler, Thompson (1967)
Dacites (6)	Japan (5), California (U.S.A.)	40—90	64	W/C	Morita (1955)
Dacites (3)	California (U.S.A.)	40—60	50	W/C	Sandell, Goldich (1943)
Dacites (7)	Calimanian Mountains (Roumania)	—	72	W/C	Savul et al. (1956a)

VII. Composition of the Earth's Crust

The composition of the upper continental crust is relatively well known as reported in Chapter 7 of Volume 1 of this handbook. Because of its minor thickness we do not consider the sedimentary cover in these computations. We only list magmatic rocks notwithstanding the large abundance of metamorphic rocks. At a first approximation, we assume that the metamorphic rocks have abundant magmatic equivalents of comparable chemical composition.

Table 30-E-8 contains the tentative proportions of the most abundant magmatic rock types as listed in Table 7—8 of Volume I of this handbook, and average zinc abundances in the major magmatic rock types as computed from data of Section 30-E.

According to more recent surveys compiled by Wedepohl (1969), there is no large difference in major elements to be expected between the upper and lower continental crust. There are not enough data on zinc in high grade metamorphic

Table 30-E-8. *Zinc abundance in the upper continental crust*

	Vol.-% of rock type	Zinc abundance in rock class (ppm)	Zinc abundance relative to rock abundance
Granites, quartz monzonites	44	48.0	21.1
Granodiorites, quartz diorites	42	52.4	22.0
Diorites	1	70	0.7
Gabbros (basalts)	13	100	13.0
Peridotites	0.5	56	0.3
Abundance in the upper continental crust:			57.1 ppm Zn

rocks (granulites, high grade mica schists etc.) to compute a reliable average. Therefore, it is suggested to use a value of 60 ppm Zn tentatively for the whole continental crust. TAYLOR (1964) who used a different approach for estimating the crustal average (in his model composed of one part granite and one part basalt) reports a value of 70 ppm Zn. If we use our zinc averages for granitic rocks and basalts we will have a value of 75 ppm Zn.

The continental crust contributes about three quarters and the oceanic crust (which is expected to be gabbroic-basaltic) only one quarter to the total earth's crust. From our data of Section 30-E, an average of 70 ppm Zn can be computed for the total crust.

Manuscript received: March 1972

30-F. Behavior in Magmatogenic Processes

I. Pegmatites

Zinc does not belong to the specific group of elements which has been commonly reported to be accumulated in certain pegmatitic melts and rocks. FERSMAN (1940) estimated the zinc concentration in average granite pegmatite melts to be as high as 400 ppm (seven times the zinc concentration of granites). But at this time, there were not many reliable minor element data on rocks available.

DE VORE (1955b) has analyzed a few micas from pegmatites besides those from numerous schists. One biotite from a pegmatite in the Littleton formation in New Hampshire contains 900 ppm Zn in contrast to about 230 ppm Zn which this author reports in biotites from 35 schist samples. His muscovites are as low in zinc as those from other rocks. WEDEPOHL (1953) found a few ppm Zn in two pegmatitic micas from Finland and Brazil, but 260 ppm Zn in a pegmatitic biotite from Haverstad (Norway). Because of the crystal chemical tendency of zinc to be incorporated in mica structures, these minerals are good indicators for the zinc concentrations in pegmatite masses.

II. Gas Transport

Zinc forms compounds of appreciable vapor pressure in magmatic vapors as is evidenced by sublimates around volcanic vents. KRAUSKOPF (1957) has computed vapor pressures of metal compounds equilibrated with the most stable solid in a model gas at 600°C of 10^3 atm H_2O, $10^{1.7}$ atm CO_2, $10^{1.5}$ atm H_2S, 10 atm HCl, and $10^{1.7}$ atm HF, in copying actual and abundant magmatic vapor compositions. $ZnCl_2$ is the most stable zinc compound (and not ZnS, ZnO or Zn) in this vapor, having a pressure of $10^{-4.7}$ atm. The vapor pressure of $ZnCl_2$ is lower than that of the chlorides of Hg, Sb, As, Bi, Pb, Sn, Mn, Fe, Cd and higher than that of the chlorides of Co, Ni, Ag and that of Cu_2S and MoO_3. The vapor pressure of zinc chloride ($10^{-4.7}$ atm) related to water (10^3 atm) gives a concentration of 50 ppb $ZnCl_2$. If the model gas does not contain sulfur, the maximum vapor pressure of $ZnCl_2$ will be much higher: 0.1 atm. This is equal to a concentration of 100 ppm $ZnCl_2$ in steam.

According to the relations between vapor pressures of zinc compounds and temperature, plotted in diagrams by HÖRBE and KNACKE (1955) (mainly on the basis of data from KUBASCHEWSKI and EVANS), the vapor pressure of zinc sulfide increases from 600° to 1,100°C in five orders of magnitude. At 1,200°C the other important volatile compound, $ZnCl_2$, will be above its temperature of sublimation.

Another compound with high vapor pressure in steam at high temperature is $Zn(OH)_2$. According to the data reported by GLEMSER et al. (1957), the partial pressure of $Zn(OH)_2+ZnO$ in steam of 1 atm and 1,300°C is 0.00102 atm. Unfortunately, we have no data on the volatility of the hydroxide in the temperature range of 600 to 700°C (of granitic intrusions), but we expect it to be much lower than at 1,300°C so that Zn will get into the ppb concentration range in steam.

From the information on vapor pressures of zinc compounds, we can draw several conclusions. We have to expect that basaltic melts would lose almost all their zinc through volatilization if its transporting agents as hydroxide and chlorine would be more than ten times higher than they were in these melts. In common basaltic melts, water is not higher than about one percent and chlorine not higher than about 1,000 ppm. At conditions of basaltic melts sulfur is mainly lost as SO_2. According to the present author, Tertiary basaltic rocks from northwest Germany have a small range of variation of zinc concentrations and large ranges of chlorine and sulfur concentrations. Because there is no correlation between zinc and chlorine, appreciable vapor losses of $ZnCl_2$ cannot be assumed to have occurred during cooling of the flows at the earth's surface.

In the temperature range of granitic intrusions, we probably can exclude $Zn(OH)_2$ transport and need only apply the two model cases of KRAUSKOPF (1957), a sulfur-containing and a sulfur-free melt. The first one, of a sulfide-containing vapor, will be much more common than the second.

If zinc concentrations in a magmatic steam of granitic intrusions are mainly controlled by vapor pressures and not by solubilities of zinc compounds in the steam, they have to be expected to be generally low. Above, we have mentioned an estimate of 50 ppb $ZnCl_2$ in a steam equilibrated with sulfide. There will not occur much ore deposition from those steams except when they accumulate zinc by leaching solid rocks. (See footnote on page 30-F-3.)

The above mentioned considerations about the sulfur control of zinc concentration in volcanic steams are confirmed by observation in nature. In the fumaroles of the Showashinzan volcano, gases high in chlorine (about 1,000 ppm) and low in sulfide contain more zinc than those low in chlorine (200 to 300 ppm) and high in sulfide. The amounts discharged from one fumarole in 1958 have been computed as being more than 500 kg Zn and the yearly production of all the fumaroles of this volcano can be estimated to be an order of magnitude larger. The computation is based on analyses with an average concentration of 2 ppm Zn in the vapor (MIZUTANI, 1970). A plot of zinc concentrations against outlet temperature of fumarolic gases indicates that there is a decrease of metals at a temperature as low as about 400°C.

Sampling of sublimates and incrustations around volcanic vents gives only qualitative or semiquantative information about the actual composition of volcanic gases in metals, because of the range of temperature and of distance from the vent at which their compounds will be precipitated.

Magnetite rich deposits of fumaroles, collected in 1919 in the Valley of Ten Thousand Smokes in Alaska, have been analyzed by ZIES (1929) as having 0.47% Zn. Zinc containing sublimation products from Vesuvius and Stromboli (Italy) are reported by ZAMBONINI (1936) and those from Japanese volcanoes by MIZUTANI *et al.* (1957). Incrustations from the andesitic volcano, White Island of New Zealand, contained a maximum of 0.3% Zn (WILSON, 1959).

III. Magmatic Fluids — Ore Deposition

It is very difficult to discriminate: a) truly magmatic fluids, b) solutions, which are mixtures of magmatic vapor, condensation products and connate waters, c) connate waters heated in the country rock through the contact of a magmatic intrusion, and d) geothermal waters or brines. Ore depositing fluids, hot brines and hydrothermal waters are additionally treated in Subsections 30-G-V, 30-H-Id and 30-I-IIIa, b of this chapter. The above mentioned different origins of hot water are not considered in numerous textbooks and publications, where at least the majority of hydrothermal vein deposits of sphalerite and galena (plus several additional ore and gangue minerals) is assumed or handled as being of purely magmatic origin.

Ore-depositing solutions probably contain zinc concentrations of at least several ppm or tens of ppm. Otherwise, too large water masses have to be saturated with respect to the mineral compounds of the country rocks, water masses, which have to pass through the channels of ore metal transport. If solutions, which tend to be saturated with respect to abundant gangue minerals (Si-, Ca-, Ba-concentrations in the X0,000 to X0 ppm range), contain less than 1 ppm Zn, the ratios of gangue to ore minerals in ore producing areas must be larger than X0 (barite) to X0,000 (quartz). From this consideration and the actual abundance ratios of gangue to ore minerals, we get limiting conditions for potential ore solutions. The theoretically derived concentration ranges of ore solutions of at least X to X00 ppm Zn are confirmed by analyses of hot brines and fluid inclusions of ore and gangue minerals reported in Subsection 30-I-IIIb. These constraints are important with respect to "primary magmatic" zinc solutions. As already mentioned, the vapor pressure of water etc. and zinc-chloride and -sulfide, in chlorine- and sulfide-containing granitic melts is expected to form steam too low in zinc to condense hydrothermal ore solutions (in the above sense).

If the process controlling zinc concentration in hydrous vapors and their condensates is not vapor pressure of zinc compounds, but solubility of zinc containing granitic minerals in hydrothermal solutions, the zinc concentration of these "primary" solutions might be higher[1]. In that case we can get information on potential concentrations from leaching experiments in silicate-oxide systems by hydrothermal solutions high in Na^+, Ca^{2+}, K^+, Cl^-, SO_4^{2-}, as ore fluids usually are. As mentioned in Subsection 30-H-Id, HEMLEY et al. (1967) have determined solubilities of ZnS at 1,000 bars, 300° to 500°C in KCl-HCl-solutions buffered by silicate mineral equilibria. The measured concentration range of soluble zinc was 0.001 to 0.1 moles per liter solution (65 to 6.500 ppm Zn). Experiments of acid decomposition of natural rocks, which give no information on the zinc concentration of the leaching solution but on the proportion leached, to primary zinc of the rocks and minerals have been published by AZZARIA (1963), TAUSON (1966), GAVRILIN et al. (1967) and WEDEPOHL (Subsection 30-G-V).

According to these papers, solutions as acid as about pH 1 to 2, leach 40 to 90 percent of the primary zinc content of granites and even more from basalts. Leaching of granites generally means leaching of biotites as their major zinc carriers.

---

[1] Note added in proof: H. D. HOLLAND (Econ. Geol. 67, 281, 1972) has proved that Zn is strongly partitioned from granitic melts into the aqueous phase proportional to the square of their chloride concentration.

Because intrusive magmatic bodies solidify from outside to inside, later condensates from inside will probably leach the outside shells of such bodies if the composition of condensates allows.

There has been some discussion about a correlation of zinc concentration in intrusive magmatic rocks and sphalerite mineralization associated with the intrusions. PUTMAN and BURNHAM (1963) studied the zinc distribution in 223 biotite samples, mainly from granitic intrusions in Arizona. They do not report a correlation between zinc and lead in intrusives and zinc and lead mineralization as obtained for copper. HAACK (1969) has investigated zinc concentrations in biotites from granitic intrusives in Central Europa. Except for a few two-mica-granites and a chloritized border granite (Harz mountains) he did not get clear indications for abnormally high or low zinc concentrations of intrusives correlated with sphalerite mineralization. PARRY and NACKOWSKI (1963) claim that Zn-, Pb- and Cu-concentrations in biotites from monzonitic intrusives in the Basin and Range stocks of Utah and Nevada are indicators of economic sphalerite-galena-copper ore deposits around the stocks. Biotites from the reported intrusives at mineralized areas are lower in zinc (300 ppm Zn) and lead than those from unmineralized occurrences of the same area. Biotites from hydrothermally altered stocks contain more copper and less lead and zinc than unaltered stocks. In contrast to these results from a relatively small area, BLAXLAND (1970) reports from much larger sampling areas of granitic intrusives in the U.S.A. and Great Britain, that it is highly unlikely that zinc concentrations in biotites (and total rocks) can be used as indicators of zinc mineralization. He explains that high zinc content of biotites from leucocratic rocks (see Table 30-D-3) reflects the decreased availability of substitution sites within the rock. GAVRILIN et al. (1967) describe correlations of albitization and sericitization of granitic rocks with hydrothermal mobilization of lead and zinc.

The reported evidences indicate leaching and reactions below the granite solidus of hydrothermal solutions with the granitic intrusives and country rocks as a potential zinc source for the formation of mineral deposits. In this case, solutions and metals can originate to a major degree from true magmatic materials. But magmatic vapors probably do not get the major proportion of their metals in the stage of magmatic melts. This evidence does not exclude the occurrence of hydrothermal zinc solutions from formation waters of pure, sedimentary, diagenetic origin. They have been described by BILLINGS, KESLER and JACKSON (1969) and by BILLINGS HITCHON and SHAW (1969) from the western Canada sedimentary basin. The authors of the former paper assume with some probability that the brines with their mean zinc concentration of 19 ppm Zn (Presqu'ile system) and 1 ppm Zn (Alberta), respectively, have leached zinc from shales of the Mackenzie basin. This assumption is based on leaching experiments by WILLIAMS (reference in BILLINGS, KESLER and JACKSON, 1969) of 82 crushed samples of Alberta shales. The leaching was by distilled water and 1 N ammonium acetate at pH 8 for 24 hours and probably much less rigorous than that by hot brines. He leached 0.6 to 53 ppm of the zinc content of the rock (on average 11 percent of the available zinc). Zinc accumulation and transport, comparable to this can be assumed for deposits of the Mississippi-Valley type.

Recent sedimentary deposits, where the brine and the sediment could be sampled, are located in several deeps of the Red Sea. Sediments contain zinc concentrations in the 0.1 to 10 percent range and probably extend to a depth of about 20 m below the surface. 104 samples of brines and 143 samples of interstitial waters from Atlantis II and Discovery Deep have zinc concentrations mainly in the 0.1 to 10 ppm Zn range (BROOKS *et al.*, 1969).

For details on the chemistry of the transport of zinc in hydrothermal solutions and brines, as chlorine or bisulfide complexes, the reader is referred to Subsection 30-H-I d.

Details and classification of commercial Pb—Zn ore deposits are a matter of special textbooks. A major classification of these deposits is on the basis of syngenetic or epigenetic deposition of the ore with a substratum, as for instance a sediment. The above mentioned Red Sea- and Mississippi-Valley-types of deposits are representatives of the syngenetic and the epigenetic classes, respectively. If ore is deposited on the wallrock of fissures or by replacement of preexisting minerals, it is called vein type or replacement type hydrothermal deposit. Not infrequently, the latter is a disseminated deposit. By far the largest amount of the world zinc production originates from epigenetic deposits.

30-G. Behavior during Weathering and Rock Alteration

Zinc, primarily occurring in the structure of silicates and oxides, goes into solution during the chemical weathering of these minerals. The concentration in weathering solutions and the distance of transport is controlled by adsorption, more than by solubility, of zinc carbonates, hydroxides and phosphates (see Section 30-H). Weathering of sulfides controlled by oxidation can support relatively high concentrations of zinc because of the high solubility of zinc sulfate. The high concentrations of this origin can be adsorbed on clay minerals, iron oxides and organic substances at a distance from the primary occurrence of sphalerite.

The low concentration of zinc in surface water indicates a restricted mobility of this element from places of rock weathering.

I. Weathering of Rocks

The behavior of zinc during basalt weathering has been investigated by the present author at different stages of decomposition of alkali olivine basalts of the Göttingen area (Table 30-G-1). The samples used for zinc analyses have been studied by Bolter (1961) on mineral and major element composition. The different stages of rock alteration are indicated by the sequence of mineral decomposition: olivine, pyroxene, plagioclase into minerals of the montmorillonite and kaolinite group and iron oxides. In general, zinc concentrations increase relatively with increasing decomposition. But this is almost a relative increase only because of the open system. To test this relative increase, the ratio Zn to Ti has been used because of the limited mobility of titanium under surface conditions. This ratio increases only slightly in the sequence of alteration. Zinc is expected to be fixed by the highly adsorbing clay minerals of the montmorillonite and kaolinite group and by freshly precipitated ferric iron oxides.

Zeissink (1971) has published zinc data, by atomic absorption procedure, of two laterite profiles on serpentinite from Queensland, Australia. The fresh rock contains 32 ppm Zn. Relative to Ti and Cr, two immobile elements, Zn is accumulated in the central zone of the laterite upto a factor of 4. Absolutely, zinc is concentrated upto a maximum of 430 ppm Zn. The main accumulation occurred in the ferruginous (30% Fe_2O_3) and in the montmorillonite zone, probably because of adsorption on ferric iron oxides and on montmorillonite.

Harriss and Adams (1966) investigated the behavior of zinc in five weathering profiles developed on granitic rocks in Oklahoma and Georgia (U.S.A.). Zinc was determined in these samples by atomic absorption. Their results are listed in Table 30-G-2 as taken from the diagrams in their publication. Nomenclature on soil horizons is the common one, indicating the unaltered parent material as D horizon. Except the two Mount Scott granite profiles with zinc decreasing in the

Table 30-G-1. *Zn concentrations in weathered Tertiary alkali olivine basalt from the Göttingen area (northwest Germany)*

Zn concentration in ppm (method: X; Wedepohl, unpublished) (ratio ppm Zn·100/percent TiO_2 in brackets).

	Locality			
	Bramburg	Backenberg	Hoher Hagen, Dransfeld	Steinberg, Meensen
Basalt, fresh	125 (0.54)	100 (0.43)	150 (0.63)	110 (0.54)
Basalt, fresh	125 (0.50)	122 (0.55)	—	142 (0.62)
Basalt, weathered, stage 1, olivine partly decomposed	—	132 (0.47)	142 (0.60)	130 (0.57)
Basalt, weathered, stage 2, pyroxene almost undecomposed	235 (0.64)	111 (0.51)	180 (0.67)	—
Basalt, weathered, stage 3, pyroxene partly decomposed	—	—	154 (0.55)	—
Basalt, weathered, stage 4, pyroxene almost completely, plagioclase partly decomposed	186 (0.60)	—	258 (0.56)	240 (0.75)
Basalt, weathered, stage 5, plagioclase almost decomposed	207 (0.63)	211 (0.54)	—	54—170 (1.13—0.59)

sequence of degree of alteration, there is a modest accumulation of this element from D to decomposed rocks (C, B). Zinc and manganese are almost parallel in their behavior. Both elements are mainly accumulated in the fine fraction of the soils with kaolinite as major mineral (range 100—300 ppm Zn). Except in the Mount Scott granite containing hornblende, the major dark mineral and host structure for zinc is biotite which is leached to a different degree in the different soil horizons. Partial decomposition of biotite, hornblende and plagioclase is characteristic for the steps of weathering.

The present author has analyzed zinc concentrations in a fresh and decomposed granite from the Harz mountains (Germany) and in a laterite profile on granite from Eklara (India) by X-ray fluorescence. The Harz granite, in which the major minerals are almost preserved during weathering, has locally lost 50 percent of its original 60 ppm zinc on leaching of biotite, or locally accumulated 30 percent relative to the primary zinc by adsorption. Both loss and accumulation of zinc have been observed during weathering of biotite from a pegmatite and a paragneiss, respectively, by Rimsaite (1967). During lateritization of the Indian granite, zinc has been lost absolutely and relative to TiO_2 in zones of kaolinite formation (upto 80 percent of the original 47 ppm Zn). Some accumulation occurred in iron oxide concretions (25 ppm Zn).

Table 30-G-2. *Zinc concentrations in 5 soil profifiles on granite and granodiorite in ppm Zn (method: A).* (After HARRISS and ADAMS, 1966)

Soil horizons (D: fresh parental material)	Tishomingo granite, Johnson County, Oklahoma (T 1)	Tishomingo granite (T 2)	Mount Scott granite, Comanche County, Oklahoma (S 1)	Mount Scott granite (S 2)	Elberton granodiorite, Elbert County, Georgia (E 1) (USA)
D	33	33	130	180	64
C_0	43	53	70		76
C_1	45	40	88	} 140	78
C_2	45	40	88		70
B	58	48	90	175	60
A	40	43	78	150	62

II. Weathering of Ore Deposits

A relatively high activity of zinc in mine waters and in natural solutions from leaching of sulfides under oxidizing conditions can produce secondary zinc minerals. Abundant supergene zinc minerals are smithsonite, hydrozincite, hemimorphite, goslarite and sauconite (or zinalsite). For some details on sauconite occurrences in Missouri ore deposits see WEDEPOHL (1953). The solubilities of the former three compounds in natural solutions at low temperatures have been computed by TAKAHASHI (1960). Each of the three compounds, hemimorphite, smithsonite and hydrozincite, can be the least soluble one at increasing pH and constant CO_2 partial pressure. A change of carbon dioxide pressure can change this sequence of solubility. Outside a halo of secondary zinc minerals around an ore deposit, which is hit by the ground-water table, clay minerals and organic substances such as peat etc., will accumulate zinc by adsorption. These secondary accumulations and higher zinc concentrations in waters and stream deposits of a certain area are the basis for geochemical prospecting of ore deposits, which is treated extensively in textbooks by HAWKES and WEBB (1962) and GINSBURG (1963), and in numerous publications.

III. Soils

Except for abnormal zinc concentrations from sulfides (as mentioned in the preceding subsection), soils usually contain the residual zinc of the weathered rocks, on which the soil has been formed. Due to the low mobility of zinc and its adsorption on clay minerals, ferric iron oxide, and organic residues, zinc is often slightly higher in soils than in the related undecomposed rocks. Zinc concentrations in soils are often incidentally increased by addition of phosphate fertilizers (containing upto 0.1% Zn). The availability of zinc for plant growth usually increases with decreasing pH. A large list of data on zinc in soils from 18 countries on all continents has been published by SWAINE (1955). Excluding high concentrations from mining areas and from peat, his data mainly range from 10 to 300 ppm Zn and can be averaged at about 70 ppm. SHACKLETTE *et al.* (1971) have published a recent survey on the chemical composition of 863 soils, sampled in a regular network over the United States. The W/C analyses gave a geometric mean of 44 ppm Zn (arithmetic mean:

54 ppm). The depth of sampling of these surficial materials was about 20 centimeters. VINOGRADOV (1959) reports an average of 50 ppm Zn from extensive sampling.

Detailed investigations about the types of occurrence of zinc in loam soils from Tennessee have been published by WHITE (1957) whose results show that about 45 percent of the zinc can be leached by reducing ferric iron oxides, and 35 percent is incorporated in mineral structures on average, in his test profiles of Tennessee soil composites. SHORT (1961) has analyzed zinc spectrographically in several samples of residual soils on granodiorite (Bighorn Mountains, Wyoming), andesite (Asheboro, North Carolina) and basalt (near Denver, Colorado). In total soil samples from different profiles, there is a slight increase of zinc, absolutely and relative to TiO_2 (maximum factor 1.5), from C-B to the A horizon. In size fractions of the soil, zinc is mainly accumulated in clay ranging from 83 to 216 ppm Zn (average of 12 samples: 125 ppm Zn which is close to our shale average of Section 30-K).

IV. Bauxites

Bauxites, developed on limestones in southern Europe, have accumulated zinc relative to limestones and even to shales. In Table 30-G-3 average data from the raw materials of an aluminium plant (VAW Lünen, Germany) are listed (for details see WEDEPOHL, 1953).

Table 30-G-3. *Zinc concentrations in bauxites from southern Europe. (Averages for several thousand tons of rocks (method: W.)* (WEDEPOHL, 1953)

Locality	ppm Zn	Ranges in ppm Zn
Eleusis (Greece)	200	160—240
Mostar (Yugoslavia)	240	200—280
Drnis, Dalmatia (Yugoslavia)	320	280—360
Istria (Yugoslavia)	160	120—200
Iszka, St. Georg (Hungary)	300	280—320
Epleny (Hungary)	300	—
Southern France	180	120—240

Bauxites have, on average, concentrated aluminium by a factor of about 3, relative to shales (clay residues of limestones). For zinc, this accumulation factor ranges between 2 and 3 indicating that only a small proportion of zinc can be lost during bauxitization, if at all. The pH of solutions during bauxitization must have been buffered by the limestone to maintain a value of about 8. At pH 8 the solubility of $Zn(OH)_2$ is low.

V. Hydrothermal Leaching of Rocks

As can expected from information compiled in Section 30-H, the effectiveness of solutions in leaching zinc from rocks depends mainly on their pH. The results of an unpublished investigation of the present author on hydrothermal decomposition of tholeiitic basalts from Iceland are plotted in Fig. 30-G-1. This investigation made

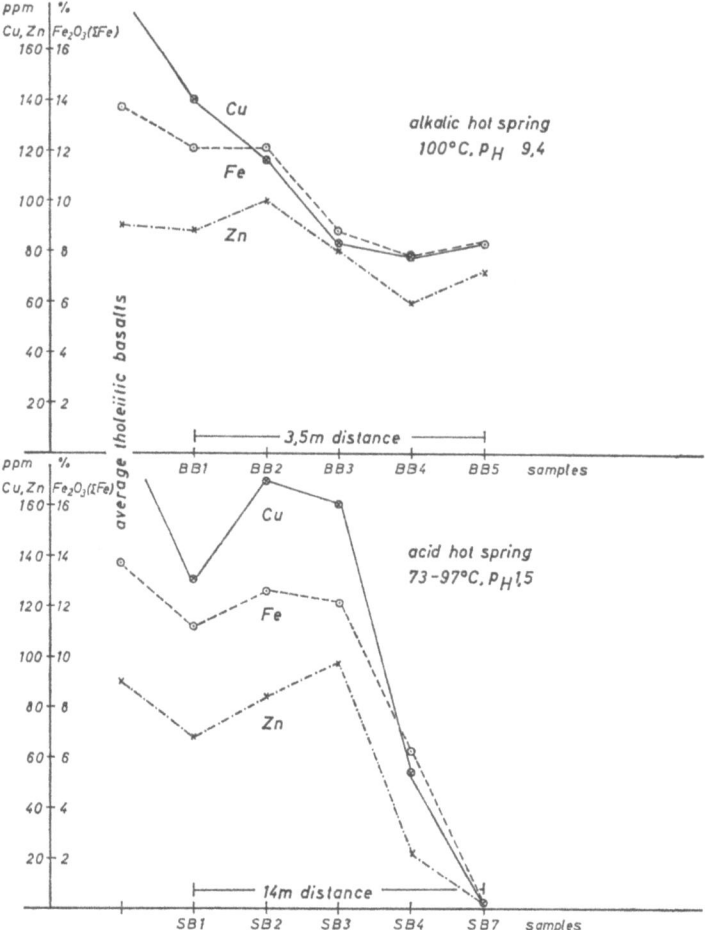

Fig. 30-G-1. Change of Zn, Cu and Fe concentrations in tholeiitic basalts from Iceland during hydrothermal decomposition by an alkalic and an acid hot spring. Increasing sample numbers mark advanced decomposition by approaching the spring. (Unpublished results of the present author on samples collected and described by SIGVALDASON, 1959)

use of samples which have been studied by SIGVALDASON (1959) for their mineral and major element composition. The samples of the two series are numbered BB 5 to BB 1 and SB 7 to SB 1; the sample number decreases with increasing distance from the hot spring. The volumes of decomposition are much larger, and the concentration of Cu, Fe and Zn in the rock of most intensive alteration are much lower in the case of acid water. The alkalic hot spring at pH 9.4 has leached almost no zinc from the rock; the ratio Zn/Ti is nearly constant in the different stages of decomposition. An alkalic hot spring water (pH 9.5) from the same Hveragerdi area is as low as 7 ppb Zn (atomic absorption analysis of the present author) comparable with many rivers. In the case of the acid spring the ratio Zn/Ti is nearly constant only in samples SB 1 to SB 3; that means that the aureole of heavy leaching has a diameter of 7 meters (SB 4 to SB 7).

ZLOBIN and PEVTSOVA (1964) present material balances of hydrothermally altered Variscan granites from Tien-Shan (U.S.S.R.). According to their investigation, no zinc was removed during propylitization and CO_2 metasomatism. But at the higher temperature of epidotization zinc was removed.

For additional information see Subsections 30-F-III, 30-H-Id and 30-I-III.

30-H. Solubilities of Compounds which Control Concentrations of Zinc in Natural Waters (I); Adsorption Processes (II)

I. Solution Chemistry

Zinc occurs in only one valence state; therefore, solution chemistry is not as complicated as that of the other 3-d elements. However, the influence of complexing in the dilute natural solutions is not yet well understood.

Of the naturally occurring inorganic anions which control solubilities by their abundance, only sulfide, phosphate and carbonate form zinc compounds of low solubility. Other compounds of comparably low solubility are the hydroxides and the oxide. Precipitation of zinc hydroxide is restricted to the pH range from 5.5 to about 10.5.

a) Sea Water

If the control of zinc in water has to be proved, sea water has its importance from its large mass. Sea water is rather constant in composition; the concentration range of controlling anions is smaller than in other natural water. However, the large ionic strength of sea water compounds has to be considered in computations of ionic concentrations. For zinc, an activity coefficient of 0.12 is used in computations for Table 30-H-1; this table contains the condensed information on the solubilities of the sulfide, hydroxide, phosphate and carbonate of zinc under sea water conditions. We have included the sulfide, notwithstanding the fact that almost all ocean water does not contain sulfide-ions. But some anoxic basins and many interstitial waters in sediments contain sulfide. Therefore zinc concentration can be controlled by the low solubility of zinc sulfide. KRAUSKOPF (1956) does not exclude that exchange between interstitial and closed basin water with open ocean water has influence on the low zinc concentration in sea water. Some effective exchange is indicated by the variation of oceanic ^{34}S with geologic time. The sulfide concentration we have chosen is that of the order of magnitude in Black Sea deep water. Similar concentrations are reported by BROOKS et al. (1968) from interstitial waters in marine sediments off the coast of southern California.

From Table 30-H-1 we can learn that neither the ions of carbonate nor hydroxide and phosphate control the concentration of zinc in normal sea water, because the latter concentration is only 2 to 5 ppm Zn. The data from this table can be used to detect natural water in which the reported anions control zinc concentrations. Except for water containing sulfide ion, the zinc concentration must be at least in the ppm range to precipitate the carbonate, hydroxide or orthophosphate (tetrahydrate: JURINAC, INOUYE, 1962) of zinc in natural waters. Even in the case of theoretical saturation, precipitation can be prevented by complexing. Ocean water zinc

Table 30-H-1. *Solubility of some zinc compounds in sea water (25° C)*

Zinc compound of low solubility	Solubility product (reference in brackets)	Concentration of anion in sea water, moles per liter	Activity of zinc in moles per liter	Activity of zinc in saturated solution: in ppm
ZnS	1.2×10^{-23} (SILLÉN, MARTELL, 1964; BROOKS et al., 1968)	10^{-4} [a] H_2S, HS^-, S^{2-} $3.3 \times 10^{-9} S^{2-}$	1.5×10^{-13}	10^{-8}
$ZnCO_3$	$10^{-10.8}$ (SILLÉN, MARTELL, 1964) $10^{-9.8}$ (CROCKET, WINCHESTER, 1966)	$10^{-4.9}$ CO_3^{2-}	3.2×10^{-5} 2.5×10^{-4}	2 16
$Zn(OH)_2$	$10^{-16.7}$ (SILLÉN, MARTELL, 1964)	10^{-7} OH^-	2.8×10^{-4}	18
$Zn_3(PO_4)_2$	10^{-32} (SILLEN, MARTELL, 1964)	$10^{-9.3} PO_4^{3-}$ $10^{-5.5} HPO_4^{2-}$	3×10^{-4}	20
$Zn_3(PO_4)_2 \times 4H_2O$	$10^{-47.3}$ (JURINAK, INOUYE, 1962)	$1.6 \times 10^{-6.6}$ H_2PO_4	1.25×10^{-4}	8

[a] Anoxic deep water of the Black Sea (order of magnitude).

can occur as $Zn(OH)_2^0$ —, $ZnCO_3^0$ —, $ZnCl^+$ —, $ZnHCO_3^+$ — complexes (ZIRINO, HEALY, 1970). At pH 8.3, zinc in sea water was calculated by the mentioned authors to be 75% $Zn(OH)_2^0$, 8% Zn^{2+}, and 4% $ZnCO_3^0$. With decreasing hydroxide ion concentration, the abundance of ionic zinc relative to complexed zinc increases. DUURSMA and SEVENHUYSEN (1966) found no conclusive evidence of a chelating influence of natural organic compounds in sea water.

BROOKS et al. (1968) report zinc concentrations in the range of 1 to 150 ppb in sulfide ion-containing interstitial waters, which theoretically could not have been more than 10^{-5} ppb Zn. Natural precipitation of sphalerite from sea water has occurred under certain conditions, as the abundance of ZnS, in concentrations upto several percent, in Permian marls of Central Europe (Kupferschiefer) and comparable deposits proves (see compilation by WEDEPOHL, 1964). The local occurrence of smithsonite in this bed seems to indicate that bottom waters or interstitial waters of the abnormal anoxic basins, with a very low rate of detrital sediment accumulation, could have contained several ppm zinc.

b) Fresh Water

Information on the behavior of Zn in fresh water can be partly derived from Subsection Ia. Data on the solubility of smithsonite, hydrozincite, hemimorphite and willemite under fresh water conditions of 25° C, 1 atm and 0.03 mole/liter CO_2, in pH ranges abundant in natural waters, have been provided by TAKAHASHI (1960). Under these conditions, and pH 8.1 to 8.3, smithsonite and hydrozincite are reported as having solubilities of 2×10^{-6} and 6×10^{-6} mole Zn per liter respectively. Solubility increases, in approaching a pH of 6.4, by a factor of about 10^2. Solubility of hemimorphite and willemite is slightly less than that of hydrozincite and slightly more than that of smithsonite at pH 7. For solubility curves as a function of pH see TAKAHASHI (1960).

Zinc hydroxide precipitation starts in a gradient of decreasing hydrogen activity at pH 6.0 to 6.3.

c) Coprecipitation of Zinc with Carbonates

Because of the scarcity of natural low temperature conditions, at which zinc carbonate will be formed, coprecipitation of zinc with calcium has some importance. Data on partition have been presented by CROCKET and WINCHESTER (1966), TSUSUE and HOLLAND (1966), DARDENNE (1967), and KITANO, TOKUYAMA and KANAMORI (1968). These data are controversial. If we compare the partition coefficients the four groups of authors reported for NaCl-free solutions and calcite, the discrepancy is obvious, notwithstanding the fact that they have been working in different ranges of temperature. In all the four papers the apparent distribution coefficient is defined as:

$$k = \frac{Ca^{2+} L}{Zn^{2+} L} \cdot \frac{Zn^{2+} \text{calcite}}{Ca^{2+} \text{calcite}} \quad \text{(L: liquid)}.$$

For NaCl-free solutions and calcite, the authors get the following values for the apparent distribution coefficient (symbol in brackets indicates the first letter of the first author of each group):

20°C:50(K); 25°C:2.9(C); 28—30°C:~3(D); 35°C:3(C); 50°C:3.2±0.2(C); 90°C:~350(T); 167°C:57±6(T); 250°C:3(T).

DARDENNE (1967) mentions from his experiments that the partition coefficient, k, increased with increasing amount of precipitated $CaCO_3$. This might be an explanation of the discrepancies among the results of the different authors.

The two groups (T and K) which have investigated NaCl-containing solutions get a remarkable decrease of the distribution coefficient with chloride concentration. For chloride concentrations close to those of sea water and calcite, KITANO et al. (1968) and TSUSUE and HOLLAND (1966) both report a distribution coefficient of 5, the former group of authors at a temperature of 20°C, the latter at 167°C. The decrease of the coefficient must be due to chlorine-complexing of zinc. Despite the controversial results in NaCl free solutions, the experiments demonstrate that the distribution coefficient for sea water precipitation of calcite is probably between 1 and 5. Otherwise, the low concentration of zinc (<20 ppm) in marine calcite (foraminifera shells) which is not altered diagenetically, could not be explained either.

If calcite of foraminifera etc. contains an average 10 ppm Zn, the accumulation of this calcite per year on the ocean floor (estimated as 2×10^9 tons at a discharge of zinc by rivers into the oceans as 2.7×10^5 tons Zn) extracts almost one tenth of the zinc supply by carbonate coprecipitation. A comparable effect can be expected from coprecipitation of zinc with apatite (Table 30-H-2).

Besides the chlorine-complexing of zinc, the influence of various organic substances (amino acids etc.) on the decrease of the partition of zinc between calcite and water has been investigated by KITANO et al. (1968).

d) Ore Depositing Fluids

The apparent increase of solubility of zinc sulfide in aqueous solutions (for the very low solubility of ZnS in pure water and sea water see Table 30-H-1) above 1 ppm, which is needed for effective zinc transport for ore deposition, has been treated by BARNES and CZAMANSKE (1967), HEMLEY et al. (1967), HELGESON (1969) and others. This transport is a problem of complexing as experimentally proved by BARNES and CZAMANSKE (1967). In H_2S saturated water (at pH $\sim$8.2 where dissociation into HS^- is large), these authors found 2,700 ppm Zn in solution (25°C; 7 atm pH_2S; ionic strength 1.4). The solution reaction: $ZnS+H_2S+HS^- \rightleftharpoons Zn(HS)_3^-$ ($K_{25°}=10^{-4}$) is nearly independent of temperature upto 2,000°C. The relative importance of the individual contributions of bisulfide and chloride complexes on the solubility of sphalerite is demonstrated by the equality of the activities of the two complexes $Zn(HS)_3^-$ and $ZnCl_2$, at an activity ratio of the anions $Cl^-/HS^- \sim 10^4$ ($a_{Cl}=1$; temperature upto 250°C). Whereas BARNES and CZAMANSKE (1967) observed increased solubilities of ZnS at neutral to weakly basic pH (and high bisulfide-H_2S concentrations), HEMLEY et al. (1967) determined solubilities at 300 to 500°C in weakly acid chloride solutions buffered by silicate minerals (muscovite) in assumed wallrock reactions. The total range of ZnS solubilities in KCl solutions (of silicate buffered systems) measured in these experiments is 65 ppm Zn (0.001 moles per liter) at 300°C to about 13,000 ppm Zn (0.2 moles per liter) at 500°C. Detailed information on chloride complexing is given by HELGESON (1969). In Fig. 21 of his report he plots a calculated solubility of ZnS (in terms of the total molality of the metal ion) in a 3 molal sodium chloride solution (pH 5), which varies from $10^{-6.15}$ (at 50°C) to $10^{-2.5}$ M Zn (at 300°C). HENNIG (1971) has measured 10^{-3} M Zn (65 ppm) dissolved from ZnS in a two molal NaCl solution at 350°C.

II. Adsorption Processes

The importance of adsorption of zinc on clay minerals has been experimentally proved by KRAUSKOPF (1956), JURINAK and THORNE (1955) and others.

Adsorption depends on temperature, pH, salinity etc. of the solution (adsorbate) and several properties such as composition, structure, grain size etc. of the adsorbent. Because of the complex system, quantitative duplication of natural processes is almost impossible. Therefore, only a few results of model experiments are reported to demonstrate the effective extraction of zinc from natural water (Table 30-H-2).

As the experiments prove, adsorption equilibria might be achieved within days or at least weeks. Zn adsorption is strongly dependent on pH. In common cases, the pH dependence is non-linear. Retention of zinc in excess of the cation-exchange capacity (above pH 6) was explained in terms of "hydroxide precipitation in the clay system" (BINGHAM et al., 1964) etc. The reaction of zinc with octahedral OH in the layer silicates was indicated by changes in the original montmorillonite infrared absorption pattern after adsorption. The high adsorption of peat could be proved by infrared as being a chelating reaction (DE MUMBRUM, JACKSON, 1956a, b). CHESTER (1965) obtained adsorption isotherms for zinc on illite with only a slight shift due to change of temperature.

Grinding of minerals increases adsorption capacity, as does the large surface of freshly precipitated iron and manganese oxides. The adsorption capacity of these

Zinc

Table 30-H-2. *Adsorption experiments*

Adsorbent	Concentration of adsorbent in suspension	Time in days/ temperature/pH	Initial Zn concentration	Final Zn concentration (ppb)	Reference
$Fe_2O_3 \cdot nH_2O$ (freshly precipitated)	30 ppm	1/18—23° C/ 7.7—8.2	sea water +2 ppm	100	1
	30 ppm	8/18—23° C/ 7.7—8.2	sea water +2 ppm	150	1
Apatite (freshly precipitated)	100 ppm	1/18—23° C/ 7.7—8.2	sea water +2 ppm	280	1
	500 ppm	1/18—23° C/ 7.7—8.2	sea water +2 ppm	30	1
Dead plankton (Oslo Fjord)	dilute	2/18—23° C/ 7.7—8.2	sea water +990 ppb	590	1
	concentrated	2/18—23° C/ 7.7—8.2	sea water +500 ppb	260	1
Peat moss	2 percent	3/18—23° C/ 7.7—8.2	sea water +2 ppm	10	1
Montmorillonite	2 percent	3/18—23° C/ 7.7—8.2	sea water +2 ppm	20	1
Illite (small size fraction)	3,200 ppm	2/26° C/ 8.2	sea water +2 ppm	1,140	2
Montmorillonite (<2 µ fraction)	1,700 ppm	7/25° C/ 7.5	0.1 N $CaCl_2$ +65 ppb	~2	3
Kaolinite (<2 µ fraction)	3,300 ppm	7/25° C/ 7.5	0.1 N $CaCl_2$ +65 ppb	~4	3

References: 1. KRAUSKOPF (1956). 2. CHESTER (1965). 3. TILLER, HODGSON (1962).

oxides decreases on aging. However, some proportion of the originally adsorbed zinc will be incorporated into the oxide lattice as analyses of iron oxides with several hundred and upto one thousand ppm Zn in red sandstone cements indicate (HARTMANN, 1963).

From Table 30-H-2, we have learnt that peat moss, montmorillonite and precipitated iron oxides adsorb zinc most effectivly. Montmorillonite suspensions are able at sea water conditions, to extract a concentration of several ppm zinc (which still could have been in equilibrium with carbonate as the least soluble zinc compound of Table 30-H-1) out of the water, to get the zinc level down to that of average sea water.

Manuscript received: March 1972

30-I. Abundance in Natural Waters and in the Atmosphere

I. Rivers, Lakes and Subsurface Fresh Water (Excluding Hydrothermal Water)

The abundance of zinc in continental water is not expected to vary much regionally due to different rock types, mainly exposed at the surface, because of similarities in average zinc concentrations. Oxidation of minor proportions of sphalerite in certain rocks can cause local anomalies. Oxidation of sphalerite from ore deposits may be traced in geochemical prospecting, using surface water among other sample types.

Three major types of pollution have to be avoided to acquire data on surface water excluding anthropogenic influences: a) acid mine water from sulfide leaching, b) industrial wastes of certain chemical and metallurgical factories, and c) dust from coal combustion.

Acid mine water can locally accumulate zinc upto the percent range of concentration. Dilution in major surface water and adsorption on clay minerals and iron oxides can decontaminate highly polluted water in distances of several kilometers from the ore bodies. The distance to which a high contamination is observable depends on the type and shape of the ore body, type of weathering, climate, local water systems etc. LEUTWEIN and WEISE (1962) and A. M. and L. E. CHERMAYAYEV (1963) and others give several details of the problem. HEIDE and SINGER (1954) have demonstrated that a local industrial zinc pollution of a medium size river is mainly decontaminated by adsorption on clay suspensions at a distance of about 15 km. HELLMANN (1970) reports that the Rhine is a polluted large river having 100 ppb Zn in solution (at Koblenz, 1969).

If one assumes an average zinc concentration of 100 ppm in coal (Table 30-L-3) and a world consumption of 3×10^9 t coal per year, the emission of zinc by local combustion is about 3×10^5 t per year (plus 1.4×10^5 t Zn from combustion of oil per year). At extreme assumptions almost all zinc dust from combustion gets into rivers and is soluble. The discharge of rivers to the sea is almost 3×10^{13} t per year; this has, at extreme conditions, 10 ppb average zinc concentration from combustion of coal. But a reasonable proportion of zinc from combustion will be precipitated in chimneys (dust containing usually more than one percent zinc) and as insoluble matter around the factories.

If one has to present an average figure for the zinc concentration in major continental surface waters (rivers etc.) one ends up at a figure close to 10 ppb Zn. TUREKIAN (1969) estimates the average for streams as 20 ppb Zn, mainly influenced by the higher figure of KONOVALOV for Russian streams.

Table 30-I-1. *Zinc concentrations in rivers, lakes and subsurface water. (Number of samples in brackets)*

Samples	Origin	Range of concentration in ppb	Average in ppb	Analytical method	Reference
Spring water (7)	Japan	0.8—7	2.4	W/C	1
Spring water (14)	Siberia (U.S.S.R.)	<0.3—3.9	⩽1.3	S	2
Streamlets (13)	Japan	1—4.1	2.2	W/C	1
Rivers (12)	Japan	1.5—9.4	5.0[a]	W/C	1
Rivers (5)	Colorado (U.S.A.)	4—30	13	W/C	3
River (40)	Columbia (U.S.A.)	2.4—37	16.2	N/R	4
Rivers (4374)	Siberia (U.S.S.R.)	0.1—5,770	13.0	S	2
Rivers	U.S.S.R.	4.2—145	39.0	S(?)	5
Rivers (polluted) (17)	Japan	12—8,200		W/C	1
River (polluted) (8)	Saale (Germany)	54—3,500		W/C	6
Lakes (31) (partially influenced by sulfide deposits and spas)	Japan	1.3—79	13.2	W/C	1
Lakes (protected against pollution) (4)	Japan	1.3—3.4	2.5	W/C	1
Lakes (440)	Maine (U.S.A.)	0.25—34	2.5	S	7
Lake (60)	Erie (U.S.A.)	small range	8	A	8
Lakes (170)	Sierra Nevada (U.S.A.)	0.3—100	1.5	W/S	9
Surface and groundwater (820)	Northwest Siberia (U.S.S.R.)		13	S	10
Groundwater (105)	U.S.A. etc.	<20—2,700	38(?)		11

[a] The same average concentration has been reported by SUGAWARA (1963) for 43 representative Japanese rivers.

References: 1. MORITA (1955). 2. UDODOV, PARILOV (1961). 3. HUFF (1948). 4. SILKER (1964). 5. KONOVALOV (1959). 6. HEIDE, SINGER (1954). 7. KLEINKOPF (1960). 8. CHAWLA, CHAU (1969). 9. BRADFORD et al. (1968). 10. KONTOROVICH et al. (1963). 11. WHITE et al. (1963).

II. Sea Water

There are not many publications reporting on zinc concentrations in sea water on the basis of modern methods. (For a list of the older data see MORITA, 1955). These data are listed in Table 30-I-2. Zinc concentrations in sea water are not very different from those in streams but slightly lower. The results of SPENCER and BREWER (1969) failed to reveal seasonal, but show regional variations. Atlantic slope waters are higher in zinc than those of the Gulf Stream and Sargasso Sea. Profiles of the former area show significant increases of zinc with depth. In the Gulf of Mexico water, zinc decreases irregularly with depth, however (SLOWEY, HOOD 1971).

Table 30-I-2. *Zinc concentrations in sea water and saline formation water. (Number of samples in brackets)*

Samples	Origin	Range in concentration in ppb	Average in ppb	Analytical method	Reference
Sea water		5—30	5	W/C	1
Sea water (off-shore) (14)	Japan	1.3—4.4	2.5	W/C	2
Sea water (coastal surface) (21)	Japan	2.8—11.7	5.6	W/C	2
Sea water (100)	North-east Atlantic	1—22	8.4	W/A	b
Sea water	Gulf of Mexico etc.	—	5	N/R	3
Sea water (232)	Sargasso Sea, Gulf of Maine	0.8—52.6	4	W/A	4
Sea water (15) (unacidified)	North-east Pacific	0.6—6.1	1.7	W/A; N/R	5
Sea water (numerous)	North-east Pacific	—	1.7	W	13
Sea water (6)	Gulf of Mexico	2.6—6.4	4	W/N/R	6
Sea water (74)	Gulf of Mexico	0.2—15	3.5	W/N/R	7
Sea water (coastal) (10)	Gulf of Mexico	2.1—10	4.2	W/N/R	7
Sea water (surface)	Black Sea	—	3.0	W/A	8,9
Sea water (deep) (130)	Black Sea	—	0.7	W/A	9
Interstitial water from marine sediments (59)	California (U.S.A.)	1.2—152	11.6	W/A	10
Formation water	Western Canada sedimentary basin	30—27,500	780	A	11
Oilfield water (73)	Louisiana, California, Mississippi (U.S.A.), Bahamas	<0.1—250	about 10	A	12

References: 1. WATTENBERG (1943). 2. MORITA (1955). 3. SLOWEY (1966). 4. SPENCER, BREWER (1969). 5. SPENCER et al. (1970). 6. RONA et al. (1962). 7. SLOWEY, HOOD (1971). 8. BREWER, SPENCER (1970). 9. SPENCER, BREWER (1971). 10. BROOKS et al. (1968). 11. BILLINGS, HITCHON and SHAW (1969). 12. SLENTZ (1971). 13. ZIRINO, HEALY (1971)[a].

[a] Limnol. Oceanogr. 16, 773 (1971).
[b] Added in proof from: RILEY, J. P., TAYLOR, D.: Deep Sea Res. 19, 307 (1972).

The difference in Black Sea surface and deep water (see Fig. 30-I-1 according to SPENCER and BREWER, 1971) is due to the difference between oxic and anoxic water; the latter contains free hydrogen sulfide to precipitate ZnS and lower the average zinc concentration.

Zinc

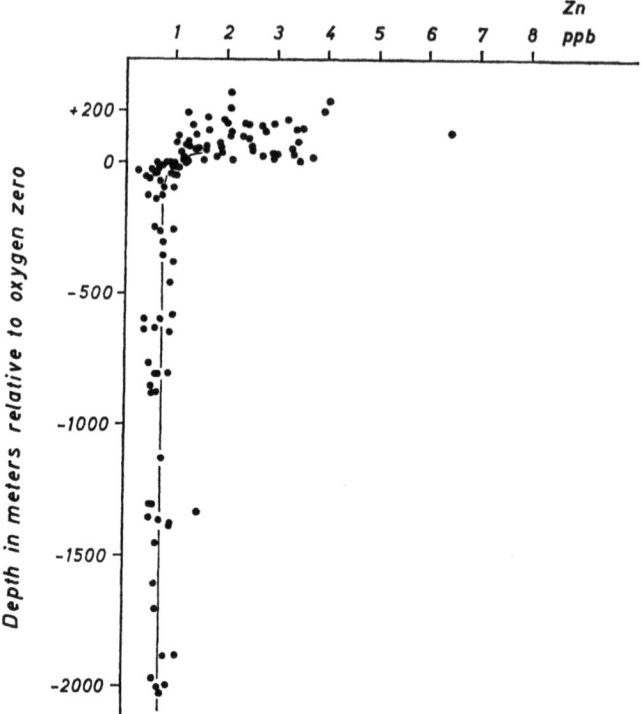

Fig. 30-I-1. Profile of the concentrations of dissolved zinc in the Black Sea. Data from 23 stations plotted with depth relative to the oxygen zero boundary. (After SPENCER and BREWER, 1971)

The average concentration of zinc in sea water is in the 2 to 10 ppb range; from this and the data of Table 30-I-1, the mean residence time of zinc in the oceans, assuming a steady state system, can be estimated to be: 2×10^3 years. This figure indicates a rapid precipitation of zinc after being discharged into the ocean, almost as rapid as Al and Fe.

Formation water with an average salinity as high as sea water, as reported by BILLINGS, HITCHON and SHAW (1969), has concentrated zinc relative to sea water by a factor of 30, so that several samples range above 1 ppm Zn. Recent marine interstitial water from southern California basins has, on average, only doubled the zinc concentration relative to sea water, whereas the results of TAGEYVA (1965) suggest that pore water of recent sediments of landlocked seas may exceed hundredfold their concentration in sea water.

III. Hydrothermal Water

a) Thermal Springs

The extensive literature on the thermal springs of the world has been recently compiled by WARING, BLANKENSHIP and BENTALL (1965). But this paper only informs about the principal constituents of the water. Isotopic evidence has confirmed that by far the largest mass of hot spring water is geothermally heated meteoric water. Thermal springs are mainly located in areas of a higher than normal thermal gradient or of volcanic activity.

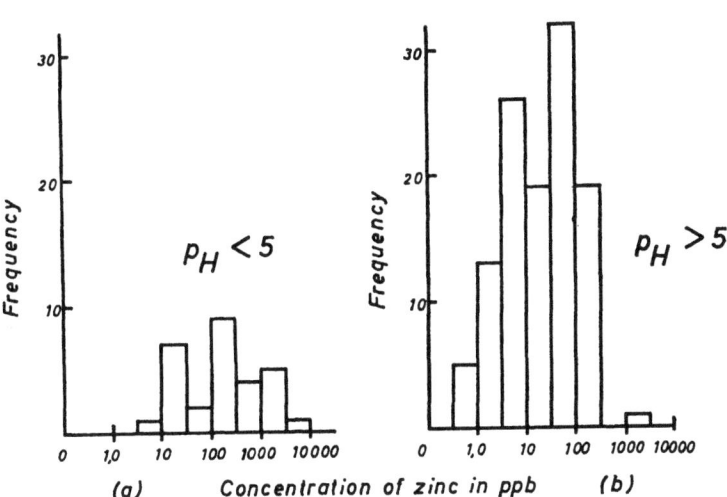

Fig. 30-I-2. Histograms on the distribution of zinc concentrations in 144 samples of thermal spring waters separated into two classes: one with pH higher than 5 and one with pH lower than 5. (After ICHIKONI, 1959)

Information on zinc in worldwide sampled hot spring water is meagre. There exists a good survey on zinc in thermal waters of Japan which are not usually high in salinity (ICHIKUNI, 1959). Fig. 30-I-2 presents histograms on the distribution of zinc concentrations in 144 samples including 24 data from the Japanese literature. The author has separated the hot spring water into two classes of pH higher and lower than 5. There is a definite trend of zinc being higher and having a larger range of variation in acid hot springs. The arithmetic mean for water of pH < 5 is 750 ppb Zn, and for water of pH > 5, 67 ppb Zn (total mean 192 ppb Zn). Leaching of country rocks by acid hot springs is probably the origin of the higher concentrations. Hot water from two wells in New Zealand, having 7 and 9 ppb Zn with an H_2S concentration of 150 and 12 ppb, fall into the low range (MERCADO, 1966/67 in: ELLIS, 1968).

The few data from additional references scattered over the literature get into the ranges of variation of Fig. 30-I-2 (GMELIN, 1956, p. 49, 50 etc.).

b) Hot Brines and Fluid Inclusions in Ore and Gangue Minerals

A few hot brines could recently be sampled, which differ from normal hot spring water in higher salinity and in concentration of several metals. SKINNER et al. (1967) report about the Salton Sea (California) geothermal brine, with zinc concentrations of 500 and 790 ppm. HELGESON (1968) mentions a fluid from the same area containing 500 ppm Zn. These brines are characterized by low sulfide, and very high chlorine concentrations (13—15 percent) and some acidity. The concentration of sulfide is still higher than that of the least soluble sulfides (Cu, Ag), so that galena and sphalerite can be precipitated in the temperature gradient. Brine samples (salinity greater than sea water), high enough in zinc (40 samples with 19 ppm, 46 samples with 1 ppm) to form commercial ore bodies of the Mississippi Valley type, have also been reported by BILLINGS, KESLER and JACKSON (1969) and BILLINGS, HITCHON and SHAW (1969) from the western Canada sedimentary basin. According to these authors, the ore bodies have probably accumulated their zinc from leaching of the shales of that basin. LEBEDEV and NIKITINA (1968) report about slightly acid, warm sodium-calcium chloride brines occurring in red beds in the Cheleken area (U.S.S.R.) having 2.3 to 5 ppm Zn (P).

Heavy metal deposits containing several percent zinc at some deeps of the bottom of the Red Sea have been precipitated from brines which could recently be sampled and analyzed for zinc (W/A) (BROOKS et al., 1969). 104 samples of brines from Atlantis II Deep and Discovery Deep and 143 samples of interstitial waters from the related sediments have zinc concentrations mainly in the 0.1 to 10 ppm range at chlorine concentrations from normal sea water to 15%. The pH of interstitial water is between 6.1 and 6.5. Extreme zinc concentrations here are 50 ppb and 400 ppm.

Fluid inclusions, high in salinity, occurring in quartz, fluorite and barite from "hydrothermal" ore deposits of the Mississippi Valley, and other types, often contain zinc concentrations in the higher ppm range. The ranges reported are: 410 to 570 ppm Zn (N/R), CZAMANSKE et al. (1963); 11 ppm to 260 ppm Zn (N/R), PUCHNER and HOLLAND (1966); 50 to 500 ppm, RYE et al. (1969); 10 to 1,040 ppm Zn (with good correlation between zinc and chlorine concentrations), PINCKNEY and HAFFTY (1970).

Zinc transport in brines in form of chloride complexes can be assumed to be an abundant process.

IV. Atmosphere

Zinc in the atmosphere occurs in dust, rain and snow and probably originates mainly from pollution. In Subsection 30-I-I we have already mentioned the high emission of 3×10^5 t Zn per year from worldwide coal and oil combustion. If this amount would be distributed on the continental precipitation (10^{14} t/year), the average zinc concentration in rain is expected to be 3 ppb. MORITA (1955) gives averages for 5 rain (2.5—5.5 ppb) and 4 snow samples (3.5—12 ppb) from Japan as having 3.6 ppb and 6.5 ppb Zn respectively. These meteoric water samples were collected in 1946 and 1947, in years of low industrial activity in Japan. SUGAWARA (1963) reports 4.2 ppb Zn for average precipitation in Japan. But around large industrial towns, zinc in rain can increase into the ppm range (GMELIN, 1956, p. 51).

Manuscript received: March 1972

30-K. Abundance in Common Sediments and Sedimentary Rocks

Due to the behavior of zinc during chemical weathering (Section 30-G) and its low solubility in natural waters (Sections 30-H and 30-I), its major transport and accumulation in the sedimentary environment is to be expected in detrital matter. But it must be considered that the detrital minerals of major importance such as quartz, muscovite and feldspars, are low in structural zinc (see Table 30-D-3). Chlorite and magnetite are better zinc carriers. Appreciable zinc transport is expected to occur in colloidal iron oxides and iron oxide coatings on other minerals. The distribution of zinc in the different rock types and environments will be handled in the following subsections in the sequence of a decreasing proportion of detrital compounds in the sediment. Data on zinc in common sedimentary rocks have been compiled by LAVERY, BARNES and WEDEPOHL (1971).

I. Psammites

a) Greywackes and Sandstones

Data on zinc in psammitic rocks are compiled in Table 30-K-1.

Greywackes are, on average, close to the mean composition of the upper continental earth's crust because they contain large proportions of chemically undecomposed materials eroded from orogenic areas. It is to be expected that the major zinc content of these rocks occurs in chlorite and iron minerals. In an investigation of Lower Triassic red sandstones, the present author got a proportion of 10 ppm Zn per every percent of Fe_2O_3 (WEDEPOHL, 1964). An iron concretion in red Triassic sandstone, with 22% Fe_2O_3, contains 760 ppm Zn. GORHAM and SWAINE (1965) report an average of 1,100 ppm Zn (S) for oxidate crusts in Recent lake sediments from the English lake district. Diagenetic reduction of ferric iron, which often occurs as hematite coating on mineral grains, mobilizes some zinc.

A composite sample of 10 quartzites with about 2% Fe_2O_3 contains 23 ppm Zn (Table 30-M-1).

b) Glacial Tills

In tills, samples from large areas with outcrops of the most abundant rock types have been collected. BAYROCK and PAWLUK (1967) have analyzed 475 Keewatin till specimens from the Alberta Plains (Canada), representing an area of 170,000 square miles or about 65% of the area of Alberta. The analytical method was emission spectrography. About 60% of the sampling area is covered by till containing 60 to 80 ppm Zn, 30% by till containing 80 to 100 ppm Zn and only 10% containing less than 60 ppm Zn. This can be averaged at about 75 ppm Zn, a value which is expected to be not far off the mean Zn concentration in the upper continental earth's crust.

Table 30-K-1. *Zinc abundance in greywackes, sandstones and sands. (Number of samples in brackets)*

Rock type	Origin	Range ppm Zn	Average ppm Zn	Analytical method	Reference
Greywackes (composite of 17)	Paleozoic, Germany	—	116	X	1
Greywackes (13)	Lower Carboniferous, Harz Mountains (Germany)	37—198	96	X	1
Greywackes, arkoses, quartz sandstones (composite of 11)	Carboniferous, Germany	—	70	X	1
Quartz sandstones, arkoses (red) (composite of 23)	Lower Triassic, Germany	—	25	X	1
Quartz sandstones (composite of 11)	Cretaceous, Germany	—	25	X	1
Quartz sandstones, arkoses (15)	Mesozoic, etc., Germany	7—135	50	X	1
Quartz sandstones, arkoses (red) (15)	Lower Permian, Lower Triassic, Germany	5—47	25	X	1
Sands (80)	Recent, coast of Honshu (Japan)	18—174	73	W/C	2
Sands (11)	Recent, coast of Korea	18—108	53	W/C	2
Sands (>70% quartz) (32)	Recent, Buzzards Bay, Atlantic (U.S.A.)	13—87	40	S	3

References: 1. WEDEPOHL (unpublished). 2. ISHIBASHI et al. (1959). 3. MOORE (1963).

II. Pelites

a) Shales and Clays

Because of their higher mean content of iron minerals, argillaceous rocks are higher in zinc than arkoses and quartz sandstones, but in the same range as greywackes. Zinc analyses of shales low in bituminous matter are compiled in Table 30-K-2, and those higher in organic carbon are listed in Table 30-K-3. Data on pelagic clays are listed in Table 30-K-4.

A general average of zinc in argillaceous rocks, low in bituminous (and carbonaceous) matter, seems to be close to 100 ppm Zn (198 samples excluding the South African and the limestone residues which have an average content of 107 ppm Zn). The present author has computed the mean and the relative standard deviation of 61 single shale samples to be 93 ± 44 ppm Zn, indicating that about 70 percent of the shales range between 50 and 130 ppm Zn. The South African argillaceous rocks investigated by HOFMEYR (1971) represent a good sampling of a reasonably large area and large time scale and contain 80 ppm Zn. The frequency distribution

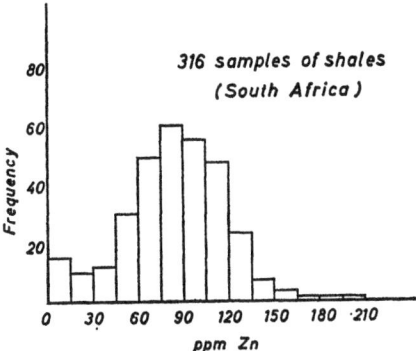

Fig. 30-K-1. Frequency distribution of zinc concentrations in South African shales ranging from the Early Precambrian Fig Tree Series (with a bituminous fraction and a mean zinc concentration of 114 ppm) to the Triassic Beaufort Series of the Karroo System. (After Hofmeyr, 1971)

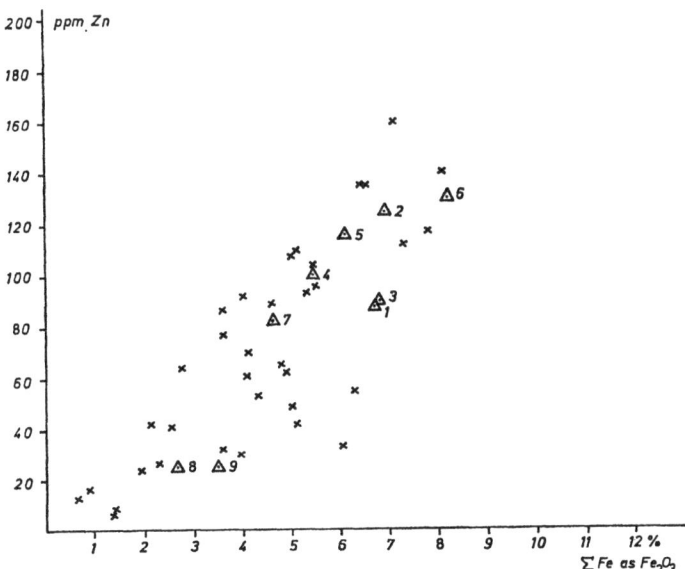

Fig. 30-K-2. Plot of zinc to total iron (Fe as Fe_2O_3) in 7 composites, 2 sets and 34 single samples of 454 sedimentary rocks listed in Tables 30-K-1 to 30-K-5 for which zinc and iron data were available. 1. 316 argillaceous rocks, South Africa (Hofmeyr, 1971; total iron data from Danchin, Ph-D thesis Capetown, 1970) 2. 36 Paleozoic shales, Europe (composite) 3. 14 Paleozoic shales, Japan (composite) 4. 10 Mesozoic shales, Japan (composite) 5. 17 Paleozoic greywackes, Germany (composite) 6. 17 Recent clays, Atlantic Ocean 7. 11 Carboniferous sandstones, Germany (composite) 8. 22 Lower Triassic sandstones, Germany (composite) 9. 11 Cretaceous sandstones, Germany (composite)

of the total of his 316 samples is plotted in Fig. 30-K-1. There is no indication of a systematic gradient of zinc content with age. Hofmeyr (1971) has analyzed the $<2\,\mu$ fraction of 33 of the South African shales with an average of 169 ppm Zn.

Zinc

This is a clear indication of a grain size effect. The grain size effect as well as the adsorption of zinc from the dissolved calcite might explain the higher zinc concentrations in acid insoluble residues of limestones. Zinc in the argillaceous rocks is probably mainly incorporated in iron minerals, chlorites and clay minerals. The few analyses of monomineralic clay mineral occurrences of Table 30-D-3 do not allow conclusions on their specific importance as zinc carriers. WHITE (1957) has investigated the occurrence of zinc in a large variety of Tennessee soils. Usually a third to one half of the zinc occurs in iron oxides, the same amount in clay mineral lattices and only 1 to 7% on exchange positions.

In Fig. 30-K-2 all clastic sediment samples for which zinc and total iron data were available to the present author are plotted. This clearly demonstrates the high degree of correlation between these two elements in sediments.

Table 30-K-2. *Zinc in shales, low in bituminous matter and nonpelagic clays.*
(Number of samples in brackets)

Rock type	Origin	Range ppm Zn	Average ppm Zn	Analytical method	Reference
Shales, marine (composite of 36)	Paleozoic, Germany, Spain, Poland	—	130	W/C	1
Shales, marine (composite of 36)	Paleozoic, Germany, Spain, Poland	—	125	X	2
Shales, marine (composite of 14)	Paleozoic, Japan	—	100	W/C	1
Shales, marine (composite of 14)	Paleozoic, Japan	—	88	X	2
Shales, marine (composite of 10)	Mesozoic, Japan	—	100	W/C	1
Shales, marine (composite of 7)	U.S.A.	—	197	X	2
Shales, marine (3)	Devonian, Rhine area (Germany)	34—140	90	X	2
Shales, marine and fresh water (25)	Upper Carboniferous, Ruhr area (sub-surface exposures) (Germany)	83—430	134	X	2
Shales, saline (4)	Upper Permian, Lower Saxony, Thuringia (Germany)	41—260	99	X	2
Shales, marine (6)	Triassic, Göttingen area (Germany)	16—106	68	X	2
Shales, marine and fresh water (3)	Cretaceous, Lower Saxony (Germany)	25—69	46	X	2
Shales, fresh water (3)	Tertiary, Meissner area (Germany), Vienna basin (Austria)	36—88	59	X	2

Table 30-K-2. (Continued)

Rock type	Origin	Range ppm Zn	Average ppm Zn	Analytical method	Reference
Shale, fresh water (1)	Pleistocene, Göttingen area (Germany)	—	125	X	2
Marls (4)	Lower Triassic, Thuringia (Germany)	60—150	105	W/C	3
Shales (184)	Early Precambrian to Triassic, South Africa	see Fig. 30-K-1	80.4	X	4
Clays, silts (<45% quartz) (20)	Recent, Buzzards Bay, Atlantic coast (U.S.A.)	28—137	64	S	5
Clay muds (29)	Recent, coast of Honshu (Japan)	14—274	114	W/C	6
Clay muds (19)	Recent, coast of Korea	2—139	63	W/C	6
Clay muds (64)	Recent, coastal areas, Japan Sea	32—780	117	W/C	7
Argillaceous (acid insoluble) residues of limestones (composite of 32)	Devonian, Germany	—	160	X	2
Argillaceous (acid insoluble) residues of limestones (composite of 45)	Jurassic, Germany	—	250	X	2
Argillaceous (acid insoluble) residues of limestones (composite of 16)	Cretaceous, Germany	—	93	X	2

References: 1. MORITA (1955). 2. WEDEPOHL (unpublished). 3. HEIDE, SINGER (1950). 4. HOFMEYR (1971). 5. MOORE (1963). 6. ISHIBASHI et al. (1959). 7. TATSUMOTO (1957).

b) Bituminous Argillaceous Rocks ("Black Shales")

Beyond zinc concentrations connected with clay minerals and oxides some additional zinc is expected to be accumulated in sediments higher in carbon fixed by the degraded organic matter, by sulfides and phosphates. However, zinc related to iron oxides might be lower in this type of rock because of the reduction and mobilization of iron and precipitation as sulfides. The low zinc concentration in tissues of the organic primary production (except plankton from the Black Sea) has been demonstrated in Table 30-L-1, but dead plankton, etc. can accumulate metals from water by adsorption (see Table 30-H-2) and by formation of chelating compounds. VINE and TOURTELOT (1970) mention in their summary report on black shales that organic fractions are locally enriched in Zn from diagenetic accumulative processes or higher than normal sea water concentrations.

Bacterial *sulfide production* in porous solutions of sediments and in bottom water needs degraded organic substances. The low sulfide is demonstrated in Table 30-H-1. Availability of sulfide ions will control the precipitation of zinc sulfide. Direct evidence of zinc sulfide precipitation in Black Sea water has been presented by SPENCER et al. (1971) by analyses of suspended matter from this water. Because of the lower solubility of iron and copper sulfides, ions of these metals can use up the available sulfide in porous solutions. Sulfide ion production in bottom water of stagnant marine basins (Black Sea, Walvis Bay, Scandinavian fjords etc.) has higher probability of zinc sulfide precipitation because of low concentrations of iron in sea water. Appreciable zinc concentrations in reducing sediments are not restricted to marine environments. GORHAM and SWAINE (1956) report an average of 800 ppm Zn (S) in 13 mud samples from English lakes.

High abundances of sphalerite in large areas of the Upper Permian marl bed "Kupferschiefer" or "Marl Slate" in Europe are a prototype of this syngenetic sedimentary sulfide precipitation (for details see: WEDEPOHL, 1964, 1971). Zinc higher than 0.5% in Kupferschiefer occurs in a broad belt of larger coastal distance than zones of copper and lead precipitation. On a paleogeographic map of metal distribution (Fig. 1 in WEDEPOHL, 1971) an area of about 2×10^4 square kilometers is covered by marl having more than 0.5% Zn. The average thickness of the bed is about one foot. Because of limited exposures, this estimate of the coverage can only be a minimum figure, totalling about 10^8 tons of zinc. The Kupferschiefer-type deposition is rare in earth's history because of a low probability of coincidence of several incidents such as large volumes of stagnant sea water under reducing conditions, oxidized sediments exposed on the continent which deliver trace metals to be accumulated and precipitated in sea water, low detrital sedimentation, and therefore limited dilution of sulfide precipitates in the sediment.

A comparable case of Recent sediments, but on a different scale of metal concentrations, is that of the Black Sea. More than half the area of the Black Sea bottom is covered by sediments containing concentrations higher than 100 ppm Zn; more than a quarter shows higher than 200 ppm Zn, and more than 10 percent, higher than 300 ppm Zn (STRACHOW et al. 1971). Higher sulfide dilution by detrital matter and lower supply of metals from the continent (surface waters have zinc in "normal" or less than normal sea water concentrations, see Fig. 30-I-1) has kept metal concentrations in the sediment more than 10 times lower than in Kupferschiefer.

Metal accumulations even smaller than in the Black Sea sediment have been observed in the organic carbon rich euxenic sediments of Walvis Bay (off southwest Africa). Zinc occurs there in the range 18 to 337 ppm Zn; the larger areas of sediment contain only 50 to 100 ppm Zn (CALVERT and PRICE, 1970) which is normal even for silty clays lacking organic matter. MANHEIM (1961) reports about minor metal accumulation in deeps of the Baltic Sea.

Comparable concentration ranges have been reported on sediments (containing 0.3 to 15.8 percent organic carbon) from the Framvaren (Norway) anoxic fjord: 68 to 247 ppm Zn (PIPER, 1971). This author discriminates between two fractions of zinc, one soluble and the other insoluble in 0.1 M HCl. Zinc in the soluble fraction increases appreciably in sediments formed under more than 15 m water by almost a

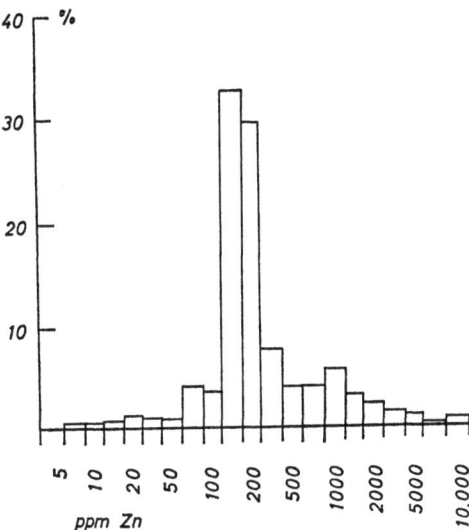

Fig. 30-K-3. Frequency distribution of zinc in 20 sets of 779 black shale samples from North America according to VINE and TOURTELOT (1970). The histogram is drawn from the sum of the percent frequency distribution of individual sets giving equal weight to each set

factor of three. The insoluble fraction is nearly constant at a mean of about 40 ppm Zn (silt from granite decomposition). The soluble fraction of 30 to 210 ppm Zn is expected to contain mainly sphalerite from the sulfide precipitation in the stagnant water.

Higher abundances of zinc in bituminous than that in "normal" argillaceous sediments can be contained (beside sphalerite) in their *phosphorite fraction*. Several authors report the following ranges of zinc concentrations in phosphorite beds and concretions: concretions in different layers of Mesozoic to Tertiary age from an area north of the Harz-Mountains (Germany), 10 to 750 ppm Zn (BENDA, 1963); phosphate rocks from the Ryuku-Islands (Japan), 40 to 6,800 ppm Zn (average 4,000 ppm Zn) (KANESHIMA, 1962); Curaçao-Guano, 400 ppm Zn (HUTCHINSON, 1950). Several hundred ppm Zn might be a mean value for marine phosphates or excretions from birds feeding on marine animals.

Information on zinc concentrations in black shales of different geologic age and origin have been compiled by WEDEPOHL (1964). A more extensive survey on 20 deposits of similar rocks from North America has been published by VINE (1966, 1969), VINE and TOURTELOT (1969, 1970) and VINE, TOURTELOT and KEITH (1969). The compilation by WEDEPOHL (1964), including some additional data, is listed in Table 30-K-3; for references see this publication. The detailed investigations by VINE et al. on North American samples from a wide variety of organic carbon producing environments (from Late Precambrian to Eocene in age) have been condensed in the report by VINE and TOURTELOT (1970). Their frequency distribution diagram of zinc in 20 sets of 729 samples (methods: S and W/C) is presented in Fig. 30-K-3. This diagram has a distinct maximum at 200 ppm Zn. It should be mentioned again that according to these authors, samples higher in zinc allow a correlation of

Table 30-K-3. *Zinc concentrations in selected black shales. (Number of samples in brackets)*

Rock type	Origin	Range ppm Zn	Average ppm Zn	Analytical method	Reference
Sapropelic clays (5.8% C; 1.6% S) (40)	Recent, Baltic Sea	30—250	110	S	1
Sapropelic clays (6.3% C; 1.5% S) (26)	Kyrkfjarden Fjord (Sweden)	—	130	S	4
Sapropelic clays and silts (8.7% C) (28)	Recent, Framvaren Fjord (Norway)	100—237	160	A	3
Sapropelic clays (2)	Recent, Kolje Fjord (Sweden)	100—110	105	X	2
Black shales (22) (0.8% C; 0.6% S)	Upper Cretaceous (Pierre), Great Plains (U.S.A.)	70—200	140	W/C	1
Black shales (27) (11.7% C; 2.7% S)	Jurassic (Lias-ε), northwest Germany	15—400	165	W/C	1
Black shales (13) (7.3% C)	Jurassic (Lias-ε), northwest and southwest Germany	22—300	110	X	2
Black shales (8)	Upper Carboniferous, Kansas, Oklahoma, Missouri (U.S.A.)	80—700	300	S	1
Black shales (6) (4.4% C)	Upper Devonian (Chattanooga), U.S.A.	75—350	160	S	1
Black shales (7) (17% C; 6.1% S)	Ordovician, Cambrian, Sweden	—	68	S	4
Black shales (2)	Silurian, Europe	24—210	117	X	2
Dark grey shales (23) (0.05% S)	Lower Precambrian (Fig Tree, Swaziland), South Africa	—	114	X	5
Black shales (779) (3.2% C)	Late Precambrian to Eocene, North America	34—1,500	300[a]	S and W/C	6

[a] For frequency distribution see Fig. 30-K-3.

References: 1. Compilation, WEDEPOHL (1964). 2. WEDEPOHL (unpublished data). 3. PIPER (1971). 4. LANDERGREN and MANHEIM (1963). 5. HOFMEYR (1971). 6. VINE and TOURTELOT (1970).

this element with the organic fraction. Besides zinc, the elements copper, nickel, chromium, molybdenum, vanadium, silver and selenium are often correlated with the organic fraction of these samples.

c) Pelagic Sediments

Compared with nearshore deposited argillaceous sediments, pelagic clays have specifically accumulated several elements such as Mn, Mo, Ni, Co, Cu, Pb, Sb etc. Zinc has a slight tendency for higher concentrations in pelagic clays than in shales as evidenced by data listed in Table 30-K-4. Some of the mentioned accumulations can be explained by diagenetic mobilization in the continental shelf regions, or by magmatism.

If one excludes the data published by TATSUMOTO (1957), which have a tendency to be high, the pelagic clays are expected to contain, on average, 130 to 150 ppm Zn, which is about 40 percent higher than the mean zinc concentration in shales. Syngenetic compounds of pelagic clays in addition to those of shales are manganese oxides (MnO_{2-x}). AHRENS et al. (1967) report an average of 600 ppm Zn (method: X) (range 172—1,409 ppm Zn) for 30 manganese nodules from the three oceans. The present author found 380 ppm Zn in 7 manganese nodules from the Pacific (method:X). If one estimates that the average pelagic clay contains almost one percent manganese and the average nodule about twenty percent manganese then an addition of about 30 ppm Zn is expected through the compound MnO_{2-x}. In the report by AHRENS et al. (1967), there is no systematic difference visible with regard to zinc concentrations among manganese nodules from the different oceans.

BOSTRÖM and PETERSON (1969) report about ferromanganoan sediments from the East Pacific Rise having as much as 315 ppm Zn in the carbonate free fraction (average of 20 samples; method: A).

Table 30-K-4. Zinc concentrations in pelagic sediments. (Number of samples in brackets)

Sediment type	Origin	Range ppm Zn	Average ppm Zn	Analytical method	Reference
Clays (17)	Atlantic Ocean	85—180	130	X	WEDEPOHL (1960)
Clays (100)	Atlantic Ocean	—	130	S	LANDERGREN and MANHEIM (1963)
Clays (450)	Atlantic and Pacific Ocean	—	85	S	LANDERGREN and MANHEIM (1963)
Clays (6)	Pacific Ocean	110—290	174	X	LEWIS and GOLDBERG (1956)
Clays (24)	Pacific Ocean	80—1,300	370	W/C	TATSUMOTO (1957)
Clays (9)	Pacific Ocean	83—288	160	X	WEDEPOHL (1960)
Calcareous oozes (5) (50% calcite)	Atlantic Ocean etc.	7.1—86.3	45.5	A	HONJO, TABUCHI (1970)
Calcareous oozes (200) (78% calcite)	Pacific Ocean	—	37	S	LANDERGREN and MANHEIM (1963)
Globigerina oozes (8)	Pacific Ocean	30—350	210	W/C	TATSUMOTO (1957)
Globigerina oozes (5)	Atlantic (4) and Pacific Ocean	18—44	32	X	WEDEPOHL (unpublished)
Pteropod oozes (6)	Gulf of Mexico	<6—312	79	X	PYLE and TIEH (1970)
Diatomaceous ooze (1)	Pacific Oceans	—	72	X	WEDEPOHL (unpublished)

III. Carbonate Rocks and Evaporite Sediments
a) Carbonate Rocks

Limestones very often contain between ten and fifteen percent clay and silt residues (insoluble in HCl). In Subsection 30-K-II, it has been demonstrated that shales and clays have, on average, about 100 ppm Zn; this will contribute 10 to 15 ppm Zn to carbonate rocks by means of 10 to 15 percent argillaceous fraction. As listed in Table 30-K-5, limestones often contain more zinc, and as reported at the end of Table 30-K-2, argillaceous residues of limestones composites, (having 14% residues) representing 93 samples, contain about twice as much zinc (192 ppm Zn) as shales and clays. This is expected to be the zinc from the carbonate fraction, adsorbed on the clay surfaces during dissolution of the limestones.

According to partition experiments on calcite and artificial sea water (Subsection 30-H-Ic), zinc concentrations in Recent marine calcite are expected to plot between 4 and 20 ppm Zn. Zinc concentrations in Recent foraminefera shells are close to 10 ppm Zn, if one excludes those with manganese oxide coating. Calcite from Recent marine clam shells, cephalopods and calcareous algae are even lower than 10 ppm Zn (1 to 4 ppm Zn, WEDEPOHL, 1955). Additional zinc can enter the carbonate fraction of limestones and dolomites during diagenesis.

GRAF (1960), in his compilation on minor elements in carbonate rocks, estimated an average of 26 ppm Zn from older references. WARREN and DELAVAULT (1961) got 18 ppm Zn in aqua regia extracts of 104 carbonate rocks of Devonian, Silurian and Ordovician age from Ontario (Canada) (if one excludes three quarries with abnormal zinc accumulation). Our mean marine limestone is close to 20 ppm Zn. Dolomites might be comparable in average zinc concentrations in contrast to some higher values listed in Table 30-K-5.

Table 30-K-5. *Zinc concentrations in carbonate rocks. (Number of samples in brackets)*

Rock type	Origin	Range ppm Zn	Average ppm Zn	Analytical method	Reference
Limestones, marine (composite of 32 samples, 12% insoluble residue, 4% dolomite)	Devonian, Germany	—	28	X	1
Limestones, marine (composite of 45 samples, 11.5% insoluble residue, 2% dolomite)	Jurassic, Germany	—	25	X	1
Limestones, marine (composite of 16 samples, 22.5% insoluble residue)	Cretaceous, Germany	—	20	X	1
Calcareous muds and reef tract (3)	Recent, Florida Bay, Key Largo (U.S.A.)	11—22	17	X	1
Limestone, marine (1)	Eocene, Roumania	—	15	W/C	2
Limestones, marine (5)	Cretaceous, northwest Germany	10—21	15	X	1

Table 30-K-5. (Continued)

Rock type	Origin	Range ppm Zn	Average ppm Zn	Analytical method	Reference
Limestones, marine (8)	Jurassic, northwest Germany	<10—90	26	X	1
Limestones, marine (6)	Triassic, northwest Germany	14—41	25	X	1
Limestones, marine (13)	Triassic, Thuringia, (Germany)	3—70	10	W/C	3
Limestones, marine (10)	Triassic, western Carpathian Mountains (Roumania)	4—52	21	W/C	2
Limestones, marine (4)	Carboniferous, New Mexico (U.S.A.)	12—55	31	W/C	4
Limestones, marine (13) (11% insoluble residue)	Paleozoic, mid-continental (U.S.A.)	<0.1—20.3	6.3	A	5
Limestones, marine (9) (15% insoluble residue)	Paleozoic, Europe	<0.1—57	18.6	A	5
Limestones (25) (6% insoluble residue)	Paleozoic, Asia, mainly Japan	<0.1—24	7.8	A	5
Limestones (296) (<10% isoluble residue)	Upper Paleozoic, Jurassic, Japan	<0.1—103	6.3	A	5
Chalks	Cretaceous, Stevns Cliff (Danmark)	12—53	21	A	5
Limestones, freshwater (travertines) (16)	Recent, western and southern Germany (except 1 sample from Thuringia, D.D.R.)	3—1,900	205 (19)[a]	X	6
Dolomites (2)	Upper Permian, south of Harz Mountains (Germany)	10—17	14	X	1
Dolomites, partially bituminous (14 subsurface samples)	Upper Permian, northwest Germany	25—85	45	X	7
Dolomites, partially bituminous (43 surface samples)	Upper Permian, northwest Germany	30—180	55	X	7
Dolomites (19)	Triassic etc., Western Carpathian Mountains (Roumania)	5—45	23	W/C	2

[a] Excluding 4 samples containing abnormally high zinc concentrations above 100 ppm.

References: 1. WEDEPOHL (unpublished). 2. SAVUL, ABABI (1956b). 3. HEIDE, SINGER (1950). 4. BARNES (1959). 5. HONJO, TABUCHI (1970). 6. SAVELLI, WEDEPOHL (1969). 7. SMYKATZ-KLOSS (1966).

Table 30-K-6. *Zinc concentrations in evaporate minerals. (Number of samples in brackets)*

Rock or mineral type	Origin	Range ppm Zn	Average ppm Zn	Analytical method	Reference
Halite	Cambrian, Siberian Platform (U.S.S.R.)	—	1.4	W/S	MOORE (1971)
Halite	Carboniferous, Utah (U.S.A.)	—	0.56	W/S	MOORE (1971)
Halite	Permian, New Mexico (U.S.A.)	—	1.1	W/S	MOORE (1971)
Halite (6 groups of samples)	Permian, south of Harz Mountains (Germany)	0.1—0.2[a]	0.3	S	HERRMANN (1958)
Halite	Jurassic, Louisiana (U.S.A.)	—	0.08	W/S	MOORE (1971)
Halite	Quaternary, California (U.S.A.)	—	2.6	W/S	MOORE (1971)
Anhydrite (3)	Permian, north-west Germany	<10—17	≤12	X	WEDEPOHL (unpublished)
Sylvite, carnallite etc. (3 groups of samples)	Permian, south of Harz Mountians (Germany)	0.3—1.45[a]	0.9	S	HERRMANN (1958)

[a] Ranges for groups of samples.

b) Evaporite Sediments

If sea water, with its 2 to 4 ppb Zn, is completely evaporated, this corresponds to a concentration by a factor of 30 in the product (mainly NaCl). Therefore, about 0.1 ppm Zn can be expected at the maximum in halite if this is a primary product of one evaporation cycle. Because of this low value, the zinc concentration in halite can be easily overestimated if samples are contaminated by detrital matter etc.(Table 30-K-6).

IV. Average Abundance in Sedimentary Rocks

Nearshore deposited, argillaceous rocks (clays, shales) are by far the most abundant sedimentary rocks. In the sandstone class, greywackes are most abundant. The proportion of shale to greywacke is rather uncertain. Because of similar average zinc concentrations in these two classes (average in shales: 120 ppm Zn; average in greywackes: 105 ppm Zn), this uncertainty has little influence on the final result.

Shales low in bituminous matter contain, on average, about 95 ppm Zn (70 percent of these shales range between 50 and 130 ppm Zn) and those higher in organic carbon have a frequency maximum at 200 ppm Zn. Pelagic clays contain, on average, 130 to 150 ppm Zn. The total portion bituminous and pelagic pelitic rocks in the total class of argillaceous deposits is expected to range between 30 and 40 percent.

The majority of sedimentary rocks contains about 100 ppm Zn with a mean at 110 ppm Zn. The less abundant sedimentary types, such as limestones (average: 20 ppm Zn), quartz sandstones and arkoses (average: about 30 ppm Zn) are low in zinc. The abundance of these two rock classes in the total column of sediments has been compiled by the present author in Chapter 8 of Volume I of this handbook. The estimates range from 5 to 29 percent for limestones (plus dolomites) and from 8 to 40 percent for a total of the sandstones; only less than half of the latter group can be quartz sandstones and arkoses. If we consider 15% limestones and 15% sandstones (excluding greywackes), the average sediment contains 85 ppm Zn. This value is 35 percent higher than that of 57 ppm Zn for the average of magmatic rocks.

The imbalance in zinc between average weathered magmatic rocks and their sink, the sediments, might be explained by an additional accumulation of zinc for degassing of the upper earth ("excess" volatile elements) in the sedimentary cover.

30-L. Biogeochemistry

Zinc is one of the common trace elements which obviously has some importance for the organic production. Our knowledge about this importance is mainly limited to the observed accumulation in special tissues, organisms and organic residues, to the occurrence of certain zinc complexes in organisms and to the behavior in nutrition experiments.

I. Accumulation in Primary Production

The organic production per year in the total mass of sea water approximately equals that on the surface of the continents. There are differences of at least a factor of ten in the production in the different types of ocean surface water (90 percent is produced on the continental shelves), resulting in a total of about 50 g C/m² year, about 70 g C/m² year and about 80 g C/m² year in the Pacific, Atlantic and Indian Oceans respectively (KOBLENTS-MISHKE et al., 1968). The vertical zoning of production in sea water is even much steeper than the lateral one, having a high rate within the upper four hundred meters of the water column. Only a very small proportion of residues from the primary production in the range of 0.0X percent is preserved in the sediment, feeding a long-time carbon cycle. More than 70% of the land surface has a higher production than 50 g C/m² year. There again is almost a factor of ten between zones of low and high productivity (excluding deserts). The average rate of long time preservation of terrestrial organic matter (in soil etc.) is unknown.

Table 30-L-1 contains information about zinc concentrations in marine organic matter, considering primary production and higher animals. Except for a few groups like acalepha, actinia (NODDACK and NODDACK, 1939) and oysters, the most common tissue of marine organisms contains 100 to 400 ppm Zn (in dry substance).

These concentrations equal 20 to 130 ppm Zn in fresh matter. Relative to sea water, the average marine primary production has accumulated zinc by a factor of about 7,500. Freshwater weeds have about the same factor of accumulation (ISHIBASHI et al., 1965). Biological cells can accumulate ions against the gradient of concentration. Abnormal zinc concentrations can be accumulated from natural waters by sulfate reducing bacteria, probably by sulfide precipitation (accumulation factors more than 1,000).

As listed in Table 30-L-2, terrestrial plants are lower in zinc than average marine organisms. Certain species can accumulate several times the average zinc concentration of terrestrial plants, which is probably close to 50 ppm (in dry substance). Abundant zinc concentrations in plant ash are in the range of 1,000 to 1,500 ppm, which indicates an accumulation by more than a factor of ten in relation to sedimentary rocks (and soils).

Plant species whose distribution is generally affected by the chemical constituents of ore deposits are called universal indicator plants. The well known universal

Table 30-L-1. *Zinc concentrations in dry substance (or ash[a]) of marine organisms. (Number of samples in brackets)*

Samples	Origin	Range in ppm	Average in ppm	Analytical method	Reference
Phytoplankton (diatoms) ash (4 species)	Black Sea	1,000—15,000[a]	—	S	1
Zooplankton ash (5 species)	Black Sea	800—20,000[a]	—	S	1
Algae	marine	120—830	—	R	2
Algae (24) (7 species)	marine, coast, England	40—136	80	W/S	3
Brown seaweeds (57) (22 species)	marine, coast, Japan	16—680	149	W/P	4
Green seaweeds (25) (11 species)	marine, coast, Japan	35—330	161	W/P	4
Red seaweeds (25) (17 species)	marine, coast, Japan	25—270	140	W/P	4
Animals (9 species)	marine, Sweden	65—1,550	454	S	5
Lamellibranchiata (12) (2 species)	marine, New Zealand	50—370	187	S	6
Lamellibranchiata (30) (23 species except oysters) soft parts	marine	66—4,400	434	W (?)	7
Oysters (6)	marine, New Zealand	850—1,500	1,100	S	6
Oysters (3) (2 species)	marine	1,000—1,400	1,200	W (?)	7
Oyster from Zn-polluted water (1)	marine		53,000	W/C	8
Gastropoda (15) (13 species)	marine	55—308	182	W (?)	7
Cephalopoda (5) (3 species)	marine	77—773	257	W (?)	7
Echinodermata	marine	3—180	33	W (?)	7

References: 1. VINOGRADOVA, KOVALSKIY (1962). 2. GUTKNECHT (1963). 3. BLACK, MITCHELL (1952). 4. ISHIBASHI et al. (1965). 5. NODDACK, NODDACK (1939). 6. BROOKS, RUMSBY (1965). 7. VINOGRADOV (1953). 8. MORITA (1955).

indicator plant of zinc is *Viola calamineria* (zinc violet). The localization of other plants such as mustards and pinks on zinc dumps may be due to acidity, sulfur etc., and not to zinc. The cress *Arabidopsis thalianum* can contain 9,000 ppm Zn in dry substance (about 6 percent in the ash) (CANNON, 1960). Due to toxity of high zinc concentrations to non-accommodated plants the following physiological and morphological changes can be expected: chlorotic leaves with green veins, white dwarfed forms; dead areas of leaf tips and roots stunted.

Table 30-L-2. *Zinc concentrations in dry substance (or ash[a]) of terrestrial organisms.*
(Number of samples in brackets)

Samples	Origin	Range in ppm	Average in ppm	Analytical method	Reference
Limnetic weeds (11) (8 species)	Japan	93—640	206	W/P	1
Fungi	Finland		150	W (?)	2
Bryophytes	U.S.A.	2—250	125	W/C	4
Ferns	Finland etc.		77	W (?)	2
Gymnosperms (417)	U.S.A. etc.		695[a] (25[b])	W (?)	3
Angiosperms (1346) (great number of species)	U.S.A. etc.		1,620[a] (57[b])	W (?)	3
Vegetables (315)	Georgia (U.S.A.)	—	38 (510[a])	W/C	8
Trees, lesser plants (449) (78 species)	British Columbia	8—120	38	W/C	5
Trees, stems (459)	Georgia (U.S.A.)	—	23.1 (800[a])	W/C	8
Trees, leaves (459)	Georgia (U.S.A.)	—	20.0 (393[a])	W/C	8
Freshwater plants (10 species)	Germany	37—104	61	W (?)	6
Insecta			400	W (?)	7
Mammalia			160	W (?)	2

[b] Converted into concentration in dry substance by assumption of average content of ash of 3.5 percent.

References: 1. ISHIBASHI et al. (1965). 2. BOWEN (1966). 3. CANNON (1960). 4. SHACKLETTE (1965). 5. WARREN (1962). 6. SEIDEL (1966). 7. SPECTOR (1956) in: BOWEN (1966). 8. SHACKLETTE et al. (1970).

II. Physiological Function

The reasonably high rates of accumulation of zinc in marine and terrestrial organisms relative to sea water and soil, respectively, indicate physiological functions of this trace element.

Analyses of the separated different types of tissues of organisms help to distinguish the zinc concentrating compounds. For mammalian tissues, TIPTON and COOK (1963; in BOWEN, 1966) give the following average zinc concentrations (in dry substance) in increasing order: skin, 13 ppm; brain, 46 ppm; lung, 62 ppm; heart, 110 ppm; liver, 130 ppm; hair, 170 ppm; muscle, 180 ppm; and kidney, 210 ppm. The kidneys are accumulators of several trace elements (Cd, Hg, Mn, Mo, Pb, Cu). According to BOWEN (1966), mammalian blood with 6.5 ppm Zn (12 ppm in red cells) is as low in zinc as blood of terrestrial invertebrates.

Zinc acts as an enzyme activator and constituent of several important metalloprotein enzymes such as flavoproteins, carbonic anhydrase (KEILIN and MANN, 1940) carboxypeptidase, alkaline phosphatase, ethanol dehydrogenase, lactic dehydrogenase, glutamic dehydrogenase (MAHLER, 1961). The influence of zinc-containing enzymes on the pancreas activity has been observed by WEITZEL et al. (1953) and others. The same group of authors (WEITZEL et al., 1954) has detected zinc as a major constituent of the tapetum lucidum part of the eyes of certain mammals. In dry substance of this tissue of fox, they got upto 12% Zn (cysteine compound). The function of a number of zinc proteins in animals has as yet to be investigated.

Zinc complexes have been observed, but not chemically characterized, in plant material. There might be replacement of Mg by zinc in the structure of chlorophyll because of its size and valency.

There is no doubt, from nutrition experiments with algae etc., that zinc is needed for autotrophic growing. This element is only slightly toxic for animals; its level of toxic concentration is much higher than that of Cd, Cu, Pb, Hg etc. CUSHING and ROSE (1970) present experimental evidence indicating dominant adsorptive uptake mechanisms of zinc by periphyton.

III. Bituminous Matter, Oil, Coal etc.

a) Bituminous sediments

Degradation of organic residues and their preservation in sediments can be connected with preservation of a high proportion of the original zinc content of the organic matter. But certain zinc complexes (porphyrins, etc.) may be formed not earlier than during diagenesis. These compounds can extract additional zinc from

Table 30-L-3. *Zinc concentrations in crude oils mainly from Chevron Oilfield Research Corporation, La Habra, California* (SLENTZ, 1971). *(Number of samples in brackets)*

Location	Age	Range in ppm	Average in ppm	Analytical method
Ventura basin, California (5)	Tertiary	7—86	28	N/R
Wilmington, California (1)	Tertiary	—	<0.5	N/R
San Joaquin Valley, California (3)	Tertiary	5—12	8	A
Louisiana (1)	Tertiary	—	4	N/R
East Texas (1)	(?)	—	160	N/R
Mississippi salt basin (2)	Cretaceous and Jurassic	<0.5—1	<1	A
Powder R. basin, Wyoming	Carboniferous	—	7	N/R
North Slope, Alaska (2)	Triassic and Cretaceous	2—7	4.5	N/R
Kenai Peninsula, Alaska (1)	Jurassic	—	6	A
Sumatra (2)	Tertiary	<1.5—4	≤2.5	N/R
El Morgan, Egypt (1)	(?)	—	3.5	N/R
Sirte, Libya (3)	Eocene and Cretaceous	13—72	49	N/R
Agha Jari, Iran (1)	Tertiary	—	<0.8	N/R
Kuwait (1)	(?)	—	<0.8	N/R

Table 30-L-4. *Concentration of zinc in peat and coal or their ashes ([a]). (Number of samples in brackets)*

Sample	Locality	Range of concentrations in ppm	Average in ppm	Analytical method	Reference
Peat (76) (1—90 percent ash)	Germany (G.D.R.)	(10[a]—1,000[a])	(163)[a]	S	1
Brown coal (35) (7—20 percent ash)	Germany (G.D.R.)	(<10[a]—500[a])	(≦27)[a]	S	1
Brown coal (6)	Victoria, Australia	70—900	300	S	2
Hard brown coal (43)	South Australia	<80—200	100	S	2
Hard coal (68)	New South Wales, Australia	<25—300	50	S	2
Hard coal (53)	Queensland, Australia	<30—1,000	200	S	2
Coal (475)	Eastern Interior Coal Prov., U.S.A.	<100—600 (7,800[a])	44 (590[a])	S	3
Coal (376)	Appalachian, U.S.A.	<100—230 (5,000[a])	7.6 (160[a])	S	4
Coal (228)	Southwestern Interior Coal Prov., U.S.A.	<100—2,100 (13,000[a])	108	S	5
Coal (221)	Northern Great Plains, U.S.A.	<200—1,000 (7,000[a])	59 (560[a])	S	6
Coal	West Virginia, U.S.A.	—1,900	(430[a])	S (?)	7
Coal (Carboniferous) (378)	Central and eastern Germany	(50[a]—2,100[a])	118 (1,370[a])	S	8
Coal (Permian) (88)	Central and eastern Germany	(150[a]—5,100[a])	300 (1,340[a])	S	8
Coal (Mesozoic) (12)	Central and eastern Germany	(300[a]—>5,000[a])	85 (1,000[a])	S	8
Coal	several areas	—10,000	200	S	9

References: 1. LEUTWEIN (1956). 2. SWAINE (1971). 3. ZUBOVIC et al. (1964). 4. ZUBOVIC et al. (1966). 5. ZUBOVIC et al. (1967). 6. ZUBOVIC et al. (1961). 7. HEADLEE, HUNTER (1955). 8. LEUTWEIN, RÖSLER (1956). 9. GOLDSCHMIDT (1937, 1954).

porous fluids. Hydrogen sulfide formed from degradation of amino compounds can be another source of secondary zinc accumulation in organic residues. SWANSON et al. (1966) present data which prove that water soluble humates can absorb about 10 percent zinc by dry weight. KRAUSKOPF (1956) did experiments showing that 40 and 99 percent of the zinc concentration of dilute solutions have been absorbed by plankton and peat-moss suspensions, respectively.

In Subsection 30-K-IIb data and information are compiled indicating higher zinc concentrations in bituminous shales than in non-bituminous pelites. Oils usually contain metal porphyrin compounds in the range of 1 to 1,000 ppm (those of vanadium, nickel and copper being most stable). The source of these pigments includes chlorophylls, hemins, enzymatic porphyrins and chlorins.

b) Oil

There are limited data on zinc concentrations in oil, which is condensed in Table 30-L-3.

From this compilation, representing crude oils from several of the important oil fields of the world, it can be seen that oil is usually low in zinc. It has not extracted larger proportions of this metal from country rocks. A gross average of a few ppm Zn in crude oil is to be expected.

c) Coal

Degradation products of higher plants are expected to be, on average, in the X0 ppm Zn range, if the original zinc concentrations of the plants are preserved during diagenesis.

Data of Table 30-L-4 indicate that coal often contains zinc in the concentration range 50 to 150 ppm. Diagenesis and alteration of plant matter into coal should have concentrated the elements for which the system was closed, by a factor of 3 to 10. If we use the factor 3, the original material plant could have contained 15 to 50 ppm Zn which seems to be reasonable in comparison with the information from Table 30-L-2. Coal ashes have generally accumulated zinc relative to the average of shales and clays by a factor of 5 to 10. Zinc often occurs as sphalerite, or associated with sulfides in coal.

Manuscript received: March 1972

30-M. Abundance in Common Metamorphic Rock Types

Because of its crystal chemical behavior, zinc in metamorphic rocks is mainly incorporated in ferrous iron and magnesium silicates and oxides as biotite, phengite, chlorite, amphibole, staurolite, garnet and magnetite (see Section 30-D). Sphalerite can locally occur in minor concentrations.

There is only a limited number of reliable zinc analyses of metamorphic rocks available as listed in Table 30-M-1. Because of their semiquantitative character, the

Table 30-M-1. *Zinc concentrations in metamorphic rocks. (Number of samples in brackets)*

Rock type	Locality	Range of concentration in ppm	Average in ppm	Analytical method	Reference
a) Serpentinites (4)	Münchberg mass, Bavaria (Germany)	57—65	62	X	1
Serpentinite (1)	Kemmletten, Alps (Switzerland)	—	70	X	1
Serpentinites (2)	Nagano, Yamaguchi (Japan)	50—95	73	W/C	12
Serpentinites, metapyroxenites	East Amisk, Saskatchewan (Canada)	—	33	X	11
b) Eclogites (2)	Münchberg mass, Bavaria (Germany)	84—105	95	X	1
Eclogite from kimberlite (1)	South Africa	—	80	X	13
Granulites, mafic (17)	Brazilian Shield	—	118	A	8
c) Amphibolites (9)	Bamble, South Norway	18—142	89	A	2
Amphibolites (20)	Dalradian (Scotland)	89—204	159	X	3
Amphibolites (16)	Northwest Adirondacks (U.S.A.)	—	153	C	4
Amphibolites (8)	Dover district, New Jersey (U.S.A.)	140—280	214	S	9
Amphibolite (1)	Wechsel, Alps (Austria)	—	77	X	1
Biotite amphibolite (1)	Toyama Prefecture (Japan)	—	140	W/C	12
Amphibolites (2)	Münchberg mass, Bavaria (Germany)	72—173	123	X	1
Meta gabbros (great number)	East Amisk, Saskatchewan (Canada)	—	48	X	11

© Springer-Verlag Berlin · Heidelberg 1972

Table 30-M-1. (Continued)

Rock type	Locality	Range of concentration in ppm	Average in ppm	Analytical method	Reference
d) Phyllites (4)	Vorderrhein area, Alps (Switzerland)	17—73	40	X	1
Mica schists (13)	Lukmanier road, Alps (Switzerland)	32—124	61	X	1
Mica schists (2)	Wechsel, Alps (Austria)	66—78	72	X	1
Mica schists (6)	Cevennes (France)	24—167	106	X	1
Mica schists (staurolite, garnet) (4)	North Idaho batholith (U.S.A.)	—	<200	S	5
Chlorite schists (29)	Dalradian (Scotland)	103—233	174	X	3
Gneisses (garnet) (2)	Lukmanier road, Alps (Switzerland)	46—121	83	X	1
Gneisses (4)	Waldviertel (Austria)	34—90	66	X	1
Gneisses (2)	Wechsel, Alps (Austria)	33—48	40	X	1
Gneiss (two micas) (1)	Verzasca, Tessin Alps (Switzerland)	—	70	X	1
Gneisses (3)	Cevennes (France)	19—161	101	X	1
Gneisses (3)	Bavarian Forest (Germany) (Austria)	62—83	71	X	6
Gneisses (3)	Black Forest (Germany)	44—74	54	X	6
Gneisses (low in mica) (6)	Spessart (Germany)	15—82	55	X	7
Gneisses (high in muscovite) (6)	Spessart (Germany)	8—22	13	X	7
Hornfels (metam. Devonian shale)	Harz Mountains (Germany)	—	153	X	1
Hornfels (metam. pelites and silts) (62)	Santa Rosa Range, Nevada (U.S.A.)	27—116	68	S	10
Granulites, intermediate (25)	Brazilian Shield	—	109	A	8
Granulites, salic (20)	Brazilian Shield	—	50	A	8
e) Marbles (4)	Austria, Yugoslavia	<5—24	≦12	X	1
f) Quartzites (composite of 10)	Germany	—	23	X	1

References: 1. WEDEPOHL (unpublished). 2. BURRELL (1965). 3. VAN DE KAMP (1970). 4. ENGEL et al. (1964). 5. HIETANEN (1969). 6. HAACK (1969). 7. OKRUSCH, RICHTER (1967). 8. SIGHINOLFI (1971). 9. COLLINS (1971). 10. WODZICKI (1971). 11. SMITH (1964). 12. MORITA (1955). 13. RIMSAITE (1971).

numerous emission spectrographic analyses of metamorphic rocks from Sweden published by LUNDEGÅRDH (1947, 1949) are not quoted in this table. They usually range higher than the more recent data from quantitative methods with higher precesion and accuracy.

The data of Table 30-M-1 are grouped in keeping with the different origin of the premetamorphic rocks. The serpentinites contain zinc concentrations comparable to those of peridotites whereas zinc in the eclogites and amphibolites is as high as in basalts and gabbros (Section 30-E).

Sandstones and limestones are usually lower in zinc than greywackes and shales. The same difference is visible in quartzites and marbles in comparison with phyllites, mica schists and gneisses. The source materials for most of the latter rocks are most probably shales, marls and greywackes. Except for the Dalradian chlorite schists (and two staurolite garnet schists from Idaho), which at least partially originate from mafic magmatics, the majority of mica schists and gneisses contains zinc in concentrations close to 65 ppm (Section 30-K). Shales low in carbonate and quartz as well as graywackes usually range higher in zinc than 65 ppm but marls have zinc concentrations comparable to the listed mica schists and gneisses. Because of the differences in average zinc concentrations between gneisses and mica schists on one side and shales and greywackes on the other side, a slight loss of zinc during metamorphism cannot be excluded. As mentioned in Section 30-N, we have investigated two profiles of rocks from Switzerland and France formed in pressure temperature gradients. They do not show any systematic change of zinc concentrations with metamorphic grade. From this it must be concluded that these rock units have behaved as closed systems with respect to zinc, even under higher amphibolite facies conditions.

Manuscript received: March 1972

30-N. Behavior in Metamorphic Reactions

If trace elements form ready volatile compounds, they can be partially lost from a rock system during thermal metamorphism (Hg for instance). For testing this behavior, the premetamorphic chemical homogeneity of the rock layer, which has undergone progressive regional metamorphisms, must be confirmed. This confirmation is a real problem and can only be given with some precaution.

A slight difference in average zinc concentrations between abundant mica schists, gneisses and similar rocks on one side and shales and greywackes on the other side, as concluded from data listed in Section 30-M, could be explained as zinc loss from pelitic or psammitic rocks during metamorphism. ENGEL and ENGEL (1960a, b) assume that measurable proportions of the original Zn concentrations of greywackes were removed during high grade (amphibolite to granulite facies) regional metamorphism of the Adirondack Mountains. In the opinion of the present author, retrograde metamorphism (replacement of biotite by chlorite-muscovite etc.) could have been effective locally in mobilization of zinc. Several additional observations are in favor of a maintenance of the original zinc concentrations during progressive metamorphism. In a Triassic profile of marls grading into phyllites, mica schists of greenschist and amphibolite facies sampled by M. Frey, of Berne, in the Swiss Alps, were obtained zinc concentrations (by X-ray method) for averages of 7 shales, 4 phyllites and 13 mica schists: 41 ppm, 40 ppm, 61 ppm Zn respectively. The relative increase of zinc must be due to a loss of carbon dioxide and water in the range of 20 to 30%. Another series of 10 mica schist and gneiss samples representing progressive metamorphism from greenschist to upper amphibolite facies was placed at our disposal by H. J. Tobschall of Mainz. They do not show any systematic change of zinc with metamorphic grade but a rather large scattering of values (mean about 100 ppm Zn), probably due to a premetamorphic heterogeneity of the sediment layers. Van de KAMP (1970) obtained evidence from greenschist and amphibolite facies, Scottish Dalradian series, that there is no systematic change of zinc concentrations with metamorphic grade.

From data in Table 30-M-1, amphibolites of probable basaltic origin are still as high in zinc as basalts and do not indicate loss of zinc. Partial melting of metamorphic rocks is connected with a mobilization of the zinc content of the latter. Zinc (as well as iron) in biotite of granites is generally more than 30% higher than zinc in biotite of gneisses and mica schists (see Section 30-D and reference: HAACK, 1969). After partial melting and zinc extraction, biotites of the residual solid rock (restite) are generally lower in zinc and higher in magnesium than biotites of mica schists and gneisses which have not yet approached temperatures of partial melting. Zinc concentration in granites generally increases with their biotite proportion (HAACK, 1969). This fact has some genetic importance and must be considered in the zinc balance of metamorphic and granitic rocks.

Salic granulites (about 70% SiO_2) as high grade metamorphic rocks, sampled by SIGHINOLFI (1971) in the Brazilian Shield, are lower in zinc (50 ppm Zn) than the average granites.

© Springer-Verlag Berlin · Heidelberg 1972

Manuscript received: March 1972

30-O. Summary (I); Economic Importance (II)

I. Geochemical Distribution and Behavior of Zinc

In peridotites as potential mantle rocks, zinc occurs in similar concentration (56 ppm Zn) as in ordinary chondrites (CH: 47 ppm Zn; CL: 60 ppm Zn). Carbonaceous chondrites (Cc_1), as well as enstatite chondrites, contain 8 times the zinc concentration of ordinary chondrites.

Basaltic and gabbroic rocks accumulate zinc relative to mantle rocks by a factor of two through partial melting. The range of variation in basaltic rocks around 100 ppm Zn is mainly controlled by scattering of the iron content and not by the magma type of basalt. Lunar basalts have ten to twenty times less zinc than terrestrial basalts.

Granitic rocks contain almost half the zinc concentration of basaltic and gabbroic rocks. There is an overlap between the ranges of variation of zinc in granites and quartz monzonites (average: 48 ppm Zn) and granodiorites and quartz diorites (average: 52.4 ppm Zn). Diorites and andesites (average: 70 ppm Zn) are almost intermediate in zinc between granitic and basaltic rocks. The zinc content of granitic rocks occurs mainly in the iron-magnesium structural positions of biotites (and/or amphibole). The major zinc content of basaltic rocks is incorporated in the magnetite structure.

Zinc has a large tendency to substitute for ferrous iron (and magnesium or other ions with about 0.8 Å radius) either in their fourfold coordinated positions or in sixfold coordinated positions having hydroxide as ligands. More than a third of all known zinc minerals contains hydroxide in structural positions.

The fraction of crustal zinc occurring in economic ore deposits is 10^{-7}. Zinc ore is mainly mined from epigenetic deposits. Zinc transport in ore-forming hydrothermal solutions is assumed to occur in chlorine and/or bisulfide complexes.

During chemical weathering and transport, zinc is accumulated in the detrital, more or less iron-rich fraction. Surface water is low in zinc (sea water: 4 ppb Zn), much lower than the solubility of zinc carbonate, hydroxide or phosphate permits. The low zinc concentration of natural water probably originates from adsorption control.

The major rock types of the sedimentary cover, in which zinc occurs in relatively high concentrations of about 100 ppm, are shales and greywackes. Zinc is generally correlated in these materials with their total iron content. Frequent values for argillaceous rocks, low and high in bituminous matter are: 95 ppm Zn and 200 ppm Zn, respectively. Pelagic clays range between 130 and 150 ppm Zn. Limestones contain, on average, 20 ppm Zn, and quartz sandstones and arkoses: 30 ppm Zn. A total average for the sediment cover of crustal rocks is 85 ppm Zn. This value is 35 percent higher than the average of magmatic rocks. The accumulation of zinc in the sedimentary environment originates primarily from chemical weathering of magmatic and metamorphic rocks and probably additionally from degassing of the upper layers of the earth.

Plant ashes have usually accumulated zinc relative to sediments by a factor of about ten. Coal ashes are often lower.

Mica schists and gneisses, as the most abundant metamorphic rocks, contain average zinc concentrations of about 65 ppm.

II. Economic Importance

Zinc came into use as metal not earlier than in the 16th century. The name has been mentioned at 1525/1526 by Paracelsus in his work: De mineralibus.

Major uses of zinc are: for galvanizing and coating iron and steel products (USA, 1956: 44% of total consumption of slab zinc), in the manufacture of brass (copper alloy with a zinc content in the range of 20 to 40 percent; U.S.A, 1956: 12% of total consumption of slab zinc), and in the zinc-base alloy metals (U.S.A. 1956: 36% of total consumption of slab zinc) for die castings, etc. In the form of rolled sheets zinc is used extensively in the manufacture of dry-cell battery cans, engraving plates etc. Zinc oxide (U.S.A. 1956: 2% of total consumption of slab zic) is consumed principally in the production of zinc pigments used in the manufacture of rubber, paints, ceramics, textiles, pharmaceuticals, chemicals etc. Zinc sulfide and zinc sulfide-cadmium sulfide mixtures constitute a major class of luminescent materials. White pigments are either ZnO or $ZnS+BaSO_4$ (lithopones). A most recent, wide use is as ZnO photoreceptor in photocopy papers; the yearly demand for this special purpose is approaching the 10^4 tons range.

In 1967 the world production of zinc ore was an equivalent of $4.66 \cdot 10^6$ tons of metallic Zn. More than 20 percent of this has been supplied by Canada (1967: $9.9 \cdot 10^5$ tons) and more than, or nearly, ten percent by the following countries each: U.S.A. (1967: $4.96 \cdot 10^5$ tons), U.S.S.R. (1967: $4.85 \cdot 10^5$ tons) and Australia (1967: $4.04 \cdot 10^5$ tons). Above the 10^5 tons production in 1967 were: Peru ($3.28 \cdot 10^5$ tons), Japan ($2.62 \cdot 10^5$ tons), Mexico ($2.13 \cdot 10^5$ tons), Poland ($1.9 \cdot 10^5$ tons), Italy ($1.2 \cdot 10^5$ tons), Congo ($1.2 \cdot 10^5$ tons), Germany (F.R.G.) ($1.1 \cdot 10^5$ tons) and Korea ($1.05 \cdot 10^5$ tons). Within the last ten years, the sequence of the main ore producers has changed much. By far, the main producer of metallic zinc is the U.S.A. ($8.34 \cdot 10^5$ tons).

The world supply in zinc is expected to be 10^8 tons. If the earth's crust contains, in dispersed form, on the average about 60 ppm Zn, this is a total of $1.5 \cdot 10^{15}$ tons Zn. The ratio of economically available to total zinc in the crust is about $1:10^7$

References: Section 30-A to 30-O

AHRENS, L. H., WILLIS, J. P., OOSTHUIZEN, C. O.: Further observations on the composition of manganese nodules, with particular reference to some of the rarer elements. Geochim. Cosmochim. Acta **31**, 2169 (1967).

ALLMANN, R.: Verfeinerung der Struktur des Zinkhydroxid-chlorids II, $Zn_5(OH)_8Cl_2 \cdot 1H_2O$. Z. Krist. **126**, 417 (1968).

ANSELL, G. B., KATZ, L.: A refinement of the crystal structure of zinc molybdenum(IV) oxide, $Zn_2Mo_3O_8$. Acta Cryst. **21**, 482 (1966).

AZZARIA, L. M.: A study of the distribution of traces of copper, lead and zinc in the minerals of a Precambrian granite. Can. Mineralogist **7**, 617 (1963).

BADALOW, S. T.: Pyrochroit, zinkhaltiger Serpentin und Allophan aus der Lagerstätte Almalyk (Usbekistan). Sapiski Wsesojusn. miner. Obsch. **87**, 698 (1958) [Russian].

BAEDECKER, P. A., SCHAUDY, R., ELZIE, J. L., KIMBERLIN, J., WASSON, J. T.: Trace element studies of rocks and soils from Oceanus Procellarum and Mare Tranquillitatis. Proc. Second Lunar Sci. Conf. **2**, 1037 (1971).

BARAGAR, W. R. A.: The geochemistry of Coppermine River basalts. Geol. Surv. Canada Paper **69,44**, 1 (1969).

BARNES, H. L.: The effect of metamorphism on metal distribution near base metal deposits. Econ. Geol. **54**, 919 (1959).

— CZAMANSKE, G. K.: Solubilities and transport of ore minerals. Chapter 8 in: BARNES, H. L. (ed.), Geochemistry of Hydrothermal Ore Deposits. New York: Holt, Rinehart & Winston 1967.

BARTIKYAN, P. M.: Native lead and zinc in the rocks of Armenia. Zapiski. Vserossiisk. Mineral. Obshchestva **95**, 99 (1965).

BARTON Jr., P. B., BETHKE, P. M., TOULMIN III, P.: Equilibrium in ore deposits. Mineral. Soc. Am. Spec. Paper **1**, 171 (1963).

— SKINNER, B. J.: Sulfide mineral stabilities. Chapter 7 in: BARNES, H. L. (ed.), Geochemistry of Hydrothermal Ore Deposits. New York: Holt, Rinehart & Winston 1967.

BATES, C. H., WHITE, W. B., ROY, R.: New high-pressure polymorph of zinc oxide. Science **137**, 993 (1962).

BAUR, W. H.: Zur Kristallchemie der Salzhydrate. Die Kristallstrukturen von $MgSO_4 \cdot 4H_2O$ (Leonhardtit) and $FeSO_4 \cdot 4H_2O$ (Rozenit). Acta Cryst. **15**, 815 (1962).

BAYER, G.: Double oxides of antimony pentoxide with spinel structure. Naturwissenschaften **48**, 46 (1961).

— Vanadates $A_3B_2V_3O_{12}$ with garnet structure. J. Am. Ceram. Soc. **48**, 600 (1965).

— HOFFMANN, W.: Über Verbindungen vom Na_xTiO_2-Typ. Z. Krist. **121**, 9 (1965).

BAYROCK, L. A., PAWLUK, S.: Trace elements in tills of Alberta. Can. J. Earth Sci. **4**, 597 (1967).

BELT, C. B.: Intrusion and ore deposition in New Mexico. Econ. Geol. **55**, 1244 (1960).

BENDA, L.: Über die Anreicherung von Uran und Thorium in Phosphoriten und Bonebeds des nördlichen Harzvorlandes. Geol. Jahrb. **80**, 313 (1963).

BILLINGS, G. K., HITCHON, B., SHAW, D. R.: Geochemistry and origin of formation waters in the Western Canada sedimentary basin 2. Alkali metals. Chem. Geol. **4**, 211 (1969).

— KESLER, S. E., JACKSON, S. A.: Relation of zinc rich formation waters, northern Alberta, to the Pine Point ore deposit. Econ. Geol. **64**, 385 (1969).

BINGHAM, F. T., PAGE, A. L., SIMS, J. R.: Retention of Cu and Zn by H-montmorillonite. Soil Sci. Soc. Am. Proc. **28**, 351 (1964).

BLACK, W. A. P., MITCHELL, R. L.: Trace elements in the common brown algae and in sea water. J. Marine Biol. Assoc. U. K. **30**, 575 (1952).

© Springer-Verlag Berlin · Heidelberg 1972

BLAXLAND, A. B.: Occurrence of zinc in granitic biotites. Master's Thesis: Washington University St. Louis, Missouri (USA) (1970).
BLIX, R., VON UBISCH, H., WICKMAN, F. E.: A search for variations in the relative abundance of the zinc isotopes in nature. Geochim. Cosmochim. Acta 11, 162 (1957).
BOLTER, E.: Über Zersetzungsprodukte von Olivin-Feldspatbasalten. Beitr. Mineral. Petrog. 8, 111 (1961).
BOSTRÖM, K., PETERSON, M. N. A.: The origin of aluminium-poor ferro-manganoan sediments in areas of high heat flow on the East Pacific Rise. Marine Geol. 7, 427 (1969).
BOWEN, H. J. M.: Trace Elements in Biochemistry. London: Academic Press 1966.
BRADFORD, G. R., BAIR, F. L., HUNSKER, V.: Trace and major element content of 170 high Sierra Lakes in California. Limnol. Oceanog. 13, 526 (1968).
BRECCIA, A., DELLONTE, S., NUCIFORA, G.: Determinazione mediante analisi per attivazione neutronica di tracce di zinco, cobalto, nichel, scandio e cromo in campioni di meteoritici di Albareto e di Alfianello. Ric. Sci., Suppl. 37, 319 (1967).
BREHLER, B.: Die Kristallstruktur des $Na_2ZnCl_4 \cdot 3H_2O$. Z. Krist. 114, 66 (1960).
— Die Kristallstruktur von $ZnBr_2$ und $ZnBr_2 \cdot 2H_2O$. Fortschr. Mineral. 39, 338 (1961).
— FOLLNER, H.: Die Kristallstruktur des α-$KZnBr_3 \cdot 2H_2O$. Naturwissenschaften 53, 177 (1966).
— HOLINSKI, R.: Die Kristallstruktur von β-$KZnBr_3 \cdot 2H_2O$. Naturwissenschaften 54, 441 (1967).
— JACOBI, H.: Die Kristallstruktur des $Li_2ZnCl_4 \cdot 2H_2O$. Naturwissenschaften 51, 11 (1964).
BREWER, P. G., SPENCER, D. W.: The distribution of some trace elements in Black Sea waters. GSA Abstracts with Programs 2, 501 (1970).
BRISI, C.: Richerche sulle melilite — Nota V. Il Camposto $2 PbO \cdot ZnO \cdot 2 SiO_2$. Ann. chim. (Rome) 54, 673 (1964).
BROOKS, R. R., KAPLAN, I. R., PETERSON, M. N. A.: Trace element composition of Red Sea geothermal brine and interstitial water. In: DEGENS, E. T., ROSS, D. A., (eds.), Hot Brines and Recent Heavy Metal Deposits in the Red Sea. Berlin-Heidelberg-New York: Springer 1969.
— PRESLEY, B. J., KAPLAN, I. R.: Trace elements in the interstitial waters of marine sediments. Geochim. Cosmochim. Acta 32, 397 (1968).
— RUMSBY, M. G.: The biogeochemistry of trace element uptake by some New Zealand bivalves. Limnol. Oceanog. 10, 521 (1965).
BROWN, H. R., SWAINE, D. J.: Inorganic constituents of Australian coals. J. Inst. Fuel 37, 422 (1964).
BROWN, P. J.: The structure of the ζ-phase in the transition metal-zinc alloy systems. Acta Cryst. 15, 608 (1962).
BRUNFELT, A. O., HEIER, K. S., STEINNES, E., SUNDVOLL, B.: Determination of 36 elements in Apollo 14 bulk fines 14163 by activation analysis. Earth Planet. Sci. Letters 11, 351 (1971).
BUNCH, T. E., KEIL, K.: Chromite and ilmenite in nonchondritic meteorites. Am. Mineralogist 56, 146 (1971).
— — SNETSINGER, K. G.: Chromite composition in relation to chemistry and texture of ordinary chondrites. Geochim. Cosmochim. Acta 31, 1569 (1967).
BURRELL, D. C.: The geochemistry and origin of amphibolites from Bamble, South Norway I. Norsk Geol. Tidsskr. 45, 21 (1965).
— Garnets from upper amphibolite lithologies of the Bamble series, Kragerö, South Norway. Norsk Geol. Tidsskr. 46, 3 (1966).
BUTLER, J. R., THOMPSON, A. J.: Cadmium and zinc in some alkali acidic rocks. Geochim. Cosmochim. Acta 31, 97 (1967).
CALVERT, S. E., PRICE, N. B.: Minor metal contents of Recent organic-rich sediments off South West Africa. Nature 227, 593 (1970).
CALVO, C.: The crystal structure and luminescence of γ-zinc orthophosphate. J. Phys. Chem. Solids 24, 141 (1963).
CAMERON, A. G. W.: A new table of abundances of the elements in the solar system. In: AHRENS, L. H. (ed.), Origin and Distribution of the Elements. Oxford: Pergamon Press 1968.

CANNON, H. L.: Botanical prospecting for ore deposits. Science **132**, 591 (1960).
CARMICHAEL, J., McDONALD, A.: The geochemistry of some natural acid glasses from the North Atlantic Tertiary volcanic province. Geochim. Cosmochim. Acta **25**, 189 (1961).
CARTER, J. L.: Mineralogy and chemistry of the earth's upper mantle based on the partial fusion-partial crystallization model. Bull. Geol. Soc. Am. **81**, 2021 (1970).
CHAWLA, V. K., CHAN, Y. K.: Trace elements in lake Erie. Proc. 12th Conf. Great Lakes Res. 760 (1969).
CHEN, JU-CHIN: Petrology and chemistry of garnet lherzolite nodules in kimberlite from South Africa. Am. Mineralogist **56**, 2088 (1971).
CHERNYAYEV, A. M., CHERNYAYEV, L. E.: Geochemistry of ground waters in the oxidized zone, Gaisk copper sulfide deposit. Geochemistry (USSR) (English Transl.) **10**, 1030 (1963).
CHESTER, R.: Adorption of zinc and cobalt on illite in sea-water. Nature **206**, 884 (1965).
CLARKE, D. B.: Tertiary basalts of Baffin Bay: possible primary magma from the mantle. Contr. Mineral. and Petrol. **25**, 203 (1970).
COCCO, G., FANFANI, L., ZANAZZI, P. F.: The crystal structure of tarbuttite, $Zn_2(OH)PO_4$. Z. Krist. **123**, 321 (1966).
COLE, H., CHAMBERS, F. W., DUNN, H. M.: Lattice parameters of Zn_3As_2. Acta Cryst. **9**, 685 (1956).
COLLINS, L. G.: Manganese and zinc in amphibolites near the Sterling Hill and Franklin Mines, New Jersey. Econ. Geol. **66**, 348 (1971).
CRAIG, D. P., WADSLEY, A. D.: Nonequivalent zinc bonds in chalcophanite. J. Chem. Phys. **22**, 346 (1954).
CROCKET, J. H., WINCHESTER, J. W.: Coprecipitation of zinc with calcium carbonate. Geochim. Cosmochim. Acta **30**, 1093 (1966).
CUSHING, C. E., ROSE, F. L.: Cycling of zinc-65 by Columbia river periphyton in a closed lotic microcosm. Limnol. Oceanog. **15**, 762 (1970).
CZAMANSKE, G. K., ROEDDER, E., BURNS, F. C.: Neutron activation analyses of fluid inclusions for copper, manganese and zinc. Science **140**, 401 (1963).
DARDENNE, M.: Experimental study of the distribution of zinc in calcium carbonates. Bull. Bur. Rech. Géol. Minières, France **5**, 75 (1967).
DESBOROUGH, G. A.: Magmatic sphalerite in Missouri basic rocks. Econ. Geol. **58**, 971 (1963).
DODGE, F. C. W., SMITH, V. C., MAYS, R. E.: Biotites from granitic rocks of the Central Sierra Nevada Batholith, California. J. Petrol. **10**, 250 (1969).
DUURSMA, E. K., SEVENHUYSEN, W.: Note on chelation and solubility of certain metals in sea water at different pH values. Neth. J. Sea Res. **3**, 95 (1966).
EL GORESY, A.: Mineralbestand und Strukturen der Graphit- und Sulfideinschlüsse in Eisenmeteoriten. Geochim. Cosmochim. Acta **29**, 1131 (1965).
— Electron microprobe analyses of coexisting sphalerite and daubreelite in the Odessa iron meteorite. Trans. Am. Geophys. Un. **48**, 159 (1967).
ELLIS, A. J.: Natural hydrothermal systems and experimental hot-water/rock interaction: Reaction with NaCl solutions and trace metal extraction. Geochim. Cosmochim. Acta **32**, 1356 (1968).
ENGEL, A. E. J., ENGEL, C. G.: Migration of elements during metamorphism in the northwest Adirondack Mountains, New York. U.S. Geol. Surv. Profess. Papers **400-B**, 465 (1960a).
— — Progressive metamorphism and granitization of the major paragneiss, northwest Adirondack Mountains, New York. Part II. Mineralogy. Bull. Geol. Soc. Am. **71**, 1 (1960b).
— — HAVENS, R. G.: Mineralogy of amphibolite interlayers in the gneiss complex, northwest Adirondack Mountains, New York. J. Geol. **72**, 131 (1964).
FERSMAN, A. E.: Pegmatites 1 [Russian]. Moscow 1940.
FILBY, R. H.: Determination of the zinc-68 zinc-64 ratio in rocks and minerals by neutron activation analysis. Anal. Chem. **36**, 1597 (1964).
FISCHER, P.: Neutronenbeugungsuntersuchung der Strukturen von $MgAl_2O_4$- und $ZnAl_2O_4$- Spinellen, in Abhängigkeit von der Vorgeschichte. Z. Krist. **124**, 275 (1967).

FLANAGAN, F. J., GWYN, M. E.: Sources of geochemical standards. Geochim. Cosmochim. Acta 31, 1211 (1967).
FLEISCHER, M.: Minor elements in some sulfide minerals. Econ. Geol. 50, 970 (1955).
FOLLNER, H., BREHLER, B.: Die Kristallstruktur des $ZnCl_2 \cdot 1^1/_3 H_2O$. (Unpublished 1969.)
FORSBERG, H.-E., NOWACKI, W.: On the crystal structure of β-ZnOHCl. Acta Chem. Scand. 13, 1049 (1959).
FRONDEL, C., ITO, J.: Hendricksite, a new species of mica. Am. Mineralogist 51, 1107 (1966a).
— — Zincian aegirine augite and jeffersonite from Franklin Furnace. Am. Mineralogist 51, 1406 (1966b).
GALLEZOT, P., WEIGEL, D., PRETTRE, M.: Structure du nitrate de nickel tetrahydrate. Acta Cryst. 22, 699 (1967).
GANAPATHY, R., KEAYS, R. R., ANDERS, E.: Apollo 12 lunar samples. Trace element analysis of a core and the uniformity of the regolith. Science 170, 533 (1970).
GATTOW, G., ZEMANN, J.: Über Doppelsulfate vom Langbeinit-Typ, $A_2^+B_2^{2+}(SO_4)_3$. Z. anorg. u. allgem. Chem. 293, 233 (1958).
GAVRILIN, R. D., PEVTSOVA, L. A.: Behavior of lead and zinc in phase and facies differentiation of magma. Geochemistry 8, 764 (1963).
— — KLASSOVA, N. S.: Lead and zinc in the differentiates of sodic granitic magma. Geochemistry International 2, 783 (1965).
— — Behavior of lead and zinc in the hydrothermal alteration of intrusive rocks. Geokhimiya 8, 954 (1967).
GELLER, S., MILLER, C. E., TREUTING, R. G.: New synthetic garnets. Acta Cryst. 13, 179 (1960).
GERASIMOWSKII, V. I., NESMEYANOVA, L. I.: Lead and zinc distribution in the rocks of the Lovozero massif. Geochemistry 7, 704 (1960).
GHOSE, S.: The crystal structure of hydrozincite, $Zn_5(OH)_6(CO_3)_2$. Acta Cryst. 17, 1051 (1964).
GIESECKE, G., PFISLER, H.: Mischkristalle des Systems $ZnSnAs_2$—InAs und des Systems $ZnGeAs_2$—InAs. Acta Cryst. 14, 1289 (1961).
GIGLIO, M.: Die Kristallstruktur von $Na_2Zn(SO_4)_2 \cdot 4H_2O$. (Zn-Blödit). Acta Cryst. 11, 789 (1958).
GINSBURG, I. I.: Grundlagen und Verfahren geochemischer Sucharbeiten auf Lagerstätten der Buntmetalle und seltenen Metalle. Berlin: Akademie Verlag 1963.
GLADKOVA, V. F., KONDRASHEV, Y. D.: Crystal structure of $ZnSeO_3 \cdot 2H_2O$. Soviet Phys. Cryst. 9, 149 (1964).
GLASS, J. J., JAHNS, R. H., STEVENS, R. E.: Helvite and danalite form New Mexico and the helvite group. Am. Mineralogist 29, 163 (1944).
GLEMSER, O., VÖLZ, H. G., MEYER, B.: Über gasförmiges Zinkhydroxyd. Z. Anorgan. Allgem. Chem. 292, 311 (1957).
GMELIN, L.: Handbuch der Anorganischen Chemie (8. Aufl.), B. 32. Zink Ergänzungsband. Weinheim: Verlag Chemie 1956.
GOLDSCHMIDT, V. M.: Geochemische Verteilungsgesetze der Elemente IX. Skrifter Norske Videnskaps-Akad. Oslo I: Nat.-Naturv. Kl. 4 (1937).
— (MUIR, A., ed.): Geochemistry. Oxford: Clarendon Press 1954.
GONCHAROVA, T., YA.: On native metallic zinc. Mem. All-Union Min. Soc. 88, 485 (1959).
GOODMAN, C. H. L.: A new group of compounds with diamond-type (chalcopyrite) structure. Nature 179, 828 (1957).
GORHAM, E., SWAINE, D. J.: The influence of oxidizing and reducing conditions upon the distribution of some elements in lake sediments. Limnol. Oceanog. 10, 268 (1965).
GOVETT, G. J. S., PANTAZIS, T. M.: Distribution of Cu, Zn, Ni and Co in the Troodos pillow lava series, Cyprus. Inst. Mining Met. Trans. 80B Bull. 771, 27 (1971).
GRAF, D. L.: Geochemistry of carbonate sediments and sedimentary carbonate rocks. Part III. Minor element distribution. Illinois State Geol. Surv. Circ. 301, 1 (1960).
GRAMSE, M.: Geochemische Untersuchungen an Titanomagnetiten tertiärer Basalte. Dipl.-thesis, University of Goettingen (Germany) (1966).

GREENLAND, L.: Fractionation of chlorine, germanium, and zinc in chondritic meteorites. J. Geophys. Res. **68**, 6507 (1963).
— The abundances of selenium, tellurium, silver, palladium, cadmium, and zinc in chondritic meteorites. Geochim. Cosmochim. Acta **31**, 849 (1967).
— GOLES, G. G.: Copper and zinc abundances in chondritic meteorites. Geochim. Cosmochim. Acta **29**, 1285 (1965).
— LOVERING, J. F.: The evolution of tektites: elemental volatilization in tektites. Geochim. Cosmochim. Acta **27**, 249 (1963).
— — Minor and trace element abundances in chondritic meteorites. Geochim. Cosmochim. Acta **29**, 821 (1965).
GUNDLACH, H., KARL, F., MÜLLER, G.: Vergleichende geochemische Untersuchungen an ost- und südalpinen Graniten, Granodioriten und Tonaliten. Contr. Mineral. and Petrol. **16**, 285 (1967).
GUTKNECHT, J.: ^{65}Zn uptake by benthic marine algae. Limnol. Oceanog. **8**, 31 (1963).
HAACK, U. K.: Spurenelemente in Biotiten aus Graniten und Gneisen. Contr. Mineral. and Petrol. **22**, 83 (1969).
HAFNER, S.: Metalloxyde mit Spinellstruktur. Schweiz. Mineral. Petrog. Mitt. **40**, 207 (1960).
HAMAGUCHI, H., SUGISITA, R., KAWAI, M., ONUMA, N.: Determination of zinc in iron meteorites by neutron activation analysis. Bunseki Kagaku **13**, 735 (1964) [Japanese].
HAMMEL, F.: The sulphates of the magnesium series. Ann. chimie (Paris) **11**, 247 (1939).
HANKE, K.: Zinktellurit, Kristallstruktur und Beziehungen zu einigen Seleniten. Naturwissenschaften **54**, 199 (1967).
— Die Kristallstruktur von $Zn_2Te_3O_8$. Naturwissenschaften **55**, 273 (1966).
HANSEN, M., ANDERKO, K.: Constitution of Binary Alloys. New York: McGraw-Hill 1958.
HARKER, R. I., HUTTA, J. J.: The stability of smithsonite. Econ. Geol. **51**, 375 (1956).
HARRISS, R. C., ADAMS, J. A. S.: Geochemical and mineralogical studies on the weathering of granitic rocks. Am. J. Sci. **264**, 146 (1966).
HARTMANN, M.: Einige geochemische Untersuchungen an Sandsteinen aus Perm und Trias. Geochim. Cosmochim. Acta **27**, 459 (1963).
HAWKES, H. E., WEBB, J. S.: Geochemistry in Mineral Exploration. Harper's Geoscience Series. New York: Harper & Row 1962.
HAYASHI, H., NAKAYAMA, N., HASEGAWA, K., MIZUKUSA, S., NOCHGUCHI, T., OGISO, S., TAKAGI, H.: Phase transition under high pressure V. Transition of Zn_2SiO_4. Rept. Govt. Ind. Res. Inst. Nagoya **14**, 389 (1965) [Japanese].
HEADLEE, A. J. W., HUNTER, R. G.: Characteristics of minable coals of West Virginia P. 5. West Va. Geol. Econ. Surv. Bull. **13-A**, 36 (1955).
HEIDE, F., SINGER, E.: Zur Geochemie des Kupfers und Zinkes. Naturwissenschaften **37**, 541 (1950).
— — Der Gehalt des Saalewassers an Kupfer und Zink. Naturwissenschaften **41**, 498 (1954).
HELGESON, H. C.: Geologic and thermodynamic characteristics of the Salton Sea geothermal system. Am. J. Sci. **266**, 129 (1968).
— Thermodynamics of hydrothermal systems at elevated temperatures and pressures. Am. J. Sci. **267**, 729 (1969).
HELLMANN, H.: Die Absorption von Schwermetallen an den Schwebstoffen des Rheins. Eine Untersuchung zur Entgiftung des Rheinwassers. Deut. Gewässerk. Mitt. **14**, 42 (1970).
HEMLEY, J. J., MEYER, C., HODGSON, C. J., THATCHER, A. B.: Sulfide solubilities in alteration-controlled systems. Science **158**, 1580 (1967).
HENNIG, W.: Löslichkeit von Zinkblende unter hydrothermalen Bedingungen im System ZnS—$NaCl$—H_2O. Neues Jahrb. Mineral. Abhandl. **116**, 1, 61 (1971).
HERRMANN, A. G.: Geochemische Untersuchungen an Kalisalzlagerstätten im Südharz. Freiberger Forschungsh. **C 43**, 1 (1958).
— WEDEPOHL, K. H.: Untersuchungen an spilitischen Gesteinen der variskischen Geosynkline in Nordwestdeutschland. Contr. Mineral. and Petrol. **29**, 255 (1970).

HIETANEN, A.: Distribution of Fe and Mg between garnet, staurolite, and biotite in aluminium rich schist in various metamorphic zones north of the Idaho batholith. Am. J. Sci. 267, 422 (1969).

HÖRBE, R., KNACKE, O.: Zusammenstellung von Dampfdruckkurven und Erläuterung einiger Anwendungen. Z. Erzbergbau Metallhüttenwesen 8, 1 (1955).

HOFMEYR, P. K.: The abundances and distribution of some trace elements in some selected South African shales. Ph.D.-Thesis, Capetown (1971).

HOLLANDER, J. M., PERLMAN, J., SEABORG, G. T.: Table of isotopes. Rev. Mod. Phys. 25, 469 (1953).

HONJO, S., TABUCHI, H.: Distribution of some minor elements in carbonate rocks; Part 1. Pacific Geol. 2, 41 (1970).

HOPPE, R., VIELHABER, E.: Neue Oxozinkate und Oxocadmate: $Na_2Zn_2O_3$, K_2ZnO_2 und N_2CdO_2. Naturwissenschaften 51, 103 (1964).

HUFF, L. C.: A sensitive field test for heavy metals in water. Econ. Geol. 43, 675 (1948).

— Abnormal Cu, Pb, and Zn of soil near metalliferous veins. Econ. Geol. 47, 517 (1952).

HUTCHINSON, G. E.: The biogeochemistry of vertebrate excretion. Bull. Am. Mus. Nat. Hist. New York 96, 474 (1950).

ICHIKUNI, M.: Teneurs en cuivre et en zinc des eaux thermales au Japon. Geochim. Cosmochim. Acta 17, 305 (1959).

IITAKA, Y.: Die Kristallstruktur von Zinkhydroxidsulfat I, $Zn(OH)_2 \cdot ZnSO_4$. Acta Cryst. 15, 559 (1962).

ISHIBASHI, M., FUJINAGA, T., YAMAMOTOTO, T., FUJITA, T., WATANABE, K.: The zinc contents of seaweeds. J. Chem. Soc. Japan, Pure Chem. Sect. (Nippon Kagaku Zasshi) 86, 728 (1965).

— UEDA, S., YAMAMOTO, Y.: Studies on the utilization of the shallow-water deposits. Records Oceanog. Works Japan, Spec. No. 3, 123 (1959).

IVANOV, V. V., PYATENKO, YU. A.: On the so-called kusterite. Mem. All.-Union Min. Soc. 88 (2), 165 (1959).

JACK, R. N., CARMICHAEL, I. S. E.: The chemical "fingerprinting" of acid volcanic rocks. In: Short contributions to California geology. California Div. Mines Geol. Spec. Rept. 100, 17 (1969).

JACOBI, H.: Die Kristallstruktur des $LiZnCl_3 \cdot 3H_2O$. (Unpublished 1969.)

JAMBOR, J. L.: Studies of basic copper and zinc carbonates. I. Synthetic zinc carbonates and their relationship to hydrozincite. Can. Mineralogist 8, 92 (1964).

— BOYLE, R. W.: Gunningite, a new zinc sulphate from the Keno hillgalena hill area, Yukon, Can. Mineralogist 7, 209 (1962).

JASMUND, K., SECK, H. A.: Geochemische Untersuchungen an Auswürflingen (Gleesiten) des Laacher-See-Gebietes. Beitr. Mineral. Petrog. 10, 275 (1964).

JURINAK, J. J., INOUYE, T. S.: Some aspects of zinc and copper phosphate formation in aqueous systems. Soil Sci. Soc. Am. Proc. 26, 144 (1962).

— THORNE, D. W.: Zinc solubility under alkaline conditions in a zinc-bentonite-system. Soil Sci. Soc. Am. Proc. 19, 446 (1955).

JUURINEN, A.: Composition and properties of staurolite. Ann. Acad. Sci. Fennicae Ser. A III (Geol. Geogr.) 47 (1956).

KAMP, P. C. VAN DE: The green beds of the Scottish Dalradian series: geochemistry, origin, and metamorphism of mafic sediments. J. Geol. 78, 281 (1970).

KANESHIMA, K.: Geochemistry of phosphate rocks from the Ryukyu Islands. Part I. J. Chem. Soc. Japan, Pure Chem. Sect. 83, 1007 (1962).

KEAYS, R. R., GANAPATHY, R., ANDERS, E.: Chemical fractionations in meteorites — IV. Abundances of fourteen trace elements in L-chondrites; implications for cosmothermometry. Geochim. Cosmochim. Acta 35, 337 (1971).

KEIL, K.: Zincian daubreelite from the Kota-Kota and St. Mark's enstatite chondrites. Am. Mineralogist 53, 491 (1968).

KEILIN, D., MANN, T.: Carbonic anhydrase purification and nature of enzyme. Biochem. J. 34, 1163 (1940).

KIESL, W., SEITNER, H., KLUGER, F., HECHT, F.: Determination of trace elements by chemical analysis and neutron activation in meteorites of the collection of the Viennese Museum of Natural History. Monatsh. Chem. **98**, 972 (1967).
— WEINKE, H. H.: Über Mangandaubréelith in den Troilitknollen des Odessa-Eisenmeteorits. Mikrochim. Acta **2**, 392 (1970).
KITANO, Y., TOKUYAMA, A., KANAMORI, N.: Measurement of the distribution coefficient of zinc and copper between carbonate precipitate and solution. J. Earth Sci. Nagoya Univ. **16**, 1 (1968).
KLEBER, W., LIEBAU, F., PIATKOWIAK, E.: Zur Struktur des Phosphophyllits, $Zn_2Fe[PO_4]_2 \cdot 4H_2O$. Acta Cryst. **14**, 795 (1961).
KLEIN, C., ITO, J.: Zincian and manganoan amphiboles from Franklin, New Jersey. Am. Mineralogist **53**, 1264 (1968).
KLEINKOPF, M. D.: Spectrographic determination of trace elements in lake waters of northern Maine. Bull. Geol. Soc. Am. **71**, 1231 (1960).
KNOX, K.: Perowskite-like fluorides. I. Structures of $KMnF_3$, $KFeF_3$, $KNiF_3$ and $KZnF_3$. Crystal field effects in the series and in $KCrF_3$ and $KCuF_3$. Acta Cryst. **14**, 583 (1961).
KOBLENTS-MISHKE, O. I., VOLKOVINSKIY, V. V., KABANOVA, Y. G.: More information about the amount of primary production of the ocean. Dokl. Acad. Nauk SSSR **183**, 1189 (1968).
KÖSTER, H. M.: Beitrag zur Geochemie der Kaoline. Proc. Intern. Clay Confer. Tokyo **1**, 273 (1969).
KOKKOROS, P. A., RENTZEPERIS, P. J.: The crystal structure of the anhydrous sulphates of copper and zinc. Acta Cryst. **11**, 361 (1958).
KONOVALOV, G. S.: Removal of microelements by the principal rivers of the USSR. Dokl. Akad. Nauk SSSR **129**, 1034 (1959). (Engl. Transl.).
KONTOROVICH, A. E., SADIKOV, M. A., SHVARTSEV, S. L.: Abundances of certain elements in the surface and ground waters of the northwestern part of the Siberian Platform. Dokl. Akad. Nauk SSSR. **149**, 168 (1963).
KORNEEVA, I. V.: The hexagonal modification of zinc selenide. Kristallografiya **6**, 5054 (1961).
KRAUSKOPF, K. B.: Factors controlling the concentrations of thirteen rare metals in seawater. Geochim. Cosmochim. Acta **9**, 1 (1956).
— The heavy metal content of magmatic vapor at 600°C. Econ. Geol. **52**, 786 (1957).
KREBS, H., LUDWIG, TH.: Die Struktur der Metallpolyphosphide des Typs $HgPbP_{14}$. Z. anorgan. u. allgem. Chem. **294**, 257 (1958).
LAIDLEY, R. A., McKAY, D. S.: Geochemical examination of obsidians from Newberry Caldera, Oregon. Contr. Mineral. and Petrol. **30**, 336 (1971).
LANDERGREN, S., MANHEIM, F. T.: Über die Abhängigkeit der Verteilung von Schwermetallen von der Fazies. Fortschr. Geol. Rheinland Westfalen **10**, 173 (1963).
LAPPE, F., NIGGLI, A., NITSCHE, R., WHITE, J. G.: The crystal structure of In_2ZnS_4. Z. Krist. **117**, 146 (1962).
LARIMER, J. W., ANDERS, E.: (1967) Preprint mentioned in: CAMERON (1968).
LAUL, J. C., CASE, D. R., WECHTER, M., SCHMIDT-BLEEK, F., LIPSCHUTZ, M. E.: An activation analysis technique for determining groups of trace elements in rocks and chondrites. J. Radioanal. Chem. **4**, 241 (1970).
LAVERY, N. G., BARNES, H. L., WEDEPOHL, K. H.: Zinc dispersion in the Wisconsin-zinc-lead district. Zinc in common sedimentary rocks. Econ. Geol. **66**, 226 (1971).
LEBEDEV, L. M., NIKITINA, I. B.: Chemical properties and ore content of hydrothermal solutions at Cheleken. Dokl. Akad. Nauk SSSR **183**, 439 (1968).
LEE, D. E., COLEMAN, R. G., BASTRON, H., SMITH, V. C.: A two-amphibole glaucophane schist in the Franciscan formation, Cazadero area, Sonoma County, California. U.S. Geol. Surv. Profess. Papers **550 C**, 148 (1966).
LEONHARDT, J., WEISS, R.: Das Kristallgitter des Kieserits $MgSO_4 \cdot H_2O$. Naturwissenschaften **44**, 338 (1957).
LEUTWEIN, F.: Untersuchungen über das Vorkommen von Spurenmetallen in Torfen und Braunkohlen. Freiberger Forschungsh. **C 30**, 28 (1956).

LEUTWEIN, F., RÖSLER, H. J.: Geochemische Untersuchungen an paläozoischen und mesozoischen Kohlen Mittel- und Ostdeutschlands. Freiberger Forschungsh. C 19, 196p (1956).
— WEISE, L.: Hydrogeochemische Untersuchungen an erzgebirgischen Gruben- und Oberflächenwässern. Geochim. Cosmochim. Acta 26, 1333 (1962).
LEWIS, G. J., GOLDBERG, E. D.: X-ray fluorescence determination of barium, titanium and zinc in sediments. Anal. Chem. 28, 1282 (1956).
LIEBAU, F.: Zur Kristallstruktur des Hopeits, $Zn_3[PO_4]_2 \cdot 4H_2O$. Acta Cryst. 18, 352 (1965).
LUNDEGÅRDH, P. H.: Some aspects to the determination and distribution of zinc. Ann. Roy. Agricult. Coll. Sweden 15, 1 (1947).
— Aspects to the geochemistry of chromium, cobalt, nickel and zinc. Sveriges Geol. Undersokn. Arsbok Ser. C Avhandl. och Uppsat. 513 43, 11, 1 (1949).
MAHLER, H. R.: In: COMAR, C. L., BRONNER, F. (eds.), Mineral Metabolism. New York and London: Academic Press 1961.
MANHEIM, F. T.: A geochemical profile in the Baltic Sea. Geochim. Cosmochim. Acta 25, 52 (1961).
MARUMO, F.: The crystal structure of nowackiite, $Cu_6Zn_3As_4S_{12}$. Z. Krist. 124, 352 (1967).
MASON, B., GRAHAM, A. L.: Minor and trace elements in meteoritic minerals. Smithsonian Contrib. Earth Sci. 3, 1 (1970).
McDONALD, W. S., CRUICKSHANK, D. W. J.: Refinement of the structure of hemimorphite. Z. Krist. 124, 180 (1967).
MELSON, W. G., THOMPSON, G., ANDEL, T. H. VAN: Volcanism and metamorphism in the Mid-Atlantic Ridge, 22°N Latitude. J. Geophys. Res. 73, 5925 (1968).
MIZUTANI, Y.: Copper and zinc in fumarolic gases of Showashinzan volcano, Hokkaido, Japan. Geochem. J. 4, 87 (1970).
— NAKAI, W., OANA, S.: Major and minor constituents of volcanic sublimates. Abstracts Symposium of Geochemistry, Sapporo (1957).
MONTGOMERY, H., LINGAFELTER, E. C.: The crystal structure of Tutton's salts. I. Zinc ammonium sulfate hexahydrate. Acta Cryst. 17, 1295 (1964).
MOORE, G. W.: Geologic significance of the minor element composition of marine salt deposits. Econ. Geol. 66, 187 (1971).
MOORE, III, J. R.: Bottom sediment studies, Buzzards Bay, Massachusetts. J. Sediment. Petrol. 33, 511 (1963).
MORITA, Y.: Distribution of copper and zinc in various phases of the earth materials. J. Earth Sci. Nagoya Univ. 3, 33 (1955).
MORRISON, G. H., GERRARD, J. T., KASHUBA, A. T., GANGADHARAM, E. V., ROTHENBERG, A. M., POTTER, N. M., MILLER, G. B., Elemental abundances of lunar soil and rocks. Proc. Apollo 11 Lunar Science Conf. 2, 1383 (1970); (Geochim. Cosmochim. Acta 34, Supplement 1).
MÜLLER, E. A.: The solar abundances p. 155 in: AHRENS, L. H. (ed.) Origin and Distribution of the Elements. Oxford: Pergamon Press 1968.
MUMBRUM, L. W. DE, JACKSON, M. L.: Copper and zinc exchange from dilute neutral solutions by soil colloidal electrolytes. Soil. Sci. 81, 353 (1956).
— — Infrared absorption evidence on exchange reaction mechanism of copper and zinc with layer silicate clays and peat. Soil Sci. Soc. Am. Proc. 20, 334 (1956).
NEUHAUS, A., CEMIČ, L.: Struktur und Mischbarkeit im System ZnS-FeS im Druckbereich bis 60 kbar. Naturwissenschaften 57, 354 (1970).
NEUMANN, H.: Notes of the mineralogy and geochemistry of zinc. Mineral. Mag. 28, 575, (1949).
NEWHOUSE, W. H.: Opaque oxides and sulphides in common igneous rocks. Bull. Geol. Soc. Am. 47, 1 (1936).
NICHIPORUK, W., CHODOS, A. A.: The concentration of vanadium, chromium, iron, cobalt, nickel, copper, zinc and arsenic in the meteoritic iron sulfide nodules. J. Geophys. Res. 64, 2451 (1959).
NICHOLLS, J., CARMICHAEL, J. S. E.: Peralkaline acid liquids. A petrological study. Contr. Mineral. and Petrol. 20, 268 (1969).

NISHIMURA, M., SANDELL, E. B.: Zinc in meteorites. Geochim. Cosmochim. Acta **28**, 1055 (1964).
NODDACK, I., NODDACK, W.: Die Häufigkeit der chemischen Elemente. Naturwissenschaften **18**, 757 (1930).
— — Die Häufigkeiten der Schwermetalle in Meerestieren. Arkiv Zool. **32-A**, 1 (1939).
NOWACKI, W., SILVERMANN, J. N.: Die Kristallstruktur von Zinkhydroxychlorid II, $Zn_5(OH)Cl_2 \cdot 1H_2O$. Z. Krist. **115**, 21 (1961).
OKRUSCH, M., RICHTER, P.: Petrographische, geochemische und mineralogische Untersuchungen zum Problem der Granitoide im mittleren Spessartkistallin. Neues Jahrb. Mineral. Abhandl. **107**, 21 (1967).
— — Zur Geochemie der Dioritgruppe. Contr. Mineral. and Petrol. **21**, 75 (1969).
OSWALD, H. R.: Zur Struktur der wasserfreien Zinkhalogenide. II. Zinkbromid und Zinkjodid. Helv. Chim. Acta **43**, 72 (1960).
PAPEZIK, V. S.: Geochemistry of some Canadian anorthosites. Geochim. Cosmochim. Acta **29**, 673 (1965).
PARRY, W. T., NACKOWSKI, M. P.: Copper, lead and zinc in biotites from Basin and Range quartz monzonites. Econ. Geol. **58**, 1126 (1963).
PASHINKIN, A. S., TISHCHENKO, G. N., KORNEEVA, I. V., RYZHENKO, B. N.: Concerning the polymorphism of some chalcogenides of zinc and cadmium. Kristallografiya (engl.) **5**, 243 (1960).
PEARRE, N. C., HEYL, A. V.: Chromite and other mineral deposits in serpentine rocks of the Piedmont Upland, Maryland, Pennsylvania and Delaware. U.S. Geol. Surv. Bull. **1082-K**, 707 (1960).
PFISTER, H.: Kristallstruktur von ternären Verbindungen der Art $A^{II}B^{IV}C^{V}_2$. Acta Cryst **16**, 153 (1963).
PINCKNEY, D. M., HAFFTY, J.: Content of zinc and copper in some fluid inclusions from the Cave-in Rock district, Southern Illinois. Econ. Geol. **65**, 451 (1970).
PIPER, D. Z.: The distribution of Co, Cr, Cu, Fe, Mn, Ni and Zn in Framvaren, a Norwegian anoxic fjord. Geochim. Cosmochim. Acta **35**, 531 (1971).
PLIETH, K., SÄNGER, G.: Die Struktur des Stranskiits, $Zn_2Cu(AsO_4)_2$. Z. Krist. **124**, 91 (1967).
PREWITT, C. T., KIRCHNER, E., PREISINGER, A.: Crystal structure of larsenite $PbZnSiO_4$. Z. Krist. **124**, 115 (1967).
PUCHNER, H. F., HOLLAND, H. D.: Studies in the Providencia area, Mexico III. Neutron activation analyses of fluid inclusions from Noche Buena. Econ. Geol. **61**, 1390 (1966).
PUTMAN, G. W., ALFORS, J. T.: Frequency distribution of minor metals in the Rocky Hill stock, Tulare County, California. Geochim. Cosmochim. Acta **31**, 431 (1967).
— BURNHAM, C. W.: Trace elements in igneous rocks, Northwestern and Central Arizona. Geochim. Cosmochim. Acta **27**, 53 (1963).
PYLE, T. E., TIEH, T. T.: Strontium, vanadium and zinc in the shells of pteropods. Limnol. Oceanog. **15**, 153 (1970).
QURASHI, M. M., BARNES, W. H.: The structures of the minerals of the descloizite and adelite groups: IV. Descloizite and conichalcite (Part II). The structure of conichalcite. Can. Mineralogist. **7**, 561 (1963).
RADER, L. F., SWADLEY, W. C., HUFFMAN, C., JR. LIPP, H. H.: New chemical determinations of zinc in basalts and rocks of similar composition. Geochim. Cosomochim. Acta **27**, 695 (1963).
— — LIPP, H. H., HUFFMANN, C.: Determination of zinc in basalts and other rocks. U.S. Geol. Surv. Profess. Papers **400-B**, 477 (1960).
RAMDOHR, P.: Die Erzmineralien in gewöhnlichen magmatischen Gesteinen. Abhandl. Preuss. Akad. Wiss. Math.-Naturw. Kl. **2**, 1 (1940).
— The opaque minerals in stony meteorites. J. Geophys. Res. **68**, 2011 (1963).
RENTZEPERIS, P. J.: The crystal structure of hodgkinsonite, $Zn_2Mn[(OH)_2/SiO_4[$. Z. Krist. **119**, 117 (1963).
RIBAR, B., ŠLJUKIĆ, M., MATKOVIC, B., GABELA, F., GIRT, E.: Crystal data for $Zn(NO_3)_2 \cdot 4H_2O$ and $Zn(NO_3)_2 \cdot 2H_2O$. Acta Cryst. **23**, 1113 (1967).

RIMSAITE, J.: Biotites intermediate between dioctahedral and trioctahedral micas p. 375 in: Clays and Clay Minerals (Proc. 15th Conf.) Oxford: Pergamon Press 1967.
— Geochemistry, mineralogy and petrology of poly-mica rocks. Proceed. XXIII. Internat. Geol. Congr. 6, 45 (1968).
— Distribution of major and minor constituents between mica and host ultrabasic rocks and between zoned mica and zoned spinel. Contr. Mineral. and Petrol. 33, 259 (1971).
RONA, E., HOOD, D. W., MUSE, L., BUGLIO, B.: Activation analysis of manganese and zinc in sea water. Limnol. Oceanog. 7, 201 (1962).
ROSS, C. S.: Sauconite — a clay mineral of the montmorillonite group. Am. Mineralogist 31, 411 (1946).
ROY, D. M., MUMPTON, F. A.: Stability of minerals in the system $ZnO-SiO_2-H_2O$. Econ. Geol. 51, 432 (1956).
RYE, R. O., HAFFTY, J.: Chemical composition of the hydrothermal fluids responsible for the lead-zinc deposits at Providencia, Zacatecas, Mexico. Econ. Geol. 64, 629 (1969).
SAALFELD, H.: Einige Strukturdaten zum Spinell 7 $ZnO \cdot Sb_2O_5$. Acta Cryst. 16, 836 (1963).
— Strukturdaten von Gahnit, $ZnAl_2O_4$. Z. Krist. 120, 476 (1964).
SANDELL, E. B., GOLDICH, S. S.: The rarer metallic constituents of some American igneous rocks. Parts I and II. I. J. Geol. 51, 99 (1943); II: J. Geol. 51, 167 (1943).
SARVER, J. F., HUMMEL, F. A.: Solid solubility and eutectic temperature in the system $Zn_2SiO_4-Mg_2SiO_4$. J. Am. Ceram. Soc. 45, 304 (1962).
SAVELLI, C.: The problem of rock assimilation by Somma-Vesuvius magma Part I. Contr. Mineral. and Petrol. 16, 328 (1967).
— WEDEPOHL, K. H.: Geochemische Untersuchungen an Sinterkalken (Travertinen) Contr. Mineral. and Petrol. 21, 238 (1969).
SAVUL, M., ABABI, V.: Cuprul, zincul si plumbul ca elemente minore in calcarele si dolomitele din Carpatii Orientalii. Acad. Rep. Populare Romine, Studii Cercetari Chim. 7, 181 (1956b).
— — NICHITA, O.: Zincul, plumbul si cuprul ca elemente minore in rocile vulcanice din Muntii Calimani. Acad. Rep. Populare Romine, Studii Cercetari Chim. 2, 89 (1956a).
SCHNERING, H. G. VON: Die Kristallstruktur des $Na[Zn(OH)_3]$. Naturwissenschaften 48, 665 (1961).
— Zur Konstitution des ε-$Zn(OH)_2$. Z. anorg. u. allgem. Chem. 330, 170 (1964).
— HOPPE, R.: Die Kristallstruktur des $SrZnO_2$. Z. anorg. u. allgem. Chem. 312, 87 (1961).
— — Zur Kenntnis des Ba_2ZnS_3. Z. anorg. u. allgem. Chem. 312, 99 (1961).
— — ZEMANN, J.: Die Kristallstruktur des $BaZnO_2$. Z. anorg. u. allgem. Chem. 305, 241 (1960).
SCOTT, S. D., BARNES, H. L.: Sphalerite geothermometry and geobarometry. Econ. Geol. 66, 653 (1971).
SEELIGER, E., MÜCKE, A.: Donathit, ein tetragonaler, Zn-reicher Mischkristall von Magnetit und Chromit. Neues Jahrb. Mineral. Monatsh. 2, 49 (1969).
SEGNIT, E. R., HOLLAND, A. E.: The system $MgO-ZnO-SiO_2$. J. Am. Ceram. Soc. 48, 409 (1965).
SEIDEL, K.: Reinigung von Gewässern durch höhere Pflanzen. Naturwissenschaften 53, 289 (1966).
SEITNER, H., KIESL, W., KLUGER, F., HECHT, F.: Wet-chemical analysis and determination of trace elements by neutron activation in meteorites. J. Radioanal. Chem. 7, 235 (1971).
SHACKLETTE, H. T.: Element content of bryophytes. U.S. Geol. Surv. Bull. 1198-D (1965).
— HAMILTON, J. C., BOERNGEN, J. G., BOWLES, J. M.: Elemental composition of surficial materials in the conterminous United States. U.S. Geol. Surv. Profess. Papers 574-D, 1 (1971).
— SAUER, H. I., MIESCH, A. T.: Geochemical environments and cardiovascular mortality rates in Georgia. U.S. Geol. Surv. Profess. Papers 574-C, 1 (1970).
SHCHEPETKIN, A. A., ZAKHAROV, R. G., ZINIGRAD, M. I., CHUFAROV, G. I.: Synthesis and mutual solubility in spinel solid solutions in Me-Ti-Fe-O systems (Me=Zn, Co, Ni and Mn). Crystallography 14, 761 (1970).
SHORT, N. M.: Geochemical variations in four residual soils. J. Geol. 69, 534 (1961).

SIGHINOLFI, G. P.: Investigations into deep crustal levels: Fractionating effects and geochemical trends related to high-grade metamorphism. Geochim. Cosmochim. Acta **35**, 1005 (1971).
SIGVALDASON, G. E.: Mineralogische Untersuchungen über Gesteinszersetzung durch postvulkanische Aktivität in Island. Beitr. Mineral. Petrogr. **6**, 405 (1959).
SILKER, W. B.: Variations in elemental concentrations in the Columbia River. Limnol. Oceanog. **9**, 540 (1964).
SILLÉN, L. G., MARTELL, A. E.: Stability Constants of Metal-Ion Complexes. The Chem. Soc. London Spec. Publ. 17, 1964.
SKINNER, B. J., WHITE, D. E., ROSE, H. J., MAYS, R. E.: Sulfides associated with the Salton Sea geothermal brine. Econ Geol. **62**, 316 (1967).
SLENTZ, L. W.: Unpublished data from Chevron Oil Co. analyses. Personal communication (1971).
SLOWEY, J. F.: Studies on the distribution of copper, manganese and zinc in the ocean using neutron activation analysis. In: The chemistry and analysis of trace metals in sea water. Final Report AEC Contract AT (40-1)-2799 Texas A+M Project 276 (1966).
— HOOD, D. W.: Copper, manganese and zinc concentrations in Gulf of Mexico waters. Geochim. Cosmochim. Acta **35**, 121 (1971).
SMALES, A. A., MAPPER, D., FOUCHÉ, K. F.: The distribution of some trace elements in iron meteorites as determined by neutron activation. Geochim. Cosmochim. Acta **31**, 673 (1967).
— — WEBB, M. S. W., WEBSTER, R. K., WILSON, J. T.: Elemental composition of lunar surface material. Proc. Apollo 11 Lunar Science Conf. **2**, 1575 (1970).
SMITH, J. R.: Distribution of nickel, copper, and zinc in bedrock of the East Amisk area Saskatschewan. Saskatschewan Res. Council Geol. Div. Rept. 6 (1964).
SMITH, J. U.: Reexamination of the crystal structure of melilite. Am. Mineralogist **38**, 643 (1953).
SMITH, P., GARCIA-BLANCO, S., RIVOIR, C.: The crystal structure of anhydrous zinc metaborate $Zn_4O(BO_2)_6$. Z. Krist. **119**, 375 (1964).
SMITH, P. L., MARTIN, J. E.: The high-pressure structures of zinc sulfide and zinc selenide. Phys. Letters **19**, 541 (1965).
SMYKATZ-KLOSS, W.: Sedimentpetrographische und chemische Untersuchungen an Karbonatgesteinen des Zechsteins. Teil II Spezieller Teil. Contrib. Mineral. and Petrol. **13**, 232 (1966).
SMYTHE, L. E., GATEHOUSE, B. M.: Polarographic determination of traces of copper, nickel, cobalt, zinc, and cadmium in rocks using rubeanic acid and 1-nitroso-2-naphtol. Anal. Chem. **27**, 901 (1955).
SPENCER, D. W., BREWER, P. G.: The distribution of copper, zinc and nickel in sea water of the Gulf of Maine and the Sargasso Sea. Geochim. Cosmochim. Acta **33**, 325 (1969).
— — Vertical advection diffusion and redox potentials as controls on the distribution of manganese and other trace metals dissolved in waters of the Black Sea. J. Geophys. Res. **76**, 5877 (1971).
— — SACHS, P. L.: Aspects of the distribution and trace element composition of suspended matter in the Black Sea. Geochim. Cosmochim. Acta **36**, 71 (1971).
— ROBERTSON, D. E., TUREKIAN, K. K., FOLSOM, T. R.: Trace element calibrations and profiles at the geosecs test station in the Northeast Pacific Ocean. J. Geophys. Res. **75**, 7688 (1970).
SPITZBERGEN, U.: The crystal structures of $BaZnO_2$, $BaCoO_2$ and $BaMnO_2$. Acta Cryst. **13**, 197 (1960).
STEIGMANN, G. A.: The crystal structures of $ZnAl_2S_4$. Acta Cryst. **23**, 142 (1967).
STRACHOW, N. M., BELOWA, I. W., GLAGOLEWA, M. A., LUBTSCHENKO, I. in: Distribution and occurrence of different elements: the surface parts of Recent Black Sea sediments. Lithology and Economic Deposits **2**, 3 (1971) [Russian].
STRUNZ, H.: Mineralogische Tabellen. Leipzig 1941 u. 1966.
SUESS, H. E., UREY, H. C.: Abundances of the elements. Rev. Mod. Phys. **28**, 53 (1956).
SÜSSE, P., BREHLER, B.: Die Kristallstruktur des $KZnCl_3 \cdot 2H_2O$. Beitr. Mineral. Petrog. **10**, 132 (1964).

Sugawara, K.: Migration of elements through phases of atmosphere and hydrosphere. Geochemical Conf. in Memory of the Centenary Birthday of Academician Vernadsky Moskow 1963.

Swaine, D. J.: The Trace Element Content of Soils. Commonwealth Bureau Soil Sci. Techn. Comm. No. 48, England 1955.

— Personal communication (1971) mainly on the basis of the paper by: Brown H. R., Swaine, D. J.: Inorganic constituents of Australian coals. Part I. J. Inst. Fuel 37, 422 (1964).

Swanson, H. E., Fuyat, R. K.: Standard x-ray diffraction powder patterns. Natl. Bur. Standards Circ. No. 539, II, 25 (1953).

Swanson, V. E., Frost, T. C., Rader, Jr., L. F., Huffmann, Jr., C. : Metal sorption by northwest Florida humate. U.S. Geol. Surv. Profess. Papers 550-C, 174 (1966).

Switzer, G.: Paradamite, a new zinc arsenate from Mexico. Science 123, 1039 (1956).

Tageyeva, N. V.: Interstitial waters of Arctic Ocean sediments. Acad. Sci. USSR Proc. 163, 207 (1965). [Engl. Transl.].

Takahashi, T.: Supergene alteration of zinc and lead deposits in limestone. Econ. Geol. 55, 1084 (1960).

Tatsumoto, M.: Chemical investigations of deep-sea deposits (No-21—26 H. Hamaguchi). Ni, Co, Cu, Zn, Sn, Ob, Be content of deep-sea deposits. J. Chem. Soc. Japan Pure Chem. Sect. (1957).

Tauson, L. V.: Factors in the distribution of the trace elements during the crystallization of magmas. Phys. Chem. Earth 6, 219 (1966).

— Geochemistry of rare elements in igneous rocks and metallogenetic specialization of magmas. In: Vinogradov, A. P. (ed.), Chemistry of the Earth's Crust 2, 248 (1964). (Israel Program for Scientific Transl. 1967.)

— Kravchenko, L. A.: Characteristics of lead and zinc distribution in minerals of Caledonian granitoids of the Susamyr batholith in Central Tian-Shan. Geochemistry 1, 78, (1956).

— Pevtsova, L. A.: On the relationships of the distribution of lead and zinc in the rocks of the Caledonian granite complex of Susamyr (Central Tian-Shan). Dokl. Akad. Nauk SSSR 103, 1069 (1955).

Taylor, H. F. W.: The dehydration of hemimorphite. Am. Mineralogist 47, 932 (1962).

Taylor, L. A., Bell, P. M., Muan, A.: The effects of pressure in the Zn_2SiO_4—Co_2SiO_4 system. Carnegie Inst. Annual Rep. Director Geophys. Lab. Yearbook 69, 194 (1971).

Taylor, S. R.: Abundance of chemical elements in the continental crust: a new table. Geochim. Cosmochim. Acta 28, 1273 (1964).

Thayer, T. P., Milton, C., Dinnin, J., Rose, H.: Zincian chromite from Outokumpu, Finland. Am. Mineralogist 49, 1178 (1964).

Theobald, P. K., Thompson, C. E.: Zinc in magnetite from alluvium and from igneous rocks associated with ore deposits. U.S. Geol. Surv. Profess. Papers 450-C, 72 (1962).

Tiller, K. G.: The distribution of trace elements during differentiation of the Mt. Wellington dolerite sill. Papers Proc. Roy. Soc. Tasmania 93, 153 (1959).

— Hodgson, J. F.: The specific sorption of cobalt and zinc by layer silicates. 9th Conf. on Clays and Clay Minerals. London-New York: Pergamon Press 1962.

Treuil, M., Varet, J., Billhot, M., Barberi, F.: Distribution of nickel, copper and zinc in the volcanic series of Erta'Ale, Ethiopia. Contr. Mineral. and Petrol. 30, 84 (1971).

Tsusue, A., Holland, H. D.: The coprecipitation of cations with $CaCO_3$-III. The coprecipitation of Zn^{2+} with calcite between 50 and 250° C. Geochim. Cosmochim. Acta 30, 439 (1966).

Turekian, K. K.: The oceans, streams and atmosphere. Chapter 10 in K. H. Wedepohl (exec. ed.) Handbook of Geochemistry, Part I. Berlin-Heidelberg-New York: Springer 1969.

Udodov, P. A., Parilov, Y. S.: Certain regularities of migration of metals in natural water. Geochemistry 8, 763 (1961) [Engl. Transl.].

Vakhrushev, V. A.: Mineralogical-geochemical zoning in the iron deposits of the Kondoma region of Gornaya Shoriya. Geochemistry 4, 471 (1959).

VINE, J. D.: Element distribution in some shelf and eugeosynclinal black shales. U.S. Geol. Surv. Bull. **1214-E** (1966).
— Element distribution in some Paleozoic black shales and associated rocks. U.S. Geol. Surv. Bull. **1214-G** (1969).
— TOURTELOT, E. B.: Geochemical investigations of some black shales and associated rocks. U.S. Geol. Surv. Bull. **1314-A** (1969).
— — Geochemistry of black shale deposits. — A summary report. Econ. Geol. **65**, 253 (1970).
— — KEITH, J. R.: Element distribution in some trough and platform types of black shale and associated rocks. U.S. Geol. Surv. Bull. **1214-H** (1969).
VINOGRADOV, A. P.: The Elementary Chemical Composition of Marine Organisms. New Haven, Conn.: Sears Foundation 1953.
— The Geochemistry of Rare and Dispersed Chemical Elements in Soils (2nd ed.) New York: Consultants Bur. Enterpr. 209 p., 1959.
— Preliminary data on lunar ground supplied by the automatic station "Luna-16". Geokhimija **3**, 261 (1971).
VINOGRADOVA, Z. A., KOVALSKIY, V. V.: Elemental composition of the Black Sea plankton. Dokl. Acad. Nauk SSSR **147**, 1458 (1962).
VORE, G. DE: The role of adsorption in the fractionation and distribution of elements. J. Geol. **63**, 159 (1955a).
— Crystal growth and the distribution of elements. J. Geol. **63**, 471 (1955b).
WAGER, L. R., MITCHELL, R. L.: The distribution of trace elements during strong fractionation of basic magma — a further study of the Skaergaard intrusion, East Greenland. Geochim. Cosmochim. Acta **1**, 129 (1951).
WALLACE, W. E.: Bonding in the zinc family metals. J. Chem. Phys. **23**, 2281 (1955).
WARING, G. A., BLANKENSHIP, R. R., BENTALL, R.: Thermal springs of the United States and other countries of the world. A summary. U.S. Geol. Surv. Profess. Papers **492** (1965).
WARREN, H. V.: Background data for biogeochemical prospecting in British Columbia. Trans. Roy. Soc. Can., Sect. III **56**, 21 (1962).
— DELAVAULT, R. E.: The lead, copper, zinc and molybdenum content of some limestones and related rocks in Southern Ontario. Econ. Geol. **56**, 1265 (1961).
WATTENBERG, H.: Ergänzung zu der Mitteilung „Zur Chemie des Meerwassers; über die in Spuren vorkommenden Elemente". Z. Anorg. Allgem. Chem. **251B**, 86 (1943).
WEBER, K.: Die Kristallstruktur des Reinerits $Zn_3[AsO_3]_2$. Habilitationsschrift T.U. Berlin 1968.
WEDEPOHL, K. H.: Untersuchungen zur Geochemie des Zinks. Geochim. Cosmochim. Acta **3**, 93 (1953).
— Schwermetallgehalte der Kalkgerüste einiger mariner Organismen. Nachr. Akad. Wiss. Goettingen Math. Phys. Chem. Abt. IIa **5**, 79 (1955).
— Spurenanalytische Untersuchungen an Tiefseetonen aus dem Atlantik. Ein Beitrag zur Deutung der geochemischen Sonderstellung von pelagischen Tonen. Geochim. Cosmochim. Acta **18**, 200 (1960).
— Untersuchungen am Kupferschiefer in Nordwestdeutschland. (Ein Beitrag zur Deutung der Genese bituminöser Sedimente.) Geochim. Cosmochim. Acta **28**, 305 (1964).
— Die Zusammensetzung der Erdkruste. Fortschr. Mineral. **46**, 145 (1969).
— "Kupferschiefer" as a prototype of syngenetic sedimentary ore deposits. Soc. Mining Geol. Japan Spec. Issue **3**, 268 (1971) (Proc. IMA IAGOD Meetings 70).
WEIGAND, P. W., RAGLAND, P. C.: Geochemistry of Mesozoic dolerite dikes from Eastern North America. Contr. Mineral. and Petrol. **29**, 195 (1970).
WEITZEL, G., FRETZDORF, A. M., STRECKER, F. J., ROESTER, U.: Zinkgehalt und Glukagoneffekt kristallisierter Insulinpräparate. Hoppe-Seylers Z. Physiol. Chem. **293**, 190 (1953).
— STRECKER, F. J., ROESTER, U., BUDDECKE, E., FRETZDORF, A. M.: Zink im Tapetum lucidum. Hoppe-Seylers Z. Physiol. Chem. **296**, 19 (1954).
WHITE, D. E., HEM, J. D., WARING, G. A.: Chemical composition of subsurface waters. Chapter F in: Data of Geochemistry (M. FLEISCHER ed.). U.S. Geol. Surv. Profess. Papers **440-F** (1963).

WHITE, J. G.: The crystal structure of the tetragonal modification of ZnP_2. Acta Cryst. **18**, 217 (1965).
WHITE, M. L.: The occurrence of zinc in soil. Econ. Geol. **52**, 645 (1957).
WILSON, S. H.: Physical and chemical investigations 1939—55 in: HAMILTON W. H., BAUMGART I. L.: White Island. New Zealand Dept. Sci. Ind. Res. Bull. **127**, 32 (1959).
WODZICKI, A.: Migration of trace elements during contact metamorphism in the Santa Rosa Range, Nevada and its bearing on the origin of ore deposits associated with granitic intrusions. Mineral. Deposita **6**, 49 (1971).
ZALKIN, A., RUBEN, H., TEMPLETON, D. H.: The crystal structure of cobalt sulfate hexahydrate. Acta Cryst. **15**, 1219 (1962).
ZAMBONINI, F.: Mineralogia Vesuviana. Torino: Rosenberg and Sellier 1936.
ZEISSINK, H. E.: Trace element behavior in two nickeliferous laterite profiles. Chem. Geol. **7**, 25 (1971).
ZIES, E. G.: The Valley of Ten Thousand Smokes. I and II. Nat. Geogr. Soc. Techn. Pap. **1**, 4, 1 (1929).
ZIRINO, A., HEALY, M. L.: Inorganic zinc complexes in seawater. Limnol. Oceanog **15**, 956 (1970).
ZLOBIN, B. I., GORSHKOVA, M. S.: Lead and zinc in alkalic rocks and their bearing on some petrological problems. Geochemistry **4**, 317 (1961).
— PEVTSOVA, L. A.: Behavior of lead and zinc in some processes of hydrothermal alteration of granitoids. Geochemistry (USSR) [English Transl.] **3**, 413 (1964).
— — KLASSOVA, N. S.: Distribution of lead and zinc and metallogenic specialization of the more calcic varieties of Variscan granitoids in the central part of the Northern Tien-Shan. Geochem. Internat. **2**, 660 (1965).
ZUBOVIC, P., SHEFFEY, N. B., STADNICHENKO, T.: Distribution of minor elements in some coals in the Western and Southwestern Regions of the Interior Coal Province. U.S. Geol. Surv. Bull. **1117-D**, (1967).
— STADNICHENKO, T., SHEFFEY, N. B.: Geochemistry of minor elements in coals of the Northern Great Plains Coal Province. U.S. Geol. Surv. Bull. **1117-A**, (1961).
— — — Distribution of minor elements in coal beds of the Eastern Interior Region. U.S. Geol. Surv. Bull. **1117-B**, (1964).
— — — Distribution of minor elements in coals of the Appalachian Region. U.S. Geol. Surv. Bull. **1117-C**, (1966).

Gallium

A   G. Gottardi   (Istituto di Mineralogia e Petrologia, Università di Modena, Italia)

B—O   J. D. Burton   (Department of Oceanography, University of Southampton, England)

and

F. Culkin   (National Institute of Oceanography, Wormley, Godalming, Surrey, England)

Introduction

The literature on the geochemistry of gallium is very extensive and several problems arise in an attempt to summarise the available information in a short account. Only a small proportion of the many determinations of gallium in rocks and minerals has been made by the most precise and accurate procedures, and analyses by emission spectrographic methods constitute a dominant part of the data given here. In Table 31-1 are summarised the results of determinations of gallium in two standard rocks by various procedures. These results show that the generally less precise spectrographic procedures give, on average, values of adequate accuracy. In view of this and also the quite low dispersion of gallium in common rocks and minerals, the average results of the less precise analyses often constitute abundance data of reasonable quality. Individual determinations or whole series of determinations by these methods are sometimes inaccurate, however, and it remains difficult to assess questions such as the extent to which differences in concentrations recorded for various areas represent genuine regional variations as distinct from systematic analytical errors. Furthermore, the low dispersion of gallium leads to an interest in relatively small variations in processes such as magmatic differentiation, and these variations tend to be obscured in many investigations through the use of imprecise methods.

Table 31-1. *Concentrations of gallium found in standard rocks G-1 and W-1 by different analytical methods*

Method	Ga (ppm)			
	range of average values found		mean and number of investigations	
	G-1	W-1	G-1	W-1
S	10 —25	8 —23	18.1 (30)	16.0 (36)
C	15 —22	14 —21.5	19.4 (3)	17.4 (4)
N/R	17.4—21.0	15.5—20.0	19.7 (4)	17.7 (5)
X	17 —20	—	18.5 (2)	24.1 (1)
Recommended value (FLEISCHER, 1969)			18	16

This table is compiled from summaries by AHRENS and FLEISCHER (1960), FLEISCHER and STEVENS (1962), and FLEISCHER (1969).

A comprehensive review of work on the geochemistry of gallium upto the end of 1955 was given by SHAW (1957). Much of the earlier work discussed there and in the review by FLEISCHER (1955) has been excluded from the present account which is primarily concerned with work published since 1950. In the tables the number of samples examined and the range of concentrations found are given wherever possible as this information appears relevant in the light of the comments above. Because it is often useful to consider the distribution of gallium in relation to that of aluminium, the weight ratio of these elements is given for many materials. These values are not necessarily based on the same number of samples as the values for gallium concentration and in a few instances were obtained by reference to theoretical compositions or analyses of analogous materials. Whenever data are quoted from a paper which has not been available to us this is indicated by the inclusion of a reference to *Chemical Abstracts* in the bibliography.

31-B. Isotopes in Nature

Two stable isotopes of gallium, ^{69}Ga and ^{71}Ga, have been found in nature. Their conventional abundances were determined by ANTKIW and DIBELER (1953) as ^{69}Ga, 60.5%, and ^{71}Ga, 39.5%, in satisfactory agreement with the results of INGHRAM et al. (1948) who obtained corresponding values of 60.2 and 39.8%. KERWIN and McELCHERAN (1956) established upper limits of 0.04% for the abundance of each of the gallium isotopes of mass numbers 66, 67, 68, 73 and 74, 0.07% for ^{70}Ga and 0.2% for ^{72}Ga.

INGHRAM et al. (1948) found no significant difference in the isotopic ratios for gallium from terrestrial and meteoritic material. There is no information on the natural fractionation of these isotopes of gallium in geological or biological processes.

31-C. Abundance in Cosmos, Meteorites, Tektites and Lunar Materials

I. Cosmos

Estimation of the solar abundance of gallium is difficult because of blending of the Ga(I) line used and other uncertainties. GOLDBERG et al. (1960) estimated log N(Ga) to be 0.86 on the scale log N(H) $=10.50$ (log N(Si) $=6.00$). ALLER (1965) revised the value to 1.01 or 1.25 depending on the f values used; this latter value is based on the data of KING et al. (1965) who gave log N(Ga) as 1.22. The estimates of the solar abundance are similar to the abundance in chondrites. Lines due to Ga(II) have been identified in the spectra of certain peculiar stars, including some "manganese" stars (BIDELMAN and CORLISS, 1962; JASCHEK et al., 1963). Marked over-abundance of gallium in a "manganese" star has been reported by BUSCOMBE and CHAMBLISS (1968).

II. Meteorites

Analyses of gallium in meteorites are summarised in Tables 31-C-1 to 31-C-3. These have been compiled from the results of the authors cited, usually by averaging all the available results for each meteorite. The classification of HEY (1966) is used with the two exceptions that the Babb's Hill specimen analysed by WASSON and KIMBERLIN (1967) has been taken as a separate ataxite from that analysed by other workers, and the Canyon Diablo (1936) octahedrite has not been treated as a separate meteorite (WASSON, 1967a).

Table 31-C-1. *Gallium in stone meteorites*

Class	Number analysed	Ga (ppm) range	Ga (ppm) mean
Aor	2	0.9— 1.1	1.0
Ap	3	1.7— 2.6	2.0
Ce_1 and Ce_2	6	11.9—19.1	15.4
CH	17	4.0— 6.8	5.4
CL	43	4.2— 8.6	5.5
CLL	2	4.6— 5	4.8
CHL	8	5.0—10.6	7.9
Cc_1	2	10.7—12.9	11.8
Cc_0	4	1.7— 9.6	7.9
Unclassified	1		4.0

This table is compiled from the results of the following workers who used the methods indicated: ONISHI and SANDELL (1956) C; GREENLAND (1965) N/R; HECHT and FENNINGER (1965) N/R; AKAIWA (1966) N/R; FOUCHÉ and SMALES (1967) N/R; SCHAUDY et al. (1967) N/R; TANDON and WASSON (1968) N/R; MASON and GRAHAM (1970) I; WASSON and BAEDECKER (1970) N/R.

Table 31-C-2. *Gallium in stony-iron meteorites*

Class	Number analysed	Ga (ppm) range	mean
P	6	14—24	21
M	1		7

This table is compiled from the results of the following workers who used the methods indicated: COBB and MORAN (1965) N/R; COBB (1967) N/R; WASSON and KIMBERLIN (1967) N/R.

Table 31-C-3. *Gallium in iron meteorites*

Class	Number analysed	Ga (ppm) range	mean
Ogg	12	6.3—96	61
Og	21	20 —94	67
Om	77	1.7—87	25
Of	34	1.6—84	15
Off	13	2.4—87	30
H	34	0.7—87	52
D	24	0.2—46	7
Others	7	<2 —51	
All irons	222		32

This table is compiled from the results of the following workers who used the methods indicated: GOLDBERG et al. (1951) N/R; LOVERING et al. (1957) S; COBB and MORAN (1965) N/R; HECHT and FENNINGER (1965) N/R; WASSON (1966, 1967a, b) N/R; WASSON and KIMBERLIN (1967) N/R; SMALES et al. (1967) N/R, C; COBB (1967) N/R; WASSON and GOLDSTEIN (1968) N/R.

The distributions of gallium concentrations in the various classes of iron meteorites are shown in Fig. 31-C-1 and 31-C-2.

Chondrites of the H and L groups show no significant difference in their average content of gallium; considerably higher concentrations occur in enstatite chondrites while the carbonaceous chondrites generally have intermediate concentrations. Of all iron meteorites, over one-third have been analysed for gallium. The distribution of the concentrations in the various classes is shown in Fig. 31-C-1 and 31-C-2; the so-called "quantized" distribution, first noted by GOLDBERG et al. (1951), emerges clearly. There appear to be at least five distinct populations of gallium concentrations and to a certain extent they can be correlated with structure. Most of the coarse octahedrites fall within the population of highest gallium concentration which has an average content of 80 to 85 ppm; the hexahedrites usually have concentrations in the range of 45 to 65 ppm; a large fraction of the medium octahedrites lie in the population with an average content of 15 to 20 ppm; a substantial part of the population with an average concentration of 2 to 2.5 ppm consists of fine octahedrites; all but one of the irons with a concentration below 1 ppm are nickel-rich ataxites. The comparatively small dispersion of some of the populations is more remarkable

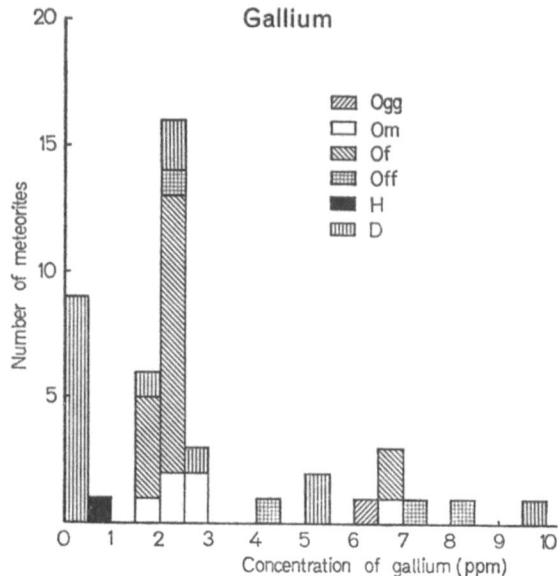

Fig. 31-C-1. Distribution of gallium concentrations in iron meteorites containing upto 10 ppm of gallium

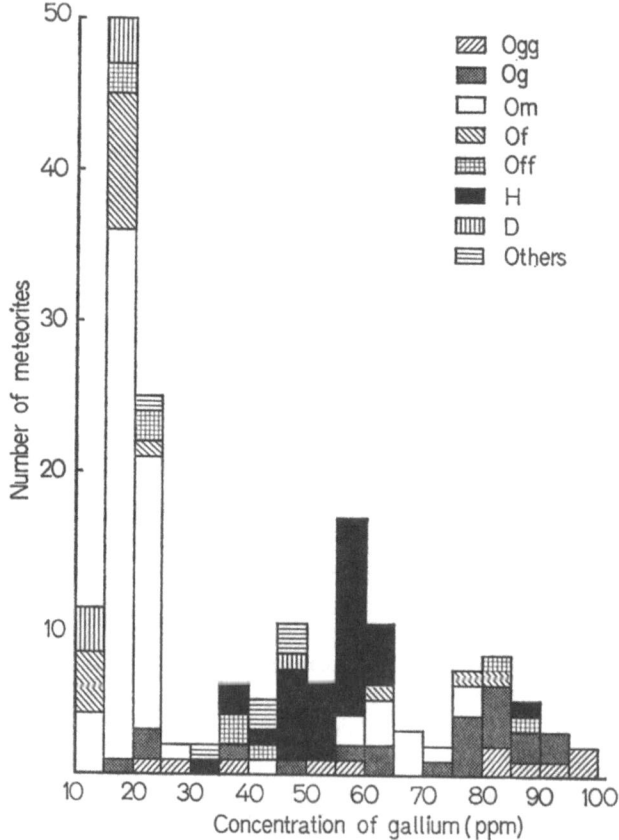

Fig. 31-C-2. Distribution of gallium concentrations in iron meteorites containing between 10 and 100 ppm of gallium

since the analytical errors of a number of separate investigations are compounded in the presentation. There is a high correlation between the concentrations of gallium and germanium in iron meteorites. A detailed discussion of the relationship of composition and structure and the significance of the distribution of gallium and other trace elements for theories of the origin of meteorites is outside the scope of this account; papers of particular relevance include those of YAVNEL (1960), SMALES et al. (1967), WASSON (1967b, 1969) and WASSON and KIMBERLIN (1967).

Results for separated phases from meteorites are given in Tables 31-C-4 and 31-C-5. The relatively high concentration of gallium in the sulfide phase of chondrites emphasises the chalcophile aspect of the behavior of gallium; the concentrations in troilite from irons are much lower. Enrichment in the metal phase is notably greater for the enstatite chondrites than for the less reduced types. In octahedrites gallium is enriched approximately ten-fold in the high-nickel taenite phase as compared with kamacite. LOVERING (1957) has discussed the factors which influence the phase distribution of gallium in meteorites.

Table 31-C-4. *Gallium in separated phases of stone and stony-iron meteorites*

Phase	Class	Number analysed	Ga (ppm) range	Ga (ppm) mean (method)	Reference
Metal	Ce_1 and Ce_2	4	54—68	61 (N/R)	COBB and MORAN (1965); FOUCHÉ and SMALES (1967)
	CH and CL	21	3.4—26	12 (N/R)	
	CH and CL	3a	3—181	16 (C)	ONISHI and SANDELL (1956)
	CLL	1		10 (I)	MASON and GRAHAM (1970)
	P	8	12—23	18 (N/R, S)	GOLDBERG et al. (1951); LOVERING et al. (1957)
	M	1		19 (S)	LOVERING et al. (1957)
Nickel-rich taenite	CH and CL	4	28—39	34 (C)	Moss et al. (1967)
Chromite	CL and CLL	2	65—69	67 (I)	MASON and GRAHAM (1970)
Non-magnetic (silicate and sulfide)	Ce_1 and Ce_2	4	0.25—5.0 b	1.7 (N/R)	FOUCHÉ and SMALES (1967)
	CH and CL	16	2.5—8.9	4.8 (N/R)	FOUCHÉ and SMALES (1967)
Silicate	Ce_2	1		0.7 (C)	Moss et al. (1967)
	CH and CL	3a	3.0—3.2	3.1 (C)	ONISHI and SANDELL (1956)
	CH and CL	4	2.4—4.8	3.7 (C)	Moss et al. (1967)

Table 31-C-4. (Continued)

Phase	Class	Number analysed	Ga (ppm) range	mean (method)	Reference
Olivine	CL and CLL	2	2—2	2 (I)	Mason and Graham (1970)
	P	2	0.6—1	0.8 (I)	Mason and Graham (1970)
Orthopyroxene	Ab	1		1 (I)	Mason and Graham (1970)
	CL and CLL	2	2—2	2 (I)	Mason and Graham (1970)
	M	1		1 (I)	Mason and Graham (1970)
Clinopyroxene	Ap	1		2 (I)	Mason and Graham (1970)
	M	1		1 (I)	Mason and Graham (1970)
Plagioclase	Ap	1		3 (I)	Mason and Graham (1970)
	CL and CLL	2	17—23	20 (I)	Mason and Graham (1970)
	M	1		8 (I)	Mason and Graham (1970)
Phosphate	CL and CLL	2	2—2	2 (I)	Mason and Graham (1970)
Sulfide	CH and CL	3[a]	9—14	11 (C)	Onishi and Sandell (1956)
	CH and CL	4	2—14	7 (C)	Moss et al. (1967)
	CL and CLL	2	5—6	5.5 (I)	Mason and Graham (1970)

[a] Composite samples.
[b] Values considerably affected by metal phase contamination.

Table 31-C-5. *Gallium in separated phases of iron meteorites*

Phase	Class	Number analysed	Ga (ppm) range	mean (method)	Reference
Kamacite	H	2	30—37	34 (S)	Nichiporuk (1958)
	Og and Om	4	19—38	26 (S)	Nichiporuk (1958)
	Of	2	<1	<1 (S)	Nichiporuk (1958)
Taenite	Og and Om	4	210—515	315 (S)	Nichiporuk (1958)
	Of	2	2—7.5	5 (S)	Nichiporuk (1958)
Troilite	Og and Om	2	0.3—0.4	0.35 (N/R)	Goldberg et al. (1951)

Table 31-C-6. *Gallium in tektites*

Description	No. analysed	Ga (ppm) range	Ga (ppm) mean (method)	Ga × 10³ / Al range	Ga × 10³ / Al mean	Reference
Various tektites	9	0.9—9.8	5.8 (C)			1
Bediasites	10	<5—16	11 (S)	<0.08—0.21	0.15	2
Martha's Vineyard and Georgia tektites	8	5.7—10	7.7 (S)	0.09—0.18	0.14	2
Indochinites	6	ca. 10—100	ca. 40	ca. 0.2—1.5	ca. 0.6	3
Phillipinites	2	6.9—7.0	7.0 (S)	0.11—0.16	0.14	4
Australites	42	4.5—12	8.3 (S)	0.06—0.21	0.14	5

References: 1. COHEN (1960). 2. CUTTITTA et al. (1967). 3. VOROBIEV (1959). 4. VOROB'YEV (1963). 5. TAYLOR (1962); TAYLOR and SACHS (1964).

Table 31-C-7. *Gallium in lunar materials*[a]

Material[b]	Number of samples	Ga (ppm) range[c]	Ga (ppm) mean	Ga × 10³ / Al range[c]	Ga × 10³ / Al mean
Apollo 11					
Type A	6	2.7—5.0	4.1	0.059—0.124	0.095
Type B	10	3—5.1	4.3	0.054—0.117	0.080
Type C	11	3.7—5.7	4.8	0.061—0.088	0.075
Type D	1	3.8—5.2	4.5	0.054—0.071	0.061
Apollo 12					
Rock 12013	1	5.9—6.4	6.1		0.097
Type AB	1		4.5		
Type B	1		2.4		
Type D	1		4.3		0.058
Core sample, Type A+D	1		2.7		
Core samples, Type D	3	5.0—5.2	5.1		

[a] This table is a summary of the results of the workers listed below who used the methods indicated.
Apollo 11: ANNELL and HELZ (1970)S; COMPSTON et al. (1970)X; GANAPATHY et al. (1970a) N/R; HASKIN et al. (1970)N/R; MORRISON et al. (1970)N/R; SMALES et al. (1970)N/R; WÄNKE et al. (1970)N/R; WASSON and BAEDECKER (1970)N/R.
Apollo 12: Rock 12013, LAUL et al. (1970)N/R; Others, GANAPATHY et al. (1970b)N/R. Aluminium contents of Apollo 12 samples were taken from the report of the LUNAR SAMPLE PRELIMINARY EXAMINATION TEAM (1970).
[b] Types A and B — mafic crystalline rocks of different textures; Type C — breccia; Type D — fines.
[c] Ranges for Apollo 11 samples are for average values calculated from the results for each rock given by the workers listed, except for Type D where the range is for the sub-samples analysed by different workers. For Rock 12013 the range is for 7 sub-samples.

III. Tektites

Determinations of gallium in tektites are summarised in Table 31-C-6. With the exception of the values reported by VOROBIEV (1959) for indochinites, tektites of various types show rather similar concentrations of gallium and a ratio of gallium to aluminium which is similar to the average ratio for terrestrial material.

IV. Lunar Materials

Results for the occurrence of gallium in lunar materials are summarised in Table 31-C-7. For the Apollo 11 materials there appears to be no significant difference on average between the Type A (fine grained) and Type B (medium grained) rocks, although a detailed examination of the data suggests that two of the Type A rocks may have distinctively low gallium contents below 3 ppm. The concentrations in the breccias and fines are similar to those in the crystalline rocks. The concentrations of gallium in the Apollo 12 crystalline rocks and breccias lie in a similar range to that of the Apollo 11 materials. The unusual gabbroic intrusion breccia, rock 12013, contains somewhat more gallium than any other lunar material examined so far. Gallium is markedly depleted in these materials in comparison with most terrestrial mafic rocks (see Section 31-E).

31-D. Abundance in Rock-forming Minerals, Gallium Minerals

I. Rock-forming Minerals

Results for the occurrence of gallium in minerals are summarised in Tables 31-D-1 to 31-D-7. Among the common rock-forming minerals gallium is generally concentrated above its crustal abundance only in some feldspars and amphiboles as well as in micas and magnetite. Its concentration is usually low in pyroxenes and is particularly low in olivine and quartz. These features reflect the close coherence of gallium and aluminium in most geochemical processes and the less close but also important relationship between gallium and iron (Fe^{3+}). The following data are useful in considering the relationships between these elements. The values are those given by AHRENS (1964) who cites two sets of electronegativity values from different sources. (For additional values of ionic radii see this handbook, Volume I, page 389).

	Ga	Al	Fe
Radius of ion (3+), Å	0.57	0.51	0.64
Electronegativity, kcal/g atom	1.82	1.47	1.64
	1.6	1.5	1.8
Ionization potential I_3, eV	30.7	28.4	30.6
Ionization potential $I_1+I_2+I_3$, eV	57.2	53.2	54.7

There are thus similarities between the ions of gallium and iron (Fe^{3+}) as well as between gallium and aluminium. The general chemical behavior of gallium, however, resembles that of aluminium very much more than that of iron.

In comparison with the aluminium-oxygen bond, the gallium-oxygen bond is slightly longer and more covalent. This suggests that gallium will be slightly excluded, relative to aluminium, in competition for lattice sites in earlier products of magmatic crystallization and so will tend to accumulate in later products of differentiation and residual materials. This effect has often been observed in studies of minerals from rocks differentiated from a common magma. An increase in the concentration of gallium and its ratio to aluminium in minerals of late crystallization was found, for example, in hornblendes, biotites and plagioclases of the Garabal Hill-Glen Fyne plutonic complex (NOCKOLDS and MITCHELL, 1948) and in other hornblendes (BORISENOK, 1959) and biotites (SEN et al., 1959). Similar trends occur in the later orthopyroxenes of the Madras charnockite series which are also richer in iron than those crystallizing earlier (HOWIE, 1955), but WILKINSON (1959) found little variation in content with differentiation in clinopyroxenes of a teschenite sill. Plagioclases of different stages of crystallization generally show little systematic

Gallium

Table 31-D-1. *Gallium in feldspars and related minerals*

Mineral and origin	No. of samples	Ga (ppm) range	Ga (ppm) mean (method)	Ga × 10³ / Al range	Ga × 10³ / Al mean	Reference
Alkali feldspar						
Granitic complexes, Ireland	21	5—12	9 (S)		0.09	1
Potassium feldspar						
S. Africa	70	10—25	(S)			2
S. California batholith	5	10—10	10 (S)	0.10—0.10	0.10	3
Charnockite series, Madras, India	9	2—15	10 (S)	0.02—0.13	0.09	4
Various rocks, U.S.S.R.		11—18	(S)		0.12	5
Lower Silesia, Poland (hydrothermal)	11	13—19	16 (S)		0.16	6
Scotland (mainly granodiorites and adamellites)	14	10—30	17 (S)	0.11—0.31	0.18	7
Orthoclase						
Alkaline rocks, U.S.S.R.	3	13—17	14 (S)			8
Metamorphic rocks, U.S.S.R.	6	15—30	23 (C, L)			9
Europe	3	9—30	18 (S, C)			10
Europe (adularia, hydrothermal)	5	9—12	10 (S)		0.11	6
Microcline						
Tedino deposit, U.S.S.R.	6	3—12	8 (S)		0.07	11
Pegmatites and parent rocks, U.S.S.R.	17	15—51	34 (S)	0.13—0.56	0.35	12
Alkaline pegmatites and rocks, Lovozero Massif, U.S.S.R.	6	40—88	63 (C, S)			13
C. Kazakhstan (hydrothermal)	6	3.5—45	27 (S)			14
Perthite						
Pegmatites, Black Hills, U.S.A.	16	30—50	35 (S)	0.27—0.48	0.33	15
Microcline-perthite						
Pegmatites, C. Kazakhstan	4	35—70	51 (S)			14
Pegmatites, Lower Silesia	20	24—46	30 (S)		0.34	6
Pegmatites, Scotland	3	20—32	26 (S)			16
Amazonite						
S.W. Africa, U.S.A., U.S.S.R.	6	48—92	74 (S)			17
Pneumatolytic pegmatites, Lower Silesia, Poland	10	34—70	44 (S)		0.45	6

Table 31-D-1. (Continued)

Mineral and origin	No. of samples	Ga (ppm) range	Ga (ppm) mean (method)	$\frac{Ga \times 10^3}{Al}$ range	$\frac{Ga \times 10^3}{Al}$ mean	Reference
Plagioclase						
S. California batholith	14	20—35	27 (S)	0.15—0.26	0.19	3
Igneous rocks, Japan	38	14—110	48 (S)			18
Charnockite series, Madras, India	6	20—45	29 (S)	0.15—0.29	0.21	4
Black Jack sill, Australia	7	15—25	19 (S)	0.09—0.16	0.12	19
Pyroxenite, U.S.S.R.		10—15	(S)			5
Other rocks, U.S.S.R.		18—25	(S)		0.22	5
Tedino deposit, U.S.S.R.	8	12—16	14 (S)		0.11	11
Metamorphic rocks, U.S.S.R.	3	18—24	21 (C,L)			9
Granite, Lower Silesia, Poland	20	26—36	31 (S)			6
Granitic complexes, Ireland	23	17—35	24 (S)		0.18	1
Igneous rocks, Scotland	9	20—50	31 (S)	0.12—0.41	0.24	7
Albite						
Acid pegmatites and parent rocks, U.S.S.R.	4	23—50	32 (S)	0.23—0.51	0.33	20
Alkaline pegmatites and rocks, U.S.S.R.	7	10—115	67 (C, S)	0.10—0.98	0.54	21
Urals, U.S.S.R.	2	36—120	78 (S)			6
Hydrothermal deposits, Europe	13	14—23	17 (S)		0.18	6
Pneumatolytic pegmatites, Lower Silesia	6	46—95	68 (S)		0.66	6
Pneumatolytic hydrothermal deposits, Lower Silesia, Poland	6	27—43	34 (S)			6
Cleavelandite						
Pegmatite, Black Hills, U.S.A.	1		70 (S)		0.67	15
Andesine, labradorite, bytownite						
Gabbro, diorite and syenite, U.S.S.R.	4	20—25	22 (S)			22
Nepheline						
Alkaline rocks and pegmatites, U.S.S.R.		34—98	(S)			23
Alkaline rocks and pegmatites, U.S.S.R.	7	70—140	100 (C, S)	0.42—0.57	0.52	21

References: 1. HALL (1967). 2. KOLBE (1965). 3. SEN et al. (1959). 4. HOWIE (1955). 5. BORISENOK (1959); BORISENOK and SAUKOV (1960); BORISENOK and TAUSON (1959). 6. WALENCZAK (1959). 7. NOCKOLDS and MITCHELL (1948). 8. BORISENOK and ZLOBIN (1959). 9. MARAKUSHEV and POLIN (1961). 10. BUTLER (1953b); BURTON et al. (1959). 11. BORISENOK and RYABCHIKOV (1962). 12. BORISENOK (1959); BORISENOK and SAUKOV (1960); GANEEV et al. (1961). 13. GERASIMOVSKII et al. (1959). 14. GANEYEV and SECHINA (1962). 15. HIGAZY (1953). 16. HITCHON (1960). 17. WALENCZAK (1959); SITNIN (1960). 18. IIDA (1961). 19. WILKINSON (1959). 20. BORISENOK and SAUKOV (1960). 21. GERASIMOVSKII et al. (1959); YEREMENKO et al. (1963). 22. BORISENOK (1959). 23. BORISENOK (1959); BORISENOK and SAUKOV (1960); BORISENOK and ZLOBIN (1959).

Gallium

Table 31-D-2. *Gallium in micas and related minerals*

Mineral and origin	No. of samples	Ga (ppm) range	Ga (ppm) mean (method)	Ga × 10³ / Al range	Ga × 10³ / Al mean	Reference
Muscovite						
Granites, S. Africa	4	37—65	(S)			1
Metamorphic rocks, U.S.A. and Europe	10	37—118	62 (S)			2
Intermediate and granitic rocks, greisens, U.S.S.R.		100—400	(S)			3
Granites and pegmatites, Azov region, U.S.S.R.	4	75—85	80 (L)	0.42—0.49	0.46	4
Tedino deposit, U.S.S.R.	34	31—124	72 (S)		0.39	5
Greisens and vein rocks, C. Kazakhstan (includes sericite)	9	27—113	68 (S)	0.17—1.00	0.51	6
Metamorphic rocks, U.S.S.R.	2	84—110	97 (C, L)		0.54	7
Biotite						
Granites, S. Africa	22	22—55	(S)			1
Metamorphic rocks, U.S.A.	17	20—71	39 (S)	0.18—0.90	0.43	2
Gabbros, granitic rocks, S. California batholith	10	15—30	22 (S)	0.16—0.40	0.25	8
Granitic rocks, syenites and nepheline syenites, U.S.S.R.	8	23—71	54 (S)			9
Granites and pegmatites, C. Kazakhstan	7	34—339	145 (S)	0.22—1.80	0.86	6
Granites, gneisses and pegmatite, Azov region, U.S.S.R.[a]	12	12—60	35 (L)	0.33—0.78	0.52	4
Tedino deposit, U.S.S.R.	5	45—146	68 (S)		0.61	5
Metamorphic rocks, U.S.S.R.	6	41—73	51 (C, L)		1.09	7
Scotland (mainly diorites, granodiorites and adamellites)	13	10—50	31 (S)	0.14—0.61	0.39	10
Pegmatites, Scotland	5	60—130	89 (S)			11
Phlogopite						
Canada	1		101 (C)			12
Altered dolomitic marbles, U.S.S.R.	3	4—15	9 (C, L)			7
Chlorite						
Various sources (mainly schists, U.S.A.)	4	35—48	39 (S)	0.29—0.40	0.35	2
Talc						
Schist, Austria	1		7.4 (S)		<0.7	2
Pyrophyllite						
Schists, U.S.A.	2	39—88	64 (S)	0.21—0.59	0.40	2

Table 31-D-3. *Gallium in amphiboles*

Mineral and origin	No. of samples	Ga (ppm) range	Ga (ppm) mean (method)	$\frac{Ga \times 10^3}{Al}$ range	$\frac{Ga \times 10^3}{Al}$ mean	Reference
Actinolite						
Galway, Ireland	1		3.6 (C)		0.09	1
Hornblende						
Various sources	10	1.5—27	15 (S)	0.14—0.67	0.34	2
Gabbros, granitic rocks, S. California batholith	12	2—20	13 (S)	0.02—0.33	0.24	3
Charnockite series, Madras, India	7	2—50	21 (S)	0.16—0.89	0.36	4
Ultramafic rocks, Urals, U.S.S.R.		12—15	(C)			5
Intermediate and granitic rocks, U.S.S.R.		10—50	(S)			6
Granites and syenite, Azov region, U.S.S.R.	3	16—21	18 (L)			7
Metamorphic rocks, U.S.S.R.	2	25—40	32 (C, L)		0.63	8
Igneous rocks and pegmatite, Scotland	9	15—30	19 (S)	0.23—0.79	0.43	9
Galway, Ireland	7	9—20	13 (C)			1
Pargasite						
Altered dolomitic marble, U.S.S.R.	1		11 (C, L)			8
Holmquistite						
Granite pegmatite, U.S.S.R.	1		31		0.50	10
Arfvedsonite						
Alkaline pegmatite, Lovozero Massif, U.S.S.R.	2	20—22	21 (C)			11

References: 1. Burton et al. (1959). 2. De Vore (1955). 3. Sen et al. (1959). 4. Howie (1955). 5. Borisenko (1963). 6. Borisenok and Saukov (1960). 7. Marchenko and Shcherbakov (1966). 8. Marakushev and Polin (1961). 9. Nockolds and Mitchell (1948); Hitchon (1960). 10. Slepnev (1962). 11. Gerasimovskii et al. (1959).

Table 31-D-2 (Continued)

References: 1. Kolbe (1965). 2. DeVore (1955). 3. Borisenok and Saukov (1960). 4. Marchenko and Shcherbakov (1966). 5. Borisenok and Ryabchikov (1962). 6. Ganeev et al. (1961). 7. Marakushev and Polin (1961). 8. Sen et al. (1959). 9. Borisenok (1959); Borisenok and Tauson (1959); Borisenok and Zlobin (1959). 10. Nockolds and Mitchell (1948). 11. Hitchon (1960). 12. Burton et al. (1959).

[a] Sericitized biotites (3) from the granites contained 85—160, mean 122 ppm Ga(L); mean Ga/Al, 1.17×10^{-3}.

Gallium

Table 31-D-4. *Gallium in pyroxenes*

Mineral and origin	No. of samples	Ga (ppm) range	Ga (ppm) mean (method)	$\dfrac{Ga \times 10^3}{Al}$ range	$\dfrac{Ga \times 10^3}{Al}$ mean	Reference
Orthopyroxene						
Pyroxenites, U.S.A. and Greenland	3	1.5—8.2	4 (S)	0.19—0.71	0.42	1
Quebec (mainly norites)	13	10—25	15 (S)	0.54—3.73	1.50	2
Charnockite series, Madras, India	13	5—30	13 (S)	0.22—2.34	0.96	3
Hypersthene						
Gabbros, norites, etc, S. California batholith	6	2—10	6 (S)	0.19—0.73	0.46	4
Peridotites and basic gneisses, Scotland	4	upto 5	(S)	0.13—0.33	0.22	5
Clinopyroxene						
Quebec (mainly norites)	11	10—20	14 (S)	0.43—1.25	0.73	2
Charnockite series, Madras, India	6	upto 10	6 (S)	0.16—0.94	0.47	3
Black Jack sill, Australia	7	12—15	14 (S)	0.49—0.79	0.64	6
Pyroxene						
Mt. Wellington sill, Tasmania	4	7—12	10 (S)			7
Mafic and ultramafic rocks, U.S.S.R.		1—6	(S)			8
Skaergaard intrusion, Greenland	6	<1—10	5 (S)	<0.05—0.47	0.25	9
Diopside						
Various sources	14	upto 13	6 (S)	0.06—0.58	0.32	1
Augite						
Gabbros, norites, etc, S. California batholith	2	4—5	4.5 (S)			4
Nepheline and calcic syenites, U.S.S.R	2	13—23	18 (S)			10
Peridotites and gneisses, Scotland	6	upto 10	(S)		0.17	5
Aegirine						
Alkaline rocks and pegmatites, U.S.S.R.	6	20—50	33 (C)			11

References: 1. DeVore (1955). 2. Philpotts (1966). 3. Howie (1955). 4. Sen et al. (1959). 5. O'Hara (1961). 6. Wilkinson (1959). 7. Tiller (1959). 8. Borisenok (1959); Borisenok and Saukov (1960). 9. Wager and Mitchell (1951). 10. Borisenok and Zlobin (1959). 11. Gerasimovskii et al. (1959); Yeremenko et al. (1963).

Table 31-D-5. *Gallium in miscellaneous silicate minerals*

Mineral and origin	No. of samples	Ga (ppm) range	Ga (ppm) mean (method)	$\dfrac{Ga \times 10^3}{Al}$ mean	Reference
Garnet					
U.S.A. and Europe (mainly schists)	21	<1—19	4 (S)	0.04	1
Gneisses, Scotland	5	5—10	6 (S)	0.05	2
Almandine					
Metamorphic rocks, U.S.S.R.	7	14—20	16 (C, L)	0.13	3
Olivine					
Black Jack sill, Australia	5	<2	<2 (S)		4
Skaergaard intrusion, Greenland	5	<1—5	4 (S)		5
Andalusite					
N. and S. America	7	45—210	83 (S)	0.25	6
Schist, Idaho, U.S.A.	1		8 (S)	0.024	7
Sillimanite					
N. America, Australia and Europe	3	33—95	66 (S)	0.20	6
Kyanite					
N. and S. America, Europe	12	12—66	38 (S)	0.11	6
Brazil	17	13—49	27 (S)	0.08	8
Epidote					
Skarns, U.S.S.R.		20—57	(S)		9
Analcite					
Black Jack sill, Australia	7	5—10	7 (S)		4

References: 1. DeVore (1955). 2. O'Hara (1961). 3. Marakushev and Polin (1961). 4. Wilkinson (1959). 5. Wager and Mitchell (1951). 6. Pearson and Shaw (1960). 7. Hietanen (1956). 8. Herz and Dutra (1964). 9. Borisenok and Saukov (1960).

variation of gallium content over a wide range of anorthite content (Howie, 1955; Sen et al., 1959; Wilkinson, 1959; Iida, 1961), although Wager and Mitchell (1951) found marked enrichment in the latest plagioclase. Hall (1967) found the gallium concentration in feldspars of granite complexes to be highest for the rocks of highest silicon content.

Enhanced concentrations of gallium in pegmatite minerals, as compared with those in minerals of associated rocks, occur to variable extents. In pegmatites associated with granitic rocks little difference in gallium content was found in feldspars by Shimer (1943) or in biotites and feldspars by Borisenok and Saukov (1960); Ganeev et al. (1961) found enhanced levels in pegmatite biotites but not in feldspars. Enrichment of gallium in pegmatite minerals has been found to occur, however, in alkaline pegmatites and this may be characteristic (Borisenok, 1959; Gerasimovskii et al., 1959; Borisenok and Saukov, 1960). Borisenok and Saukov (1960)

Table 31-D-6. *Gallium in oxide minerals*[a]

Mineral and origin	No. of samples	Ga (ppm) range	mean (method)	Reference
Quartz[b]				
Various sources	5	0.008—2.0	1.0 (C)	BURTON et al. (1959)
Granitic rocks, U.S.S.R.	5	2—6	3.4 (S, C)	BORISENOK and TAUSON (1959); SITNIN (1960); VERSHKOVSKAYA and SALTYKOVA (1961)
Detrital in soil, N. Kazakhstan	ca. 200	upto 10	(S)	DOBROVOL'SKII (1959)
Magnetite				
S. California batholith	12	50—120	88 (S)	SEN et al. (1959)
Charnockite series, Madras, India	14	10—120	57 (S)	HOWIE (1955)
Granitoids, Azov region, U.S.S.R.	13	20—120	60 (L)	MARCHENKO and SHCHERBAKOV (1966)
Titanomagnetite				
Black Jack sill, Australia	7	70—80	77 (S)	WILKINSON (1959)

[a] Results for bauxite are given in Table 31-G-1.
[b] See also Table 31-D-7.

emphasise the occurrence of gallium enrichment in situations where volatile components were important in the processes of formation. The importance of this factor is stressed by the valuable study of feldspars by WALENCZAK (1959) who found that the concentration of gallium and its ratio to aluminium were highest in feldspars of pneumatolytic pegmatites and that gallium is lower in feldspars of later hydrothermal stages (see Table 31-D-1). WALENCZAK (1966a, b) has also studied the occurrence of gallium in various types of quartz (Table 31-D-7). The content is usually very low in quartz but remarkably enhanced concentrations occur in amethysts and gallium enters here independently of aluminium. In mica pegmatites, BORISENOK and RYABCHIKOV (1962) found that muscovites of earlier crystallization were richer in gallium than later muscovites. Some residual phases, such as analcite and pollucite, are surprisingly low in gallium and while predictions based on the properties of gallium and aluminium already discussed are often broadly confirmed, more complex factors enter into many situations. For example, the chalcophile behavior of gallium may dominate in hydrothermal processes (see Section 31-F). The impoverishment of gallium in certain high temperature minerals such as cryolite and cryolithonite may be due to the contrasting behavior of gallium and aluminium in the presence of fluoride as the elements show differences in volatility of the fluorides and the mobility of their complex fluorides in acid solutions (BORISENOK and SAUKOV, 1960). Low levels of gallium in topaz may be explained by similar effects.

Table 31-D-7. *Gallium in quartz from central Europe*[a]

Origin and type	Number of samples	Ga (ppm)		$\dfrac{Ga \times 10^3}{Al}$	
		range	mean	range	mean
Geodes in melaphyres					
Colourless quartz	4	0.02—0.08	0.04	0.18— 3.6	1.5
Amethyst	4	0.11—0.35	0.23	6 —18	13
Metamorphic rocks					
Colourless quartz	8	0.05— 2	0.46	0.11— 1.4	0.74
Amethyst	4	12 —22	18	75 —185	132
Hydrothermal veins in andesites					
Colourless quartz	7	0.08— 0.80	0.41	0.17— 4.6	1.9
Amethyst	7	4.4 —16	7.7	36 — 87	64
Granites and pegmatites					
Quartz	24	0.02—0.11[b]	0.06	0.09— 0.28[b]	0.19
Milky quartz[c]	7	0.06— 0.45	0.27	0.20— 0.71	0.37
Amethyst	7	17 —25	21	94 —280	182

[a] Summary of results (S, C) of WALENCZAK (1966a, b).
[b] Average values from different sources.
[c] Crystallizing before amethyst.

Studies of co-existing minerals in a variety of rocks show that the concentration of gallium and its ratio to aluminium in plagioclase is usually at least twice that in potassium feldspar (SEN et al., 1959; WALENCZAK, 1959; BORISENOK and SAUKOV, 1960). Orthopyroxenes have ratios of gallium to aluminium which are about twice those for co-existing clinopyroxenes (HOWIE, 1955; PHILPOTTS, 1966). The results of DEVORE (1955) show that garnets are generally depleted in gallium relative to the common rock-forming minerals; the concentration of gallium in garnet is about three to four times lower than in co-existing plagioclase in basic gneisses (O'HARA, 1961).

Biotites from a variety of rocks have higher ratios of gallium to aluminium than associated plagioclases and this has been attributed to the readier substitution of gallium for aluminium when the latter is in six-fold rather than four-fold coordination (BELL, 1955; BORISENOK and TAUSON, 1959). BORISENOK and ZLOBIN (1959) suggest that this is also the reason that there is a lower amount of gallium in late crystallizing nepheline in alkaline rocks as compared with co-existing biotite, despite the high aluminium content of the nepheline; they emphasize the importance of the same factor in influencing the gallium content of the hornblende and augite. The opposite view has been put forward by MARAKUSHEV and POLIN (1961) to explain the distribution of gallium between muscovite and biotite; biotite shows lower concentrations of gallium but higher ratios of gallium to aluminium than co-existing muscovite. Certainly it appears that preferential substitution for six-coordinated aluminium is not an over-riding factor. Hornblende, while relatively poor in aluminium, contains more six-coordinated aluminium than does biotite, yet for co-existing pairs the biotite is generally considerably richer in gallium and sometimes also shows the higher ratio of gallium to aluminium (DeVore, 1955; Moxham, 1965).

A relationship with iron (Fe^{3+}) is shown by the distribution of gallium in certain minerals. High levels of gallium commonly occur in magnetite. The gallium content of magnetites remained fairly constant during differentiation of basic magma in the Skaergaard intrusion (Wager and Mitchell, 1951) but rose in the magnetites of later differentiating rocks of the Madras charnockite series (Howie, 1955), and the Mt. Wellington dolerite sill (Tiller, 1959). In teschenites, Wilkinson (1959) found that titanomagnetites contained about four times as much gallium as the associated plagioclases. Iron oxides other than magnetite do not show notable concentrations of gallium. The views have been advanced that substitution for iron (Fe^{3+}) is an important factor leading to enhanced concentrations of gallium in biotite (Yeremenko et al., 1963) and aegirine (Borisenok and Saukov, 1960). Substitution of gallium for iron (Fe^{2+}) and magnesium has been advanced as an explanation for unusually high ratios of gallium to aluminium in aluminium deficient minerals such as hedenbergite and anthophyllite (Bell, 1955) and diopside and olivine (Borisenko, 1963). There is, however, no other evidence of coherence of gallium with these ions and an alternative explanation may lie in the readier entry of gallium into silicate tetrahedra on account of its somewhat greater electronegativity (Borisenok, 1959).

Among the less abundant minerals notable concentrations of gallium have been found in spinel and crysoberyl.

II. Gallium Minerals

The only two known independent minerals of gallium are the very rare sulfide of tetragonal structure, gallite ($CuGaS_2$), which has been found in two ore deposits in South-West and Central Africa (Strunz et al., 1958) and the hydroxide, soehngeite ($Ga(OH)_3$) (Strunz, 1965). Gallite occurs as grains or aggregates in association with other sulfides and this probably accounts for earlier findings of unusually high amounts of gallium in germanite.

Revised manuscript received: October 1970

31-E. Abundance in Common Igneous Rock Types

Results for the occurrence of gallium in igneous rocks are summarised in Tables 31-E-1 to 31-E-4. The main feature of the distribution of gallium in igneous rocks is its rather uniform occurrence in most mafic, intermediate and granitic rocks. The low dispersion is related to the role of feldspars as the main carriers of gallium in these rocks. Tendencies for increases in gallium concentration in late crystallizing minerals tend to be minor (see Section 31-D) and are offset by the low gallium content of quartz. In addition to the feldspars, micas contribute significantly to the gallium content and magnetite can also be important.

Variation in gallium content beyond the rather narrow range of the commoner rocks is limited mainly to ultramafic and alkaline types. Almost all ultramafic rocks are poor in gallium because of the low levels of the element in olivine and augite and the small amounts of feldspars and micas in the rocks. Furthermore the plagioclases present in ultramafic rocks contain less gallium than those of similar composition in mafic rocks (BORISENOK, 1959). Some ultramafic rocks, however, have gallium concentrations more typical of later differentiates. This applies to some of

Table 31-E-1. *Gallium in ultramafic rocks*

Type and origin	No. of samples	Ga (ppm) range	Ga (ppm) mean (method)	$\dfrac{Ga \times 10^3}{Al}$ range	$\dfrac{Ga \times 10^3}{Al}$ mean	Reference
Dunite						
U.S.S.R.	10		1 (S)		0.11	1
Urals, U.S.S.R.	25	upto 5	(C)			2
Garabal Hill — Glen Fyne complex, Scotland	1		<1 (S)			3
Insch complex, Scotland	4	0.5—0.5	0.5 (S)	0.01—0.03	0.02	4
Peridotite						
Central Africa	2	<1—5	(S)	<0.05—0.14		5
U.S.S.R.	20		3 (S)		0.11	1
Urals, U.S.S.R.	6	3—6	4.5 (C)			2
Scotland	1		10.5 (C)		0.38	6
Garabal Hill — Glen Fyne complex, Scotland	1		<1 (S)			3
Insch complex, Scotland	1		1 (S)		0.06	4

Gallium

Table 31-E-1 (Continued)

Type and origin	No. of samples	Ga (ppm) range	Ga (ppm) mean (method)	$\text{Ga} \times 10^3 / \text{Al}$ range	$\text{Ga} \times 10^3 / \text{Al}$ mean	Reference
Harzburgite						
Unstated	1		4 (L)		0.15	7
Norite-harzburgite, Rhodesia	1		1.3 (C)			6
Lherzolite						
S. Africa	1		1.0 (C)			6
Kimberlite						
Basutoland	14	<3—30	<9 (S)	0.19—0.41		8
U.S.S.R.	10		1 (S)		0.10	1
Pyroxenite						
Central Africa	4	10—30	20	0.39—0.97	0.68	5
Charnockite series, Madras, India	2	5—5	5 (S)	0.11—0.19	0.15	9
U.S.S.R.	20		4 (S)		0.10	1
Urals, U.S.S.R.	71	3—28	(C)	0.21—0.78	0.48	2
Garabal Hill — Glen Fyne complex, Scotland	1		5 (S)		0.17	3
Troctolite						
Insch complex, Scotland	2	1—1	1 (S)	0.02—0.03	0.02	4
Serpentinite						
U.S.S.R.	15		1 (S)			1
Lower Silesia, Poland	2	2—7.5	5 (S)	0.30—0.41	0.36	10
Various ultramafic types						
N. America	6[a]		1.5 (L)		ca. 0.15	7
U.S.S.R.		1—3[b]	2 (S)	0.10—0.11[b]	0.10	11
Ireland	6	7—38	18 (S)	0.20—0.59	0.36	12

Note: Data on some rare ultramafic rocks of Central Africa are given by HIGAZY (1954a).

References: 1. BORISENOK (1959). 2. BORISENKO (1963). 3. NOCKOLDS and MITCHELL (1948). 4. READ and HAQ (1963). 5. HIGAZY (1954a). 6. BURTON et al. (1959). 7. SANDELL (1949). 8. DAWSON (1962). 9. HOWIE (1955). 10. WALENCZAK and PENDIAS (1958). 11. BORISENOK and SAUKOV (1960). 12. EVANS (1964).

[a] Analysed as composite sample.
[b] Range given is that of regional averages.

Gallium

Table 31-E-2. *Gallium in mafic rocks*

Type and origin	No. of samples	Ga (ppm) range	Ga (ppm) mean (method)	$\frac{Ga \times 10^3}{Al}$ range	$\frac{Ga \times 10^3}{Al}$ mean	Reference
Gabbro and related rocks						
Gabbro, Paresis complex, S. W. Africa	2	35—37	36 (S)	0.40—0.42	0.41	1
Gabbro-diorite, California, U.S.A.	4[a]		17 (L)		0.17	2
Anorthosite, U.S.A.	3	19—21	20 (L)	0.12—0.15	0.13	2
Norite and jotunite, S. Quebec, Canada	12	10—30	19 (S)	0.10—0.37	0.23	3
Olivine and quartz gabbros, Falkland Islands	3	20—30	25 (S)	0.19—0.30	0.26	4
Gabbroic rock, U.S.S.R.	97		14 (S)		0.20	5
Gabbro, Scotland	10	10—22	17 (S)	0.12—0.28	0.21	6
Gabbro-picrite, Skaergaard intrusion, E. Greenland	2	5—10	8 (S)		0.22	7
Gabbro and ferrogabbro, Skaergaard intrusion, E. Greenland	20	10—30	18 (S)	0.13—0.61	0.25	7
Gabbro, Basistoppen sheet, E. Greenland	5	23—35	30 (S)			8
Basalt and related rocks						
Basalt, Paresis complex, S. W. Africa	3	27—29	28 (S)	0.34—0.44	0.37	1
Basalt, Kenya and Hawaii	5	12—19	17 (C)	0.25—0.31	0.29	9
Basaltic rocks, N. America and Hawaii	139		17 (S)		0.20	10
Basalt, Hawaii	26	13—30	22 (S)	0.24—0.44	0.32	11
Basalt, Michigan, U.S.A. (composite sample of Greenstone flow)	1		17.5 (L)		0.20	2
Basalt ("diabase"), Minnesota, U.S.A.	2	15.5—19	17 (L)	0.14—0.24	0.19	2
Basalt ("diabase"), Ontario, Canada	57	12—26	17 (S)			12
Tholeiitic basalt ("dolerite",) Great Lake sheet, Tasmania	34	6—31	14 (S)	0.09—0.51	0.18	13
Basalt ("dolerite"), Mt. Wellington sill, Tasmania	20	16—30	22 (S)			14
Basaltic rocks, U.S.S.R.	23		18 (S)		0.21	5
Basalt and spilite, Lower Silesia, Poland	5	16—20	18 (S)	0.21—0.30	0.26	15
Olivine alkali basalt, N. W. Germany	10		15 (S)			16
Spilite, Switzerland	5	6—18	13 (S)	0.08—0.19	0.13	17
Basalt, E. Greenland	3	20—20	20 (S)	0.27—0.28	0.28	7
Basalt, Sea floor, Gulf of Aden	7	15—15	15 (X)	0.16—0.18	0.18	18

References: 1. SIEDNER (1965). 2. SANDELL (1949). 3. PHILPOTTS (1966). 4. ADIE (1955). 5. BORISENOK (1959). 6. NOCKOLDS and MITCHELL (1948); READ and HAQ (1963). 7. WAGER and MITCHELL (1951). 8. DOUGLAS (1964). 9. BURTON et al. (1959). 10. RADER et al. (1963). 11. WAGER and MITCHELL (1953); NOCKOLDS and ALLEN (1956). 12. FAIRBAIRN et al. (1953). 13. GREENLAND and LOVERING (1966). 14. TILLER (1959). 15. WALENCZAK and PENDIAS (1958). 16. WEDEPOHL (1961). 17. AMSTUTZ (1953). 18. CANN (1970).

[a] Analysed as composite sample.

Table 31-E-3. *Gallium in intermediate and granitic rocks*

Type and origin	No. of samples	Ga (ppm) range	Ga (ppm) mean (method)	Ga × 10³ / Al range	Ga × 10³ / Al mean	Reference
Diorite						
U.S.A.	2[a]		15 (L)		0.18	1
Quartz diorite, Falkland Islands	7	20—30	24 (S)	0.21—0.29	0.26	2
U.S.S.R.	60		16.5 (S)		0.18	3
Dzhida complex, U.S.S.R.	8	6—13	11 (S)	0.08—0.13	0.11	4
Scotland	10	15—25	18 (S)	0.16—0.24	0.20	5
Olivine diorite and related rocks, Kûngnât complex, S. Greenland	6	12—29	20 (S)	0.13—0.47	0.23	6
Syenite						
Paresis complex, S. W. Africa	3	31—37	33 (S)	0.34—0.38	0.36	7
S. Africa	2	15—17	16 (S)			8
U.S.A.	2[a]		22 (L)		0.25	1
U.S.S.R.	76		18 (S)		0.20	9
Insch complex, Scotland	6	13—20	17 (S)	0.13—0.20	0.18	10
Kûngnât complex, S. Greenland	18	16—40	24 (S)	0.20—0.50	0.31	6
Andesite						
Hawaii	5	20—25	21 (S, C)	0.21—0.29	0.24	11
U.S.S.R.	13		14 (S)		0.18	3
Trachyte						
N. W. Germany	4		17 (S)			12
Trachyte and mugearite, Skye, Scotland	3	20—25	23 (S)			13
Tonalite						
Tanzania (Standard T-1)	1		21 (S)		0.24	14
Susamyr batholith, U.S.S.R.	1		20 (S)		0.21	15
Monzonite						
U.S.S.R.	10		23 (S)		0.30	3
Mangerite						
S. Quebec, Canada	9	15—25	22 (S)	0.19—0.37	0.29	16
Granitic rocks						
Various regions (rocks with > 65% SiO$_2$)	140	1—50	16 (S)			17
Various regions	64[b]	15—18.5	16 (L)	0.19—0.30	0.23	1
Granite, various sources	29	10.5—43	18.5 (C)			18
"Younger acidic" rocks, Africa	62	15—100	35 (S)		0.53	19[c]

[a] Includes composite.
[b] Analysed as 7 composite samples.
[c] The data incorporate additional information supplied by J. M. ROOKE.

Gallium

Table 31-E-3 (Continued)

Type and origin	No. of samples	Ga (ppm) range	mean (method)	Ga × 10³ / Al range	mean	Reference
"Older acidic" rocks, Africa	68	5—85	24 (S)		0.32	19c
Granite, Nigeria	38	18—60	34 (C)	0.26—0.83	0.48	20
Granite and other "acidic" rocks, S. Africa	10	10—22	18 (S)			8
Granite, Canada	32	13—22	17 (S)			21
Granitic rocks, Brazil	39	7—24	16 (S)	0.21—0.37		22
Granite, Japan	8	16—20	18 (L)			23
Granite, New South Wales, Australia	35	12—23	(S)			24
Granite and other "acidic" rocks, U.S.S.R.	355	15—23 d	19 (S)	0.20—0.32 d	0.26	25
Granite and pegmatite, Azov Region, U.S.S.R.	14	13—18	15 (L)	0.15—0.28	0.20	26
Granodiorite, adamellite and granite, Dzhida complex, U.S.S.R.	21	11—39	19 (S)	0.13—0.57	0.26	4
Granodiorite and granite, Lower Silesia, Poland	15	14—18	15 (S)	0.19—0.25	0.21	27
Granitic rocks, S. W. England and Haig Fras	5	19—22	21	0.24—0.28	0.27	28
Granodiorite, adamellite, granite and aplite, Scotland	15	15—25	16 (S)	0.15—0.38	0.19	5
Feldspar rhyolite, Paresis complex, S. W. Africa	5	16—22	19 (S)	0.24—0.27	0.26	7
Comendite, Paresis complex, S. W. Africa	9	25—38	32 (S)	0.40—0.59	0.49	7

References: 1. SANDELL (1949). 2. ADIE (1955). 3. BORISENOK (1959). 4. PETROVA and KUSAKINA (1965). 5. NOCKOLDS and MITCHELL (1948). 6. UPTON (1960). 7. SIEDNER (1965). 8. EDGE and AHRENS (1962). 9. BORISENOK (1959); PETROVA and KUSAKINA (1965). 10. READ and HAQ (1963). 11. WAGER and MITCHELL (1953); BURTON et al. (1959). 12. WEDEPOHL (1961). 13. WAGER and MITCHELL (1953). 14. THOMAS (1963). 15. BORISENOK and TAUSON (1959). 16. PHILPOTTS (1966). 17. FLEISCHER (1955). 18. BURTON et al. (1959). 19. ROOKE (1964). 20. BOWDEN (1964). 21. AHRENS (1954). 22. HERZ and DUTRA (1960). 23. NISHIKAWA (1958). 24. KOLBE (1965). 25. BORISENOK (1959); BORISENOK and SAUKOV (1960). 26. MARCHENKO and SHCHERBAKOV (1966). 27. WALENCZAK and PENDIAS (1958). 28. EXLEY (1966).

d Range given is that for regional or formation averages.

the kimberlites examined by DAWSON (1962) and some of the pyroxenites analysed by BORISENKO (1963). In the pyroxenites wide variations in gallium content were clearly related to the variations in the minerals present, the higher gallium levels being due to increased amounts of amphibole and titanomagnetite. Enriched levels of gallium often occur in alkaline rocks although this is not an invariable character-

Table 31-E-4. *Gallium in alkaline rocks*

Type and origin	No. of samples	Ga (ppm) range	Ga (ppm) mean (method)	$\frac{Ga \times 10^3}{Al}$ range	$\frac{Ga \times 10^3}{Al}$ mean	Reference
Alkali syenite, U.S.S.R.	15		24 (S)		0.30	1
Nepheline syenite, Arkansas, U.S.A.	3	20—20	20 (S)			2
Nepheline syenite, U.S.S.R.	70		47 (S)		0.43	1
Nepheline syenite, Lovozero Massif, U.S.S.R.	9	30—50	40 (C)	0.35—0.46	0.39	3
Nepheline syenite, S. Greenland	6	20—50	32 (S)	0.22—0.47	0.32	4
Foyaite, U.S.S.R.	4		24 (S)			1
Foyaite, Lovozero Massif, U.S.S.R.	5	60—76	70 (C)	0.77—0.79	0.78	3
Phonolite, Africa	4[a]		22 (L)		0.23	5
Trachyphonolite, Lower Silesia, Poland	1		19 (S)		0.19	6

Note: Extensive data for rarer alkaline rock types are given by GERASIMOVSKII *et al.* (1959) and YEREMENKO *et al.* (1963). GERASIMOVSKIY *et al.* (1968) give a range of values of 10—140 ppm (S) for nepheline syenites of various types from many areas. They conclude that, excluding certain unusually high values, the mean content of gallium in rocks of the nepheline syenite type is 36 ppm.

References: 1. BORISENOK (1959). 2. GORDON and MURATA (1952). 3. GERASIMOVSKII *et al.* (1959). 4. UPTON (1964). 5. SANDELL (1949). 6. WALENCZAK and PENDIAS (1958).

[a] Analysed as composite sample.

istic. The increase in gallium content is especially related to the presence of nepheline, sodalite and cancrinite and is not characteristic of the more mafic types (BORISENOK, 1959).

In studies of related differentiates a tendency has often been noted for the concentration of gallium and its ratio to aluminium to increase during magmatic crystallization (HIGAZY, 1952b; AMSTUTZ, 1953; PATTERSON, 1952, 1953; GERASIMOVSKII *et al.*, 1959; UPTON, 1960; PETROVA and KUSAKINA, 1965; SIEDNER, 1965; DIETRICH and HEIER, 1967). Some aspects of this trend were discussed in Section 31-D. The tendency for the ratio to rise, however, is frequently slight and sometimes undetectable or only detectable in extreme differentiates (NOCKOLDS and MITCHELL, 1948; WAGER and MITCHELL, 1951; COATS, 1952; NOCKOLDS and ALLEN, 1953; ADIE, 1955; HOWIE, 1955; BORISENOK and TAUSON, 1959; BORISENOK and ZLOBIN, 1959; WILKINSON, 1959; BORISENOK and SAUKOV, 1960; DOUGLAS, 1964). In some instances a slight reversal of the general trend is apparent (NOCKOLDS and ALLEN, 1954). Analytical limitations make interpretation difficult in many cases where the trends are not pronounced. PETROVA and KUSAKINA (1965) note that the tendency for gallium to concentrate in late differentiates is more marked with magma of high alkali content and this is generally borne out by the available observations. The advantages of using the ratio $Ga/Al+Fe^{3+}$ in studying the trends of gallium during differentiation have been stressed (BOWDEN, 1964).

BOWDEN (1964) has discussed the difficulty of distinguishing the effects of differentiation and albitization as factors producing high gallium contents in riebeckite granites. High ratios of gallium to aluminium in younger granites of some African provinces were found by ROOKE (1964) to be related to some degree of albitization. The effects of albitization are further discussed in Section 31-N.

On the basis of those investigations in which accurate analyses have been made on a range of samples, the following estimates of the average concentration of gallium may be made: basalts, 17 ppm; granites, 18.5 ppm. These values are satisfactorily consistent with the abundance value for igneous rocks of 18.5 ppm, derived by HORN and ADAMS (1966), and a rounded value of 18 ppm appears at present to be the best estimate of the crustal abundance of gallium.

31-F. Behavior in Magmatogenic Processes

As indicated by the discussion of pegmatite minerals in Section 31-D, there is sometimes enhancement of the concentration of gallium and its ratio to aluminium in pegmatites (BORISENOK and SAUKOV, 1960; GINZBOURG, 1960). This is a particular characteristic of alkaline pegmatites and in other types increased levels of gallium are often not observed, for example in the mica pegmatites investigated by BORISENOK and RYABCHIKOV (1962). Changes in gallium content associated with greisenization and albitization are discussed in Section 31-N.

In hydrothermal processes the chalcophile behavior of gallium is clearly shown by its concentration in sulfide ores (see Table 31-F-1). The main affinity of gallium in these reactions is with zinc and it is in sphalerite that unusually high concentrations of gallium have been widely recorded. The content of gallium in sphalerite is related to both the geological province and the temperature of formation. High gallium concentrations are usually characteristic of low temperature deposits (GABRIELSON, 1945; MORRIS and BREWER, 1954; RIGAULT, 1956; BORISENOK and SAUKOV, 1960). The results of TROSHIN (1962) suggest that the ratio of gallium to indium may be a more characteristic indicator of the type of the deposit than the gallium concentration alone.

While temperature of formation has generally been considered to be an important factor in determining the concentration of gallium in an ore body, FRYKLUND and FLETCHER (1956) have suggested that the concentrations of gallium and most other minor elements cannot act as indicators of temperature of formation because they do not attain equilibrium values; these workers have also criticised the criteria generally used to establish the temperature of formation.

It is difficult to decide whether gallium in sphalerite is substituted in the lattice or not (MORRIS and BREWER, 1954; RIGAULT, 1956, 1957; TROSHIN and TROSHINA, 1965). The hypotheses concerning possible modes of gallium substitution in the sphalerite lattice are discussed by IVANITSKII and GVARAMADZE (1960) who conclude that isomorphic substitution of Ga for Zn in ZnS is most probable. The main evidence advanced in support of the view that the association involves a mechanism other than lattice substitution is the ease with which gallium is dispersed when sphalerite is oxidized (see below).

The features of the distribution of gallium in residual products and the striking departures that may occur from the predicted trend of gallium enrichment are often explicable in terms of the conflicting lithophile and chalcophile tendencies of the element. The latter are particularly important in low temperature hydrothermal conditions especially where associated phases have little affinity for gallium, such as in quartz-fluorite veins (VERSHKOVSKAYA and FABRIKOVA, 1957). This question is extensively considered in the review by VERSHKOVSKAYA (1966) of gallium geochemistry. Among the competing processes, reactions with country rocks must be

Table 31-F-1. *Gallium in sulfide minerals*

Mineral and origin	No. of samples	Ga (ppm) range	mean (method)	Reference
Galena				
Spain and England	2	0.08—0.12	0.10 (C)	1
Sphalerite				
New Mexico and Utah, U.S.A.	187	upto 150	(S)	2
Idaho, U.S.A.	59	upto 60	11 (S)	3
N. Peru	7	upto 5	(S)	4
Naugarzan and Takob deposits, U.S.S.R.	100	10—900	200 (C)	5
Khrustal'noe deposit, U.S.S.R.		2.5—6.4		6
Transbaikaliya, U.S.S.R.,				
low temperature	22	18—200	60 (S)	7
medium temperature	33	0.4—32	7 (S)	7
high temperature	11	0.3—14	4 (S)	7
Europe	71	<6—1,000	(S)	8
N. Alps (lime deposits)	ca. 200	<10—1,600	85 (S)	9
Sweden	64	upto 500	45 (S)	10
Chalcopyrite				
Khrustal'noe deposit, U.S.S.R.		2.5—13		6
Georgia, U.S.S.R.	34		17 (S)	11
Pyrite				
Various sources	3	0.1—0.7	0.5 (C)	1

Note: Results for many rarer sulfide minerals are given by BURTON *et al.* (1959).

References: 1. BURTON *et al.* (1959). 2. ROSE (1967). 3. FRYKLUND and FLETCHER (1956). 4. SIMONS (1955). 5. VERSHKOVSKAYA and SALTYKOVA (1961). 6. VERTSKOVSKAYA (1963). 7. TROSHIN (1962). 8. RIGAULT (1956). 9. FRUTH (1966). 10. GABRIELSON (1945). 11. IVANITSKII and GVARAMADZE (1960).

considered. GARMASH and VLASOVA (1963) examined deposits in which ore formation was accompanied by intense chloritization of country rocks with increase in their gallium content; the extent to which gallium was concentrated in sphalerite was influenced by the amount of sericite formed at the same time. The entry of substantial amounts of gallium into altered country rocks has been observed in other situations (ISHIKAWA *et al.*, 1962; EFENDIEV *et al.*, 1965). KULIKOVA (1964) and MEITUV (1962) have contrasted the dominant tendency of gallium to concentrate in sphalerite occurring in association with carbonate rocks with its tendency to dispersion when sulfide ores are deposited in silicate country rocks. Diminished reactivity with country rocks is probably one factor which favours entry of gallium into low temperature sphalerite (VERSHKOVSKAYA and SALTYKOVA, 1961; TROSHIN, 1962).

The dispersion of gallium during the oxidation of sulfide ores has been noted by several workers (BORISENOK and SAUKOV, 1960; KULIKOVA, 1964, 1966). In Central

Asian deposits KULIKOVA (1966) found that the concentrations of gallium were generally low in the zinc and lead minerals of the oxidized zone; the commonest occurrences were in plumbojarosite (2—50 ppm) and hydrous iron oxides such as hydrogoethite (1—44 ppm). The mobility of gallium in the oxidation process was reflected also in its enhancement in some samples of limestones and skarns from surrounding rocks, the concentration reaching 375 ppm in one sample of limonitized limestone. The importance of precipitation as the hydroxide with other hydroxides such as that of iron, as a factor affecting the distribution of gallium in the oxidized zone, is emphasised by the experimental studies of RAZENKOV and GALAKTIONOVA (1962).

31-G. Behavior during Weathering and Alteration of Rocks

Results are given in Table 31-G-1 for the occurrence of gallium in soils and related materials. Gallium, like aluminium, is enriched in the products of intense weathering; it is more mobile than aluminium, however, and the ratio of gallium to aluminium commonly decreases in the residual materials in the weathering processes. This tendency is shown by the results of Butler (1953b, 1954a) for weathering of a number of rock types and has been observed in the formation of bauxite from andesite (Wolfenden, 1965) and to a lesser degree in laterite formation (Burton et al., 1959). However, Chowdhury et al. (1965) found variable trends in the ratio in bauxite formation and Gordon and Murata (1952) found the ratio to be increased in the formation of bauxite from nepheline syenite. Evidence of variable trends in the ratio during the weathering of a variety of rocks is given by Ronov and Migdisov (1965). These contrasting trends may be associated with differences in the rate of breakdown of gallium-bearing minerals in relation to the rate of formation of residual structures in which gallium becomes incorporated.

Although there is some separation of gallium and aluminium in weathering processes, the coherence between the elements remains a major factor in determining the distribution of gallium, and high correlations of their concentrations in the clays and bauxites of individual formations have been found (Migdisov and Borisenok, 1963; Ronov and Migdisov, 1965). Burton et al. (1959) noted the lack of coherence between gallium and iron (Fe^{3+}) in a laterite profile, but Turton et al. (1962) found the elements to be associated in a similar situation and Lavrenchuk and Tenyakov (1962, 1963b) found preferential association of gallium with magnetic iron minerals in bauxite ores. Wolfenden (1965) found that bauxites occurring under reducing conditions were richer in gallium than similar deposits in non-reducing environments; he suggested that gallium might be precipitated under reducing conditions. In this connection it is notable that Gorham and Swaine (1965) found slightly enhanced levels of gallium in reduced lake muds compared with oxidized muds.

The general tendency of gallium to be concentrated in residual materials, even though generally to a lesser degree than aluminium, is reflected in the fact that in many soil profiles the concentration of gallium is positively correlated with the amount of clay present (McKenzie, 1957; LeRiche and Weir, 1963; Giles, 1964). The usual trend in soil profiles is for an increase in gallium content with depth (Butler, 1954b; Connor et al., 1957; McKenzie, 1957; Dobrovol'skii, 1959; LeRiche and Weir, 1963; Oertel and Giles, 1963, 1964; Giles, 1964). A pronounced opposite trend appears to have been found only in the gravel fraction of some gravelly laterites (Turton et al., 1962). Trends with depth in the ratio of gallium to aluminium are too variable for generalizations to be made.

Table 31-G-1. *Gallium in laterites, bauxites and soils*

Type and origin	No. of samples	Ga (ppm) range	Ga (ppm) mean (method)	$\frac{Ga \times 10^3}{Al}$ range	$\frac{Ga \times 10^3}{Al}$ mean	Reference
Laterite, N. Nigeria	11	22—27	25 (C)	0.29—0.32	0.31	1
Soil, U.S.A.	28	1—25	10 (S)	0.08—0.51	0.20	2
Kaolinitic clay, Malaysia	5	20—25	23 (S)	0.13—0.15	0.14	3
Soil, S. and S. W. Australia	81	6—77	35 (S)			4
Soil, Queensland	320	<5—49	22 (S)	0.14—0.90	0.40	5a
Soil, New Zealand and S. W. Pacific Is.	134	6.4—39[b]	26 (S)	0.15—0.43[b]	0.30	6
Soil, N. Kazakhstan	ca. 200		43 (S)			7
Soil, Britain	62	8—70	33 (S)			8
Bauxite and bauxitic clay, Arkansas, U.S.A.	14	50—100	86 (S)			9
Bauxite, India	28	5—122	60 (C)	0.12—4.3	2.0	10
Bauxite, Malaysia	14	22—47	33 (S)	0.07—0.18	0.12	3
Bauxite, Russian platform	9[c]		54 (S)		0.22	11
Bauxitic clay, Kazakhstan	135	20—87	44 (C)		0.25	12
Bauxite, Kazakhstan	6	45—70	54 (C)	0.16—0.31	0.23	13
Bauxite, various sources (mainly U.S.S.R.)[d]	2,298	10—170	(C)			14

References: 1. BURTON et al. (1959). 2. CONNOR et al. (1957); PRINCE (1957b). 3. WOLFENDEN (1965). 4. MCKENZIE (1957); TURTON et al. (1961, 1962). 5. GILES (1959, 1964); OERTEL and GILES (1959). 6. WELLS (1960). 7. DOBROVOL'SKII (1959). 8. BUTLER (1954b); SWAINE and MITCHELL (1960). 9. GORDON and MURATA (1952). 10. CHOWDHURY et al. (1965). 11. MIGDISOV and BORISENOK (1963). 12. LAVRENCHUK and TENYAKOV (1963a). 13. LAVRENCHUK and TENYAKOV (1962, 1963b). 14. TENYAKOV (1968).

[a] The data incorporate additional values supplied by J. B. GILES. The concentration values refer to the total mineral matter but are approximately the same on a dry weight basis.

[b] Range given is that of averages for soils on different rock types and of different organic content.

[c] Composite samples.

[d] These samples represent more than 70 deposits. On the basis of these results and those of other workers, TENYAKOV (1968) estimates the world average concentration of gallium in bauxite as 52 ppm.

31-H. Solubilities of Compounds which Control Concentrations of the Element in Natural Waters; Valence States in Natural Environments

LATIMER (1952) gives a value of $\varepsilon°$ of $ca.$ $+0.65$ volt for the reaction

$$Ga^{2+} = Ga^{3+} + e$$

The element exists as Ga^{3+} in natural environments and the most important factor affecting its behavior in the cycle of weathering and sedimentation is the low solubility of the hydroxide. According to LATIMER (1952), the values of the relevant equilibrium constants are:

$$Ga(OH)_{3(s)} = Ga^{3+} + 3OH^- \quad K = ca.\ 5 \cdot 10^{-37}$$

$$H_3GaO_{3(s)} = H_2GaO_3^- + H^+ \quad K = ca.\ 1 \cdot 10^{-15}$$

Gallium hydroxide is a stronger acid and a weaker base than is aluminium hydroxide. It is completely precipitated at a pH of 3.4 as compared with 4.1 for aluminium hydroxide. This difference may be important in allowing some separation of gallium and aluminium in processes such as the oxidation of sulfide ores (RAZENKOV and GALAKTIONOVA, 1962).

31-I. Abundance in Natural Waters

The small amount of information available on the occurrence of gallium in natural waters is summarised in Table 31-I-1. Gallium is only slightly mobile in most natural waters. The useful study of HEIDE and KÖDDERITZSCH (1964) showed the ratio of gallium to aluminium in filtered Saale river water to average 1.4×10^{-4} over 12 months; this value is similar to the average ratio for the lithosphere. The corresponding ratio in sea water is probably not less than 3×10^{-3} (BURTON et al., 1959). From these limited data it appears that gallium may accumulate in sea water relative to aluminium, as was predicted by GOLDSCHMIDT (1954).

Table 31-I-1. *Gallium in natural waters*

Type and origin	No. of samples	Ga (ppb) range	mean (method)	Reference
River water				
Saale and Elbe, Germany	8	0.05—0.14	0.09 (C)	HEIDE and KÖDDERITZSCH (1964)
Saale, Germany (12 month mean)	—		0.07 (C)	HEIDE and KÖDDERITZSCH (1964)
Sea water				
N. Atlantic, Irish Sea, English Channel	9	0.02—0.05	0.03 (C)	BURTON et al. (1959)
N.W. Pacific	2	0.015—0.02	0.02 (L)	ISHIBASHI et al. (1961)
Groundwater				
Siberian platform, U.S.S.R.	130		0.27	KONTOROVICH et al. (1963)
Oil-field water				
Apsheron Peninsula, U.S.S.R.	15	0.3—0.8	0.5 (C)	EFENDIYEV and SHIK (1965)
Subsurface water				
Bulgaria	—	0.01—100[a]	(S)	PENTCHEVA (1964)
Hot-spring water				
Bath, England	1		0.63 (C)	RILEY (1961)
U.S.S.R.	—	upto 2	—	KRAINOV (1961)
Japan:				
acid (pH < 4)	⎫	0.52—72	⎫	UZUMASA and NASU (1960)
neutral (pH 4—8)	⎬ 98	0.11—5.0	⎬ 2.5 (L)	UZUMASA and NASU (1960)
alkaline (pH > 8)	⎭	0.19—5.4	⎭	UZUMASA and NASU (1960)

[a] High concentrations were associated with nitrogenous thermal waters. PENTCHEVA (1965) reports a concentration of 1.8 ppb Ga (S) in a sample of saline underground water in N. Bulgaria. SILVEY (1967) detected gallium in only 5 of 72 samples of spring waters of California but the concentrations reported for those samples ranged from 5 to 13 ppb Ga (S).

31-K. Abundance in Common Sediments and Sedimentary Rock Types

Information on the occurrence of gallium in the main types of sedimentary rocks is summarised in Table 31-K-1. Results for modern sediments are given in Table 31-K-2.

Particular interest is attached to gallium as a diagnostic element for the salinity of depositional environments, its concentration being generally higher in fresh water than in marine argillaceous deposits (DEGENS et al., 1957, 1958; MIGDISOV and BORISENOK, 1963; SHIMADA, 1963), although TOURTELOT (1964) found a definite reduction in gallium content only in shales deposited in an *offshore* marine environment. The gallium content is not by itself an adequate indicator of depositional environment, but is of value in association with those of other indicator elements such as rubidium and boron. POTTER et al. (1963) found that gallium was more abundant in marine argillaceous sediments than in freshwater deposits, contrary to the above findings.

There is little differentiation of gallium and aluminium in argillaceous sediments as compared with crustal material, the slightly higher concentration of gallium in clays and shales corresponding to an increased aluminium content. Near-shore and pelagic marine clays show similar concentrations of gallium. HIRST (1962) noted the persistence of the coherence of gallium and aluminium from sand to clay in the sediments of the Gulf of Paria. He suggests that kaolinite may be an important carrier of gallium in weathering and transportation; this view is supported by the conclusions of DEGENS et al. (1957). NICHOLLS and LORING (1962) suggested that in Carboniferous

Table 31-K-1. *Gallium in sedimentary rocks*

Type and origin	No. of samples	Ga (ppm) range	Ga (ppm) mean (method)	$\frac{Ga \times 10^3}{Al}$ range	$\frac{Ga \times 10^3}{Al}$ mean	Reference
Arenaceous sedimentary rocks						
Greywacke, N. America and Europe	75	upto 36	9 (S)			1
Arkose, New Zealand	8	8—20	13 (S)	0.09—0.21	0.15	2
Sand and silt, Russian platform	421[a]		8		0.21	3
Lias sand, S. England	35	3.3—22	11 (S)	0.18—0.37	0.25	4
Sandstone and greywacke, British Isles	6	0.8—15	7.3 (C)			5

[a] Composites, comprising 6217 samples.

Table 31-K-1. (Continued)

Type and origin	No. of samples	Ga (ppm) range	Ga (ppm) mean (method)	$\frac{Ga \times 10^3}{Al}$ range	$\frac{Ga \times 10^3}{Al}$ mean	Reference
Argillaceous sedimentary rocks						
Fresh and brackish water shale, W. Pennsylvania, U.S.A.	45	2—55	16 (S)			6
Marine shale, W. Pennsylvania, U.S.A.	30	<2—40	7 (S)			6
Freshwater clay and shale, U.S.A.	13	9—25	16 (S)			7
Marine clay and shale, U.S.A.	20	10—48	25 (S)			7
Shale and claystone, Western U.S.A.	107	6—24	18 (S)			8
Shale, low-grade pelite and slate, New Hampshire, U.S.A.	13	9—35	21 (S)	0.17—0.35	0.27	9
Marlstone and bentonite, Western U.S.A.	16		17 (S)			10
Shale, Japan	1[b]		14 (S)			11
Clay and shale, Russian platform	511[c]		18		0.21	2
Shale, Europe	1[d]		23 (S)			11
Shale and mudstone, British Isles	64	3—40	22 (S, C)	0.11—0.44	0.21	12
Black shale, Nod Glas sediments, N. Wales	4		15 (S)			13
Carbonate sedimentary rocks						
Dolomite and aragonite-magnesite, Utah, U.S.A.	4	2—4	3 (S)	0.31—0.43	0.35	14
Limestone, Japan	3		<1 (L)			15
Limestone and dolomite (some impure), Scotland	276	<3—30	<3 (S)			16
Limestone and dolomite, British Isles	6	0.02—0.10	0.06 (C)			5
Siliceous and phosphatic sedimentary rocks						
Chert, Wales	1		2.4 (C)			5
Phosphorite, Phosphoria formation, U.S.A.	60	upto 10	<10 (S)	0.46—1.18	0.76	17
Phosphorite, Nod Glas sediments, N. Wales	6	3—5	3 (S)			13

References: 1. WEBER and MIDDLETON (1961). 2. McLAUGHLIN (1955). 3. RONOV and MIGDISOV (1965). 4. CAMERON (1957). 5. BURTON et al. (1959). 6. DEGENS et al. (1957, 1958). 7. POTTER et al. (1963). 8. TOURTELOT (1964). 9. SHAW (1954, 1956). 10. TOURTELOT et al. (1960). 11. WEDEPOHL (1960). 12. SPENCER (1957); BURTON et al. (1959); NICHOLLS and LORING (1962); CURTIS (1969). 13. CAVE (1965). 14. GRAF et al. (1961). 15. NISHIKAWA (1958). 16. MUIR et al. (1956). 17. GULBRANDSEN (1966).

[b] Composite of 14 samples.
[c] Composites, comprising 9452 samples.
[d] Composite of 36 samples.

Gallium

Table 31-K-2. *Gallium in recent sediments*

Type and origin	No. of samples	Ga (ppm) range	Ga (ppm) mean (method)	$\frac{Ga \times 10^3}{Al}$ range	$\frac{Ga \times 10^3}{Al}$ mean	Reference
Marine sediments:						
Argillaceous deposits						
Pelagic clay, Pacific Ocean	21	10—31	19 (S)	0.11—0.36	0.20	1
Pelagic clay, Atlantic Ocean	16	15—39	21 (S)			2
Pelagic clay and mud	18	14—28[a]	19 (C)	0.17—0.28	0.22	3
Clay	14	12—29	20 (S)			4
Calcareous deposits						
Globigerina ooze, Pacific Ocean	1		13[b] (S)		0.33	1
Calcareous ooze	7	3—20	10 (C)	0.07—0.92	0.36	5
Siliceous deposits						
Diatomaceous sediment[c], Pacific Ocean	2	16—17	16 (S)	0.28—0.33	0.30	1
Diatomaceous and radiolarian ooze, S. Atlantic Ocean	7	11—16	14 (C)			6
Ferro-manganese concretions and other deposits						
Pacific Ocean[d]	3	16—22	19 (C)	0.58—1.5	0.97	6
Pacific Ocean	54	2—30	10 (X)		0.35	7
Iron-rich sediment, Atlantis II Deep, Red Sea	1		20 (S)		ca. 3	8
Glauconite		9—12	(L)			9
Fresh water sediments:						
Argillaceous deposits, N. America	19	6—26	14 (S)			4
Oxidate crusts, English Lakes	8	ca. 1—25	13 (S)			10
Mud and glacial clay, English Lakes	16	10—25	17 (S)			10

[a] Includes the average value of 11 sections from one core.
[b] 34 ppm on $CaCO_3$ — free basis; a sample of calcareous ooze contained 9 ppm on this basis.
[c] With volcanic ash.
[d] Results are for acid-soluble fraction.

References: 1. GOLDBERG and ARRHENIUS (1958). 2. WEDEPOHL (1960). 3. WAKEEL and RILEY (1961). 4. POTTER et al. (1963). 5. BURTON et al. (1959); WAKEEL and RILEY (1961). 6. BURTON et al. (1959). 7. MERO (1962). 8. MILLER et al. (1966). 9. ISHIBASHI et al. (1961). 10. GORHAM and SWAINE (1965).

shales and mudstones gallium was carried mainly in illite. The low levels of gallium in sandstones accord with the known depletion of the element in quartz.

In marine ferromanganese concretions the ratio of gallium to aluminium is greater than that in crustal material. This reflects the authigenic nature of most of the material of these concretions and the relatively high ratio of gallium to aluminium in sea water.

Data on carbonate rocks have been reviewed by GRAF (1960). Reported values range from the very low concentrations found by BURTON et al. (1959) to the values for impure types comparable with those for argillaceous deposits. There is abundant evidence that the gallium in carbonate rocks is present almost entirely in clay impurities. It appears that the clay fractions of the carbonate rocks contain enhanced levels of gallium as compared with the content of corresponding fractions of dominantly argillaceous deposits (DEGENS et al., 1958; TOURTELOT et al., 1960).

The most reliable values for the average abundance of gallium in the main sedimentary types are those computed by HORN and ADAMS (1966). Their values (rounded) are: shale, 23; sandstone, 6; carbonate rock, 2.5; pelagic clay, 19; pelagic carbonate, 12 ppm.

Revised manuscript received: October 1970

31-L. Biogeochemistry

Information on the occurrence of gallium in organisms and related materials is summarised in Table 31-L-1. No organism has been found which accumulates gallium to a striking extent. So far as can be judged from these very limited data, the ratio of gallium to aluminium in organisms is not very different from the average ratio in crustal material. CULKIN and RILEY (1958) suggest that in the marine animals that they analysed this was due to derivation of the elements from mud rather than from sea water. In land plants there appears to be a tendency for the ratio to be somewhat higher than in the soils from which the elements are derived (PRINCE, 1957a, b; SHACKLETTE, 1965).

The role of gallium in the growth of micro-organisms has been studied (see STEINBERG, 1945) but earlier suggestions that it is an essential nutrient have not been confirmed (BERTRAND, 1954). Some investigations have been made of the metabolism of gallium in vertebrates. Experimental studies (DUDLEY and LEVINE, 1949; DUDLEY and MADDOX, 1949; DUDLEY et al., 1949) show that inhaled gallium is deposited in the lungs and injected gallium appears primarily in the liver, kidneys and skeletal tissues; uptake from the gut is small. Analyses by BELOZEROV (1966 b) of natural levels in human tissues show the concentrations to be highest in spleen, liver and kidney and lowest in lung and bone. Lung was the only tissue, however, in which KOCH et al. (1956) were able to detect gallium.

Data on the occurrence of gallium in *carbonaceous sediments* are given in Table 31-L-2. These deposits have received much attention because the high concentrations of the element found in coal ashes and flue dusts from coal-gas production are of economic interest. Evidence on the way in which gallium is incorporated in coal is not conclusive. The importance of its association with organic matter has been emphasised by HORTON and AUBREY (1950), RYCZEK (1959) and ZUBOVIC (1966a). Other workers (INAGAKI and YAMAGUCHI, 1958; IDZIKOWSKI, 1959) consider, however, that gallium is not concentrated in organic material and DALTON and PRINGLE (1962) found it to be predominantly associated with the lighter alumino-silicate minerals with some in the organic fraction but little or none in the carbonates, pyrites or heavier clay minerals. Several workers (NICHOLLS and LORING, 1962; ZUBOVIC, 1966b) have stressed that the concentrations of gallium may reflect adsorption and redistribution subsequent to the initial formation of the deposits. ZUBOVIC (1966a) considers that gallium may be bonded to degraded lignin. HYDEN (1961) suggests that gallium in crude oil is present as an organometallic complex.

Gallium

Table 31-L-1. *Gallium in biological and related materials*

Material and origin	No. of samples	Ga (ppm) range	Ga (ppm) mean (method)	$\frac{Ga \times 10^3}{Al}$ range	$\frac{Ga \times 10^3}{Al}$ mean	Reference
Land plants						
Field corn (leaves), New Jersey, U.S.A.	25	0.02—0.32[a]	0.08 (S)	0.21—0.94	0.56	1
Ragweed (*Ambrosia artemisiifolia*), New Jersey, U.S.A.	5	0.21—1.8[a]	0.73 (S)			2
Tree and shrub foliage, U.S.A.	43	upto 0.06[a]	(S)			3
Lichens, ferns, grasses, trees, etc., Finland	—	<3—60[b]	(S)			4
Bryophytes, U.S.A.	38	upto 30[b]	18 (S)		0.43	5
Artemisia scoparia, U.S.S.R.	—	—	10 (S)			6
Marine algae, Irish Sea	14	0.01—0.56[a]	0.15 (C)			7
Human tissues						
(20 organs)	—	upto 6[b]	(S)			8
(10 organs)	—	1.7—29	(C)			9
Human blood	1	—	0.28 (C)			9
Human hair	ca. 130	20—140	ca. 85 (N/R)			10
Rat blood	4		5×10^{-4} (N/R)			11
Marine invertebrates, Irish Sea	21	0.007—0.93[a]	0.14 (C)	0.035—0.63	0.25	7
Mollusc shells, Irish Sea	6	0.008—0.07[a]	0.02 (C)	0.10—0.47	0.32	7
Ascidians	—	upto 3[b]	(S)			12
Planktonic organisms, Black Sea	—	upto 20[b]				13
Foodstuffs, Sverdlovsk region, U.S.S.R	—	0.13—0.66				14
Tap water, Sverdlovsk region, U.S.S.R.	1		0.006			14

References: 1. PRINCE (1957a, b). 2. PRINCE (1957a). 3. HANNA and GRANT (1962). 4. LOUNAMAA (1956). 5. SHACKLETTE (1965). 6. MAKSUDOV et al. (1962). 7. CULKIN and RILEY (1958). 8. KOCH et al. (1956). 9. BELOZEROV (1966b). 10. PERKONS and JERVIS (1966). 11. BOWEN (1959). 12. KOVAL'SKII et al. (1962). 13. VINOGRADOVA and KOVAL'SKIY (1964). 14. BELOZEROV (1966a).

[a] In dry material.
[b] In ash.

Table 31-L-2. *Gallium in carbonaceous sediments*

Material and origin	No. of samples	Ga (ppm) in coal or oil range	Ga (ppm) in coal or oil mean (method)	Ga (ppm) in ash range	Ga (ppm) in ash mean (method)	Reference
Coal						
Sub-bituminous coal, Nigeria	4	30—70	45			1
W. Virginia seams, U.S.A.	596			8—1,000	18 (S)	2
Interior coal province, U.S.A.		upto 6				3
Lignite, U.S.A.	2			20—36	28 (S)	4
Thermally metamorphosed coal, Falkland Islands	7			30—60	40 (S)	5
Japan		3.6—15.4	(S)	35—75	(S)	6
Coal, lignite and peat, Bulgaria	14	1—10	4 (L)			7
Coal blend, Bulgaria	8	7—21	16 (L)	20—44	31 (L)	7
Poland				upto 350		8
Upper Silesia, Poland	637	10—500	(S)			9
Norway	34			<10—135	40 (S)	10
Germany	ca. 1,000	upto 700	(S)			11
Italy	10			30—150	50 (L)	12
England (samples representing whole seams)	26	2—14	5.5 (C)	20—282	83 (C)	13
England (individual samples)	75	<1—29	8.7 (C)	9—282	43 (C)	13
Wales	4	6—25	15 (S)			14
Crude Oil						
Western U.S.A.	102			<1.5 to 3,000 (S)		15
Emba region, U.S.S.R.	16	0.003—0.042	0.015 (C)			16
U.S.S.R.		0.0005—0.003			3.8	17

References: 1. SIMPSON (1954). 2. HEADLEE and HUNTER (1953). 3. ZUBOVIC (1966b). 4. O'NEIL and SUHR (1960). 5. BROWN and TAYLOR (1960). 6. INAGAKI and YAMAGUCHI (1958). 7. BEKYAROVA and ROUSCHEV (1963). 8. RYCZEK (1959). 9. IDZIKOWSKI (1959). 10. BUTLER (1953a). 11. LEUTWEIN and RÖSLER (1956). 12. BERTETTI (1955). 13. DALTON and PRINGLE (1962). 14. NICHOLLS and LORING (1962). 15. HYDEN (1961). 16. KOTOVA and VIKTOROVA (1965). 17. NURIEV and EFENDIEV (1966).

31-M. Abundance in Common Metamorphic Rock Types

Data on the abundance of gallium in metamorphic rocks are given in Table 31-M-1.

Table 31-M-1. *Gallium in metamorphic rocks*

Type and origin	No. of samples	Ga (ppm) range	Ga (ppm) mean (method)	$\frac{Ga \times 10^3}{Al}$ range	$\frac{Ga \times 10^3}{Al}$ mean	Reference
Hornfels, Ireland	11	23—92	48 (S)	0.14—0.60	0.33	1
Skarn, Soviet Armenia	259	10—50	(S)			2
Skarn, U.S.S.R.		5—24	(S)	0.10—0.20		3
Metasomatic skarn, U.S.S.R.		1—10	(S)			3
Schist and skarn, Ireland	7	22—37	33 (S)	0.33—0.51	0.43	4
Schist						
Black Hills, U.S.A.	2	35—40	38 (S)	0.45—0.49	0.47	5
Ireland	17	12—43	26 (S)	0.07—0.30	0.19	1
Garabal Hill — Glen Fyne complex, Scotland	11	10—25	18 (S)			6
Schist and gneiss						
New Hampshire, U.S.A.	50	6—33	18 (S)	0.15—0.25	0.19	7
Scotland	5	25—75	50 (S)	0.30—0.49	0.42	8
Gneiss						
Tedino deposit, U.S.S.R.	7	14—22	18 (S)			9
Lower Silesia, Poland	4	13—20	16 (S)	0.19—0.25	0.22	10
Scotland — Ultrabasic	2	<2	<2 (S)			11
Scotland — Basic	6	<2—15	12 (S)	0.13—0.21	0.18	11
Skaergaard metamorphic complex, E. Greenland	7	20—35	25 (S)			12
Amphibolite						
U.S.S.R. (mainly gabbro amphibolites)	18		18 (S)		0.21	13
Scotland	3	10—40	22 (S)			14
Skaergaard metamorphic complex, E. Greenland	2	15—25	20 (S)			12
Lower Silesia, Poland	5	17.5—20	19 (S)	0.20—0.28	0.24	10
Granulite, Lower Silesia	1		13.5 (S)		0.20	10
Eclogite, Lower Silesia	1		18 (S)		0.25	10

References: 1. EVANS (1964). 2. NESTERENKO et al. (1958). 3. BORISENOK and SAUKOV (1960). 4. HIGAZY (1952a). 5. HIGAZY (1953). 6. NOCKOLDS and MITCHELL (1948). 7. SHAW (1954). 8. HIGAZY (1954b). 9. BORISENOK and RYABCHIKOV (1962). 10. WALENCZAK and PENDIAS (1958). 11. O'HARA (1961). 12. WAGER and MITCHELL (1951). 13. BORISENOK (1959). 14. HITCHON (1960).

31-N. Behavior in Metamorphic Reactions

The data given in Section 31-M show that both the concentration of gallium and the ratio of gallium to aluminium in the common metamorphic rocks are in general similar to those in unmetamorphosed material. In a number of detailed studies, however, increases in both quantities have been found to occur in some rocks showing various degrees of metamorphic reaction. BORISENOK and SAUKOV (1960) noted such increases in the formation of vesuvianite skarn, especially at the pneumatolytic stage; epidote from the last mineralization was particularly enriched in gallium. According to NESTERENKO et al. (1958), gallium also enters garnet during skarn formation. Some evidence of an increase in the ratio of gallium to aluminium in metasomatic change is provided by the results of HIGAZY (1952a) who suggests that, in the formation of lepidomelane skarns, gallium may substitute for iron (Fe^{3+}) as well as for aluminium. EVANS (1964) in a study of rocks from Connemara found that the concentration of gallium and its ratio to aluminium rose with increasing intensity of metamorphism and he also suggested that the relationship between gallium and iron (Fe^{3+}) might be an important factor; the highest concentrations of gallium occurred in xenoliths which were rich in magnetite.

A number of investigations (SITNIN, 1960; SEVEROV and VERSHKOVSKAYA, 1961; GANEEV et al., 1961; GANEYEV and SECHINA, 1962; KUTS, 1966; MISHCHENKO et al., 1966) have been made of the changes in the concentration of gallium and its ratio to aluminium during the alteration of granitic rocks by greisenization and albitization. High concentrations of gallium, upto 170 ppm, have been found in albitized granites, the increase being largely due to high levels of gallium in the albite. The degree of enrichment of gallium is related to the degree of sodium metasomatism. This has been attributed to the mobility in alkaline solutions of gallium which can thus replace aluminium in sodic plagioclases. As a result albitized rocks may show higher concentrations of gallium in the feldspars than in micas and amphiboles which tend to have the higher concentrations in unmodified rocks. The effects observed in granites are less marked in the alteration of sedimentary rocks. The results of SLEPNEV (1962, 1964) suggest a more complex relationship in the albitization of granite pegmatites. In these rocks the gallium contents were lower in strongly albitized rocks than in related rocks showing less albitization; intensity of greisenization appeared to have no appreciable effect. The reduction of gallium in the most albitized rocks was probably associated with its dispersion in holmquistite in the contact zone. FOMIN et al. (1965) considered that albitization was the cause of increased levels of gallium in mariupolite compared with those in foyaite.

According to SITNIN (1960) gallium becomes depleted relative to aluminium in greisen minerals, contrary to the data given by BORISENOK and SAUKOV (1960). The results of GANEEV et al. (1961) show that in conversion of biotite to muscovite the content of gallium is little changed. The topaz of quartz-topaz greisens is very poor in gallium as is general for this mineral.

31-O. Relations to Other Elements; Economic Importance

The distribution of gallium is dominantly attributable to its close relationship with aluminium in magmatic processes, and in the cycle of weathering and sedimentation. It shows a significant, although less important, association with iron (Fe^{3+}) in magmatic differentiation. The only other relationship of clearly established significance is that with zinc under certain hydrothermal conditions where the chalcophile tendencies of gallium become dominant. These associations are reflected in the fact that gallium is recovered commercially as a by-product of the processing of bauxite and sphalerite. The production of gallium is on a small-scale as its present uses are rather limited. A summary of technological aspects of the element is given by EILERSTEN (1965). The element and its compounds have become of considerable importance in semiconductor technology. The low melting point (29.8°C) of the metal has led to its use in wide range thermometry and as a sealing material for glass joints. Other uses are in selenium rectifiers, vapour lamps and dental alloys.

© Springer-Verlag Berlin · Heidelberg 1972

Revised manuscript received: October 1970

References: Sections 31-B to 31-O

ADIE, R. J.: The petrology of Graham Land II. The Andean granite-gabbro intrusive suite. Falkland Islands Dependencies Surv. Sci. Repts No. 12, 39 pp. (1955).

AHRENS, L. H.: The lognormal distribution of the elements. (A fundamental law of geochemistry and its subsidiary). Geochim. Cosmochim. Acta **5**, 49 (1954).

— The significance of the chemical bond for controlling the geochemical distribution of the elements. Part 1. Phys. Chem. Earth **5**, 1 (1964).

— FLEISCHER, M.: Second report on a cooperative investigation of the composition of two silicate rocks. Part 4. Report on trace constituents in granite G-1 and diabase W-1. U.S. Geol. Surv. Bull. **1113**, 83 (1960).

AKAIWA, H.: Abundances of selenium, tellurium, and indium in meteorites. J. Geophys. Res. **71**, 1919 (1966).

ALLER, L. H.: The abundance of elements in the solar atmosphere. Advan. Astron. Astrophys. **3**, 1 (1965).

AMSTUTZ, G. C.: Geochemistry of Swiss lavas. Geochim. Cosmochim. Acta **3**, 157 (1953).

ANNELL, C. S., HELZ, A. W.: Emission spectrographic determination of trace elements in lunar samples from Apollo 11. Proc. Apollo 11 [Eleven] Lunar Sci. Conf. **2**, 991 (1970).

ANTKIW, S., DIBELER, V. H.: Mass spectrum of gallium vapor. J. Chem. Phys. **21**, 1890 (1953).

BEKYAROVA, E. E., ROUSCHEV, D. D.: The germanium, gallium, and molybdenum content of some solid fuels and their industrial products. Fuel **42**, 359 (1963).

BELL, C. K.: Some aspects of the geochemistry of gallium. Bull. Geol. Soc. Am. **66**, 1529 (1955).

BELOZEROV, E. S.: The content of gallium in some food products. Vopr. Pitaniya **25**, 85 (1966a); Chem. Abstr. **65**, 1291c (1966).

— Distribution of gallium in healthy humans. Byul. Eksperim. Biol. i. Med. **62**, No. 10, 61 (1966b); Chem. Abstr. **66**, 790 u (1967).

BERTETTI, J.: Sulla presenza del gallio e del germanio in alcuni carboni fossili Italiani. Atti Accad. Ligure Sci. e Lettere **11**, 53 (1955).

BERTRAND, D.: Le gallium peut il être considéré comme un oligo-élément indispensable pour l'*Aspergillus niger*? Compt. Rend. **239**, 1704 (1954).

BIDELMAN, W. P., CORLISS, C. H.: Identification of Ga II lines in stellar spectra. Astrophys. J. **135**, 968 (1962).

BORISENKO, L. F.: Some characteristics of the distribution of gallium in ultramafic rocks. Geochemistry (USSR) (English Transl.) 778 (1963).

BORISENOK, L. A.: Distribution of gallium in the rocks of the Soviet Union. Geochemistry (USSR) (English Transl.) 52 (1959).

— RYABCHIKOV, I. D.: Gallium in the minerals of mica pegmatites of the Tedino deposit. Geochemistry (USSR) (English Transl.) 68 (1962).

— SAUKOV, A. A.: Geochemical cycle of gallium. Intern. Geol. Congr., 21st, Copenhagen, 1960, Rept Session, Norden Part 1, 96 (1960).

— TAUSON, L. V.: Geochemistry of gallium in the granitoids of the Susamyr batholith (Central Tien Shan). Geochemistry (USSR) (English Transl.) 178 (1959).

— ZLOBIN, B. I.: Gallium in the alkalic rocks of the Sandyk Mountains massif (Northern Kirgiziya). Geochemistry (USSR) (English Transl.) 612 (1959).

BOWDEN, P.: Gallium in younger granites of Northern Nigeria. Geochim. Cosmochim. Acta **28**, 1981 (1964).

BOWEN, H. J. M.: The determination of gallium and molybdenum in biological material by activation analysis. Intern. J. Appl. Radiation Isotopes **5**, 227 (1959).

BROWN, H. R., TAYLOR, G. H.: Metamorphosed coal from the Theron Mountains. Trans-Antarctic Exped. 1955—1958, Sci. Rept. No. 12, 1 (1960).
BURTON, J. D., CULKIN, F., RILEY, J. P.: The abundances of gallium and germanium in terrestrial materials. Geochim. Cosmochim. Acta 16, 151 (1959).
BUSCOMBE, W., CHAMBLISS, C. R.: Spectral analysis of the manganese star HD 1909. Monthly Notices Roy. Astron. Soc. 140, 369 (1968).
BUTLER, J. R.: Geochemical affinities of some coals from Svalbard. Kong. Ind.-, Handverk-, Skipsfartdept., Norsk Polarinst., Skrifter No. 96, 1 (1953a).
— The geochemistry and mineralogy of rock weathering (1) The Lizard area, Cornwall. Geochim. Cosmochim. Acta 4, 157 (1953b).
— The geochemistry and mineralogy of rock weathering (2) The Nordmarka area, Oslo. Geochim. Cosmochim. Acta 6, 268 (1954a).
— Trace-element distribution in some Lancashire soils. J. Soil Sci. 5, 156 (1954b).
CAMERON, E. M.: A geochemical study of certain Jurassic sands. Ph.D. Thesis, University of Manchester (1957).
CANN, J. R.: Petrology of basalts dredged from the Gulf of Aden. Deep-Sea Res. Oceanogr. Abstr. 17, 477 (1970).
CAVE, R.: The Nod Glas sediments of Caradoc age in North Wales. Geol. J. 4, 279 (1965).
CHOWDHURY, A. N., CHAKRABORTY, S. C., BOSE, B. B.: Geochemistry of gallium in bauxite from India. Econ. Geol. 60, 1052 (1965).
COATS, R. R.: Magmatic differentiation in Tertiary and Quaternary volcanic rocks from Adak and Kanaga Islands, Aleutian Islands, Alaska. Bull. Geol. Soc. Am. 63, 485 (1952).
COBB, J. C.: A trace-element study of iron meteorites. J. Geophys. Res. 72, 1329 (1967).
— MORAN, G.: Gallium concentrations in the metal phases of various meteorites. J. Geophys. Res. 70, 5309 (1965).
COHEN, A. J.: Trace element relationships and terrestrial origin of tektites. Nature 188, 653 (1960).
COMPSTON, W., CHAPPELL, B. W., ARRIENS, P. A., VERNON, M. J.: The chemistry and age of Apollo 11 lunar material. Proc. Apollo 11 [Eleven] Lunar Sci. Conf. 2, 1007 (1970).
CONNOR, J., SHIMP, N. F., TEDROW, J. C. F.: A spectrographic study of the distribution of trace elements in some podzolic soils. Soil Sci. 83, 65 (1957).
CULKIN, F., RILEY, J. P.: The occurrence of gallium in marine organisms. J. Marine Biol. Assoc. U. K. 37, 607 (1958).
CURTIS, C. D.: Trace element distribution in some British Carboniferous sediments. Geochim. Cosmochim. Acta 33, 519 (1969).
CUTTITTA, F., CLARKE, R. S., JR., CARRON, M. K., ANNELL, C. S.: Martha's Vineyard and selected Georgia tektites: New chemical data. J. Geophys. Res. 72, 1343 (1967).
DALTON, I. M., PRINGLE, W. J. S.: The gallium content of some Midland coals. Fuel 41, 41 (1962).
DAWSON, J. B.: Basutoland kimberlites. Bull. Geol. Soc. Am. 73, 545 (1962).
DEGENS, E. T., WILLIAMS, E. G., KEITH, M. L.: Environmental studies of Carboniferous sediments. Part I: Geochemical criteria for differentiating marine from fresh-water shales. Bull. Am. Assoc. Petrol. Geologists 41, 2427 (1957).
— — — Environmental studies of Carboniferous sediments. Part II. Application of geochemical criteria. Bull. Am. Assoc. Petrol. Geologists 42, 981 (1958).
DEVORE, G. W.: Crystal growth and the distribution of elements. J. Geol. 63, 471 (1955).
DIETRICH, R. V., HEIER, K. S.: Differentiation of quartz-bearing syenite (nordmarkite) and riebeckitic arfvedsonite granite (ekerite) of the Oslo Series. Geochim. Cosmochim. Acta 31, 275 (1967).
DOBROVOL'SKII, V. V.: Epigenesis in the Quaternary deposits of Northern Kazakhstan. Geochemistry (USSR) (English Transl.) 220 (1959).
DOUGLAS, J. A. V.: Geological investigations in East Greenland. Part VII. The Basistoppen sheet. A differentiated basic intrusion into the upper part of the Skaergaard complex, East Greenland. Medd. Groenland 164, No. 5, 1 (1964).
DUDLEY, H. C., LEVINE, M. D.: Studies of the toxic action of gallium. J. Pharmacol. Exp. Therap. 95, 487 (1949).

DUDLEY, H. C., MADDOX, G. E.: Deposition of radio gallium (Ga^{72}) in skeletal tissues. J. Pharmacol. Exp. Therap. **96**, 224 (1949).
— — LA RUE, H. C.: Studies of the metabolism of gallium. J. Pharmacol. Exp. Therap. **96**, 135 (1949).
EDGE, R. A., AHRENS, L. H.: Studies on the trace element content of some South African rocks. Trans. Geol. Soc. S. Africa **65**, 113 (1962).
EFENDIYEV, G. KH., SHIK, E. I.: Occurrence of gallium in oil field waters. Geochemistry Intern. **2**, 242 (1965).
EFENDIEV, G. KH., NOVRUZOV, N. A., ABDULLAEVA, R. S.: Geochemistry of Ga in pyrite-polymetallic and Cu pyrrhotite type deposits. Dokl. Akad. Nauk Azerb. SSR **21**, No. 11, 12 (1965); Chem. Abstr. **64**, 17288 g (1966).
EILERSTEN, D. E.: Gallium. In: Mineral facts and problems. U.S. Dept. Interior, Bureau of Mines, Bull. 630 (1965).
EVANS, B. W.: Fractionation of elements in the pelitic hornfelses of the Cashel-Lough Wheelaun intrusion, Connemara, Eire. Geochim. Cosmochim. Acta **28**, 127 (1964).
EXLEY, C. S.: The granitic rocks of Haig Fras. Nature **210**, 365 (1966).
FAIRBAIRN, H. W., AHRENS, L. H., GORFINKLE, L. G.: Minor element content of Ontario diabase. Geochim. Cosmochim. Acta **3**, 34 (1953).
FLEISCHER, M.: Estimates of the abundances of some chemical elements and their reliability. Geol. Soc. Am., Spec. Papers **62**, 145 (1955).
— U.S. Geological Survey standards. I. Additional data on rocks G-1 and W-1, 1965—1967. Geochim. Cosmochim. Acta **33**, 65 (1969).
— STEVENS, R. E.: Summary of new data on rock samples G-1 and W-1. Geochim. Cosmochim. Acta **26**, 525 (1962).
FOMIN, O. B., KUTS, V. P., ORLOVA, L. A.: Accumulation of gallium in rocks of the Oktyabr and Elanchik massifs. Dopovidi Akad. Nauk Ukr. RSR, 78 (1965); Chem. Abstr. **62**, 10251e (1965).
FOUCHÉ, K. F., SMALES, A. A.: The distribution of trace elements in chondritic meteorites. 1. Gallium, germanium and indium. Chem. Geol. **2**, 5 (1967).
FRUTH, I.: Spurengehalte der Zinkblenden verschiedener Pb-Zn-Vorkommen in den nördlichen Kalkalpen. Chem. Erde **25**, 105 (1966).
FRYKLUND, V. C., JR., FLETCHER, J. D.: Geochemistry of sphalerite from the Star Mine, Coeur d'Alene district, Idaho. Econ. Geol. **51**, 223 (1956).
GABRIELSON, O.: Studier över elementfördelningen i zinkbländen från svenska fyndorter. Sveriges Geol. Undersokn. Arsbok, Ser. C: Avhandl. och Uppsat. No. 468, 52 pp (1945).
GANAPATHY, R., KEAYS, R. R., LAUL, J. C., ANDERS, E.: Trace elements in Apollo 11 lunar rocks: Implications for meteorite influx and origin of moon. Proc. Apollo 11 [Eleven] Lunar Sci. Conf. **2**, 1117 (1970a).
— — ANDERS, E.: Apollo 12 lunar samples: Trace element analysis of a core and the uniformity of the regolith. Science **170**, 533 (1970b).
GANEYEV, I. G., SECHINA, N. P.: Geochemical characteristics of albitized granites. Geochemistry (USSR) (English Transl.) 158 (1962).
GANEEV, I. G., PACHADZHANOV, D. N., BORISENOK, L. A.: Geochemistry of gallium, tin and some other elements in the process of greisenization. Geochemistry (USSR) (English Transl.) 830 (1961).
GARMASH, A. A., VLASOVA, N. K., Geochemistry of gallium during formation of pyrite-polymetallic ores. Tr. Inst. Mineralog., Geokhim. i Kristallokhim. Redkikh Elementov, Akad. Nauk SSSR. **10**, 184 (1963); Chem. Abstr. **61**, 2841 g (1964).
GERASIMOVSKIY, V. I., BORISENOK, L. A., TUZOVA, A. M.: Geochemistry of gallium in nepheline syenites. Geochemistry Intern. **5**, 985 (1968).
GERASIMOVSKII, V. I., TUZOVA, A. M., BORISENOK, L. A., RASSKAZOVA, V. S.: Gallium in the rocks of the Lovozero alkali massif. Geochemistry (USSR) (English Transl.) 550 (1959).
GILES, J. B.: Trace element content of some Queensland surface soils. Australia, Commonwealth Sci. Ind. Res. Organ., Div. Soils, Div. Rept. No. 1/59, 22 pp. (1959).
— Trace element profiles of some grey brown soils of heavy texture with acid substrate from the Brigalow Lands, Queensland. Australia, Commonwealth Sci. Ind. Res. Organ., Div. Soils, Div. Rept. No. 12/62, 44 pp. (1964).

GINZBOURG, A. I.: Specific geochemical features of the pegmatitic process. Intern. Geol. Congr., 21st, Copenhagen, 1960, Rept. Session, Norden Part 17, 111 (1960).
GOLDBERG, E. D., ARRHENIUS, G. O. S.: Chemistry of Pacific pelagic sediments. Geochim. Cosmochim. Acta 13, 153 (1958).
GOLDBERG, E., UCHIYAMA, A., BROWN, H.: The distribution of nickel, cobalt, gallium, palladium and gold in iron meteorites. Geochim. Cosmochim. Acta 2, 1 (1951).
GOLDBERG, L., MÜLLER, E. A., ALLER, L. H.: The abundances of the elements in the solar atmsophere. Astrophys. J., Suppl. Ser. 5, Suppl. No. 45, 1 (1960).
GOLDSCHMIDT, V. M.: Geochemistry. Oxford: Clarendon Press 1954.
GORDON, M., JR., MURATA, K. J.: Minor elements in Arkansas bauxite. Econ. Geol. 47, 169 (1952).
GORHAM, E., SWAINE, D. J.: The influence of oxidizing and reducing conditions upon the distribution of some elements in lake sediments. Limnol. Oceanog. 10, 268 (1965).
GRAF, D. L.: Geochemistry of carbonate sediments and sedimentary carbonate rocks. Part III. Minor element distribution. Illinois State Geol. Surv., Circ. 301, 71 pp. (1960).
— EARDLEY, A. J., SHIMP, N. F.: A preliminary report on magnesium carbonate formation in glacial Lake Bonneville. J. Geol. 69, 219 (1961).
GREENLAND, L.: Gallium in chondritic meteorites. J. Geophys. Res. 70, 3813 (1965).
— LOVERING, J. F.: Fractionation of fluorine, chlorine and other trace elements during fractionation of a tholeiitic magma. Geochim. Cosmochim. Acta 30, 963 (1966).
GULBRANDSEN, R. A.: Chemical composition of phosphorites of the Phosphoria Formation. Geochim. Cosmochim. Acta 30, 769 (1966).
HALL, A.: The distribution of some major and trace elements in feldspars from the Rosses and Ardara granite complexes, Donegal, Ireland. Geochim. Cosmochim. Acta 31, 835 (1967).
HANNA, W. J., GRANT, C. L.: Spectrochemical analysis of the foliage of certain trees and ornamentals for 23 elements. Bull. Torrey Botan. Club 89, 293 (1962).
HASKIN, L. A., ALLEN, R. O., HELMKE, P. A., PASTER, T. P., ANDERSON, M. R., KOROTEV, R. L., ZWEIFEL, K. A.: Rare earths and other trace elements in Apollo 11 lunar samples. Proc. Apollo 11 [Eleven] Lunar Sci. Conf. 2, 1213 (1970).
HEADLEE, A. J. W., HUNTER, R. G.: Elements in coal ash and their industrial significance. Ind. Eng. Chem. 45, 548 (1953).
HECHT, F., FENNINGER, H.: Determination of trace elements in meteorites. Progr. Oceanog. 3, 145 (1965).
HEIDE, F., KÖDDERITZSCH, H.: Der Galliumgehalt des Saale- und Elbewassers. Naturwissenschaften 51, 104 (1964).
HERZ, N., DUTRA, C. V.: Minor element abundance in a part of the Brazilian shield. Geochim. Cosmochim. Acta 21, 81 (1960).
— — Geochemistry of some kyanites from Brazil. Am. Mineralogist 49, 1290 (1964).
HEY, M. H.: Catalogue of Meteorites. London: Trustees of the British Museum (Natural History) 1966.
HIETANEN, A.: Kyanite, andalusite, and sillimanite in the schist in Boehls Butte Quadrangle, Idaho. Am. Mineralogist 41, 1 (1956).
HIGAZY, R. A.: Behaviour of the trace elements in a front of metasomatic-metamorphism in the Dalradian of Co. Donegal. Geochim. Cosmochim. Acta 2, 170 (1952a).
— The distribution and significance of the trace elements in the Braefoot Outer Sill, Fife. Trans. Geol. Soc. Edinburgh 15, 150 (1952b).
— Observations on the distribution of trace elements in the perthite pegmatites of the Black Hills, South Dakota. Am. Mineralogist 38, 172 (1953).
— Trace elements of volcanic ultrabasic potassic rocks of Southwestern Uganda and adjoining part of the Belgian Congo. Bull. Geol. Soc. Am. 65, 39 (1954a).
— A geochemical study of the regional metamorphic zones of the Scottish Highlands. Congr. Geol. Intern. Compt. Rend., 19e, Algiers, 1952 Fasc XV, 415 (1954b).
HIRST, D. M.: The geochemistry of modern sediments from the Gulf of Paria. II. The location and distribution of trace elements. Geochim. Cosmochim. Acta 26, 1147 (1962).

HITCHON, B.: The geochemistry, mineralogy, and origin of pegmatites from three Scottish Pre-Cambrian metamorphic complexes. Intern. Geol. Congr., 21st., Copenhagen, 1960, Rept. Session, Norden Part 17, 36 (1960).
HORN, M. K., ADAMS, J. A. S.: Computer-derived geochemical balances and element abundances. Geochim. Cosmochim. Acta **30**, 279 (1966).
HORTON, L., AUBREY, K. V.: The distribution of minor elements in vitrain: three vitrains from the Barnsley seam. J. Soc. Chem. Ind. **69**, Suppl. Issue No. 1, S41 (1950).
HOWIE, R. A.: The geochemistry of the charnockite series of Madras, India. Trans. Roy. Soc. Edinburgh **62**, 725 (1955).
HYDEN, H. J.: Uranium and other metals in crude oils. B. Distribution of uranium and other metals in crude oils. U.S. Geol. Surv. Bull. **1100**, 17 (1961).
IDZIKOWSKI, A.: O wystepowaniu niektorych mikroelementow w weglach kamiennych warstw rudzkich i siodlowych na Gornym Slasku. Arch. Mineral. **23**, 271 (1959).
IIDA, C.: Trace elements in minerals and rocks of the Izu-Hakone region, Japan. Part II. Plagioclase. J. Earth Sci., Nagoya Univ. **9**, 14 (1961).
INAGAKI, M., YAMAGUCHI, T.: Gallium in Japanese coals. Tanken **9**, 161 (1958); Chem. Abstr. **53**, 684 g (1959).
INGHRAM, M. G., HESS, D. C., JR., BROWN, H. S., GOLDBERG, E.: On the isotopic composition of meteoritic and terrestrial gallium. Phys. Rev. **74**, 343 (1948).
ISHIBASHI, M., SHIGEMATSU, T., NISHIKAWA, Y., HIRAKI, K.: Gallium content of sea water, marine organisms, sea sediments, etc. Nippon Kagaku Zasshi **82**, 1141 (1961); Chem. Abstr. **56**, 5767a (1962).
ISHIKAWA, H., KURODA, R., SUDO, T.: Minor elements in some altered zones of "Kuroko" (black ore) deposits in Japan. Econ. Geol. **57**, 785 (1962).
IVANITSKII, T. V., GVARAMADZE, N. D.: Content and distribution of certain dispersed elements in the principal sulfides of the lead-zinc and polymetallic deposits of Georgia. Geochemistry (USSR) (English Transl.) 167 (1960).
JASCHEK, M., JASCHEK, C., LAVAGNINO, C. J.: On the presence of Ga II in peculiar stars. Publ. Astron. Soc. Pacific **75**, 15 (1963).
KERWIN, L., MCELCHERAN, D. E.: Some upper limits of isotopic abundance — II: Ne, Cl, and Ga. Can. J. Phys. **34**, 1497 (1956).
KING, R. B., LAWRENCE, G. M., LINK, J. K.: Solar abundances of gallium, silver, and indium. Astrophys. J. **142**, 386 (1965).
KOCH, H. J., JR., SMITH, E. R., SHIMP, N. F., CONNOR, J.: Analysis of trace elements in human tissues. I. Normal tissues. Cancer **9**, 499 (1956).
KOLBE, P.: The use of an oxygen jet in the spectrochemical determination of trace amounts of Pb, Tl, Ga, Cu and Sn in some silicate rocks and minerals. Geochim. Cosmochim. Acta **29**, 153 (1965).
KONTOROVICH, A. E., SADIKOV, M. A., SHVARTSEV, S. L.: Abundances of certain elements in the surface and ground waters of the northwestern part of the Siberian platform. Dokl. Akad. Nauk SSSR: Earth Sci. Sect. (English Transl.) **149**, 168 (1963).
KOTOVA, A. V., VIKTOROVA, M. YE.: Gallium and germanium content in crude oils of the Emba region. Geochemistry Intern. **2**, 1024 (1965).
KOVAL'SKII, V. V., REZAEVA, L. T., KOL'TSOV, G. V.: Concentration of trace elements in the organism and blood cells of Ascidia. Dokl.-Biol. Sci. Sect. (English Transl.) **147**, 1225 (1962).
KRAINOV, S. R.: New data on mineral waters of the maritime territory. Byul. Nauchn.-Tekhn. Inform. Min. Geol. i Okhrany Nedr SSSR. **2**, 19 (1961); Chem. Abstr. **57**, 7029h (1962).
KULIKOVA, M. F.: Gallium in the oxidized zones of some polymetallic deposits of Eastern Transbaikaliya. Dokl.: Earth Sci. Sect. (English Transl.) **143**, 134 (1964).
— Geochemistry of gallium and indium in the oxidized zone of lead-zinc deposits in Soviet Central Asia. Geochemistry Intern. **3**, 982 (1966).
KUTS, V. P.: Distribution of gallium in granites of the Azov area. Dopovidi Akad. Nauk Ukr. RSR, 788 (1966); Chem. Abstr. **65**, 16710d (1966).
LATIMER, W. M.: The Oxidation States of the Elements and Their Potentials in Aqueous Solutions. Second Edition, New York: Prentice-Hall 1952.

LAUL, J. C., KEAYS, R. R., GANAPATHY, R., ANDERS, E.: Abundance of 14 trace elements in lunar rock 12013, 10. Earth Planet. Sci. Lett. 9, 211 (1970).
LAVRENCHUK, V. N., TENYAKOV, V. A.: Distribution of gallium in bauxites. Geochemistry (USSR) (English Transl.) 862 (1962).
— — On the average gallium content in clays. Dokl. Akad. Nauk SSSR: Earth Sci. Sect. (English Transl.) 151, 176 (1963a).
— — Gallium balance in bauxites. Dokl. Akad. Nauk SSSR 151, 1430 (1963b); Chem. Abstr. 59, 15069f (1963).
LERICHE, H. H., WEIR, A. H.: A method of studying trace elements in soil fractions. J. Soil Sci. 14, 225 (1963).
LEUTWEIN, F., RÖSLER, H. J.: Geochemische Untersuchungen an paläozoischen und mesozoischen Kohlen Mittel- und Ostdeutschlands. Freiberger Forschungsh. C 19, 1 (1956).
LOUNAMAA, J.: Trace elements in plants growing wild on different rocks in Finland. A semi-quantitative spectrographic survey. Ann. Botan. Soc. Zoo. Botan. Fennicae "Vanamo" 29, No. 4, 196 pp (1956).
LOVERING, J. F.: The geochemical behaviour of elements in meteorites. J. Proc. Roy. Soc. N.S. Wales 91, 149 (1957).
— NICHIPORUK, W., CHODOS, A., BROWN, H.: Distribution of gallium, germanium, cobalt, chromium, and copper in iron and stony-iron meteorites in relation to nickel content and structure. Geochim. Cosmochim. Acta 11, 263 (1957).
LUNAR SAMPLE PRELIMINARY EXAMINATION TEAM: Preliminary examination of lunar samples from Apollo 12. Science 167, 1325 (1970).
MAKSUDOV, N. KH., POGORELKO, I. P., YULDASHEV, P. KH.: A chemical investigation of Artemisia scoparia. Uzbeksk. Khim. Zh. 6, No. 5, 84 (1962); Chem. Abstr. 58, 10514h (1963).
MARAKUSHEV, A. A., POLIN, YU. K.: Distribution of gallium in minerals of Archean metamorphic rocks of the Aldan Shield. Geochemistry (USSR) (English Transl.), 204 (1961).
MARCHENKO, YE. YA., SHCHERBAKOV, V. P.: Notes on gallium distribution in the granitoids of the Azov Sea region. Geochemistry Intern. 3, 1102 (1966).
MASON, B., GRAHAM, A. L.: Minor and trace elements in meteoritic minerals. Smithsonian Contrib. Earth Sci. 3 (1970).
MCKENZIE, R. M.: The distribution of trace elements in some South Australian red-brown earths. Australian J. Agr. Res. 8, 246 (1957).
MCLAUGHLIN, R. J. W.: Geochemical changes due to weathering under varying climatic conditions. Geochim. Cosmochim. Acta 8, 109 (1955).
MEITUV, G. M.: Geochemistry of rare elements in lead-zinc deposits of the Klichkinsk district (Eastern Transbaikal Region). Geochemistry (USSR) (English Transl.) 694 (1962).
MERO, J. L.: Ocean-floor manganese nodules. Econ. Geol. 57, 747 (1962).
MIGDISOV, A. A., BORISENOK, L. A.: Geochemistry of gallium in sedimentation under humid conditions. Geochemistry (USSR) (English Transl.) 1113 (1963).
MISHCHENKO, V. S., KUTS, V. P., ORLOVA, L. A.: The geochemistry of gallium in high-temperature postmagmatic processes. Geochemistry Intern. 3, 330 (1966).
MILLER, A. R., DENSMORE, C. D., DEGENS, E. T., HATHAWAY, J. C., MANHEIM, F. T., MCFARLIN, P. F., POCKLINGTON, R., JOKELA, A.: Hot brines and recent iron deposits in deeps of the Red Sea. Geochim. Cosmochim. Acta 30, 341 (1966).
MORRIS, D. F. C., BREWER, F. M.: The occurrence of Ga in blende. Geochim. Cosmochim. Acta 5, 134 (1954).
MORRISON, G. H., GERARD, J. T., KASHUBA, A. T., GANGADHARAN, E. V., ROTHENBERG, A. M., POTTER, N. M., MILLER, G. B.: Elemental abundances of lunar soil and rocks. Proc. Apollo 11 [Eleven] Lunar Sci. Conf. 2, 1383 (1970).
MOSS, A. A., HEY, M. H., ELLIOTT, C. J., EASTON, A. J.: Methods for the chemical analysis of meteorites: II. The major and some minor constituents of chondrites. Mineral. Mag. 36, 101 (1967).
MOXHAM, R. L.: Distribution of minor elements in coexisting hornblendes and biotites. Can. Mineralogist 8, 204 (1965).

Muir, A., Hardie, H. G. M., Mitchell, R. L., Phemister, J.: The limestones of Scotland. Chemical analyses and petrography. Geol. Surv. Gt Brit., Mem. Geol. Surv., Spec. Rep. Mineral Resources Gt Brit. 37 (1956).

Nesterenko, G. V., Studenikova, Z. V., Savinova, E. N.: Rare and dispersed elements in skarns of Tyrny-Auz (Soviet Armenia). Geochemistry (USSR) (English Transl.) 287 (1958).

Nicholls, G. D., Loring, D. H.: The geochemistry of some British Carboniferous sediments. Geochim. Cosmochim. Acta 26, 181 (1962).

Nichiporuk, W.: Variations in the content of nickel, gallium, germanium, cobalt, copper and chromium in the kamacite and taenite phases of iron meteorites. Geochim. Cosmochim. Acta 13, 233 (1958).

Nishikawa, Y.: Gallium content of rocks and minerals of Japan. Nippon Kagaku Zasshi 79, 351 (1958); Chem. Abstr. 52, 13564d (1958).

Nockolds, S. R., Allen, R.: The geochemistry of some igneous rock series. Geochim. Cosmochim. Acta 4, 105 (1953).

— — The geochemistry of some igneous rock series: Part II. Geochim. Cosmochim. Acta 5, 245 (1954).

— — The geochemistry of some igneous rock series: Part III. Geochim. Cosmochim. Acta 9, 34 (1956).

— Mitchell, R. L.: The geochemistry of some Caledonian plutonic rocks: A study in the relationship between the major and trace elements of igneous rocks and their minerals. Trans. Roy. Soc. Edinburgh 61, 533 (1948).

Nuriev, A. I., Efendiev, G. Kh.: Thallium and gallium contents in stratal waters, crude oils, and surrounding rocks. Azerb. Khim. Zh. 112 (1966); Chem. Abstr. 65, 15098h (1966).

Oertel, A. C., Giles, J. B.: Trace element profiles of some Queensland soils. Australia, Commonwealth Sci. Ind. Res. Organ., Div. Soils, Div. Rept. No. 8/58, 28 pp. (1959).

— — Trace element contents of some Queensland soils. Australian J. Soil Res. 1, 215 (1963).

— — A study of some Brigalow soils based on trace-element profiles. Australian J. Soil Res. 2, 162 (1964).

O'Hara, M. J.: Zoned ultrabasic and basic gneiss masses in the early Lewisian metamorphic complex at Scourie, Scotland. J. Petrol. 2, 248 (1961).

O'Neil, R. L., Suhr, N. H.: Determination of trace elements in lignite ashes. Appl. Spectry 14, 45 (1960).

Onishi, H., Sandell, E. B.: Gallium in chondrites. Geochim. Cosmochim. Acta 9, 78 (1956).

Patterson, E. M.: A petrochemical study of the Tertiary lavas of northeast Ireland. Geochim. Cosmochim. Acta 2, 283 (1952).

— Petrochemical data for some acid intrusive rocks from the Mourne Mountains and Slieve Gullion. Proc. Roy. Irish Acad., Sect. B 55, 171 (1953).

Pearson, G. R., Shaw, D. M.: Trace elements in kyanite, sillimanite and andalusite. Am. Mineralogist 45, 808 (1960).

Pentcheva, E.: Sur le critère microchimique caractérisant les eaux thermales azotées de Bulgarie. Compt. Rend. Acad. Bulgare Sci. 17, 1021 (1964).

— Sur la distribution des éléments rares et disperses dans des eaux salées souterraines bulgares. Compt. Rend. Acad. Bulgare Sci. 18, 149 (1965).

Perkons, A. K., Jervis, R. E.: Trace elements in human head hair. J. Forensic Sci. 11, 50 (1966).

Petrova, Z. I., Kusakina, L. V.: Gallium distribution in Paleozoic granitic rocks of the Dzhida Area (West Trans'Baykal Region). Geochemistry Intern. 2, 924 (1965).

Philpotts, A. R.: Origin of the anorthite-mangerite rocks in southern Quebec. J. Petrol. 7, 1 (1966).

Potter, P. E., Shimp, N. F., Witters, J.: Trace elements in marine and fresh-water argillaceous sediments. Geochim. Cosmochim. Acta 27, 669 (1963).

PRINCE, A. L.: Influence of soil types on the mineral composition of corn tissues as determined spectrographically. Soil Sci. **83**, 399 (1957a).
— Trace element delivering capacity of 10 New Jersey soil types as measured by spectrographic analyses of soils and mature corn leaves. Soil Sci. **84**, 413 (1957b).
RADER, L. F., SWADLEY, W. C., HUFFMAN, C., JR., LIPP, H. H.: New chemical determinations of zinc in basalts and rocks of similar composition. Geochim. Cosmochim. Acta **27**, 695 (1963).
RAZENKOV, N. I., GALAKTIONOVA, G. F.: On the mode of occurrence of gallium in the oxidized zone of sulfide deposits. Geochemistry (USSR) (English Transl.), 104 (1962).
READ, H. H., HAQ, B. T.: The distribution of trace-elements in the dunite-syenite differentiated series of the Insch Complex, Aberdeenshire. Proc. Geologists' Assoc. (Engl.) **74**, 203 (1963).
RIGAULT, G.: Gallio e indio nella blenda. Periodico Mineral. (Rome) **25**, 43 (1956).
— Relazioni tra la struttura della blenda e il contenuto in gallio e indio. Nuovo Cimento **5**, 579 (1957).
RILEY, J. P.: Composition of mineral water from the hot spring at Bath. J. Appl. Chem. (London) **11**, 190 (1961).
RONOV, A. B., MIGDISOV, A. A.: Principal features of the geochemistry of hydrolyzate elements in weathering and sedimentation. Geochemistry Intern. **2**, 92 (1965).
ROOKE, J. M.: Element distribution in some acid igneous rocks of Africa. Geochim. Cosmochim. Acta **28**, 1187 (1964).
ROSE, A. W.: Trace elements in sulfide minerals from the Central district, New Mexico and the Bingham district, Utah. Geochim. Cosmochim. Acta **31**, 547 (1967).
RYCZEK, M.: Occurrence of germanium and gallium in Polish coal and methods for their concentration. Przeglad Gorniczy **15**, 420 (1959); Chem. Abstr. **59**, 15077 g (1963).
SANDELL, E. B.: The Ga content of igneous rocks. Am. J. Sci. **247**, 40 (1949).
SCHAUDY, R., KIESL, W., HECHT, F.: Activation analytical determination of elements in meteorites. Chem. Geol. **2**, 279 (1967).
SEN, N., NOCKOLDS, S. R., ALLEN, R.: Trace elements in minerals from rocks of the S. Californian batholith. Geochim. Cosmochim. Acta **16**, 58 (1959).
SEVEROV, E. A., VERSHKOVSKAYA, O. V.: The behavior of gallium during the process of albitization of granitoids. Dokl. Akad. Nauk SSSR.: Earth Sci. Sect. (English Transl.) **135**, 1096 (1961).
SHACKLETTE, H. T.: Element content of bryophytes. U.S. Geol. Surv. Bull. **1198-D**, 1 (1965).
SHAW, D. M.: Trace elements in pelitic rocks. Part I. Variation during metamorphism. Bull. Geol. Soc. Am. **65**, 1151 (1954).
— Geochemistry of pelitic rocks. Part III. Major elements and general geochemistry. Bull. Geol. Soc. Am. **67**, 919 (1956).
— The geochemistry of gallium, indium, thallium—A review. Phys. Chem. Earth **2**, 164 (1957).
SHIMADA, I.: Extractable organic constituents in sediments, with reference to the relation between composition of the chromatographic fractions in extracts and depositional environment. Sci. Rept. Tohaku Univ., Ser. III **8**, 421 (1963).
SHIMER, J. A.: Spectrographic analysis of New England granites and pegmatites. Bull. Geol. Soc. Am. **54**, 1049 (1943).
SIEDNER, G.: Geochemical features of a strongly fractionated alkali igneous suite. Geochim. Cosmochim. Acta **29**, 113 (1965).
SILVEY, W. D.: Occurrence of selected minor elements in the waters of California. U.S. Geol. Surv., Water Supply Papers **1535-L** (1967).
SIMONS, F. S.: The lead-zinc veins of the Chilete mining district in Northern Peru. Econ. Geol. **50**, 399 (1955).
SIMPSON, A.: The Nigerian coalfield. The geology of parts of Onitsha, Owerri, and Benue Provinces. Geol. Surv. Nigeria, Bull. **24**, 85 pp. (1954).
SITNIN, A. A.: Distribution of rare elements in amazonite granites of the Etyka massif (Eastern Transbaikaliya). Geochemistry (USSR) (English Transl.) 361 (1960).

SLEPNEV, YU. S.: Gallium content in the granite pegmatites of the Sayan Mountains. Geochemistry (USSR) (English Transl.) 742 (1962).
— Geochemical characteristics of the rare metal granitic pegmatites of the Sayan Mountains. Geochemistry Intern. 1, 221 (1964).
SMALES, A. A., MAPPER, D., FOUCHÉ, K. F.: The distribution of some trace elements in iron meteorites, as determined by neutron activation. Geochim. Cosmochim. Acta 31, 673 (1967).
— — WEBB, M. S. W., WEBSTER, R. K., WILSON, J. D.: Elemental composition of lunar surface material. Proc. Apollo 11 [Eleven] Lunar Sci. Conf. 2, 1575 (1970).
SPENCER, D. W.: The geochemistry of a Lower Silurian graptolite band: a study of the geochemical variation occurring at a single geological horizon. Ph. D. Thesis, University of Manchester (1957).
STEINBERG, R. A.: Use of microorganisms to determine essentiality of minor elements. Soil Sci. 60, 185 (1945).
STRUNZ, H.: Söhngeit, Ga(OH)$_3$, ein neues Mineral. Naturwissenschaften 52, 493 (1965).
— GEIER, B. H., SEELIGER, E.: Gallit, CuGaS$_2$, das erste selbständige Galliummineral, und seine Verbreitung in den Erzen der Tsumeb- und Kipushi-Mine. Neues Jahrb. Mineral., Monatsh., 241 (1958).
SWAINE, D. J., MITCHELL, R. L.: Trace-element distribution in soil profiles. J. Soil Sci. 11, 347 (1960).
TANDON, S. N., WASSON, J. T.: Gallium, germanium, indium and iridium variations in a suite of L-group chondrites. Geochim. Cosmochim. Acta 32, 1087 (1968).
TAYLOR, S. R.: The chemical composition of australites. Geochim. Cosmochim. Acta 26, 685 (1962).
— SACHS, M.: Geochemical evidence for the origin of australites. Geochim. Cosmochim. Acta 28, 235 (1964).
TENYAKOV, V. A.: Abundance of Ga in bauxite (as related to the calculation of its new average content). Dokl. Akad. Nauk SSSR.: Earth Sci. Sect. (English Transl.) 181, 205 (1968).
THOMAS, W. K. L.: Standard geochemical sample T-1. Suppl. 1. Part 1, Chemical analyses of T-1. Geol. Surv. Div., Ministry Commerce Ind., Tanzania 1 (1963).
TILLER, K. G.: The distribution of trace elements during differentiation of the Mt. Wellington dolerite sill. Papers Proc. Roy. Soc. Tasmania 93, 153 (1959).
TOURTELOT, H. A.: Minor-element composition and organic carbon content of marine and nonmarine shales of late Cretaceous age in the western interior of the United States. Geochim. Cosmochim. Acta 28, 1579 (1964).
— SCHULTZ, L. G., GILL, J. R.: Stratigraphic variations in mineralogy and chemical composition of the Pierre Shale in South Dakota and adjacent parts of North Dakota, Nebraska, Wyoming, and Montana. U.S. Geol. Surv. Profess. Paper 400-B, B 447 (1960).
TROSHIN, YU. P.: Gallium-indium ratio in sphalerites of Transbaikaliya. Geochemistry (USSR) (English Transl.) 378 (1962).
— TROSHINA, G. M.: On the stability of distribution of trace elements during the formation of polymetallic deposits. Geochemistry Intern. 2, 42 (1965).
TURTON, A. G., MARSH, N. L., MCKENZIE, R. M.: Morphological, chemical, physical, clay mineralogy and minor element data on lateritic profiles from South Western Australia. Australia, Commonwealth Sci. Ind. Res. Organ., Div. Soils, Div. Rept. No. 8/61, 56 pp. (1961).
— — — MULCAHY, M. J.: The chemistry and mineralogy of lateritic soils in the southwest of Western Australia. Australia, Commonwealth Sci. Ind. Res. Organ., Soil Publ. No. 20 (1962).
UPTON, B. G. J.: The alkaline igneous complex of Kûngnât Fjeld, South Greenland. Medd. Groenland 123, No. 4, 1 (1960).
— The geology of Tugtutôq and neighbouring islands, South Greenland. Part IV. The nepheline syenites of the Hviddal composite dyke. Medd. Groenland 169, No. 3, 53 (1964).

Uzumasa, Y., Nasu, Y.: Chemical investigations of hot springs in Japan. LVII. Gallium in hot springs. Nippon Kagaku Zasshi **81**, 732 (1960); Chem. Abstr. **55**, 3883a (1961).

Vershkovskaya, O. V.: Gallium. In: Geochemistry and Mineralogy of Rare Elements and Genetic Types of Their Deposits (ed. by K. A. Vlasov), vol. I. Geochemistry of Rare Elements, p. 437. Jerusalem: Israel Program for Scientific Translations 1966.

— Fabrikova, E. A.: Gallium in sphalerite. Geochemistry (USSR) (English Transl.) 377 (1957).

— Saltykova, V. S.: Gallium in rocks containing fluorite-sulfide mineralization. Geochemistry (USSR) (English Transl.) 464 (1961).

Vertskovskaya, O. V.: Gallium in rocks and minerals of the Khrustal'noe deposit. Tr. Inst. Mineralog., Geokhim. i Kristallokhim. Redkikh Elementov, Akad. Nauk SSSR **10**, 201 (1963); Chem. Abstr. **61**, 1656 g (1964).

Vinogradova, Z. A., Koval'skiy, V. V.: Elemental composition of the Black Sea plankton. Dokl. Akad. Nauk SSSR.: Earth Sci. Sect. (English Transl.) **147**, 217 (1964).

Vorobiev, G. G.: Composition of tektites. I. Indochinites. Meteoritika **17**, 64 (1959); Chem. Abstr. **55**, 6300b (1961).

Vorob'yev, G. G.: A study of tektite composition. Philippinites (Risalites). Meteoritica (English Transl.) **23**, 98 (1963).

Wänke, H., Rieder, R., Baddenhausen, H., Spettel, B., Teschke, F., Quijano-Rico, M. Balacescu, A.: Major and trace elements in lunar material. Proc. Apollo 11 [Eleven] Lunar Sci. Conf. **2**, 1719 (1970).

Wager, L. R., Mitchell, R. L.: The distribution of trace elements during strong fractionation of basic magma — a further study of the Skaergaard intrus on, East Greenland. Geochim. Cosmochim. Acta **1**, 129 (1951).

— — Trace elements in a suite of Hawaiian lavas. Geochim. Cosm him. Acta **3**, 217 (1953).

Wakeel, S. K. El., Riley, J. P.: Chemical and mineralogical studies of deep-sea sediments. Geochim. Cosmochim. Acta **25**, 110 (1961).

Walenczak, Z.: Gallium content in feldspars of the granites and pegmatites of Karkonosze and Strzegom (Lower Silesia). Bull. Acad. Polon. Sci., Ser. Sci. Chim., Geol. Geograph. **7**, 595 (1959).

— Anomalous Ga content in amethysts. Bull. Acad. Polon. Sci., Ser. Sci. Chim., Geol. Geograph. **14**, 61 (1966a).

— Geochemistry of Ga and Al in quartz. Bull. Acad. Polon. Sci., Ser. Sci. Chim., Geol. Geograph. **14**, 69 (1966b).

— Pendias, H.: Gallium in igneous and metamorphic rocks of Lower Silesia. Bull. Acad. Polon. Sci., Ser. Sci. Chim., Geol. Geograph. **6**, 75 (1958).

Wasson, J. T.: Butler, Missouri: Fe meteorite with extremely high Ge content. Science **153**, 976 (1966).

— Concentrations of Ni, Ga, and Ge in a series of Canyon Diablo and Odessa meteorite specimens. J. Geophys. Res. **72**, 721 (1967a).

— The chemical classification of iron meteorites: I. A study of iron meteorites with low concentrations of gallium and germanium. Geochim. Cosmochim. Acta **31**, 161 (1967b).

— The chemical classification of iron meteorites — III. Hexahedrites and other irons with germanium concentrations between 80 and 200 ppm. Geochim. Cosmochim. Acta **33**, 859 (1969).

— Baedecker, P. A.: Ga, Ge, In, Ir and Au in lunar, terrestrial and meteoritic basalts. Proc. Apollo 11 [Eleven] Lunar Sci. Conf. **2**, 1741 (1970).

— Goldstein, J. I.: The North Chilean hexahedrites: Variations in composition and structure. Geochim. Cosmochim. Acta **32**, 329 (1968).

— Kimberlin, J.: The chemical classification of iron meteorites — II. Irons and pallasites with germanium concentrations between 8 and 100 ppm. Geochim. Cosmochim. Acta **31**, 2065 (1967).

Weber, J. N., Middleton, G. V.: Geochemistry of the turbidites of the Nomanskill and Charny formations. II. Distribution of trace elements. Geochim. Cosmochim. Acta **22**, 244 (1961).

WEDEPOHL, K. H.: Spurenanalytische Untersuchungen an Tiefseetonen aus dem Atlantik. Ein Beitrag zur Deutung der geochemischen Sonderstellung von pelagischen Tonen. Geochim. Cosmochim. Acta 18, 200 (1960).
— Geochemische und petrographische Untersuchungen an einigen jungen Eruptivgesteinen Nordwestdeutschlands. Fortschr. Mineral. 39, 142 (1961).
WELLS, N.: Total elements in topsoils from igneous rocks: An extension of geochemistry. J. Soil Sci. 11, 409 (1960).
WILKINSON, J. F. G.: The geochemistry of a differentiated teschenite sill near Gunnedah, New South Wales. Geochim. Cosmochim. Acta 16, 123 (1959).
WOLFENDEN, E. B.: Geochemical behaviour of trace elements during bauxite formation in Sarawak, Malaysia. Geochim. Cosmochim. Acta 29, 1051 (1965).
YAVNEL, A. A.: The relation between the structure of iron meteorites and their chemical composition. Soviet Phys. "Doklady" (English Transl.) 5, 217 (1960).
YEREMENKO, G. K., WALTER, A. A., KLIMENCHUK, V. I.: On the distribution of gallium in alkalic rocks, Azov region. Geochemistry (USSR) (English Transl.) 145 (1963).
ZUBOVIC, P.: Physicochemical properties of certain minor elements as controlling factors in their distribution in coal. Advan. Chem. Ser. 55, 221 (1966a).
— Minor element distribution in coal samples of the interior coal province. Advan. Chem. Ser. 55, 232 (1966b).

Revised manuscript received: October 1970

Selenium 34

A   R. Fischer   (Institut für Mineralogie und Kristallographie der Universität, A-1010 Wien, Austria)

B—O   F. Leutwein   (Centre de Recherches Pétrographiques et Géochimiques, F-54 Vandœuvre-Nancy, France)

34-B. Isotopes in Nature

I. Average Composition

Selenium has 6 stable isotopes; their relative abundances are given in the following Table 34-B-1 (BAINBRIDGE and NIER, 1950).

Table 34-B-1

Se	%
74	0.87
76	9.02
77	7.58
78	23.52
80	49.82
82	9.19

No radioactive process is known which forms a selenium-isotope as a stable end-product. Several artificial selenium-isotopes are known, all of very short life. Only ^{79}Se (a β-emitter, giving ^{79}Br half life, period $6.5 \cdot 10^4$ a) could perhaps be detected as a natural fission-product of uranium minerals (RANKAMA, 1954).

II. Isotopic Fractionation under Natural Conditions

KROUSE and THODE (1962) studied the isotopic differentiation of selenium isotopes, covering a rather large field of mass-units. The experimental measurement of the fractionation coefficient of ^{74}Se to ^{82}Se is, in the experimental sense, not favorable because ^{74}Se is about 10 times rarer than the heaviest isotope ^{82}Se. Therefore the authors decided to study the pair ^{76}Se/^{82}Se, the mass-difference here being only ~7%, but both isotopes have nearly the same natural abundance. For mass-spectrometric work they used SeF$_6$, prepared in a monel-metall line; the inlet-system to the mass-spectrometer was of the same metal. The most frequent ion obtained in the source of the mass-spectrometer is SeF$_5^+$. Measurements were made with a double collector for ^{76}SeF$_5$ and ^{82}SeF$_5$, using the same technique as for sulfur analyses. Table 34-B-2 gives their results on natural materials. The δ ^{82}Se is calculated in the same way as the δ ^{34}S values, taking the Canon Diablo-selenium as reference (a short description of their chemical procedure is given in the publication).

These results show that the enrichments of the heavy isotopes of sulfur and selenium have the same trends. Quite typical is the depletion of ^{34}S in the native sulfur of the Shirane-Volcano, δ^{34}S $= -5\%$ (SAKAI and NAGASAWA, 1958), but further research is highly desirable.

Table 34-B-2. $\delta\,^{82}Se$-*values from different natural materials.* (After KROUSE and THODE, 1962)

Locality	Material	^{82}Se
Canon Diablo, USA	Troilite	0.000
Noranda, Ontario, Canada	Massive sulfides	$+0.5 \pm 0.5$
Flin Flon, Manitoba, Canada	Massive sulfides	$+0.5 \pm 0.5$
Sudbury, Ontario	Massive sulfides	$+0.5 \pm 0.5$
Mt. Lyell, Australia	Mesothermal sulfides	-1.0 ± 0.5
MG-Wingen, N. S. Wales	Native Se under a pyrite-bed	-1.0 ± 0.5
Beaverlodge lake, Sask., Canada	Umangite in a pitchblende vein	0.00 ± 0.1
	Chalcomenite	0.5 ± 0.1
County Meath, Ireland,	Seleniferous soil	$+9.0 \pm 0.5$
SW of the USA	Astragalus bisulcatus	$+2.0 \pm 0.5$
locality unknown	Astragalus pattersoni	-11 ± 1.0
Volcano Shirane, Japan	Native sulfur, seleniferous	-2.0 ± 1.0
	Refined selenium (commercial)	$+0.5 \times 0.1$
	commercial SeF_6	$+0.5 \times 0.5$

Revised manuscript received: June 1971

34-C. Abundance in Cosmos, Meteorites, Lunar Materials and Tektites

I. Cosmic Abundance

Since the most important spectral-lines of Se are in the ultra-violet, this element has not yet been detected by direct observation in solar or stellar spectra.

CAMERON (1968) estimated the cosmic abundance by analysing type I carbonaceous chondrites. (Numerous scientists admit that meteorites of this type represent specimens of solar material except for gases of high volatility). Relative to an abundance of $Si = 10^6$, CAMERON found the following data:

$$^{16}S \quad 5.06 \cdot 10^5$$
$$^{34}Se \quad 70.1$$
$$^{52}Te \quad 6.76$$

Thus, the cosmic abundance of selenium is about seven thousand times lower than that of sulfur.

II. Abundance in Meteorites

Concerning the abundance of selenium in iron meteorites and troilite, some older data is reported by GOLDSCHMIDT and STROCK (1935).

Table 34-C-1. *Selenium in iron meteorites and troilite.* (After GOLDSCHMIDT and STROCK, 1935)

Material	Locality	Method	Se ppm
Iron	Canon Diablo, Arizona (USA)	W/C	4.6
Troilite	Canon Diablo, Arizona (USA)		91
Iron	Corisotillo (Chile)	W/C	1.5
Troilite	Corisotillo (Chile)		132

VINOGRADOV (1962) cites for stony meteorites (chondrites) 10 ppm, a proportion of 1.98 Se relative to $1 \cdot 10^6$ Si.

The abundances of selenium (and other trace-elements) in chondrites have recently been studied by DU FRESNE (1960), SCHINDEWOLF (1960), GREENLAND (1967), and KIESL *et al.* (1970). DU FRESNE (1960) reports an average value of 7.9 ppm Se (W/C) for 22 ordinary and enstatite chondrites, which range from 5.3 (Bjurbole) to 15.1 ppm Se (Indarch). The average is close to 9.8 ppm Se (N/R) as published by SCHINDEWOLF (1960) for 4 chondrites. The same author mentions the achondrites Johnstown and Nuevo Laredo as being very low in selenium (7 and 1.6 ppb). The results by GREENLAND (1967) are listed in Table 34-C-2. KIESL *et al.* (1970) report an average value of 10 ppm Se (N/R) for 16 chondrites.

© Springer-Verlag Berlin · Heidelberg 1972

Table 34-C-2. *Se and Te in chondritic meteorites.* (GREENLAND, 1967) (*N/R-method*)

Meteorite	ppm Se	ppm Te
Carbonaceous chondrites type I		
Ivuna	26	1.0
Orgueil	27	0.94
Carbonaceous chondrites type II		
Mighei	14	1.2
Murray	10	3.3
Carbonaceous chrondrites type III		
Chainpur	9.0	1.4
Felix	11	0.44
Lancé	9.3	0.15
Mokoia	9.6	0.79
Vigarano	6.5	0.82
Warrenton	11	0.90
Enstatite chondrites type I		
Abee	11	3.9
Indarch	34	2.6
Enstatite chondrites type II		
Daniel's Kuil	19	0.21
Hvittis	10	0.79
Khairpur	24	0.74
Pillistfer	13	0.18
Ordinary chondrites		
Allegan (CH)	6.1	—
Bruderheim (CL)	6.4	0.76
Ehole (CH)	6.4	0.93
Mocs (CL)	7.3	0.89
Pantar (CH)	4.8	0.61
Peace River (CL)	6.4	0.82
Walters (CL)	8.2	0.60

Some Se/S and Se/Te ratios in chondrite groups are listed in Table 34-C-3.

Table 34-C-3. *Some trace-element atomic ratios in several meteorite types.* (After GREENLAND, 1967)

Ratio	Carbonaceous Ch. Type I	Carbonaceous Ch. Type II	Carbonaceous Ch. Type III	Enstatite Ch. Type I	Enstatite Ch. Type II	Ordinary Chondrites	Cosmos (SUESS and UREY, 1956)
$Se/10^5$ S.	18	15	16	16	20	12	18
Se/Te	44	8.4	19	11	57	14	14

Data on the abundance of selenium in *tektites* are not available.

III. Lunar Materials

Our information on the content of selenium in lunar materials is not yet very complete. We have information only on the selenium content in some total rock specimens from the Apollo 11 landing site "Mare Tranquillitatis". The rock and soil specimens are grouped in several textural classes: fine-grained vesicular type A ("vesicular basalt"), coarse-grained crystalline type B ("gabbro"), the lunar breccia C, and type D ("soil") which is the fine material. From each type, there are several samples (numbered with 5 digits), and from each sample, several batches (2-digit-numbers) have been distributed to various laboratories. Selenium-analyses have been made by neutron activation methods. (HASKIN et al., 1970; TUREKIAN and KHARKAR, 1970) and by mass spectrometry using spark-source excitation (MORRISON et al., 1970). They are listed in Table 34-C-4.

Table 34-C-4. *Selenium content in lunar materials.* (After HASKIN et al., "H"; TUREKIAN et al. "T"; and MORRISON et al. "M")

Rock-type	Sample and batch number	ppm Se	Reference
A	10022—38	0.7 ± 0.5	H
	10020—26	0.4	M
	10044—27	0.23	T
	10049—33	0.20	T
	10057—79	0.12	T
B	10045—25	0.8 ± 0.4	H
	10046—23	0.4	M
	10020—27	0.25	T
	10062—25	0.23	T
	10056—21	0.28	T
C	10048—47	1.6 ± 0.8	H
	10060—13	0.9	M
	10021—33	0.17	T
	10046—24	0.28	T
D	10084—52	0.8 ± 0.6	H
	10084—55	0.2	M
	10084—57	0.39	T

To allow a comparison with terrestrial materials, we cite here some analyses of sulfur from lunar material published by KAPLAN et al. (1970): the "soil" type D has 720 ± 40, the breccia-type C 1,100 ± 20, and the fine-grained rock of basaltic type A has 2,240 ± 40 ppm S.

The selenium analyses show a wide scattering. This may be due to analytical errors, but it is also not certain whether the "total rock" samples are indeed sufficiently homogeneous. For the "soil" samples the following ratios were obtained: a S/Se ratio of 1:7,000, and for the A-type rocks 1:10,000 to 1:20,000 — relations which are not very different from terrestrial data for comparable mafic rocks.

34-D. Abundance in Rock-Forming Minerals (I), Selenium Minerals (II)[1]

I. Rock Forming Minerals

In natural environments, selenium has little lithophile tendency and silicates of selenium are not known. SINDEEVA (1964) gave the following compilation of contents of selenium in minerals (W/C methods) (cited from VLASSOV, 1964).

Table 34-D-1. *Ranges of selenium contents in minerals of hypogene origin*

			ppm
Galena	0—20%	Pentlandite	27— 67
Molybdenite	0—1,000 ppm	Millerite	5— 10
Volcanic sulfur	0.x—5.18%	Stibnite	0— 28
Linneite	0.x—4.69%	Sphalerite	1— 120
Chalcopyrite	0—1,000 ppm	Bismuthinite	150—8,000
Pyrite	0—3%	Arsenopyrite	1— 144
Pyrrhotite	1—60 ppm	Cinnabar	0—4,000
		Bornite	100— 150
		Marcasite	3— 80

FLEISCHER (1955) published some semi-quantitative data of selenium contents in sulfide minerals of various origins (mostly by spectrographic methods) as listed in Table 34-D-2.

Table 34-D-2. *Semi-quantitative data on selenium in sulfide-minerals.* (FLEISCHER, 1955)

Sphalerite
  Very low content in 41 samples; 36 range between 50 and less than 10 ppm Se; 1 sample had 900 ppm, none was free of this element
Pyrite
  One sample had 300 ppm, 80 range between 10 and 50 ppm; in 3 samples Se had not been detected
Chalcopyrite
  One sample had 2,100 ppm, 22 range between 10 and 50 ppm; none was free of this element
Marcasite
  One sample had 11 ppm, 7 samples had much smaller concentrations than 10 ppm; 1 had no Se at all
Pyrrhotite
  One had 63 ppm, 19 range between 5 and 50 ppm
Arsenopyrite
  3 samples had 42, 47 and 57 ppm
Galena
  PbS forms solid solutions with PbSe; Swedish galenas range mostly between 50 and 1,000 ppm, max. 1.5% Se; in general, galenas have only 0 to 15 ppm of Se

---

[1] For a bibliography on the geology of selenium see LUTTRELL (1959).

II. Selenium Minerals

Table 34-D-3. *Selenium minerals*

1. Native selenium:
Native selenium, Se (trig. and monoclinic).

2. Selenides and intermetallic compounds:

Naumannite	Ag_2Se (below 133° C orthorhombic, above cubic)
Aguilarite	Ag_4SeS
Eucairite	AgCuSe
Crookesite	$(Cu, Tl, Ag)_2Se$
Clausthalite	PbSe
Tiemannite	HgSe
Onofrite	Hg(SeS)
Achavalite	FeSe
Ferroselite	$FeSe_2$
Cadmoselite	CdSe (wurtzite—type)
Eskebornite	Fe_3CuSe_4
Permingeatite	Cu_3SbS_4
Klockmannite	CuSe
Berzelianite	$Cu_{(2-a)}Se$
Umangite	$Cu_{(4-a)}Se$
Guanajuatite	$Bi_2(Se, S)_3$
Laitakariite	$Bi_8(S, Se)_7(?)$
Jeromite	$As(S, Se)_2$
Stilleite	ZnSe
Trogtalite	$CoSe_2$
Hastite	$(Co, Fe) Se_2$
Bornhardtite	Co_3Se_4
Freboldite	CoSe
Tyrrellite	$(Cu, Co, Ni)_3 Se_4$
Blockite	$NiSe_2$

3. Sulfosalts:

Weibullite	$PbBi_2(S, Se)_4(?)$
Platynite	$PbBi_2(Se, S)_3(?)$
Wittite	$Pb_5Bi_6(Se, S)_{14}(?)$

4. Oxides:

Selenolite	SeO_2.

5. Selenates and selenites:

Kerstenite	$PbSeO_4 \cdot 2 H_2O(?)$
Chalcomenite	$CuSeO_3 \cdot 2 H_2O$
Ahlfeldite	$NiSeO_3 \cdot 2 H_2O$.

GONI and GUILLEMIN (1953) found that our knowledge of natural selenates and selenites is — due to the scarcity of those minerals — rather limited and that further studies are necessary.

It should be noted that polymorphism is rather frequent in the group of selenium minerals, but the phase relations have not always been sufficiently studied.

Selenium forms a homogenous series of mixed crystals with sulfur. In spite of the far greater abundance of sulfur, selenium contents from traces to 0.2% in native sulfur occur very frequently, and similarly, native selenium often contains sulfur.

Another well studied isomorphous series is that between PbS and PbSe. The latter compound (clausthalite) is indeed the most abundant selenium mineral. Under natural conditions, this lead selenide has only traces of Te, though the PbTe (altaite) is a rare, but well known tellurium mineral of perfect crystallographic analogy to PbS or PbSe. Experimental results on the partition of selenium between galena and sphalerite, and galena and chalopyrite are reported in Section 34-F (BETHKE and BARTON, 1971 etc.).

It should be noted that selenium-gold minerals do not exist. Though selenides occur often with gold and gold-minerals, the existence of gold selenides has not yet been confirmed.

34-E. Abundance in Common Igneous Rock Types

As pointed out previously, selenium has strong chalcophile tendencies under endogenic conditions. It is not surprising that its content in igneous rocks is low and not yet well known. Selenium probably accompanies the accessory sulfide minerals of these rocks. TUREKIAN and WEDEPOHL (1961) gave 0.05 ppm as normal content for all magmatic rocks, ultrabasic, basaltic, granitic and syenitic types. RANKAMA (1950) gives an average of 0.09 ppm for all igneous rocks. The most complete information is given by SINDEEVA (1964), see Table 34-E-1.

Table 34-E-1. *Selenium content in various magmatic rocks mainly from the USSR* (According to SINDEEVA, 1964, methods: Wand W/C; except four basalt samples see footnotes

Rocks	Region	Stratigraphic age	Se ppm
Biotite-granite (medium-grained), composite sample	Talitskii—Massif (Altai)	Variscan	0.1
Alaskite-granite, composite sample	Kuraminskii Mountains (Central Asia)	Post-lower Permian	0.1
Biotite-granite (medium-grained)	Great Canyon Massif (Magadan Dept.)	Cretaceous	0.2
Granodiorite	Kapchagai Massif (Magadan Dept.)	Cretaceous	0.2
Gabbro	Norilsk	Variscan	0.1
Basalt, composite sample,	Tuvinstaya Autonom. Region	Lower Devonian	0.1
Basalt, tholeiitic [b]	Pacific (2), Atlantic (1)	Recent	0.15—0.18
Basalt [a] BCR-1	Columbia river (USA)		0.11
Quartz-porphyrites and their tuffs, composite sample	Tuvinskaya Autonom. Region	idem	0.1
Andesitic gabbro	Norilsk	Variscan	0.1
Olivine-gabbro	Kundai Tunguska	Lower Triassic	0.15
Monzonite	Kadzharan (Caucasia)	Miocene	0.14
Miaskite, composite sample	Vishnerye Gory-Urals	Devonian	0.21
Nepheline syenite, composite sample	Tataralkaline Massif. Krasnoyarskii Rayon	Postcambrian	0.1

[a] HASKIN *et al.* (1970) N/R.
[b] LAUI *et al.* Geochim. Cosmochim. Acta **36**, 329 (1972).

Selenium

SINDEEVA (1964) gives the following data for the average content of Se in various magmatic rocks of the USSR (Table 34-E-2).

Table 34-E-2

Rock-type	Occurrence in USSR in % of area	Number of samples and specimens	Se ppm (W/C)
Salic intrusives and extrusives	62.2	5	0.14
Mafic and ultramafic intrusives and extrusives	37.4	10	0.13
Alkalic rocks	0.4	3	0.10
Total:	100.0	18	Average: 0.14

On the Devonian spilites (diabases) of the eastern Harz-mountains (East-Germany), LEUTWEIN and STARKE (1957) found 0.54 ppm Se in the normal rock. In a fresh spilite (diabase) with metasomatic hematite-ore, the selenium-content was 1.26 ppm (W method).

In several international standard rock samples, BRUNFELT and STEINNES (1967) found by N/R methods the following selenium concentrations (Table 34-E-3):

Table 34-E-3. *Content of selenium in some standard rocks (ppm).* [According to BRUNFELT and STEINNES, 1967, (N/R)]

Basalt ("diabase") W-1	Basalt BCR-1	Nepheline Syenite STM 1	Granite G-2	Granodiorite GSP-1	Andesite AGV-1	Peridotite PCC-1	Dunite DTS-1
0.098	0.093	0.012	0.005	0.055	0.007	0.023	0.003
0.104	0.112	0.008		0.063	0.009	0.021	0.004
0.114							
0.123							
0.110	0.103	0.010	0.005	0.059	0.008	0.022	0.004

34-F. Behavior during Magmatic Processes, Pegmatites, Gas Transport and Ore Deposition

I. Deposits of Magmatic Origin

Selenium with its chalcophile character and its high fugacity is not accumulated under the thermal conditions of magmatic melts nor in the early crystallizations of magnetite, chromite, etc. However, because of the similarity of the ionic radii of Se^{2-} and S^{2-} as well as the EK-coefficients for their lattice energy (FERSMANN, 1939), this element is always found in sulfides, even those of magmatic origin. GOLDSCHMIDT and HEFTER (1933) and GOLDSCHMIDT and STROCK (1935) found a Se:S-ratio of 1:7,000 in primary pyrrhotine and pentlandite of basic magmatic rocks and in magmatic pyrite and chalcopyrite ratios from 1:10,000 to 1:20,000. These ratios are not very far from the ratio of Se:S in the earth's crust. Under high temperatures and strong reducing conditions, a separation of S from Se is evidently possible. The close diadochy decreases if the oxygen fugacity increases — and under strong oxidizing conditions, both elements will show strongly different behavior.

Under magmatic and hot hydrothermal conditions, selenium enters especially pyrite, bornite, pentlandite and chalcopyrite lattices. Examples of deposits of this type are Sudbury (Canada), Norilsk (USSR) and Boliden (Sweden). The Se content in these ores is very low (3—30 ppm, and exceptionally 100 ppm Se) and selenium minerals are absent. (Only Te forms its own minerals in microscopic dimensions.) The metallurgical treatments of ores of this type give by-products sufficiently rich in Se and Te to permit a special separation (fly dusts and electrolytic sludges for Se). The production of ores from mines of this genetic type is very important — the production of Se (and Te) from ores of this type represents about 80% of the total world production of these elements. Compared to these deposits, the production of specific selenium-deposits is nearly insignificant.

Table 34-F-1. *Selenium content in minerals from the Norilsk and Boliden-deposits.* (After SINDEEVA, 1964, and GRIP and WIRSTAM, 1970)

Mineral	Norilsk ppm Se	Boliden ppm Se
Pyrrhotite	36—140	26— 70
Chalcopyrite	50—200	700—1,400
Pentlandite	67	
Millerite	10	
Pyrite	tr.	40— 30
Bornite	100—150	
Arsenopyrite		400—1,600
Sphalerite		300— 900

© Springer-Verlag Berlin · Heidelberg 1972

Analyses of the selenium-content in typical minerals from Norilsk (USSR) have been published by SINDEEVA (1964), and from Boliden (Sweden) by GRIP and WIRSTAM (1970), see Table 34-F-1 (indications on the analytical methods are not given).

II. Deposits of Pegmatitic and Pneumatolytic Genesis

In the group of pegmatitic deposits, sulfides are rare and information on their selenium content is missing. In the greisen-type deposits with cassiterite, molybdenite, wolframite and even in wolframite-molybdenite-quartz-veins, selenium has been found. The pneumatolytic formations are characterized by strong acid and slightly oxidizing conditions. Trivalent iron (hematite) is very common, in addition to the sulfides, molybdenite is very stable; pyrite and arsenopyrite may be present. Here, Se enters the molybdenite lattice; in cassiterite and wolframite it has not been found. CHRUSČOV et al. (1960) found 30 to 240 ppm of Se in molybdenites of different deposits; TISCHENDORF (1965) gave a systematic report on the selenium content of the Sn—Mo—W deposits of the Erzgebirge and found 50 to 200 ppm of Se W/C method) in the molybdenites. In all molybdenites a slight enrichment of Se appears relative to S. The Se content is independent of the genetic temperatures and depends merely on the Eh-pH conditions. KVAČEK and TRDLIČKA (1970) studied molybdenites in greisen-type and pneumatolytic veins from the Bohemian Massif (W/C method). Se contents vary from 100 to 180 ppm and the Se:S ratio from 1:3900 to 1:2100. In hydrothermal metallic veins they found 66 to 300 ppm Se, and Se:S ratios ranging from 1:5700 to 1:1300.

III. Hydrothermal Veins

In the typical quartz-gold-veins, tellurium is often found in very rare minerals (tetradymite), but selenium, although in content about 10 times higher, is always present in pyrite, chalcopyrite and eventually arsenopyrite, but not as its own minerals. In lead zinc deposits, Se enters the galena lattice and to a lesser extent in sphalerite, but even here, the copper minerals are richer in this element. Experimental studies of the partition coefficients of selenium between galena and sphalerite have been made by HALBIG (1970) and by BETHKE and BARTON (1971). They found experimentally that the partition coefficients K in the systems $Pb \cdot Zn \cdot S \cdot Se$, $Pb \cdot Cu \cdot Fe \cdot S \cdot Se$ and $Zn \cdot Cu \cdot Fe \cdot S \cdot Se$ depend mainly on the temperature of

Table 34-F-2. *Experimental results of the partition of Se among galena and sphalerite and galena and chalcopyrite.* (After BETHKE and BARTON, 1971)

Distribution coefficient	Temperature range in °C	$\log K = \dfrac{-\Delta H - \Delta VP}{2.3\,RT} + C$ in weight percent
$K_{Se}^{gn-sp} = \dfrac{PbSe_{gn} \cdot ZnS_{sp}}{PbS_{gn} \cdot ZnSe_{sp}}$	600—890	$\log K = \dfrac{2850 + 0.0027P}{T_{(°K)}} - 1.33$
$K_{Se}^{gn-ccp} = \dfrac{PbSe_{gn}}{Cu_{0.526}Fe_{0.526}Se_{ccp}}$	300—595	$\log K = \dfrac{3410}{T_{(°K)}} - 3.10$

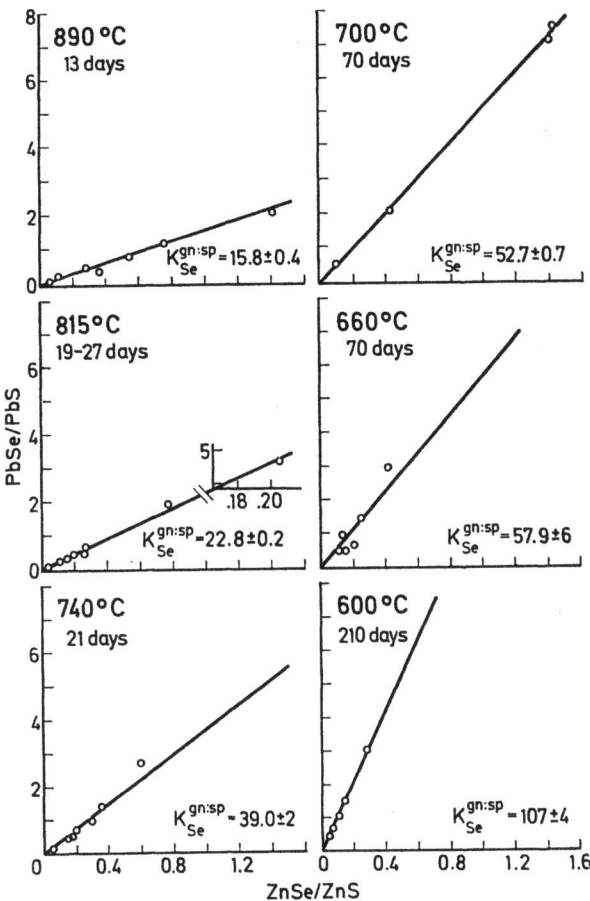

Fig. 34-F-1. Plots of mole fraction ratios PbSe/PbS *vs* ZnSe/ZnS calculated from experimental data for 6 temperatures (after BETHKE and BARTON, 1971)

co-precipitation of galena and sphalerite, galena and chalcopyrite or sphalerite and chalcopyrite (Table 34-F-2 and Fig. 34-F-1).

In other hydrothermal deposits of the vein-type, Se is evidently missing or of no importance and it is only necessary to mention some very rare, even unique deposits where selenium appears in its own minerals.

SINDEEVA (1964) found in the north-eastern part of the USSR (Seychemskoye and other areas) veins of cobalt-minerals in very complex parageneses, mostly associated with differentiated granite-intrusions, (cobalt-arsenides with hessite, clausthalite, proustite, etc...). Here, the mean Se-content of the ores varies from 1 to 20 ppm Se, always in the form of selenium minerals.

A unique vein type in Devonian shales represented the now exhausted selenium deposit of Pacajake in Bolivia. In a barite-siderite-hematite-calcite vein blockite, naumannite, clausthalite and others were found (BLOCK and AHLFELD, 1937).

The only gold-selenium deposit so far described was the famous gold-mine of Redshang — Lebong (Sumatra). ZWIERZYCKI (1937) found that the main mineral was pyrite associated with native gold, berzelianite, aguilarite, stromeyerite, etc. The average content of Se in the ore was 160 ppm. The veins were situated in a tectonic zone between Miocene bituminous limestone and andesites and dacites.

The gold-selenium deposits of Western Germany (Korbach-Waldeck) and the complex selenide occurences of Tilkerode in the Harz-Mountains (East-Germany) have been studied by several authors (RAMDOHR, 1931 and 1955; LEUTWEIN and STARKE, 1957; TISCHENDORF and UNGETHÜM, 1964). In the classic literature, these veins are described as telethermal epigenetic hydrothermal types. However, LEUTWEIN and STARKE (1957) and TISCHENDORF and UNGETHÜM (1964) came to the conclusion that the metal content of these deposits, especially the selenium, has been leached out from Se-bearing host-rocks under favorable Eh-pH conditions, so they are of typical lateral-secretionary origin in the sense of von Sandberger. It was in veins of this type that RAMDOHR (1955) discovered the new minerals, eskebornite, trogthalite, hastite and bornhardtite. The main Se-mineral is clausthalite. More interesting than these unique deposits is the occurence of clausthalite in uranium-deposits, especially associated with massive ore-shoots of pitchblende with pyrite, hematite and sulfides. This is an interesting example of slightly oxidizing conditions, produced by the reduction of hexavalent, soluble uranium to the insoluble tetravalent state. Under such conditions, few sulfides are stable, e.g., molybdenite, pyrite, chalcopyrite, clausthalite. Galena needs more reducing conditions and appears here — if ever — as a later formation, eventually with other sulfosalts, such as proustite.

IV. Hydrothermal Pyrite and Chalcopyrite Deposits of Magmatic or Sedimentary Origin

In magmatic-hydrothermal pyrite and pyrite-copper deposits, lenses as well as vein-types, Se is always present in trace amounts (mostly less than 20 ppm). These occurences are often of great economic value as sulfur or copper-deposits. Their important production may consequently give a considerable selenium production too, as mentioned above for the magmatic group. However, in this hydrothermal group, their genesis is not always clear. Most of them appear to be of submarine exhalative origin — others are perhaps of pure sedimentary biogenic origin. Mixed types are also well known (e.g., Meggen or Rammelsberg in Germany). Often, they have been subjected to a later metamorphism as in the Norwegian deposits. GOLDSCHMIDT and STROCK (1935) found that sedimentary sulfides have a Se:S ratio of 1:50,000 to 1:100,000, but sulfides of doubtless magmatic origin (the question of intrusive or volcanic was not discussed) have Se:Se ratios from 1:10,000 to 1:20,000. This problem needs further study, see for example LOFTUS-HILLS and SOLOMON (1967) and VOROBJEW (1969).

V. Volcanic Exhalations

It is well known that H_2Se and SeO_2 appear in volcanic gases and that Se is found in sulfur of volcanic origin. It scontent may increase from traces upto 5% Se. Other minerals in this association are gypsum and, rarely, realgar. GOLDSCHMIDT and STROCK (1935) mention that native sulfur of sedimentary origin is poor in Se, and that of volcanic origin is far richer. Recently, SREBRODOL'SKIY and SIDEL'-NIKOVA (1970) published data from the volcanoes of the Kuriles; a resumé of their results is given in Table 34-F-3, (with no information on analytical methods).

Table 34-F-3. *Selenium content in native sulfur of volcanic origin.* (From SREBRODOL'SKIY and SIDEL'NIKOVA, 1970)

Volcano (Kuriles)	Range of variation of Se content	Number of analyses
Ebeko	20—1,040 ppm	8
Kuntomintar	14— 760 ppm	5
Mendeleyew	1— 2 ppm	2
Golovnina	40 ppm	1

Revised manuscript received: June 1971

34-G. Behavior during Weathering and Alteration of Rocks

A very characteristic feature of selenium is that under oxidizing conditions selenium-bearing sulfides and selenides are rapidly oxidized. The resulting selenite ions are very stable and able to migrate until they are adsorbed on iron hydroxides and soil minerals. There they are, in contrast to sulfate ions, nearly quantitatively fixed. Under the same conditions sulfides are oxidized to sulfuric acid or sulfates — most of them are very soluble. The migrating sulfuric acid may react with sulfides to form H_2S which precipitates sulfides. If selenite ions participate in this migration, there will eventually be mixtures of cementative selenides plus native selenium. In general, once formed, selenite ions will be fixed by iron oxide minerals, and in this way, they will stay in the "iron cap" of the deposits as well as gold. The result is that the iron cap may well be an exploitable deposit of gold and of selenium, even if the primary ore is not. In contrast to tellurides, especially tetradymite, selenides in this zone are oxidized and absorbed, so that selenium-bearing placers, contrary to tellurium, are not known.

The existence of selenates is the object of controversy. As a general rule, the oxidation potential necessary for their formation will not be reached. It is so high, that water begins to dissociate before SeO_4^{2-} compounds may be formed. Therefore if they exist, selenates will only exist under extremely dry, alkaline and very high oxidation conditions (see Section 34-H). The formation of native selenium is by far more probable.

In about 200 to 300 years, the selenium leached from a bituminous, copper-rich marl dump of a mine has accumulated upto detectable levels in the underlying soil, but only traces of copper, and no enrichment of S, have occurred (LEUTWEIN-STARCKE, 1957).

Native sulfur of bacterial origin is, as previously noted, poor in Se. But native sulfur resulting from the oxidation of sulfide ores may well have a Se content upto 1% or even more.

In the oxidized uranium ores of sedimentary or of other origin, especially in the carnotite-uranium ores of Colorado (U.S.A.), selenium has been always detected. For geochemical prospecting, it has been useful to follow the selenium indications (in plants — see Section 34-L). Selenium, once adsorbed in soils, is far less mobile than uranium, which is easely leached and found only in deeper zones.

34-H. Solubilities of Compounds which Control Selenium Concentrations in Natural Waters; Valence States in Natural Environments; Adsorption

Selenium, located under sulfur in the periodic system, has analogous compounds and states of oxidation, but their thermodynamic stability is quite different from those of the lighter and far more frequent sulfur compounds. The ionic and atomic radii are from Tables 12-7 and 12-8 of volume 1 of this handbook:

$$S^{2-} \quad 1.74 \text{ Å} \quad\quad Se^{2-} \quad 1.91 \text{ Å}$$
$$S \quad\quad 1.27 \text{ Å} \quad\quad Se \quad\quad 1.40 \text{ Å}$$
$$S^{6+} \quad 0.34 \text{ Å} \quad\quad Se^{+6} \quad 0.3\text{—}0.4 \text{ Å}$$

The dissociation of H_2S occurs in two stages:

$$H_2S \rightleftharpoons HS^- + H^+ \quad \text{and} \quad HS^- \rightleftharpoons H^+ + S^{2-}.$$

The analogous dissociation of hydrogenselenide and the constants (K) of dissociation are:

$$H_2Se \rightarrow H^+ + HSe^- \quad \text{and} \quad HSe^- \rightleftharpoons H^+ + Se^{2-},$$
$$K_1 = 1.30 \times 10^{-4}, \quad K_2 = 1 \times 10^{-11}.$$

The products of solubility of several sulfides and selenides, after TISCHENDORF and UNGETHÜM (1964), are listed in the following table.

Table 34-H-1. *Solubility constants (K_L) of some sulfides and analogous selenides.* (After TISCHENDORF and UNGETHÜM, 1964)

HgSe, tiemannite	10^{-59}	HgS, zinnabar	$1.88 \cdot 10^{-53}$
Ag$_2$Se, haumannite	?	Ag$_2$S, argentite	$6.31 \cdot 10^{-50}$
Cu$_2$Se, berzelianite	?	Cu$_2$S, chalcosite	$7.25 \cdot 10^{-49}$
CuSe, klockmannite	10^{-49}	Cu$_2$S, covellite	$3.98 \cdot 10^{-36}$
PbSe, clausthalite	10^{-38}	PbS, galena	$7 \cdot 10^{-28}$
ZnSe, stilleite	10^{-31}	ZnS, sphalerite	$1.52 \cdot 10^{-24}$
FeSe, achavalite	10^{-26}	FeS, pyrrhotite	$5.13 \cdot 10^{-18}$

The data on sulfides are from CZAMANSKE (1959), that on tiemannite is reported after BJERRUM et al. (1964); all other data are after LATIMER (1952).

The geochemical behavior of selenium in aqueous systems is demonstrated in pH-Eh diagrams. DELAHAY et al. (1952) published such a diagram for selenium. Later, TISCHENDORF and UNGETHÜM (1964), studying the conditions of the formation of clausthalite in presence of galena, gave a more detailed diagram of this type. DELAHAY et al. (1959) calculated a diagram for Σ Se = 1, at 25°C. For geochemical purposes, the diagram of TISCHENDORF and UNGETHÜM (1964) is more useful, see

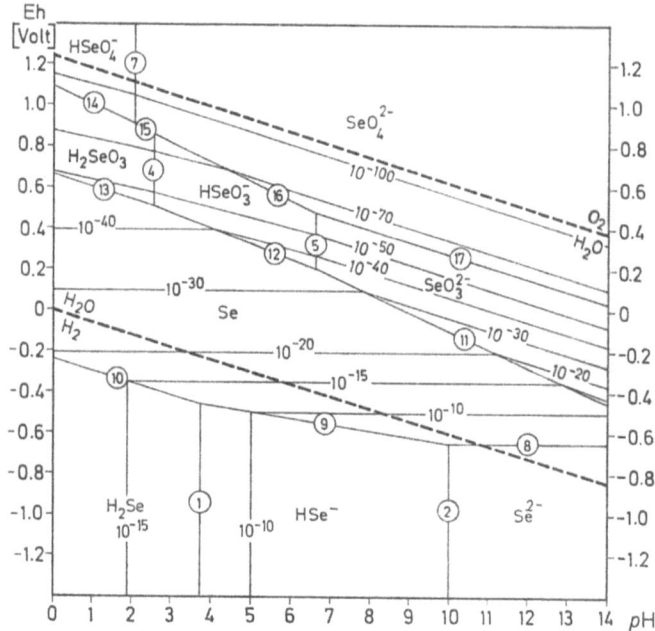

Fig. 34-H-1. Fields of predominance of selenium-bearing compounds for $\Sigma\,Se=10^{-5}$, 25°C and 1 atm. total pressure. The thin lines give the activities of the selenide-ions. (TISCHENDORF and UNGETHÜM, 1964)

Fig. 34-H-1. DYACHKOVA and KHODAKOWSKIY (1968) published analogous diagrams for 15°, 150° and 300°C, and 1, 5 and 85 atm. (see Fig. 34-H-2).

The field of predominance of native selenium is far more extended than that of native sulfur. We should expect (GOLDSCHMIDT has already mentioned this problem) that native selenium in tiny crystals should be rather abundant in oxidized sediments, but until now, observations of this were unreported. Fig. 34-H-2 shows that HS⁻-ions are active upto pH 13 and S^{2-} ion activity is limited to the field of extremly basic conditions. In contrast to this, the field of predominance of H_2Se is less and that of Se^{2-} more extended. From Fig. 34-H-1, it is seen that the field of predominance of SeO_4^{2-} ions is restricted to extremely oxidized conditions, so that the existence of selenates is rather improbable. If selenates exist, they could only exist in hot arid climates, where solutions normally have a strong alkaline character.

Under reducing conditions, we have a rather good camouflage of Se in sulfides. Under slightly oxidizing conditions in greisens, skarns and in those hydrothermal veins where Fe^{3+} and SO_4^{2-} ions are present as well as Fe^{2+} and HS⁻, a separation of Se may be expected. Under such conditions, selenium as a selenide is less mobile than sulfur as a sulfate-ion. Under stronger oxidizing conditions, selenides are oxidized to stable selenites (under natural conditions sulfites are quite unstable and are rapidly oxidized to sulfate-ions). Selenites of alkalies are soluble, but those of iron are not. Selenites are rapidly and nearly completely fixed by iron hydroxides and oxides. The separation of selenium from sulfur in the exogenic cycle is based on the greater mobility of sulfate ions. The problem of the equilibria of sulfate and sulfide and their influence on the distribution of selenium ions has been studied

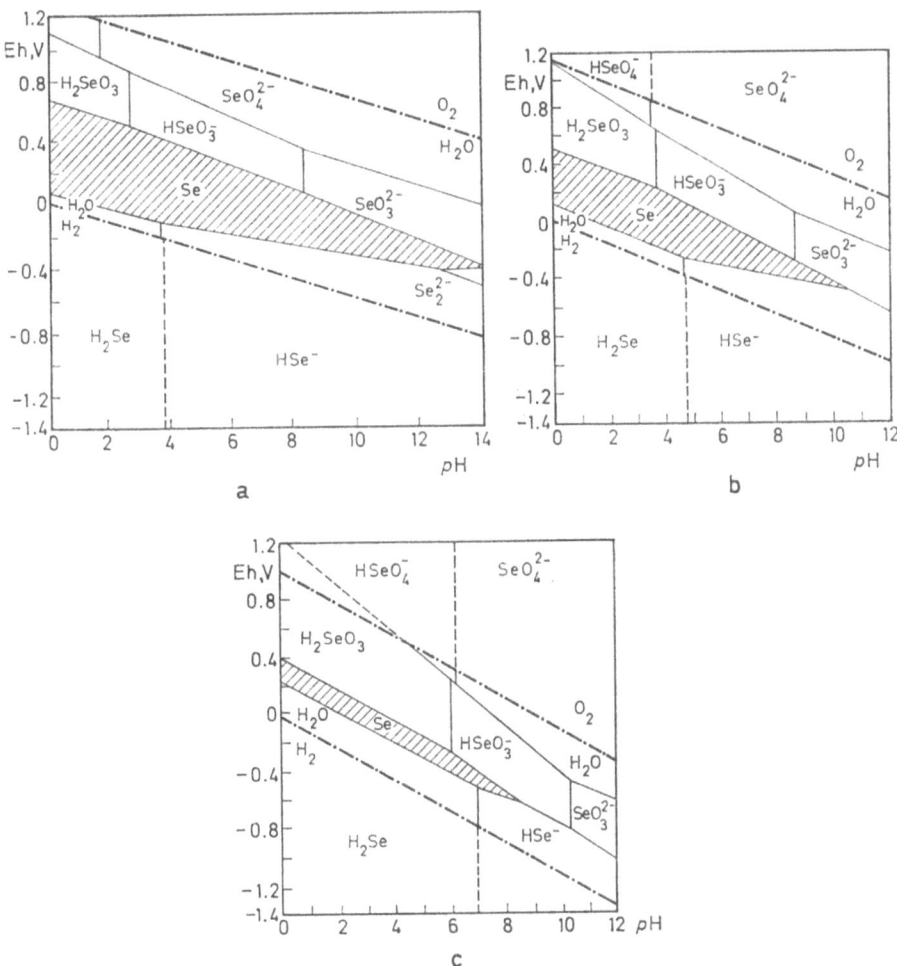

Fig. 34-H-2a—c. Thermodynamic equilibria in the system H_2O-Se at $\Sigma\, Se = 10^{-6}$, a: 25°C, P=1 atm.; b: 150°C, P=5 atm.; c: 300°C, P 85 atm. (DYACHKOVA and KHODAKOVSKIY, 1968)

recently by BADALOV et al. (1970). These authors studied the selenium content of anhydrite and coexisting pyrite, chalcopyrite and enargite, by fluorometric analyses. The selenium content of the sulfides ranged from 30 to 220 ppm (average 88 ppm) while that of coexisting anhydrite varied between 0.8 and 3 ppm (mean 1.5 ppm).

From these observations it is not surprising that Se has a short residence time in the oceans. Contrary to sulfur, selenium is accumulated in marine clay sediments mostly by adsorption (GEERING et al., 1968) in near contrast to sulfur, which is predominantly either precipitated as sulfides in the upper layers of the sediments (bacteriogenic), or precipitated mainly as gypsum or even anhydrite when it reaches the salinar-facies in the oxidized state. Selenium has rarely been found in these parageneses.

Revised manuscript received: June 1971

34-I. Abundance in Natural Waters

The information on the hydrogeochemistry of selenium is incomplete. In ocean water, WATTENBERG (1943) found 0.004 ppb (35°/$_{00}$ salinity), SCHUTZ and TUREKIAN (1965) found 0.09 ppb Se.

Data on the selenium content of rivers are rare. BYERS (1937) mentions that draining waters of selenium-rich soils are often seleniferous: 0.1 to 0.2 ppm. In the Colorado river at Topock, Arizona, he found 0.03 ppm, in the Guanajuato river (Mexico) at about 400 m from mines 0.2 ppm Se. LEUTWEIN (1943) (W+S methods) found in the waste water of the smelter of Halsbrücke/Freiberg (Saxonia), especially after slag-granulation, 0.1—0.3 ppm Se. This water enters the Mulde river, only 1 km from the smelter, and no selenium was detectable in the water, even though this river has just passed through the famous Freiberg ore district (lead-zinc-silver-ores with only traces of Se). Here, selenium is evidently rapidly adsorbed on the muds of the river, just as is the case for lead ions (LEUTWEIN, unpublished results).

Recently, KHARKAR et al. (1968) published some data on selenium transport of rivers not affected by draining waters from abnormally selenium rich regions. The analyses were made by neutron activation and Table 34-I-1 gives some average values.

Table 34-I-1. *Selenium content of some rivers.* (KHARKAR, TUREKIAN and BERTINE, 1968)

River	Locality	Date	Se ppb
Mississippi	Minneapolis	July 1965	0.114
Susquehanna	Marietta/Pa	June 1966	0.325
Rhone	Avignon/France	June 1966	0.15
Amazon	Santarem, Brazil	Jan. 1965	0.21

Such data are very valuable, but not yet sufficient for a detailed scheme of the cycle of selenium — this would only be possible if the selenium content of different types of rivers are studied by several analyses per year. Several authors state that selenium is mostly supplied by the rivers as a dissolved element, with a relatively low fraction in the suspended clay minerals.

Selenium in wells and mineral springs has been observed. The highest content, 0.2 ppm Se was found by TABOURY (1909) in the mineral springs of La Roche Posay (Dept. Vienne, France).

SREBRODOL'SKIY and SIDEL'NIKOVA (1970) report that oxygenated water of desert areas contain 10^{-9} to 10^{-7} ppm Se. Sodium rich thermal waters of Central Asia may have upto $6 \cdot 10^{-3}$ ppm of Se. H_2S-rich waters from a sulfur deposit of Carpathia have 100 to 250 ppm H_2S and 0.2 ppm H_2Se.

The problem of the so-called venomous springs (or wells) of South Dakota (U.S.A.), has been studied by BYERS (1937). These waters have $4 \cdot 10^{-3}$ ppb and are definitely not dangerous. But the grass around these sources has accumulated selenium and this is the real cause for the observed poisoning of cattle.

Revised manuscript received: June 1971

34-K. Abundance in Common Sediments and Sedimentary Rock Types

I. Major Rock Types

During weathering processes, selenium will be oxidized from selenide to selenite. Calcium and sodium selenites are soluble; iron-selenites are nearly insoluble. Selenite-ions in hydrous solutions are very rapidly adsorbed by iron hydroxides.

As RANKAMA (1950) remarks, selenium is mainly found in clay sediments and the degree of enrichment will be still higher in oxide rich deposits. The processes of removal of Se from ocean waters are — as for arsenic — a very important "depoisoning" of the oceans, especially in the biological sense, for contrary to sulfur, selenium is hostile to most organisms.

SINDEEVA (1964) notes that fresh oceanic sediments from the Bering-Sea, the Arctic, the Caribbean and the Gulf of California always have a selenium content of about 1 ppm, with a maximum content of 5 ppm.

VOROBJEW (1969) recently studied the mode of selenium occurrence in marine Paleocene to Oligocene black schists of Western Central Asia. Selenium is normally in good correlation with the content of organic carbon and iron of the sediments, but an abnormal enrichment of selenium for the Paleogene is reported which is ascribed to the activity of submarine volcanism at this epoch (see Table 34-K-1).

Table 34-K-1. *Selenium content of normal (Mesozoic, Neogene) and abnormal (Paleogene) deposits of West Central Asia.* (After V. P. VOROBJEW, 1969), (No information about the analytical method)

Type of environment of deposition	Selenium content in ppm	
	Mesozoic and Neogene	Paleogene
Continental	2	0.2
Marine		
Littoral	4	12
Neritic	7	40
Pelagic	3	66
Gross average	5	35

TUREKIAN and WEDEPOHL (1961) report, mostly from older analyses (GOLDSCHMIDT and STROCK, 1935; MINAMI, 1934/35), that shales have an average content of 0.6 ppm, and sandstones and limestones 0.05 to 0.08 ppm Se. Because of the rapid adsorption of Se in marine clastic sediments, almost no selenium traces are found in evaporites. However, these deposits concentrate considerable amounts of sulfur.

Selenium

Intensive analytical work has been carried out on the Cretaceous series of the central parts of the United States and the following Table 34-K-2 is compiled from data published by WILLIAMS and BYERS (1935) and BYERS (1937).

Table 34-K-2. *Selenium in Cretaceous sedimentary rocks of the United States.* (After WILLIAMS and BYERS, 1935; and BYERS, 1937)

	ppm
Limestones of Niobrara	0.4—30
Overlying Pierre-shales	trace to 100
Alabama-schists	0.15—0.6
Gypsum-bearing sands of New-Mexico	0.2
Gypsum-bearing schists of Nebraska, New Mexico and Kansas	1—10
Cretaceous schists of South Dakota	3—6
Niobrara-series	wide range upto 3
Overlying Smoky Hill series	3—6
Maximum content (at Custer, S. Dakota)	30
Same stratigraphic horizon at Nebraska	upto 100

The following Table 34-K-3, gives some data from literature.

Table-K-3. *Se contents of sedimentary rocks*

	ppm Se	Analytical method	Reference
Composite sample of 10 Mesozoic shales (Japan)	0.38	W/C	MINAMI (1935)
Composite sample of 14 Paleozoic shales from Hondo (Japan)	0.24	W/C	MINAMI (1935)
36 European Paleozoic shales	0.12	W/C	MINAMI (1935)
Culmian shales, Lehesten (Germany)	0.1	W/C	GOLDSCHMIDT and STROCK (1935)
Separated pyrite of these shales	26	W/C	GOLDSCHMIDT and STROCK (1935)
Devonian shales, Bernau, Black Forest (Germany)	0.4	W/S	LEUTWEIN (1938)
Culmian shales from Sulzburg, Baden (Germany)	0.5	W/S	LEUTWEIN (1938)
Gotlandian lydites, Thuringia (black)	2.2	W/S	LEUTWEIN (1938)
Gotlandian lydites, Thuringia (red)	2.5	W/S	LEUTWEIN (1938)
Black shales of Korbach (Germany)	5	W/S	LEUTWEIN (1938)
Silurian shales, Harz Mountains (Germany)	2	W/S	LEUTWEIN and STARKE (1957)
Arkoses of Rotliegend age, near Wettin (Germany)	0.2	W/S	LEUTWEIN and STARKE
Dictyonema shales of the Esthonian SSR (3 samples)	0.3, 0.7 and 9	W/C	SINDEEVA (1964)

The Kupferschiefer is a typical euxinic marine sediment. NODDACK and NODDACK (1930) found 300 ppm Se, WAGENMANN (see RÖSLER and LANGE, 1965) mentions only 20 ppm. The latter analysis seems to be more reliable.

LEUTWEIN and STARKE (1957) studied a profile of the Kupferschiefer of a fresh outcrop at Dobis near Wettin. The selenium distribution in a profile is listed in Table 34-K-4.

Table-K-4. *Selenium content in a profile of the Upper Permian Kupferschiefer.* (LEUTWEIN and STARKE, 1957.) The zero level is the conglomerate at the base in contact with the shale

Se ppm	Level in centimeters relative to a marker bed	Rock-type
0.60	−120 to −30	Conglomerate of porphyry
1.45	− 30 to 0	Conglomerate
15.00	0 to 16	Cupriferous marl
5.05	+ 16 to 29	Cupriferous marl
0.8	+ 29 to 65	Limestone

It is not known whether Se is incorporated in copper minerals (bornite), or whether it forms copper selenides and native selenium, both in submicroscopic dimensions.

II. Native Sulfur Deposits

Recently, an interesting study of selenium in native sulfur has been published by SREBRODOL'SKIY and SIDEL'NIKOVA (1970). They reported about 30 analyses of sulfur of different origin. Unfortunately no information on the analytical method is given. Their analyses are made on pure, isolated sulfur, well separated from the host-rock. Volcanic sulfur contains 1.6 to 1,040 ppm Se, sedimentary sulfur only 0.1 to 10 ppm. Only in sublimated native sulfur from burning coal of Ravatskoje (Tadzhik, S.S.R.) did the authors find selenium contents as high as 480 and 1,300 ppm Se.

Se content in some sedimentary sulfur deposits of Carpathia (Rozdol, Yazov, etc.) and of Central Asia (Shorsu, Gaurdak) is in the range from 0.66 to 9.9 ppm, average 1.97 ppm.

GOLDSCHMIDT and HEFTER (1933) found in native sulfur from Sicely (Italy) 0.9 ppm Se (W/C, methods). However, in the sulfur of Texas and Louisiana, the same authors could not detect selenium. They found concentrations in native sulfur of a gypsum quarry at Weenzen (Lower Saxony, Germany) upto 1 ppm Se.

Revised manuscript received: June 1971

34-L. Biogeochemistry

Information on the influence of selenide or selenite ions on bacteria and algae is not yet available. It seems possible that some of these organisms tolerate traces of selenium, but evidently no group exists for which selenium is as essential an element as sulfur. As a general rule, selenium, even in very low concentrations, is toxic for living matter with the exception of several plants which may grow even if considerable quantities of selenium occur in a soil, and some species are known for which selenium ions seem to be a necessary stimulator. For more details see SHRIFT (1964).

I. Soils

In soils, selenium has been found in several regions of the earth. The higher contents were detected in some areas of the United States and in Canada (Saskatchewan). Selenium-bearing soils have been reported in Mexico and in Ireland. SINDEEVA (1964) mentions that they probably exist in West China, Turkestan and Tibet. VINOGRADOV (1954) studied this problem in several regions of the Russian plains covered by different types of soils from podsol, chernosjem to subtropical types in their different horizons. He never found more than 0.01 ppm Se in the soils; often, selenium was not detectable. From this, it follows that the selenium content of the so-called selenium soils may be more than 1,000 times greater than the world-wide normal content.

II. Plants

Calcium and sodium selenites are very soluble, ferric selenites and adsorbed selenite ions on hydroxides of iron are far less soluble. It is well possible, that selenium exists in the upper humic horizons also in the form of organic compounds, albumins, amino-acids, etc., as a result of the decomposition of plants (see also ROSENFELD and BEATH, 1964). Evidently most plants tolerate a certain selenium level in soils, since selenium is less toxic for plants than for animals. There are some families of plants which are characteristic for higher selenium content in soils and these may well serve as selenium indicators. In the United States, *Astragalus pectinatus* and *Aplipappus Fremontii* are such indicator plants. *Astragalus, Cruciferae* and several composites are able to collect selenium in their leafs. In Germany (Harz-mountains), LEUTWEIN and STARKE (1957) tried to find selenium traces in plants accumulated on selenium-bearing soils (0.1 to 11 ppm Se). However, in all plants studied, the selenium content was below the limit of detection (≤ 2 ppm). The following were analyzed: *Vicia silvatica, Fagus silvatica, Carpinus betulus* (leafs), *Sieglingia decumbens, Euphorbia cyparissias, Astragalus glycophyllus* (whole plants). Indeed, adaption to selenium soils is only limited to certain species and not common to the whole families. VINOGRADOV (1965) has also studied several plants (Table 34-L-1).

Table 34-L-1. *Selenium content of plants from central Russia (in percent of dry plant).*
(After VINOGRADOV, 1965)

Astragalus	bisulcatus	upto 0.31%
Astragalus	pectinatus	0.18—0.22%
Astragalus	grayi	0.02—0.06%
Astragalus	succulentus	upto 0.01%
Oonopsis	condensata	0.17—0.21%
Oonopsis	engelmannii	upto 0.02%
Cruciferae	pinnata	0.06—0.1%
Cruciferae	stanleya	upto 0.01%

III. Animals and Se Diseases

BYERS (1937) reports an example of selenium poisoning from Iraputo, Mexico. Grass on selenium-accumulating soils had a selenium content of 0.3 ppm. In the milk of the cows, 0.06 ppm Se (for dry substance) has been detected, and is responsible for a chronic disease, erroneously attributed to Hg poisoning. Systematic research in selenium-rich regions of Dakota gave no clear indication of the toxicity for the population. It seems that a certain adaption is possible. The organism accumulates selenium in urine. Especially for the agriculture of the United States, the problem of selenium accumulation in special soils is rather important because of "alkalosis", a dangerous cattle disease. It is characterized by severe deformations of horns and hooves. In the sulfur-rich organic substances of the horns, sulfur is replaced by the toxic selenium. This disease does not always appear on selenium-rich soils. Selenium need only be present in soluble form, accessible to the plants. Cattle suffering from "alkalosis" had concentrations of 0.33 to 1.7 ppm Se in urine. A systematic investigation of the animal population in this region gave, in 92% of all analyzed cases, concentrations of 0.02 to 1.33 ppm Se in the urine, but no toxicity or disease could be cited. In plants (cereals), the proteins (gluten and S-aminoacids) accumulate Se. It seems that the famous "alkalosis", indeed a selenium desease, is typical for cattle (bovides) only.

IV. Coals

GOLDSCHMIDT and STROCK (1935) state that selenium is a trace element in coals, cokes and dusts of their combustion. Traces of selenium were found in the ashes of anthracitic and other coals of eastern Germany (Upper Carboniferous as well as in lignites of Tertiary age). It is certain that the main part of the selenium has been lost during calcination. Pyrites from coal deposits are selenium-bearing (10 to 30 ppm). The problem has not been studied in detail, but the higher selenium contents were found in coals with higher contents of ashes and sulfur (pyrite-rich coals). However, the existence of organic selenium compounds cannot be excluded, though the main part of selenium is evidently present in the sulfides of coals. The higher selenium values were found in the uranium-bearing coal of early Permian age from Freital near Dresden. A pyrite concentrate from these coals has 100 ppm Se (W). The ashes calcinated at low temperature gave 3 to 10 ppm (W/S) so it seems improbable that the whole selenium content of coals is concentrated in pyrites.

It is very probable, that the plants of earlier geologic periods, Carboniferous to Tertiary, accumulated selenium. There is no doubt that the sulfides, in which Se is actually concentrated, are of diagenetic origin.

Revised manuscript received: June 1971

34-M and 34-N. Abundance in Common Metamorphic Rock Types and Behavior in Metamorphic Reactions

Data on the selenium content of gneisses and other metamorphic rocks are not available. Selenium is a chalcophile element under endogenic conditions and its properties allow only a very limited incorporation in silicates. In metamorphic pyrite deposits, selenium contents are frequent, as well as those of non-metamorphic pyrite and pyrite-chalcopyrite deposits. The metamorphism evidently does not affect or mobilize selenium. GOLDSCHMIDT and HEFTER (1933) report Se data on 4 Norwegian metamorphic pyrite-chalcopyrite deposits. The separated pyrites from Varaldsö have 30 to 36 ppm, from Orkla 4 ppm, from Bakke 5 ppm and from Bossmo 29 ppm of selenium. GRIP and WIRSTAM (1970) gave the following data for the Boliden deposit: (Table 34-M and N-1, method not given).

Table 34-M and N-1. *Mean selenium content in minerals from Boliden, Sweden* (GRIP and WIRSTAM, 1970)

Pyrite	4— 30 ppm Se
Chalcopyrite	70—1400 ppm Se
Sphalerite	30— 90 ppm Se
Arsenopyrite	40—1600 ppm Se
Pyrrhotite	2— 7 ppm Se

The mean selenium content of arsenopyrite ore of this mine during 1933 to 1943 was 1,500 ppm. LEUTWEIN (1939, unpublished) analysed the pyrite ore from the metamorphic deposit at Bayerland near Waldsassen (Bayrischer Wald), which is probably of sedimentary origin, and found only traces of Se: 0.5 to 1 ppm (W/C method, limit of detection: 0.5 ppm). Analyses (by the same method) on amphibolite, ortho- and paragneiss from the Kinzigtal (Black Forest) indicated less than 1 ppm Se, but the pyrites isolated from several amphibolites of the Black Forest yielded upto 6 ppm of Se. Pyrites from an amphibolite (meta-diorite) of the Odenwald (Germany) gave only 2 ppm Se.

34-O. Relations to Other Elements; Economic Importance[1]

The rare element selenium is located in the periodic system under the abundant element sulfur. The oxidation potential of selenium is so different from that of sulfur, that separation of these elements is not uncommon except under reducing conditions. Selenites are far more stable than sulfites and the oxidation potential of selenates is so high, that they may only exist under extremely oxidizing conditions in hot arid climates.

The main sources for selenium production are pyrite deposits. In chalcopyrite, pentlandite and molybdenum ores, selenium is often present. Selenium minerals are rare; the most abundant, though economically unimportant, is PbSe, clausthalite. Deposits of selenium minerals are even rarer than those of tellurium.

In the exogenic cycle, selenium is often separated from sulfur. Selenites are easily adsorbed on ferric iron oxides of soils. In the oceans, selenium is rapidly fixed in oxides and hydroxides of iron and manganese in clay sediments. Volcanism may be a source of selenium transport too, but without economic importance.

The annual world production of selenium is not very high but there is an increasing interest in this element, in very pure form, by the electrotechnical and electronics industries. The use of selenium salts in the glass industries, either in form of traces for decoloration of slightly iron-containing ordinary glasses, or in higher contents for red rubi-glasses is much older. The main utilization is actually in the electrical industries. The world's largest producer of Se, the United States, imports about 20% of its needs. Information on the world production from 1910 to 1958 (excluding the production of the USSR) is listed in Table 34-O-1.

The production of the USSR must be important and probably ranges between the annual production of the USA and Canada.

Table 34-O-1. *Production of Se (in tons)*. (After SINDEEVA, 1964)

Years	Total world (except USSR)	USA	Canada
1910—1915	6	6	—
1916—1920	27	27	—
1921—1925	54	54	—
1926—1930	no data	no data	—
1931—1935	183	114	69
1936—1940	277	113	134
1941—1945	425	245	180
1946—1950	460	213	180
1951—1955	615	331	153
1956	850	502	230
1957	640	482	146
1958	600	302	183

[1] For a bibliography on the geology of selenium see LUTTRELL (1959).

References: Sections 34-B to 34-O

BADALOV, S. T., BELOPOLSKAYA, T. J., PRIKHIDKO, P. L., TURESEBEKOV, A.: Geochemistry of selenium in sulfate-sulfide mineral parageneses. Geochemistry International **7**, 799 (1970).

BAINBRIGE, K. T., NIER, A. O.: Relative isotopic abundances of the elements. Preliminary Report Nucl. Sci. Ser., Nat. Research Council U.S., Washington, D. C. **9** (1950).

BETHKE, P. M., BARTON, P. B.: Distribution of some minor elements between coexisting sulfide minerals. Econ. Geol. **66**, 140 (1971).

BJERRUM, J., SCHWARZENBACH, G., SILLÉN, L. G.: Stability Constants of Metal-ion Complexes, Part II: Inorganic Ligands. London: The Chemical Society 1964.

BLOCK, H., AHLFELD, F.: Die Selenerzlagerstätte Pasajake. Bolivia. Z. Prakt. Geol. **45**, 16 (1937).

BRUNFELT, A. O., STEINNES, E.: Determination of selenium in standard rocks by neutron activation analysis. Geochim. Cosmochim. Acta **31**, 283 (1967).

BYERS, H. G.: Selenium in Mexico. Ind. Eng. Chem. **29**, 1200 (1937).

CAMERON, A. G. W.: A new table of abundances of the elements in the solar system. In: L. H. AHRENS (ed.), Origin and Distribution of the Elements. London: Pergamon Press 1968.

CHRUSČOV, N. A., KRUGLOVA, V. G., PENSIONEROVA, V. M.: Die Verteilung von Rhenium, Selen und Tellur in den Molybdänlagerstätten der Sowjetunion [Russ.]. Mineralische Rohstoffe **1**, 86 (1960).

COLEMAN, R. C., DELEVAUX, M.: Occurrence of selenium in sulfides from some sedimentary rocks of the western United States. Econ. Geol. **52**, 499 (1957).

CZAMANSKE, G. K.: Sulfide solubility in aqueous solutions. Econ. Geol. **54**, 57 (1959).

DELAHAY, P., POURBAIX, P., VAN RYSSELBERGHE, P.: Diagrammes d'équilibre potentiel pH de quelques éléments. C. R. Réunion du Comité Thermodynamique Cinét. Electrochim. Milano, 1952.

DU FRESNE, A.: Selenium and tellurium in meteorites. Geochim. Cosmochim. Acta **20**, 141 (1960).

DYACHKOVA, J. B., KHODAKOVSKIY, J. L.: Thermodynamic equilibria in the systems S—H_2O, Se—H_2O and Te—H_2O in the 25—300° temperature range and their geochemical interpretations. Geokhimiya **5**, 1108 (1968).

FERSMANN, A. E.: Geokhimija. Acad. Sci., Moscow 1939.

FLEISCHER, M.: Minor elements in some sulfide minerals. Econ. Geol. **50**, 970 (1955).

GEERING, H. R., CARY, E. E., JONES, L. H. P., ALLAWAY, W. H.: Solubility and redox criteria for the possible forms of selenium in soils. Soil Sci. Soc. Am. Proc. **32**, 35 (1968).

GOLDSCHMIDT, V. M., HEFTER, O.: Zur Geochemie des Selens. Nachr. Ges. Wiss. Göttingen, Math.-Physik. Kl., III, **35**, IV, 245 (1933).

— STROCK, L.: Zur Geochemie des Selens. II. Nachr. Ges. Wiss. Göttingen, Math.-Physik. Kl., IV, N.F. **1**, 123 (1935).

GONI, J., GUILLEMIN, C.: Données nouvelles sur les sélénites et séléniates naturels. Bull. Soc. Franç. Minéral. **76**, 422 (1953).

GREENLAND, L.: The abundances of Se, Te, Ag, Pd, Ce and Zn in chondritic meteorites. Geochim. Cosmochim. Acta **31**, 849 (1967).

GRIP, E., WIRSTAM, A.: The Boliden sulphide deposit — a review of geo-investigation carried out during the lifetime of the Boliden mine Sweden, 1924—1967. Sveriges Geologiska Undersökning Ser. C, **651**, 68 (1970).

© Springer-Verlag Berlin · Heidelberg 1972

HALBIG, J. B.: Trace element studies in synthetic sulfide systems; the solubility of thallium in sphalerite and the partition of selenium between sphalerite and galena. Diss. Abstr. Int. 30, 7, 3240B (1970).

HASKIN, L. A., ALLEN, R. O., HELMKE, P. A., PASTER, T. P., ANDERSON, M. R., KOROTEV, R. L., ZWEIFEL, K. A.: Rare earth and other trace elements in Apollo 11 lunar samples. Proceedings of the Apollo 11 Lunar Science Conference. Geochim. Cosmochim. Acta, Supplement I, 2, 1213 (1970).

KAPLAN, J. R., SMITH, J. W., RUTH, E.: Carbon and sulfur concentration and isotopic composition in Apollo 11 lunar samples. Proceedings of the Apollo 11 Lunar Science Conference. Geochim. Cosmochim. Acta, Supplement I, 2, 1317 (1970).

KHARKAR, D. P., TUREKIAN, K. K., BERTINE, K. K.: Stream supply of dissolved Ag, Mo, Sb, Se, Cr, Co, Rb and Cs in the oceans. Geochim. Cosmochim, Acta 32, 285 (1968).

KIESL, W., GRASS, F., BÖCKL, R., PONTA, U.: Cosmochemical abundances of trace elements in meteorites. I. Determination of Se, Te, Tl, Sr, Ba and Ta in chondrites. J. Radioanal. Chem. 6, 447 (1970).

KROUSE, H. R., THODE, H. G.: Thermodynamic properties and geochemistry of isotopic compounds of selenium. Can. J. Chem. 40, 367 (1962).

KVAČEK, M., TRDLIČKA, Z.: Die Rhenium und Selengehalte in einigen Molybdäniten der böhmischen Masse. Acta Univ. Carol. Geologica 2, 69 (1970).

LATIMER, W. M.: Oxidation States of the Elements and their Potentials in Aqueous Solutions. New York: Prentice Hall 1952.

LEUTWEIN, F.: Unpublished results, 1938, 1939, 1943.

— STARKE, R.: Über die Möglichkeit der geochemischen Prospektion auf Selen, untersucht am Beispiel des Kupferschiefers und des Tilkeröder Erzbezirks. Geologie 6, 349 (1957).

LOFTUS-HILLS, G., SOLOMON, M.: Cobalt, nickel and selenium in sulfides as indicators of ore genesis. Mineral. Dep. 2, 228 (1967).

LUTTRELL, G. W.: Annotated bibliography on the geology of selenium. U.S. Geol. Surv. Bull. 1019 M, 863 (1959).

MINAMI, E.: Selen-Gehalte von europäischen und japanischen Tonschiefern. Nachr. Gött. Ges. Wiss. IV, NF 1, 155 (1934/1935).

MORRISON, G. H., GERARD, J. T., KASHUBA, A. TH., GANGADHARAM, E. V.: Elemental abundances of lunar soil and rocks. Proceedings of the Apollo 11 Lunar Science Conference. Geochim. Cosmochim. Acta, Supplement I, 2, 1383 (1970).

NODDACK, I., NODDACK, W.: Die Häufigkeit der chemischen Elemente. Naturwissenschaften 18, 757 (1930).

RAMDOHR, P.: Die Goldlagerstätte des Eisenbergs bei Korbach. Abt. prakt. Geol. Bergwirtschaftslehre, Berlin, Bd. 2, H. 1 (1931).

— Die Erzmineralien und ihre Verwachsungen, 2. Edition. Berlin: Akademie Verlag 1955.

RANKAMA, K.: Isotope Geology. London: Pergamon Press 1954.

— SAHAMA, T. G.: Geochemistry. Chicago: University Press 1950.

RÖSLER, H. J., LANGE, H.: Geochemische Tabellen. Leipzig: 1965.

ROSENFELD, I., BEATH, O. A.: Selenium. New York: Academic Press 1964.

SAKAI, H., NAGASAWA, H.: Fractionation of sulfur isotopes in volcanic gases. Geochim. Cosmochim. Acta 15, 32 (1958).

SCHINDEWOLF, U.: Selenium and tellurium content of stony meteorites by neutron activation. Geochim. Cosmochim. Acta 19, 134 (1960).

SCHUTZ, D. F., TUREKIAN, K. K.: The investigation of the geographical and vertical distribution of several trace elements in sea water using neutron activation analysis. Geochim. Cosmochim. Acta 29, 259 (1965).

SHRIFT, A.: A selenium cycle in nature? Nature 201, 1304 (1964).

SINDEEVA, N. D.: Mineralogy and Types of Deposits of Selenium and Tellurium. (Engl. Transl., Ed. E. Ingerson). New York: Interscience Publishers 1964.

SREBRODOL'SKIY, B. I., SIDEL'NIKOVA, V. D.: Selenium in native sulfur. Geochemistry 8, 803 (1970).

SUESS, H. E., UREY, H. C.: Abundances of the elements. Rev. Mod. Phys. 28, 53 (1956).

TABOURY, F.: Sur la présence du sélénium dans les eaux minérales de la Roche-Posay (Vienne). Bull. Soc. Chim. 5, 865 (1909).

TISCHENDORF, G.: Über den Selengehalt erzgebirgischer Molybdänite. Freiberger Forschungsh. C. 186, 261 (1965).

— UNGETHÜM, H.: Über die Bildungsbedingungen von Clausthalit-Galenit und Bemerkungen zur Selenverteilung im Galenit in Abhängigkeit vom Redoxpotential und vom pH-Wert. Chemie der Erde 23, 279 (1964).

TUREKIAN, K. K., KHARKAR, D. P.: Neutron activation analysis of milligram quantities of Apollo 11 lunar rocks and soil. Proceedings of the Apollo 11 Lunar Science Conference, Geochim. Cosmochim. Acta Suppl. 1, 2 (1970).

— WEDEPOHL, K. H.: Distribution of the elements in some major units of the earth crust. Bull. Geol. Soc. Am. 72, 175 (1961).

VALENSI, G.: Contribution au diagramme potentiel pH du soufre. C. R. II. Réunion du C.I.T.C.E., Milano (1951).

VINOGRADOV, A. P.: Geochemie seltener und nur in Spuren vorhandener chemischer Elemente im Boden. [German transl. of the Russian original.] Berlin: Akademie Verlag 1954.

— Atomic abundance of elements in the sun and stony meteorites. Geokhimiya 1, 329 (1962).

— Oligo-elemente in biologischen Objekten (I, Se, Li, Al, Mn). Agrochimie [Russ.] 8, 20 (1965).

VLASSOV, K. A.: Geochemistry, Mineralogy and Genetic Types of Deposits of Rare Elements, vol. 1, Geochemistry of the Rare Elements. Moscow: 1964. [Russian Edition.]

VOROBJEW, V. P.: Two modes of origin of sedimentary selenium concentrations. Dokl. Acad. Nauk, Earth-Science Series 186, 218 (1969).

WATTENBERG, H.: Zur Chemie des Meerwassers. Z. anorg. allg. Chemie 251, 71 (1943).

WILLIAMS, K. T., BYERS, H. G.: Selenium in deep sea deposits. Ind. Eng. Chem. 13, 353 (1935).

ZWIERZYCKI, J.: De geologie van de goudertsafsetting Redjang Lebong en de kansen van verdere exploratie. Rapp. Betr. Geol. Onderz. in Opdr. v. d. Directie der Mijnb. Mij. Redjang-Lebong Uitger. op hare Concessieterreinen in de Res. Benkoelen, Batavia 1936/37 [Neues Jahrb. Miner. Geol. Paläontol. Referate II, 720 (1937)].

Revised manuscript received: June 1971

Rubidium 37

A	G. Cocco, L. Fanfani and P. F. Zanazzi	(Istituto di Mineralogia dell'Università di Perugia, Perugia, Italy)
B—E, G, I—N	K. S. Heier and	(Mineralogisk-Geologisk Museum, Oslo, Norway)
	G. K. Billings	(Department of Geosciences, New Mexico Institute of Mining and Technology, Socorro, New Mexico, U.S.A.)

37-A. Crystal Chemistry

Rubidium, the fourth member of the alkali metal family, has the electronic configuration $1s^2\, 2s^2\, 2p^6\, 3s^2\, 3p^6\, 3d^{10}\, 4s^2\, 4p^6\, 5s^1$. After the first ionization, the rubidium ion attains the configuration of the noble gas krypton. This explains the chemical behavior of Rb and its tendency to achieve the oxidation state $+1$. Rb^+ has an ionic radius very similar to that of K^+, the preceding alkali metal in the Periodic Table; its usually accepted value (c. f., this handbook, Vol. I, Table 12-8, p. 389) is 1.49 Å, only 11% larger than that of K^+ (1.33 Å), and therefore the crystal chemistry of these two elements is very similar. In nature Rb does not form minerals of its own, but it is dispersed especially in K-minerals, where it is concentrated in the later crystallizates. The highest rubidium contents are observed in amazonites, microclines, muscovites, lepidolites, zinnwaldites and biotites of special occurrences. The similarity of ionic radii associates Rb^+ with Tl^+ (ionic radius 1.49 Å), for example in muscovite and microcline. In pollucite and rhodizite, rubidium seems to replace cesium (ionic radius 1.65 Å) to some extent; however, the crystallochemical properties of this element differ considerably from rubidium.

Rb^+ exhibits an ionic radius similar to that of the complex cation H_3O^+; this substitution has been observed in synthetic compounds belonging to the metatorbernite group (SCHULTE, 1965).

As for the other alkali metals, the present crystallochemical review on Rb considers only halogenides and oxygen-containing compounds. Reliable information on Rb-halogen and Rb-oxygen bond lengths are listed in Table 37-A-1.

I. Halogen Compounds

The simple Rb halides have the "NaCl" structure type. Rb-halogen distances are: Rb—F, 2.815 Å; Rb—Cl, 3.290 Å; Rb—Br, 3.444 Å; Rb—I, 3.671 Å (after SYSIÖ, 1969). $RbHF_2$ as the analogous K compound, is known in two forms; the form stable at room temperature is tetragonal and becomes cubic above 180° C; both structures are of a fluorite-like type (KRUH et al., SR 1956, 215). In $RbLiF_2$, Rb^+ ions are surrounded by eight fluorine atoms with distances ranging from 2.78 to 3.16 Å (average value, 2.95 Å; BURNS and BUSING, 1965). Several double fluorides have been obtained: $RbCaF_3$ and $RbMnF_3$ are cubic, $RbZnF_3$ is tetragonal, $RbMgF_3$ is monoclinic, but all the structures belong to the various modifications of the perowskite type (LUDEKENS and WELCH, SR 1952, 166; HOPPE et al., SR 1961, 308). $RbCoF_3$ has the same arrangement and is cubic above and tetragonal below 101°K (NOUET et al., 1969). $RbNiF_3$ is isotypic with the hexagonal modification of $BaTiO_3$: the mean $Rb^{[12]}$—F value is 2.93 Å (BABEL, 1969). Rb_3AlF_6 has tetragonal symmetry at room temperature. At 375°C it undergoes a transition in a cubic phase with a behavior closely analogous to that of Cs_3AlF_6 (HOLM, 1965). $RbBF_4$ has a baryte-like atomic arrangement (HOARD and BLAIR, SB 1935, 419); each Rb^+ is

Table 37-A-1

		Range Å	Mean Å	Reference
RbF	Rb[6]—F	2.815	2.82	Sysiö (1969)
RbLiF$_2$	Rb[8]—F	2.78—3.16	2.95	Burns and Busing (1965)
RbNiF$_3$	Rb$_I$[12]—F	2.922—3.043	2.98	Babel (1969)
	Rb$_{II}$[12]—F	2.796—2.942	2.90	
RbBF$_4$	Rb[12]—F	2.92—3.37	3.07	Clark and Lynton (1969)
α-RbFeF$_4$	Rb[6]—F	2.61—3.14	2.78	Tressaud et al. (1969)
β-RbFeF$_4$	Rb[8]—F	2.97	2.97	Tressaud et al. (1969)
γ-Rb$_2$BeF$_4$	Rb[6]—F	2.87—3.07	2.97	Mustafaev et al. (1966)
	Rb$_{II}$[8]—F	2.61—2.96	2.85	
RbBe$_2$F$_5$	Rb[6]—F	2.82—3.02	2.91	Iljukhin and Belov (SR 1961, 303)
RbPaF$_6$	Rb[10]—F	2.81—3.17	2.96	Burns et al. (1968)
RbCl	Rb[6]—Cl	3.290	3.29	Sysiö (1969)
RbNiCl$_3$	Rb[12]—Cl	3.480—3.640	3.56	Asmussen et al. (1969)
α-RbMnCl$_3$·2H$_2$O	Rb[8]—Cl	3.377—3.573	3.48	Jensen (1967)
β-RbMnCl$_3$·2H$_2$O	Rb[8]—Cl	3.313—3.622	3.43	Jensen (1967)
RbBr	Rb[6]—Br	3.444	3.44	Sysiö (1969)
RbI	Rb[6]—I	3.671	3.67	Sysiö (1969)
RbAg$_4$I$_5$	Rb[6]—I	3.62	3.62	Bradley and Greene (1967)
Rb$_2$O	Rb[4]—O	2.92	2.92	Helms and Klemm (SB 1939, 85)
Rb$_2$O$_2$	Rb[6]—O	2.85	2.85	Föppl (SR 1957, 234)
Rb$_2$Ti$_6$O$_{13}$	Rb[8]—O	2.77—3.14	2.92	Andersson and Wadsley (1962)
Rb$_x$Mn$_x$Ti$_{2-x}$O$_4$	Rb[8]—O	2.93—3.20	3.06	Reid et al. (1968)
RbPO$_3$	Rb[7]—O	2.90—3.20	2.99	Corbridge (SR 1956, 300)
β-Rb$_2$[SO$_4$]	Rb$_I$[10]—O	2.90—3.23	3.10	Wyckoff (1965)
	Rb$_{II}$[9]—O	2.82—3.22	2.98	
RbAl[SO$_4$]$_2$·12H$_2$O	Rb[6]—O	3.07	3.07	Larson and Cromer (1967)
Rb(AmO$_2$)[CO$_3$]	Rb[12]—O	3.03—3.10	3.06	Ellinger and Zachariasen (SR 1954, 510)
Rb(UO$_2$)[NO$_3$]$_3$ by neutron diffraction	Rb[14]—O	2.94—3.29	3.19	Barclay et al. (1965)

surrounded by twelve F atoms with distances in the range 2.92—3.37 Å (calculated from Clark and Lynton, 1969).

RbFeF$_4$ exhibits a reversible polymorphic transformation at 650°C; in the orthorhombic α-form, each Rb⁺ has six F neighbors with distances from 2.61 to 3.14 Å (the average value is 2.78 Å). In the tetragonal β-form, the coordination polyhedron around Rb⁺ is a square prism with Rb—F distances of 2.97 Å (Tressaud et al., 1969).

Rb$_2$BeF$_4$ shows three polymorphic modifications. In the γ-form Rb is in six- and eight-fold coordination with Rb—F distances from 2.61 to 3.05 Å (Mustafaev et al., 1966). In RbBe$_2$F$_5$ there are brucite-type layers of Rb octahedra, the Rb[6]—F mean distance is 2.91 Å (Iljukhin and Belov, SR 1961, 303). In contrast to RbNbF$_6$ and RbTaF$_6$ (Cox, SR 1956, 224) which have the KOsF$_6$ structure, in RbPaF$_6$ the atomic arrangement resembles that of K_2ZrF_6 (BURNS et al., 1968); Rb^+ has ten nearest fluorine atoms in the range 2.81—3.17 Å.

The hexagonal atomic arrangements of $RbNiCl_3$ and $RbNiBr_3$ are related to that of $BaNiO_3$; the average $Rb^{[12]}$—Cl distance is 3.56 Å (ASMUSSEN et al., 1969). About the same bond lengths are found in isostructural $RbCoCl_3$ (ENGBERG and SOLING, 1967). The α- and β-forms of $RbMnCl_3 \cdot 2H_2O$ are orthorhombic and triclinic respectively. In both structures the Rb^+ ions have eight Cl^- ions as nearest neighbors with average bond lengths of 3.48 and 3.43 Å (JENSEN, 1967). In $RbAg_4I_5$, Rb ions are surrounded by six I^- ions; the distance Rb—I is 3.62 Å (BRADLEY and GREENE, 1967).

II. Oxygen Compounds

$Rb_2^{[4]}O^{[8]}$ has the anti-fluorite arrangement with Rb—O distances of 2.92 Å (HELMS and KLEMM, SB 1939, 85). Rb_2O_2 is orthorhombic; the alkali-cation is irregularly coordinated by six oxygen atoms with Rb—O distances of 2.85 Å (FÖPPL, SR 1957, 234). $RbPaO_3$ has the perovskite structure (KELLER, 1965). $Rb_2Ti_6O_{13}$ is isomorphous with $Na_2Ti_6O_{13}$ and $K_2Ti_6O_{13}$; Rb^+ is in eightfold coordination with a $Rb^{[8]}$—O mean distance of 2.92 Å (ANDERSSON and WADSLEY, 1962). In the non-stoichiometric compound $Rb_xMn_xTi_{2-x}O_4$ with $0.60 < x < 0.80$, Rb^+ is surrounded by eight oxygen atoms at distances up to 3.20 Å. Two additional oxygen atoms at 3.56 Å are probably unbonded (REID et al., 1968). In $^1_\infty Rb[PO_3]$, Rb^+ ions are coordinated to seven oxygen atoms in a range from 2.90 to 3.20 Å (mean value of $Rb^{[7]}$—O is 2.99 Å; CORBRIDGE, SR 1956, 300). At room temperature Rb_2SO_4 is orthorhombic with the β-K_2SO_4 arrangement. Coordination around the two non-equivalent Rb^+ ions is ten- and nine-fold with average distances of 3.10 and 2.98 Å respectively (calculated from WYCKOFF, 1965). A hexagonal form occurs at temperatures above 657°C. This modification is isotypic with the high-temperature form of cesium and potassium sulphates (TABRIZI et al., 1968). Rb_2CrO_4 is isotypic with the orthorhombic modification of Rb_2SO_4; average values of Rb—O distances are 3.19 and 3.09 Å (SMITH and COLBY, SR 1940, 158). Many double rubidium sulphates with the chemical formula $Rb_2M_2^{2+}[SO_4]_3$, where M^{2+} can be Ca, Cd, Co, Fe^{2+}, Mg, Mn^{2+} or Ni, have the langbeinite atomic arrangement (GATTOW and ZEMANN, 1958).

According to LEDSHAM and STEEPLE (1968), $RbAl[SO_4]_2 \cdot 12H_2O$ and $RbCr[SO_4]_2 \cdot 12H_2O$ belong to the same class of α-alums as Tl- and K-alums, in contrast to Cs alums which have been classified as β-alums. In the aluminum alum Rb^+ is surrounded by six water molecules at 3.07 Å (LARSON and CROMER, 1967). Rb_2MnO_4 is orthorhombic and isotypic with the analogous compounds of potassium and cesium (DUQUENOY, 1969). $Rb_2Cr_2O_7$, which is known at room temperature in triclinic and monoclinic modifications isotypic with those of $K_2Cr_2O_7$, undergoes a phase change on heating at 330°C (ERDEY et al., 1965).

The crystal structure of $RbAmO_2[CO_3]$ is isotypic with that of $KPuO_2[CO_3]$; the mean value of $Rb^{[12]}$—O distances is 3.06 Å (ELLINGER and ZACHARIASEN, SR 1954, 510). In $RbUO_2[NO_3]_3$ (BARCLAY et al., 1965), the Rb^+ ion has a rather large coordination number; 14 oxygen atoms surround the alkali cation with distances from 2.94 to 3.29 Å.

References: Section 37-A

ANDERSON, S., WADSLEY, A. D.: The structures of $Na_2Ti_6O_{13}$ and $Rb_2Ti_6O_{13}$ and the alkali metal titanates. Acta Cryst. **15**, 194 (1962).

ASMUSSEN, R. W., LARSEN, T. K., SOLING, H.: The crystal structure of $RbNiCl_3$ and $RbNiBr_3$. Acta Chem. Scand. **23**, 2055 (1969).

BABEL, D.: Die Kristallstrukturen der hexagonalen Fluorperowskite. Z. Anorg. Allgem. Chem. **369**, 117 (1969).

BARCLAY, G. A., SABINE, T. M., TAYLOR, J. C.: The crystal structure of rubidium uranyl nitrate: A neutron-diffraction study. Acta Cryst. **19**, 205 (1965).

BRADLEY, J. N., GREENE, P. D.: Relationship of structure and ionic mobility in solid MAg_4I_5. Trans. Faraday Soc. **63**, 2516 (1967).

BURNS, J. H., BUSING, W. R.: Crystal structures of rubidium lithium fluoride, $RbLiF_2$, and cesium lithium fluoride, $CsLiF_2$. Inorg. Chem. **4**, 1510 (1965).

— LEVY, H. A., KELLER, O. L. JR.: The crystal structure of rubidium hexafluoroprotoactinate (V), $RbPaF_6$. Acta Cryst. B **24**, 1675 (1968).

CLARK, M. J. R., LYNTON, H.: Crystal structures of potassium, ammonium, rubidium and cesium tetrafluoborates. Can. J. Chem. **47**, 2579 (1969).

DUQUENOY, G.: Synthèses entre solides à partir d'un supraoxyde alcalin-manganates de potassium, rubidium ou cesium. Compt. Rend. C **268**, 828 (1969).

ENGBERG, A., SOLING, H.: On the crystal structures of $RbCoCl_3$ and Rb_3CoCl_5. Acta Chem. Scand. **21**, 168 (1967).

ERDEY, L., LIPTAY, G., BIDLO, G.: Polymorphous transformation of rubidium dichromate. J. Inorg. Nucl. Chem. **27**, 2451 (1965).

GATTOW, G., ZEMANN, J.: Double sulfates of the langbeinite type, $A_2^I B_2^{II}(SO_4)_3$. Z. Anorg. Allgem. Chem. **293**, 233 (1958).

HOLM, J. L.: Phase transitions and structure of the high-temperature phases of some compounds of the cryolite family. Acta Chem. Scand. **19**, 261 (1965).

JENSEN, S. J.: The crystal structures of α- and of β-$RbMnCl_3 \cdot 2H_2O$. Acta Chem. Scand. **21**, 889 (1967).

KELLER, C.: Ternary or quaternary protactinium oxides with perovskite structure. J. Inorg. Nucl. Chem. **27**, 321 (1965).

LARSON, A. C., CROMER, D. T.: Refinement of the alum structures. III. X-ray study of the α-alums, K, Rb and $NH_4Al(SO_4)_2 \cdot 12H_2O$. Acta Cryst. **22**, 793 (1967).

LEDSHAM, A. H. C., STEEPLE, H.: The classification of the chromium alums. Acta Cryst. B **25**, 398 (1969).

MUSTAFAEV, N. M., ILYUKHIN, V. V., BELOV, N. V.: The crystal structures of K and Rb orthofluoroberyllates and their relationship to compounds of general formula M_2BX_4. Soviet Phys.-Cryst. **10**, 676 (1966).

NOUET, J., KLEINBERGER, R., DE KOUCHKOVSKY, R.: Etude radiocristallographique à basse température de la perowskite fluorée $RbCoF_3$. Compt. Rend. B **269**, 986 (1969).

REID, A. F., MUMME, W. G., WADSLEY, A. D.: A new class of compound $M_x^+ A_x^{3+} Ti_{2-x}O_4$ ($0.60 < x < 0.80$) typified by $Rb_xMn_xTi_{2-x}O_4$. Acta Cryst. B **24**, 1228 (1968).

SCHULTE, E.: Zur Kenntnis der Uranglimmer. Neues Jahrb. Mineral. Monatsh. 242 (1965).

SYSIÖ, P. A.: On the additivity of crystal radii in alkali halides. Acta Cryst. B **25**, 2374 (1969).

© Springer-Verlag Berlin · Heidelberg 1972

TABRIZI, D., GAULTIER, M., PANNETIER, G.: Analyse radiocristallographique des formes "basse" (β) et "haute" (α) temperature de sulfate de césium Cs_2SO_4. Bull. Soc. Chim. France 935 (1968).

TRESSAUD, A., GALY, J., PORTIER, J.: Structure cristalline des variétés basse et haute température du fluoferrite de rubidium $RbFeF_4$. Bull. Soc. Franc. Minéral. Crist. **92**, 335 (1969).

WYCKOFF, R. W. G.: Crystal Structures (second ed.) vol. 3. New York and London: Interscience Publisher 1965.

Revised manuscript received: June 1970

Strontium 38

A   K. Fischer   (Lehrstuhl für Kristallographie, Univ. des Saarlandes, Saarbrücken, Germany)

B   Z. E. Peterman and C. Hedge   (United States Geological Survey, Denver, Col. 80225, U.S.A.)

C—O   K. K. Turekian   (Department of Geology, Yale University, New Haven, Conn. 06520, U.S.A.)

38-A. Crystal Chemistry
I. General

At present only about 25 strontium minerals are known. In most of them, Sr is bound to oxygen (or OH, H_2O) exclusively and its valency is always two. The ionic radius of Sr^{2+} is between those of Ca^{2+} and Ba^{2+} and is somewhat smaller than the Pb^{2+} radius. Therefore, in a number of minerals and artificial compounds, Sr is replaced by these atoms and also replaces them; for example, celestite $Sr[SO_4]$ is isotypic with barite $Ba[SO_4]$, anglesite $Pb[SO_4]$ and baryto-celestite, $(Sr, Ba)[SO_4]$. Strontianite $Sr[CO_3]$ is isotypic with witherite $Ba[CO_3]$, cerussite $Pb[CO_3]$, and aragonite $Ca[CO_3]$ (but not with calcite which forms an isomorphous series with carbonates of smaller cations). Belovite, $Sr_5(OH)[PO_4]_3$, belongs to the apatite series, and in natural brewsterite, $(Sr, Ba, Ca)[Al_2Si_6O_{16}] \cdot 5H_2O$, Sr is partially replaced by Ba and Ca. Synthetic $SrCu[Si_4O_{10}]$ and $Sr_2[Al^{[4]}SiO_7]$ are members of the gillespite and melilite groups, respectively. In addition, $Eu(OH)_2 \cdot H_2O$ is isostructural with both $Sr(OH)_2 \cdot H_2O$ and $Ba(OH)_2 \cdot H_2O$ (BÄRNIGHAUSEN, 1966).

In Table 38-A-1, a few minerals and selected artificial compounds with fairly or well known crystal structures are listed. The coordination numbers of Sr^{2+} vary from 6 to 12 (or possibly even more, cf. $Sr[B_4O_7]$). A coordination number of 9 appears to be rather stable in a coordination polyhedron which can be described as a trigonal prism with 6 atoms at the corners and 3 additional atoms above the centers of its prismatic faces (cf. also computations of DUNITZ and ORGEL, 1960). Besides a cubic coordination polyhedron in SrF_2, other polyhedrons with a coordination number of 8 are listed in Table 38-A-1. The polyhedrons for higher coordination numbers are in most cases irregular. They appear to be dominated by the anions of the crystal structures which provide the stronger binding forces. The reported Sr—(O, OH, H_2O) distances range from 2.5 to 3.25 Å.

II. Crystal Chemistry of Some Minerals and Selected Artificial Compounds

In $Sr(OH)_2 \cdot H_2O$, the eight-fold coordination of Sr by oxygen atoms is dominated by the Sr—(OH, H_2O) bonds (BÄRNIGHAUSEN and WEIDLEIN, 1967). The coordination polyhedron can be described in two different ways: as a distorted trigonal prism of (OH) plus two H_2O molecules with slightly longer distances (see Table 38-A-1) or as a distorted quadratic antiprism (as also reported for $Sr(OH)_2 \cdot 8H_2O$ and $Sr[Cr_2O_7]$). A well-defined coordination number of 9 has been found for two independent Sr positions in $Sr[B(OH)_4]_2$, and for a single Sr in $SrH[AsO_4] \cdot H_2O$ (Sr-haidingerite) and in $Sr[VO_3]_2 \cdot 4H_2O$. The coordination polyhedron is very similar in all the three compounds and no systematic difference of Sr—O distances between the atoms at the corners of the trigonal prism and those

Table 38-A-1. *Coordination of Sr in some crystal structures*

Mineral or artificial compound	Ligands and their distances to Sr in Å		Coordination polyhedron	References
$Sr^{[6]}O^{[6]}$	6O	2.58	regular octahedron	1
$Sr^{[8]}(OH)_2 \cdot H_2O$	6OH	2.59—2.69	distorted trigonal prism	2
	$2H_2O$	2.70; 2.74	above top of two side faces of this prism	
$Sr^{[8]}(OH)_2 \cdot 8H_2O$	$8H_2O$	2.60	quadratic antiprism	3
$Sr^{[12]}Ti^{[6]}O_3$	8O	2.76	cube-octahedron	4
$Sr^{[8]}F_2^{[4]}$	8F	2.60	cube	5
$Sr^{[9]}[B^{[4]}(OH)_4]_2$	Sr_1: 9OH	2.53—2.97	distorted trigonal prism with 3OH on top of prismatic faces	6
	Sr_2: 9OH	2.52—2.84		
$Sr^{[9]}[B_4^{[4]}O_7]$	9O	2.52—2.84	see text	7
	(6O	3.04—3.20)		
$Sr^{[10]}[B_3^{[3]}B_3^{[4]}O_9 \cdot (OH)_2] \cdot 3H_2O$ tunellite	6O	2.64—2.90	irregular, see text	8
	$4H_2O$	2.61—2.98		
$Sr^{[12]}[SO_4]$ celestite	10O	2.48—2.99	see text	9
	2O	3.25		
$Sr^{[8]}[Cr_2O_7]$	8O	2.53—2.66	distorted quadratic antiprism	10
$Sr^{[12]}Sr_2^{[10]}[PO_4]_2$	6+6O	2.63; 3.10	distorted icosahedron, see text	11
	10O	2.48—2.72		
$Sr^{[9]}H[AsO_4] \cdot H_2O$ (Sr-haidingerite)	7O	2.57—3.12	distorted trigonal prism with "centered" side faces	12
	$2H_2O$	2.68; 2.69		
$Sr^{[9]}[V^{[4]}O_3]_2 \cdot 4H_2O$	$6H_2O$	2.62; 2.73	distorted trigonal prism on top of the prismatic faces	13
	3O	2.54—2.74		

References: 1. GERLACH, SB 1913—1928, 118. 2. BÄRNIGHAUSEN and WEIDLEIN (1967). 3. SMITH (SR 1954, 520), PREISINGER (*ibid.*). 4. DONNAY *et al.* (1963). 5. VAN ARKEL (SB 1913—1928, 186). 6. KRAVCHENKO (1965), cf. KUTSCHABSKY (1965). 7. PERLOFF and BLOCK (1966); cf. WITZMANN and BEULICH (1965), KROGH-MOE (1964). 8. CLARK (1964). 9. GARSKE and PEACOR (1965). 10. WILHELMI (1966). 11. ZACHARIASEN (SR 1947—1948, 388). 12. BINAS (1966). 13. SEDLACEK and DORNBERGER-SCHIFF (1965).

on top of the prismatic faces could be detected (cf., however, $Sr(OH)_2 \cdot H_2O$). The coordination number in $Sr[B_4O_7]$ is not well defined; in addition to the 9 oxygen atoms at a distance of 2.84 Å or less (forming an irregular polyhedron, point symmetry m), three oxygen atoms occur at distances of 3.04 and 3.05 Å, two oxygen atoms at 3.15 Å and one oxygen atom at 3.20 Å. A coordination number of 10 has been reported for tunellite, $Sr[B_6O_9(OH)_2] \cdot 3H_2O$. The coordination polyhedron, which fills "holes" in the borate sheets, has no symmetry and exhibits no simple geometrical form. For information on the two different coordination polyhedrons of $Sr_3[PO_4]_2$, which is isostructural with $Ba_3[PO_4]_2$, see subsection 56-A-II. The (idealized) 12-fold coordination in celestite consists of a five-membered ring on one side and a parallel six-membered ring on the other side, the latter having one oxygen on top of its center. For additional information on the 10- or 12-fold coordination, see subsection 56-A-II (barite). In the zeolite brewsterite, $(Sr, Ba, Ca)^{[9]}[Al_2Si_6O_{16}] \cdot 5H_2O$, the coordination polyhedron must not be typical for Sr because in its usual cation position substantial proportions of Ba and Ca occur (PERROTTA and SMITH, 1964).

Revised manuscript received: July 1970

References: Section 38-A

BÄRNINGHAUSEN, H.: Gitterkonstanten und Raumgruppe der isotypen Verbindungen Eu(OH)$_2$ · H$_2$O, Sr(OH)$_2$ · H$_2$O und Ba(OH)$_2$ · H$_2$O. Z. Anorg. Allgem. Chem. **342**, 233 (1966).
— WEIDLEIN, J.: Die Kristallstruktur von Strontiumhydroxid-Monohydrat. Acta Cryst. **22**, 252 (1967).
BINAS, H.: Die Struktur des Sr-Haidingerits, SrHAsO$_4$ · H$_2$O. Z. Anorg. Allgem. Chem. **347**, 140 (1966).
CLARK, J. R.: The crystal structure of tunellite, SrB$_6$O$_9$(OH)$_2$ · 3H$_2$O. Am. Mineralogist **49**, 1549 (1964).
DONNAY, J. D. H., DONNAY, G., COX, E. G., KENNARD, O., KING, M. V. (eds.): Crystal Data Determinative Tables, 2nd ed. Am. Cryst. Assoc., Monograph No. 5 (1963).
DUNITZ, J. D., ORGEL, L. E.: Stereochemistry of ionic solids. Advan. Inorg. Chem. Radiochem. **2**, 1 (1960)
GARSKE, D., PEACOR, D. R.: Refinement of the structure of celestite SrSO$_4$. Z. Krist. **121**, 204 (1965).
KRAVCHENKO, V. B.: The crystal structure of the monoclinic modification of SrB$_2$O$_4$ · 4H$_2$O = Sr[B(OH)$_4$]$_2$. Zh. Strukt. Khim. **6**, 835 (1965) (translated).
KROGH-MOE, J.: The crystal structure of strontium diborate, SrO · 2B$_2$O$_3$. Acta Chem. Scand. **18**, 2055 (1964).
KUTSCHABSKY, L.: Zur Kristallstruktur des Sr[B(OH)$_4$]$_2$. Z. Chem. **5**, 110 (1965).
PERLOFF, A., BLOCK, S.: The crystal structure of the strontium and lead tetraborates, SrO · 2B$_2$O$_3$ and PbO · 2B$_2$O$_3$. Acta Cryst. **20**, 274 (1966).
PERROTTA, A. J., SMITH, J. V.: The crystal structure of brewsterite. Acta Cryst. **17**, 857 (1964).
SEDLACEK, P., DORNBERGER-SCHIFF, K.: Das Strukturprinzip des Strontiummetavandat, Sr(VO$_3$)$_2$ · 4H$_2$O. Acta Cryst. **18**, 407 (1965).
WILHELMI, K.-A.: The crystal structure of strontium dichromate, SrCr$_2$O$_7$. Arkiv Kemi **26**, 149 (1966).
WITZMANN, H., BEULICH, W.: Beitrag zur Struktur wasserfreier Strontiumborate. Naturwissenschaften **52**, 157 (1965).

© Springer-Verlag Berlin · Heidelberg 1972

Revised manuscript received: July 1970

Cadmium 48

A   B. Brehler   (Mineralogisch-Kristallographisches Institut, Technische Universität, Clausthal-Zellerfeld 1, Germany)

B—M   H. Wakita   (Department of Chemistry, Oregon State University, Corvallis, Oregon, U.S.A.)

and

O   R. A. Schmitt   (Department of Chemistry, Oregon State University, Corvallis, Oregon, U.S.A.)

48-A. Crystal Chemistry

In nature cadmium is exclusively known as occurring in compounds. In some cases it replaces other elements in their minerals, especially zinc. The Cd-minerals (greenockite, β-CdS; monteponite, CdO; otavite, $CdCO_3$; hawleyite, α-CdS) originate mainly from weathering of Cd-bearing zinc minerals.

In all known compounds the valency of Cd is two. The ionic radius of Cd^{2+} has been found to be 1.03 Å (GOLDSCHMIDT, 1926; see Chapter 12 of Volume I of this handbook) and 0.97 Å (PAULING, 1927; AHRENS, 1952; see Chapter 12 of Volume I of this handbook); PAULING and HUGGINS (1934) derived the tetrahedral covalent radius as 1.48 Å.

The observed coordination numbers of Cd in its compounds are usually four and six, in a few cases also five, seven, eight, nine and twelve. Cd-compounds are often isotypic with the corresponding compounds of Zn^{2+}, Mg^{2+}, Fe^{2+}, Co^{2+}, Ni^{2+} and in some cases of Ca^{2+}. Some cadmium chalcogen compounds and silicates have applications in electronics.

Elementary cadmium crystallizes in a very distorted hexagonal close-packing; the axial ratio c/a is 1.885 (STENZEL and WEERTS, SB 1928—32, 164), instead of 1.633 as in the ideal hexagonal close-packings. The distance from each atom to its six neighbors in (00.1) is 2.96 Å, to the other six neighbors 3.28 Å (HULL and DAVEY, SB 1913—28, 42).

In the following lists of known crystal-structures of cadmium compounds, the following are omitted: all organic compounds, the large number of alloys, and all compounds whose number of cadmium atoms is smaller than a quarter of the total number of cations.

In Table 48-A-1 the halogenides and hydroxides of cadmium are listed.

In CdO the cadmium atoms are surrounded by six oxygens. The sulfide, selenide and telluride have fourfold coordinations; they can form both zincblende and wurtzite structures.

From the lattice energies calculated by DÄWERITZ (1971) from the Born-Landé equation, it can be concluded that the stability of the zincblende structure increases in the order CdS-CdSe-CdTe. In Table 48-A-2 the cadmium chalcogenides are listed.

In all structurally known double halogenides, Cd is octahedrally surrounded. $KCdF_3$ (BRISI, SR 1952, 166), $RbCdF_3$ (COUSSEINS and PINA PEREZ, 1968), $CsCdCl_3$ and $CsCdBr_3$ (NÁRAY-SZABÓ, SR 1947—48, 454) crystallize in the perovskite-type. $KCdCl_3$ (BRANDENBERGER, SR 1947—48, 438) and $RbCdCl_3$ (MAC GILLAVRY, NIJEVELD, DIERDORP and KARSTEN, SB 1939, 115) are isotypic to $(NH_4)CdCl_3$ (BRASSEUR and PAULING, SB 1938, 79). Rb_2CdCl_4 (SEIFERT and KOKNAT, 1968) and some other compounds are isotypic to K_2NiF_4.

© Springer-Verlag Berlin · Heidelberg 1972

Table 48-A-1. *Cadmium halogenides and hydroxides*

Compound	Structure-type	Cd-coordination	Reference
$Cd^{[8]}F_2^{[4]}$	fluorite	8 F Cd—F: 2.33 Å	KOLDERUP (SB 1913—28, 188); HAENDLER and BERNARD (SR 1951, 144)
$^2_\infty Cd^{[6]}Cl_2^{[3\,py]}$	$CdCl_2$	6 Cl Cd—Cl: 2.74 Å	PAULING (SB 1913—28, 774)
$^2_\infty Cd^{[6]}Br_2^{[3\,py]}$	$CdCl_2$	6 Br	FERRARI and GIORGIO (SB 1928—32, 246)
$^2_\infty Cd^{[6]}Br_2^{[3\,py]}$	"Wechsel-struktur"[a,b]	6 Br	BIJVOET and NIEUWENKAMP (SB 1933—35, 280); HÄGG, KIESSLING and LINDÉN (SR 1947—48, 490)
$^2_\infty Cd^{[6]}I_2^{[3py]c}$	CdI_2 CdI_2 (2. Modif.)	6 I Cd—I: 2.99 Å	BOZORTH (SB 1913—28, 189); HASSEL (SB 1933—35, 282)
$^2_\infty Cd^{[6]}(OH)_2$	CdI_2[d]	6 OH	NATTA (SB 1913—28, 195 and 775)
$\gamma\text{-}Cd^{[6]}(OH)_2$[e]		6 OH Cd—(OH): 2.16—2.39 Å	FEITKNECHT (SR 1957, 239); GLEMSER, HAUSCHILD and RICHERT (SR 1957, 239); DE WOLFF (1966, 1967)
$^2_\infty Cd^{[6]}OHCl$[f]	CdOHCl	3 Cl + 3 OH Cd—Cl: 2.69 Å (3×) Cd—OH: 2.34 Å (3×)	HOARD and GRENKO (SB 1933—35, 373)

[a] This structure is a random layer lattice of C6 and C19 type layers. With heating, it transforms rapidly into the C19 type (HÄGG, KIESSLING, LINDÉN, SR 1947—48, 490).

[b] A C6 type in some samples with enclosed 6—8 layers of the C27 type was found by PINSKER and LAPIDUS (SR 1947—48, 490) and PINSKER (SR 1947—48, 491). An X-ray study of the polytypes was given by MITCHELL (1962). Dislocations in $CdBr_2$ are discused by AGRAWAL and TRIGUNAYAT (1970).

[c] More than 100 polytypes of CdI_2 are known (e.g., MITCHELL, SR 1956, 252; JAIN, CHADHA and TRIGUNAYAT, 1970; PRASAD and SRIVASTAVA, 1970). In the system $CdBr_2$—CdI_2, one intermediate phase with the composition CdBrI appears. It was stated by HÄGG and LINDÉN (SR 1947—48, 490) that CdBrI posses a "Wechselstruktur"; it cannot be decided whether the I and Br ions form separate layers, or are divided statistically.

[d] During the precipitation of $Cd(OH)_2$, the "α-Form" will grow first, which is a random layer structure (FEITKNECHT, SB 1938, 89). MITCHELL (1966) studied 50 crystals of $Cd(OH)_2$ by the Weissenberg method; he observed no evidence for structural polytypism. All crystals were shown to be type 2H, $[(A\,\gamma\,B)]_n$.

[e] In this structure, the $Cd(OH)_6$-octahedra occur in pairs, sharing a face parallel to (010). All octahedra share edges with neighbors in the c-direction. Finally, the infinite double strings thus formed share corners with four neighboring strings.

[f] There are several hydroxyhalogenides known i.e., $CdCl_{0.67}(OH)_{1.33}$; $CdBr_{0.6}(OH)_{1.4}$; $CdI_{0.5}(OH)_{1.5}$ (FEITKNECHT et al., SB 1937, 49; SR 1942—44, 217; SR 1945—46, 171). They have layer structures with replacement of (OH) and halogen atoms.

Table 48-A-2. *The chalcogenides of cadmium*

Compound	Structure-type	Cd-coordination	Reference
$Cd^{[6]}O^{[6]}$ [a] monteponite	NaCl	6 O Cd—O: 2.35 Å	Davey and Hoffmann (SB 1913—28, 120)
$\alpha\text{-}Cd^{[4]}S^{[4]}$ [b] hawleyite	sphalerite	4 S Cd—S: 2.52 Å	Ulrich and Zachariasen (SB 1913—28, 129); Traill and Boyle (1955) [c]
$\beta\text{-}Cd^{[4]}S^{[4]}$ [b, d] greenockite	wurtzite	4 S	Bragg (SB 1913—28, 129); Smith (SR 1955, 403) (lattice dimensions)
$\alpha\text{-}Cd^{[4]}Se^{[4]}$ [e]	sphalerite	4 Se Cd—Se: 2.6 Å	Goldschmidt (SB 1913—28, 136)
$\beta\text{-}Cd^{[4]}Se^{[4]}$ cadmoselite	wurtzite	4 Se	Goldschmidt (SB 1913—28, 136)
$Cd^{[4]}Te^{[4]}(I)$ [f]	sphalerite	4 Te Cd—Te: 2.80 Å	Zachariasen (SB 1913—28, 136)
$Cd^{[4]}Te^{[4]}(II)$	wurtzite	4 Te	Semiletov (SR 1955, 81)
High pressure forms:			
$Cd^{[6]}S^{[6]}$ (β-CdS, 18 kb)	NaCl	Cd—S: 2.78 Å	Kabalkina and Troitskaya (1963)
$Cd^{[6]}Se^{[6]}$ (> 32 kb)	NaCl		Mariano and Warekois (1963); Onodera (1969)
$Cd^{[6]}Te^{[6]}$ (high pressure)	NaCl	Cd—Te: 2.98 Å	Smith and Martin (1963); Onodera (1969)
$Cd^{[4+2]}Te^{[4+2]}$ (> 90 kb)	analog to β-Sn		Smith and Martin (1963)

[a] CdO forms solid solutions with CaO (Natta and Passerini, SB 1928—32, 227).

[b] Both α-CdS and β-CdS form solid solutions with HgS (Rittner and Schulman, SR 1942—44, 282).

[c] Isometric CdS has been found as a mineral. CdS was precipitated from salt solutions under several different conditions by Hartmann (1966).

[d] β-CdS forms a continuous series of solid solutions with ZnS (Ballentyne and Roy, SR 1961, 251). There are also some other mixed sulfides (Cd, Zn, Fe, Mn)S (Skinner and Bethke, SR 1961, 251; Skinner, SR 1961, 252). β-CdS also forms solid solutions with CdSe and CdTe (Vitrikhovskij and Mizeckaja, SR 1959, 76).

[e] CdSe forms solid solutions with ZnSe, InAs and In_2Se_3 (Gorjunova, Komović and Frank-Kameneckij, SR 1955, 80).

[f] Cubic CdTe forms solid solutions with ZnTe (Kolomiec and Malkova, SR 1958, 71) and HgTe (Gavriščak, SR 1960, 83; Wolley and Ray, SR 1960, 234).

In oxyspinels Cd^{2+} has the coordination numbers four (e.g., $CdFe_2O_4$, Gorter, SR 1950, 242) and six (e.g., $CdIn_2O_4$, Skribljak, Dasgupta and Biswas, SR 1959, 377).

Table 48-A-3 shows the cadmium silicates of known structure.

Table 48-A-3. *Cadmium silicates*

Compound	Remarks	Reference
$Cd_3^{[6]}O[SiO_4]$	similar to $Ca_3O[SiO_4]$	Dent Glasser and Glasser (1964); Dent Glasser (1965)
$Cd_3^{[8]}Al_2^{[6]}[SiO_4]_3$	garnet structure	Gentile and Roy (1960)
$Cd_3^{[8]}V_2^{[6]}[SiO_4]_3$	garnet structure	Mill (1964)
$Cd_2^{[6]}[SiO_4]$	related to $Na_2SO_4(V)$ Cd—O: 2.15—2.35 Å (4×) 2.61 Å (2×)	Dent Glasser and Glasser (1964)
$Li_2^{[4]}Cd^{[4]}[SiO_4]$ [a]	related to Li_3PO_4	Tarte et Cahay (1970)
$Cd^{[6]}[SiO_3]$	related to β-$CaSiO_3$	Dent Glasser and Glasser (1964)
$Na_4^{[4]}Cd_2^{[6]}[Si_3O_{10}]$	Cd—O: 2.13—2.27 Å	Simonov, Egorov-Tismenkov and Belov (1968)

[a] The compounds of Zn, Mg, Mn, Co and Fe are isostructural.

Table 48-A-4. *Complex cadmium oxygen compounds*

Compound	Remarks	Reference
	Coordination number 4	
α-$Cd^{[4]}SO_4$	isotypic with $HgSO_4$ Cd—O: 2.11 Å, 2.18 Å (2×) 2.21 Å	Kokkoros and Rentze-Peris (1963)
$Cd^{[4]}[B^{[4]}B^{[3]}O_7]$	Cd—O: 2.18—2.22 Å	Ihara and Krogh-Moe (1966)
	Coordination number 5	
$Ba^{[7]}Cd^{[5]}O_2$	(trig. bipyramidal) Cd—O: 2.23 Å (2×), 2.29 Å (2×), 2.62 Å	Schnering (1962)
	Coordination number 6	
$Cd^{[6]}[CO_3]$ [a] otavite	calcite type	Zachariasen (SB 1913—28, 338); Ramdohr and Strunz (1941)
$Cd^{[6]}Ti^{[6]}O_3$	ilmenite type	Barth and Posnjak (SB 1933—35, 379)
$Cd_2^{[6]}Mn_3^{[6]}O_8$	isotypic with Mn_5O_8 Cd—O: 2.16—2.52 Å av. 2.35 Å	Oswald, Wampetich (1970)
β-$Cd^{[6]}[SO_4]$ [b]	isotypic with $CrVO_4$ Cd—O: 2.23 Å	Coing-Boyat (SR 1961, 441)

[a] $CdCO_3$ forms a continuous series of solid solutions with $MnCO_3$, and shows limited solubility in series with $CaCO_3$ (Oftedahl and Vegard, SR 1947—48, 494). Phase relations in the system $CaCO_3$—$CdCO_3$ were investigated by Chang and Brice (1971).

[b] Isotypic with the Mg, Ni, Co(II), Mn(II) sulfates.

Table 48-A-4 (Continued)

Compound	Remarks	Reference
α-$Cd^{[6]}[CrO_4]$ (low temperature)	isotypic with $CrVO_4$	Brandt (SR 1942—44, 181)
β-$Cd^{[6]}[CrO_4]$ (high temperature)	isotypic with α-$MnMoO_4$	Müller, White and Roy (1969)
β-$Cd^{[6]}[U^{[6]}O_4]$		Kovba, Polunina, Ippolitova, Simanov and Spicyn (SR 1961, 403)
$Cd^{[6]}[Sb_2^{[6]}O_6]$ $Cd^{[6]}[As_2^{[6]}O_6]$	isotypic with $PbSb_2O_6$	Magnéli (SR 1940—41, 156)
$Cd^{[6]}Ta_2^{[6]}O_6$	columbite type	Ismailzade (SR 1958, 316)
$A_2^I Cd_2^{[6]}[SO_4]_3$ $A^I = (NH_4)$, K, Rb, Tl	langbeinite type	Gattow and Zemann (1958)
$Cd_3^{[6]}[Cl\|B_7O_{13}]^d$ cadmium borate	Cd: 4O+2Cl	Jona (SR 1959, 502)
$Cd^{[6]}Sn^{[6]}(OH)_6$	6 OH isotypic with stottite	Giglio and Novales (1964)

Coordination number 7

$Cd_2^{[7]}SnO_4$	isotypic with Sr_2PbO_4	Trömel (1967, 1969)

Coordination number 8

$Cd^{[8]}[MoO_4]$	scheelite type	Broch (SB 1928—32, 450) Swanson, Gilfrich and Cook (SR 1956, 451)
$(CdU_2)^{[8]}O_6$	fluorite type^e	Rüdorff, Kemmler and Leutner (1962)
$Cd_2^{[8]}[Nb_2^{[6]}O_7]^f$	pyrochlore type	Byström (SR 1945—46, 117)
$Cd_2^{[8]}[Ta_2^{[6]}O_7]$	pyrochlore type Cd—O: 2.50 Å (6×) 2.25 Å (2×)	Cook and Jaffe (SR 1952, 254); Aleshin and Roy (1962)
$Cd_2^{[8]}[Re_2^{[6]}O_7]$	pyrochlore type Cd—O: 2.66 Å, 2.21 Å	Donohue, Longo, Rosenstein and Katz (1965)
$Cd_3^{[8]}Fe_2^{[6]}[SnO_4]_3$	garnet type	Geller, Bozorth, Gilleo and Miller (SR 1960, 350)
$Cd_3^{[8]}X_2^{[6]}[GeO_4]_3$ X = Fe, Al, Cr, Se, Ga	garnet type	Strunz, Freigang and Contag (SR 1960, 456); Tauber, Whinfrey and Banks (SR 1961, 415)
$Cd_2^{[6]}Bi_2O_4Br_2$(I, II)	Cd—O: 2.28 Å (4×) isotypic Cd—Br: 3.44 Å (4×)	Sillén (SR 1939, 159)
$Cd_2^{[6]}Bi_2O_4I_2$	Cd—O: 2.32 Å Cd—I: 3.52 Å	Sillén (SR 1947—48, 307)

^{c} Oxygen in distorted cubic close packing. The metal distribution is closely related to that of the spinels. Isotypic with the Ni, Co, Cu, Zn compounds.

^{d} See also Fouassier, Levasseur, Joubert, Muller, Hagenmuller (1970).

^{e} The cations are statistically distributed.

^{f} $Cd_2Nb_2O_7$ forms solid solutions with $Sr_2Nb_2O_7$ and $NaBiNb_2O_7$ (Ismailzade, SR 1958, 371) and with $NaNbO_3$ (Lewis and White, SR 1956, 298). $Cd_2MM'O_5F_2$ (with M = Ta(V) or Nb(V) and M' = Fe(III), Cr(III), Al, Ga, Se, In) crystallizes also in the pyrochlore type (Pannetier and Lucas, 1970).

Table 48-A-4 (Continued)

Compound	Remarks	Reference
	Coordination number 12	
$Cd^{[12]}Ti^{[6]}O_3$	perovskite type (monoclinic distorted)^g	BARTH and POSNJAK (SB 1933—35, 379); LEBEDEV, VENEVSTSEV and SHDANOV (1970)
	Coordination numbers 4 and 6	
$X^{[6]}X^{[4]}[Mo_3^{[6]}O_8]$ X = Cd, Mg, Mn, Co, Ni, Zn		MAC CARROLL, KATZ and WARD (SR 1957, 333)
	Coordination numbers 5 and 6	
$Cd^{[6]}Cd_2^{[5]}[AsO_4]$	$Cd^{[6]}$—O: 2.19—2.50 Å $Cd^{[5]}$—O: 2.11—2.39 Å	ENGEL and KLEE (1970)
	Coordination numbers 6 and 8	
$Cd_2^{[8]}Cd^{[6]}Ge^{[6]}[GeO_4]_3$ (high pressure phase)	garnet like structures $Cd^{[8]}(I)$—O: 2.45 Å (av.) $Cd^{[8]}(II)$—O: 2.41 Å (av.) $Cd^{[6]}$—O: 2.24 Å (av.)	PREWITT and SLEIGHT (1969)
$Cd^{[8]}Cd^{[6]}[Sb_2^{[6]}O_7]$	isotypic with weberite	BYSTRÖM (SR 1945—46, 117)
	Coordination numbers 7 and 9	
$Cd_2^{[6+3]}Cd_3^{[7]}(OH)[PO_4]_3$ $Cd_2^{[6+3]}Cd_3^{[7]}(Y)[XO_4]_3$ X = P, As, V Y = F, Cl, Br	apatite type Cd (I): 9 O, Cd (II): 6 O + 1 (OH)	KLEMENT and ZUREDA (SR 1940—41, 206); KLEMENT and HASELBECK (1965); ENGEL (1968 and 1970)

^{g} Some other cadmium compounds (e.g., $Cd^{[12]}Sn^{[6]}O_3$) crystallize also in the perovskite structure (NÁRAY-SZABÓ, SR 1947—48, 454).

In its complex oxygen compounds, in most cases cadmium has coordination numbers 6, 8 and 12, and in a few cases also 4, 5, 7 and 9. In Table 48-A-4 some complex oxygen compounds are listed.

Table 48-A-5 gives some information about the structurally known hydrated complex compounds of cadmium.

There are some binary cadmium chalcogene compounds. Some are isotypic with the normal spinel type e. g., $Cd^{[4]}In_2^{[6]}S_4$. A large group has the thiogallate structure (HAHN, FRANK, KLINGLER, STÖRGER and STÖRGER, SR 1955, 414) e. g., $Cd^{[4]}Al^{[4]}S_4$ and $Cd^{[4]}Ga^{[4]}S_4$.

Some high pressure modifications have been found by RANGE, BECKER and WEISS (1968a, b and 1969). $Cd^{[4]}Al_2^{[6]}S_4$ crystallizes at 400° C and at a pressure above 48 kb in the normal spinel type. At different temperatures and pressures four modifications of $CdIn_2S_4$ are known. One of these crystallizes in a NaCl defect structure, the other one with a fourfold coordination of cadmium.

Table 48-A-5. *Hydrated complex compounds of cadmium*

Compound	Cd-coordination	Reference
$Cd^{[6]}(H_2O)[SO_4]$	$2\,OH_2 + 4\,O$ Cd—O: 2.21—2.36 Å	Brégeault and Herpin (1970)
$Cd^{[6]}(H_2O)_2[SeO_4]$	$2\,OH_2 + 4\,O$	Kokkoros (SR 1940—41, 196); Hiriyana and Sakuri (SR 1949, 251)
$Cd^{[6]}(H_2O)_2[SO_4] \cdot {}^2/_3\,H_2O$	$2\,OH_2 + 4\,O$	Lipson (SB 1936, 182)
$Cd^{[4+4]}(H_2O)_3[H_3IO_6]$	$2\,OH_2 + 2\,O$ $(+2\,OH_2 + 2\,O)$ Cd—O: 2.23 Å (av.) 2× —OH_2: 2.27 Å; 2.41 Å Cd—OH_2: 2.50 Å (av.) 2× —O: 2.64 Å; 2.79 Å	Braibanti, Tiripicchio, Bigoli and Pelling-Helli (1970)

$Cd^{[6]}(H_2O)_4[NO_3]_2$	$4\,OH_2 + 4\,O$
	Cd—OH_2: 2.26 Å (2×)
	         2.33 Å (2×)
	Cd—O: 2.44 Å (2×)
	         2.59 Å (4×)
		Matković and Ribar (1963);
		Matković, Ribar, Zelenko
		and Peterson (1966)

$(NH_4)_2Cd^{[6]}(H_2O)_6[SO_4]_2$	$6\,OH_2$ Cd—OH_2: 2.30 Å (4×) 2.24 Å (2×)	Hofmann (SB 1928—32, 436); Montgomery and Linga- felter (1966)
$Cd^{[6]}(H_2O)_6[ClO_4]_2$ $Cd^{[6]}(H_2O)_6[BF_4]_2$	$6\,OH_2$ $6\,OH_2$	West (SB 1933—35, 452); Moss, Russel and Sharp (SR 1961, 304)

Some arsenide and phosphide compounds of cadmium and the elements of the fourth group of the periodic system crystallize in the chalcopyrite type, e.g., $Cd^{[4]}\,Ge^{[4]}As$ (Goodman, SR 1957, 38), and $Cd^{[4]}Sn^{[4]}P_2$ (Buehler and Wernick, 1971).

Revised manuscript received: February 1972

References: Section 48-A

AGRAWAL, V. K., TRIGUNAYAT, G. C.: Arcing phenomenon in single crystals of cadmium bromide. Acta Cryst. A **26**, 426 (1970).

ALESHIN, E., ROY, R.: Crystal chemistry of pyrochlore. J. Am. Ceram. Soc. **45**, 18 (1962).

BRAIBANTI, A., TIRIPICCHIO, A., BIGOLI, F., PELLINGHELLI, M. A.: Crystal and molecular structure of cadmium trihydrogenhexaoxoiodate. (VII) trihydrate. Acta Cryst. B **26**, 1069 (1970).

BRÉGEAULT, J.-M., HERPIN, P.: Étude structurale du sulfate de cadmium monohydraté $CdSO_4 \cdot H_2O$. Bull. Soc. Franç. Minéral. Cristall. **93**, 37 (1970).

BUEHLER, E., WERNICK, J. H.: Concerning growth of single crystals of the II-IV-V diamond-like compounds $ZnSiP_2$, $CdSiP_2$, $ZnGeP_2$, and $CdZnP_2$ and standard enthalpies of formation for $ZnSiP_2$ and $CdSiP_2$. J. Crystal Growth **8**, 324 (1971).

CHANG, L. L. Y., BRICE, W. R.: Subsolidus phase relations in the system calcium carbonate-cadmium carbonate. Am. Mineralogist **56**, 338 (1971).

COUSSEINS, J.-C., PINA PEREZ, C.: Fluorures doubles de cadmium et d'éléments metalliques monovalents. Rev. Chim. Minerale **5**, 147 (1968).

DÄWERITZ, L.: Relative stability of zincblende and wurtzite structure in $A^{II}B^{IV}$ compounds. Kristall u. Technik **6**, 101 (1971).

DENT GLASSER, L. S.: Silicates M_3SiO_5. II. Relationships between Sr_3SiO_5, Cd_3SiO_5 and Ca_3SiO_5. Acta Cryst. **18**, 455 (1965).

— GLASSER, F. P.: The preparation and crystal data of the cadmium silicates $CdSiO_3$, Cd_2SiO_4, and Cd_3SiO_5. Inorg. Chem. **3**, 1228 (1964).

DONOHUE, P. C., LONGO, J. M., ROSENSTEIN, R. D., KATZ, L.: The preparation and structure of cadmium rhenium oxide, $Cd_2Re_2O_7$. Inorg. Chem. **4**, 1152 (1965).

ENGEL, G.: Einige Apatite des Cadmiums. Z. Anorg. Allgem. Chem. **362**, 273 (1968).

— Cadmiumapatite sowie die Verbindungen Cd_2XO_4F mit X = P, As und V. Z. Anorg. Allgem. Chem. **378**, 49 (1970).

— KLEE, W. E.: Die Kristallstruktur des Cadmiumorthoarsenates. Z. Krist. **132**, 332 (1970).

FOUASSIER, C., LEVASSEUR, A., JOUBERT, J. C., MULLER, J., HAGENMULLER, P.: Les systèmes B_2O_3—MO—MS (M = Mg, Mn, Fe, Cd) et sodalites M—S (M = Co, Zn). Z. Anorg. Allgem. Chem. **375**, 202 (1970).

GATTOW, G., ZEMANN, J.: Über Doppelsulfate vom Langbeinit-Typ, $A_2^+B_2^{2+}(SO_4)_3$. Z. Anorg. Allgem. Chem. **293**, 233 (1958).

GENTILE, A. L., ROY, R.: Isomorphism and crystalline solubility in the garnet family. Am. Mineralogist **45**, 701 (1960).

GIGLIO, M., NOVALES, H.: $ZnSn(OH)_6$ und $CdSn(OH)_6$. Naturwissenschaften **51**, 56 (1964).

HARTMANN, H.: Über die Polymorphie unterschiedlich präparierter Cadmiumsulfide. Kristall u. Technik **1**, 267 (1966).

IHARA, M., KROGH-MOE, J.: The crystal structure of cadmium diborate, $CdO \cdot 2B_2O_3$. Acta Cryst. **20**, 132 (1966).

JAIN, R. K., CHADHA, G. K., TRIGUNAYAT, G. C.: Crystal structures of four new polytypes of cadmium iodide. Acta Cryst. B **26**, 1785 (1970).

KABALKINA, S. S., TROITSKAYA, Z. V.: Cadmium sulfide structures at high pressures up to 90 kilobars. Dokl. Akad. Nauk SSSR **151**, 1068 (1963).

KLEMENT, R., HASELBECK, H.: Apatite und Wagnerite zweiwertiger Metalle. Z. Anorg. Allgem. Chem. **336**, 113 (1965).

© Springer-Verlag Berlin · Heidelberg 1972

KOKKOROS, P. A., RENTZEPERIS, P. J.: The crystal structure of the anhydrous cadmium and mercuric sulfates. Z. Krist. 119, 234 (1963).

LEBEDEV, V. M., VENEVTSEV, YU. N., SHDANOV, G. S.: High-temperature x-ray investigation of the perovskite modification of $CdTiO_3$. Soviet Phys.-Cryst. 15, 318 (1970).

MARIANO, A. N., WAREKOIS, E. P.: High-pressure phases of some compounds of group II—VI. Science 142, 672 (1963).

MATKOVIĆ, B., RIBAR, B.: The crystal structure of cadmium nitrate tetrahydrate. Croat. Chem. Acta 35, 147 (1963).

— — ZELENKO, B., PETERSON, S. W.: Refinement of the structure of $Cd(NO_3)_2 \cdot 4H_2O$. Acta Cryst. 21, 719 (1966).

MILL, B. V.: Hydrothermal synthesis of garnets containing V^{3+}, In^{3+} and Sc^{3+}. Dokl. Akad. Nauk SSSR 156, 914 (1964).

MITCHELL, R. S.: Single crystal x-ray study of structural polytypism in cadmium bromide. Z. Krist. 117, 309 (1962).

— Comments on an x-ray study of cadmium hydroxide single crystals. Z. Krist. 123, 459 (1966).

MONTGOMERY, H., LINGAFELTER, E. C.: The crystal structure of tutton's salts. IV. Cadmium ammonium sulfate hexahydrate. Acta Cryst. 20, 728 (1966).

MÜLLER, O., WHITE, W. B., ROY, R.: X-ray diffraction study of the chromates of nickel, magnesium and cadmium. Z. Krist. 130, 112 (1969).

ONODERA, A.: High pressure transition in cadmium selenide. Rev. Phys. Chem. Japan 39, 65 (1969).

OSWALD, H. R., WAMPETICH, M. J.: Die Kristallstrukturen von Mn_5O_8 und $Cd_2Mn_3O_8$. Helv. Chim. Acta 50, 2023 (1970).

PANNETIER, J., LUCAS, J.: Niobates et tantalates de cadmium oxyfluores contenant de cations trivalents. Compt. Rend., Ser. C 270, 1013 (1970).

PAULING, L., HUGGINS, M. L.: Covalent radii of atoms and interatomic distances in crystals containing electron-pair-bonds. Z. Krist. (A) 87, 205 (1934).

PRASAD, R., SRIVASTAVA, O. N.: A new hexagonal polytype 32H of cadmium iodide, its structure and growth. Z. Krist. 131, 376 (1970).

PREWITT, C. T., SLEIGHT, A. W.: Garnet-like structures of high-pressure cadmium germanate and calcium germanate. Science 163, 386 (1969).

RAMDOHR, P., STRUNZ, H.: Isomorphie von Otavit mit Kalkspat. Zentr. Mineralogie A, 97/98 (1941).

RANGE, K.-J., BECKER, W., WEISS, A.: Über Hochdruckphasen des $CdAl_2S_4$, $HgAl_2S_4$, $ZnAl_2Se_4$, $CdAl_2Se_4$ und $HgAl_2Se_4$ mit Spinellstruktur. Z. Naturforsch. 23b, 1009 (1968a).

— — — Über Hochdruckphasen des $CdIn_2Te_4$ und $HgIn_2Te_4$ mit NaCl-Struktur. Z. Naturforsch. 23b, 1261 (1968b).

— — — Das Verhalten von $CdIn_2Se_4$ bei hohen Drucken. Z. Naturforsch. 24b, 1654 (1969).

RÜDORFF, W., KEMMLER, S., LEUTNER, H.: Ternäre Uran (V)-oxyde [1]. Angew. Chem. 74, 429 (1962).

SCHNERING, H. G. VON: Die Kristallstruktur des $BaCdO_2$. Z. Anorg. Allgem. Chem. 314, 144 (1962).

SEIFERT, H. J., KOKNAT, F. W.: Neue Alkalichlorocadmate (II) in den Systemen $CsCl/CdCl_2$ und $RbCl/CdCl_2$. Z. Anorg. Allgem. Chem. 357, 314 (1968).

SIMONOV, M. A., EGOROV-TISMENKO, YU. K., BELOV, N. V.: New trisilicate radical $[Si_3O_{10}]$ in the structure of $Na_4Cd_2[Si_3O_{10}]$. Dokl. Acad. Nauk SSSR 179, 1329 (1968).

SMITH, P. L., MARTIN, J. E.: High pressure structures of cadmium sulfide and cadmium telluride. Phys. Letters 6, 42 (1963).

TARTE, P., CAHAY, R.: Synthesis and structure of a new family of $Li_2X(II)SiO_4$ and $Li_2X(II)GeO_4$ compounds structurally related to lithium phosphate, [X(II) = magnesium, zinc, manganese, cobalt, iron and cadmium]. Compt. Rend. Ser. C 271, 777 (1970).

TRAILL, R. J., BOYLE, R. W.: Hawleyite, isometric cadmium sulphide, a new mineral. Am. Mineralogist **40**, 555 (1955).

TRÖMEL, M.: Kristallstrukturdaten für Ca_2SnO_4 und Cd_2SnO_4. Naturwissenschaften **54**, 17 (1967).

— Die Kristallstruktur der Verbindungen vom Sr_2PbO_4-Typ. Z. Anorgan. Allgem. Chem. **371**, 237 (1969).

WOLFF, P. M. DE: The crystal structure of γ-$Cd(OH)_2$. Acta Cryst. **21**, 432 (1966).

— The crystal structure of γ-$Cd(OH)_2$. Corrigenda. Acta Cryst. **22**, 441 (1967).

Revised manuscript received: February 1972

Indium 49

B—M, T. A. LINN, JR. (Kennecott Research Center, Salt Lake City,
O   Utah 84111, U.S.A.)

and

R. A. SCHMITT (Department of Chemistry, Oregon State University, Corvallis, Oregon 97331, U.S.A.)

49-B. Isotopes in Nature

Indium (atomic number 49) consists of two isotopes of mass numbers 113 and 115. The natural isotopic abundances of ^{113}In and ^{115}In are 4.33 and 95.67 atomic percent respectively (SUNDERMAN and TOWNLEY, 1960; LEDERER et al., 1967). Indium-115 is naturally radioactive, decaying with a 5×10^{14}-year half-life to stable ^{115}Sn.

49-C. Abundance in Cosmos, Meteorites and Tektites

I. Cosmic Abundance

The cosmic abundance of indium has been based on solar abundance and the available data are summarized in Tables 49-C-1 and 49-C-2.

Table 49-C-1. *Indium cosmic abundances relative to hydrogen*

Indium abundance[a]	Reference
0.75	ALLER (1961)
1.16	GOLDBERG et al. (1960)
1.28	ALLER (1965)
1.45	ALLER (1965)
1.51	ALLER (1968)
1.68	GREVESSE et al. (1968)

Atomic abundances based on silicon (Si $= 10^6$) are given in Table 49-C-2.

Table 49-C-2. *Indium cosmic abundances relative to silicon*

Indium abundance[b]	Reference
0.11	SUESS AND UREY (1956)
0.084	VINOGRADOV (1956)
0.098	AHRENS and TAYLOR (1961)
0.22	CAMERON (1968)
0.80	UREY (1967)

The indium content of sunspot spectra was reported as 0.75 ± 0.2 (log N_{In} based on log $N_H = 12.0$) by DUBOV and KHROMOVA (1966). A cosmic mass abundance of 0.24 ppb indium was given by GREEN (1959).

II. Meteorites

The indium content of iron meteorites are shown in Table 49-C-3, and the abundances of indium in the various classes of stony meteorites are given in Table 49-C-4.

---

[a] log N_{In} based on log $N_H = 12.0$.
[b] log N_{In} based on log $N_{Si} = 6.0$.

Table 49-C-3. *Indium in iron meteorites*

Material	No. of samples	In (ppb) range	mean	Method	Reference
Hexahedrite (H)	5	0.5—10	3	(N/R)	Smales et al. (1967)
Hexahedrite (H)	2	0.36—0.39	0.38	(N/R)	Tandon and Wasson (1967a)
Ataxite, Ni-poor (Da)	4	4—10	10	(N/R)	Smales et al. (1967)
Octahedrite, coarsest (Ogg)	4	4—11	9	(N/R)	Smales et al. (1967)
Octahedrite, coarsest (Ogg)	1	—	10.6	(N/R)	Tandon and Wasson (1967a)
Octahedrite, coarse (Og)	10	1—14	11	(N/R)	Smales et al. (1967)
Octahedrite, medium (Om)	20	0.5—30	12	(N/R)	Smales et al. (1967)
Octahedrite, medium (Om)	2	14.8—16.1	15.5	(N/R)	Tandon and Wasson (1967a)
Octahedrite, fine (Of)	10	0.3—<10	—		Smales et al. (1967)
Octahedrite, finest (Off)	4	1.1—<10	—		Smales et al. (1967)
Ataxite, Ni-rich (Dr)	6	0.4—41	14	(N/R)	Smales et al. (1967)

Table 49-C-4. *Indium in achondrites and chondrites*

Meteorite class	No. determined	In (ppb) range	mean	Method	Reference
Achondrites:					
Ae	2	0.34—0.40	0.37±0.04	(N/R)	Schmitt and Smith (1968)
Ap	2	0.24—1.0	0.5 ±0.1	(N/R)	Schmitt and Smith (1968)
Chondrites:					
Ce	4	4—76	35	(N/R)	Fouché and Smales (1967)
Ce_1	1	135	135	(N/R)	Akaiwa (1966)
Ce_1	2	90—93	92	(N/R)	Schmitt and Smith (1968)
Ce_2	3	0.24—2.9	1.1	(N/R)	Schmitt and Smith (1968)
CH	9	0.5—8.2	2.3	(N/R)	Fouché and Smales (1967)
CH	3	0.1—2.1	0.8	(N/R)	Schmitt and Smith (1968)
CH	1	0.5	0.5	(N/R)	Tandon and Wasson (1967b)
CL	9	0.5—1.3	0.7	(N/R)	Fouché and Smales (1967)
CL	26	0.05—55	6	(N/R)	Tandon and Wasson (1968)
CL	5	0.3—14	5	(N/R)	Schmitt and Smith (1968)
CL	10	0.07—26	7	(N/R)	Keays, Ganapathy and Anders (1971)
CLL	1	74	74	(N/R)	Schmitt and Smith (1968)
CHL	3	23—32	27	(N/R)	Schmitt and Smith (1968)
CHL	1	31	31	(N/R)	Akaiwa (1966)
CHL	1	27	27	(N/R)	Wakita and Schmitt (1970)
Cc_1	2	64—80	75	(N/R)	Schmitt and Smith (1968)
Cc_1	2	82—88	85	(N/R)	Fouché and Smales (1967)
Cc_1	1	115	115	(N/R)	Akaiwa (1966)
Cc_2	2	47—64	55	(N/R)	Schmitt and Smith (1968)
Cc_2	2	46—49	48	(N/R)	Fouché and Smales (1967)
Cc_2	2	23—50	36	(N/R)	Akaiwa (1966)

LARIMER and ANDERS (1967) reported the averages of indium content in various types of chondritic meteorites. Their data are summarized in Table 49-C-5.

Table 49-C-5. *Indium in chondritic meteorites*

Chondrite type	In (ppb)
Carbonaceous, I	220
Carbonaceous, II	100
Carbonaceous, III	43
Ordinary	0.4
Enstatite, I	140
Enstatite, II	3

TAYLOR and SACHS (1960) attempted to determine indium in tektites, but found less than their detection limit (1 ppm In) in 14 australites. No other data are available concerning indium in tektites.

49-D. Abundance in Rock-forming Minerals and Ore Minerals; Indium Minerals

I. Rock-forming and Ore Minerals

Indium is rare in the common minerals, being either an inclusion in the principal mineral or occupying a lattice site in the mineral crystal. Tables 49-D-1 and 49-D-2 list abundances of indium in the principal rock-forming minerals and the principal ore-forming minerals.

Table 49-D-1. *Indium in rock-forming minerals*

Material	No. of samples	In (ppb) range	In (ppb) mean	Method	Reference
Quartz from alaskite, Yakutiya (U.S.S.R.)	—	—	10	(C)	1
Plagioclases and K-feldspars	9	—	<20	(S)	2
Feldspars from granite, Yakutiya (U.S.S.R.)	—	—	15	(C)	1
Feldspars (plagioclases) from gabbro, Skaergaard intrusion (East Greenland)	2	3.0—3.4	3.2	(N/R)	3
Feldspars (plagioclases) from gabbro, Skaergaard intrusion (East Greenland)	2	—	4±1	(N/R)	4
Albite from alaskite, Yakutiya (U.S.S.R.)	—	—	20	(C)	1
Tourmalines, Deputaskoe (U.S.S.R.)	3	3,000—13,000	7,000	(C)	5
Chlorite, Deputaskoe (U.S.S.R.)	1	—	8,000	(C)	5
Muscovites (Finland and U.S.A.)	2	<20—4,500	3,000	(S)	6
Muscovites from granite, Yakutiya (U.S.S.R.)	2	50—80	65	(C)	1
Biotites from granite, Saxony (Eastern Germany)	2	250—320	290	(S)	7
Biotites	2	—	<20		6
Biotites from granites, Yakutiya (U.S.S.R.)	2	490—1,800	1,100	(C)	1
Augite (diopside)	—	—	100±4	(N/R)	8
Pyroxenes from gabbros, Skaergaard intrusion (East Greenland)	9	170—1,080	700	(N/R)	3
Pyroxenes from gabbros, lamprophyres (Greenland, U.S.A., Finland, and Canada)	4	<20—310	91	(S)	6

© Springer-Verlag Berlin · Heidelberg 1972

Table 49-D-1 (continued)

Material	No. of samples	In (ppb) range	In (ppb) mean	Method	Reference
Olivines, Skaergaard intrusion (East Greenland)	2	55—57	56	(N/R)	3
Serpentinites	4	—	15	(S)	7
Amphiboles (U.S.A.)	2	<20—5,800	3,000	(S)	6
Garnets from gneisses (Greenland and U.S.A.)	3	26—180	100	(S)	6
Epidote	1	—	590	(S)	6
Cassiterites (U.S.S.R.)	9	16—30 ppm	—	(C)	9
Cassiterites (U.S.S.R.)	186	10—100 ppm	—		9
Cassiterite	1	<1 ppm	—		10
Cassiterites	72	10—1,000 ppm	—		11
Cassiterites	—	X—100 ppm	—	(S)	12
Magnetites from gabbro, Skaergaard intrusion (East Greenland)	2	150—160	160	(N/R)	3
Magnetites from lamprophyre	2	71—120	95	(S)	6
Magnetites from gabbro, Skargaard intrusion (East Greenland)	2	—	90±5	(N/R)	13
Magnetite-ilmenites from gabbro, Skaergaard intrusion (East Greenland)	2	—	150±4	(N/R)	13
Ilmenites from gabbro, Skaaergaard intrusion (East Greenland)	2	280—290	290	(N/R)	3
Ilmenites	2	—	194±4	(N/R)	13
Pyrrhotites, Deputaskoe (U.S.S.R.)	6	1—50 ppm	28 ppm	(C)	5
Pyrite, Deputaskoe (U.S.S.R.)	1	—	2.5 ppm	(C)	5
Siderites, Deputaskoe (U.S.S.R.)	2	—	4 ppm	(C)	5
Mangano-siderites, Deputaskoe (U.S.S.R.)	2	2—6 ppm	4 ppm	(C)	5
Calcites, North Kantau (U.S.S.R.)	3	320—1,800	—	(P)	14
Aragonites, Karkhona (U.S.S.R.)	6	220—1,600	—	(P)	14

X: less than limit of detection.

References: 1. IVANOV (1963). 2. SHAW (1957). 3. WAGER et al. (1958). 4. SMALES et al. (1957). 5. IVANOV and ROZBIANSKAYA (1961). 6. SHAW (1952). 7. VOLAND (1969). 8. ONUMA et al. (1968). 9. IVANOV and LIZUNOV (1960). 10. AHRENS (1948). 11. ITZIKSON and RUSANOV (1946). 12. BREWER and BAKER (1936). 13. SMALES et al. (1957). 14. KULIKOVA (1966).

Indium concentrates in sulfide minerals, especially those having tetrahedral coordination about the principal metal ion. Note the especially high concentrations in cassiterites, sphalerites, stannites, and chalcopyrites.

Table 49-D-2. *Indium in ore-forming minerals*

Material	No. of samples	In (ppm or ppb) range	mean	Method	Reference
Wolframites, Deputaskoe (U.S.S.R.)	2	3—50 ppm	26 ppm	(C)	1
Wolframites, Northern Yakutiya (U.S.S.R.)	17	10—100 ppm	—		1
Sphalerites, (U.S.S.R.)	7	200 ppb to 100 ppm	—	(P)	2
Sphalerites, Eastern Transbaikaliya (U.S.S.R.)	—	10—5,000 ppm	—		2
Sphalerites, Eastern Transbaikaliya (U.S.S.R.)	48	30—3,300 ppm	—	(P)	3
Sphalerites, Eastern Transbaikaliya (U.S.S.R.)	9	—	20 ppm	(P)	3
Sphalerites, Karamazar and Chatkal (U.S.S.R.)	51	70—330 ppm	—		4
Sphalerites, Deputaskoe (U.S.S.R.)	12	20—3,100 ppm	800 ppm	(C)	1
Sphalerites, Northern Yakutiya (U.S.S.R.)	85	10—4,700 ppm	—		1
Sphalerites	24	X—800 ppm	—		5
Sphalerites	—	70—120 ppm	—		6
Sphalerites	75	10—700 ppm	—		7
Chalcopyrites, Eastern Transbaikaliya (U.S.S.R.)	11	5—100 ppm	—	(P)	3
Chalcopyrites, Karamazar and Chatkal (U.S.S.R.)	10	30—315 ppm	—		4
Chalcopyrites, Deputaskoe (U.S.S.R.)	5	570—1,500 ppm	—	(C)	1
Chalcopyrites, Northern Yakutiya (U.S.S.R.)	58	50—750 ppm	—		8
Chalcoyprite	1	—	4.7 ppm	(S)	9
Stannites, Deputaskoe (U.S.S.R.)	2	800—900 ppm	850 ppm	(C)	1
Stannites, Northern Yakutiya (U.S.S.R.)	27	50—1,500 ppm	—		1
Stannites	2	—	> 100 ppm		5
Galenas, Kansay (U.S.S.R.)	10	0.3—70 ppm	—	(P)	2
Galenas, Eastern Transbaikaliya (U.S.S.R.)	—	10—50 ppm	—		10
Galenas, Eastern Transbaikaliya (U.S.S.R.)	11	1—34 ppm	—	(P)	3
Galenas, Karamazar and Chatkal (U.S.S.R.)	25	—	5 ppm		4
Galenas, Deputaskoe (U.S.S.R.)	6	5—10 ppm	5.3 ppm	(C)	1
Galena	1	—	910 ppb	(S)	9

X = less than limit of detection.

References: 1. IVANOV and ROZBIANSKAYA (1961). 2. KULIKOVA (1966). 3. MEITUV (1962). 4. BADALOV (1961). 5. ANDERSON (1953). 6. CAMBI and MALETESTA (1936). 7. OFTEDAL (1940). 8. IVANOV and LIZUNOV (1959). 9. SHAW (1952). 10. KULIKOVA (1962).

II. Indium Minerals

Although rare in occurrence, three indium minerals have been reported (FLEISCHER, 1966) and are listed in Table 49-D-3.

Table 49-D-3. *Indium minerals*

Mineral	Formula	Reference
Dzhalindite (or Jalindite)	$In(OH)_3$	GENKIN and MURAV'EVA (1963)
Indite	$FeIn_2S_4$	GENKIN and MURAV'EVA (1963)
Roquesite	$CuInS_2$	PICOT and PIERROT (1963)

GENKIN and MURAV'EVA (1963) described the more recently discovered indium minerals, which were found in cassiterite from tin ore deposits in the far eastern USSR. Roquesite was reported by PICOT and PIERROT (1963), who determined the mineral composition from electron probe measurements.

49-E. Abundance in Common Igneous Rocks

The indium contents of the principal igneous rock types are given in Tables 49-E-1 through 49-E-4.

Table 49-E-1. *Indium in ultramafic rocks*

Material	No. of samples	In (ppb) range	In (ppb) mean	Method	Reference
Dunites, average	—	—	13	(—)	1
Dunites, Urals, Tagil'sk massif (U.S.S.R.)	1	—	20	(C)	2
Dunites, Canwell glacier, Alaska (U.S.A.)	1	—	6	(N/R)	3
Peridotites (Alaska; Venezuela; Puerto Rico)	3	5—33	15	(N/R)	3
Peridotites, Miyamori, Iwate, and Toba, Mie (Japan)	2	23—26	25	(N/R)	4
Peridotites, average	—	—	13	(—)	1
Peridotites, Kola Peninsula, Monchegora (U.S.S.R.)	(composite)	—	40	(C)	2
Pyroxenites, average	—	—	13	(—)	1
Pyroxenites, Urals, Tagil'sk massif (U.S.S.R.)	1	—	50	(C)	2
Pyroxenites, Kola Peninsula, Monchegora (U.S.S.R.)	(composite)	—	60	(C)	2
Garnet-pyroxenites, Kakanui (New Zealand)	1	—	14±2	(N/R)	3

References: 1. VINOGRADOV (1962). 2. IVANOV (1963). 3. SCHMITT and SMITH (1968). 4. HAMAGUCHI et al. (1967).

Table 49-E-2. *Indium in mafic rocks*

Material	No. of samples	In (ppb) range	In (ppb) mean	Method	Reference
Basalt, Rhön (Germany)	1	—	180	(S)	1
Basalts, Thüringer Wald (East Germany)	6	40—210	93	(S)	1
Basalts, Aleutian Islands and Greenland	3	130—320	—	(S)	2
Basalts (Japan)	(composite of 14)	—	83 ± 4	(N/R)	3
Basalts	(composite of > 200)	—	67 ± 6	(N/R)	4
Basalts (East Pacific Rise; Guadalupe Island)	4	65—94	76	(N/R)	4
Basalts	—	—	60 ± 3	(N/R)	5
Spilites, Elbe Valley (East Germany)	3	15—50	27	(S)	1
Quartz-spilites, Elbe Valley (East Germany)	6	100—170	127	(S)	1
Olivine basalt (1921), Kilauea (Hawaii)	1	—	84 ± 4	(N/R)	3
Olivine-basalts (dolerites), Thüringer Wald (East Germany)	3	35—56	47	(S)	1
Olivine-basalt, Rhön (Germany)	1	—	80	(S)	1
Nepheline-basalts, Lausitz and Erzgebirge (East Germany)	4	25—40	34	(S)	1
Leucite-basalts, Lausitz and Erzgebirge (East Germany)	2	40—60	50	(S)	1
Gabbros (Finland, Canada, and U.S.A.)	6	—	15	(S)	2
Basalts and gabbros, average	—	—	220	(—)	6
Gabbros, Urals (U.S.S.R.)	(composite of 3)	40—90	60	(C)	7
Gabbros, Skaergaard intrusion (East Greenland)	18	50—62	—	(N/R)	8
Lamprophyres (East Germany)	4	40—80	59	(S)	1
Lamprophyres (Canada and Finland)	2	17—71	44	(S)	2
Eclogites	6	—	51	(S)	1

References: 1. VOLAND (1969). 2. SHAW (1952). 3. HAMAGUCHI *et al.* (1967). 4. SCHMITT and SMITH (1968). 5. ONUMA *et al.* (1968). 6. VINOGRADOV (1962). 7. IVANOV (1963). 8. WAGER *et al.* (1958).

Indium

Table 49-E-3. *Indium in intermediate rocks*

Material	No. of samples	In (ppb) range	In (ppb) mean	Method	Reference
Diorites, Yakutiya (U.S.S.R.)	—	—	50	(C)	IVANOV (1963)
Quartz-diorites, Yakutiya (U.S.S.R.)	1	—	<50	(C)	IVANOV (1963)
Hornblende-diorites, Yakutiya (U.S.S.R.)	(composite)	—	30	(C)	IVANOV (1963)
Hornblendite, Kaalamo (Finland)	—	—	53	(S)	SHAW (1952)
Andesites (Japan)	(composite)	—	50±4	(N/R)	HAMAGUCHI et al. (1967)
Diorites, Yakutiya (U.S.S.R.)	—	—	96	(C)	IVANOV (1963)
Monzonites, Yakutiya (U.S.S.R.)	1	—	130	(C)	IVANOV (1963)
Syenites, Yakutiya (U.S.S.R.)	4	50—380	139	(C)	IVANOV (1963)
Nepheline-syenites (U.S.S.R.)	3	30—36	33	(S)	VOLAND (1969)
Phonolites, Lausitz, Erzgebirge (East Germany)	2	40—45	43	(S)	VOLAND (1969)

Table 49-E-4. *Indium in granitic rocks*

Material	No. of samples	In (ppb) range	In (ppb) mean	Method	Reference
Granodiorites, Mongolia	4	45—80	60	(S)	1
Granodiorites, Lausitz (East Germany)	3	35—85	53	(S)	1
Granodiorites, Urals, Kasakhstan, Tuva and Yakutiya (U.S.S.R.)	8	<20—140	94	(C)	2
Granodiorites, Kitashidara, Aichi (Japan)	1	—	27±1	(N/R)	3
"Granites" (<60% SiO$_2$)	(composite of 85)	—	83±8	(N/R)	4
Granites (60—70% SiO$_2$)	(composite of 191)	—	67±6	(N/R)	4
Granites (>70% SiO$_2$)	(composite of 213)	—	57±6	(N/R)	4
Granites (Japan)	(composite of 20)	—	42±3	(N/R)	3
Granites, Shimoina, Nagana (Japan)	—	—	47±3	(N/R)	3
Granites, Yakutiya (U.S.S.R.)	4	20—87	47	(C)	2

Indium

Table 49-E-4 (continued)

Material	No. of samples	In (ppb) range	In (ppb) mean	Method	Reference
Granites (East Germany)	11	13—22	18	(S)	1
Granites, Far East (U.S.S.R.)	(composite of 18)	—	290	(C)	2
Granites, Eastern Transbaikaliya (U.S.S.R.)	(composite of 8)	—	140	(C)	2
Granites Eastern Kasakhstan (U.S.S.R.)	(composite of 9)	—	530	(C)	2
Granites, Ukraine (U.S.S.R.)	(composite of 8)	—	200	(C)	2
Granites, Gornyi Altai (U.S.S.R.)	(composite of 18)	—	160	(C)	2
Granites, Western Tuva (U.S.S.R.)	(composite of 71)	—	20	(C)	2
Granites, Northern Caucasus	(composite of 18)	—	130	(C)	2
Granites, Urals	(composite of 96)	—	20	(C)	2
Granites, Lausitz (East Germany)	3	18—45	28	(S)	1
Granites, Thuringia (East Germany)	8	5—50	23	(S)	1
Granites, Niederbobritz (East Germany)	4	11—60	30	(S)	1
Granites, Saxony (East Germany)	3	17—60	34	(S)	1
Granites, Western Erzgebirge (East Germany)	20	14—250	42	(S)	1
Granites	10	<20—2,000	260	(S)	5
"Tin-Granites", Erzgebirge (East Germany)	19	13—340	170	(S)	1
Granite-gneiss, Yakutiya (U.S.S.R.)	1	—	210	(C)	2
Alaskites, Yakutiya (U.S.S.R.)	3	24—50	36	(C)	2
Aplites (granitic), Yakutiya (U.S.S.R.)	3	X—33	—	(C)	2
Aplitic granites	27	—	18	(S)	1
Rhyolites (quartz-porphyries), Erzgebirge (East Germany)	35	11—85	43	(S)	1
Rhyolites (quartz-porphyries), Northwest Saxony (East Germany)	2	58—70	64	(S)	1
Rhyolites (pyroxene-poor quartz porphyries), Northwest Saxony (East Germany)	3	95—230	230	(S)	1
Pyroxene-rich quartz porphyry, Northwest Saxony (East Germany)	1	—	640	(S)	1

X = less than limit of detection.

References: 1. VOLAND (1969). 2. IVANOV (1963). 3. HAMAGUCHI et al. (1967). 4. SCHMITT and SMITH (1968). 5. SHAW (1952).

IVANOV (1963) and VOLAND (1969) noted that the indium content in granites decreased with decrease in total iron content (Figs. 49-E-1 and 49-E-2), and thus the Fe/In ratio remained relatively constant in spite of considerable varaition in indium content.

Some additional data on indium concentrations in igneous rocks of special interest are given in Tables 49-E-5 and 49-E-6.

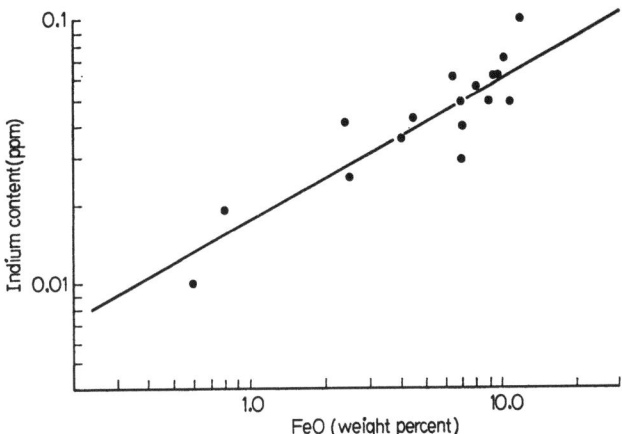

Fig. 49-E-1. Correlation between indium and FeO content of ultramafic and mafic rocks (VOLAND, 1969)

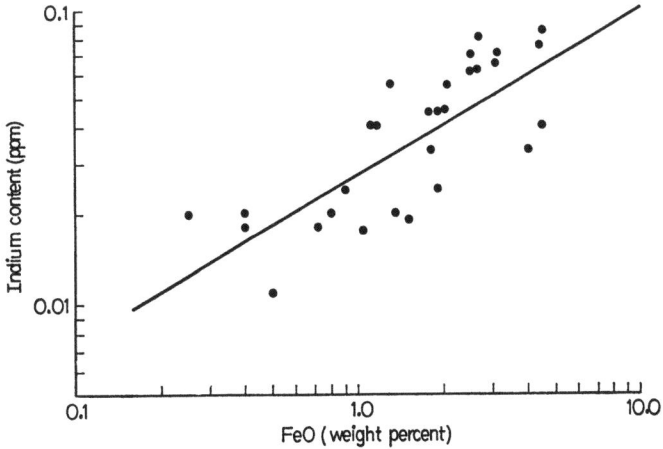

Fig. 49-E-2. Correlation between indium and FeO content of granitic rocks of norma evolution (VOLAND, 1969)

Indium

Table 49-E-5. *Indium in rocks of the Skaergaard intrusion, East Greenland*

Material	Collection No.	No. of samples	In (ppb) range	In (ppb) mean	Method	Reference
Chilled marginal gabbro, east margin	EG-1825	7	59—62	61±2	(N/R)	1
Chilled marginal gabbro	EG-1825	9	52—62	61	(N/R)	2
Chilled marginal gabbro, south margin	EG-4507	11	50—58	54±3	(N/R)	1
Chilled marginal gabbro	EG-4507	8	50—58	54	(N/R)	2
Fayalite ferrogabbro	EG-4327	7	150—174	160±10	(N/R)	1
Fayalite ferrogabbro	EG-4327	7	151—170	170	(N/R)	2
Fayalite ferrogabbro	EG-4328	8	164—194	180±20	(N/R)	1
Fayalite ferrogabbro	EG-4328	12	170—194	180	(N/R)	2
Acid granophyre	EG-3058	2	89—93	91±2	(N/R)	1
Acid granophyre	EG-3058	2	89—93	91	(N/R)	2
Hypersthene olivine gabbro	EG-5086	2	59—62	60±2	(N/R)	1
Hypersthene olivine gabbro	EG-5086	2	59—62	60	(N/R)	2
Hortonolite ferrogabbro	EG-5181	3	77—83	79	(N/R)	2

References: 1. SMALES et al. (1957). 2. WAGER et al. (1958).

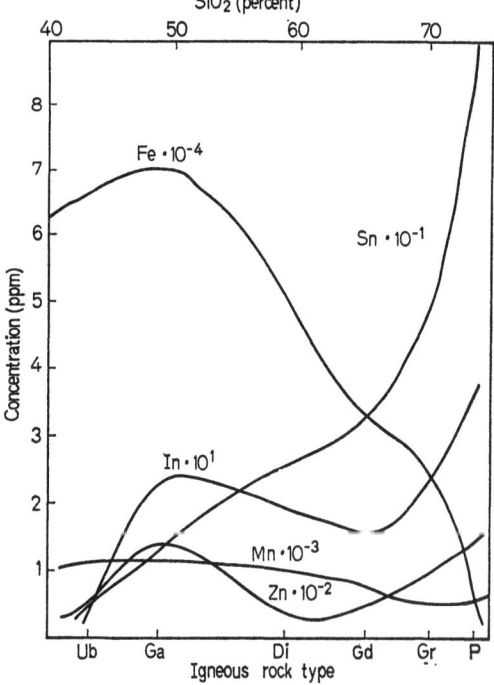

Fig. 49-E-3. Variation diagram of In, Fe, Sn, Mn and Zn (SMITH, 1963). *Ub* Ultramafic; *Ga* Gabbro; *Di* Diorite; *Gd* Granodiorite; *Gr* Granite; *P* Pegmatite

Table 49-E-6. *Indium in U.S.G.S. silicate rock standards*

Material	No. of samples	In (ppb) range	In (ppb) mean	Method	Reference
DTS-1 dunite	2	2.6—2.8	2.7	(N/R)	1
PCC-1 peridotite	2	4.1—4.5	4.3	(N/R)	1
BCR-1 basalt	2	107—110	109	(N/R)	1
W-1 basalt ("diabase")	2	67.4—67.5	67	(N/R)	1
W-1 basalt ("diabase")	—	55—59	57	(N/R)	2
W-1 basalt ("diabase")	—	—	42	(S)	3
W-1 basalt ("diabase")	7	61—70	64 ± 4	(N/R)	4
W-1 basalt ("diabase")	—	—	50	(S)	5
W-1 basalt ("diabase")	—	—	55	(N/R)	6
W-1 basalt ("diabase")	—	—	<200	(mass. spec.)	7
W-1 basalt ("diabase")	—	—	25	(C)	8
AGV-1 andesite	2	42.6—44.5	44	(N/R)	1
STM-1 nepheline-syenite	2	85.2—87.9	87	(N/R)	1
GSP-1 granodiorite	2	49.7—57.1	53	(N/R)	1
G-1 granite	2	23.5—24.5	24	(N/R)	1
G-1 granite	—	25—27	26	(N/R)	2
G-1 granite	—	—	42	(S)	3
G-1 granite	8	24—29	25 ± 2	(N/R)	4
G-1 granite	—	—	25	(N/R)	6
G-1 granite	—	—	<200	(mass spec.)	7
G-1 granite	—	—	20	(C)	8
G-1 granite	—	—	22	(S)	5
G-2 granite	2	27.0—29.1	28	(N/R)	1

References: 1. JOHANSEN and STEINNES (1966). 2. HAMAGUCHI *et al.* (1967). 3. BROOKS and AHRENS (1961). 4. SMALES *et al.* (1957). 5. VOLAND (1969). 6. PIERCE and PECK (1961). 7. BROWN and WOLSTENHOLME (1964). 8. IVANOV and CHOLODOV (1966).

Indium remains in magma until the latest stages of crystallization. Some concentration of indium at earlier stages (in basalts and gabbros) has been observed, and the trend in the abundance data is shown by the variation diagram for indium (Fig. 49-E-3). Distribution of indium abundances in common magmatic rocks is presented in frequency diagrams (Fig. 49-E-4 and Fig. 49-E-5).

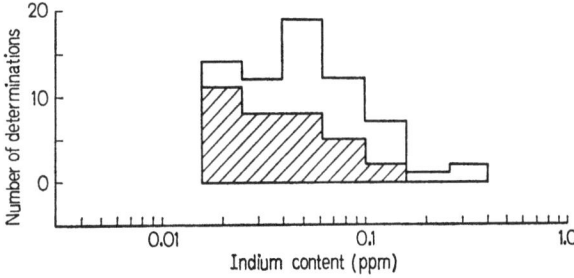

Fig. 49-E-4. Frequencies of distribution of indium in mafic and granitic volcanic rocks; 62 values (IVANOV, 1963). ☐ Mafic; ▨ Granitic

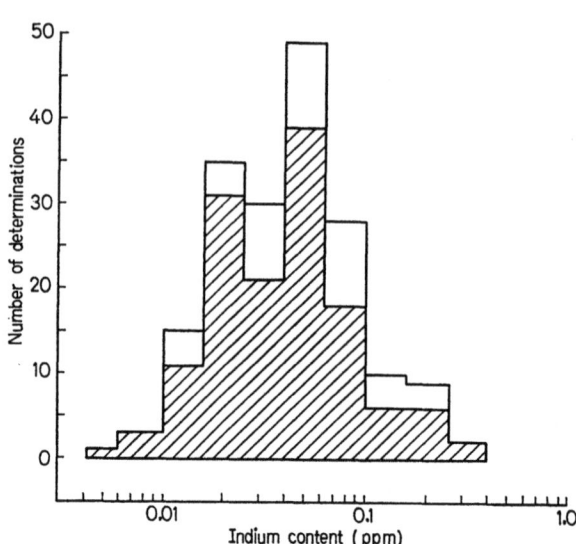

Fig. 49-E-5. Frequencies of distribution of indium in mafic and granitic volcanic rocks; 179 values (VOLAND, 1969). ☐ Mafic; ▨ Granitic

Revised manuscript received: October 1970

49-F. Behavior in Magmatogenic Processes

Among the principal ore-forming elements in the early stages of ore deposition, indium closely resembles iron, manganese and tin (IVANOV and LIZUNOV, 1959). Indium is concentrated in silicates with high iron content, and in later stages of sulfide formation, indium concentrates in the sulfides of copper and zinc and in iron-containing sulfides and carbonates (IVANOV and ROZBIANSKAYA, 1961).

Indium has strongly chalcophile properties and is concentrated by sulfide minerals, especially in products of hydrothermal and pneumatolytic processes (RANKAMA and SAHAMA, 1950[1]). Indium is also enriched in pegmatites containing tin, especially cassiterite (SHAW, 1952). Considerable amounts of indium are found in sphalerites which form at intermediate temperatures and, in general, sphalerites are the most important ore-minerals of indium (OFTEDAHL, 1940).

The variation diagram for indium in the Skaergaard intrusion follows that of manganese (WAGER et al., 1957). Variation diagrams for Fe, Mn, Sn, and Zn are compared with that for In in Fig. 49-E-3 (after SMITH, 1963).

---

[1] Editor's note: The accumulation of In by pneumatolytic processes is confirmed by a concentration of 1450 ppb In in 7 greisen samples from the Erzgebirge area (Saxony, East Germany) (VOLAND, 1969). Indium is generally high in rocks of tin provinces (VOLAND, 1969).

Revised manuscript received: October 1970

49-G. Behavior during Weathering and Alteration of Rocks

During ordinary weathering, indium is oxidized to In^{3+} (SHAW, 1952) which disperses following the Fe^{3+} and Mn^{4+} ions, and usually precipitates under conditions which form the hydrous oxides (KULIKOVA, 1962). Relatively high indium contents were observed in limonite, and electrodialysis experiments showed that indium was present both as free hydroxide and as hydroxide bonded in the structure of hydrous iron oxide (KULIKOVA, 1964).

On oxidation, the primary sulfide ores are depleted of indium by dissolution in slightly acidic or alkaline environments. Such depletion is accelerated where complex ions can form as in the presence of sulfate, chloride or organic acid anions, e.g., humates, carbonates, oxalates, etc. (KULIKOVA, 1966).

Deposition of indium usually occurs in the oxidates following iron and manganese, in some hydrolysate sediments (SHAW, 1957), and in bauxites with aluminium (GOLDSCHMIDT, 1958). Indium found in some marine sediments may have accumulated due to the insolubility of borates of trivalent indium (KULIKOVA, 1962).

49-H. Solubilities of Compounds which Control Indium Concentrations in Natural Waters; Adsorption Processes; Valence States in Natural Environments

In neutral aqueous solutions indium forms a number of insoluble compounds. Nearly all are soluble in acidic solutions or in concentrated solutions of the associated anion. Table 49-H-1 lists four compounds of some interest in natural environments.

Table 49-H-1. *Insoluble compounds of indium*

Compound	Solubility product (K_{sp})
$In(OH)_3$	1.3×10^{-34}
In_2S_3	(10^{-40})a
$In_2(CO_3)_3$	(10^{-15})a
$InPO_4$	(10^{-25})a

a Estimated values.

Indium forms a number of complex ions in aqueous solutions, especially with halide ions. Low concentrations of hydroxide ion permit indium complexation due to incomplete hydrolysis. Several organic anions (e.g., oxalate ion) are capable of complexing indium (SUNDERMAN and TOWNLEY, 1960).

The electrochemical half-cell (reduction) potentials of indium species in aqueous solution are shown in Table 49-H-2 (after LATIMER, 1952).

Table 49-H-2. *Standard potentials for indium*

Half-cell reactions	$\varepsilon°$ (vs. N.H.E.)
$In^{3+} + 3e^- = In$	-0.342
$In(OH)_3(s) + 3e^- = 3OH^- + In^{3+}$	-1.0

The compounds of trivalent indium are the most stable, and only the trivalent ion is stable in aqueous solutions. Compounds of lower valent indium undergo oxidation-reduction in the presence of water to form In^{3+} and elemental indium (BUSEV, 1962).

Revised manuscript received: October 1970

© Springer-Verlag Berlin · Heidelberg 1972

49-I. Abundance in Natural Waters

There are very few data on indium concentrations in natural waters. GOLDBERG (1961) assumed that sea water contains less than 20 ppb indium. Earlier information indicated that no indium was detectable in sea water (GOLDSCHMIDT, 1958), and SHAW (1952) reported no detectable amounts of indium in ocean salts.

CHOW and SNYDER (1969) recently reported measurements of the indium content of Pacific Ocean water at various depths below the surface, and at depths in core material from Mohole experiments. The indium profile to a depth of 3571 meters indicated a uniform indium concentration of 0.004 ppb (micrograms per liter of sea water) by isotope dilution mass spectrometry. MATTHEWS and RILEY (1970) report a range of about 0.0003 to 0.0001 ppb In of decreasing indium concentration with depth from an Atlantic (sea water) sampling station (neutron activation analysis). Core samples at depths from 33.3 to 149.2 meters below the sea floor contained from 40 to 20 ppb indium. The concentration of indium in sea water appears to be controlled by pH through the solubility product of indium hydroxide and the hydroxo-indium complex $[In(OH)_4^-]$.

49-K. Abundance in Common Sediments and Sedimentary Rocks

Table 49-K-1. *Indium in sediments and sedimentary rocks*

Material	No. of samples	In (ppb) range	In (ppb) mean	Method	Reference
Sediments					
Pelagic clays	6	—	73	(S)	SHAW (1957)
Pelagic clays Atlantic (4) Pacific (2)	6	<20—280	74	(S)	SHAW (1952)
Pelagic clay, Atlantic	1	—	74	(N)	MATTHEWS and RILEY (1970)
Kaolines	12	—	69	(S)	VOLAND (1969)
Kaolines	2	<20—120	70	(S)	SHAW (1952)
Siliceous ooze, Pacific	1	—	<20	(S)	SHAW (1952)
Diatom ooze, Atlantic	1	—	120	(N)	MATTHEWS and RILEY (1970)
Globigerina ooze, Pacific	1	—	<20	(S)	SHAW (1952)
Globigerina ooze, Atlantic	1	—	23	(N)	MATTHEWS and RILEY (1970)
Sedimentary rocks					
Shales, argillites, etc. (U.S.A., Finland, U.S.S.R.)	7	<20—180	57	(S)	SHAW (1952)
Shales	16	X—230	54	(S)	SHAW (1957)
Shales, Paleozoic (Europe)	(composite of 36)	—	48±5	(N/R)	SCHMITT and SMITH (1968)
Shales (North America)	(composite of 40)	—	70±7	(N/R)	SCHMITT and SMITH (1968)
Shale, Thuringia (East Germany)	1	—	70	(S)	VOLAND (1969)
Slates (U.S.A.)	2	—	37	(S)	SHAW (1952)
Shales (East Germany)	16	—	62	(S)	VOLAND (1969)
Cherts (East Germany)	3	—	10	(S)	VOLAND (1969)
Greywackes (U.S.A. and Canada)	3	33—230	104	(S)	SHAW (1952)
Carbonaceous sedimentary rocks	3	—	72	(S)	SHAW (1957)
Limestone, Harz (Germany)	1	—	9	(S)	VOLAND (1969)
Bauxite	—	—	170	(S)	SHAW (1952)

X = less than limit of detection.

VINOGRADOV (1962) reported an average of 50 ppb indium in sedimentary rocks (clays and shales) which is comparable with the data listed in Table 49-K-1 for sediments.

49-L. Biogeochemistry

The data on indium in biological materials and organisms are extremely sparse in the present literature. Certain coal ashes were reported to contain upto 2 ppm indium (GOLDSCHMIDT, 1937). KOCH and ROESMER (1962) reported 16 ppb indium (N/R) in dried mammalian muscle. Indium in mammalian blood was studied by WOLSTENHOLME (1964) who found less than 70 ppb indium (spark source mass spectroscopy) in dried blood plasma.

49-M. Abundance in Common Metamorphic Rocks

The literature is lacking much data on indium in metamorphic rocks. Those that are available are summarized in Table 49-M-1.

Table 49-M-1. *Indium in metamorphic rocks*

Material	No. of samples	In (ppb) range	In (ppb) mean	Method	Reference
Phyllites and schists, New York State (U.S.A.)	11	<20—460	110	(S)	Shaw (1952)
Gneisses, Wyoming (U.S.A.), Greenland	11	<20—1,900	495	(S)	Shaw (1952)
Ortho-gneisses, Erzgebirge (East Germany)	9	—	56	(S)	Voland (1969)
Para-gneisses, Erzgebirge (East Germany)	22	—	60	(S)	Voland (1969)
Gneisses, Erzgebirge (East Germany)	6	—	110	(S)	Voland (1969)
Mica-schists, etc., Erzgebirge (East Germany)	29	—	104	(S)	Voland (1969)
Amphibolites (East Germany)	14	—	110	(S)	Voland (1969)
Skarns	8	—	3,800	(S)	Voland (1969)
Granulites, Saxony (East Germany)	3	—	23	(S)	Voland (1969)
Eclogites (East Germany)	6	—	51	(S)	Voland (1969)

Because several sets of data in Table 49-M-1 are on gneisses and mica-schists from a tin province (Erzgebirge), values higher than 100 ppb In are probably abnormal.

Revised manuscript received: October 1970

© Springer-Verlag Berlin · Heidelberg 1972

49-O. Concluding Notes

Table 49-O-1 summarizes the ranges and average abundances of indium in different geochemical materials. Since many of the data in this table are only approximate, these values should be treated with caution in correlative studies.

Table 49-O-1. *Abundances of indium in various cosmic and terrestrial matter*

Type of Matter	In (ppb)	
	range	average
Iron meteorites	0.3—41	10
Carbonaceous chondrites	46—112	70
Enstatite chondrites	0.2—150	70
Ordinary chondrites	0.05—55	0.5
Igneous rocks		
Ultramafic rocks	5—60	20
Basalts and gabbros	15—320	70
Granites	10—2,000	50
Shales	30—230	60
Pelagic clays	<20—280	70

SHAW (1957) has reviewed the geochemistry of Ga, In and Tl, all members of Group III of the periodic table of the chemical elements. All three elements are geochemically dispersed in that they form separate minerals only in very rare cases and usually occur in combination with minerals of other elements. Their distributions in nature are obviously determined by their geochemical affinities, which are dependent on some chemical properties such as ionic radii, electronegativity, electronic orbitals for complexing, etc. Elements, with similar behavior and affinity, which govern the uptake of In are principally Fe, Zn, Sn, Pb and Cu. The chalcophilic property of In greatly exceeds the siderophilic and lithophilic tendencies.

Although In has been observed in iron meteorites by SMALES *et al.* (1967) using the sensitivite N/R technique (see Table 49-C-3), the possibility of In incorporation in small sulfide inclusions cannot be ruled out. Since siderophilic properties have been observed for both Ga and Ge in iron meteorites (WASSON, 1967) and also for Sn in iron meteorites (GOLDSCHMIDT, 1958), it may be reasonably postulated, from the general periodic behavior of chemical elements, that In also has siderophilic properties. Indium appears to behave similarly to As, Sb and Ag in iron meteorites (SMALES *et al.*, 1967).

A close similarity for the ionic radii of In^{3+} (0.81 Å) and Fe^{2+} (0.74 Å) may be partially responsible for In^{3+}—Fe^{2+} diadochism in igneous rocks and minerals

© Springer-Verlag Berlin · Heidelberg 1972

(SHAW, 1952, 1957). BREWER and BAKER (1936) suggested that In tends to associate with Ag, Cd, Sn and Sb; the first two elements, Ag and Cd, immediately precede In in the Periodic Table and they terminate the second transition series in sequence and progressively add successive electrons to the $5s\,5p$ subshell; thus some similar chemical properties might be expected for these elements.

ANDERSON (1953) has discussed the occurrence of In in sulfides and sulphosalts. From a study of sulfide structural characteristics he concluded that: a) sulfides, the best collectors of In, show a tetrahedral coordination; b) the In—S tetrahedral bond length is greater than the bond lengths of Zn—S, Cu—S, and Fe—S; moreover, the In—S bond length approaches most closely the Sn—S length in stannite, a very efficient collector of In; and c) indium rejection occurs largely in minerals with small metal-sulfur bond lengths.

Indium achieved economic importance in 1924, some 60 years after its discovery. The chief commercial source of indium is the processing of the zinc mineral sphalerite and other zinc-lead ores. Producers are found in the United States, Canada, Japan, and several European countries. The United States, the chief consumer, uses about 20 tons per year and has estimated reserves of 320 tons. The principal uses of indium are in the manufacture of electronic semiconductor devices, sleeve bearings, special alloys for dental casts, fusible safety plugs and links, and solders (SCHROEDER, 1965). Recent research has been directed toward specialized uses of indium in the field of electronics.

References: 49-B to 49-M, 49-O

AHRENS, L. H.: Evidence of geological age against decay of Sn-115 to In-115 by electron capture. Nature **162**, 413 (1948).
— TAYLOR, S. R.: Spectrochemical Analysis, 2nd ed. Reading, Massachusetts: Addison-Wesley Publishing Co., Inc. 1961.
AKAIWA, H.: Abundances of selenium, tellurium, and indium in meteorites. J. Geophys. Res. **71**, 1919 (1966).
ALLER, L. H.: The Abundance of the Elements. New York: Interscience Publishers, Inc. 1961.
— The abundance of elements in the solar atmosphere. Advances in Astronomy and Astrophysics. New York: Academic Press 1965.
— The chemical composition of the sun and the solar system. Proc. Australian Astron. Soc. **1**, 133 (1968).
ANDERSON, J. S.: Observations on the geochemistry of indium. Geochim. Cosmochim. Acta **4**, 225 (1953).
BADALOV, S. T.: The geochemical relationship between indium and silver in zinc-silver-lead deposits. Geochemistry (English Transl.). **10**, 1005 (1961).
BREWER, F. M., BAKER, E.: Arc spectrographic determination of indium in minerals and the association of indium with tin and silver. J. Chem. Soc. 1286 (1936).
BROOKS, R. R., AHRENS, L. H.: The determination of indium and thallium in G-1, W-1 and other silicate rocks by a new technique. Geochim. Cosmochim. Acta **23**, 145 (1961).
BROWN, R., WOLSTENHOLME. W. A.: Analysis of geological samples by spark source mass spectrometry. Nature **201**, 598 (1964).
BUSEV, A. I.: The Analytical Chemistry of Indium. Oxford: Pergamon Press 1962.
CAMBI, L., MALESTESTA, L.: Germanium, gallium, and indium in Sardinian sphalerite. Rend. Inst. Lombardo Sci. **69**, 369 (1936).
CAMERON, A. G. W.: A new table of abundances of the elements in the solar system. Proc. I.A.G.C. Symposium on the Origin and Distribution of the Elements, Paris (L. H. AHREN., ed.) Oxford: Pergamon Press 1968.
CHOW, T. J., SNYDER, C. B.: Indium content of sea water. Earth and Planet. Sci. Lett. **7**, 221 (1970).
DUBOV, E. E., KHROMOVA, T. P.: Determination of the content of certain elements in the sun from sunspot spectra. III. The indium abundance. Izv. Krym. Astrofiz. Observ. **35**, 186 (1966).
FLEISCHER, M.: Index of new mineral names, discredited minerals, and changes of mineralogical nomenclature in volumes 1—50 of the American Mineralogist. Am. Mineralogist **51**, 1248 (1966).
FOUCHÉ, K. F., SMALES, A. A.: The distribution of trace elements in chondritic meteorites. I. Gallium, germanium and indium. Chem. Geol. **2**, 5 (1967).
GENKIN, A. D., MURAV'EVA, I. V.: Indite and jalindite, new indium minerals. Zap. Vses. Mineralog. Obshchestva **92**, 445 (1963).
GOLDBERG, E. D.: Marine geochemistry. Ann. Rev. Phys. Chem. **12**, 24 (1961).
GOLDBERG, L., MÜLLER, E. A., ALLER, L. H.: Abundances of the elements in the solar atmosphere. Astrophys. J., Suppl. Ser. **5**, 1 (1960).
GOLDSCHMIDT, V. M.: The principles of distribution of chemical elements in minerals and rocks. J. Chem. Soc. 655 (1937).
— Geochemistry. Oxford: Oxford University Press 1958.
GREVESSE, N., BLANQUET, G., BOURY, A.: Proc. I.A.G.C. Symposium on the Origin and Distribution of the Elements, Paris (L. H. AHRENS, ed.). Oxford: Pergamon Press 1968.

© Springer-Verlag Berlin · Heidelberg 1972

GREEN, J.: Geochemical table of the elements for 1959. Bull. Geol. Soc. Am. **70**, 1127 (1959).
HAMAGUCHI, G., TOMURA, K., ONUMA, N., HIGUCHI, H., SUDA, K.: Determination of indium in rocks by neutron activation analysis. Bunseki Kagaku **16**, 1233 (1967).
IRVING, H., VAN R. SMIT, J., SALMON, L.: Determination of indium in cylindrite by neutron activation analysis and other methods. Analyst **82**, 549 (1957).
ITZIKSON, M. I., RUSANOV, A. K.: Indium in the tin-ore deposits of the Far East. Dokl. Akad. Nauk SSSR **53**, 631 (1946).
IVANOV, V. V.: Indium in some igneous rocks of the USSR. Geochemistry (English Transl.) **12**, 115 (1963).
— CHOLODOV, V. N.: Indij. Sammelwerk: Metally v. osadočnych t olščach Nauka, Moskva 187 (1966).
— LIZUNOV, N. V.: Indium in some tin deposits of Yakutia. Geochemistry (English Transl.) **4**, 416 (1959).
— — On certain pecularities of the distribution of indium in endogenetic ore deposits. Geochemistry (English Transl.) **1**, 53 (1960).
— ROZBIANSKAYA, A. A.: Geochemistry of indium in cassiterite-silicate-sulfide ores. Geochemistry (English Transl.) **1**, 71 (1961).
JOHANSEN, O., STEINNES, E.: Determination of indium in standard rocks by neutron activation analysis. Talanta **13**, 1177 (1966).
KEAYS, R. R., GANAPATHY, R., ANDERS, E.: Chemical fractionations in meteorites. IV. Abundances of fourteen trace elements in L-chondrites; implications for cosmothermometry. Geochim. Cosmochim. Acta. **35**, 337 (1971).
KOCH, R. C., ROESMER, J.: Application of activation analysis to the determination of trace-element concentrations in meat. J. Food Sci. **27**, 309 (1962).
KULIKOVA, M. F.: Behavior of indium in the oxidized zones of some polymetallic deposits of Eastern Transbaikaliya. Geochemistry (English Transl.) **7**, 707 (1962).
— Distribution of trace elements in the oxidized zone of some lead-zinc deposits of Middle Asia. Geochemistry (English Transl.) **5**, 981 (1964).
— Geochemistry of gallium and indium in the oxidized zone of lead-zinc deposits in Soviet Central Asia. Geochemistry (English Transl.) **5**, 982 (1966).
LARIMER, J. W., ANDERS, E.: Chemical fractionations in meteorites. II. Abundance patterns and their interpretation. Geochim. Cosmochim. Acta **31**, 1239 (1967).
LATIMER, W. M.: Oxidation States of the Elements and their Potentials in Aqueous Solutions, 2nd ed. New Jersey: Prentice-Hall, Inc. 1952.
LEDERER, C. M., HOLLANDER, J. M., PERLMAN, I.: Table of Isotopes, 6th ed. New York: John Wiley & Sons, Inc. 1967.
MATTHEWS, A. D., RILEY, J. P.: Occurence of indium in seawater and some marine sediments. Nature **225**, 1242 (1970).
MEITUV, G. M.: Geochemistry of rare elements in the lead-zinc deposit of the Klichkinskii region (Eastern Transbaikaliya). Geochemistry (English Transl.) **7**, 694 (1962).
OFTEDAL, I.: Untersuchungen über die Nebenbestandteile von Erzmineralen norwegischer zinkblendeführender Vorkommen. Skr. Norske Vid.-Akad. Oslo 1. Mat.-Naturv. Kl. No. 8 (1940).
ONUMA, N., HIGUCHI, H., WAKITA, H., NAGASAWA, H.: Trace element partition between two pyroxenes and the host lava. Earth Planet. Sci. Lett. **5**, 47 (1968).
PICOT, P., PIERROT, R.: Roquesite. Bull. Soc. Franc. Mineral. Crist. **86**, 7 (1963).
PIERCE, T. B., PECK, P. F.: A rapid method for determining indium by neutron activation. Analyst **86**, 580 (1961).
PREUSS, E.: Beiträge zur spektranalytischen Methodik. II. Bestimmung von Zn, Cd, Hg, In, Tl, Ge, Sn, Sb, Pb und Bi durch fraktionierte Destillation. Z. Angew. Min. **3**, 8 (1940).
RANKAMA, K., SAHAMA, T. G.: Geochemistry. Chicago: Chicago University Press 1950.
SCHINDEWOLF, U., WAHLGREN, M.: The rhodium, silver and indium content of some chondritic meteorites. Geochim. Cosmochim. Acta **18**, 36 (1960).
SCHMITT, R. A., SMITH, R. H.: Indium abundances in chondritic and achondritic meteorites, and in terrestrial rocks. Proc. I.A.G.C. Symposium on the Origin and Distribution of the Elements, Paris (L. H. AHRENS, ed.). Oxford: Pergamon Press 1968.
SCHROEDER, H. J.: Indium, mineral facts and problems. U.S. Bur. Mines Bull. **630** (1965).

SHAW, D. M.: The geochemistry of indium. Geochim. Cosmochim. Acta 2, 185 (1952).
— The geochemistry of gallium, indium, thallium — A review. Phys. Chem. Earth 2, 164 (1957).
SMALES, A. A., MAPPER, D., FOUCHÉ, K. F.: The distribution of some trace elements in iron meteorites, as determined by neutron activation. Geochim. Cosmochim. Acta 31, 673 (1967).
— VAN R. SMIT, J., IRVING, H. M.: Determination of indium in rocks and minerals by radioactivation. Analyst 82, 539 (1957).
SMITH, F. G.: Physical Geochemistry. Reading, Massachusetts: Addison-Wesley Publishing Co., Inc. 1963.
SUESS, H., UREY, H. C.: Abundances of the elements. Rev. Mod. Phys. 28, 53 (1956).
SUNDERMAN, D. N., TOWNLEY, C. W.: The radiochemistry of indium. NAS-NS-3014, U.S. At. En. Comm., 1960.
TANDON, S. N., WASSON, J. T.: Neutron activation determination of indium in meteorites. Radiochim. Acta 8, 184 (1967a).
— — Indium variations in a petrologic suite of L-group chondrites. Science 158, 259 (1967b).
— — Gallium, germanium, indium and iridium variations in a suite of L-group chondrites. Geochim. Cosmochim. Acta 32, 1087 (1968).
TAYLOR, S. R.: Abundance of chemical elements in the continental crust: A new table. Geochim. Cosmochim. Acta 28, 1273 (1964).
— SACHS, M.: Trace elements in australites. Nature 188, 387 (1960).
UREY, H. C.: The abundance of the elements with special reference to the problem of the iron abundance. Quart. J. Roy. Astron. Soc. 8, 23 (1967).
VINOGRADOV, A. P.: Regularity of distribution of elements in the earth's crust. Geokhimiya 1, 6 (1956).
— Average contents of chemical elements in the principal types of igneous rocks of the earth's crust. Geochemistry (English Transl.). 7, 641 (1962).
VOLAND, B.: Die Verteilung des Indiums in Eruptivgesteinen. Ein Beitrag zur Geochemie des Indiums. Freiberger Forschungsh. C 246, 67 (1969).
WAGER, L. R. VAN R. SMIT, J., IRVING, H.: Indium content of rocks and minerals from the Skaergaard intrusion, East Greenland. Geochim. Cosmochim. Acta 13, 81 (1958).
WAKITA, H., SCHMITT, R. A.: Rare earth and other elemental abundances in the Allende meteorite. Nature 227, 478 (1970).
WASSON, J. T.: The chemical classification of iron meteorites. I. A study of iron meteorites with low concentrations of gallium and germanium. Geochim. Cosmochim. Acta 31, 161 (1967).
WOLSTENHOLME, W. A.: Analysis of dried blood plasma by spark source mass spectrometry. Nature 203, 1284 (1964).

Revised manuscript received: October 1970

Tellurium **52**

A   J. Zemann   (Mineralogisches Institut der Universität Wien, Austria)

B—K, O   F. Leutwein   (Centre de Recherches Pétrographiques et Géochimiques, 54-Vandoeuvre-Nancy, France)

52-B. Isotopes in Nature

Tellurium (atomic number 52) has 8 stable isotopes whose relative abundances (BAINBRIDGE and NIER, 1950) are given in the following table:

Table 52-B-1. *Isotopes of Te and their relative abundances*

Te isotope	%
120	0.089
122	2.46
123	0.87
124	4.61
125	6.99
126	18.71
128	31.79
130	34.49

If the antimony isotope ^{123}Sb (relative abundance: 42.75%) were radioactive, forming ^{123}Te under β-decay, the half-life would be extremely long, more than 10^{12} to 10^{14} years (RANKAMA, 1954). INGHRAM and REYNOLDS (1950) found that ^{130}Te is unstable, forming ^{130}Xe at the end of two β-decay processes — but the half-life-time is about 10^{21} a.

52-C. Abundance in Cosmos, Meteorites and Lunar Samples

I. Cosmic Abundance

Direct spectral observations of tellurium in solar or stellar atmospheres are lacking. CAMERON (1967) calculated the cosmic abundance of this element by using carbonaceous chondrites of type 1 (for comparison, the values for S and Se are also given). Relative to $Si = 10^6$, he found:

$$S = 5.06 \cdot 10^5$$
$$Se = 70.1$$
$$Te = 6.76$$

In these meteorites, tellurium is about ten times rarer than selenium. According to SUESS and UREY (1956), the ratio of Te/S in the cosmos is 13×10^6 that of Se/Te 14.

II. Abundance in Meteorites

In the older investigations on tellurium in meteorites by NODDACK and NODDACK (1933), a value (17 ppm Te) has only been reported for troilite. GOLDSCHMIDT has calculated the average concentration of tellurium in meteorites as 0.73 ppm. Recent investigations on the abundance of tellurium in meteorites with reliable and sensitive methods have been published by DU FRESNE (1960), SCHINDEWOLF (1960), GOLES and ANDERS (1960, 1962), REED (1963) and GREENLAND (1967).

The average tellurium concentration in 20 ordinary and enstatite chondrites has been reported by DU FRESNE (1960) (using a spectrophotometric method) to be 1.34 ppm Te (7.9 ppm Se). SCHINDEWOLF's mean value for 4 ordinary chondrites (using neutron activation) is lower, 0.61 ppm Te; REED (1963) obtained a value of 0.64 ppm for 11 chondrites and KIESL et al. (1970) a value of 0.81 ppm Te for 16 chondrites by neutron activation.

In 1962, GOLES and ANDERS (1962) published 23 analyses of meteorites by neutron activation. The range for Te in 7 bronzite and hypersthene chondrites is upto 0.51 ppm; in three enstatite chondrites, they found 2.1 ppm and in two carbonaceous chondrites 1.9 ppm. For iron meteorites, the values cover a wide range of 17—90 ppm Te, whereas in five troilites 1.2—5.0 ppm Te were found.

In 1967, GREENLAND published a very complete study on several trace elements in chondritic meteorites and these results are given in the following two tables; the analyses were made by neutron activation.

To obtain a better comparison with stellar spectrographic data, Table 52-C-2 gives the results calculated in atom-percentage for $Si = 10^6$.

Table 52-C-1. *Analytical results (data in ppm) on some chondritic meteorites by N/R.*
(After GREENLAND, 1967)

Meteorite	Se	Te
Carbonaceous type I		
Ivuna	26	1.0
Orgueil	27	0.94
Carbonaceous type II		
Mighei	14	1.2
Murray	10	3.3
Carbonaceous type III		
Chainpur	9.0	1.4
Felix	11	0.44
Lancé	9.3	0.15
Mokoia	9.6	0.70
Vigarano	6.5	0.82
Warrenton	11	0.90
Enstatite type A		
Abee	11	3.9
Indarch	34	2.6
Enstatite type B		
Daniei's Kuil	19	0.21
Hvittis	10	0.79
Khairpur	24	0.74
Pillistfer	13	0.18
Hypersthene-bronzite		
Allegan (H)	6.1	—
Bruderheim (L)	6.4	0.76
Ehole (H)	6.4	0.93
Mocs (L)	7.3	0.89
Pantar (H)	4.8	0.61
Peace River (L)	6.4	0.82
Walters (L)	8.2	0.60

(H) indicates a high-iron olivine bronzite chondrite and (L) indicates a low-iron olivine hypersthene chondrite.

Table 52-C-2. *Atomic abundance relative to* $Si = 10^6$ (GREENLAND, 1967)

Material	Se	Te
Carbonaceous type I chondrites	89	2.0
Carbonaceous type II chondrites	32	3.8
Carbonaceous type III chondrites	21	1.1
Enstatite type A chondrites	48	4.2
Enstatite type B chondrites	32	0.56
Ordinary chondrites	13	0.96
Cosmic (SUESS and UREY, 1956)	68	4.7

III. Lunar Samples

Only a limited amount of data has been published on tellurium in lunar rocks. GANAPATHY et al. (1970) report 8 to 13 ppm Te in four Apollo 11 samples, and 72 and 73 ppm Te in two lunar breccias from the same sampling locality (by neutron activation analysis). Lunar fines and breccias are mainly enriched in relatively volatile elements. VINOGRADOV (1971) has published three values of 150 to 200 ppm Te on lunar regolith from the Russian automatic station Luna 16 (mass spectrometric analysis).

Revised manuscript received: January 1971

52-D. Abundance in Rock-Forming Minerals; Tellurium Minerals

In natural environments, tellurium has few lithophile tendencies; it is mainly restricted to sulfides and to low-temperature supergene minerals. In rock-forming silicates it has not yet been detected probably because of the limit of sensitivity of the methods applied.

The ionic radii of tellurium in its various ionization states are very different from those of sulfur and even of selenium. It forms anions in minerals (tellurides, tellurites and tellurates), and its redox potentials are widely different from those of the analogous sulfur components. Therefore, it is not surprising that there exists a rather large number (about 40) of independent minerals of this element. The tendency for isomorphic replacement of the far more abundant sulfur in its compounds is smaller for Te than for Se.

Some tellurium minerals, e.g. sylvanite, were exploited a long time before the detection of tellurium (1798) in important gold ores. Tellurium minerals are rare and are often only of microscopic size. Because of the grain size and the presence of intergrowths, analyses made without a microprobe are often of minor value as a basis for mineral formulae. STUMPFL (1970) found by microprobe, that sylvanite has often more gold than its formula indicates. He also states, that a correct distinction of krennerite and calaverite is only possible by microprobe.

The supergene tellurium minerals tellurites and tellurates have a high proportion of ionic bonding. In the group of hypogene Te minerals, metallic luster, proportions of metallic bonds (often accompanied by infra-red transparency and photoelectric effects) are frequent. Some minerals are typical intermetallic compounds.

In the following table tellurium minerals are listed mainly after SINDEEVA (1964).

The ionic radius of tellurium and the physico-chemical properties of H_2Te, H_2TeO_3, H_2TeO_4 are so different from those of the far more abundant sulfur compounds that there is only a very limited camouflage or diadochy to be expected. There are far more tellurium than selenium minerals. There is, surprising, a great number of gold-tellurium minerals (8), followed by silver, bismuth, copper and mercury ores. Only one lead telluride is known, but no zinc, cadmium, manganese or tin-tellurium compounds are known as natural minerals at this time.

Phase Equilibria

Selenium and tellurium form homogeneous melts in all proportions and below the solidus, form homogeneous mixed crystals in the hexagonal rhombohedral form of selenium. Te and S melts are completely miscible; in the solid state miscibility is rather limited — in β-sulfur maximum contents of 2% Te and in α-sulfur only 0.5% Te have been reported.

Table 52-D-1. *Tellurium minerals*

1. Native tellurium

Native tellurium. Often contains traces of iron and gold, and no traces of selenium.

Seleniferous tellurium. Under laboratory conditions, Se and Te are able to form homogeneous mixed crystals, though the atomic radii are rather different. The natural mineral usually has a higher Te content ($^2/_3$ Te, $^1/_3$ Se).

2. Tellurides

Hessite	Ag_2Te
Empressite	$AgTe_3$
Petzite	Ag_3AuTe_2
Stützite	$Ag_{(5-\alpha)}Te_3$
Muthmannite	$(Au, Ag)Te$
Calaverite	$AuTe_2$
Krennerite	$(Au, Ag)Te_2$
Sylvanite	$(Au, Ag)Te_4$
Montbrayite	Au_2Te_3
Altaite	$PbTe$
Coloradoite	$HgTe$
Frohbergite	$FeTe_2$
Melonite	$NiTe_2$
Tellurobismuthite	Bi_2Te_3
Wehrlite	$Bi_{(2+\alpha)}Te_{(3-\alpha)}$
Hedleyite	Bi_7Te_3
Tetradymite	Bi_2Te_2S
Csiklovaite	Bi_2TeS_2
Joseïte A	$Bi_{(4+\alpha)}Te_{(1-\alpha)}S_2$
Joseïte B	$Bi_{(4+\alpha)}Te_{(2-\alpha)}S$
Gruenlingite	Bi_4TeS_3
Oruetite	Bi_8TeS_4 perhaps
Rickardite	$Cu_{(4-\alpha)}Te_2$
Weissite	Cu_2Te
Nagyagite	$Pb_5Au_{0.7}Sb_{1.1}(Te_{2.1}, S_{5.4})_{7.5}$

3. Sulfosalts

Goldfieldite	$Cu_6Sb_2(S, Te)_3$
Arsenotellurite	$Te_2As_2S_7$

The existence of neither minerals is yet confirmed on the base of very detailed studies.

4. Oxides

Tellurite	TeO_2 (orthorhomb.)
Paratellurite	TeO_2 (tetrag.)

5. Tellurites and tellurates

Montanite	$Bi_2TeO_4(OH)_4$	(?)	
Teineite	$CuTeO_3 \cdot 2H_2O$		
Emmonsite	$Fe_2(TeO_3)_3 \cdot 2H_2O$		
Mackayite	$Fe(OH)(Te_2O_5)$		
Blakeite	$Fe_2(TeO_3)_3$	(?)	
Schmitterite	$UO_2 TeO_3$		
Magnolite	Hg_2TeO_4	(?)	chemical and crystallographic data lacking
Ferrotellurite	$FeTeO_4$	(?)	
Lead tellurate	$PbTeO_4$	(?)	(DOMEYKO, 1875), possibly identical with a mineral later described as dunhamite (FAIRBANKS, 1946)
Denningite	$(Mn, Ca, Zn)Te_2O_5$		
Spiroffite	$(Mn, Zn, Ca)_2Te_3O_8$		
Moctezumite	$PbUTe_2O_8$		
Poughite	$Fe_2[TeO_3]_2[SO_4] \cdot 3H_2O$		

Revised manuscript received: January 1971

52-E. Abundance in Common Igneous Rock Types

The average content of tellurium in the earth's crust — the clarke — is mainly based on extrapolations or on analyses of composite samples of "igneous rocks"; rather contraversial results have been obtained. NODDACK and NODDACK (1930) report 0.8 ppm Se and 0.01 ppm Te as average values. More recently, VINOGRADOV (1954) made the following estimate of the order of magnitude of the crustal abundance: 0.01 ppm selenium and 0.001 ppm tellurium. SINDEEVA (1964) analyzed a large number of different rock-types for Se and Te — but the Te-contents were always under the limit of detection (0.01 ppm)[1]. The tellurium content of the whole continental earth's crust ($\sim$16 km) was estimated by SCHTSCHERBINA (1937) to be 2×10^{11} t.

TUREKIAN and WEDEPOHL (1961) give 0.002 ppm as average content of all magmatic rocks of the upper continental crust.

---

[1] GANAPATHY et al. (1970) analyzed a pyroxene gabbro from the Adirondacks and obtained a value of 10 ppbTe(N/R).

52-F. Behavior in Magmatogenic Processes

The presence of traces of tellurium and of gaseous H_2Te in volcanic exhalations is often mentioned; in addition, Te might be transported as gaseous $TeO(OH)_2$ (GLEMSER *et al.* 1964). Volcanic sulfur as well as exhalation-crusts from the Liparian islands can contain traces of this element. In general, traces of 8—15 ppm, and rarely even 200—250 ppm in seleniferous volcanic sulfur are mentioned.

Tellurium has not been detected in pegmatites, but in hydrothermal sulfides, even of higher origin, and in skarn deposits, tellurium minerals are well known (SINDEEVA, 1964).

In sulfides from high temperature quartz, wolframite cassiterite veins, contents of selenium and traces of tellurium have been reported.

Tellurium has not been detected in tin and molybdenum greisens, but hydrothermal copper-molybdenum deposits (characterized by several ore-forming phases) often contain tellurium, especially in the younger molybdenite, either in isomorphous form or as tellurium minerals. The average content of 107 molybdenites from the Kadzharan deposit, USSR, is reported to be 33 ppm Te (KARAMYAN, 1962).

The deposits of magmatic pyrrhotite-pentlandite with chalcopyrite may have considerable amounts of Se and Te — though the concentrations are usually low. The recovery of these elements from this type of ore is relatively easy, because Se and Te accumulate in dusts or anodic slugs of the copper electrolysis plants.

Another important group of telluriferous deposits are the sedimentary pyrite masses of high temperature sources. Deposits are known where the amount of tellurium equals that of selenium or exceeds it. In general, pyrite, pyrrhotine and chalcopyrite are the richest, and sphalerite the poorest hosts of Te and Se. Selenium minerals are in general absent. Tellurium mostly appears in the form of fine-grained separate minerals — tetradymite, altaite, hessite and others. As a general tendency, younger deposits, of Mesozoic or Tertiary age, are richer in these elements than older ones (SINDEEVA, 1964).

From a mineralogical point of view, the most interesting tellurium deposits are the sulfide-bearing gold veins. (From the economical point of view, the two deposits cited before are by far the more important). They may have very different ages and they are not always of subvolcanic origin. Deposits of this type are not very abundant, but those from Roumania (here subvolcanic), Calgoorlie (Australia), Cripple Creek (Nevada), Beresovskoye in the Central Ural, Dzhelambet (Kazakstan), Kirovskoye in the Amur-region (for detailed description, see SINDEEVA, 1964) and from the early Precambrian gold belt of Ontario-Quebec (Canada) are known.

Kirovskoye is a vein deposit with a complex mineral composition. The gangue is quartz with rare carbonates. The common sulfides are pyrite, arseno-pyrite, bismuthinite, chalcopyrite, sphalerite, galena, tetraedrite, pyrrotite and gold. Rarer minerals are boulangerite, jamesonite, scheelite and molybdenite and tellurides of gold, silver and bismuth. This evidently represents mineralization in several phases of catathermal to mesothermal character. The host rocks are granitic gneisses and sediments of Jurassic age, which are cut by a post-middle Jurassic granodiorite and its differentiates. A table of the Se and Te contents is given by SINDEEVA (1964).

Table 52-F-1. *Contents of selenium and tellurium (by W, C) in minerals of the Kirovskoye deposit.* (After SINDEEVA, 1964)

Minerals	Percent Te	Percent Se
Bismuthinite	0.3275—0.2600	0.3275—0.2800
Arsenopyrite	0.0240—0.0001	0.019—0.0006
Chalcopyrite	0.0016—0.0008	0.003—0.0008
Pyrite	upto 0.0066	upto 0.0030
Jamesonite	upto 0.0018	upto 0.0005

Table 52-F-2. *Contents of tellurium (ppm) in sulfide-minerals of some pyrite deposits of the Ural.* After VLASSOV (1969). (Analytical method not stated)

Ore deposit	Pyrite	Chalcopyrite	Sphalerite	Tetraedrite
III — Internationala	6—220	20—387	<1—9	n.d.
Krasnogwardeskoë	6— 39	nil—236	>1	20—860
Lewicha	tr —50	9—6,600	nil—1	n.d.
Sibaï	1— 32	88	nil—7	n.d.
Karabasch	2—1,350	400	n.d.	1—900

Table 52-F-3. *Content of tellurium (ppm) in sulfide minerals of some copper-molybdenum deposits of the USSR.* After VLASSOV (1964), (W, C)

Ore deposit	Pyrite	Chalcopyrite	Molybdenite	Sphalerite	Galena
Almalyk	5—510	5—20	10—12	5—30	50—520
Kadscharan	5—12	22	30	—	—
Dasstakert	22	9—31	n.d.	n.d.	40

It is interesting to note that the relative enrichment of Te widely exceeds that of selenium; no selenium-minerals have been found in this deposit.

The very important group of hydrothermal lead-zinc deposits (of vein type) is of no significance as regards the Se and Te contents. Traces of both elements are often found here — more often Te than Se — but the absolute quantities are low. At Freiberg/Saxony, the author found traces of 2 to 20 ppm of Te especially in tetraedrite and the bismuthiferous galena; selenium was present in even lower concentrations (unpublished spectrographic data).

Tellurium minerals are formed from high to low thermal conditions. Tetradymite is mostly a high thermal mineral, appearing even in skarn deposits. Sylvanite and nagyagite are known only from mesothermal veins.

SCHTSCHERBINA (1937) inferred from microscopic observations as well as from thermodynamic calculations that there exists a series of different thermodynamic affinities of certain metals for tellurium in preference to sulfur:

$$Cu-Pb-Ni-Bi-Hg-Ag-Au$$

(sequence of increasing affinity). Gold tellurides may exist in the presence of sulfides of all other metals of this sequence. Melonite (NiTe) may only co-exist with sulfides of Cu, Pb and all other tellurides, and not with sulfides of the metals to the right of nickel. Copper-tellurides may co-exist with all tellurides of this sequence, but not in presence of the sulfides of these metals. (It must be noted that this rule holds under favorable thermodynamic conditions and only for minerals of the same paragenesis *sensu stricto*.) From SCHTSCHERBINA's conclusions, it follows that $Ag_2Te + PbS$ (hessite + galena) is a stable paragenesis, but not $PbTe + Ag_2S$ (altaite + argentite), which is in good agreement with microscopic observations.

52-G. Behavior during Weathering and Alteration of Rocks

In the oxidation zone of ore deposits, tellurium shows little tendency for migration. The tellurides are — in contrast to sulfides and selenides — stable even at rather high oxidation potentials. The oxidation potential for hexavalent sulfur (SO_4^{2-}) is under natural conditions very easily attained. No Te^{6+}-minerals are well known; they are to be expected only under exceptional climatic conditions. The stable oxidized compounds are mostly tellurites. Because of their stability tellurides, even under oxidising conditions, accumulate in placerdeposits together with gold and other heavy minerals. Examples of such deposits are reported by SINDEEVA (1964). In the Angora region in Siberia, several important gold-tellurium placers are known but are not at present exploited. Here the heavy minerals are: gold, scheelite, cassiterite, bismuthinite, native silver, pyrite, chalcopyrite. Tellurium minerals are hessite and joseite. This list shows, that only the special climatic conditions of this tundra region may allow the existence of readily oxidizable sulfides. It is interesting that the primary sources of these placers are not yet known.

The natural migration of Te may only occur in zones of a high oxidation potential and of leaching above the so-called "iron cap", especially in arid regions. Tellurium is fixed under such conditions either, as native tellurium or selenotellurium or, as in deposits of Colorado and Montana (USA), it is oxidized to TeO_2 which has a strong amphoteric character and may migrate under acid and alkaline conditions. So tetradymite gives the bismuth tellurite montanite ($Bi_2TeO_4 \cdot 2H_2O$) and coloradoite gives magnolite (Hg_2TeO_4). Finally sulfuric acid replaces tellurium in the metallic compounds and we find such iron tellurites as emmonsite and mackayite.

52-H. Solubilities of Compounds which Control Concentration of Tellurium in Waters

Under earth's surface conditions, tellurium may exist as native tellurium and its field of existence under various pH-Eh conditions is more extensive than that of native sulfur. Tellurates (TeO_4^{2-}) are only to be expected in strong alkaline solutions under extremely oxidizing conditions. The most abundant form of Te will be the tetravalent state (TeO_3^{2-}). The behavior of tellurium-ions plotted in a pH-Eh diagram (Fig. 52-H-1) has recently been investigated by DYACHKOVA and KHODAKOVSKIY (1968) and the following solubility products were found:

	25° C	150° C	300° C
PbTe	38.60	31.1	29.2
Ag_2Te	64.00	47.2	39.2
ZnTe	25.87	22.9	23.4
CdTe	34.40	28.9	28.1

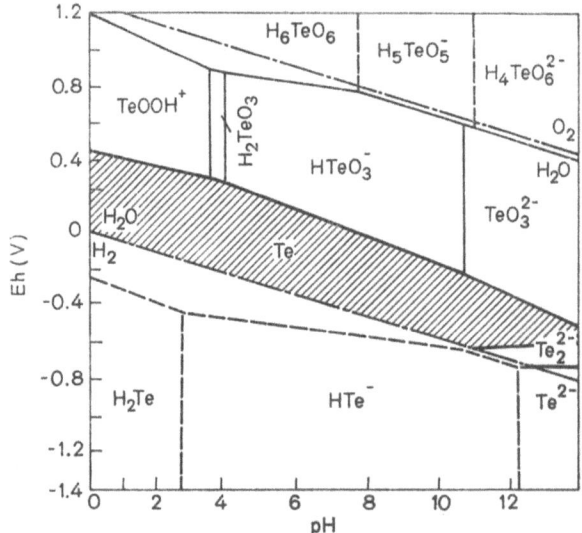

Fig. 52-H-1a

Fig. 52-H-1a—c. Stability fields of various compounds of tellurium in relation to Eh and pH at Σ Te = 10^{-7} at: a = 25° C, 1 atm.; b = 150° C, 5 atm.; c = 300° C, 85 atm. After DYACHKOVA and KHODAKORSKIY (1968)

Tellurium

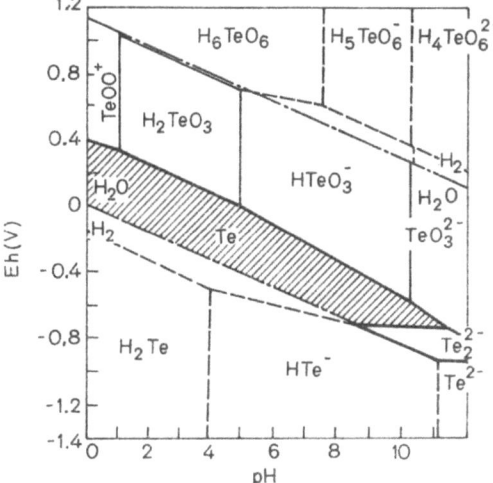

Fig. 52-H-1 b

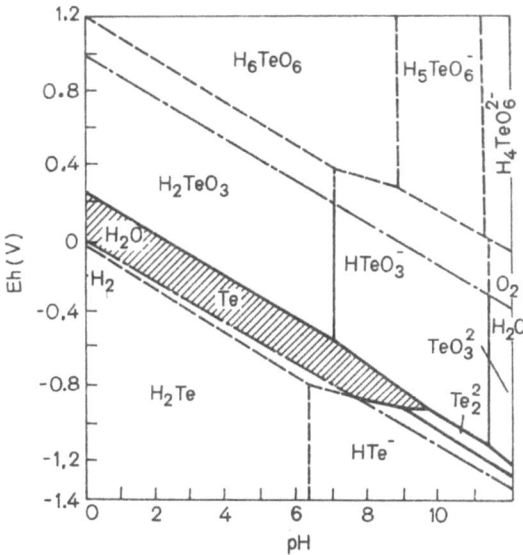

Fig. 52-H-1c

52-I. Abundance in Natural Waters

The tellurium content of ocean and fresh waters is not yet known. SINDEEVA (1964) mentions that in the waters of an abandoned mine in Siberia, 0.4 ppm Te was found (W, C). Unpublished data of the author (1966) on 6 samples of Elbe river waters upstream from Hamburg and seawater from near Helgoland showed only in the waters of the outflow of the Hamburg harbour 0.05 ppm Te (but 1 ppm Se); in all other samples the Te content was less than 0.01 ppm Te (W, C).

52-K. Abundance in Common Sediments, Sedimentary Rock Types and Sedimentary Ore Deposits

Adequate data on the tellurium content of rocks of the exogenic cycle are virtually lacking. In graywackes and shales its content is less than 0.1—1 ppm Te. LAKIN et al. (1963) report a range of 5—125 ppm Te (average: 48 ppm Te) in 12 manganese nodules from the Pacific and Indian Oceans. In samples of several sedimentary pyrite-nodules, SINDEEVA (1964) reported 2 ppm Te; in sedimentary iron, manganese and phosphorite-ores, SINDEEVA detected no tellurium. However, in the sedimentary sulfur deposits of Vetroyamovskoye, there exists 30 ppm Se and 6 ppm Te. In a systematic analysis of salt minerals (sylvinite, carnallite from Solikamsk, szaybelyite, ulexite, colemanite and hydroborazite from Inder), from potassium deposits of Ukrainia, no Te could be detected (limit of detection: 0.5 ppm).

In the Kupferschiefer marls from an euxinic environment (Eisleben) NODDACK and NODDACK (1936) found 0.02 ppm Te by X-ray analysis.

In contrast to selenium, tellurium has not yet been found in soils, nor plants, coal or lignite-ashes. [Unpublished data of LEUTWEIN (1948) from coals and soils of Eastern Germany — limit of detection 0.1—1 ppm (S-W method).]

52-O. Relations to Other Elements; Economic Importance, etc.

Many additional analytical investigations involving Te are required as the geochemical cycle of this element is only little known. However, it does appear that the cosmic abundance is higher than the abundance in the earth's crust. The chemical behavior and ionic radii of S, Se and Te are so different, that a camouflage of Te in the far more abundant sulfur minerals can not at present be detected. Though very rare (crustal abundance about 0.01 ppm), Te forms a rather large number of separate minerals but the crystallochemical relationships between Te and Se are not close and the ratio of Se to Te in different rocks is not well known. In pyrite, pyrrhotite, chalcopyrite and pentlandite ore deposits the content of these elements is very low, but nevertheless such deposits provide an estimated 80% of the world's production. Here the ratio of Se to Te is about 10:1.

Tellurium is an important element for the electrotechnical and electronics industries but technical applications in metallurgy and chemical industries appear to be less important. The famous telluride deposits of gold, silver, and other metals are of only minor economic importance. Both metals are accumulated during the metallurgical processes in the copper and pyrite smelters as by-products. For tellurium the main sources are flue-dust and anode-slugs of electrolysis plants but complete information on the world annual production is not available. The main producers are the United States and Canada and for 1958, SINDEEVA (1964) gives the respective annual amounts produced as 77 tons and 20 tons respectively.

© Springer-Verlag Berlin · Heidelberg 1972

Revised manuscript received: January 1971

References: Sections 52-B to 52-K, 52-O

BAINBRIDGE, K. T., NIER, A. O.: Relative isotopic abundances of the elements. Preliminary Report No. 9, Nucl. Sci. Series, Nat. Research Council U.S., Washington, D. C. (1950).

BOULADON, J., PICOT, P.: Sur les minéralisations en cuivre des ophiolites de Corse, des Alpes françaises et de Ligurie. Bull. du B.R.G.M., p. 23 (1968).

CHUKROV, F. V.: Discovery of a bismuth telluride in Manka Deposit (S. Altai). Notes of the All-Russian Min. Soc., No. 2, 76 (1947).

DOMEYKO, I.: Über die Entdeckung von Tellur-Mineralien in Chile. Compt. Rend. 81 (1875).

DU FRESNE, A.: Selenium and tellurium in meteorites. Geochim. Cosmochim. Acta 20, 141 (1960).

DYACKKOVA, I. B., KHODAKOVSKIY, J. L.: Thermodynamic equilibria in the systems $S-H_2O$, $Se-H_2O$, $Te-H_2O$ in the 25—300° temperature range and their geochemical interpretation (Geokhimija, 1968). Cited after Geochemistry International 5, 1108 (1968).

FAIRBANKS, E. E.: The punched card identification of ore minerals. Econ. Geol. 41, 761 (1946).

FERSMAN, A. E.: Geochemie. Vol. 4. Leningrad: Staatl. Wiss. Tech. Verlag chemischer Literatur 1939 (Russ.)

GANAPATHY, R., KEAYS, R. R., LAUL, J. C., ANDERS, E.: Trace elements in Apollo 11 lunar rocks: Implication for meteorite influx and origin of moon. Geochim. Cosmochim. Acta, Proceedings of the Apollo 11 Lunar Science Conference, Part 2, Suppl. 1, Vol. 34 (1970).

GARRELS, R. M.: Mineral Equilibria. New York: Harpers Geoscience Series 1960.

GLEMSER, v. O., HAESELER, R. v., MÜLLER, A.: Über gasförmiges $TeO(OH)_2$. Z. Anorg. Allgem. Chem. 329, 53 (1964).

GMELIN's Handbuch der anorganischen Chemie, VIII. ed., Syst. No. 11, Tellurium. Weinheim: Verlag Chemie 1940.

GOLDSCHMIDT, V. M.: Geochemische Verteilungsgesetze und kosmische Häufigkeit der Elemente. Naturwissenschaften 18, 999 (1930).

— BARTH, T., HOLMSEN, D., LUNDE, G., THOMASSEN, L., ULRICH, F., ZACHARIASSEN, W.: Geochemische Verteilungsgesetze der Elemente I—IX. Skrifter Norske Vidensskaps-Akad. Oslo, I. Math. Naturw. Kl. 1923—1937.

GOLES, G. G., ANDERS, F.: I content of meteorites and their $^{129}I-^{129}X$ ages. J. Geophys. Res. 65, 4181 (1960).

— — Abundance of iodine, tellurium and selenium in meteorites. Geochim. Cosmochim. Acta 26, 723 (1962).

GREENLAND, L.: The abundances of Se, Te, Ag, Pd, Ce and Zn in chondritic meteorites. Geochim. Cosmochim. Acta 31, 849 (1967).

INGHRAM, M. G., REYNOLDS, I. H.: Double β-decay of ^{130}Te. Phys. Rev. 78, 822 (1950).

KARAMYAN, K. A.: Correlation between rhenium, selenium and tellurium in the molybdenites of the Kadzkaran copper-molybdenum deposit. Geochemistry 2, 194 (1962).

KIESL, W., GRASS, F., BÖCKL, R., PONTA, U.: Cosmochemical abundances of trace elements in meteorites I. J. Radioanal. Chem. 6, 447 (1970).

LAKIN, H. W., THOMPSON, C. E., DAVIDSON, D. F.: Tellurium content of marine manganese oxides and other manganese oxides. Science 142, 1568 (1963).

NODDACK, I., NODDACK, W.: Die Häufigkeit der chemischen Elemente. Naturwissenschaften 18, 757 (1930).

— — Die geochemischen Verteilungskoeffizienten der Elemente. Svensk. Kem. Tidskr. 46, 173 (1934).

© Springer-Verlag Berlin · Heidelberg 1972

RAMDOHR, P.: Die Erzmineralien und ihre Verwachsungen, 2 ed. Berlin: Akademie-Verlag 1955.
RANKAMA, K.: Isotope Geology. London: Pergamon Press 1954.
— SAHAMA, T.: Geochemistry. Chicago: The University of Chicago Press 1950.
REED, G. W.: Heavy elements in the Pantar meteorite. J. Geophys. Res. **68**, 3531 (1963).
SCHINDEWOLF, U.: Selenium and tellurium content of stony meteorites by neutron activation. Geochim. Cosmochim. Acta **19**, 134 (1960).
SCHTSCHERBINA, W. W.: Principal features of the geochemistry of tellurium [Russ.]. Bull. Acad. USSR, Ser. Geol. 980 (1937).
SINDEEVA, N. D.: Mineralogy and Types of Deposits of Selenium and Tellurium. Engl. transl. (ed. E. Ingerson.) New York: Interscience Publishers 1964.
STUMPFL, E. F.: New electronprobe and optical data on gold-tellurides. Am. Mineralogist **55**, 808 (1970).
SUESS, H. E., UREY, H. C.: Abundances of the elements. Rev. Mod. Phys. **28**, 53 (1956).
TUREKIAN, K. K., WEDEPOHL, K. H.: Distribution of the elements in some major units of the earth's crust. Geol. Soc. Am. Bull. **72**, 175 (1961).
VINOGRADOV, A. P.: Geochemie seltener und nur in Spuren vorhandener chemischer Elemente im Boden (1954). German translation (Acad. Verlag, Berlin) of the Russian original, 1950.
— Trace-elements and research problems: microelements in plants of various systematic position. Agrochimija, No. 8, 2—30 (1965) [Russ.]. (Mikroelementy i zadatchi na'uky.)
— Introduction to the Geochemistry of the Oceans. Ed. Nauka (Sciences) Moscow 1967 [Russ.].
— Preliminary data on lunar ground supplied by the automatic station "Luna 16". Geochimija **3**, 261 (1971).
VLASSOV, V. A.: Geochemistry and Mineralogy of the Genetic Types of Deposits of Rare Elements, vol. 1, chapt. Tellurium, p. 586. Moskow 1964 [Russ.]

Cesium 55

A   G. Cocco, L. Fanfani and P. F. Zanazzi   (Istituto di Mineralogia dell'Università di Perugia, Perugia, Italy)

B—O   K. S. Heier   (Mineralogisk-Geologisk Museum, Oslo, Norway)

and

G. K. Billings   (Department of Geosciences, New Mexico Institute of Mining and Technology, Socorro, New Mexico, U.S.A.)

55-A. Crystal Chemistry

The electronic structure of the cesium atom is $1s^2\,2s^2\,2p^6\,3s^2\,3p^6\,3d^{10}\,4s^2\,4p^6\,4d^{10}\,5s^2\,5p^6\,6s^1$. As cesium is the heaviest stable alkali metal, it exhibits a high reactivity within the alkali-metal group. In all its compounds cesium occurs as a monovalent cation which has the stable electronic configuration of the noble gas xenon. The usually accepted ionic radius of Cs+ is 1.65 Å (see Table 12-8 of Vol. I of this handbook). This value places cesium closest to rubidium and monovalent thallium but, because of the lack of Rb-minerals and of the different geochemical behavior of Tl, the larger concentrations of Cs occur in K-minerals. Isomorphism between these two elements, however, is inhibited by the relative difference of the ionic radii (24%) so that K-minerals, sometimes containing significant concentrations of Rb, generally accomodate less than 100 ppm of cesium (see section 55-D). This is partly the effect of a limited supply of Cs. Larger Cs-contents are observed in some beryl varieties — in morganite (upto 7.2%), in rhodizite (upto 7.1%) and in avogadrite (upto 7.0%). Pollucite, with a Cs-content ranging from 22% to 36%, represents the only independent Cs-mineral.

The present crystallochemical review is confined to the halogen- and oxygen-containing compounds. In Table 55-A-1 there are listed several structures of Cs-compounds from which reliable information concerning Cs coordination can be obtained.

Table 55-A-1

		Range, Å	Mean, Å	Reference
CsF	$Cs^{[6]}$—F	3.00	3.00	after Sysiö (1969)
CsLiF$_2$	$Cs^{[6+2]}$—6F	2.96—3.15	3.07	Burns and
	—2F	3.50—3.53		Busing (1965)
CsCoF$_3$	$Cs_I^{[12]}$—F	3.10—3.14	3.12	Babel (1969)
	$Cs_{II}^{[12]}$—F	3.11—3.30	3.17	
CsNiF$_3$	$Cs_I^{[12]}$—F	3.13—3.32	3.22	Babel (1969)
	$Cs_{II}^{[12]}$—F	3.23	3.23	
CsBeF$_3$	$Cs^{[8]}$—F	2.96—3.40	3.13	Steinfink and Brunton (1968)
Cs[UF$_6$]	$Cs^{[12]}$—F	3.10—3.15	3.12	Rosenzweig (1967)
Cs$_4$[Mg$_3$F$_{10}$]	$Cs_I^{[11]}$—F	3.05—3.47	3.22	Steinfink and
	$Cs_{II}^{[10]}$—F	2.94—3.19	3.09	Brunton (1969)
CsCl (low temp.)	$Cs^{[8]}$—Cl	3.57	3.57	after Sysiö (1969)
CsCl (high temp.)	$Cs^{[6]}$—Cl	3.54	3.54	after Poyhonen and Ruuskanen (1964)
CsCl · $\tfrac{1}{3}$H$_3$OHCl$_2$	$Cs^{[9]}$—Cl	3.44—3.69	3.59	Schroeder and Ibers (1968)

© Springer-Verlag Berlin · Heidelberg 1972

Cesium

Table 55-A-1 (continued)

		Range, Å	Mean, Å	Reference
$CsCoCl_3$	$Cs^{[12]}$—Cl	3.60—3.75	3.68	Soling (1968)
$Cs_2[PuCl_6]$	$Cs^{[12]}$—Cl	3.70—3.72	3.71	Zachariasen (SR 1948, 413)
$Cs_2[UCl_6]$	$Cs^{[12]}$—Cl	3.62—3.76	3.70	Siegel (SR 1956, 249)
$Cs_2[ThCl_6]$	$Cs^{[12]}$—Cl	3.63—3.82	3.74	Siegel (SR 1956, 249)
$Cs_2[BkCl_6]$	$Cs_I^{[12]}$—Cl	3.59—3.88	3.74	Morss and Fuger (1969)
	$Cs_{II}^{[12]}$—Cl	3.53—3.90	3.73	
$CsNa[BkCl_6]$	$Cs^{[12]}$—Cl	3.82	3.82	Morss and Fuger (1969)
$Cs_3[Cr_2Cl_9]$	$Cs_I^{[12]}$—Cl	3.61—3.79	3.66	Wessel and Ijdo (SR 1957, 277)
	$Cs_{II}^{[12]}$—Cl	3.59—3.61	3.60	
$Cs_3[Re_3Cl_{12}]$	$Cs_I^{[12]}$—Cl	3.44—3.87	3.67	Bertrand et al. (1963)
	$Cs_{II}^{[11]}$—Cl	3.29—3.89	3.65	
	$Cs_{III}^{[10]}$—Cl	3.38—3.77	3.62	
CsBr	$Cs^{[8]}$—Br	3.72	3.72	after Sysiö (1969)
$CsBr \cdot \frac{1}{3}H_3OHBr_2$	$Cs^{[9]}$—Br	3.56—3.81	3.72	Schroeder and Ibers (1968)
Cs_2ZnBr_4	$Cs_I^{[11]}$—Br	3.72—4.27	4.02	Morosin and Lingafelter (SR 1959, 317)
	$Cs_{II}^{[9]}$—Br	3.58—4.02	3.72	
$Cs_2[TeBr_6]$	$Cs^{[12]}$—Br	3.89	3.89	Das and Brown (1966)
CsI	$Cs^{[8]}$—I	3.96	3.96	after Sysiö (1969)
$Cs_3[Bi_2I_9]$	$Cs_I^{[12]}$—I	4.21—4.24	4.22	Lindqvist (1968)
	$Cs_{II}^{[12]}$—I	4.20—4.31	4.26	
$CsMnCl_3 \cdot 2H_2O$	$Cs^{[12]}$—8 Cl	3.48—3.67	3.60	Jensen et al. (1962)
	—4 O	3.78—3.84	3.81	
$Cs_2MnCl_4 \cdot 2H_2O$	$Cs^{[10]}$—8 Cl	3.35—3.70	3.54	Jensen (1964)
	—2 O	3.56—3.62	3.59	
$Cs_2RuCl_5 \cdot H_2O$	$Cs_I^{[11]}$—9 Cl	3.41—3.85	3.54	Hopkins et al. (1966)
	—2 O	3.72	3.72	
	$Cs_{II}^{[7]}$—6 Cl	3.62—3.70	3.65	
	—1 O	4.15		
$Cs_2[(UO_2)Cl_4]$	$Cs^{[11]}$—8 Cl	3.49—3.73	3.60	Hall et al. (1966)
	—3 O	3.57—3.94	3.82	
$Cs_{0.9}[(UO_2)OCl_{0.9}]$	$Cs^{[8]}$—3 Cl	3.50	3.50	Allpress et al. (1963)
	—5 O	2.92—3.22	3.10	
$Cs[PO_2F_2]$	$Cs^{[12]}$—6 F	3.26—3.62	3.52	Trotter and Whitlow (1967)
	—6 O	3.07—3.29	3.17	
$CsAl[SO_4]_2 \cdot 12H_2O$	$Cs^{[12]}$—O	3.37—3.45	3.41	Cromer et al. (1966)
$^3_\infty Cs[B^{[4]}B_3^{[3]}O_{14}]$	$Cs^{[10]}$—O	3.14—3.40	3.30	Krogh-Moe and Ihara (1967)
$Cs[V_3^{[5+1]}O_8]$	$Cs^{[12]}$—O	2.99—3.68	3.30	Evans and Block (1966)
$^2_\infty Cs_2[(U^{[7]}O_2)_2V_2^{[5]}O_8]$	$Cs^{[6+5]}$—6 O	3.01—3.28	3.16	Appleman and Evans (1965)
	—5 O	3.47—3.60	3.52	
$Cs_{0.7}Na_{0.3}[AlSi_2O_6] \cdot 0.3 H_2O$ pollucite	$Cs^{[12]}$—O	3.40—3.57	3.48	Newnham (1967)

I. Halogen Compounds

CsF has the "NaCl" structure type, $Cs^{[6]}$—F bond lengths being 3.004 Å (Sysiö, 1969). The radius ratio for CsCl = 0.93 allows 8-fold coordination but is so near to the ratio for 6-fold coordination that cesium chloride is dimorphous. At room temperature the cesium chloride structure is characterized by a body-centered lattice: each Cs atom is at the center of a cube of Cl atoms with $Cs^{[8]}$—Cl distances of 3.571 Å (Sysiö, 1969). At 445°C the CsCl structure changes into the NaCl arrangement with $Cs^{[6]}$—Cl bonds (at 469°) of 3.537 Å (calculated from Poyhonen and Ruuskanen, 1964). CsBr and CsI show only one modification corresponding to the arrangement in CsCl which is stable at room temperature; $Cs^{[8]}$-halogen distances are 3.720 Å and 3.956 Å respectively (Sysiö, 1969).

In $CsCl \cdot \frac{1}{3}H_3OHCl_2$ and in $CsBr \cdot \frac{1}{3}H_3OHBr_2$, cesium is 9-coordinated; average $Cs^{[9]}$—Cl and $Cs^{[9]}$—Br bond lengths are 3.59 and 3.72 Å respectively (Schroeder and Ibers, 1968). $CsHF_2$, together with the K- and Rb-bifluoride, exists in an α phase at room temperature and in a β phase at higher temperature. The transition temperature for the cesium term is 61°C (Kruh et al., SR 1956, 215). In $CsLiF_2$, isostructural with $RbLiF_2$, 8 fluorine atoms in an asymmetric array surround Cs+ ions. Six of them are at an average distance of 3.07 Å while the other ones are at a distance of 3.50 and 3.53 Å from the cation (Burns and Busing, 1965).

A large number of $CsMeX_3$ compounds — where Me represents a divalent cation and X an element of the halogen series — have a more or less distorted perowskite arrangement. Among these double halogenides, which often exhibit more than one modification, are $CsCaF_3$, $CsZnF_3$, $CsMgF_3$, $CsCoF_3$, $CsNiF_3$, $CsMnF_3$, $CsFeF_3$, $CsCdCl_3$, $CsHgCl_3$, $CsPbCl_3$, $CsCdBr_3$, $CsHgBr_3$ and $CsPbBr_3$. In $CsCoF_3$ (structure-type: $BaRuO_3$) and in $CsNiF_3$ (structure type: $BaNiO_3$), Cs is in 12-fold coordination with Cs—F distances upto 3.32 Å (Babel, 1969). The crystal structure of $CsCoCl_3$ is of the $CsNiCl_3$-type; Cs+ ions are 12-coordinated with an average $Cs^{[12]}$—Cl distance of 3.68 Å (Soling, 1968). $Cs_2AgAuCl_6$ and $Cs_2Au_2Cl_6$, ordinarily cubic, become tetragonal after heating to 350° for several days and their structures are characterized by superlattices on the perowskite arrangement (Elliott and Pauling, SB 1938, 126). In $CsCuCl_3$ cesium and chlorine atoms form a hexagonal close-packing, the average $Cs^{[12]}$—Cl bond lengths being 3.63 Å (Schlueter et al., 1966). The same arrangement has been ascribed to $CsNiCl_3$ and $CsVCl_3$.

There are three structural modifications of $CsBeF_3$ stable in ranges of increasing temperature (α-$CsBeF_3$ > 360° > β-$CsBeF_3$ > 140° > γ-$CsBeF_3$). A recent investigation on crystals probably belonging to the β-form shows an 8-fold coordination polyhedron around Cs with Cs—F distances in the range 2.96—3.40 Å (Steinfink and Brunton, 1968). $CsBF_4$ is isostructural with $RbBF_4$. The atomic arrangement is baryte-like and the coordination around Cs is 12-fold (Clark and Lynton, 1969). Cs_2ZnBr_4 has essentially the same structure as Cs_2ZnCl_4, Cs_2CuCl_4 and Cs_2CuBr_4. The coordination around the two different Cs+ ions is 11-fold and 9-fold with Cs—Br distances from 3.581 to 4.266 Å (Morosin and Lingafelter, SR 1959, 317). $Cs_3[AlF_6]$ has a behavior closely similar to that of $Rb_3[AlF_6]$; this compound, which is tetragonal at room temperature, undergoes a transition in a cubic phase after heating (Holm, 1965).

$Cs[UF_6]$ is isostructural with $K[OsF_6]$. The Cs atom has 12 neighbors, 6 at 3.101 Å and 6 at 3.147 Å (ROSENZWEIG, 1967). $Cs[PuF_6]$, $Cs[NpF_6]$ and many other compounds with a general chemical formula of $Cs[Me^V F_6]$ show an arrangement closely related to $Cs[UF_6]$. The structures of many cubic cesium halogenide complexes, $Cs_2[Me^{IV} F_6]$, such as $Cs_2[CrF_6]$, $Cs_2[GeF_6]$ and $Cs_2[GeCl_6]$, can be referred to that of $K_2[PtCl_6]$. In $Cs_2[TeBr_6]$ which belongs to this group of compounds, $Cs^{[12]}$—Br bond lengths are 3.89 Å, in good agreement with the sum of the ionic radii corrected for a 12-coordination (DAS and BROWN, 1966). According to ZACHARIASEN (SR 1948, 413), $Cs_2[PuCl_6]$ is trigonal and cesium is bonded to twelve Cl atoms with a $Cs^{[12]}$—Cl distance of 3.71 Å. $Cs_2[UCl_6]$ and one form of $Cs_2[ThCl_6]$ are isostructural with it. The average $Cs^{[12]}$—Cl distances are 3.70 Å and 3.74 Å respectively (SIEGEL, SR 1956, 249). $Cs_2[BkCl_6]$ is not isomorphous with $Cs_2[PuCl_6]$. Its symmetry is hexagonal and its structure is typified by $Rb_2[MnF_6]$. Average $Cs_I^{[12]}$—Cl is 3.74 Å and $Cs_{II}^{[12]}$—Cl is 3.73 Å (MORSS and FUGER, 1969).

Another class of compounds, with the chemical formula $Cs_3[Me_2^{III} X_9]$, crystallizes in the hexagonal system. In $Cs_3[Bi_2I_9]$, coordination around the two different Cs atoms is 12-fold with distances from 4.205 to 4.308 Å (LINDQVIST, 1968). In $Cs_3[Cr_2Cl_9]$, Cs—Cl distances range from 3.59 to 3.79 Å (WESSEL and IJDO, SR 1957, 277).

II. Oxygen Compounds

Cs_2O crystallizes in the hexagonal system with a cubic packing of cations and the anti-$CdCl_2$ arrangement; the Cs—O bond length is 2.86 Å (TSAI et al., SR 1956, 260).

$CsIO_3$ has a perowskite-type structure (NARAY-SZABO, SR 1947, 454) while $CsVO_3$ is orthorhombic (CALVO, SR 1958, 433) and $CsNO_3$ rhombohedral (FERRONI et al., SR 1957, 352). The crystal structure of the modification of $CsIO_4$ stable at room temperature appears to be a relatively slight distortion of the $CaWO_4$ arrangement (BEINTEMA, SB 1937; 89); $CsReO_4$ and $CsTcO_4$ are isostructural with it. Cs_2SO_4 is dimorphic; the orthorhombic low-temperature form and the hexagonal high-temperature form are isotypic with the corresponding modifications of K and Rb sulfates (TABRIZI et al., 1968).

Because of the large size of the cation, $CsAl[SO_4]_2 \cdot 12H_2O$ belongs to the β alum series. In this structure cesium can accomodate 12 oxygen neighbors of 12 water molecules at a mean distance of 3.41 Å (CROMER et al., 1966). $CsCr[SO_4]_2 \cdot 12H_2O$ also has the β structure (LEDSHAM and STEEPLE, 1968).

In $Cs[PO_2F_2]$, isomorphous with K- and Rb-difluorophosphate, Cs^+ ions are surrounded by 6 oxygen and by 6 fluorine atoms with distances upto 3.62 Å; this compound has the $Ba[SO_4]$ structure (TROTTER and WHITLOW, 1967). In $Cs[B_9O_{14}]$, each cesium atom is coordinated by 10 oxygen atoms with distances ranging from 3.14 Å to 3.40 Å (KROGH-MOE and IHARA, 1967). In CsV_3O_8, the alkali-cation has a 12-fold coordination with average $Cs^{[12]}$—O bond length of 3.30 Å (EVANS and BLOCK, 1966). A structural investigation on $Cs_2[(UO_2)_2V_2O_8]$, the cesium analogue of anhydrous carnotite, has been performed by APPLEMAN and EVANS (1966); in this structure Cs^+ ions connect the sheets built up by U and V coordination polyhedra; there are 11 Cs—O bonds in the range 3.01—3.60 Å.

Few data concerning Cs-minerals are available. The crystal structure of pollucite, determined by NEWNHAM (1967) on a sample with the approximate chemical composition $Cs_{0.7}Na_{0.3}AlSi_2O_6 \cdot 0.3\ H_2O$, shows cesium coordinated to a distorted close-packed arrangement of twelve oxygen atoms, six at 3.40 Å and six at 3.57 Å. The atomic arrangement of rhodizite has been determined assuming the ideal formula of the mineral to be $CsBe_4B_{12}Al_4O_{28}$, although the sample analyzed by FRONDEL and ITO (1965) reveals large amounts of K and Rb and a sum of $Cs+Rb+K<1 \cdot$ A $Cs^{[12]}$—O distance of 3.243 Å is reported. However uncertainties in the chemical formula allow only an approximate description of the atomic arrangement (TAXER and BUERGER, 1967).

References: Section 55-A

ALLPRESS, J. G., WADSLEY, A. D.: The crystal structure of caesium uranyl oxychloride $Cs_xUO_2OCl_x$ (x approximately 0.9). Acta Cryst. **17**, 41 (1964).

APPLEMAN, D. E., EVANS, H. T., JR.: The crystal structures of synthetic anhydrous carnotite, $K_2(UO_2)_2V_2O_8$, and its cesium analogue, $Cs_2(UO_2)_2V_2O_8$. Am. Mineralogist **50**, 825 (1965).

BABEL, D.: Die Kristallstrukturen der hexagonalen Fluorperowskite. Z. Anorg. Allgem. Chem. **369**, 117 (1969).

BERTRAND, J. A., COTTON, F. A., DOLLASE, W. A.: The crystal structure of cesium dodecachlorotrirhenate (III), a compound with a new type of metal atom cluster. Inorg. Chem. **2**, 1166 (1963).

BURNS, J. H., BUSING, W. R.: Crystal structures of rubidium lithium fluoride, $RbLiF_2$, and cesium lithium fluoride, $CsLiF_2$. Inorg. Chem. **4**, 1510 (1965).

CLARK, M. J. R., LYNTON, H.: Crystal structures of potassium, ammonium, rubidium and cesium tetrafluoborates. Canad. J. Chem. **47**, 2579 (1969).

CROMER, D. T., KAY, M. I., LARSON, A. C.: Refinement of the alum structures. I. X-ray and neutron diffraction study of $CsAl(SO_4)_2 \cdot 12H_2O$, a β alum. Acta Cryst. **21**, 383 (1966).

DAS, A. K., BROWN, I. D.: A refinement of the crystal structures of $(NH_4)_2TeBr_6$ and Cs_2TeBr_6. Canad. J. Chem. **44**, 939 (1966).

EVANS, H. T., BLOCK, S.: The crystal structures of potassium and cesium trivanadates. Inorg. Chem. **5**, 1808 (1966).

FRONDEL, C., ITO, J.: Composition of rhodizite. Tschermaks Mineral. Petrog. Mitt. **10**, 409 (1965).

HALL, D., RAE, A. D., WATERS, T. N.: The crystal structure of dicaesium tetrachlorodioxouranium (VI). Acta Cryst. **20**, 160 (1966).

HOLM, J. L.: Phase transitions and structure of the high-temperature phases of some compounds of the cryolite family. Acta Chem. Scand. **19**, 261 (1965).

HOPKINS, T. E., ZALKIN, A., TEMPLETON, D. H., ADAMSON, M. G.: The crystal structure of cesium aquopentachlororuthenate. Inorg. Chem. **5**, 1431 (1966).

JENSEN, S. J.: The crystal structure of $Cs_2MnCl_4 \cdot 2H_2O$ and $Rb_2MnCl_4 \cdot 2H_2O$. Acta Chem. Scand. **18**, 2085 (1964).

— ANDERSEN, P., RASMUSSEN, S. E.: The crystal structure of $CsMnCl_3 \cdot 2H_2O$. Acta Chem. Scand. **16**, 1890 (1962).

KROGH-MOE, J., IHARA, M.: The crystal structure of caesium enneaborate, $Cs_2O \cdot 9B_2O_3$. Acta Cryst. **23**, 427 (1967).

LEDSHAM, A. H. C., STEEPLE, H.: The crystal structures of sodium chromium alum and caesium chromium alum. Acta Cryst. B **24**, 1287 (1968).

LINDQVIST, O.: The crystal structure of caesium bismuth iodide, $Cs_3Bi_2I_9$. Acta Chem. Scand. **22**, 2943 (1968).

MORSS, L. R., FUGER, J.: Preparation and crystal structures of dicesium sodium berkelium hexachloride. Inorg. Chem. **8**, 1433 (1969).

NEWNHAM, R. E.: Crystal structure and optical properties of pollucite. Am. Mineralogist **52**, 1515 (1967).

POYHONEN, J., RUUSKANEN, A.: X-ray investigation of the transition of CsCl at 469°. Ann. Acad. Sci. Fennicae, Ser. A VI **146**, 12 (1964).

ROSENZWEIG, A.: The crystal structure of $CsUF_6$. Acta Cryst. **23**, 865 (1967).

SCHLUETER, A. W., JACOBSON, R. A., RUNDLE, R. E.: A redetermination of the crystal structure of $CsCuCl_3$. Inorg. Chem. **5**, 277 (1966).

© Springer-Verlag Berlin · Heidelberg 1972

SCHROEDER, W., IBERS, J. A.: The bihalide ions ClHCl⁻ and BrHBr⁻: Crystal structures of cesium chloride — $\frac{1}{3}$ — (hydronium bichloride) and cesium bromide — $\frac{1}{3}$ — (hydronium bibromide). Inorg. Chem. 7, 594 (1968).

SOLING, H.: The crystal structure and magnetic susceptibility of $CsCoCl_3$. Acta Chem. Scand. 22, 2793 (1968).

STEINFINK, H., BRUNTON, G. D.: The crystal structure of $CsBeF_3$. Acta Cryst. B 24, 807 (1968).

— — The crystal structure of $Cs_4Mg_3F_{10}$. Inorg. Chem. 8, 1665 (1969).

SYSIÖ, P. A.: On the additivity of crystal radii in alkali halides. Acta Cryst. B 25, 2374 (1969).

TABRIZI, D., GAULTIER, M., PANNETIER, G.: Analyse radiocristallographique des formes "basse" (β) et "haute" (α) temperature de sulfate de césium. Bull. Soc. Chim. Franç. 935 (1968).

TAXER, K. J., BUERGER, M. J.: The crystal structure of rhodizite. Z. Krist. 125, 423 (1967).

TROTTER, J., WHITLOW, S. H.: The structures of caesium and rubidium difluorophosphates. J. Chem. Soc. (A) 1383 (1967).

Revised manuscript received: August 1970

Barium 56

A   K. Fischer   (Lehrstuhl für Kristallographie, Universität des Saarlandes, Saarbrücken, Germany)

B—O   H. Puchelt   (Mineralogisch-Petrographisches Institut der Universität, Tübingen, Germany)

56-A. Crystal Chemistry
I. General

In the crystal structures of barium minerals, Ba is surrounded by O, OH, H_2O or a halogen ion as nearest neighbors. Its valency is two and the bond type is predominantly ionic. Of the divalent positive ions, Ba^{2+} has the largest ionic radius (except Ra^{2+}). Isostructural replacement of, or by, other large cations such as Pb^{2+} and Sr^{2+} is frequently observed. Less common is replacement of K^+ and of the smaller ion Ca^{2+}. A few examples of crystal chemical relations which are not listed in the tables below are the following synthetic compounds: $BaFe_{12}O_{19}$, $BaAl_{12}O_{19}$, $SrFe_{12}O_{19}$, $SrAl_{12}O_{19}$ etc. in the magnetoplumbite series; pandaite with Ba partially replacing Ca in the pyrochlore group; isotypy of $K[NO_3]$ and witherite $Ba[CO_3]$; Ca—Ba-mimetisite; heinrichite and meta-uranocircite in the torbernite and meta-torbernite group.

The coordination of Ba has been reviewed by MANOHAR and RAMASESHAN (1964). The coordination number ranges from 6 to 12 for O, OH, H_2O. The existence of a large variety of coordination polyhedra is illustrated in Tables 51-A-1, 51-A-2, and 51-A-3 in which crystal structures of Ba minerals and a few selected synthetic compounds are listed. In general, oxygen atoms with distances larger than 3.3 Å are not considered as sharing one coordination polyhedron (exception: barite). This limitation appears to be justified as in the majority of the structures the distance of the "next-nearest" O neighbors is substantially larger than the range of distances attributed to the first-order coordination. In other cases, however, the distances show a broad and rather uniform range of variation (e.g. Ba_3 in taramellite) so that it is difficult to define the coordination number. The type of coordination polyhedron of Ba apparently has a minor influence on the energy balance of these structures. Nevertheless it can control the structure type as has been demonstrated for silicates by LIEBAU (1962, 1968).

The following averages for atomic distances were obtained for the different coordination numbers indicated:

Coordination number	Atomic distance, Å	
6	2.7_6	(BaO, $BaZnO_2$, benitoite)
7	2.7_8	(paracelsian, $Ba[GeO_3]$ high)
8	2.8_1	($BaNiO_2$, $Ba(OH)_2 \cdot 8H_2O$, $BaUO_4$, gillespite, taramellite)
9	2.8_5	($Ba[S_5O_6] \cdot 2H_2O$, $Ba[B(OH)_4]_2$, $Ba[B_4O_7]$, celsian, sanbornite)
10	2.8_7	(BaO_2, $BaTi_4O_9$, $Ba[B_4O_7]$, $Ba_3[PO_4]_2$)
12	2.9_3	(psilomelane, high $BaTiO_3$, nitrobarite, barite, norsethite, $Ba_3[PO_4]_2$, "hexagonal" $Ba[Al_2Si_2O_8]$, α-$BaO_2 \cdot 2H_2O_2$)

These data demonstrate a general increase of distance with increasing coordination number. This can be considered only as a first approximation because of the restricted selection of minerals and artificial compounds; (for some other compounds used for the computation of the averages, which are not listed in Tables 56-A-1, 56-A-2 and 56-A-3, see MANOHAR and RAMASESHAN, 1964).

II. Crystal Chemistry of Some Minerals and Synthetic Compounds

a) Oxides, Halides

The structure of psilomelane $(Ba, H_2O)_2Mn_5O_{10}$ consists of MnO_6 octahedra and $(Ba, H_2O)O_{12}$ cubo-octahedra (only oxygen atoms with distances ≤ 3.16 Å are considered). Ba and H_2O occupy the same equipoint (partial ordering assumed), probably because of very similar distances of Ba—O and O—H_2O atoms and molecules (H-bridges?). The crystal structure of hollandite $BaMn_8O_{16}$ has the same coordination polyhedron for Ba (BYSTRÖM and BYSTRÖM, SR **13**, 1950, 186). In synthetic oxides and hydroxides almost all coordination numbers from 6 to 12 occur;

Table 56-A-1. *Coordination of Ba in some oxides and halides*

Mineral or synthetic compound	Coordinated atoms, distances to Ba in Å		Coordination polyhedron	References
$Ba^{[6]}O^{[6]}$	6O	2.76	octahedron	1
$Ba^{[6]}Zn^{[4]}O_2$	6O (2O	2.64—2.97 3.36)	octahedron, severely distorted	2
$Ba^{[8]}Ni^{[4]}O_2$	8O	2.80; 2.84	quadratic prism, severely distorted	3
$Ba^{[8]}(OH)_2 \cdot 8H_2O$	8OH	2.69—2.77	distorted quadratic antiprism	4
$Ba^{[8]}U^{[6]}O_4$	8O	2.71—2.99	irregular	5
$Ba^{[10]}O_2$	8O 2O	2.79 2.68	tetragonal prism with tetragonal pyramids on two base faces	6
$Ba^{[10]}Ti_4^{[6]}O_9$	10O	2.81—3.09	pentagonal prism	7
psilomelane, $(Ba, H_2O)_2^{[12]}Mn_5^{[6]}O_{10}$	8O 4H_2O	2.85—3.16 2.78; 2.88	distorted cubo-octahedron	8
$Ba^{[12]}Ti^{[6]}O_3$ (high-temp. phase)	12O	2.83	cubo-octahedron	9
$Ba^{[8]}F_2^{[4]}$	8F	2.68	cube	10
$Ba^{[9]}Cl_2 \cdot 2H_2O$	7Cl 2H_2O	3.14—3.38 2.76	distorted trigonal prism with 2Cl and 1H_2O on top of prismatic faces	11

References: 1. GERLACH (SB 1913—1928, 119). 2. v. SCHNERING, HOPPE and ZEMANN (SR 24, 1960, 454). 3. LANDER (SR 15, 1951, 180); for structure, see LANDER (1951). 4. MANOHAR and RAMASESHAN (1964). 5. SAMSON and SILLÉN (SR 11, 1947—48, 441). 6. ABRAHAMS and KALNAJS (SR 18, 1954, 364). 7. TEMPLETON and DAUBEN (SR 24, 1960, 326). 8. WADSLEY (SR 17, 1953, 429). 9. GOLDSCHMIDT (SB 1, 1913—1928, 333). 10. DAVEY (SB 1, 1913—1928, 187). 11. JENSEN (SR 10, 1945—1946, 95).

halides and hydrated halides are known with coordination numbers of 8 and 9. A selection of these compounds (especially oxides and hydroxides with coordination polyhedrons not described for minerals in this section) is included in Table 56-A-1.

b) Nitrates, Carbonates, Borates

The BaO_{12} icosahedron found in nitrobarite $Ba[NO_3]_2$ was also observed in $Ba[ClO_4]_2 \cdot 3H_2O$, $\alpha\text{-}BaO_2 \cdot 2H_2O_2$, $Ba[SiF_6]$ and $Ba[GeF_6]$ (MANOHAR and RAMASESHAN, 1964). In the carbonate minerals, Ba replaces Ca in the aragonite series (witherite and solid solutions) with a coordination polyhedron which can be described as a distorted cube with one of its edges elongated for housing the 9th O atom above its centre; the same coordination polyhedron is also reported for sanbornite $Ba_2[Si_4O_{10}]$. Barytocalcite $BaCa[CO_3]_2$ has an independent structure also with a coordination number of 9 (two distances of 3.28 and 3.30 Å are not included). In norsethite, $BaMg[CO_3]_2$, Ba has a $6+6$ coordination forming a distorted ditrigonal prism (LIPPMANN, 1968). No barium borate mineral structures are known at present. Of the synthetic borates, $Ba[B(OH)_4]_2$ is isostructural with the corresponding Sr borate (see subsection 38-A-II), whereas $Ba[B_4O_7]$ is not (coordination numbers 9 and 10).

c) Sulfates, Phosphates

Barite $Ba[SO_4]$ is isostructural with anglesite $Pb[SO_4]$ and celestite $Sr[SO_4]$. (Solid solutions are known as angleso-barite, baryto-celestite and calcio-barite.)

Table 56-A-2. *Coordination of Ba in some oxo-salts*

Mineral or synthetic compound	Coordinated atoms, distances to Ba in Å		Coordination polyhedron	References
Nitrobarite, $Ba^{[12]}[NO_3]_2$	$6+6O$	2.86; 2.95	distorted icosahedron	1
Witherite, $Ba^{[9]}[CO_3]$	9O	ca. 2.8	see text	2
Barytocalcite, $Ba^{[9]}Ca^{[6]}[CO_3]_2$	9O (2O	2.56—2.99 3.28; 3.30)		3
Norsethite, $Ba^{[12]}Mg^{[6]}[CO_3]_2$	$6+6O$	2.72; 3.18	distorted ditrigonal prism	4
$Ba^{[9]}[B(OH)_4]_2$	Ba_1: 9OH Ba_2: 9OH	2.73—2.90 2.77—2.99	trigonal prism with centered faces	5
$Ba^{[9]}Ba^{[10]}[B_4^{[3]}B_4^{[4]}O_{14}]$	Ba_1: 9O Ba_2: 10O	2.61—3.08 2.71—3.12	irregular	6
Barite, $Ba^{[2]}[SO_4]$	10O (2O	2.76—3.08 3.30)	see text	7
$Ba^{[12]}Ba_2^{[10]}[PO_4]_2$	Ba_1: $6+6O$	2.80; 3.23	distorted icosahedron	8
	Ba_2: 10O	2.71—2.83	see text	

References: 1. VEGARD and BILBERG (SB **2**, 1928—1932, 386); cf. LUTZ (SR **24**, 1960, 421). 2. COLBY and LA COSTE (SB **3**, 1933—35, 407). 3. ALM (SR **24**, 1960, 425). 4. LIPPMANN (1968). 5. KRAVCHENKO (1965). BLOCK and PERLOFF (1965). 7. SAHL (1963); see also COLVILLE and STAUDHAMMER (1967). 8. ZACHARIASEN (SR **11**, 1947—48, 388).

The coordination polyhedron based on a coordination number of 12 consists of two nearly parallel rings of 5 and 6 oxygens on both sides of the Ba atom, with a 12th O on top of the 6-ring (for an alternative description, see MANOHAR and RAMASESHAN, 1964). In this coordination polyhedron, 2O with distances of 3.30 Å are included. Omitting them from the 5-ring would lead to a rather one-sided coordination polyhedron (SAHL, 1963). $Ba_3[PO_4]_2$, isostructural with $Sr_3[PO_4]_2$, has two coordination polyhedra for Ba with coordination numbers of 12 and 10; the latter one consists of a distorted "hexagonal" pyramid (symmetry 3m) slightly above the Ba atom and a triangle below.

d) Silicates

$Ba_2^{[6]}[SiO_4]$ and $Sr_2^{[6]}[SiO_4]$ are isostructural with olivine (O'DANIEL and TSCHEISCHWILI, SR **9**, 1942—44, 261). Another silicate with isolated tetrahedra, $BaO \cdot SiO_2 \cdot 6H_2O$, has two different coordination polyhedra with coordination numbers of 8 and 10 which share faces with each other but have no common corner with the $[SiO_4]$ tetrahedron. The ten-fold coordination can be described as a pentagonal pyramid above a square. In benitoite $BaTi[Si_3O_9]$ (isomorphous to pabstite $BaSn[Si_3O_9]$), the coordination polyhedron is severely distorted. (The next 6O atoms beyond the nearest ligands are not considered because of their distance of 3.43 Å.) Taramellite $Ba_2(Fe, Ti, Mg)_2[Si_4O_{12}](OH)_2$, with rings of $4[SiO_4]$ tetrahedra, has three different coordination polyhedra for Ba with a coordination number of 6, 6 and 7 (?). The last coordination polyhedron can also be described as having a coordination number of 9 (two additional O's at a distance of 3.12 Å) or with a coordination number of 11 (two additional oxygen atoms at a distance of 3.21 Å). $Ba[GeO_3]$, with a germanate chain of the "Zweierketten"-type, is one of the few examples for a coordination number of 7. (Higher condensed silicate chains with Ba as cations have been investigated by KATSCHER and LIEBAU (1965, 1966)). In the sheet silicates, small percentages of Ba have been found in some muscovites. The mineral anandite $(Ba, K)(Fe, Mg)_3[(Si, Al, Fe)_4O_{10}](O, OH)_2$ was recently described as a member of the trioctahedral mica family (PATTIARATCHI, SAARI and SAHAMA, 1967). Other Ba silicates with single sheet structures do not have the usual plane silicate sheet of the mica type and do not contain OH or H_2O (LIEBAU, 1968). Sanbornite, $Ba[Si_2O_5]$, with undulating sheets built of 6-membered rings has the same coordination polyhedron for a coordination number of 9 as described for witherite. In the structure of gillespite $BaFe[Si_4O_{10}]$ (with folded sheets of 4- and 8-membered rings of $[SiO_4]$ tetrahedra), the coordination polyhedron of Ba can be derived from a cube whose faces are each divided into 2 triangles to form a triangular dodecahedron of symmetry $\bar{4}$ (cf. also cuprorivaite: PABST, 1959; MAZZI and PABST, 1962). Ba replaces Sr or Ca in synthetic compounds of the melilite series $(Ba_2[Fe^{[4]}Si_2O_7]$, $Ba_2[Mn^{[4]}Si_2O_7]$, cf. WYCKOFF, 1968, p. 225—227)[1]. For the non feldspar form of $Ba[Al_2Si_2O_8]$, containing double sheets, a hexagonal (or pseudo-hexagonal) prism has been described as the coordination polyhedron (on the orthorhombic distortion, cf. TAKÉUCHI, SR **22**, 1958, 501). The same coordination polyhedron has been found in cymrite $Ba[AlSi_3O_8(OH)]$. In the Ba-feldspar celsian $Ba[Al_2Si_2O_8]$, the coordination polyhedron is similar to that in witherite (if a tenth oxygen atom at a

---

[1] The classification of tetrahedral structures follows ZOLTAI's suggestion (1960).

Table 56-A-3. *Coordination of Ba in some silicates*

Mineral or synthetic compound	Coordinated atoms, distances to Ba in Å			Coordination polyhedron	Ref.
BaO · SiO$_2$ · 6H$_2$O	Ba$_1$:	8OH, H$_2$O	2.83—2.90	distorted quadratic antiprism	1
	Ba$_2$:	10OH, H$_2$O	2.83—2.90	see text	
Benitoite, Ba$^{[6]}$Ti$^{[6]}$[Si$_3$O$_9$]		6O	2.77	heavily distorted octahedron or trigonal prism	2
Taramellite, Ba$_2$(Fe, Ti, Mg)$_{\frac{1}{2}}^{[6]}$(OH)$_2$[Si$_4$O$_{12}$]	Ba$_1$:	6O	2.83; 2.90	distorted hexagon	3
	Ba$_2$:	6O	2.73; 2.75	distorted trigonal prism	
	Ba$_3$:	7O	2.71—3.00	see text	
		(2+2O	3.12; 3.21)		
$_\infty^1$Ba$^{[7]}$[Ge$^{[4]}$O$_3$] (high)		7O	2.62—2.94	irregular	4
Sanbornite, $_\infty^2$Ba$^{[9]}$[Si$_2$O$_5$]		7O	2.74—2.94	see text	5
		2O	3.14		
$_\infty^2$Gillespite, Ba$^{[8]}$Fe$^{[4\,p2]}$[Si$_4$O$_{10}$]		8O	2.73; 2.98	heavily distorted cube or trigonal dodecahedron	6
Ba$^{[12]}$[Al$_2$Si$_2$O$_8$] ("hexagonal celsian")		12O	~2.89	hexagonal prism	7
Cymrite, Ba$^{[12]}$[AlSi$_3$O$_8$(OH)]		12O (2×½ H$_2$O	3.05 3.39)	hexagonal prism	8
Celsian, Ba$^{[9]}$[Al$_2$Si$_2$O$_8$]		9O (1O	2.67—3.14 3.42)	see text	9
Paracelsian, Ba$^{[7]}$[Al$_2$Si$_2$O$_8$]		7O (2O	2.73—2.83 3.32; 3.37)	see text	10
Barylite, Ba$^{[12\,?]}$[Be$_2^{[4]}$Si$_2$O$_7$]		11O (1O	2.82—3.12 3.34)	distorted cubo-octahedron (minus one?)	11
Harmotome, Ba$^{[10]}$[Al$_2$Si$_6$O$_{16}$] · 6H$_2$O		8O, H$_2$O 2O	2.77—3.08 3.26	see text	12

References: 1. HÖHNE and DORNBERGER-SCHIFF (SR **26**, 1961, 513). 2. FISCHER (1969). 3. MAZZI and ROSSI (1965). 4. HILMER (1962). 5. DOUGLASS (SR **22**, 1958, 490). 6. PABST (SR **9**, 1942—44, 249); cf. MAZZI and PABST (1962), PABST (SR **23**, 1959, 486). 7. MATSUMOTO (SR **15**, 1951, 304); cf. TAKÉUCHI (SR **22**, 1958, 501). 8. KASHAEV (1966). 9. NEWNHAM and MEGAW (SR **24**, 1960, 491). 10. BAKAKIN and BELOV (SR **24**, 1960, 492); cf. SMITH (SR **17**, 1953, 556). 11. ABRASHEW, ILYUKHIN and BELOV (1964). 12. SADANAGA, MARUMO and TAKÉUCHI (SR **26**, 1961, 532).

distance of 3.42 Å is not counted), with one surprisingly short distance of 2.67 Å (see NEWNHAM and MEGAW, SR **24**, 1960, 491). In the framework structure of paracelsian (with the same chemical composition as celsian), which has a structure closely related to danburite Ca[B$_2^{[4]}$Si$_2$O$_8$], a similar coordination polyhedron with a coordination number of 9 has been described, including two distances larger than 3.3 Å. Omitting these latter two oxygen atoms, the coordination polyhedron consists of a distorted rectangle with Ba nearly in its plane plus two oxygen atoms forming a gable-roof on top of it and the seventh oxygen below its centre. The distorted cubo-octahedra around Ba in barylite $Ba[Be_2Si_2O_7]$ (with one distance larger than 3.3 Å), form chains by linked triangles. Two crystal structures of natural Ba-zeolites, edingtonite $Ba[Al_2Si_3O_{10}] \cdot 4H_2O$ (TAYLOR and JACKSON, SB **3**, 1933 to 1935, 529) and harmotome $Ba[Al_2Si_6O_{16}] \cdot 6H_2O$ are known. In harmotome, Ba is surrounded by ten O or H_2O in a coordination polyhedron similar to that of $BaO \cdot SiO_2 \cdot 6H_2O$. Two of the 6 distances from Ba to framework oxygen atoms are rather long (3.26 Å).

56-B. Isotopes in Nature

Natural Ba is a mixture of seven stable isotopes (Table 56-B-1). The atomic weight is 137.34 (Atomic Weight Conference of the IUPAC, 1961, Butterworth Scientific Publications, 1962), calculated on the basis that $^{12}C = 12.00000000$.

Table 56-B-1. *Stable Ba isotopes in nature*
(STROMINGER et al., 1958; MATTAUCH et al., 1965; EUGSTER et al., 1969)

Stable isotopes	% in natural mixtures	Isotope masses $^{12}C = 12.00000000$
^{130}Ba	0.1056 ± 0.0002	129.9062
^{132}Ba	0.1012 ± 0.0002	131.9051
^{134}Ba	2.417 ± 0.003	133.9046
^{135}Ba	6.592 ± 0.002	134.9056
^{136}Ba	7.853 ± 0.004	135.9043
^{137}Ba	11.232 ± 0.004	136.9055
^{138}Ba	71.699 ± 0.007	137.9050

Measurements (I) made by UMEMOTO (1962) indicate an enrichment of the lighter Ba isotopes in the Pasamonte and Nuevo Laredo achondrites and Bruderheim chondrite with a maximum of 2.5% for ^{130}Ba in the Bruderheim achondrite. This potential accumulation decreases almost linearly to mass ^{138}Ba.

Recent investigations of Ba isotope distribution in meteorites (EUGSTER et al., 1969) showed that differences in isotopic composition between meteoritic and terrestrial samples are always <0.1% for all isotopes.

A very small amount of the short-lived ^{140}Ba occurs in nature as a fission product of ^{238}U (KURODA and EDWARDS, 1957; HEYDEGGER and KURODA, 1959).

56-C. Abundance in Cosmos, Meteorites, Tektites, and Lunar Samples

I. Cosmos

Ba has been detected spectroscopically in stars of all spectral classes except classes O and B. These classes correspond to such high temperatures that all resonance transitions of the highly ionized atoms fall into the ultraviolet range, which is inaccessible for terrestrial observation. Ba abundances are normally given as atomic ratios per 10^{12} H atoms or 10^6 Si atoms, or as ratios of Ba atoms/Fe atoms.

Atmospheres of main sequence stars always seem to contain Ba in similar concentrations. In Table 56-C-1, the sun is given as one example. Pronounced differences of Ba concentrations have been reported for two other groups of stars which do not belong to the main sequence. Ba II stars, which are considered to be carbon stars with higher temperatures than N stars (GORDON, 1968), exhibit an overabundance of s process elements, including Ba (BIDELMAN and KEENAN, 1951; BURBIDGE and BURBIDGE, 1957; WARNER, 1965). FUJITA and TSUJI (1965) found s process

Table 56-C-1. *Barium abundance in stars*

Type of star	Name	log Ba for log H = 12.00	Reference
High velocity stars:			
F 6 IV-V	γ Serpentis	2.09	KEGEL (1962)
Subdwarf	HD 140283	0.11	BASCHEK (1962)
Horizontal branch	HD 161817	0.94	KODAIRA (1964)
Subdwarf	HD 19445	0.35	ALLER, GREENSTEIN (1960)
Subdwarf	HD 122563	−2.65	WALLERSTEIN, PARKER, GREENSTEIN, HELFER, ALLER (1963)
Main sequence stars:			
G 2 V	sun	2.10	GOLDBERG et al. (1960)
Other stars			
G 8 III	ε Virginis	2.01	CAYREL, CAYREL (1966)
		$(\log Ba/\log Fe)_{star}$ $-(\log Ba/\log Fe)_{sun}$	
Main sequence stars:			
G V	β Com	0.06 ± 0.20	HELFER et al. (1963)
G V	99 Her A	0.11 ± 0.20	HELFER et al. (1963)
G V	85 Peg A	-0.10 ± 0.20	HELFER et al. (1963)
Other stars (subgiant):			
G IV	ζ Her	0.02 ± 0.15	HELFER et al. (1963)

elements in N star Y CVn to be overabundant by factors of 10^2 to 10^3 when compared to main sequence stars. Strong enhancement of Ba lines in N stars has been detected by GORDON (1968) and UTSUMI (1967). A certain relationship of BaII stars to S type stars was observed (WARNER, 1965) whereas R stars show normal Ba abundances. CLAYTON (1964) has developed the idea that in a special type of s process nucleosynthesis, the elements heavier than Zr are favored and this process is assumed to work in BaII stars.

Population II stars which are characterized by their high velocity show higher atomic ratios, H/Ba, than main sequence stars. Nevertheless, the abundance ratios of elements heavier than carbon are similar to main sequence stars (UNSÖLD, 1967). BURBIDGE and BURBIDGE (1957) found that the subdwarfs HD 106223 and λ Bootis are depleted in Ba by factors of 20 to 30.

ALLER (1961) gives the logarithms of the ratio, number of Ba atoms in the sun/number of Ba atoms in the star, in the subdwarf for three typical subdwarfs: HD 140283, $\log(N_{sun}/N_{star})=2.59$; HD 19445, $\log(N_{sun}/N_{star})=2.15$; HD 219617, $\log(N_{sun}/N_{star})=1.40$. Spectral measurements for the CH stars HD 26 and HD 626 (WALLERSTEIN and GREENSTEIN, 1964) and comparison to the G 8III star ε Virginis showed them to be Ba deficient by a factor of 30 to 50. HELFER et al. (1963) gave the logarithms of atomic ratios $(N_{Ba}/N_{Fe})_{star}-(N_{Ba}/N_{Fe})_{sun}$ for four αG stars: ζ Herculis ($+0.02\pm0.15$), β Comae Berenicis ($+0.06\pm0.20$), 99 Herculis ($+0.11\pm0.20$), 85 Pegasi (-0.10 ± 0.20). Logarithms of atomic abundances of Ba in cosmos are published by ALLEN (1963) as 2.11 for H=12. The weight ratio is 4.25. SUESS and UREY (1956) and CAMERON (1959) calculated the ratio of Ba atoms per 10^6 Si atoms as 3.66; in the sun the same ratio is 3.978 (ALLER, 1961).

II. Meteorites

The published data are summarized in Tables 56-C-2 and 3. Distribution patterns of Ba are plotted in Figures 56-C-1a through d. Generally meteorite "finds" show higher Ba values than "falls", indicating a probable terrestrial contamination of the "finds". Thus a log of the normal distribution of Ba can be observed only in chondrite falls, whereas the "finds" of the same class exhibit a random distribution.

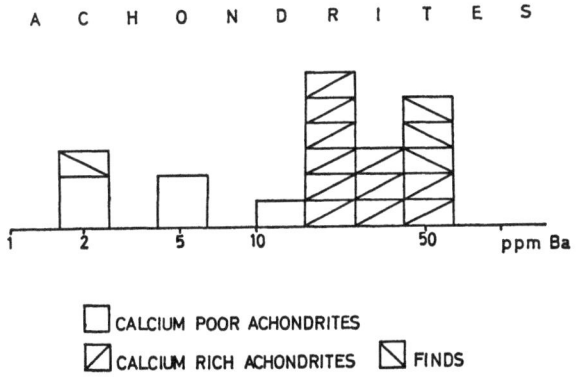

Fig. 56-C-1a. Distribution of Barium in achondrites

Barium

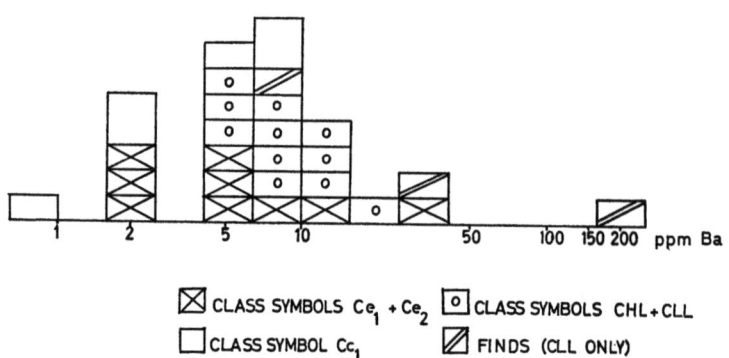

Fig. 56-C-1b. Distribution of Barium in carbonaceous and enstatite chondrites etc.

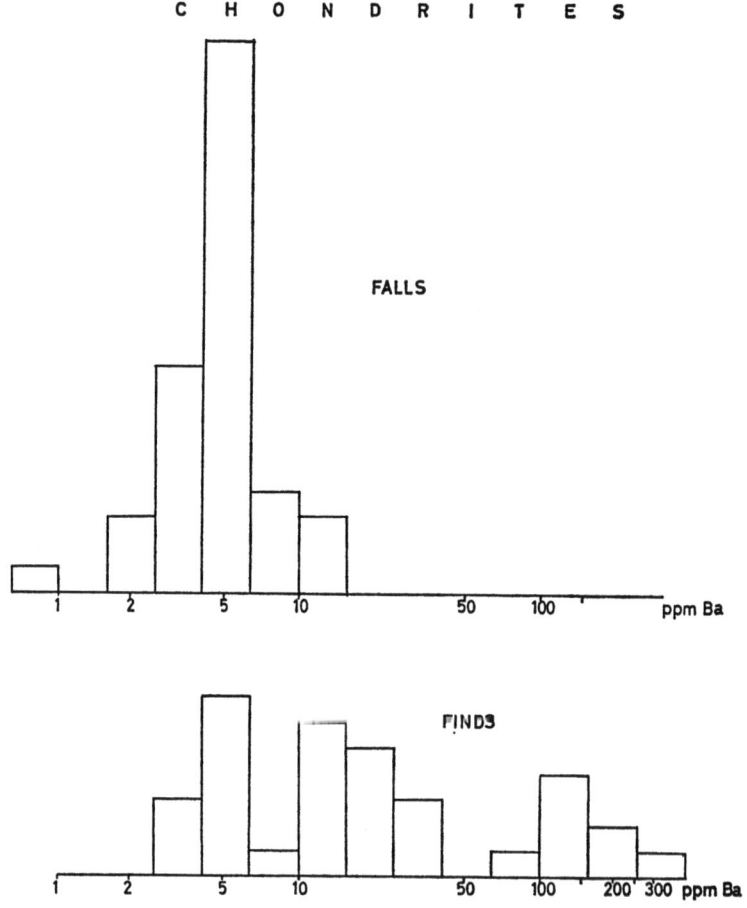

Fig. 56-C-1c. Distribution of Barium in ordinary chondrites, Class symbol CH

Barium

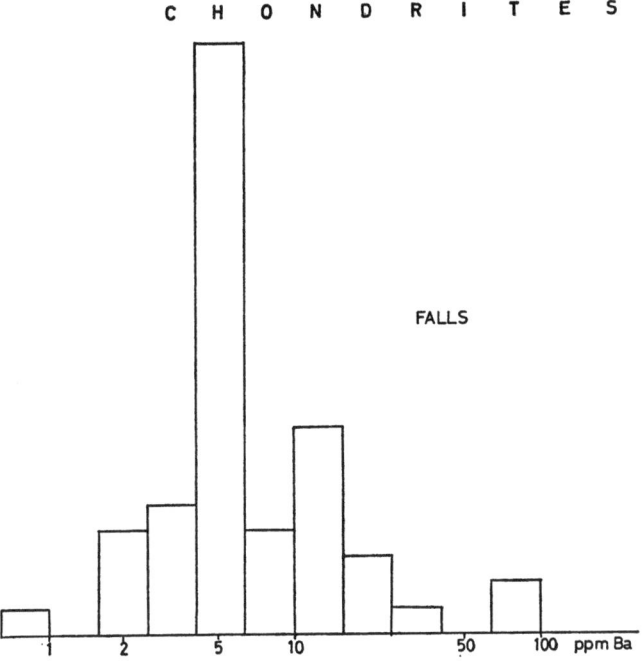

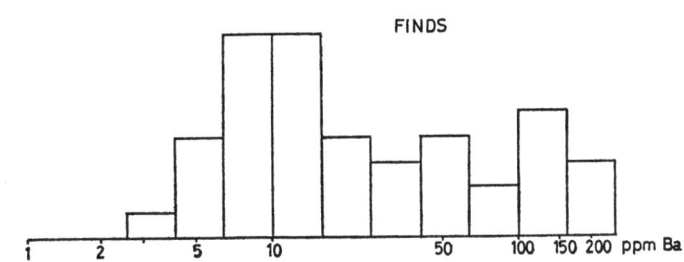

Fig. 56-C-1 d. Distribution of Barium in ordinary chondrites, Class symbol CL

Calcium rich achondrites are also high in Ba, whereas Ca-poor achondrites show low Ba values, close to chondrites. In calculations of Ba averages in meteorites, the finds have to be omitted. The few data available for nonsilicate meteorite phases (all <9 ppm) indicate that Ba is definitely a lithophilic element. Table 56-C-4 lists the mean concentrations of Ba in individual meteorite groups. Numbers of analyses used are given in brackets. Class CH and CL meteorites are by far the most frequent stony meteorites and by using their data, the Ba average for chondrites is calculated as 6.3 ppm.

Table 56-C-2. *Barium in stony meteorites*

Name of meteorite	Ba content ppm	Reference	Name of meteorite	Ba content ppm	Reference
Calcium poor achondrites			St. Marks	34.6	6
Ae (falls):			*Ce₂ (falls):*		
Cumberland Falls	14	10	Hvittis	5.9	6
Norton County	2	10	Khairpur	6.3	6
Ab (falls):			Pillistfer	5.5	6
Johnstown	2.5	10	*CH (finds):*		
	5	12	Acme	120	10
Shalka	4	10	Alamagordo	26	10
			Aurora	20	10
Calcium rich achondrites			Cavour	4	10
			Colby (Kansas)	170	10
Aor (falls):			Coldwater	10	10
Bununu	18.5	11	Cook	20.3	6
Ab (falls):			Coolidge	32	10
Juvinas	9—27	2	Covert	115	10
	26	1	Dimboola	12.8	6
	30.2	15	Estacado	6	12
Moore County	17	1	Farley	290	10
	22	5	Gladstone	17	10
Pasamonte	26	1	Hugoton	200	10
	28.6—29.8	15	Kaldoonera Hill	16.8	6
	38	5	Kansas City	4	10
	28.2	3	Kissij	5	8
Sioux County	20	1	Marsland	3	10
	25	5	Modoc	3.6	7
	27.2	15		5	10
Stannern	62	1		3.8	13
	48	2	Morland	5	10
	53	15	Morven	10.1	6
Bereba	28.6	15	Nardoo	12.9	6
Jonzac	29—29.3	15	Orlovka	18	10
Angra dos Reis	21.5	15	Petropavlovka	4	10
Ap finds):			Plainview	10	10
Binda	2	1	Ransom	28	10
Nuevo Laredo	40	1		7	12
	46	7	Seibert	125	10
	44	13	Texline	13	10
	39.3	3	Tulia	120	10
	39.3	15	Wilmot	82	10
			CH (falls)		
Chondrites			Alexandrowsky	10	10
			Allegan	4.6	6
Ce₁ (falls):				4	10
Abee	11.8	6	Aviles	<1	2
	1.8	13	Barbotan	1—3	2
	1.6	14	Beardsley	5	10
	2.41	3		3	13
	2.25	15	Beaver Creek	5	12
Indarch	6.4	6	Cangas de Onis	5	12
	1.9	13	Erxleben	1—3	2

Table 56-C-2. (Continued)

Name of meteorite	Ba content ppm	Reference	Name of meteorite	Ba content ppm	Reference
Forest City	3.37	3	Hayes Center	32	12
	3.7	7		30	10
	4	10	Herimitage Plains	52.4	6
	9	12	Kingfisher	5	10
	3.3	13	Kulnine	21.2	6
	3.30—3.37	15	Ladder Creek	94	10
Forest Vale	4.3	6	Lake Brown	8.4	6
Hessle	8	12	La Lande	210	10
Ichkala	4	10	Long Island	100	12
Kernouve	6	12		190	10
Kesen	4	10	Loongana (Forest Lake)	13.1	6
Lumpkin	6	12
Monroe	5	12	McKinney	6	10
Mount Browne	7.9	6	Melrose	155	10
	4	10	Nardoo (2)	15.5	6
Nanjemoy	13	10	Ness County (1894)	20	10
Ochansk (1)	5	10	Otis	7	10
Ochansk (2)	2	10	Potter	155	10
Olmedilla de Alarcon	5	10	Rawlinna	15	10
Pantar	4	4	Roy	72	10
	3	4	Rush Creek	5	10
	5	10	Silverton (N.S. Wales)	10.3	6
	3.2	13
	2.9	13	Tryon	190	10
Pultusk	3	10	Vincent	15.8	6
	7	12	Waconda	7	12
Richardton	3.2	7	Yalgoo	18.2	6
	4	10	CL (falls):		
Uberaba	4	10	Alfianello	3	10
Weston	6	10	Bjurböle	<1	2
Yatoor	6	10		8	12
Zhovtnevyi	11.3	6		5	10
	6	10		14.2	6
CL (finds):			Bruderheim	3.37; 3.4	3
Accalana	16.0	6		4.3	14
Adelie Land	11.3	6		3.37	15
Arriba	53	10	Chateau Renard	6	10
Asson	8	12	Chantonnay	1 to 3	2
Barratta	6.6	6	Colby (Wisconsin)	4.5	10
	7	12	Dhurmsala	6	10
Beenham	34	10	Elenovka	5	10
Berdyansk	7	10		10.5	6
Bluff	3	10	Farmington	9.1	6
Brisco	150	10	Holbrook	2.7—9	2
Cadell	5.8	6		5.8	3
Cocunda	10.0	6		12.2	6
Cook	40.6	6		4	7
Coolamon	46.5	6		26	10
De Nova	115	10		9	12
Goodland	10	10		3.6	13
Harrisonville	7	10

Table 56-C-2. (Continued)

Name of meteorite	Ba content ppm	Reference	Name of meteorite	Ba content ppm	Reference
Homestead	11.0	12	Tané	5	10
	5.1	6	Tenham	15.8	6
Khohar	13.8	6	Tennasilm	10	12
Knyahinya	1—3	2	CLL (falls):		
	5	10	Ensisheim (light)	6.9	13
Krasnoi-Ugol	5	10	(dark)	7.1	13
Kuleschovka	3.5	10	Mangwendi	12.3	6
Kunashak	4	10	Olivenza	6	10
(light)	4.8	9	Savtschenkoje	4	10
(dark)	5.2	9	CLL (finds):		
L'Aigle	2.7—9	2	Kelly	165	10
Leedey	3.76; 3.85	3	Lake Labyrinth	9.9	6
	3.64—3.76	15	Shaw	26	10
Marion	3	10	CHL:		
Maziba	4	10	Felix	4	10
Mocs	21.7	6	Karoonda	11.9	6
	5	10	Lancé	21.8	6
	6	12		8.2	13
Narellan	7.6	6	Mokoia	9.5	6
New Concord	4	10	Warrenton	10.5	6
Ni Kolskoje	2	10	Cc_1:		
Olivenza	16.4	6	Orgueil	9.8	6
Parnallee	3.5	10		<1	12
Pavlograd	6.5	10		2.4	13
Perpeti	8.6	6	Hessle	8	12
Pervomaisky	5	10	Mighei	2.5	12
Saint Michel	13.3	6	Murray County	4	10
	4	10			
Saratov	22	10			
Stavropol	2.5	10			

References (methods in brackets): 1. DUKE and SILVER, 1967 (?); 2. VON ENGELHARDT, 1936 (S); 3. EUGSTER et al., 1969 (I); 4. FREDRIKSON and KEIL, 1963 (S); 5. GAST, 1965 (I); 6. GREENLAND and LOVERING, 1965 (S); 7. HAMAGUCHI et al., 1957 (N/R); 8. HEY, 1966 (?); 9. LAVRUKHINA et al., 1966 (N/R); 10. MOORE and BROWN, 1963 (S); 11. PHILPOTTS et al., 1967 (I); 12. PINSON et al., 1953 (S); 13. REED, 1963 (N/R); 14. SHIMA and HONDA, 1967 (N/R); 15. TERA et al., 1970 (I).

Table 56-C-3. *Barium in stony irons and irons*

Name of meteorite	Ba content ppm	Reference	Name of meteorite	Ba content ppm	Reference
M (fall):			Om (find):		
Estherville	5	1	Ni poor ataxite		
M (find):			Toluca (troilite)	<0.1	13
Pallasite olivine	7	12	El Taco	2.37	3
Og (find):			Octahedrite, Weekeroo Station	8.70	3
Canon Diablo (troilite)	0.4	13			
	<0.006	13			
	<0.3	13			

References: see Table 56-C-2.

Table 56-C-4. *Average Ba concentrations of stony meteorites*

		Falls	Finds
Ae	Enstatite achondrites	8 (2)	
Ab	Bronzite achondrites	3.8 (3)	
Ap	Pigeonite-plagioclase achondrites (eucrites)	30.0 (11)	34.2 (5)
Ce_1	Enstatite chondrites	8.6 (7)	
Ce_2	Enstatite chondrites	5.9 (3)	
CH	High iron (H)-group chondrites	4.9 (42)	43 (33)
CL	Low iron (L)-group chondrites	7.2 (53)	47.1 (42)

All calculations of Ba averages in meteorites suffer from one or both of the following uncertainties: inhomogeneity of the samples and difficulties in analytical methods. The first point was demonstrated by MOORE and BROWN (1963), when they analyzed different parts of the Holbrook chondrite, which was a fall. They got a range of 8 to 110 ppm Ba for the different parts of the specimen.

III. Tektites

Ba concentrations in tektites are reported to be in the range of 300 to 7,700 ppm (Table 56-C-5, Fig. 56-C-2). Older data which, are generally lower, are reviewed by GMELIN (1960). All the more recent papers show Ba values to be much higher in tektites than in any meteoritic material. Whereas most investigators found Ba con-

Table 56-C-5. *Barium in tektites and other natural glasses*

Locality	Ba concentration range	mean	No. of anal.	Method	Reference
Africa					
Ivory Coast	648—665	657	2	I	SCHNETZLER et al. (1967)
Asia					
Indomalaysia	300—320	310	2	S	PINSON et al. (1953)
Philippines	420		1	S	PINSON et al. (1953)
Indochina	900—2,000	1,300	6	S	VOROB'EV (1959)
Australia					
Australia	340—420	375	6	S	CHAO (1963)
	540—800	620	43	S	TAYLOR and SACHS (1964)
	540—800	630	24	S	TAYLOR (1962)
Europe					
Czechoslovakia	2,000—7,700	3,600	10	S	VOROB'EV (1960)
North America					
Texas	370—1,100	610	21	S	CHAO (1963)
	380—1,100	580	10	S	CUTTITTA et al. (1967)
Georgia	340—715	566	7	S	CUTTITTA et al. (1967)
Martha's Vineyard	390		1	S	CUTTITTA et al. (1967)

Barium

Table 56-C-5. (Continued)

Locality	Ba concentration		No. of anal.	Method	Reference
	range	mean			
Natural glasses					
Tasmania					
Darwin Glass	290—360	340	8	S	Taylor and Solomon (1964)
Darwin Glass, dark	550		1	M	Chapman et al. (1967)
Darwin Glass, light	300		1	M	Chapman et al. (1967)
Australia					
Macedon, dark	400		1	M	Chapman et al. (1967)
Macedon, light	200		1	M	Chapman et al. (1967)
Henbury impact glass	600—700	650	2	S	Taylor and Kolbe (1964)
Africa					
Bosumtwi Crater	533—624	579	2	I	Schnetzler et al. (1967)

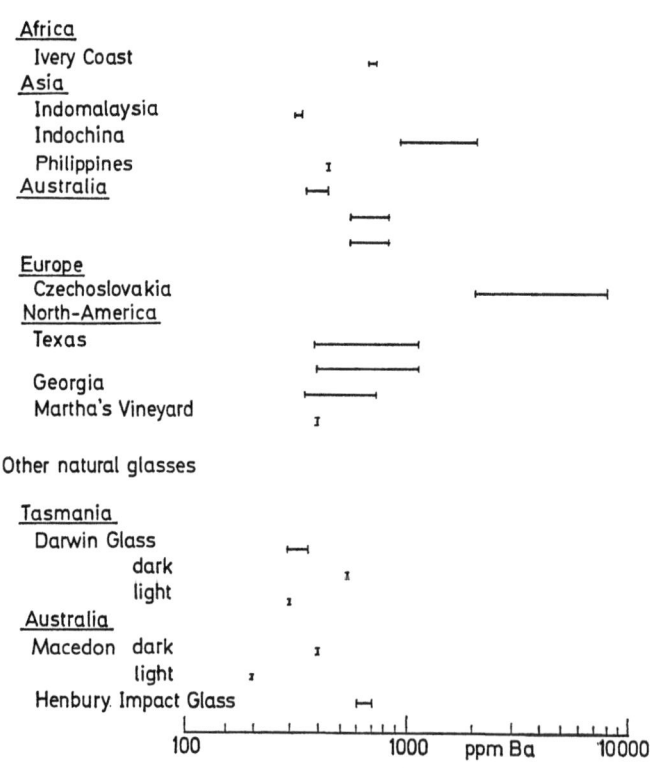

Fig. 56-C-2. Barium in tektites and other natural glasses

centrations in the range of 330 to 1,100 ppm, the values of VOROB'EV (1959, 1960) both for moldavites and indochinites are higher by a factor as great as ten, compared with tektite values from the same areas investigated by other scientists. PREUSS (1935) gives a mean of 450 ppm Ba for 36 tektites, 14 of which were moldavites. BOUSKA and POVONDRA (1964) published semiquantitative spectrographic data for moldavites showing that six out of seven samples had 100 to 1,000 ppm Ba, and the other had less than 100 ppm Ba. Thus, it cannot be excluded that VOROB'EV's values are too high due to a systematic error, especially since no simple parent material can be imagined which would provide so much Ba. TAYLOR (1965, 1966) could relate australites by their Ba concentration and the content of other elements to Henbury impact glass and Henbury greywacke as parent materials. Similar investigations of SCHNETZLER et al. (1967) showed the consistency of Ba and RE concentrations in Ivory Coast tektites and Bosumtwi Crater impact glasses and strong similarities to Bosumtwi phyllites. Calculations by TAYLOR (1962) made it obvious that no meteorite splash could affect the Ba concentrations of the resulting impact glasses and tektites.

IV. Lunar Samples

Considerable effort has been expended on analytical investigations of lunar samples from Apollo 11 and 12 missions. From data of the individual investigators, averages were calculated for the different types of material according to the classification established by the *Lunar Sample Preliminary Examination Team* (1969, 1970), (Table 56-C-6).

The overall unweighted average for Apollo 11 material calculated from the means of the particular batches (33) is 240 ppm Ba, $s=76$. The range of data for

Table 56-C-6. *Barium in lunar rocks*

Sample type	Barium content (ppm)		No. of analyses
	arith. mean	stand. dev.	
Apollo 11 (Mare Tranquillitatis):			
Type A: fine grained vesicular rocks	251	130	37
Type B: medium grained vuggy igneous rocks	176	72	23
Type C: breccia	230	66	37
Type D: "lunar soil" (fines)	169	40	16
Apollo 12 (Ocean of Storms)			
Basaltic rocks	72	50	13
Breccia	420	210	3
Fines	586	189	7
Sample 12013	3,088	2,256	24

References for Apollo 11 samples (methods in brackets): ANNEL and HELZ (1970) (S); BROWN et al. (1970) (X); COMPSTON et al. (1970) (X); GAST et al. (1970) (I); GOLES et al. (1970) (N/R); HASKIN et al. (1970) (N/R); MAXWELL et al. (1970) (S); MORRISON et al. (1970) (N/R, M); MURTHY et al. (1970) (I); PHILPOTTS and SCHNETZLER (1970) (I); SMALES et al. (1970) (I); TAYLOR et al. (1970) (S); TERA et al. (1970) (I); WAKITA et al. (1970) (N/R); WÄNKE et al. (1970) (N/R).

Reference for Apollo 12 samples: DRAKE et al. (1970) (M); HUBBARD et al. (1970) (I); HUBBARD et al. (1971) (I); LSPET (1970) (S); MAXWELL and WIIK (1971) (S); SCHNETZLER et al. (1970) (I); WAKITA and SCHMITT (1970) (N/R).

total rock analyses covers 10 to 370 ppm Ba, but values of upto 55,900 ppm and 24,400 ppm have been found (M) in interstitial K rich (9.4 and 10.9% K_2O) phases from a 4 mm fragment from lunar soil (10085-LR-1) (ALBEE and CHODOS, 1970). These K rich phases consist predominantly of glass and extremely fine-grained crystalline material. Their chemical composition approaches but does not reach K feldspar composition. The K rich phase, which is present in about 6 vol.-% in sample 10085-LR-1, may also be responsible for Ba, Rb, and K concentrations in other samples, where a close relationship between these elements was observed. For several samples, a grouping to separate high K, Rb, Ba material from low K, Rb, Ba material has been tried. A few samples do not fit this separation. Still, the K/Ba ratios are rather constant in general. Inhomogeneities of Ba content in different chips of one sample have been reported, especially by WAKITA et al., N/R, (1970). Different chips of their batch 10019 differ by as much as 210 ppm in their Ba content (130 and 340 ppm). Analyses (N/R) of size fractions of lunar soil showed a Ba enrichment in the finer fraction (WAKITA et al., 1970).

Enrichment of Ba in moon rocks from Mare Tranquillitatis is 26 to 110 times (GAST et al., 1970) compared to chondritic abundances. Within a single rock, concentration factors are similar for the "incompatable elements" Ba, U, Th, Zr, and REE except Eu.

Several individual minerals have been analyzed for Ba. The results are given in Table 56-C-7.

Apollo 12 samples from the Ocean of Storms show lower Ba content in the basaltic rocks (average 72 ppm), but higher values for breccia (average 420 ppm) and for "fines" (average 586 ppm). A most unusual rock is sample 12013 with 61% SiO_2, 2% K_2O and an average of 3088 ppm Ba. Microprobe investigations of individual points in alkali feldspars of this sample gave upto 8% BaO and 13.1% K_2O, i.e., compositions of celsian-orthoclase solid solution series (*Lunatic Asylum*, 1970). In rock 12013, Ba is enriched upto more than 2,000 times compared to chondrites (HUBBARD et al., 1970).

Table 56-C-7. *Ba content of individual minerals from Apollo 11 and 12 samples*

Mineral	Lunar sample	Ba concentration ppm	Method	Reference
Apollo 11:				
Plagioclase	10085	<1; <1; 15; 1,500	M	ANDERSEN et al. (1970)
Plagioclase	10044—24	271	I	PHILPOTTS et al. (1970)
Plagioclase	10062—29	70.1	I	PHILPOTTS et al. (1970)
Clinopyroxene	10085	<1; 45	M	ANDERSEN et al. (1970)
Pyroxene	10044—24	93.7	I	PHILPOTTS et al. (1970)
Pyroxene	10062—29	30.9	I	PHILPOTTS et al. (1970)
Ilmenite		81.2	I	MURTHY et al. (1970)
Apollo 12:				
Plagioclase	12013	910	M	DRAKE et al. (1970)
Alk. feldspar	12013	10650	M	DRAKE et al. (1970)
Alk. feldspar	12013—10 (40)	895; 910; 8,680; 20,400	M	LUNATIC ASYLUM (1970)

Revised manuscript received: September 1971

56-D. Abundance in Rock-Forming Minerals (I) and Barium Minerals (II)

I. Rock-Forming Minerals

In igneous rocks of the earth's crust Ba usually does not form minerals of its own, but is distributed among a number of silicate structures, mainly feldspars and micas. The most important substitution is for potassium due to the nearly identical ion sizes, even with the somewhat more covalent character of the Ba—O bond. Substitution for Ca is observed in plagioclases, pyroxenes and amphiboles. Apatite and calcite are the most important rock forming non-silicates containing Ba.

a) Feldspars

K feldspars are the most important Ba carriers. Investigations by ROY (1965, 1967) and GAY and ROY (1968) with synthetic members have shown that a continuous series of solid solutions exist at high temperatures between K feldspars and celsian. In the subsolidus region, two gaps of miscibility seem to exist, one close to the microcline composition and the other between hyalophane and celsian (Fig. 56-D-1).

The system $BaO-Al_2O_3-SiO_2$ was recently investigated by LIN and FOSTER (1968, 1969).

In nature, K feldspars with BaO >2% are rare. They are mostly restricted to alpine type fissures (adularia) and deposits of manganese oxide.

Ba concentrations upto 9.5% in feldspars of the orthoclase-celsian series are reported from the alkalic rock complex, Magnet Cove (ERICKSON and BLADE, 1963) and from phonolites in South West Germany (WEISKIRCHNER, 1969). Ba distribution in K feldspars has been investigated by many authors. As a general pattern, it was found that in magmatic sequences the early crystallized K feldspars show the highest Ba concentrations, whereas microclines from pegmatites were low (S) in this element (SHIMER, 1943; BRAY, 1942). Metasomatic alteration of a granite body (Black Forest, Germany) was discussed as a cause of abundant K feldspar phenocrysts with a high and rather homogeneous Ba distribution by EMMERMANN (1968), (X).

Distribution of Ba abundances in K feldspars is shown for igneous rocks, pegmatites, and alpine fissures in Fig. 56-D-2 using the data of Table 56-D-1. The histograms clearly demonstrate that, in general, average pegmatitic K feldspars contain less Ba than those from granitic rocks.

Ba may enter the plagioclase structure depending on composition in the An-Ab series, temperature and pressure, of competing elements. DUCHESNE (1968) found a correlation between potassium and Ba concentrations. Many indications exist supporting the view that all the Ba in feldspars can be plotted in the four component system Ab, An, Or, Cn. Plots of Ba concentrations for different members

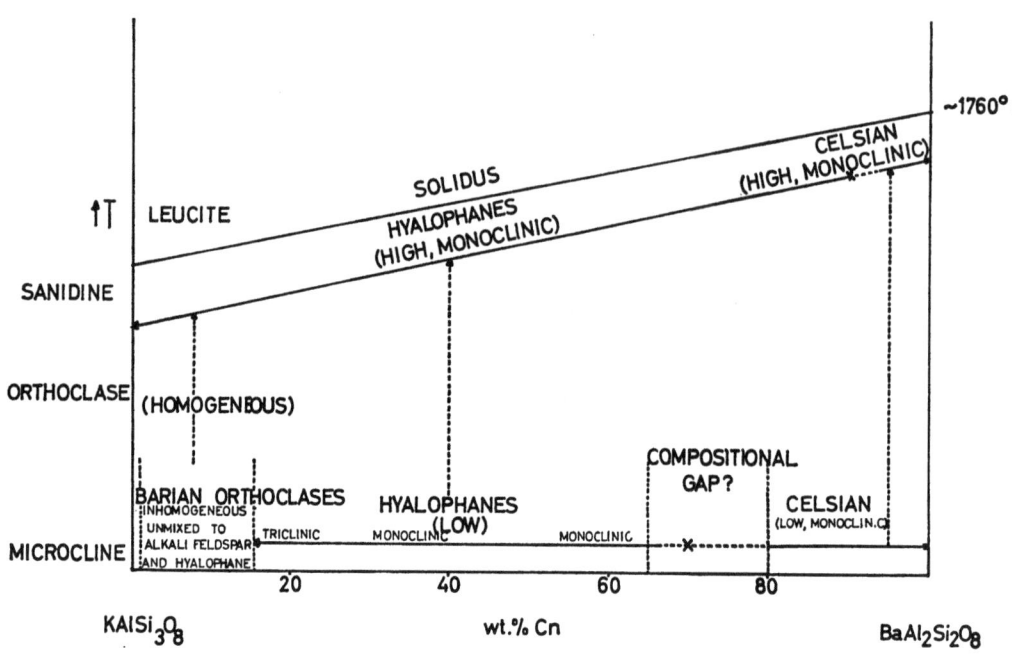

Fig. 56-D-1. Phases in the system KAlSi$_3$O$_8$—BaAl$_2$Si$_2$O$_8$ (GAY and ROY, 1968)

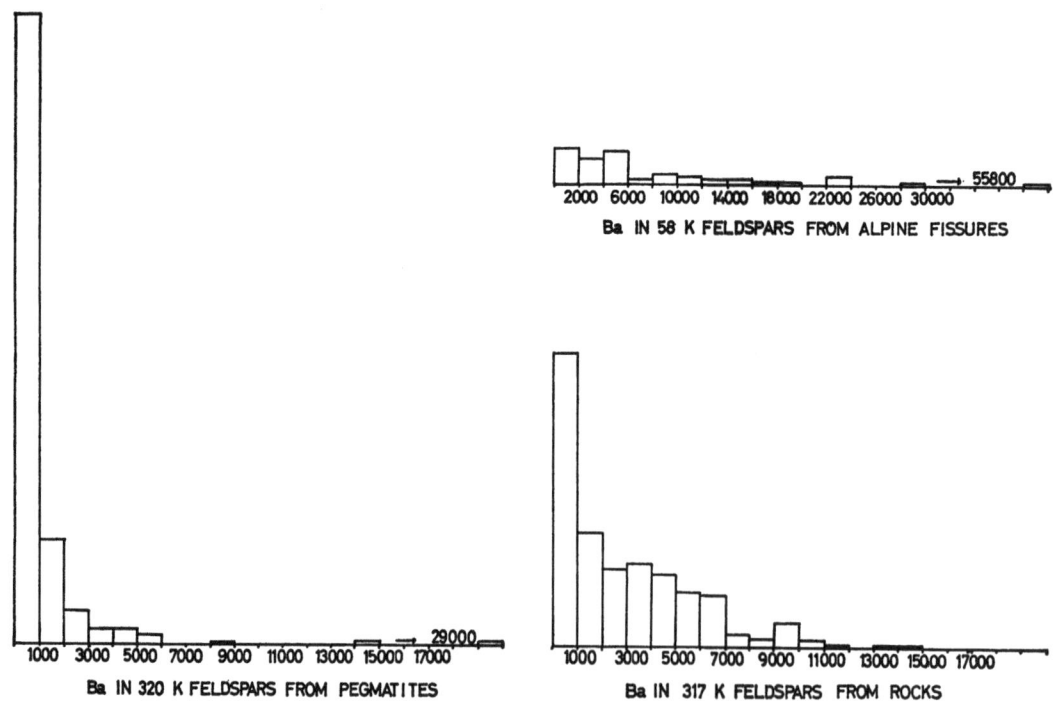

Fig. 56-D-2. Distribution of Ba concentrations in K feldspars (see Table 56-D-1 for references)

Table 56-D-1. *Barium concentrations in K feldspars*

Rock type	No. of analyses	Barium concentration range ppm	arith. mean	Method	Reference
Pegmatites	7	79— 1,130	270	S	Bray (1942)
	38	19— 450	100	X	Correia Neves (1964)
	6	3— 470	132	S	von Engelhardt (1936)
	5	200— 6,000	1,600	S	Erickson and Blade (1963)
	44	50— 3,400	710	S	Heier and Taylor (1959)
	17	20— 10,000	911	S	Higazy (1953)
	15	100— 1,700		S	Higazy (1949)
	5	435— 5,620	2,408	S	Hitchon (1960)
	9	190— 4,750	1,290	X	Markart and Preisinger (1964)
	11	150— 600	355	S	Oftedal (1961)
	40	150— 1,000	436	S	Oftedal (1962)
	44	<10— 9,000	1,660	S	Oftedal (1958)
	16	171— 10,525	4,581	S	Pirani and Simboli (1963)
	64	76— 397	251	S	Shcherba et al. (1964)
	11	27— 3,550	495	S	Taylor et al. (1960)
	4	236—>6,000		S	Townend (1966)
		<10— 10,525	863		*average for 320 samples*
Alpine fissures	13	220— 55,800	7,730	N	Rybach and Nissen (1967)
	3	2,400— 3,270	2,690	X	Markart and Preisinger (1964)
	15	2,600— 29,000	14,050	S	Weibel (1957)
	33	630— 15,300	5,000	S	Weibel and Meyer (1957)
	1	1,970	1,970	S	Hewlett (1959)
		220— 55,800	7,481		*average for 65 samples*
Igneous rocks					
Granite	11	162— 2,340	1,150	S	Bray (1942)
Granodiorite	4	322— 1,440	810	S	Bray (1942)
Quartzmonzonite	3	1,080— 1,260	1,170	S	Bray (1942)
Trachyte	2	5,100— 8,600	6,850	S	Carmichael (1965)
Granite	10	108— 361	201	S	Emiliani and Vespignani (1964)
Granite[a]	150	800— 5,400	4,700	X	Emmermann (1968)
Granite-larvikite	5	90— 14,300	3,920	S	von Engelhardt (1936)
Alkalic rock	4	900— 4,000	2,475	S	Erickson and Blade (1963)
Monzonite, granite	25	130— 12,000		S	Heier (1960)
Granite, gneiss	41	1,000— 9,500	3,840	S	Heier and Taylor (1959)
Granite, granodiorite, gneiss	12	90— 5,480	2,765	S	Herz and Dutra (1966)
Trachyte	10	360— 13,600	3,370	S	Hewlett (1959)
Granite	3	420— 2,700	1,550	S	Hewlett (1959)
Charnockite	4	2,500— 5,000		S	Howie (1955)
Monzonitic accumulate[a]	16	1,070— 12,100	5,000	X	Jasmund and Seck (1964)
Granite, gneiss	19	364— 4,620	2,220	X	Markart and Preisinger (1964)

Table 56-D-1. (Continued)

Rock type	No. of analyses	Barium concentration range ppm	arith. mean	Method	Reference
Alkalic rock	4	630— 5,900	314	X	Perchuck and Ryabchikov (1968)
Granite	70	26— 9,480	2,124	X	Rhodes (1969)
Granite	9	4,350— 4,770	4,560	X	Richter (1966)
Quartz, monzonitic	29	3— 3,000	990	S	Rogers (1958)
Granite	4	835— 1,020	950	S	Scharbert (1966)
Granite-tonalite	5	2,000— 3,000	2,500	S	Sen et al. (1959)
Granite	91	45— 392	176	S	Shcherba et al. (1964)
Granodiorite	14		836	S	Shcherba et al. (1964)
Adamellite	3		460	S	Shcherba et al. (1964)
Syenite	4		162	S	Shcherba et al. (1964)
Monzonite	2		155	S	Shcherba et al. (1964)
Syenodiorite	6		595	S	Shcherba et al. (1964)
Different rocks	32	<200— 18,000	3,559	M	Smith and Ribbe (1966)
Different rocks[a]	23	<200— 6,200	1,183	M	Smith and Ribbe (1966)
Granite	12	277—>6,000		S	Townend (1966)
Granite	7	766— 1,526	1,240	X	White (1966)
		3— 18,000	2,626		average for 598 samples

[a] Sanidine.

of the plagioclase solid solution series show that the pure components contain less Ba than the intermediate plagioclases (Table 56-D-2, Fig. 56-D-3). This may be due to the fact that the pure end members are mostly very late or secondary crystals which formed from a Ba-poor liquid. Distribution of Ba between coexisting K feldspars and plagioclases is discussed as a potential geothermometer by Heier (1960, 1961). Barth (1961) showed that a straight line relation exists between log ratios of Ba in coexisting feldspars and log inverse absolute temperatures. He deduced that below 250° C, plagioclase is the preferred host mineral for Ba. Iiyama (1968) studied Ba distribution between potassium feldspars and plagioclases experimentally by hydrothermal runs at 600° C and 1,000 bars. His results contradict the observations in natural rocks as he finds higher Ba incorporation in the plagioclases than in the coexisting K feldspars. Rudert (1970) experimentally found an incorporation of Ba in albite at 930° C/1 kb, upto 30 weight % of celsian molecules; Bruno and Gazzoni (1970) substituted large amounts of Ca in the modifications of $BaAl_2Si_2O_8$. In the hexagonal modification synthesized by solid state reaction at 1,200° C, Ca replaces Ba upto 37% (atomic fraction); in the hexagonal modification obtained by crystallization of a melt, the replacement is limited to 25%; the same value was found for the monoclinic modification by heating the mentioned modifications to 1,450° C.

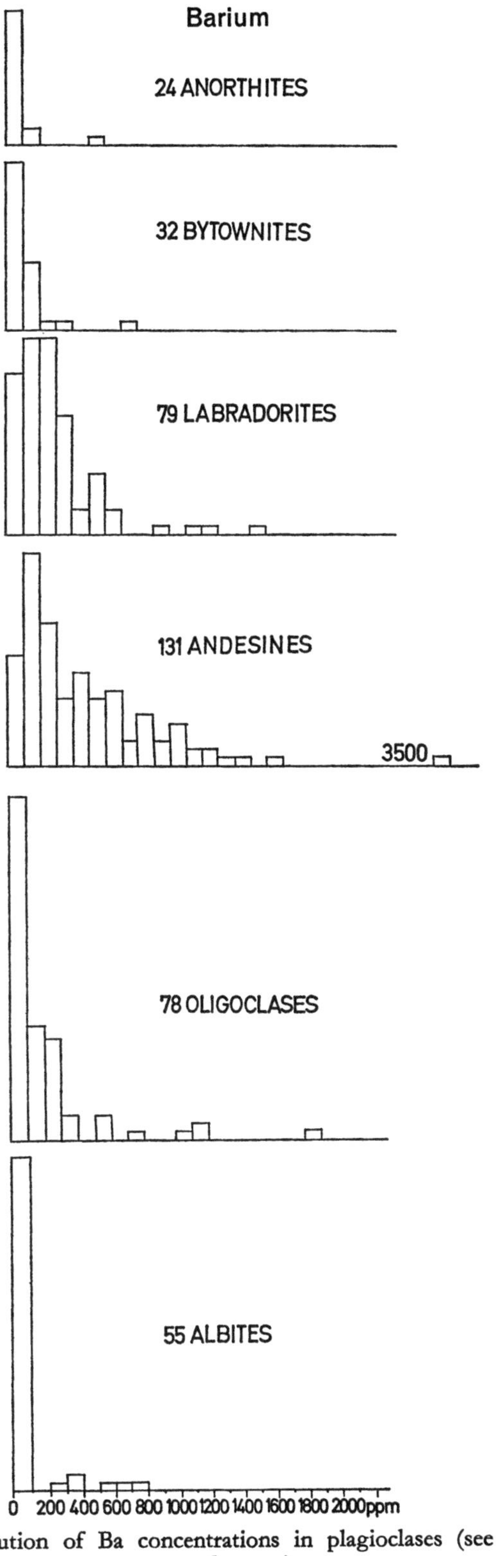

Fig. 56-D-3. Distribution of Ba concentrations in plagioclases (see Table 56-D-2 for references)

Table 56-D-2. *Barium concentrations in plagioclases from igneous and metamorphic rocks and rocks of hydrothermal origin*

Rock type	No. of analyses	Barium concentration range ppm		arith. mean	Method	Reference
Anorthites from:						
Basalt	1	60		60	S	Byers (1961)
Pumice	1	10		10	S	Coats (1952)
Anorthosite, gabbro and others	13	<100—	500	<217	M	Corlett and Ribbe (1967)
Metam. rock (amphibolite facies)	3	<45		<45	S	Sen (1960)
Metam. rock (granulite facies)	2	<45—	80	<60	S	Sen (1960)
Dacite-andesite	3	<45—	80	<63	S	Sen (1960)
Norite	1	10		10	S	Sen et al. (1959)
Bytownites from:						
Anorthosite, norite and others	12	<100—	100	<100	M	Corlett and Ribbe (1967)
Gabbro-anorthosite	9	40—	320	130	S	Emmons (1952)
Basalt	2	22—	270	146	S	Muir et al. (1964)
Metam. rock (amphibolite facies)	7	<45—	54	<46	S	Sen (1960)
Gabbro	2	80+	85	82	S	Sen et al. (1959)
Labradorites from:						
Anorthosite, gabbro and others	38	<100—	300	<195	M	Corlett and Ribbe (1967)
Basalt	9	200—	800	400	S	Cornwall and Rose (1957)
Gabbro-anorthosite	9	130—	580	347	S	Emmons (1952)
Anorthosite	2	85+	160	122	S	Papezik (1965)
Metam. rock (amphibolite facies)	3	115—	150	160	S	Sen (1960)
Metam. rock (granulite facies)	4	270—	480	295	S	Sen (1960)
Norite	3	45—	130	75	S	Sen et al. (1959)
Gabbro	2	50—	80	65	S	Wagner and Mitchell (1951)
Teschenite-basalt, gabbro	9	80— 1,500		451	S	Wilkinson (1959)
Andesines from:						
Anorthosite	9	630— 3,500		1,250	M	Anderson (1966)
Granite, granodiorite	6	370— 1,000		310	S	Bray (1942)
Gneiss, anorthosite and others	19	<100—	600	<177	M	Corlett and Ribbe (1967)
Granodiorite-anorthosite	5	90—	330	224	S	Emmons (1952)
Monzonitic accumulates	12	440— 1,020		735	X	Jasmund and Seck (1964)
Gneiss	11	900—18,000		6,100	S	Oftedal (1958)
Pegmatite	12	10— 1,430		300	S	Oftedal (1958)
Anorthosite	10	150— 1,075		384	S	Papezik (1965)

Table 56-D-2. (Continued)

Rock type	No. of analyses	Barium concentration range ppm	Barium concentration arith. mean	Method	Reference
Metam. rock (amphibolite facies)	15	<45— 990	<224	S	SEN (1960)
Metam. rock (granulite facies)	12	45— 700	380	S	SEN (1960)
Dacite-andesite	4	450— 870	560	S	SEN (1960)
Granite	2	250+ 320	285	S	SEN (1960)
Tonalite, granodiorite	6	80— 255	174	S	SEN et al. (1959)
Gabbro	6	200— 600	390	S	WAGER and MITCHELL (1951)
Teschenite, gabbro	2	350+ 600	475	S	WILKINSON (1959)

Oligoclases from:

Rock type	No. of analyses	range ppm	arith. mean	Method	Reference
Anorthosite	1	3,480	3,480	M	ANDERSON (1966)
Pegmatite, granite and others	43	<100— 1,100	<134	M	CORLETT and RIBBE (1967)
Basalt	2	700+ 1,000	850	S	CORNWALL and ROSE (1957)
Granite	4	70— 380	132	S	EMMONS (1952)
Pegmatite	1	365	365	S	HITCHON (1960)
Granite	2	150— 268	209	S	PIRANI and SIMBOLI (1963)
Granite	4	180— 250	210	S	SCHARBERT (1966)
Metam. rock (amphibolite facies)	5	45— 540	205	S	SEN (1960)
Metam. rock (granulite facies)	7	<90— 1,170	<425	S	SEN (1960)
Rhyolite-rhyodacite	2	45— 1,800	922	S	SEN (1960)
Granite	5	63— 380	155	S	SEN (1960)
Granite, granodiorite	2	120+ 235	175	S	SEN et al. (1959)

Albites from:

Rock type	No. of analyses	range ppm	arith. mean	Method	Reference
Pegmatite, alpine fissure	36	<100— 650	<145	M	CORLETT and RIBBE (1967)
Basalt	1	700	700	S	CORNWALL and ROSE (1957)
Pegmatite	6	54— 80	69	X	CORREIA NEVES (1964)
Pegmatite	1	1	1	S	EMMONS (1952)
Pegmatite	1	2	2	S	VON ENGELHARDT (1936)
Pegmatite	1	5	5	S	HIGAZY (1953)
Metam. rock (greenschist facies)	1	<45	<45	S	SEN (1960)
Different rocks	2	<200	<200	M	SMITH and RIBBE (1966)
Pegmatite	5	12— 49	~24	S	TAYLOR et al. (1960)
Granophyre	1	30	30	S	WAGER and MITCHELL (1951)

Anorthoclases from:

Rock type	No. of analyses	range ppm	arith. mean	Method	Reference
Basalt, trachyte	2	1,000+ 1,100	1,050	S	WILKINSON (1962)
Phonolite	1	2,160	2,160	S	HEWLETT (1959)

b) Micas and Other Important Rock Forming Minerals

Micas are the next important Ba carriers in rocks. Ranges of Ba concentrations in these and other minerals are given in Table 56-D-3 and in Fig. 56-D-4. Ba is a concentrated in biotites as in muscovites. Only lepidolites have much lower Ba values. Generally micas from pegmatites show lower Ba values than those from the adjacent country rocks (TAKUBO and TATEKAWA, 1954), but each pegmatite may have its own "Ba level". Extensive material on Ba contents, mainly from pegmatite muscovites, was published by HEINRICH et al. (1956). These authors found Ba concentrations from 1 to 9,900 ppm in 162 samples, predominantly pegmatite muscovites. They calculated an average of 1,020 ppm Ba.

PETROV et al. (1965) found that the relative concentrations of Ba in biotites increase with metamorphic grade. In rocks free of K feldspars, biotite is often the main Ba carrier, sometimes together with pyroxenes and amphiboles. In skarns

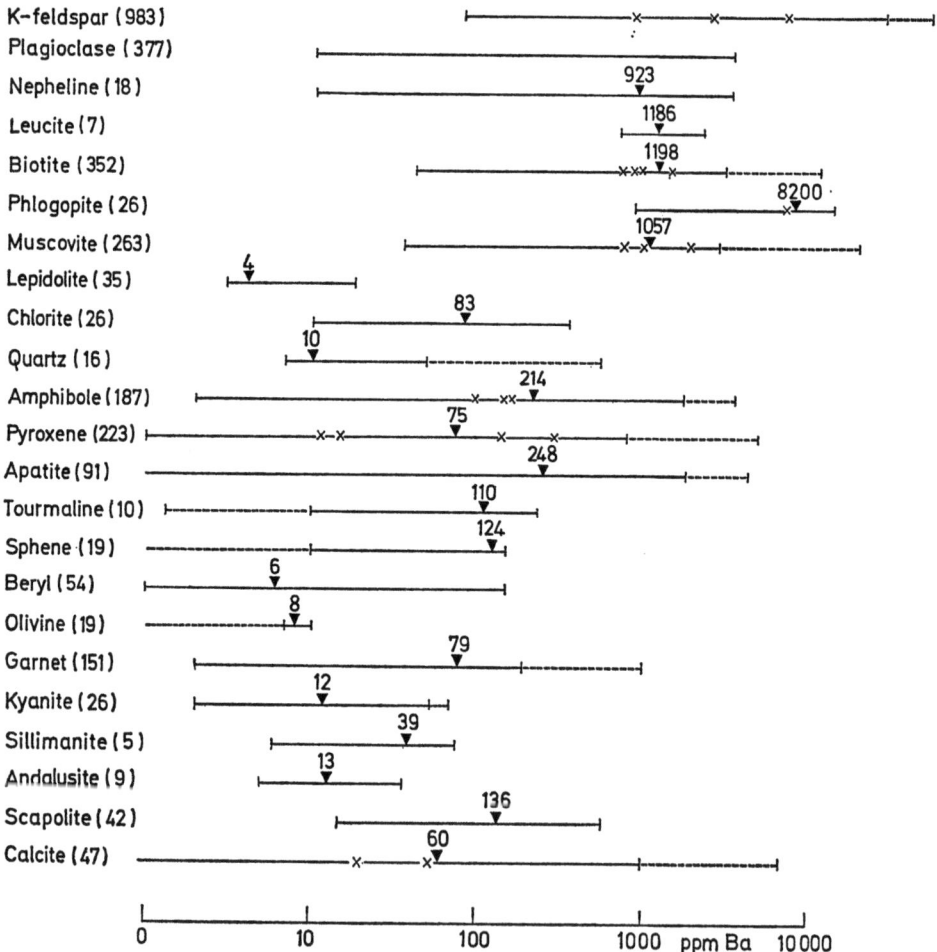

Fig. 56-D-4. Ba concentrations in rock forming minerals (see Table 56-D-4 for references). ▼ Indicates total averages. × Indicates group averages

Table 56-D-3. *Barium concentrations in rock forming minerals of igneous and metamorphic rocks*

Rock type	No. of ana-lyses	Barium concentration range ppm	arith. mean	Method	Reference
Nepheline					
Alkalic rock	3	<45— 3,320	1,270	S	von Eckermann (1952)
Alkalic rock	5	100— 3,000	1,200	S	Erickson and Blade (1963)
Pegmatite	6	<10— 20	20	S	Oftedal (1962)
Alkalic rock	4	<100— 3,400	1,670	X	Perchuck and Ryabchikov (1968)
		<10— 3,400	923		average for 18 samples
Leucite					
Leucite-basanite	2	1,160— 1,250	1,200	S	von Engelhardt (1936)
Leucitophyre-leucite basanite	5	710— 2,250	1,180	S	Henderson (1965)
		710— 2,250	1,186		average for 7 samples
Biotite					
Granite, granodiorite	12	67— 720	294	S	Bray (1942)
Pegmatite	3	103— 470	241	S	Bray (1942)
Schist	18	600— 1,100	790	S	Butler (1967)
Metapelite	5	500— 2,000	1,500	S	Card (1964)
Rhyolite	8	1,430— 6,730	5,200	M	Carmichael (1967a)
Granitic rock	34	240— 5,000	1,405	S	Dodge et al. (1969)
Alkalic rock (carbonatite)	8	1,700— 6,360	4,065	S	von Eckermann (1952)
Granite	2	84+ 100	90	S	Emiliani and Vespignani (1964)
Paragneiss	25	300— 2,200	930	S	Engel and Engel (1960)
Alkalic rock	4	900—12,000	4,500	S	Erickson and Blade (1963)
Quartzdiorite-granite	13	400— 2,700	1,830	S	Haslam (1968)
Charnockite	2	900+ 2,300	1,600	S	Heier (1960)
Pegmatite	7	45— 2,320	717	S	Hitchon (1960)
Gneiss, schist, amphibolite	35	100— 1,200	750	S	Hunzicker (1966)
Gneiss, amphibolite	8	900— 3,180	~2,400	S	Kretz (1959)
Gneiss, schist	20	120— 2,450	810	S	Moxham (1965)
Pegmatite	3	200— 400	300	S	Oftedal (1962)
Quartzmonzonite, granodiorite	24	450— 8,500	1,212	X	Rimsaite (1964)
Gneiss, schist	17	900— 2,700	808	X	Rimsaite (1964)
Metaarkose	7	472— 1,086	780	S	Schwarcz (1966)
Granite, norite	10	500— 2,500	1,822	S	Sen et al. (1959)
Pegmatite, wallrock	48	<150— 2,000	873	S	Stern (1966)
Metamorphic rock (staurolite zone)	22	42— 1,410	800	S	Turekian and Phinney (1962)
Migmatite, granite	8	240— 1,335	880	X	White (1966)
Basalt	1	2,500	2,500	S	Wilkinson (1962)
Cordierite-biotite, gneiss	8	60— 530	290	S	Wynne-Edwards and Hay (1962)
		42— 8,500	1,198		average for 352 samples

Barium

Table 56-D-3. (Continued)

Rock type	No. of analyses	Barium concentration range ppm	arith. mean	Method	Reference
Phlogopite					
Wyomingite, orendite	12	3,670—10,100	7,350	M	CARMICHAEL (1967b)
Wolgidite, wyomingite, fitzroyite	9	3,140—14,300	12,000	M	CARMICHAEL (1967b)
Alkalic rock	2	1,000+ 4,000	2,500	S	ERICKSON and BLADE (1963)
Lamprophyre, basalt	3	2,700— 4,850	4,016	X	RIMSAITE (1964)
		1,000—14,300	8,202		average for 26 samples
Muscovite					
Granite	7	108— 900	490	S	BRAY (1942)
Pegmatite	5	23— 360	125	S	BRAY (1942)
Schist	22	1,400— 2,900	1,930	S	BUTLER (1967)
Granite	2	1,000+ 1,100	1,050	S	EMILIANI and VESPIGNANI (1964)
Pegmatite	168	2— 9,800	1,020	S	HEINRICH et al. (1953)
Pegmatite	1	36	36	S	HEINRICH (1967)
Pegmatite	13		160	S	HITCHON (1960)
Gneiss, schist	14	110— 1,200	850	S	HUNZICKER (1966)
Alpine fissure	2	230+ 300	265	S	SCHWANDER et al. (1968)
Metamorphic rock	1	21,400	21,400	W	SNETSINGER (1966)
Pegmatite	28	180— 1,650	797	S	STERN (1966)
		2—21,400	1,057		average for 263 samples
Lepidolite					
Pegmatite	26	2— 9	3	S	HEINRICH et al. (1953)
Pegmatite	9	3— 18	7	S	HEINRICH (1967)
		2— 18	4		average for 35 samples
Chlorite					
Basalt	11	20— 300	120	S	CORNWALL and ROSE (1957)
Granite	3	10— 50	25	S	EMILIANI and VESPIGNANI (1964)
Schist	11	17— 148 (2,100)	36	S	GRESENS (1967)
Garnet, chlorite, schist	1	358	358	S	MOHR (1956)
		10— 2,100	83		average for 26 samples
Quartz					
Granite	4	280— 550	416	S	BRAY (1942)
Pegmatite	11	30— 50	11	S	HITCHON (1960)
Basic intrusion	1	7	7	S	WAGER and MITCHELL (1951)
		7— 550	112		average for 16 samples
Amphibole					
Metamorphic rock (amphibolite)	3	<50— 200	100	S	CARD (1964)
Wolgidite, orendite	4	1,800— 5,400	3,700	M	CARMICHAEL (1967b)

Table 56-D-3. (Continued)

Rock type	No. of analyses	Barium concentration range ppm	Barium concentration arith. mean	Method	Reference
Schist, glaucophane	9	2— 120	20	S	COLEMAN and PAPIKE (1968)
Schist, riebeckite	4	<4— 140	74	S	COLEMAN and PAPIKE (1968)
Schist, actinolite	2	12+ 410	211	S	COLEMAN and PAPIKE (1968)
Granite	17	7— 50	30	S	DODGE et al. (1968)
Granodiorite, quartz-diorite	3	15— 95	60	S	DODGE et al. (1968)
Metamorphic rock (amphibolite)	24	24— 320	100	S	ENGEL and ENGEL (1962)
Diorite, porphyr	15	100— 300	150	S	ENGEL (1959)
Schist	13	44— 104	66	S	GRESENS (1967)
Quartzdiorite	6	10— 36	20	S	HASLAM (1968)
Granite	4	10— 36	20	S	HASLAM (1968)
Rhyodacite	1	180	180	S	HASLAM (1968)
Gneiss, schist	8	<10— 195 (1,200)	80	S	HUNZICKER (1966)
Monzonitic accumulate	11	90— 1,780	755	X	JASMUND and SECK (1964)
Gneiss, amphibolite	16	<90— 300	160	S	KRETZ (1959)
Amphibolite	9	<90— 360	230	S	KRETZ (1960)
Gneiss, schist	20	10— 485	150	S	MOXHAM (1965)
Nepheline-syenite, pegmatite	2	10+ 20	15	S	OFTEDAL (1962)
Ophiolite	1	27	27	S	PLAS and HÜGI (1961)
Tonalite, granodiorite	12	10— 80	40	S	SEN et al. (1959)
Essexite	2	25— 150	90	S	SIMPSON (1954)
Basalt	1	100	100	S	WILKINSON (1962)
		2— 5,400	214		average for 187 samples
Pyroxene					
Basic layered rock	14	12— 22	18	S	ATKINS (1969)
Basalt	1	60	60	S	BYERS (1961)
Basalt etc.	8	30— 800	195	S	CORNWALL and ROSE (1957)
Alkalic rock	9	<45— 540	160	S	VON ECKERMANN (1952)
Alkalic rock	32	<10— 5,000	300	S	ERICKSON and BLADE (1963)
Peridotite	2	10— 18	14	S	GREEN (1964)
Quartzdiorite	4	5— 27	16	S	HASLAM (1968)
Granite	1	10	10	S	ISHIOKA (1967)
Metamorphic rock (skarn)	10	<90— 360	<145	S	KRETZ (1960)
Charnockite	20	<5— 22	≤11	S	LEELANANDAM (1967)
Basalt, trachyte	2	10+ 50	30	S	LEMAITRE (1962)
Metamorphic rock (skarn)	38	<1— 44	12	S	MOXHAM (1960)
Basalt	8	5— 45	21	S	MUIR et al. (1964)
Kimberlite	9	<10— 150	33	S	NIXON et al. (1963)

Table 56-D-3. (Continued)

Rock type	No. of analyses	Barium concentrations range ppm		arith. mean	Method	Reference
Pegmatite	1	10		10	S	OFTEDAL (1962)
Olivine basalt	1	1.5		1.5	N/R	ONUMA et al. (1968)
Metamorphic rock (skarn)	38	1—	65	15	S	SHAW et al. (1963)
Syenite, gabbro, essexite	5	<10—	30	<21	S	SIMPSON (1954)
Gabbro	9	<5—	60	16	S	WAGER and MITCHELL (1951)
Teschenite, basalt, gabbro	11	<5—	60	11	S	WILKINSON (1959)
		<1—	5,000	75		average for 223 samples

from the Greenville province, KRETZ (1960) reports that the Ba content decreases in the following sequence: biotite > amphibole > pyroxene. The structure of chain silicates, where Ba occupies Ca positions, seems to accept more Ba at higher temperature of formation; ENGEL and ENGEL (1962) found an increasing Ba content (9, 18, 86, 106 ppm) in hornblendes from metamorphic rocks formed at 400, 500, 525, and 625° C.

Quartz has a rather low Ba content, and it is not certain whether contamination by other minerals was avoided in all analyses published.

From minerals of metamorphic origin, garnets exhibit the highest Ba values. Kyanites, sillimanites, and andalusites are much lower and always contain less than 100 ppm Ba.

Calcite from alkalic rocks in the form of carbonatites is somewhat enriched in Ba. Crystals from hydrothermal veins generally have very low Ba concentrations. The very rare Ca-Ba-carbonates, alstonite, barytocalcite, and benstonite are of no importance in rock formation and do not form solid solution series with the end members.

Depending on type and environment of formation, zeolites contain different amounts of Ba in their structures. Upto 300 ppm Ba were found in analcime and natrolite (ERICKSON and BLADE, 1963; WILKINSON, 1959). In stilbite, thomsonite, chabasite, gmelinite, and gismandine, concentrations between 600 and 5,200 ppm Ba were reported (Hoss and ROY, 1960). Phillipsites, or constituents of the series phillipsite-harmotome, which occur in manganese nodules and in larger amounts in certain deep sea sediments, may be extremely enriched in Ba, upto 44,500 ppm (SHKABARA, 1950; Hoss and ROY, 1960).

c) Barium Partition between Mineral and Host Rock and Between Coexisting Minerals

During the first stages of differentiation of basaltic magmas, Ba is enriched in the liquid phase. With progressing crystallization Ba is incorporated, especially in K feldspars and micas, which extract this element from the melt. Pegmatitic stages of differentiation series are consequently often impoverished in Ba.

Barium 56-D-13

Table 56-D-4. *Barium distribution between minerals and total rock*

Pair		No. of pairs	Barium concentration range		Distribution coefficient Ba_{min}/Ba_{rock}		Method	Reference
mineral	rock		mineral ppm	rock ppm				
Sanidine	Trachyte	7	140—6,900	120—1,800	1.17—	8.95	S	BERLIN and HENDERSON (1969)
Plagioclase	Trachyte	5	560—3,500	770—1,100	0.72—	3.98	S	BERLIN and HENDERSON (1969)
Anorthoclase	Phonolite	2	2,600—3,000	1,100—1,400	2.14—	2.37	S	BERLIN and HENDERSON (1969)
Biotite	Dacite, rhyolite trachyte	8	1,400—6,700	400—1,100	1.6 —	15	M	CARMICHAEL (1967a)
Garnet	Eclogite	11	150— 165	<10—1,140	0.13—>16.3		S	HAHN-WEINHEIMER and LÜCKE (1963)
Amphibole	Rhyodacite	1	180	580	0.31		S	HASLAM (1968)
Biotite	Metaarkose	7	472—1,086	1,190—1,730	0.29—	0.65	S	HASLAM (1968)
Pseudoleucite	Juvite, tinguaite	7	50—4,000	50—6,500	0.23—	1.00	X	HENDERSON (1965)
Leucite	Leucite-porphyre, leucite-basanite	5	710—2,250	1,950—2,250	0.36—	0.81	X	HENDERSON (1965)
Biotite	Amphibolite	9	575—1,150	270— 760	1.22—	4.60	S	HUNZICKER (1966)
Biotite	Gneiss	19	290—1,150	250—1,050	0.40—	2.44	S	HUNZICKER (1966)
Muscovite	Amphibolites	5	1,000—1,200	270— 760	1.33—	4.08	S	HUNZICKER (1966)
Muscovite	Gneiss	9	390—1,100	310—1,050	0.85—	3.55	S	HUNZICKER (1966)
Plagioclase	Granulite	12	110— 475	160—2,500	0.11—	3.37	S	SEN (1960)
Orthoclase	Granitic gneiss	5	3,330—5,380	706— 731	4.67—	7.38	X	WHITE (1966)
Biotite	Granitic gneiss	2	847—1,110	706— 730	1.20—	1.52	X	WHITE (1966)
Biotite	Biotite gneiss	8	60— 530	103— 380	0.41—	3.95	S	WYNNE-EDWARDS *et al.* (1962)

Table 56-D-5. Barium distribution between coexisting minerals

Pairs mineral I	mineral II	No. of pairs	Rock type	Range mineral I ppm Ba	Range mineral II ppm Ba	Range of distrib. coeffic.	Method	Reference
Biotite	Muscovite	5	Gneiss, migmatite	1,250 [a]	630 [a]	1.98	X	1
		12	Schist, paragneiss	1,525 [a]	1,970 [a]	0.80	X	1
		18	Schist	600 — 1,100	1,400 — 2,900	0.31— 0.56	S	2
Biotite	Orthoclase	7	Migmatite, gneiss	240 — 1,335	1,176 — 6,718	0.20— 0.25	X	3
Biotite	Garnet	8	Paragneiss	350 — 2,200	35 — 1,100	2.00—15.0	S	4
		22	Metam. sequence (42)	580 — 1,410	11 — 38	21.0 —75.8	S	5
Biotite	Hornblende	5	Metamorphic rock	1,000 — 3,000	15 — 60	33 —68	S	6
		20	Gneiss, schist	120 — 2,450	10 — 485	2.50—60	S	7
Biotite (phlogopite)	Pyroxene	2	Phonolite, rhyo-dacite	2,480 + 4,670	21.3 + 55.0	116 +85	I	8
Biotite	Ilmenite	7	Metaarkose	472 — 1,086	7 — 132	0.01 — 0.28	S	9
Alk. feldsp.	Plagioclase	4	Trachyte	2,700 — 6,900	560 — 3,500	0.93 —10.8	S	10
Sanidine	Plagioclase	12	Monz. cumulates	900 —12,100	394 — 1,165	2.30 —15.9	X	11
Sanidine	Amphibole	11	Monz. cumulates	900 —12,100	90 — 1,770	1.50 —22.1	X	11
Orthoclase	Nepheline	3	Alkalic rock	990 — 2,300	100 — 3,400	0.38 —25.0	X	12
Plagiocl.	Amphibole	10	Monz. cumulates	395 — 1,170	415 — 1,780	0.51 — 1.69	X	11
Plagiocl.	Clinopyroxene	3	Oceanite, andesite	23 — 58.2	1.26— 10.2	4.75 —18.7	I	8
Pyroxene	Amphibole	5	Skarn	2.6— 26	12 —>500	<0.016— 2.00	S	13
Pyroxene	Scapolite	19	Skarn	1 — 31	14 — 210	0.013— 0.20	S	13
Clinopyrox.	Olivine	2 (6)	Ankoranite, alkali basalt	4.9— 175	1.9 — 5.1	2.59 —34.5	I	8

Reference: 1. RIMSAITE (1964); 2. BUTLER (1967); 3. WHITE (1966); 4. ENGEL and ENGEL (1960); 5. TUREKIAN and PHINNEY (1962); 6. HIETANEN (1971); 7. MOXHAM (1965); 8. PHILPOTTS and SCHNETZLER (1970a); 9. SCHWARCZ (1966); 10. BERLIN and HENDERSON (1969); 11. JASMUND and SECK (1964); 12. PERCHUCK and RYABCHIKOV (1968); 13. SHAW et al. (1963).

[a] average.

The degree of relative incorporation of an element into a specific mineral is given by the ratio:

$$D = \frac{\text{concentration of Ba in the phenocryst}}{\text{concentration of Ba in the melt}}.$$

This distribution coefficient can be calculated if equilibrium phenocrysts are compared with total ground mass, for instance, of volcanic rocks. If the amount of phenocrysts is low, the Ba content of the total rock can be used in the denominator of the above given equation. Distribution coefficients can also be calculated for metamorphic rocks, which constitute equilibrium mineral assemblages.

A number of these coefficients have been calculated from published data (Table 56-D-4). It must be mentioned that the distribution coefficients depend on several factors: pressure, temperature, the amount of Ba available, presence of competing elements, etc.

BERLIN and HENDERSON (1969) determined Ba in sanidine and plagioclase crystals, and the embedding ground mass of trachytes and phonolites. They could reconstruct, by Ba and Sr determinations, the sequence of crystallization (sanidine or plagioclase first) and even detected in one case indications for secondary redistribution.

PHILPOTTS and SCHNETZLER (1970a), analyzing (I) phenocrysts and a matrix of basic to intermediate volcanic rocks, calculated partition coefficients (Ba concentration in mineral/Ba concentration in matrix) for: plagioclases, $D = 0.0537$ to 0.589 (9 pairs); K feldspars, $D = 6.12$; clinopyroxenes, $D = 0.0129$ to 0.388 (10 pairs); orthopyroxenes, $D = 0.121$ and 0.141; micas (biotite-phlogopite), $D = 1.09$, 6.36 15.3; hornblendes, $D = 0.0996$, 0.417, 0.731; garnet, $D = 0.0172$; and olivines, $D = 0.00864$ and 0.0112. Only K feldspars and micas concentrate Ba relative to the matrix. They observed a strong coherence of Ba and K in almost all phenocrysts. The D-value of the K/Ba ratio is within the range of 0.5 to 2.0 in nearly all their examples. Since this K/Ba ratio is found to be extremely consistent in basic rock types, these authors recommend it as an "indexing criterion" for solar systems.

Ba distribution coefficients between coexisting minerals (Table 56-D-5) give some idea of the relative importance of minerals for the Ba content of a rock. It must be kept in mind that analyses of mineral phases isolated from a certain rock do not necessarily represent equilibrium conditions. Ba concentration in a definite mineral is affected by pressure, temperature, Ba availability, structure and position in the sequence of separation from the melt, and other parameters.

A useful index of fractionation is the ratio Ba/Rb. Ba has a tendency to be captured in early K minerals, whereas Rb is enriched in the residual melts due to its smaller charge and larger ion size. TAYLOR and HEIER (1960) report a variation of the Ba/Rb ratio in feldspars from gneisses, granites and pegmatites from 54 to 0.04.

II. Barium Minerals

Because of their abundance, the most important Ba minerals and consequently the main Ba carriers in the earth's crust are: in igneous rocks, K feldspars, which contain a certain percentage of celsian molecules; and in sedimentary rocks and hydrothermal deposits, barite. Under special conditions, a large number of well defined Ba minerals can form. Sometimes they are reported from only one locality. In Table 56-D-6 the nonsilicates and silicates are listed. Criteria for the presentation in these lists have been: (1) reported by STRUNZ (1966), and/or (2) approved by the Commission on New Minerals of IMA.

Table 56-D-6. *Barium minerals*

Mineral	Formula
Oxides	
Billietite	$(BaO \cdot 6UO_3) \cdot 11H_2O$
Hollandite	$Ba_2Mn_8O_{16}$
Pandaite	$(Ba, Sr, Ca)(Nb, Ti, Ta)_2O_6 \cdot H_2O$
Priderite	$(K, Ba)_{1.3}(Ti, Fe)_8O_{16}$
Psilomelane	$(Ba, H_2O)Mn_5O_{10}$
Rijkeboerite	$Ba_{1-x}(Ta, Nb)_2O_5(H_2O)$
Todorokite	$(Mn, Mg, Ca, Ba, Na, K)_2Mn_5O_{12} \cdot 3H_2O$
Carbonates	
Alstonite (Ba-aragonite)	$BaCa[CO_3]_2$
Barytocalcite	$BaCa[CO_3]_2$
Benstonite	$(Ca, Mg, Mn)_7(Ba, Sr)_6[CO_3]_{13}$
Burbankite	$(Na, Ca, Sr, Ba, Ce)_6[CO_3]_5$
Carbocernaite	$(Ca, RE, Na, Sr, Ba)[CO_3]$
Ewaldite	$Ba(Ca, RE, Na, K, Sr, U, \square)[CO_3]_2$
Huanghoite	$BaCe[CO_3]_2F$
Kordylite	$Ba(Ce, La, Nd)_2F_2/[CO_3]_3$
Mckelveyite	$Na_2Ba_4CaY_2[CO_3]_9 \cdot H_2O$
Norsethite	$BaMg[CO_3]_2$
Stenonite	$(Sr, Ba, Na)_2Al[CO_3]F$
Witherite	$BaCO_3$
Nitrate	
Nitrobarite	$Ba[NO_3]_2$
Sulfate[a]	
Barite	$BaSO_4$
Selenite	
Guilleminite	$Ba(UO_2)_3[SeO_3]_2(OH)_4 \cdot 3H_2O$
Phosphates, Arsenates, Vanadates	
Babefphite	$Be_5Ba_4[PO_4]_4O\ F_4 \cdot 0.35H_2O$
Bergenite[b]	$Ba(UO_2)_4[PO_4]_2(OH)_4 \cdot 8H_2O$
Dussertite	$BaFe^{3+}_3H[AsO_4]_2(OH)_6$
Ferrazite[c]	$(Pb, Ba)_3[PO_4]_2 \cdot 8H_2O$
Francevillite	$(Ba, Pb)(UO_2)_2[VO_4]_2 \cdot 8H_2O$
Gamagarite	$Ba_4(Fe, Mn)_2U_4O_{15}(OH)_2$
Gorceixite	$BaAl_3(OH)_5[PO_4]_2 \cdot H_2O$
Heinrichite	$Ba(UO_2)_2[AsO_4]_2 \cdot 10H_2O$
Metaankoleite	$(K, Ba)(UO_2)_2[PO_4]_2 \cdot 6H_2O$
Metaheinrichite	$Ba(UO_2)_2[AsO_4]_2 \cdot 8H_2O$
Metauranocircite I	$Ba(UO_2)_2[PO_4]_2 \cdot 8H_2O$
Metauranocircite II	$Ba(UO_2)_2[PO_4]_2 \cdot 6H_2O$
Strontiumapatite[b]	$(Sr, Ba)_6(Ca, RE, Mg, Na)_4[PO_4]_6(F, OH)_2$
Uranocircite I	$Ba(UO_2)_2[PO_4]_2 \cdot 12H_2O$
Uranocircite II	$Ba(UO_2)_2[PO_4]_2 \cdot 10H_2O$
Vesignietite	$BaCu_3[VO_4]_2(OH)_2$
Weilerite[d]	$BaAl_3H_{0-1}[AsO_4, SO_4]_2(OH)_{7-6}$

Table 56-D-6. (Continued)

Mineral	Formula
Nesosilicates	
Bariumuranophane	$BaH[UO_2/SiO_4]_2 \cdot 5H_2O$
Garrelsite	$(Ba, Ca)_4H_6Si_2B_6O_{20}$
Sorosilicates	
Bafertisite	$BaFe_2TiSi_2O_9$
Barylite	$BaBe_2Si_2O_7$
Hyalotektite	$(Pb, Ca, Ba)_4B[Si_6O_{17}](F, OH)$
Innelite	$Ba_2(Na, K, Mn, Ti)_2Ti(O, OH, F)_2[(S, Si)O_4/Si_2O_7]$
Labuntsovite	$(K, Ba, Na)(Ti, Nb)(Si, Al)_2(O, OH)_7 \cdot H_2O$
Nenandkevichite	$(Na, K, Ca, Ba)(Nb, Ti)[Si_2O_7] \cdot 2H_2O$
Shcherbakovite	$(K, Na, Ba)_3(Ti, Nb)_2[Si_2O_7]_2$
Yoshimuraite	$(Ba, Sr)_2(Mn, Fe, Mg)_2(Ti, Fe)(OH, Cl)_2[(S, P, Si)O_4/Si_2O_7]$
Ring silicates	
Armenite	$BaCa_2Al_6Si_8O_{28} \cdot 2H_2O$
Baotite	$Ba_4(Ti, Nb)_8Si_4O_{28}Cl$
Benitoite	$BaTiSi_3O_9$
Cappelenite	$(Ba, Ca, Ce, Na)_3(Y, Ce, La)_6[BO_3]_6[Si_3O_9]$
Muirite	$Ba_{10}Ca_2MnTiSi_{10}O_{30}(OH, Cl, F)_{10}$
Papstite	$Ba(Sn, Ti)Si_3O_9$
Taramellite	$Ba_2(Fe^{3+}, Ti, Fe^{2+})_2(OH)_2[Si_4O_{12}]$
Traskite	$Ba_9Fe_2Ti_2Si_{12}O_{36}(OH, Cl, F)_6 \cdot 6H_2O$
Verplanckite	$Ba_2(Mn, Fe, Ti)Si_2O_6(O, OH, Cl, F)_2 \cdot 3H_2O$
Chain silicates	
Batisite	$Na_2BaTi_2[Si_2O_7]_2$
Krauskopfite	$BaSi_2O_5 \cdot 3H_2O$
Walstromite	$BaCa_2Si_3O_9$
Sheet silicates	
Anandite	$(Ba, K)(Fe, Mg)_3(Si, Al, Fe)_4O_{10}(O, OH)_2$
Bariumphlogopite	$(K, Ba)Mg_3(F, OH)_2[AlSi_3O_{10}]$
Barium-vanadium-muscovite	$(K, Na, Ba)(Al, Ti, V, Mg)_2(OH)_2[AlSi_3O_{10}]$
Gillespite	$BaFe[Si_4O_{10}]$
Oellacherite	$(K, Ba)(Al, Mg)_2 (OH, F)_2[AlSi_3O_{10}]$
Sanbornite	$Ba_2[Si_4O_{10}]$
Tectosilicates (without zeolites)	
Banalsite	$BaNa_2[Al_2Si_2O_8]$
Celsian	$Ba[Al_2Si_2O_8]$
Paracelsian	$Ba[Al_2Si_2O_8]$
Bariumalbite	solid solution Ab-Or-Ce
Bariumplagioclase	solid solution Ab-An-Ce
Calciocelsian	solid solution Ce-An
Hyalophane	solid solution Or-Ce
Bariumsanidine	solid solution
Cymrite	$BaAlSi_3O_8(OH)$
Wenkite	$(Ba,Ca)_9[SO_4]_2Al_9Si_{12}O_{42}(OH)_5$

Table 56-D-6. (Continued)

Mineral	Formula
Zeolites	
Barium heulandite	$(Ca,Ba)[Al_2Si_7O_{18}] \cdot 6H_2O$
Brewsterite	$(Sr, Ba, Ca)_2\ Al_4Si_{12}O_{32} \cdot 10H_2O$
Edingtonite	$Ba\ Al_2Si_3O_{10} \cdot 3H_2O$
Harmotome	$Ba_2\ Al_4Si_{12}O_{32} \cdot 12H_2O$
Wellsite	solid solution phillipsite-harmotome
Unclassified silicates	
Fresnoite	$Ba_2TiSi_2O_8$
Joaquinite	$NaBa(Ti, Fe)_3Si_4O_{15}$
Leukosphenite	$BaNa_4(TiO)_2[Si_2O_5]_5$
Macdonaldite	$BaCa_4Si_{15}O_{35} \cdot 11H_2O$
Tienshanite	$Na_2BaMnTiB_2Si_6O_{20}$

Calciobarite (Ca), Baritocelestite, Celestobarite (Sr), Baritoanglesite, Anglesobarite, Hokutolite, Weisbachite (Pb), Radiobarite (Ra). Reviews on composition, occurence, and crystallographic and physical properties are given by DANA (1951), HINTZE (1930, 1938, 1960). X-ray evidence for the existence of a barite-celestite isomorphous series has been obtained by SABINE and YOUNG (1954).

BOSTRÖM *et al.* (1968) studied subsolidus phase relations and lattice constants in the system $BaSO_4$-$SrSO_4$-$PbSO_4$.

[a] Barite forms more or less continuously, solid solutions with Ca, Sr, Ra, and Pb sulfates. In keeping with respective compositions, different names are used for the members of the series:

[b] No decisive vote of the IMA new mineral commission.

[c] Doubtful mineral species.

[d] Not yet approved by IMA new mineral commission.

56-E. Abundance in Common Igneous Rock Types

A survey of Ba concentrations of rocks published by v. ENGELHARDT (1936) showed that Ba content in igneous rock series normally increases with increasing SiO_2 concentration. Recent data of Ba concentrations in common igneous rocks are summarized in Tables 56-E-1 and 2. The grouping follows the outlines of WEDEPOHL in volume 1 of this handbook. The listing contains the range of individual values, the arithmetic means and the standard deviation (s) of these means. Ba distribution in the main rock types is briefly discussed, as well as its bearing on genetic interpretations.

I. Ultramafic Rocks

The most important ultramafic rocks in the development of magmatic series are dunites and peridotites. From the available data, mostly spectrographic, an average of 8.8 ppm for dunite and of 25 ppm for peridotite was calculated. In their table of elemental distribution in the earth's crust, TUREKIAN and WEDEPOHL (1961) preferred a neutron activation value of 0.4 ppm Ba for dunite rather than a spectrographic determination average of 6 ppm.

All data for these rocks suffer from potential contamination (see discrepancy between finds and falls of chondrites) and from analytical difficulties occurring close to the detection limit of Ba.

Pyroxenites contain a little more Ba (average 23 ppm); biotite pyroxenites have concentrations upto 3,200 ppm Ba (HIGAZY, 1954), as do kimberlites, where phlogopite is the main Ba carrier. Very high values are also reported for carbonatites (average 3,520 ppm) which sometimes even contain barite.

II. Gabbroic and Basaltic Rocks

Gabbroic rocks of intrusive occurence (average Ba content 246 ppm) closely resemble continental tholeiitic basalts (average Ba content 246 ppm).

Basalts can be divided into three groups according to their trace element concentration (Fig. 56-E-1). The averages for Ba are as follows:

Oceanic tholeiitic basalts	14.5 ppm Ba
Tholeiitic basalts of continents and oceanic islands	246 ppm Ba
Alkali basalts	613 ppm Ba.

All trace element data for oceanic tholeiitic basalts indicate a general uniformity for widely separated parts of the oceans. MUIR *et al.* (1964), however, determined Ba from the rift zone of the Midatlantic Ridge, at 45° N, and found 55 to 220 ppm (S). The continental tholeiitic basalts contain more Si, K, Ba, Cs, Pb, and Rb than oceanic tholeiitic basalts (GAST, 1960).

© Springer-Verlag Berlin · Heidelberg 1972

Table 56-E-1. *Barium in intrusive rocks*

Rock type	No. of localities or No. of mean values used	No. of individual values	Range of individual values in ppm	Arith. mean of means grouped by locality in ppm	s[a]	References[b]
Alkali granite	3	19	22— 2,100	857		11, 12, 73
Granite	28	608	22— 3,000	732	453	6, 17, 21, 24, 26, 31, 32, 49, 50, 54, 55, 57, 62, 67, 72, 73, 75, 80, 83, 85, 87, 89, 93, 106, 121, 126
Granodiorite	7	53	400— 1,815	888		6, 73, 83, 87, 124, 131
Quartzdiorite	5	18	150— 1,250	811		87, 90, 131
Quartzmonzonite	3	18	233— 6,000	1,605		12, 87, 124
Alkali syenite	4	10	300— 2,070	1,067		21, 34, 75, 109
Syenite	9	30	230—18,000	2,753		24, 34, 44, 58, 63, 97, 100, 109, 111
Monzogabbro	1	4	7,500—14,000	10,360		100
Diorite	3	25	126— 1,150	714		75, 87, 90
Gabbro	10	148	5— 1,250	246	228	4, 21, 75, 87, 91, 100, 101, 109, 111, 127
Anorthosite	3	66	65— 475	171		75, 91
Norite	4	18	5— 310	78		75, 87, 99
Nepheline syenite	12	524	10— 6,300	1,427	1,206	27, 34, 35, 43, 44, 58, 59, 61, 75, 111, 125
Essexite	3	8	850— 2,800	1,598		64, 88, 111
Teschenite	1	3	650— 2,000	1,500		63
Ijolite, melteigite, jacupyrangite	5	66	88— 2,430	973		27, 35, 43, 44, 75
Dunite	9	18	0.3— 40	8.8		34, 39, 56, 75, 98, 100
Peridotite	6	9	1— 70	38		38, 39, 64, 75, 98
Pyroxenite	2	13	5.5— 67	23		75, 98
Kimberlite	4	35	20— 2,860	847		23, 29, 53, 68
Carbonatite	9	215	88—54,000	3,799		23, 27—29, 35, 41, 46, 64, 69, 105, 123

[a] s is the standard deviation of the mean values of the different authors, standard deviations are given only when 10 or more mean values were available.

[b] For list of references see Page 56-E-4.

Table 56-E-2. *Barium in volcanic rocks*

Rock type	No. of localities or No. of mean values used	No. of individual values	Range of individual values in ppm	Arith. mean of means grouped by locality in ppm	s[a]	References[b]
Alkali rhyolite	8	126	1— 700	118		25, 30, 36, 86, 102, 103, 109
Rhyolite	20	153	5— 3,650	1,127	632	8, 9, 18, 25, 26, 30, 74, 79, 88, 89, 92, 103, 108, 109, 112, 121
Rhyodacite	3	12	550— 1,800	1,210		18, 30, 89
Dacite	7	22	150— 1,250	629		87, 89, 118, 122
Quartz latite	1			550		112
Alkali trachyte	2	2	800+ 1,500	1,150		88, 128
Trachyte	10	30	20— 3,000	1,177	667	3, 36, 71, 88, 92, 107, 128
Latite minette	3	4	137— 2,500	1,379		18, 88, 112
Latite andesite	2	7	300— 2,250	841		88, 112
Andesite	23	185	80— 2,700	703	475	2, 3, 8, 39, 45, 51, 74, 79, 88, 89, 108, 110, 112, 115, 118, 120, 122, 128
Tholeiitic basalt	51	555	20— 1,160	246	197	3, 5, 8, 10, 13—16, 19, 20, 22, 37, 48, 51, 52, 60, 65, 74, 76, 77, 81, 87—89, 92, 94—96, 99, 113, 114, 117, 119, 127, 128
Alkali basalt	16	100	30— 1,350	613	223	1, 3, 33, 51, 66, 76, 78, 81, 88, 116, 117, 129, 130, 132
Oceanic tholeiite	41	41	3— 46	14.5	11.3	33, 42, 70, 82, 84
Phonolite	5	42	10— 2,000	999		27, 44, 71, 88, 104
Nepheline basanite Nepheline tephrite Leucite basanite Leucite tephrite	8	95	250— 9,360	1,976		7, 40, 64, 88, 104, 107, 130, 133
Nephelinite	2	5	600— 5,900	3,444		64, 130
Ankaratrite	1	2	1,800+ 4,000	2,900		64
Leucitite	1	9	1,800— 7,000	3,510		64
Melilitite	1	2	2,200+10,000	6,100		64
Alnöite	2	22	450— 6,930	1,890		27—29, 47

[a] s is the standard deviation of the mean values of the different authors, standard deviations are given only when 10 or more mean values were available.
[b] For list of references see Page 56-E-4.

References (methods in brackets): 1. BAKER, 1969 (X); 2. BAKER, 1968 (S); 3. BAKER et al., 1964 (S); 4. BARAGAR, 1960 (S); 5. BARTEL et al., 1963 (S); 6. BRÄUER, 1965; 7. BROWN and CARMICHAEL, 1969 (X); 8. BYERS, JR., 1961 (S); 9. CARMICHAEL and MCDONALD, 1961 (S); 10. CLARKE, 1970; 11. CLIFFORD et al., 1962 (S); 12. CLIFFORD et al., 1969 (S); 13. COATS, 1952 (S); 14. COATS, 1953 (S); 15. COATS, 1959 (S); 16. COATS et al., 1961 (S); 17. COCCO, 1953 (S); 18. CORNWALL, 1962 (S); 19. CORNWALL and ROSE JR., 1957 (S); 20. COX and HORNUNG, 1966 (S, X); 21. COX et al., 1965 (S); 22. COX et al., 1967 (S); 23. DAWSON, 1962 (S); 24. DIETRICH and HEIER, 1967; 25. DIXON et al., 1968 (S); 26. DUNHAM, 1968 (S); 27. v. ECKERMANN, 1948 (S); 28. v. ECKERMANN, 1966 (S); 29. v. ECKERMANN, 1967 (S); 30. EL-HINNAWI, 1969 (S); 31. EMILIANI and VESPIGNANI, 1964 (S); 32. EMMERMANN, 1968 (X); 33. ENGEL et al., 1965 (S); 34. v. ENGELHARDT, 1936 (S); 35. ERICKSON and BLADE, 1963 (S); 36. EWART et al., 1968 (S); 37. FAIRBAIRN et al., 1953 (S); 38. FISHER and ENGEL, 1969 (S); 39. FLANAGAN, 1969 (S, N/R, X); 40. FORNASERI et al., 1963 (W); 41. GARSON, 1967; 42. GAST, 1965; 43. GERASIMOVSKII, 1966 (S); 44. GERASIMOVSKII and BELYAEV, 1963 (S); 45. GIUSCA and IONESCU, 1965; 46. GOLD, 1963; 47. GOLD, 1967; 48. GREENLAND and LOVERING, 1966 (S); 49. GROHMANN and SCHROLL, 1966 (S); 50. GROUT, 1935 (W); 51. GUNN, 1965 (X); 52. GUNN, 1966 (X); 53. HAHN-WEINHEIMER, 1959 (S); 54. HAHN-WEINHEIMER and ACKERMANN, 1967 (X); 55. HALL, 1967 (X); 56. HAMAGUCHI et al., 1957 (N/R); 57. HEIER, 1960 (S); 58. HEIER, 1964 (S); 59. HEIER, 1965 (S); 60. HEIER et al., 1966 (S); 61. HENDERSON, 1965 (S); 62. HERZ and DUTRA, 1960 (S); 63. HIGAZY, 1952 (S); 64. HIGAZY, 1954 (S); 65. HOTZ, 1953 (S); 66. HUCKENHOLZ, 1969 (X); 67. HÜGI and SWAINE, 1963 (S); 68. JANSE, 1962; 69. JOHNSON, 1961 (S); 70. KAY et al., 1970 (I); 71. KING and SUTHERLAND, 1967; 72. KOLBE, 1964; 73. KOLBE and TAYLOR, 1966 (S); 74. KUNO et al., 1957 (S); 75. LIEBENBERG, 1960 (S); 76. LIPMAN, 1969 (S); 77. MACDONALD and EATON, 1964 (S); 78. LE MAITRE, 1962 (S); 79. MARKHININ et al., 1964 (S); 80. MARMO and SIIVOLA, 1966 (S); 81. MATHIAS, 1957 (S); 82. MELSON et al., 1968 (S); 83. MOENKE, 1960 (S); 84. MUIR et al., 1964 (S); 85. MUKHERJEE, 1968 (S); 86. NOBLE and HAFFTY, 1969 (S); 87. NOCKOLDS and ALLEN, 1953 (S); 88. NOCKOLDS and ALLEN, 1954 (S); 89. NOCKOLDS and ALLEN, 1956 (S); 90. OKRUSCH and RICHTER, 1969 (X); 91. PAPEZIK, 1965 (S); 92. PATTERSON, 1951 (S); 93. PATTERSON, 1953 (S); 94. PATTERSON and SWAINE, 1955 (S); 95. PATTERSON et al., 1955 (S); 96. PECK et al., 1966 (S); 97. PECORA, 1962 (S); 98. PINSON et al., 1953 (S); 99. PRINZ, 1964 (S); 100. READ and HAQ, 1963 (S); 101. READ et al., 1965 (X); 102. RENFREW et al., 1966 (S); 103. RENFREW et al., 1968 (S); 104. RIDLEY, 1970 (S, X); 105. RUSSELL et al., 1954 (S); 106. SAHAMA, 1945 (S); 107. SAVELLI, 1967 (X); 108. SHELTON, 1955; 109. SIEDNER, 1965 (S); 110. SIEGERS et al., 1969 (X); 111. SIMPSON, 1954 (S); 112. SINHA and TIWARI, 1964 (S); 113. SINHA and KARKARE, 1964a (S); 114. SINHA and KARKARE, 1964b (S); 115. SMITH, 1964 (S); 116. SMITH and CARMICHAEL, 1969 (X); 117. SNAVELY JR. et al., 1968 (S); 118. STARITSIN, 1964 (S); 119. STARK and TRACEY, 1963 (S); 120. TAYLOR and WHITE, 1966 (S); 121. TAYLOR et al., 1968 (S); 122. TAYLOR et al., 1969 (S); 123. TEMPLE and GROGAN, 1965 (X); 124. TOWNEND, 1966 (S); 125. VLASOV et al., 1966 (S); 126. VOLBORTH, 1962 (W); 127. WAGER and MITCHELL, 1951 (S); 128. WAGER and MITCHELL, 1953 (S); 129. WEDEPOHL, 1954 (S); 130. WEDEPOHL, 1961 (S); 131. WEIBEL, 1960 (S); 132. WILKINSON, 1959 (S); 133. WILKINSON, 1968.

Differences in Ba concentration among tholeiitic basalts of particular regions were found in South Africa, where Rhodesian tholeiites show unusually high mean Ba values upto 1,020 ppm (Cox et al., 1967), while the basalts of Basutoland and Swaziland are of normal tholeiitic geochemistry. GREENLAND and LOVERING (1966) studied differentiation within a tholeiitic sill in Tasmania (Australia). Ba concentrations increase here, from the bottom to the 1735 ft-high top of the flow, from 160 to 500 ppm. Elements with similar enrichment trends are F and Ga, whereas nonparallel behavior was found for Ni, Co, Cr and Sc.

Barium

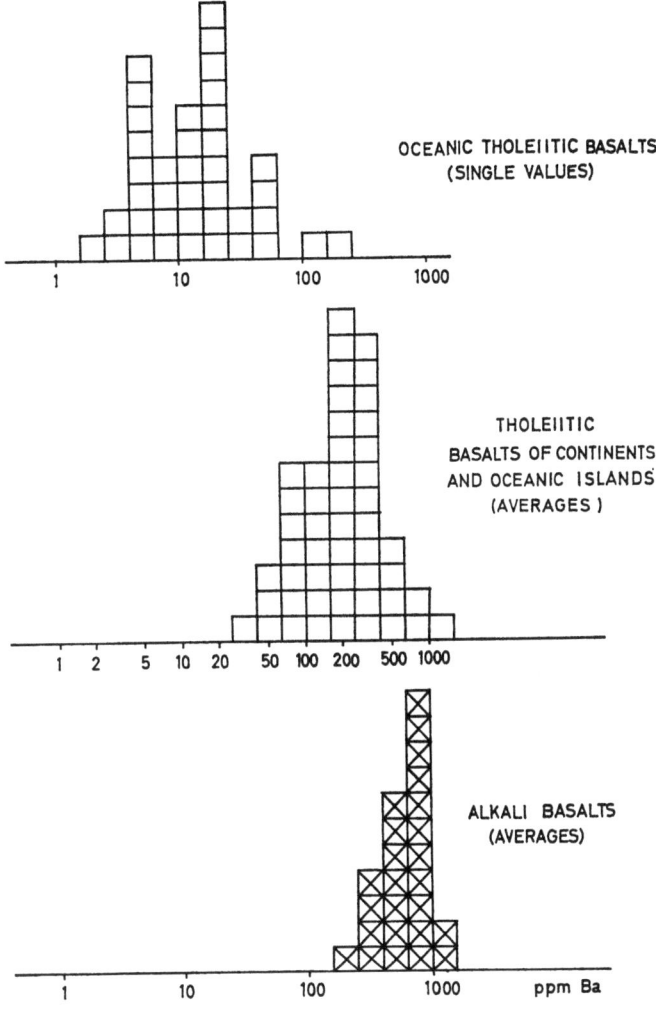

Fig. 56-E-1. Barium distribution in basaltic rocks

ENGEL et al. (1965) gave the ratio of the masses of alkali olivine basalts to tholeiitic basalts as 2:98. Consequently, an average of 253 ppm Ba is to be assumed for average continental basaltic rock. PRINZ (1967), in his summary of trace element data for basalts, found an arithmetic mean for all basalts of 303 ppm and a geometric mean of 220 ppm for 253 analyses. From the wide range of Ba concentrations within similar petrographic types and from similar Ba values for different petrographic types within any region, it is indicated that differences in the initial abundance of Ba in basaltic magmas exist (PRINZ, 1967). The generation of such differences is attributed to different degrees of partial melting, mantle inhomogeneity or wall rock reactions (JAMIESON and CLARKE, 1970).

Anorthosites and norites are both lower in Ba than tholeiitic basalts. The normally observed relationship between K and Ba is not found in Canadian anorthosites. In these rocks, which are rich in Ca, the Ba-Ca diadochy is superimposed on the more common Ba-K relation.

III. Granitic Rocks

a) High Ca Content

Granitic rocks with a high Ca content, as in granodiorites and quartz-diorites, are usually high in Ba. Averages are 888 ppm and 811 ppm, respectively; within particular areas the Ba content may vary considerably. An average for both rock types, calculated with the percentages of WEDEPOHL (1969), is 873 ppm.

b) Low Ca Content

Granitic rocks with low Ca content, which are represented by granites and quartzmonzonites, show extremely different Ba values (Fig. 56-E-2). This may be

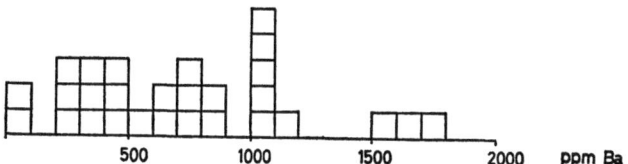

Fig. 56-E-2. Ba distribution in granites (28 local means, see Table 56-E-1)

partly due to origin and to differences in the respective processes of formation. Metasomatical alterations may also have changed the Ba content in some granites (cf. EMMERMANN, 1969). The granite average from 28 regional mean values is 732 ppm. Within certain granite bodies, the Ba concentration may exhibit a narrow spread, while the means are low or high. Gradational change of Ba content was also observed (EMMERMANN, 1969). For three quartzmonzonite areas, an average of 1,605 ppm was calculated.

The effusive equivalents of granitic rocks contain more Ba (average 1,127 ppm) than granites. The single quartz latite value, however, is much lower than the average quartzmonzonite content.

IV. Intermediate Rocks

Intermediate rocks, such as syenites and trachytes, are strongly enriched in Ba. While the syenite average is 2,753 ppm, the effusive equivalent averages only 1,177 ppm. TUREKIAN and WEDEPOHL (1961) gave a value of 1,600 ppm, which was calculated from data of v. ENGELHARDT (1936) and SAHAMA (1945).

V. Alkalic Rocks

Alkalic rocks are all considerably enriched in Ba. All averages are higher than 1,000 ppm, with nepheline-syenite showing 1,427 ppm, phonolite 1,000 ppm and nepheline-basanite 1,976 ppm Ba. In rocks of the Lovozero alkali massif (USSR), Ba is enriched upto 3,300 ppm (VLASOV et al., 1966). The Kola alkali complex (USSR) contains upto 1,600 ppm Ba in its nepheline-syenites and upto 1,350 ppm Ba in the foyaites (GERASIMOVSKII and BELYAEV, 1963).

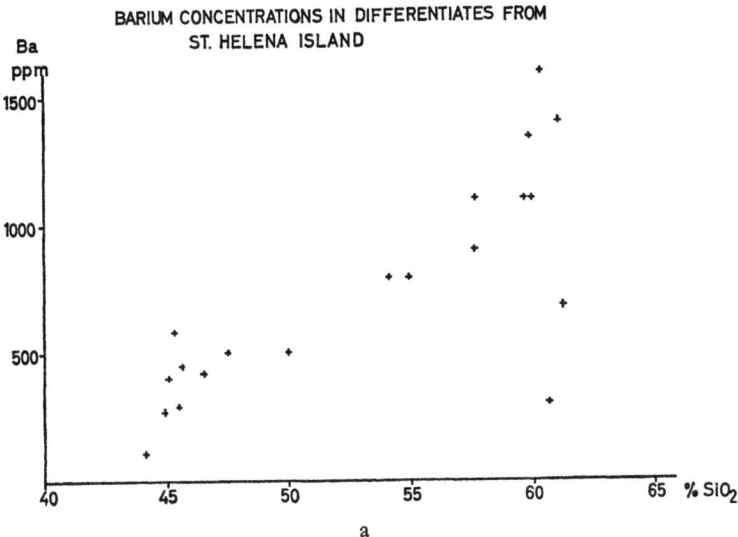

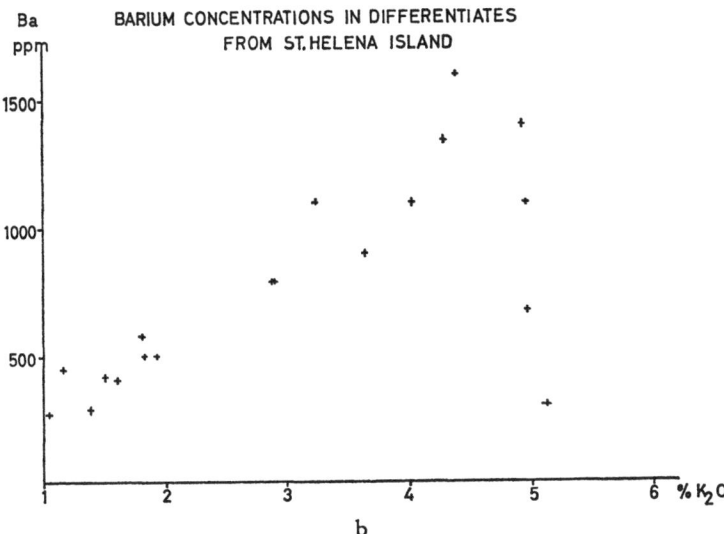

Fig. 56-E-3. Correlation of Ba concentration and SiO_2 (upper plot) and K_2O content in a differentiation series (BAKER, 1969)

Many investigators (v. ENGELHARDT, 1936; WAGER and MITCHELL, 1951; NOCKOLDS and ALLEN, 1953, 1954, 1956; WILKINSON, 1959; MARKHININ et al., 1964; P. E. BAKER, 1968) found that Ba concentrations increase during progressing differentiation. One example of correlation between concentrations of Ba and SiO_2 and K_2O (as parameter of differentiation) is given for volcanic rocks of Saint Helena island, South Atlantic (Fig. 56-E-3).

Sometimes Sr shows the same trend, while a nonparallel behavior is observed for Ca, V, Cu, Co, Ni, and Sc. In several instances the relation between Ba and other elements is not as simple; plotting versus differentiation, solidification, mafic index or Larsen factor provides a better insight into differentiation behavior.

In series which proceed far in differentiation, Ba concentrations normally pass a maximum. Since Ba substitutes mostly for K, this indicates that a distribution coefficient $\left(\frac{Ba}{K}\right)_{crystal} \Big/ \left(\frac{Ba}{K}\right)_{melt}$, larger than unity is effective in acid magmas for at least one major mineral — mostly potash feldspar or mica.

This type of Ba distribution prevails in the example given in Fig. 56-E-3, in the Scottish Caledonian rocks, and the East Central Sierra Nevada rocks analyzed by NOCKOLDS and ALLEN (1953). As a consequence of this differentiation behavior, Ba is always very low in pegmatites of truly magmatic origin (cf. Table 56-D-1 and v. ENGELHARDT, 1936). A trend of K/Ba ratios to increase from 26 to 36 in a tholeiitic sequence was found by GUNN (1965).

Data for averages of Ba concentration in igneous rocks have been subject to changes in the last year due to the fact that more and better data have been published. Consequently, the mean calculated Ba content in igneous crustal rocks has changed, since better insight into the relative abundance of rock types was obtained. With the Ba concentrations of Table 56-E-1, one arrives at a Ba average of 728 ppm for the upper continental crust of the earth, using the data for the relative amounts of intrusive rocks by WEDEPOHL (1969, handbook; Tables 7—8). Examples for Ba distribution between minerals of certain rocks are given in Table 56-E-3.

Compilation and interpretation of Ba data from literature suffer somewhat from the difficulties in analytical determination. This is shown graphically by the Ba values for the standard rocks G1 and W1 (Fig. 56-E-4). The considerable spread of values — even with the same analytical method — cannot be explained by too coarse grain size of the reference samples (KLEEMANN, 1967), since similar observations were made with the much more finely ground new standard rocks of the USGS (FLANAGAN, 1969). As the individual authors use their own different values for internal reference, difficulties arise when results from different sources are compared. This must be considered when the data from these paragraphs are evaluated.

Table 56-E-3. *Barium distribution*

Rock type	Total rock Ba ppm	Quartz Ba ppm	Ortho- clase Ba ppm	Plagio- clase Ba ppm	Biotite Ba ppm
Hortonolite ferro gabbro	50			200	
Olivine norite	5			10	
Quartz biotite norite	300			110[a]	1,600
Granodiorite	1,000		3,000	90[a]	2,500
Granite	1,000		2,500	125[a]	2,000
Predazzo granite	70		108	65	100
Adamellite porphyrite	600			275	1,250
Adamellite	110		225		120
Sphen pyroxenite	500	100	900		
Teschenite	275			350	

[a] Quartz and plagioclase.

Barium

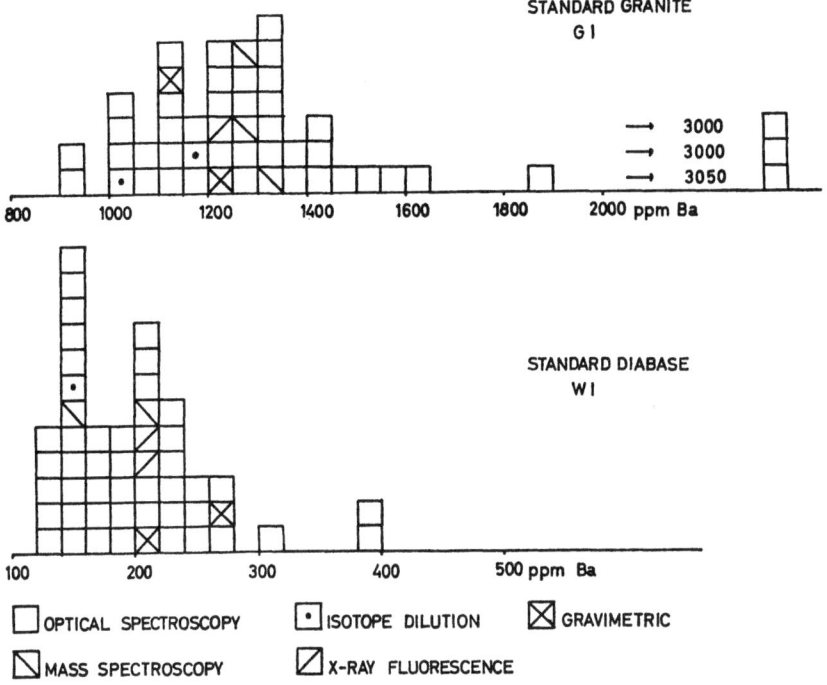

Fig. 56-E-4. Ba determinations on standard rocks (FAIRBAIRN et al., 1951; STEVENS et al., 1960; FLEISCHER and STEVENS, 1962; FLEISCHER, 1965, 1969; FLANAGAN, 1969)

in rocks

Muskovite Ba ppm	Amphibole	Pyroxene Ba ppm	Olivine Ba ppm	Method	References
		10	10	S	WAGER and MITCHELL (1951)
	15			S	SEN et al. (1959)
	5	5		S	SEN et al. (1959)
				S	SEN et al. (1959)
				S	SEN et al. (1959)
				S	EMILIANI and VESPIGNANI (1964)
	120	35			WILKINSON et al. (1964)
10				S	BUTLER (1953a)
		20		S	ERICKSON and BLADE (1963)
		10		S	WILKINSON (1959)

Revised manuscript received: September 1971

56-F. Behavior during Processes Connected with Magmatism

I. Pegmatites

Pegmatites of magmatic origin generally contain less Ba than their embedding wallrocks of magmatic or metamorphic origin. This feature is most apparent in the main Ba carriers, feldspar and mica. By analytical studies of alkali feldspars from the South Norwegian Precambrian basement complex, HEIER and TAYLOR (1959) found that the concentration of Ba in potash feldspars decreases with increasing differentiation. The K feldspars of large pegmatites in the surveyed area always contain less Ba than the host rock. This general trend was diagramatically shown by HEIER (1962) and can also be seen in Fig. 56-D-2, for which additional analyses were used. TAYLOR and HEIER (1960) stressed the importance of the Ba/Rb ratio in feldspars for judging the degree of fractionation. Since Rb is more discriminated against in the K feldspar structure than Ba, Rb is continuously enriched in the fluid during crystallization. Under the assumption of constant distribution coefficients for both elements, this leads to the highest Rb values in the last crystallization. Low Ba values in granite-pegmatite feldspars are also reported by v. ENGELHARDT (1936), OFTEDAL (1958), and TAKUBO and TATEKAWA (1954). Within a pegmatite body, OFTEDAL (1959) found the younger microcline to be poorer in Ba than the older one. Similar Ba impoverishment was found in nepheline-syenite pegmatites (v. ENGELHARDT, 1936; OFTEDAL, 1962) as compared to the lardalite and larvikite from which they are supposed to be derived. Biotite from granite pegmatite also is impoverished in Ba compared to the biotite from the mother granite (TAKUBO and TATEKAWA, 1954).

Ba behavior is different in the small pegmatite bodies which sometimes form by metamorphic processes. In the plagioclase gneiss area of Justøy, Norway, small pegmatitic veins occur with high Ba values. HITCHON (1960) investigated pegmatites of three metamorphic complexes in Scotland. While two complexes showed the usual Ba relation between pegmatite and country rock, pegmatites of the third complex (Laxfordian) were markedly enriched in Ba in all minerals (microcline-perthite 3,785 and 5,620 ppm; oligoclase 365 ppm; biotite 900 ppm, 760 ppm, 1,100 ppm and 2,320 ppm Ba). In these pegmatites, the Ba content of the individual minerals increases towards the core of the respective body. It is possible that in the case of Laxfordian pegmatites metasomatic processes caused a later Ba enrichment in the pegmatite minerals.

II. Metasomatism, Wallrock Alteration, Greisenization

Metasomatic changes of Ba content of rocks occur sometimes with emplacement of pegmatites. Evidence for a large scale metasomatic Ba addition in a rock body was found for the Albtal granite, Germany, by EMMERMANN (1968, 1969). This author investigated the distinct differences in the distribution patterns of Ba in the two occurring K feldspar generations and reached the conclusion that the potash feldspar megacrysts (mean Ba content 4,600 ppm) had grown during a postmagmatic stage from a metasomatic, Ba rich fluid, whereas the groundmass K feldspar (mean Ba content 1,600 ppm) represents the first generation.

Metasomatic alterations in granodiorite, dacite, and gabbro related to serpentinization of ultramafic rocks in the Western United States, generally resulted in a distinct Ba impoverishment. In a few cases, small intermediate zones with Ba enrichment were observed (COLEMAN, 1967).

LUR'YE (1963), analyzing spectrographically the silicic wallrocks of Zambarah zinc-lead ore deposit, Central Asia, found a pronounced decrease in the concentration of Ba and Sr towards the veins. The Ba content drops from 3,000 to 6,000 ppm, in fresh rock containing abundant K feldspar, to 300 ppm as a maximum in completely sericitized, feldspar free rocks close to the ore. He concludes that all barite and its Sr content originates from the feldspar decomposition in the wall rocks. Granitic wallrocks of hydrothermal veins in the Black Forest, Germany, were analysed for Ba distribution by DEGENS (1956). Extensive studies of wallrock alterations were carried out by TOOKER (1963) in the Front Range Mineral Belt, Colorado, USA. His spectrographic analyses showed that Ba, as with other large ions, normally tends to be removed veinward from all Precambrian and Tertiary metamorphic and igneous rocks he investigated. During alteration in the rock, pH drops along with the decrease of K, Ba etc.

Greisenization normally proceeds with the removal of Ba from the affected rock. In a mixture of samples from 24 greisen, v. ENGELHARDT (1936) found an abundance of 160 ppm Ba. This value is far below the content in the respective unaltered igneous silicic rocks. SOLOMON (1966) reports Ba removal from granites (430 ppm and 250 ppm) by greisenization (Ba in the altered rocks: 160 ppm and 105 ppm respectively) in the North Pennine ore field, Great Britain.

III. Ore Deposition

Probably, magmatic-hydrothermal fluids originally do not contain any significant amount of Ba, but obtain this element by leaching suitable wallrocks. The importance of this mechanism, mentioned in the previous paragraph, was already stressed by v. ENGELHARDT (1936). Another mechanism working in sedimentary environments is dissolution of barite from the sediment by bacterial sulfate reduction in suitable diagenetic environment (PUCHELT, 1967). As a third way of producing Ba containing fluids, certain metamorphic reactions may give off Ba because structures which had incorporated high Ba concentrations become unstable.

The most common Ba ore is barite. It is deposited by fluids with a high oxidation potential, where sulfur is present as sulfate. Suitable conditions of this kind occur close to the earth's surface or in subsurface areas where sulfate solutions mix with reduced Ba containing waters (cf. Subsection 56-I-IV). In solution, Ba migrates to the region of sulfate stability and thus is often bound to a narrow zone close to the earth's surface.

Modes of barite formation and possibilities of $BaSO_4$ transport in solution are discussed by PUCHELT (1967). Barite contains, in all cases, certain amounts of isomorphous Sr. Concentrations of this element in marine barites (upto 3.36%) are summarized by CHURCH (1970). STARKE (1964) analysed a large number of vein barites which contain upto 12% $SrSO_4$. The conditions necessary for witherite formation from ore forming fluids are not often fulfilled. Calculations for $BaCO_3$ formation, at 250°C and CO_2 fugacities of 0.1, 1.0 and 100 atm., were carried out by HOLLAND (1965) for varying sulfur and oxygen fugacities.

Products of hydrothermal activities are also the alpine fissure type minerals of which adularia are sometimes most apparently enriched in Ba (cf. Table 56-D-1).

IV. Volcanic Exhalations, Gas Transport

The possibility of Ba transport through hydrous gas phases was demonstrated by the experiments of STRÜBEL (1967). Results are discussed in chapter 56-H. NABOKO (1945) found traces of Ba in fumarole sublimates (mostly NaCl, KCl and NH_4Cl) of Klyucherskoy volcano, USSR. MINGUZZI (1948) determined traces of Ba in fumarole products of Vesuvius, Italy.

Revised manuscript received: September 1971

56-G. Behavior during Weathering and Alteration of Rocks

Experimental weathering of K feldspar in distilled water (PUCHELT, 1967) showed that Ba is preferentially released from this silicate structure into the solution. The weight ratio K_2O/BaO, being 18.9 in the mineral, is much lower in the weathering solution (8.1).

In the naturally occurring weathering series biotite — hydrobiotite — vermiculite, BOETTCHER (1966) observed, by spectrographic analysis, a decrease of the BaO content from 4,500 ppm to 300 ppm.

ROSENQUIST (1939) leached a granite powder (0.11% BaO) with distilled water. In the residue of the weathering solution BaO was enriched to 0.91%.

Extensive studies on different rocks and their weathering products were carried out by BUTLER (1953b, 1954) (Table 56-G-1). In three types, Ba is enriched in the silt and clay fraction of the weathered material, while in a fourth example even the weathering residue is leached with respect to Ba.

Table 56-G-1. *Barium distribution during rock weathering* (BUTLER, 1953b, 1954)

Rock type	Ba content (ppm)			Reference (Analytical method)
	fresh rock	from weathered rock		
		silt fraction	Ca saturated clay fraction	
Granite	110	180	500	BUTLER, 1953b (S)
Hornblende schist	10	190	500	BUTLER, 1953b (S)
Quartz free syenite	1,000	300	870	BUTLER, 1954 (S)
Hypersthene monzonite	1,000	45	15	BUTLER, 1954 (S)

Both Ba increase and Ba decrease have been observed in the weathering products in studies by a great number of investigators. Among the factors which influence Ba behavior in this process are: climate; type of clay minerals, which form during the decomposition; amount and kind of organic material present; and sulfur or sulfate content. Since barium sulfate is a compound of very low solubility, this last factor predominates in sequences originally rich in sulfides.

A survey of Ba content of soils was published by SWAINE (1955). Most soils contain 100 to 3,000 ppm Ba of which only trace amounts can normally be extracted by means of $1N\ NH_4$ acetate. The highest value which SWAINE reports (3.3% Ba) comes from Tennessee, USA, from areas where barite had been mined. Older literature is listed in GMELIN (1960). Under special arid conditions of weathering in deserts, varnishes form, which always show Ba enrichment (ENGEL and SHARP, 1958) (S).

Revised manuscript received: September 1971

© Springer-Verlag Berlin · Heidelberg 1972

56-H. Solubilities of Compounds which Control Concentrations of Barium in Natural Waters (I), Adsorption Processes (II)

I. Solubilities

Only two Ba compounds exist which can control the Ba content of natural waters: $BaSO_4$ — barite, and $BaCO_3$ — witherite.

a) Solubility of $BaSO_4$

$BaSO_4$ is the least soluble and most abundant Ba mineral in the earth's crust. Its solubility in water upto 100°C has been repeatedly determined. Some of the available data are given in Fig. 56-H-1. STRÜBEL (1967) published data for $BaSO_4$ solubility in the hydrothermal range upto 600°C. At this temperature and 1,084 bars, he obtained a solubility of 9.61 ± 1.95 mg $BaSO_4$/1,000 g H_2O.

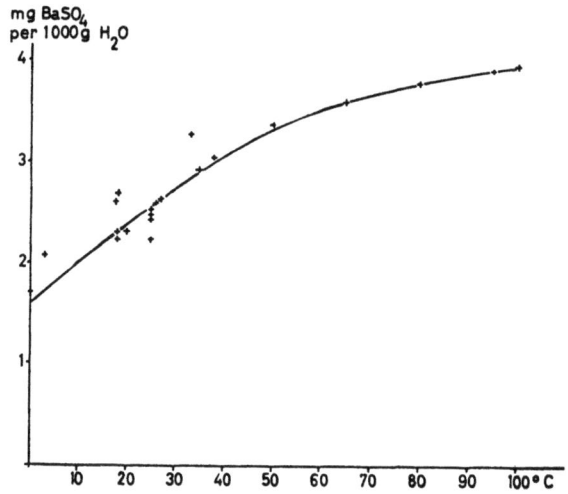

Fig. 56-H-1. Solubility of barium sulfate in distilled water (KOHLRAUSCH, 1908; MELCHER, 1910; NEUMANN, 1933; ROSSEINSKY, 1958; TEMPLETON, 1960; BURTON et al., 1968)

Electrolytes considerably increase the $BaSO_4$ solubility. TEMPLETON (1960) has investigated sodium chloride influence upto 95°C with solutions upto 5 molal NaCl. PUCHELT (1967) radiochemically determined $BaSO_4$ solubility at 25 and 50°C in upto 6.08 molal NaCl solutions. Experiments upto 350°C were carried out by UCHAMEYSHVILI et al. (1966) with 0.25 N, 1.0 N and 2.0 N sodium chloride solutions. Investigations upto 600°C with upto 2 molal (or 11.69% ?) NaCl solutions were

© Springer-Verlag Berlin · Heidelberg 1972

performed by STRÜBEL (1967). Solubility data upto the boiling point are plotted versus NaCl molality in Fig. 56-H-2. For 600°C, 1,990 bars and 2N(?) NaCl solutions, STRÜBEL reports a solubility of 971 mg/kg H_2O. In the hydrothermal range, $BaSO_4$ solubility sensitively increases with pressure. STRÜBEL's investigations show that an area of retrograde $BaSO_4$ solubility exists between 350 and 450°C.

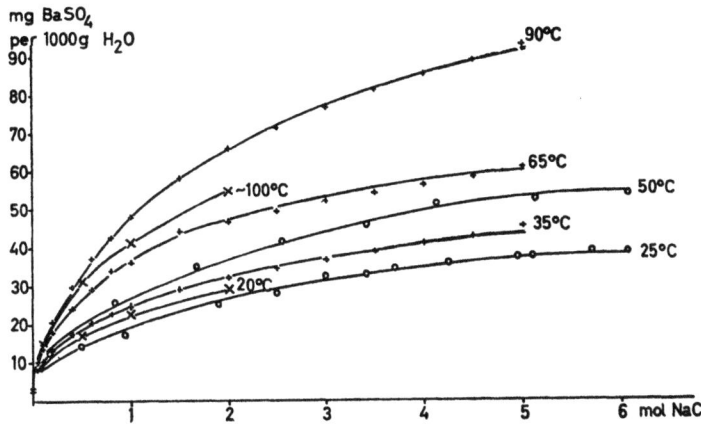

Fig. 56-H-2. Solubility of barium sulfate in NaCl solution. + TEMPLETON, 1960; × STRÜBEL, 1967; ○ PUCHELT, 1967

Influences of other electrolytes on $BaSO_4$ solubility in aqueous solutions were studied by: NEUMANN (1933), KCl, KNO_3, $MgCl_2$, $Mg(NO_3)_2$, $LaCl_3$, $La(NO_3)_3$; UCHAMEISHVILI et al. (1966), KCl, $MgCl_2$, $CaCl_2$; and PUCHELT (1967), KCl, $MgCl_2$, $CaCl_2$. COLLINS and ZELINSKI (1966) investigated the effect of synthetic brines containing NaCl, $MgCl_2$, $CaCl_2$, and $NaHCO_3^-$.

The results of PUCHELT are:

Solution	Maximum solubility found at		$BaSO_4$ solubility mg/1,000 g H_2O
	temperature	ionic strength	
KCl	25° C	5.0	40.8
KCl	50° C	5.0	54.0
$CaCl_2$	25° C	5.5	42.4
$CaCl_2$	50° C	5.5	59.8
$MgCl_2$	25° C	6.0	47.2
$MgCl_2$	50° C	6.0	71.6

UCHAMEISHVILI et al. (1966) found a strong increase in $BaSO_4$ solubility in $CaCl_2$ solutions with temperatures between 100 and 255°C. In $MgCl_2$ solutions, too, a barium sulfate solubility higher than that in NaCl and KCl solutions was observed by these investigators.

Barium sulfate solubility in sea water was calculated by CHOW and GOLDBERG (1960), and experimentally studied at one atmosphere pressure by PUCHELT (1967), who also made investigations regarding the kinetics of $BaSO_4$ precipitation in sea water. The solubility of 89 µg $BaSO_4$/l (at 20°C), found by PUCHELT agrees well with the value of 87.9 µg $BaSO_4$/l (at 25°C) of CHOW and GOLDBERG. Recently, BURTON et al. (1968) obtained a mean value of 81 µg/l. Close to saturation, complete equilibrium is reached slowly. Despite seeding, PUCHELT (1967) needed about 80 days. He also made experiments to study the influence of salinity of sea water on solubility and covered the range upto 87.5‰ salinity. HANOR (1969) and CHURCH (1970) calculated the effect of aqueous complexing and presence of Sr, Ca, and K on the solubility of $BaSO_4$—$SrSO_4$ mixed crystals in sea water.

Pressure increases the solubility of $BaSO_4$. In pure water the solubility product is, by a factor of 5.4, larger at 1 kilobar than at 1 atm. pressure. In a sodium chloride solution of 0.727 molality the same ratio is only 4.2.

b) Solubility of $BaCO_3$

Barium carbonate solubility depends largely on the CO_2 partial pressure of equilibrium atmosphere. At 25°C and 1 atm. CO_2 pressure, GARRELS et al. (1960) determined a dissociation constant of $10^{-8.64}$ for witherite. The solubility product of ^{14}C-labeled $BaCO_3$ in basic aqueous solutions at 25°C was found to be $4.0 \cdot 10^{-10} \pm 0.5 \cdot 10^{-10}$ by BACCANARI et al. (1968).

Increase in carbon dioxide pressure causes an increase in $BaCO_3$ solubility. This effect is much smaller at higher temperatures than at lower temperature (MALININ, 1963), (Fig. 56-H-3).

TOWNLEY et al. (1937) showed that LiCl, NaCl, and KCl at 25°C and 40°C, increased $BaCO_3$ solubility according to their respective concentration. In all concentrations investigated (upto 3 molal) LiCl produced the strongest increase in solubility. The effect of KCl in the same molality was the least.

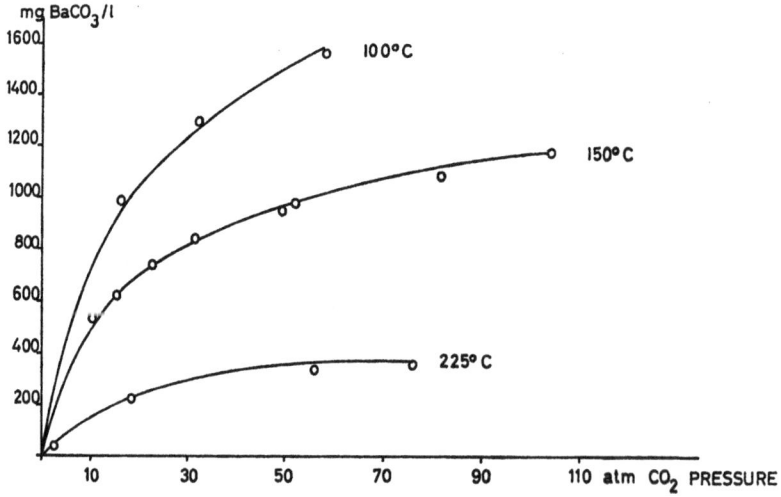

Fig. 56-H-3. Solubility of $BaCO_3$ in water with increasing CO_2 pressure (MALININ, 1963)

II. Adsorption Processes

Ba is adsorbed from solutions by clays, hydroxides, and organic matter. In addition to $BaSO_4$ solubility, these processes control the amount of Ba present in natural waters.

a) Clays

Ba adsorption on the sodium charged form of standard montmorillonite, illite, and kaolinite, at 25°C was studied by PUCHELT (1967) for pure and electrolyte containing solutions. Ba adsorption decreases with ionic strength of the exchange solution. In rivers, the ratio of Ba which is adsorbed by suspended matter depends on the type of suspension and the concentration of ions competing for adsorption sites.

Ba exchange on vermiculite and bentonite was investigated by LEVI and SCHIEWER (1965). Bentonite adsorbs Ba more strongly than NH_4^+, Mg^{++}, and Ca^{++} (KOMLEV et al., 1965). CARLSON and OVERSTREET (1967) found a high adsorption of incompletely dissociated Ba hydroxide by bentonite at pH 6. The heat of exchange of Ba ions on bentonite with H^+, Na^+, and K^+ were calorimetrically determined by TADZIEV and MUKSINOV (1967). GANGULY and MUKHERJEE (1951) investigated Ba exchange on bentonite, kaolinite, illite, and mica.

b) Hydroxides

Ba adsorption by hydrous ferric oxide was investigated by DUVAL and KURBATOV (1952). PUCHELT (1967) studied, experimentally, Ba adsorption on γ MnO(OH) and found that NaCl concentrations upto 3.5% do not influence the amount of Ba adsorption. γ MnO(OH) can adsorb as much as 20% (by weight) of its Mn content of Ba. These results probably can explain the Ba content of deep sea manganese nodules. PUCHELT also observed that γ MnO(OH) adsorbs more than 85% of the Ba concentration which exists in equilibrium with a $BaSO_4$ precipitate. Adsorption of Ba ions on silica gels from acid solutions depends on time of exchange, specific surface, and pore size of the gel (KIRICHENKO et al., 1965).

c) Organic Substances

BEL'KEVICH et al. (1966) equilibrated 0.05 to 0.2N earth alkali solutions with the H-form of peat. They found Ba to have the strongest tendency to substitute for H in peat. Adsorption of ^{137}Ba by coal humic acid was studied by MATSUMARA and ISHIYAMA (1966). PUCHELT (1967) observed that bacteria may extract Ba from solutions, but it is not yet clear, whether this happens by adsorption or incorporation.

56-I. Abundance in Natural Waters

I. Springs and Fresh Water Wells

Only very few spring and fresh waters are free of sulfate. Thus the solubility product of $BaSO_4$ is the limiting factor for the Ba concentration. As spring waters normally have only low amounts of dissolved solids and moderate temperatures, no considerable increase of $BaSO_4$ over the distilled water solubility is to be expected. These waters originate normally from rain water which had only a limited time for equilibration in sediments and soils. The Ba content is mainly controlled by the solution of Ba compounds (mostly barite), and exchange of Ba from silicate structures. Several analyses are published for water which served medical purposes. A survey of older data (methods: W) is given by DELKESKAMP (1900, 1902); some values published before 1949 are tabulated by GMELIN (1960). PUCHELT (1967) surveyed the more recent literature grouping the waters in hydrocarbonate, chloride, sulfide and sulfate types according to their prevailing anion.

Table 56-I-1. *Ba concentrations in European spring waters.* (PUCHELT, 1967)

Type of water	No. of springs	Ba range ppb	Arith. mean ppb	Standard deviation
Hydrocarbonate	16	4—22,900	1,757	5,672
Chloride	22	12— 9,500	1,340	2,681
Sulfide	3	150— 750		
Sulfate	9	1— 230		

In very few cases of spring water, higher values were observed than were expected from the $BaSO_4$ solubility product. Their existence is explained by supersaturation which sometimes occurs for a short time after adding sulfate to Ba solutions.

Drinking water from fresh water wells was analysed in the USA by DURFOR and BECKER (1964). For 10 wells from all parts of the country they found 4.6 to 34 ppb Ba (S).

Additional new data for ground and spring waters were published for: South Africa (KENT, 1949; KENT and RUSSELL, 1949), Bulgaria (PENCHEV et al., 1958, 1960), Czechoslovakia (RUBESKA and MIKSOVSKY, 1963), Finland (WILSKA, 1952), Germany (FRICKE, 1968), Hungary (STRAUB, 1950), Japan (IKEDA, 1955a/b; ICHIKUNI, 1966; IWASAKI et al., 1963), and the Soviet Union (BABINETS and RADKO, 1956; GRUSHKO and SHIPITSYN, 1948; KONTOROVICH et al., 1963; OSTROUMOV and RUSSKIKH, 1965; SHINKARENKO, 1948; YUSUROVA, 1957). The highest value reported comes from a Japanese warm spring of sodium chloride type and is 62 ppm Ba.

II. Rivers and Lakes

Only North American rivers have been extensively analysed for Ba. The investigations of DURUM et al. (1960) and DURUM and HAFFTY (1961) cover long periods of time, climatic conditions and discharge for a number of rivers. In all cases they found, over a year, a considerable and complex variance of Ba content. One example is given in Table 56-I-2.

Table 56-I-2. *Variation of Ba concentration in Mississippi water near Baton Rouge, Lousiana, U.S.A.*

Date of sampling	Run-off m³/sec	Dissolved solids ppm	Ba ppb
May 10, 1958	24.550	160	78
Oct. 14, 1958	7.400	223	127
March 13, 1959	18.600	184	72
May 18, 1959	11.800	255	84

DURUM et al. (1960) covered 30% of the total run-off of the North American rivers with their analyses, and DURUM and HAFFTY (1963) found a median Ba concentration of 45 ppb from the available data. Ba/Sr ratios (by weight) vary between 0.2 and 3.9 (PUCHELT, 1967), but give a geometric mean of 0.87 for North American rivers (DURUM and HAFFTY, 1963), which may be compared with $0.78 \cdot 10^{-3}$ in the oceans. A number of rivers and lakes being used as drinking water resources have been analysed for Ba by DURFOR and BECKER (1964). By spectrographic analyses, they determined a range of Ba content in rivers between 3.1 and 340 ppb (arithmetic mean: 75.7 ppb). Lakes and brackish water in North America and Europe ranged from 3 to 140 ppb Ba (DURFOR and BECKER, 1964; BROWN et al., 1962; WILSKA, 1952; LANDERGREN and MANHEIM, 1963). Concentration ranges for several rivers are plotted in Fig. 56-I-1.

Local variations of Ba content due to rock composition of the drained area were found by MILLER (1961) in New Mexico and BROWN et al. (1962) in Alaska. They observed the highest Ba values from regions with sediments (sandstone, slates), less from granites, and obtained the lowest means from quarzite. The distribution of Ba along a river was studied by LEUTWEIN and WEISE (1962) for the Mulde in Germany. They observed values of 5 to 100 ppb Ba in true solution in the river itself but upto 730 ppb in certain adjoining creeks which drained mining areas. According to these authors, in the upper part of the river 70% and more of the Ba is transported in true solution as the ion. On flowing into the Elbe after 245 km, only 20% of the Ba is still in the ionic form, 80% having been adsorbed onto clays and organic matter. In regions with extreme sulfate concentrations, Ba content is low in accord with the $BaSO_4$ solubility product. TUREKIAN et al. (1967) describe the Ba variation of the Neuse river (North Carolina, USA). Ba concentration decreases in the upper part of the river, (16 to 5.7 ppb) in slate and granite areas, but increases regularly downstream to 22 ppb in slate, schist, sand and limestone

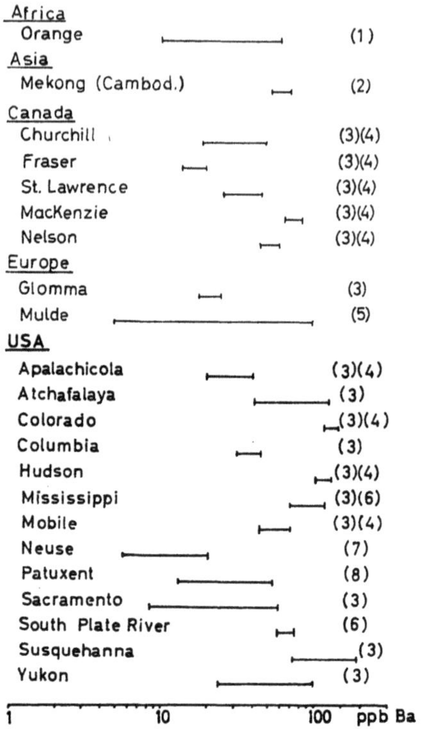

Fig. 56-I-1. Ba concentration ranges of rivers (all data obtained by spectrographic methods). 1. DEVILLIERS (1962); 2. DURUM and HAFFTY (1963); 3. DURUM et al. (1960); 4. DURUM and HAFFTY (1961); 5. LEUTWEIN and WEISE (1962); 6. DURFOR and BECKER (1964); 7. TUREKIAN et al. (1967); 8. HEIDEL and FRENIER (1965)

areas. Where the petrographic composition influences the drainage waters of a certain area distinctly, no general conclusion can be drawn from trace element data from large river basins regarding the origin of the particular trace element.

III. Oceans

From all oceans, Ba determinations are available in surface to bottom profiles. The concentration ranges upto 78 ppb Ba and the estimated mean is about 20 ppb (TUREKIAN and JOHNSON, 1966). Analytical data are compiled in Table 56-I-3.

In general, the Ba concentration seems to be lower in the Atlantic than in the Pacific. In most cases the surface layers are depleted in Ba. Special features have been observed for the distribution of Ba in the Pacific close to the equator; CHOW and GOLDBERG (1960) have found a steady increase of Ba concentration with depth in the respective profiles (Fig. 56-I-2a). They explained this by high biological activity in the surface layers of this region, incorporation or adsorption on organic matter, and a downward transport with the organic debris. They found Ba to resemble radium in distribution with depth. WOHLGEMUTH and BROECKER (1970) could also correlate Ba content with concentrations of other bio-important elements and found parallels with Ra distribution. These authors sampled from very deep

Table 56-I-3. Ba concentrations in the oceans

Locality	Maximum sampling depth (m)	No. of samples	Ba range ppb	Method	References
Pacific Ocean					
Central part, close to equator	4,752	18	10 —63	I	Chow and Goldberg (1960)
	4,350	4	19 —33	N/R	Turekian and Johnson (1966)
Philippine Sea	4,000	10	11 —33		Turekian and Johnson (1966)
Antarctic	5,120	125	8 —56		Turekian and Johnson (1966)
South East Pacific	1,500	8	9 —17		Turekian and Johnson (1966)
South West Pacific	5,000	5	19 —78		Turekian and Johnson (1966)
East Pacific	4,580	13	8.5 —31.2	I	Wolgemuth and Broecker (1970)
East Pacific	4,000	13	6.1 —23.5	I	Wolgemuth (1970)
Indian Ocean					
Central part	surface	1	14	N/R	Turekian and Johnson (1966)
West Indian Ocean	400	3	21 —46		Turekian and Johnson (1966)
South Indian Ocean	4,500	3	10 —15	N/R	Bolter et al. (1964)
Atlantic Ocean					
Long Island Sound	39	60	9 —32 (65)	N/R	Turekian and Johnson (1966)
Caribbean, Gulf of Mexico	3,019	17	5 —23		Turekian and Johnson (1966)
South Atlantic	5,000	3	15 —21		Turekian and Johnson (1966)
North Atlantic	5,061	21	9 —31		Turekian and Johnson (1966)
	4,100	3	12.9 —13.1	N/R	Bolter et al. (1964)
	3,000	8	12 —18	I	Chow and Patterson (1966)
	2,098	16	0.04—22.8	F	Andersen and Hume (1968)
Equatorial	4,387	35	0.80—37.5		Andersen and Hume (1968)
English Channel	surface	1	6.3	N/R	Bowen (1956)
Caribbean	4,729	20	7 —23	S	Szabo and Joensuu (1967)
Puerto Rico Trench and off Barbados	7,540	28	7.9 —19.1	I	Wolgemuth and Broecker (1970)

ocean regions but did not find a marked Ba increase towards the sea floor. Turekian and Johnson (1966) observed, in some places, a maximum of Ba concentration in depths of 600 to 1,200 m (Fig. 56-I-2b) which, however, coincides with the region of lowest Ba sulfate solubility (35 µg/l) in agreement with the interaction of tem-

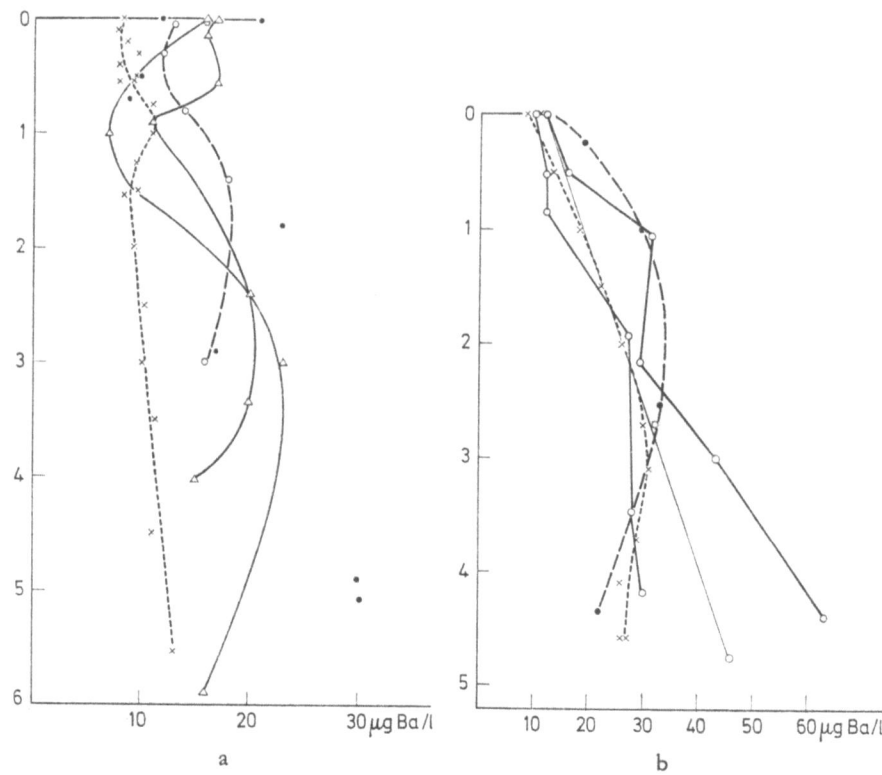

Fig. 56-I-2a and b. Barium distribution in the oceans. a Barium profiles in the Atlantic. × WOLGEMUTH and BROECKER, 1970; ○ CHOW and PATTERSON, 1966; ● TUREKIAN and JOHNSON, 1966; △ SZABO and JOENSUU, 1967. b Barium profiles in the Pacific. × WOLGE-MUTH and BROECKER, 1970; ○ CHOW and GOLDBERG, 1960; TUREKIAN and JOHNSON, 1966

perature decrease and pressure increase (CHOW and GOLDBERG, 1960). One possible explanation for these high values could be that microcrystals of barite have contaminated the samples. SZABO and JOENSUU (1967) found in profiles in the Caribbean the lowest concentration in the depths of about 1,000 m (Fig. 56-I-2a). In two areas, the equatorial Pacific and the Atlantic off southwest Africa, high values for Ba were found both in the sediments and in sea water, whereas in other places no correlation could be detected.

From the Ba supply of the streams, saturation of Ba sulfate should almost be reached in the oceans. From stream discharge ($3.6 \cdot 10^4$ km^3/yr) and Ba concentration of 45 ppb (DURUM and HAFFTY, 1963), ocean-mass of $1,372 \cdot 10^6$ km^3, and Ba content of 20 µg/l (TUREKIAN and JOHNSON, 1966), the residence time of Ba in the sea can be calculated to be $17 \cdot 10^3$ years. Using a sedimentation rate of 0.05 g SiO$_2$/yr for diatomaceous ooze (based on ^{32}Si), a Ba content of 6,000 ppm and a mean ocean depth of 5,000 m, TUREKIAN and JOHNSON computed a residence time of only 33 years above this particular sediment. Possibly this low value and the observed increase of Ba content with depth indicate an additional supply of Ba from some volcanic source on the ocean floor. With an average of 20 µg/l Ba, the total content of the oceans is $27.4 \cdot 10^9$ tons of Ba.

Table 56-I-4. *Barium concentration in formation waters*

Period	Country	Total No. of samples	No. of samples >1 ppm Ba	Mean Ba conc. of samples with Ba >1 ppm	Maximum Ba conc. ppm	Analytical method	Reference
Precambrian	U.S.A.	1	1	4	4	S	White (1965)
Ordovician	U.S.A.	4	3	61	100	S	McGrain and Thomas (1951)
Silurian	U.S.A.		1	320	320	S	McGrain and Thomas (1951)
Devonian	Canada		2	2	2	S	White (1965)
	Germany	2	1		1.1	W	Fresenius (see Michel 1963)
	U.S.A.	37	21	218	1,140	S	Poth (1962)
		13	7	205	700	W, S	Price et al. (1937), White et al. (1963)
		69	12	828	2,000	S	Hoskins (1947)
	U.S.S.R.	73	67		971	S	Kozin (1964)
Carboniferous (Mississippian, (Pennsylvanian)	Belgium	1	1	347	347	S	Camerman (1951)
	Germany	4	3	1,007	1,260	S	Jakobshagen and Münnich (1964)
		249	46	1,006	2,860	W	Michel (1963)
		225	69	832	2,806	W, S	Puchelt (1964), (1967)
		152	38	798	2,000	W, S	Wasserwirtschaftsstelle (in Puchelt, 1967)
	Great Britain	38	22		5,100		Anderson (1945), Gibson (1963)
	U.S.A.	152	112	447	5,530	W, S	Price et al. (1937)
		77	34	265	1,080	S	Poth (1962), Hoskins (1947)
		72			600		Collins (1969)
		36	20	236	1,980	S	McGrain and Thomas (1951)
	U.S.S.R.	111	≥36		190	S	Kozin (1964)
Permian	Germany		4	11	21	S	Herrmann (1961)
	U.S.A.		3	5.2	9.6	S	White et al. (1963)
	U.S.S.R.	48	≥5		73	S	Kozin (1964)
Jurassic	Germany	26	7	14.8	30	S, W	Puchelt (1967)
Cretaceous	Sweden	8	2	1	1	S	Assarson (1948)
	U.S.A.		24	14	72		Buckley et al. (1958), White et al. (1963)
Tertiary	Japan	3	1	59	59	S	Bailey et al. (1961)
	U.S.A.		15	37	150	S	Bailey et al. (1961), White (1965)
Quarternary	Poland	6	2	3.6	4		Dowgiallo (1965)
	U.S.A.	2	2	16	28	S	White et al. (1963)

IV. Formation Waters

Frequently, Ba was discovered in formation waters which had lost their initial sulfate content through bacterial activity during diagenesis. As these bacteria require a reducing environment and organic substances to live on, their areas of activity and thus high Ba concentrations in formation waters, are always connected with occurences of organic matter (oil, bitumen, coal, or gas). Ba has been found in such waters from beds of all geological ages from all over the world. The Ba concentration may reach 5,500 ppm, but no correlation of the Ba concentration with any other parameter of the solutions (dissolved solids, Sr, Ca, K concentration) could be found. Often the Ba/Sr weight ratio is larger than unity. An extensive survey of data is given by PUCHELT (1967). A more condensed compilation was prepared for this chapter (Table 56-I-4).

In the Soviet Union, Ba concentrations in formation waters have been successfully used for a correlation of stratigraphic horizons (KOZIN, 1964; NIKOLAEV et al., 1960). Ba has been assayed in formation waters of a few areas, especially in connection with oil field investigations (AKHUNDOV and SAPPO, 1960; DODONOV et al., 1949; KATCHENKO and FLEGONTOVA, 1955, 1956; KAVEEV and VASIL'EV, 1956; KOROLEV, 1938; KUKABAEV and SYDYKOV, 1962; SKROBOV and SMIRNOV, 1939; SUKHAREV, 1961; TIMASHEVA, 1963; VAROV and ROMM, 1942).

Mixing of Ba-containing formation waters with sulfate waters is often the reason for a scale formation in oil wells (GATES and CARAWAY, 1965; TEMPLETON, 1960) and mines (PATTEISKY, 1954; ANDERSON, 1945). PUCHELT (1967) presented evidence for the abundant formation of certain types of barite deposits from those waters.

V. Brines

Geothermal brines were tapped by deep wells near the Salton Sea, California, USA, an area characterized by rhyolites and Tertiary sediments. With a total of 319,000 ppm evaporation residue (180°C), 200 ppm Ba were reported for a 1963 sample (WHITE, 1965) whereas 1966 samples from two wells gave 235 and 250 ppm Ba (SKINNER et al., 1967). A hot brine from the Atlantis II Deep in the Red Sea containing more than 300 g dissolved solids per liter had almost 1,100 ppb Ba (MILLER et al., 1966).

56-K. Abundance in Common Sediments and Sedimentary Rock Types

A compilation of data on the Ba distribution in recent and fossil sediments has been published by PUCHELT (1967). An abbreviated review, but with the important new data included, is given below.

I. Recent Sediments

a) Deep Sea Sediments

Deep sea clays were often analysed for Ba in recent years. Generally, they are enriched in Ba compared to shales. WEDEPOHL (1960) found a definite difference between Atlantic clays (average: 750 ppm Ba) and Pacific clays (average: 4,000 ppm Ba). Since the rate of Ba deposition is about the same in both oceans, he assumed the difference to be caused by a lower rate of detrital accumulation in the Pacific. GOLDBERG and ARRHENIUS (1958) found Ba enrichment in sediments under equatorial waters and related this observation to the high biological productivity in the surface layers of the waters. Several planktonic organisms are known to accumulate Ba in their tests which carry the Ba bottomward after death. They may cause a Ba enrichment in layers close to the bottom, where they dissolve, and may generate local $BaSO_4$ precipitations (PUCHELT, 1967; BRONGERSMA-SANDERS, 1967; cf. Table 56-L-2). Barium is not deposited homogeneously on the sea floor. The zones of high biological activity as well as the ocean ridge systems usually have higher Ba concentrations (TUREKIAN, 1968) than normal deep sea sediments. The origin of Ba in deep sea sediments is a matter of discussion.

BOSTRÖM and PETERSON (1966) determined upto 3.1% Ba in cores from the flanks of the East Pacific Rise. From additional data for other elements and data for the heat flow, it can be concluded that volcanic activity adds several of the enriched elements. TUREKIAN (1968) concluded from data of Ba supply to the oceans by streams, from average Ba content of clays (shales), from the model of Ba enrichment by plankton and from some additional information, that a volcanic or hydrothermal Ba supply need not be assumed to explain the observed Ba data. Ba concentrations in deep sea matter do not often correlate with any of the major constituents. It has to be concluded that Ba adsorption on clays is of less importance. The main Ba carrier is most likely barite. Locally, manganese oxides and phillipsite will accumulate Ba. Calcium carbonate of organic origin is usually very low in Ba (<100 ppm, often only 10 to 30 ppm). In order to reach comparable data for clay sediments, TUREKIAN and TAUSCH (1964) have calculated their Ba determination of Atlantic cores on calcium carbonate free basis. Because the original data are not given, only these corrected values are included in Table 56-K-1. These authors found higher Ba values in the South Atlantic than in the North Atlantic. The areas

Table 56-K-1. *Ba concentrations in deep sea sediments*

Origin	No. of samples	Ba content ppm	Method	References
a) Deep sea clays				
Atlantic				
Area north of equator	1	200	S	v. Engelhardt (1936)
South Atlantic	2	590	S	v. Engelhardt (1936)
Area north of 20° S	37	596[a]	S	Ericson et al. (1961)
Area north of 10° N	62	1,100[a]	S	Turekian and Tausch (1964)
Area south of 10° N	63	2,000[a]	S	Turekian and Tausch (1964)
Area north of equator	5	910	S	El Wakeel and Riley (1961a)
North American trench, south part	15	725	X	Wedepohl (1960)
Kap Verde Trench	3	470	X	Wedepohl (1960)
Mid Atlantic Ridge	1 (9)	4,400[a]	N/R	Turekian (1968)
Average of Atlantic clays	189	1,260		
Pacific and Indian Ocean				
Area south of equator	8	1,150	S	Goldberg and Arrhenius (1958)
Area north of equator	5	5,700	S	Goldberg and Arrhenius (1958)
Baja Californian seamounts	3	5,800	S	Goldberg and Arrhenius (1958)
North Pacific	2	2,450	S	Goldberg and Arrhenius (1958)
Area north of equator	4	6,100	W ?	Grim et al. (1949)
Trench NNW Sixty Mile Bank	2	2,700	W ?	Grim et al. (1949)
Indian Ocean	18	610	S	Katchenko and Flegontova (1964)
Area north of equator	8	2,120	S	El Wakeel and Riley (1961a)
Indian Ocean equator area	2	1,330	S	El Wakeel and Riley (1961a)
Area north of equator	20	8,050	S	Young (1954)
Area south of equator	1	300	X	Wedepohl (1960)
From all over the Pacific	9	6,700	X	Wedepohl (1960)
Average of Pacific clays	82	4,160		
b) Deep sea carbonates[b]				
Atlantic and Mediterranean				
Caribbean Sea	2	210	S	Ericson and Wollin (1956)
Mid Atlantic Ridge	1 (9)	689	N/R	Turekian (1968)
Various places	4	190	S	Turekian and Wedepohl (1961)
North Atlantic (with clay)	4	840	S	El Wakeel and Riley (1961a)
South Atlantic off Africa	1	1,900	S	El Wakeel and Riley (1961a)
Mediterranean	2	1,600	S	El Wakeel and Riley (1961a)
Pacific and Indian Ocean				
Equatorial area, East Pacific (manganiferous)	2	5,000	S	Goldberg and Arrhenius (1958)

Table 56-K-1. (Continued)

Origin	No. of samples	Ba content ppm	Method	References
Equatorial area, Central Pacific	1	680	S	EL WAKEEL and RILEY (1961a)
North equator area, Central Pacific	3	540	S	YOUNG (1954)
c) Deep sea siliceous muds				
Atlantic				
Atlantic off African coast	1	700	S	EL WAKEEL and RILEY (1961a)
Equatorial area, Central Pacific	3	3,470	S	EL WAKEEL and RILEY (1961a)
Central northern Pacific	6	10,400	S	YOUNG (1954)
Central equatorial Pacific	3	8,100	S	YOUNG (1954)

[a] Deep-sea clays calculated on $CaCO_3$ free basis.
[b] Raw analyses of carbonate rich cores but not necessarily indicative of the pure carbonate fraction.

closer to the continents are normally lower than the Central Ocean areas, but just off the coast of Africa (20 to 25°S), an area with values of >4,000 ppm Ba was detected.

TUREKIAN (1968) has analyzed samples from different depths in a deep sea core for Ba. Concentrations range between 1,700 and 6,700 ppm Ba (calculated $CaCO_3$ and salt free); they indicate changes in the rate of Ba deposition within the last 30,000 years. Accumulation rates for Ba reported by TUREKIAN (1968) vary from $<90\ \mu g/cm^2$ per 1,000 years to $790\ \mu g/cm^2$ per 1,000 years within the special core and are about $1,000\ \mu g/cm^2$ per 1,000 years in the Antarctic. In Table 56-K-1, averages for Ba are calculated using the "reduced $CaCO_3$ free" data. This means that the actual data may be lower.

Deep sea carbonates. Foraminifera ooze high in carbonate contains low Ba concentrations (10 to 30 ppm, Table 56-L-2). The barium content in carbonate sediments is either due to $BaSO_4$ or to manganese oxides or clay. Since all these sources may be active at the same time and do not work coherently, a recalculation to "pure carbonates" is very difficult. TUREKIAN and TAUSCH (1964) extrapolated deep-sea cores in the North Atlantic to 100% $CaCO_3$ and got 10 to 30 ppm Ba for the pure carbonate. The available data are given in the table. The value of 190 ppm Ba by TUREKIAN and WEDEPOHL (1961) for deep sea carbonates derived from 4 globigerina oozes from Atlantic cores is as yet the best information.

Manganese nodules cover wide areas of the deep sea bottom of all oceans and contain upto 20,000 ppm Ba. A survey by PUCHELT (1967) of published literature shows that the Ba means are 4,500 ppm, 5,200 ppm and 3,700 ppm Ba for nodules from the Pacific, Atlantic, and Indian Oceans, respectively. In manganese nodules, Ba is either adsorbed, incorporated in acid soluble compounds (zeolites), or occurs as barite (ARRHENIUS, 1963).

Siliceous sediments occurring in deep areas of the oceans where carbonates are no longer stable can locally contain more than 1% Ba.

b) Shallow Water Sediments

The barium content of near shore and shelf sediments is influenced by the amount and kind of detrital matter and the barium content of the rivers. A review of literature by PUCHELT (1967) shows that clay sediments of these areas are generally higher in Ba than sand and silt fractions. Clays from the Mississippi delta are especially high in Ba.

Three studies of reef carbonates (STEHLI and HOWER, 1961; SENAKOLIS, 1964; FRIEDMAN, 1968) demonstrated that reef debris, reef material and oolitic muds contain only limited amounts of Ba. STEHLI and HOWER (1961) found a range from 10 to 61 ppm in 59 samples and an average of 18.4 ppm Ba. FRIEDMAN (1968) obtained spectrographically, 18 to 62 ppm Ba in corals with encrusting coralline algae, and 15.5 to 68 ppm Ba in carbonate sands from reef aprons. Carbonate sands with admixed terrigenous debris showed 130 to 280 ppm Ba. In the analysed samples, Ba concentrations change parallel to the "insoluble residue". Similar observations are reported by SENAKOLIS (1964). Within the internal parts of the reef, average Ba content in clastic limestone was 3.1 ppm, peripheral parts of the reef had 436 ppm Ba.

II. Barium in Consolidated Sediments

a) Sandstones, Cherts, Graywackes

Pure quartz sandstones are very low in Ba, but since most sandstones contain considerable amounts of feldspars, these minerals are the most important Ba carriers besides micas, which occasionally occur. PETTIJOHN (1963) calculated the proportion of sandstone types to be 34% quartzite, 26% graywacke, 25% subgraywacke, and 15% arkose. This combination has an average K_2O content of 1.3%, to which a Ba content should be proportional. Comparison of the K/Ba ratios in low Ca granites and in sandstones and graywackes gives additional support to this proportionality. Single sandstones vary widely in Ba content, since even barite concentration (as cement) occurs locally (cf. PUCHELT, 1967). If barite is a sandstone constituent, normally the weight ratio Ba/Sr is greater than 10, for barite generally contains much less than 10% $SrSO_4$. The Ba content of sandstones and graywackes ranges from 5 to 900 ppm. An average composition calculated from the European, Russian, and American sandstone and graywacke is 316 ppm Ba. This value is subject to changes when the individual data can be properly weighted. Nevertheless it is a more realistic value than the X0 ppm guess of TUREKIAN and WEDEPOHL (1961). Cherts constitute a special group in silica sediments and always exhibit higher Ba means.

Table 56-K-2. *Barium in quartz sandstones, cherts, and graywackes*

Locality	No. of samples	Barium concentration range ppm	group mean ppm	total mean ppm	Method	References
Quartz sandstone						
Germany	73	50—810		406		
			134		X	Bertsch (1964)
			313		X	Wedepohl (1961)
			770		S	Zurlo (1963) see Puchelt (1967)
Sunda Islands	9	5—900	150		S	v. Tongeren (1938)
U.S.A.	289			280		Shoemaker et al. (1958)
			1,800		S	Young (1954)
U.S.S.R.	399	230—820		249		
			340		S	Babina and Kotorovich (1966) see Puchelt (1967)
			223		S	Lebedev (1967) see Puchelt (1967)
			140		S	Litvin (1961)
			250		S	Litvin (1963)
			290		S	Sinkarenko (1948)
Chert						
C.S.S.R.	2	300— 500		400	S	Leutwein (1957)
Finland	46	360— 630		495	S	Sahama (1945)
Germany	311	30—1,000		440	S	Leutwein (1957)
					S	Prashnowsky (1957)
Indonesia	22	50—1,900		420		Audlee-Charles (1965)
Different places		350		350		Maxwell (1953) see Puchelt (1967)
Graywacke and arcose						
Africa				258	S	Danchin (1970) see Puchelt (1967)
Europe	51	189—670		370		
			290		S	v. Engelhardt (1936)
			360		S	Klein (1935)
			447		S	Kuenen (1941)
			389		S	Rivalenti and Sighinolfi (1969)
			270		X	Wedepohl (1961)
			335			Westermann (1961)
			500		S	Zurlo (1963) see Puchelt (1967)
North America	95	30—830		330		
					S	Macpherson (1958)
					S	Weber and Middleton (1961)
Indonesia					S	McLaughlin (1955)
New Zealand	12	30—480		252	S	v. Tongeren (1938)

b) Shales

Ba averages for shales reported in the literature vary from 250 to 800 ppm (PUCHELT, 1967; VINOGRADOV, 1956). The average of this paper is 546 ppm (s for the 25 means used: 212). Individual samples gave values from 10 to 5,000 ppm. From the data summarized in Table 56-K-3, it can be observed that especially low values have been found in shales of the Dnieper-Donets depression (Russia) and the west Siberian depression (LITVIN, 1961, 1963; TOLKACHEV, 1968). If these low

Table 56-K-3. *Barium in shales*

Locality	No. of samples[a]	Barium concentration range ppm	Barium concentration mean ppm	Method	References
Africa:	2	270— 450	360		JUNNER and JAMES (1947)
	25 (323)	394—1,004[b]	681	S	DANCHIN (1970)
Asia:					
Japan	1 (14)		540	X	WEDEPOHL (1960)
U.S.S.R.	8 (521)	150— 370[b]	270	S	BABINA and KOTOROVICH (1966)
	6 (920)	260— 520[b]	394		LEBEDEV (1963)
	19	30— 450	83		LITVIN (1961)
	6 (32)	140— 230	188		LITVIN (1963)
	1		360		SINKARENKO (1948)
	5 (185)	140— 220	182	S	TOLKACHEV (1968)
Europe:	1 (36)		800	X	WEDEPOHL (1960)
Finland	17 (105)	9—2,700	654	S	SAHAMA (1945)
Germany	3	480— 540	513	S	HEIDE and CHRIST (1953)
	66	50—5,000	750	S	LEUTWEIN (1951)
	3 (9)	700— 900	824	S	PRASHNOWSKY (1957)
	3	390— 730[b]	527	S	ZURLO (1963)
Great Britain	24	210—2,240	739	S	MOHR (1959)
	6	330—1,050	555	S	NICHOLLS and LORING (1962)
	27	325—1,280	723	S	SPENCER (1966)
Sweden			500	S	LANDERGREN and MANHEIM (1963)
	9	450—2,150	866		LARSSON (1932)
North America:	33	250—1,000	526	S	DEGENS et al. (1957)
	31	100— 750	393	S	FENNER and HAGNER (1967)
	41	10—1,020	470	S	MACPHERSON (1958)
	26	190— 610	359	S	MURRAY (1954)
	15		580		SHAW (1954, 1957)
	17		720		TOURTELOT (1957)
	3 (?)		750	S	YOUNG (1954)
Pacific Islands:	2	500—1,050	775	S	EL WAKEEL and RILEY (1961b)
	4	20—1,800	840	S	v. TONGEREN (1938)

[a] No. of individual samples used for bulk samples, or for means, are given in parantheses.

[b] Range of means, not of individual samples.

values for Russia are omitted from averaging, the mean is 628 ppm with a standard deviation of 157 ppm.

Ba does not seem to be an environmental indicator for shales. VINE (1966) and LEBEDEV (1967) found a Ba increase in shales from fresh water to marine environment, while MURRAY (1954) reports opposite observations from Indiana and Illinois, USA. In general, shales have higher Ba contents than graywackes or sandstones, but locally, siltstones or sandstones may be higher in Ba (ALANOV, 1963).

The mode of Ba binding in shales is complex. Several indications exist which point to a correlation of Ba with mica; parallelism with the amount of illite present has been found (FENNER and HAGNER, 1967); and $BaSO_4$ was shown to be another possible carrier. Black shales often contain more Ba than normal shales thus suggesting a connection of Ba with organic matter. While certain shales retained their Ba content of deposition, others gained or lost some. Redistribution in diagenetic processes is possible.

c) Carbonate Rocks

Literature on Ba in carbonate rocks is summarized by GRAF (1960) and PUCHELT (1967). The Ba content of carbonate rocks varies from 1 to 10,000 ppm (cf. PUCHELT, 1967). Using the literature cited in Table 56-K-4, an average Ba concentration for carbonate rocks of 90 ppm was calculated. This value is within the range reported by GRAF (1960) (150 ± 110 ppm). TUREKIAN and WEDEPOHL (1961) based their average limestone value (10 ppm) on the Ba content of modern molluscan shells. This value seems to be too low even for average carbonate, excluding detrital material. Means calculated by area as listed in PUCHELT (1967) are plotted in Fig. 56-K-1.

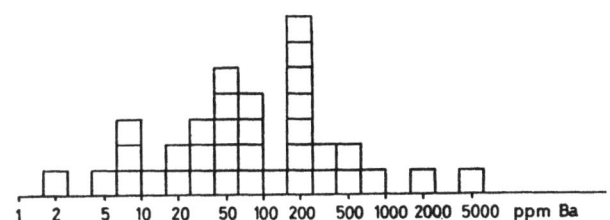

Fig. 56-K-1. Frequency distribution of Ba concentrations in carbonates (PUCHELT, 1967)

High average Ba values in relation to the overall mean are reported for 10 Ordovician dolomites from Missouri, USA (620 ppm, KELLER et al., 1950), Cretaceous limestones, USA (900 and 1,800 ppm, YOUNG, 1954), 91 Pennsylvanian limestones from Illinois, USA (260 ppm, OSTROM, 1957) and 6 limestones from Africa (1,330 ppm, JUNNER and JAMES, 1947).

Ba in carbonate rocks originates mainly from 3 sources or processes:
1. detrital clay material,
2. redistribution during diagenetic processes, in which $BaSO_4$ can be precipitated,
3. incorporation of Ba in carbonate minerals.

In most cases processes 1 and 2 are quantitatively more important. Reference is made here to the difference between "pure" carbonates within recent reefs and carbonate sands from their peripheral parts (cf. chapter 56-K-I). In carbonate sands with detrital clay, FRIEDMAN (1968) observed an increase of Ba concentrations parallel to "insoluble residue". Fossil carbonate sediments exhibit similar features (VINOGRADOV et al., 1952).

During evaporation of sea water, Ba is precipitated as barite which usually occurs disseminated in the calcium carbonate (PUCHELT, 1967). Diagenetic alterations may cause local BaSO₄ concentrations. Thus CAMERON (1966) found between 1 and 6,100 ppm Ba in a core from carbonate rocks within a distance of 80 feet. This diagenetic barite usually can be recognized under the microscope. Manganese-containing carbonate sediments are sometimes enriched in Ba (MOHR and ALLEN, 1965).

Table 56-K-4. *Barium in carbonate rocks*

Locality	No. of samples[a]	Barium concentration range ppm	mean ppm	Method	References
Asia:					
Indonesia	4	9— 450	220	S	v. TONGEREN (1938) [b]
U.S.S.R.	8	30— 240	65	S	GORLITSKY and KALYAEV (1962)
				S	LEBEDEV (1967)
					LITVIN (1963)
	59 (2,973)	1— 250	36	S	RONOV (1956)
	198	12— 690	47		SINKARENKO (1948)
					VINOGRADOV and RONOV (1956)
Europe:					
Germany	41 (131)	1— 300	62	S	v. ENGELHARDT (1936)
				S	HEIDE and CHRIST (1953)
				S	PRASHNOWSKY (1957)
	125	15— 403	92	X	BAJOR (1965)
Great Britain	183	<5— 8,000	220	S	MUIR et al. (1956)
Roumania	15	10— 300	100	S	IMREH and ECATERINE (1965)
Scandinavia	5	3— 330	78	S	v. ENGELHARDT (1936)
				S	HENRIQUES (1964)
				S	SAHAMA (1945)
North America:					
U.S.A.	10	200— 2,000	620	S	KELLER et al. (1950) [b]
	420	2—10,000	106		CANNON (1955)
				S	LAMAR and THOMPSON (1956)
					MOORE (1960)
				S	OSTROM (1957)
				S	RUNNELS and SCHLEICHER (1956)

[a] Number of individual samples used for bulk samples, or for means, are given in parantheses.
[b] Data not included in calculation of average.

Revised manuscript received: September 1972

56-L. Biogeochemistry

Barium is present in recent and fossil plants, in animals and fuels. Ba accumulation in plants and animals was found by several investigators, but there is no evidence that this element is physiologically necessary. Ba is moderately toxic for plants and slightly toxic for mammals. Reviews of barium biogeochemistry are published by PUCHELT (1967) and BOWEN (1966), who also makes reference to earlier compilations.

BOWEN (1966) summarizes the available literature for Ba content in dry weights as follows:

marine plants	30 ppm
land plants	14 ppm
marine animals	0.2—3 ppm (higher in hard tissues)
land animals	0.75 ppm

I. Plants

In ash of marine plants, Ba varies over a wide range (Table 56-L-1). Coccoliths, which form the main constituents of marine carbonates, contain 10 to 30 ppm Ba in ash (TUREKIAN and TAUSCH, 1964). Ash of phytoplankton species from the Black Sea (*Chaetoceros Curvistus* and *Rhizosolenia Calcar Avis*) is very high in Ba (4,000 and 20,000—30,000 ppm, respectively; VINOGRADOVA and KOVAL'SKIY, 1962). Since these diatoms are very abundant in surface waters in summer, they form a considerable Ba enrichment, which may contribute to the Ba content of pelagic sediments. The tests of rhizosolenia and chaetoceros are very delicate and under marine conditions not stable. Thus they are subject to dissolution and are not to be found in sediments, although they certainly serve as barium conveyors to the sea floor (BRONGERSMA-SANDERS, 1967).

BOWEN (1966) reports concentration factors (ppm Ba in fresh organisms/ppm Ba in sea water) for plankton and brown algae to be 120 and 260, respectively.

Algae from the coast of Great Britain show seasonal variance in their Ba content in ash (upto 900 ppm Ba; BLACK and MITCHELL, 1952).

For terrestrial plants, an extensive study exists for bryophytes (SHACKLETTE, 1965). Some of the species investigated concentrate Ba considerably. The highest enrichment factor for Ba in bryophyte ash versus soil is 2,000. Equisetum (horsetail) ash was analysed by CANNON et al. (1968) and BOROVIK-ROMANOVA (1939). Ba content in ash of this plant (70 to 4,500 ppm) resembles approximately the concentration in the respective substrata.

Phanerogames reabsorb distinct amounts of Ba from the soil. Fir and spruce have 500 to 6,200 ppm Ba in ash with the highest concentrations in twigs (LOTSPEICH and MARKWARD, 1963). Black walnut, hickory, and red-ash leaves contain 870 to 2,570 ppm. Relative Ba enrichment was reported for oaks, which gave upto 2.30% Ba in the ash of twigs (BLOSS and STEINER, 1960). The average for Ba in ash of legumes is 1,420 ppm (CANNON, 1964).

Table 56-L-1. *Barium in plants*

Plant	Ba in dry tissue ppm	Ba in ash ppm	References
Schizophyta			
Bacteria	—62,000		Foerster and Foster (1966)
Phycophyta			
Coccoliths		10 — 30	Turekian and Tausch (1964)
Diatoms		20 —30,000	Vinogradova and Koval'skiy (1962)
Brown algae	0.4—120 ⌀ 31	270 — 900	Puchelt (1967)[a]
Red algae	50	0.6— 5.6	Puchelt (1967)[a]
Bryophyta	5—200 ⌀ 150	200 —50,000	Puchelt (1967)[a] Bowen (1966)[a]
Pteridophyta			
Equisitinae		30 — 4,500	Cannon et al. (1968)
Ferns	8		Bowen (1966)[a]
Spermatophyta			
Conifers		10 — 100	Puchelt (1967)[a]
		500 — 6,200	Lotspeich and Markward (1963)
Angiosperms	14		Bowen (1966)[a]
Deciduous trees		10 — 2,700	Puchelt (1967)[a]
		—23,000	Bloss and Steiner (1960) Robinson et al. (1950)
Legumes		average: 1,420	Cannon (1964)

[a] Compilations.

II. Animals

Organisms are only important for trace element geochemistry, if they occur in large amounts. Zooplankton (especially crustacees) from the Black Sea shows Ba accumulation upto 2,000 ppm in ash (Vinogradova and Koval'skiy, 1962). Since these animals constitute about 80% of the planktonic population in the survey area, they may also contribute to the Ba content of marine sediments.

Protozoan skeletons and shells consisting of $CaCO_3$ or SiO_2 contain upto 270 ppm Ba. They are the source for Ba in pelagic globigerina and radiolarian oozes. From investigations of Arrhenius (1963) it must be concluded that in Recent planktonic foraminifera, Ba is mostly bound to organic matter (upto 700 ppm in ash). In Table 56-L-2 the radiolaria Acantharia and the rhizopod Xenophyophora, which concentrate $BaSO_4$ in their skeletons, are listed. After death these skeletons are dissolved, but during this process they transport Ba towards the sediment. No fossil skeletons of these species are reported.

Goldberg and Arrhenius (1958) assume a certain Ba accumulation by digestion of benthonic organisms, since they found Ba enrichment in fecal pellets from the sea floor.

Table 56-L-2. *Barium in animals.* (Compilations by BOWEN, 1966; PUCHELT, 1967)

	Ba in dry tissue ppm	Ba in ash ppm	Ba in hard tissue ppm	Material in hard tissue
Protozoa[a]	10—270			CaCO$_3$
Foraminifera			180	CaCO$_3$
			500	SiO$_2$
Coelanterates	11—450		8.6—35[b]	CaCO$_3$
Corals				
Ctenophora		40—2,000		
Echinodermata	20—50[c]		35	CaCO$_3$
Annelida (Vermes)		17—50[d]		
Tentaculata				
Bryozoes		12—2,000[e]		
Brachiopodes				
Mollusca	3		<1—90[b]	CaCO$_3$
Lamellibranchiates	4—75	4—500	4—75[f]	
Gastropodes			4—50[g]	
Cephalopodes \} Scaphopoda	7.2+20			
Arthopoda		15—800		
Mammalia	2.3		6.9	apatite

[a] Radiolaria Acantharia contains 5,400 ppm Ba (ARRHENIUS, 1963), BaSO$_4$ is the hard tissue in the rhizopod Xenophyophora (VINOGRADOV, 1953).

[b] Including data by FRIEDMAN (1969).

[c] Upto 5,000 ppm Ba were found in dry tissue of *Asterias Linkii* from the Barents Sea.

[d] High values (1,000 to 1,500 ppm) are reported from Black Sea plankton (VINOGRADOVA and KOVALSKIY, 1962).

[e] Including data by SCHOPF and MANHEIM (1967).

[f] With certain species of Anodonta, Pecten, Astarte and Tellina, higher Ba concentrations (upto 500 ppm) were found.

[g] Some samples of Helix species, Littorina species, and Neptuna species contain upto 500 ppm Ba in their shells.

Much support for Ba in molluscan shells has been published (cf. PUCHELT, 1967). LEUTWEIN (1963) and PILKEY (1963) found a distinct Ba enrichment in Recent mollusc shells from fresh and brackish water environments. Obviously, the structure of the shells (calcite or aragonite) is of less importance as a cause of Ba incorporation than Ba content in the water and the environment. A reconstruction of palaeoenvironments from Ba concentrations of unaltered fossil shells has been attempted by PROKOF'EV (1964), FRIEDMAN (1967) and others. TUREKIAN and ARMSTRONG (1960, 1961), investigating Recent and fossil shells, found Ba to be much higher in molluscan shells of the Fox Hill Formation (Cretaceous), South Dakota, than in Recent species. According to their studies diagenetic alterations might considerably shift the initial trace element composition, even with only slightly altered mineralogy of the shell. Animals do not contain appreciable Ba concentrations. A few data from BOWEN (1966) are included in Table 56-L-2. Additional references of detailed investigations are compiled by GMELIN (1960).

Table 56-L-3. *Barium in fuels*

Locality	Stratigraphy	Ba content in ash ppm	Method	References
Brown coal				
Australia	Permian-Tertiary	upto 800[a]	S	Swaine (1967)
				Brown and Swaine (1964)
Czechoslovakia	Tertiary	100— 1,000		Honek and Jirele (1965), see Puchelt (1967)
Germany	Tertiary	700— 7,600	X	Pietzner and Wolf (1964)
	Tertiary	200— 2,800	S	Rösler and Lange (1965), see Puchelt (1967)
U.S.A.	Tertiary	100—10,000	S	Breger *et al.* (1955), see Puchelt (1967)
			S	Brewer and Ryerson (1935), see Puchelt (1967)
	Cretaceous	100— 1,000	S	Deul and Annel (1956)
U.S.S.R.		140— 2,730		Tkachev *et al.* (1965)
Hard coal				
Australia	Permian	1,000—10,000	S	Clarke and Swaine (1962)
Canada		20— 2,200	S	Hawley (1955)
Finland		360— 1,600	S	Lokka (1943)
Germany	Carboniferous	100—27,000	S	v. Engelhardt (1936)
			S	Thilo (1934), see Puchelt (1967)
			S	Leutwein and Rösler (1956), see Puchelt (1967)
			S	Radmacher (1965), see Puchelt (1967)
	Permian	100— 500	S	Leutwein and Rösler (1956), see Puchelt (1967)
	Triassic, Jurassic	100	S	Leutwein and Rösler (1956), see Puchelt (1967)
Great Britain	Carboniferous	90— 710	S	Nicholls and Loring (1962), see Puchelt (1967)
			S	Gibson (1963)
New Zealand		250— 8,400	S	Brown and Taylor (1960)
			S	Coal Res. Company (1949)
Norway		average: 4,000	S	Butler (1953)
U.S.A.		270—22,000	S	Headlee and Hunter (1953), see Puchelt (1967)
Oils and bitumina				
Germany	Triassic—Cretaceous	—10,000	S	Heide (1938), see Puchelt (1967)
U.S.A.	Cambrian—Tertiary	X—X 00,000	S	Bell (1960), see Puchelt (1967)
			S	Erickson *et al.* (1954), see Puchelt (1967)
			S	Hyden (1961)
U.S.S.R.	Devonian—Tertiary	100— 3,000		Katchenkov (1951), see Puchelt (1967)

[a] Ba concentration in dried coal.

III. Fuels (Including Coal)

Fuels of all geologic ages contain Ba in their ashes; often in amounts considerably above the earth's crust mean (Table 56-L-3).

From Ba data in Recent plants it can be deduced that at least part of the barium originates from living plants. During diagenetic alteration, humic acids may absorb additional Ba from the involved solutions. Extremely high Ba values of coal ashes (upto 4.76%), as reported from Great Britain (REYNOLDS, 1939), may be caused by a secondary $BaSO_4$ mineralization. No general trend of Ba concentration with maturity of coal could be observed. In one instance, LEUTWEIN (1966) paralleled Ba content with the amount of clays in a brown coal profile. ERSHOV (1958) carried out several electrodialysis experiments on coal samples and concluded that Ba—in addition to other elements—was present either as soluble minerals or in weakly absorbed form. It is not bound as strongly as Ge, which could not be extracted by this procedure.

It is to be assumed that Ba is incorporated in certain metallic-organic compounds in oil, but no investigations on the specific types have been published.

Revised manuscript received: September 1971

56-M. Abundance in Common Metamorphic Rock Types

Barium concentrations in metamorphic rocks exhibit a large variation within each type (Table 56-M-1). They vary as widely as do values for all igneous and sedimentary rocks. Consequently, no meaningful Ba averages can be calculated. The only exception seems to be eclogites which form under special pressure conditions. Comparison of Ba concentrations in typical metamorphic minerals such as sillimanite, staurolite, garnet etc., show that these structures have no appreciable tolerance for Ba.

Data on Ba distribution between coexisting minerals in metamorphic rocks are included in Table 56-D-5.

Table 56-M-1. *Barium in metamorphic rocks*

Rock type and locality	No. of samples	Barium concentration range ppm	Barium concentration mean ppm	Method	Reference
Gneiss, Randesund, Norway	8	<100— 1,060	<605	X	BALL (1966)
Slightly altered gneiss, Adirondacks, U.S.A.	10	220— 1,400	610	S	ENGEL and ENGEL (1958)
"Granitized" gneiss, Adirondacks, U.S.A.	18	125— 2,600	980	S	ENGEL and ENGEL (1958)
Gneiss, Montana, U.S.A.	7	1,800— 3,800	2,580	S	FOSTER (1962)
Gneiss, Langøy, Norway	5	823— 1,300	1,050	S	HEIER (1960)
Gneiss, Lewisian, Inverness-shire, Scotland	4	300— 920	520	S	LAMBERT (1964)
Basic gneiss, Scotland	8	20— 300	110	S	O'HARA (1961)
Gneiss, S. W. Finland	40	<340— 654	≦430	S	PARRAS (1958)
Glaucophane schist, California, U.S.A.	11	7— 300	92	S	COLEMAN and LEE (1963)
Pelitic schist, Connemara, Eire	16	550— 1,850	1,300	S	EVANS (1964)
Sillimanite schist, Montana, U.S.A.	2	1,300+ 1,900	1,600	S	FOSTER (1962)
Schist, Moine, Inverness-shire, Scotland	5	800— 1,440	970	S	LAMBERT (1964)
Phyllite, Finland	174	90— 1,500	552	S	LONKA (1967)
Quartz-albite-biotite schist, New Zealand	8	450— 1,500	690	S	TAYLOR (1955)
Greenschist, New Zealand	6	10— 125	34	S	TAYLOR (1955)

Table 56-M-1. (Continued)

Rock type and locality	No. of samples	Barium concentration range ppm	Barium concentration mean ppm	Method	Reference
Hornfels, Connemara, Eire	11	670— 2,000	1,210	S	Evans (1964)
Hornfels, Palaman district, India	7		160	S	Ghose (1966)
Amphibolite, Randesund, Norway	17	260— 680	450	X	Ball (1966)
Amphibolite, Brazil	20	21— 270	67	S	Barros-Gomes et al. (1964)
Amphibolite, Adirondacks, U.S.A.	16	42— 140	84	S	Engel and Engel (1962)
Sericitite and biotite-amphibolite, Adirondacks, U.S.A	11	140— 1,650	220	S	Engel and Engel (1962)
Amphibolite, Australian shield	61	<610—> 990	817	X	Lambert and Heier (1968)
Amphibolite, S. W. Finland	20		251	S	Parras (1958)
Granulite, Australian shield	89	<420—>1,090	720	X	Lambert and Heier (1968)
Metabasite, Saxonia, Germany	9	100— 125	110	S	Mathé (1969)
Charnockite, Finland	29		570	S	Parras (1958)
Paracharnockite, Finland	24		672	S	Parras (1958)
Eclogite, Naustdal, Norway	1		<5	S	Binns (1967)
Eclogite, Nordfjord, W. Norway	6	<10— 30	≦20	S	Bryhni et al. (1969)
Eclogite, California, U.S.A.	2	15+ 300	160	S	Coleman and Lee (1963)
Eclogite, different places	5	5.6—136	56	I	Griffin and Murthy (1968)
Eclogitic rock, Münchberg, Germany	18	<100— 355	≦190	S	Hahn-Weinheimer (1959)

56-N. Behavior in Metamorphic Reactions

Only very few investigations exist on Ba behavior under metamorphism. LONKA (1967), analyzing Precambrian phyllites of Finland, observed no differences in Ba concentration between phyllites of lower and higher degree of metamorphism. Trace element data including Ba concentrations have been used by TAYLOR (1955) to discuss the origin of New Zealand metamorphic rocks under the assumption of isochemical metamorphism.

In the Adirondacks, New York, ENGEL and ENGEL (1958) studied progressive metamorphism and granitization of the major paragneiss. They found Ba to decrease with increasing metamorphism, while the Ba content in biotites increased (580, 717, 888, 1,766 ppm). Granitized gneisses of this area generally showed much higher Ba values than normal gneisses. TUREKIAN and PHINNEY (1962) could not detect characteristic changes of Ba content in garnets and coexisting biotites in a metamorphic sequence from Nova Scotia.

A special feature of transport during metamorphism is skarn formation. HIGAZY (1952) observed an increase of Ba content from epidiorite, 90 ppm (S), to biotite-epidiorite, 270 ppm, to biotite skarn, 910 ppm. The subsequent alteration to lepidomelan-skarn (720 ppm) and chlorite-skarn (270 ppm) caused distinct decreases in Ba. NESTERENKO et al. (1958) found Ba to be depleted from biotite hornfels when this rock was altered to pyroxene-garnet-skarn.

56-O. Relations to Other Elements, Crustal Distribution, Economic Importance etc.

I. Inter-element Relationships

In igneous rocks, Ba generally substitutes for K in silicate structures. A certain correlation of Ba and Ca in igneous rocks with low K was demonstrated in sections 56-D and E. In the sedimentary cycle, Ba preferentially occurs as barite, in clays and in feldspar. Presence of barite is dependant on sulfate abundance, which in turn requires suitable redox conditions. In sediments, including evaporites, the correlation Ba-K is much less pronounced than in igneous rocks. The substitution Ba-Ca observed in few carbonate minerals is of less general importance.

II. Distribution in the Earth's Crust

Details discussed in the preceding sections are summarized in the following table:

Table 56-O-1. *Abundance of Ba in important masses of the earth's crust. (Means calculated on the basis of* WEDEPOHL'S, 1969, *data on the abundance of rock units)*

Igneous intrusive rocks (mean)	728 ppm
Gabbroic rocks	246 ppm
Granites	732 ppm
Granodiorites and quartzdiorites	873 ppm
Diorites	714 ppm
Consolidated sediments (mean)	538 ppm
Sandstones (including graywackes)	316 ppm
Shales	628 ppm[a]
Carbonate rocks	90 ppm
Sea water	0.020 ppm

[a] 546 ppm, if the Russian shales low in Ba are included.

A discrepancy exists between the Ba means for consolidated sediments and magmatic rocks. Part of the Ba missing in the fossil sediments is bound in pelagic clays (mean Ba content 2,000 to 3,000 ppm; cf. Table 56-K-1) which constitute at least 10% of the total sediment mass (WEDEPOHL, 1969). But part of the discrepancy may be due to the fact that the individual data used for the averages could not be weighted properly for the compilation.

III. Technical and Economic Importance

Barite and witherite are the only barium minerals of economic interest. Descriptions of deposits and their production are given by GMELIN (1960), BROBST (1970) and others. World production of barite has been almost constant since 1964 at about 4 million short tons per year (ORR, 1970; for detailed information see U.S. Geol. Surv. Bulletin 1321).

Barite is used for drilling muds in oil and gas geology (consuming about 75% of the world's production) and for radiation-shielding concrete. Chemically treated $BaSO_4$ of fine particle size is an important filling material in the rubber, paper, and fabric industries. It is one of the components of the white pigment "lithopone". Ba compounds are used in glass and enamel production as flows and for glasses with special optical properties. Barium chloride serves as a rat poison and insecticide; barium chlorate causes the green colour of pyrotechnics; barium titanate is a ferroelectric substance used in the electro-industry; certain barium compounds are the active substances of fluorescent screens.

Reviews on barium compounds and their industrial uses are given by STÖHR and FLASCH (1953), in GMELIN (1960), in RÖMPP (1966) and in KIRK and OTHMER (1958). A bibliography on barium chemistry was compiled by SCHWIND (1952).

Acknowledgements

I am very grateful to Dr. MICHAEL FLEISCHER, U.S.G.S., Washington, D.C., for making available to me his literature surveys and giving many valuable suggestions. Thanks are also due to Professor WEDEPOHL for his many suggestions for improvement.

Revised manuscript received: September 1971

References: Section 56-A to 56-O

ABRASHEV, K. K., ILYUKHIN, V. V., BELOV, N. V.: Crystal structure of barylite, $BaBe_2Si_2O_7$. Kristallografija 9, 691 (1964), translated.

AKHUNDOV, A. R., SAPPO, P. V.: Distribution of some trace elements in formation water of the pay stratum in the Balakhany-Sabunchi-Ramaninsk oil pool. Azerb. Neft. Khoz. 39, 9 (1960).

ALANOV, A.: Geochemistry of the Lower Cretaceous deposits in the Gaurdak Region. Izv. Akad. Nauk. Turkm. SSR, Ser. Fiz.-Tekhn. Khim. i Geol. Nauk 116 (1963).

ALBEE, A. L., CHODOS, A. A.: Microprobe investigations on Apollo 11 samples. Proceedings of the Apollo 11 Lunar Science Conference 1, 135 (1970).

ALLEN, C. W.: Astrophysical Quantities. London: University of London, Athlone Press 1963.

ALLER, L. H.: The Abundance of the Elements. Interscience Monographs and Texts in Physics and Astronomy 7 (1961).

— GREENSTEIN, J. L.: The abundances of the elements in G-type subdwarfs. Astrophys. J., Suppl. 5, 139 (1960).

ANDERSEN, C. A., HINTHORNE, J. R., FREDRIKSSON, K.: Ion microprobe analysis of lunar material from Apollo 11. Proceedings of the Apollo 11 Lunar Science Conference 1, 159 (1970).

ANDERSEN, N. R., HUME, D. N.: Strontium and barium content of sea water. Advan. Chem. Ser. 73, 296 (1968).

ANDERSON, A. T., Jr.: Mineralogy of the Labrieville anorthosite, Quebec. Am. Mineralogist 51, 1671 (1966).

ANDERSON, W.: On the chloride waters of Great Britain. Geol. Mag. 82, 267 (1945).

ANNELL, C. S., HELZ, A. W.: Emission spectrographic determination of trace elements in lunar samples from Apollo 11. Proceedings of the Apollo 11 Lunar Science Conference 2, 991 (1970).

ARRHENIUS, G.: Pelagic sediments. In: The Sea. Ideas and Observations on Progress in the Study of the Seas, vol. III, ed. by M. N. HILL. New York-London: Interscience Publ. 1963.

ASSARSSON, G.: On the winning of salt from the brines in southern Sweden. Sveriges Geol. Undersökn., Ser. C 501 (1948).

ATKINS, F. B.: Pyroxenes of the Bushveld intrusion, South Africa. J. Petrol. 10, 222 (1969).

BABINA, N. M., KOTOROVICH, A. E.: Alkali and alkaline-earth metals in sedimentary rocks of the Western Siberia Lowland. Geochem. International, 508 (1966).

BABINETS, A. E., RAD'KO, N. I.: Microelements in the mineral waters of the southern slopes of the Soviet Carpathians. Geol. Zh., Akad. Nauk Ukrain. RSR 16, 21 (1956).

BACCANARI, D. P., BUCKMAN, B. A., YEVITZ, M. M., SWAIN, H. A., Jr.: The solubility of carbon-14-labelled barium carbonate in aqueous systems. Talanta 15, 416 (1968).

BAILEY, E. H., SNAVELY, P. D., Jr., WHITE, D. E.: Chemical analyses of brines and crude oil, Cymric field, Kern County, California. U.S. Geol. Surv., Profess. Papers 424-D, 306 (1961).

BAJOR, M.: Geochemistry of Tertiary lacustrine beds of the Steinheim basin in Steinheim on Albuch (Württemberg). Jahresh. Geol. Landesamtes Baden-Württemberg 7, 355 (1965).

BAKER, I.: Petrology of the volcanic rocks of St. Helena Island, South Atlantic. Bull. Geol. Soc. Am. 80, 1283 (1969).

BAKER, P. E.: Petrology of Mt. Misery Volcano, St. Kitts, West Indies. Lithos 1, 124 (1968).

— GASS, I. G., HARRIS, P. G., LEMAITRE, R. W.: The volcanological report of the Royal Society expedition to Tristan da Cunha, 1962. Phil. Trans. Roy. Soc. London 256, 439 (1964).

© Springer-Verlag Berlin · Heidelberg 1972

BALL, T. K.: The geochemistry of the Randesund Gneisses. Norsk. Geol. Tidskr. 46, 379 (1966).
BARAGAR, R. A.: Petrology of basaltic rocks in part of the Labrador Trough. Bull. Geol. Soc. Am. 71, 1589 (1960).
BARKER, F.: Sapphirine-bearing rock, Val Codera, Italy. Am. Mineralogist 49, 146 (1964).
BARROS GOMES, C. DE, SANTINI, P., DUTRA, C. V.: Petrochemistry of a Precambrian amphibolite from the Jaraguá area, Sao Paulo, Brazil. J. Geol. 72, 664 (1964).
BARTEL, A. J., FENNELLY, E. J., HUFFMAN, C., Jr., RADER, L. F., Jr.: Some new data on the arsenic content of basalt. U.S. Geol. Surv. Profess. Papers 475-B, B20 (1963).
BARTH, T. F. W.: The feldspar lattices as solvents of foreign ions. Instituto Lucas Mallada, Cursillos y Conferencias 8 (1961).
BASCHEK, B.: Häufigkeitsbestimmung für Kohlenstoff aus CH-Banden im Unterzwerg HD 140283 und in der Sonne. Z. Astrophys. 56, 207 (1962).
BEL'KEVICH, P. I., CHISTOVA, L. R., STROGONOVA, L. F.: Effect of temperature and form of ion and anion on the ion-exchange equilibrium in peat. Vestsi Akad. Nauk Belarusk. SSR, Ser. Khim. Nauk 4, 29 (1966).
BERLIN, R., HENDERSON, C. M. B.: The distribution of Sr and Ba between the alkali feldspar, plagioclase and groundmass phases of porphyritic trachytes and phonolites. Geochim. Cosmochim. Acta 33, 247 (1969).
BERTSCH, W.: Barium in Buntsandsteinen. Master's Thesis, University of Tübingen, 26 pp. 1964.
BIDELMAN, W. P., KEENAN, P. C.: The Ba II stars. Astrophys. J. 114, 473 (1951).
BINNS, R. A.: Barroisite — bearing eclogite from Naustdal, Sogu og Fjordane, Norway. J. Petrol. 8, 349 (1967).
BLACK, W. A. P., MITCHELL, R. L.: Trace elements in the common brown algae and in sea water. J. Marine Biol. Assoc. U. K. 30, 575 (1952).
BLOCK, A., PERLOFF, S.: The crystal structure of barium tetraborate, $BaO \cdot 2B_2O_3$. Acta Cryst. 19, 297—300 (1965).
BLOSS, F. D., STEINER, R. L.: Biogeochemical prospecting for manganese in north-east Tennessee. Bull. Geol. Soc. Am. 71, 1053 (1960).
BOETTCHER, A. L.: Vermiculite, hydrobiotite, and biotite in the Rainy Creek igneous complex near Libby, Montana. Clay Minerals Bull. 6, 283 (1966).
BOLTER, E., TUREKIAN, K. K., SCHUTZ, D. F.: The distribution of rubidium, cesium and barium in the oceans. Geochim. Cosmochim. Acta 28, 1459 (1964).
BOROVIK-ROMANOVA, M. F.: Spectroscopic determination of barium in the ash of plants. Tr. Biogeokhim. Lab. Akad. Nauk SSSR 5, 175 (1939).
BOSTRÖM, K., PETERSON, M. N. A.: Precipitates from hydrothermal exhalations on the east Pacific Rise. Econ. Geol. 61, 1258 (1966).
— FRAZER, J., BLANKENBURG, J.: Subsolidus phase relations and lattice constants in the system $BaSO_4$—$SrSO_4$—$PbSO_4$. Arkiv Mineral. Geol. 4, 477 (1968).
BOUSKA, V., POVONDRA, P.: Correlation of some physical and chemical properties of moldavites. Geochim. Cosmochim. Acta 28, 783 (1964).
BOWEN, H. J. M.: Strontium and barium in sea water and in marine organisms. J. Marine Biol. Assoc. U.K. 35, 451 (1956).
— Trace Elements in Biochemistry. London-New York: Academic Press 1966.
BRÄUER, H.: Spurenelementgehalte in Graniten Thüringens und Sachsens. Ph. D. Thesis, Bergakademie Freiberg, 1965.
BRAY, J. M.: Spectroscopic distribution of minor elements in igneous rocks from Jamestown, Colorado. Bull. Geol. Soc. Am. 53, 765 (1942).
BROBST, D. A.: Barite: world production, reserves and future prospects. Bull. U.S. Geol. Surv. 1321, 46 pp. (1970).
BRONGERSMA-SANDERS, M.: Barium in pelagic sediments and in diatoms. Proc. Koninkl. Ned. Akad. Wetenschap., Ser. B 70, 93 (1967).
BROWN, F. H., CARMICHAEL, I. S. A.: Quaternary volcanoes of the Lake Rudolf region: 1. The basanite-tephrite series of the Korath Range. Lithos 2, 239 (1969).

BROWN, G. M., EMELEUS, C. H., HOLLAND, J. G., PHILLIPS, R.: Mineralogical, chemical and petrological features of Apollo 11 rocks and their relationship to igneous processes. Proceedings of the Apollo 11 Lunar Science Conference 1, 195 (1970).

BROWN, H. R., SWAINE, D. J.: Inorganic constituents of Australian coals. I. Nature and mode of occurrence. J. Inst. Fuel 37, 422 (1964).

— TAYLOR, G. H.: Metamorphosed coal from the Theron Mountains (Falkland Islands Dependencies). Trans-Antarctic Expedition 1955—58, Sci. Rept. 12, 1 (1960).

BROWN, J., GRANT, C. L., UGOLINI, F. C., TEDROW, J. C. F.: Mineral composition of some drainage waters from arctic Alaska. J. Geophys. Res. 67, 2447 (1962).

BRUNO, E., GAZZONI, G.: On the system $Ba(Al_2Si_2O_8)$-$Ca(Al_2Si_2O_8)$ I. Ca-Ba substitution in polymorphic modifications of $BaAl_2Si_2O_8$. Contr. Mineral. and Petrol. 25, 144 (1970).

BRYHNI, I., BOLLINGBERG, H. J., GRAFF, P. R.: Eclogites in quartzofeldspathic gneisses of Nordfjord, West Norway. Norsk Geol. Tidsskr. 49, 193 (1969).

BUCKLEY, S. E., HOCOTT, C. R., TAGGART, M. S., Jr.: Distribution of dissolved hydrocarbons in subsurface waters. In: L. G. Weeks (ed.), Habitat of Oil. Tulsa: Am. Assoc. Petrol. Geol. 1958.

BURBIDGE, E. M., BURBIDGE, G. R.: Chemical composition of the Ba II star HD 46407 and its bearing on element synthesis in stars. Astrophys. J. 126, 357 (1957).

BURTON, J. D., MARSHALL, N. J., PHILLIPS, A. J.: Solubility of barium sulfate in sea water. Nature 217, 834 (1968).

BUTLER, B. C. M.: Chemical study of minerals from the Moine Schists of the Ardnamurchan Area, Argyllshire, Scotland. J. Petrol. 8, 233 (1967).

BUTLER, J. R.: Geochemical affinities of some coals from Svalbard (Spitzbergen). Kong. Ind.-, Handverk, Skips-fartdept., Norsk Polarinst., Skrifter 96, 1 (1953a).

— The geochemistry and mineralogy of rock weathering. I. The Lizard area, Cornwall. Geochim. Cosmochim. Acta 4, 157 (1953b).

— The geochemistry and mineralogy of rock weathering. II. The Nordmarka area, Oslo. Geochim. Cosmochim. Acta 6, 268 (1954).

BYERS, F. M., Jr.: Petrology of three volcanic suites, Umnak and Bogoslof Islands, Aleutian Islands, Alaska. Bull. Geol. Soc. Am. 72, 93 (1961).

CAMERMAN, C.: Composition d'une eau à forte salure du bassin houiller de Charleroi. Bull. Soc. Belge Géol. 60, 361 (1951).

CAMERON, A. G. V.: A revised table of abundance of the elements. Astrophys. J. 129, 676 (1959).

CAMERON, E. M.: Evaluation of sampling and analytical methods for the regional geochemical study of a subsurface carbonate formation. J. Sediment. Petrol. 36, 755 (1966).

CANNON, H. L.: Geochemistry of rocks and related soils and vegetation in the Yellow Cat area, Grand County, Utah. U.S. Geol. Surv. Bull. 1176 (1964).

— SHACKLETTE, H. T., BASTRON, H.: Metal absorption by equisetum (Horsetail). U.S. Geol. Surv. Bull. 1278-A (1968).

CARD K. D.: Metamorphism in the Agnew Lake area, Sudbury District, Ontario, Canada. Bull. Geol. Soc. Am. 75, 1011 (1964).

CARLOSN, R. M., OVERSTREET, R.: The ion exchange behavior of the alkaline earth metals. Soil Sci. 103 213 (1967).

CARMICHAEL, J. S. E.: Trachytes and their feldspars phenocrysts. Mineral. Mag. 34, 107 (1965).

— The iron-titanium oxides of salic volcanic rocks and their associated ferromagnesian silicates. Contr. Mineral. and Petrol. 14, 36 (1967a).

— The mineralogy and petrology of the volcanic rocks from the Leucite Hills, Wyoming. Contr. Mineral. and Petrol. 15, 24 (1967b).

— McDONALD, A.: The geochemistry of some natural acid glasses from the North Atlantic Tertiary volcanic province. Geochim. Cosmochim. Acta 25, 189 (1961).

CAYREL, R., CAYREL DE STROBEL, G.: Abundance determination from stellar spectra. Ann. Rev. Astronomy Astrophys. 4, 1 (1966).

CHAO, E. C. T.: The petrographic and chemical characteristics of tektites. In: J. O'KEEFE (ed.), Tektites. Chicago: University Chicago Press 1963.

CHAPMAN, D. R., KEIL, K., ANNELL, C.: Comparison of Macedon and Darwin glass. Geochim. Cosmochim. Acta **31**, 1595 (1967).
CHOW, T. J., GOLDBERG, E. D.: On the marine geochemistry of barium. Geochim. Cosmochim. Acta **20**, 192 (1960).
— PATTERSON, C. C.: Concentration profiles of barium and lead in Atlantic waters off Bermuda. Earth Planet. Sci. Lett. **1**, 397 (1966).
CHURCH, T. M.: Marine barite. Ph. D. Thesis, University of California, San Diego, 1970.
CLARKE, D. B.: Tertiary basalts of Baffin Bay: possible primary magma from the mantle. Contr. Mineral. and Petrol. **25**, 203 (1970).
CLARKE, M. C., SWAINE, D. J.: Trace elements in coal. I. New South Wales coals. II. Origin, mode of occurrence, and economic importance. Fuel Research Tech. Comm. **45**, 115 (1962).
CLAYTON, D. D.: Implications of the solar-system abundances near atomic weight 90. J. Geophys. Res. **69**, 5081 (1964).
CLIFFORD, T. N., NICOLAYSEN, L. O., BURGER, A. J.: Petrology and age of the pre-Otavi basement granite at Franzfontein, Northern South West Africa. J. Petrol. **3**, 244 (1962).
— ROOKE, J. M., ALLSOPP, H. L.: Petrochemistry and age of the Franzfontein granitic rocks of Northern South West Africa. Geochim. Cosmochim. Acta **33**, 973 (1969).
Coal Research Committee: D.S.I.R. Wellington N. Z., Coal Report No. **249** (1949).
COATS, R. R.: Magmatic differentiation in Tertiary and Quarternary volcanic rocks from Adak and Kanaga Islands, Aleutian Islands, Alaska. Bull. Geol. Soc. Am. **63**, 485 (1952).
— Geology of Buldir Island, Aleutian Islands, Alaska. Bull. U.S. Geol. Surv. **989-A**, 1 (1953).
— Geologic reconnaissance of Semisopochnoi Island, Western Aleutian Islands, Alaska. Bull. U.S. Geol. Surv. **1028-O**, 477 (1959).
— NELSON, W. H., LEWIS, R. Q., POWERS, H. A.: Geologic reconnaissance of Kiska Island, Aleutian Islands, Alaska. Bull. U.S. Geol. Surv. **1028-R**, 568 (1961).
COCCO, G.: Genesis of the Elba granite rocks: geochemistry of strontium and barium. Rend. Soc. Mineral. Ital. **9**, 48 (1953).
COLEMAN, R. G.: Low temperature reaction zones and alpine ultramafic rocks of California, Oregon and Washington. Bull. Geol. Soc. Am. **1247** (1967).
— LEE, D. E.: Glaucophane bearing metamorphic rock types of the Cazadero Area, California. J. Petrol. **4**, 260 (1963).
— PAPIKE, J. J.: Alkali amphiboles from the blueschists of Cazadero, California. J. Petrol. **9**, 105 (1968).
COLLINS, A. G.: Chemistry of some Anadarko basin brines containing high concentrations of iodide. Chem. Geol. **4**, 169 (1969).
— ZELINSKI, W. P.: The solubilities of barium and strontium sulfates in oil field brines. Am. Chem. Soc. Div. Water Air Waste Chem., Preprints **6**, 7 (1966).
COLVILLE, A., STAUDHAMMER, K.: A refinement of the structure of barite. Am. Mineralogist **52**, 1877—1880 (1967).
COMPSTON, W., CHAPPELL, B. W., ARRIENS, P. A., VERNON, M. J.: The chemistry and age of Apollo 11 lunar material. Proceedings of the Apollo 11 Lunar Science Conference **2**, 1007 (1970).
CORLETT, M., RIBBE, P. H.: Electron probe microanalysis of minor elements — plagioclase feldspars. Schweiz. Mineral. Petrog. Mitt. **47**, 317 (1967).
CORNWALL, H. R.: Calderas and associated volcanic rocks near Beatty, Nye County, Nevada. Geol. Soc. Am. Petrologic Studies, Buddington volume, 357 (1962).
ROSE, H. J., Jr.: Minor elements in Keweenawan lavas, Michigan. Geochim. Cosmochim. Acta **12**, 209 (1957).
CORREIA NEVES, J. M.: Genese des zonar gebauten Beryllpegmatits von Venturinha (Viseu, Portugal) in geochemischer Sicht. Beitr. Mineral. Petrog. **10**, 357 (1964).
COX, K. G., HORNUNG, G.: The petrology of the Karroo basalts of Basutoland. Am. Mineralogist **51**, 1414 (1966).
— JOHNSON, R. L., MONKMAN, L. J., STILLMAN, C., VAIL, J. R., WOOD, D. N.: The geology of the Nuanetsi igneous province. Philos. Trans. Roy. Soc. London **257**, 71 (1965).

Cox, K. G., Macdonald, R., Hornung, G.: Geochemical and petrographic provinces in the Karroo basalts of Southern Africa. Am. Mineralogist 52, 1451 (1967).

Cuttitta, F., Clarke, R. S., Jr., Carron, M. K., Annell, C. S.: Martha's Vineyard and selected Georgia tektites: new chemical data. J. Geophys. Res. 72, 1343 (1967).

Dana, J. D., Dana, E. S.: In: C. Palache, H. Berman, Frondel, C.: The System of Mineralogy, 7th ed. New York: John Wiley & Sons 1951 and 1952.

Danchin, R. V.: Aspects of the geochemistry of some selected south African fine grained sediments. Ph. D. Thesis, University of Cape Town, 1970.

Dawson, J. B.: Basutoland kimberlites. Bull. Geol. Soc. Am. 73, 545 (1962).

Degens, E. T.: Geochemical investigations into the country rock of the fluorite-barite-Co-Ni-Bi-Ag-Cl- ore veins of the middle Black Forest. Glückauf 92, 842 (1956).

— Williams, E. G., Keith, M. L.: Environmental studies of Carboniferous sediments, part 1: Geochemical criteria for differentiating marine and fresh water shales. Bull. Am. Ass. Petrol. Geologists 41, 2427 (1957).

Delkeskamp, R.: Die Schwerspathvorkommnisse in der Wetterau und Rheinhessen und ihre Entstehung. Notizbl. geol. Landesanst. Hessen, IV. F. 21, 47 (1900).

— Die weite Verbreitung des Baryums in Gesteinen und Mineralquellen und die sich hieraus ergebenden Beweismittel für die Anwendbarkeit der Lateralsecretions- und Thermaltheorie auf die Genesis der Schwerspathgänge. Z. prakt. Geol. 10, 117 (1902).

DeVilliers, P. R.: The chemical composition of the water of the Orange River at Vioolsdrif, Cape Province. Rep. Suid-Afrika, Dept. Mynwese Ann. Geol. Opname 1, 197 (1962).

Dietrich, R. V., Heier, K. S.: Differentiation of quartz-bearing syenite (nordmarkite) and riebeckitic-arfvedsonite granite (ekerite) of the Oslo series. Geochim. Cosmochim. Acta 31, 275 (1967).

Dixon, J. E., Cann, J. R., Renfrew, C.: Obsidian and the origins of trade. Sci. Am. 218, 38 (1968).

Dodge, F. C. W., Papike, J. J., Mays, R. E.: Hornblendes from granitic rocks of the Central Sierra Nevada Batholith, California. J. Petrol. 9, 378 (1968).

— Smith, V. C., Mays, R. E.: Biotites from granitic rocks of the Central Sierra Nevada Batholith, California. J. Petrol. 10, 250 (1969).

Dodonov, Ya. Ya., Eferova, L. V., Kolosova, V. S.: Über die Salze der Erdalkalimetalle in den Wässern der Bohrungen von der Gaslagerstätte Saratov. Dokl. Akad. Nauk SSSR 65, 887 (1949).

Dowgiallo, J.: The occurrence of brines within the Koobrzeg Unit, their genesis and relation to tectonics. Bull. Acad. Polon. Sci. Sér. Sci. Géol. Géograph. 13, 305 (1965).

Drake, M. J., McCallum, I. S., McKay, G. A., Weill, D. F.: Mineralogy and petrology of Apollo 12 sample no. 12013: a progress report. Earth Planet. Sci. Lett. 9, 103 (1970).

Duchesne, J. C.: Strontium vs. calcium and barium vs. potassium relations in plagioclases from southern Rogaland anorthosites. Ann. Soc. Geol. Belg. Bull. 90, 643 (1968).

Duke, M. B., Silver, L. T.: Petrology of eucrites, howardites and mesosiderites. Geochim. Cosmochim. Acta 31, 1637 (1967).

Dunham, A. C.: The felsites, granophyres, explosion breccias, and tuffisites of the north east margin of the Tertiary igneous complex of Rhum, Inverness-shire. Quart. J. Geol. Soc. London 123, 327 (1968).

Durfor, G. N., Becker, E.: Public water supplies for the 100 largest cities in the United States 1962. U.S. Water Supply Papers 1812 (1964).

Durum, W. H., Haffty, J.: Occurrence of minor elements in water. U.S. Geol. Surv. Circular 445 (1961).

— — Implications of the minor element content of some major streams of the world. Geochim. Cosmochim. Acta 27, 1 (1963).

— Heidel, S. G., Tison, L. J.: World wide runoff of dissolved solids. Int. Ass. Sci. Hydrology Gen. Assembly Helsinki 1960, Commission of Surface Waters, Publ. 51, 618 (1960).

Duval, J. E., Kurbatov, M. H.: The adsorption of cobalt and barium ions by hydrous ferric oxide at equilibrium. J. Phys. Chem. 56, 982 (1952).

ECKERMANN, H. VON: The distribution of barium in the alkaline rocks and fenites of Alnö Island (Sweden). Intern. Geol. Congr. Rept. 18th Session Gt. Britain, Part II, 46 (1948).
— The distribution of barium and strontium in the rocks and minerals of the syenitic and alkaline rocks of Alnö island. Arkiv Mineral. Geol. 1, 367 (1952).
— The strontium and barium contents of the Alnö carbonatites. Min. Soc. of India, IMA volume, 106 (1966).
— The strontium and barium contents of the Alnö sovites and the alkaline, carbonatitic, and ultrabasic rocks. Arkiv Mineral. Geol. 4, 417 (1967).
EL-HINNAWI, E. E.: Trace element distribution in Chilean ignimbrites. Contr. Mineral. and Petrol. 24, 50 (1969).
EMILIANI, F., VESPIGNANI-BALZANI, G. C.: Sr and Ba distribution in the Predazzo granite. Mineral. Petrogr. Acta 10, 81 (1964).
EMMERMANN, R.: Differentiation und Metasomatose des Albtalgranits (Südschwarzwald). Neues Jahrb. Mineral. Abhandl. 109, 94 (1968).
— Genetic relations between two generations of K feldspar in a granite pluton. Neues Jahrb. Mineral. Abhandl. 111, 289 (1969).
EMMONS, R. C.: Selected petrogenetic relationships of plagioclase. Chemical analyses. Geol. Soc. Am. Mem. 52, 11 (1952).
ENGEL, A. E. J., ENGEL, C. G.: Progressive metamorphism and granitization of the major paragneiss, northwest Adirondack Mountains, New York. Bull. Geol. Soc. Am. 69, 1369 (1958).
— — Progressive metamorphism and granitization of the major paragneiss, northwest Adirondack Mountains, New York. Bull. Geol. Soc. Am. 71, 1 (1960).
— — Hornblendes formed during progressive metamorphism of amphibolites, northwest Adirondack Mountains, New York. Bull. Geol. Soc. Am. 73, 1499 (1962).
— HAVENS, R. G.: Chemical characteristics of oceanic basalts and the upper mantle. Bull. Geol. Soc. Am. 76, 719 (1965).
ENGEL, C. G.: Igneous rocks and constituent hornblendes of the Henry Mountains, Utah. Bull. Geol. Soc. Am. 70, 951 (1959).
— SHARP, R. P.: Chemical data on desert varnish. Bull. Geol. Soc. Am. 69, 487 (1958).
ENGELHARDT, W. VON: Die Geochemie des Barium. Chem. Erde 10, 187 (1936).
ERICKSON, R. L., BLADE, L. V.: Geochemistry and petrology of the alkalic igneous complex at Magnet Cove, Arkansas. U.S. Geol. Surv., Profess. Papers. 425 (1963).
ERICSON, D. B., EWING, M., WOLLIN, G., HEEZEN, B. C.: Atlantic deep-sea sediment cores. Bull. Geol. Soc. Am. 72, 193 (1961).
— WOLLIN, G.: Correlation of six cores from the equatorial Atlantic and the Caribbean. Deep-Sea Res. 3, 104 (1956).
ERSHOV, V. M.: Relation of germanium to the organic matter in fossil coals. Geochemistry, 763 (1958).
EUGSTER, O., TERA, F., WASSERBURG, G. J.: Isotopic analyses of barium in meteorites and terrestrial samples. J. Geophys. Res. 74, 3897 (1969).
EVANS, B. W.: Fractionation of elements in the pelitic hornfelses of the Cashel-Lough Wheelaun intrusion, Connemara, Eire. Geochim. Cosmochim. Acta 28, 127 (1964).
EWART, A., TAYLOR, S. R., CAPP, A. C.: Geochemistry of the pantellerit of Mayor Island, New Zealand. Contr. Mineral. and Petrol. 17, 116 (1968).
FAIRBAIRN, H. W., AHRENS, L. H., GORFINKLE, L. G.: Minor element content of Ontario diabase. Geochim. Cosmochim. Acta 3, 34 (1953).
— SCHLECHT, W. G., STEVENS, R. E., DENNEN, W. H., AHRENS, L. H., CHAYES, F.: A cooperative investigation of precision and accuracy in chemical, spectrochemical and modal analysis of silicate rocks. Bull. U.S. Geol. Surv. 980 (1951).
FENNER, P., HAGNER, A. F.: Correlation of variations in trace elements and mineralogy of the Esopus Formation, Kingston, New York. Geochim. Cosmochim. Acta 31, 237 (1967).
FISCHER, K.: Verfeinernug der Kristallstruktur von Benitoit $BaTi[Si_3O_9]$. Z. Krist. 129, 222 (1969).
FISHER, R. L., ENGEL, C. G.: Ultramafic and basaltic rocks dredged from the near shore flank of the Tonga trench. Bull. Geol. Soc. Am. 80, 1373 (1969).

FLANAGAN, F. J.: U.S. Geological Survey standards. II. First compilation of data for the new U.S.G.S. rocks. Geochim. Cosmochim. Acta 33, 81 (1969).
FLEISCHER, M.: Summary of new data on rock samples G-1 and W-1, 1962—1965. Geochim. Cosmochim. Acta 29, 1263 (1965).
— U.S. Geological Survey standards. — I. Additional data on rocks G-1 and W-1, 1965—1967. Geochim. Cosmochim. Acta 33, 65 (1969).
— STEVENS, R. E.: Summary of new data on rock samples G-1 and W-1. Geochim. Cosmochim. Acta 26, 525 (1962).
FOERSTER, H. F., FOSTER, J. W.: Endotrophic calcium, strontium, and barium spores of bacillus megaterium and bacillus cereus. J. Bacteriol. 91, 1333 (1966).
FORNASERI, M., SCHERILLO, A., VENTRIGLIA, U.: La regione vulcanica dei Colli Albani. Consiglio Nazionale delle Ricerche Centro di Mineralogia e Petrografia Aziende Tipografiche Eredi Dott. G. Bardi (1963).
FOSTER, R. J.: Precambrian corundum bearing rocks, Madison Range, southwestern Montana. Bull. Geol. Soc. Am. 73, 131 (1962).
FREDRIKSON, K., KEIL, K.: The light-dark structure in the Pantar and Kapoeta stone meteorites. Geochim. Cosmochim. Acta 27, 717 (1963).
FRICKE, K.: Neue hydrogeologische Untersuchungen und Neubohrungen im Heilquellengebiet von Bad Hermannsborn, NRW. Das Gas- Wasserfach 109, 251 (1968).
FRIEDMAN, G. M.: Obtaining paleoenvironmental information. U. S. Patent Office No. 3, 343, 917 (1967).
— Geology and geochemistry of reefs, carbonate sediments, and waters, Gulf of Aqaba (Elat), Red Sea. J. Sediment. Petrol. 38, 895 (1968).
— Trace elements as possible environmental indicators in carbonate sediments. Soc. Econ. Paleontologists Mineralogists, Spec. Publ. 14, 193 (1969).
FUJITA, Y., TSUJI, I.: Spectrophotometry of Y Canum Venaticorum. Publ. Dominion Astrophys. Obs. 12, 339 (1965).
GANGULY, A. K., MUKHERJEE, S. K.: The cation exchange behaviour of heteroionic and homoionic clays of silicate minerals. J. Phys. and Colloid Chem. 55, 1429 (1951).
GARRELS, R. M., THOMPSON, M. E., SIEVER, R.: Stability of some carbonates at 25° C and one atmosphere total pressure. Am. J. Sci. 258, 402 (1960).
GARSON, M. S.: Carbonatites in Malawi, pp. 33—77. In: O. F. TUTTLE, GITTINS, J. (eds.), Carbonatites. New York: Interscience Publishers 1967.
GAST, P. W.: Limitations on the composition of the upper mantle. J. Geophys. Res. 65, 1287 (1960).
— Terrestrial ratio of potassium to rubidium and the composition of the earth's mantle. Science 147, 858 (1965).
— HUBBARD, N. J., WIESMANN, H.: Chemical composition and petrogenesis of basalts from Tranquillity Base. Proceedings of the Apollo 11 Lunar Science Conference 2, 1143 (1970).
GATES, G. L., CARAWAY, W. H.: Oil-well scale formation in water flood operations using ocean brines, Wilmington, California. U.S. Bur. Mines, Rept. Invest. 6658 (1965).
GAY, P., ROY, N. N.: The mineralogy of the potassium-barium feldspar series. III. Subsolidus relationships. Mineral. Mag. 36, 914 (1968).
GERASIMOVSKII, V. I.: Geochemical features of agpaitic nepheline-syenites, p. 104. In: VINOGRADOV (ed.), Chemistry of the Earth's Crust, I. Jerusalem: 1966.
— BELYAEV, Yu. I: Content of manganese, barium and strontium in alkaline rocks of the Kola Peninsula. Geochemistry, 1161 (1963).
GHOSE, N. C.: Behaviour of trace elements during thermal metamorphism and (oblique) or granitization of the metasediments and basic igneous rocks. Geol. Rundschau 55, 608 (1966).
GIBSON, J.: Personal communication (1963).
GIUSCA, D., IONESCU, J.: Geochemistry of volcanic rocks in the Gutai Mountains. Probl. Geokhim., Akad. Nauk SSSR, Inst. Geokhim. i Analit. Kkim., 436 (1965).
GMELIN: Gmelins Handbuch der anorganischen Chemie, 8th edit. Barium Ergänzungsband. Weinheim: Verlag Chemie 1960.

GOLD, D. P.: Average chemical composition of carbonatites. Ec. Geol. 58, 988 (1963).
— Alkaline ultrabasic rocks in the Montreal Area, Quebec, p. 288. In: WYLLIE (ed.), Ultramafic and Related Rocks. New York: John Wiley & Sons 1967.
GOLDBERG, E. D., ARRHENIUS, G. O. S.: Chemistry of Pacific pelagic sediments. Geochim. Cosmochim. Acta 13, 153 (1958).
GOLDBERG, L., MÜLLER, E. A., ALLER, L. H.: The abundances of the elements in the solar atmosphere. Astrophys. J., Suppl. 5, 1 (1960).
GOLES, G. G., RANDLE, K., OSAWA, M., SCHMITT, R. A., WAKITA, H., EHMANN, W. D., MORGAN, J. W.: Elemental abundances by instrumental activation analyses in chips from 27 lunar rocks. Proceedings of the Apollo 11 Lunar Science Conference, 2, 1165 (1970).
GORDON, C. P.: The Ba II and N stars as a temperature sequence. Astrophys. J. 153, 915 (1968).
GRAF, D. L.: Geochemistry of carbonate sediments and sedimentary carbonate rocks. III. Minor element distribution. Illinois State Geol. Surv. Circ. 301 (1960).
GREEN, D. H.: The petrogenesis of the high-temperature peridotite intrusion in the Lizard area, Cornwall (England). J. Petrol. 5, 134 (1964).
GREENLAND, L. P., LOVERING, J. F.: Minor and trace element abundances in the chondritic meteorites. Geochim. Cosmochim. Acta 29, 821 (1965).
— — Fractionation of fluorine, chlorine, and other trace elements during differentiation of a tholeiitic magma. Geochim. Cosmochim. Acta 30, 963 (1966).
GRESENS, R. L.: Tectonic-hydrothermal pegmatites. II. An example. Contr. Mineral. and Petrol. 16, 1 (1967).
GRIFFIN, W. L., MURTHY, V. R.: Abundances of K, Rb, Sr, and Ba in ultramafic rocks and minerals. Earth Planet. Sci. Lett. 4, 497 (1968).
GRIM, R. E., DIETZ, R. S., BRADLEY, W. F.: Clay mineral composition of some sediments from the Pacific Ocean off the Californian coast and the Gulf of California. Bull. Geol. Soc. Am. 60, 1785 (1949).
GROHMANN, H., SCHROLL, E.: Seltene Elemente in Granitoiden der südlichen Böhmischen Masse. Tschermaks Mineral. Petrog. Mitt. 11, 348 (1966).
GROSS, E.: Pabstite, the tin analogue of benitoite. Am. Mineralogist 50, 1164 (1965).
GROUT, F. F.: The compositions of some African granitoid rocks. J. Geol. 43, 281 (1935).
GRUSHKO, Ya. M., SHIPITSYN, S. A.: Toxic substances in the drinking waters of Irkutsk from spectral analyses. Gigiena i Sanit. 13, 4 (1948).
GUNN, B. M.: K/Rb and K/Ba ratios in Antarctic and New Zealand tholeiites and alkali basalts. J. Geophys. Res. 70, 6241 (1965).
— Modal and element variation in Antarctic tholeiites. Geochim. Cosmochim. Acta 30, 881 (1966).
HAHN-WEINHEIMER, P.: Geochemische Untersuchungen an den ultrabasischen und basischen Gesteinen der Münchberger Gneismasse (Fichtelgebirge). Neues Jahrb. Mineral., Abhandl. 92, 203 (1959).
— LUECKE, W.: Garnets from the eclogites of the Muenchberger gneiss massif (N. E. Bavaria). Can. Mineralogist 7, 764 (1963).
— ACKERMANN, H.: Geochemical investigation of differentiated granite plutons of the Southern Black Forest — II. The zoning of the Malsburg granite pluton as indicated by the elements titanium, zirconium, phosphorus, strontium, barium, rubidium, potassium, and sodium. Geochim. Cosmochim. Acta 31, 2197 (1967).
HALL, A.: The variation of some trace elements in the Rosses granite complex, Donegal. Geol. Mag. 104, 99 (1967).
HAMAGUCHI, H., REED, G. W., TURKEWITSCH, A.: Uranium and barium in stone meteorites. Geochim. Cosmochim. Acta 12, 337 (1957).
HANOR, J. S.: Barite saturation in sea water. Geochim. Cosmochim. Acta 33, 894 (1969).
HASKIN, L. A., ALLEN, R. O., HELMKE, P. A., PASTER, T. P., ANDERSON, M. R., KOROTEV, R. L., ZWEIFEL, K. A.: Rare earths and other trace elements in Apollo 11 lunar samples. Proceedings of the Apollo 11 Lunar Science Conference 2, 1213 (1970).
HASLAM, H. W.: The crystallization of intermediate and acid magmas at Ben Nevis, Scotland. J. Petrol. 9, 84 (1968).

Hawley, J. E.: Germanium content of some Nova Scotian coals. Ec. Geol. 50, 517 (1955).
Heide, F., Christ, W.: Zur Geochemie des Strontiums und Bariums. Chem. Erde 16, 327 (1953).
Heidel, S. G., Frenier, W. W.: Chemical quality of water and trace elements in the Patuxent River Basin. Maryland Geol. Surv. Rept. Invest. No. 1, 1 (1965).
Heier, K. S.: Petrology and geochemistry of high-grade metamorphic and igneous rocks on Langøy, Northern Norway. Norsk Geol. Undersökning 207 (1960).
— The amphibolite-granulite facies transition reflected in the mineralogy of potassium feldspars. Instituto Lucas Mallada, Cursillos y Conferencias fasc. 8 (1961).
— Trace elements in feldspars — a review. Norsk Geol. Tidsskr. 42, 415 (1962).
— Geochemistry of the nepheline-syenite on Stjernøy, North Norway. Norsk Geol. Tidsskr. 44, 205 (1964).
— A geochemical comparison of the Blue Mountain (Ontario, Canada), and Stjernøy (Finnmark, North Norway) nepheline syenites. Norsk Geol. Tidsskr. 45, 41 (1965).
— Taylor, S. R.: Distribution of Ca, Sr and Ba in southern Norwegian pre-Cambrian alkali feldspars. Geochim. Cosmochim. Acta 17, 286 (1959).
— Chappell, B. W., Arriens, P. A., Morgans, J. W.: The geochemistry of four Icelandic basalts. Norsk. Geol. Tidsskr. 46, 427 (1966).
Heinrich, E. W. M.: Micas of the Brown Derby pegmatites, Gunnison County, Colorado. Am. Mineralogist 52, 1110 (1967).
— Levinson, A. A., Levandowski, D. W., Hewitt, C. H.: Studies in the natural history of micas. Univ. Mich., Final Rept. Project M 978 (1953).
— — — Geochemistry of muscovite and related micas. 20th Internat. Geological Congr. Mexico 1956.
Helfer, H. L., Wallerstein, G., Greenstein, J. L.: Metal abundance in the sub-giant Hercules and three other dG stars. Astrophys. J. 138, 97 (1963).
Henderson, C. M. B.: Minor element chemistry of leucite and pseudoleucite. Mineral. Mag. 35, 596 (1965).
Herrmann, A. G.: Über das Vorkommen einiger Spurenelemente in Salzlösungen aus dem deutschen Zechstein. Kali Steinsalz 3, 209 (1961).
Herz, N., Dutra, C. V.: Minor element abundance in a part of the Brazilian shield. Geochim. Cosmochim. Acta 21, 81 (1960).
— — Geochemistry of some kyanites from Brazil. Am. Mineralogist 49, 1290 (1964).
— — Trace elements in alkali feldspars, Quadrilátero ferrifero, Minas Gerais, Brazil. Am. Mineralogist 51, 1593 (1966).
Hewlett, C. G.: Optical properties of potassic feldspars. Bull. Geol. Soc. Am. 70, 511 (1959).
Hey, M. H.: Catalogue of Meteorites, 3rd edit. British Museum 1966.
Heydegger, H. R., Kuroda, P. K.: Natural occurrence of the short-lived barium and strontium isotopes. J. Inorg. Nucl. Chem. 12, 12 (1969).
Hietanen, A.: Distribution of elements in biotite-hornblende pairs and in an orthopyroxene-clinopyroxene pair from zoned plutons, Northern Sierra Nevada, California. Contr. Mineral. and Petrol. 30, 161 (1971).
Higazy, R. A.: Petrogenesis of perthite pegmatites in the Black Hills, South Dakota. J. Geol. 57, 555 (1949).
— Behaviour of the trace elements in a front of metasomatic-metamorphism in the Dalradian of Co. Donegal. Geochim. Cosmochim. Acta 2, 170 (1952).
— Observations on the distribution of trace elements in the perthite pegmatites of the Black Hills, South Dakota. Am. Mineralogist 38, 172 (1953).
— Trace elements of volcanic ultrabasic potassic rocks of south-western Uganda and adjoining parts of the Belgian Congo. Bull. Geol. Soc. Am. 65, 39 (1954).
Hilmer, W.: Die Struktur der Hochtemperaturform des $BaGeO_3$. Acta Cryst. 15, 1101 (1962).
Hintze, C.: Handbuch der Mineralogie, Bd. I, 3. Abt., 1. Hälfte. Berlin: Walter de Gruyter 1930.
— Handbuch der Mineralogie, Bd. I, 3. Abt., 2. Hälfte. Berlin: Walter de Gruyter 1930.
— Handbuch der Mineralogie. Ergänzungsband I. Berlin: Walter de Gruyter 1938.
— Handbuch der Mineralogie. Ergänzungsband II. Berlin: Walter de Gruyter 1960.

HITCHON, B.: The geochemistry, mineralogy, and origin of pegmatites of three Scottish Precambrian metamorphic complexes. Intern. Geol. Congr. Rept. 21st session Copenhagen XVII, 36 (1960).

HOLLAND, H. D.: Some applications of thermochemical data to problems of ore deposits. II. Mineral assemblages and the composition of ore-forming fluids. Ec. Geol. 60, 1101 (1965).

HOSKINS, H. A.: Analyses of West Virginia brines. State of West Virginia Geol. and Ec. Surv. Rept. Invest. 1 (1947).

HOSS, H., ROY, R.: On natural phillipsite, gismondite, harmotome, chabasite, gmelinite. Beitr. Mineral. Petrog. 7, 389 (1960).

HOTZ, P. E.: Petrology of granophyre in diabase near Dillsburg, Pa. Bull. Geol. Soc. Am. 64, 675 (1953).

HOWIE, R. A.: The geochemistry of the charnockite series of Madras, India. Trans. Roy. Soc. Edinburgh 62, 725 (1953—55).

HUBBARD, N. J., GAST, P. W., WIESMANN, H.: Rare earth, alkaline and alkali metal and $^{87/86}$Sr data for subsamples of lunar sample 12013. Earth Planet. Sci. Lett. 9, 181 (1970).

— MEYER, C., Jr., GAST, P. W.: The composition and derivation of Apollo 12 soils. Earth Planet. Sci. Lett. 10, 341 (1971).

HUCKENHOLZ, G.: Personal communication (1969).

HÜGI, TH., SWAINE, D. J.: The geochemistry of some Swiss granites. J. Proc. Roy. Soc. New S. Wales 96, 65 (1963).

HUNZICKER, J. C.: Zur Geologie und Geochemie des Gebietes zwischen Valle Antigoro (Provincia di Novara) und Valle di Campo (Kt. Tessin). Schweiz. Mineral. Petrog. Mitt. 46, 473 (1966).

HYDEN, H. J.: Uranium and other metals in crude oils. II. Distribution of uranium and other metals in crude oils. U.S. Geol. Surv. Bull. 1100-B, 17 (1961).

ICHIKUNI, I. M.: Barium content of barite-forming hydrothermal waters as exemplified by Tamagawa Hot Spring waters. Bull. Chem. Soc. Japan 39, 898 (1966).

IIJAMA, J. T.: Etude experimentale de la distribution d'élément en traces entre deux feldspaths. Feldspath potassique et plagioclase coexistants. I. Distribution de Rb, Cs, Sr et Ba à 600° C. Bull. Soc. Franc. Mineral. Crist. 91, 130 (1968).

IKEDA, N.: Chemical studies on the hot springs of Nasu. VII. Spectrochemical determination of rarer element contents of Motoyu spring. J. Chem. Soc. Japan, Pure Chem. Sect. 76, 711 (1955a).

— Chemical studies on the hot springs of Nasu. VIII. The spring water of the Motoyu. J. Chem. Soc. Japan, Pure Chem. Sect. 76, 713 (1955b).

IMREH, J., ECATERINE, J.: Geochemical study of some Eocene limestones from the Transylvanian Basin, Romania. Bull. Serv. Carte Geol. Alsace Lorraine 18, 277 (1965).

ISHIOKA, K.: A clinopyroxene granitic rock from Myogase, Japan. Geochem. J. 1, 95 (1967).

IWASAKI, I., KATSURA, T., TARUTANI, T., OZAWA, T., YOSHIDA, M.: Geochemical studies of Tamagawa Hot Spring. Geochem. Tamagawa Hot Springs, 7 (1963).

JAKOBSHAGEN, V., MÜNNICH, K. O.: ^{14}C-Altersbestimmung und andere Isotopenuntersuchungen an Thermalsolen des Ruhrkarbons. Neues Jahrb. Geol. Palaeontol. Monatsh. 9, 561 (1964).

JAMIESON, B. G., CLARKE, D. B.: Potassium and associated elements in tholeiitic basalts. J. Petrol. 11, 183 (1970).

JANSE, A. J. A.: Monticellite-peridotite from Mt Brukkaros, South West Africa. Univ. of Leeds Res. Inst. African Geol. 86th Ann. Rept., 21 (1962—1963).

JASMUND, K., SECK, H. A.: Geochemische Untersuchungen an Auswürflingen (Gleesiten) des Laacher-See-Gebietes. Beitr. Mineral. Petrog. 10, 275 (1964).

JOHNSON, R. L.: The geology of the Dorowa and Shawa carbonatite complexes Southern Rhodesia. Trans. Geol. Soc. S. Africa 64, 101 (1961).

JUNNER, N. R., JAMES, W. T.: Chemical analyses of gold coast rocks, ores, and minerals. Gold Coast Geol. Surv. Bull. 15, 1 (1947).

KASHAEV, A.: Crystal structure of cymrite. Dokl. Akad. Nauk SSSR 169, 201 (1966).

KATCHENKOV, S. M., FLEGONTOVA, E. I.: Trace elements in the Devonian (petroleum) deposits of the Volga-Ural region, as found by spectral analysis. Tr. Vses. Neft. Nauchn.-Issled. Geologorazved. Inst., Geol. Sbornik 83, 466 (1955).
— — Distribution of chemical elements in sedimentary rocks, waters, and in the ashes of the crudes of the Gronznyi-Dagestan region. Tr. Vses. Neft. Nauchn.-Issled. Geologorazved. Inst., Geol. Sbornik 95, 481 (1956).
— — Trace elements in bottom muds of the Indian Ocean. Tr. Vses. Neft. Nauchn.-Issled. Geologorazved. Inst. 227, 202 (1964).
KATSCHER, H., LIEBAU, F.: Über die Kristallstruktur von $Ba_3Si_3O_8$, ein Silikat mit Dreifachketten. Naturwissenschaften 52, 512 (1965).
— — Triple, quadruple and quintruple chains in barium silicates. Acta Cryst. 21, Suppl., A 58 (1966).
KAVEEV, M. S., VASIL'EV, B. V.: Hydrogeology of petroleum formations in Devonian deposits of southeast Tatar. Tr. Soveshch. Probl. Neftegaz. Uralo-Volzhskoi Oblasti, Tr. Soveshchaniya 10—15 Maya, 337 (1954, 1956).
KAY, R., HUBBARD, N. J., GAST, P. W.: Chemical characteristics and origin of oceanic ridge volcanic rocks. J. Geophys. Res. 75, 1585 (1970).
KEGEL, W. H.: Die Atmosphäre des F 6 IV–V-Sternes γ Serpentis. Z. Astrophys. 55, 221 (1962).
KELLER, W. D., KLEMME, A. W., PICKETT, E. E.: Detailed survey of the chemical composition of rock layers in an agricultural limestone quarry. Ec. Geol. 45, 461 (1950).
KENT, L. E.: The thermal waters of the Union of South Africa and South West Africa. Trans. Proc. Geol. Soc. S. Africa 52, 231 (1949).
— RUSSELL, H. D.: The warm spring on Buffelshoek, near Thabazimbi, Transvaal. Trans. Roy. Soc. S. Africa 32, 161 (1949).
KING, B. C., SUTHERLAND, D. S.: The carbonatite complexes of Eastern Uganda, p. 73. In: TUTTLE, O. F., GITTINS, V. (eds.), Carbonatites. New York: Interscience Publishers 1967.
KIRICHENKO, L. F., CHERTOV, V. M., VYSOTSKII, Z. Z., STRAZHESKO, D. N.: Sorption of cations from acid solutions on silica gels obtained by the hydrothermal method. Dokl. Akad. Nauk SSSR 164, 618 (1965).
KIRK, R. E., OTHMER, D. F.: Encyclopedia of Chemical Technology. New York: Interscience Publishers 1963.
KLEEMAN, A. W.: Sampling error in the chemical analysis of rocks. J. Geol. Soc. Australia 14, 43 (1967).
KODAIRA, K.: Atmosphärenstruktur und chemische Zusammensetzung des Schnelläufers HD 161817. Z. Astrophys. 59, 139 (1964).
KOHLRAUSCH, F.: Über gesättigte wäßrige Lösungen schwerlöslicher Salze II. Teil: Die gelösten Mengen mit ihrem Temperaturgang. Z. Physik. Chem. 64, 129 (1908).
KOLBE, P.: Geochemistry of some African and Australian granitic rocks. Ph. D. Thesis, Australian National University 1964.
— TAYLOR, S. R.: Major and trace element relationships in granodiorites and granites from Australia and South Africa. Contr. Mineral. and Petrol. 12, 202 (1966).
KOMLEV, O. I., TIKHA, N. I., ZHERDEVA, N. I.: Tricationic exchange on Transcarpathian bentonite. Visn. L'viv. Derzh. Univ., Ser. Khim. 8, 54 (1965).
KONTOROVICH, A. E., SADIKOV, M. A., SHVARTSEV, S. L.: Abundances of certain elements in the surface and ground waters of the northwestern part of the Siberian platform. Dokl. Akad. Nauk SSSR 149, 168 (1963).
KOROLEV, A.: The water content of the petroleum deposit of Kala and the chemical classification of this water. Azerb. Neft. Khoz. 15, 35 (1938).
KOZIN, A. N.: Barium in formation waters of oil pools in the Volga area near Kuibyshev. Geokhimiya 9, 937 (1964).
KRAVCHENKO, V. B.: Crystal structure of $BaB_2O_4 \cdot 4H_2O \equiv Ba[B(OH)_4]_2$. Zh. Strukt. Khim. 6, 724 (1965), translated.
KRETZ, R.: Chemical study of garnet, biotite, and hornblende from gneisses of southwestern Quebec, with analysis on distribution of elements in coexisting minerals. J. Geol. 67, 371 (1959).
— The distribution of certain elements among coexisting calcic pyroxenes, calcic amphiboles and biotites in skarns. Geochim. Cosmochim. Acta 20, 161 (1960).

KUKABAEV, B., SYDYKOV, ZH.: Hydrochemistry of subsurface waters of the Permian-Triassic deposits in the southwestern part of the Ural and Emba interfluve. Izv. Akad. Nauk Kaz. SSR, Ser. Geol. 4, 58 (1962).
KUNO, H., YAMASAKI, K., IIDA, CH., NAGASHIMA, K.: Differentiation of Hawaiian magmas. Japan. J. Geol. Geography, Trans. 28, 179 (1957).
KURODA, P. K., EDWARDS, R. R.: Radiochemical measurement of the natural fission rate of uranium and the natural occurrence of ^{140}Ba. J. Inorg. Nucl. Chem. 3, 345 (1957).
LAMBERT, I. B., HEIER, K. S.: Geochemical investigations of deep-seated rocks in the Australian shield. Lithos 1, 30 (1968).
LAMBERT, R. ST.: The relationship of the Moine schists and Lewisian gneisses near Mallaigmore, Inverness-shire. Proc. Geologists' Assoc. 75, 1 (1964).
LANDER, J.: The crystal structures of NiO · 3BaO, NiO · BaO, BaNiO and intermediate faces with composition near $Ba_2Ni_2O_5$; with a note on NiO. Acta Cryst. 4, 148 (1951).
LANDERGEEN, ST., MANHEIM, F. T.: Über die Abhängigkeit der Verteilung von Schwermetallen von der Fazies. Fortschr. Geol. Rheinld. u. Westf. 10, 173 (1963).
LARSSON, W.: Chemical analyses of Swedish rocks. Bull. Geol. Inst. Univ. Upsala 24, 47 (1932).
LAVRUKHINA, A. K., KOLESOV, G. M., KALICHEVA, I. S., AKOL'ZINA, L. D.: Activation analysis for Ce, Eu, Sc, Ba, U, and P in the dark and light varieties of the Kunashak and Pervomayskiy Poselok chondrites. Geochem. Int., 217 (1966).
LEBEDEV, B. A.: Trace elements in Jurassic and Lower Cretaceous formations of the Caspian depression. Dokl. Akad. Nauk SSSR 173, 192 (1967a).
— Comparison of trace element contents in marine and fresh water clays. Geochem. Int., 821 (1967b).
LEELANANDAM, C.: Chemical study of pyroxenes from the charnockitic rocks of Kondagalli (Andrah Pradesh), India, with emphasis on the distribution of elements in coexisting pyroxenes. Mineral. Mag. 36, 153 (1967).
LEMAITRE, R. W.: Petrology of volcanic rocks, Gough Island, South Atlantic. Bull. Geol. Soc. Am. 73, 1309 (1962).
LEUTWEIN, F.: Geochemische Untersuchungen an Alaun- und Kieselschiefern Thüringens. Arch. Lagerstättenforsch. 82, 1 (1951).
— Spurenelemente in rezenten Cardien verschiedener Fundorte. Fortschr. Geol. Rheinld. u. Westf. 10, 283 (1963).
— Geochemische Charakteristica mariner Einflüsse in Torfen und anderen quarternären Sedimenten. Geol. Rundschau 55, 97 (1966).
— WEISE, L.: Hydrogeochemische Untersuchungen an erzgebirgischen Gruben- und Oberflächenwässern. Geochim. Cosmochim. Acta 26, 1333 (1962).
LEVI, H. W., SCHIEWER, E.: Bariumaustausch an Vermikulit und Bentonit. Z. Anorg. Allgem. Chem. 337, 105 (1965).
LIEBAU, F.: Die Systematik der Silikate. Naturwissenschaften 49, 481 (1962).
— Ein Beitrag zur Kristallchemie der Schichtsilikate. Acta Cryst. B 24, 690 (1968).
LIEBENBERG, C. J.: The trace elements of the rocks of the Bushveld igneous complex. Publikasies Univ. Pretoria, Nuwe Reeks 12 (1960).
LIN, H. C., FOSTER, W. R.: Studies in the system $BaO-Al_2O_3-SiO_2$. I. The polymorphism of celsian. Am. Mineralogist 53, 134 (1968).
— — Studies in the system $BaO-Al_2O_3-SiO_2$. III. The binary system sanbornite-celsian Mineral. Mag. 37, 459 (1969).
LIPMAN, P. W.: Alkalic and tholeiitic volcanism related to the Rio Grande depression, Southern Colorado and Northern New Mexico. Bull. Geol. Soc. Am. 80, 1343 (1969).
LIPPMANN, F.: Die Kristallstruktur des Norsethit, $BaMg(CO_3)_2$. Tschermak's Min. Petr. Mitteil. 12, 299 (1968).
LITVIN, I. I.: Minor elements in Lower Cretaceous rocks of the Dnieper-Donets depression. Dokl. Akad. Nauk SSSR 139, 450 (1961).
LITVIN, S. V.: Trace elements in the Upper Carboniferous sedimentary rocks of the Donets Basin and in northeastern part of the Dnieper-Donets syncline. Dokl. Akad. Nauk SSSR 152, 1453 (1963).

LONKA, A.: Trace elements in the Finnish Precambrian phyllites as indicators of salinity at the time of sedimentation. Bull. Comm. Géol. Finlande **228** (1967).
LOTSPEICH, F. B., MARKWARD, E. L.: Minor elements in bedrock soil, and vegetation at an outcrop of the Phosphoria Formation on Snowdrift Mountain, southeastern Idaho. U.S. Geol. Surv. Bull. **1181**-F, F1 (1963).
LSPET (Lunar Sample Preliminary Examination Team): Preliminary examination of lunar samples from Apollo 11. Science **165**, 1211 (1969).
— Preliminary examination of lunar samples from Apollo 12. Science **167**, 1325 (1970).
Lunatic Asylum: Mineralogic and isotopic investigations on lunar rock 12013. Earth Planet. Sci. Lett. **9**, 137 (1970).
LURYE, L. M.: Migration of barium and strontium during country rock metasomatism in the Zambarak ore field. Dokl. Akad. Nauk SSSR **149**, 1167 (Engl. translation 185) (1963).
MACDONALD, G. A., EATON, J. P.: Hawaiian volcanoes during 1955. Bull. U.S. Geol. Surv. **1171** (1964).
MACPHERSON, H. G.: A chemical and petrographic study of Pre-Cambrian sediments. Geochim. Cosmochim. Acta **14**, 73 (1958).
MALININ, S. D.: An experimental investigation of the solubility of calcite and witherite under hydrothermal conditions. Geochemistry, 650 (1963).
MANOHAR, H., RAMASESHAN, S.: Crystal coordination of the barium ion. Proc. Ind. Acad. Sci. **60**, 317 (1964).
— — The crystal structure of barium hydroxide octahydrate Ba (OH)$_2 \cdot$ 8H$_2$O. Z. Kirst. **119**, 357 (1964).
MARKART, H., PREISINGER, A.: Zur Bestimmung der Feldspate in Gesteinen. Tschermaks Mineral. Petrog. Mitt., 3. F., 315 (1964/65).
MARKHININ, E. K., SAPOZHNIKOVA, A. M., STRATULA, D. S.: Barium in the volcanic rocks of Kamchatka and the Kurile islands. Geochemistry, 933 (1964).
MARMO, V., SIIVOLA, J.: The barium content of some granites of Finland. Bull. Comm. Géol. Finlande **222**, 169 (1966).
MATHÉ, G.: Die Metabasite des sächsischen Granulitgebirges. Freiberger Forschungsh. C **251** (1969).
MATHIAS, M.: The geochemistry of the Messum Igneous Complex, South-West Africa. Geochim. Cosmochim. Acta **12**, 29 (1957).
MATSUMURA, T., ISHIYAMA, T.: Adsorption of radioactive materials by coal humic acid. Ann. Rept. Radiat. Center Osaka Prefect **7**, 14 (1966).
MATTAUCH, J. H. E., THIELE, W., WAPSTRA, A. H.: 1964 atomic mass table. Nucl. Phys. **67**, 1 (1965).
MAXWELL, J., PECK, L. C., WIIK, H. B.: Chemical composition of Apollo 11 lunar samples 10017, 10020, 10072 and 10084. Proceedings of the Apollo 11 Lunar Science Conference **2**, 1369 (1970).
— WIIK, H. B.: Chemical composition of Apollo 12 lunar samples 12004, 12033, 12051, 12052 and 12065. Earth Planet. Sci. Lett. **10**, 285 (1971).
MAZZI, F., PABST, A.: Re-examination of cuprorivaite. Am. Mineralogist **47**, 409 (1962).
— ROSSI, G.: The crystal structure of taramellite Ba$_2$(Fe, Ti, Mg)$_2$H$_2$[O$_2$(Si$_4$O$_{12}$)]. Z. Krist. **121**, 243 (1965).
MCGRAIN, P., THOMAS, G. R.: Preliminary report on the natural brines of Eastern Kentucky. Kentucky Geol. Surv. Rept. Invest., Ser. 9 **3**, 1 (1951).
MELCHER, A. C.: The solubility of silver chloride, barium sulphate and calcium sulphate at high temperatures. Am. Chem. Soc. **32**, 50 (1910).
MELSON, W. G., THOMSON, G., ANDEL, T. H. VAN: Volcanism and metamorphism in the Mid-Atlantic Ridge, 22°N latitude. J. Geophys. Res. **73**, 5925 (1968).
MICHEL, G.: Untersuchungen über die Tieflage der Grenze Süßwasser-Salzwasser im nördlichen Rheinland und anschließenden Teilen Westfalens, zugleich ein Beitrag zur Hydrogeologie und Chemie des tiefen Grundwassers. Forschungsber. Landes Nordrhein-Westfalen **1239** (1963).
— Zur Mineralisation des tiefen Grundwassers in Nordrhein-Westfalen. J. Hydrology **3**, 73 (1965).

MILLER, A. R., DENSMORE, C. D., DEGENS, E. T., HATHAWAY, J. C., MANHEIM, F. T., MCFARLIN, P. F., POCKLINGTON, R., JOKELA, A.: Hot brines and recent iron deposits in deeps of the Red Sea. Geochim. Cosmochim. Acta 30, 341 (1966).

MILLER, J. P.: Solutes in small streams draining single rock types, Sangre de Cristo Range, New Mexico. U.S. Geol. Surv. Water Supply 1535-F (1961).

MINGUIZZI, C.: Spectrography of some products of Vesuvius fumaroles. Rend. Soc. Mineral. Ital. 5, 60 (1948).

MOENKE, H.: Trace elements in Variscan and pre-Variscan German granites. A spectrochemical analysis of granitic rocks of different age. Chemie Erde 20, 227 (1960).

MOHR, P. A.: Trace element distribution in a garnet-chlorite-rock from Foel Ddu, near Harlech, Merionethshire. Mineral. Mag. 31, 319 (1956).

— A geochemical study of the shales of the Lower Cambrian manganese shale group of the Harlech Dome, North Wales. Geochim. Cosmochim. Acta 17, 186 (1959).

— ALLEN, R.: Further considerations on the deposition of the Middle Cambrian manganese carbonate beds of Wales and New Foundland. Geol. Mag. 102, 328 (1965).

MOORE, C. B., BROWN, B.: Barium in stony meteorites. J. Geophys. Res. 68, 4293 (1963).

MORRISON, G. H., GERARD, J. T., KASHUBA, A. T., GANGADHARAM, E. V., ROTHENBERG, A. M., POTTER, N. M., MILLER, G. B.: Elemental abundances of lunar soil and rocks. Proceedings of the Apollo 11 Lunar Science Conference 2, 1383 (1970).

MOXHAM, R. L.: Minor element distribution in some metamorphic pyroxenes. Can. Mineralogist 6, 522 (1960).

— Distribution of minor elements in coexisting hornblendes and biotites. Can. Mineralogist 8, 204 (1965).

MUIR, A., HARDIE, H. G. M., MITCHELL, R. L., PHEMISTER, J.: Limestones of Scotland: chemical analyses and petrography. Mem. Geol. Surv. Gt. Brit. Mineral Resources Gt. Brit. 37, 1 (1956).

MUIR, T. D., TILLEY, C. E., SCOON, J. H.: Basalts from the Northern Part of the rift zone of the Mid-Atlantic Ridge. J. Petrol. 5, 409 (1964).

MUKHERJEE, B.: Genetic significance of trace elements in certain rocks of Singhlburn, India. Mineral. Mag. 36, 661 (1968).

MURRAY, H. H.: Genesis of clay minerals in some Pennsylvanian shales of Indiana and Illinois. Proc. 2nd Nat. Conf. Clays and Clay Minerals 1953, 47 (1954).

MURTHY, V. R., EVENSEN, N. M., COSCIO, M. R.: Distribution of K, Rb, Sr and Ba and Rb-Sr isotopic relations in Apollo 11 lunar samples. Proceedings of the Apollo 11 Lunar Science Conference 2, 1393 (1970).

NABOKO, S. I.: The sublimates of the Klyuchevskoy volcano. Bull. Acad. Sci. USSR, Geol. Ser. 1, 50 (1945).

NESTERENKO, G. V., STUDENIKOVA, Z. V., SAVINOVA, E. N.: Rare and disseminated elements in skarns of Tyrny-Auz (Soviet Armenia). Geochemistry, 287 (1958).

NEUMANN, E. W.: Solubility relations of barium sulfate in aqueous solutions of strong electrolytes. J. Am. Chem. Soc. 55, 879 (1933).

NICHOLLS, G. D., LORING, D. H.: The geochemistry of some British Carboniferous sediments. Geochim. Cosmochim. Acta 26, 181 (1962).

NIKOLAEV, V. M., SOKIRKO, L. E., PESTOVA, N. M.: Spectral analysis for correlation of formation waters in Mesozoic deposits of the Eastern Caucasus Region. Tr. Groznensk. Neft. Nauchn.-Issled. Inst. 8, 200 (1960).

NIXON, P. H., KNORRING, O. VON, ROOKE, J. M.: Kimberlites and associated inclusions of Basutoland a mineralogical and geochemical study. Am. Mineralogist 48, 1090 (1963).

NOBLE, D. C., HAFFTY, J.: Minor-element and revised major-element contents of some Mediterranean pantellerites and comendites. J. Petrol. 10, 502 (1969).

NOCKOLDS, S. R., ALLEN, R.: The geochemistry of some igneous rock series. I. Calcalkalic rocks. Geochim. Cosmochim. Acta 4, 105 (1953).

— — The geochemistry of some igneous rock series. II. Alkalic rocks. Geochim. Cosmochim. Acta 5, 245 (1954).

— — The geochemistry of some igneous rock series. III. Tholeiitic rocks. Geochim. Cosmochim. Acta 9, 34 (1956).

OFTEDAL, I.: On the development of granite pegmatite in gneiss areas. Norsk Geol. Tidsskr. 38, 231 (1958).
— Distribution of Ba and Sr in microcline in sections across a granite pegmatite band in gneiss. Norsk Geol. Tidsskr. 39, 343 (1959).
— Remarks on the variable contents of Ba and Sr in microcline from a single pegmatite body. Norsk Geol. Tidsskr. 41, 271 (1961).
— Contribution to the geochemistry of nephelinesyenitic pegmatite in the Langesundsfjord area. Norsk Geol. Tidsskr. 42, 167 (1962).
O'HARA, M. J.: Zoned ultrabasic and basic gneiss masses in the early Lewisian metamorphic complex at Scourie, Scotland. J. Petrol. 2, 248 (1961).
OKRUSCH, M., RICHTER, P.: Zur Geochemie der Dioritgruppe. Vergleichende Untersuchungen an Gesteinen des Bayerischen Waldes, des Spessarts und des Odenwaldes (Süd-Deutschland). Contr. Mineral. and Petrol. 21, 75 (1969).
ONUMA, N., HIGUCHI, H., WAKITA, H., NAGASAWA, H.: Trace element partition between two pyroxenes and the host lava. Earth Planet. Sci. Lett. 5, 47 (1968).
ORR, A. R. D.: Barytes. Mining Annual Review, London 108 (1970).
OSTROM, M. E.: Trace elements in Illinois Pennsylvanian limestones. Illinois State Geol. Surv. Circ. 243, 1 (1957).
OSTROUMOV, V. M., RUSSKIKH, A. M.: Sarazon spring. Izv. Altai. Geogr. Obshchest. SSSR 6, 45 (1965).
PABST, A.: Structures of some tetragonal sheet silicates. Acta Cryst. 12, 733 (1959).
PAPEZIK, V. S.: Geochemistry of some Canadian anorthosites. Geochim. Cosmochim. Acta 29, 673 (1965).
PARRAS, K.: The charnockites in the light of a highly metamorphic rock complex in south-western Finland. Bull. Comm. Géol. Finlande 181 (1958).
PATTEISKY, K.: Die thermalen Solen des Ruhrgebietes und ihre juvenilen Quellgase. Glückauf 90, 1334, 1508 (1954).
PATTERSON, E. M.: A petrochemical study of the Tertiary lavas of north-east Ireland. Geochim. Cosmochim. Acta 2, 283 (1951).
— Petrochemical data for some acid intrusive rocks from the Mourne Mountains and Slieve Gullion. Proc. Roy. Irish Acad., Sect B 55, 171 (1953).
— MITCHELL, W. A., SWAINE, D. J.: Tertiary volcanic succession in the western part of the Antrim Plateau. Proc. Roy. Irish Acad., Sect. B 57, 155 (1955).
— SWAINE, D. J.: A petrochemical study of Tertiary tholeiitic basalts: the middle lavas of the Antrim Plateau. Geochim. Cosmochim. Acta 8, 173 (1955).
PATTIARATCHI, D. B., SAARI, E., SAHAMA, TH. G.: Anandite, a new barium iron silicate from Wilagedera, North Western Province, Ceylon. Mineral. Mag. 36, 1 (1967).
PECK, D. L., WRIGHT, T. L., MOORE, J. G.: Crystallization of tholeiitic basalt in Alae Lava Lake, Hawaii. Bull. Volcanol. 29, 629 (1966).
PECORA, W. T.: Carbonatite problem in the Bearpaw Mountains, Montana. Bull. Geol. Soc. Am., Buddington Volume 83 (1962).
PENCHEV, N. P., PENCHEVA, E. N., BONCHEV, P. R.: Spectrographic investigation of the trace elements in Bulgarian mineral waters. Compt. Rend. Acad. Bulgare Sci. 11, 375 (1958).
— — — Spectrographic investigations of the microcomponents of Bulgarian mineral waters. II. Compt. Rend. Acad. Bulgare Sci. 13, 55 (1960).
PERCHUCK, L. L., RYABCHIKOV, I. D.: Mineral equilibria in the system nepheline-alkali feldspar-plagioclase and their petrological significance. J. Petrol. 9, 123 (1968).
PETROV, V. P., PREDOVSKII, A. A., SERGEEV, A. S., GALIBIN, V. A.: Some peculiarities in the distribution of trace elements in biotites from crystalline schists and gneisses in the Northern Ladoga Region. Vestn. Leningr. Univ. Ser. Geol. i Geogr. 4, 5 (1965).
PETTIJOHN, F. J.: Data of Geochemistry, 6th ed. U.S. Geol. Surv., Profess. Papers 440-S (1963).
PHILPOTTS, J. A., SCHNETZLER, C. C.: Phenocryst-matrix partition coefficients for K, Rb, Sr and Ba, with applications to anorthosite and basalt genesis. Geochim. Cosmochim. Acta 34, 307 (1970a).

PHILPOTTS, J. A., SCHNETZLER, C. C.: Apollo 11 lunar samples: K, Rb, Sr, Ba and rare-earth concentrations in some rocks and separated phases. Proceedings of the Apollo 11 Lunar Science Conference 2, 1471 (1970b).
— — THOMAS, H. H.: Rare-earth and barium abundances in the Bununu howardite. Earth Planet. Sci. Lett. 2, 19 (1967).
PIETZNER, H., WOLF, M.: Geochemical study of brown coal ashes, and coal-petrographic study of brown coals from the Lower Rhine district. Fortschr. Geol. Rheinland Westfalen 12, 517 (1964).
PILKEY, O. H.: Trace elements in Recent marine mollusk shells. Thesis, Ann Arbor, Michigan, 1963.
PINSON, W. H., AHRENS, L. H., FRANCK, M. L.: The abundance of Li, Sc, Sr, Ba and Zr in chondrites and some ultramafic rocks. Geochim. Cosmochim. Acta 4, 251 (1953).
PIRANI, R., SIMBOLI, G.: Genesis of Sardinian granite of the Buddoso platea. Geochemistry and structure of the chief mineralogical components. I. K-feldspar. Mineral. Petrogr. Acta 9, 179 (1963a).
— — Plagioclase and the use of the 2 coexisting feldspars in geothermal determinations. Mineral. Petrogr. Acta 9, 211 (1963b).
PLAS, L. VAN DER, HÜGI, T.: A ferrian sodium-amphibole from Vals, Switzerland. Schweiz. Mineral. Petrog. Mitt. 41, 371 (1961).
POTH, CH. N.: The occurrence of brine in Western Pennsylvania. Pennsylv. Bull. Geol. Surv., 4th Ser. M 47 (1962).
PRASHNOWSKY, A. A.: Sedimentpetrographische und geochemische Untersuchungen im südlichen Rheinischen Schiefergebirge. Neues Jahrb. Geol. Palaeontol. Abhandl. 105, 47 (1957).
PREUSS, E.: Spektralanalytische Untersuchung der Tektite. Chem. Erde 9, 365 (1935).
PRICE, P. H., HARE, C. E., MCCUE, J. B., HOSKINS, H. A.: Salt brines of West Virginia. West Virg. Geol. Surv. Rept. 8 (1937).
PRINZ, M.: Geologic evolution of the Beartooth Mountains, Montana, and Wyoming. Part 5. Mafic Dike Swarms of the Southern Beartooth Mountains. Bull. Geol. Soc. Am. 75, 1217 (1964).
— Geochemistry of basaltic rocks: trace elements. In: HESS, H. H., POLDERVAART, A. (ed.), Basalts. The Poldervaart Treatise on Rocks of Basaltic Composition, vol. I, New York: Interscience Publishers 1967.
PROKOF'EV, V. A.: Elemental chemical composition of shells of the Paleozoic brachiopods from spectral analysis data. Geokhimiya, 75 (1964).
PUCHELT, H.: Zur Geochemie des Grubenwassers im Ruhrgebiet. Z. Deutsch. Geol. Ges. 116, 167 (1964).
— Zur Geochemie des Bariums im exogenen Zyklus. Sitzungsber. Heidelb. Akad. Wiss. Math.-nat. Kl. 4. Abh. (1967).
READ, H. H., HAQ, B. T.: The distribution of trace elements in the dunite-syenite differentiated series of the Insch Complex. Aberdeenshire. Proc. Geologists' Assoc. 74, 203 (1963).
— SADASHIVAIAH, M. S., HAQ, B. T.: The hyperstene-gabbro of the Insch Complex, Aberdeenshire. Proc. Geologists' Assoc. 76, 1 (1965).
REED, G. W.: Heavy elements in the Pantar meteorite. J. Geophys. Res. 68, 3531 (1963).
RENFREW, C., DIXON, J. E., CANN, J. R.: Obsidian and early cultural contact in the Near East. Proc. Prehistoric Soc. 32, 30 (1966).
— — — Further analysis of near eastern obsidians. Proc. Prehistoric Soc. 34, 319 (1968).
REYNOLDS, F. M.: Bemerkungen über das Vorkommen von Ba in der Kohle. J. Soc. Chem. Ind. 58, 64 (1939).
RHODES, J. M.: On the chemistry of potassium feldspars in granitic rocks. Chem. Geol. 4, 373 (1969).
RICHTER, W.: Die Feldspate des Granites von Eisenkappel (Kärnten) und seines Randporphyres. Tschermaks Mineral. Petrog. Mitt., 3. F. 11, 439 (1966).
RIDLEY, W. I.: The petrology of the Las Canadas volcanoes, Tenerife, Canary Islands. Contr. Mineral. and Petrol. 26, 124 (1970).

Rimsaite, J. H. Y.: On micas from magmatic and metamorphic rocks. Beitr. Mineral. Petrog. 10, 152 (1964).

Rivalenti, G., Sighinolfi, G. P.: Geochemical study of graywackes as a possible starting material of para-amphibolites. Contr. Mineral. and Petrol. 23, 173 (1969).

Robinson, W. O., Whetstone, R. R., Edgington, G.: Barium in soils and plants. U.S. Dept. Agr. Tech. Bull. 1013 (1950).

Römpp, H.: Chemie Lexikon, vol. I. Stuttgart: Francksche Verlagsbuchhandlung 1966.

Rogers, J. J. W.: Textural and spectrochemical studies of the White Tank quartz monzonite, California. Bull. Geol. Soc. Am. 69, 449 (1958).

Rosenqvist, I. Th.: Note on leaching of granite with special reference to lead, radium and barium. Norsk Geol. Tidsskr. 19, 110 (1939).

Rosseinsky, D. R.: The solubilities of sparingly soluble salts in water. Trans. Faraday Soc. 54, 116 (1958).

Roy, N. N.: The mineralogy of the potassium-barium feldspar series. I. The determination of the optical properties of natural members. Mineral. Mag. 35, 508 (1965).

— II. Studies on hydrothermally synthesized members. Mineral. Mag. 36, 43 (1967).

Rubeska, I., Miksovsky, M.: Strontium and barium in plain and mineral waters of the Karlovy Vary and Teplice areas and their determination. Vestn. Ustredneko Ustavu Geol. 38, 153 (1963).

Rudert, V.: Phasenbeziehungen im System $NaAlSi_3O_8$-$BaAl_2Si_2O_8$-H_2O. Fortschr. Mineral. 48, 86 (1970).

Russell, H. D., Hiemstra, S. A., Groeneveld, D.: The mineralogy and petrology of the carbonatite at Loolelop, eastern Transvaal. Trans. Proc. Geol. Soc. S. Africa 58, 197 (1954).

Rybach, L., Nissen, H. U.: Zerstörungsfreie Simultanbestimmung von Na, K und Ba in Adular mittels Neutronenaktivierung. Schweiz. Mineral. Petrog. Mitt. 47, 189 (1967).

Sabine, P. A., Young, B. R.: Cell size and composition of the baryte-celestite isomorphous series. Acta Cryst. 7, 630 (1954).

Sahama, Th., G.: Spurenelemente der Gesteine im südlichen Finnisch-Lappland. Bull. Comm. Géol. Finlande 135 (1945a).

— On the chemistry of east Fennoscandian Rapakivi granites. Bull. Comm. Géol. Finlande 136, 15 (1945b).

Sahl, K.: Die Verfeinerung der Kristallstrukturen von $PbCl_2$ (Cotunnit), $BaCl_2$, $PbSO_4$ (Anglesit) und $BaSO_4$ (Baryt). Beitr. Mineral. Petrog. 9, 111 (1963).

Savelli, C.: The problem of rock assimilation by Somma-Vesuvius magma. I. Composition of Somma and Vesuvius lavas. Contr. Mineral. and Petrol. 16, 328 (1967).

Scharbert, S.: Mineralbestand und Genesis des Eisgarner Granits im niederösterreichischen Waldviertel. Tschermaks Mineral. Petrog. Mitt. 3. F. 11, 388 (1966).

Schnetzler, C. C., Philpotts, J. A., Bottino, M. L.: Li, K, Rb, Sr, Ba and rare-earth concentrations, and Rb-Sr age of lunar rock 12013. Earth Planet. Sci. Lett. 9, 185 (1970).

— — Thomas, H. H.: Rare earth and barium abundances in Ivory Coast tektites and rocks from the Bosumtwi Crater area, Ghana. Geochim. Cosmochim. Acta 31, 1987 (1967).

Schopf, T. J. M., Manheim, F. T.: Chemical composition of Ectoprocta (Bryozoa). J. Paleontol. 41, 1197 (1967).

Schwander, H., Hunzicker, J., Stern, W.: Zur Mineralchemie in Hellglimmern in den Tessiner Alpen. Schweiz. Mineral. Petrog. Mitt. 48, 357 (1968).

Schwarcz, H. P.: Chemical and mineralogic variations in an arkosic quartzite during progressive regional metamorphism. Bull. Geol. Soc. Am. 77, 509 (1966).

Schwind, S. B.: Barium. A bibliography of unclassified literature. U.S. Atomic Energy Comm. Nat. Sci. Foundation Wash. TID-369 (1952).

Sen, N., Nockolds, S. R., Allen, R.: Trace elements in minerals from rocks of the S. Californian batholith. Geochim. Cosmochim. Acta 16, 58 (1959).

Sen, S. K.: Some aspects of the distribution of barium, strontium, iron, and titanium in plagioclase feldspars. J. Geol. 68, 638 (1960).

Senakolis, A. F.: Lithology and geochemistry of the Cambrian carbonate rocks in some sections of the Sayan-Altai territory. Materialy po Geol. i Polezn. Iskop. Zapadn. Sibiri 193 (1964).

SHACKLETTE, H. T.: Element content of bryophytes. Contr. Geochem. Prosp. Minerals, Geol. Surv. Bull. **1198-D** (1965).
SHAW, D. M.: Trace elements in pelitic rocks I, II. Bull. Geol. Soc. Am. 65, 1151, 1167 (1954).
— Some aspects of the determination of barium in silicate rocks. Spectrochim. Acta 10, 125 (1957).
— MOXHAM, R. L., FILBY, R. H., LAPKOWSKY, W. W.: The petrology and geochemistry of some Grenville skarns. II. Geochemistry. Can. Mineralogist 7, 578 (1963).
SHCHERBA, G. N., GUKOVA, V. D., KUDRYASHOV, A. V., SENCHILO, N. P.: Alkali feldspars from feldspar-quartz veinlets in molybdenum-tungsten deposits, Kazakhstan. IV. Quartz and feldspar-quartz veins and veinlets. Geochem. Int. 1, 141 (1964).
SHELTON, J. S.: Glendova volcanic rocks, Los Angeles Basin, Cal. Bull. Geol. Soc. Am. 66, 45 (1955).
SHIMA, M., HONDA, M.: Distributions of alkali, alkaline earth and rare earth elements in component minerals of chondrites. Geochim. Cosmochim. Acta 31, 1995 (1967).
SHIMER, J. A.: Spectrographic analysis of New England granites and pegmatites. Bull. Geol. Soc. Am. 54, 1049 (1943).
SHINKARENKO, A. L.: The gas component and content of microelements in mineral springs of the Caucasian mineral waters. Tr. Lab. Gidrogeol. Probl. 3, 253 (1948).
SHKABARA, M. N.: On the mineralogy of wellsite. Zentr. Mineralogie, 180 (1950).
SIEDNER, G.: Geochemical features of a strongly fractionated alkali igneous suite. Geochim. Cosmochim. Acta 29, 113 (1965).
SIEGERS, A., PICHLER, H., ZEIL, W.: Trace element abundances in the "andesite" formation of northern Chile. Geochim. Cosmochim. Acta 33, 882 (1969).
SIMPSON, E. S. W.: The Okonjeje igneous complex, southwest Africa. Trans. Proc. Geol. Soc. S. Africa 57, 125 (1954).
SINHA, R. C., KARKARE, S. G.: Distribution and behavior of trace elements in some of the Deccan basalts. Geol. Soc. India Bull. 1, 21 (1964a).
— — Geochemistry of Deccan basalts: a study of the behaviour of major and trace elements in the basaltic flows of India. Intern. Geol. Congress, Rep. 21, Session India, Part VII, 85 (1964b).
— TIWARI, B. D.: Geochemistry of the volcanic rocks of Pavagarh. Intern. Geol. Congr. Rep. 21, Session India, Part VII, 104 (1964).
SKINNER, B. J., WHITE, D. E., ROSE, H. J., MAYS, R. E.: Sulfides associated with the Salton Sea geothermal brine. Ec. Geol. 62, 316 (1967).
SKROBOV, A. A., SMIRNOV, V. J.: Die natürlichen Mineralwässer des nördlichen Gebiets. Nördl. Geol. Verw. Archangelsk (1939).
SMALES, A. A., WEBB, M. S. W., WEBSTER, R. K., WILSON, J. D.: Elemental composition of lunar surface material. Proceedings of the Apollo 11 Lunar Science Conference 2, 1575 (1970).
SMITH, A. L., CARMICHAEL, I. S. E.: Quaternary trachybasalts from southeastern California. Am. Mineralogist 54, 909 (1969).
SMITH, G. J.: Geology and volcanic petrology of the Lava Mountains, San Bernardino County, California. U.S. Geol. Surv. Profess. Papers **457** (1964).
SMITH, J. V., RIBBE, P. H.: X-ray-emission microanalysis of rock-forming minerals. III. Alkali feldspars. J. Geol. 74, 197 (1966).
SNAVELY, P. D., Jr., MACLEOD, N. S., WAGNER, H. C.: Tholeiitic and alkalic basalts of the Eocene Siletz river. Am. J. Sci. 266, 454 (1968).
SNETSINGER, K. G.: Barium-vanadium muscovite and vanadium tourmaline from Mariposa County, California. Am. Mineralogist 51, 1623 (1966).
SOLOMON, M.: Origin of barite in the North Pennine ore field. Inst. Mining Met., Trans. 75, 230 (1966).
SPENCER, D.: Factors affecting element distribution in a Silurian graptolite band. Chem. Geol. 1, 221 (1966).
STARITSIN, F. V.: Stratigraphic subdivision of Mesozoic extrusives of eastern Transbaikaliya by trace elements. Geochemistry 83 (1964).

Stark, J. T., Tracey, J. T.: Petrology of volcanic rocks of Guam. U.S. Geol. Surv. Profess. Papers **403-C** (1963).

Starke, R.: Die Strontiumgehalte der Baryte. Freiberger Forschungsh. C **150** (1964).

Stehli, F. G., Hower, J.: Mineralogy and early diagenesis of carbonate sediments. J. Sediment. Petrol. **31**, 358 (1961).

Stern, W. B.: Zur Mineralchemie von Glimmern aus Tessiner Pegmatiten. Schwei z Mineral. Petrog. Mitt. **46**, 137 (1966).

Stevens, R. E., Fleischer, M., Niles, W. W., Chodos, A. A., Filby, H., Leininger, R. K., Flanagan, F. J.: Second report on a cooperative investigation of the composition of two silicate rocks. Bull. U.S. Geol. Surv. **1113** (1960).

Stöhr, H., Flasch, H.: Barium und Bariumverbindungen. In: Ullmanns Encyklopädie der technischen Chemie, vol. IV, München-Berlin: Urban & Schwarzenberg 1953.

Straub, J.: Chemical composition of mineral waters of Transylvania: their rare constituents and the biochemical importance. Magy. Allami Foldt Int. Evkonyve **39** (1950).

Strominger, D., Hollander, J. M., Seaborg, G. T.: Table of isotopes. Rev. Mod. Phys. **30** (1958).

Strübel, G.: Zur Kenntnis und genetischen Bedeutung des Systems $BaSO_4$-NaCl-H_2O. Neues Jahrb. Mineral. Monatsh., 223 (1967).

Strunz, H.: Mineralogische Tabellen, 4. Aufl. Leipzig: Akademische Verlagsgesellschaft Geest & Bortig 1966.

Suess, H. E., Urey, H. C.: Abundances of the elements. Rev. Mod. Phys. **28**, 53 (1956).

Sukharev, G. H.: The composition of connate waters in Mesozoic deposits in the Caucasus region and their connection with the prospecting for oil and gas. Geol. Nefti i Gaza **5**, 17 (1961).

Swaine, D. J.: The trace-element content of soils. Commonwealth Agr. Bur. Tech. Com. **48**, 16 (1955).

— Inorganic constituents in Australian coals. Mitt. Naturforsch. Ges. Bern (N. F.) **24**, 49 (1967).

Szabo, B. J., Joensuu, O.: Emission spectrographic determination of barium in sea water using a cation-exchange concentration procedure. Environ. Sci. Technol. **1**, 499 (1967).

Tadzhiev, F. Kh., Muksinov, T. Kh.: Change in the thermodynamic functions of some ion-exchange reactions on Oglanly bentonite. Uzbeksk. Khim. Zh. **11**, 49 (1967).

Takubo, J., Tatekawa, M.: Distribution of minor elements in pegmatites and granites from Oku-Tango District, Kyoto Prefecture, Japan, I. Distribution of barium and strontium. Kobutsugaku Zasshi **1**, 301 (1954).

Taylor, S. R.: The origin of some New Zealand metamorphic rocks as shown by their major and trace element composition. Geochim. Cosmochim. Acta **8**, 182 (1955).

— Consequences for tektite composition of an origin by meteoritic splash. Geochim. Cosmochim. Acta **26**, 915 (1962).

— Similarity in composition between Henbury impact glass and australites. Geochim. Cosmochim. Acta **29**, 599 (1965).

— The application of trace element data to problems of petrology. Phys. Chem. Earth **6**, 133 (1965).

— Australites, Henbury impact glass and subgreywacke: a comparison of the abundances of 51 elements. Geochim. Cosmochim. Acta **30**, 1121 (1966).

— Capp, A. C., Graham, A. L.: Trace element abundances in andesites. II. Saipan, Bougainville and Fiji. Contr. Mineral. and Petrol. **23**, 1 (1969).

— Ewart, A., Capp, A. C.: Leucogranites and rhyolites: trace element evidence for fractional crystallisation and partial melting. Lithos **1**, 179 (1968).

— Heier, K. S.: The petrological significance of trace element variations in feldspars. Proc. XXI. Inter. Geol. Congr. Norden **47** (1960).

— Johnson, P. H., Martin, R., Bennett, D., Allen, J., Nance, W.: Preliminary chemical analyses of Apollo 11 lunar samples. Proceedings of the Apollo 11 Lunar Science Conference **2**, 1627 (1970).

— Kolbe, P.: Henbury impact glass: parent material and behaviour of volatile elements during melting. Nature **203**, 390 (1964).

TAYLOR, S. R., SACHS, M.: Geochemical evidence for the origin of australites. Geochim. Cosmochim. Acta 28, 235 (1964).
— SOLOMON, M.: The geochemistry of Darwin glass. Geochim. Cosmochim. Acta 28, 471 (1964).
— — SVERDRUP, T. L.: Contributions to the mineralogy of Norway. V. Trace element variations in three generations of feldspars from the Landsverk I pegmatite Evje, Southern Norway. Norsk Geol. Tidsskr. 40, 133 (1960).
— WHITE, A. J. R.: Trace element abundances in andesites. Bull. Volcanol. 29, 177 (1966).
TEMPLE, A. K., GROGAN, R. M.: Carbonatite and related alkalic rocks at Powder Horn, Colorado. Ec. Geology 60, 672 (1965).
TEMPLETON, CH. C.: Solubility of barium sulfate in sodium chloride solutions from 25° to 95° C. J. Chem. Eng. Data 5, 514 (1960).
TERA, F., EUGSTER, O., BURNETT, D. S., WASSERBURG, G. J.: Comparative study of Li, Na, K, Rb, Cs, Ca, Sr and Ba abundances in achondrites and in Apollo 11 lunar samples. Proceedings of the Apollo 11 Lunar Science Conference 2, 1637 (1970).
TIMASHEVA, E. E.: Distribution of trace elements in formation waters of the Carpathian syncline and in southwestern boundaries of the Russian Platform. Tr. Ukr. Nauchn.-Issled. Geol. Razved. Inst. 3, 344 (1963).
TKACHEV, Yu. A., SKIBA, N. S., BONDARENKO, C. F.: Distributions of strontium and barium in coals of Kirgizia. Litol., Geokhim. i Polezn. Iskop. Osad. Obrazov. Tyan-Shanya, Akad. Nauk Kirg. SSR, Inst. Geol., 24 (1965).
TOLKACHEV, M. V.: Alkaline-earth elements in sedimentary rocks of the west Siberian depression. Geokhimiya 4, 489 (1968).
TOOKER, E. W.: Altered wallrocks in the central part of the Front Range mineral belt, Gilpin and clear Creek Counties, Colorado. U.S. Geol. Surv. Profess. Papers 439 (1963).
TOURTELOT, H. A.: Chemical composition of the Pierre shale and equivalent rocks of late Cretaceous age, Great Plains region. Bull. Geol. Soc. Am. 68, 1806 (1957).
TOWNEND, R.: The geology of some granite plutons from Western Connemara, Co. Galway. Roy. Irish Acad. Proc. 65, 157 (1966).
TOWNLEY, R. N., WHITNEY, W. B., FELSING, W. A.: The solubilities of barium and strontium carbonates in aqueous solutions of some alkali chloride. J. Am. Chem. Soc. 59, 631 (1937).
TUREKIAN, K. K.: Deep-sea deposition of barium, cobalt, and silver. Geochim. Cosmochim. Acta 32, 603 (1968).
— ARMSTRONG, R. L.: Magnesium, strontium, and barium concentrations and calcite-aragonite ratios of some Recent molluscan shells. J. Marine Res. 18, 133 (1960).
— — Chemical and mineralogical composition of fossil molluscan shells from the Fox Hills formation, South Dakota. Bull. Geol. Soc. Am. 72, 1817 (1961).
— HARRISS, R. C., JOHNSON, D. G.: The variations of Si, Cl, Na, Ca, Sr, Ba, Co, and Ag in the Neuse River, North Carolina. Limnol. Oceanog. 12, 702 (1967).
— JOHNSON, D. J.: The barium distribution in seawater. Geochim. Cosmochim. Acta 30, 1153 (1966).
— PHINNEY, W. C.: The distribution of Ni, Co, Cr, Cu, Ba, and Sr between biotite-garnet pairs in a metamorphic sequence. Am. Mineralogist 47, 1434 (1962).
— TAUSCH, E. H.: Barium in deep-sea sediments of the Atlantic Ocean. Nature 201, 696 (1964).
— WEDEPOHL, K. H.: Distribution of the elements in some major units of the earth's crust. Bull. Geol. Soc. Am. 72, 175 (1961).
UCHAMEISHVILI, N. E., MALININ, S. D., KHITAROV, N. J.: Solubility of barite in concentrated solutions of chlorides of some metals at elevated temperatures in relation to the genesis of barite deposits. Geochemistry, 951 (1966).
UMEMOTO, S.: Isotopic composition of barium and cerium in stone meteorites. J. Geophys. Res. 67, 375 (1962).
UNSÖLD, A.: Der neue Kosmos. Heidelberger Taschenbücher 16/17 (1967).
UTSUMI, K.: Spectral analysis of some carbon stars in the visual region. Publ. Astron. Soc. Japan 19, 342 (1967).

Varov, A. A., Romm, I. F.: Occurrence of strontium and barium in oilfield waters of the Ural-Volga-Region. Dokl. Akad. Nauk. SSSR **35**, 114 (1942).

Vine, J. D.: Elemental distribution in some shelf and eugeosynclinal black shales. U.S. Geol. Surv. Bull. **1214-E** (1966).

Vinogradov, A. P.: The Elementary Chemical Composition of Marine Organisms. New Haven: Memoir Sears Foundation for Marine Research, No. II, Yale University 1953.

— Regularity of distribution of chemical elements in the earth crust. Geochemistry, 1 (1956).

— Ronov, A. B.: Composition of sedimentary rocks of the Russian platform in relation to the history of their tectonic movements. Geochemistry, 533 (1956).

— — Ratynsky, V. M.: Changes of chemical composition of carbonate rocks of the Russian Platform. Akad. Nauk SSSR Izv. Geol. Ser. **1**, 33 (1952).

Vinogradova, Z. A., Koval'skiy, V. V.: Elemental composition of the Black Sea plankton. Dokl. Akad. Nauk SSSR **147**, 1458 (1962).

Vlasov, K. A., Kuzmenko, M. Z., Es'kova, E. M.: The Lovozero Alkali Massif. Edinburgh-London: Oliver & Boyd Publishers 1966.

Volborth, A.: Rapakivi-type granites in the Precambrian complex of Gold Butte, Clark County, Nevada. Bull. Geol. Soc. Am. **73**, 813 (1962).

Vorob'ev, G. G.: Composition of tektites. I. Indochinites. Meteoritika **17**, 64 (1959).

— Composition of tektites. II. Moldavites. Meteoritika **18**, 35 (1960).

Wänke, H., Rieder, R., Baddenhausen, H., Spettel, B., Teschke, F., Quijano-Rico, M., Balacescu, A.: Major and trace elements in lunar material. Proceedings of the Apollo 11 Lunar Science Conference **2**, 1719 (1970).

Wager, L. R., Mitchell: R. L. The distribution of trace elements during strong fractionation of basic magma — a further study of the Skaergaard intrusion, East Greenland. Geochim. Cosmochim. Acta **1**, 129 (1951).

— — Trace elements in a suite of Hawaiian lavas. Geochim. Cosmochim. Acta **3**, 217 (1953).

Wakeel, S. K. El, Riley, J. P.: Chemical and mineralogical studies of deep-sea sediments. Geochim. Cosmochim. Acta **25**, 110 (1961b).

— — Chemical and mineralogical studies of fossil red clays from Timor. Geochim. Cosmochim. Acta **24**, 260 (1961a).

Wakita, H., Schmitt, R. A.: Elemental abundances in seven fragments from lunar rock 12013. Earth Planet. Sci. Lett. **9**, 169 (1970).

— — Rey, P.: Elemental abundances of major, minor and trace elements in Apollo 11 lunar rocks, soil and core samples. Proceedings of the Apollo 11 Lunar Science Conference **2**, 1685 (1970).

Wallerstein, G., Greenstein, J. L.: Chemical composition of two CH stars, HD 26 and HD 626. Astrophys. J. **139**, 1163 (1964).

— — Parker, R., Helfer, H. L., Aller, L. H.: Red giants with extreme metal deficiencies. Astrophys. J. **137**, 280 (1963).

Warner, B.: The barium stars. Monthly notices of the Roy. Astron. Soc. **129**, 263 (1965).

Wedepohl, K. H.: Der trachydoleritische Basalt (Olivin-Andesin-Basalt) des Backenberges bei Güntersen, westlich von Göttingen. Heidelberger Beitr. Mineral. Petrog. **4**, 217 (1954).

— Spurenanalytische Untersuchungen an Tiefseetonen aus dem Atlantik. Geochim. Cosmochim. Acta **18**, 200 (1960).

— Written communication August 1961.

— Composition and abundance of common igneous rocks, p. 227. In: K. H. Wedepohl, Handbook of Geochemistry, I. Berlin-Heidelberg-New York: Springer 1969.

Weibel, M.: Zum Chemismus alpiner Adulare (II). Schweiz. Mineral. Petrog. Mitt. **37**, 545 (1957).

— Chemismus und Mineralzusammensetzung von Gesteinen des nördlichen Bergeller Massivs. Schweiz. Mineral. Petrog. Mitt. **40**, 69 (1960).

— Meyer, F.: Zum Chemismus alpiner Adulare (I). Schweiz. Mineral. Petrog. Mitt. **37**, 153 (1957).

Weiskirchner, W.: Personal communication 1969.

WHITE, A. J. R.: Genesis of migmatites from the Palmer region of south Australia. Chem. Geol. **1**, 165 (1966).
WHITE, D. E.: Saline waters of sedimentary rocks. In: Fluids in subsurface environments. Am. Ass. Petr. Geol. Memoir **4**, 342 (1965).
— HEM, J. D., WARING, G. A.: Data of Geochemistry. 6th Edition. Chapter F.: Chemical composition of subsurface waters. Geol. Surv. Profess. Papers **440-F** (1963).
WHITE, R. W.: Ultramafic inclusions in basaltic rocks from Hawaii. Contr. Mineral. and Petrol. **12**, 245 (1966).
WILKINSON, J. F. G.: The geochemistry of a differentiated teschenite sill near Gunnedah, New South Wales. Geochim. Cosmochim. Acta **16**, 123 (1959).
— Mineralogical, geochemical, and petrogenetic aspects of an analcite-basalt from the New England district of New South Wales. J. Petrol. **3**, 192 (1962).
— Analcimes from some potassic igneous rocks and aspects of analcime rich igneous assemblages. Contr. Mineral. and Petrol. **18**, 252 (1968).
— VERNON, R. H., SHAW, S. E.: The petrology of an adamellite-porphyrite from the New England Bathylith (New S. Wales). J. Petrol. **5**, 461 (1964).
WILSKA, S.: Trace elements in Finnish ground and mine waters: a spectroanalytical study. Ann. Acad. Sci. Fennicae Ser. A II. Chem. **46**, 7 (1952).
WOLGEMUTH, K.: Barium analyses from the first geosecs test cruise. J. Geophys. Res. **75**, 7686 (1970).
— BROECKER, W. S.: Barium in sea water. Earth Planet. Sci. Lett. **8**, 372 (1970).
WYCKOFF, R.: Crystal Structures, 2nd ed., vol. 4. New York and London: Interscience Publishers 1968.
WYNNE-EDWARDS, H. R., HAY, P. W.: Coexisting cordierite and garnet in regionally metamorphosed rocks from the Westport area, Ontario. Can. Mineralogist **7**, 453 (1962).
YOUNG, E. J.: Studies of trace elements in sediments. Thesis, Mass. Inst. Techn. 1954.
YUSUROVA, S. M.: Geochemistry of the mineral water of thermal springs of Tadzhikistan (rare elements in thermal waters of Tadzhikistan). Dokl. Akad. Nauk Tadzh. SSR **21**, 19 (1957).
ZOLTAI, T.: Classification of silicates and other minerals with tetrahedral structures. Am. Mineralogist **45**, 960 (1960).
ZURLO, P.: Geochemische Untersuchungen im Unterkarbon von Doberlug und im Oberkarbon von Zwickau. Geologie **9**, 1112 (1963).

Platinum 44 – 45 – 46
76 – 77 – **78**

A   E. Parthé   (Laboratoire de Cristallographie aux Rayons X, Université de Genève, Genève, Suisse)

B—G, K, M, O   J. H. Crocket   (Department of Geology, McMaster, University, Hamilton, Ontario, Canada)

78-A. Crystal Chemistry

The term "platinum metals" is generally applied to the following six elements: ruthenium (Ru), rhodium (Rh), palladium (Pd), osmium (Os), iridium (Ir) and platinum (Pt). The first three with a density $\varrho \sim 12$ g/cm³ are often called the light platinum metals, the remaining three with $\varrho \sim 22$ g/cm³ the heavy platinum metals. All these noble metals crystallize in either the cubic close-packed Cu (A1) structure type or in the hexagonal close-packed Mg (A3) structure type. Crystal structure data together with other characteristics of these elements are summarized in Table 78-A-1.

Table 78-A-1. *Crystal structure data and other important data for platinum metals*

Element	Atomic No.	Atomic wt.	Electronic ground-state configuration	Melting point (°C)	ϱ (g/cm³)	Crystal structure type	Metallic atom radius (Å)	Reference[a]
Ru	44	101.07	[Kr] $4d^7 5s$	2,250	12.30	A3	1.32_5	1
Rh	45	102.905	[Kr] $4d^8 5s$	1,960	12.4	A1	1.34_5	1
Pd	46	106.4	[Kr] $4d^{10}$	1,552	11.97	A1	1.37_5	2
Os	76	190.2	[Xe] $4f^{14} 5d^6 6s^2$	3,000	22.48[b]	A3	1.33_7	3
Ir	77	192.2	[Xe] $4f^{14} 5d^7 6s^2$	2,443	22.421	A1	1.35_7	2
Pt	78	195.09	[Xe] $4f^{14} 5d^9 6s$	1,769	21.45[b]	A1	1.38_7	4

[a] Reference for crystal structure: 1. HULL and DAVEY, SB 1913—1928, 69. 2. HULL and DAVEY, SB 1913—1928, 70. 3. HULL, SB 1913—1928, 70. 4. HULL and DAVEY, SB 1913—1928, 71.
[b] at 20° C

The phase diagrams of binary systems containing only platinum metals are simple. Elements with A1 structure form a complete series of solid solutions at least at high temperatures. Similarly Os and Ru with A3 structure are fully miscible in the solid state. Systems between elements with A3 structure and elements with A1 structure are of the eutectic or peritectic type, usually with wide solid solution ranges. A presentation of the known alloying behavior of the platinum metals with themselves is shown in Fig. 78-A-1. The minerals of the platinum metals found in placers may contain all six platinum metals in varying proportions. It is possible to distinguish between two groups of minerals depending on the crystal structure. *Iridosmine* (nevyanskite and sysertskite) has A3 structure. *Native platinum, platiniri-*

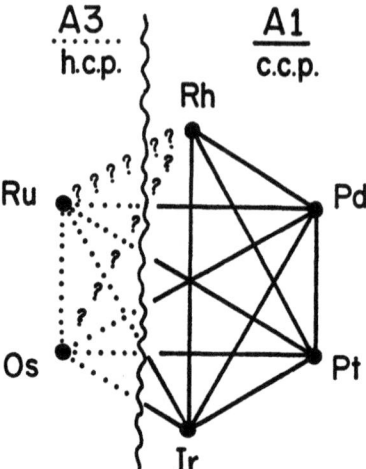

Fig. 78-A-1. The alloying behavior of the platinum metals with themselves. For the most part exact data on solubility limits in A3—A1 systems are missing. The solubility limits shown in the drawing are schematic

dium, osmiridium and aurosmirid (Au—Os—Ir) are minerals of the platinum metals with A1 structure.

The structures of the platinum metals and their alloys have been interpreted in terms of the Engel-Brewer theory (BREWER, 1965; BREWER, 1967; HUME-ROTHERY, 1968). This theory assumes that the crystal structures of the elements are based on definite electronic states of the atoms which are determined solely by the number of s and p electrons. The A2 (W), A3 (Mg) and A1 (Cu) structure types are regarded as resulting from the electron configurations $d^{n-1} s$, $d^{n-2} s p$ and $d^{n-3} s p^2$ respectively. An element will select that particular crystal structure for which the net cohesion is a maximum; the latter factor is the difference between the promotion energy (needed to create one of the specific electron configurations) and the bonding energy (which becomes available if the promoted atoms combine). All unpaired electrons of the promoted atoms, s, p and d, take part in the crystal bonding, but the d electrons play no part in determining the type of crystal structure. The electron configurations of the platinum metals are:

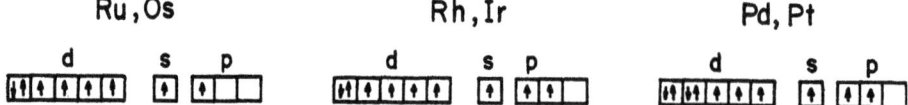

The promotion energies to create the $d^6 s p^2$ and $d^7 s p^2$ configurations are quite high; however, this is compensated by the increase in bonding energy due to the newly available unpaired d electrons.

The Engel-Brewer theory can explain the stability and structures of certain intermetallic compounds of the platinum metals. In particular, combinations between platinum metals and transition metals from the left side on the periodic table like

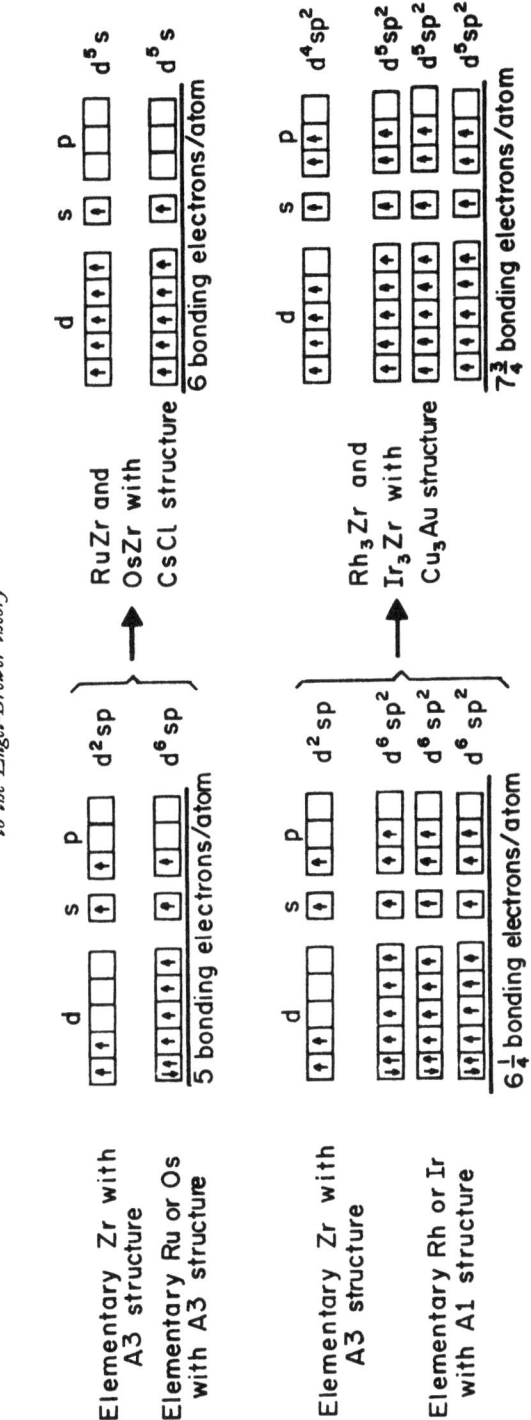

Table 78-A-2. Electron configurations for the elements in intermetallic platinum metal compounds and their components in the unalloyed state according to the Engel-Brewer theory

Ti, Zr and Hf, should produce very stable compounds. The differences between the transition metals from the left and right sides of the periodic table can be summarized in terms of incomplete use of available bonding orbitals and full use of available electrons on the left side and incomplete use of electrons and full use of bonding orbitals on the right side. A mixture of metals from both sides will allow electron transfer from the platinum metals with a surplus of electrons to the metals with low-lying vacant orbitals to make fullest use of all electrons and all orbitals. As an example one may consider intermetallic compounds of the platinum metals with Zr. RuZr and OsZr have the CsCl structure, an A2 derivative structure. Rh_3Zr and Ir_3Zr crystallize with the Cu_3Au structure, an A1 derivative structure. The electron configurations of the pure elements and of the elements in these intermetallic compounds are given in Table 78-A-2. The increase in the average number of bonding electrons per atom explains the unusual high stability of these compounds.

It should be mentioned that the Engel-Brewer correlation, especially the correlation between specific electron configuration and crystal structure has been the subject of heated discussions and has been attacked on theoretical grounds. However, this correlation as a whole has been surprisingly successful in explaining many observations for which no other explanations are available.

Compounds of the platinum metals which are found in nature can be grouped in metallic and non-metallic minerals as shown in Table 78-A-3.

The CuAu structure type of potarite, PdHg, corresponds to an ordered derivative of the A1 (Cu) structure, while all other metallic minerals for which the structure is known except froodite have the NiAs structure or a variant of it like the CdI_2 structure. A satisfactory crystal chemical interpretation of these metallic structures is unfortunately not possible at this time.

In the lower part of Table 78-A-3 are given the non-metallic semiconducting, diamagnetic chalcogenide and pnictide minerals of the platinum metals. These ionic-covalent compounds satisfy simultaneously the electronic conditions for covalent bonding according to the valence bond theory and the conditions for ionic bonding according to the electrostatic crystal-field theory and the general theory for valence compounds. Thus, two canonical formulae can be written for each compound, one being a covalent formula, the other an ionic formula. The valence bond theory specifies that six octahedral bonding orbitals (possibly also somewhat distorted) are formed if the transition metal atom can form a d^2sp^3 hybrid or four square planar orbitals with a dsp^2 hybrid. The theory shows further that the chalcogen and P, As, Sb atoms will form four tetrahedral orbitals with sp^3 hybridization. If all the electrons of the transition elements which participate in the hybridization are shifted momentarily to the chalcogen or P, As, Sb atoms, one obtains the limiting ionic formula. The electrostatic crystal-field theory specifies then that the diamagnetic transition metal ions must have a d^6 configuration for regular octahedral coordination, a d^4 configuration for a stretched octahedral or rhombic bi-pyramidal coordination or a d^8 configuration for a square planar coordination. Further the general valence compound equation states that the anions will use all the provided electrons to fill their octet shells either by forming isolated anions or anion dumb bells or anion chains. Using this information one can calculate as shown in Table 78-A-4 possible compositions for semiconducting platinum metal chalcogenides and pnictides assuming particular cation coordinations and anion nettings.

Table 78-A-3. *Compounds of the platinum metals found in nature*

Mineral	Formula	Structure	Reference
Metallic minerals			
Potarite	PdHg	tetragonal, AuCu (L1$_0$) type	TERADA and CAGLE, SR 1960, 181; BITTNER and NOWOTNY, SR 1952, 115
Froodite	PdBi$_2$	monoclinic, α PdBi$_2$ type	HAWLEY and BERRY, SR 1958, 52
Niggliite	Pt(Sn, Te)	hexagonal, NiAs (B8) type	GROENEVELD MEIJER, SR 1955, 420
Arsenopalladinite	Pd$_3$As	hexagonal, does not agree with artificial Pd$_3$As which has tetragonal Fe$_3$P type	
Michenerite	PdBi$_2$	cubic pyrite, does not agree with β PdBi$_2$ which has tetragonal MoSi$_2$ (C11) type	HAWLEY and BERRY, SR 1958, 52
Stibiopalladinite	Pd$_3$Sb	unknown structure	
Stannopalladinite	Pd$_3$Sn$_2$	hexagonal "filled up" NiAs (B8) type	NOWOTNY, SCHUBERT and DETTINGER, SR 1947—1948, 173
Kotulskite	Pd(Te, Bi)$_{1-2}$	hexagonal "partly empty" CdI$_2$ (C6) type	
Monchëite	(Pt, Pd)(Te, Bi)$_2$	hexagonal CdI$_2$ (C6) type	
Semiconducting, diamagnetic minerals			
Cooperite	PtS	tetragonal PtS (B17) type	BANNISTER and HEY, SB 1928 bis 1932, 9, 234
Braggite	(Pt, Pd, Ni)S	tetragonal PdS (B34) type	GASKELL, SB 1937, 3, 41
Vysotskite	(Pd, Ni)S	tetragonal PdS (B34) type	
Laurite	RuS$_2$	cubic pyrite (C2) type	OFTEDAL, SB 1913—1928, 780
Sperrylithe	PtAs$_2$	cubic pyrite (C2) type	THOMASSEN, SB 1913—1928, 781
Geversite	PtSb$_2$	cubic pyrite (C2) type	THOMASSEN, SB 1913—1928, 781

In Table 78-A-5 structure types are shown of all known semiconducting and diamagnetic chalcogenides and pnictides of the platinum metals. In addition to the cation coordinations and anion nettings considered in Table 78-A-4, more complicated arrangements are also found. However, in all cases there is excellent agree-

Table 78-A-4. *Possible compositions for semiconducting platinum metal chalcogenides and pnictides with prescribed cation coordination and anion netting*

Large numerals are used for the number of valence electrons, including here in the case of the platinum metals all d electrons. Small numerals are conventional chemical composition parameters.

T Elements: 8: Ru or Os; 9: Rh or Ir; 10: Pd or Pt.
X Elements: 5: P, As or Sb; 6: O, S, Se or Te.

Formulae for which examples are known are enclosed in rectangles

CATION COORDINATION

d electron energy levels and their occupation with electrons for T metal ions in different coordinations.

ANION NETTING	spherical coordination	regular octahedr. coordination d^6	stretched rhombic or octahedr. bipyramid d^4	square planar coordination d^8
|X| isolated anions 8 electrons/anion		$9_2 6_3$	impossible	$10\ 6$
|X—X| anions dumb bells 7 electrons/anion		86_2 $10\ 5_2$ $9\ 56$	85_2	impossible
—X—X—X— chains or rings 6 electrons/anion		95_3	impossible	86_2 $10\ 5_2$ $9\ 56$

Table 78-A-5. *Crystal structure types, electron configuration and ionic formulae of the semiconducting and diamagnetic chalcogenides and pnictides of the platinum metals*

Structural features	Structure type	Atoms per cell and space group	Electron distribution for covalent formula	Ionic formula	Examples	References [a]
			TX Compounds			
square planar T coord., tetrahedral X coord., $\boxed{X}$ only	Pd[4]S[4] cooperite (B17 type)	4 P4$_2$/mmc	d^8 / $ds p^2$ hybrid / sp^3 hybrid	$Pt^{2+}[\underline{S}]^{2-}$	*10 6*: PdO, PtO, PtS	1.
same; differences only in second nearest coordination	Pd[4]S[4] (B34 type)	16 P4$_2$/m	same	$Pd^{2+}[\underline{S}]^{2-}$	*10 6*: PdS, PdSe	2.
			T$_2$X$_3$ Compounds			
octahedral T coord., tetrahedral X coord., $\boxed{X}$ only	Rh$_2^{[6]}$S$_3^{[4]}$	20 Pbcn	d^6 / $d^2 s p^3$ hybrids / sp^3 hybrids	$Rh_2^{3+}[\underline{S}]_3^{2-}$	*9 6$_3$*: Rh$_2$S$_3$, Rh$_2$Se$_3$, Ir$_2$S$_3$	3.
			TX$_2$ Compounds			
octahedral T coord., tetrahedral X coord., $\overline{X-X}$ dumb bells	Fe[6]S$_2^{[3+1]}$ pyrite (C2 type) or Ni[6+3]Sb[3+1][3+1] ullmannite ≡ ternary ordered pyrite type	12 Pa3 12 P2$_1$3	$d^6 s p^3$ hybrid / sp^3 hybrid	$Ru^{2+}[\underline{S}-\underline{S}]^{2-}$ $Rh^{3+}[\underline{P}-\underline{S}]^{2-}$ $Pt^{4+}[\underline{P}-\underline{P}]^{4-}$	*8 6$_2$*: RuS$_2$, RuSe$_2$, RuTe$_2$, OsS$_2$, OsSe$_2$, OsTe$_2$ *9 5 6*: RhPS, RhAsS, RhSbS, RhBiS, RhPSe, RhAsSe, RhSbSe, RhBiSe, RhAsTe, RhSbTe, RhBiTe, IrPS, IrAsS, IrSbS, IrBiS, IrBiSe, IrAsSe, IrSbSe, IrBiSe, IrAsTe, IrSbTe, IrBiTe *10 5$_2$*: PdAs$_2$, PdSb$_2$, PtP$_2$, PtAs$_2$, PtSb$_2$	4.

Table 78-A-5. (Continued)

Structural features	Structure type	Atoms per cell and space group	Electron distribution for covalent formula T X (d s p / s p)	Ionic formula	Examples	References[a]
same; differences only in second nearest coordination	Fe[6]S[3+1] marcasite (C18 type)	6 Pnnm	same	same		5.
rhombic bipyramidal T coordination, tetrahedral X coord., $\overline{X}$–$\overline{X}$ dumb bells	Fe[6]As[3+1] loellingite similar to marcasite (C18 type) but with changed axial ratios ($c/a \sim 0.55$ instead of 0.74 and $c/b \sim 0.48$ instead of ~ 0.62)	6 Pnnm	d^2sp^3 hybrid / sp^3 hybrids	$Ru^{4+}[\overline{P}-\overline{P}]^{4-}$	85_2: RuP_2, $RuAs_2$, $RuSb_2$, OsP_2, $OsAs_2$, $OsSb_2$	6.
square planar T coord., tetrahedral X coord., $\overline{X}$–$\overline{X}$ zig-zag chains	Pd[4]Pt[3+1]	12 C2/c	dsp^2 hybrid / sp^3 hybrids	$Pd^{2+}[-\overline{P}-\overline{P}-]^{2-}$	$10\,5_2$: PdP_2	7.
octahedral T coord., every X atom with only 3 tetrahedral neighbors, $\boxed{X}$ only, layer structure	Cd[6]Cl[3] (C6 type)	3 P$\overline{3}$m1	d^2sp^3 hybrid / imperfect sp^3 hybrids with one non-bonding orbital each	$Pt^{4+}[\boxed{S}]_2^{2-}$	$10\,6_2$: PtS_2, $PtSe_2$	8.
square planar T coordination, every X atom with only 3 tetrahedral neighbors, $\overline{X}$–$\overline{X}$ dumb bells, layer structure	Pd[4]Se[3+1]	12 Pbca	dsp^2 hybrid / imperfect sp^3 hybrids with one non-bonding orbital each	$Pd^{2+}[\overline{Se}-\overline{Se}]^{2-}$	$10\,6_2$: PdS_2, $PdSe_2$	9.

octahedral T coord., half of X atoms with only 3 tetrahedral neighbors, others with four; half of the X are isolated $\underline{X}$, half form dumb bells $\overline{X}-\overline{X}$	$Ir[^6]Se_3^{[s+1]}Se^{[3]}$	24 Pnma	d^{10} sp^3 hybrid	$Ir_3^{4+}[\underline{Se}]\frac{1}{3}[\overline{Se}-\overline{Se}]^{3-}$	$96_3:$ RhSe$_2$, IrS$_2$, IrSe$_2$	10.
deformed octahedral T coord, tetrahedral X coordination, T–T dumb bells, $\overline{X}-\overline{X}$ dumb bells	$Co[^{4+1}]As_2^{[s+1]}$ or $Fe[^{s+s+1}]As^{[s+1]}S^{[s+1]}$ arseno pyrite ≡ ternary ordered $CoAs_2$ type	12 P2$_1$/c 12 P2$_1$/c	d^s sp^3 hybrid d^s sp^3 hybrid d electron bond between two transition metal atoms	$[Rh-Rh]^{8+}[\overline{P}-\overline{P}]_2^{4-}$ imperfect sp^3 hybrid with one non-bonding orbital $[Ru-Ru]^{6+}[\overline{P}-\overline{S}]_2^{3-}$ sp^3 hybrids	$95_2:$ RhP$_2$, RhAs$_2$, RhSb$_2$, RhBi$_2$, IrP$_2$, IrAs$_2$, IrSb$_2$, IrBi$_2$ $8\,56:$ RuPS, RuAsS, RuSbS, RuPSe, RuAsSe, RuSbSe, RuAsTe, RuSbTe, OsPSe, OsAsSe, OsSbSe, OsBiSe, OsAsTe, OsSbTe, OsPS, OsAsS, OsSbS, OsBiTe	11.

TX_3 Compounds

octahedral T coord., deformed tetrahedral X coord, somewhat deformed ring $\begin{array}{cc}|\overline{X} & \overline{X}|\\ |\overline{X} & \overline{X}|\end{array}$	$Co[^6]As_3^{[s+1]}$ skutterudite (DO$_2$ type)	32 Im3	d^6 sp^3 hybrids	$Rh^{3+}[-\overline{P}-\overline{P}-\overline{P}-\overline{P}-]^{3-}$	$95_3:$ RhP$_3$, RhAs$_3$, RhSb$_3$, IrP$_3$, IrAs$_3$, IrSb$_3$	12.
octahedral T coord., on the average every X atom with only 3 tetrahedral neighbors, $\overline{X}-\overline{X}$ dumb bells, ordered pyrite defect structure	$Ir[^6]Se_3^{[s+1]}$	32 P$\overline{1}$	d^6 sp^3 hybrid imperfect sp^3 hybrids with one non-bonding orbital each	$Ir_{0.046}^{3+}\square_{0.33}[\overline{Se}-\overline{Se}]^{3-}$	$96_3:$ IrSe$_3$	13.

[a] References for structure types: 1. BANNISTER and HEY, SB 1928—1932, 234. 2. GASKELL, SB 1937, 3, 41. 3. PARTHÉ, HOHNKE and HULLIGER (1967). 4. BRAGG, SB 1913—1928, 150, 215; TAKEUCHI, SB 1957, 34. 5. BUERGER, SB 1928—1932, 272. 6. HOLSETH and KJEKSHUS (1968); BUERGER, SB 1928—1932, 273. 7. ZACHARIASEN (1963). 8. BOZORTH, SB 1913—1928, 161, 189. 9. GRØNVOLD and RØST, SB 1957, 161. 10. BARRICELLI, SB 1958, 145. 11. DARMON and WINTENBERGER (1966). 12. OFTEDAL, SB 1913—1928, 232, 250; RUNDQVIST and ERSSON (1968). 13. PARTHÉ and HOHNKE (1970).

ment between the observed structural features and the number of valence electrons of the compounds as required by the rules given above. This can be verified by studying the limiting covalent and ionic formulae listed in Table 78-A-5. Drawings for all the different chalcogenide and pnictide structures can be found in a recent review article by HULLIGER (1968). These drawings may be examined to determine in detail the structural features of these compounds.

Revised manuscript received: March 1970

References: Section 44-45-46, 76-77-78-A

BREWER, L.: Predictions of high temperature metallic phase diagrams. In: ZACKAY, V. F. (ed.), High Strength Materials, p. 12—103. New York: John Wiley 1965.
— A most striking confirmation of the Engel metallic correlation. Acta Met. **15**, 553 (1967).
DARMON, R., WINTENBERGER, M.: Structure cristalline de $CoAs_2$. Bull. Soc. Franc. Mineral. Crist. **89**, 213 (1966).
HOLSETH, H., KJEKSHUS, A.: Compounds with the marcasite type crystal structure. I. Compositions of the binary pnictides. Acta Chem. Scand. **22**, 3273 (1968). II. On the crystal structures of the binary pnictides. Acta Chem. Scand. **22**, 3284 (1968).
HULLIGER, H.: Crystal chemistry of the chalcogenides and pnictides of the transition elements. Structure and Bonding **4**, 83 (1968).
HUME-ROTHERY, W.: The Engel-Brewer theories of metals and alloys. Progr. Mater. Sci. **13**, 229 (1968).
PARTHÉ, E., HOHNKE, D.: The ordered vacancy arrangement in pyrite defect structures. In: EYRING, L., O'KEEFE, M. (eds.), The Chemistry of Extended Defects in Nonmetallic Solids, p. 220—222. Amsterdam: North-Holland Publishing Company 1970.
— — HULLIGER, F.: A new structure type with octahedron pairs for Rh_2S_3, Rh_2Se_3 and Ir_2S_3. Acta Cryst. **23**, 832 (1967).
RUNDQVIST, S., ERSSON, N. O.: Structure and bonding in skutterudite-type phosphides. Arch. Kemi **30**, 103 (1968).
ZACHARIASEN, W. H.: The crystal structure of palladium diphosphide. Acta Cryst. **16**, 1253 (1963).

Revised manuscript received: March 1970

© Springer-Verlag Berlin · Heidelberg 1972

Gold

A	R. Allmann	(Fachbereich Geowissenschaften der Universität, Marburg a. d. Lahn, Germany)
B—O	J. H. Crocket	(Department of Geology, McMaster University, Hamilton, Ontario, Canada)

79-A. Crystal Chemistry

Like copper and silver, gold belongs to group 1b of the periodic table and has a single s electron outside a completed d shell. In spite of the similarity in electronic structures there are, however, only few resemblances between Cu, Ag and Au. Gold is strongly siderophilic and somewhat chalcophilic, whereas copper and silver are mainly chalcophilic. This means that gold occurs in nature mainly as the metal and as various alloys, especially with silver. Its chalcophilic character is limited to the formation of several gold tellurides (see Table 79-A-1).

The metallic radii of gold and silver are nearly equal, 1.44 Å, as calculated from the lattice constants of their ccp cell: $a=4.0783$ Å[1] for Au and $a=4.0856$ Å for Ag. The metallic radius of copper is about 11% smaller: $r=1.28$ Å from $a=3.6153$ Å. Because of their similar metallic radii, gold and silver form a continuous solid solution series and native gold always contains some amount of silver, commonly 10—15% by weight. Copper is mostly present too, but in smaller amounts of 0.1 to 0.5%. The purest gold was found in the Kalgoorlie district of Western Australia, having a fineness of 999.1 [$=1,000 \cdot$ Au/(Au+Ag)] (PALACHE et al., 1944, p. 91).

The maximum high temperature solid solubilities of some metals in gold are as follows (after WISE, 1964):

100%	Ag, Cu, Ni, Pd, Pt	10.9—7.7%	Mn, Ta, Co, In
46%	Fe	5.2—1.2%	V, Ga, Sn, Mg, Al, Ti, Ge
21.5—~19%	Cd, Cr, Hg	<1%	As, Bi, Ca, Mo, Pb, Pr, Rh, Sb, Th, Tl, U
13%	Zn		

At lower temperatures some of these alloys form ordered phases, e.g. the following Au—Cu phases are known: Au_3Cu (cubic, $a=3.98$ Å, Au—Au and Au—Cu distance $= 2.81$ Å); CuAu (tetragonal, $a=3.98$, $c=3.72$ Å, Au—Au $=$ Cu—Cu $=2.81$ Å, Au—Cu $=2.73$ Å); and Cu_3Au (cubic, $a=3.71$ Å, Cu—Cu $=$ Cu—Au $=2.65$ Å). The latter two have been found in nature as auricupride (RAMDOHR, 1967).

Several gold tellurides are known, most of them belonging to the system Au—Ag—Te (CAPRI, 1965). In this system, the naturally occurring compounds montbrayite (Au_2Te_3), empressite (AgTe), and muthmannite (Ag, Au)Te could not be synthesized at the temperatures used ($\geqq 290°$ C). In the related ditellurides calaverite, krennerite and sylvanite, Au and Ag have a distorted octahedral coordination with an average Au—Te distance of 2.89 Å. These minerals may be considered as consisting of the ions Au^{3+}, Ag^{3+} and Te^{2-} with about 0.5 additional Te—Te bonds per Te atom (TUNELL and PAULING, 1952). The metallic and ionic radii of AHRENS and PAULING (from STRUNZ, 1970, p. 28/30) do not fit very well with the experimental values: $r(Au^{3+})+r(Te^{2-})=0.85+2.21=3.06$ Å. According to SHANNON and PREWITT (1969), $r(Au^{3+})$ is $=0.70$ Å for a square planar coordination of Au^{3+}. This value yields a better fit: Au^{3+}—$Te^{2-}=0.70+2.21=2.91$ Å (found 2.89 Å).

---

[1] All cited lattice constants from STRUNZ (1970).

Gold

Table 79-A-1. *Gold-containing minerals*

Name	Formula	Wt.-% Au	Density	Comments (references, distances in Å, etc.)
a) Varieties of native gold (mostly after PALACHE et al., 1944)				
Gold	(Au, Ag)	>80	19.3 (pure Ag)	M. P. 1,062.4° C
Electrum	(Au, Ag)	50—80		argentian gold
Silver	(Ag, Au)	0—50	10.5 (pure Ag)	M. P. 960.6° C, aurian silver (küstelite)
Porpezite	(Au, Pd)	90—95		palladian gold
Rhodite	(Au, Rh)	57—66	15.5—16.8	rhodian gold, not confirmed
Auricupride	(Cu, Au) but also Cu_3Au and $CuAu$ (RAMDOHR, 1967) = cuproauride			
Aurosmiride	(Ir, Os, Au)	19.3	20	aurian osmiridium, probably a mixture
b) Intermetallic compounds				
Gold-amalgam[c]	Au_2Hg_3(?)	34.2—41.6	15.5	(PALACHE et al., 1944, p. 105)
Maldonite	Au_2Bi	64.5—65.1	15.5	(JURRIAANSE, 1935, SB 3, 315) artificial Au_2Bi: cubic, $a = 7.98$ Å, Cu_2Mg-type, $Dx = 15.7$, Au—Au = 2.82 Å, Au—Bi = 3.30 Å, Bi—Bi = 3.45 Å
Aurostibite	$AuSb_2$	43.5—50.9	9.89 (Dx)	(GRAHAM and KAIMAN, 1952, SR 15, 15) cubic, $a = 6.66$ Å, pyrit-type, Au—Sb = 2.79 Å, Sb—Sb = 2.63 Å
c) Gold tellurides, structure determined (under remarks: Au—Te distances in Å)				
Calaverite[a]	$AuTe_2$	39.2—42.8[b]	9.0—9.39[b]	2.68 (2), 2.97 (4)
Krennerite[a]	$(Au, Ag)Te_2$	30.7—43.9	8.35—8.62	2.66 (2), 2.99 (4); {2.63, 2.67, 2.87, 2.95, 3.15, 3.22; {2.63, 2.66, 2.94 (?), 3.04 (2)
Sylvanite[a]	$AuAgTe_4$	24.25—29.99	8.07—8.16	{2.67 (2), 2.75 (2), 3.25 (2)
Petzite[c]	Ag_3AuTe_2	19.0—25.2	8.7—9.02	2.53 (2) (FRUEH, SR 1959, 154)
Hessite	Ag_2Te	4.73	8.24—8.45	Ag—Te: 2.85—3.04 (see 47-A)

Table 79-A-1. (Continued)

Name	Formula	Wt.-% Au	Density	Comments(references, distances in Å, etc.)
	d) Gold tellurides, structure unknown			
Montbrayite[c]	Au_2Te_3	38.6—44.3	9.94	PEACOCK and THOMPSON (1946), SR **10**, 55
Muthmannite	(Ag, Au)Te?	22.9—31	5.6	PALACHE *et al.* (1944), p. 260
Kostovite	$CuAuTe_4$	25.2	?	TERZIEV (1966)
	e) Complex gold minerals			
Nagyagite	$AuTe_2 \cdot 6Pb(S, Te)$?	7.4—10.2	7.35—7.46	PALACHE *et al.* (1944), p. 168 (formula after STRUNZ, 1970, p. 136)
Aurobismuthinite	$(Bi, Au, Ag)_5S_6$?	12.3	?	PALACHE *et al.* (1944), p. 278

[a] TUNELL and PAULING (1952).
[b] All gold contents and densities after PALACHE *et al.* (1944).
[c] Note added in proof: Gold-amalgam is a mixture of Au_6Hg_5 and Au_3Hg [LINDAHL, T.: Acta Chem. Scand. **24**, 946—952 (1970)]. Fischesserite, Ag_3AuSe_2, the first gold-containing selenide, is isostructural with petzite [JOHAN, Z., PICOT, P., PIERROT, R., KVAČEK, M.: Bull. Soc. Fr. Minéral. Cristallogr. **94**, 381—384 (1971)]. Montbrayite is only stable in presence of Sb [BACHECHI, F.: Amer. Mineralog. **57**, 146—154 (1972)].

Petzite, Ag_3AuTe_2, contains linear Te—Au—Te groups with a very short Au—Te distance of 2.53 Å (FRUEH, SR 1959, 154), whereas the sum of the ionic radii $r(Au^{1+}) + r(Te^{2-}) = 1.37 + 2.21 = 3.58$ Å. Therefore this Au—Te bond is more intermetallic than ionic: $r(Au) + r(Te) = 1.44 + 1.37 = 2.81$ Å (for coordination number 12, but in petzite Au has only a two-fold coordination). The silver atoms in petzite have a distorted tetrahedral surrounding as in hessite, Ag_2Te, (average Ag—Te = 2.94 Å; FRUEH, SR 1959, 240). Au may replace some Ag in hessite.

All other minerals contain only very minor amounts of Au. Older analyses tend to be significantly higher in gold than more recent analyses (including those by neutron activation) of the same minerals (for a compilation of analyses upto 1967, see JONES and FLEISCHER, 1969). Roughly one may state that the gold content of sulfides is in the ppm region and of oxides and silicates in the ppb region. Nevertheless gold commonly occurs in sulfide minerals, but largely, if not entirely, as the free metal; it is uncertain whether any gold occurs in these minerals in true isomorphous substitution.

References: Section 79-A

CABRI, L. J.: Phase relations in the Au—Ag—Te system and their mineralogical significance. Econ. Geol. **60**, 1569 (1965).

JONES, R. S., FLEISCHER, M.: Gold in minerals and the composition of native gold. U.S. Geol. Surv. Circ. **612** (1969).

PALACHE, C., BERMAN, H., FRONDEL, C.: Dana's System of Mineralogy, vol. 1 (7th ed.). New York: John Wiley & Sons, 1944.

RAMDOHR, P.: A widespread mineral association, connected with serpentinization. Neues Jahrb. Mineral., Abhandl. **107**, 241 (1967).

SHANNON, R. D., PREWITT, C. T.: Effective ionic radii in oxides and fluorides. Acta Cryst. B **25**, 925 (1969).

STRUNZ, H.: Mineralogische Tabellen (5. Aufl.). Leipzig: Akad. Verlagsges. Geest & Portig 1970.

TERZIEV, G.: Kostovite, a gold-copper telluride from Bulgaria. Am. Mineralogist **51**, 29 (1966).

TUNELL, G., PAULING, L.: The atomic arrangement and bonds in the gold-silver ditellurides. Acta Cryst. **5**, 375 (1952).

WISE, E. M.: Gold: Recovery, Properties and Applications. New York: D. Van Nostrand Co. Inc. 1964.

Manuscript received: June 1971

© Springer-Verlag Berlin · Heidelberg 1972

Thallium

A	K. Sahl	(Institut für Mineralogie, Ruhr-Universität, Bochum, Germany)
B—O	C. A. R. de Albuquerque and D. M. Shaw	(Department of Geology, McMaster University, Hamilton, Ontario, Canada)

81-B. Isotopes in Nature

The two thallium stable isotopes (^{203}Tl and ^{205}Tl) were discovered by SCHÜLER and KEYSTON (1931) and ASTON (1931) and the relative abundances of these isotopes are 29.50% (^{203}Tl) and 70.50% (^{205}Tl) (MORRIS and KILLICK, 1960). OSTIC (1967) gives a value of 2.378±0.005 for the (^{205}Tl/^{203}Tl) ratio.

GOLUBCHINA et al. (1957) found that the isotopic composition of thallium does not vary with geological processes (Table 81-B-1) and this seems to be confirmed by more recent investigations (ANDERS and STEVENS, 1960; OSTIC, 1967).

Determinations of the isotopic compositions of terrestrial and meteoritic thallium agree to within 1% (ANDERS and STEVENS, 1960; OSTIC, 1967), and therefore it can be assumed that no radiogenic excess of ^{205}Tl is present in meteorites (Table 81-B-2).

The radionuclides ^{204}Tl and ^{206}Tl may be obtained on irradiation of ^{203}Tl and ^{205}Tl with thermal neutrons (MORRIS and KILLICK, 1960) and they decay to ^{204}Hg and ^{204}Pb respectively. The half-life of ^{206}Tl is 4.19 minutes and that of ^{204}Tl is 4.1 years. The latter nuclide, due to its longer life, is used for the radiometric determination of thallium.

Table 81-B-1. *Isotopic composition of thallium in rocks of different age.*
(After GOLUBCHINA et al., 1957)

Rock	Locality	Age (m. y.)	^{205}Tl/^{203}Tl [a]
Granite	Zmeinogorsk, Altai	230	2.458 + 0.025
Granite	Zmeinogorsk, Altai	230	2.453 + 0.025
Diorite porphyry (dyke)	Zmeinogorsk, Altai	230	2.448 + 0.025
Granite	Leninogorsk, Altai	230	2.455 + 0.025
Granite	Leninogorsk, Altai	230	2.443 + 0.025
Granite	Kemenskii Massif, Eastern Siberia	1,000	2.452 ± 0.025
Granodiorite	Shakthtama, Eastern Transbaikalia	150	2.458 + 0.025

[a] The values of this ratio are, apparently, too high, although for the study of the possible variation of the ^{205}Tl/^{203}Tl ratio, these figures may be considered without correction.

Table 81-B-2. *Isotopic compositions of terrestrial and meteoritic thallium*

Terrestrial thallium		$^{205}Tl/^{203}Tl$	Reference
Normal thallium		2.390	BAINBRIDGE and NIER (1950) (in ANDERS and STEVENS, 1960)
Normal thallium[a]		2.367 ± 0.020	ANDERS and STEVENS (1960)
Weighed average (7 determinations)		2.378 ± 0.005	OSTIC (1967)
Ivigtut galena		2.380 ± 0.003 (I)	OSTIC (1967)

Meteoritic thallium	Class	$^{205}Tl/^{203}Tl$	Reference
Iron meteorites			
Canyon Diablo (Metal)	Og	2.372 ± 0.003	OSTIC (1967)
Canyon Diablo (Troilite)	Og	2.375 ± 0.002	OSTIC (1967)
Canyon Diablo (Troilite)[b]	Og	2.373 ± 0.020	ANDERS and STEVENS (1960)
Toluca (Troilite)[b]	Om	2.376 ± 0.020	ANDERS and STEVENS (1960)
Chondrites			
Enstatite chondrites:			
Abee[b]	Ce_1	2.387 ± 0.020	ANDERS and STEVENS (1960)
Ordinary chondrites (high-iron group):			
Beardsley[b]	CH	2.387 ± 0.020	ANDERS and STEVENS (1960)
Plainview	CH	2.382 ± 0.003	OSTIC (1967)
Richardton[b]	CH	2.404 ± 0.040	ANDERS and STEVENS (1960)
Carbonaceous chondrites:			
Mighei[b]	Cc_2	2.381 ± 0.030 (I)	ANDERS and STEVENS (1960)

[a] These values are not corrected for mass discrimination in the instrument and therefore the ratio is systematically low.

[b] The values given in this table are the adjusted figures (OSTIC and KOHMAN, 1968) for the difference in mass spectrometers (2.378/2.367).

81-C. Abundances in Cosmos, Meteorites, Tektites, Earth and Lunar Samples

I. Cosmos

The estimates of the "cosmic" abundance of thallium are based on the contents of meteoritic and terrestrial matter and therefore are valid only for the solar system.

There is considerable discrepancy between the estimates given by different investigators (Table 81-C-1), due in part to the variable contents of this element in the different types of meteorite, on which these estimates are largely based.

Table 81-C-1. *Atomic abundance of thallium* $(Si = 10^6)$

Tl	^{203}Tl	Reference
0.108		NODDACK and NODDACK (1934)
0.170		GOLDSCHMIDT (1937) (modified by SUESS and UREY, 1956)
0.110		UREY (1952)
0.108	0.03190	SUESS and UREY (1956)
0.001 [a]	0.00030 [b]	EHMAN and HUIZENGA (1959)
0.00044 [a]	0.00013	REED et al. (1958), in EHMAN and HUIZENGA (1959)
0.182 [c]	0.0537	CAMERON (1968)

[a] Values obtained from the abundance of ^{203}Tl (^{205}Tl/^{203}Tl = 2.378).
[b] Based on the determinations on five meteorites.
[c] Based on the abundances in Type I carbonaceous chondrites.

II. Meteorites

The thallium contents of meteorites are given in Table 81-C-2 and average abundances or range of variations, according to several investigators, for different groups of meteorites are listed in Table 81-C-3.

In iron meteorites, thallium is concentrated in the troilite phase relative to the metal phase.

Enstatite chondrites and carbonaceous chondrites are enriched in thallium relative to ordinary chondrites and achondrites by factors of 120 and 170 respectively.

Values of Rb/Tl for meteorites vary widely, from about 10 for carbonaceous chondrites to more than 10,000 for some chondrites (Fig. 81-C-1).

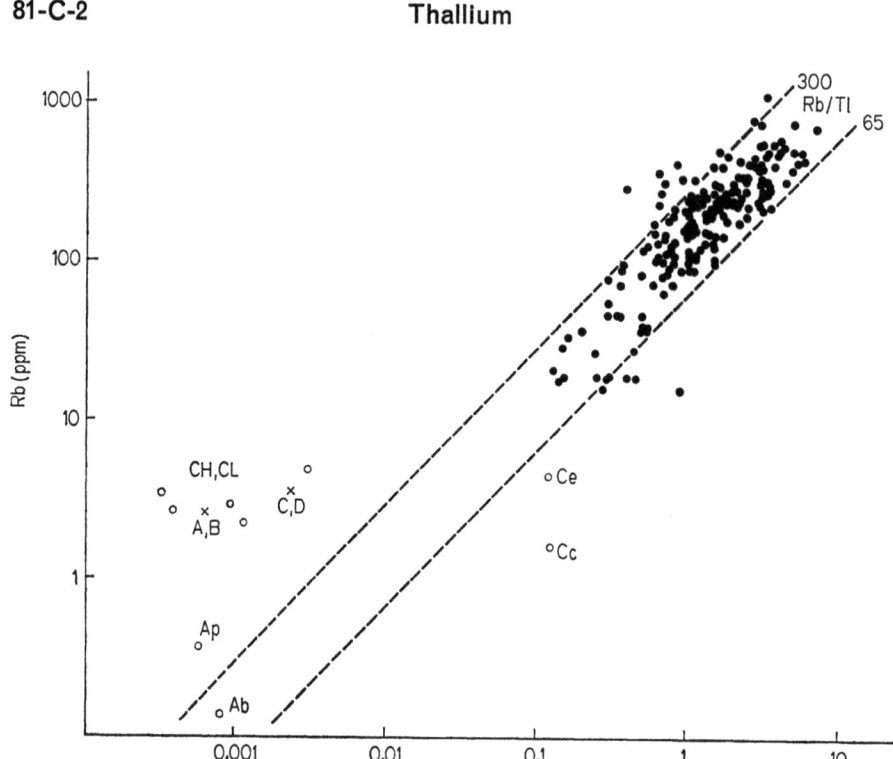

Fig. 81-C-1. *Relationship between thallium and rubidium in igneous rocks, lunar rocks and meteorites.*
○ = *meteorites*, ● = *igneous rocks*, × = *lunar rocks*

Table 81-C-2. *Thallium contents of meteorites*

Meteorite	Class	Tl (ppb)		Reference
Iron meteorites				
Canyon Diablo (Metal)	Og	0.25	(W)	El-Badry et al. (1960)
Canyon Diablo (Metal)	Og	5.8, 7.8	(N/R)	Tandon (1967)
Canyon Diablo (Troilite)	Og	1.5	(W)	El-Badry et al. (1960)
Canyon Diablo (Troilite)	Og	74, 65	(N/R)	Tandon (1967)
Canyon Diablo (Troilite)	Og	13, 7.8	(N/R)	Reed et al. (1960)
Odessa (Metal)	Og	2.4, 0.33	(N/R)	Tandon (1967)
Toluca (Troilite)	Om	198	(N/R)	Reed et al. (1960)
Campo del Cielo (Metal)	H	0.13, 0.21	(N/R)	Tandon (1967)
Sikhote Alin (Metal)	II	0.48, 0.26	(N/R)	Tandon (1967)
Average metal		1.35		
Average troilite		113		
Achondrites				
Nuevo Laredo	Ap	0.75, 0.57	(N/R)	Reed et al. (1960)
Johnstown[a]	Ab	0.98, 0.61	(N/R)	Ehman and Huizenga (1959)
Average		0.75		

Table 81-C-2. (Continued)

Meteorite	Class	Tl (ppb)		Reference
Chondrites				
Enstatite chondrites				
Abee	Ce_1	71, 96	(N/R)	REED et al. (1960)
Abee	Ce_1	150	(I)	ANDERS and STEVENS (1960)
Indarch	Ce_1	125	(N/R)	REED et al. (1960)
Average		120		
Ordinary chondrites				
Beardsley[b]	CH	0.36	(N/R)	REED et al. (1960)
Beardsley	CH	5.8	(I)	ANDERS and STEVENS (1960)
Beardsley[a]	CH	1.9, 4.6	(N/R)	EHMAN and HUIZENGA (1959)
Forest City[a]	CH	0.51, 0.35	(N/R)	EHMAN and HUIZENGA (1959)
Forest City	CH	0.51, 0.39, 0.14	(N/R)	REED et al. (1960)
Plainview	CH	0.7	(W)	EL-BADRY et al. (1960)
Richardton	CH	0.94	(I)	ANDERS and STEVENS (1960)
Average		1.3		
Holbrook	CL	0.45, 0.56, 3.8, 0.42	(N/R)	REED et al. (1960)
Holbrook[a]	CL	0.91, 1.18	(N/R)	EHMAN and HUIZENGA (1959)
Modoc[a]	CL	0.32	(N/R)	EHMAN and HUIZENGA (1959)
Modoc	CL	0.83, 0.14, 0.23, 0.03,	(N/R)	REED et al. (1960)
Average		0.75		
Carbonaceous chondrites				
Orgueil	Cc_1	141	(N/R)	REED et al. (1960)
Mighei	Cc_2	97	(N/R)	REED et al. (1960)
Mighei	Cc_2	140	(I)	ANDERS and STEVENS (1960)
Average		130		

[a] The figures given were calculated from the abundances of ^{203}Tl using the ratio $(^{205}Tl/^{203}Tl) = 2.378$.

[b] Low figure probably due to leaching of the meteorite by ground waters (ANDERS and STEVENS, 1961).

III. Tektites

Although the geochemistry of tektites is fairly well known, very few analyses are available which include thallium determinations. It is noteworthy that thallium seems to occur in very small amounts in tektites (Table 81-C-4).

Table 81-C-3. Average thallium abundances in meteorites (ppb)

Iron meteorites metal	troilite	Enstatite chondrites	Ordinary chondrites	Carbonaceous chondrites	Meteorites	Achondrites	Feldspathic achondrites	Reference
—	—	—	—	—	1.35	—	—	Ehman and Huizenga (1959)
—	10–200	0–140	0.4	70–140	—	—	—	Reed et al. (1960)
—	—	—	1.0	140	—	—	—	Taylor (1964b)
—	100[a]	6[b]	1.0	97[c]	—	0.7[d]	0.8[e]	Vinogradov (1965)
1.35 (3)	113 (2)	120 (2)	1.3 (4) H-group 0.75 (2) L-group	130 (2)	—	0.75 (2)	—	This work

[a] Based on the value given by Nichiporuk.
[b] Based on determinations on the Abee meteorite.
[c] Based on determinations on the Mighei meteorite.
[d] Based on determinations on the Johnstown meteorite.
[e] Based on determinations on the Nuevo Laredo meteorite.

Table 81-C-4. *Thallium contents of tektites*

Rock		Tl (ppm)	Locality	Reference
Australite	(S) (5)	0.1	Australia	TAYLOR (1966b)
Henbury impact glass	(S)	0.4	Henbury, Australia	TAYLOR (1966b)

IV. Earth

The thallium abundance in the earth's crust has been estimated by different authors and the results are summarized in Table 81-C-5.

The thallium content of the earth is, according to BROOKS and AHRENS (1961a), 0.004 ppm.

Table 81-C-5. *Thallium abundance in the Earth's crust*

Tl (ppm)	Reference
0.85×10^{-3}	CLARKE and WASHINGTON (1924)
0.1	NODDACK and NODDACK (1930)
0.3	GOLDSCHMIDT (1937)
3.0	AHRENS (1948)
0.6	RANKAMA and SAHAMA (1950)
1.3	SHAW (1952, 1957)
1.7	VINOGRADOV (1956)
1.0	MASON (1958)
0.7[a]	BROOKS and AHRENS (1961a)
1.0[a]	VINOGRADOV (1962)
0.45[b]	TAYLOR (1964a, 1966a)
0.55[b]	TAYLOR and WHITE (1967)
0.8	This paper

[a] Based on the (acid rock/basic rock) ratio of 2:1.
[b] Continental crust.

V. Lunar Samples

Thallium was detected in lunar samples collected by the U.S.A. Apollo 11 astronauts. The average Tl contents of 5 Types A and B samples and 8 Types C and D were determined to be 0.63 and 2.44 ppb (KEAYS et al., 1970) with Rb/Tl ratios of 4,231 and 1,454 respectively.

81-D. Abundances in Rock-Forming Minerals and Thallium Minerals

I. Thallium Minerals

The thallium minerals are of very rare occurrence in nature, and are only formed during epithermal stages of hydrothermal processes or under supergene conditions (VLASOV, 1966). These minerals and their chemical composition are listed in Table 81-D-1.

Table 81-D-1. *Thallium minerals*

Mineral	Chemical composition	Paragenesis
Lorandite	$TlAsS_2$	Realgar, orpiment, colloform pyrite
Picotpaulite	$TlFe_2S_3$	
Chalkothallite	Cu_3TlS_2	
Vrbaite	$Hg_3Tl_4As_8Sb_2S_{20}$	Realgar, orpiment, colloform pyrite
Hutchinsonite	$(Pb, Tl)_2(Cu, Ag)As_5S_{10}$	Sulfides etc of As, Zn, Pb, Fe
Bukovite	$Cu_{3+a}Tl_2FeS_{4-a}$	
Wallisite	$PbTlCuAs_2S_5$	
Hatchite	$PbTlAgAs_2S_5$	
Crookesite	$(Cu, Tl, Ag)_2Se$	Selenides
Avicennite	Tl_2O_3	Hematite, calcite

II. Minerals from Ultramafic, Mafic and Intermediate Rocks

Data on the thallium contents of minerals from these rocks are scarce (Table 81-D-2).

Phlogopite from a dunite pipe from Transvaal, South Africa (AHRENS, 1945) contains 3.0 ppm Tl.

The distribution of thallium in minerals from gabbro and ferrodiorite from the Skaergaard Intrusion (Table 81-D-2) shows some irregularities. In the gabbro, thallium is concentrated in magnetite while in the ferrodiorite the highest content of this element is found in olivine. The magnetite from a lamprophyre, associated with a carbonatite complex, is richer in thallium than the augite from the same rock (Table 81-D-2).

Table 81-D-2. *Thallium contents of minerals from mafic rocks*

Mineral	Tl (ppm)				
	1	2	3	4	5
Plagioclase	0.1	0.1	—	—	0.017, 0.029
Olivine	—	0.8	—	—	
Pyroxene	0.5	0.2	0.044	0.1	
Ilmenite	1.6	—	—	—	
Magnetite	2.2	0.05	0.36	—	
Method	(N/R)	(N/R)	(S)	(S)	(S)

1 = Gabbro (MZ), Skaergaard Intrusion (East Greenland) (ADAMS, 1961, in WAGER and BROWN, 1967). 2 = Ferrodiorite (UZ), Skaergaard Intrusion (East Greenland) (ADAMS, 1961, in WAGER and BROWN, 1967). 3 = Lamprophyre, Seabrook Lake, Ontario (SHAW, 1952). 4 = Gabbro Pegmatite, Harzburg, Germany (OTTEMAN, 1940). 5 = Stillwater Complex, Montana, U.S.A. (SHAW, 1952).

III. Minerals from Granitic Rocks

Very few data are available on the thallium distribution in co-existing minerals of granitic rocks (Table 81-D-3). This element is concentrated in biotite, as can be expected from its crystal chemistry (Table 81-D-8). Relatively high amounts of thallium occur in the potassium feldspar and in amphibole and it may be noted that the amount in the latter mineral may exceed that of the co-existing potassium feldspar.

In biotite, thallium occurs generally in the range 3—15 ppm, while the amounts in potassium feldspar and possibly muscovite are mostly in the range 1—6 ppm (Tables 81-D-3; 81-D-8).

IV. Minerals from Alkalic Rocks

The available data for thallium in minerals from alkalic rocks (Table 81-D-4) show that the content of this element in both biotite and potassium feldspar is in the same range as in granitic rocks. Two leucite analyses are included in Table 81-D-8.

V. Minerals from Granite Pegmatites

Data on the thallium contents of potassium feldspar from granites and related pegmatites show a thallium enrichment in the feldspar from the pegmatites (Tables 81-D-3, 6, 7).

Large variations are observed for the thallium contents of potassium feldspars from pegmatites, where it occurs mostly in the range 0.5—50 ppm, although occasionally contents upto 150 ppm have been found. A marked enrichment of thallium is observed in strongly differentiated pegmatites, where contents upto 600 ppm have been reported in the potassium feldspar. Amazonite is also particularly rich in thallium, with amounts in the range 6—100 ppm (Table 81-D-6, 7).

Table 81-D-3. *Thallium contents of minerals from granitic rocks*

Mineral	Method	Tl (ppm) range	Tl (ppm) mean	Rock	Locality	Ref.
Quartz	(?)		0.1	Granodiorite	Almalyk area, UzbekS.S.R.	1
Potassium feldspar	(?)		1.4	Granodiorite	Almalyk area, UzbekS.S.R.	1
	(C)		2.4	Porphyritic adamellite	Susamyr batholith, U.S.S.R.	2
	(S) (2)	4.5—4.8	4.6	Granite	Dzirul Massif, U.S.S.R.	3
	(S) (3)	0.7—1.1	0.9	Granite	E. Transvaal	4
	(S) (70)	0.9—5.0		"Granite"	South Africa	5
	(S) (26)	1.3—30		Granite	S. Norway	6
	(S)		8.2	Farsundite	S. Norway	6
	(S)		1.0	Granite	Langøy, N. Norway	7
Amazonite	(C)		44	Amazonite granite	E. Transbaikalia, U.S.S.R.	8
Albite	(C)		10	Amazonite granite	E. Transbaikalia, U.S.S.R.	8
Oligoclase	(?)		0.7	Granodiorite	Almalyk area, UzbekS.S.R.	1
Biotite	(?)		8.4	Granodiorite	Almalyk area, UzbekS.S.R.	1
	(S) (4)	3.0—3.6	3.2	Biotite quartz-diorite	Dzirul Massif, U.S.S.R.	3
	(S) (3)	3.7—4.3	4.1	Granite (Hercynian)	Dzirul Massif, U.S.S.R.	3
	(S) (2)	3.0—3.8	3.4	Granite (Mesozoic)	Dzirul Massif, U.S.S.R.	3
	(C) (4)	5.0—7.2	6.0	Biotite granite	Tyrny-Auz, U.S.S.R.	9
	(C) (3)	5.2—6.0	—	"Granite"	Susamyr batholith, U.S.S.R.	2
	(N/R) (4)	2.6—12.4	—	Granite	Amo Complex, N. Nigeria	10
	(S) (22)	5—16	—	"Granite"	South Africa	5
Phlogopite	(S)		4.4	Graphic granite	Transvaal	4
Zinnwaldite	(C)		48	Amazonite granite	E. Transbaikalia, U.S.S.R.	8
Muscovite	(S) (5)	1.5—3.7		Granite	South Africa	4,11
Hornblende	(C)		0.8	Porphyritic adamellite	Susamyr batholith, U.S.S.R.	2
	(?)		2.6	Granodiorite	Almalyk area, UzbekS.S.R.	1
Magnetite	(?)		0.4	Granodiorite	Almalyk area, UzbekS.S.R.	1

References: 1. BADALOV and RABINOVICH (1966); 2. TAUSON and BUSAEV (1957); 3. ODIKADZE (1967); 4. AHRENS (1945); 5. KOLBE (1965); 6. HEIER and TAYLOR (1959); 7. HEIER (1960); 8. SITNIN (1960); 9. VOSKRESENSKAYA (1959); 10. BUTLER (1962); 11. KOLBE and TAYLOR (1966b).

Table 81-D-4. *Thallium contents of minerals from the alkalic rocks of the Mt. Sandyk Massif, U.S.S.R.* (According to Zlobin, 1958)

Mineral	Tl (ppm)		
	alkaline-earth syenite (Phase I)	hornblende alkalic syenite (Phase II)	nepheline syenite (Phase III)
Orthoclase	1.2	2.5	3.4
Biotite	3.2	7.6	5.3
Hornblende	—	0.7	—
Rock	1.1	2.2	2.7

Table 81-D-5. *Thallium contents of minerals from a differentiated granitic pegmatite, Mongolian Altai.* (According to Solodov, 1962)

Mineral	Tl (ppm) (C), pegmatite zones									
	I	II	III	IV	V	VI	VII	VIII	IX	X
Microline	18	10	25	—	—	—	—	—	61	—
Muscovite	—	15	—	21	—	—	—	—	—	—
Gilbertite	—	—	—	—	—	28	22	—	—	—
Lepidolite	—	—	—	—	—	—	—	72	—	—
Beryl	—	4	—	4	6	7	—	—	—	—
Albite	—	4	—	—	—	—	—	—	—	—
Spodumene	—	—	—	—	—	4	—	—	—	—

Zones defined in the pegmatite (from the margin to the centre): I = Graphic granite; II = Fine grained albite; III = Blocky microcline I; IV = Quartz-muscovite; V = Cleavelandite-spodumene; VI = Quartz-spodumene; VII = Fine platy albite; VIII = Scaly muscovite; IX = Blocky microcline II; X = Blocky quartz.

Muscovite is relatively rich in thallium, particularly late muscovite in replacement zones of pegmatites, which may contain upto 280 ppm. The thallium contents of this mineral vary, in general, between 2 and 60 ppm.

Lepidolite, lithium-rich biotite and pollucite contain appreciable amounts of thallium.

Other minerals from pegmatites contain very low amounts of thallium (Table 81-D-5, 7).

VI. Minerals from Syenite Pegmatites

The potassium feldspars from the Palabora pegmatites (E. Transvaal) contain thallium in the range 0.4—1.0 ppm (Ahrens, 1945), which is notably lower than in the same mineral from granitic pegmatites. Similarly, the thallium contents of phlogopite from the same pegmatite are low ($<$1—1.9 ppm).

VII. Minerals from Alkalic Pegmatites

Gerasimovsky and Rasskazova (1962) investigated the distribution of thallium in the minerals from pegmatites related to the alkalic rocks of the Lovozero Massif.

Table 81-D-6. *Thallium contents*

	Tl (ppm)				
	1	2	3	4	5
Microcline	48	84	1.5–3.2	—	0.5–3.1
Amazonite	—	—	19	10—30	—
Albite	—	—	—	3.0	—
Cleavelandite	—	6.8	—	—	—
Muscovite	145	47	2.0—37	6.0—6.7	10
Lepidolite	105—210	60—95	—	95—220	—
Pollucite	—	18	—	105	—
Method	(S)	(S)	(S)	(S)	(S)

Locality: 1. Harding Mine Pegmatite, Dixon, New Mexico (AHRENS, 1948); 2. Custer Mt. Pegmatites, Black Hills, S. Dakota (AHRENS, 1948; SHAW, 1952); 3. Cape Province, South Africa (AHRENS, 1945); 4. Old granite, S. W. Africa (AHRENS, 1945; AHRENS, 1948); 5. Old granite, Natal (AHRENS, 1945); 6. Old granite, E. Transvaal (AHRENS, 1945);

Table 81-D-7. *Thallium contents of minerals from granitic pegmatites*

Mineral	Method	Tl (ppm) range	mean	Locality	Ref.
Potassium feldspar	(S) (4)	6.7—7.7	7.2	Small pegmatites, S. Norway	1
	(S) (38)	2—140	19.5	Large pegmatites, S. Norway	1
	(S) (8)	380—610	500	Differentiated vein, Kola Peninsula, U.S.S.R.	2
	(W?) (4)	5.1—15	8.0	Karaoba Massif, U.S.S.R.	3
	(S) (8)	0.0—16	9.8	Dzirul Massif, U.S.S.R.	4
	(C)		70	Albite pegmatite, Altai, Mongolia	5
	(S) (3)	1.0—1.5	1.2	Bushveld granite	6
	(S) (10)	3.2—20	6.5	Johannesburg, S. Africa	6
Amazonite	(S)		50	Honeydew, Transvaal	6
	(S)		35	Madagascar	7
	(S) (2)		10	Labrador, Ontario	7
	(S)		77	Norway	7
	(S) (2)	6—50	—	Colorado	7
Oligoclase (An$_{11}$)	(S)		1.0	Snarum, Norway	8
Muscovite	(W)		18	Replacement zone, Sayan Mts., U.S.S.R.	9
	(S) (2)	260—280	270	Replacement zone, Kola Peninsula, U.S.S.R.	2
Lepidolite	(S)		18	Old granite, Mozambique	6
	(S) (2)	70—95	—	Connecticut	7

of minerals from several granitic pegmatites

	Tl (ppm)				
	6	7	8	9	10
Microcline	2.6—3.7	—	9—29	5—17	12—200
Amazonite	—	15	33—99	—	—
Albite	0.4	—	0.7—1.1	3	—
Cleavelandite	—	—	1.2—1.4	—	—
Muscovite	3.3—26	2.0—58	—	4—11	82—140
Lepidolite	50—58	—	—	—	170—520
Pollucite	—	—	—	—	160—440
Method	(S)	(S)	(S)	(W)	(S)

7. Namaqualand (AHRENS, 1945); 8. Landsverk, Evje, S. Norway (TAYLOR et al., 1960); 9. Sayan Mts., U.S.S.R. (SLEPNEV, 1961); 10. Kola Peninsula, U.S.S.R. (BOROVIK-ROMANOVA and SOSEDKO, 1960).

Table 81-D-7. (Continued)

Mineral	Method	Tl (ppm)		Locality	Ref.
		range	mean		
	(S) (2)	175—200	185	Pala, California	7
	(S) (2)		155	Maine	7
	(S) (2)	230—270	250	Brown Derby, Colorado	7
	(S)		95	Copper Mt., Wyoming	7
	(S) (3)	115—165	—	S. E. Manitoba	7
	(S)		95	Lipowka, Urals	7
	(S)		125	Moravia	7
Biotite (lithium-rich)	(S)		80	Strickland Quarry, Connecticut	7
	(S)		385	King's Mt., N. Carolina	7
Zinnwaldite	(S) (2)	62—72	—	Saxony, Erzgebirge (Germany)	6
Beryl	(S)		39	Manjak, Madagascar	7
Rhodizite	(S)		95	Manjak, Madagascar	7
Pollucite	(S) (2)	13—18	—	Greenwood, Norway (Maine)	7
	(S)		77	Varuträsk, Sweden	7
Tourmaline	(W)		4	Sayan Mts., U.S.S.R.	9
Garnet	(W)		4	Sayan Mts., U.S.S.R.	9
Triphylite	(C)		4	Altai, Mongolia	5
Apatite	(S)		0.23	Black Hills, S. Dakota	10

List of References: 1. HEIER and TAYLOR (1959); 2. BOROVIK-ROMANOVA and SOSEDKO (1960); 3. GANEYEV and SECHINA (1962); 4. ODIKADZE (1967); 5. SOLODOV (1962); 6. AHRENS (1945); 7. AHRENS (1948); 8. HEIER (1962); 9. SLEPNEV (1961); 10. SHAW (1952).

Table 81-D-8. *Thallium contents of various rock-forming minerals*

Mineral	Method	Tl (ppm)	Locality	Reference
Zircon	(S)	0.36	Mt. Titanium, Georgia, U.S.A.	Shaw (1952)
Titanite	(S)	0.1	Eganville, Renfrew Co., Ontario	Shaw (1952)
Garnet	(S)	0.019	Adirondacks, New York	Shaw (1952)
Garnet	(S)	0.009	N. Stromfjord, Greenland	Shaw (1952)
Almandine	(S)	0.007	Koirinkoiski, Kitila, K.F.S.S.R.	Shaw (1952)
Beryl	(S)	0.40	Lyndoch Tp., Renfrew Co., Ontario	Shaw (1952)
Tremolite	(S)	0.65	Lee, Massachussets	Shaw (1952)
Biotite G-24	(S)	27.8	Australia	Kolbe (1965)
Biotite G-22	(C)	12.2	Australia	Kolbe (1965)
Biotite F-234	(C)	15.9	Australia	Kolbe (1965)
Lepidolite GMM-P	(C)	83	Australia	Kolbe (1965)
Nepheline	(S)	7.6		Shaw (1952)
Leucite	(S)	65	Civita Castellano, Italy	Shaw (1952)
Leucite	(F)	15	Italy	Fornaseri and Penta (1963)
Opal	(S)	0.043	Grand Manan Is., New Brunswick	Shaw (1952)
Asbecasite	(C?)	0.90%	Binnatal, Kt. Wallis, Switzerland	Graeser (1966)

Table 81-D-9. *Thallium contents of potassium feldspars from metamorphic rocks*

Rock (Number of samples)		Tl (ppm) range	Tl (ppm) mean	Locality	Reference
Monzonitic granulite	(4)	tr. −0.9	0.7	Langøy, N. Norway	Heier (1960)
Monzonitic and granodioritic granulite (banded series)	(2)	0.8—1.0	0.9	Langøy, N. Norway	Heier (1960)
Monzonitic gneisses (retrograde metamorphosed)	(2)	0.9—1.0	1.0	Langøy, N. Norway	Heier (1960)
Granodioritic charnockite	(2)	0.8—0.9	0.9	Langøy, N. Norway	Heier (1960)
"Granodioritic" gneiss (amphibolite facies)	(5)	0.6—0.9	0.8	Langøy, N. Norway	Heier (1960)
Augen gneiss	(2)		0.8	Langøy, N. Norway	Heier (1960)
Granitic-pegmatitic vein in gneiss			0.7	Langøy, N. Norway	Heier (1960)
Gneiss granite			5.4	S. Norway	Heier and Taylor (1959)
Granitic gneiss			5.8	S. Norway	Heier and Taylor (1959)
Gneiss	(3)	5.3—12	8.2	S. Norway	Heier and Taylor (1959)
Augen gneiss	(2) (S)	1.9—7.9	4.9	S. Norway	Heier and Taylor (1959)

This element is concentrated in the K—Na feldspar (2.78 ppm), nepheline contains 1.3 ppm while aegirine shows a low content of thallium —0.44 ppm.

HEIER (1966) found that the thallium concentration in potassium feldspar, nepheline and albite from the pegmatites related to the nepheline syenites of the Sternjøy Complex (Northern Norway), to be less than 0.5 ppm.

VIII. Minerals from Metamorphic Rocks

The thallium contents of potassium feldspars from gneisses from the South of Norway (Table 81-D-9) are in the same range as in the same mineral from granitic rocks.

The potassium feldspars from granulite facies rocks of different composition (Table 81-D-9), on the other hand, are particularly low in thallium, with contents of this element not exceeding 1.0 ppm.

Revised manuscript received: March 1970

81-E. Abundance in Common Igneous Rock Types

The results of thallium determinations for some international standard rocks are given in Table 81-E-1 and average thallium contents and atomic abundances in igneous rocks, according to different authors, are summarized in Table 81-E-7.

I. Ultramafic Rocks

The thallium contents of these rocks are typically low, mostly in the range 0.07—0.30 ppm (Table 81-E-2). The mineralogical composition of the rocks has, apparently, very little influence on the thallium content, as dunites, peridotites and pyroxenites show the same range of variation in the amount of this element.

II. Mafic Rocks

The thallium contents of mafic rocks are variable, mostly in the range 0.05—0.7 ppm (Table 81-E-3) and no systematic differences are found between plutonic and volcanic rocks. Dolerites also contain thallium varying in the same range. The contents of anorthosites are lower than those of other basic rocks.

Enrichment of thallium in mafic rocks is observed in the chilled margins of intrusions, e.g. the Skaergaard gabbro and the Jagersfontein Mine dolerite (BROOKS and AHRENS, 1961a) (Table 81-E-3).

III. Intermediate Rocks

A definite increase in the thallium contents is observed in these rocks relative to mafic and ultramafic rocks. Most values are in the range 0.15—1.4 ppm and the higher figures are found more frequently in the more acid types (syenite, monzonite and dacite) (Table 81-E-4).

IV. Granitic-Rhyolitic Rocks

1. Granitic Rocks

Most granitic rocks contain thallium in the range 0.6—3.5 ppm (Table 81-E-5) showing that these rocks are enriched in this element relative to the more mafic rocks.

Granitic rocks of different age from South West Africa show enrichment in thallium toward the more modern types (SIEDNER, 1968). This would be the result of the processes of partial melting of former granitic rocks. The data available for the thallium contents of granitic rocks of different ages do not show that this element has become enriched in the more modern rocks.

Table 81-E-1. Thallium contents of international standard rocks

Tl (ppm)											Method	Reference
G-1	W-1	Sy-1	T-1	G-2	GSP-1	AGV-1	PCC-1	DTS-1	BCR-1	Sulphide Ore-1		
1.3	0.05	—	—	—	—	—	—	—	—	—	(W/S)	Brooks et al. (1960)
1.3	0.17	—	—	—	—	—	—	—	—	—	(N/R)	Morris and Killick (1960)
1.27	0.21	—	—	—	—	—	—	—	—	—	(W/S)	Brooks and Ahrens (1961b)
1.3	0.10	—	—	—	—	—	—	—	—	—	(N/R)	Vincent and Adams (1961) (in Fleischer and Stevens, 1962)
1.5	—	—	—	—	—	—	—	—	—	—	(N/R)	Butler (1962)
0.4	—	—	—	—	—	—	—	—	—	—	(I)	Brown and Wolstenholme (1964)
1.3	0.2	1.8	—	—	—	—	—	—	—	—	(S)	Taylor and Kolbe (1964)
—	—	1.7	—	—	—	—	—	—	—	—	(S)	Kolbe (1965)
1.3	0.14	—	—	—	—	—	—	—	—	—	(I)	Taylor (1965a)
—	0.185	—	—	—	—	—	—	—	—	—	(I)	Taylor (1965b)
1.0	0.1	—	0.5	—	—	—	—	—	—	—	(S)	Schroll and Weninger (1965)
—	—	1.8	—	—	—	—	—	—	—	1.8	(S)	Webber (1965)
1.2	0.12	1.3	—	0.85	0.71	0.27	—	—	—	2.90	(S)	Tennant (1967)
0.66	<0.05	0.86	—	—	—	—	<0.05	<0.05	0.36	0.2	(P)	Wahler (1968)
—	—	1.4	—	—	—	—	—	—	—	—	(S)	Champ (1968) (in Sine et al., 1969)
1.3	—	—	—	—	—	—	—	—	—	—	(S)	Champ (1968) (in Fleischer, 1969)
—	—	—	—	1.3	1.6	1.6	—	—	—	—	(S)	Champ (1968) (in Flanagan, 1969)
1.3	0.13	—	—	—	—	—	—	—	—	—		Recommended value (Fleischer, 1969)

Table 81-E-2. *Thallium contents of ultramafic rocks*

Rock	Method	Tl (ppm) range	Tl (ppm) mean	Locality	Reference
Dunite	(S)		0.11	Khabozero, Kola Peninsula	Shaw (1952)
Dunite	(C)		0.13	Kola Peninsula	Voskresenskaya (1961)
Serpentinized dunite	(C) (5)		0.07	Cheremshanka, Urals	Voskresenskaya (1961)
Peridotite	(N/R)		1.04	Insch Mass, Aberdeenshire, Scotland	Morris and Killick (1960)
Peridotite	(S)		0.13	St. George, Garden Cove, Aleutian Is.	Shaw (1952)
Peridotite	(C) (10)		0.07	Kola Peninsula	Voskresenskaya (1961)
Serpentinized peridotite	(C) (5)		0.13	Kola Peninsula	Voskresenskaya (1961)
Lherzolite	(W/S)		0.12	Lac Lherz, Pyrenees, France	Brooks et al. (1960)
Pyroxenite	(C) (10)		0.27	Eastern Sayan Mt., U.S.S.R.	Voskresenskaya (1961)
Feldspathic pyroxenite	(C) (10)		0.07	Kola Peninsula	Voskresenskaya (1961)
Serpentinite	(C) (4)	0.3—0.5	0.4	Urusht Complex, Greater Caucasus	Voskresenskaya (1961)

Table 81-E-3. *Thallium contents of mafic rocks and lamprophyres*

Rock	Method	Tl (ppm) range	Tl (ppm) mean	Locality	Ref.
Plutonic rocks					
Gabbro-norite	(C) (10)		0.13	Northern Karelia, U.S.S.R.	1
Gabbro (composite)	(S) (11)		0.3	Germany	2
Gabbro	(S)		0.16	Sudbury, Ontario	3
Gabbro	(S)		0.058	Duluth, Minnesota	3
Gabbro	(S)		0.060	Moskuvaara, Finland	3
Gabbro	(R)		0.6	Kohyama, Yamaguchi, Japan	4
Gabbro	(C)		0.2	Degtyarka, Urals	1
Gabbro	(W/S)		0.18	Paresis Complex, S. W. Africa	5
Hornblende gabbro	(S)		0.46	Salem, Massachusetts	3
Gabbro-diorite	(C) (5)		0.3	Central Kazakhstan, U.S.S.R.	1
Gabbro-diorite	(C) (4)		0.3	Caucasus, U.S.S.R.	1
Olivine gabbro	(N/R)		0.048	Insch Mass, Scotland	6
Syenogabbro	(N/R)		0.046	Insch Mass, Scotland	6
Olivine gabbro (LZ)	(N/R) (2)		0.03	Skaergaard Intrusion, East Greenland	7

Table 81-E-3. (Continued)

Rock	Method	Tl (ppm) range	Tl (ppm) mean	Locality	Ref.
Olivine-free gabbro (MZ)	(N/R) (2)	0.01—0.02	—	Skaergaard Intrusion, East Greenland	7
Chilled marginal gabbro	(N/R)		0.05	Skaergaard Intrusion, East Greenland	7
Ferrodiorite (UZ)	(N/R) (4)	0.01—0.04	0.025	Skaergaard Intrusion, East Greenland	7
Anorthosite	(S)		0.029	Stillwater Complex, Montana	3
Anorthosite	(S)		tr.	Egersund, Norway	3

Doleritic rocks

Rock	Method	Tl (ppm) range	Tl (ppm) mean	Locality	Ref.
Basalt ("diabase")	(C) (7)		0.22	Northern Caucasus	1
Basalt ("diabase")	(C) (4)	0.2—0.4	0.3	Greater Caucasus	1
Basalt (dolerite sill) (chilled margin)	(W/S) (12)	0.02—0.15	0.06	Jagersfontein Mine,	8
			1.10	Orange Free State,	8
Basalt (dolerite sill)	(W/S) (5)	0.03—0.45	0.12	South Africa	8
Basalt (dolerite)	(W/S)		0.54	Kloof Nek, Cape Town, S. Africa	9
Olivine basalt ("diabase porphyry")	(C) (13)		0.06	Northern Caucasus	1
Olivine basalt ("olivine diabase")	(S)		0.073	Säppi, Luvia, Finland	3
Basalt pegmatite ("diabase-pegmatite")	(S)		0.47	Säppi, Luvia, Finland	3

Volcanic rocks

Rock	Method	Tl (ppm) range	Tl (ppm) mean	Locality	Ref.
Basalt	(S) (3)	0.019—0.19	—	Aleutian Islands	3
Basalt	(R) (3)	0.6—0.9	—	Mt. Mihara, Japan	4
Basalt (Caledonian)	(C) (6)	0.2—0.3	0.2	Urusht Complex, Greater Caucasus	1
Basalt (Hercynian)	(C) (6)	0.2—0.4	0.3	Khudes, Greater Caucasus	1
Basalt (Alpine)	(C) (7)	0.1—0.4	0.22	Mt. Indyuk, Greater Caucasus	1
Basalt	(C) (10)		0.3	Northern Caucasus	1
Amygdaloidal basalt	(W/S)		0.07	Paresis, S. W. Africa	8
Spilite	(C) (9)	0.0—0.3	0.2	Khudes, Urusht, Greater Caucasus	1

Lamprophyres

Rock	Method	Tl (ppm) range	Tl (ppm) mean	Locality	Ref.
Lamprophyre	(S)		0.4	Seabrook Lake, Ontario	3
Lamprophyre	(S)		0.12	Åvo Brändö, Finland	3

List of References: 1. Voskresenskaya (1961a); 2. Preuss (1940); 3. Shaw (1952); 4. Ishimori and Takashima (1955); 5. Brooks and Ahrens (1961b); 6. Morris and Killick (1960); 7. Adams (1961) in Wager and Brown (1967); 8. Brooks and Ahrens (1961a); 9. Brooks et al. (1960).

Table 81-E-4. *Thallium contents of intermediate and syenitic rocks*

Rock	Method	Tl (ppm) range	Tl (ppm) mean	Locality	Ref.
Gabbro diorite and diorite	(C) (4)	0.6—0.8	0.7	Susamyr Batholith, U.S.S.R.	1
Diorite	(C) (2)	0.12—0.5	—	Greater Caucasus	2
Diorite	(?) (8)		1.1	Almalyk, UzbekS.S.R.	3
Diorite	(S)		0.3	Jersey City, N. J.	4
Diorite and monzonite	(?)		0.6	Shakhtaminski Massif, U.S.S.R.	5
Syenite-diorite	(C) (8)	0.11—0.4	—	Greater Caucasus	2
Syenodiorite	(?) (14)		1.3	Almalyk, UzbekS.S.R.	3
Monzonite	(S)		1.7	San Juan Co., Colorado	4
Syenite	(S) (2)	1.1—1.3	—	Marquis Tp., Ontario	4
Syenite	(N/R)		0.29	Insch Mass, Scotland	6
Biotite quartz-diorite	(C)		0.5	Teberda River, Greater Caucasus	2
Biotite quartz-diorite	(S) (21)	0.0—9.0	1.7	Dzirul Massif, Georgia, U.S.S.R.	7
Tonalite	(C) (7)	0.5—0.8	0.7	Greater Caucasus	2
Tonalite	(C) (9)	0.2—0.4	0.3	Malaya Laba Massif, Caucasus	8
Tonalite	(S) (2)	0.29—0.50	—	Ontario	4
Volcanic rocks					
Basaltic andesite	(C) (5)	0.2—0.6	0.4	Armenia, U.S.S.R.	2
Basaltic andesite	(C) (2)	0.4—0.5	—	S. Osetiya, Georgia, U.S.S.R.	2
Basaltic andesite	(C) (5)	0.2—0.4	0.3	Elbrus, Greater Caucasus	2
Andesite	(C)		1.1	Elbrus, Greater Caucasus	2
Andesite	(C) (8)	0.0—0.13	0.06	Mt. Indyuk, Greater Caucasus	2
Andesite	(C) (6)	0.0—0.4	0.06	Khudes, Greater Caucasus	2
Andesite	(C)		0.6	Gudzhareti, Georgia, U.S.S.R.	2
Andesite	(R) (8)	0.6—1.4	—	Mt. Asama, Mt. Yake, Japan	9
Andesite (mafic)	(S) (3)	0.14—0.90	0.43	Hahajima, Japan	9
Andesite	(S) (2)	0.16—0.50	0.33	Tongariro National Park, New Zealand	10
Andesite	(S) (2)	0.15—0.27	0.21	Asama volcano, Izu, Japan	10
Andesite-dacite	(C) (19)	0.2—1.1	0.7	Elbrus, Greater Caucasus	2
Andesite-dacite	(C) (5)	0.2—0.8	0.5	Kazbek, U.S.S.R.	2
Andesite-dacite	(C)		0.5	S. Osetiya, U.S.S.R.	2
Dacite	(S)		0.51	St. George, Aleutian Islands	4
Dacite	(R)	1.1—1.4	1.3	Japan	9
Dacite pumice tuff	(C)		1.8	Elbrus, Greater Caucasus	2
Trachyte	(C) (5)	0.13—0.33	0.23	Mt. Indyuk, Greater Caucasus	2
Keratophyres					
Keratophyre	(C) (4)		0.5	Urusht, Greater Caucasus	2
Keratophyre	(C) (2)		0.3	Buron, Northern Caucasus	2

List of References: 1. Tauson and Busaev (1957); 2. Voskresenskaya (1961a); 3. Badalov and Rabinovich (1966); 4. Shaw (1952); 5. Tauson (1968); 6. Morris and Killick (1960); 7. Odikadze (1967); 8. Demin and Khitarov (1958); 9. Ishimori and Takashima (1955); 10. Taylor and White (1967).

Table 81-E-5. *Thallium contents of granitic and rhyolitic rocks*

Rock	Method	Tl (ppm) range	Tl (ppm) mean	Locality	Ref.
Granitic rocks					
Granite (Precambrian)	(S) (6)	0.4—6.4	2.9	Finland	1
Rapakivi granite	(S) (3)	2.8—4.0	3.3	Finland	1
Rapakivi granite	(C) (4)	1.4—1.7	—	Ukraine, Karelia, Dresden	2
Granite	(S) (7)	0.36—0.77	0.61	Huab, S. W. Africa	3
Granite	(S) (8)	0.95—1.60	1.22	Damara, S. W. Africa	3
Granite	(S) (34)	1.1—3.1	1.8	Cape, South Africa	4
Gneissic granite	(S) (4)	1.0—1.7	1.3	Snowy Mountains, Australia	4
Granodiorite	(S) (20)	0.6—1.6	1.1	Snowy Mountains, Australia	4
Leucogranite	(S) (8)	1.5—3.2	2.1	Snowy Mountains, Australia	4
Granite	(S) (4)	1.0—2.1	1.5	Harz, Germany	5
Biotite granite	(N/R) (3)	2.7—4.0	3.2	Cornwall, S. W. England	6
Muscovite-biotite granite	(N/R)		6.9	Cornwall, S. W. England	6
Granite	(W?) (5)	3.0—5.8	—	Karaoba, C. Kazakhstan	7
Aplite	(W?)		5.9	Karaoba, C. Kazakhstan	7
Granite	(C) (2)	0.6—0.7	0.6	Greater Caucasus	2
Alaskite	(C)		0.3	Greater Caucasus	2
Microcline granite	(C) (17)	0.7—1.7	1.1	Greater Caucasus	2
Two-mica granite	(C) (38)	0.5—4.0	1.4	Greater Caucasus	2
Biotite granite	(C) (14)	0.7—1.3	0.9	Greater Caucasus	2
Porphyritic granite	(C) (87)	0.5—2.5	1.2	Greater Caucasus	2
Alaskite	(C) (3)	0.6—0.9	0.8	Greater Caucasus	2
"Granodiorite"	(R) (9)	0.4—1.1	0.9	Japan	8
"Granodiorite" (Mesozoic)	(C) (4)	0.3—1.0	—	Greater Caucasus	2
Leucocratic granite	(C) (13)	0.1—0.4	0.2	Tyrny-Auz, Greater Caucasus	2
Porphyritic biotite granite	(C) (15)	0.9—1.8	1.3	Tyrny-Auz, Greater Caucasus	2
Granite	(R) (4)	0.6—1.3	1.0	Japan	8
Granodiorite	(R) (4)	0.9—2.0	1.4	Japan	8
Mesozoic granite	(S) (13)	0.8—3.6	1.99	S. W. Africa	3
Granodiorite and granite	(C) (18)	1.2—1.8	1.3	Susamyr Batholith, U.S.S.R.	9
Leucocratic granite	(C) (8)	1.9—2.8	2.5	Susamyr Batholith, U.S.S.R.	9
Aplite	(C) (4)	4.2—6.3	4.7	Susamyr Batholith, U.S.S.R.	9
Granite	(C) (15)	0.8—2.4	1.4	Musht Complex, N. Caucasus	10
Aplite	(C) (6)	1.7—2.3	2.0	Musht Complex, N. Caucasus	10
Granite	(S) (28)	0.0—4.0	2.4	Dzirul Massif, Georgia, U.S.S.R.	11
Aplite	(S) (8)	0.0—5.0	1.1	Dzirul Massif, Georgia U.S.S.R.	11
Granite	(C) (10)		0.6	Malaya Laba Massif, Caucasus	12

Table 81-E-5. (Continued)

Rock	Method	Tl (ppm) range	Tl (ppm) mean	Locality	Ref.
Alaskite	(C) (2)		0.7	Malaya Laba Massif, Caucasus	12
Granodiorite	(?)		1.3	Shakhtaminski, E. Transbaikaliya	13
Granite	(?)		1.9	Shakhtaminski, E. Transbaikaliya	13
Leucocratic granite	(C) (6)	0.15—0.5	0.3	Turgoyak Pluton, Urals	14
Biotite granite	(C) (12)	0.5—1.0	0.8	Turgoyak Pluton, Urals	14
Granodiorite	(C) (9)	0.25—0.6	0.7	Turgoyak Pluton, Urals	14
Biotite granite	(N/R) (16)	3.0—4.9	3.9	Perak, Malaya	6
Biotite granite	(C) (2)		0.8	Baksan, U.S.S.R.	15
Muscovite-biotite granite	(C) (4)	0.9—1.3	1.1	Baksan, U.S.S.R.	15
Granophyre	(N/R) (2)	0.25—0.35	—	Skaergaard Intrusion, E. Greenland	16
Volcanic rocks					
Rhyolite	(R) (2)	1.0—1.3	—	Japan	8
Rhyolite-dacite	(C) (5)	0.10—0.25	0.16	Mt. Indyuk, Caucasus	2
Quartz porphyry	(C) (4)	0.13—0.20	0.17	Mt. Indyuk, Caucasus	2
Rhyolite-dacite	(C) (6)	0.6—1.0	0.8	Elbrus, Caucasus	2
Rhyolite	(C) (25)	0.6—1.7	1.3	Elbrus, Caucasus	2
Obsidian	(R)		1.2	Mt. Asama, Japan	8
Obsidian	(S)		1.4	Lake Co., Oregon	1
Obsidian	(S)		5.6	Millard Co., Utah	1
Vitrophyre	(C)		2.2	Elbrus, Caucasus	2
Rhyolite and ignimbrite	(S)	0.7—1.6	1.1	Taupo, New Zealand	17
Hypabyssal rocks					
Granodiorite porphyry	(?) (24)		1.2	Almalyk, UzbekS.S.R.	18
Granite porphyry	(R)		1.6	Miyazaki, Japan	8
Granite porphyry	(C) (6)	0.7—0.9	0.8	Greater Caucasus	2
Granite porphyry	(C) (38)		5.3	Northern Caucasus	2, 19
Granosyenite porphyry	(C) (96)		3.2	Northern Caucasus	2, 19
Quartz-syenite porphyry	(C) (40)		3.2	Northern Caucasus	2, 19
Quartz porphyry	(C) (6)	1.0—1.8	1.3	Malaya Laba, Greater Caucasus	2
Quartz albitophyre	(C) (4)	0.7—0.9	0.8	Malaya Laba, Greater Caucasus	2
Quartz albitophyre	(C) (2)	0.7—1.2	1.0	Kvaisa, Greater Caucasus	2
Quartz albitophyre	(C)		1.2	Kvaisinisk, U.S.S.R.	15

List of References: 1. SHAW (1952); 2. VOSKRESENSKAYA (1961a); 3. SIEDNER (1968); 4. KOLBE and TAYLOR (1966b); 5. OTTEMAN (1940); 6. BUTLER (1962); 7. GANEYEV and SECHINA (1962); 8. ISHIMORI and TAKASHIMA (1955); 9. TAUSON and BUSAEV (1957); 10. VOSKRESENSKAYA and SU SHOU-T'IEN (1961); 11. ODIKADZE (1967); 12. DEMIN and KHITAROV (1958); 13. TAUSON (1968); 14. KOGARKO (1959); 15. VOSKRESENSKAYA (1959); 16. ADAMS (1961) (*in* WAGER and BROWN, 1967); 17. EWART *et al.* (1968); 18. BADALOV and RABINOVICH (1966); 19. VOSKRESENSKAYA *et al.* (1962b).

Table 81-E-6. *Thallium contents of alkalic rocks and carbonatites*

Rock	Method	Tl (ppm) range	Tl (ppm) mean	Locality	Ref.

Plutonic rocks

Rock	Method	range	mean	Locality	Ref.
Ijolite	(S)		1.4	Iivaara, Finland	1
Mariupolite	(C)		1.0	Zhdanov, Ukraine	2
Nepheline syenite (phase I)	(C) (5)	0.80—1.40	1.18	Lovozero Massif, U.S.S.R.	3, 4
Urtite, foyaite, lujavrite (phase II)	(C) (15)	0.53—1.45	0.88	Lovozero Massif, U.S.S.R.	3, 4
Lujavrite, melteijite, juvite, syenite (phase III)	(C) (14)	0.55—1.70	0.66	Lovozero Massif, U.S.S.R.	3, 4
Monchiquite (phase IV)	(C)		0.55	Lovozero Massif, U.S.S.R.	3, 4
"Monzonite" (phase I)	(C) (3)	1.0—1.3	—	Mt. Sandyk Massif, U.S.S.R.	5, 6
Syenite (phase I)	(C) (10)	1.0—2.0	—	Mt. Sandyk Massif, U.S.S.R.	5, 6
Alkali syenite (phase II)	(C) (6)	1.6—2.3	2.0	Mt. Sandyk Massif, U.S.S.R.	5, 6
Nepheline syenite (phase II)	(C) (5)	1.5—4.4	2.9	Mt. Sandyk Massif, U.S.S.R.	5, 6
Urtite	(C)	0.40—0.6	—	Khibina Massif, U.S.S.R.	3, 6
Foyaite	(C) (2)	0.45—0.8	—	Khibina Massif, U.S.S.R.	3, 6
Khibinite	(C) (4)	0.45—0.95	—	Khibina Massif, U.S.S.R.	3, 6
Rischorrite	(C)		0.66	Khibina Massif, U.S.S.R.	3, 6
Miaskite	(C)		0.5	Sakhariok Massif, U.S.S.R.	3, 6
Ijolite-urtite	(C)		1.0	Osernaya Varaka Massif, U.S.S.R.	3, 6
Nepheline syenite	(S)		2.2	Methuen Tp., Ontario	1
Nepheline syenite	(C) (5)		0.6	Vishnevyye Gory, Urals	2
Nepheline syenite	(C) (5)		0.2	Keivy, Kola Peninsula	2
Nepheline syenite	(C) (25)		1.2	Yenisei Range, U.S.S.R.	2
Nepheline syenite	(C) (10)		2.7	Central Kazakhstan, U.S.S.R.	2
Nepheline syenite	(W/S)		0.59	Paresis Complex, S.W. Africa	7
Foyaite	(W/S) (2)	0.06—0.11	0.09	Paresis Complex, S.W. Africa	8
Syenite	(W/S) (3)	0.06—0.22	—	Ondura Korume Complex, S.W. Africa	8
Ekerite	(S) (2)	1.7—1.8	—	Oslo Series, Norway	9
"Granite"	(N/R) (9)	1.3—5.0	3.0	Liruei Complex, Nigeria	10
"Granite"	(N/R) (14)	0.7—3.6	1.7	Amo Complex, Nigeria	10
"Granite"	(N/R) (6)	1.1—2.2	1.8	Shere, Northern Nigeria	10
Granite	(N/R) (4)	2.8—3.1	3.0	Ropp, Northern Nigeria	10
Granite	(N/R) (3)	1.5—1.6	1.6	Kaleri, Northern Nigeria	10

Table 81-E-6. (Continued)

Rock	Method	Tl (ppm) range	Tl (ppm) mean	Locality	Ref.
Volcanic rocks					
Leucitite	(F)		5.2	Capo di Bove, Rome	11
Phonolite	(S)		7.2	Cripple Creek, Colorado	1
Rhyolite	(N/R) (6)	1.5—2.5	2.0	Liruei Complex, Nigeria	10
Rhyolite	(W/S)		0.05	Paresis Complex, S. W. Africa	8
Riebeckite rhyolite	(W/S)		0.89	Paresis Complex, S. W. Africa	8
Na-rhyolite	(W/S)		0.34	Paresis Complex, S. W. Africa	8
Pantellerite	(S)		0.93	Mayor Island	12
Tephrite	(S)		0.16	Rheinprussen	1
Leucitophyre	(C)		0.3	Armenia	2
Leucite-basalt	(S)		0.26	Rheinpreussen	1
Carbonatites					
Micaceous carbonatite	(W/S)		0.19	Otjiwarongo, S. W. Africa	8
Dolomitic carbonatite	(W/S)		0.03	Otjiwarongo, S. W. Africa	8

List of References: 1. SHAW (1952); 2. VOSKRESENSKAYA (1961a); 3. GERASIMOVSKII and RASSKAZOVA (1962); 4. GERASIMOVSKII et al. (1968). 5. ZLOBIN (1958); 6. ZLOBIN and LEBEDEV (1960); 7. BROOKS and AHRENS (1961b); 8. BROOKS and AHRENS (1961a); 9. DIETRICH et al. (1965); 10. BUTLER (1962); 11. FORNASERI and PENTA (1963); 12. EWART et al. (1968).

2. Volcanic and Hypabyssal Rocks

Much less information exists for these rocks (Table 81-E-5). Rhyolites and obsidians show contents of thallium mostly in the range 1.0—1.6 ppm, similar to those of granitic rocks.

Quartz-syenite and granosyenite porphyries of the Caucasus show similar contents of thallium, while granite porphyries from the same region are richer in thallium. It may be added that these rocks are relatively rich in thallium, in particular those of Neogene age.

V. Alkalic Rocks

The thallium contents of alkalic rocks are generally low (Table 81-E-6), although nepheline syenites, phonolites and leucitites may contain upto 2.7, 7.2 and 5.2 ppm respectively of this element.

VI. Series of Differentiation

The evidence on the variation of the thallium contents in series of differentiation is inconclusive.

Rocks ranging from gabbro-diorite or diorite to granite may show higher thallium contents in the later differentiates (e.g. Susamyr Batholith, U.S.S.R.,

Tables 81-E-4, 81-E-5). The same can be observed in series of alkalic rocks (Mt. Sandyk Massif, Northern Kirgiziya, U.S.S.R., Table 81-E-6).

On the other hand, the alkalic rocks of the Lovozero Massif, Kola Peninsula, show a different distribution of thallium, as the amount of this element decreases towards the late differentiates (Table 81-E-6).

No trends are apparent for the variation of the thallium contents of the rocks of the Ring Complexes of the Younger Granites of Northern Nigeria (BUTLER, 1962).

An important role on the distribution of thallium may be played by the vapour phase during the crystallization of the magmas, as pointed out by DIETRICH et al. (1965) and TAUSON (1968). This may, in part, explain the differences on the distribution of thallium in different series of differentiation.

The geochemistry of thallium is dominated by a close coherence with potassium and rubidium. In igneous rocks the values of K/Tl and Rb/Tl fall in the ranges 4,000—112,000 and 19—590 (see Fig. 81-C-1), with averages close to 30,000 and 150 respectively.

Table 81-E-7. *Average thallium contents and atomic abundance of thallium in igneous rocks*

Ultra-mafic rocks	Mafic rocks	Inter-mediate rocks	Granitic-rhyolitic rocks	Syenite	Alkalic rocks	Reference
Thallium contents (ppm)						
0.06	0.12	0.15	3.2	1.4	—	SHAW (1952, 1957)
0.06	0.21	—	0.72 (high-Ca) 2.3 (low-Ca)	1.4	—	TUREKIAN and WEDEPOHL (1961)
—	0.11	—	0.73	—	—	BROOKS and AHRENS (1961a)
0.01	0.2	0.5	1.5	—	—	VINOGRADOV (1962)
0.05	0.1	—	0.75	0.3	—	TAYLOR (1964a, 1966a)
0.1	0.27	0.83	1.6	—	1.5	IVANOV (1966)
—	0.1	—	1.0	—	—	TAYLOR and WHITE (1967)
0.05	0.18	0.55	1.7	—	1.2	This paper
Atomic abundance (atoms /10^6 Si atoms)						
0.0072	0.114	0.264	0.636	—	—	VINOGRADOV (1962)
0.033	0.11	0.30	0.73	—	0.73	This paper

81-F. Behavior in Magmatogenic Processes

I. Metasomatic and Hydrothermal Processes

Enrichment of thallium is generally observed in rocks affected by metasomatic processes.

Potassium metasomatism is believed to be responsible for the high thallium content of the amazonite granite of the Etyka Massif (U.S.S.R.) (SITNIN, 1960), which contains 13 ppm of this element.

Table 81-F-1. *Thallium contents of rocks modified by metasomatic or hydrothermal processes*

Rock	Method	Tl (ppm) range	Tl (ppm) mean	Locality	Ref.
Two-mica granite	(C) (7)	0.5—1.2	0.9	Kuban River, Ullu-Kam, Greater Caucasus	1
Greisenized granite	(C) (5)	1.1—2.0	1.5	Kuban River, Ullu-Kam, Greater Caucasus	1
Amazonite granite	(C)		13	Etyka Massif, U.S.S.R.	2
Greisenized contact rocks	(C) (3)	17—26		Etyka Massif, U.S.S.R.	2
Siltstone, 200 m from contact	(C)		5	Etyka Massif, U.S.S.R.	2
Leucocratic granite (unaltered)	(C) (4)	1.4—1.7	1.5	Musht, N. Caucasus	3
Leucocratic granite (fault zone)	(C) (5)	2.0—4.0	3.2	Musht, N. Caucasus	3
Silicified and microclinized granite	(C) (3)	1.2—1.8	1.4	Musht, N. Caucasus	3
Same, fault zone	(C) (2)	2.1—2.9	2.5	Musht, N. Caucasus	3
Tourmalinized coarse-grained granite	(N/R) (2)	3.7—5.2	—	Cornwall, S. W. England	4
Granite porphyry (unaltered)	(C) (6)		0.27	Tyrny-Auz, Greater Caucasus	5
Same, hydrothermally altered	(C) (2)		0.90	Tyrny-Auz, Greater Caucasus	5
Granosyenite porphyry (unaltered)	(C) (81)		3.3	Northern Caucasus	5
Same, opalized, kaolinized	(C) (5)		3.7	Northern Caucasus	5
Same, pyritized, limonitized	(C) (5)		9.8	Northern Caucasus	5
Quartz albitophyre (unaltered)	(C)		1.2	Kvaisinsk, Greater Caucasus	5
Same, albitized, sericitized	(C)		0.7	Kvaisinsk, Greater Caucasus	5

List of References: 1. VOSKRESENSKAYA (1961a); 2. SITNIN (1960); 3. VOSKRESENSKAYA and SU SHOU T'IEN (1961); 4. BUTLER (1962); 5. VOSKRESENSKAYA (1959).

The Musht granites do not show much variation in their thallium contents, even when the silicified and microclinized rocks are taken into account (Table 81-F-1), but a marked enrichment of thallium is observed in rocks occurring in the fault zones.

Hydrothermal alteration frequently leads to thallium enrichment. The kaolinized and opalized granosyenite prophyries of the Greater Caucasus (VOSKRESENSKAYA

Table 81-F-2. *Thallium contents of rocks modified by contact-metamorphic and metasomatic processes*

Rock	Method	Tl (ppm) range	Tl (ppm) mean	Locality	Ref.
Limestone (beyond the boundaries of the Lachin-Khana ore deposit)	(C) (10)		n.f.	Ugam Range, Western Tien Shan	1
Limestone (deposit area)	(C) (8)		3	Ugam Range, Western Tien Shan	1
Marble (near the ore bodies)	(C) (12)		6	Ugam Range, Western Tien Shan	1
Porphyritic biotite granite	(C) (14)		1.2	Tyrny-Auz, Greater Caucasus	2
Same, at contact with biotite hornfels	(C)		1.8	Tyrny-Auz, Greater Caucasus	2
Same, at limestone contact	(C) (3)	0.7—0.9	0.8	Tyrny-Auz, Greater Caucasus	2
Limestone	(C)		0.4	Tyrny-Auz, Greater Caucasus	2
Marble	(C)		0.34	Tyrny-Auz, Greater Caucasus	2
Granite porphyry	(C) (38)		5.3	Northern Caucasus	3
Same, contact with marls	(C) (4)		6.2	Northern Caucasus	3
Marl (contact with granite prophyry)	(C) (3)		8.8	Northern Caucasus	3
Granosyenite porphyry	(C) (81)		3.3	Northern Caucasus	3
Same, at contact with clay, marl, argillite	(C) (11)		5.5	Northern Caucasus	3
Same, at contact with carbonate rocks	(C) (4)		4.4	Northern Caucasus	3
Maikop clay (at contact)	(C)		5.6	Northern Caucasus	3
Same, unaltered	(C) (2)		0.7	Northern Caucasus	3
Argillaceous marl (at contact)	(C) (4)		10	Northern Caucasus	3
Same, unaltered	(C) (2)		0.8	Northern Caucasus	3
Hornfels (sandstone)	(C)		1.6	Northern Caucasus	3
Sandstone (unaltered)	(C) (2)		1.2	Northern Caucasus	3
Quartz syenite porphyry	(C) (27)		3.0	Northern Caucasus	3
Same, at contact with marls (argillaceous)	(C) (2)		4.5	Northern Caucasus	3
Same, at contact with limestone	(C) (2)		2.7	Northern Caucasus	3
Skarn (marl)	(C) (7)		7.2	Northern Caucasus	3
Skarn (limestone)	(C) (5)		2.8	Northern Caucasus	3

List of References: 1. DUNIN-BARKOVSKAYA (1961); 2. VOSKRESENSKAYA (1959); 3. VOSKRESENSKAYA et al. (1962b).

et al., 1962b) contain slightly higher amounts of thallium than the unaltered rocks, while a three fold enrichment of this element is observed in the pyritized and limonitized porphyries from the same region.

On the other hand, during albitization thallium may be removed. The albitized and sericitized quartz albitophyre of the Kvaisinsk Deposit region shows a decrease in thallium content relative to the unaltered rock (Table 81-F-1).

II. Contact Metamorphic and Metasomatic Processes

Enrichment of thallium in the country rocks surrounding intrusions of magmatic rocks and hydrothermal mineral deposits has been observed in different regions (Table 81-F-2 and Table 81-K-4).

The magmatic rocks themselves may also have higher thallium contents near the contact, although this enrichment is not so marked as in the country rock.

In several instances, thallium enrichment is related to the formation of greisen, and micas seem to be the main thallium carriers.

The mineralogical assemblage of the country rock is, apparently, an important factor conditioning the distribution of thallium. Limestones show a much more marked enrichment of thallium than sandstones or quartzites, if these rocks are not converted to greisen.

Table 81-F-3. *Thallium contents of the ores of the Lachin-Khana polymetallic deposit* (after DUNIN-BARKOVSKAYA, 1961) (Tl ppm, C, S)

Mineral	Stage I	Stage II	Stage III	Stage IV	Stage V
Marcasite	(5) 170	(16) 2,110	—	(4) 40	—
Pyrite	—	(3) 350	(6) 2	(5) 34	—
Sphalerite	—	(6) 55	—	—	(2) 6
Galena	—	—	—	(7) 31	—
Tetrahedrite	—	—	—	(3) 2	—

Stage I: Chalcedony-Iron Sulfide. Stage II: Iron Sulfide-Sphalerite. Stage III: Pyrite-Fluorite. Stage IV: Barite-Quartz. Stage V: Sphalerite.

Table 81-F-4. *Thallium contents of ore minerals from the Musht Region polymetallic deposit, Caucasus.* (According to VOSKRESENSKAYA, 1961b)

Mineral		Tl (ppm) (C)	
		range	mean
Sphalerite			
1st generation	(2)	1.3—1.6	1.5
2nd generation	(?)	1.3—2.2	1.9
3rd generation	(3)	4.0—8.0	5.7
Galena	(17)	0.4—1.6	0.95
Pyrite	(3)	0.8—2.0	1.4
Marcasite			
Radiated	(2)	22—80	50
Colloform	(7)	170—300	220
From oxidised zone	(2)	10—20	15
Siderite			19.5

Table 81-F-5. *Thallium contents in ores from mineral deposits*

Mineral	Method	Tl (ppm) range	Tl (ppm) mean	Deposit	Ref.
Sulfides					
Galena	(S)		0.27	Sakkijarvi, Finland	2
Galena	(?) (19)	1.0—7.8	3.2	Verkhnyaya Kvaisa, Caucasus	5
Galena	(S) (29)		53	V. Kvaisa, G. Caucasus	8
Galena					
Coarse-grained	(C) (3)	0.0—2.6	0.9	Zolotoi Kurgan polymetallic deposit, Caucasus	6
Fine-grained	(C) (4)	74—112	51	Zolotoi Kurgan polymetallic deposit, Caucasus	6
Galena	(S) (18)	<1—150	32	Lead-zinc deposit (Replacement deposit), Kootenay, B.C.	12
Galena	(S) (23)		<5	Lead-zinc deposit (Fissure deposit), Kootenay, B.C.	12
Galena	(W) (7)	10—5,000	—	Polymetallic deposits, E. Transbaikaliya, U.S.S.R.	10
Galena	(W) (15)		33	Polymetallic deposits (in carbonate rocks), E. Transbaikaliya, U.S.S.R.	14
Galena	(W) (30)		7	Polymetallic deposits (aluminosilicate gangue minerals), E. Transbaikaliya, U.S.S.R.	14
Galena	(C) (37)	0.0—16	—	Polymetallic deposits, Caucasus, U.S.S.R.	6
Galena	(C) (5)	14.5—38	22.5	Buron iron sulphide deposit, Caucasus, U.S.S.R.	6
Sphalerite	(C) (24)	0.0—16	—	Polymetallic deposits, Caucasus, U.S.S.R.	6
Sphalerite	(C) (2)	3.4—8.0	5.7	Urup iron sulphide deposit, Caucasus, U.S.S.R.	6
Sphalerite	(?) (5)	100—1,000	—	Belgium, Westphalia, Silesia	3
Sphalerite	(W)		700	Hanaoka, Akita Pref., Japan	1
Sphalerite	(S) (38)		19	V. Kvaisa, G. Caucasus	8
Sphalerite	(S)		9.3	Sukhumi, Dusheti region, Caucasus	8
Sphalerite	(?) (10)	1.6—9.0	5.8	Verkhnyaya Kvaisa, Caucasus	5
Pyrite	(W)		70	Hanaoka, Akita Pref., Japan	1
Pyrite (Colloform)	(S)		210	Sukhumi, Dusheti region, Caucasus	8
Pyrite	(C) (6)	0.6—3.0	—	Polymetallic deposits, Caucasus, U.S.S.R.	6
Pyrite	(C) (9)	0.0—2.0	—	Iron sulfide deposits, Caucasus, U.S.S.R.	6

Table 81-F-5. (Continued)

Mineral	Method	Tl (ppm) range	Tl (ppm) mean	Deposit	Ref.
Pyrite	(C) (1)		3.0	Iron sulfide-polymetallic deposit, Khudes, Caucasus	6
Pyrite	(C) (2)		—	Iron sulfide-polymetallic deposit, Beskes, Caucasus	6
Pyrite	(C) (1)		0.6	Mercury-arsenic deposit, Tsess, Caucasus	6
Marcasite	(W)		1,250	Hanaoka, Akita Pref., Japan	1
Marcasite	(W)		850	Mori, Yamagata Pref., Japan	1
Marcasite	(S)		0.24	Veules, Normandy	2
Marcasite	(C) (3)	56—64	59	Verkhnyaya Kvaisa, Caucasus	5
Marcasite	(C) (3)	56—64	60	E. Kvaisa polymetallic deposit, Caucasus, U.S.S.R.	6
Marcasite[a]	(S) (14)		55	V. Kvaisa, G. Caucasus	8
Marcasite[a]	(S) (10)		140	Khvamli, G. Caucasus	8
Marcasite[a]	(S) (24)		450	Tkhmori, G. Caucasus	8
Marcasite	(?)	1,000—9,000	—	Daraiso lead-zinc deposit, U.S.S.R.	9
Pyrrhotite	(S)		0.074	Creighton Mine, Sudbury, Ontario	2
Pyrrhotite	(C) (5)		—	Polymetallic and iron sulfide-polymetallic deposits, Caucasus	6
Chalcopyrite	(S)		0.028	Creighton Mine, Sudbury, Ontario	2
Chalcopyrite	(C) (2)		0.8	Sadon polymetallic deposit, Caucasus, U.S.S.R.	6
Chalcopyrite	(C) (3)		—	Iron sulfide deposits, Caucasus, U.S.S.R.	6
Molybdenite	(S)		0.008	Temagami, Ontario	2
Cinnabar	(C)		—	Mercury-arsenic deposit, Akhei, Caucasus	6
Cinnabar	(C)		0.4	Mercury-arsenic deposit, Tibs, Caucasus	6
Stibnite	(C)		1.8	Mercury-arsenic deposit, Tibs, Caucasus	6
Stibnite	(C) (2)	0.2—0.3	0.25	Ferberite-stibnite deposit, Zopikhto, Caucasus	6
Ferberite	(C)		—	Ferberite-stibnite deposit, Notsaro, Caucasus	6
Berthierite	(C)		10	Ferberite-stibnite deposit, Zopikhto, Caucasus	6
Realgar	(C)		—	Mercury-arsenic deposit, Kodis-Dziri, Caucasus	6
Enargite	(W)		560	Shimokita, Aomon Pref., Japan	1
Geocrinite	(?)	100—5,000		Eastern Transbaikaliya, U.S.S.R.	10

[a] Contains pyrite as impurity.

Table 81-F-5. (Continued)

Mineral	Method	Tl (ppm) range	Tl (ppm) mean	Deposit	Ref.
Meneghinite	(?)	100—1,000		Eastern Transbaikaliya, U.S.S.R.	10
Boulangerite	(?)	10—1,000		Eastern Transbaikaliya, U.S.S.R.	10
Jamesonite	(?)	10—100		Eastern Transbaikaliya, U.S.S.R.	10
Thallium-jordanite	(?)		1.56%	Lead-zinc deposit, Upper Silesia	9

Oxides

Mineral	Method	Tl (ppm) range	Tl (ppm) mean	Deposit	Ref.
Braunite	(C) (3)	1.2—4.5	—	Eastern Kradzhal, U.S.S.R.	7
Braunite	(C) (8)	0.0—14	—	Dzhezda deposit, Central Kazakhstan	4
Pyrolusite	(C) (2)	2.4—5.0	—	Eastern Kradzhal, U.S.S.R.	7
Psilomelane	(C) (25)	20—300	154	Dzhesda deposit, Central Kazakhstan	7
Psilomelane	(W)	320—1,000		Eastern Transbaikaliya, U.S.S.R.	10
Manganite[b]	(C)		3.8	Dzhezda deposit, Central Kazakhstan	4
Goethite	(?)		100	Daraiso lead-zinc deposit, U.S.S.R.	9
Hydrogoethite, goethite	(?)	5—14		Sarykan deposit, Middle Asia, U.S.S.R.	11
Hydrogoethite	(?)	2—4		Kumyshkan deposit, Middle Asia, U.S.S.R.	11
Hydrous iron oxides	(?)	3—85		Kurgashinkan deposit, Middle Asia, U.S.S.R.	11
Limonite	(W)		6	Polymetallic deposits, E. Transbaikaliya, U.S.S.R.	10
Mn-rich limonite	(W)	10—50		Polymetallic deposits, E. Transbaikaliya, U.S.S.R.	10
Mn-hydroxides	(C, S) (2)	40—95	68	Lachin-Khana polymetallic deposit, U.S.S.R.	13
Iron oxides	(C, S) (18)	tr.—18	4	Lachin Khana polymetallic deposit, U.S.S.R.	13

Sulfates, carbonates etc.

Mineral	Method	Tl (ppm) range	Tl (ppm) mean	Deposit	Ref.
Thallium-jarosite		1.75—2.04%		Daraiso lead-zinc deposit, U.S.S.R.	9
Plumbojarosite	(?)		1.5	Sarykan lead-zinc deposit, Middle Asia	11
Plumbojarosite	(?)	3.7—4.7		Kurgashinkan lead-zinc deposit, Middle Asia	11
Plumbojarosite	(?)		—	Kurgash deposit, Middle Asia	11
Plumbojarosite	(C, S) (3)	11—44	27	Lachin-Khana polymetallic deposit, U.S.S.R.	13

[b] Contains a little braunite as impurity.

Table 81-F-5. (Continued)

Mineral	Method	Tl (ppm) range	Tl (ppm) mean	Deposit	Ref.
Jarosite	(C, S) (11)	1—40	19	Lachin-Khana polymetallic deposit, U.S.S.R.	13
Melanterite	(?)		100	Daraiso lead-zinc deposit, U.S.S.R.	9
Colourless cerussite	(W) (4)	n.f.-tr.		Polymetallic deposits, E. Transbaikaliya, U.S.S.R.	10
Cerussite	(?)		tr.	Kurgash deposit, Middle Asia	11
Black cerussite	(W) (6)	0—40	—	Polymetallic deposits, E. Transbaikaliya, U.S.S.R.	10
Black cerussite	(?)		2.0	Sarykan lead-zinc deposit, Middle Asia	11
Black cerussite	(?)	10—20		Kumyshkan lead-zinc deposit, Middle Asia	11
Anglesite	(?)		tr.	Kurgash deposit, Middle Asia	11
Mimetite	(?)		41	Kurgashinkan lead-zinc deposit, Middle Asia	11
Alunite	(C, S)		2	Lachin-Khana polymetallic deposit, U.S.S.R.	13
Chalcanthite	(C, S)		2	Lachin-Khana polymetallic deposit, U.S.S.R.	13
Cuprohalloysite	(C, S)		12	Lachin-Khana polymetallic deposit, U.S.S.R.	13

List of References: 1. MURAKAMI et al. (1950); 2. SHAW (1952); 3. STOIBER (1940) (in SHAW, 1957); 4. VOSKRESENSKAYA and USEVICH (1957); 5. VOSKRESENSKAYA and KARPOVA (1958); 6. VOSKRESENSKAYA (1961b); 7. VOSKRESENSKAYA and SOBOLEVA (1961); 8. IVANITSKII and GVARAMADZE (1960); 9. MOGAROVSKII (1961); 10. KULIKOVA (1962); 11. KULIKOVA (1964); 12. SINCLAIR (1964); 13. DUNIN-BARKOVSKAYA (1961); 14. MEITUV (1962).

III. Thallium in the Processes of Ore Deposition

The most detailed studies on the behaviour of thallium in ore deposition have been made on the hydrothermal deposits of the Caucasus region.

The ore minerals from polymetallic deposits show the highest concentrations of thallium (Table 81-F-3) and, in the Caucasus region, these deposits are related to acid intrusives (quartz albitophyres, quartz syenite and granite porphyries).

Enrichment of thallium is accompanied by concentration of volatiles such as arsenic, antimony and silver.

The critical factor governing the distribution of thallium in mineral deposits appears to be the sorption of this element in the sulfide gels, with isomorphism playing a secondary role. This explains the concentration of thallium in sphalerite rather than in galena (Tables 81-F-4, 81-F-5). Ore minerals which crystallized from colloform solutions show a strong enrichment of this element (Table 81-F-4).

IVANOV et al. (1956) (in VOSKRESENSKAYA and KARPOVA, 1958) point out that when sphalerite and galena crystallize from true solutions, thallium is concentrated in galena, its distribution being governed by the crystallochemical laws with thallium replacing lead in galena.

The mode of occurrence of thallium in sulfides is not well known. VOSKRESENSKAYA and KARPOVA (1958) found a correlation between thallium and arsenic in galena and an antipathetic behavior of arsenic and antimony in the case of a compensation replacement of lead by thallium; however, IVANOV et al. (1956) (in VOSKRESENSKAYA and KARPOVA, 1958) suggest the substitutions Tl—Bi and Tl—Sb for 2 Pb. Most sulfides are low in thallium (Table 81-F-5) although sulfides of lead and antimony or arsenic and lead may contain appreciable amounts of this element, in particular jordanite.

Gangue minerals do not show any appreciable enrichment in thallium (Table 81-F-6).

Table 81-F-6. *Thallium contents in gangue minerals of polymetallic ore deposits of the U.S.S.R.*

Mineral	Method	Tl (ppm) range	Tl (ppm) mean	Deposit	Ref.
Garnet	(C) (7)		n.f.	Zolotoi Kurgan, Caucasus	1
Datolite	(C)		n.f.	Zolotoi Kurgan, Caucasus	1
Axinite	(W) (2)	1—7	—	Klichkinskii, E. Transbaikaliya	2
Muscovite	(W) (5)	14—20	—	Klichkinskii, E. Transbaikaliya	2
Phlogopite	(W) (4)	10—20	—	Klichkinskii, E. Transbaikaliya	2
Sericite	(W) (12)	10—27	—	Klichkinskii, E. Transbaikaliya	2
Chlorite	(C) (2)		0.6	Sadon, Caucasus	1
Serpentine	(W) (3)		n.f.	Klichkinskii, E. Transbaikaliya	2
Apophylite	(C) (2)		1.0	Zolotoi Kurgan, Caucasus	1
Potassium feldspar	(W) (3)	10—12	—	Klichkinskii, E. Transbaikaliya	2
Albite	(W)		1	Klichkinskii, E. Transbaikaliya	2
Quartz	(C, S)		1	Lachin-Khana	3
Analcime	(C)		1.5	Zolotoi Kurgan, Caucasus	1
Zeolites	(C) (3)		n.f.	Zolotoi Kurgan, Caucasus	1
Barite	(C) (7)		n.f.	Greater Caucasus	1
Barite	(C, S)		1	Lachin-Khana	3
Calcite	(C, S)		2	Lachin-Khana	3
Calcite	(C) (2)		n.f.	Sadon, Caucasus	1

List of References: 1. VOSKRESENSKAYA (1961b); 2. MEITUV (1962); 3. DUNIN-BARKOVSKAYA (1961).

IV. Discussion

It is apparent from the evidence on the thallium distribution in magmatic, contact metamorphic and metasomatic rocks that this element is highly mobile.

VOSKRESENSKAYA (1959) suggested that thallium migrates from the central parts of the intrusive bodies to the endocontact zone and to the invaded rocks as halide or fluoride.

The possibility that thallium may form a monochloride in magmatic melts is advocated by ZLOBIN (1958).

Therefore, it seems that in magmatogenic processes the high volatility of the thallium halides plays a major role in the distribution of this element.

Revised manuscript received: March 1970

81-G. Behavior during Weathering and Alteration of Rocks

I. Rocks

Data on the thallium contents of unaltered and weathered rocks are given in Table 81-G-1.

The altered granites of the Musht Region, Northern Caucasus (VOSKRESENSKAYA and SU SHOU T'IEN, 1961), and of Perak, Malaya (BUTLER, 1962) do not show much variation in their thallium contents when compared to the unaltered rocks. Similarly, the clay derived from the Bosahan adamellite, Cornwall (BUTLER, 1953), contains the same amount of thallium as the unaltered rock.

The thallium contents of the weathered mantle of felsic and intermediate igneous rocks, in Middle Asia (VOSKRESENSKAYA et al., 1962a) are similar to the contents of unaltered, comparable rocks.

Table 81-G-1. *Thallium contents of weathered rocks*

Rock	Method	Tl (ppm) range	Tl (ppm) mean	Locality	Ref.
Adamellite	(S)		3.0	Bosahan, Cornwall	1
Ca-saturated clay from adamellite	(S)		3.0	Bosahan, Cornwall	1
Granite	(S) (3)	1—2	—	Cornwall, S. W. England	1
Unaltered rhyolitic welded tuffs and sericitized and kaolinized volcanic rocks	(C) (30)		1.0	Ugam Range, W. Tien Shan	2
Kaolinized welded tuffs with finely dispersed pyrite	(C) (100)	1—3	—	Ugam Range, W. Tien Shan	2
Weathered mantle of felsic and intermediate igneous rocks	(C) (4)	0.3—1.5	—	Middle Asia	3
Biotite granite (unaltered)	(C) (7)	0.8—1.3	1.1	Musht, Northern Caucasus	4
Same, altered	(C) (7)	1.1—1.5	1.4	Musht, Northern Caucasus	4
Biotite granite (unaltered)	(N/R) (6)	3.0—4.9	3.9	Perak, Malaya	5
Same, kaolinized, chloritized	(N/R) (2)	5.0—5.7	5.4	Perak, Malaya	5
Same, weathered	(N/R)		2.9	Perak, Malaya	5
Muscovite granite, kaolinized	(N/R)		4.1	Cornwall, S. W. England	5

List of References: 1. BUTLER (1953); 2. DUNIN-BARKOVSKAYA (1961); 3. VOSKRESENSKAYA et al. (1962a); 4. VOSKRESENSKAYA and SU SHOU T'IEN (1961); 5. BUTLER (1962).

Concentration of thallium in weathering products is included in KURBANAYEV's study of the Maykain Deposit Area, Northern Kazakhstan (Table 81-G-2). A successive enrichment of thallium is observed in the weathered mantle, clay and loam and silt derived from granodiorite, andesite and shale.

It is apparent from these studies that thallium is largely retained during the weathering processes, and, probably as a result of this, enrichment of this element may be observed in the products of weathering.

Table 81-G-2. *Thallium contents in the bedrock and in weathering products of the Maykain Deposit Area, Northern Kazakhstan.* (After KURBANAYEV, 1966)

Bedrock	Weathered rock and shingle	Weathered mantle	Clays	Loam and silt
Granodiorite	0.2 (125)	0.26 (22)	0.3 (104)	0.33 (73)
Andesite	0.22 (43)	0.03 (13)	0.32 (28)	0.77 (14)
Silty shale	0.04 (121)	0.10 (115)	0.13 (122)	0.28 (79)

Thallium contents in ppm (S, W) and number of analysed samples in parentheses.

II. Oxidized Zone of Ore Deposits

Thallium tends to disperse during oxidation of sulphide ores and many oxidized minerals are very poor in this element. Therefore, the oxidized zone of ore deposits is poorer in thallium than the sulphide zone (Table 81-F-5).

In some instances, minerals from the oxidized zone of ore deposits show appreciable enrichment in thallium. Mimetite may concentrate thallium, probably due to the formation of the nearly insoluble compounds $TlAsO_3$ and $TlAsO_4$.

Of much more importance as thallium carriers are jarosite, which may form thallium-rich varieties, and manganese oxides, especially psilomelane.

Thallium seems to precipitate together with manganese, probably as Tl_2O_3, in strong oxidizing environments. According to VOSKRESENSKAYA and USEVICH (1957), thallium is preferentially associated with Mn^{4+} and, therefore, the thallium contents of the manganese minerals are not a function of the Mn-content but of the contents of the different oxides of the latter element.

It seems possible that the observed enrichment of thallium in jarosite is a consequence of the substitution of thallium for potassium of the structure of this mineral.

81-H. Solubilities of Compounds which Control Concentrations in Natural Waters; Adsorption Processes; Valence States in Natural Environments

The concentration of thallium in natural waters is practically unknown and the processes involved in the transport of thallium constitute a problem which needs investigation.

Thallium was detected in one out of seven samples of mine waters from the Lachin-Khana lead-zinc deposit (DUNIN-BARKOVSKAYA, 1961), in which a concentration of 0.002 ppm thallium was found.

It has been mentioned above (Section 81-G) that thallium is largely retained during the process of weathering of rocks. This element is also adsorbed in clay minerals (SHAW, 1957) owing to the fact that thallium is a large electropositive ion. It seems likely that thallium is transported in solution as a univalent cation. In strongly oxidizing conditions, thallium may be removed from solution, with formation of Tl^{3+} by precipitation with manganese and/or iron.

Adsorption processes appear to play an important role in the concentration of thallium in mineral deposits as colloform ores are particularly rich in this element. VOSKRESENSKAYA and KARPOVA (1958) explained the absence of thallium in the oxidized ores of mineral deposits as being due to the high solubility of its sulfates and carbonates.

In natural environments, thallium seems to occur in general as a univalent cation, and only in strongly oxidizing conditions will it occur as trivalent thallium as in the oxidation zones of mineral deposits. The coherence between thallium on one hand, and potassium and rubidium on the other hand, its adsorption in clays and the enrichment of thallium in sedimentary rocks formed in reducing conditions, strongly suggest that this element is present in most natural environments as univalent thallium.

81-I. Abundance in Natural Waters

Thallium is below the detection limits of the methods used for its determination in natural waters, and in particular in ocean water. MANHEIM (1965) (in AHRENS et al., 1967) found that the thallium content of sea-water is less than 0.01 ppb.

Waters from hot springs and drillholes in thermal areas along the margin of the Taupo Volcanic Zone (New Zealand) contain upto 7 ppb of thallium (WEISSBERG, 1969). They ascent through fissures and silicic volcanic rocks and, at a depth of 1 km, the waters are slightly alkaline (pH 6—7).

The precipitates formed from these waters are amorphous sulfides containing gold, silver, arsenic, antimony and mercury. They contain thallium in variable amounts, from less than 1 ppm to 0.5%, according to the same author, and only 0.4% of the thallium in the waters is deposited as precipitate.

81-K. Abundance in Common Sediments and Sedimentary Rock Types

The thallium contents of sedimentary rocks are generally in the range 0.1 to 2.0 ppm. The highest values for the concentration of this element are found in sandstones (1—3 ppm) and shales; argillaceous rocks and coal may be relatively rich in thallium, with contents upto 2.2—3.0 ppm (Table 81-K-1).

Table 81-K-1. *Thallium contents of sedimentary rocks*

Rock	Method	Tl (ppm) range	Tl (ppm) mean	Locality	Ref.
Sandstone (composite)	(S) (23)		2.0	Triassic, Germany	1
Sandstone	(S)		1.3	Huvitas, Finland	2
Sandstone	(C) (2)		1.2	Caucasus, U.S.S.R.	3
Siltstone (argillaceous)	(?) (5)		0.68		4
Arkose	(S)		1.0	Fair Haven, Connecticut	2
Greywacke	(S) (3)	0.13—0.21	—	North America	2
Subgreywacke	(S) (5)		0.6	Henbury, Australia	5
Clay	(C, S) (3)	0.24—0.70	—	Finland, U.S.S.R., U.S.A.	2, 3
Argillite	(S)		1.4	Michigan	2
Argillaceous rock	(S) (323)		0.36		6
Slate	(S)		0.81	Eagle Bridge, New York	2
Shale (composite)	(S) (36)		2.0	Paleozoic, Germany	1
Shale	(S, W/S) (4)	0.12—0.50	—	New York, South Africa	2, 7, 8
Shale (Dictyonema)	(S)		3.1	Tallinn, Estonia	2
Phyllite	(S)		0.79	Finland	2
Shale	(?) (7)		1.3		4
Carbonaceous schist	(S)		0.60	Manitoba	2
Schungite	(S)		0.11	Finland	2
Anthraxolite	(S)		2.8	Ontario	2
Porcellanite	(S)		0.84	Michigan	2
Carbonaceous marl	(C) (2)		0.8	Caucasus, U.S.S.R.	3
Argillaceous limestone	(C) (2)		0.7	Caucasus, U.S.S.R.	3
Argillaceous marl	(C) (2)		2.2	Caucasus, U.S.S.R.	3
Loess	(S)		0.31	Illinois	2
Laterite	(S)		0.10	Madeira	2
Ocean mud (depth 500 m)	(W/S)		0.13	Antarctica	7

© Springer-Verlag Berlin · Heidelberg 1972

Table 81-K-1. (Continued)

Rock	Method	Tl (ppm) range	Tl (ppm) mean	Locality	Ref.
Eluvium-alluvium	(S, W) (230)		0.1	N. Kazakhstan	10
Clay (from parting in coal)	(C) (5)	0.3—1.0	—	Middle Asia	3
Sandstone (from parting in coal)	(C) (2)	1.0—3.0	2.0	Middle Asia	3
Coal	(C) (9)	0.3—1.5	0.85	Middle Asia	3
Coal	(C) (7)	tr. — 2.3	—	Northern Caucasus	3

List of References: 1. Preuss (1940), in Shaw (1957); 2. Shaw (1952); 3. Voskresenskaya et al. (1962a); 4. Erlank (1965), in Kolbe and Taylor (1966a); 5. Taylor (1966b); 6. Canney (1953); 7. Brooks et al. (1960); 8. Brooks and Ahrens (1961a); 9. Brooks and Ahrens (1961b); 10. Kurbanayev (1966).

Table 81-K-2. *Thallium contents of deep-sea sediments*

Rock	Method	Tl (ppm) range	Tl (ppm) mean	Reference
Iron-manganese-rich ooze	(S)		0.3	Shaw (1952)
Globigerina ooze	(S)		0.16	Shaw (1952)
Siliceous ooze	(S)		—	Shaw (1952)
Clay	(S) (6)	0.16—1.5	0.66	Shaw (1952)
Manganese nodules	(S) (19)	nf. — 180	70	Willis and Ahrens (1962)
Manganese nodules	(S) (31)	2—164	140	Ahrens et al. (1967)
Manganese nodules	(?)		100	Manheim (1965) in Ahrens et al. (1967)

Deep-sea sediments contain usually low amounts of thallium (Table 81-K-2), with the exception of manganese nodules which may show concentrations of this element as high as 600 ppm and an average concentration of about 100 ppm (Manheim, 1965, in Ahrens et al., 1967).

The average thallium contents of sedimentary rocks, according to different authors, are given in Table 81-K-5.

Argillaceous rocks with high carbonaceous content have contents of thallium about twice the value for normal pelitic rocks (Shaw, 1952, 1957). Thallium will be concentrated owing to the reducing conditions in which such rocks were formed.

Thallium occurs in the sulfide phases in coal rather than in the coal itself (Table 81-K-3, 81-K-4). According to Voskresenskaya et al. (1962a), thallium is not firmly held in the organic mass of the coal as it forms soluble compounds of the humate and fulvate types with organic components in coal.

Pyrite from the coal is richer in thallium than the same mineral from the enclosing sedimentary rocks. Voskresenskaya (1968) pointed out that thallium must have been delivered to the coal basins during sedimentation or early diagenesis,

Table 81-K-3. *Thallium contents of sedimentary rocks from coal basins of the U.S.S.R.* (According to Voskresenskaya, 1968)

Locality	Tl (ppm) (S, C)					
	Coal	Ash-rich coal or dull coal	Carbonaceous clay	Clay	Sandstone	Coal with sulfides
Middle Asia Basin (Jurassic)						
Soguty	0.5 (3)	—	0.8 (3)	—	n.f.	135
Kavak	n.f.	0.4 (2)	—	0.9 (2)	n.f.	—
Tashkumyr	0.1 (3)	1.0 (3)	—	—	—	—
Dzhergalan	0.5 (4)	—	—	—	0.5	—
Kara-Kiche	0.8 (4)	1.9 (2)	1.0 (3)	—	—	—
Aksay	0.6 (3)	—	1.2	—	0.5	—
Moscow Basin (Carboniferous)						
Kimovo	n.f. (2)	—	—	0.7 (2)	—	—
Dnepr Basin (Paleogene)						
Baydakov	n.f. (2)	—	—	—	n.f. (8)	—
Transcarpathian Basin (Neogene)						
Suvorovsk	—	0.4	—	—	—	4.0

Table 81-K-4. *Thallium contents of minerals from sedimentary rocks and hydrothermally altered or contact metamorphosed sedimentary rocks*

Mineral		Tl (ppm) (C)		Rock	Locality	Ref.
		range	mean			
Pyrite	(2)		2.0	Sandstone	Northern Caucasus	2
Pyrite	(2)	2.0—2.2	2.1	Argillite	Caucasus	2
Pyrite	(9)	0.3—2.2	1.1	Sedimentary rocks (unaltered)	Greater Caucasus	1
Pyrite	(7)	2.8—270	50	Sedimentary rocks (hydrothermally alt.)	Greater Caucasus	1
Pyrite	(3)		n.f.	Marl (unaltered)	Greater Caucasus	1
Pyrite	(7)		1.0	Marl	Caucasus	2
Pyrite			18.5	Skarn	Greater Caucasus	1
Pyrite			2.0	Limestone	Suvorovsk, Moscow basin	3
Pyrite			25	Barite vein in coal	Middle Asia basin	3
Pyrite	(7)	1.0—30	10	Clay and sandstone	Middle Asia basin	3
Pyrite	(30)	1.0—200	34	Coal	Middle Asia basin	3
Pyrite	(18)	0.0—20	4.3	Clay and sandstone	Moscow basin	3

Table 81-K-4. (Continued)

Mineral		Tl (ppm) range	(C) mean	Rock	Locality	Ref.
Pyrite	(14)	0.0—13	3.3	Coal	Moscow basin	3
Pyrite	(5)	4—23	11	Coal	Transcarpathian basin	3
Pyrite	(16)	0.0—0.6	—	Clay and sandstone	Dnepr basin	3
Pyrite	(23)	0.0—0.5	0.5	Coal	Dnepr basin	3
Pyrite			1.0	Tuffite	Transcarpathian basin	3
Sphalerite			4.0	Sandstone	Northern Caucasus	2
Galena			2.6	Sandstone	Caucasus	2
Magnetite	(3)	157—300	219	Skarn	Greater Caucasus	1
Magnetite			0.6	Spotted marl	Greater Caucasus	1
Ilsemannite	(2)	12—62	37	Coal	Middle Asia basin	2
Barite		1.0—3.0	—	Clay partin in coal	Middle Asia basin	2
Hydrous iron oxides			—	Coal	Middle Asia basin	2
Sulfur			0.7	Coal	Middle Asia basin	2

List of References: 1. VOSKRESENSKAYA (1961b); 2. VOSKRESENSKAYA et al. (1962a); 3. VOSKRESENSKAYA (1968).

Table 81-K-5. *Average thallium contents in sedimentary rocks (ppm Tl)*

Shales	Sandstones	Carbonates	Deep-sea sediments		Greywackes	Manganese nodules	Reference
			carbonate	clay			
0.69	0.82	0.05	—	—	—	—	SHAW (1957)
1.4	0.82	0.0x	0.16	0.8	—	—	TUREKIAN and WEDEPOHL (1961)
0.3	—	—	—	—	0.3	—	TAYLOR (1966a)
1.55	1.54	0.065	0.18	0.75	—	—	HORN and ADAMS (1966)
0.9	1.0 (greywacke excluded)	0.05	0.16	0.6	0.3	100	This paper

as the contents of this element do not depend on the petrographic characteristics of the coal and the contents in pyrite are uniform throughout the vertical sections of the coal beds.

According to this author the concentration of thallium in the basins is governed by the distance to the source area of the sediments and, more important, by the composition of the rock in the source area.

The leading role in the concentration of thallium in the sedimentary cycle is played by its chalcophile tendency and sorption (VOSKRESENSKAYA et al., 1962a).

The possibility of thallium being adsorbed in clay minerals to a considerable extent was pointed out by SHAW (1957) and, according to CANNEY (1952), the relative degree of adsorption of the large univalent ions in clay is $Cs > Rb > K > Tl$.

The association of thallium with manganese in the sedimentary cycle is a consequence of the oxidizing environment precipitation (SHAW, 1957). The same seems to be valid for iron also. This explanation was offered by different authors for the presence of thallium in some ore minerals as well.

The coherence of thallium and rubidium appears to extend to sedimentary rocks. Taking the Tl values in Table 81-K-5 and Rb figures from Chapter 37, the average Rb/Tl ratios for shales and sandstones approximate to 140 and 40 respectively.

81-L. Biogeochemistry

The behavior of thallium in biogeochemical processes is not known in detail.

This element has been identified, according to GOLDSCHMIDT (1954), in land plants and in seaweed.

The dry matter of marine animals may contain upto 0.03 ppm Tl.

It seems possible that decaying organic matter will concentrate thallium and it has been shown that this element is enriched in sedimentary rocks containing carbonaceous matter (SHAW, 1952). Among the carbonaceous samples of argillaceous rocks analysed by SHAW (1952) was an Archean schist containing a carbonaceous aggregate believed to be a fossil. This sample contains 0.79 ppm Tl, a value above the average for such rocks (Table 81-K-1).

The thallium contents of coal from the Northern Caucasus and Middle Asia regions (VOSKRESENSKAYA et al., 1962a) are mostly in the range 0.3—2.3 ppm (Table 81-K-3).

81-M—N. Abundance in Common Metamorphic Rock Types and Behavior in Metamorphic Reactions

The thallium distribution in metamorphic rocks has not yet been thoroughly investigated, and the variation of the contents of this element with grade of metamorphism is practically unknown.

Table 81-M—N-1. *Thallium contents of metamorphic rocks*

Rock	Method	Tl (ppm) range	Tl (ppm) mean	Locality	Ref.
Phyllite	(S) (4)	0.56—0.66	0.6	Clove Quadr., New York	1
Phyllite	(S)		0.27	Moores Mills, New York	1
Schist	(W)		2.0	Sayan Mountains, U.S.S.R.	2
Sericite-bearing quartzose rock	(S, W) (102)		0.02	Maykan, N. Kazakhstan	3
Quartz-sericite schist	(S,W) (81)		0.16	Maykan, N. Kazakhstan	3
Garnet schist	(S) (3)	0.61—1.3	1.0	Clove Quadr., New York	1
Biotite-hornblende schist	(S)		0.35	Danbury Quadr., New York	1
Sillimanite schist					
Dark layer	(S)		1.9	Clove Quadr., New York	1
Light layer	(S)		0.74	Clove Quadr., New York	1
Gneiss	(R) (3)	0.6—0.9	0.8	Oki Island, Japan	4
Gneiss	(W/S)		2.2	Lützow-Holm Bay, Antarctica	5
Noritic gneiss	(S)		tr.	Albany Co., Wyoming	1
Dioritic gneiss	(S)		0.42	Albany Co., Wyoming	1
"Granodioritic" gneiss	(S) (3)	0.74—0.9	0.81	Albany Co., Wyoming	1
Granodioritic gneiss	(S)		0.72	Egedesminde, Greenland	1
Enderbitic gneiss	(S)		0.1	N. Isortog, Greenland	1
Garnet gneiss	(S) (2)	0.13—0.37	0.25	N. Isortog, Greenland	1
Granulite gneiss	(S)		0.026	N. Isortog, Greenland	1
Amphibolite	(S)		0.40	Impilahti, K.F.S.S.R.	1
Amphibolite	(W) (2)	2—3	—	Sayan Mountains, U.S.S.R.	2
Eclogite	(W/S) (2)		0.03	Nodule, Kimberley, S. Africa	6
Eclogite	(W/S)		0.56	Nodule, Kimberley, S. Africa	6

List of References: 1. SHAW (1952); 2. SLEPNEV (1961); 3. KURBANAYEV (1966); 4. ISHIMORI AND TAKASHIMA (1955); 5. BROOKS et al. (1960); 6. BROOKS and AHRENS (1961a).

Quartzose rocks contain very little thallium (0.02 ppm), according to KURBANAYEV (1966), while the quartz-sericite schists from the same region contain 0.16 ppm (Table 81-M—N-1).

Slates and phyllites contain thallium in the range 0.27—0.66 ppm and higher grade schists from the same region (New York State) show contents of this element in the range 0.35—1.3 ppm (SHAW, 1952). According to this author, the dark layers of a silimanite schist are enriched in thallium relative to the light layers by a factor of about 2.5 (Table 81-M—N-1).

The thallium contents of gneissic rocks are variable, and apparently depend mostly on the mineralogical composition of the gneiss. Granodioritic gneisses are richer in this element than dioritic or noritic gneisses by a factor of at least 2. Enderbitic gneisses are also poor in thallium (0.10 ppm) and a sample of granulite contains only 0.026 ppm thallium (Table 81-M—N-1). These figures may indicate that thallium is depleted in granulite facies rocks, confirming the observations made on the potassium feldspars from such rocks (Section 81-D, Table 81-D-9).

The scanty data on amphibolites show the thallium contents to vary in the range 0.43—3 ppm (Table 81-M—N-1). This upper limit seems to be very high for this type of rock from that predicted on theoretical grounds.

Eclogites occuring as nodules in ultrabasic rocks from South Africa (BROOKS and AHRENS, 1961a) contain thallium in the range 0.03—0.56 ppm, and therefore no conclusions can be drawn for the abundance of thallium in these rocks until more data are available.

81-O. Sensitivity Limits for the Determination of Thallium by Different Methods

Method		Sensitivity	Reference
D-C arc emission spectrography	(S)	1.00 ppm	AHRENS and TAYLOR (1961)
Emission spectrography: double-arc method	(S)	0.05 ppm	SHAW (1952)
X-ray spectrometry	(X)	20.00 ppm	JENKINS and DE VRIES (1967)
Neutron activation	(N/R)	0.01 ppm	MORRIS and KILLICK (1960)
Neutron activation	(N/R)	0.03 ppb[a]	REED et al. (1960)
Isotope dilution	(I)	1.00 ppb[a]	ANDERS and STEVENS (1960)
Radiometric	(R)	0.30 ppm[a]	ISHIMORI and TAKASHIMA (1955)
Flame photometry	(F)	0.20 ppm	FORNASERI and PENTA (1963)
Atomic fluorescence spectrometry	(A)	0.10 ppm	MANNING and HENEAGE (1968)
Atomic absorption spectrometry	(A)	0.20 ppm	SLAVIN et al. (1964)
Atomic absorption: sampling boat technique	(A)	0.025 ppm	KAHN et al. (1968)
Chemical/spectrographic	(W/S)	0.01—0.05 ppm[a]	BROOKS et al. (1960)
Colorimetric	(C)	0.01 ppm[a]	VOSKRESENSKAYA (1961a)
Polarographic	(P)	0.05 ppm[a]	WAHLER (1968)

[a] Figure based on the results given, which may be taken as a limit.

© Springer-Verlag Berlin · Heidelberg 1972

Revised manuscript received: March 1970

References: Sections 81-B to 81-O

AHRENS, L. H.: The geochemical relationship between thallium and rubidium in minerals of igneous origin. Trans. Geol. Soc. S. Africa **48**, 207 (1945).
— The unique association of thallium and rubidium in minerals. J. Geol. **56**, 578 (1948).
— TAYLOR, S. R.: Spectrochemical Analysis, 2nd ed. Reading, Mass., U.S.A.: Addison-Wesely Pub. Co. Inc. 1961.
— WILLIS, J. P., OOSTHUIZEN, C. O.: Further observations on the composition of manganese nodules, with particular reference to some of the rarer elements. Geochim. Cosmochim. Acta **31**, 2169 (1967).
ANDERS, E., STEVENS, C. M.: Search for extinct lead 205 in meteorites. J. Geophys. Res. **65**, 3043 (1960).
ASTON, F. W.: Constitution of thallium and uranium. Nature **128**, 725 (1931).
BADALOV, S. T., RABINOVICH, A. V.: The geochemistry of indium and thallium in the Karamazar Ore region (the Uzbek and Tadzhik Republics). Geochem. Int. 1095 (1966).
BOROVIK-ROMANOVA, T. F., SOSEDKO, A. F.: Relation between thallium and rubidium in minerals from pegmatite veins of the Kola Peninsula. Geochem. Int. 34 (1960).
BROOKS, R. R., AHRENS, L. H.: Some observations on the distribution of thallium, cadmium and bismuth in silicate rocks and the significance of covalency on their degree of association with other elements. Geochim. Cosmochim. Acta **23**, 100 (1961a).
— — The determination of indium and thallium in G-1, W-1 and other silicate rocks by a new technique. Geochim. Cosmochim. Acta **23**, 145 (1961b).
— TAYLOR, S. R.: The determination of trace elements in silicate rocks by a combined spectrochemical-anion exchange technique. Geochim. Cosmochim. Acta **18**, 162—175 (1960).
BROWN, R., WOLSTENHOLME, W. A.: Analysis of geological samples by spark source mass spectrometry. Nature **201**, 598 (1964).
BUTLER, J. R.: The geochemistry and mineralogy of rock weathering: the Lizard Area, Cornwall. Geochim. Cosmochim. Acta **4**, 157 (1953).
— Tl in some igneous rocks. Geochem. Int. 599 (1962).
CAMERON, E. G. W.: A new table of abundances of the elements in the solar system. In: Origin and Distribution of the Elements (ed. L. H. AHRENS), **30**, 125. Pergamon Press 1968.
CANNEY, F. C.: Some aspects of the geochemistry of potassium, rubidium, caesium and thallium in sediments. Am. Mineralogist **38**, 331 (1953).
DAWSON, K. R., MAXWELL, J. A., PARSONS, D. E.: A description of the meteorite which fell near Abee, Alberta, Canada. Geochim. Cosmochim. Acta **21**, 127 (1960).
DEMIN, A. M., KHITAROV, D. N.: Geochemistry of potassium, rubidium and thallium in application to problems of petrology. Geochem. Int. 721 (1958).
DIETRICH, R. V., HEIER, K. S.: Differentiation of quartz-bearing syenite (nordmarkite) and riebeckitic-arfvedsonite granite (ekerite) of the Oslo series. Geochim. Cosmochim. Acta **31**, 275 (1967).
— TAYLOR, S. R.: Studies on the igneous rock complex of the Oslo Region. XX. Petrology and geochemistry of ekerite. Skrifter Norske Videnskaps-Akad. Oslo, I. Mat.-Naturv. Kl., Ny Ser. No. 19 (1965).
DUNIN-BARKOVSKAYA, E. A.: Thallium in the ores and minerals of Lachin Khana (Western Tien Shan, Ugam Range). Geochem. Int. 740 (1961).

EDWARDS, G.: Sodium and potassium in meteorites. Geochim. Cosmochim. Acta **8**, 285 (1955).
— UREY, H. C.: Determination of alkali metals in meteorites by a distillation process. Geochim. Cosmochim. Acta **7**, 154 (1955).
EHMANN, W. D., HUIZENGA, J. R.: Bismuth, thallium and mercury in some meteorites by activation analysis. Geochim. Cosmochim. Acta **17**, 125 (1959).
EL-BADRY, H. M., HODGE, E. S., BAER, W. K., KOHMAN T. P.: Thallium in meteorites. U.S.A.E.C. Rep. NYO-8920 (1960).
EWART, A., TAYLOR, S. R., CAPP, A. C.: Geochemistry of the pantellerites of Mayor Island, New Zealand. Contr. Mineral. and Petrol. **17**, 2, 116 (1968).
FLANAGAN, F. J.: U.S. geological survey standards. II. First compilation of data for the new U.S.G.S. rocks. Geochim. Cosmochim. Acta **33**, 80 (1969).
FLEISCHER, M.: Summary of new data on rock samples G-1 and W-1, 1962—1965. Geochim. Cosmochim. Acta **29**, 1263 (1965).
— U.S. geological survey standards. I. Additional data on rocks G-1 and W-1, 1965—1967. Geochim. Cosmochim. Acta **33**, 65 (1969).
— STEVENS, R. E.: Summary of new data on rock samples G-1 and W-1. Geochim. Cosmochim. Acta **26**, 525 (1962).
FORNASERI, M., PENTA, A.: Determinazione del tallio nei silicati per fotometria di fiamma. Metallurgia Italiana **55**, 8, 437 (1963).
GANEYEV, I. G., SECHINA, N. P.: Geochemical characteristics of albitized granites. Geochem. Int. 158 (1962).
GAST, P. W.: Alkali metals in stone meteorites. Geochim. Cosmochim. Acta **19**, 1 (1960).
— Limitations on the composition of the upper mantle. J. Geophys. Res. **65**, 1287 (1960).
GERASIMOVSKII, V. I., RASSKAZOVA, V. S.: Distribution of thallium in nepheline syenites of the Lovozero massif (Kola Peninsula). Geochem. Int. 275 (1962).
— VOLKOV, V. P., KOGARKO, L. N., POLYAKOV, A. I., SAPRYKINA, T. V., BALASHOV, YU. A.: The Geochemistry of the Lovozero Alkaline Massif. Pt. I. Geology and Petrology; Pt. 2. Geochemistry. Toronto: University of Toronto Press 1968.
GOLDSCHMIDT, V. M.: Geochemistry. Oxford: Clarendon Press 1954.
GOLUBCHINA, M. N., RABINOVICH, A. V., MURTAZINA, T. M.: Isotopic composition of thallium in igneous rocks. Geochem. Int. 231 (1957).
GRAESER, S.: Asbecasit und Cafarsit, zwei neue Mineralien aus dem Binnatal (Kt. Wallis). Schweiz. Mineral. Petrog. Mitt. **46**, 367 (1966).
HEIER, K. S.: Petrology and geochemistry of high-grade metamorphic and igneous rocks on Langøy, Northern Norway. Norg. Geol. Undersokelse 207 (1960).
— Trace elements in feldspars — a review. Norsk Geol. Tidsskr. **42**, 415 (1962).
— Some crystallochemical relations of nephelines and feldspars on Stjernoy, North Norway. J. Petrol. **7**, 95 (1966).
— TAYLOR, S. R.: Distribution of Li, Na, K, Rb, Cs, Pb and Tl in Southern Norwegian pre-Cambrian alkali feldspars. Geochim. Cosmochim. Acta **15**, 284 (1959).
HIGAZY, R. A.: Petrogenesis of perthite pegmatites in the Black Hills, South Dakota. J. Geol. **1**, 555 (1949).
— Observations on the distribution of trace elements in the perthite pegmatites of the Black Hills, South Dakota. Am. Mineralogist **38**, 172 (1953).
HORN, M. K., ADAMS, J. A. S.: Computer-derived geochemical balances and element abundances. Geochim. Cosmochim. Acta **30**, 279 (1966).
ISHIMORI, T., TAKASHIMA, Y.: Thallium content of Japanese igneous rocks. Mem. Fac. Sci. Kyushu Univ., Ser. C (Chem.) **2** (2), 65 (1955).
IVANITSKII, T. V., GVARAMADZE, N. D.: Content and distribution of certain dispersed elements in the principal sulfides of the lead-zinc and polymetallic deposits of Georgia. Geochem. Int. 167 (1960).
IVANOV, V. V.: Geochemistry of the Rare Elements Ga, Ge, Gd, In and Tl in Hydrothermal Deposits. Moscow: Nedra 1966.

Jenkins, R., DeVries, J. L.: Practical X-ray spectrometry. Berlin-Heidelberg-New York: Springer 1967.
Kahn, H. L., Peterson, G. E., Schallis, J. E.: Atomic absorption micro-sampling with the "sampling boat" technique. Atomic Absorption Newsletter **7**, 35 (1968).
Keays, R. R., Ganapathy, R., Laul, J. C., Anders, E., Herzog, G. F., Jeffery, P. M.: Trace elements and radioactivity in lunar rocks: implications for meteorite infall, solar-wind flux, and formation conditions of the moon. Science **167**, 490 (1970).
Kogarko, L. N.: Distribution of alkali elements and thallium in granitic rocks of the Turgoyak massif (Central Urals). Geochem. Int. 557 (1959).
Kohman, T. P.: Extinct natural radioactivity: possibilities and potentialities. Ann. N.Y. Acad. Sci. **62**, 503 (1955).
Kolbe, P.: The use of an oxygen jet in the spectrochemical determination of trace amounts of Pb, Tl, Ga, Cu and Sn in some silicate rocks and minerals. Geochim. Cosmochim. Acta **29**, 153 (1965).
— Taylor, S. R.: Geochemical investigation of the granitic rocks of the Snowy Mountains area, New South Wales. J. Geol. Soc. Australia **13**, 1 (1966a).
— — Major and trace element relationships in granodiorites and granites from Australia and South Africa. Contr. Mineral. and Petrol. **12**, 202 (1966b).
Kulikova, M. F.: Trace elements in the oxidized zone of polymetallic deposits of Eastern Transbaikaliya. Geochem. Int. 176 (1962).
— Distribution of trace elements in the oxidized zone of some lead-zinc deposits of Middle Asia. Geochem. Int. 977 (1964).
Kurbanayev, M. S.: Thallium in the dispersion aureoles of the Maykain gold-barite polymetallic deposit. Geochem. Int. 568 (1966).
Larimer, J. W., Anders, E.: Chemical fractionations in meteorites. II. Abundance patterns and their interpretation. Geochim. Cosmochim. Acta **31**, 1239 (1967).
Manning, D. C., Heneage, P.: Element detection limits by atomic fluorescence spectroscopy. Atomic Absorption Newsletter **7**, 80 (1968).
Mason, B.: Principles of Geochemistry (2nd. ed.). New York: John Wiley & Sons, Inc. 1958.
— Wiik, H. B.: The Holbrook, Arizona, chondrite. Geochim. Cosmochim. Acta **21**, 276 (1961).
— — The composition of the Forest City, Tennasilm, Weston, and Geidam meteorites. Am. Mus. Novitates 2220 (1965).
Meituv, G. M.: Geochemistry of rare elements in the lead-zinc deposit of the Klichkimskii region (Eastern Transbaikaliya). Geochem. Int. 694 (1962).
Mogarovskii, V. V.: Geochemistry of thallium in the oxidized zone of the Daraiso sulfide deposit (Middle Asia). Geochem. Int. 848 (1961).
— Correlation between thallium and zinc content in the Daraiso sulfide deposit (Middle Asia). Geochem. Int. 734 (1962).
Morris, D. F. C., Killick, R. A.: The determination of silver and thallium in rocks neutron-activation analysis. Talanta **4**, 51 (1960).
Müller, E. A.: The solar abundances. In: Origin and Distribution of the Elements (ed. L. H. Ahrens), **30**, 155. Oxford: Pergamon Press 1968.
Nockolds, S. R.: Average chemical compositions of some igneous rocks. Bull. Geol. Soc. Am. **65**, 1007 (1954).
— The behaviour of some elements during fractional crystallization of magma. Geochim. Cosmochim. Acta **30**, 267 (1966).
Noddack, I., Noddack, W.: Die Häufigkeit der chemischen Elemente. Naturwissenschaften **18**, 35, 757 (1930).
— Die geochemischen Verteilungskoeffizienten der Elemente. Svensk Kem. Tidskr. **46**, 173 (1934).
Odikadze, G. L.: Some geochemical characteristics of potassium, rubidium and thallium in the granitoids of the Dzirul crystalline massif (Western Georgia). Geochem. Int. 58 (1967).
Ostic, R. G.: Isotopic composition of thallium in meteorites. U.S.A.E.C. Rep. NYO-844-71, 45 (1967).

OSTIC, R. G., KOHMAN, T. P.: Interpretation of meteoritic thallium isotope ratios. U.S.A.E.C. Rep. NYO-844-75, 31—37 (1968).

OTTEMANN, J.: Untersuchungen zur Verteilung von Spurenelementen, insbesondere Zinn, in Tiefengesteinen und einigen gesteinsbildenden Mineralien des Harzes. Z. Angew. Mineral. 3, 2, 142 (1940).

POLDERVAART, A.: Chemistry of the earth's crust. In: Crust of the Earth (a symposium) (A. POLDERVAART, ed.). G.S.A. Spec. Paper 62, 119 (1955).

RANKAMA, K., SAHAMA, TH. G.: Geochemistry. Chicago: Chicago Univ. Press 1950.

REED, G. W., KIGOSHI, K., TURKEVICH, A.: Determinations of concentrations of heavy elements in meteorites by activation analysis. Geochim. Cosmochim. Acta 20, 122 (1960).

SCHROLL, E., WENINGER, M.: Eine empfindliche spektrochemische Analysenmethode zur Bestimmung von Germanium und Zinn unter Verwendung sulfidierender thermochemischer Reagenzien. Mikrochim. Acta 378 (1965).

SCHÜLER, H., KEYSTON, J. E.: Über einen Isotopenverschiebungseffekt der Hyperfeinstrukturtherme von Thallium. Z. Phys. 69, 1 (1931).

SHAW, D. M.: The geochemistry of thallium. Geochim. Cosmochim. Acta 2, 118 (1952).

— The geochemistry of gallium, indium, thallium — a review. Phys. Chem. Earth 2, 164 (1957).

SHIMA, M., HONDA, M.: Distributions of alkali, alkaline earth and rare earth elements in component minerals of chondrites. Geochim. Cosmochim. Acta 31, 1995 (1967).

SIEDNER, G.: Distribution of alkali metals and thallium in some South-West African granites. Geochim. Cosmochim. Acta 32, 1303 (1968).

SINCLAIR, A. J.: A lead isotope study of mineral deposits in the Kootenay Arc. Ph. D. thesis, University of British Columbia (1964).

SINE, N. M., TAYLOR, W. O., WEBBER, G. R., LEWIS, C. L.: Third report of analytical data for CAAS sulphide ore and syenite rock standards. Geochim. Cosmochim. Acta 33, 121 (1969).

SITNIN, A. A.: Distribution of rare elements in amazonite granites of the Etyka massif, Eastern Transbaikalia. Geochem. Int. 361 (1960).

SLAVIN, W., SPRAGUE, S., MANNING, D. C.: Detection limits in analytical atomic absorption spectrophotometry. Atomic Absorption Newsletter, No. 18 (1964).

SLEPNEV, Y. S.: The ratio of Tl to Rb, Cs and K in the schists, granites and rare metal pegmatites of the Sayan Mountains. Geochem. Int. 382 (1961).

SMALES, A. A., HUGHES, T. C., MAPPER, D., MCINNES, C. A. J., WEBSTER, R. W.: The determination of rubidium and cesium in stony meteorites by neutron activation analysis and by mass spectrometry. Geochim. Cosmochim. Acta 28, 209 (1964).

SOLODOV, N. A.: Distribution of thallium among the minerals of a zoned pegmatite. Geochem. Int. 738 (1962).

SUESS, H. E., UREY, H. C.: Abundances of the elements. Rev. Mod. Phys. 28, 53 (1956).

TANDON, S. N.: Mercury, thallium and bismuth in metal and troilite phases of iron meteorites by neutron activation analysis. U.S.A.E.C. Rep. No. NYO-844-71, 35 (1967).

TAUSON, L. V.: Distribution regularities of trace elements in granitoid intrusions of the batholith and hypabyssal types. In: Origin and Distribution of the Elements (ed. L. H. AHRENS), 30, 629. Oxford: Pergamon Press 1968.

— BUZAEV, N. N.: Geochemistry of thallium in granitoids of the Susamyr batholith, Central Tian-Shan. Geochem. Int. 703 (1957).

— STAVROV, O. D.: The geochemistry of rubidium in granitoids. Geochem. Int. 819 (1957).

TAYLOR, S. R.: Abundances of chemical elements in the continental crust: a new table. Geochim. Cosmochim. Acta 28, 1273 (1964a).

— Trace element abundances and the chondritic earth model. Geochim. Cosmochim. Acta 28, 1989 (1964b).

— Geochemical application of spark source mass spectrography. Nature 205, 34 (1965).

— Geochemical analysis by spark source mass spectrography. Geochim. Cosmochim. Acta 29, 1243 (1965).

TAYLOR, S. R.: The application of trace element data to problems in petrology. Phys. Chem. Earth 6, 133 (1966a).
— Australites, Henbury impact glass and subgreywacke: a comparison of the abundances of 51 elements. Geochim. Cosmochim. Acta 30, 1121 (1966b).
— Geochemistry of andesites. In: Origin and Distribution of the Elements (ed. L. H. AHRENS), 30, 559. Oxford: Pergamon Press 1968.
— EWART, A., CAPP, A. C.: Leucogranites and rhyolites: trace element evidence for fractional crystallization and partial melting. Lithos 1, 2, 179 (1968).
— HEIER, K. S., SVERDRIEP, T. L.: Contributions to the mineralogy of Norway. No. 5. Trace element variations in three generations of feldspars from the Landsverk 1 pegmatite. Norsk. Geol. Tidssk. 40, 133 (1960).
— KOLBE, P.: Geochemical standards. Geochim. Cosmochim. Acta 28, 447 (1964).
— WHITE, A. J. R.: Trace element abundances in andesites. Bull. Vulcan. 29, 174 (1967).
TENNANT, W. C.: General spectrographic techniques for the determination of volatile trace elements in silicates. Appl. Spectry. 21, 282 (1967).
TUREKIAN, K. K., WEDEPOHL, K. H.: Distribution of the elements in some major units of the earth's crust. Bull. Geol. Soc. Am. 72, 175 (1961).
UREY, H. C.: A review of atomic abundances in chondrites and the origin of meteorites. Rev. Geophys. 2, 1 (1964).
— CRAIG, H.: The composition of the stone meteorites and the origin of the meteorites. Geochim. Cosmochim. Acta 4, 36 (1953).
VINOGRADOV, A. P.: Regularity on distribution of chemical elements in the earth's crust. Geochem. Int. 1 (1956).
— Average contents of chemical elements in the principal types of igneous rocks of the earth's crust. Geochem. Int. 641 (1962).
— Meteoritic matter. Geochem. Int. 947 (1965).
VLASOV, K. A. (ed.): Geochemistry and Mineralogy of Rare Elements and Genetic Types of Their Deposits. Vol. I: Geochemistry of Rare Elements. Israel Program for Scientific Translations 1966.
VOSKRESENSKAYA, N. T.: Geochemistry of thallium and rubidium in igneous rocks. Geochem. Int. 599 (1959).
— Thallium content in the igneous rocks of the Greater Caucasus and some other regions of the U.S.S.R. Geochem. Int. 609 (1961a).
— Thallium in some hydrothermal deposits of the Greater Caucasus. Geochem. Int. 731 (1961b).
— Thallium in coal. Geochem. Int. 158 (1968).
— FEL'DMAN, V. I.: Behaviour of rubidium and thallium during autometasomatism of alkalic granitoids of the Taidut massif (Central Transbaikaliya). Geochem. Int. (abstract), 56 (1964).
—KARPOVA, I. S.: Thallium in ore minerals of the Verkhnyaya Kvaisa. Geochem. Int. 552 (1958).
— SHOU-T'IEN, S.: Geochemistry of alkali metals and Tl in the Musht granites (North-ern Caucasus). Geochem. Int. 534 (1961).
— SOBOLEVA, L. T.: Once more on thallium in manganese minerals. Geochem. Int. 309 (1961).
— TIMOFEYEVA, N. V., TOPKHANA, M.: Thallium in some minerals and ores of sedimentary origin. Geochem. Int. 851 (1962a).
— TITKOVA, N. F., SHULYAKOVSKAYA, N. S., TS'UI-YING, C.: Geochemistry of thallium, rubidium and lithium in the magmatic process. Geochem. Int. 282 (1962b).
— USEVICH, T. D.: The occurrence of thallium in manganese minerals. Geochem. Int. 710 (1957).
WAGER, L. R., BROWN, G. M.: Layered Igneous Rocks. San Francisco: W. H. Freeman & Co. 1967.
WAHLER, W.: Pulse-polarographische Bestimmung der Spurenelemente Zn, Cd, In, Tl, Pb und Bi in 37 geochemischen Referenzproben nach Voranreicherung durch selektive Verdampfung. Neues Jahrb. Mineral., Abhandl. 108, 36 (1968).

Webber, G. R.: Second report of analytical data for CAAS syenite and sulphide standards. Geochim. Cosmochim. Acta 29, 229 (1965).

Weissberg, B. G.: Gold-silver ore-grade precipitates from New Zealand thermal waters. Econ. Geol. 64, 95 (1969).

Wiik, H. B.: The chemical composition of some stony meteorites. Geochim. Cosmochim. Acta 9, 279 (1956).

Willis, J. P., Ahrens, L. H.: Some investigations on the composition of manganese nodules with particular reference to certain trace elements. Geochim. Cosmochim. Acta 26, 751 (1962).

Zlobin, B. I.: Geochemistry of Tl in alkalic rocks. Geochem. Int. 560 (1958).

— Lebedev, V. I.: Geochemical relationship of Li, Na, K, Rb and Tl in alkalic magma and its petrogenetic significance. Geochem. Int. 101 (1960).

Revised manuscript received: March 1970

Lead 82

A   K. Sahl   (Institut für Mineralogie, Ruhr-Universität, Bochum, Germany)

B   B. R. Doe   (U.S. Geological Survey, Denver, Colorado, U.S.A.)

C—O   K. H. Wedepohl   (Geochemisches Institut der Universität, Göttingen, Germany)

82-B. Isotopes in Nature[1]

I. Introduction

The geochemistry of lead isotopes is intimately associated with that of uranium and thorium and in consequence, the preparation of separate chapters on uranium, thorium and lead has necessitated a subdivision of topics involving these three elements. This section contains a discussion of U—Th—Pb dating, observed values of $^{238}U/^{204}Pb$ and $^{232}Th/^{204}Pb$ ratios in various rock types, some consideration of values of the Th/U ratio predicted from lead isotopic compositions, and some geological applications of radioactive lead that occurs as an intermediate daughter in the uranium and thorium decay chains. Chapter 92 (Uranium) includes a discussion of both uranium and thorium in connection with heat production in the earth, autoradiographic studies, fractionation of thorium and uranium during weathering and sedimentation, and the utilization of nuclear energy. Chapter 90 (Thorium) includes discussions of age determinations by means of radioactive disequilibria of short-lived uranium and thorium isotopes in the uranium and thorium decay chains, Th/U ratios in igneous rocks, and the general relationship of thorium and uranium to igneous petrology.

a) General

Lead has four stable isotopes:

Symbol	Mass (BHANOT et al., 1960)	Typical extreme abundances (%)	
		Southeast Missouri lead ore (BROWN, 1962)	Cœur d'Alene district lead ore Idaho (LONG et al., 1960)
^{204}Pb	204.038	~1.28	1.38
^{206}Pb	206.040	~26.73	23.65
^{207}Pb	207.042	~20.51	22.40
^{208}Pb	208.043	~51.48	52.51

The isotopic abundances of the lead isotopes vary considerably because three of them are generated by radioactive decay: ^{206}Pb by decay of ^{238}U, ^{207}Pb by decay of ^{235}U, and ^{208}Pb by decay of ^{232}Th, ^{204}Pb has no long-lived radioactive parent. Shown above, therefore, are only the approximate average compositions of the two isotopically extreme major lead-producing deposits — the Cœur d'Alene in Idaho and the southeast Missouri lead belt — both in the U.S.A. The modes of decay

---

[1] Publication authorized by the Director, U.S. Geological Survey.

are complex, each parent-daughter series involving many relatively short-lived radioactive intermediate daughters before the stable lead isotope is reached.

Variations in lead isotopic composition arise directly or indirectly according to the equations in Table 82-B-1 or according to mixtures of systems that have had different values of U/Pb and Th/Pb for long periods of time. The relevant decay constants for the equations are:

$\lambda(^{238}U) = 1.54 \times 10^{-10} \text{yr}^{-1}$ (KOVARIK and ADAMS, 1955)
$\lambda(^{235}U) = 9.71 \times 10^{-10} \text{yr}^{-1}$ (FLEMING et al., 1952)
$\lambda(^{232}Th) = 4.99 \times 10^{-11} \text{yr}^{-1}$ (KOVARIK and ADAMS, 1938; PICCIOTTO and WILGAIN, 1956).

Of utmost importance is the fact that the present abundance ratio of $^{238}U/^{235}U$ is nearly constant in nature:

$$\left(\frac{^{238}U}{^{235}U}\right) = 137.8 \quad \text{(INGHRAM, 1946)}.$$

The values of the constants given above are those recommended in the careful evaluation by ALDRICH and WETHERILL (1958) and differ somewhat from those used by STIEFF et al. (1959) in compiling their tables for age calculations. Except for $\lambda(^{232}Th)$, however, the differences are within the uncertainties of the constants. For $\lambda(^{232}Th)$, STEIGER and WASSERBURG (1966) show that the constant recommended by ALDRICH and WETHERILL (1958) gives ages in better agreement with other methods than does the method used by STIEFF et al. (1959).

Physico-chemical variations such as those found for hydrogen-deuterium, carbon, oxygen, and sulfur are very small for heavy isotopes such as lead. There is no evidence as yet of any variations in lead isotopic composition caused by anything other than radioactive decay or mixing of leads of different isotopic compositions.

This chapter is in large part the condensation of a monograph on lead isotopes (DOE, 1970). Those who wish to see expanded discussions on the topics covered in this chapter are referred to this monograph.

b) Isotopic Measurements

Mass Spectrometry. By far the most popular method of analyzing lead isotope abundances is by mass spectrometry, preferably without an electron multiplier in the ion collector. The most precise method (0.05 percent of a ratio) appears to be that involving the use of a thermal ionization source (triple filament) (CATANZARO, 1967); however, about 500 micrograms of lead are required. Ratios are nearly absolute (<0.1% correction). This method is not yet very popular because of the large sample size required and because most mass spectrometers require modification for use in this manner.

Perhaps most data are generated using a surface emission technique where PbS and NH_4NO_3 are mixed on a single filament. This method has been evaluated by DOE et al. (1967) and although as little as 10 micrograms of lead may be analyzed, the reproducibility is not as good as that quoted above (about 0.05 percent per mass unit separation at best) and corrections for mass spectrometer fractionation are necessary.

Lead

Table 82-B-1. *Equations and parameters used in lead isotope calculations*

1. Basic equation: $-\dfrac{dN}{dt} = \lambda N$

   N = number of radioactive atoms
   t = time
   λ = constant of proportionality (decay constant)

2. Working equations (atomic ratios):

   a) $^{206}Pb/^{238}U$ age

   $$\left(\frac{^{206}Pb}{^{204}Pb}\right)_{observed} - \left(\frac{^{206}Pb}{^{204}Pb}\right)_{initial} = \left(\frac{^{238}U}{^{204}Pb}\right)_{observed} \cdot (e^{\lambda_8 T} - 1)$$

   b) $^{207}Pb/^{235}U$ age

   $$\left(\frac{^{207}Pb}{^{204}Pb}\right)_{observed} - \left(\frac{^{207}Pb}{^{204}Pb}\right)_{initial} = \left(\frac{^{235}U}{^{204}Pb}\right)_{observed} \cdot (e^{\lambda_5 T} - 1)$$

   c) $^{208}Pb/^{232}Th$ age

   $$\left(\frac{^{208}Pb}{^{204}Pb}\right)_{observed} - \left(\frac{^{208}Pb}{^{204}Pb}\right)_{initial} = \left(\frac{^{232}Th}{^{204}Pb}\right)_{observed} \cdot (e^{\lambda_2 T} - 1)$$

   d) Pb—Pb age or isochron age

   $$\frac{\left(\frac{^{207}Pb}{^{204}Pb}\right)_{observed} - \left(\frac{^{207}Pb}{^{204}Pb}\right)_{initial}}{\left(\frac{^{206}Pb}{^{204}Pb}\right)_{observed} - \left(\frac{^{206}Pb}{^{204}Pb}\right)_{initial}} = \left(\frac{^{235}U}{^{238}U}\right)_{observed} \cdot \frac{(e^{\lambda_5 T}-1)}{(e^{\lambda_8 T}-1)}$$

   e) Explanation of abbreviated symbols
   T = geologic age.
   λ = decay constant; subscript is the last mass number of the radioactive parent.
   Initial ratio is that ratio which the phase contained when it was formed.

3. Working equations for common lead calculations (atomic ratios)

Ratio at geologic time (t)	Ratio when earth formed	Ratio in earth today	Decay from time earth was formed (T) to the geologic time of interest (t)

Primary growth equations

a) $\left(\dfrac{^{206}Pb}{^{204}Pb}\right)_t = \left(\dfrac{^{206}Pb}{^{204}Pb}\right)_T + \left(\dfrac{^{238}U}{^{204}Pb}\right)_n \cdot (e^{\lambda_8 T} - e^{\lambda_8 t})$.

b) $\left(\dfrac{^{207}Pb}{^{204}Pb}\right)_t = \left(\dfrac{^{207}Pb}{^{204}Pb}\right)_T + \left(\dfrac{^{235}U}{^{204}Pb}\right)_n \cdot (e^{\lambda_5 T} - e^{\lambda_5 t})$.

c) $\left(\dfrac{^{208}Pb}{^{204}Pb}\right)_t = \left(\dfrac{^{208}Pb}{^{208}Pb}\right)_T + \left(\dfrac{^{232}Th}{^{204}Pb}\right)_n \cdot (e^{\lambda_2 T} - e^{\lambda_2 t})$.

Primary isochron equation [sometimes referred to as the Gerling-Houtermans-Holmes equation (GERLING, 1942; HOUTERMANS, 1946; HOLMES, 1946)]

d) $\dfrac{\left(\frac{^{207}Pb}{^{204}Pb}\right)_t - \left(\frac{^{207}Pb}{^{204}Pb}\right)_T}{\left(\frac{^{206}Pb}{^{204}Pb}\right)_t - \left(\frac{^{206}Pb}{^{206}Pb}\right)_T} = \left(\dfrac{^{235}U}{^{238}U}\right)_n \left(\dfrac{e^{\lambda_5 T} - e^{\lambda_5 t}}{e^{\lambda_8 T} - e^{\lambda_8 t}}\right)$.

Table 82-B-1. (Continued)

Secondary growth equations

	Normal growth stage $\left(\frac{^{20 \times}\text{Pb}}{^{204}\text{Pb}}\right)_t$, [X=6,7,8]	Secondary growth
e)	$\left(\frac{^{206}\text{Pb}}{^{204}\text{Pb}}\right)_{t'} = \left(\frac{^{206}\text{Pb}}{^{204}\text{Pb}}\right)_T + \left(\frac{^{238}\text{U}}{^{204}\text{Pb}}\right)_n \cdot (e^{\lambda_8 T} - e^{\lambda_8 t})$	$+ \left(\frac{^{238}\text{U}}{^{204}\text{Pb}}\right)_n'' \cdot (e^{\lambda_8 t} - e^{\lambda_8 t'})$.
f)	$\left(\frac{^{207}\text{Pb}}{^{204}\text{Pb}}\right)_{t'} = \left(\frac{^{207}\text{Pb}}{^{204}\text{Pb}}\right)_T + \left(\frac{^{238}\text{U}}{^{204}\text{Pb}}\right)_n \cdot (e^{\lambda_5 T} - e^{\lambda_5 t})$	$+ \left(\frac{^{235}\text{U}}{^{204}\text{Pb}}\right)_n' \cdot (e^{\lambda_5 t} - e^{\lambda_5 t'})$.
g)	$\left(\frac{^{208}\text{Pb}}{^{204}\text{Pb}}\right)_{t'} = \left(\frac{^{208}\text{Pb}}{^{204}\text{Pb}}\right)_T + \left(\frac{^{238}\text{U}}{^{204}\text{Pb}}\right)_n \cdot (e^{\lambda_2 T} - e^{\lambda_2 t})$	$+ \left(\frac{^{232}\text{Th}}{^{204}\text{Pb}}\right)_n' \cdot (e^{\lambda_2 t} - e^{\lambda_2 t'})$,

where $\left(\frac{^{238}\text{U}}{^{204}\text{Pb}}\right)_n'' = \left(\frac{^{235}\text{U}}{^{204}\text{Pb}}\right)_n'' = \left(\frac{^{232}\text{Th}}{^{204}\text{Pb}}\right)_n'' = 0$.

h) Secondary isochron:

$$\frac{\left(\frac{^{207}\text{Pb}}{^{204}\text{Pb}}\right)_{t'} - \left(\frac{^{207}\text{Pb}}{^{204}\text{Pb}}\right)_t}{\left(\frac{^{206}\text{Pb}}{^{204}\text{Pb}}\right)_{t'} - \left(\frac{^{206}\text{Pb}}{^{206}\text{Pb}}\right)_t} = \left(\frac{^{235}\text{U}}{^{238}\text{U}}\right)_n \cdot \left(\frac{e^{\lambda_5 t} - e^{\lambda_5 t'}}{e^{\lambda_8 t} - e^{\lambda_8 t'}}\right).$$

i) Explanation of symbols, their equivalents in other conventions, and present estimates of key values with references to data.

	Canadian convention	Swiss convention	Value	Reference
$\left(\frac{^{206}\text{Pb}}{^{204}\text{Pb}}\right)_t$	X	α	$t=0$; 18.66	1
			18.51	2
$\left(\frac{^{206}\text{Pb}}{^{204}\text{Pb}}\right)_T$	a_0	α_0	9.56	3
			9.54	4
			9.346	5
$\left(\frac{^{238}\text{U}}{^{204}\text{Pb}}\right)_n$	137.8 V	μ_0	8.99	1
			8.7	2
			9.09	6
$\left(\frac{^{207}\text{Pb}}{^{204}\text{Pb}}\right)_t$	Y	β	$t=0$; 15.79	1
			15.72	2
$\left(\frac{^{207}\text{Pb}}{^{204}\text{Pb}}\right)_T$	b_0	β_0	10.42	3
			10.27	4
			10.218	5
$\left(\frac{^{235}\text{U}}{^{204}\text{Pb}}\right)_n$	V	$\frac{\mu_0}{137.8}$		
$\left(\frac{^{208}\text{Pb}}{^{204}\text{Pb}}\right)_t$	Z	γ	$t=0$; 39.06	1
			38.44	2
$\left(\frac{^{208}\text{Pb}}{^{204}\text{Pb}}\right)_T$	c_0	γ_0	30.0	1
			29.7	3
			29.46	4
			28.96	5
$\left(\frac{^{232}\text{Th}}{^{204}\text{Pb}}\right)_n$	W	$\mu \cdot k$	35.55	1
			34.8	2
			38.3	6
$\left(\frac{^{232}\text{Th}}{^{238}\text{U}}\right)$	$\frac{W}{V}$	k	3.92	1
			4.0	2
			3.8	3
			4.21	6

References: 1. KANASEWICH (1968a); 2. DOE (1962b); 3. MURTHY and PATTERSON (1962); 4. COW and PATTERSON (1961), corrected according to CHOW and PATTERSON (1962b); 5. OVERSBY (1969); 6. STACEY et al. (1969).

Below 10 micrograms of lead a refractory high temperature emittor is used, the most common type being silica gel with or without other high temperature additives such as $ZrSiO_4$ or H_2PO_4 (AKISHIN et al., 1957; CAMERON et al., 1969). The analytical uncertainties involved are not yet fully evaluated; however, TATSUMOTO and ROSHOLT (1970) have analyzed 0.2 microgram of lead.

Perhaps the greatest volume of precise published lead isotope data uses the gaseous lead tetramethyl method with electron bombardment ionization. This method has been compared with the thermal emission method by STACEY et al. (1969). Aside from the large sample size required, the tetramethyl method has several defects: the high toxicity of the compound being analyzed, problems of instrumental clean-up to prevent contamination between analyses, and corrections for instrumental fractionation of the data due to the gas leak. The more versatile thermal emission method now appears to be replacing the tetramethyl method.

Activation Analysis. Although it is possible to measure relative abundances of ^{204}Pb, ^{206}Pb, and ^{208}Pb by various kinds of activation analyses, the surface emission method with refractory emittors requires less speed of analysis, is probably more accurate, and measures lead quantities down to the contamination level involved in activation work. The activation method remains more of a novelty.

II. U—Th—Pb Dating

a) General

Three independent ages may be obtained in the U—Th—Pb system: $^{206}Pb/^{238}U$, $^{207}Pb/^{235}U$ or $^{207}Pb/^{206}Pb$, and $^{208}Pb/^{232}Th$. Emphasis has been placed on U—Pb dating because the value $^{238}U/^{235}U$ is a physical constant that permits internal treatment of the data not found in any other dating system. This treatment helps to eliminate the assumption that the phase being dated has remained closed to changes in the parent-daughter system. The theoretical systematics are expressed in Fig. 82-B-1. WETHERILL (1956a, b) showed that a phase, which is subject to no lead loss or uranium gain (a closed system), will have $^{207}Pb/^{235}U$ ages equal to $^{206}Pb/^{238}U$ ages and that the data will lie along a curved line called *concordia*. In addition, he showed that phases subject to lead loss or uranium gain during a period of time that is short compared with the age of the phase (episodic bulk daughter losses or parent gains), recently or in the distant past, will have data that will lie along a straight line called *discordia*. The lower intersection of discordia with concordia represents the time of the episodic event and the upper intersection represents the age of the phase. NICOLAYSEN (1957) suggested that diffusion of lead out of a phase might take place at a constant rate over the entire history of the phase (continuous diffusion), and TILTON (1960) showed that continuous diffusion with bulk lead loss or uranium gain also closely approaches a straight line over much of its length. WASSERBURG (1963) considered the possible effects of a diffusion constant that varies with time, such as might be caused by radiation damage; WETHERILL (1963) discussed most other aspects of the theory of U—Pb behavior, such as continuous diffusion with superimposed episodic loss and the theory of uranium loss and bulk radiogenic daughter gain. The U—Th—Pb system has intermediate radioactive daughters, including the gas radon. If there is radon loss over a long period of

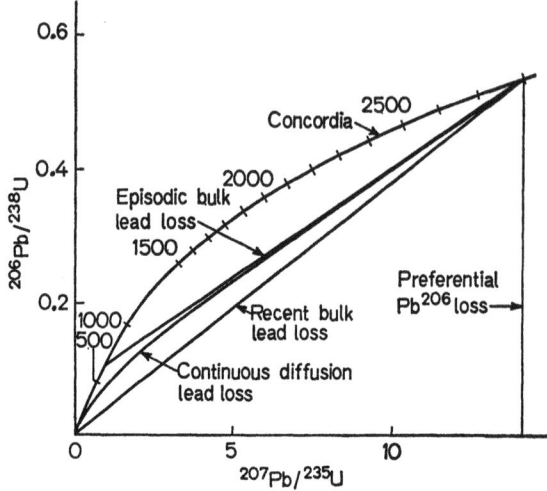

Fig. 82-B-1. Systematics of U—Pb dating. (After CATANZARO and KULP, 1964)

time, the calculated ages may be affected. If the system is subject to radon leakage, the result is reflected in preferential loss of ^{206}Pb, because the radon in the ^{238}U chain has a much longer half-life (3.8 days) than that in the ^{235}U chain (3.9 seconds). The longer half-life allows a longer time for radon escape.

Though considerable data have been obtained on Th—Pb dating, this dating has received less emphasis because of the lack of a companion radioactive parent isotope to go with ^{232}Th, so that Th—Pb dating lacks the protection against parent and daughter losses or gains inherent in the U—Pb dating method. The theoretical behavior of Th—Pb dating in terms of a U—Th—Pb concordia diagram is, however, given by STEIGER and WASSERBURG (1966).

Most U—Th—Pb dating has been done on uraninite, pitchblende, and other uranium minerals. Uraninite under some conditions gives the most concordant ages, but it is not a common accessory mineral in most rock types. Emphasis, therefore, will here be placed on the more common accessories — zircon, sphene, phosphates (particularly monazite) — and on whole-rock dating. Zircon has been subjected to the most thorough and rigorous laboratory and field investigation and is therefore chosen for first discussion.

b) Zircon (ZrSiO$_4$)

The isotopic ages found in zircon (selected data given in Table 82-B-2) are generally discordant with ^{206}Pb/^{238}U age < ^{207}Pb/^{235}U age < ^{207}Pb/^{206}Pb age. This relationship of ages is often referred to as *normal discordance*. One such example of zircons with discordant ages is the study by SILVER and DEUTSCH (1963) which was particularly useful in that the data lie along an episodic lead-loss line, and it was the first study that verified the episodic lead-loss hypothesis. The data do not lie along continuous diffusion lines and, in fact, no clear example has yet been found to support the continuous diffusion hypothesis. A good example of episodic lead loss (and uranium gain) is given in the study of DAVIS *et al.* (1968) presented in

Table 82-B-2. *U—Th—Pb data for selected minerals*

Geologic unit	Locality	Element concentration (ppm)			Atomic ratios		
		U	Th	Pb	$^{206}Pb/^{204}Pb$	$^{207}Pb/^{204}Pb$	$^{208}Pb/^{204}Pb$
		A. Zircons (from CATANZARO and KULP, 1964)					
Fresh granite gneiss	Little Belt Mountains, Montana	2,227	1,421	261	229	36.6	59.1
		549	217	138	694	90.1	126.9
		1,417	276	216	658	90.1	95.2
		553	159	115	1,042	127.2	121.3
		710	346	123	629	81.4	106.7
		1,047	338	176	303	46.4	74.2
		B. Sphenes (OOSTHUYZEN and BURGER, 1965)					
Syenite	South Africa	63.8	230	32.1	121.4	24.47	149.5
		—	—	—	28.49[a]	16.70[a]	50.48[a]
		C. Apatite (OOSTHUYZEN and BURGER, 1965)					
Syenite	South Africa	20.7	—	28	38.16	17.60	83.58
		D. Monazite (CATANZARO and KULP, 1964; HEIMLICH and BANKS, 1968)					
Granite gneiss	Little Belt Mountains, U.S.A.	0.213	8.84	0.802	312.5	50.2	35.4
	Bighorn Mountains, U.S.A.	0.0830	—	0.998[b]	206.3	53.8	4,521

[a] Acid leach of sphene.
[b] radiogenic lead only.

Figs. 82-B-2 and 82-B-3. The zircons are from the Front Range of Colorado from Precambrian rocks adjacent to Tertiary igneous stocks; in this case, the episodic event was intrusion of the Tertiary stocks. Hydrothermal experiments on a metamict Ceylon zircon in 2 molal NaCl at 500° C and 1,000 bars (PIDGEON et al., 1966) showed a very rapid episodic lead loss of approximately 50 percent of the lead in 11 hours. Uranium content was little affected. Apparently weathering can also cause discordance in U—Pb ages of zircon (STERN et al., 1966). The conclusion from these studies is that zircon ages are usually discordant and that this discordance may be caused rapidly by geological processes not usually considered to be strong. Also important is the fact that usually only part of the lead is lost. If the Wetherill concordia diagram is used, reliable mineral ages can be obtained.

The mechanism which causes the discordant ages in zircon is not yet known. The most promising approach seems to be consideration of the formation of domains rich in H_2O, heavy trace elements (including uranium and thorium), and yttrium, such as discussed by GRÜNENFELDER (1963), GRÜNENFELDER and EBERHARD (1969),

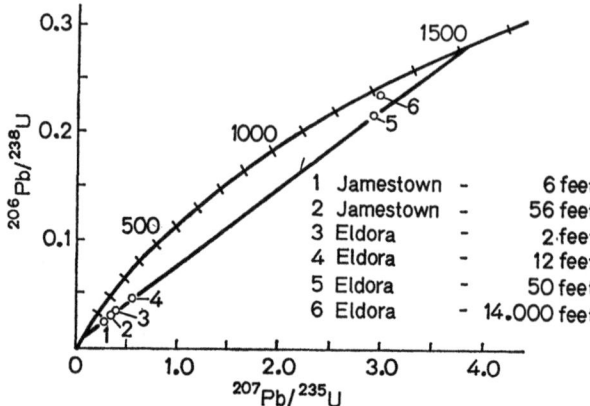

Fig. 82-B-2. Concordia diagram for zircons from the Precambrian Idaho Springs Formation adjacent to the Tertiary Eldora stock and from the Precambrian Silver Plume Granite adjacent to the Tertiary Jamestown stock, Colorado (DAVIS et al., 1968)

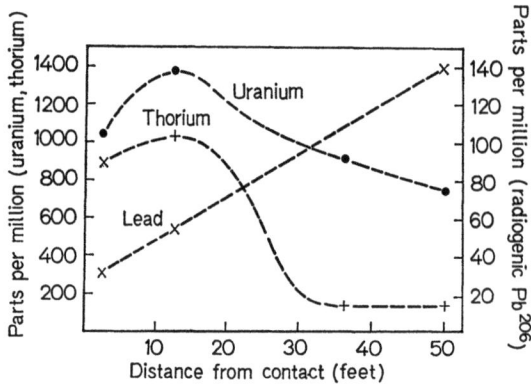

Fig. 82-B-3. Change in concentration of uranium, thorium, and lead in zircon with distance from the Eldora stock, Colorado (DAVIS et al., 1968)

and STEIGER and WASSERBURG (1966). Whether this formation of domains is a continuous process or is assisted by catastrophism has not yet been determined; support for the importance of GRÜNENFELDER's "multiphase" approach to the dating of zircons has been found by STEIGER and WASSERBURG (1969). They have found examples where the secondary age (lower intersection of the discordia line) determined when an episodic model is assumed, disagrees with that expected from other information. Their argument is convincing that some zircons behave like two phases — one that is nearly concordant and another that is highly discordant in a manner expected in models involving some sort of continuous diffusion.

Th—Pb ages on zircons usually, but not always, range from the $^{206}Pb/^{238}U$ age to the $^{207}Pb/^{206}Pb$ age. The Th—Pb ages are rigorously interpretable in some cases by use of the U—Th—Pb concordia diagram (STEIGER and WASSERBURG, 1966) but are generally unreliable for dating work.

c) Sphene (CaTiSiO$_5$)

As sphene is one of the most widespread radioactive accessory minerals in igneous and metamorphic rocks (it is particularly abundant in amphibolite), the lack of more extensive dating of this mineral is surprising (some representative data are given in Table 82-B-2). The first analyses are reported from as early as 1952; however, real interest had to await publication of the study by OOSTHUYZEN and BURGER (1965) who analyzed co-existing sphene, zircon and apatite (Fig. 82-B-4), and found the sphene to be more concordant than the zircon. In addition to the concordia

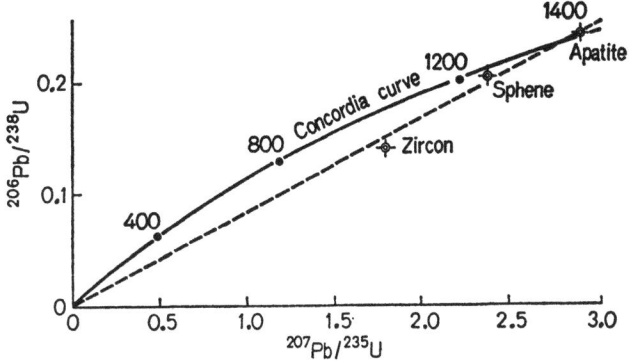

Fig. 82-B-4. Concordia diagram for Leeuwfontein Syenite, South Africa, showing isotopic relations of zircon, sphene, and apatite. (From OOSTHUYZEN and BURGER, 1965)

relationship ($t=1,380$ m.y.), the Th—Pb age of the sphene was determined to be 1,190 m.y., whereas that of the zircon was only 705 m.y. The Th—Pb age on the sphene is therefore also more concordant than that of the zircon.

The major evaluation study by TILTON and GRÜNENFELDER (1968) indicates that sphenes will usually give concordant ages but they fall at either end of a discordia line. This characteristic is illustrated by the analyses of TILTON and GRÜNENFELDER on granite-gneiss from along the French River in Ontario. The whole-rock Rb—Sr isochron age of this gneiss is given by HART et al. (1967) as 1,725 m.y., whereas Rb—Sr mineral ages such as those for biotite reflect the age of the so-called Grenville orogeny at about 1,000 m.y. Two sphene fractions were a little discordant but gave $^{207}Pb/^{206}Pb$ ages of 1,060 and 1,100 m.y. In contrast, the zircon (KROGH, in TILTON and GRÜNENFELDER, 1968) again demonstrated the difficulty of completely resetting the zircon in nature as it has a $^{207}Pb/^{206}Pb$ age of 1,470 m.y.

Little laboratory experimental work has been done on sphene. OOSTHUYZEN and BURGER (1965) did analyze the isotopic composition of the acid leach solution of their sphene (Table 82-B-2) and found it to be somewhat radiogenic but no more so than acid leachings of zircon. The effect of this leaching on the sphene has not yet been completely evaluated.

d) Phosphates

Monazite (CePO$_4$). Next to uranium minerals and zircon, monazite receives the greatest emphasis today. Although monazite usually has normal discordance of

^{206}Pb/^{238}U age < ^{207}Pb/^{235}Pb age < ^{207}Pb/^{206}Pb age (strongly so in some specimens), some examples are known to have strong *reverse discordance* where

$$^{206}Pb/^{238}U \text{ age} > {}^{207}Pb/^{235}U \text{ age} > {}^{207}Pb/^{206}Pb \text{ age}.$$

(For one such example see the analytical data of ZYKOV et al., 1964, Table 82-B-2). Why this reverse discordance occurs in occasional monazites is not understood but the theoretical base provided by WETHERILL (1963) allows distinction of continuous diffusion of uranium from episodic loss of uranium.

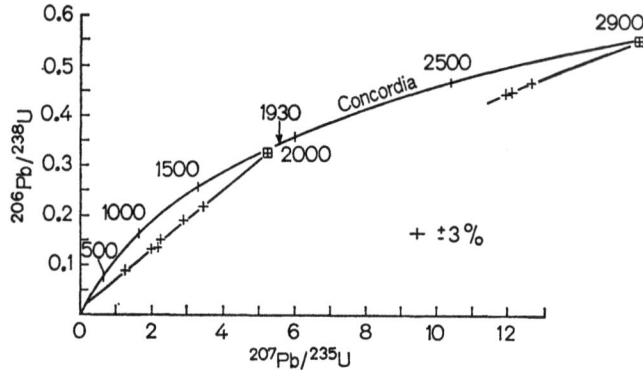

Fig. 82-B-5. Concordia diagram showing monazite and zircon relations. Discordia line intersecting concordia at 1,930 m.y. is from CATANZARO and KULP (1964), Little Belt Mountains, Montana; the discordia line intersecting concordia at 2,900 m.y. is from HEIMLICH and BANKS (1968), Bighorn Mountains, Wyoming. Crosses are plottings for zircons; boxes are plottings for monazites

Much emphasis has been placed on acid-leaching experiments (BURGER et al., 1967, and papers cited therein) which have shown that some uranium and uranogenic lead are leached relatively easily in the laboratory and that Th and ^{208}Pb are less easily leached. BURGER et al. (1967) indicate that leaching in nature does not reflect the results of the laboratory experiments and the U—Pb system on these monazites gives the best age estimate. They also point out that it is apparent from modern detailed studies where independent evidence of age exists, that the Th—Pb age is usually too low; these studies suggest the likelihood of episodic losses in nature also. Examples of monazites with reverse discordance, which have well-established zircon isochrons, are needed to better evaluate this conclusion.

Examples are also known where monazite is nearly concordant (Fig. 82-B-5, Table 82-B-2). In the Rocky Mountain region, monazite would be a reasonable dating tool because it is relatively abundant and appears to have a tendency to be concordant.

Little rigorous evaluation has been made on the effect of metamorphism on monazite so that the ability of this mineral to retain some isotopic identity through periods of metamorphism has not yet been estimated.

Apatite $Ca_5[PO_4]_3(OH, F)$. Apatite has seldom been used or investigated for dating but is of possible interest. A comparison of apatite with zircon and sphene (data in Table 82-B-2) is given in Fig. 82-B-4 (from OOSTHUYZEN and BURGER, 1965). the apatite is nearly concordant but has slight reverse discordance which, because the radiogenic enrichment is not large, might be due to the common lead correction.

e) Whole Rock

Whole-rock dating is difficult because of the lack of radiogenic enrichment in the lead, particularly ^{207}Pb.

In a series of papers, SOBOTOVICH and his co-workers report success in dating very old rocks by the ^{207}Pb—^{206}Pb isochron method (Fig. 82-B-6). In the study of the 3,000-m.y.-old granites and gneisses of the Taromskoe Quarry in the Ukraine

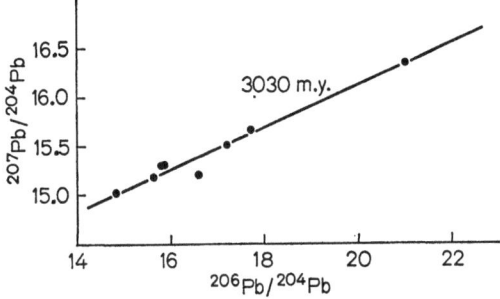

Fig. 82-B-6. Whole-rock plot of ^{207}Pb/^{204}Pb versus ^{206}Pb/^{204}Pb for rocks of granitic composition from the Taromskoe Quarry, Dnjeper Region, Ukraine (SOBOTOVICH et al., 1963)

(SOBOTOVICH et al., 1963), the Pb—Pb isochron was somewhat older than K—Ar ages on micas which are known to be easily affected by metamorphism. The determined values of uranium relative to lead were found to be too great for the observed lead isotopic compositions so that the ^{238}U/^{204}Pb—^{206}Pb/^{204}Pb age is much younger than the Pb—Pb age. Data on a rock sample as presented by ZARTMAN (1965) showed an opposite effect of far too little uranium. ZARTMAN's other rocks appeared to have a fairly good U—Pb balance. Apparently the disturbances in U—Pb are recent. ROSHOLT and PETERMAN (1969) have found a dramatic recent depletion of uranium relative to lead, but not thorium, in the Precambrian basement granites of the Cenozoic sandstone uranium ores of Wyoming.

III. Common Lead

a) General

"Common lead" is any lead from a phase with a low value of U/Pb and/or Th/Pb such that no significant radiogenic lead has been generated *in situ* since the phase formed. Such phases are galena and other sulfides such as pyrite, feldspars (in particular K-feldspar), micas, and most abundant rock types of Cenozoic age. Data on common lead are used in determining ages and, more important, in the solution of genetic problems. Presented here, therefore, are data on those phases relevant to common lead applications even if sufficient radiogenic lead has been generated to markedly alter the initial isotopic composition of the lead such as in Precambrian crystalline rocks.

b) Meteorites, the Moon, and Tektites

Studies of lead isotopes in meteorites have been very fundamental to all of earth science; the first reliable age of meteorites (4,500 m.y.) was a lead isochron age (PATTERSON, 1956; MURTHY and PATTERSON, 1962) (Fig. 82-B-7). Strictly speaking, the present day lead isotopic composition of most meteorites does not fit the definition of common lead because it has been drastically changed since the meteorites formed. Lead from the troilite phase of meteorites does appear to be primordial and is the least radiogenic lead known. These values taken from the evaluations of MURTHY and PATTERSON (1962) and of OVERSBY (1970) are included in Table 82-B-1. These numbers are a very important parameter in most common lead calculations and, therefore, all meteorites are considered together here. The values determined by OVERSBY (1970) have been corrected for mass spectrometer bias, and the attempt has been made to eliminate samples that have been contaminated by terrestrial lead. STACEY et al. (1969) have pointed out that there is little change in derived values such as age of the earth or primary $^{238}U/^{204}Pb$ if all samples are in absolute ratios. Care should be taken to make sure that all samples are in absolute ratios when using the values of primordial lead recommended by OVERSBY (1970).

The line that best fits the data in Fig. 82-B-7 is a straight line and lead from meteorites form a physical example of a primary isochron or Gerling-Houtermans-Holmes equation (Eq. 3d in Table 82-B-1) indicating that the meteorites were formed 4,550 million years ago (PATTERSON, 1956). From the slope of the line in the plot of $^{208}Pb/^{204}Pb$ versus $^{206}Pb/^{204}Pb$, the value of $^{232}Th/^{238}U$ may be calculated to be 3.8 (MURTHY and PATTERSON, 1962).

The very recent results (TATSUMOTO and ROSHOLT, 1970; SILVER, 1970) on the samples returned by Apollo 11 from the Mare Tranquillitatis on the moon have values of $^{207}Pb/^{204}Pb$ of 96-590 which are much greater than those of meteorites or the earth. These large ratios for the moon materials signify depletion of lead relative to uranium a long time ago when ^{235}U was still abundant. These data on the fines and breccia from the moon lie very near the extension of the primary isochron of meteorites and the calculated U—Pb and Th—Pb ages support the age signified by the isochron. If the fines and breccia do not contain lead that has primarily been added to the moon from some foreign source, then the age of the moon must be close to 4,600 m.y., similar to that of meteorites. This age is not necessarily the age at which the observed values of U/Pb and Th/Pb were established.

The lead in the lunar volcanic rocks also give nearly concordant ages between about 3,700 and 4,200 m.y. The lead isotope ages therefore infer that the moon was tectonically active for at least a billion years after its formation.

Meteorites have large variations in values of $^{207}Pb/^{204}Pb$ — from 12 to 35 — and those of the moon are even greater — from 96 to 590 — whereas for the earth this ratio is rather uniform, ranging mainly between 15 and 16. TILTON (1958) used the difference in the nature of $^{207}Pb/^{204}Pb$ between meteorites and the earth to try to determine whether tektites are terrestrial or extraterrestrial in origin. All tektites are found to have a terrestrial type of $^{207}Pb/^{204}Pb$ value (such as included in the cross-hatched area in Fig. 83-B-7). Thus arguments based on lead isotopes suggest a derivation of tektites from a terrestrial source rather than from a meteoritic or lunar source. WAMPLER et al. (1969), through analyses of tektites and rocks of

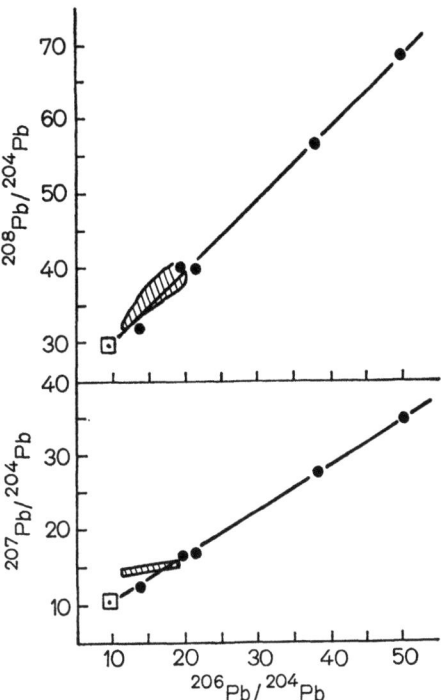

Fig. 82-B-7. A plot of ^{207}Pb/^{204}Pb and ^{208}Pb/^{204}Pb versus ^{206}Pb/^{204}Pb of troilite primordial lead (□) and stone meteorite leads (●) as selected by MURTHY and PATTERSON (1962). The solid line in the lower plot is a primary isochron with a slope of 0.59 and the slope of the solid line in the upper plot represents a Th/U of 3.8. Cross-hatched areas enclose terrestrial leads

possible impact crater sources, have shown that the lead isotopes support the growing body of data relating the Ivory Coast tektites to the Bosumtwi crater. They also show that moldavites are compatible with derivation from Ries crater and suggest that other tektites may be ultimately derived from pelagic sediments.

c) General Aspects for the Earth

Age of Earth. All leads from terrestrial rocks, except those from uranium and thorium ores, are enclosed in the cross-hatched areas of Fig. 82-B-7. The proximity of terrestrial lead to the primary meteorite isochron has been used by PATTERSON (PATTERSON, 1956; MURTHY and PATTERSON, 1962) in estimating the age of the earth at 4,550 m.y. as well as for estimating the isotopic composition of terrestrial primordial lead. Some studies on terrestrial materials suggest that the earth may be as much as 200 m.y. older than the estimate from meteorites (TILTON and STEIGER, 1969); however, if the earth is even that much older, either there must have also been differentiation of U relative to lead in the early history (PATTERSON and TATSUMOTO, 1964) or the values of primordial lead for the earth and for meteorites must differ (DOE et al., 1965).

Table 82-B-3. Data on selected stratiform ores whose lead approaches single stage lead isotope development

Location	Host rock	No. of samples in average	$^{206}Pb/^{204}Pb$	$^{207}Pb/^{204}Pb$	$^{208}Pb/^{204}Pb$	Reference number
Bathurst, New Brunswick, Canada	Middle Ordovician metamorphosed rocks	3	18.291	15.781	38.526	1
	Middle Ordovician metamorphosed rocks	1	18.20	15.69	38.24	2
	Middle Ordovician metamorphosed rocks	1	18.204[a]	15.655[a]	38.122[a]	3
Balmat, New York	Marbles about 1,100 m.y. in age	1	17.019	15.677	36.857	4
	Marbles about 1,100 m.y. in age	1	16.96	15.56	36.60	2
	Marbles about 1,100 m.y. in age	1	16.935[a]	15.505[a]	36.423[a]	3
Broken Hill, Australia	Metamorphosed rocks approximately 1,600—1,700 m.y. old	6	16.12	15.54	36.06	5
	Metamorphosed rocks approximately 1,600—1,700 m.y. old	1	16.07	15.52	36.03	2
	Metamorphosed rocks approximately 1,600—1,700 m.y. old	1	16.007[a]	15.397[a]	35.675[a]	3
Geneva Lake, Ontario, Canada	Precambrian metamorphosed rocks	2	14.08	15.02	34.10	6
	Precambrian metamorphosed rocks	1	14.002[a]	14.870[a]	33.716[a]	3
Manitouwadge Lake, Ontario, Canada	Archean metamorphosed rocks approximately 2,700—2,800 m.y. old	2	13.34	14.56	33.27	6
	Archean metamorphosed rocks approximately 2,700—2,800 m.y. old	4	13.400	14.565	33.545	1
	Archean metamorphosed rocks approximately 2,700—2,800 m.y. old	1	13.25	14.42	33.18	7
	Archean metamorphosed rocks approximately 2,700—2,800 m.y. old	1	13.211[a]	14.401[a]	33.069[a]	3

References: 1. Osiric et al. (1967); 2. Doe (1962); 3. Stacey et al. (1969); 4. Reynolds and Russell (1968); 5. Kollar et al. (1960); 6. Russell and Farquhar (1960); 7. Doe et al. (1965).

[a] Absolute ratio.

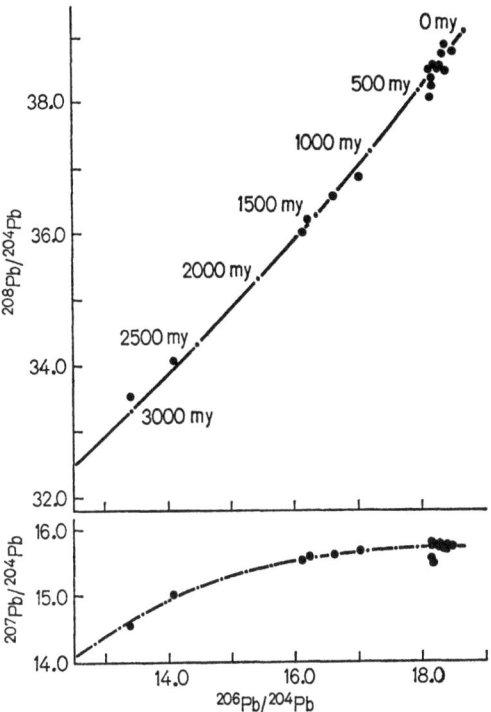

Fig. 82-B-8. $^{207}Pb/^{204}Pb$ and $^{208}Pb/^{204}Pb$ versus $^{206}Pb/^{204}Pb$ for stratiform ores which appear to have formed at approximately the same time as the enclosing sediments and approach conditions of single-stage lead isotope growth. Parameters of the growth curves are those of KANASEWICH (1968a) and are given in Table 82-B-1 and I. Introduction

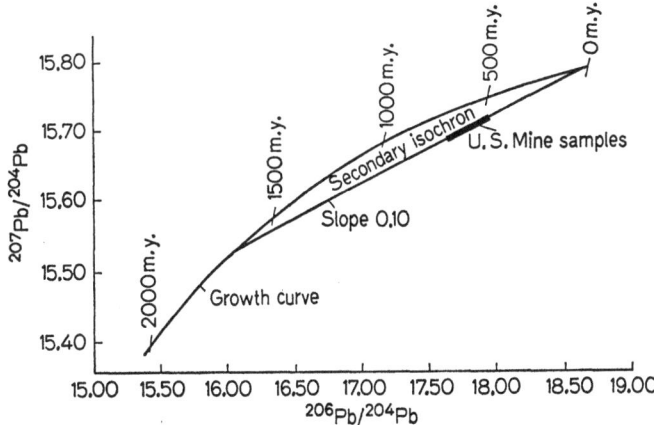

Fig. 82-B-9. Secondary isochron for galenas from the U.S. Mine (STACEY et al., 1967)

Primary or single stage growth curves. A physical example where the lead isotope data forms a close approach to primary growth curves (Eqs. 3a, b and c, in Table 82-B-1) is found for the lead isotope data of *conformable ore deposits* (STANTON and RUSSELL, 1959). These ore deposits are tabular (grossly conformable with the country rock), have ill-defined margins, and are pyrite-rich; many occur in rocks bearing organic matter and most are in metamorphic terranes. STANTON has interpreted the terrane containing the deposits to be of the island-arc and related tectonic classes. The data for some of these deposits are given in Table 82-B-3 and are shown in Fig. 82-B-8. The source of the lead in such deposits is uncertain.

Secondary or two stage growth curves. One particularly good example of a secondary isochron is lead accompanying mineralization in Utah at the U.S. Mine (STACEY et al., 1967), covering only a small range in $^{206}Pb/^{204}Pb$ and $^{207}Pb/^{204}Pb$ ratios (Fig. 82-B-9). Later studies (STACEY et al., 1968) in Utah also showed that the lead isotopic composition has economic implications in that the largest sulfide ores in Utah have lead isotope ratios nearest those of possibly related igneous rocks.

Because the development of lead isotope ratios has occurred in two or more stages, the observed ratios may be greater than they would be in a single stage development because the lead has evolved in an integrated system with $^{238}U/^{204}Pb >$ about 9 over a period of time $\leq 3,500$ m.y. Such lead is *radiogenic* (i.e., the ratios are greater) in comparison with a single stage lead of that age. Special radiogenic leads having future model lead ages are generally referred to as J-type or Joplin type. If the lead isotope ratios are less than they would be in a single stage development because of an integrated $^{238}U/^{204}Pb <$ about 9, the term *unradiogenic* lead relative to a single stage lead of that age will be used. Unradiogenic leads are sometimes called B-type or Bleiberg type. Leads from the Bleiberg deposits do not fit to the isotopic definition, however, and the use of the term B-type should be discontinued.

d) Observed Values of $^{238}U/^{204}Pb$ and $^{232}Th/^{204}Pb$ in Rocks

The lead isotopic composition of igneous rocks and ores places constraints on the observed atomic values of $^{238}U/^{204}Pb$ and $^{232}Th/^{204}Pb$ in the source rocks. Knowledge of the values of these ratios observed in various rock types is therefore desirable in radiogenic tracer studies (Fig. 82-B-10).

e) Whole-rock Studies, Precambrian and Paleozoic

Knowledge of the isotopic composition of lead in Precambrian and Paleozoic upper crustal rocks is important in answering many questions of genesis, such as the source of igneous rocks and ores. Fig. 82-B-10 shows a wide spectrum of present day isotopic compositions in these rocks (for references to data see DOE, 1970). All North American data are enclosed in the area shown and tend to be radiogenic. Lines are drawn on the diagram to show the approximate position of a terrestrial lead that lies on the meteorite geochron (a model lead of zero age). The rocks of the Altai Mountains in Asia (SOBOTOVICH and GRASHCHENKO, 1965) also appear to be radiogenic.

Lead

Group	^{232}Th/^{204}Pb observed ($\mu \cdot k$)	General Description	No. of samples	No. with $\mu \cdot k$ above 36
7	(11.4)⊢─(41)─⊢(54)─────┤(194)	Clastic sediments and crystalline metamorphic rocks below granulite facies	8	5
6	(18)⊢────(65)↕(105)──────────↑(585)	Silicic volcanics and intrusives	23	21
5	(6.5)⊢─(36)↕(48)──────────┤(175)	Intermediate volcanics and intrusives	51	26
4	(2.4)⊢(34)↕(50)	Xenocryst bearing subalkaline basalts	9	3
3	(21)⊢─(66)↕(78)──────┤(143)	Alkaline and peralkaline basalts, gabbro sills, and kimberlite	34	30
2	(1.7)⊢(15)↕(26)────┤(94)	Subalkaline basalts	39	10
1	(1.0)⊢(18)↕(28)──┤(76)	Mafic inclusions in pipes	5	0

0   50   100   150   200

Group	^{238}U/^{204}Pb observed (μ)	General Description	No. of samples	No. with μ below 10
7	(10.3)⊢(20.0)↕(22.3)─┤(40.3)	Clastic sediments and crystalline metamorphic rocks below granulite facies	8	0
6	(3.0)⊢──(23.5)↕(24.9)──────↑(86.5)	Silicic volcanics and intrusives	43	7
5	(3.8)⊢(10)↕(13.8)──────┤(50.9)	Intermediate volcanics and intrusives	55	21
4	(7.1)⊢(9.0)↕(9.2)↕(16.8)	Xenocryst bearing subalkaline basalts	9	7
3	(7.0)⊢(14.5)↕(18.5)──┤(35.6)	Alkaline and peralkaline basalts, gabbro sills, and kimberlite	34	5
2	(6.0)⊢(8.9)──────┤(35.6)	Subalkaline basalts	44	31
1	(0.4)⊢(4.0)↕(6.1)┤(18.7)	Mafic inclusions in pipes	5	4

0   12   24   36   48   60

Fig. 82-B-10. Extremes, means (long vertical bar), and fiftieth percentiles (arrows) for various rock types for which lead isotope data are available. Numbers in parentheses give values for each indicator. (Compilation and figure are from Doe, 1970)

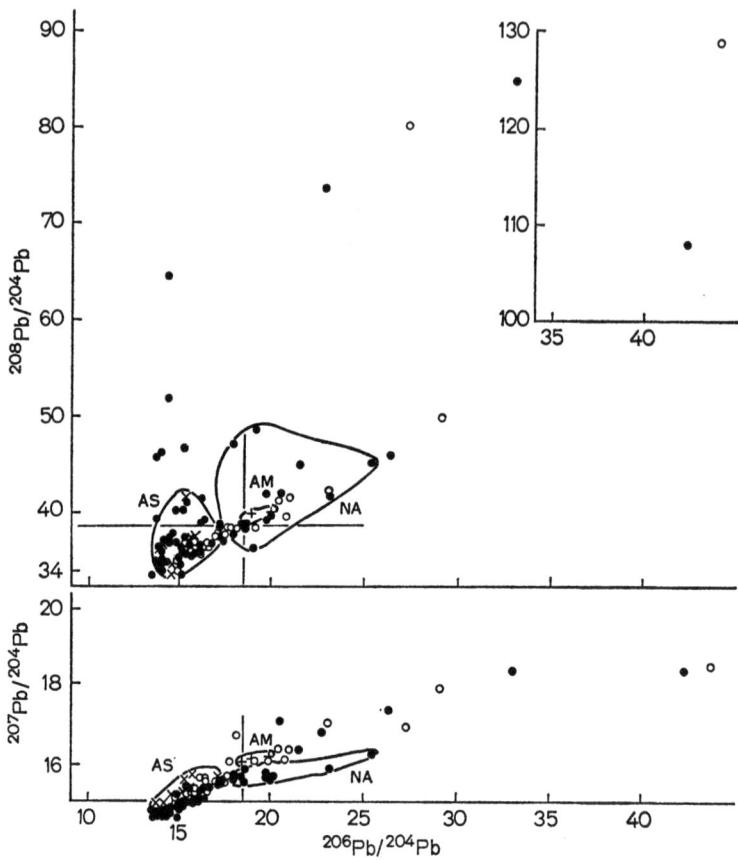

Fig. 82-B-11. $^{207}Pb/^{204}Pb$ and $^{208}Pb/^{204}Pb$ versus $^{206}Pb/^{204}Pb$ for whole-rock samples of Precambrian and Paleozoic crystalline rocks and composites: ○ Ukrainian Shield; ● Baltic Shield and Scotland; × Aldan Shield (AS); + Altai Mountains (AM); ● North America (NA). (Compilation is from DOE, 1970)

Fig. 82-B-12. $^{207}Pb/^{204}Pb$ and $^{208}Pb/^{204}Pb$ versus $^{206}Pb/^{204}Pb$ for sediments of Cenozoic age from various localities

Area or symbol	Description
A	Pelagic sediments of the Red Sea and basins of the Atlantic, Antarctic, and Indian Oceans except for the Gulf of Aqaba, HCl soluble lead
∧	Sediment from the Gulf of Aqaba, HCl soluble lead
P	Pelagic sediments from the Pacific Ocean basin, HCl soluble lead
ST-GC	Calcareous clastic sediments from the Salton Sea Trough and a manganese nodule from the Gulf of California:
○	HCl soluble lead
●	residue lead
B	Sediments from the Baltic Sea:
×	HCl soluble lead

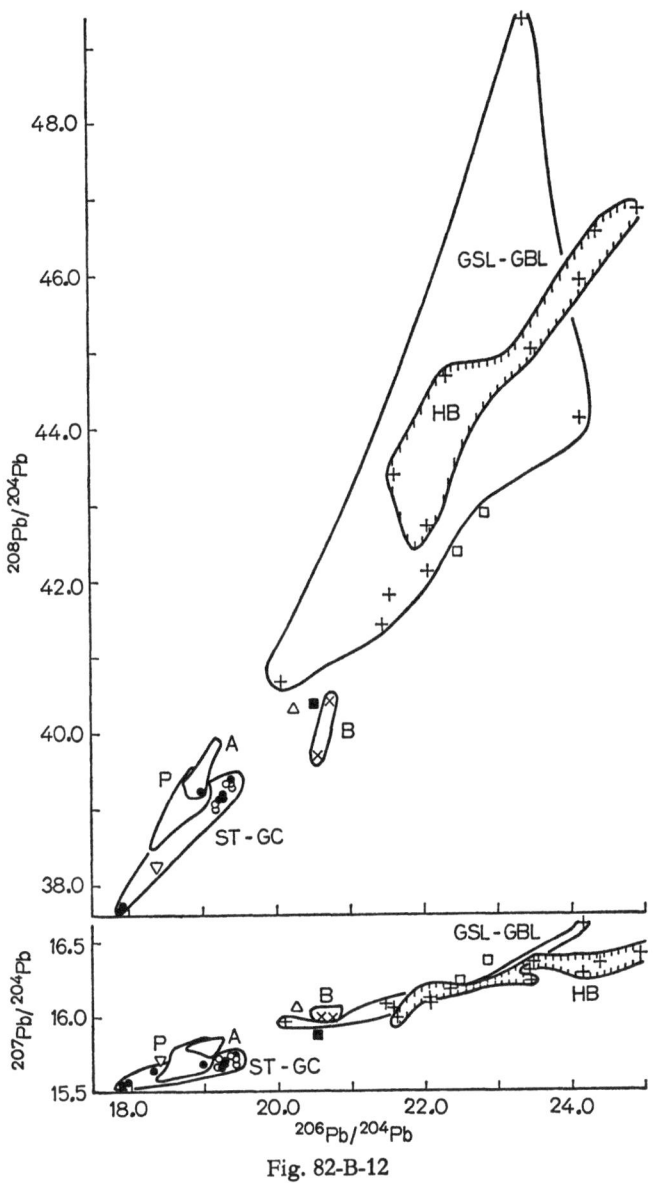

Fig. 82-B-12

□	Sediments from Lake Superior in the Canadian Shield: HCl and water soluble lead
■	residue lead
HB	Sediments from Hudson Bay in the Canadian Shield, HCl soluble lead
GSL-GBL	Sediments from Great Bear Lake and Great Slave Lake in the Canadian Shield: HCl soluble lead
×	
▽	Sediments from the Mediterranian Sea, HCl soluble lead

Such radiogenic rocks are not necessarily characteristic for large segments of the earth's crust as may be seen from data (Fig. 82-B-11) on the Precambrian crystalline rocks of the Baikalian block of the Aldan Shiled in Asia (SOBOTOVICH et al., 1965).

Data from the Baltic and Ukrainian Shield areas range widely, which makes averages difficult to determine. The leads in rocks of the Ukrainian Shield are also probably unradiogenic today. The Baltic Shield is more problematical although many of the rocks are unradiogenic.

A detailed study of high grade Precambrian metamorphic rock from the Lewisian basement of northwest Scotland has recently become available (MOORBATH et al., 1969). All present day values of $^{206}Pb/^{204}Pb$ are strongly nonradiogenic — in the range 13.5 to 16.5 — regardless of rock type. Rocks of granulite facies metamorphism about 2,600 m.y. ago and those in the transition zone to an area of amphibolite facies metamorphism about 1,400 to 1,600 m.y. ago, also have nonradiogenic values of $^{208}Pb/^{204}Pb$ — in the range 33.9 to 37.5. This range also includes many samples from the region of subsequent amphibolite facies metamorphism; however, in 40 percent of these samples, $^{208}Pb/^{204}Pb$ is greater than 40 and must be considered radiogenic if the ratios, which were determined on leads volatilized from the sample, are nearly the same as those for the whole rocks. Significant thorium but apparently little uranium may have been added during the amphibolite facies metamorphism to account for these isotopic relationships.

f) Cenozoic Sediments, Whole-rock and HCl-soluble Lead

Lead isotopic composition of the main sediment types is more a function of age of source areas for the sediment than of rock type because the source rocks contain significant quantities of U and Th relative to lead (Fig. 82-B-12). Young sediments in restricted basins draining Precambrian terranes, such as Hudson Bay, Great Slave Lake, Great Bear Lake, Lake Superior, and the Baltic Sea (CHOW, 1965; HART and TILTON, 1966), tend to be highly radiogenic and variable in lead isotope ratios.

On the other hand, sediments, well-mixed because they are far from their source, tend to be isotopically rather uniform even if the detritus is Precambrian, but even on these rocks the lead is still somewhat radiogenic (DOE et al., 1966).

Pelagic sediments from the ocean basins (CHOW and PATTERSON, 1962a; CHOW and TATSUMOTO, 1964; WAMPLER and KULP, 1964; CHOW, 1968) fall in restricted ranges of lead isotopic composition (Fig. 82-B-12), although the means from the Atlantic, Indian and Antarctic Oceans and the Red Sea are somewhat more radiogenic than those of the Pacific Ocean. CHOW and PATTERSON (1962a) have concluded that this relationship reflects a younger source terrane around the Pacific than the terranes that surround the other areas.

g) Cenozoic and Mesozoic Igneous Rocks

Ocean Basins. Data on abyssal basalts from oceanic ridges and rises (TATSUMOTO, 1966) and basalts from islands in "open ocean" structures such as Hawaii [see DOE (1970) for a compilation of references] are given in Table 82-B-4 and are shown in enclosed areas along with all Hawaiian data in Fig. 82-B-13. The isotope ratios

Table 82-B-4. *Isotopic composition of lead in abyssal basalts and in some island volcanics of the East Pacific Rise* (TATSUMOTO, 1966[a])

Description	$^{206}Pb/^{204}Pb$	$^{207}Pb/^{204}Pb$	$^{208}Pb/^{204}Pb$
Abyssal basalts			
Mid-Atlantic Ridge			
High alumina tholeiite (22° 40′ S; 13° 16′ W)[b]	18.47	15.54	38.01
High alumina tholeiite (5° 47′ S; 11° 25′ W)[b]	17.82	15.54	37.52
High alumina tholeiite (9° 39′ N; 40° 27′ W)[b]	18.82	15.68	38.65
East Pacific Rise			
High alumina tholeiite (7° 47′ S; 108° 10′ W)	18.19	15.54	37.93
High alumina tholeiite (12° 52′ S; 110° 57′ W)[c]	18.24	15.53	38.03
High alumina tholeiite (18° 25′ S; 113° 20′ W)	18.50	15.58	38.34
Island volcanics, East Pacific Rise			
Easter Island			
Obsidian, Mount Ourito	19.31	15.66	39.15
Andesine andesite	19.25	15.58	38.94
Alkali basalt	19.30	15.73	39.46
Tholeiite with alkalic affinities	19.28	15.67	39.16

[a] Lead extraction by volatilization through melting. Mass spectrometric analysis uses the surface emission mode of ionization of $PbS-NH_4NO_3$ from a rhenium filament with Faraday cup collection of ions. Data are not corrected for fractionation effects in the ionization process.
[b] Average of two analyses.
[c] Average of three analyses.

are clearly not uniform, although the average of all samples would approach a modern lead within the limits of error that such models may be calculated. The age of origin for the sources of abyssal basalts may be quite old, of the order of 1,000 m.y., and there may be little differentiation of uranium relative to lead, as TATSUMOTO has demonstrated.

Island volcanics from oceanic ridges and rises have lead isotope ratios [see DOE (1970) for a compilation of references, data on Fig. 82-B-13; selected data on Table 82-B-4] which are in general more radiogenic than those of abyssal basalts and Hawaiian volcanics (DOE, 1968).

This relationship is particularly evident on the plot of $^{208}Pb/^{204}Pb$ *versus* $^{206}Pb/^{204}Pb$ where the only overlap is with Iceland Group I volcanics of WELKE *et al.* (1968).

Island arc data may be rather similar to island volcanic data from oceanic ridges and rises, although $^{207}Pb/^{204}Pb$ and $^{208}Pb/^{204}Pb$ tend to be greater in the island arc data. Values of $^{206}Pb/^{204}Pb$ and especially $^{208}Pb/^{204}Pb$ are distinctly unradiogenic for volcanics from Japan relative to most data from oceanic ridges and rises. ARMSTRONG (1968) and TATSUMOTO and KNIGHT (1969) suggest that much of the isotopic characteristics for lead in volcanics from an island arc terrane is due to contamination of the mantle in the magma-generating zone with dragged down oceanic crust. The initial data of COOPER and RICHARDS (1969b) on New Zealand andesites is rather similar to those from eastern Japan.

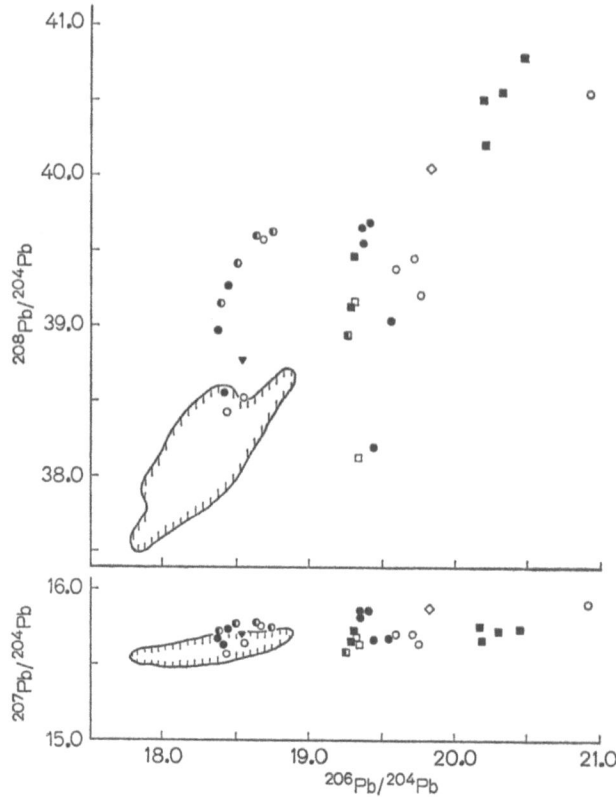

Fig. 82-B-13. ^{207}Pb/^{204}Pb and ^{208}Pb/^{204}Pb versus ^{206}Pb/^{204}Pb for volcanics from islands and a volcanogenic sediment from the East-Pacific Rise (■ basalt, ◨ andesite or trachyandesite, □ others), Mid-Atlantic Ridge (● basalt, ◐ andesite or trachyandesite, ○ others), Vema Seamount, S.E. Atlantic (◇ phonolite), and Reunion Islands, Indian Ocean (▼ olivine basalt). For comparison, the enclosed areas are superimposed which include all the isotopic data on abyssal basalts and volcanics of the Hawaiian Islands. (Compilation is from DOE, 1970)

Continental igneous rocks. Various kinds of igneous rocks from the continents (Fig. 82-B-14) range widely in lead isotope ratios [for a compilation of references see DOE (1970)]. Data for California, Oregon, Washington, Alaska and Nevada rocks appear similar to those for oceanic ridges and rises (DOE, 1968) and are compatible with the recent conclusion, made on the basis of other evidence, that the East Pacific Rise extends under this region (YEATS, 1968). Selected data from Oregon and Colorado are given in Table 82-B-5. In North America, the unradiogenic leads are concentrated in the Rocky Mountain region and farther east with Utah an isotopic transition zone between Nevada and the Rocky Mountain region (Fig. 82-B-14). Basalts from the Rocky Mountain region are rather similar in lead isotopic composition to silicic igneous rocks, even if obviously contaminated basalts are eliminated (Table 82-B-5). These data suggest that both rock types come from a similar age source with similar values of U/Pb and Th/Pb (DOE, 1968). Contrary

Table 82-B-5. *Selected data on continental igneous rocks from North America.*
(Abbreviations: prim. = primitive; cont. = contaminated)

Sample locality	Rock type	$^{206}Pb/^{204}Pb$	$^{207}Pb/^{204}Pb$	$^{208}Pb/^{205}Pb$	Reference
		Oregon			
Northwest coast	Tholeiitic basalt[a]	18.88	15.76	39.25	2
Marys Peak gabbroic sill	Aplite dike	19.11	15.73	39.30	2
	Pegmatite	19.25	15.75	39.26	2
	Granophyric diorite	19.22	15.74	39.30	2
	Basalt, chilled border	19.31	15.75	39.34	2
Widow Creek	Alkalic basalt[a]	19.34	15.70	39.32	2
Coffin Butte	Tholeiitic basalt	18.98	15.61	38.74	2
		San Juan Mountains, Colorado			
Creede area	Silicic volcanic	18.74	15.63	38.09	1
Wagon Wheel Gap area	Silicic volcanic	18.36	15.58	37.68	1
La Jara Reservoir area	Hinsdale Formation				
	Basalt (prim.)	18.33	15.55	37.67	3
	Basalts (cont.)[a]	17.94	15.54	37.48	3
	Basalt (cont.)	18.38	15.62	38.26	1

References: 1. DOE (1967); after dissolution by HF—HClO$_4$ (and volatilization of lead in the case of the basalts) mass spectrometric analysis was made from PbS—NH$_4$NO$_3$ by the surface emission mode of ionization from rhenium filament with Faraday cup collection of ions. Data are normalized to agree with Ta-1 filament material to bring ratios within stated analytical uncertainties of Tilton's 1:1 lead isotope gravimetric standard. 2. TATSUMOTO and SNAVELY (1969); lead extraction by volatilization below the melting point with other analytical factors as given in Table 32-B-4. No normalization of data has been made. 3. DOE et al. (1969); basalts were treated by volatilization extraction with one checked by HF—HClO^4 dissolution. All data are normalized to agree with Ta-1 filament material with is within the error limits indicated by the Tilton 1:1 lead isotope gravimetric standard.

[a] Average of 2 samples.

to the isotopic relations in the Rocky Mountain region, in Scotland the mafic igneous rocks statistically appear to be more radiogenic than the silicic igneous rocks of Cenozoic age (MOORBATH and WELKE (1968). The silicic igneous rocks, which are completely overlapped in their lead isotope ratios by those from whole-rock samples from the nearby Lewisian basement (MOORBATH et al., 1969), might be interpreted as being derived from the Lewisisan basement with the mafic igneous rocks coming from the mantle and contaminated to varying degrees with Lewisian basement.

Ultramafic rocks. Only the data for the kimberlite from South Africa (LOVERING and TATSUMOTO, 1968) and peridotitic portions of a layered mafic intrusion in Scotland (MOORBATH and WELKE, 1968) can be considered reliable whole-rock values at this time. In view of the great importance of data on these kinds of rocks and the difficult analytical problems, only those data considered reliable whole-rock or mineral values are included in Table 82-B-6.

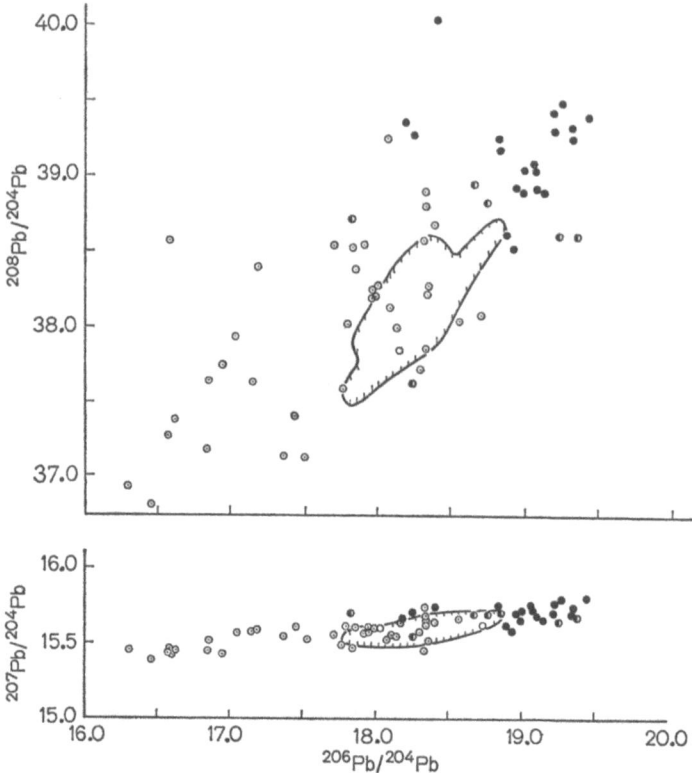

Fig. 82-B-14. $^{207}Pb/^{204}Pb$ and $^{208}Pb/^{204}Pb$ versus $^{206}Pb/^{204}Pb$ for Cenozoic and Mesozoic igneous rocks of western North America: ● California, Oregon, Washington, British Columbia, Alaska, and Nevada; ◐ Utah; ○ New Mexico, Colorado, Wyoming, Idaho Montana, South Dakota, and Texas. Enclosed areas superimposed for comparison include all the isotope data on abyssal basalts and volcanics of the Hawaiian Islands. (Compilation is from DOE, 1970)

Table 82-B-6- *Isotopic composition of lead in ultrabasic and related rocks (no normalization to data)*

Sample type	$^{206}Pb/^{204}Pb$	$^{207}Pb/^{204}Pb$	$^{208}Pb/^{204}Pb$	Reference
Kimberley pipe, South Africa				
Kimberlite (initial ratios)[a]	19.51	15.82	39.32	1
Kimberlite (present-day ratios)	19.77	15.83	39.66	
Cuillins layered intrusion, Scotland				
Gabbro, Zone I (leucocractic band)	17.84	15.69	37.97	2
Gabbro, Zone I (melanocractic band)	17.77	15.67	37.84	2
Gabbro, Zone IV	17.60	15.55	38.00	2
Peridotite, main body	18.13	15.66	38.44	2
Sligachan area, Scotland				
Peridotite dike in basalt	17.86	15.58	38.02	2

References: 1. LOVERING and TATSUMOTO (1968); 2. MOORBATH and WELKE (1968).
[a] Corrected for 174×10^6 yrs. of decay of uranium and thorium.

Lead isotope data are still valuable even when the lead is only partially extracted and its lead isotopic composition not equal to that of the whole-rock. COOPER and GREEN (1969) have volatilized lead from lherzolite inclusions and host alkali basalt from western Victoria, Australia. Although only a preliminary report is available, MANTON and TATSUMOTO (1969) find somewhat similar results to those of COOPER and GREEN between eclogite inclusions and host kimberlite from the Roberts Victor Mine in South Africa. Both the lherzolites from Australia and the eclogites from South Africa are calculated to come from sources 2,000 to 2,500 m.y. old.

h) Paleozoic and Precambrian Igneous Rocks

Igneous rocks older than Mesozoic are somewhat difficult to work on as the corrections for *in situ* radioactive decay of U and Th are large and the values of U/Pb and Th/Pb often have undergone a recent disturbance. The more successful approach is to analyze the lead in a phase with low values of U/Pb and Th/Pb such as K-feldspars. Such data for rocks about, 1,000 m.y. old are given in Fig. 82-B-15

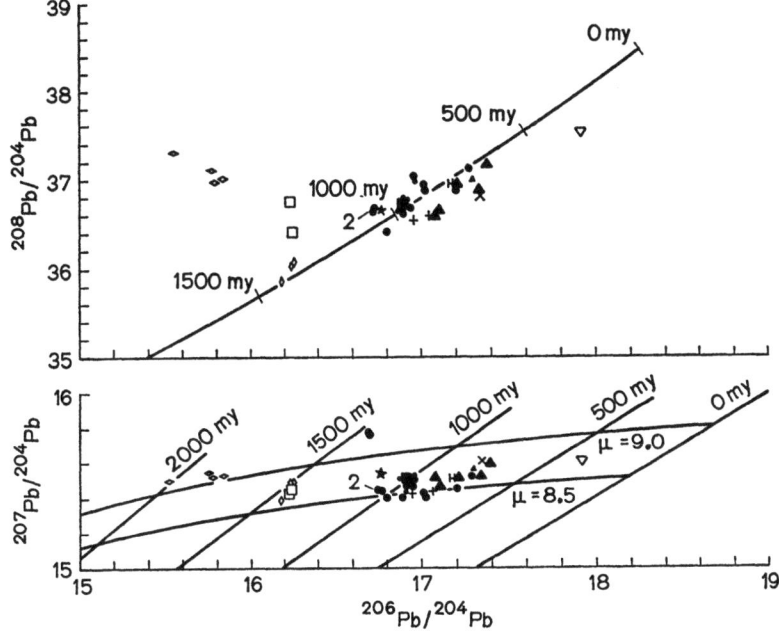

Fig. 82-B-15. Lead isotope relationships for feldspar leads from 1,000-m.y.-old igneous rocks. Curved lines on the plot of $^{207}Pb/^{204}Pb$ versus $^{206}Pb/^{204}Pb$ are the single stage growth curves of $^{238}U/^{204}Pb$ for values of 9.0 and 8.5, and the heavy sloping lines are the isochrons for the labeled ages. The normal growth curve involving $^{208}Pb/^{204}Pb$ and $^{206}Pb/^{204}Pb$ is given in that plot with positions of model leads indicated for various ages. The plot is from ZARTMAN and WASSERBURG (1969) and includes their data from: ● Texas suite, ▼ Western Adirondack Mountains, V.Y., ▽ Eastern Adirondack Mountains, N.Y., + Westport. Ontario, * Leg Pond, Newfoundland, □ ■ Port Cartier-Mt. Reed, Quebec, H Herefoss, Norway, · Pikes Peak , Colorado, × Duluth, Minnesota, ◇ Mellen.Wisconsin, ◇ Gold Butte, Nevada; as well as the following data: ● Llano,Texas (ZARTMAN, 1965b), ▲ Shenandoah National Park, Virginia (DOE et al., 1965), ▲ Balmat, New York (DOE, 1962), · Pikes Peak, Colorado (DOE, 1967)

Table 82-B-7. *Selected feldspar lead data on Paleozoic and Precambrian rocks (corrections for decay of U and Th are less than 2 percent)*

Sample locality	$^{206}Pb/^{204}Pb$	$^{207}Pb/^{204}Pb$	$^{208}Pb/^{204}Pb$	Ages (m.y.) isochron model	$^{208}Pb/^{204}Pb$	accepted	Reference
Some K-feldspars in which model lead ages are in good agreement with accepted ages							
Pikes Peak batholith, Colorado	16.91	15.49	36.71	980	980	1,050	1
Commercial pegmatite, Wyoming	14.62	15.27	34.24	2,530	2,410	2,450	2
Microcline granite, Kola Peninsula, Baltic Shield	13.65	14.59	33.50	2,750	2,660	2,800	3
Some K-feldspars in which model lead ages are older than accepted ages							
Pegmatite, Pitkyaranta, Baltic Shield	14.53	15.09	34.43	2,450	2,220	1,900	3
Rapakivi granite, Nevada[a]	16.21	15.44	35.99	1,450	1,350	1,100	4
Granite, Wisconsin[b]	15.73	15.51	37.08	1,870	770	1,100	4
Some K-feldspars in which model lead ages are younger than accepted ages							
Pegmatite in Baltimore Gneiss, Maryland	18.60	15.75	38.65	−80	20	450	5
	18.50[c]	15.62[c]	38.17[c]				
Charnockite, Sabarov, Ukraine	15.97	15.27	36.05	1,450	1,260	2,200	3
Feldspars with at least one ratio reasonably unaffected by metamorphism							
Granite gneiss, Pitkyaranta, Baltic Shield	14.35	15.13	33.51	2,600	2,320	2,600 (1,800)[d]	3
Baltimore gneiss, Maryland	17.42	15.52	40.01	610	−890	1,100 (300—450)[d]	5
	18.10	15.53	37.09	70	770		

References: 1. Doe (1967); 2. Catanzaro and Gast (1960); 3. Tugarinov *et al.* (1963a); 4. Zartman and Wasserburg (1969); 5. Doe *et al.* (1965).

[a] Average of 3 samples. [b] Average of 4 samples. [c] Muscovite.
[d] Age in ( ) is the age of metamorphism. The other age is the accepted age of the rock.

(ZARTMAN and WASSERBURG, 1969). Selected data are given in Table 82-B-7. Radiogenic leads appear to be the rule even of most granitic rocks found in terranes that are much older than the granitic rock. Leads unradiogenic for their age as in Mesozoic and Cenozoic igneous rocks have been much more difficult to find. Existence of older equivalents of the Rocky Mountain type seems likely. The ones that have been found are small, shallow intrusions (ZARTMAN and WASSERBURG, 1969) and in a pegmatite from the Baltic Shield at Pitkyaranta in Southwest Karelia (TUGARINOV et al., 1963). Though many old granite whole-rock samples of the Aldan Shild are unradiogenic in their present-day ratios, TUGARINOV et al. (1963) give no data that demonstrates that any are very unradiogenic in their initial ratios. Apparently, most older granitic rocks form in a terrane somewhat similar to that of the present Pacific coast states of the United States or oceanic ridges or rises.

i) Ore Genesis

Magmatogenic Ores. Ores in which the isotopic composition of ore lead is the same as that in the presumably related igneous source at the time of crystallization are good candidates for a magmatogenic origin. These ores are usually isotopically uniform with $^{206}Pb/^{204}Pb < 19.5$ and $^{208}Pb/^{204}Pb < 39$. Examples are the ores at Butte, Montana (DOE et al., 1968), Bingham Canyon, Utah (STACEY et al., 1969), and the Nelson batholith, British Columbia (REYNOLDS and SINCLAIR, 1971). It may be, however, that in cases such as the Boulder batholith the lead was derived from the igneous rock during the wall rock alteration rather than being a differentiate of magma at depth. Both processes would give the same lead isotopic composition.

Indirect Magmatogenic Ores. Many ores thought to be derived in the igneous process have lead isotopic compositions different from the likely igneous sources of the region or of even larger areas. The lead is usually more radiogenic than the igneous rocks such as the ores of the Milford region, Utah, where $^{206}Pb/^{204}Pb$ varies between 18.6 and 20.2 for the ores, but the ratio for the igneous rock is 18.4 to 18.8 (STACEY et al., 1969), although such deposits around the Nelson batholith are less radiogenic (REYNOLDS and SINCLAIR, 1971). The isotopic data demand that a large component of the ore lead is derived from sources outside the magma and is lead probably generated by radioactive decay of U and Th from rocks of a much older terrane. The facts that the igneous rock has usually the least radiogenic lead and that the ores form a continuum of more radiogenic values suggest that the igneous rock supplied lead as well as heat during ore-formation and that there was a magmatic component of lead in the ore.

Deposits Formed by Metamorphism of Much Older Rocks. Lead isotope data clearly support a metamorphic origin for some deposits of this kind, although well-documented studies are few for this little-emphasized category. Greater variability in isotope ratios may be encountered. The ratios may range from those expected for the time of formation of the source rocks to those much more radiogenic than expected for the age of metamorphism. A Pb—Pb age may be calculated to determine the age of the source terrane if the metamorphic age is known or *vice versa*; the results may be confirmed by other dating methods. The Thackaringa deposits (RUSSELL et al., 1961) in Australia are the best example of this type.

Lateral Secretion Through Much Older Rocks. Many important stratiform ores, such as those of the Mississippi Valley type, do not have any close association with either

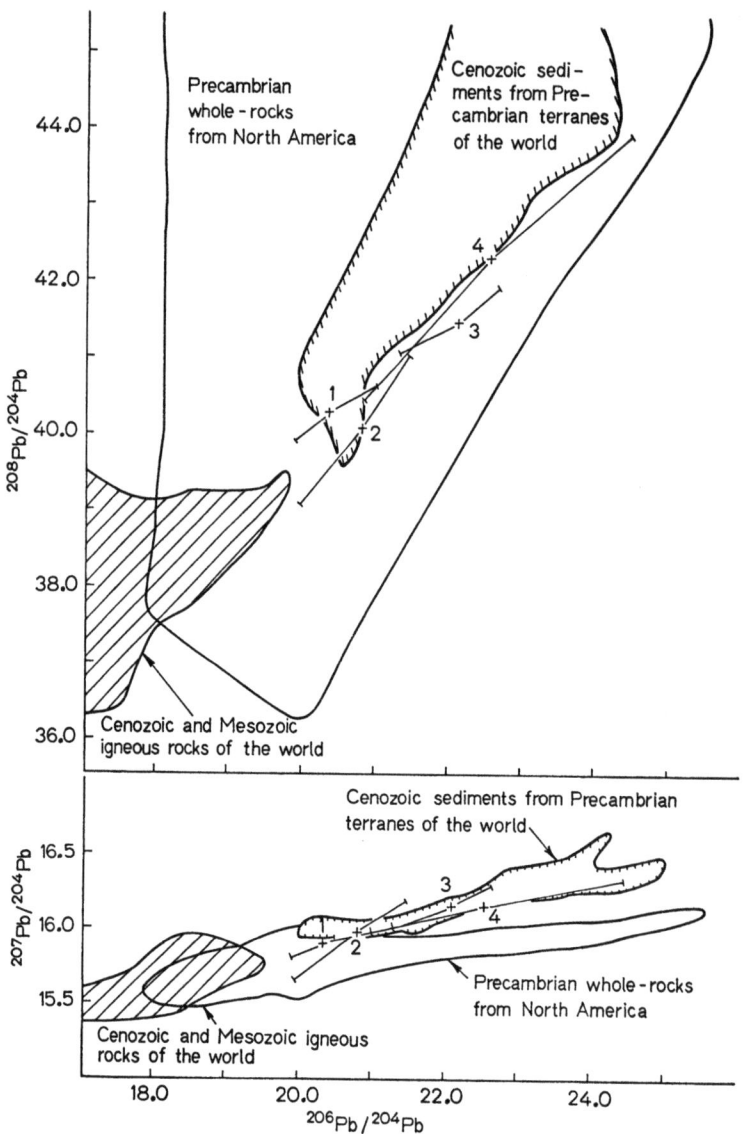

Fig. 82-B-16. $^{207}Pb/^{204}Pb$ and $^{208}Pb/^{204}Pb$ versus $^{206}Pb/^{204}Pb$ for extremes and means of four Mississippi Valley districts: **1** Illinois-Kentucky fluorspar district (HEYL et al., 1966); **2** Southeast Missouri lead belt (BROWN, 1967); **3** Tri-State zinc district (RUSSEL and FARQUHAR, 1960); **4** Wisconsin-Illinois-Iowa lead-zinc district (HEYL et al., 1966). For comparison, enclosed areas include the kinds of data as labeled. (Figure is from DOE, 1970)

volcanism or metamorphism. (Some lead isotope data are shown on Fig. 82-B-16.) These ores are characterized by isotopic heterogeneity, although apparently not to the extent of the indirect magmatogenic ores. The lead isotope data nicely fit a lateral secretion mechanism for these deposits, perhaps with the source of the lead being the basement rocks, as suggested by KANASEWICH (1962) and HEYL et al. (1966), because of the similarity of the Pb—Pb age to that accepted for the basement. These ores characteristically are highly radiogenic with isotope ratios greater than those initially present in any igneous rock of any kind, any where in the world.

Deposits of Uncertain Origin. Most other deposits carry lead isotopic compositions of a more equivocal nature. An important isotopic grouping are those ores listed in Table 82-B-3 which, in general, are a large grouping of isotopically-uniform, conformable or stratiform ores including some of the major mineral deposits of the world. The nearly "model" characteristics of the lead isotope ratios in these deposits in fact adds to the confusion as to genesis more than helping to solve this problem. The one possible source that isotopically is a good candidate is pelagic sediments, as was first noted by BROWN (1965). ARMSTRONG (1968) feels the mechanism of getting the lead from such sediments is through downwarping of oceanic crust along the Benioff zone off continental margins and island arcs. The original interpretation of STANTON and RUSSELL (1959) used a volcanic-sedimentary origin in Island arcs. In contrast, KANASEWICH (1968b) has proposed that many of these ores are related to zones of pull-apart and upwelling of fresh mantle such as, for examples, the Red Sea and California's Salton Sea basin. Heavy metal-rich chloride brines are found in these localities. OSTIC et al. (1967), while allowing for a mantle origin, interprets some as being derived by metamorphism of sediments soon after they are deposited (such as Bathurst, New Brunswick). WEDEPOHL (1964) has made a convincing case of syngenetic derivation from a sandstone shoreline for the Kupferschiefer Marl Slate. Many derivations may be involved, but the common tie of the noted isotopic characteristics with most conformable or stratiform deposits remains an interesting puzzle.

j) Isotopic Composition of Lead in Natural Waters

Oceans. The isotopic composition of lead has not yet been measured directly in ocean water because of the low lead content; however, the isotopic composition of HCl-soluble lead in manganese nodules is commonly accepted to represent that of lead in the oceans (Table 82-B-8). The lead isotopic compositions of manganese nodules are also not much different from those in the pelagic sediments.

Brines. The chloride-rich brines that carry great concentrations of heavy metals are a relatively new discovery and may be the ore fluid for some of the categories mentioned above. Several vein and stratiform ores near the Red Sea could well be related to the Red Sea geothermal brines for example. COOPER and RICHARDS (1969a) suggest that the variation in lead isotopic composition of acid-soluble lead in the deeper sediments may be due to contamination of the lead from the slightly acidic brine with lead from calcareous pelagic sediments. The new data available so far (Table 82-B-9) show the lead in the brines to be somewhat radiogenic (DOE et al., 1966; DELEVAUX et al., 1967).

IV. Radioactive Lead Isotopes

There are four radioactive isotopes of lead: ^{210}Pb and ^{214}Pb in the ^{238}U decay chain, ^{211}Pb in the ^{235}U decay chain, and ^{212}Pb in the ^{232}Th decay chain. All except ^{210}Pb have half-lives of less than 12 hours. Such short half-lives are of limited use-

Table 82-B-8. *Isotopic composition of lead in natural waters and in the atmosphere*

Locality (reference)	Material	No. samples	$^{206}Pb/^{204}Pb$	$^{207}Pb/^{204}Pb$	$^{208}Pb/^{204}Pb$
Arctic Ocean water (1)	Jurassic Mn nodule	1	18.78	15.84	39.33
Atlantic Ocean water (1)					
N.E. Atlantic, area F in (1)	Cenozoic Mn nodules	3	19.06	15.82	39.58
N.W. Atlantic, area G in (1)	Cenozoic Mn nodules	7	19.13	15.82	39.66
Pacific Ocean water					
West Pacific (1), area A in (1)	Cenozoic Mn nodules	2	18.55	15.67	39.00
West Pacific (2), area A in (1)	Cenozoic Mn nodules	4	18.51	15.62	38.80
North Pacific (1), area B in (1)	Cenozoic Mn nodules	8	18.81	15.79	39.28
East Pacific (1), area C in (1)	Cenozoic Mn nodules	10	18.85	15.78	39.29
S.E. Pacific (1), area D in (1)	Cenozoic Mn nodules	4	18.99	15.88	39.47
Gulf of California (1)	Cenozoic Mn nodules	1	19.20	15.73	38.98

References: 1. CHOW and PATTERSON (1962a); 2. CHOW and TATSUMOTO (1964).

Tabke 82-B-9. *Isotopic composition of lead in brines, and precipitate from brines*

| Region | Lead in | No. of samples | Isotopic ratios [a] | | | Reference |
			$^{206}Pb/^{204}Pb$	$^{207}Pb/^{204}Pb$	$^{208}Pb/^{204}Pb$	
Salton Sea Trough geothermal area, California	Chloride brine	4 (avg.)	19.13	15.67	39.07	2
Red Sea, Africa (Atlantic II deep)	Chloride brine	1	18.72	15.68	38.52	1
Red Sea, Africa	Surface precip.	1	18.68	15.75	38.82	3
	Deeper precip. from hot spots	12 (avg.)	18.92	15.77	39.08	3
Red Sea, Africa	Pelagic sediments	9 (avg.)	18.95	15.79	39.32	4

References: 1. DELEVAUX et al. (1967); 2. DOE et al. (1966); 3. COOPER and RICHARDS (1969); 4. CHOW (1968).

[a] All samples were analyzed by surface emission mass spectrometry. All precipitate and sediment data are on acid-soluble lead only and are not whole-rock values.

fulness in geology; therefore, ^{210}Pb with a half-life of 22.0 ± 0.5 yr. (HOUTERMANS et al., 1964; RAMTHUN, 1964; GUTEN et al., 1967) has received the major interest. The analytical techniques for ^{210}Pb are given by GOLDBERG (1963).

The ^{210}Pb has been found useful in studies of volcanic rocks where it is often found in disequilibrium with ^{238}U in Recent volcanics (SOMAYAJULU et al., 1966; OVERSBY and GAST, 1968) and in fumarolic matter (SOMAYAJULU et al., 1966; HOUTERMANS et al., 1964). These studies show promise of dating Recent volcanism within the last 100 yrs. or so, where it may occur free of precursors.

The ^{210}Pb isotope is also useful as an atmospheric tracer [JAWOROWSKI, 1966; for a compilation of references see DOE (1970)]. It is of particular interest in the dating of snow (CROZAZ and LANGWAY, 1966; GOLDBERG, 1963) and other precipitation as again the ^{210}Pb occurs free of precursors. Some studies have used ^{210}Pb/^{206}Pb in dating of uranium minerals (KULP et al., 1953).

GOLDBERG (1963) has shown that the ^{210}Pb content shows a rapid depletion between the source regions of rivers (2 to 4 dpm/l) and downstream locations (≤ 0.1 dpm/l). The reason for the rapid depletion is not known. Major rivers in regions of major enrichments of uranium have high ^{210}Pb contents (14.7 dpm/l for Colorado River water at Grand Junction, Colorado). GOLDBERG (1963) has shown that the turnover rate of ^{210}Pb in the oceans is very rapid, and he attributes this to biological causes.

Revised manuscript received: July 1970

References: Section 82-B

AKISHIN, P. A., NIKITIN, O. T., PANCHENKOV, G. M.: A new effective ion emittor for the isotopic lead analysis. Geokhimiya No. 5, 429 (1967).

ALDRICH, L. T., WETHERILL, G. W.: Geochronology by radioactive decay. In: Annual Review of Nuclear Science (Ed. E. G. SEGRÈ). Natl. Research Council, Natl. Acad. Sci., Stanford, Calif., Ann. Revs. **8**, 257 (1958).

ARMSTRONG, R. L.: A model for the evolution of strontium and lead isotopes in a dynamic earth. Rev. Geophys. **6**, 175 (1968).

BHANOT, V. B., JOHNSON, W. H., JR., NIER, A. O.: Atomic masses in the heavy mass region. Phys. Rev. **120**, 235 (1960).

BROWN, J. S.: Ore leads and isotopes. Econ. Geol. **57**, 673 (1962).

— Oceanic lead isotopes and ore genesis. Econ. Geol. **60**, 47 (1965).

— Isotopic zoning of lead and sulfur in southeast Missouri. In: Genesis of Stratiform Lead-zinc-fluorite Deposits (Mississippi Valley Type Deposits). A symposium, New York, 1966. Econ. Geol. Mon. **3**, 410 (1967).

BURGER, A. J., NICOLAYSEN, L. O., AHRENS, L. H.: Controlled leaching of monazites. J. Geophys. Res. **72**, 3585 (1967).

CAMERON, A. E., SMITH, D. H., WALKER, R. L.: Mass spectrometry of nanogram-size samples of lead. Anal. Chem. **41**, 525 (1969).

CATANZARO, E. J.: Triple-filament method for solid-sample lead isotope analysis. J. Geophys. Res. **72**, 1325 (1967).

— GAST, P. W.: Isotopic composition of lead in pegmatitic feldspars. Geochim. Cosmochim. Acta **19**, 113 (1960).

— KULP, J. L.: Discordant zircons from the Little Belt (Montana), Bear tooth (Montana) and Santa Catalina (Arizona) Mountains. Geochim. Cosmochim. Acta **28**, 87 (1964).

— MURPHY, T. J., SHIELDS, W. H., GARNER, E. L.: Absolute isotopic abundance ratios of common, equal-atom and radiogenic lead isotope standards. J. Res. Natl. Bur. St, A. Phys. and Chem. **72**, 261 (1968).

CHOW, T. J.: Radiogenic leads of the Canadian and Baltic Shield regions. In: Symposium on Marine Geochemistry, 1964. Rhode Island Univ. Narragansett Marine Lab. Occasional Publ. **3**, 169 (1965).

— Lead isotopes of the Red Sea region. Earth Planet. Sci. Lett. **5**, 143 (1968).

— PATTERSON, C. C.: On the primordial lead of the Canyon Diablo meteorite. Geokhimiya, **12**, 1124 (1961).

— — The occurrence and significance of lead isotopes in pelagic sediments. Geochim. Cosmochim. Acta **26**, 263 (1962a).

— — Correction to "Lead isotopes in manganese nodules". Geochim. Cosmochim. Acta **26**, 973 (1962b).

— Tatsumoto, M.: Isotopic composition of lead in the sediments near Japan Trench. In: Recent Researches in the Fields of Hydrosphere, Atmosphere, and Nuclear Geochemistry, Chapt. 10. Tokyo: Maruzen Co. Ltd. 1964.

COOPER, J. A., GREEN, D. H.: Lead isotope measurements on lherzolite inclusions and host basanites from western Victoria, Australia. Earth Planet. Sci. Lett. **6**, 69 (1969).

— RICHARDS, J. R.: Lead isotope measurements on sediments from Atlantis II and Discovery deep areas. In: Hot Brines and Recent Heavy Metal Deposits in the Red Sea, 600 p. Berlin-Heidelberg-New York: Springer 1969a.

— — Lead isotope measurements on volcanics and associated galenas from Coromandel-TeAroha region, New Zealand. Geochem. Jour. (Japan) **3**, 1 (1969b).

Lead

CROZAZ, G., LANGWAY, J. R.: Dating Greenland firn-ice cores with Pb-210. Earth Planet. Sci. Lett. **1**, 194 (1966).

DAVIS, G. L., HART, S. R., TILTON, G. R.: Some effects of contact metamorphism on zircon ages. Earth Planet. Sci. Lett. **5**, 27 (1968).

DELEVAUX, M. H., DOE, B. R., BROWN, G. F.: Preliminary lead isotope investigations of brine from the Red Sea, galena from the Kingdom of Saudi Arabia, and galena from the United Arab Republic (Egypt). Earth Planet. Sci. Lett. **3**, 139 (1967).

DOE, B. R.: Relationships of lead isotopes among granites, pegmatites and sulfide ores near Balmat, New York. J. Geophys. Res. **67**, 2895 (1962).

— The bearing of lead isotopes on the source of granitic magma. J. Petrol. **8**, 51 (1967).

— Lead Isotopes. 37 p. Berlin-Heidelberg-New York: Springer 1970.

— Lead and strontium isotope studies of Cenozoic volcanic rocks in the Rocky Mountain region — a summary. Quart. Colo. School Mines **63**, 149 (1968).

— HEDGE, C. E., WHITE, D. E.: Preliminary investigation of the source of lead and strontium in deep geothermal brines underlaying the Salton Sea geothermal area. Econ. Geol. **61**, 462 (1966).

— LIPMAN, P. W., HEDGE, C. E., KURASAWA, H.: Primitive and contaminated basalts from the Southern Rocky Mountains, U.S.A. Contr. Mineral. and Petrol. **21**, 142 (1969).

— TATSUMOTO, M., DELEVAUX, M. H., PETERMAN, Z. E.: Isotope dilution determination of five elements in G-2 (granite), with a discussion of the analysis of lead. U.S. Geol. Surv. Profess. Papers **575B**, B 170 (1967)

— TILLING, R. J., HEDGE, C. E., KLEPPER, M. R.: Lead and strontium isotope studies of the Boulder batholith, southwestern Montana. Econ. Geol. **63**, 884 (1968).

— TILTON, G. R., HOPSON, C. A.: Lead isotopes in feldspars from selected granitic rocks associated with regional metamorphism. J. Geophys. Res. **70**, 1947 (1965).

GERLING, E. K.: Age of the earth according to radioactivity data. Compt. Rend. Acad. Sci. U.R.S.S. **34**, 259 (1942).

FLEMING, E. H., JR., GHIORSO, A., CUNNINGHAM, B. B.: The specific alpha-activities and half-lives of U^{234}, U^{235}, and U^{236}. Phys. Rev. **88**, 642 (1952).

GOLDBERG, E. D.: Geochronology with lead-210. In: Symposium on Radioactive Dating; Athens 1962. Proc. Vienna Internal. Atomic Energy Agency p. 121 (1963).

GRÜNENFELDER, M. H.: Heterogeneity of accessory zircons and the petrographic significance of their uranium decay age. Pt. 1: Zircons of granodiorite gneisses of Acquacalda (Lukmanier Pass). [In German.] Schweiz. Mineral. Petrog. Mitt. **43**, 235 (1963).

— EBERHARD, E.: Phase unmixing in zircons: A possible cause for discordant U/Pb age patterns. In: Geochronology. Switzerland 1969 Meeting, Zürich-Bern, Aug. 29-Sept. (1969).

GUTEN, H. R., WYTTENBACK, A., DULAKAS, H.: Half-life of ^{210}Pb (RaD). J. Inorg. and Nucl. Chem. **29**, 2826 (1967).

HART, S. R., KROGH, T. E., DAVIS, G. L., ALDRICH, L. T., MUNIZAGA, F.: Rb/Sr geochronology of granitic rocks southeast of Sudbury, Ontario. Carnegie Inst. Wash. Year Book 1965—1966, **65**, 59 (1967).

— TILTON, G. R.: The isotope geochemistry of strontium and lead in Lake Superior sediments and water (p. 127—137). In: The Earth Beneath the Continents. Am. Geophys. Union Geophys. Mon., Ser. 10, 127 (1966).

HEIMLICH, R. A., BANKS, P. O.: Radiometric age determinations, Bighorn Mountains, Wyoming. Am. J. Sci. **266**, 180 (1968).

HEYL, A. V., DELEVAUX, M. H., ZARTMAN, R. E., BROCK, M. R.: Isotopic study of galenas from the upper Mississippi Valley, the Illinois Kentucky, and some Appalachian Valley mineral districts. Econ. Geol. **61**, 933 (1966).

HOLMES, A.: An estimate of the age of the earth. Nature **157**, 680 (1946).

HOUTERMANS, F. G.: The isotope frequency in natural lead and the age of uranium. Naturwissenschaften **33**, 185 (1946).

— EBERHARDT, A., FERRARA, G.: Lead of volcanic origin. Chap. 18 (p. 233—243). In: Isotopic and Cosmic Chemistry, Chapt. 18. Amsterdam: North-Holland Pub. Co. 1964.

INGRAM, M. G.: Manhattan Project, Tech. Ser., National Nuclear Energy Ser., Div. II, vol. 14, Chapt. 5, p. 35. New York: McGraw-Hill Book Co. 1946.

JAWOROWSKI, Z.: Temporal and geographical distribution of radium D (lead-210). Nature 212, 886 (1966).
KANASEWICH, E. R.: Approximate age of tectonic activity using anomalous lead isotopes. Roy. Astron. Soc. Geophysics J. 7, 158 (1962).
— The interpretation of lead isotopes and their geological significance. In: Radiometric Dating for Geologists (E. J. HAMILTON and R. M. FARQUHAR, eds.), p. 137. New York: Interscience Publ. 1968 (a).
— Precambrain rift; genesis of strata-bound ore deposits. Science 161, 1002 (1968b).
KOLLAR, F., RUSSELL, R. D., ULRYCH, T. J.: Precision intercomparisons of lead isotope ratios, Broken Hill and Mount Isa. Nature 187, 754 (1960).
KOVARIK, A. F., ADAMS, N. N., JR.: The disintegration constant of thorium and the branching ratio of thorium C. Phys. Rev. 54, 413 (1938).
— — Redetermination of the disintegration constant of U^{238}. Phys. Rev. 98, 46 (1955).
KULP, J. L., BROCKER, W. S., ECKELMAN, W. R.: Age determination of uranium minerals by the Pb^{210} method. Nucleonics 11, 19 (1953).
LONG, A., SILVERMAN, A. J., KULP, J. L.: Isotopic composition of lead and Precambrian mineralization of the Cœur d'Alene district, Idaho. Econ. Geol. 55, 645 (1960).
LOVERING, J. F., TATSUMOTO, M.: Lead isotopes and the origin of granulite and eclogite inclusions in deep-seated pipes. Earth Planet. Sci. Lett. 4, 350 (1968).
MANTON, W. J., TATSUMOTO, M.: Isotopic composition of lead and strontium in eclogites and related rocks: Annual Report Geosci. Div., 1967—1968, Southwest Center Adv. Studies, p. 19—20, 22 (1969).
MOORBATH, S., WELKE, H.: Lead isotope studies on igneous rocks from the Isle of Skye, northwest Scotland. Earth Planet. Sci. Lett. 5, 217 (1968).
— — GALE, N. H.: The significance of lead isotope studies in ancient, high-grade metamorphic basement complexes, as exemplified by the Lewisian rocks of northwest Scotland. Earth Planet. Sci. Lett. 6, 245 (1969).
MURTHY, V. R., PATTERSON, C. C.: Primary isochron of zero age for meteorites and the earth. J. Geophys. Res. 67, 1161 (1962).
NICOLAYSEN, L. O.: Solid diffusion in radio-active minerals and the measurement of absolute age. Geochim. Cosmochim. Acta 11, 41 (1957).
OOSTHUYZEN, E. J., BURGER, A. J.: Radiometric dating of intrusives associated with the Waterberg system. S. Africa Geol. Survey Annals 3, pt. 2, 87 (1965).
OSTIC, R. G., RUSSELL, R. D., STANTON, R. L.: Additional measurements of the isotopic composition of lead from stratiform deposits. Can. J. Earth Sci. 4, 245 (1967).
OVERSBY, V. M.: The isotopic composition of lead in iron meteorites. Geochim. Cosmochim. Acta 34, 77 (1970).
— GAST, P. W.: Lead isotopic compositions and uranium decay series disequilibrium in Recent volcanic rocks. Earth Planet. Sci. Lett. 5, 199 (1968).
PATTERSON, C. C.: The Pb^{207}/Pb^{206} ages of some stone meteorites. Geochem. Cosmochim. Acta 10, 230 (1956).
— TATSUMOTO, M.: The significance of lead isotopes in detrital feldspar with respect to chemical differentation within the Earth's mantle. Geochim. Cosmochim. Acta 28, 1 (1964).
PICCIOTTO, E. E., WILGAIN, S.: Confirmation of the half-lifes of thorium-232. [In French.] Nuovo Cimento 4, 1525 (1956).
PIDGEON, R. T., O'NEIL, J. R., SILVER, L. T.: Uranium and lead isotopic stability in a metamict zircon under experimental hydrothermal conditions. Science 154, 1538 (1966).
RAMTHUN, H.: New calorimetric determination of the half-life period of RaD (^{210}Pb). Z. Naturforsch. 19a, 1064 (1964).
REYNOLDS, P. H., RUSSELL, R. D.: Isotopic composition of lead from Balmat, New York. Can. J. Earth Sci. 5, 1239 (1968).
— SINCLAIR, A. J.: Rock and ore-lead isotopes from the Nelson batholith and the Kootenay Arc, British Columbia, Canada. Econ. Geol. 66, 259 (1971).
ROSHOLT, J. N., PETERMAN, Z. E.: Uranium, thorium, lead systematics in the Granite Mountains, Wyoming. In: Geol. Soc. Am., Abstr. with Programs 1969, 1 pt. 5, 70 (1969).

Lead

Russell, R. D., Farquhar, R. M.: Lead Isotopes in Geology. New York: Interscience Publ. 1960.
— Ulrych, T. J., Kollar, F.: Anomalous leads from Broken Hill, Australia. J. Geophys. Res. 66, 1495 (1961).
Silver, L. T.: Uranium-thorium-lead isotope relations in some lunar materials collected by Apollo 11. Science 167, 468 (1970).
— Deutsch, S.: Uranium-lead isotopic variations in zircons — a case study. J. Geol. 71, 721 (1963).
Sobotovich, E. V., Grashenko, S. M.: Isotopic compositions of Recent lead as age criterion for isolated igneous rock samples. [In Russian.] Akad. Nauk SSSR Izv. Serv. Geol. No. 4, 3 (1965).
— — Lovtsyus, A. V.: Rock age of Taromskoe quarry according to data of the lead isochronous method. Akad. Nauk SSSR Kom. Opredeleniyu Absolyut. Vozrasta Geol. Formatsii Trudy, sess. 11, 353 (1963).
— — — Age of the rocks of the Sharyzhalagay series (Baikal block). [In Russian.] Akad Nauk SSSR Izv. Ser. Geol., No. 9, 38 (1965).
Somayajulu, B. L. K., Tatsumoto, M., Rosholt, J. N., Knight, R. J.: Disequilibrium of the ^{238}U series in basalt. Earth Planet. Sci. Lett. 1, 387 (1966).
Stacey, J. S., Delevaux, M. E., Ulrych, T. J.: Some triple-filament lead isotope ratio measurements and an absolute growth curve for single stage leads. Earth Planet. Sci. Lett. 6, 15 (1969).
— Moore, W. J., Rubright, R. D.: Precision measurement of lead isotope ratios — Preliminary analyses from the U.S. Mine, Bingham Canyon, Utah. Earth Planet. Sci. Lett. 2, 489 (1967).
— Zartman, R. E., Nkomo, I. T.: A lead isotope study of galenas and selected feldspars from mining districts in Utah. Econ. Geol. 63, 796 (1968).
Stanton, R. L., Russell, R. D.: Anomalous leads and the emplacement of lead sulfide ores. Econ. Geol. 54, 588 (1959).
Steiger, R. H., Wasserburg, G. J.: Systematics in the Pb^{208}—Th^{232}, Pb^{207}—U^{235}, and Pb^{206}—U^{238} systems. J. Geophys. Res. 71, 6065 (1966).
— — Comparative U—Th—Pb systematics in 2.7×10^9 yr. plutons of different geologic histories. Geochim. Cosmochim. Acta 33, 1213 (1969).
Stern, T. W., Goldich, S. S., Newell, M. F.: Effects of weathering on the U—Pb ages of zircon from the Morton Gneiss, Minnesota. Earth Planet. Sci. Lett. 1, 369 (1966).
Stieff, L. R., Stern, T. W., Seiki, Oshiro, Senftle, F. E.: Tables for the calculation of lead isotope ages. U.S. Geol. Surv. Profess. Papers 334-A, 1 (1959).
Tatsumoto, M.: Genetic relations of oceanic basalts as indicated by lead isotopes. Science 153, 1094 (1966).
— Knight, R. J.: Isotopic composition of lead in volcanic rocks from central Honshu — with regard to basalt genesis. Geochem. J. 3, 53 (1969).
— Rosholt, J. N.: The age of the moon: An isotopic study of uranium-thorium-lead systematics of lunar samples. Science 167, 461 (1970).
— Snavely, P. D., Jr.: Isotopic composition of lead in rocks of the Coast Range, Oregon and Washington. J. Geophys. Res. 74, 1087 (1969).
Tilton, G. R.: Volume diffusion as a mechanism for discordant ages. J. Geophys. Res. 65, 2933 (1960).
— Isotopic composition of lead from tektites. Geochim. Cosmochim. Acta 14, 323 (1958).
— Grünenfelder, M. H.: Sphene, uranium-lead ages. Science 159, 1458 (1968).
— Steiger, R. H.: Mineral ages and isotopic composition of primary lead at Manitouwadge, Ontario. J. Geophys. Res. 74, 2118 (1969).
Tugarinov, A. I., Gavrilova, J. K., Bedrinov, V. P.: Evolution of the isotopic composition of lead in Precambrain granitic rocks. Vopr. Prikladn. Radiogeol. 228 (1963).
Wampler, J. M., Kulp, J. L.: An isotopic study of lead in sedimentary pyrite. Geochim. Cosmochim. Acta 28, 1419 (1964).
— Smith, D. H., Cameron, A. E.: Isotopic comparison of lead in tektites with lead in earth materials. Geochim. Cosmochim. Acta 33, 1045 (1969).

WASSERBURG, G. J.: Diffusion processes in lead-uranium systems. J. Geophys. Res. 68, 4823 (1963).
WEDEPOHL, K. H.: Investigation of the Kupferschiefer in northwest Germany; a contribution to the explanation of the genesis of bituminous sediments. [In German]. Geochim. Cosmochim. Acta 28, 305 (1964).
WELKE, H., MOORBATH, S., CUMMING, G. L., SIGURDSSON, H.: Lead isotope studies on igneous rocks from Iceland. Earth Planet. Sci. Lett. 4, 221 (1968).
WETHERILL, G. W.: An interpretation of the Rhodesia and Witwatersrand age patterns. Geochim. Cosmochim. Acta 9, 290 (1956a).
— Discordant uranium-lead ages. Am. Geophys. Union Trans. 37, 320 (1956b).
— Discordant uranium-lead ages — Pt. 2: Discordant ages resulting from diffusion of lead and uranium. J. Geophys. Res. 68, 2957 (1963).
YEATS, R. S.: Southern California structure, sea floor spreading, and history of the Pacific Basin. Geol. Soc. Am. Bull. 79, 1969 (1968).
ZARTMAN, R. E.: The isotopic compositions of lead in microclines from the Llano uplift, Texas. J. Geophys. Res. 70, 965 (1965).
— WASSERBURG, G. J.: The isotopic composition of lead in potassium feldspars from some 1.0-b.y. old North American igneous rocks. Geochim. Cosmochim. Acta 33, 901 (1969).
ZYKOV, S. J., TUGARINOV, A. J., BEL'KOV, J. V., BIBIKOVA, E. V.: The age of the most ancient formations of the Kola Peninsula. Geokhymiya No. 4, 307 (1964).

Revised manuscript received: July 1970

Bismuth 83

A	V. Kupčík	(Mineralogisch-Kristallographisches Institut der Universität Göttingen, Germany)
B—M, O	L. H. Ahrens and A. J. Erlank	(Department of Geochemistry, University of Cape Town, Rondebosch, Cape, South Africa)

83-A. Crystal Chemistry

Bismuth is crystallo-chemically very closely related to arsenic and antimony, as well as to lead. This follows from its position in the periodic table (group Vb), from the common electronic structure of Bi^{+3} and Pb^{+2}, and from the common geochemical occurrence. Bismuth is found mainly in sulfidic ore deposits and their oxidation zones. It occurs in rocks only as a trace element. The main groups of bismuth minerals include:

a) metallic bismuth and alloys (rare),
b) sulfides, selenides, tellurides, and sulfosalts,
c) oxides, oxosalts and halogen-oxosalts.

Bismuth, in contrast to arsenic and antimony, is trivalent in all its minerals. The ionic radius of Bi^{3+} is 1.20 Å according to WYCKOFF, and 0.96 Å according to AHRENS. The covalent radius (nowhere given precisely) is about 1.40 Å; the electron structure is $6\ s^2p^3$. The electro-positive character of Bi is the strongest within group Vb, but is only slightly stronger than that of antimony.

I. Metallic Bi and Alloys

Metallic bismuth occurs frequently in sulfidic ores. It is, like antimony and arsenic, trigonal of space group R3m. In the crystal structure (CUCKA and BARRET, 1962), each bismuth atom is coordinated by six neighbors arranged in a trigonal antiprism ($Bi^{[3+3]}$, Bi—Bi distances: 3×3.07 Å and 3×3.53 Å).

In alloy-like compounds, bismuth can, e.g., replace upto $\sim 10\%$ of the Ag atoms in chilenite (DANA, 1868), or it can form ordered structures, e.g., maldonite, $Au_2^{[6\ Au+6\ Bi]}\ Bi_k^{[12\ Au+4\ Bi]}$ (JURRIAANSE, SB 1933—1935, 315).

II. Sulfides, Selenides, Tellurides, and Sulfosalts

The structures of the S, Se, and Te compounds of Bi may be divided into two groups:

a) Bi_2S_3, with a small isomorphic addition of Se (upto a ratio of $S:Se=2:1$). This has a chain structure comparable with antimonite, $\frac{1}{\infty}[Sb_2^{[3+2]}S_3]$ (cf. ŠĆAVNIČAR, SB 1960, 51).[1]

b) A trigonal layer structure, which occurs with higher Se content in the presence of Te. In this structure Bi is coordinated by S, Se, or Te in the form of a trigonal antiprism. The structures of other compounds of the type $Bi_n(S, Se, Te)_m$ are most probably also layer structures (see Fig. 83-A-1 for the structure of tetradymite, $\frac{2}{\infty}[Bi^{[3+3]}]Te_2S$; HARKER, SB 1933—1935, 287).

---

[1] A very flat trigonal antiprism has been found in artificial $Bi^{[3+3]}(Bi_2S_3)_9I_3$, MIEHE and KUPČÍK (1971).

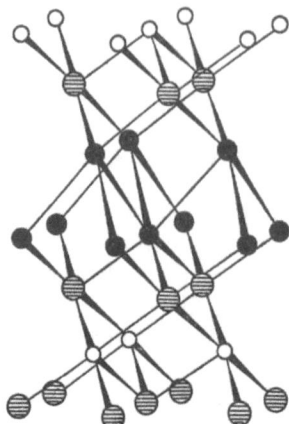

Fig. 83-A-1. Crystal structure of tetradymite, Bi_2Te_2S. Large hatched circles: Bi; medium black circles: Te; small white circles: S

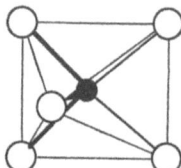

Fig. 83-A-2. Typical $[BiS_5]$ coordination pyramid

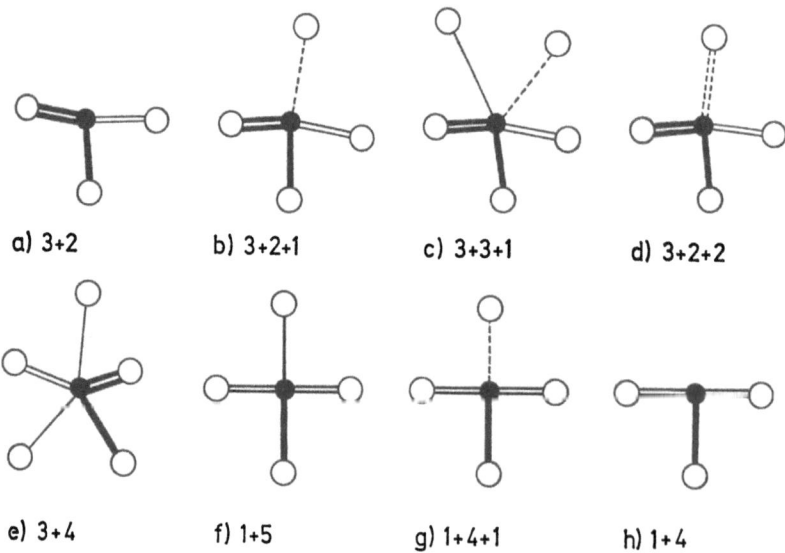

a) 3+2   b) 3+2+1   c) 3+3+1   d) 3+2+2

e) 3+4   f) 1+5   g) 1+4+1   h) 1+4

Fig. 83-A-3a—h. Different $[BiS_n]$ coordination polyhedra. See text for explanation. Small black circles: Bi; large white circles: S

Table 83-A-1

Chemical formula Total	Structural	Name	Shape of the coordination polyhedra and some interatomic distances (Å)	References
Bi_2S_3	$^2_\infty[Bi^{3+3} Bi^{[3+2+1]} S_3]$	bismuthinite	$c\begin{cases}1\times 2.69\\2\times 2.67\\1\times 3.05\\2\times 3.07\end{cases} + d\begin{cases}1\times 2.54\\2\times 2.76\\2\times 2.97\end{cases}$	Kupčík- and Veselá-Nováková (1970)
$CuBiS_2$	$Cu^{[4]}_\infty[Bi^{[3+3]}_{\frac{1}{2}} S_4]$	emplectite	$a\begin{cases}1\times 2.55\\2\times 2.66\\2\times 3.16\end{cases}$	Hofmann SB 1933—35, 393; Kupčík (1965)
$Cu_4Bi_6S_{11}$	a	hodrushite	$4\times d + 1\times b + 1$ octahedron	Kupčík and Makovický (1968)
$PbBi_2S_4$	$Pb^{[5]}_{\frac{3}{\infty}}[Bi^{[3+2+1]}_{\frac{1}{2}} S_4]$	galenobismuthite	$b\begin{cases}1\times 2.63\\2\times 2.73\\2\times 2.99\\1\times 3.12\end{cases} + e\begin{cases}2\times 2.76\\1\times 2.79\\1\times 3.00\\2\times 3.02\\1\times 3.10\end{cases}$	Iitaka and Nowacki (1962)
$Pb_2Bi_2S_5$	a	cosalite	$3\times b + 1\times g$	Weitz and Hellner, SR 1960, 63
$CuPbBiS_3$	$Cu^{[4]} Pb^{[5]}_\infty[Bi^{[3+2]} S_3]$	aikinite	$b\begin{cases}1\times 2.66\\2\times 2.76\\2\times 2.95\\1\times 3.16\end{cases}$	Ohmasa and Nowacki (1970)
$Pb_3Bi_2S_6$	$Pb^{[6+2]}_{\frac{2}{\infty}}[Pb^{[6]}_{\frac{1}{2}} Bi^{[1+5]}_{\frac{1}{2}} S_6]$	phase III	$f\begin{cases}1\times 2.80\\4\times 2.90\\1\times 2.92\end{cases}$	Otto and Strunz (1968)
$(Cu, Fe)Pb_6(Bi, Sb)_7S_{17.5}$	a	kobellite	$2\times a;\ 2\times b;\ b;\ g;\ b$	Miehe (1971)

a Structural formula too complicated, therefore omitted.

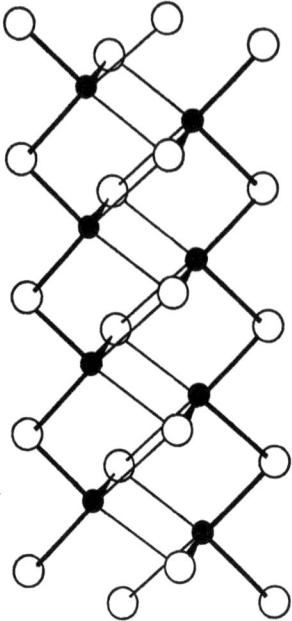

Fig. 83-A-4. Typical [Bi$_2$S$_4$] double chain in emplectite, CuBiS$_2$

Bismuthinite, Bi$_2$S$_3$ (KUPČÍK, 1970), with its irregular coordination of sulfur atoms around bismuth, is representative of the whole group of Bi sulfosalts. The common coordination polyhedron is a tetragonal pyramid (Fig. 83-A-2) where the bismuth atom is located slightly above or below the base. One or two additional atoms are often located at the back side of the pyramid. There is only one known exception: galenobismuthite, PbBi$_2$S$_4$ (IITAKA and NOWACKI, 1962), see Fig. 83-A-3e.

The Bi—S valences can be divided according to bond character and strength into the following groups:

a) Bi—S distances from 2.52 to 2.67 Å. These are p^3 bonds approximately rectangular to each other. The vector to the top of the coordination pyramid belongs always to this group of relatively short bonds (see Fig. 83-A-3; heavy lines).

b) Bi—S distances above 2.80 Å. These are hybrid p-d bonds (which mostly lie within the base of the pyramid; medium lines).

c) Bi—S distances from 2.94 to 3.17 Å. These are pure d-γ bonds lying within the base of the pyramid (light lines).

d) Bi—S distances above 3.27 Å. These are weak d-γ bonds (effect of the inner s-electron pair; broken lines).

In every structure there are several slightly differently coordinated Bi atoms. The coordination polyhedra polymerize by edge-sharing infinite single chains ∞[BiS$_2$], or further to double or multiple chains, e.g., ∞[Bi$_2$S$_4$], ∞[Bi$_4$S$_6$] (see also Fig. 83-A-4).

Isomorphic replacement of As by Bi is unknown and of Sb by Bi possible but, for geochemical reasons, rare. A typical example for a wide-range substitution is the structure of kobellite (Cu, Fe)Pb$_6$(Bi, Sb)$_7$S$_{17.5}$ (MIEHE, 1971).

The occurrence of a regular coordination (octahedron with Bi atom in the center) must be interpreted statistically. In such a case Bi may be isomorphically replaced by Pb as in hodrushite (Pb, Bi, Ag)Cu$_4$Bi$_5$S$_{11}$ (KUPČÍK and MAKOVICKÝ, 1968).

Replacement of Bi by Pb may also occur when each Bi atom replaced by a Pb results in an additional Cu atom occupying tetrahedral interstices of the bismuthinite structure. This explains the isomorphism of the bismuthinite-aikinite series, (Bi$_{2-x}$, Pb$_x$)Cu$_x$S$_3$ with $x=0$ to 1 (OHMASA and NOWACKI, 1970). It is assumed that the intermediate members have an ordered Bi—Pb, Cu distribution, which leads to multiple lattice constants and therefore to different mineral types. Data are given in Table 83-A-1.

III. Oxidic Compounds

In contrast to Bi sulfides and sulfosalts, which are very similar to the corresponding Sb compounds and in which Bi always has the same sulfur coordination, the oxidic compounds of Bi are much more complicated. E. g., α- and β-Bi$_2$O$_3$ are not related crystallo-chemically to the corresponding Sb compounds. The coordination around

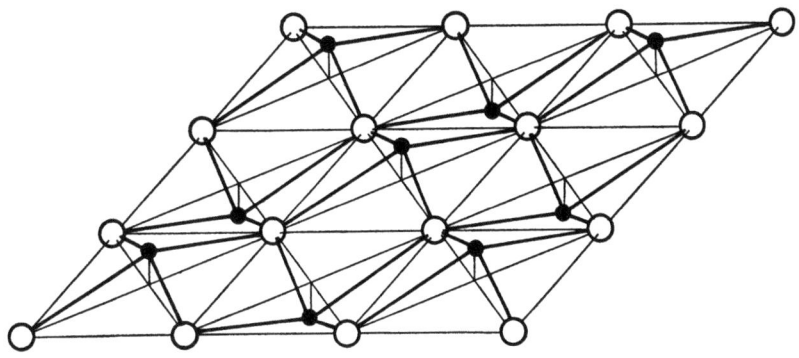

Fig. 83-A-5. $\overset{2}{\infty}$ [BiO] nets formed from linked [BiO$_4$] pyramids in kettnerite, CaBi[OF|CO$_3$]. Small black circles: Bi; large white circles: O

Bi is very irregular (except in bismuthite and β-Bi$_2$O$_3$) and is difficult to describe in terms of common polyhedra types, though two frequently occurring basic arrangements may be distinguished:

a) $p^3 d^{2-3}$ bonding (similar to sulfides) in a deformed and more or less incomplete octahedral coordination with 3 short bonds (tetragonal pyramid with Bi occupying the basis in α-Bi$_2$O$_3$ or trigonal antiprism in eulytite, Bi$_4$Si$_3$O$_{12}$);

b) extremely flat tetragonal pyramid with Bi occupying the apex.

Schematically, bond lengths can be related to the kind of hybridization:

a) ~2.09—2.30 Å   pure p^3 bonding
b) ~2.30—2.40 Å $p^3 d^n$ hybrid bonding
c) ~2.50—2.65 Å   pure d bonding
d) >2.70 Å   weak d bonding with repulsion effects from inert electron pair

Table 83-A-2

Chemical formula	Name	Bi–O interatomic distances [Å]	Shape of the coordination polyhedra	Reference
α-Bi$_2$O$_3$	bismite	Bi$_1$ 2.08; 2.17; 2.21; 2.54; 2.63 B$_2$ 2.14; 2.22; 2.29; 2.48; 2.54; 2.80	Tetragonal pyramid (3+2) Trigonal antiprism	Malmros (1970)
β-Bi$_2^{[6]}$O$_3$	—	6×2.39	Trigonal antiprism	Sillén SR 1937, 69
Bi$_2^{[2]}$O$_2$[CO$_3$]	bismutite	4×2.29	Tetragonal pyramid	Lagerkrantz and Sillén, SR 1947–1948, 308
Bi$^{[4+4]}$[PO$_4$]·0.5 H$_2$O	—	2×2.31; 2×2.33; 2×2.56;	Quadr.-planar + 2×2	Mooney and Slater (1962)
Bi$^{[2+2+1]}$Mo$^{[2+4]}$O$_6$	koechlinite	2×2.25; 2×2.39; 2.53	Tetragonal pyramid + 1	Zemann SR 1956, 449
Bi$^{[4+4]}$V$^{[4]}$O$_4$	pucherite	2×2.20; 2×2.31; 2×2.54; 2×2.73	Tetragonal prisma	Qurashi and Barnes SR 1953, 503
CaBi$_2^{[4]}$O$_2$[CO$_3$]$_2$	beyerite	4×2.29	Tetragonal pyramid	Lagerkrantz and Sillén, SR 1947–1948, 308
Bi$_4^{[3+3]}$[SiO$_4$]$_3$	eulytite	3×2.15; 3×2.62	Trigonal antiprism	Segal et al. (1966)
Bi$^{[4]}$OCl	bismoclite	4×2.32	Tetragonal pyramid	Sillén, SR 1947–1948, 312
CaBi$^{[4]}$[OF/CO$_3$]	kettnerite	4×2.29 + 4 from CO$_3$ undefined	Tetragonal pyramid	Syneček and Žák, SR 1960, 426

Data are given in Table 83-A-2.

There are several different ways of linking the coordination polyhedra, and the following types have been found:

a) isolated [BiO$_8$] groups in eulytite, Bi$_4$[SiO$_4$]$_3$;

b) $^2_\infty$[BiO] nets built of BiO$_4$ pyramids sharing edges (see Fig. 83-A-5) in bismoclite, BiOCl, bismutite, Bi$_2$O$_2$[CO$_3$], and kettnerite, CaBi[OF|CO$_3$];

c) probably $^1_\infty$[BiO$_2$] chains with edge-sharing in koechlinite, BiMoO$_6$;

d) three-dimensional arrays $^3_\infty$[Bi$_2$O$_3$], resulting from polyhedra sharing edges and corners in bismite, α-Bi$_2$O$_3$.

Isomorphous replacement of Bi by other elements is not known in oxidic minerals.

References: Section 83-A

CUCKA, P., BARRET, C. S.: The crystal structure of Bi and solid solutions of Pb, Sn, Sb and Te in Bi. Acta Cryst. **15**, 865 (1962).

DANA in: STRUNZ, H. Mineralogische Tabellen (4. ed.), p. 100. Leipzig: Akad. Verlagsgesellschaft 1966.

IITAKA, Y., NOWACKI, W.: A redetermination of the crystal structure of galenobismutite. $PbBi_2S_4$. Acta Cryst. **15**, 691 (1962).

KUPČÍK, V.: Struktur des Emplektit $CuBiS_2$. Referate der 8. Diskussionstagung der Sektion für Kristallkunde der DMG, Marburg, 16 (1965).

— MAKOVICKÝ, E.: Die Kristallstruktur des Minerals $(Pb, Ag, Bi)Cu_4Bi_5S_{11}$. Neues Jahrb. Mineral., Monatsh. **236** (1968).

— VESELÁ-NOVÁKOVÁ, L.: Zur Kristallstruktur des Bismuthinits, Bi_2S_3. Mineral. Petrog. Mitt. **14**, 55 (1970).

MALMROS, G.: The crystal structure of α-Bi_2O_3. Acta Cryst. **24**, 384 (1970).

MIEHE, G.: Crystal structure of kobellite. Nature Physical Science **231**, 133 (1971).

— KUPČÍK, V.: Die Kristallstruktur des $Bi(Bi_2S_3)_9J_3$. Naturwissenschaften **58**, 219 (1971).

MOONEY-SLATER, R. C. L.: Polymorphic forms of bismuth phosphate. Z. Krist. **117**, 317 (1962).

OHMASA, M., NOWACKI, W.: Note on the space group and on the structure of aikinite derivatives. Neues Jahrb. Mineral., Monatsh. **158** (1970).

— — A redetermination of the crystal structure of aikinite $[BiS_2|S|Cu^{IV}Pb^{VII}]$. Z. Krist. **132**, 7 (1970).

OTTO, H. H., STRUNZ, H.: Zur Kristallchemie synthetischer Blei-Wismut-Spießglanze. Neues Jahrb. Mineral., Abhandl. **108**, 1 (1968).

SEGAL, D. J., SANTORO, R. P., NEWNHAM, R. E.: Neutron-diffraction study of $Bi_4Si_3O_{12}$. Z. Krist. **123**, 73 (1966).

Manuscript received June 1972

© Springer-Verlag Berlin · Heidelberg 1972

MIX
Papier aus verantwortungsvollen Quellen
Paper from responsible sources
FSC® C105338

If you have any concerns about our products,
you can contact us on
ProductSafety@springernature.com

In case Publisher is established outside the EU,
the EU authorized representative is:
Springer Nature Customer Service Center GmbH
Europaplatz 3, 69115 Heidelberg, Germany

Printed by Libri Plureos GmbH
in Hamburg, Germany